RSMeans

Residential Cost Data

23rd Annual Edition

- *Square Foot Costs*
- *Systems Costs*
- *Unit Costs*

2004

Senior Editor
Robert W. Mewis, CCC

Contributing Editors
Barbara Balboni
Robert A. Bastoni
John H. Chiang, PE
Robert J. Kuchta
Robert C. McNichols
Melville J. Mossman, PE
John J. Moylan
Jeannene D. Murphy
Stephen C. Plotner
Michael J. Regan
Eugene R. Spencer
Marshall J. Stetson
Phillip R. Waier, PE

Senior Engineering Operations Manager
John H. Ferguson, PE

President
Curtis B. Allen

Vice President, Product Management
Roger J. Grant

Sales Director
John M. Shea

Production Manager
Michael Kokernak

Production Coordinator
Marion E. Schofield

Technical Support
Thomas J. Dion
Jonathan Forgit
Mary Lou Geary
Gary L. Hoitt
Paula Reale-Camelio
Robin Richardson
Kathryn S. Rodriguez
Sheryl A. Rose

Book & Cover Design
Norman R. Forgit

First Printing

Foreword

RSMeans is a product line of Reed Construction Data, a leading provider of construction information, products, and services in North America and globally. Reed Construction Data's project information products include more than 100 regional editions, national construction data, sales leads, and local plan rooms in major business centers. Reed Construction Data's PlansDirect provides surveys, plans and specifications. The First Source suite of products consists of *First Source for Products,* SPEC-DATA™, MANU-SPEC™, CADBlocks, Manufacturer Catalogs and First Source Exchange (www.firstsourceexchange.com) for the selection of nationally available building products. Reed Construction Data also publishes ProFile, a database of more than 20,000 U.S. architectural firms. RSMeans provides construction cost data, training, and consulting services in print, CD-ROM and online. Reed Construction Data, headquartered in Atlanta, is owned by Reed Business Information (www.cahners.com), a leading provider of critical information and marketing solutions to business professionals in the media, manufacturing, electronics, construction and retail industries. Its market-leading properties include more than 135 business-to-business publications, over 125 Webzines and Web portals, as well as online services, custom publishing, directories, research and direct-marketing lists. Reed Business Information is a member of the Reed Elsevier plc group (NYSE: RUK and ENL)—a world-leading publisher and information provider operating in the science and medical, legal, education and business-to-business industry sectors.

Our Mission

Since 1942, RSMeans has been actively engaged in construction cost publishing and consulting throughout North America.

Today, over 50 years after the company began, our primary objective remains the same: to provide you, the construction and facilities professional, with the most current and comprehensive construction cost data possible.

Whether you are a contractor, an owner, an architect, an engineer, a facilities manager, or anyone else who needs a fast and reliable construction cost estimate, you'll find this publication to be a highly useful and necessary tool.

Today, with the constant flow of new construction methods and materials, it's difficult to find the time to look at and evaluate all the different construction cost possibilities. In addition, because labor and material costs keep changing, last year's cost information is not a reliable basis for today's estimate or budget.

That's why so many construction professionals turn to RSMeans. We keep track of the costs for you, along with a wide range of other key information, from city cost indexes . . . to productivity rates . . . to crew composition . . . to contractor's overhead and profit rates.

RSMeans performs these functions by collecting data from all facets of the industry, and organizing it in a format that is instantly accessible to you. From the preliminary budget to the detailed unit price estimate, you'll find the data in this book useful for all phases of construction cost determination.

The Staff, the Organization, and Our Services

When you purchase one of RSMeans' publications, you are in effect hiring the services of a full-time staff of construction and engineering professionals.

Our thoroughly experienced and highly qualified staff works daily at collecting, analyzing, and disseminating comprehensive cost information for your needs. These staff members have years of practical construction experience and engineering training prior to joining the firm. As a result, you can count on them not only for the cost figures, but also for additional background reference information that will help you create a realistic estimate.

The RSMeans organization is always prepared to help you solve construction problems through its five major divisions: Construction and Cost Data Publishing, Electronic Products and Services, Consulting Services, Insurance Services, and Educational Services.

Besides a full array of construction cost estimating books, Means also publishes a number of other reference works for the construction industry. Subjects include construction estimating and project and business management; special topics such as HVAC, roofing, plumbing, and hazardous waste remediation; and a library of facility management references.

In addition, you can access all of our construction cost data through your computer with RSMeans CostWorks 2004 CD-ROM, an electronic tool that offers over 50,000 lines of Means detailed construction cost data, along with assembly and whole building cost data. You can also access Means cost information from our Web site at www.rsmeans.com

What's more, you can increase your knowledge and improve your construction estimating and management performance with a Means Construction Seminar or In-House Training Program. These two-day seminar programs offer unparalleled opportunities for everyone in your organization to get updated on a wide variety of construction-related issues.

RSMeans also is a worldwide provider of construction cost management and analysis services for commercial and government owners and of claims and valuation services for insurers.

In short, RSMeans can provide you with the tools and expertise for constructing accurate and dependable construction estimates and budgets in a variety of ways.

Robert Snow Means Established a Tradition of Quality That Continues Today

Robert Snow Means spent years building his company, making certain he always delivered a quality product.

Today, at RSMeans, we do more than talk about the quality of our data and the usefulness of our books. We stand behind all of our data, from historical cost indexes... to construction materials and techniques... to current costs.

If you have any questions about our products or services, please call us toll-free at 1-800-334-3509. Our customer service representatives will be happy to assist you or visit our Web site at www.rsmeans.com

Table of Contents

How the Book Is Built: An Overview

A Powerful Construction Tool

You have in your hands one of the most powerful construction tools available today. A successful project is built on the foundation of an accurate and dependable estimate. This book will enable you to construct just such an estimate.

For the casual user the book is designed to be:

- quickly and easily understood so you can get right to your estimate
- filled with valuable information so you can understand the necessary factors that go into the cost estimate

For the regular user, the book is designed to be:

- a handy desk reference that can be quickly referred to for key costs
- a comprehensive, fully reliable source of current construction costs and productivity rates, so you'll be prepared to estimate any project
- a source book for preliminary project cost, product selections, and alternate materials and methods

To meet all of these requirements we have organized the book into the following clearly defined sections.

Square Foot Cost Section

This section lists Square Foot costs for typical residential construction projects. The organizational format used divides the projects into basic building classes. These classes are defined at the beginning of the section. The individual projects are further divided into ten common components of construction. An outline of a typical page layout, an explanation of Square Foot prices, and a Table of Contents are located at the beginning of the section.

Assemblies Cost Section

This section uses an "Assemblies" (sometimes referred to as "systems") format grouping all the functional elements of a building into 9 construction divisions.

At the top of each "Assembly" cost table is an illustration, a brief description, and the design criteria used to develop the cost. Each of the components and its contributing cost to the system is shown.

Material: These cost figures include a standard 10% markup for "handling". They are national average material costs as of January of the current year and include delivery to the job site.

Installation: The installation costs include labor and equipment, plus a markup for the installing contractor's overhead and profit.

For a complete breakdown and explanation of a typical "Assemblies" page, see "How To Use Assemblies Cost Tables" at the beginning of the Assembly Section.

Unit Price Section

All cost data has been divided into the 16 divisions according to the MasterFormat system of classification and numbering as developed by the Construction Specifications Institute (CSI) and Construction Specifications Canada (CSC). For a listing of these divisions and an outline of their subdivisions, see the Unit Price Section Table of Contents.

Estimating tips are included at the beginning of each division.

Reference Section

This section includes information on Reference Numbers, Crew Listings, Location Factors, and a listing of Abbreviations. It is visually identified by a vertical gray bar on the edge of pages.

Reference Numbers: At the beginning of selected major classifications in the Unit Price Section are "reference numbers" shown in bold squares. These numbers refer you to related information in the Reference Section.

In this section, you'll find reference tables, explanations, and estimating information that support how we develop the unit price data. Also included are alternate pricing methods, technical data, and estimating procedures, along with information on design and economy in construction. You'll also find helpful tips on what to expect and what to avoid when estimating and constructing your project.

It is recommended that you refer to the Reference Section if a "reference number" appears within the section you are estimating.

Crew Listings: This section lists all the crews referenced in the book. For the purposes of this book, a crew is composed of more than one trade classification and/or the addition of power equipment to any trade classification. Power equipment is included in the cost of the crew. Costs are shown both with bare labor rates and with the installing contractor's overhead and profit added. For each, the total crew cost per eight-hour day and the composite cost per labor-hour are listed.

Location Factors: Costs vary depending upon regional economy. You can adjust the "national average" costs in this book to over 930 major cities throughout the U.S. and Canada by using the data in this section.

Abbreviations: A listing of the abbreviations used throughout this book, along with the terms they represent, is included.

Index

A comprehensive listing of all terms and subjects in this book to help you find what you need quickly when you are not sure where it falls in MasterFormat.

The Scope of This Book

This book is designed to be as comprehensive and as easy to use as possible. To that end we have made certain assumptions and limited its scope in three key ways:

1. We have established material prices based on a "national average."
2. We have computed labor costs based on a 7 major region average of open shop wage rates.
3. We have targeted the data for projects of a certain size range.

Project Size

This book is intended for use by those involved primarily in Residential construction costing less than $750,000. This includes the construction of homes, row houses, townhouses, condominiums and apartments.

With reasonable exercise of judgment the figures can be used for any building work. For other types of projects, such as repair and remodeling or commercial buildings, consult the appropriate MEANS publication for more information.

How to Use the Book: The Details

What's Behind the Numbers? The Development of Cost Data

The staff at RSMeans continuously monitors developments in the construction industry in order to ensure reliable, thorough and up-to-date cost information.

While *overall* construction costs may vary relative to general economic conditions, price fluctuations within the industry are dependent upon many factors. Individual price variations may, in fact, be opposite to overall economic trends. Therefore, costs are continually monitored and complete updates are published yearly. Also, new items are frequently added in response to changes in materials and methods.

Costs—$ (U.S.)

All costs represent U.S. national averages and are given in U.S. dollars. The Means Location Factors can be used to adjust costs to a particular location. The Location Factors for Canada can be used to adjust U.S. national averages to local costs in Canadian dollars.

Material Costs

The RSMeans staff contacts manufacturers, dealers, distributors, and contractors all across the U.S. and Canada to determine national average material costs. If you have access to current material costs for your specific location, you may wish to make adjustments to reflect differences from the national average. Included within material costs are fasteners for a normal installation. RSMeans engineers use manufacturers' recommendations, written specifications and/or standard construction practice for size and spacing of fasteners. Adjustments to material costs may be required for your specific application or location. Material costs do not include sales tax.

Labor Costs

Labor costs are based on the average of open shop wages from across the U.S. for the current year. Rates along with overhead and profit markups are listed on the inside back cover of this book.

- If wage rates in your area vary from those used in this book, or if rate increases are expected within a given year, labor costs should be adjusted accordingly.

Labor costs reflect productivity based on actual working conditions. These figures include time spent during a normal workday on tasks other than actual installation, such as material receiving and handling, mobilization at site, site movement, breaks, and cleanup.

Productivity data is developed over an extended period so as not to be influenced by abnormal variations and reflects a typical average.

Equipment Costs

Equipment costs include not only rental, but also operating costs for equipment under normal use. The operating costs include parts and labor for routine servicing such as repair and replacement of pumps, filters and worn lines. Normal operating expendables such as fuel, lubricants, tires and electricity (where applicable) are also included. Extaordinary operating expendables with highly variable wear patterns such as diamond bits and blades are excluded. These costs are included under materials. Equipment rental rates are obtained from industry sources throughout North America—contractors, suppliers, dealers, manufacturers, and distributors.

Crew Equipment Cost/Day— The power equipment required for each crew is included in the crew cost. The daily cost for crew equipment is based on dividing the weekly bare rental rate by 5 (number of working days per week), and then adding the hourly operating cost times 8 (hours per day). This "Crew Equipment Cost/Day" is listed in Subdivision 01590.

Mobilization/Demobilization—The cost to move construction equipment from an equipment yard or rental company to the job site and back again is not included in equipment costs. Mobilization (to the site) and demobilization (from the site) costs can be found in Section 02305-250. If a piece of equipment is already at the job site, it is not appropriate to utilize mob/demob costs again in an estimate.

Factors Affecting Costs

Costs can vary depending upon a number of variables. Here's how we have handled the main factors affecting costs.

Quality— The prices for materials and the workmanship upon which productivity is based represent sound construction work. They are also in line with U.S. government specifications.

Overtime— We have made no allowance for overtime. If you anticipate premium time or work beyond normal working hours, be sure to make an appropriate adjustment to your labor costs.

Productivity— The productivity, daily output, and labor-hour figures for each line item are based on working an eight-hour day in daylight hours in moderate temperatures. For work that extends beyond normal work hours or is performed under adverse conditions, productivity may decrease. (See the section in "How To Use the Unit Price Pages" for more on productivity.)

Size of Project— The size, scope of work, and type of construction project will have a significant impact on cost. Economies of scale can reduce costs for large projects. Unit costs can often run higher for small projects. Costs in this book are intended for the size and type of project as previously described in "How the Book Is Built: An Overview." Costs for projects of a significantly different size or type should be adjusted accordingly.

Location— Material prices in this book are for metropolitan areas. However, in dense urban areas, traffic and site storage limitations may increase costs. Beyond a 20-mile radius of large cities, extra trucking or transportation charges may also increase the material costs slightly. On the other hand, lower wage rates may be in effect. Be sure to consider both these factors when preparing an estimate, particularly if the job site is located in a central city or remote rural location.

In addition, highly specialized subcontract items may require travel and per diem expenses for mechanics.

Other factors—

- season of year
- contractor management
- weather conditions
- local union restrictions
- building code requirements
- availability of:
    - adequate energy
    - skilled labor
    - building materials
- owner's special requirements/restrictions
- safety requirements
- environmental considerations

General Conditions—The "Square Foot" and "Assemblies" sections of this book use costs that include the installing contractor's overhead and profit (O&P). The Unit Price Section presents cost data in two ways: Bare Costs and Total Cost including O&P (Overhead and Profit). General Conditions, when applicable, should also be added to the Total Cost including O&P. The costs for General Conditions are listed in Division 1 of the Unit Price Section and the Reference Section of this book. General Conditions for the *Installing Contractor* may range from 0% to 10% of the Total Cost including O&P. For the *General* or *Prime Contractor*, costs for General Conditions may range from 5% to 15% of the Total Cost including O&P, with a figure of 10% as the most typical allowance.

Overhead & Profit—Total Cost including O&P for the *Installing Contractor* is shown in the last column on both the Unit Price and the Assemblies pages of this book. This figure is the sum of the bare material cost plus 10% for profit, the base labor cost plus overhead and profit, and the bare equipment cost plus 10% for profit. Details for the calculation of Overhead and Profit on labor are shown on the inside back cover and in the Reference Section of this book. (See the "How to Use the Unit Price Pages" for an example of this calculation.)

Unpredictable Factors—General business conditions influence "in-place" costs of all items. Substitute materials and construction methods may have to be employed. These may affect the installed cost and/or life cycle costs. Such factors may be difficult to evaluate and cannot necessarily be predicted on the basis of the job's location in a particular section of the country. Thus, where these factors apply, you may find significant, but unavoidable cost variations for which you will have to apply a measure of judgment to your estimate.

Rounding of Costs

In general, all unit prices in excess of $5.00 have been rounded to make them easier to use and still maintain adequate precision of the results. The rounding rules we have chosen are in the following table.

Prices from ...	Rounded to the nearest ...
$.01 to $5.00	$.01
$5.01 to $20.00	$.05
$20.01 to $100.00	$.50
$100.01 to $300.00	$1.00
$300.01 to $1,000.00	$5.00
$1,000.01 to $10,000.00	$25.00
$10,000.01 to $50,000.00	$100.00
$50,000.01 and above	$500.00

Final Checklist

Estimating can be a straightforward process provided you remember the basics. Here's a checklist of some of the items you should remember to do before completing your estimate.

Did you remember to . . .

- factor in the Location Factor for your locale
- take into consideration which items have been marked up and by how much
- mark up the entire estimate sufficiently for your purposes
- read the background information on techniques and technical matters that could impact your project time span and cost
- include all components of your project in the final estimate
- double check your figures to be sure of your accuracy
- call RSMeans if you have any questions about your estimate or the data you've found in our publications

Remember, RSMeans stands behind its publications. If you have any questions about your estimate . . . about the costs you've used from our books . . . or even about the technical aspects of the job that may affect your estimate, feel free to call the RSMeans editors at 1-800-334-3509.

Square Foot Cost Section

Table of Contents

Introduction to Square Foot Cost Section

The Square Foot Cost Section of this manual contains costs per square foot for four classes of construction in seven building types. Costs are listed for various exterior wall materials which are typical of the class and building type. There are cost tables for Wings and Ells with modification tables to adjust the base cost of each class of building. Non-standard items can easily be added to the standard structures.

Cost estimating for a residence is a three-step process:

(1) Identification
(2) Listing dimensions
(3) Calculations

Guidelines and a sample cost estimating procedure are shown on the following pages.

Identification

To properly identify a residential building, the class of construction, type, and exterior wall material must be determined. Page 8 has drawings and guidelines for determining the class of construction. There are also detailed specifications and additional drawings at the beginning of each set of tables to further aid in proper building class and type identification.

Sketches for eight types of residential buildings and their configurations are shown on pages 10 and 11. Definitions of living area are next to each sketch. Sketches and definitions of garage types are on page 12.

Living Area

Base cost tables are prepared as costs per square foot of living area. The living area of a residence is that area which is suitable and normally designed for full time living. It does not include basement recreation rooms or finished attics, although these areas are often considered full time living areas by the owners.

Living area is calculated from the exterior dimensions without the need to adjust for exterior wall thickness. When calculating the living area of a 1-1/2 story, two story, three story or tri-level residence, overhangs and other differences in size and shape between floors must be considered.

Only the floor area with a ceiling height of six feet or more in a 1-1/2 story residence is considered living area. In bi-levels and tri-levels, the areas that are below grade are considered living area, even when these areas may not be completely finished.

Base Tables and Modifications

Base cost tables show the base cost per square foot without a basement, with one full bath and one full kitchen for economy and average homes and an additional half bath for custom and luxury models. Adjustments for finished and unfinished basements are part of the base cost tables. Adjustments for multi family residences, additional bathrooms, townhouses, alternative roofs, and air conditioning and heating systems are listed in Modifications, Adjustments and Alternatives tables below the base cost tables.

Costs for other modifications, adjustments and alternatives, including garages, breezeways and site improvements, are on pages 93 to 96.

Listing of Dimensions

To use this section of the manual only the dimensions used to calculate the horizontal area of the building and additions, modifications, adjustments and alternatives are needed. The dimensions, normally the length and width, can come from drawings or field measurements. For ease in calculation, consider measuring in tenths of feet, i.e., 9 ft. 6 in. = 9.5 ft., 9 ft. 4 in. = 9.3 ft.

In all cases, make a sketch of the building. Any protrusions or other variations in shape should be noted on the sketch with dimensions.

Calculations

The calculations portion of the estimate is a two-step activity:

(1) The selection of appropriate costs from the tables
(2) Computations

Selection of Appropriate Costs

To select the appropriate cost from the base tables, the following information is needed:

(1) Class of construction
 (a) Economy
 (b) Average
 (c) Custom
 (d) Luxury
(2) Type of residence
 (a) One story
 (b) 1-1/2 story
 (c) 2 story
 (d) 3 story
 (e) Bi-level
 (f) Tri-level
(3) Occupancy
 (a) One family
 (b) Two family
 (c) Three family
(4) Building configuration
 (a) Detached
 (b) Town/Row house
 (c) Semi-detached
(5) Exterior wall construction
 (a) Wood frame
 (b) Brick veneer
 (c) Solid masonry
(6) Living areas

Modifications, adjustments and alternatives are classified by class, type and size.

Computations

The computation process should take the following sequence:

(1) Multiply the base cost by the area
(2) Add or subtract the modifications, adjustments, alternatives
(3) Apply the location modifier

When selecting costs, interpolate or use the cost that most nearly matches the structure under study. This applies to size, exterior wall construction and class.

How to Use the Residential Square Foot Cost Pages

The following is a detailed explanation of a sample entry in the Residential Square Foot Cost Section. Each bold number below corresponds to the item being described on the facing page with the appropriate component of the sample entry following in parenthesis.

Prices listed are costs that include overhead and profit of the installing contractor. Total model costs include an additional mark-up for General Contractors' overhead and profit, and fees specific to class of construction.

Base cost per square foot of living area

Exterior Wall	1000	1200	1400	1600	1800	2000	2200	2600	3000	3400	3800
						Living Area					
Wood Siding - Wood Frame	92.85	84.30	80.40	77.75	75.10	**71.95**	70.05	66.25	62.50	60.80	59.35
Brick Veneer - Wood Frame	100.60	91.50	87.15	84.20	81.15	77.85	[illegible]	71.30	67.10	65.25	63.55
Stucco on Wood Frame	92.45	83.95	80.10	77.50	74.80	71.70	[illegible]	66.00	62.20	60.60	59.15
Solid Masonry	112.20	102.20	97.10	93.75	90.15	86.50	83.80	78.75	74.00	71.80	69.75
Finished Basement, Add	14.35	14.25	13.90	13.60	13.30	13.15	[illegible]	12.50	12.15	12.00	11.75
Unfinished Basement, Add	5.35	5.10	4.90	4.75	4.60	4.45	[illegible]	4.05	3.90	3.80	3.65

Modifications

Add to the total cost

Upgrade Kitchen Cabinets	$ + 2240
Solid Surface Countertops	+ 896
Full Bath - including plumbing, wall and floor finishes	+ 4060
Half Bath - including plumbing, wall and floor finishes	+ 2583
One Car Attached Garage	+ 8958
One Car Detached Garage	+ 11,757
Fireplace & Chimney	+ 4365

Adjustments

For multi family - add to total cost

Additional Kitchen	$ + 4228
Additional Bath	+ 4060
Additional Entry & Exit	+ 1236
Separate Heating	+ 1237
Separate Electric	+ 1487

For Townhouse/Rowhouse - Multiply cost per square foot by

Inner Unit	.90
End Unit	.95

Alternatives

Add to or deduct from the cost per square foot of living area

Cedar Shake Roof	+ 1.16
Clay Tile Roof	+ 2.04
Slate Roof	+ 3.89
Upgrade Walls to Skim Coat Plaster	+ .38
Upgrade Ceilings to Textured Finish	+ .39
Air Conditioning, in Heating Ductwork	+ 1.67
In Separate Ductwork	+ 4.32
Heating Systems, Hot Water	+ 1.38
Heat Pump	+ 2.15
Electric Heat	– .52
Not Heated	– 2.46

Additional upgrades or components

Kitchen Cabinets & Countertops	Page 93
Bathroom Vanities	94
Fireplaces & Chimneys	94
Windows, Skylights & Dormers	94
Appliances	95
Breezeways & Porches	95
Finished Attic	95
Garages	96
Site Improvements	96
Wings & Ells	56

1 Class of Construction (Average)

The class of construction depends upon the design and specifications of the plan. The four classes are economy, average, custom and luxury.

2 Type of Residence (2 Story)

The building type describes the number of stories or levels in the model. The seven building types are 1 story, 1-1/2 story, 2 story, 2-1/2 story, 3 story, bi-level and tri-level.

3 Specification Highlights (Hot Air Heat)

These specifications include information concerning the components of the model, including the number of baths, roofing types, HVAC systems, and materials and workmanship. If the components listed are not appropriate, modifications can be made by consulting the information shown lower on the page or the Assemblies Section.

4 Exterior Wall System (Wood Siding - Wood Frame)

This section includes the types of exterior wall systems and the structural frame used. The exterior wall systems shown are typical of the class of construction and the building type shown.

5 Living Areas (2000 SF)

The living area is that area of the residence which is suitable and normally designed for full time living. It does not include basement recreation rooms or finished attics. Living area is calculated from the exterior dimensions without the need to adjust for exterior wall thickness. When calculating the living area of a 1-1/2 story, 2 story, 3 story or tri-level residence, overhangs and other differences in size and shape between floors must be considered. Only the floor area with a ceiling height of six feet or more in a 1-1/2 story residence is considered living area. In bi-levels and tri-levels, the areas that are below grade are considered living area, even when these areas may not be completely finished. A range of various living areas for the residential model are shown to aid in selection of values from the matrix.

6 Base Costs per Square Foot of Living Area ($71.95)

Base cost tables show the cost per square foot of living area without a basement, with one full bath and one full kitchen for economy and average homes and an additional half bath for custom and luxury models. When selecting costs, interpolate or use the cost that most nearly matches the residence under consideration for size, exterior wall system and class of construction. Prices listed are costs that include overhead and profit of the installing contractor, a general contractor markup and an allowance for plans that vary by class of construction. For additional information on contractor overhead and architectural fees, see the Reference Section.

7 Basement Types (Finished)

The two types of basements are finished or unfinished. The specifications and components for both are shown on the Building Classes page in the Introduction to this section.

8 Additional Costs for Basements ($13.15)

These values indicate the additional cost per square foot of living area for either a finished or an unfinished basement.

9 Modifications and Adjustments (Solid Surface Countertops $896)

Modifications and Adjustments are costs added to or subtracted from the total cost of the residence. The total cost of the residence is equal to the cost per square foot of living area times the living area. Typical modifications and adjustments include kitchens, baths, garages and fireplaces.

10 Multiplier for Townhouse/ Rowhouse (Inner Unit .90)

The multipliers shown adjust the base costs per square foot of living area for the common wall condition encountered in townhouses or rowhouses.

11 Alternatives (Skim Coat Plaster $.38)

Alternatives are costs added to or subtracted from the base cost per square foot of living area. Typical alternatives include variations in kitchens, baths, roofing, air conditioning and heating systems.

12 Additional Upgrades or Components (Wings & Ells page 56)

Costs for additional upgrades or components, including wings or ells, breezeways, porches, finished attics and site improvements, are shown in other locations in the Residential Square Foot Cost Section.

How to Use the Square Foot Cost Pages *(Continued)*

2

1

Average 2 Story

Living Area - 2000 S.F.
Perimeter - 135 L.F.

		Labor-Hours	Cost Per Square Foot Of Living Area		
			Mat.	Labor	Total
1 Site Work	Site preparation for slab; 4' deep trench excavation for foundation wall.	.034		.52	.52
2 Foundation	Continuous reinforced concrete footing 8" deep x 18" wide; dampproofed and insulated reinforced concrete foundation wall, 8" thick, 4' deep, 4" concrete slab on 4" crushed stone base and polyethylene vapor barrier, trowel finish.	.066	2.12	3.01	5.13
3 Framing	Exterior walls - 2" x 4" wood studs, 16" O.C.; 1/2" plywood sheathing; 2" x 6" rafters 16" O.C. with 1/2" plywood sheathing, 4 in 12 pitch; 2" x 6" ceiling joists 16" O.C.; 2" x 8" floor joists 16" O.C. with 5/8" plywood subfloor; 1/2" plywood subfloor on 1" x 2" wood sleepers 16" O.C.	.131	5.41	6.64	12.05
4 Exterior Walls	Beveled wood siding and #15 felt building paper on insulated wood frame walls; 6" attic insulation; double hung windows; 3 flush solid core wood exterior doors with storms.	.111	8.18	4.46	12.64
5 Roofing	25 year asphalt shingles; #15 felt building paper; aluminum gutters, downspouts, drip edge and flashings.	.024	.47	.90	1.37
6 Interiors	Walls and ceilings, 1/2" taped and finished drywall, primed and painted with 2 coats; painted baseboard and trim, finished hardwood floor 40%, carpet with 1/2" underlayment 40%, vinyl tile with 1/2" underlayment 15%, ceramic tile with 1/2" underlayment 5%; hollow core and louvered interior doors.	.232	10.87	10.92	21.79
7 Specialties	Average grade kitchen cabinets - 14 L.F. wall and base with plastic laminate counter top and kitchen sink; 40 gallon electric water heater.	.021	1.22	.54	1.76
8 Mechanical	1 lavatory, white, wall hung; 1 water closet, white; 1 bathtub with shower; enameled steel, white; gas fired warm air heat.	.060	2.15	2.13	4.28
9 Electrical	200 Amp. service; romex wiring; incandescent lighting fixtures, switches, receptacles.	.039	.75	1.24	1.99
10 Overhead	Contractor's overhead and profit and plans.		5.28	5.14	10.42
		Total	36.45	35.50	**71.95**

3 4 5 6 7 8 9

1 *Specifications*

The parameters for an example dwelling from the previous pages are listed here. Included are the square foot dimensions of the proposed building. Living Area takes into account the number of floors and other factors needed to define a building's total square footage. Perimeter dimensions are defined in terms of linear feet.

2 *Building Type*

This is a sketch of a cross section view through the dwelling. It is shown to help define the living area for the building type. For more information, see the Building Types pages in the Introduction.

3 *Components (4 Exterior Walls)*

This page contains the ten components needed to develop the complete square foot cost of the typical dwelling specified. All components are defined with a description of the materials and/or task involved. Use cost figures from each component to estimate the cost per square foot of that section of the project. The components listed on this page are typical of all sizes of residences from the facing page. Specific quantities of components required would vary with the size of the dwelling and the exterior wall system.

4 *Labor-Hours (.111)*

Use this column to determine the number of labor-hours needed to perform a task. This figure will give the builder labor-hours per square foot of building. The total labor-hours per component is determined by multiplying the living area times the labor-hours listed on that line.

5 *Materials (8.18)*

This column gives the amount needed to develop the cost of materials. The figures given here are not bare costs. Ten percent has been added to bare material cost for profit.

6 *Installation (4.46)*

Installation includes labor and equipment costs. The labor rates included here incorporate the total overhead and profit costs for the installing contractor. The average mark-up used to create these figures is 72.0% over and above bare labor costs. The equipment rates include 10% for profit.

7 *Line Totals (12.64)*

The extreme right-hand column lists the sum of two figures. Use this total to determine the sum of material cost plus installation cost. The result is a convenient total cost for each of the ten components.

8 *Overhead*

The costs in components 1 through 9 include overhead and profit for the installing contractor. Item 10 is overhead and profit for the general contractor. This is typically a percentage mark-up of all other costs. The amount depends on size and type of dwelling, building class and economic conditions. An allowance for plans or design has been included where appropriate.

9 *Bottom Line Total (71.95)*

This figure is the complete square foot cost for the construction project and equals the sum of total material and total labor costs. To determine total project cost, multiply the bottom line total times the living area.

Building Classes

Economy Class

An economy class residence is usually built from stock plans. The materials and workmanship are sufficient to satisfy building codes. Low construction cost is more important than distinctive features. The overall shape of the foundation and structure is seldom other than square or rectangular.

An unfinished basement includes 7′ high 8″ thick foundation wall composed of either concrete block or cast-in-place concrete.

Included in the finished basement cost are inexpensive paneling or drywall as the interior finish on the foundation walls, a low cost sponge backed carpeting adhered to the concrete floor, a drywall ceiling, and overhead lighting.

Custom Class

A custom class residence is usually built from plans and specifications with enough features to give the building a distinction of design. Materials and workmanship are generally above average with obvious attention given to construction details. Construction normally exceeds building code requirements.

An unfinished basement includes a 7′-6″ high 10″ thick cast-in-place concrete foundation wall or a 7′-6″ high 12″ thick concrete block foundation wall.

A finished basement includes painted drywall on insulated 2″ x 4″ wood furring as the interior finish to the concrete walls, a suspended ceiling, carpeting adhered to the concrete floor, overhead lighting and heating.

Average Class

An average class residence is a simple design and built from standard plans. Materials and workmanship are average, but often exceed minimum building codes. There are frequently special features that give the residence some distinctive characteristics.

An unfinished basement includes 7′-6″ high 8″ thick foundation wall composed of either cast-in-place concrete or concrete block.

Included in the finished basement are plywood paneling or drywall on furring that is fastened to the foundation walls, sponge backed carpeting adhered to the concrete floor, a suspended ceiling, overhead lighting and heating.

Luxury Class

A luxury class residence is built from an architect's plan for a specific owner. It is unique both in design and workmanship. There are many special features, and construction usually exceeds all building codes. It is obvious that primary attention is placed on the owner's comfort and pleasure. Construction is supervised by an architect.

An unfinished basement includes 8′ high 12″ thick foundation wall that is composed of cast-in-place concrete or concrete block.

A finished basement includes painted drywall on 2″ x 4″ wood furring as the interior finish, suspended ceiling, tackless carpet on wood subfloor with sleepers, overhead lighting and heating.

Configurations

Detached House

This category of residence is a free-standing separate building with or without an attached garage. It has four complete walls.

Semi-Detached House

This category of residence has two living units side-by-side. The common wall is a fireproof wall. Semi-detached residences can be treated as a row house with two end units. Semi-detached residences can be any of the building types.

Town/Row House

This category of residence has a number of attached units made up of inner units and end units. The units are joined by common walls. The inner units have only two exterior walls. The common walls are fireproof. The end units have three walls and a common wall. Town houses/row houses can be any of the building types.

Building Types

One Story

This is an example of a one-story dwelling. The living area of this type of residence is confined to the ground floor. The headroom in the attic is usually too low for use as a living area.

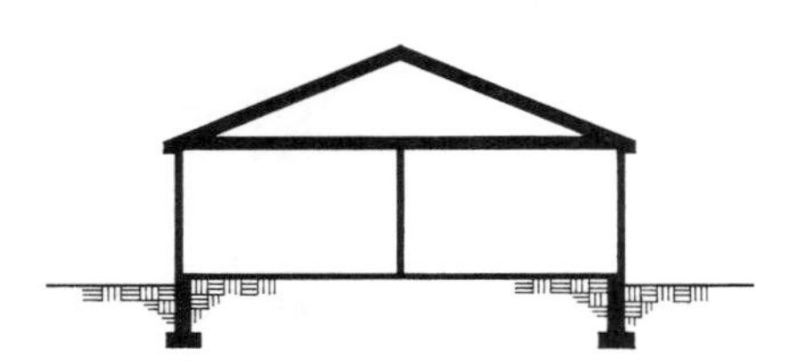

One-and-a-half Story

The living area in the upper level of this type of residence is 50% to 90% of the ground floor. This is made possible by a combination of this design's high-peaked roof and/or dormers. Only the upper level area with a ceiling height of 6 feet or more is considered living area. The living area of this residence is the sum of the ground floor area plus the area on the second level with a ceiling height of 6 feet or more.

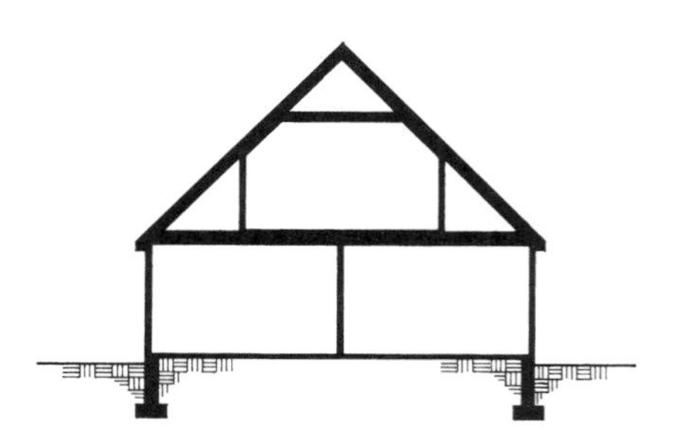

One Story with Finished Attic

The main living area in this type of residence is the ground floor. The upper level or attic area has sufficient headroom for comfortable use as a living area. This is made possible by a high peaked roof. The living area in the attic is less than 50% of the ground floor. The living area of this type of residence is the ground floor area only. The finished attic is considered an adjustment.

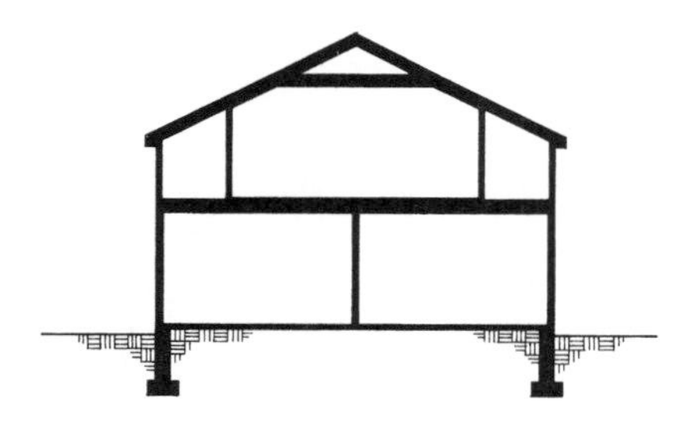

Two Story

This type of residence has a second floor or upper level area which is equal or nearly equal to the ground floor area. The upper level of this type of residence can range from 90% to 110% of the ground floor area, depending on setbacks or overhangs. The living area is the sum of the ground floor area and the upper level floor area.

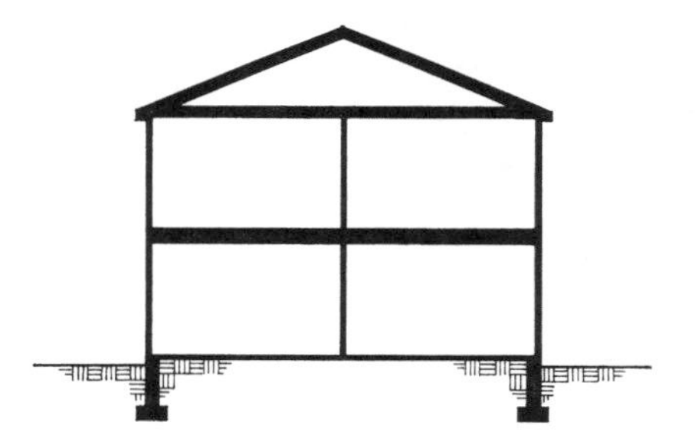

Two-and-one-half Story

This type of residence has two levels of equal or nearly equal area and a third level which has a living area that is 50% to 90% of the ground floor. This is made possible by a high peaked roof, extended wall heights and/or dormers. Only the upper level area with a ceiling height of 6 feet or more is considered living area. The living area of this residence is the sum of the ground floor area, the second floor area and the area on the third level with a ceiling height of 6 feet or more.

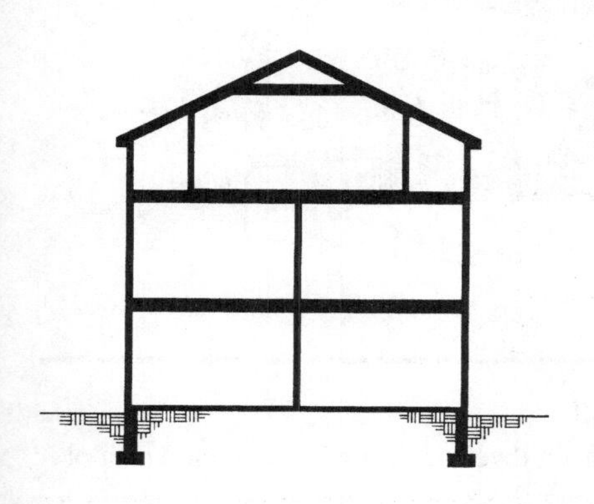

Bi-level

This type of residence has two living areas, one above the other. One area is about 4 feet below grade and the second is about 4 feet above grade. Both areas are equal in size. The lower level in this type of residence is originally designed and built to serve as a living area and not as a basement. Both levels have full ceiling heights. The living area is the sum of the lower level area and the upper level area.

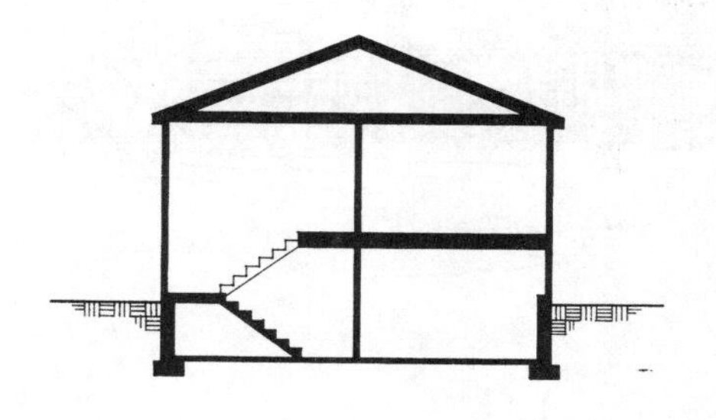

Three Story

This type of residence has three levels which are equal or nearly equal. As in the 2 story residence, the second and third floor areas may vary slightly depending on setbacks or overhangs. The living area is the sum of the ground floor area and the two upper level floor areas.

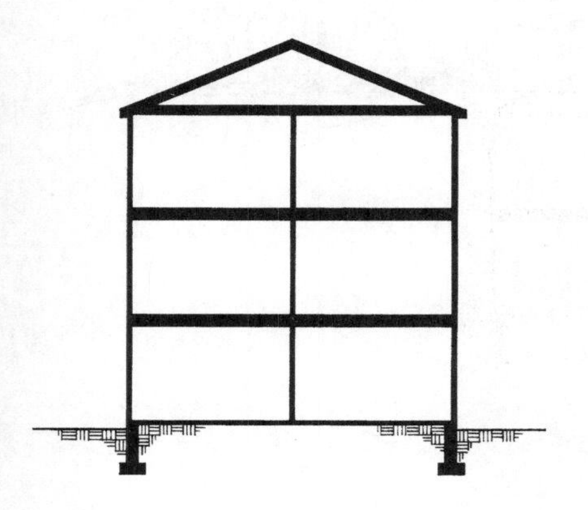

Tri-level

This type of residence has three levels of living area, one at grade level, one about 4 feet below grade and one about 4 feet above grade. All levels are originally designed to serve as living areas. All levels have full ceiling heights. The living area is a sum of the areas of each of the three levels.

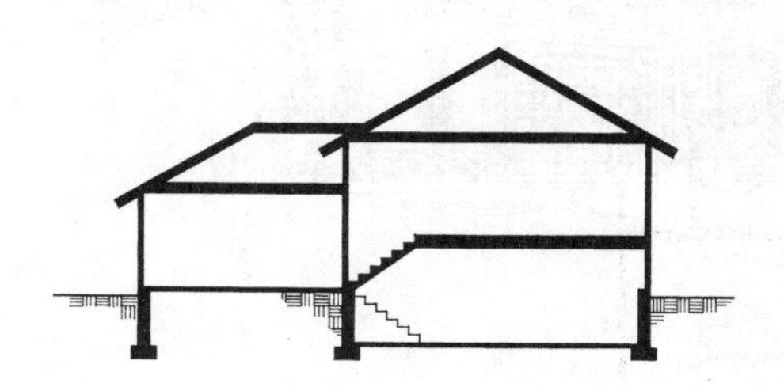

Garage Types

Attached Garage

Shares a common wall with the dwelling. Access is typically through a door between dwelling and garage.

Basement Garage

Constructed under the roof of the dwelling but below the living area.

Built-In Garage

Constructed under the second floor living space and above basement level of dwelling. Reduces gross square feet of living area.

Detached Garage

Constructed apart from the main dwelling. Shares no common area or wall with the dwelling.

Building Components

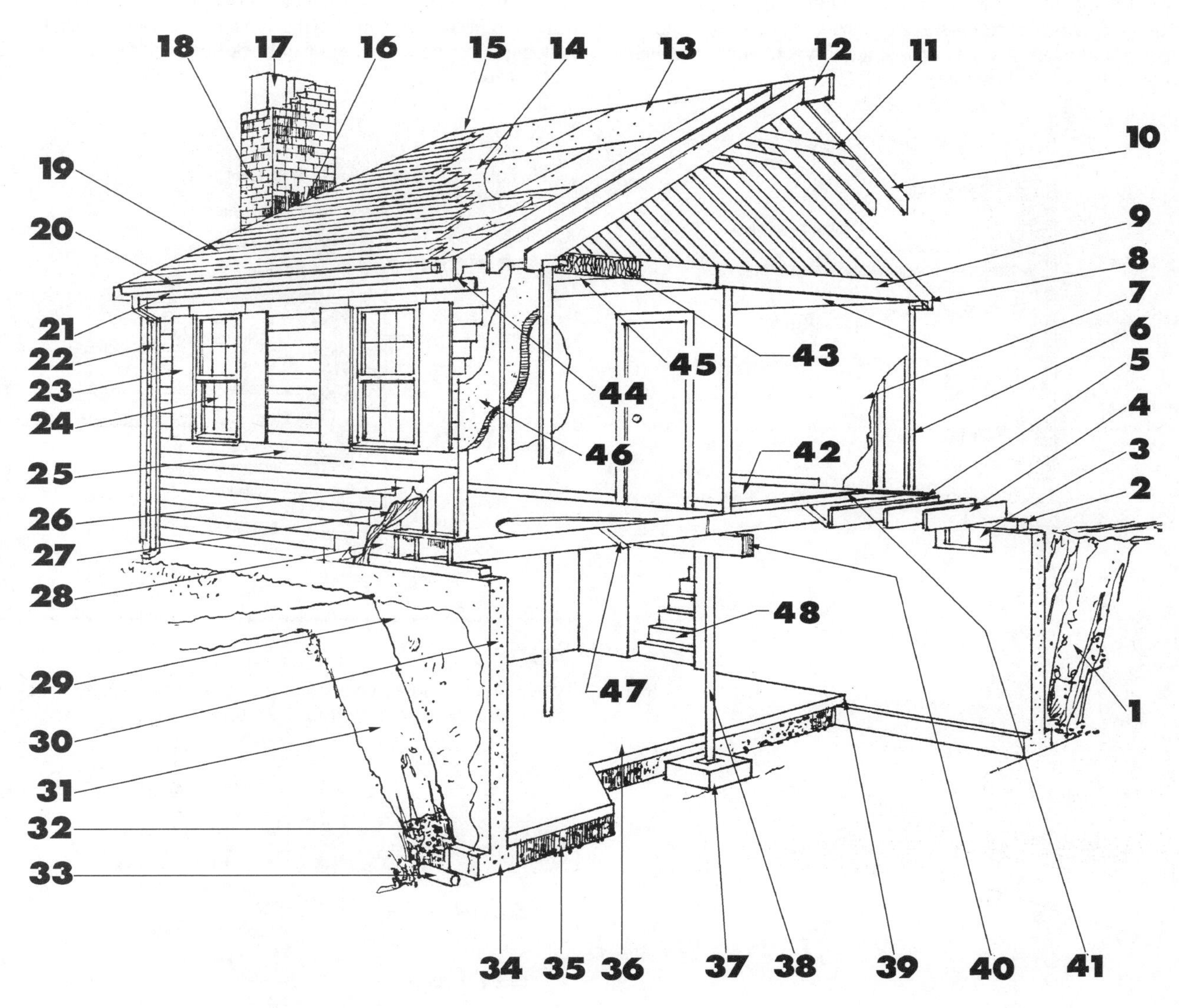

1. Excavation
2. Sill Plate
3. Basement Window
4. Floor Joist
5. Shoe Plate
6. Studs
7. Drywall
8. Plate
9. Ceiling Joists
10. Rafters
11. Collar Ties
12. Ridge Rafter
13. Roof Sheathing
14. Roof Felt
15. Roof Shingles
16. Flashing
17. Flue Lining
18. Chimney
19. Roof Shingles
20. Gutter
21. Fascia
22. Downspout
23. Shutter
24. Window
25. Wall Shingles
26. Building Paper
27. Wall Sheathing
28. Fire Stop
29. Dampproofing
30. Foundation Wall
31. Backfill
32. Drainage Stone
33. Drainage Tile
34. Wall Footing
35. Gravel
36. Concrete Slab
37. Column Footing
38. Pipe Column
39. Expansion Joint
40. Girder
41. Sub-floor
42. Finish Floor
43. Attic Insulation
44. Soffit
45. Ceiling Strapping
46. Wall Insulation
47. Cross Bridging
48. Bulkhead Stairs

Exterior Wall Construction

Typical Frame Construction

Typical wood frame construction consists of wood studs with insulation between them. A typical exterior surface is made up of sheathing, building paper and exterior siding consisting of wood, vinyl, aluminum or stucco over the wood sheathing.

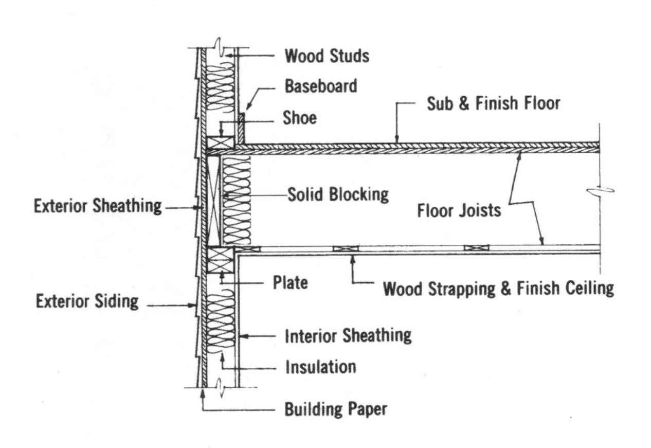

Brick Veneer

Typical brick veneer construction consists of wood studs with insulation between them. A typical exterior surface is sheathing, building paper and an exterior of brick tied to the sheathing, with metal strips.

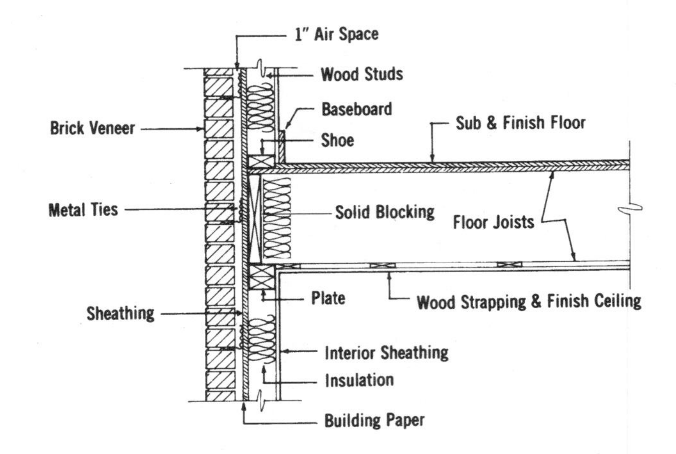

Stone

Typical solid masonry construction consists of a stone or block wall covered on the exterior with brick, stone or other masonry.

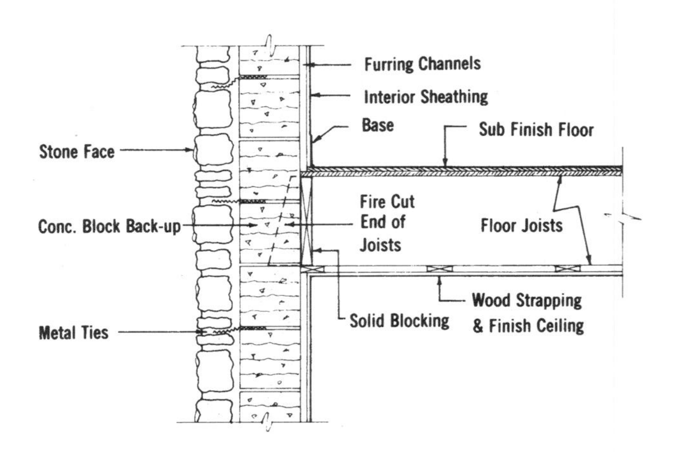

Residential Cost Estimate Worksheets

Worksheet Instructions

The residential cost estimate worksheet can be used as an outline for developing a residential construction or replacement cost. It is also useful for insurance appraisals. The design of the worksheet helps eliminate errors and omissions. To use the worksheet, follow the example below.

1. Fill out the owner's name, residence address, the estimator or appraiser's name, some type of project identifying number or code, and the date.
2. Determine from the plans, specifications, owner's description, photographs or any other means possible the class of construction. The models in this book use economy, average, custom and luxury as classes. Fill in the appropriate box.
3. Fill in the appropriate box for the residence type, configuration, occupancy, and exterior wall. If you require clarification, the pages preceding this worksheet describe each of these.
4. Next, the living area of the residence must be established. The heated or air conditioned space of the residence, not including the basement, should be measured. It is easiest to break the structure up into separate components as shown in the example. The main house (A), a one-and-one-half story wing (B), and a one story wing (C). The breezeway (D), garage (E) and open covered porch (F) will be treated differently. Data entry blocks for the living area are included on the worksheet for your use. Keep each level of each component separate, and fill out the blocks as shown.
5. By using the information on the worksheet, find the model, wing or ell in the following square foot cost pages that best matches the class, type, exterior finish and size of the residence being estimated. Use the *modifications, adjustments, and alternatives* to determine the adjusted cost per square foot of living area for each component.
6. For each component, multiply the cost per square foot by the living area square footage. If the residence is a townhouse/rowhouse, a multiplier should be applied based upon the configuration.
7. The second page of the residential cost estimate worksheet has space for the additional components of a house. The cost for additional bathrooms, finished attic space, breezeways, porches, fireplaces, appliance or cabinet upgrades, and garages should be added on this page. The information for each of these components is found with the model being used, or in the *modifications, adjustments, and alternatives* pages.
8. Add the total from page one of the estimate worksheet and the items listed on page two. The sum is the adjusted total building cost.
9. Depending on the use of the final estimated cost, one of the remaining two boxes should be filled out. Any additional items or exclusions should be added or subtracted at this time. The data contained in this book is a national average. Construction costs are different throughout the country. To allow for this difference, a location factor based upon the first three digits of the residence's zip code must be applied. The location factor is a multiplier that increases or decreases the adjusted total building cost. Find the appropriate location factor and calculate the local cost. If depreciation is a concern, a dollar figure should be subtracted at this point.
10. No residence will match a model exactly. Many differences will be found. At this level of estimating, a variation of plus or minus 10% should be expected.

Adjustments Instructions

No residence matches a model exactly in shape, material, or specifications.

The common differences are:

1. Two or more exterior wall systems
   a. Partial basement
   b. Partly finished basement
2. Specifications or features that are between two classes
3. Crawl space instead of a basement

EXAMPLES

Below are quick examples. See pages 17-19 for complete examples of cost adjustments for these differences:

1. Residence "A" is an average one-story structure with 1,600 S.F. of living area and no basement. Three walls are wood siding on wood frame, and the fourth wall is brick veneer on wood frame. The brick veneer wall is 35% of the exterior wall area.

   Use page 38 to calculate the Base Cost per S.F. of Living Area.

   Wood Siding for 1,600 S.F. = $73.15 per S.F.
   Brick Veneer for 1,600 S.F. = $86.45 per S.F.

   .65 ($73.15) + .35 ($86.15) = $77.69 per S.F. of Living Area.

2. a. Residence "B" is the same as Residence "A"; However, it has an unfinished basement under 50% of the building. To adjust the $77.69 per S.F. of living area for this partial basement, use page 38.

   $77.69 + .50 ($7.35) = $81.37 per S.F. of Living Area.

   b. Residence "C" is the same as Residence "A"; However, it has a full basement under the entire building. 640 S.F. or 40% of the basement area is finished.

   Using Page 38:

   $77.69 + .40 ($21.05) + .60 ($7.35) = $90.52 per S.F. of Living Area.

3. When specifications or features of a building are between classes, estimate the percent deviation, and use two tables to calculate the cost per S.F.

   A two-story residence with wood siding and 1,800 S.F. of living area has features 30% better than Average, but 70% less than Custom.

   From pages 42 and 64:

   Custom 1,800 S.F. Base Cost = $93.05 per S.F.
   Average 1,800 S.F. Base Cost = $75.10 per S.F.
   DIFFERENCE = $17.95 per S.F.

   Cost is $75.10 + .30 ($17.95) = $80.49 per S.F. of Living Area.

4. To add the cost of a crawl space, use the cost of an unfinished basement as a maximum. For specific costs of components to be added or deducted, such as vapor barrier, underdrain, and floor, see the "Assemblies" section, pages 101 to 281.

Model Residence Example

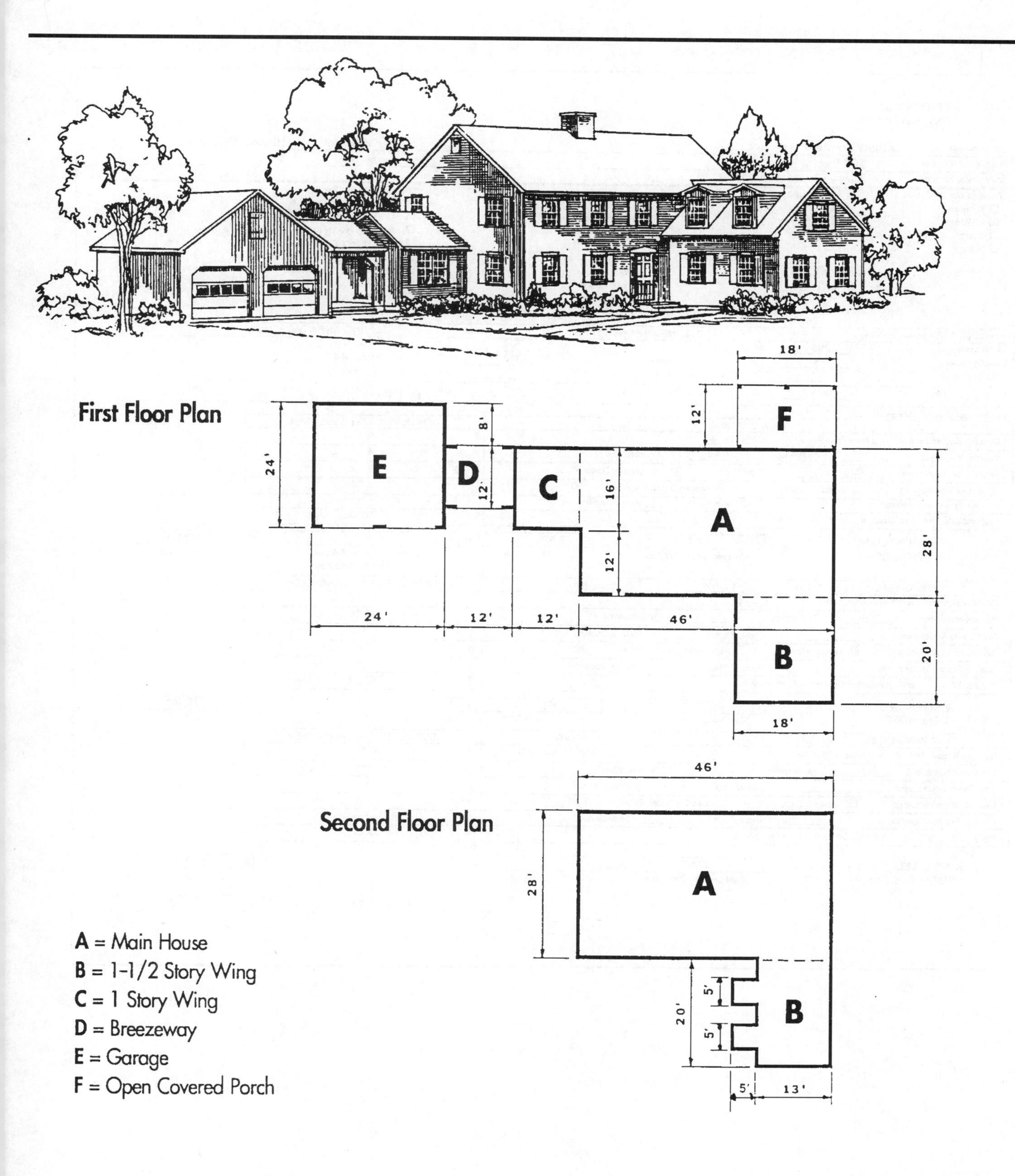

RESIDENTIAL COST ESTIMATE

OWNERS NAME:	Albert Westenberg	APPRAISER:	Nicole Wojtowicz
RESIDENCE ADDRESS:	300 Sygiel Road	PROJECT:	# 55
CITY, STATE, ZIP CODE:	Three Rivers, MA 01080	DATE:	Jan. 2, 2004

CLASS OF CONSTRUCTION	RESIDENCE TYPE	CONFIGURATION	EXTERIOR WALL SYSTEM
☐ ECONOMY	☐ 1 STORY	☑ DETACHED	☑ WOOD SIDING - WOOD FRAME
☑ AVERAGE	☐ 11/2 STORY	☐ TOWN/ROW HOUSE	☐ BRICK VENEER - WOOD FRAME
☐ CUSTOM	☑ 2 STORY	☐ SEMI-DETACHED	☐ STUCCO ON WOOD FRAME
☐ LUXURY	☐ 2 1/2 STORY		☐ PAINTED CONCRETE BLOCK
	☐ 3 STORY	OCCUPANCY	☐ SOLID MASONRY (AVERAGE & CUSTOM)
	☐ BI-LEVEL	☑ ONE FAMILY	☐ STONE VENEER - WOOD FRAME
	☐ TRI-LEVEL	☐ TWO FAMILY	☐ SOLID BRICK (LUXURY)
		☐ THREE FAMILY	☐ SOLID STONE (LUXURY)
		☐ OTHER	

* LIVING AREA (Main Building)		* LIVING AREA (Wing or Ell) (B)		* LIVING AREA (Wing or Ell) (C)	
First Level	1288 S.F.	First Level	360 S.F.	First Level	192 S.F.
Second level	1288 S.F.	Second level	310 S.F.	Second level	S.F.
Third Level	S.F.	Third Level	S.F.	Third Level	S.F.
Total	2576 S.F.	Total	670 S.F.	Total	192 S.F.

* Basement Area is not part of living area.

MAIN BUILDING		COSTS PER S.F. LIVING AREA
Cost per Square Foot of Living Area, from Page	30	$ 70.05
Basement Addition: ____ % Finished,	100 % Unfinished	+ 4.35
Roof Cover Adjustment: Cedar Shake	Type, Page 30 (Add or Deduct)	( ) 1.16
Central Air Conditioning: ☐ Separate Ducts	☑ Heating Ducts, Page 30	+ 1.67
Heating System Adjustment: ____	Type, Page ____ (Add or Deduct)	( ) ____
Main Building: Adjusted Cost per S.F. of Living Area		$ 77.23

MAIN BUILDING TOTAL COST: $ 77.23 /S.F. (Cost per S.F. Living Area) x 2,576 S.F. (Living Area) x —— (Town/Row House Multiplier (Use 1 for Detached)) = $ 198,944 TOTAL COST

WING OR ELL (B) 1 - 1/2 STORY		COSTS PER S.F. LIVING AREA
Cost per Square Foot of Living Area, from Page	37	$ 61.50
Basement Addition: 100 % Finished,	____ % Unfinished	+ 19.05
Roof Cover Adjustment: ____	Type, Page ____ (Add or Deduct)	( ) —
Central Air Conditioning: ☐ Separate Ducts	☑ Heating Ducts, Page 30	+ 1.67
Heating System Adjustment: ____	Type, Page ____ (Add or Deduct)	( ) —
Wing or Ell (B): Adjusted Cost per S.F. of Living Area		$ 82.22

WING OR ELL (B) TOTAL COST: $ 82.22 /S.F. (Cost per S.F. Living Area) x 670 S.F. (Living Area) = $ 55,087 TOTAL COST

WING OR ELL (C) 1 STORY		COSTS PER S.F. LIVING AREA
Cost per Square Foot of Living Area, from Page	30 (WOOD SIDING)	$ 93.10
Basement Addition: ____ % Finished,	____ % Unfinished	+ —
Roof Cover Adjustment: ____	Type, Page ____ (Add or Deduct)	( ) —
Central Air Conditioning: ☐ Separate Ducts	☐ Heating Ducts, Page ____	+ —
Heating System Adjustment: ____	Type, Page ____ (Add or Deduct)	( ) —
Wing or Ell (C) Adjusted Cost per S.F. of Living Area		$ 93.10

WING OR ELL (C) TOTAL COST: $ 93.10 /S.F. (Cost per S.F. Living Area) x 192 S.F. (Living Area) = $ 17,875 TOTAL COST

TOTAL THIS PAGE: 271,906

Page 1 of 2

RESIDENTIAL
COST ESTIMATE

	QUANTITY		UNIT COST		
Total Page 1				$	271,906
Additional Bathrooms: 2 Full, 1 Half 2 @ 4060 1 @ 2583				+	10,703
Finished Attic: N/A Ft. x ____ Ft.		S.F.			
Breezeway: ☑ Open ☐ closed 12 Ft. x 12 Ft.	144	S.F.	17.00	+	2,448
Covered Porch: ☑ Open ☐ Enclosed 18 Ft. x 12 Ft.	216	S.F.	25.25	+	5,454
Fireplace: ☑ Interior Chimney ☐ Exterior Chimney ☑ No. of Flues (2) ☑ Additional Fireplaces 1 - 2nd Story				+	7,360
Appliances:				+	—
Kitchen Cabinets Adjustments: (±)					—
☑ Garage ☐ Carport: 2 Car(s) Description Wood, Attached (±)					15,194
Miscellaneous:				+	

ADJUSTED TOTAL BUILDING COST $ 313,065

REPLACEMENT COST		
ADJUSTED TOTAL BUILDING COST	$	313,065
Site Improvements		
(A) Paving & Sidewalks	$	
(B) Landscaping	$	
(C) Fences	$	
(D) Swimming Pools	$	
(E) Miscellaneous	$	
TOTAL	$	313,065
Location Factor	x	1.028
Location Replacement Cost	$	321,831
Depreciation	-$	32,183
LOCAL DEPRECIATED COST	$	289,648

INSURANCE COST		
ADJUSTED TOTAL BUILDING COST	$	
Insurance Exclusions		
(A) Footings, sitework, Underground Piping	-$	
(B) Architects Fees	-$	
Total Building Cost Less Exclusion	$	
Location Factor	x	
LOCAL INSURABLE REPLACEMENT COST	$	

SKETCH AND ADDITIONAL CALCULATIONS

Page 2 of 2

Economy Class

1 Story

1-1/2 Story

2 Story

Bi-Level

Tri-Level

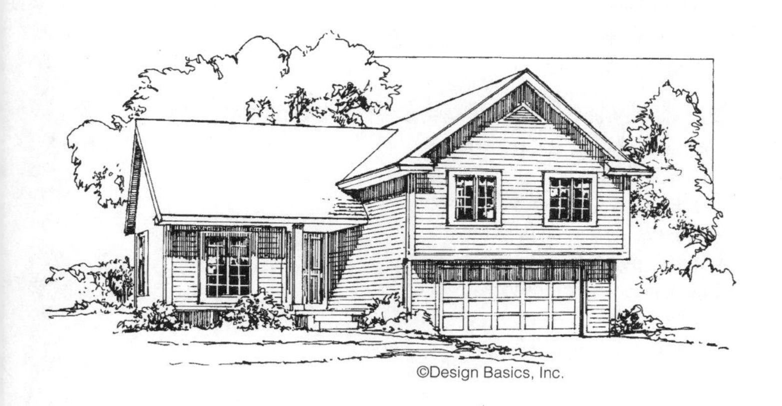

- **Mass produced from stock plans**
- **Single family — 1 full bath, 1 kitchen**
- **No basement**
- **Asphalt shingles on roof**
- **Hot air heat**
- **Drywall interior finishes**
- **Materials and workmanship are sufficient to meet codes**

Note: The illustration shown may contain some optional components (for example: garages and/or fireplaces) whose costs are shown in the modifications, adjustments, & alternatives below or at the end of the square foot section.

Base cost per square foot of living area

Exterior Wall	Living Area 600	800	1000	1200	1400	1600	1800	2000	2400	2800	3200
Wood Siding - Wood Frame	82.60	75.10	69.35	64.55	60.35	57.75	56.50	54.80	51.05	48.40	46.65
Brick Veneer - Wood Frame	89.15	80.90	74.65	69.30	64.75	61.85	60.40	58.50	54.45	51.50	49.50
Stucco on Wood Frame	79.95	72.65	67.15	62.60	58.60	56.05	54.90	53.25	49.70	47.10	45.45
Painted Concrete Block	83.70	76.10	70.25	65.35	61.10	58.40	57.15	55.40	51.65	48.90	47.10
Finished Basement, Add	20.05	19.10	18.40	17.75	17.20	16.85	16.60	16.30	15.90	15.55	15.25
Unfinished Basement, Add	8.65	7.90	7.35	6.90	6.45	6.15	6.05	5.80	5.45	5.20	4.95

Modifications

Add to the total cost

Upgrade Kitchen Cabinets	$ + 360
Solid Surface Countertops	+ 679
Full Bath - including plumbing, wall and floor finishes	+ 3235
Half Bath - including plumbing, wall and floor finishes	+ 2058
One Car Attached Garage	+ 8261
One Car Detached Garage	+ 10,644
Fireplace & Chimney	+ 3925

Adjustments

For multi family - add to total cost

Additional Kitchen	$ + 2142
Additional Bath	+ 3235
Additional Entry & Exit	+ 1236
Separate Heating	+ 1237
Separate Electric	+ 800

For Townhouse/Rowhouse - Multiply cost per square foot by

Inner Unit	.95
End Unit	.97

Alternatives

Add to or deduct from the cost per square foot of living area

Composition Roll Roofing	– .62
Cedar Shake Roof	+ 2.61
Upgrade Walls and Ceilings to Skim Coat Plaster	+ .58
Upgrade Ceilings to Textured Finish	+ .39
Air Conditioning, in Heating Ductwork	+ 2.72
In Separate Ductwork	+ 5.26
Heating Systems, Hot Water	+ 1.45
Heat Pump	+ 1.77
Electric Heat	– 1.25
Not Heated	– 3.16

Additional upgrades or components

Kitchen Cabinets & Countertops	Page 93
Bathroom Vanities	94
Fireplaces & Chimneys	94
Windows, Skylights & Dormers	94
Appliances	95
Breezeways & Porches	95
Finished Attic	95
Garages	96
Site Improvements	96
Wings & Ells	34

Important: See the Reference Section for Location Factors (to adjust for your city) and Estimating Forms

Economy 1 Story

Living Area - 1200 S.F.
Perimeter - 146 L.F.

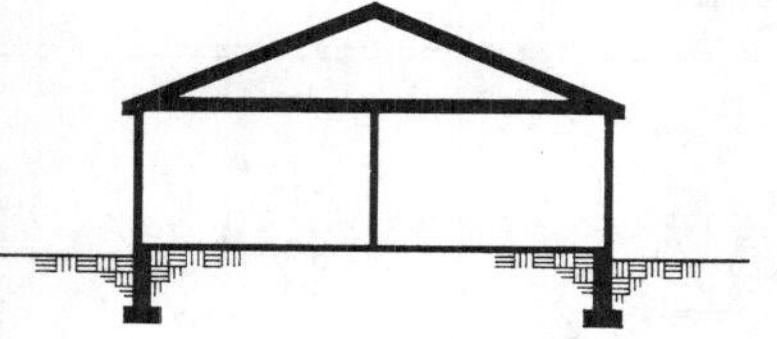

		Labor-Hours	Cost Per Square Foot Of Living Area		
			Mat.	Labor	Total
1 Site Work	Site preparation for slab; 4' deep trench excavation for foundation wall.	.060		.85	.85
2 Foundation	Continuous reinforced concrete footing, 8" deep x 18" wide; dampproofed and insulated 8" thick reinforced concrete block foundation wall, 4' deep; 4" concrete slab on 4" crushed stone base and polyethylene vapor barrier, trowel finish.	.131	3.85	5.38	9.23
3 Framing	Exterior walls - 2" x 4" wood studs, 16" O.C.; 1/2" insulation board sheathing; wood truss roof frame, 24" O.C. with 1/2" plywood sheathing, 4 in 12 pitch.	.098	3.60	4.62	8.22
4 Exterior Walls	Metal lath reinforced stucco exterior on insulated wood frame walls; 6" attic insulation; sliding sash wood windows; 2 flush solid core wood exterior doors with storms.	.110	4.58	4.80	9.38
5 Roofing	20 year asphalt shingles; #15 felt building paper; aluminum gutters, downspouts, drip edge and flashings.	.047	.91	1.75	2.66
6 Interiors	Walls and ceilings, 1/2" taped and finished drywall, primed and painted with 2 coats; painted baseboard and trim; rubber backed carpeting 80%, asphalt tile 20%; hollow core wood interior doors.	.243	6.27	9.03	15.30
7 Specialties	Economy grade kitchen cabinets - 6 L.F. wall and base with plastic laminate counter top and kitchen sink; 30 gallon electric water heater.	.004	1.23	.68	1.91
8 Mechanical	1 lavatory, white, wall hung; 1 water closet, white; 1 bathtub, enameled steel, white; gas fired warm air heat.	.086	2.72	2.49	5.21
9 Electrical	100 Amp. service; romex wiring; incandescent lighting fixtures, switches, receptacles.	.036	.57	1.10	1.67
10 Overhead	Contractor's overhead and profit.		3.57	4.60	8.17
		Total	27.30	35.30	**62.60**

- **Mass produced from stock plans**
- **Single family — 1 full bath, 1 kitchen**
- **No basement**
- **Asphalt shingles on roof**
- **Hot air heat**
- **Drywall interior finishes**
- **Materials and workmanship are sufficient to meet codes**

Note: The illustration shown may contain some optional components (for example: garages and/or fireplaces) whose costs are shown in the modifications, adjustments, & alternatives below or at the end of the square foot section.

Base cost per square foot of living area

Exterior Wall	600	800	1000	1200	1400	1600	1800	2000	2400	2800	3200
	Living Area										
Wood Siding - Wood Frame	95.95	80.20	71.50	67.60	64.80	60.50	58.45	56.20	51.60	49.85	48.00
Brick Veneer - Wood Frame	105.10	86.80	77.60	73.35	70.25	65.45	63.15	60.65	55.55	53.60	51.50
Stucco on Wood Frame	92.25	77.45	69.00	65.25	62.55	58.45	56.45	54.35	50.00	48.25	46.60
Painted Concrete Block	97.50	81.30	72.55	68.60	65.70	61.25	59.20	56.90	52.30	50.50	48.60
Finished Basement, Add	15.15	12.95	12.50	12.20	11.95	11.50	11.30	11.05	10.60	10.45	10.20
Unfinished Basement, Add	7.40	5.75	5.40	5.15	4.95	4.60	4.45	4.30	3.95	3.80	3.60

Modifications

Add to the total cost

Upgrade Kitchen Cabinets	$ + 360
Solid Surface Countertops	+ 679
Full Bath - including plumbing, wall and floor finishes	+ 3235
Half Bath - including plumbing, wall and floor finishes	+ 2058
One Car Attached Garage	+ 8261
One Car Detached Garage	+ 10,644
Fireplace & Chimney	+ 3925

Adjustments

For multi family - add to total cost

Additional Kitchen	$ + 2142
Additional Bath	+ 3235
Additional Entry & Exit	+ 1236
Separate Heating	+ 1237
Separate Electric	+ 800

For Townhouse/Rowhouse - Multiply cost per square foot by

Inner Unit	.95
End Unit	.97

Alternatives

Add to or deduct from the cost per square foot of living area

Composition Roll Roofing	− .43
Cedar Shake Roof	+ 1.89
Upgrade Walls and Ceilings to Skim Coat Plaster	+ .70
Upgrade Ceilings to Textured Finish	+ .39
Air Conditioning, in Heating Ductwork	+ 2.03
In Separate Ductwork	+ 4.62
Heating Systems, Hot Water	+ 1.39
Heat Pump	+ 1.95
Electric Heat	− .99
Not Heated	− 2.91

Additional upgrades or components

Kitchen Cabinets & Countertops	Page 93
Bathroom Vanities	94
Fireplaces & Chimneys	94
Windows, Skylights & Dormers	94
Appliances	95
Breezeways & Porches	95
Finished Attic	95
Garages	96
Site Improvements	96
Wings & Ells	34

Economy 1-1/2 Story

Living Area - 1600 S.F.
Perimeter - 135 L.F.

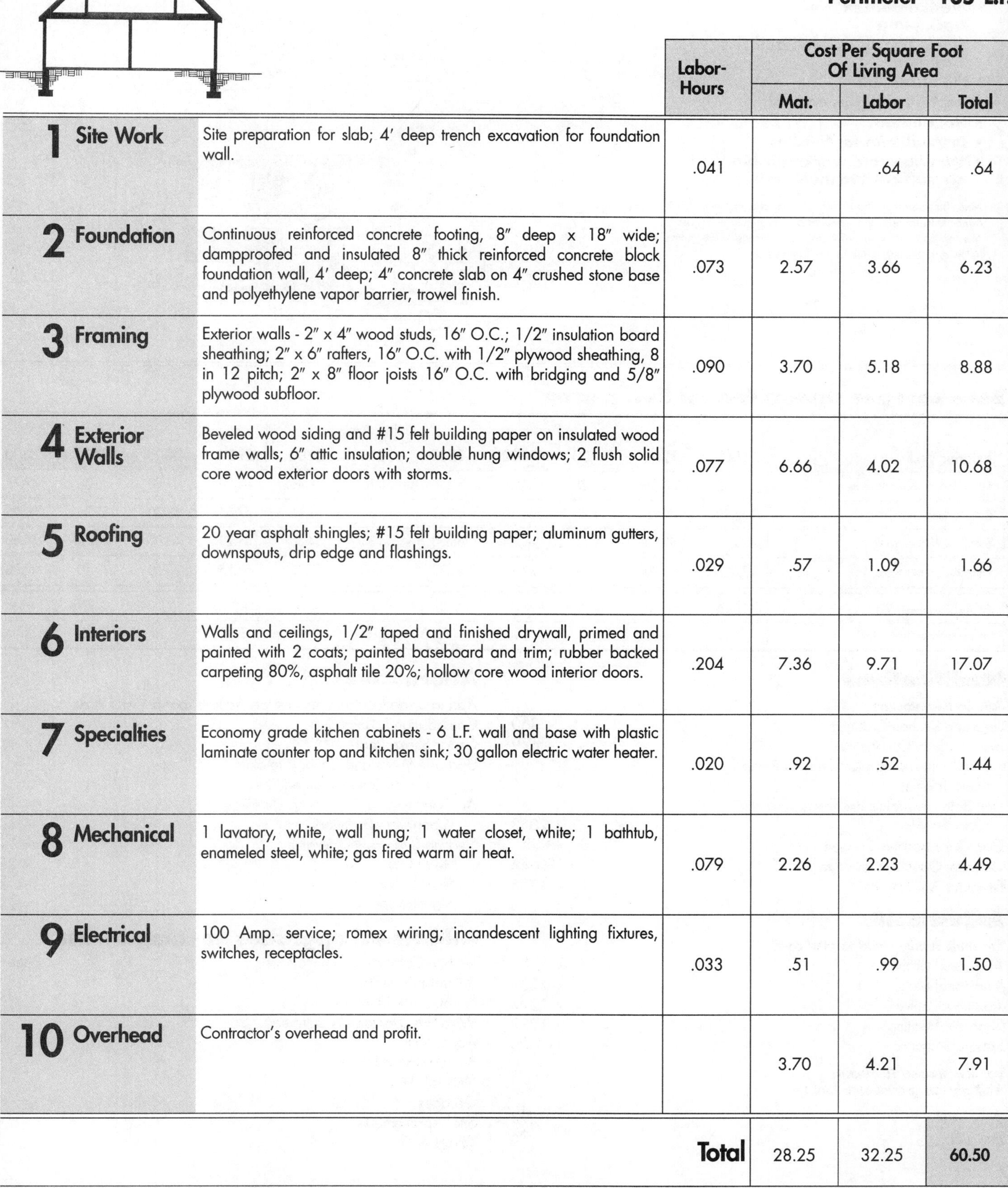

		Labor-Hours	Cost Per Square Foot Of Living Area		
			Mat.	Labor	Total
1 Site Work	Site preparation for slab; 4′ deep trench excavation for foundation wall.	.041		.64	.64
2 Foundation	Continuous reinforced concrete footing, 8″ deep x 18″ wide; dampproofed and insulated 8″ thick reinforced concrete block foundation wall, 4′ deep; 4″ concrete slab on 4″ crushed stone base and polyethylene vapor barrier, trowel finish.	.073	2.57	3.66	6.23
3 Framing	Exterior walls - 2″ x 4″ wood studs, 16″ O.C.; 1/2″ insulation board sheathing; 2″ x 6″ rafters, 16″ O.C. with 1/2″ plywood sheathing, 8 in 12 pitch; 2″ x 8″ floor joists 16″ O.C. with bridging and 5/8″ plywood subfloor.	.090	3.70	5.18	8.88
4 Exterior Walls	Beveled wood siding and #15 felt building paper on insulated wood frame walls; 6″ attic insulation; double hung windows; 2 flush solid core wood exterior doors with storms.	.077	6.66	4.02	10.68
5 Roofing	20 year asphalt shingles; #15 felt building paper; aluminum gutters, downspouts, drip edge and flashings.	.029	.57	1.09	1.66
6 Interiors	Walls and ceilings, 1/2″ taped and finished drywall, primed and painted with 2 coats; painted baseboard and trim; rubber backed carpeting 80%, asphalt tile 20%; hollow core wood interior doors.	.204	7.36	9.71	17.07
7 Specialties	Economy grade kitchen cabinets - 6 L.F. wall and base with plastic laminate counter top and kitchen sink; 30 gallon electric water heater.	.020	.92	.52	1.44
8 Mechanical	1 lavatory, white, wall hung; 1 water closet, white; 1 bathtub, enameled steel, white; gas fired warm air heat.	.079	2.26	2.23	4.49
9 Electrical	100 Amp. service; romex wiring; incandescent lighting fixtures, switches, receptacles.	.033	.51	.99	1.50
10 Overhead	Contractor's overhead and profit.		3.70	4.21	7.91
		Total	28.25	32.25	**60.50**

- **Mass produced from stock plans**
- **Single family — 1 full bath, 1 kitchen**
- **No basement**
- **Asphalt shingles on roof**
- **Hot air heat**
- **Drywall interior finishes**
- **Materials and workmanship are sufficient to meet codes**

Note: The illustration shown may contain some optional components (for example: garages and/or fireplaces) whose costs are shown in the modifications, adjustments, & alternatives below or at the end of the square foot section.

Base cost per square foot of living area

Exterior Wall	Living Area 1000	1200	1400	1600	1800	2000	2200	2600	3000	3400	3800
Wood Siding - Wood Frame	76.05	68.60	65.25	63.05	60.80	58.10	56.40	53.10	49.80	48.40	47.15
Brick Veneer - Wood Frame	83.10	75.15	71.40	68.90	66.30	63.35	61.40	57.65	54.00	52.35	50.95
Stucco on Wood Frame	73.15	65.90	62.75	60.65	58.50	55.85	54.30	51.20	48.05	46.75	45.60
Painted Concrete Block	77.25	69.70	66.30	64.00	61.70	59.00	57.20	53.85	50.50	49.10	47.80
Finished Basement, Add	10.40	10.10	9.80	9.60	9.40	9.25	9.10	8.75	8.55	8.40	8.25
Unfinished Basement, Add	4.60	4.35	4.15	4.00	3.85	3.75	3.60	3.35	3.15	3.05	2.95

Modifications

Add to the total cost

Upgrade Kitchen Cabinets	$ + 360
Solid Surface Countertops	+ 679
Full Bath - including plumbing, wall and floor finishes	+ 3235
Half Bath - including plumbing, wall and floor finishes	+ 2058
One Car Attached Garage	+ 8261
One Car Detached Garage	+ 10,644
Fireplace & Chimney	+ 4335

Adjustments

For multi family - add to total cost

Additional Kitchen	$ + 2142
Additional Bath	+ 3235
Additional Entry & Exit	+ 1236
Separate Heating	+ 1237
Separate Electric	+ 800

For Townhouse/Rowhouse - Multiply cost per square foot by

Inner Unit	.93
End Unit	.96

Alternatives

Add to or deduct from the cost per square foot of living area

Composition Roll Roofing	– .31
Cedar Shake Roof	+ 1.31
Upgrade Walls and Ceilings to Skim Coat Plaster	+ .66
Upgrade Ceilings to Textured Finish	+ .39
Air Conditioning, in Heating Ductwork	+ 1.64
In Separate Ductwork	+ 4.20
Heating Systems, Hot Water	+ 1.35
Heat Pump	+ 2.07
Electric Heat	– .82
Not Heated	– 2.75

Additional upgrades or components

Kitchen Cabinets & Countertops	Page 93
Bathroom Vanities	94
Fireplaces & Chimneys	94
Windows, Skylights & Dormers	94
Appliances	95
Breezeways & Porches	95
Finished Attic	95
Garages	96
Site Improvements	96
Wings & Ells	34

Economy 2 Story

Living Area - 2000 S.F.
Perimeter - 135 L.F.

		Labor-Hours	Cost Per Square Foot Of Living Area		
			Mat.	Labor	Total
1 Site Work	Site preparation for slab; 4′ deep trench excavation for foundation wall.	.034		.51	.51
2 Foundation	Continuous reinforced concrete footing, 8″ deep x 18″ wide; dampproofed and insulated 8″ thick reinforced concrete block foundation wall, 4′ deep; 4″ concrete slab on 4″ crushed stone base and polyethylene vapor barrier, trowel finish.	.069	2.06	2.93	4.99
3 Framing	Exterior walls - 2″ x 4″ wood studs, 16″ O.C.; 1/2″ insulation board sheathing; wood truss roof frame, 24″ O.C. with 1/2″ plywood sheathing, 4 in 12 pitch; 2″ x 8″ floor joists 16″ O.C. with bridging and 5/8″ plywood subfloor.	.112	3.79	5.43	9.22
4 Exterior Walls	Beveled wood siding and #15 felt building paper on insulated wood frame walls; 6″ attic insulation; double hung windows; 2 flush solid core wood exterior doors with storms.	.107	6.77	4.12	10.89
5 Roofing	20 year asphalt shingles; #15 felt building paper; aluminum gutters, downspouts, drip edge and flashings.	.024	.46	.88	1.34
6 Interiors	Walls and ceilings, 1/2″ taped and finished drywall, primed and painted with 2 coats; painted baseboard and trim; rubber backed carpeting 80%, asphalt tile 20%; hollow core wood interior doors.	.219	7.22	9.71	16.93
7 Specialties	Economy grade kitchen cabinets - 6 L.F. wall and base with plastic laminate counter top and kitchen sink; 30 gallon electric water heater.	.017	.74	.41	1.15
8 Mechanical	1 lavatory, white, wall hung; 1 water closet, white; 1 bathtub, enameled steel, white; gas fired warm air heat.	.061	1.99	2.08	4.07
9 Electrical	100 Amp. service; romex wiring; incandescent lighting fixtures; switches, receptacles.	.030	.48	.92	1.40
10 Overhead	Contractor's overhead and profit.		3.54	4.06	7.60
		Total	27.05	31.05	**58.10**

- **Mass produced from stock plans**
- **Single family — 1 full bath, 1 kitchen**
- **No basement**
- **Asphalt shingles on roof**
- **Hot air heat**
- **Drywall interior finishes**
- **Materials and workmanship are sufficient to meet codes**

Note: The illustration shown may contain some optional components (for example: garages and/or fireplaces) whose costs are shown in the modifications, adjustments, & alternatives below or at the end of the square foot section.

Base cost per square foot of living area

Exterior Wall	Living Area 1000	1200	1400	1600	1800	2000	2200	2600	3000	3400	3800
Wood Siding - Wood Frame	71.00	63.85	60.90	58.85	56.85	**54.25**	52.75	49.80	46.75	45.55	44.45
Brick Veneer - Wood Frame	76.25	68.80	65.45	63.25	61.00	58.20	56.55	53.20	49.90	48.50	47.30
Stucco on Wood Frame	68.75	61.90	59.00	57.05	55.15	52.60	51.20	48.40	45.45	44.30	43.25
Painted Concrete Block	71.90	64.70	61.65	59.55	57.55	54.90	53.40	50.40	47.25	46.00	44.85
Finished Basement, Add	10.40	10.10	9.80	9.60	9.40	9.25	9.10	8.75	8.55	8.40	8.25
Unfinished Basement, Add	4.60	4.35	4.15	4.00	3.85	3.75	3.60	3.35	3.15	3.05	2.95

Modifications

Add to the total cost

Upgrade Kitchen Cabinets	$ + 360
Solid Surface Countertops	+ 679
Full Bath - including plumbing, wall and floor finishes	+ 3235
Half Bath - including plumbing, wall and floor finishes	+ 2058
One Car Attached Garage	+ 8261
One Car Detached Garage	+ 10,644
Fireplace & Chimney	+ 3925

Adjustments

For multi family - add to total cost

Additional Kitchen	$ + 2142
Additional Bath	+ 3235
Additional Entry & Exit	+ 1236
Separate Heating	+ 1237
Separate Electric	+ 800

For Townhouse/Rowhouse - Multiply cost per square foot by

Inner Unit	.94
End Unit	.97

Alternatives

Add to or deduct from the cost per square foot of living area

Composition Roll Roofing	− .31
Cedar Shake Roof	+ 2.61
Upgrade Walls and Ceilings to Skim Coat Plaster	+ .62
Upgrade Ceilings to Textured Finish	+ .39
Air Conditioning, in Heating Ductwork	+ 1.64
In Separate Ductwork	+ 4.20
Heating Systems, Hot Water	+ 1.35
Heat Pump	+ 2.07
Electric Heat	− .82
Not Heated	− 2.75

Additional upgrades or components

Kitchen Cabinets & Countertops	Page 93
Bathroom Vanities	94
Fireplaces & Chimneys	94
Windows, Skylights & Dormers	94
Appliances	95
Breezeways & Porches	95
Finished Attic	95
Garages	96
Site Improvements	96
Wings & Ells	34

Important: See the Reference Section for Location Factors (to adjust for your city) and Estimating Forms

Economy Bi-Level

Living Area - 2000 S.F.
Perimeter - 135 L.F.

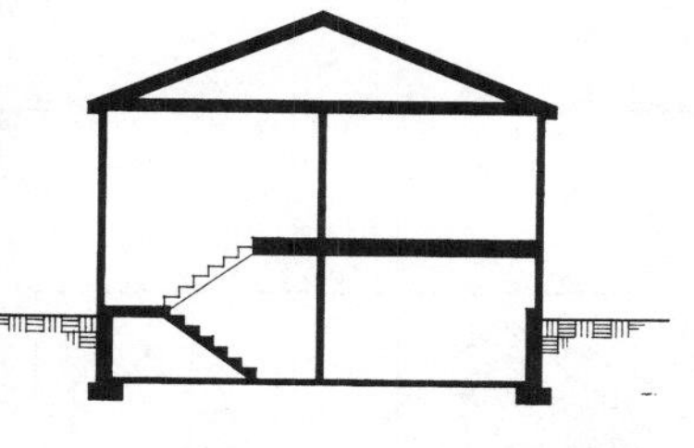

		Labor-Hours	Cost Per Square Foot Of Living Area		
			Mat.	Labor	Total
1 Site Work	Excavation for lower level, 4′ deep. Site preparation for slab.	.029		.51	.51
2 Foundation	Continuous reinforced concrete footing, 8″ deep x 18″ wide; dampproofed and insulated 8″ thick reinforced concrete block foundation wall, 4′ deep; 4″ concrete slab on 4″ crushed stone base and polyethylene vapor barrier, trowel finish.	.069	2.06	2.93	4.99
3 Framing	Exterior walls - 2″ x 4″ wood studs, 16″ O.C.; 1/2″ insulation board sheathing; wood truss roof frame, 24″ O.C. with 1/2″ plywood sheathing, 4 in 12 pitch; 2″ x 8″ floor joists 16″ O.C. with bridging and 5/8″ plywood subfloor.	.107	3.57	5.11	8.68
4 Exterior Walls	Beveled wood siding and #15 felt building paper on insulated wood frame walls; 6″ attic insulation; double hung windows; 2 flush solid core wood exterior doors with storms.	.089	5.34	3.22	8.56
5 Roofing	20 year asphalt shingles; #15 felt building paper; aluminum gutters, downspouts, drip edge and flashings.	.024	.46	.88	1.34
6 Interiors	Walls and ceilings, 1/2″ taped and finished drywall, primed and painted with 2 coats; painted baseboard and trim, rubber backed carpeting 80%, asphalt tile 20%; hollow core wood interior doors.	.213	7.07	9.40	16.47
7 Specialties	Economy grade kitchen cabinets - 6 L.F. wall and base with plastic laminate counter top and kitchen sink; 30 gallon electric water heater.	.018	.74	.41	1.15
8 Mechanical	1 lavatory, white, wall hung; 1 water closet, white; 1 bathtub, enameled steel, white; gas fired warm air heat.	.061	1.99	2.08	4.07
9 Electrical	100 Amp. service; romex wiring; incandescent lighting fixtures; switches, receptacles.	.030	.48	.92	1.40
10 Overhead	Contractor's overhead and profit.		3.24	3.84	7.08
	Total		24.95	29.30	**54.25**

- **Mass produced from stock plans**
- **Single family — 1 full bath, 1 kitchen**
- **No basement**
- **Asphalt shingles on roof**
- **Hot air heat**
- **Drywall interior finishes**
- **Materials and workmanship are sufficient to meet codes**

Note: The illustration shown may contain some optional components (for example: garages and/or fireplaces) whose costs are shown in the modifications, adjustments, & alternatives below or at the end of the square foot section.

Base cost per square foot of living area

Exterior Wall	Living Area 1200	1500	1800	2000	2200	2400	2800	3200	3600	4000	4400
Wood Siding - Wood Frame	65.40	60.05	56.10	54.50	52.10	50.35	48.85	46.85	44.40	43.70	41.85
Brick Veneer - Wood Frame	70.25	64.50	60.05	58.30	55.75	53.70	52.10	49.90	47.20	46.40	44.40
Stucco on Wood Frame	63.35	58.30	54.45	52.95	50.65	48.90	47.50	45.60	43.25	42.55	40.75
Solid Masonry	66.15	60.80	56.75	55.15	52.75	50.85	49.40	47.30	44.90	44.15	42.30
Finished Basement, Add*	12.75	12.25	11.80	11.65	11.45	11.20	11.05	10.85	10.60	10.50	10.30
Unfinished Basement, Add*	5.30	4.90	4.60	4.45	4.30	4.15	4.00	3.80	3.60	3.55	3.45

*Basement under middle level only.

Modifications

Add to the total cost

Upgrade Kitchen Cabinets	$ + 360
Solid Surface Countertops	+ 679
Full Bath - including plumbing, wall and floor finishes	+ 3235
Half Bath - including plumbing, wall and floor finishes	+ 2058
One Car Attached Garage	+ 8261
One Car Detached Garage	+ 10,644
Fireplace & Chimney	+ 3925

Adjustments

For multi family - add to total cost

Additional Kitchen	$ + 2142
Additional Bath	+ 3235
Additional Entry & Exit	+ 1236
Separate Heating	+ 1237
Separate Electric	+ 800

For Townhouse/Rowhouse - Multiply cost per square foot by

Inner Unit	.93
End Unit	.96

Alternatives

Add to or deduct from the cost per square foot of living area

Composition Roll Roofing	– .43
Cedar Shake Roof	+ 1.89
Upgrade Walls and Ceilings to Skim Coat Plaster	+ .55
Upgrade Ceilings to Textured Finish	+ .39
Air Conditioning, in Heating Ductwork	+ 1.36
In Separate Ductwork	+ 3.92
Heating Systems, Hot Water	+ 1.29
Heat Pump	+ 2.15
Electric Heat	– .73
Not Heated	– 2.66

Additional upgrades or components

Important: See the Reference Section for Location Factors (to adjust for your city) and Estimating Forms

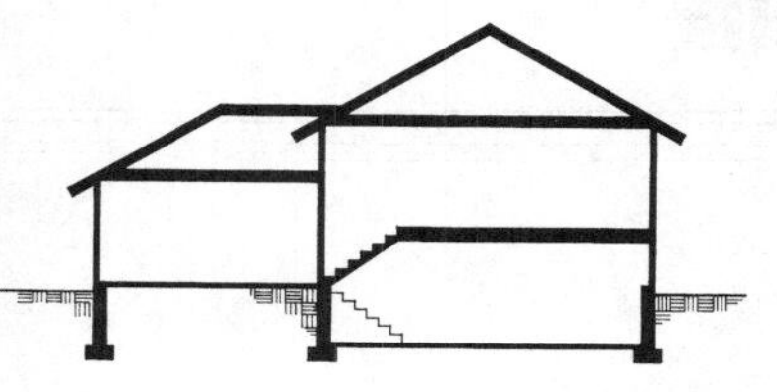

Economy Tri-Level

Living Area - 2400 S.F.
Perimeter - 163 L.F.

SQUARE FOOT COSTS

		Labor-Hours	Cost Per Square Foot Of Living Area		
			Mat.	Labor	Total
1 Site Work	Site preparation for slab; 4′ deep trench excavation for foundation wall, excavation for lower level, 4′ deep.	.027		.43	.43
2 Foundation	Continuous reinforced concrete footing, 8″ deep x 18″ wide; dampproofed and insulated 8″ thick reinforced concrete block foundation wall, 4′ deep; 4″ concrete slab on 4″ crushed stone base and polyethylene vapor barrier, trowel finish.	.071	2.33	3.19	5.52
3 Framing	Exterior walls - 2″ x 4″ wood studs, 16″ O.C.; 1/2″ insulation board sheathing; wood truss roof frame, 24″ O.C. with 1/2″ plywood sheathing, 4 in 12 pitch; 2″ x 8″ floor joists 16″ O.C. with bridging and 5/8″ plywood subfloor.	.094	3.37	4.60	7.97
4 Exterior Walls	Beveled wood siding and #15 felt building paper on insulated wood frame walls; 6″ attic insulation; double hung windows; 2 flush solid core wood exterior doors with storms.	.081	4.63	2.79	7.42
5 Roofing	20 year asphalt shingles; #15 felt building paper; aluminum gutters, downspouts, drip edge and flashings.	.032	.61	1.17	1.78
6 Interiors	Walls and ceilings, 1/2″ taped and finished drywall, primed and painted with 2 coats; painted baseboard and trim, rubber backed carpeting 80%, asphalt tile 20%; hollow core wood interior doors.	.177	6.21	8.33	14.54
7 Specialties	Economy grade kitchen cabinets - 6 L.F. wall and base with plastic laminate counter top and kitchen sink; 30 gallon electric water heater.	.014	.62	.35	.97
8 Mechanical	1 lavatory, white, wall hung; 1 water closet, white; 1 bathtub, enameled steel, white; gas fired warm air heat.	.057	1.81	1.98	3.79
9 Electrical	100 Amp. service; romex wiring; incandescent lighting fixtures, switches, receptacles.	.029	.45	.88	1.33
10 Overhead	Contractor's overhead and profit.		3.02	3.58	6.60
	Total		23.05	27.30	**50.35**

1 Story — Base cost per square foot of living area

Exterior Wall	Living Area 50	100	200	300	400	500	600	700
Wood Siding - Wood Frame	115.35	87.85	75.80	63.50	59.60	57.25	55.70	56.30
Brick Veneer - Wood Frame	131.85	99.60	85.65	70.05	65.50	62.75	60.90	61.30
Stucco on Wood Frame	108.50	82.90	71.75	60.85	57.20	55.00	53.50	54.20
Painted Concrete Block	118.10	89.80	77.40	64.60	60.60	58.15	56.60	57.10
Finished Basement, Add	32.70	26.65	24.15	19.95	19.15	18.60	18.25	18.00
Unfinished Basement, Add	18.40	13.70	11.75	8.55	7.90	7.55	7.25	7.10

1-1/2 Story — Base cost per square foot of living area

Exterior Wall	Living Area 100	200	300	400	500	600	700	800
Wood Siding - Wood Frame	90.55	72.55	61.55	55.40	52.05	50.50	48.50	47.85
Brick Veneer - Wood Frame	105.35	84.30	71.35	63.05	59.15	57.20	54.80	54.15
Stucco on Wood Frame	84.55	67.70	57.50	52.25	49.20	47.75	45.90	45.30
Painted Concrete Block	93.00	74.50	63.20	56.70	53.25	51.65	49.55	48.95
Finished Basement, Add	21.80	19.30	17.60	15.80	15.25	14.95	14.60	14.60
Unfinished Basement, Add	11.35	9.40	8.10	6.70	6.30	6.05	5.80	5.80

2 Story — Base cost per square foot of living area

Exterior Wall	Living Area 100	200	400	600	800	1000	1200	1400
Wood Siding - Wood Frame	92.10	68.30	57.90	48.20	44.80	42.75	41.40	42.15
Brick Veneer - Wood Frame	108.60	80.10	67.70	54.75	50.70	48.25	46.60	47.15
Stucco on Wood Frame	85.30	63.45	53.80	45.50	42.40	40.50	39.20	40.05
Painted Concrete Block	94.85	70.30	59.55	49.30	45.80	43.70	42.30	42.95
Finished Basement, Add	16.40	13.35	12.10	10.00	9.60	9.35	9.15	9.05
Unfinished Basement, Add	9.20	6.85	5.95	4.25	3.95	3.80	3.65	3.55

Base costs do not include bathroom or kitchen facilities. Use Modifications/Adjustments/Alternatives on pages 93-96 where appropriate.

Average Class

1 Story

1-1/2 Story

2 Story

2-1/2 Story

Bi-Level

Tri-Level

- **Simple design from standard plans**
- **Single family — 1 full bath, 1 kitchen**
- **No basement**
- **Asphalt shingles on roof**
- **Hot air heat**
- **Drywall interior finishes**
- **Materials and workmanship are average**

Note: The illustration shown may contain some optional components (for example: garages and/or fireplaces) whose costs are shown in the modifications, adjustments, & alternatives below or at the end of the square foot section.

Base cost per square foot of living area

Exterior Wall	Living Area 600	800	1000	1200	1400	1600	1800	2000	2400	2800	3200
Wood Siding - Wood Frame	102.70	93.60	86.65	81.05	76.25	73.15	71.55	69.50	65.25	62.20	60.15
Brick Veneer - Wood Frame	118.70	108.80	101.30	95.10	89.85	86.45	84.65	82.35	77.80	74.40	72.10
Stucco on Wood Frame	111.15	102.05	95.20	89.60	84.85	81.70	80.10	78.10	73.90	70.80	68.75
Solid Masonry	129.45	118.35	109.95	103.00	96.95	93.10	91.10	88.45	83.30	79.50	76.85
Finished Basement, Add	24.85	24.25	23.25	22.30	21.55	21.05	20.85	20.40	19.80	19.30	18.90
Unfinished Basement, Add	9.95	9.20	8.65	8.15	7.65	7.35	7.20	7.00	6.65	6.35	6.15

Modifications

Add to the total cost

Upgrade Kitchen Cabinets	$ + 2240
Solid Surface Countertops	+ 896
Full Bath - including plumbing, wall and floor finishes	+ 4060
Half Bath - including plumbing, wall and floor finishes	+ 2583
One Car Attached Garage	+ 8958
One Car Detached Garage	+ 11,757
Fireplace & Chimney	+ 3910

Adjustments

For multi family - add to total cost

Additional Kitchen	$ + 4228
Additional Bath	+ 4060
Additional Entry & Exit	+ 1236
Separate Heating	+ 1237
Separate Electric	+ 1487

For Townhouse/Rowhouse - Multiply cost per square foot by

Inner Unit	.92
End Unit	.96

Alternatives

Add to or deduct from the cost per square foot of living area

Cedar Shake Roof	+ 2.32
Clay Tile Roof	+ 4.08
Slate Roof	+ 7.77
Upgrade Walls to Skim Coat Plaster	+ .32
Upgrade Ceilings to Textured Finish	+ .39
Air Conditioning, in Heating Ductwork	+ 2.80
In Separate Ductwork	+ 5.46
Heating Systems, Hot Water	+ 1.49
Heat Pump	+ 1.84
Electric Heat	– .69
Not Heated	– 2.63

Additional upgrades or components

Kitchen Cabinets & Countertops	Page 93
Bathroom Vanities	94
Fireplaces & Chimneys	94
Windows, Skylights & Dormers	94
Appliances	95
Breezeways & Porches	95
Finished Attic	95
Garages	96
Site Improvements	96
Wings & Ells	56

Average 1 Story

Living Area - 1600 S.F.
Perimeter - 163 L.F.

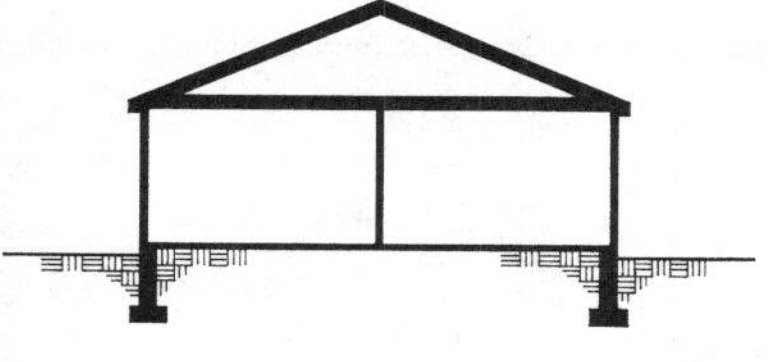

			Labor-Hours	Cost Per Square Foot Of Living Area		
				Mat.	Labor	Total
1	Site Work	Site preparation for slab; 4′ deep trench excavation for foundation wall.	.048		.66	.66
2	Foundation	Continuous reinforced concrete footing 8″ deep x 18″ wide; dampproofed and insulated reinforced concrete foundation wall, 8″ thick, 4′ deep; 4″ concrete slab on 4″ crushed stone base and polyethylene vapor barrier, trowel finish.	.113	3.59	4.90	8.49
3	Framing	Exterior walls - 2″ x 4″ wood studs, 16″ O.C.; 1/2″ plywood sheathing; 2″ x 6″ rafters 16″ O.C. with 1/2″ plywood sheathing, 4 in 12 pitch; 2″ x 6″ ceiling joists 16″ O.C.; 1/2″ plywood subfloor on 1″ x 2″ wood sleepers 16″ O.C.	.136	4.96	6.88	11.84
4	Exterior Walls	Beveled wood siding and #15 felt building paper on insulated wood frame walls; 6″ attic insulation; double hung windows; 3 flush solid core wood exterior doors with storms.	.098	6.93	3.76	10.69
5	Roofing	25 year asphalt shingles; #15 felt building paper; aluminum gutters, downspouts, drip edge and flashings.	.047	.94	1.80	2.74
6	Interiors	Walls and ceilings, 1/2″ taped and finished drywall, primed and painted with 2 coats; painted baseboard and trim, finished hardwood floor 40%, carpet with 1/2″ underlayment 40%, vinyl tile with 1/2″ underlayment 15%, ceramic tile with 1/2″ underlayment 5%; hollow core and louvered interior doors.	.251	9.32	9.64	18.96
7	Specialties	Average grade kitchen cabinets - 14 L.F. wall and base with plastic laminate counter top and kitchen sink; 40 gallon electric water heater.	.009	1.54	.67	2.21
8	Mechanical	1 lavatory, white, wall hung; 1 water closet, white; 1 bathtub with shower, enameled steel, white; gas fired warm air heat.	.098	2.45	2.29	4.74
9	Electrical	200 Amp. service; romex wiring; incandescent lighting fixtures, switches, receptacles.	.041	.83	1.35	2.18
10	Overhead	Contractor's overhead and profit and plans.		5.19	5.45	10.64
			Total	35.75	37.40	**73.15**

- **Simple design from standard plans**
- **Single family — 1 full bath, 1 kitchen**
- **No basement**
- **Asphalt shingles on roof**
- **Hot air heat**
- **Drywall interior finishes**
- **Materials and workmanship are average**

Note: The illustration shown may contain some optional components (for example: garages and/or fireplaces) whose costs are shown in the modifications, adjustments, & alternatives below or at the end of the square foot section.

Base cost per square foot of living area

	Living Area										
Exterior Wall	**600**	**800**	**1000**	**1200**	**1400**	**1600**	**1800**	**2000**	**2400**	**2800**	**3200**
Wood Siding - Wood Frame	116.40	98.10	88.00	83.45	80.10	75.15	**72.70**	70.10	64.90	62.80	60.65
Brick Veneer - Wood Frame	126.45	105.35	94.75	89.75	86.15	80.60	77.90	75.00	69.25	66.90	64.45
Stucco on Wood Frame	115.90	97.65	87.70	83.15	79.80	74.85	72.40	69.80	64.70	62.55	60.45
Solid Masonry	141.35	116.20	104.75	99.10	95.05	88.70	85.60	82.35	75.65	73.05	70.10
Finished Basement, Add	20.05	17.60	17.00	16.55	16.25	15.65	15.35	15.05	14.45	14.15	13.85
Unfinished Basement, Add	8.35	6.60	6.30	6.00	5.80	5.50	5.30	5.15	4.80	4.65	4.45

Modifications

Add to the total cost

Upgrade Kitchen Cabinets	$ + 2240
Solid Surface Countertops	+ 896
Full Bath - including plumbing, wall and floor finishes	+ 4060
Half Bath - including plumbing, wall and floor finishes	+ 2583
One Car Attached Garage	+ 8958
One Car Detached Garage	+ 11,757
Fireplace & Chimney	+ 3910

Adjustments

For multi family - add to total cost

Additional Kitchen	$ + 4228
Additional Bath	+ 4060
Additional Entry & Exit	+ 1236
Separate Heating	+ 1237
Separate Electric	+ 1487

For Townhouse/Rowhouse - Multiply cost per square foot by

Inner Unit	.92
End Unit	.96

Alternatives

Add to or deduct from the cost per square foot of living area

Cedar Shake Roof	+ 1.69
Clay Tile Roof	+ 2.95
Slate Roof	+ 5.61
Upgrade Walls to Skim Coat Plaster	+ .38
Upgrade Ceilings to Textured Finish	+ .39
Air Conditioning, in Heating Ductwork	+ 2.10
In Separate Ductwork	+ 4.75
Heating Systems, Hot Water	+ 1.41
Heat Pump	+ 2.04
Electric Heat	– .61
Not Heated	– 2.54

Additional upgrades or components

Kitchen Cabinets & Countertops	Page 93
Bathroom Vanities	94
Fireplaces & Chimneys	94
Windows, Skylights & Dormers	94
Appliances	95
Breezeways & Porches	95
Finished Attic	95
Garages	96
Site Improvements	96
Wings & Ells	56

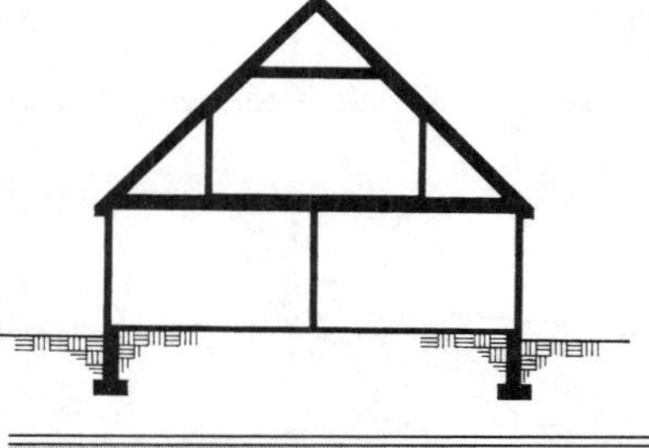

Average 1-1/2 Story

Living Area - 1800 S.F.
Perimeter - 144 L.F.

		Labor-Hours	Cost Per Square Foot Of Living Area		
			Mat.	Labor	Total
1 Site Work	Site preparation for slab; 4′ deep trench excavation for foundation wall.	.037		.59	.59
2 Foundation	Continuous reinforced concrete footing 8″ deep x 18″ wide; dampproofed and insulated reinforced concrete foundation wall, 8″ thick, 4′ deep; 4″ concrete slab on 4″ crushed stone base and polyethylene vapor barrier, trowel finish.	.073	2.54	3.59	6.13
3 Framing	Exterior walls - 2″ x 4″ wood studs, 16″ O.C.; 1/2″ plywood sheathing; 2″ x 6″ rafters 16″ O.C. with 1/2″ plywood sheathing, 8 in 12 pitch; 2″ x 8″ floor joists 16″ O.C. with 5/8″ plywood subfloor; 1/2″ plywood subfloor on 1″ x 2″ wood sleepers 16″ O.C.	.098	5.21	6.43	11.64
4 Exterior Walls	Beveled wood siding and #15 felt building paper on insulated wood frame walls; 6″ attic insulation; double hung windows; 3 flush solid core wood exterior doors with storms.	.078	7.53	4.10	11.63
5 Roofing	25 year asphalt shingles; #15 felt building paper; aluminum gutters, downspouts, drip edge and flashings.	.029	.59	1.12	1.71
6 Interiors	Walls and ceilings, 1/2″ taped and finished drywall, primed and painted with 2 coats; painted baseboard and trim, finished hardwood floor 40%, carpet with 1/2″ underlayment 40%, vinyl tile with 1/2″ underlayment 15%, ceramic tile with 1/2″ underlayment 5%; hollow core and louvered interior doors.	.225	10.98	10.93	21.91
7 Specialties	Average grade kitchen cabinets - 14 L.F. wall and base with plastic laminate counter top and kitchen sink; 40 gallon electric water heater.	.022	1.36	.59	1.95
8 Mechanical	1 lavatory, white, wall hung; 1 water closet, white; 1 bathtub with shower, enameled steel, white; gas fired warm air heat.	.049	2.29	2.21	4.50
9 Electrical	200 Amp. service; romex wiring; incandescent lighting fixtures, switches, receptacles.	.039	.79	1.29	2.08
10 Overhead	Contractor's overhead and profit and plans.		5.31	5.25	10.56
		Total	36.60	36.10	**72.70**

- **Simple design from standard plans**
- **Single family — 1 full bath, 1 kitchen**
- **No basement**
- **Asphalt shingles on roof**
- **Hot air heat**
- **Drywall interior finishes**
- **Materials and workmanship are average**

Note: The illustration shown may contain some optional components (for example: garages and/or fireplaces) whose costs are shown in the modifications, adjustments, & alternatives below or at the end of the square foot section.

Base cost per square foot of living area

Exterior Wall	Living Area 1000	1200	1400	1600	1800	2000	2200	2600	3000	3400	3800
Wood Siding - Wood Frame	92.85	84.30	80.40	77.75	75.10	**71.95**	70.05	66.25	62.50	60.80	59.35
Brick Veneer - Wood Frame	100.60	91.50	87.15	84.20	81.15	77.85	75.60	71.30	67.10	65.25	63.55
Stucco on Wood Frame	92.45	83.95	80.10	77.50	74.80	71.70	69.80	66.00	62.20	60.60	59.15
Solid Masonry	112.20	102.20	97.10	93.75	90.15	86.50	83.80	78.75	74.00	71.80	69.75
Finished Basement, Add	14.35	14.25	13.90	13.60	13.30	13.15	12.90	12.50	12.15	12.00	11.75
Unfinished Basement, Add	5.35	5.10	4.90	4.75	4.60	4.45	4.35	4.05	3.90	3.80	3.65

Modifications

Add to the total cost

Upgrade Kitchen Cabinets	$ + 2240
Solid Surface Countertops	+ 896
Full Bath - including plumbing, wall and floor finishes	+ 4060
Half Bath - including plumbing, wall and floor finishes	+ 2583
One Car Attached Garage	+ 8958
One Car Detached Garage	+ 11,757
Fireplace & Chimney	+ 4365

Adjustments

For multi family - add to total cost

Additional Kitchen	$ + 4228
Additional Bath	+ 4060
Additional Entry & Exit	+ 1236
Separate Heating	+ 1237
Separate Electric	+ 1487

For Townhouse/Rowhouse - Multiply cost per square foot by

Inner Unit	.90
End Unit	.95

Alternatives

Add to or deduct from the cost per square foot of living area

Cedar Shake Roof	+ 1.16
Clay Tile Roof	+ 2.04
Slate Roof	+ 3.89
Upgrade Walls to Skim Coat Plaster	+ .38
Upgrade Ceilings to Textured Finish	+ .39
Air Conditioning, in Heating Ductwork	+ 1.67
In Separate Ductwork	+ 4.32
Heating Systems, Hot Water	+ 1.38
Heat Pump	+ 2.15
Electric Heat	– .52
Not Heated	– 2.46

Additional upgrades or components

Kitchen Cabinets & Countertops	Page 93
Bathroom Vanities	94
Fireplaces & Chimneys	94
Windows, Skylights & Dormers	94
Appliances	95
Breezeways & Porches	95
Finished Attic	95
Garages	96
Site Improvements	96
Wings & Ells	56

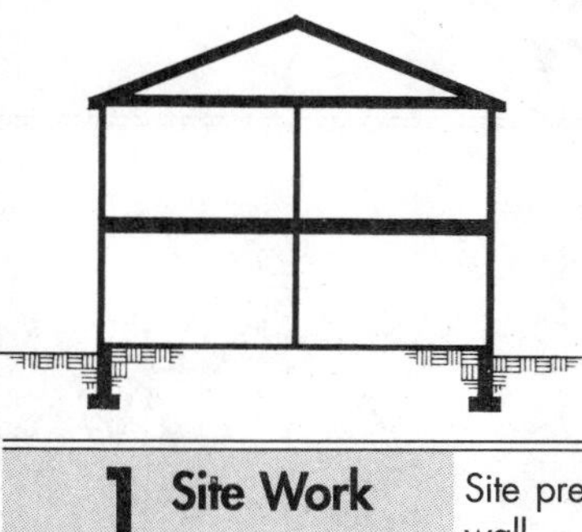

Average 2 Story

Living Area - 2000 S.F.
Perimeter - 135 L.F.

			Labor-Hours	Cost Per Square Foot Of Living Area		
				Mat.	Labor	Total
1	Site Work	Site preparation for slab; 4′ deep trench excavation for foundation wall.	.034		.52	.52
2	Foundation	Continuous reinforced concrete footing 8″ deep x 18″ wide; dampproofed and insulated reinforced concrete foundation wall, 8″ thick, 4′ deep, 4″ concrete slab on 4″ crushed stone base and polyethylene vapor barrier, trowel finish.	.066	2.12	3.01	5.13
3	Framing	Exterior walls - 2″ x 4″ wood studs, 16″ O.C.; 1/2″ plywood sheathing; 2″ x 6″ rafters 16″ O.C. with 1/2″ plywood sheathing, 4 in 12 pitch; 2″ x 6″ ceiling joists 16″ O.C.; 2″ x 8″ floor joists 16″ O.C. with 5/8″ plywood subfloor; 1/2″ plywood subfloor on 1″ x 2″ wood sleepers 16″ O.C.	.131	5.41	6.64	12.05
4	Exterior Walls	Beveled wood siding and #15 felt building paper on insulated wood frame walls; 6″ attic insulation; double hung windows; 3 flush solid core wood exterior doors with storms.	.111	8.18	4.46	12.64
5	Roofing	25 year asphalt shingles; #15 felt building paper; aluminum gutters, downspouts, drip edge and flashings.	.024	.47	.90	1.37
6	Interiors	Walls and ceilings, 1/2″ taped and finished drywall, primed and painted with 2 coats; painted baseboard and trim, finished hardwood floor 40%, carpet with 1/2″ underlayment 40%, vinyl tile with 1/2″ underlayment 15%, ceramic tile with 1/2″ underlayment 5%; hollow core and louvered interior doors.	.232	10.87	10.92	21.79
7	Specialties	Average grade kitchen cabinets - 14 L.F. wall and base with plastic laminate counter top and kitchen sink; 40 gallon electric water heater.	.021	1.22	.54	1.76
8	Mechanical	1 lavatory, white, wall hung; 1 water closet, white; 1 bathtub with shower; enameled steel, white; gas fired warm air heat.	.060	2.15	2.13	4.28
9	Electrical	200 Amp. service; romex wiring; incandescent lighting fixtures, switches, receptacles.	.039	.75	1.24	1.99
10	Overhead	Contractor's overhead and profit and plans.		5.28	5.14	10.42
			Total	36.45	35.50	**71.95**

- **Simple design from standard plans**
- **Single family — 1 full bath, 1 kitchen**
- **No basement**
- **Asphalt shingles on roof**
- **Hot air heat**
- **Drywall interior finishes**
- **Materials and workmanship are average**

Note: The illustration shown may contain some optional components (for example: garages and/or fireplaces) whose costs are shown in the modifications, adjustments, & alternatives below or at the end of the square foot section.

Base cost per square foot of living area

Exterior Wall	Living Area 1200	1400	1600	1800	2000	2400	2800	3200	3600	4000	4400
Wood Siding - Wood Frame	92.90	87.45	80.00	78.55	75.90	71.60	68.05	64.65	62.90	59.65	58.60
Brick Veneer - Wood Frame	101.15	94.95	86.90	85.50	82.35	77.45	73.65	69.75	67.70	64.10	62.95
Stucco on Wood Frame	92.45	87.05	79.65	78.20	75.55	71.30	67.75	64.45	62.65	59.45	58.40
Solid Masonry	113.45	106.05	97.15	95.75	91.95	86.15	81.95	77.30	74.85	70.75	69.35
Finished Basement, Add	12.20	12.00	11.55	11.55	11.30	10.90	10.75	10.35	10.20	9.95	9.90
Unfinished Basement, Add	4.45	4.15	3.95	3.95	3.75	3.50	3.40	3.20	3.10	3.00	2.95

Modifications

Add to the total cost

Upgrade Kitchen Cabinets	$ + 2240
Solid Surface Countertops	+ 896
Full Bath - including plumbing, wall and floor finishes	+ 4060
Half Bath - including plumbing, wall and floor finishes	+ 2583
One Car Attached Garage	+ 8958
One Car Detached Garage	+ 11,757
Fireplace & Chimney	+ 4960

Adjustments

For multi family - add to total cost

Additional Kitchen	$ + 4228
Additional Bath	+ 4060
Additional Entry & Exit	+ 1236
Separate Heating	+ 1237
Separate Electric	+ 1487

For Townhouse/Rowhouse - Multiply cost per square foot by

Inner Unit	.90
End Unit	.95

Alternatives

Add to or deduct from the cost per square foot of living area

Cedar Shake Roof	+ 1.01
Clay Tile Roof	+ 1.77
Slate Roof	+ 3.37
Upgrade Walls to Skim Coat Plaster	+ .38
Upgrade Ceilings to Textured Finish	+ .39
Air Conditioning, in Heating Ductwork	+ 1.53
In Separate Ductwork	+ 4.17
Heating Systems, Hot Water	+ 1.24
Heat Pump	+ 2.20
Electric Heat	– .95
Not Heated	– 2.87

Additional upgrades or components

Kitchen Cabinets & Countertops	Page 93
Bathroom Vanities	94
Fireplaces & Chimneys	94
Windows, Skylights & Dormers	94
Appliances	95
Breezeways & Porches	95
Finished Attic	95
Garages	96
Site Improvements	96
Wings & Ells	56

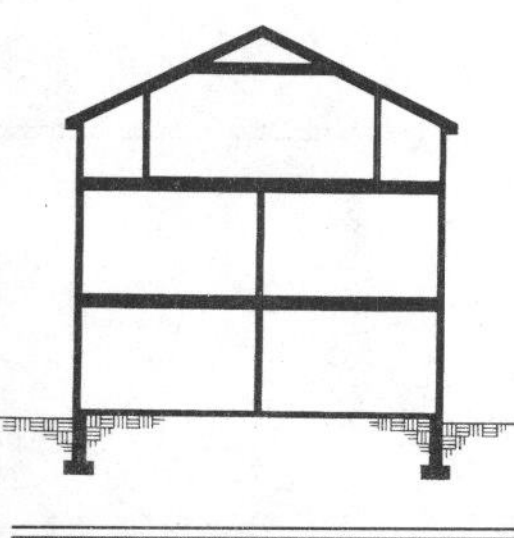

Average 2-1/2 Story

Living Area - 3200 S.F.
Perimeter - 150 L.F.

		Labor-Hours	Cost Per Square Foot Of Living Area		
			Mat.	Labor	Total
1 Site Work	Site preparation for slab; 4′ deep trench excavation for foundation wall.	.046		.33	.33
2 Foundation	Continuous reinforced concrete footing 8″ deep x 18″ wide, dampproofed and insulated reinforced concrete foundation wall, 8″ thick, 4′ deep; 4″ concrete slab on 4″ crushed stone base and polyethylene vapor barrier, trowel finish.	.061	1.54	2.14	3.68
3 Framing	Exterior walls - 2″ x 4″ wood studs, 16″ O.C.; 1/2″ plywood sheathing; 2″ x 6″ rafters 16″ O.C. with 1/2″ plywood sheathing, 4 in 12 pitch; 2″ x 6″ ceiling joists 16″ O.C.; 2″ x 8″ floor joists 16″ O.C. with 5/8″ plywood subfloor; 1/2″ plywood subfloor on 1″ x 2″ wood sleepers 16″ O.C.	.127	5.38	6.51	11.89
4 Exterior Walls	Beveled wood siding and #15 felt building paper on insulated wood frame walls; 6″ attic insulation; double hung windows; 3 flush solid core wood exterior doors with storms.	.136	6.88	3.76	10.64
5 Roofing	25 year asphalt shingles; #15 felt building paper; aluminum gutters, downspouts, drip edge and flashings.	.018	.36	.69	1.05
6 Interiors	Walls and ceilings, 1/2″ taped and finished drywall, primed and painted with 2 coats; painted baseboard and trim, finished hardwood floor 40%, carpet with 1/2″ underlayment 40%, vinyl tile with 1/2″ underlayment 15%, ceramic tile with 1/2″ underlayment 5%; hollow core and louvered interior doors.	.286	10.72	10.58	21.30
7 Specialties	Average grade kitchen cabinets - 14 L.F. wall and base with plastic laminate counter top and kitchen sink; 40 gallon electric water heater.	.030	.75	.32	1.07
8 Mechanical	1 lavatory, white, wall hung; 1 water closet, white; 1 bathtub with shower, enameled steel, white; gas fired warm air heat.	.072	1.71	1.89	3.60
9 Electrical	200 Amp. service; romex wiring; incandescent lighting fixtures, switches, receptacles.	.046	.63	1.09	1.72
10 Overhead	Contractor's overhead and profit and plans.		4.73	4.64	9.37
		Total	32.70	31.95	**64.65**

- **Simple design from standard plans**
- **Single family — 1 full bath, 1 kitchen**
- **No basement**
- **Asphalt shingles on roof**
- **Hot air heat**
- **Drywall interior finishes**
- **Materials and workmanship are average**

Note: The illustration shown may contain some optional components (for example: garages and/or fireplaces) whose costs are shown in the modifications, adjustments, & alternatives below or at the end of the square foot section.

Base cost per square foot of living area

	Living Area										
Exterior Wall	1500	1800	2100	2500	3000	3500	4000	4500	5000	5500	6000
Wood Siding - Wood Frame	86.35	78.50	75.15	72.50	**67.40**	65.25	62.20	58.80	57.70	56.50	55.25
Brick Veneer - Wood Frame	94.10	85.70	81.85	78.90	73.20	70.80	67.20	63.45	62.20	60.80	59.30
Stucco on Wood Frame	85.95	78.10	74.80	72.15	67.10	64.95	61.90	58.55	57.45	56.30	55.05
Solid Masonry	105.65	96.40	91.85	88.40	81.95	78.95	74.65	70.35	68.85	67.20	65.40
Finished Basement, Add	10.50	10.45	10.20	10.00	9.70	9.55	9.20	9.05	8.90	8.85	8.75
Unfinished Basement, Add	3.60	3.45	3.30	3.20	3.00	2.90	2.75	2.60	2.60	2.50	2.45

Modifications

Add to the total cost

Upgrade Kitchen Cabinets	$ + 2240
Solid Surface Countertops	+ 896
Full Bath - including plumbing, wall and floor finishes	+ 4060
Half Bath - including plumbing, wall and floor finishes	+ 2583
One Car Attached Garage	+ 8958
One Car Detached Garage	+ 11,757
Fireplace & Chimney	+ 4960

Adjustments

For multi family - add to total cost

Additional Kitchen	$ + 4228
Additional Bath	+ 4060
Additional Entry & Exit	+ 1236
Separate Heating	+ 1237
Separate Electric	+ 1487

For Townhouse/Rowhouse - Multiply cost per square foot by

Inner Unit	.88
End Unit	.94

Alternatives

Add to or deduct from the cost per square foot of living area

Cedar Shake Roof	+ .77
Clay Tile Roof	+ 1.36
Slate Roof	+ 2.59
Upgrade Walls to Skim Coat Plaster	+ .38
Upgrade Ceilings to Textured Finish	+ .39
Air Conditioning, in Heating Ductwork	+ 1.53
In Separate Ductwork	+ 4.17
Heating Systems, Hot Water	+ 1.24
Heat Pump	+ 2.20
Electric Heat	– .74
Not Heated	– 2.67

Additional upgrades or components

Kitchen Cabinets & Countertops	Page 93
Bathroom Vanities	94
Fireplaces & Chimneys	94
Windows, Skylights & Dormers	94
Appliances	95
Breezeways & Porches	95
Finished Attic	95
Garages	96
Site Improvements	96
Wings & Ells	56

Average 3 Story

Living Area - 3000 S.F.
Perimeter - 135 L.F.

		Labor-Hours	Cost Per Square Foot Of Living Area		
			Mat.	Labor	Total
1 Site Work	Site preparation for slab; 4′ deep trench excavation for foundation wall.	.038		.35	.35
2 Foundation	Continuous reinforced concrete footing 8″ deep x 18″ wide, dampproofed and insulated reinforced concrete foundation wall, 8″ thick, 4′ deep; 4″ concrete slab on 4″ crushed stone base and polyethylene vapor barrier, trowel finish.	.053	1.41	2.01	3.42
3 Framing	Exterior walls - 2″ x 4″ wood studs, 16″ O.C.; 1/2″ plywood sheathing; 2″ x 6″ rafters 16″ O.C. with 1/2″ plywood sheathing, 4 in 12 pitch; 2″ x 6″ ceiling joists 16″ O.C.; 2″ x 8″ floor joists 16″ O.C. with 5/8″ plywood subfloor; 1/2″ plywood subfloor on 1″ x 2″ wood sleepers 16″ O.C.	.128	5.51	6.70	12.21
4 Exterior Walls	Horizontal beveled wood siding; #15 felt building paper; 3-1/2″ batt insulation; wood double hung windows; 3 flush solid core wood exterior doors; storms and screens.	.139	7.82	4.28	12.10
5 Roofing	25 year asphalt shingles; #15 felt building paper; aluminum gutters, downspouts, drip edge and flashings.	.014	.31	.60	.91
6 Interiors	Walls and ceilings, 1/2″ taped and finished drywall, primed and painted with 2 coats; painted baseboard and trim, finished hardwood floor 40%, carpet with 1/2″ underlayment 40%, vinyl tile with 1/2″ underlayment 15%, ceramic tile with 1/2″ underlayment 5%; hollow core and louvered interior doors.	.280	11.10	10.96	22.06
7 Specialties	Average grade kitchen cabinets - 14 L.F. wall and base with plastic laminate counter top and kitchen sink; 40 gallon electric water heater.	.025	.80	.34	1.14
8 Mechanical	1 lavatory, white, wall hung; 1 water closet, white; 1 bathtub with shower, enameled steel, white; gas fired warm air heat.	.065	1.75	1.91	3.66
9 Electrical	200 Amp. service; romex wiring; incandescent lighting fixtures, switches, receptacles.	.042	.64	1.11	1.75
10 Overhead	Contractor's overhead and profit and plans.		5.01	4.79	9.80
		Total	34.35	33.05	**67.40**

RESIDENTIAL | Average | Bi-Level

- Simple design from standard plans
- Single family — 1 full bath, 1 kitchen
- No basement
- Asphalt shingles on roof
- Hot air heat
- Drywall interior finishes
- Materials and workmanship are average

Note: The illustration shown may contain some optional components (for example: garages and/or fireplaces) whose costs are shown in the modifications, adjustments, & alternatives below or at the end of the square foot section.

Base cost per square foot of living area

	Living Area										
Exterior Wall	**1000**	**1200**	**1400**	**1600**	**1800**	**2000**	**2200**	**2600**	**3000**	**3400**	**3800**
Wood Siding - Wood Frame	86.95	78.85	75.35	72.90	70.50	**67.55**	65.85	62.45	58.90	57.50	56.15
Brick Veneer - Wood Frame	92.80	84.25	80.35	77.75	75.05	71.95	70.05	66.20	62.40	60.80	59.30
Stucco on Wood Frame	86.65	78.65	75.00	72.65	70.25	67.35	65.70	62.20	58.70	57.25	55.95
Solid Masonry	101.45	92.25	87.90	84.90	81.80	78.45	76.20	71.80	67.60	65.65	63.95
Finished Basement, Add	14.35	14.25	13.90	13.60	13.30	13.15	12.90	12.50	12.15	12.00	11.75
Unfinished Basement, Add	5.35	5.10	4.90	4.75	4.60	4.45	4.35	4.05	3.90	3.80	3.65

Modifications

Add to the total cost

Upgrade Kitchen Cabinets	$ + 2240
Solid Surface Countertops	+ 896
Full Bath - including plumbing, wall and floor finishes	+ 4060
Half Bath - including plumbing, wall and floor finishes	+ 2583
One Car Attached Garage	+ 8958
One Car Detached Garage	+ 11,757
Fireplace & Chimney	+ 3910

Adjustments

For multi family - add to total cost

Additional Kitchen	$ + 4228
Additional Bath	+ 4060
Additional Entry & Exit	+ 1236
Separate Heating	+ 1237
Separate Electric	+ 1487

For Townhouse/Rowhouse - Multiply cost per square foot by

Inner Unit	.91
End Unit	.96

Alternatives

Add to or deduct from the cost per square foot of living area

Cedar Shake Roof	+ 2.32
Clay Tile Roof	+ 4.08
Slate Roof	+ 3.89
Upgrade Walls to Skim Coat Plaster	+ .38
Upgrade Ceilings to Textured Finish	+ .39
Air Conditioning, in Heating Ductwork	+ 1.67
In Separate Ductwork	+ 4.32
Heating Systems, Hot Water	+ 1.38
Heat Pump	+ 2.15
Electric Heat	– .52
Not Heated	– 2.46

Additional upgrades or components

Kitchen Cabinets & Countertops	Page 93
Bathroom Vanities	94
Fireplaces & Chimneys	94
Windows, Skylights & Dormers	94
Appliances	95
Breezeways & Porches	95
Finished Attic	95
Garages	96
Site Improvements	96
Wings & Ells	56

Important: See the Reference Section for Location Factors (to adjust for your city) and Estimating Forms

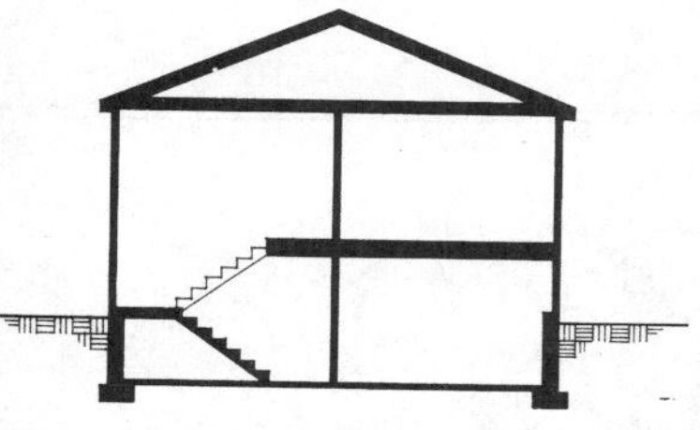

Average Bi-Level

Living Area - 2000 S.F.
Perimeter - 135 L.F.

SQUARE FOOT COSTS

		Labor-Hours	Cost Per Square Foot Of Living Area		
			Mat.	Labor	Total
1 Site Work	Excavation for lower level, 4′ deep. Site preparation for slab.	.029		.52	.52
2 Foundation	Continuous reinforced concrete footing 8″ deep x 18″ wide, dampproofed and insulated reinforced concrete foundation wall, 8″ thick, 4′ deep; 4″ concrete slab on 4″ crushed stone base and polyethylene vapor barrier, trowel finish.	.066	2.12	3.01	5.13
3 Framing	Exterior walls - 2″ x 4″ wood studs, 16″ O.C.; 1/2″ plywood sheathing; 2″ x 6″ rafters 16″ O.C. with 1/2″ plywood sheathing, 4 in 12 pitch; 2″ x 6″ ceiling joists 16″ O.C.; 2″ x 8″ floor joists 16″ O.C. with 5/8″ plywood subfloor; 1/2″ plywood subfloor on 1″ x 2″ wood sleepers 16″ O.C.	.118	5.19	6.31	11.50
4 Exterior Walls	Horizontal beveled wood siding; #15 felt building paper; 3-1/2″ batt insulation; wood double hung windows; 3 flush solid core wood exterior doors; storms and screens.	.091	6.41	3.47	9.88
5 Roofing	25 year asphalt shingles; #15 felt building paper; aluminum gutters, downspouts, drip edge and flashings.	.024	.47	.90	1.37
6 Interiors	Walls and ceilings, 1/2″ taped and finished drywall, primed and painted with 2 coats; painted baseboard and trim, finished hardwood floor 40%, carpet with 1/2″ underlayment 40%, vinyl tile with 1/2″ underlayment 15%, ceramic tile with 1/2″ underlayment 5%; hollow core and louvered interior doors.	.217	10.71	10.61	21.32
7 Specialties	Average grade kitchen cabinets - 14 L.F. wall and base with plastic laminate counter top and kitchen sink; 40 gallon electric water heater.	.021	1.22	.54	1.76
8 Mechanical	1 lavatory, white, wall hung; 1 water closet, white; 1 bathtub with shower, enameled steel, white; gas fired warm air heat.	.061	2.15	2.13	4.28
9 Electrical	200 Amp. service; romex wiring; incandescent lighting fixtures, switches, receptacles.	.039	.75	1.24	1.99
10 Overhead	Contractor's overhead and profit and plans.		4.93	4.87	9.80
		Total	33.95	33.60	**67.55**

- **Simple design from standard plans**
- **Single family — 1 full bath, 1 kitchen**
- **No basement**
- **Asphalt shingles on roof**
- **Hot air heat**
- **Drywall interior finishes**
- **Materials and workmanship are average**

Note: The illustration shown may contain some optional components (for example: garages and/or fireplaces) whose costs are shown in the modifications, adjustments, & alternatives below or at the end of the square foot section.

Base cost per square foot of living area

Exterior Wall	Living Area 1200	1500	1800	2100	2400	2700	3000	3400	3800	4200	4600
Wood Siding - Wood Frame	82.45	76.30	71.80	67.70	65.15	63.60	62.00	60.45	57.85	55.80	54.75
Brick Veneer - Wood Frame	87.85	81.15	76.15	71.65	68.90	67.20	65.40	63.65	60.90	58.65	57.50
Stucco on Wood Frame	82.15	76.10	71.50	67.45	64.90	63.40	61.80	60.25	57.70	55.60	54.60
Solid Masonry	95.80	88.45	82.65	77.60	74.40	72.60	70.45	68.55	65.40	62.85	61.60
Finished Basement, Add*	16.70	16.45	15.85	15.35	15.00	14.85	14.55	14.40	14.10	13.85	13.75
Unfinished Basement, Add*	6.25	5.85	5.45	5.15	5.00	4.85	4.70	4.60	4.45	4.30	4.25

*Basement under middle level only.

Modifications

Add to the total cost

Upgrade Kitchen Cabinets	$ + 2240
Solid Surface Countertops	+ 896
Full Bath - including plumbing, wall and floor finishes	+ 4060
Half Bath - including plumbing, wall and floor finishes	+ 2583
One Car Attached Garage	+ 8958
One Car Detached Garage	+ 11,757
Fireplace & Chimney	+ 4365

Adjustments

For multi family - add to total cost

Additional Kitchen	$ + 4228
Additional Bath	+ 4060
Additional Entry & Exit	+ 1236
Separate Heating	+ 1237
Separate Electric	+ 1487

For Townhouse/Rowhouse - Multiply cost per square foot by

Inner Unit	.90
End Unit	.95

Alternatives

Add to or deduct from the cost per square foot of living area

Cedar Shake Roof	+ 1.69
Clay Tile Roof	+ 2.95
Slate Roof	+ 5.61
Upgrade Walls to Skim Coat Plaster	+ .39
Upgrade Ceilings to Textured Finish	+ .39
Air Conditioning, in Heating Ductwork	+ 1.40
In Separate Ductwork	+ 4.05
Heating Systems, Hot Water	+ 1.33
Heat Pump	+ 2.23
Electric Heat	− .43
Not Heated	− 2.37

Additional upgrades or components

Kitchen Cabinets & Countertops	Page 93
Bathroom Vanities	94
Fireplaces & Chimneys	94
Windows, Skylights & Dormers	94
Appliances	95
Breezeways & Porches	95
Finished Attic	95
Garages	96
Site Improvements	96
Wings & Ells	56

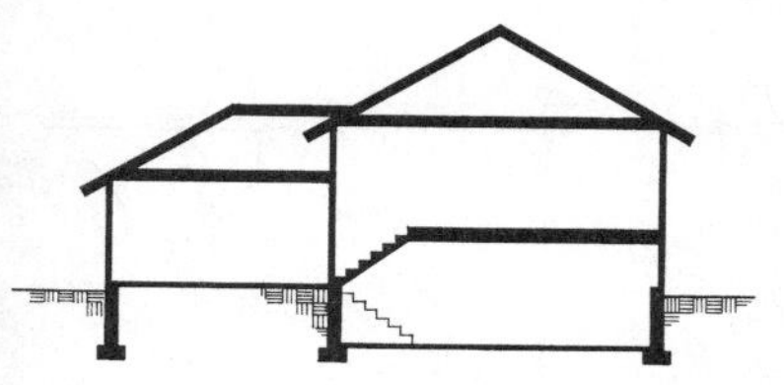

Average Tri-Level

Living Area - 2400 S.F.
Perimeter - 163 L.F.

			Labor-Hours	Cost Per Square Foot Of Living Area		
				Mat.	Labor	Total
1	Site Work	Site preparation for slab; 4′ deep trench excavation for foundation wall, excavation for lower level, 4′ deep.	.029		.44	.44
2	Foundation	Continuous reinforced concrete footing 8″ deep x 18″ wide; dampproofed and insulated reinforced concrete foundation wall, 8″ thick, 4′ deep; 4″ concrete slab on 4″ crushed stone base and polyethylene vapor barrier, trowel finish.	.080	2.40	3.28	5.68
3	Framing	Exterior walls - 2″ x 4″ wood studs, 16″ O.C.; 1/2″ plywood sheathing; 2″ x 6″ rafters 16″ O.C. with 1/2″ plywood sheathing, 4 in 12 pitch; 2″ x 6″ ceiling joists 16″ O.C.; 2″ x 8″ floor joists 16″ O.C. with 5/8″ plywood subfloor; 1/2″ plywood subfloor on 1″ x 2″ wood sleepers 16″ O.C.	.124	4.89	5.93	10.82
4	Exterior Walls	Horizontal beveled wood siding: #15 felt building paper; 3-1/2″ batt insulation; wood double hung windows; 3 flush solid core wood exterior doors; storms and screens.	.083	5.56	3.01	8.57
5	Roofing	25 year asphalt shingles; #15 felt building paper; aluminum gutters, downspouts, drip edge and flashings.	.032	.63	1.20	1.83
6	Interiors	Walls and ceilings, 1/2″ taped and finished drywall, primed and painted with 2 coats; painted baseboard and trim, finished hardwood floor 40%, carpet with 1/2″ underlayment 40%, vinyl tile with 1/2″ underlayment 15%, ceramic tile with 1/2″ underlayment 5%; hollow core and louvered interior doors.	.186	10.72	10.37	21.09
7	Specialties	Average grade kitchen cabinets - 14 L.F. wall and base with plastic laminate counter top and kitchen sink; 40 gallon electric water heater.	.012	1.02	.44	1.46
8	Mechanical	1 lavatory, white, wall hung; 1 water closet, white; 1 bathtub with shower, enameled steel, white; gas fired warm air heat.	.059	1.89	2.03	3.92
9	Electrical	200 Amp. service; romex wiring; incandescent lighting fixtures, switches, receptacles.	.036	.70	1.17	1.87
10	Overhead	Contractor's overhead and profit and plans.		4.74	4.73	9.47
			Total	32.55	32.60	**65.15**

- **Post and beam frame**
- **Log exterior walls**
- **Simple design from standard plans**
- **Single family — 1 full bath, 1 kitchen**
- **No basement**
- **Asphalt shingles on roof**
- **Hot air heat**
- **Drywall interior finishes**
- **Materials and workmanship are average**

Note: The illustration shown may contain some optional components (for example: garages and/or fireplaces) whose costs are shown in the modifications, adjustments, & alternatives below or at the end of the square foot section.

Base cost per square foot of living area

Exterior Wall	Living Area 600	800	1000	1200	1400	1600	1800	2000	2400	2800	3200
6" Log - Solid Wall	101.70	92.00	84.30	78.30	73.50	69.30	67.30	65.90	61.40	57.90	56.00
8" Log - Solid Wall	101.00	91.20	83.60	77.70	72.90	68.80	66.90	65.40	61.00	57.30	55.50
Finished Basement, Add	23.40	22.00	21.00	20.00	19.40	18.90	18.60	18.20	17.60	17.20	16.90
Unfinished Basement, Add	10.00	9.10	8.50	7.80	7.40	7.10	6.90	6.70	6.30	6.00	5.80

Modifications

Add to the total cost

Upgrade Kitchen Cabinets	$ + 2240
Solid Surface Countertops	+ 896
Full Bath - including plumbing, wall and floor finishes	+ 4060
Half Bath - including plumbing, wall and floor finishes	+ 2583
One Car Attached Garage	+ 8958
One Car Detached Garage	+ 11,757
Fireplace & Chimney	+ 3910

Adjustments

For multi family - add to total cost

Additional Kitchen	$ + 4228
Additional Bath	+ 4060
Additional Entry & Exit	+ 1236
Separate Heating	+ 1237
Separate Electric	+ 1487

For Townhouse/Rowhouse - Multiply cost per square foot by

Inner Unit	.92
End Unit	.96

Alternatives

Add to or deduct from the cost per square foot of living area

Cedar Shake Roof	+ 2.32
Air Conditioning, in Heating Ductwork	+ 2.80
In Separate Ductwork	+ 5.46
Heating Systems, Hot Water	+ 1.49
Heat Pump	+ 1.83
Electric Heat	– .71
Not Heated	– 2.63

Additional upgrades or components

Kitchen Cabinets & Countertops	Page 93
Bathroom Vanities	94
Fireplaces & Chimneys	94
Windows, Skylights & Dormers	94
Appliances	95
Breezeways & Porches	95
Finished Attic	95
Garages	96
Site Improvements	96
Wings & Ells	56

Solid Wall 1 Story

Living Area - 1600 S.F.
Perimeter - 163 L.F.

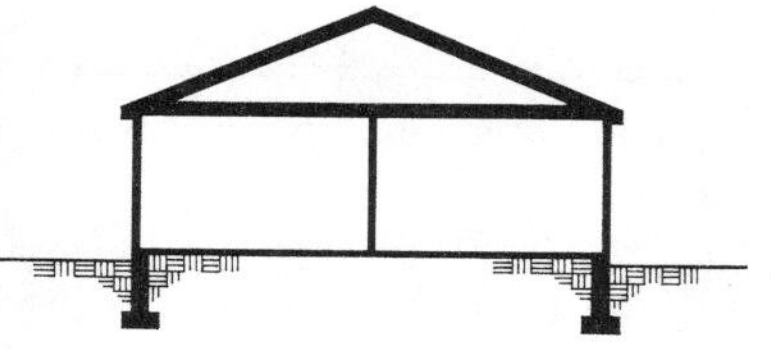

		Labor-Hours	Cost Per Square Foot Of Living Area		
			Mat.	Labor	Total
1 Site Work	Site preparation for slab; 4′ deep trench excavation for foundation wall.	.048		.82	.82
2 Foundation	Continuous reinforced concrete footing 8″ deep x 18″ wide; dampproofed and insulated reinforced concrete foundation wall, 8″ thick, 4′ deep; 4″ concrete slab on 4″ crushed stone base and polyethylene vapor barrier, trowel finish.	.113	3.82	4.77	8.59
3 Framing	Exterior walls - Precut traditional log home. Handicrafted white cedar or pine logs. Delivery included.	.201	11.42	8.10	19.52
4 Exterior Walls					
5 Roofing	25 year asphalt shingles; #15 felt building paper; aluminum gutters, downspouts, drip edge and flashings.	.047	.95	1.66	2.61
6 Interiors	Walls and ceilings, 1/2″ taped and finished drywall, primed and painted with 2 coats; painted baseboard and trim, finished hardwood floor 40%, carpet with 1/2″ underlayment 40%, vinyl tile with 1/2″ underlayment 15%, ceramic tile with 1/2″ underlayment 5%; hollow core and louvered interior doors.	.232	8.31	8.15	16.46
7 Specialties	Average grade kitchen cabinets - 14 L.F. wall and base with plastic laminate counter top and kitchen sink; 40 gallon electric water heater.	.009	1.73	.58	2.31
8 Mechanical	1 lavatory, white, wall hung; 1 water closet, white; 1 bathtub with shower, enameled steel, white; gas fired warm air heat.	.098	2.72	2.12	4.84
9 Electrical	200 Amp. service; romex wiring; incandescent lighting fixtures, switches, receptacles.	.041	.87	1.27	2.14
10 Overhead	Contractor's overhead and profit and plans.		7.03	4.98	12.01
		Total	36.85	32.45	**69.30**

RESIDENTIAL | Solid Wall | 2 Story

- Post and beam frame
- Log exterior walls
- Simple design from standard plans
- Single family — 1 full bath, 1 kitchen
- No basement
- Asphalt shingles on roof
- Hot air heat
- Drywall interior finishes
- Materials and workmanship are average

Note: The illustration shown may contain some optional components (for example: garages and/or fireplaces) whose costs are shown in the modifications, adjustments, & alternatives below or at the end of the square foot section.

Base cost per square foot of living area

Exterior Wall	Living Area 1000	1200	1400	1600	1800	2000	2200	2600	3000	3400	3800
6" Log-Solid	101.90	84.00	80.10	77.30	74.10	71.20	69.10	65.00	60.30	58.70	57.10
8" Log-Solid	97.80	87.70	83.40	80.70	77.50	74.10	72.10	67.70	63.10	61.40	59.50
Finished Basement, Add	13.50	13.10	12.60	12.40	12.10	11.80	11.60	11.20	10.90	10.80	10.60
Unfinished Basement, Add	5.50	5.10	4.80	4.60	4.50	4.20	4.20	3.90	3.70	3.60	3.50

Modifications

Add to the total cost

Upgrade Kitchen Cabinets	$ + 2206
Solid Surface Countertops	+ 896
Full Bath - including plumbing, wall and floor finishes	+ 3861
Half Bath - including plumbing, wall and floor finishes	+ 2405
One Car Attached Garage	+ 8424
One Car Detached Garage	+ 11,040
Fireplace & Chimney	+ 4160

Adjustments

For multi family - add to total cost

Additional Kitchen	$ + 4061
Additional Bath	+ 3861
Additional Entry & Exit	+ 1128
Separate Heating	+ 1211
Separate Electric	+ 1410

For Townhouse/Rowhouse - Multiply cost per square foot by

Inner Unit	.92
End Unit	.96

Alternatives

Add to or deduct from the cost per square foot of living area

Cedar Shake Roof	+ 1.50
Air Conditioning, in Heating Ductwork	+ 1.56
In Separate Ductwork	+ 4.00
Heating Systems, Hot Water	+ 1.34
Heat Pump	+ 2.05
Electric Heat	– .50
Not Heated	– 2.46

Additional upgrades or components

Kitchen Cabinets & Countertops	Page 93
Bathroom Vanities	94
Fireplaces & Chimneys	94
Windows, Skylights & Dormers	94
Appliances	95
Breezeways & Porches	95
Finished Attic	95
Garages	96
Site Improvements	96
Wings & Ells	56

Solid Wall 2 Story

Living Area - 2000 S.F.
Perimeter - 135 L.F.

		Labor-Hours	Cost Per Square Foot Of Living Area		
			Mat.	Labor	Total
1 Site Work	Site preparation for slab; 4′ deep trench excavation for foundation wall.	.034		0.60	0.60
2 Foundation	Continuous reinforced concrete footing 8″ deep x 18″ wide; dampproofed and insulated reinforced concrete foundation wall, 8″ thick, 4′ deep; 4″ concrete slab on 4″ crushed stone base and polyethylene vapor barrier, trowel finish.	.066	2.29	2.90	5.19
3 Framing	Exterior walls - Precut traditional log home. Handicrafted white cedar or pine logs. Delivery included.	.232	13.89	9.22	23.11
4 Exterior Walls					
5 Roofing	25 year asphalt shingles; #15 felt building paper; aluminum gutters, downspouts, drip edge and flashings.	.024	.47	.86	1.33
6 Interiors	Walls and ceilings, 1/2″ taped and finished drywall, primed and painted with 2 coats; painted baseboard and trim, finished hardwood floor 40%, carpet with 1/2″ underlayment 40%, vinyl tile with 1/2″ underlayment 15%, ceramic tile with 1/2″ underlayment 5%; hollow core and louvered interior doors.	.225	9.74	9.56	19.30
7 Specialties	Average grade kitchen cabinets - 14 L.F. wall and base with plastic laminate counter top and kitchen sink; 40 gallon electric water heater.	.021	1.36	.48	1.84
8 Mechanical	1 lavatory, white, wall hung; 1 water closet, white; 1 bathtub with shower, enameled steel, white; gas fired warm air heat.	.060	2.40	1.96	4.36
9 Electrical	200 Amp. service; romex wiring; incandescent lighting fixtures, switches, receptacles.	.039	.79	1.15	1.94
10 Overhead	Contractor's overhead and profit and plans.		6.88	6.65	13.53
		Total	37.82	33.38	**71.20**

1 Story — Base cost per square foot of living area

Exterior Wall	Living Area 50	100	200	300	400	500	600	700
Wood Siding - Wood Frame	138.05	106.75	93.10	79.40	74.95	72.25	70.45	71.20
Brick Veneer - Wood Frame	147.40	110.90	95.10	77.75	72.60	69.50	67.40	67.90
Stucco on Wood Frame	136.95	105.85	92.35	78.80	74.45	71.80	69.95	70.75
Solid Masonry	183.15	139.00	119.95	97.30	91.05	87.25	85.95	86.05
Finished Basement, Add	41.90	34.70	31.20	25.40	24.25	23.55	23.05	22.75
Unfinished Basement, Add	19.95	15.15	13.15	9.85	9.20	8.80	8.50	8.35

1-1/2 Story — Base cost per square foot of living area

Exterior Wall	Living Area 100	200	300	400	500	600	700	800
Wood Siding - Wood Frame	111.20	89.55	76.60	69.60	65.65	63.85	61.50	60.70
Brick Veneer - Wood Frame	164.15	120.85	99.65	87.15	80.75	77.30	73.65	72.20
Stucco on Wood Frame	147.10	107.20	88.30	78.30	72.60	69.60	66.35	65.00
Solid Masonry	188.20	140.10	115.70	99.70	92.30	88.25	83.95	82.40
Finished Basement, Add	28.25	25.50	23.20	20.60	19.95	19.50	19.05	19.00
Unfinished Basement, Add	12.55	10.55	9.20	7.75	7.35	7.10	6.85	6.80

2 Story — Base cost per square foot of living area

Exterior Wall	Living Area 100	200	400	600	800	1000	1200	1400
Wood Siding - Wood Frame	110.60	83.25	71.20	60.20	56.30	54.00	52.30	53.25
Brick Veneer - Wood Frame	167.20	115.45	91.65	73.80	67.60	63.85	61.30	61.55
Stucco on Wood Frame	148.15	101.80	80.30	66.20	60.80	57.50	55.25	55.75
Solid Masonry	194.20	134.70	107.65	84.55	77.25	72.85	69.85	69.85
Finished Basement, Add	22.40	18.85	17.10	14.20	13.60	13.30	13.05	12.90
Unfinished Basement, Add	10.10	7.70	6.70	5.00	4.70	4.55	4.40	4.30

Base costs do not include bathroom or kitchen facilities. Use Modifications/Adjustments/Alternatives on pages 93-96 where appropriate.

Important: See the Reference Section for Location Factors (to adjust for your city) and Estimating Forms

Custom Class

1 Story

1-1/2 Story

2 Story

2-1/2 Story

Bi-Level

Tri-Level

- **A distinct residence from designer's plans**
- **Single family — 1 full bath, 1 half bath, 1 kitchen**
- **No basement**
- **Asphalt shingles on roof**
- **Forced hot air heat/air conditioning**
- **Drywall interior finishes**
- **Materials and workmanship are above average**

Note: The illustration shown may contain some optional components (for example: garages and/or fireplaces) whose costs are shown in the modifications, adjustments, & alternatives below or at the end of the square foot section.

Base cost per square foot of living area

Exterior Wall	Living Area 800	1000	1200	1400	1600	1800	2000	2400	2800	3200	3600
Wood Siding - Wood Frame	123.65	112.80	103.95	96.70	91.75	89.25	86.05	80.00	75.65	72.60	69.25
Brick Veneer - Wood Frame	139.20	127.75	118.30	110.60	105.35	102.65	99.25	92.90	88.20	84.95	81.35
Stone Veneer - Wood Frame	143.30	131.45	121.60	113.60	108.20	105.45	101.85	95.25	90.35	87.00	83.25
Solid Masonry	145.80	133.80	123.70	115.50	110.00	107.15	103.45	96.75	91.75	88.25	84.35
Finished Basement, Add	39.55	39.35	37.55	36.15	35.20	34.65	33.85	32.85	32.05	31.35	30.75
Unfinished Basement, Add	16.20	15.45	14.70	14.10	13.75	13.55	13.20	12.75	12.40	12.10	11.80

Modifications

Add to the total cost

Upgrade Kitchen Cabinets	$ + 656
Solid Surface Countertops	+ 1280
Full Bath - including plumbing, wall and floor finishes	+ 4811
Half Bath - including plumbing, wall and floor finishes	+ 3061
Two Car Attached Garage	+ 18,506
Two Car Detached Garage	+ 20,976
Fireplace & Chimney	+ 4070

Adjustments

For multi family - add to total cost

Additional Kitchen	$ + 8917
Additional Full Bath & Half Bath	+ 7872
Additional Entry & Exit	+ 1236
Separate Heating & Air Conditioning	+ 4555
Separate Electric	+ 1487

For Townhouse/Rowhouse - Multiply cost per square foot by

Inner Unit	.90
End Unit	.95

Alternatives

Add to or deduct from the cost per square foot of living area

Cedar Shake Roof	+ 1.99
Clay Tile Roof	+ 3.75
Slate Roof	+ 7.43
Upgrade Ceilings to Textured Finish	+ .39
Air Conditioning, in Heating Ductwork	Base System
Heating Systems, Hot Water	+ 1.52
Heat Pump	+ 1.82
Electric Heat	– 2.04
Not Heated	– 3.30

Additional upgrades or components

Kitchen Cabinets & Countertops	Page 93
Bathroom Vanities	94
Fireplaces & Chimneys	94
Windows, Skylights & Dormers	94
Appliances	95
Breezeways & Porches	95
Finished Attic	95
Garages	96
Site Improvements	96
Wings & Ells	74

Custom 1 Story

Living Area - 2400 S.F.
Perimeter - 207 L.F.

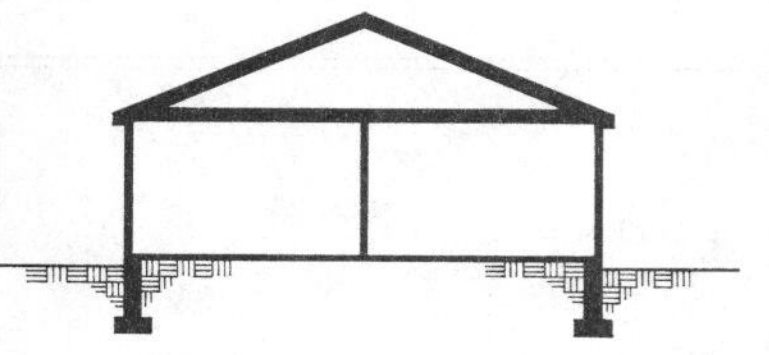

		Labor-Hours	Cost Per Square Foot Of Living Area		
			Mat.	Labor	Total
1 Site Work	Site preparation for slab; 4′ deep trench excavation for foundation wall.	.028		.52	.52
2 Foundation	Continuous reinforced concrete footing 8″ deep x 18″ wide; dampproofed and insulated reinforced concrete foundation wall, 8″ thick, 4′ deep; 4″ concrete slab on 4″ crushed stone base and polyethylene vapor barrier, trowel finish.	.113	4.02	5.38	9.40
3 Framing	Exterior walls - 2″ x 6″ wood studs, 16″ O.C.; 1/2″ plywood sheathing; 2″ x 8″ rafters 16″ O.C. with 1/2″ plywood sheathing, 4 in 12 pitch; 2″ x 6″ ceiling joists 16″ O.C.; 5/8″ plywood subfloor on 1″ x 3″ wood sleepers 16″ O.C.	.190	3.17	4.98	8.15
4 Exterior Walls	Horizontal beveled wood siding; #15 felt building paper; 6″ batt insulation; wood double hung windows; 3 solid core wood exterior doors; storms and screens.	.085	7.00	2.52	9.52
5 Roofing	30 year asphalt shingles; #15 felt building paper; aluminum gutters, downspouts and drip edge; copper flashings.	.082	2.81	2.61	5.42
6 Interiors	Walls and ceilings - 5/8″ drywall, skim coat plaster, painted with primer and 2 coats; hardwood baseboard and trim, sanded and finished; hardwood floor 70%, ceramic tile with underlayment 20%, vinyl tile with underlayment 10%; wood panel interior doors, primed and painted with 2 coats.	.292	11.73	9.42	21.15
7 Specialties	Custom grade kitchen cabinets - 20 L.F. wall and base with plastic laminate counter top and kitchen sink; 4 L.F. bathroom vanity; 75 gallon electric water heater, medicine cabinet.	.019	3.08	.67	3.75
8 Mechanical	Gas fired warm air heat/air conditioning; one full bath including: bathtub, corner shower, built in lavatory and water closet; one 1/2 bath including: built in lavatory and water closet.	.092	4.34	2.34	6.68
9 Electrical	200 Amp. service; romex wiring; fluorescent and incandescent lighting fixtures, switches, receptacles.	.039	.78	1.33	2.11
10 Overhead	Contractor's overhead and profit and design.		7.37	5.93	13.30
		Total	44.30	35.70	**80.00**

- **A distinct residence from designer's plans**
- **Single family — 1 full bath, 1 half bath, 1 kitchen**
- **No basement**
- **Asphalt shingles on roof**
- **Forced hot air heat/air conditioning**
- **Drywall interior finishes**
- **Materials and workmanship are above average**

Note: The illustration shown may contain some optional components (for example: garages and/or fireplaces) whose costs are shown in the modifications, adjustments, & alternatives below or at the end of the square foot section.

Base cost per square foot of living area

Exterior Wall	Living Area 1000	1200	1400	1600	1800	2000	2400	2800	3200	3600	4000
Wood Siding - Wood Frame	113.50	106.10	100.80	94.25	90.60	86.85	79.80	76.65	73.80	71.40	68.20
Brick Veneer - Wood Frame	120.20	112.40	106.80	99.70	95.70	91.75	84.10	80.70	77.55	75.05	71.60
Stone Veneer - Wood Frame	124.45	116.40	110.60	103.15	99.00	94.85	86.85	83.30	80.00	77.40	73.75
Solid Masonry	127.15	118.90	113.05	105.30	101.05	96.80	88.55	84.95	81.45	78.85	75.15
Finished Basement, Add	26.35	26.40	25.65	24.60	24.00	23.40	22.25	21.80	21.15	20.90	20.40
Unfinished Basement, Add	10.75	10.40	10.10	9.65	9.40	9.20	8.70	8.55	8.25	8.15	7.90

Modifications

Add to the total cost

Upgrade Kitchen Cabinets	$ + 656
Solid Surface Countertops	+ 1280
Full Bath - including plumbing, wall and floor finishes	+ 4811
Half Bath - including plumbing, wall and floor finishes	+ 3061
Two Car Attached Garage	+ 18,506
Two Car Detached Garage	+ 20,976
Fireplace & Chimney	+ 4595

Adjustments

For multi family - add to total cost

Additional Kitchen	$ + 8917
Additional Full Bath & Half Bath	+ 7872
Additional Entry & Exit	+ 1236
Separate Heating & Air Conditioning	+ 4555
Separate Electric	+ 1487

For Townhouse/Rowhouse - Multiply cost per square foot by

Inner Unit	.90
End Unit	.95

Alternatives

Add to or deduct from the cost per square foot of living area

Cedar Shake Roof	+ 1.44
Clay Tile Roof	+ 2.71
Slate Roof	+ 5.37
Upgrade Ceilings to Textured Finish	+ .39
Air Conditioning, in Heating Ductwork	Base System
Heating Systems, Hot Water	+ 1.44
Heat Pump	+ 1.92
Electric Heat	– 1.77
Not Heated	– 3.04

Additional upgrades or components

Kitchen Cabinets & Countertops	Page 93
Bathroom Vanities	94
Fireplaces & Chimneys	94
Windows, Skylights & Dormers	94
Appliances	95
Breezeways & Porches	95
Finished Attic	95
Garages	96
Site Improvements	96
Wings & Ells	74

Custom 1-1/2 Story

Living Area - 2800 S.F.
Perimeter - 175 L.F.

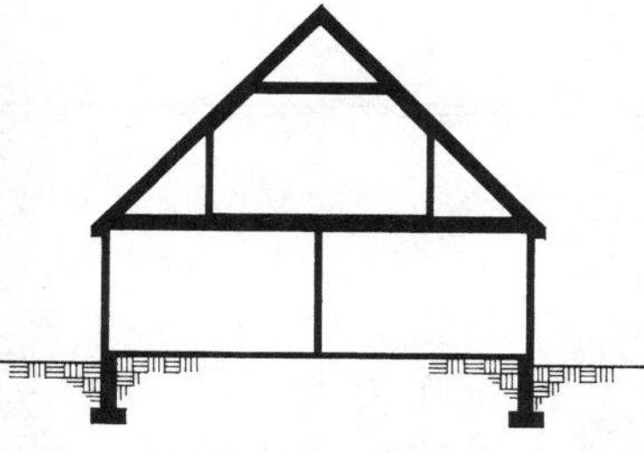

		Labor-Hours	Cost Per Square Foot Of Living Area		
			Mat.	Labor	Total
1 Site Work	Site preparation for slab; 4′ deep trench excavation for foundation wall.	.028		.45	.45
2 Foundation	Continuous reinforced concrete footing 8″ deep x 18″ wide; dampproofed and insulated reinforced concrete foundation wall, 8″ thick, 4′ deep; 4″ concrete slab on 4″ crushed stone base and polyethylene vapor barrier, trowel finish.	.065	2.80	3.83	6.63
3 Framing	Exterior walls - 2″ x 6″ wood studs, 16″ O.C.; 1/2″ plywood sheathing; 2″ x 8″ rafters 16″ O.C. with 1/2″ plywood sheathing, 8 in 12 pitch; 2″ x 10″ floor joists 16″ O.C. with 5/8″ plywood subfloor; 5/8″ plywood subfloor on 1″ x 3″ wood sleepers 16″ O.C.	.192	4.09	5.27	9.36
4 Exterior Walls	Horizontal beveled wood siding; #15 felt building paper; 6″ batt insulation; wood double hung windows; 3 solid core wood exterior doors; storms and screens.	.064	6.99	2.56	9.55
5 Roofing	30 year asphalt shingles; #15 felt building paper; aluminum gutters, downspouts and drip edge; copper flashings.	.048	1.76	1.63	3.39
6 Interiors	Walls and ceilings - 5/8″ drywall, skim coat plaster, painted with primer and 2 coats; hardwood baseboard and trim, sanded and finished; hardwood floor 70%, ceramic tile with underlayment 20%, vinyl tile with underlayment 10%; wood panel interior doors, primed and painted with 2 coats.	.259	12.97	10.35	23.32
7 Specialties	Custom grade kitchen cabinets - 20 L.F. wall and base with plastic laminate counter top and kitchen sink; 4 L.F. bathroom vanity; 75 gallon electric water heater, medicine cabinet.	.030	2.63	.56	3.19
8 Mechanical	Gas fired warm air heat/air conditioning; one full bath including: bathtub, corner shower, built in lavatory and water closet; one 1/2 bath including: built in lavatory and water closet.	.084	3.75	2.19	5.94
9 Electrical	200 Amp. service; romex wiring; fluorescent and incandescent lighting fixtures, switches, receptacles.	.038	.74	1.27	2.01
10 Overhead	Contractor's overhead and profit and design.		7.17	5.64	12.81
		Total	42.90	33.75	**76.65**

- **A distinct residence from designer's plans**
- **Single family — 1 full bath, 1 half bath, 1 kitchen**
- **No basement**
- **Asphalt shingles on roof**
- **Forced hot air heat/air conditioning**
- **Drywall interior finishes**
- **Materials and workmanship are above average**

Note: The illustration shown may contain some optional components (for example: garages and/or fireplaces) whose costs are shown in the modifications, adjustments, & alternatives below or at the end of the square foot section.

Base cost per square foot of living area

	Living Area										
Exterior Wall	**1200**	**1400**	**1600**	**1800**	**2000**	**2400**	**2800**	**3200**	**3600**	**4000**	**4400**
Wood Siding - Wood Frame	107.00	100.95	96.65	93.05	88.80	82.85	**77.65**	74.30	72.35	70.25	68.50
Brick Veneer - Wood Frame	114.15	107.70	103.05	99.00	94.55	88.05	82.40	78.75	76.60	74.35	72.40
Stone Veneer - Wood Frame	118.70	111.95	107.15	102.85	98.25	91.35	85.45	81.55	79.35	76.90	74.90
Solid Masonry	121.55	114.65	109.65	105.25	100.55	93.45	87.30	83.35	81.10	78.50	76.45
Finished Basement, Add	21.25	21.30	20.70	20.15	19.65	18.80	18.10	17.65	17.35	16.95	16.75
Unfinished Basement, Add	8.60	8.30	8.15	7.90	7.75	7.35	7.05	6.90	6.75	6.60	6.50

Modifications

Add to the total cost

Upgrade Kitchen Cabinets	$ + 656
Solid Surface Countertops	+ 1280
Full Bath - including plumbing, wall and floor finishes	+ 4811
Half Bath - including plumbing, wall and floor finishes	+ 3061
Two Car Attached Garage	+ 18,506
Two Car Detached Garage	+ 20,976
Fireplace & Chimney	+ 4595

Adjustments

For multi family - add to total cost

Additional Kitchen	$ + 8917
Additional Full Bath & Half Bath	+ 7872
Additional Entry & Exit	+ 1236
Separate Heating & Air Conditioning	+ 4555
Separate Electric	+ 1487

For Townhouse/Rowhouse - Multiply cost per square foot by

Inner Unit	.87
End Unit	.93

Alternatives

Add to or deduct from the cost per square foot of living area

Cedar Shake Roof	+ 1.00
Clay Tile Roof	+ 1.88
Slate Roof	+ 3.72
Upgrade Ceilings to Textured Finish	+ .39
Air Conditioning, in Heating Ductwork	Base System
Heating Systems, Hot Water	+ 1.41
Heat Pump	+ 2.14
Electric Heat	– 1.77
Not Heated	– 2.88

Additional upgrades or components

Kitchen Cabinets & Countertops	Page 93
Bathroom Vanities	94
Fireplaces & Chimneys	94
Windows, Skylights & Dormers	94
Appliances	95
Breezeways & Porches	95
Finished Attic	95
Garages	96
Site Improvements	96
Wings & Ells	74

Important: See the Reference Section for Location Factors (to adjust for your city) and Estimating Forms

Custom 2 Story

Living Area - 2800 S.F.
Perimeter - 156 L.F.

		Labor-Hours	Cost Per Square Foot Of Living Area		
			Mat.	Labor	Total
1 Site Work	Site preparation for slab; 4′ deep trench excavation for foundation wall.	.024		.45	.45
2 Foundation	Continuous reinforced concrete footing 8″ deep x 18″ wide; dampproofed and insulated reinforced concrete foundation wall, 8″ thick, 4′ deep; 4″ concrete slab on 4″ crushed stone base and polyethylene vapor barrier, trowel finish.	.058	2.37	3.30	5.67
3 Framing	Exterior walls - 2″ x 6″ wood studs, 16″ O.C.; 1/2″ plywood sheathing; 2″ x 8″ rafters 16″ O.C. with 1/2″ plywood sheathing, 6 in 12 pitch; 2″ x 8″ ceiling joists 16″ O.C.; 2″ x 10″ floor joists 16″ O.C. with 5/8″ plywood subfloor; 5/8″ plywood subfloor on 1″ x 3″ wood sleepers 16″ O.C.	.159	4.49	5.42	9.91
4 Exterior Walls	Horizontal beveled wood siding; #15 felt building paper; 6″ batt insulation; wood double hung windows; 3 solid core wood exterior doors; storms and screens.	.091	7.94	2.94	10.88
5 Roofing	30 year asphalt shingles; #15 felt building paper; aluminum gutters, downspouts and drip edge; copper flashings.	.042	1.41	1.31	2.72
6 Interiors	Walls and ceilings - 5/8″ drywall, skim coat plaster, painted with primer and 2 coats; hardwood baseboard and trim, sanded and finished; hardwood floor 70%, ceramic tile with underlayment 20%, vinyl tile with underlayment 10%; wood panel interior doors, primed and painted with 2 coats.	.271	13.19	10.59	23.78
7 Specialties	Custom grade kitchen cabinets - 20 L.F. wall and base with plastic laminate counter top and kitchen sink; 4 L.F. bathroom vanity; 75 gallon electric water heater, medicine cabinet.	.028	2.63	.56	3.19
8 Mechanical	Gas fired warm air heat/air conditioning; one full bath including: bathtub, corner shower; built in lavatory and water closet; one 1/2 bath including: built in lavatory and water closet.	.078	3.87	2.23	6.10
9 Electrical	200 Amp. service; romex wiring; fluorescent and incandescent lighting fixtures, switches, receptacles.	.038	.74	1.27	2.01
10 Overhead	Contractor's overhead and profit and design.		7.31	5.63	12.94
		Total	43.95	33.70	**77.65**

- **A distinct residence from designer's plans**
- **Single family — 1 full bath, 1 half bath, 1 kitchen**
- **No basement**
- **Asphalt shingles on roof**
- **Forced hot air heat/air conditioning**
- **Drywall interior finishes**
- **Materials and workmanship are above average**

Note: The illustration shown may contain some optional components (for example: garages and/or fireplaces) whose costs are shown in the modifications, adjustments, & alternatives below or at the end of the square foot section.

Base cost per square foot of living area

	Living Area										
Exterior Wall	1500	1800	2100	2400	2800	3200	3600	4000	4500	5000	5500
Wood Siding - Wood Frame	106.50	96.20	90.20	86.70	82.05	77.45	75.00	71.00	69.00	67.20	65.30
Brick Veneer - Wood Frame	114.00	103.00	96.35	92.55	87.60	82.50	79.75	75.50	73.20	71.20	69.10
Stone Veneer - Wood Frame	118.70	107.40	100.30	96.20	91.15	85.70	82.85	78.30	75.85	73.75	71.50
Solid Masonry	121.70	110.15	102.65	98.55	93.35	87.75	84.70	80.05	77.55	75.30	73.00
Finished Basement, Add	16.80	16.75	15.85	15.40	15.00	14.45	14.10	13.70	13.40	13.20	12.95
Unfinished Basement, Add	6.90	6.60	6.25	6.05	5.95	5.65	5.50	5.35	5.25	5.10	5.05

Modifications

Add to the total cost

Upgrade Kitchen Cabinets	$ + 656
Solid Surface Countertops	+ 1280
Full Bath - including plumbing, wall and floor finishes	+ 4811
Half Bath - including plumbing, wall and floor finishes	+ 3061
Two Car Attached Garage	+ 18,506
Two Car Detached Garage	+ 20,976
Fireplace & Chimney	+ 5190

Adjustments

For multi family - add to total cost

Additional Kitchen	$ + 8917
Additional Full Bath & Half Bath	+ 7872
Additional Entry & Exit	+ 1236
Separate Heating & Air Conditioning	+ 4555
Separate Electric	+ 1487

For Townhouse/Rowhouse - Multiply cost per square foot by

Inner Unit	.87
End Unit	.94

Alternatives

Add to or deduct from the cost per square foot of living area

Cedar Shake Roof	+ .86
Clay Tile Roof	+ 1.62
Slate Roof	+ 3.22
Upgrade Ceilings to Textured Finish	+ .39
Air Conditioning, in Heating Ductwork	Base System
Heating Systems, Hot Water	+ 1.27
Heat Pump	+ 2.20
Electric Heat	– 3.15
Not Heated	– 2.88

Additional upgrades or components

Kitchen Cabinets & Countertops	Page 93
Bathroom Vanities	94
Fireplaces & Chimneys	94
Windows, Skylights & Dormers	94
Appliances	95
Breezeways & Porches	95
Finished Attic	95
Garages	96
Site Improvements	96
Wings & Ells	74

Important: See the Reference Section for Location Factors (to adjust for your city) and Estimating Forms

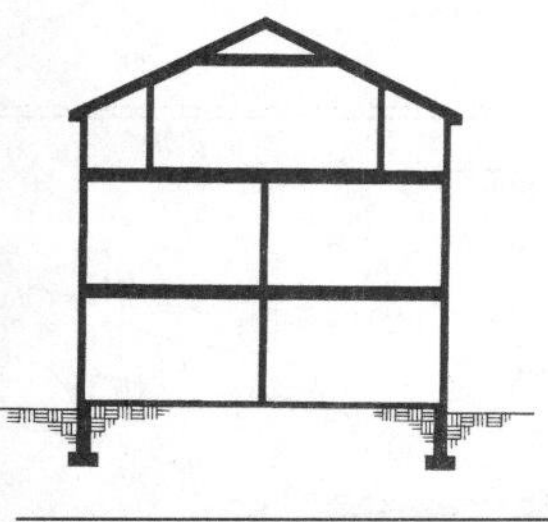

Custom 2-1/2 Story

Living Area - 3200 S.F.
Perimeter - 150 L.F.

			Labor-Hours	Cost Per Square Foot Of Living Area		
				Mat.	Labor	Total
1	Site Work	Site preparation for slab; 4′ deep trench excavation for foundation wall.	.048		.40	.40
2	Foundation	Continuous reinforced concrete footing 8″ deep x 18″ wide; dampproofed and insulated reinforced concrete foundation wall, 8″ thick, 4′ deep; 4″ concrete slab on 4″ crushed stone base and polyethylene vapor barrier, trowel finish.	.063	1.96	2.74	4.70
3	Framing	Exterior walls - 2″ x 6″ wood studs, 16″ O.C.; 1/2″ plywood sheathing; 2″ x 8″ rafters 16″ O.C. with 1/2″ plywood sheathing, 6 in 12 pitch; 2″ x 8″ ceiling joists 16″ O.C.; 2″ x 10″ floor joists 16″ O.C. with 5/8″ plywood subfloor; 5/8″ plywood subfloor on 1″ x 3″ wood sleepers 16″ O.C.	.177	4.86	5.65	10.51
4	Exterior Walls	Horizontal beveled wood siding; #15 felt building paper; 6″ batt insulation; wood double hung windows; 3 solid core wood exterior doors; storms and screens.	.134	8.22	3.04	11.26
5	Roofing	30 year asphalt shingles; #15 felt building paper; aluminum gutters, downspouts and drip edge; copper flashings.	.032	1.08	1.00	2.08
6	Interiors	Walls and ceilings - 5/8″ drywall, skim coat plaster, painted with primer and 2 coats; hardwood baseboard and trim, sanded and finished; hardwood floor 70%, ceramic tile with underlayment 20%, vinyl tile with underlayment 10%; wood panel interior doors, primed and painted with 2 coats.	.354	13.92	11.28	25.20
7	Specialties	Custom grade kitchen cabinets - 20 L.F. wall and base with plastic laminate counter top and kitchen sink; 4 L.F. bathroom vanity; 75 gallon electric water heater, medicine cabinet.	.053	2.32	.49	2.81
8	Mechanical	Gas fired warm air heat/air conditioning; one full bath including: bathtub, corner shower; built in lavatory and water closet; one 1/2 bath including: built in lavatory and water closet.	.104	3.53	2.12	5.65
9	Electrical	200 Amp. service; romex wiring; fluorescent and incandescent lighting fixtures, switches, receptacles.	.048	.71	1.24	1.95
10	Overhead	Contractor's overhead and profit and design.		7.30	5.59	12.89
			Total	43.90	33.55	**77.45**

- **A distinct residence from designer's plans**
- **Single family — 1 full bath, 1 half bath, 1 kitchen**
- **No basement**
- **Asphalt shingles on roof**
- **Forced hot air heat/air conditioning**
- **Drywall interior finishes**
- **Materials and workmanship are above average**

Note: The illustration shown may contain some optional components (for example: garages and/or fireplaces) whose costs are shown in the modifications, adjustments, & alternatives below or at the end of the square foot section.

Base cost per square foot of living area

Exterior Wall	Living Area 1500	1800	2100	2500	3000	3500	4000	4500	5000	5500	6000
Wood Siding - Wood Frame	106.10	95.90	91.05	87.05	80.80	77.60	73.75	69.55	68.10	66.55	64.95
Brick Veneer - Wood Frame	113.85	103.05	97.80	93.50	86.60	83.15	78.70	74.15	72.55	70.85	69.00
Stone Veneer - Wood Frame	118.80	107.60	102.05	97.55	90.30	86.60	81.85	77.10	75.40	73.60	71.60
Solid Masonry	121.90	110.45	104.70	100.10	92.60	88.80	83.85	78.95	77.15	75.25	73.15
Finished Basement, Add	14.85	14.80	14.10	13.75	13.10	12.70	12.25	11.90	11.70	11.55	11.30
Unfinished Basement, Add	6.00	5.75	5.50	5.40	5.15	5.00	4.80	4.60	4.60	4.50	4.40

Modifications

Add to the total cost

Upgrade Kitchen Cabinets	$ + 656
Solid Surface Countertops	+ 1280
Full Bath - including plumbing, wall and floor finishes	+ 4811
Half Bath - including plumbing, wall and floor finishes	+ 3061
Two Car Attached Garage	+ 18,506
Two Car Detached Garage	+ 20,976
Fireplace & Chimney	+ 5190

Adjustments

For multi family - add to total cost

Additional Kitchen	$ + 8917
Additional Full Bath & Half Bath	+ 7872
Additional Entry & Exit	+ 1236
Separate Heating & Air Conditioning	+ 4555
Separate Electric	+ 1487

For Townhouse/Rowhouse - Multiply cost per square foot by

Inner Unit	.85
End Unit	.93

Alternatives

Add to or deduct from the cost per square foot of living area

Cedar Shake Roof	+ .66
Clay Tile Roof	+ 1.25
Slate Roof	+ 2.48
Upgrade Ceilings to Textured Finish	+ .39
Air Conditioning, in Heating Ductwork	Base System
Heating Systems, Hot Water	+ 1.27
Heat Pump	+ 2.20
Electric Heat	– 3.15
Not Heated	– 2.78

Additional upgrades or components

Important: See the Reference Section for Location Factors (to adjust for your city) and Estimating Forms

Custom 3 Story

Living Area - 3000 S.F.
Perimeter - 135 L.F.

		Labor-Hours	Cost Per Square Foot Of Living Area		
			Mat.	Labor	Total
1 Site Work	Site preparation for slab; 4′ deep trench excavation for foundation wall.	.048		.42	.42
2 Foundation	Continuous reinforced concrete footing 8″ deep x 18″ wide; dampproofed and insulated reinforced concrete foundation wall, 8″ thick, 4′ deep; 4″ concrete slab on 4″ crushed stone base and polyethylene vapor barrier, trowel finish.	.060	1.83	2.59	4.42
3 Framing	Exterior walls - 2″ x 6″ wood studs, 16″ O.C.; 1/2″ plywood sheathing; 2″ x 8″ rafters 16″ O.C. with 1/2″ plywood sheathing, 6 in 12 pitch; 2″ x 8″ ceiling joists 16″ O.C.; 2″ x 10″ floor joists 16″ O.C. with 5/8″ plywood subfloor; 5/8″ plywood subfloor on 1″ x 3″ wood sleepers 16″ O.C.	.191	5.12	5.83	10.95
4 Exterior Walls	Horizontal beveled wood siding; #15 felt building paper; 6″ batt insulation; wood double hung windows; 3 solid core wood exterior doors; storms and screens.	.150	9.39	3.47	12.86
5 Roofing	30 year asphalt shingles; #15 felt building paper; aluminum gutters, downspouts and drip edge; copper flashings.	.028	.94	.87	1.81
6 Interiors	Walls and ceilings - 5/8″ drywall, skim coat plaster, painted with primer and 2 coats; hardwood baseboard and trim, sanded and finished; hardwood floor 70%, ceramic tile with underlayment 20%, vinyl tile with underlayment 10%; wood panel interior doors, primed and painted with 2 coats.	.409	14.37	11.66	26.03
7 Specialties	Custon grade kitchen cabinets - 20 L.F. wall and base with plastic laminate counter top and kitchen sink; 4 L.F. bathroom vanity; 75 gallon electric water heater, medicine cabinet.	.053	2.47	.53	3.00
8 Mechanical	Gas fired warm air heat/air conditioning; one full bath including: bathtub, corner shower; built in lavatory and water closet; one 1/2 bath including: built in lavatory and water closet.	.105	3.69	2.18	5.87
9 Electrical	200 Amp. service; romex wiring; fluorescent and incandescent lighting fixtures, switches, receptacles.	.048	.72	1.26	1.98
10 Overhead	Contractor's overhead and profit and design.		7.72	5.74	13.46
		Total	46.25	34.55	**80.80**

- **A distinct residence from designer's plans**
- **Single family — 1 full bath, 1 half bath, 1 kitchen**
- **No basement**
- **Asphalt shingles on roof**
- **Forced hot air heat/air conditioning**
- **Drywall interior finishes**
- **Materials and workmanship are above average**

Note: The illustration shown may contain some optional components (for example: garages and/or fireplaces) whose costs are shown in the modifications, adjustments, & alternatives below or at the end of the square foot section.

Base cost per square foot of living area

Exterior Wall	Living Area 1200	1400	1600	1800	2000	2400	2800	3200	3600	4000	4400
Wood Siding - Wood Frame	101.15	95.50	91.40	87.95	84.00	78.50	73.75	70.70	68.85	66.95	65.35
Brick Veneer - Wood Frame	106.50	100.50	96.20	92.50	88.35	82.45	77.25	74.05	72.10	69.95	68.25
Stone Veneer - Wood Frame	109.90	103.70	99.30	95.35	91.10	84.90	79.60	76.15	74.10	71.90	70.15
Solid Masonry	112.05	105.75	101.15	97.20	92.85	86.45	81.00	77.50	75.40	73.15	71.25
Finished Basement, Add	21.25	21.30	20.70	20.15	19.65	18.80	18.10	17.65	17.35	16.95	16.75
Unfinished Basement, Add	8.60	8.30	8.15	7.90	7.75	7.35	7.05	6.90	6.75	6.60	6.50

Modifications

Add to the total cost

Upgrade Kitchen Cabinets	$ + 656
Solid Surface Countertops	+ 1280
Full Bath - including plumbing, wall and floor finishes	+ 4811
Half Bath - including plumbing, wall and floor finishes	+ 3061
Two Car Attached Garage	+ 18,506
Two Car Detached Garage	+ 20,976
Fireplace & Chimney	+ 4070

Adjustments

For multi family - add to total cost

Additional Kitchen	$ + 8917
Additional Full Bath & Half Bath	+ 7872
Additional Entry & Exit	+ 1236
Separate Heating & Air Conditioning	+ 4555
Separate Electric	+ 1487

For Townhouse/Rowhouse - Multiply cost per square foot by

Inner Unit	.89
End Unit	.95

Alternatives

Add to or deduct from the cost per square foot of living area

Cedar Shake Roof	+ 1.99
Clay Tile Roof	+ 3.75
Slate Roof	+ 3.72
Upgrade Ceilings to Textured Finish	+ .39
Air Conditioning, in Heating Ductwork	Base System
Heating Systems, Hot Water	+ 1.41
Heat Pump	+ 2.14
Electric Heat	− 1.78
Not Heated	− 2.78

Additional upgrades or components

Kitchen Cabinets & Countertops	Page 93
Bathroom Vanities	94
Fireplaces & Chimneys	94
Windows, Skylights & Dormers	94
Appliances	95
Breezeways & Porches	95
Finished Attic	95
Garages	96
Site Improvements	96
Wings & Ells	74

Important: See the Reference Section for Location Factors (to adjust for your city) and Estimating Forms

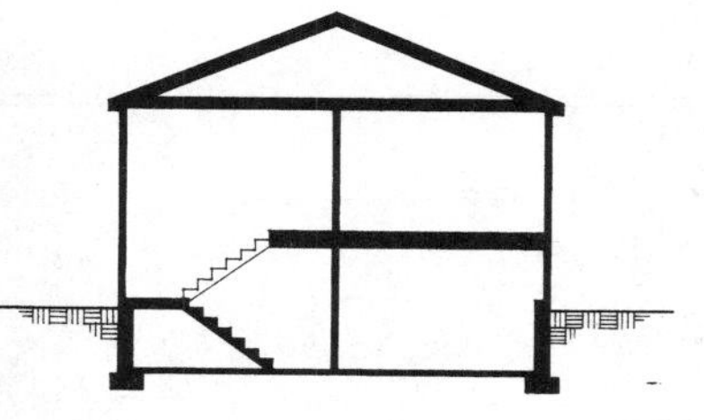

Custom Bi-Level

Living Area - 2800 S.F.
Perimeter - 156 L.F.

		Labor-Hours	Cost Per Square Foot Of Living Area		
			Mat.	Labor	Total
1 Site Work	Excavation for lower level, 4′ deep. Site preparation for slab.	.024		.45	.45
2 Foundation	Continuous reinforced concrete footing 8″ deep x 18″ wide; dampproofed and insulated reinforced concrete foundation wall, 8″ thick, 4′ deep; 4″ concrete slab on 4″ crushed stone base and polyethylene vapor barrier, trowel finish.	.058	2.37	3.30	5.67
3 Framing	Exterior walls - 2″ x 6″ wood studs, 16″ O.C.; 1/2″ plywood sheathing; 2″ x 8″ rafters 16″ O.C. with 1/2″ plywood sheathing, 6 in 12 pitch; 2″ x 8″ ceiling joists 16″ O.C.; 2″ x 10″ floor joists 16″ O.C. with 5/8″ plywood subfloor; 5/8″ plywood subfloor on 1″ x 3″ wood sleepers 16″ O.C.	.147	4.26	5.17	9.43
4 Exterior Walls	Horizontal beveled wood siding; #15 felt building paper; 6″ batt insulation; wood double hung windows; 3 solid core wood exterior doors; storms and screens.	.079	6.22	2.26	8.48
5 Roofing	30 year asphalt shingles; #15 felt building paper; aluminum gutters, downspouts and drip edge; copper flashings.	.033	1.41	1.31	2.72
6 Interiors	Walls and ceilings - 5/8″ drywall, skim coat plaster, painted with primer and 2 coats; hardwood baseboard and trim, sanded and finished; hardwood floor 70%, ceramic tile with underlayment 20%, vinyl tile with underlayment 10%; wood panel interior doors, primed and painted with 2 coats.	.257	13.07	10.32	23.39
7 Specialties	Custom grade kitchen cabinets - 20 L.F. wall and base with plastic laminate counter top and kitchen sink; 4 L.F. bathroom vanity; 75 gallon electric water heater, medicine cabinet.	.028	2.63	.56	3.19
8 Mechanical	Gas fired warm air heat/air conditioning; one full bath including: bathtub, corner shower, built in lavatory and water closet; one 1/2 bath including: built in lavatory and water closet.	.078	3.87	2.23	6.10
9 Electrical	200 Amp. service; romex wiring; fluorescent and incandescent lighting fixtures, switches, receptacles.	.038	.74	1.27	2.01
10 Overhead	Contractor's overhead and profit and design.		6.93	5.38	12.31
		Total	41.50	32.25	**73.75**

- **A distinct residence from designer's plans**
- **Single family — 1 full bath, 1 half bath, 1 kitchen**
- **No basement**
- **Asphalt shingles on roof**
- **Forced hot air heat/air conditioning**
- **Drywall interior finishes**
- **Materials and workmanship are above average**

Note: The illustration shown may contain some optional components (for example: garages and/or fireplaces) whose costs are shown in the modifications, adjustments, & alternatives below or at the end of the square foot section.

Base cost per square foot of living area

Exterior Wall	Living Area 1200	1500	1800	2100	2400	2800	3200	3600	4000	4500	5000
Wood Siding - Wood Frame	103.90	94.90	88.10	82.30	78.60	75.85	72.60	69.00	67.55	64.25	62.45
Brick Veneer - Wood Frame	109.20	99.75	92.45	86.30	82.30	79.40	75.95	72.10	70.60	67.00	65.10
Stone Veneer - Wood Frame	112.60	102.85	95.20	88.80	84.70	81.70	78.00	74.00	72.45	68.75	66.75
Solid Masonry	114.70	104.75	96.95	90.40	86.20	83.10	79.30	75.30	73.65	69.85	67.75
Finished Basement, Add*	26.40	26.25	25.05	24.10	23.50	23.00	22.45	21.90	21.65	21.15	20.80
Unfinished Basement, Add*	10.80	10.30	9.80	9.40	9.20	8.95	8.75	8.50	8.40	8.15	8.00

*Basement under middle level only.

Modifications

Add to the total cost

Upgrade Kitchen Cabinets	$ + 656
Solid Surface Countertops	+ 1280
Full Bath - including plumbing, wall and floor finishes	+ 4811
Half Bath - including plumbing, wall and floor finishes	+ 3061
Two Car Attached Garage	+ 18,506
Two Car Detached Garage	+ 20,976
Fireplace & Chimney	+ 4595

Adjustments

For multi family - add to total cost

Additional Kitchen	$ + 8917
Additional Full Bath & Half Bath	+ 7872
Additional Entry & Exit	+ 1236
Separate Heating & Air Conditioning	+ 4555
Separate Electric	+ 1487

For Townhouse/Rowhouse - Multiply cost per square foot by

Inner Unit	.87
End Unit	.94

Alternatives

Add to or deduct from the cost per square foot of living area

Cedar Shake Roof	+ 1.44
Clay Tile Roof	+ 2.71
Slate Roof	+ 5.37
Upgrade Ceilings to Textured Finish	+ .39
Air Conditioning, in Heating Ductwork	Base System
Heating Systems, Hot Water	+ 1.36
Heat Pump	+ 2.22
Electric Heat	− 1.59
Not Heated	− 2.78

Additional upgrades or components

Kitchen Cabinets & Countertops	Page 93
Bathroom Vanities	94
Fireplaces & Chimneys	94
Windows, Skylights & Dormers	94
Appliances	95
Breezeways & Porches	95
Finished Attic	95
Garages	96
Site Improvements	96
Wings & Ells	74

Important: See the Reference Section for Location Factors (to adjust for your city) and Estimating Forms

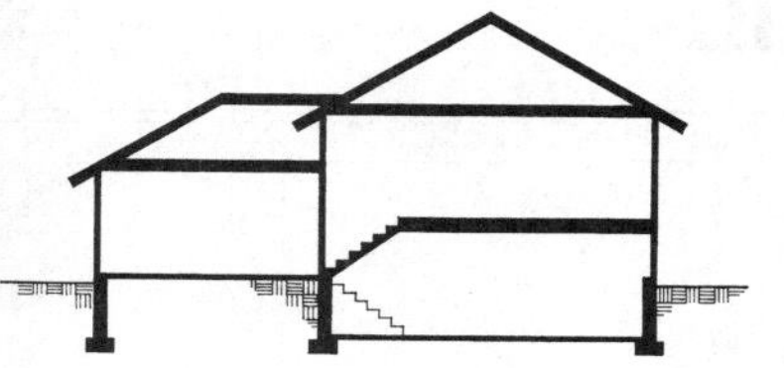

Custom Tri-Level

Living Area - 3200 S.F.
Perimeter - 198 L.F.

		Labor-Hours	Cost Per Square Foot Of Living Area		
			Mat.	Labor	Total
1 Site Work	Site preparation for slab; 4′ deep trench excavation for foundation wall, excavation for lower level, 4′ deep.	.023		.40	.40
2 Foundation	Continuous reinforced concrete footing 8″ deep x 18″ wide; dampproofed and insulated reinforced concrete foundation wall, 8″ thick, 4′ deep; 4″ concrete slab on 4″ crushed stone base and polyethylene vapor barrier, trowel finish.	.073	2.81	3.80	6.61
3 Framing	Exterior walls - 2″ x 6″ wood studs, 16″ O.C.; 1/2″ plywood sheathing; 2″ x 8″ rafters 16″ O.C. with 1/2″ plywood sheathing, 6 in 12 pitch; 2″ x 8″ ceiling joists 16″ O.C.; 2″ x 10″ floor joists 16″ O.C. with 5/8″ plywood subfloor; 5/8″ plywood subfloor on 1″ x 3″ wood sleepers 16″ O.C.	.162	3.87	5.09	8.96
4 Exterior Walls	Horizontal beveled wood siding; #15 felt building paper; 6″ batt insulation; wood double hung windows; 3 solid core wood exterior doors; storms and screens.	.076	6.02	2.19	8.21
5 Roofing	30 year asphalt shingles; #15 felt building paper; aluminum gutters, downspouts and drip edge; copper flashings.	.045	1.87	1.75	3.62
6 Interiors	Walls and ceilings - 5/8″ drywall, skim coat plaster, painted with primer and 2 coats; hardwood baseboard and trim, sanded and finished; hardwood floor 70%, ceramic tile with underlayment 20%, vinyl tile with underlayment 10%; wood panel interior doors, primed and painted with 2 coats.	.242	12.46	9.85	22.31
7 Specialties	Custom grade kitchen cabinets - 20 L.F. wall and base with plastic laminate counter top and kitchen sink; 4 L.F. bathroom vanity; 75 gallon electric water heater, medicine cabinet.	.026	2.32	.49	2.81
8 Mechanical	Gas fired warm air heat/air conditioning; one full bath including: bathtub, corner shower, built in lavatory and water closet; one 1/2 bath including: built in lavatory and water closet.	.073	3.53	2.12	5.65
9 Electrical	200 Amp. service; romex wiring; fluorescent and incandescent lighting fixtures, switches, receptacles.	.036	.71	1.24	1.95
10 Overhead	Contractor's overhead and profit and design.		6.71	5.37	12.08
		Total	40.30	32.30	**72.60**

1 Story — Base cost per square foot of living area

Exterior Wall	Living Area 50	100	200	300	400	500	600	700
Wood Siding - Wood Frame	162.85	126.25	110.35	93.75	88.55	85.40	83.30	84.05
Brick Veneer - Wood Frame	180.90	139.15	121.05	100.90	94.95	91.45	89.00	89.60
Stone Veneer - Wood Frame	192.40	147.35	127.90	105.45	99.10	95.25	92.65	93.10
Solid Masonry	199.60	152.55	132.20	108.30	101.65	97.65	94.95	95.30
Finished Basement, Add	62.10	53.05	48.05	39.75	38.10	37.10	36.45	35.95
Unfinished Basement, Add	32.10	25.00	21.95	17.30	16.35	15.80	15.40	15.15

1-1/2 Story — Base cost per square foot of living area

Exterior Wall	Living Area 100	200	300	400	500	600	700	800
Wood Siding - Wood Frame	129.80	106.05	91.70	83.35	79.00	77.05	74.35	73.55
Brick Veneer - Wood Frame	145.85	118.95	102.45	91.75	86.75	84.35	81.25	80.35
Stone Veneer - Wood Frame	156.15	127.15	109.25	97.05	91.65	89.00	85.65	84.70
Solid Masonry	162.60	132.35	113.60	100.45	94.80	91.85	88.40	87.45
Finished Basement, Add	41.55	38.45	35.10	31.45	30.45	29.85	29.20	29.15
Unfinished Basement, Add	20.65	17.65	15.70	13.70	13.05	12.70	12.30	12.30

2 Story — Base cost per square foot of living area

Exterior Wall	Living Area 100	200	400	600	800	1000	1200	1400
Wood Siding - Wood Frame	129.80	98.75	85.20	72.45	68.00	65.35	63.55	64.55
Brick Veneer - Wood Frame	147.80	111.65	95.90	79.60	74.50	71.40	69.25	70.05
Stone Veneer - Wood Frame	159.30	119.85	102.70	84.15	78.55	75.20	72.90	73.55
Solid Masonry	166.50	125.05	107.05	87.00	81.10	77.60	75.20	75.75
Finished Basement, Add	31.15	26.50	24.05	19.95	19.10	18.60	18.25	18.00
Unfinished Basement, Add	16.05	12.50	11.00	8.65	8.20	7.90	7.70	7.60

Base costs do not include bathroom or kitchen facilities. Use Modifications/Adjustments/Alternatives on pages 93-96 where appropriate.

Luxury Class

1 Story

1-1/2 Story

2 Story

2-1/2 Story

Bi-Level

Tri-Level

- **Unique residence built from an architect's plan**
- **Single family — 1 full bath, 1 half bath, 1 kitchen**
- **No basement**
- **Cedar shakes on roof**
- **Forced hot air heat/air conditioning**
- **Double drywall interior**
- **Many special features**
- **Extraordinary materials and workmanship**

Note: The illustration shown may contain some optional components (for example: garages and/or fireplaces) whose costs are shown in the modifications, adjustments, & alternatives below or at the end of the square foot section.

Base cost per square foot of living area

Exterior Wall	Living Area 1000	1200	1400	1600	1800	2000	2400	2800	3200	3600	4000
Wood Siding - Wood Frame	142.25	131.90	123.45	117.65	114.50	110.75	103.75	**98.70**	95.15	91.35	88.25
Brick Veneer - Wood Frame	148.95	137.95	128.90	122.75	119.45	115.40	108.00	102.60	98.75	94.70	91.40
Solid Brick	159.80	147.75	137.75	131.10	127.50	123.00	114.90	108.95	104.60	100.25	96.60
Solid Stone	159.25	147.15	137.30	130.60	127.05	122.60	114.55	108.65	104.35	99.90	96.25
Finished Basement, Add	38.90	41.75	39.95	38.85	38.15	37.25	35.95	34.95	34.05	33.35	32.70
Unfinished Basement, Add	16.80	15.85	15.15	14.65	14.40	14.00	13.45	13.00	12.60	12.25	11.95

Modifications

Add to the total cost

Upgrade Kitchen Cabinets	$ + 866
Solid Surface Countertops	+ 1536
Full Bath - including plumbing, wall and floor finishes	+ 5581
Half Bath - including plumbing, wall and floor finishes	+ 3551
Two Car Attached Garage	+ 21,366
Two Car Detached Garage	+ 24,051
Fireplace & Chimney	+ 5705

Adjustments

For multi family - add to total cost

Additional Kitchen	$ + 11,470
Additional Full Bath & Half Bath	+ 9132
Additional Entry & Exit	+ 1854
Separate Heating & Air Conditioning	+ 4555
Separate Electric	+ 1487

For Townhouse/Rowhouse - Multiply cost per square foot by

Inner Unit	.90
End Unit	.95

Alternatives

Add to or deduct from the cost per square foot of living area

Heavyweight Asphalt Shingles	– 1.99
Clay Tile Roof	+ 1.76
Slate Roof	+ 5.65
Upgrade Ceilings to Textured Finish	+ .39
Air Conditioning, in Heating Ductwork	Base System
Heating Systems, Hot Water	+ 1.62
Heat Pump	+ 1.96
Electric Heat	– 1.77
Not Heated	– 3.57

Additional upgrades or components

Kitchen Cabinets & Countertops	Page 93
Bathroom Vanities	94
Fireplaces & Chimneys	94
Windows, Skylights & Dormers	94
Appliances	95
Breezeways & Porches	95
Finished Attic	95
Garages	96
Site Improvements	96
Wings & Ells	92

Luxury 1 Story

Living Area - 2800 S.F.
Perimeter - 219 L.F.

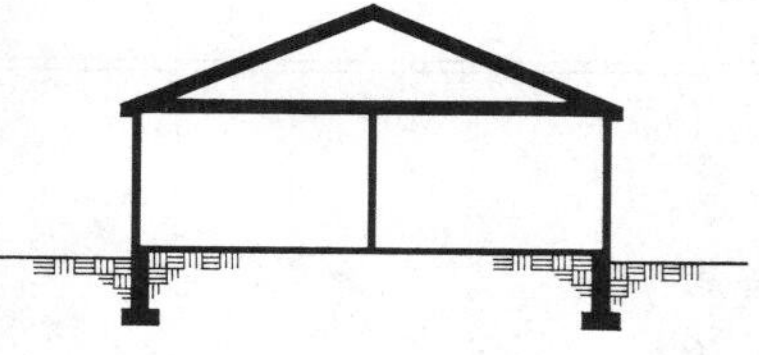

		Labor-Hours	Cost Per Square Foot Of Living Area		
			Mat.	Labor	Total
1 Site Work	Site preparation for slab; 4′ deep trench excavation for foundation wall.	.028		.49	.49
2 Foundation	Continuous reinforced concrete footing 8″ deep x 18″ wide; dampproofed and insulated reinforced concrete foundation wall, 12″ thick, 4′ deep; 4″ concrete slab on 4″ crushed stone base and polyethylene vapor barrier, trowel finish.	.098	4.78	5.58	10.36
3 Framing	Exterior walls - 2″ x 6″ wood studs, 16″ O.C.; 5/8″ plywood sheathing; 2″ x 10″ rafters 16″ O.C. with 5/8″ plywood sheathing, 6 in 12 pitch; 2″ x 8″ ceiling joists 16″ O.C.; 5/8″ plywood subfloor on 1″ x 3″ wood sleepers 16″ O.C.	.260	9.27	10.21	19.48
4 Exterior Walls	Horizontal beveled wood siding; #15 felt building paper; 6″ batt insulation; wood double hung windows; 3 solid core wood exterior doors; storms and screens.	.204	6.72	2.46	9.18
5 Roofing	Red cedar shingles; #15 felt building paper; aluminum gutters, downspouts and drip edge; copper flashings.	.082	3.29	3.05	6.34
6 Interiors	Walls and ceilings - 5/8″ drywall, skim coat plaster, painted with primer and 2 coats; hardwood baseboard and trim, sanded and finished; hardwood floor 70%, ceramic tile with underlayment 20%, vinyl tile with underlayment 10%; wood panel interior doors, primed and painted with 2 coats.	.287	9.94	10.36	20.30
7 Specialties	Luxury grade kitchen cabinets - 25 L.F. wall and base with plastic laminate counter top and kitchen sink; 6 L.F. bathroom vanity; 75 gallon electric water heater; medicine cabinet.	.052	3.43	.70	4.13
8 Mechanical	Gas fired warm air heat/air conditioning; one full bath including: bathtub, corner shower; built in lavatory and water closet; one 1/2 bath including: built in lavatory and water closet.	.078	4.43	2.47	6.90
9 Electrical	200 Amp. service; romex wiring; fluorescent and incandescent lighting fixtures; intercom, switches, receptacles.	.044	.88	1.53	2.41
10 Overhead	Contractor's overhead and profit and architect's fees.		10.26	8.85	19.11
	Total		53.00	45.70	**98.70**

- **Unique residence built from an architect's plan**
- **Single family — 1 full bath, 1 half bath, 1 kitchen**
- **No basement**
- **Cedar shakes on roof**
- **Forced hot air heat/air conditioning**
- **Double drywall interior**
- **Many special features**
- **Extraordinary materials and workmanship**

Note: The illustration shown may contain some optional components (for example: garages and/or fireplaces) whose costs are shown in the modifications, adjustments, & alternatives below or at the end of the square foot section.

Base cost per square foot of living area

Exterior Wall	Living Area 1000	1200	1400	1600	1800	2000	2400	2800	3200	3600	4000
Wood Siding - Wood Frame	132.10	123.30	116.90	109.20	104.85	100.40	92.20	**88.45**	85.05	82.35	78.65
Brick Veneer - Wood Frame	139.80	130.55	123.75	115.40	110.75	106.05	97.10	93.20	89.35	86.55	82.55
Solid Brick	152.30	142.25	134.90	125.60	120.40	115.20	105.20	100.80	96.40	93.35	88.85
Solid Stone	151.65	141.55	134.30	125.05	119.90	114.65	104.85	100.40	96.05	93.00	88.60
Finished Basement, Add	27.50	29.75	28.80	27.50	26.70	26.05	24.70	23.95	23.25	22.90	22.30
Unfinished Basement, Add	11.80	11.40	11.05	10.45	10.20	9.90	9.35	9.05	8.75	8.60	8.35

Modifications

Add to the total cost

Upgrade Kitchen Cabinets	$ + 866
Solid Surface Countertops	+ 1536
Full Bath - including plumbing, wall and floor finishes	+ 5581
Half Bath - including plumbing, wall and floor finishes	+ 3551
Two Car Attached Garage	+ 21,366
Two Car Detached Garage	+ 24,051
Fireplace & Chimney	+ 5705

Adjustments

For multi family - add to total cost

Additional Kitchen	$ + 11,470
Additional Full Bath & Half Bath	+ 9132
Additional Entry & Exit	+ 1854
Separate Heating & Air Conditioning	+ 4555
Separate Electric	+ 1487

For Townhouse/Rowhouse - Multiply cost per square foot by

Inner Unit	.90
End Unit	.95

Alternatives

Add to or deduct from the cost per square foot of living area

Heavyweight Asphalt Shingles	– 1.44
Clay Tile Roof	+ 1.27
Slate Roof	+ 4.14
Upgrade Ceilings to Textured Finish	+ .39
Air Conditioning, in Heating Ductwork	Base System
Heating Systems, Hot Water	+ 1.55
Heat Pump	+ 2.17
Electric Heat	– 1.77
Not Heated	– 3.29

Additional upgrades or components

Kitchen Cabinets & Countertops	Page 93
Bathroom Vanities	94
Fireplaces & Chimneys	94
Windows, Skylights & Dormers	94
Appliances	95
Breezeways & Porches	95
Finished Attic	95
Garages	96
Site Improvements	96
Wings & Ells	92

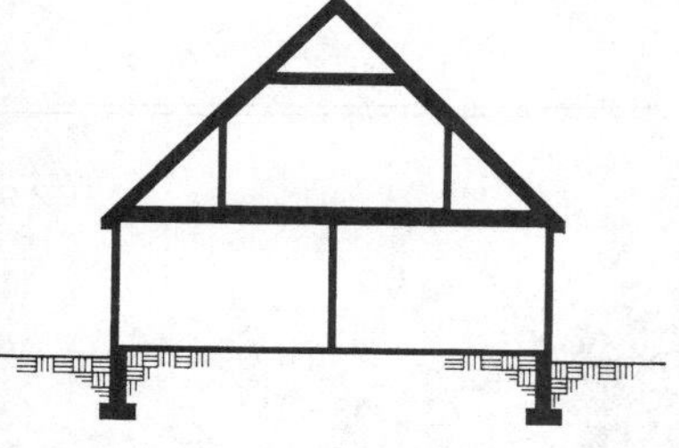

Luxury 1-1/2 Story

Living Area - 2800 S.F.
Perimeter - 175 L.F.

		Labor-Hours	Cost Per Square Foot Of Living Area		
			Mat.	Labor	Total
1 Site Work	Site preparation for slab; 4' deep trench excavation for foundation wall.	.025		.49	.49
2 Foundation	Continuous reinforced concrete footing 8" deep x 18" wide; dampproofed and insulated reinforced concrete foundation wall, 12" thick, 4' deep; 4" concrete slab on 4" crushed stone base and polyethylene vapor barrier, trowel finish.	.066	3.42	4.21	7.63
3 Framing	Exterior walls - 2" x 6" wood studs, 16" O.C.; 5/8" plywood sheathing; 2" x 10" rafters 16" O.C. with 5/8" plywood sheathing, 8 in 12 pitch; 2" x 8" ceiling joists 16" O.C.; 2" x 12" floor joists 16" O.C. with 5/8" plywood subfloor; 5/8" plywood subfloor on 1" x 3" wood sleepers 16" O.C.	.189	5.61	6.65	12.26
4 Exterior Walls	Horizontal beveled wood siding; #15 felt building paper; 6" batt insulation; wood double hung windows; 3 solid core wood exterior doors; storms and screens.	.174	7.38	2.75	10.13
5 Roofing	Red cedar shingles; #15 felt building paper; aluminum gutters, downspouts and drip edge; copper flashings.	.065	2.06	1.91	3.97
6 Interiors	Walls and ceilings - 5/8" drywall, skim coat plaster, painted with primer and 2 coats; hardwood baseboard and trim, sanded and finished; hardwood floor 70%, ceramic tile with underlayment 20%, vinyl tile with underlayment 10%; wood panel interior doors, primed and painted with 2 coats.	.260	11.60	11.82	23.42
7 Specialties	Luxury grade kitchen cabinets - 25 L.F. wall and base with plastic laminate counter top and kitchen sink; 6 L.F. bathroom vanity; 75 gallon electric water heater; medicine cabinet.	.062	3.43	.70	4.13
8 Mechanical	Gas fired warm air heat/air conditioning; one full bath including: bathtub, corner shower; built in lavatory and water closet; one 1/2 bath including: built in lavatory and water closet.	.080	4.43	2.47	6.90
9 Electrical	200 Amp. service; romex wiring; fluorescent and incandescent lighting fixtures; intercom, switches, receptacles.	.044	.88	1.53	2.41
10 Overhead	Contractor's overhead and profit and architect's fees.		9.29	7.82	17.11
		Total	48.10	40.35	**88.45**

RESIDENTIAL | Luxury | 2 Story

- **Unique residence built from an architect's plan**
- **Single family — 1 full bath, 1 half bath, 1 kitchen**
- **No basement**
- **Cedar shakes on roof**
- **Forced hot air heat/air conditioning**
- **Double drywall interior**
- **Many special features**
- **Extraordinary materials and workmanship**

Note: The illustration shown may contain some optional components (for example: garages and/or fireplaces) whose costs are shown in the modifications, adjustments, & alternatives below or at the end of the square foot section.

Base cost per square foot of living area

	Living Area										
Exterior Wall	**1200**	**1400**	**1600**	**1800**	**2000**	**2400**	**2800**	**3200**	**3600**	**4000**	**4400**
Wood Siding - Wood Frame	122.95	115.80	110.75	106.35	101.40	94.50	88.60	84.60	82.35	79.90	77.75
Brick Veneer - Wood Frame	131.10	123.50	118.05	113.25	108.05	100.55	94.00	89.70	87.25	84.55	82.20
Solid Brick	144.50	135.95	130.00	124.50	118.90	110.35	102.90	98.05	95.30	92.15	89.50
Solid Stone	143.80	135.30	129.35	123.90	118.35	109.75	102.45	97.55	94.85	91.75	89.15
Finished Basement, Add	22.10	24.00	23.30	22.55	22.05	20.95	20.05	19.45	19.15	18.70	18.40
Unfinished Basement, Add	9.45	9.10	8.85	8.60	8.35	7.95	7.60	7.35	7.20	7.00	6.90

Modifications

Add to the total cost

Upgrade Kitchen Cabinets	$ + 866
Solid Surface Countertops	+ 1536
Full Bath - including plumbing, wall and floor finishes	+ 5581
Half Bath - including plumbing, wall and floor finishes	+ 3551
Two Car Attached Garage	+ 21,366
Two Car Detached Garage	+ 24,051
Fireplace & Chimney	+ 6255

Adjustments

For multi family - add to total cost

Additional Kitchen	$ + 11,470
Additional Full Bath & Half Bath	+ 9132
Additional Entry & Exit	+ 1854
Separate Heating & Air Conditioning	+ 4555
Separate Electric	+ 1487

For Townhouse/Rowhouse - Multiply cost per square foot by

Inner Unit	.86
End Unit	.93

Alternatives

Add to or deduct from the cost per square foot of living area

Heavyweight Asphalt Shingles	– 1.00
Clay Tile Roof	+ .88
Slate Roof	+ 2.83
Upgrade Ceilings to Textured Finish	+ .39
Air Conditioning, in Heating Ductwork	Base System
Heating Systems, Hot Water	+ 1.51
Heat Pump	+ 2.29
Electric Heat	– 1.59
Not Heated	– 3.11

Additional upgrades or components

Kitchen Cabinets & Countertops	Page 93
Bathroom Vanities	94
Fireplaces & Chimneys	94
Windows, Skylights & Dormers	94
Appliances	95
Breezeways & Porches	95
Finished Attic	95
Garages	96
Site Improvements	96
Wings & Ells	92

Important: See the Reference Section for Location Factors (to adjust for your city) and Estimating Forms

Luxury 2 Story

Living Area - 3200 S.F.
Perimeter - 163 L.F.

		Labor-Hours	Cost Per Square Foot Of Living Area		
			Mat.	Labor	Total
1 Site Work	Site preparation for slab; 4′ deep trench excavation for foundation wall.	.024		.43	.43
2 Foundation	Continuous reinforced concrete footing 8″ deep x 18″ wide; dampproofed and insulated reinforced concrete foundation wall, 12″ thick, 4′ deep; 4″ concrete slab on 4″ crushed stone base and polyethylene vapor barrier, trowel finish.	.058	2.78	3.42	6.20
3 Framing	Exterior walls - 2″ x 6″ wood studs, 16″ O.C.; 5/8″ plywood sheathing; 2″ x 10″ rafters 16″ O.C. with 5/8″ plywood sheathing, 6 in 12 pitch; 2″ x 8″ ceiling joists 16″ O.C.; 2″ x 12″ floor joists 16″ O.C. with 5/8″ plywood subfloor; 5/8″ plywood subfloor on 1″ x 3″ wood sleepers 16″ O.C.	.193	5.64	6.58	12.22
4 Exterior Walls	Horizontal beveled wood siding, #15 felt building paper; 6″ batt insulation; wood double hung windows; 3 solid core wood exterior doors; storms and screens.	.247	7.77	2.90	10.67
5 Roofing	Red cedar shingles; #15 felt building paper; aluminum gutters, downspouts and drip edge; copper flashings.	.049	1.64	1.53	3.17
6 Interiors	Walls and ceilings - 5/8″ drywall, skim coat plaster, painted with primer and 2 coats; hardwood baseboard and trim, sanded and finished; hardwood floor 70%, ceramic tile with underlayment 20%, vinyl tile with underlayment 10%; wood panel interior doors, primed and painted with 2 coats.	.252	11.44	11.75	23.19
7 Specialties	Luxury grade kitchen cabinets - 25 L.F. wall and base with plastic laminate counter top and kitchen sink; 6 L.F. bathroom vanity; 75 gallon electric water heater; medicine cabinet.	.057	3.01	.61	3.62
8 Mechanical	Gas fired warm air heat/air conditioning; one full bath including: bathtub, corner shower; built in lavatory and water closet; one 1/2 bath including: built in lavatory and water closet.	.071	4.04	2.34	6.38
9 Electrical	200 Amp. service; romex wiring; fluorescent and incandescent lighting fixtures; intercom, switches, receptacles.	.042	.85	1.49	2.34
10 Overhead	Contractor's overhead and profit and architect's fee.		8.93	7.45	16.38
		Total	46.10	38.50	**84.60**

SQUARE FOOT COSTS

- **Unique residence built from an architect's plan**
- **Single family — 1 full bath, 1 half bath, 1 kitchen**
- **No basement**
- **Cedar shakes on roof**
- **Forced hot air heat/air conditioning**
- **Double drywall interior**
- **Many special features**
- **Extraordinary materials and workmanship**

Note: The illustration shown may contain some optional components (for example: garages and/or fireplaces) whose costs are shown in the modifications, adjustments, & alternatives below or at the end of the square foot section.

Base cost per square foot of living area

Exterior Wall	Living Area 1500	1800	2100	2500	3000	3500	4000	4500	5000	5500	6000
Wood Siding - Wood Frame	121.00	109.05	102.15	97.00	90.10	84.90	79.95	77.50	75.40	73.15	70.90
Brick Veneer - Wood Frame	129.55	116.90	109.20	103.65	96.15	90.40	85.00	82.30	80.00	77.55	75.10
Solid Brick	143.50	129.75	120.65	114.55	106.00	99.35	93.30	90.10	87.50	84.65	81.90
Solid Stone	142.75	129.05	120.05	113.95	105.40	98.85	92.90	89.75	87.05	84.30	81.55
Finished Basement, Add	17.65	18.95	17.80	17.20	16.40	15.65	15.20	14.85	14.50	14.25	13.95
Unfinished Basement, Add	7.60	7.25	6.85	6.60	6.25	5.90	5.75	5.60	5.45	5.35	5.20

Modifications

Add to the total cost

Upgrade Kitchen Cabinets	$ + 866
Solid Surface Countertops	+ 1536
Full Bath - including plumbing, wall and floor finishes	+ 5581
Half Bath - including plumbing, wall and floor finishes	+ 3551
Two Car Attached Garage	+ 21,366
Two Car Detached Garage	+ 24,051
Fireplace & Chimney	+ 6255

Adjustments

For multi family - add to total cost

Additional Kitchen	$ + 11,470
Additional Full Bath & Half Bath	+ 9132
Additional Entry & Exit	+ 1854
Separate Heating & Air Conditioning	+ 4555
Separate Electric	+ 1487

For Townhouse/Rowhouse - Multiply cost per square foot by

Inner Unit	.86
End Unit	.93

Alternatives

Add to or deduct from the cost per square foot of living area

Heavyweight Asphalt Shingles	– .86
Clay Tile Roof	+ .76
Slate Roof	+ 2.47
Upgrade Ceilings to Textured Finish	+ .39
Air Conditioning, in Heating Ductwork	Base System
Heating Systems, Hot Water	+ 1.36
Heat Pump	+ 2.36
Electric Heat	– 3.15
Not Heated	– 3.11

Additional upgrades or components

Kitchen Cabinets & Countertops	Page 93
Bathroom Vanities	94
Fireplaces & Chimneys	94
Windows, Skylights & Dormers	94
Appliances	95
Breezeways & Porches	95
Finished Attic	95
Garages	96
Site Improvements	96
Wings & Ells	92

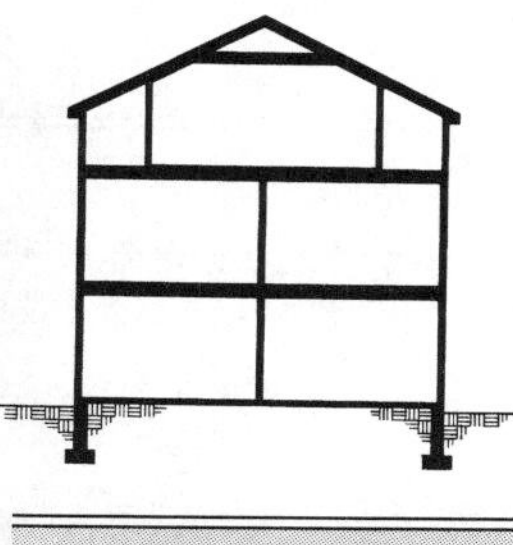

Luxury 2-1/2 Story

Living Area - 3000 S.F.
Perimeter - 148 L.F.

		Labor-Hours	Cost Per Square Foot Of Living Area		
			Mat.	Labor	Total
1 Site Work	Site preparation for slab; 4′ deep trench excavation for foundation wall.	.055		.45	.45
2 Foundation	Continuous reinforced concrete footing 8″ deep x 18″ wide; dampproofed and insulated reinforced concrete foundation wall, 12″ thick, 4′ deep; 4″ concrete slab on 4″ crushed stone base and polyethylene vapor barrier, trowel finish.	.067	2.43	3.12	5.55
3 Framing	Exterior walls - 2″ x 6″ wood studs, 16″ O.C.; 5/8″ plywood sheathing; 2″ x 10″ rafters 16″ O.C. with 5/8″ plywood sheathing, 8 in 12 pitch; 2″ x 8″ ceiling joists 16″ O.C.; 2″ x 12″ floor joists 16″ O.C. with 5/8″ plywood subfloor; 5/8″ plywood subfloor on 1″ x 3″ wood sleepers 16″ O.C.	.209	5.79	6.78	12.57
4 Exterior Walls	Horizontal beveled wood siding; #15 felt building paper; 6″ batt insulation; wood double hung windows; 3 solid core wood exterior doors; storms and screens.	.405	9.09	3.44	12.53
5 Roofing	Red cedar shingles; #15 felt building paper; aluminum gutters, downspouts and drip edge; copper flashings.	.039	1.27	1.18	2.45
6 Interiors	Walls and ceilings - 5/8″ drywall, skim coat plaster, painted with primer and 2 coats; hardwood baseboard and trim, sanded and finished; hardwood floor 70%, ceramic tile with underlayment 20%, vinyl tile with underlayment 10%; wood panel interior doors, primed and painted with 2 coats.	.341	13.05	13.18	26.23
7 Specialties	Luxury grade kitchen cabinets - 25 L.F. wall and base with plastic laminate counter top and kitchen sink; 6 L.F. bathroom vanity; 75 gallon electric water heater; medicine cabinet.	.119	3.21	.66	3.87
8 Mechanical	Gas fired warm air heat/air conditioning; one full bath including: bathtub, corner shower; built in lavatory and water closet; one 1/2 bath including: built in lavatory and water closet.	.103	4.23	2.42	6.65
9 Electrical	200 Amp. service; romex wiring; fluorescent and incandescent lighting fixtures; intercom, switches, receptacles.	.054	.86	1.52	2.38
10 Overhead	Contractor's overhead and profit and architect's fee.		9.57	7.85	17.42
		Total	49.50	40.60	**90.10**

RESIDENTIAL | Luxury | 3 Story

- Unique residence built from an architect's plan
- Single family — 1 full bath, 1 half bath, 1 kitchen
- No basement
- Cedar shakes on roof
- Forced hot air heat/air conditioning
- Double drywall interior
- Many special features
- Extraordinary materials and workmanship

Note: The illustration shown may contain some optional components (for example: garages and/or fireplaces) whose costs are shown in the modifications, adjustments, & alternatives below or at the end of the square foot section.

Base cost per square foot of living area

	Living Area										
Exterior Wall	1500	1800	2100	2500	3000	3500	4000	4500	5000	5500	6000
Wood Siding - Wood Frame	120.20	108.40	102.75	98.00	90.70	87.10	82.65	77.80	76.05	74.25	72.45
Brick Veneer - Wood Frame	129.05	116.60	110.45	105.30	97.40	93.40	88.35	83.10	81.15	79.15	77.05
Solid Brick	143.50	129.95	122.90	117.20	108.25	103.60	97.65	91.70	89.50	87.20	84.60
Solid Stone	142.70	129.30	122.25	116.60	107.70	103.10	97.15	91.20	89.10	86.75	84.20
Finished Basement, Add	15.50	16.65	15.95	15.40	14.65	14.20	13.60	13.10	12.95	12.70	12.40
Unfinished Basement, Add	6.60	6.30	6.05	5.90	5.60	5.40	5.20	4.95	4.85	4.80	4.65

Modifications

Add to the total cost

Upgrade Kitchen Cabinets	$ + 866
Solid Surface Countertops	+ 1536
Full Bath - including plumbing, wall and floor finishes	+ 5581
Half Bath - including plumbing, wall and floor finishes	+ 3551
Two Car Attached Garage	+ 21,366
Two Car Detached Garage	+ 24,051
Fireplace & Chimney	+ 6840

Adjustments

For multi family - add to total cost

Additional Kitchen	$ + 11,470
Additional Full Bath & Half Bath	+ 9132
Additional Entry & Exit	+ 1854
Separate Heating & Air Conditioning	+ 4555
Separate Electric	+ 1487

For Townhouse/Rowhouse - Multiply cost per square foot by

Inner Unit	.84
End Unit	.92

Alternatives

Add to or deduct from the cost per square foot of living area

Heavyweight Asphalt Shingles	− .66
Clay Tile Roof	+ .59
Slate Roof	+ 1.88
Upgrade Ceilings to Textured Finish	+ .39
Air Conditioning, in Heating Ductwork	Base System
Heating Systems, Hot Water	+ 1.36
Heat Pump	+ 2.36
Electric Heat	− 3.15
Not Heated	− 3.01

Additional upgrades or components

Kitchen Cabinets & Countertops	Page 93
Bathroom Vanities	94
Fireplaces & Chimneys	94
Windows, Skylights & Dormers	94
Appliances	95
Breezeways & Porches	95
Finished Attic	95
Garages	96
Site Improvements	96
Wings & Ells	92

Important: See the Reference Section for Location Factors (to adjust for your city) and Estimating Forms

Luxury 3 Story

Living Area - 3000 S.F.
Perimeter - 135 L.F.

		Labor-Hours	Cost Per Square Foot Of Living Area		
			Mat.	Labor	Total
1 Site Work	Site preparation for slab; 4' deep trench excavation for foundation wall.	.055		.45	.45
2 Foundation	Continuous reinforced concrete footing 8" deep x 18" wide; dampproofed and insulated reinforced concrete foundation wall, 12" thick, 4' deep; 4" concrete slab on 4" crushed stone base and polyethylene vapor barrier, trowel finish.	.063	2.19	2.83	5.02
3 Framing	Exterior walls - 2" x 6" wood studs, 16" O.C.; 5/8" plywood sheathing; 2" x 10" rafters 16" O.C. with 5/8" plywood sheathing, 6 in 12 pitch; 2" x 8" ceiling joists 16" O.C.; 2" x 12" floor joists 16" O.C. with 5/8" plywood subfloor; 5/8" plywood subfloor on 1" x 3" wood sleepers 16" O.C.	.225	5.89	6.84	12.73
4 Exterior Walls	Horizontal beveled wood siding; #15 felt building paper; 6" batt insulation; wood double hung windows; 3 solid core wood exterior doors; storms and screens.	.454	9.89	3.72	13.61
5 Roofing	Red cedar shingles; #15 felt building paper; aluminum gutters, downspouts and drip edge; copper flashings.	.034	1.10	1.02	2.12
6 Interiors	Walls and ceilings - 5/8" drywall, skim coat plaster, painted with primer and 2 coats; hardwood baseboard and trim, sanded and finished; hardwood floor 70%, ceramic tile with underlayment 20%, vinyl tile with underlayment 10%; wood panel interior doors, primed and painted with 2 coats.	.390	13.09	13.26	26.35
7 Specialties	Luxury grade kitchen cabinets - 25 L.F. wall and base with plastic laminate counter top and kitchen sink; 6 L.F. bathroom vanity; 75 gallon electric water heater; medicine cabinet.	.119	3.21	.66	3.87
8 Mechanical	Gas fired warm air heat/air conditioning; one full bath including: bathtub, corner shower; built in lavatory and water closet; one 1/2 bath including: built in lavatory and water closet.	.103	4.23	2.42	6.65
9 Electrical	200 Amp. service; romex wiring; fluorescent and incandescent lighting fixtures; intercom, switches, receptacles.	.053	.86	1.52	2.38
10 Overhead	Contractor's overhead and profit and architect's fees.		9.69	7.83	17.52
		Total	50.15	40.55	**90.70**

- **Unique residence built from an architect's plan**
- **Single family — 1 full bath, 1 half bath, 1 kitchen**
- **No basement**
- **Cedar shakes on roof**
- **Forced hot air heat/air conditioning**
- **Double drywall interior**
- **Many special features**
- **Extraordinary materials and workmanship**

Note: The illustration shown may contain some optional components (for example: garages and/or fireplaces) whose costs are shown in the modifications, adjustments, & alternatives below or at the end of the square foot section.

Base cost per square foot of living area

Exterior Wall	Living Area 1200	1400	1600	1800	2000	2400	2800	3200	3600	4000	4400
Wood Siding - Wood Frame	116.30	109.65	104.85	100.65	96.05	89.60	84.10	80.55	78.35	76.20	74.15
Brick Veneer - Wood Frame	122.45	115.45	110.35	105.85	101.00	94.05	88.20	84.35	82.10	79.65	77.50
Solid Brick	132.50	124.75	119.25	114.30	109.15	101.40	94.85	90.65	88.10	85.30	83.00
Solid Stone	131.95	124.30	118.80	113.85	108.70	101.05	94.50	90.30	87.80	85.00	82.70
Finished Basement, Add	22.10	24.00	23.30	22.55	22.05	20.95	20.05	19.45	19.15	18.70	18.40
Unfinished Basement, Add	9.45	9.10	8.85	8.60	8.35	7.95	7.60	7.35	7.20	7.00	6.90

Modifications

Add to the total cost

Upgrade Kitchen Cabinets	$ + 866
Solid Surface Countertops	+ 1536
Full Bath - including plumbing, wall and floor finishes	+ 5581
Half Bath - including plumbing, wall and floor finishes	+ 3551
Two Car Attached Garage	+ 21,366
Two Car Detached Garage	+ 24,051
Fireplace & Chimney	+ 5705

Adjustments

For multi family - add to total cost

Additional Kitchen	$ + 11,470
Additional Full Bath & Half Bath	+ 9132
Additional Entry & Exit	+ 1854
Separate Heating & Air Conditioning	+ 4555
Separate Electric	+ 1487

For Townhouse/Rowhouse - Multiply cost per square foot by

Inner Unit	.89
End Unit	.94

Alternatives

Add to or deduct from the cost per square foot of living area

Heavyweight Asphalt Shingles	– 1.99
Clay Tile Roof	+ 1.76
Slate Roof	+ 5.65
Upgrade Ceilings to Textured Finish	+ .39
Air Conditioning, in Heating Ductwork	Base System
Heating Systems, Hot Water	+ 1.51
Heat Pump	+ 2.29
Electric Heat	– 1.59
Not Heated	– 3.11

Additional upgrades or components

Kitchen Cabinets & Countertops	Page 93
Bathroom Vanities	94
Fireplaces & Chimneys	94
Windows, Skylights & Dormers	94
Appliances	95
Breezeways & Porches	95
Finished Attic	95
Garages	96
Site Improvements	96
Wings & Ells	92

Luxury Bi-Level

Living Area - 3200 S.F.
Perimeter - 163 L.F.

		Labor-Hours	Cost Per Square Foot Of Living Area		
			Mat.	Labor	Total
1 Site Work	Excavation for lower level, 4′ deep. Site preparation for slab.	.024		.43	.43
2 Foundation	Continuous reinforced concrete footing 8″ deep x 18″ wide; dampproofed and insulated reinforced concrete foundation wall, 12″ thick, 4′ deep; 4″ concrete slab on 4″ crushed stone base and polyethylene vapor barrier, trowel finish.	.058	2.78	3.42	6.20
3 Framing	Exterior walls - 2″ x 6″ wood studs, 16″ O.C.; 5/8″ plywood sheathing; 2″ x 10″ rafters 16″ O.C. with 5/8″ plywood sheathing, 6 in 12 pitch; 2″ x 8″ ceiling joists 16″ O.C.; 2″ x 12″ floor joists 16″ O.C. with 5/8″ plywood subfloor; 5/8″ plywood subfloor on 1″ x 3″ wood sleepers 16″ O.C.	.232	5.38	6.25	11.63
4 Exterior Walls	Horizontal beveled wood siding; #15 felt building paper; 6″ batt insulation; wood double hung windows; 3 solid core wood exterior doors; storms and screens.	.185	6.10	2.29	8.39
5 Roofing	Red cedar shingles: #15 felt building paper; aluminum gutters, downspouts and drip edge; copper flashings.	.042	1.64	1.53	3.17
6 Interiors	Walls and ceilings - 5/8″ drywall, skim coat plaster, painted with primer and 2 coats; hardwood baseboard and trim, sanded and finished; hardwood floor 70%, ceramic tile with underlayment 20%; vinyl tile with underlayment 10%; wood panel interior doors, primed and painted with 2 coats.	.238	11.31	11.48	22.79
7 Specialties	Luxury grade kitchen cabinets - 25 L.F. wall and base with plastic laminate counter top and kitchen sink; 6 L.F. bathroom vanity; 75 gallon electric water heater; medicine cabinet.	.056	3.01	.61	3.62
8 Mechanical	Gas fired warm air heat/air conditioning; one full bath including: bathtub, corner shower; built in lavatory and water closet; one 1/2 bath including: built in lavatory and water closet.	.071	4.04	2.34	6.38
9 Electrical	200 Amp. service; romex wiring; fluorescent and incandescent lighting fixtures; intercom, switches, receptacles.	.042	.85	1.49	2.34
10 Overhead	Contractor's overhead and profit and architect's fees.		8.44	7.16	15.60
		Total	43.55	37.00	**80.55**

- **Unique residence built from an architect's plan**
- **Single family — 1 full bath, 1 half bath, 1 kitchen**
- **No basement**
- **Cedar shakes on roof**
- **Forced hot air heat/air conditioning**
- **Double drywall interior**
- **Many special features**
- **Extraordinary materials and workmanship**

Note: The illustration shown may contain some optional components (for example: garages and/or fireplaces) whose costs are shown in the modifications, adjustments, & alternatives below or at the end of the square foot section.

Base cost per square foot of living area

Exterior Wall	Living Area 1500	1800	2100	2400	2800	3200	3600	4000	4500	5000	5500
Wood Siding - Wood Frame	109.95	101.95	95.25	90.90	87.65	83.80	79.75	78.00	74.15	72.05	69.70
Brick Veneer - Wood Frame	115.45	106.90	99.80	95.15	91.70	87.60	83.25	81.45	77.25	75.00	72.55
Solid Brick	124.50	115.05	107.20	102.10	98.35	93.75	89.00	87.00	82.40	79.85	77.05
Solid Stone	124.05	114.60	106.80	101.70	98.05	93.40	88.70	86.70	82.15	79.60	76.80
Finished Basement, Add*	25.95	27.90	26.70	25.95	25.40	24.65	24.00	23.70	23.05	22.60	22.25
Unfinished Basement, Add*	11.20	10.60	10.05	9.80	9.60	9.25	8.95	8.80	8.55	8.35	8.20

*Basement under middle level only.

Modifications

Add to the total cost

Upgrade Kitchen Cabinets	$ + 866
Solid Surface Countertops	+ 1536
Full Bath - including plumbing, wall and floor finishes	+ 5581
Half Bath - including plumbing, wall and floor finishes	+ 3551
Two Car Attached Garage	+ 21,366
Two Car Detached Garage	+ 24,051
Fireplace & Chimney	+ 6255

Adjustments

For multi family - add to total cost

Additional Kitchen	$ + 11,470
Additional Full Bath & Half Bath	+ 9132
Additional Entry & Exit	+ 1854
Separate Heating & Air Conditioning	+ 4555
Separate Electric	+ 1487

For Townhouse/Rowhouse - Multiply cost per square foot by

Inner Unit	.86
End Unit	.93

Alternatives

Add to or deduct from the cost per square foot of living area

Heavyweight Asphalt Shingles	– 1.44
Clay Tile Roof	+ 1.27
Slate Roof	+ 4.14
Upgrade Ceilings to Textured Finish	+ .39
Air Conditioning, in Heating Ductwork	Base System
Heating Systems, Hot Water	+ 1.46
Heat Pump	+ 2.38
Electric Heat	– 1.41
Not Heated	– 3.01

Additional upgrades or components

Kitchen Cabinets & Countertops	Page 93
Bathroom Vanities	94
Fireplaces & Chimneys	94
Windows, Skylights & Dormers	94
Appliances	95
Breezeways & Porches	95
Finished Attic	95
Garages	96
Site Improvements	96
Wings & Ells	92

Luxury Tri-Level

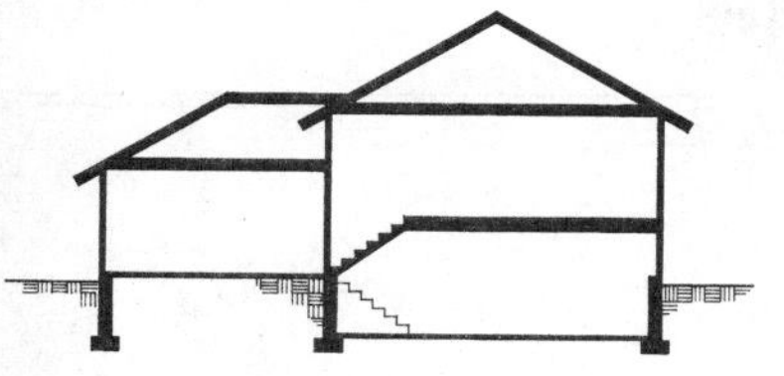

Living Area - 3600 S.F.
Perimeter - 207 L.F.

		Labor-Hours	Cost Per Square Foot Of Living Area		
			Mat.	Labor	Total
1 Site Work	Site preparation for slab; 4′ deep trench excavation for foundation wall, excavation for lower level, 4′ deep.	.021		.37	.37
2 Foundation	Continuous reinforced concrete footing 8″ deep x 18″ wide; dampproofed and insulated reinforced concrete foundation wall, 12″ thick, 4′ deep; 4″ concrete slab on 4″ crushed stone base and polyethylene vapor barrier, trowel finish.	.109	3.33	3.96	7.29
3 Framing	Exterior walls - 2″ x 6″ wood studs, 16″ O.C.; 5/8″ plywood sheathing; 2″ x 10″ rafters 16″ O.C. with 5/8″ plywood sheathing, 6 in 12 pitch; 2″ x 8″ ceiling joists 16″ O.C.; 2″ x 12″ floor joists 16″ O.C. with 5/8″ plywood subfloor; 5/8″ plywood subfloor on 1″ x 3″ wood sleepers 16″ O.C.	.204	5.32	6.28	11.60
4 Exterior Walls	Horizontal beveled wood siding; #15 felt building paper; 6″ batt insulation; wood double hung windows; 3 solid core wood exterior doors; storms and screens.	.181	5.85	2.16	8.01
5 Roofing	Red cedar shingles; #15 felt building paper; aluminum gutters, downspouts and drip edge; copper flashings.	.056	2.19	2.03	4.22
6 Interiors	Walls and ceilings - 5/8″ drywall, skim coat plaster, painted with primer and 2 coats; hardwood baseboard and trim, sanded and finished; hardwood floor 70%, ceramic tile with underlayment 20%, vinyl tile with underlayment 10%; wood panel interior doors, primed and painted with 2 coats.	.217	10.56	10.73	21.29
7 Specialties	Luxury grade kitchen cabinets - 25 L.F. wall and base with plastic laminate counter top and kitchen sink; 6 L.F. bathroom vanity; 75 gallon electric water heater; medicine cabinet.	.048	2.68	.55	3.23
8 Mechanical	Gas fired warm air heat/air conditioning; one full bath including: bathtub, corner shower; built in lavatory and water closet; one 1/2 bath including: built in lavatory and water closet.	.057	3.73	2.28	6.01
9 Electrical	200 Amp. service; romex wiring; fluorescent and incandescent lighting fixtures; intercom, switches, receptacles.	.039	.81	1.46	2.27
10 Overhead	Contractor's overhead and profit and architect's fees.		8.28	7.18	15.46
		Total	42.75	37.00	**79.75**

1 Story — Base cost per square foot of living area

Exterior Wall	Living Area 50	100	200	300	400	500	600	700
Wood Siding - Wood Frame	181.85	140.75	122.80	104.05	98.30	94.75	92.40	93.25
Brick Veneer - Wood Frame	202.60	155.55	135.15	112.30	105.70	101.70	98.95	99.55
Solid Brick	236.25	179.55	155.20	125.60	117.70	112.90	109.60	109.85
Solid Stone	234.45	178.35	154.10	124.95	117.05	112.25	109.05	109.30
Finished Basement, Add	68.15	61.80	55.50	44.95	42.80	41.55	40.70	40.15
Unfinished Basement, Add	35.80	27.75	24.40	18.80	17.75	17.05	16.60	16.25

1-1/2 Story — Base cost per square foot of living area

Exterior Wall	Living Area 100	200	300	400	500	600	700	800
Wood Siding - Wood Frame	144.65	117.95	101.75	92.40	87.50	85.20	82.20	81.30
Brick Veneer - Wood Frame	163.10	132.75	114.15	102.00	96.35	93.60	90.15	89.15
Solid Brick	193.15	156.80	134.15	117.65	110.80	107.25	103.05	101.95
Solid Stone	191.60	155.55	133.05	116.80	110.05	106.50	102.30	101.30
Finished Basement, Add	44.90	44.45	40.25	35.55	34.30	33.45	32.70	32.55
Unfinished Basement, Add	22.85	19.55	17.30	14.85	14.15	13.75	13.30	13.25

2 Story — Base cost per square foot of living area

Exterior Wall	Living Area 100	200	400	600	800	1000	1200	1400
Wood Siding - Wood Frame	142.65	107.85	92.50	78.05	73.15	70.15	68.15	69.15
Brick Veneer - Wood Frame	163.35	122.65	104.80	86.30	80.50	77.05	74.65	75.55
Solid Brick	197.05	146.65	124.90	99.65	92.50	88.30	85.35	85.85
Solid Stone	195.25	145.40	123.80	98.90	91.90	87.70	84.85	85.30
Finished Basement, Add	34.15	30.95	27.80	22.50	21.50	20.85	20.45	20.10
Unfinished Basement, Add	17.90	13.90	12.20	9.40	8.85	8.50	8.30	8.10

Base costs do not include bathroom or kitchen facilities. Use Modifications/Adjustments/Alternatives on pages 93-96 where appropriate.

Kitchen cabinets - Base units, hardwood *(Cost per Unit)*

	Economy	Average	Custom	Luxury
24" deep, 35" high,				
One top drawer,				
One door below				
12" wide	$124	$165	$220	$290
15" wide	162	216	285	380
18" wide	176	234	310	410
21" wide	183	244	325	425
24" wide	207	276	365	485
Four drawers				
12" wide	266	355	470	620
15" wide	210	280	370	490
18" wide	233	310	410	545
24" wide	255	340	450	595
Two top drawers,				
Two doors below				
27" wide	229	305	405	535
30" wide	244	325	430	570
33" wide	251	335	445	585
36" wide	259	345	460	605
42" wide	278	370	490	650
48" wide	296	395	525	690
Range or sink base				
(Cost per unit)				
Two doors below				
30" wide	205	273	365	480
33" wide	218	291	385	510
36" wide	229	305	405	535
42" wide	244	325	430	570
48" wide	255	340	450	595
Corner Base Cabinet				
(Cost per unit)				
36" wide	326	435	580	760
Lazy Susan *(Cost per unit)*				
With revolving door	323	430	570	755

Kitchen cabinets - Wall cabinets, hardwood *(Cost per Unit)*

	Economy	Average	Custom	Luxury
12" deep, 2 doors				
12" high				
30" wide	$134	$179	$240	$ 315
36" wide	153	204	270	355
15" high				
30" wide	140	186	245	325
33" wide	155	206	275	360
36" wide	157	209	280	365
24" high				
30" wide	170	226	300	395
36" wide	186	248	330	435
42" wide	203	271	360	475
30" high, 1 door				
12" wide	122	163	215	285
15" wide	137	182	240	320
18" wide	147	196	260	345
24" wide	164	218	290	380
30" high, 2 doors				
27" wide	198	264	350	460
30" wide	193	257	340	450
36" wide	218	290	385	510
42" wide	236	315	420	550
48" wide	263	350	465	615
Corner wall, 30" high				
24" wide	141	188	250	330
30" wide	165	220	295	385
36" wide	177	236	315	415
Broom Closet				
84" high, 24" deep				
18" wide	356	475	630	830
Oven Cabinet				
84" high, 24" deep				
27" wide	506	675	900	1180

Kitchen countertops *(Cost per L.F.)*

	Economy	Average	Custom	Luxury
Solid Surface				
24" wide, no backsplash	$86	$114	$150	$200
with backsplash	93	124	165	215
Stock plastic laminate, 24" wide				
with backsplash	15	20	25	35
Custom plastic laminate, no splash				
7/8" thick, alum. molding	21	29	40	50
1-1/4" thick, no splash	26	35	45	60
Marble				
1/2" - 3/4" thick w/splash	40	53	70	110
Maple, laminated				
1-1/2" thick w/splash	60	80	105	140
Stainless steel				
(per S.F.)	87	116	155	205
Cutting blocks, recessed				
16" x 20" x 1" (each)	80	106	140	185

Vanity bases *(Cost per Unit)*

	Economy	Average	Custom	Luxury
2 door, 30" high, 21" deep				
24" wide	$164	$218	$290	$380
30" wide	190	253	335	445
36" wide	248	330	440	580
48" wide	296	395	525	690

Solid surface vanity tops *(Cost Each)*

	Economy	Average	Custom	Luxury
Center bowl				
22" x 25"	$284	$307	$331	$358
22" x 31"	325	351	379	409
22" x 37"	370	400	432	466
22" x 49"	460	497	537	579

Fireplaces & Chimneys *(Cost per Unit)*

	1-1/2 Story	2 Story	3 Story
Economy (prefab metal)			
Exterior chimney & 1 fireplace	$3925	$4335	$4750
Interior chimney & 1 fireplace	3755	4180	4380
Average (masonry)			
Exterior chimney & 1 fireplace	3910	4365	4960
Interior chimney & 1 fireplace	3665	4110	4475
For more than 1 flue, add	285	480	800
For more than 1 fireplace, add	2770	2770	2770
Custom (masonry)			
Exterior chimney & 1 fireplace	4070	4595	5190
Interior chimney & 1 fireplace	3820	4320	4655
For more than 1 flue, add	320	550	750
For more than 1 fireplace, add	2925	2925	2925
Luxury (masonry)			
Exterior chimney & 1 fireplace	5705	6255	6840
Interior chimney & 1 fireplace	5445	5955	6300
For more than 1 flue, add	470	785	1100
For more than 1 fireplace, add	4495	4495	4495

Windows and Skylights *(Cost Each)*

	Economy	Average	Custom	Luxury
Fixed Picture Windows				
3'-6" x 4'-0"	$ 403	$ 435	$ 470	$ 508
4'-0" x 6'-0"	656	708	765	826
5'-0" x 6'-0"	844	912	985	1064
6'-0" x 6'-0"	857	926	1000	1080
Bay/Bow Windows				
8'-0" x 5'-0"	1136	1227	1325	1431
10'-0" x 5'-0"	1586	1713	1850	1998
10'-0" x 6'-0"	1693	1829	1975	2133
12'-0" x 6'-0"	1993	2153	2325	2511
Palladian Windows				
3'-2" x 6'-4"		1667	1800	1944
4'-0" x 6'-0"		1667	1800	1944
5'-5" x 6'-10"		2060	2225	2403
8'-0" x 6'-0"		2546	2750	2970
Skylights				
46" x 21-1/2"	346	374	477	515
46" x 28"	382	413	468	505
57" x 44"	476	514	634	685

Dormers *(Cost/S.F. of plan area)*

	Economy	Average	Custom	Luxury
Framing and Roofing Only				
Gable dormer, 2" x 6" roof frame	$20	$23	$25	$40
2" x 8" roof frame	21	24	26	42
Shed dormer, 2" x 6" roof frame	13	14	16	26
2" x 8" roof frame	14	15	17	27
2" x 10" roof frame	15	17	18	28

Appliances (Cost per Unit)

	Economy	Average	Custom	Luxury
Range				
30″ free standing, 1 oven	$ 310	$1005	$1353	$1700
2 oven	1675	1838	1919	2000
30″ built-in, 1 oven	550	1163	1469	1775
2 oven	1175	1550	1738	1925
21″ free standing				
1 oven	320	368	391	415
Counter Top Ranges				
4 burner standard	251	426	513	600
As above with griddle	535	678	749	820
Microwave Oven	169	385	492	600
Combination Range, Refrigerator, Sink				
30″ wide	1025	1525	1775	2025
60″ wide	2950	3393	3614	3835
72″ wide	3350	3853	4104	4355
Comb. Range, Refrig., Sink, Microwave Oven & Ice Maker	4969	5715	6087	6460
Compactor				
4 to 1 compaction	520	570	595	620
Deep Freeze				
15 to 23 C.F.	475	565	610	655
30 C.F.	885	993	1046	1100
Dehumidifier, portable, auto.				
15 pint	164	189	201	213
30 pint	183	211	224	238
Washing Machine, automatic	415	945	1210	1475
Water Heater				
Electric, glass lined				
30 gal.	365	448	489	530
80 gal.	745	960	1068	1175
Water Heater, Gas, glass lined				
30 gal.	515	638	699	760
50 gal.	770	960	1055	1150
Water Softener, automatic				
30 grains/gal.	610	702	747	793
100 grains/gal.	830	955	1017	1079
Dishwasher, built-in				
2 cycles	375	455	495	535
4 or more cycles	395	443	466	490
Dryer, automatic	475	788	944	1100
Garage Door Opener	320	385	418	450
Garbage Disposal	90	147	175	203
Heater, Electric, built-in				
1250 watt ceiling type	163	202	222	241
1250 watt wall type	195	228	244	260
Wall type w/blower				
1500 watt	221	306	348	390
3000 watt	390	429	449	468
Hood For Range, 2 speed				
30″ wide	121	456	623	790
42″ wide	325	545	655	765
Humidifier, portable				
7 gal. per day	164	189	201	213
15 gal. per day	197	227	241	256
Ice Maker, automatic				
13 lb. per day	445	512	546	579
51 lb. per day	1275	1467	1562	1658
Refrigerator, no frost				
10-12 C.F.	530	678	751	825
14-16 C.F.	540	580	600	620
18-20 C.F.	620	835	943	1050
21-29 C.F.	775	1600	2013	2425
Sump Pump, 1/3 H.P.	210	290	330	370

Breezeway (Cost per S.F.)

Class	Type	Area (S.F.)			
		50	100	150	200
Economy	Open	$ 17.85	$15.20	$12.75	$12.50
	Enclosed	85.60	66.15	54.90	48.10
Average	Open	22.10	19.45	17.00	15.45
	Enclosed	93.85	69.90	57.20	50.35
Custom	Open	31.50	27.70	24.10	22.10
	Enclosed	129.40	96.25	78.70	69.15
Luxury	Open	32.50	28.50	25.70	24.90
	Enclosed	130.95	97.10	78.55	70.35

Porches (Cost per S.F.)

Class	Type	Area (S.F.)				
		25	50	100	200	300
Economy	Open	$ 51.15	$ 34.25	$26.75	$22.65	$19.35
	Enclosed	102.25	71.25	53.85	42.00	36.00
Average	Open	62.10	39.55	30.35	25.25	25.25
	Enclosed	122.70	83.15	62.90	48.70	41.25
Custom	Open	80.45	53.35	40.20	35.10	31.45
	Enclosed	159.90	109.25	83.15	64.70	55.75
Luxury	Open	86.10	56.30	41.55	37.35	33.25
	Enclosed	168.50	118.35	87.85	68.20	58.75

Finished attic (Cost per S.F.)

Class	Area (S.F.)				
	400	500	600	800	1000
Economy	$13.70	$13.20	$12.65	$12.45	$12.00
Average	21.05	20.50	20.05	19.75	19.25
Custom	25.65	25.05	24.55	24.10	23.65
Luxury	32.30	31.55	30.80	30.10	29.60

Alarm system (Cost per System)

	Burglar Alarm	Smoke Detector
Economy	$ 360	$ 54
Average	410	66
Custom	697	138
Luxury	1050	170

Sauna, prefabricated
(Cost per unit, including heater and controls—7′ high)

Size	Cost
6′ x 4′	$ 4200
6′ x 5′	4675
6′ x 6′	5000
6′ x 9′	6200
8′ x 10′	8175
8′ x 12′	9575
10′ x 12′	10200

Garages *

(Costs include exterior wall systems comparable with the quality of the residence. Included in the cost is an allowance for one personnel door, manual overhead door(s) and electrical fixture.)

Class	Type									
	Detached			Attached			Built-in		Basement	
	One Car	Two Car	Three Car	One Car	Two Car	Three Car	One Car	Two Car	One Car	Two Car
Economy										
Wood	$10,644	$16,205	$21,767	$ 8261	$14,218	$19,779	$-1335	$-2670	$1130	$1488
Masonry	14,779	21,380	27,981	10,849	17,845	24,446	-1843	-3685		
Average										
Wood	11,757	17,598	23,440	8958	15,194	21,035	-1471	-2943	1285	1797
Masonry	14,907	21,540	28,174	10,929	17,957	24,590	-1858	-3172		
Custom										
Wood	13,650	20,976	28,301	10,722	18,506	25,832	-2964	-3019	1978	3183
Masonry	16,723	24,821	32,920	12,645	21,202	29,300	-3341	-3774		
Luxury										
Wood	15,403	24,051	32,699	12,260	21,366	30,014	-3034	-3160	2667	4246
Masonry	19,496	29,174	38,851	14,821	24,957	34,634	-3537	-4165		

*See the Introduction to this section for definitions of garage types.

Swimming pools *(Cost per S.F.)*

Residential (includes equipment)	
In-ground	$ 19.10 - 47.50
Deck equipment	1.30
Paint pool, preparation & 3 coats (epoxy)	2.93
Rubber base paint	2.72
Pool Cover	.70
Swimming Pool Heaters (Cost per unit)	
(not including wiring, external piping, base or pad)	
Gas	
155 MBH	$ 2800
190 MBH	3775
500 MBH	9200
Electric	
15 KW 7200 gallon pool	2075
24 KW 9600 gallon pool	2725
54 KW 24,000 gallon pool	4100

Wood and coal stoves

Wood Only	
Free Standing (minimum)	$ 1325
Fireplace Insert (minimum)	1336
Coal Only	
Free Standing	$ 1509
Fireplace Insert	1652
Wood and Coal	
Free Standing	$ 3104
Fireplace Insert	3180

Sidewalks *(Cost per S.F.)*

Concrete, 3000 psi with wire mesh	4" thick	$ 2.71
	5" thick	3.29
	6" thick	3.69
Precast concrete patio blocks (natural)	2" thick	8.15
Precast concrete patio blocks (colors)	2" thick	8.65
Flagstone, bluestone	1" thick	12.00
Flagstone, bluestone	1-1/2" thick	15.80
Slate (natural, irregular)	3/4" thick	11.75
Slate (random rectangular)	1/2" thick	17.60
Seeding		
Fine grading & seeding includes lime, fertilizer & seed	per S.Y.	1.88
Lawn Sprinkler System	per S.F.	.72

Fencing *(Cost per L.F.)*

Chain Link, 4' high, galvanized	$ 11.45
Gate, 4' high (each)	126.00
Cedar Picket, 3' high, 2 rail	9.95
Gate (each)	127.00
3 Rail, 4' high	11.10
Gate (each)	137.00
Cedar Stockade, 3 Rail, 6' high	11.30
Gate (each)	136.00
Board & Battens, 2 sides 6' high, pine	16.45
6' high, cedar	23.50
No. 1 Cedar, basketweave, 6' high	13.05
Gate, 6' high (each)	157.00

Carport *(Cost per S.F.)*

Economy	$ 6.44
Average	9.76
Custom	14.55
Luxury	16.52

Assemblies Section

Table of Contents

How to Use the Assemblies Cost Tables

The following is a detailed explanation of a sample Assemblies Cost Table. Included are an illustration and accompanying system descriptions. Additionally, related systems and price sheets may be included. Next to each bold number below is the item being described with the appropriate component of the sample entry following in parenthesis. General contractors should add an additional markup to the figures shown in the Assemblies section. Note: Throughout this section, the words assembly and system are used interchangeably.

System Identification (3 Framing 12)

Each Assemblies section has been assigned a unique identification number, component category, system number and system description.

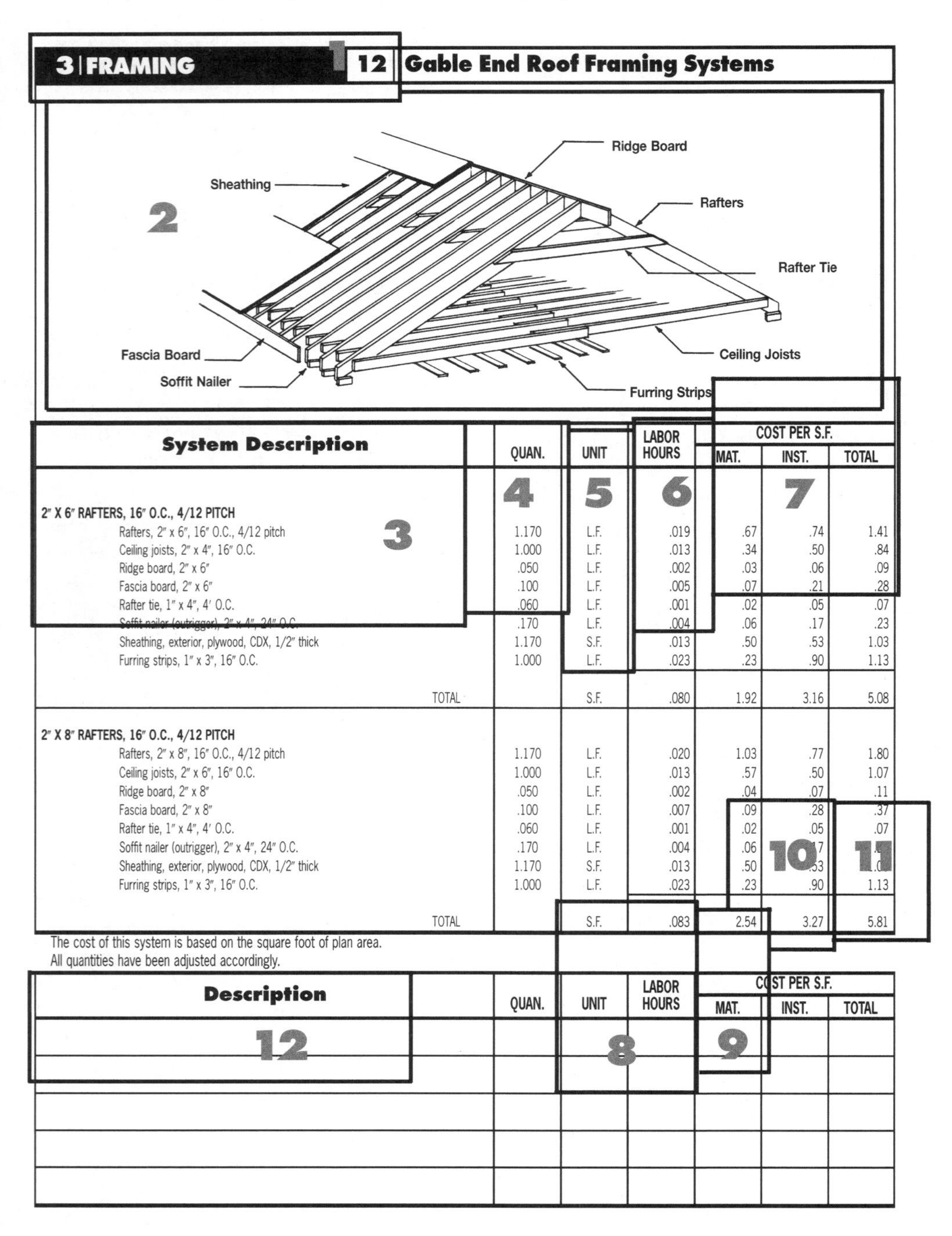

System Description	QUAN.	UNIT	LABOR HOURS	COST PER S.F. MAT.	INST.	TOTAL
2" X 6" RAFTERS, 16" O.C., 4/12 PITCH						
Rafters, 2" x 6", 16" O.C., 4/12 pitch	1.170	L.F.	.019	.67	.74	1.41
Ceiling joists, 2" x 4", 16" O.C.	1.000	L.F.	.013	.34	.50	.84
Ridge board, 2" x 6"	.050	L.F.	.002	.03	.06	.09
Fascia board, 2" x 6"	.100	L.F.	.005	.07	.21	.28
Rafter tie, 1" x 4", 4' O.C.	.060	L.F.	.001	.02	.05	.07
Soffit nailer (outrigger), 2" x 4", 24" O.C.	.170	L.F.	.004	.06	.17	.23
Sheathing, exterior, plywood, CDX, 1/2" thick	1.170	S.F.	.013	.50	.53	1.03
Furring strips, 1" x 3", 16" O.C.	1.000	L.F.	.023	.23	.90	1.13
TOTAL		S.F.	.080	1.92	3.16	5.08
2" X 8" RAFTERS, 16" O.C., 4/12 PITCH						
Rafters, 2" x 8", 16" O.C., 4/12 pitch	1.170	L.F.	.020	1.03	.77	1.80
Ceiling joists, 2" x 6", 16" O.C.	1.000	L.F.	.013	.57	.50	1.07
Ridge board, 2" x 8"	.050	L.F.	.002	.04	.07	.11
Fascia board, 2" x 8"	.100	L.F.	.007	.09	.28	.37
Rafter tie, 1" x 4", 4' O.C.	.060	L.F.	.001	.02	.05	.07
Soffit nailer (outrigger), 2" x 4", 24" O.C.	.170	L.F.	.004	.06	[illegible]	[illegible]
Sheathing, exterior, plywood, CDX, 1/2" thick	1.170	S.F.	.013	.50	[illegible]	[illegible]
Furring strips, 1" x 3", 16" O.C.	1.000	L.F.	.023	.23	.90	1.13
TOTAL		S.F.	.083	2.54	3.27	5.81

The cost of this system is based on the square foot of plan area.
All quantities have been adjusted accordingly.

Description	QUAN.	UNIT	LABOR HOURS	COST PER S.F. MAT.	INST.	TOTAL

2 *Illustration*

At the top of most assembly pages is an illustration with individual components labeled. Elements involved in the total system function are shown.

3 *System Description (2" x 6" Rafters, 16" O.C., 4/12 Pitch)*

The components of a typical system are listed separately to show what has been included in the development of the total system price. Each page includes a brief outline of any special conditions to be used when pricing a system. Alternative components can be found on the opposite page. Simply insert any chosen new element into the chart to develop a custom system.

4 *Quantities for Each Component*

Each material in a system is shown with the quantity required for the system unit. For example, there are 1.170 L.F. of rafter per S.F. of plan area.

5 *Unit of Measure for Each Component*

The abbreviated designation indicates the unit of measure, as defined by industry standards, upon which the individual component has been priced. In this example, items are priced by the linear foot (L.F.) or the square foot (S.F.).

6 *Labor Hours*

This is the amount of time it takes to install the quantity of the individual component.

7 *Unit of Measure (Cost per S.F.)*

In the three right-hand columns, each cost figure is adjusted to agree with the unit of measure for the entire system. In this case, cost per S.F. is the common unit of measure.

8 *Labor Hours (.083)*

The labor hours column shows the amount of time necessary to install the system per the unit of measure. For example, it takes .083 labor hours to install one square foot (plan area) of this roof framing system.

9 *Materials (2.54)*

This column contains the material cost of each element. These cost figures include 10% for profit.

10 *Installation (3.27)*

This column contains labor and equipment costs. Labor rates include bare cost and the installing contractor's overhead and profit. On the average, the labor cost will be 69.5% over the bare labor cost. Equipment costs include 10% for profit.

11 *Totals (5.81)*

The figure in this column is the sum of the material and installation costs.

12 *Work Sheet*

Using the selective price sheet on the page opposite each system, it is possible to create estimates with alternative items for any number of systems.

Division 1
Site Work

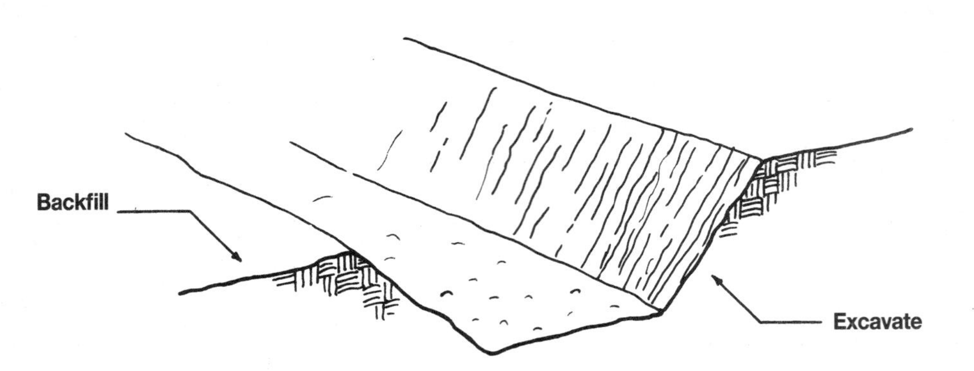

System Description	QUAN.	UNIT	LABOR HOURS	COST EACH		
				MAT.	INST.	TOTAL
BUILDING, 24′ X 38′, 4′ DEEP						
Cut & chip light tres to 6″ diam.	.190	Acre	9.120		489.25	489.25
Excavate, backhoe	174.000	C.Y.	2.319		247.08	247.08
Backfill, dozer, 4″ lifts, no compaction	87.000	C.Y.	.580		90.48	90.48
Rough grade, dozer, 30′ from building	87.000	C.Y.	.580		90.48	90.48
TOTAL		Ea.	12.599		917.29	917.29
BUILDING, 26′ X 46′, 4′ DEEP						
Cut & chip light tres to 6″ diam.	.210	Acre	10.080		540.75	540.75
Excavate, backhoe	201.000	C.Y.	2.679		285.42	285.42
Backfill, dozer, 4″ lifts, no compaction	100.000	C.Y.	.667		104	104
Rough grade, dozer, 30′ from building	100.000	C.Y.	.667		104	104
TOTAL		Ea.	14.093		1,034.17	1,034.17
BUILDING, 26′ X 60′, 4′ DEEP						
Cut & chip light tres to 6″ diam.	.240	Acre	11.520		618	618
Excavate, backhoe	240.000	C.Y.	3.199		340.80	340.80
Backfill, dozer, 4″ lifts, no compaction	120.000	C.Y.	.800		124.80	124.80
Rough grade, dozer, 30′ from building	120.000	C.Y.	.800		124.80	124.80
TOTAL		Ea.	16.319		1,208.40	1,208.40
BUILDING, 30′ X 66′, 4′ DEEP						
Cut & chip light tres to 6″ diam.	.260	Acre	12.480		669.50	669.50
Excavate, backhoe	268.000	C.Y.	3.572		380.56	380.56
Backfill, dozer, 4″ lifts, no compaction	134.000	C.Y.	.894		139.36	139.36
Rough grade, dozer, 30′ from building	134.000	C.Y.	.894		139.36	139.36
TOTAL		Ea.	17.840		1,328.78	1,328.78

The costs in this system are on a cost each basis.
Quantities are based on 1′-0″ clearance on each side of footing.

Description	QUAN.	UNIT	LABOR HOURS	COST EACH		
				MAT.	INST.	TOTAL

Important: See the Reference Section for critical supporting data - Reference Nos., Crews & Location Factors

Footing Excavation Price Sheet	QUAN.	UNIT	LABOR HOURS	COST EACH		
				MAT.	INST.	TOTAL
Clear and grub, medium brush, 30′ from building, 24′ x 38′	.190	Acre	9.120		490	490
26′ x 46′	.210	Acre	10.080		540	540
26′ x 60′	.240	Acre	11.520		620	620
30′ x 66′	.260	Acre	12.480		670	670
Light trees, to 6″ dia. cut & chip, 24′ x 38′	.190	Acre	9.120		490	490
26′ x 46′	.210	Acre	10.080		540	540
26′ x 60′	.240	Acre	11.520		620	620
30′ x 66′	.260	Acre	12.480		670	670
Medium trees, to 10″ dia. cut & chip, 24′ x 38′	.190	Acre	13.029		700	700
26′ x 46′	.210	Acre	14.400		770	770
26′ x 60′	.240	Acre	16.457		880	880
30′ x 66′	.260	Acre	17.829		960	960
Excavation, footing, 24′ x 38′, 2′ deep	68.000	C.Y.	.906		96.50	96.50
4′ deep	174.000	C.Y.	2.319		247	247
8′ deep	384.000	C.Y.	5.119		545	545
26′ x 46′, 2′ deep	79.000	C.Y.	1.053		113	113
4′ deep	201.000	C.Y.	2.679		286	286
8′ deep	404.000	C.Y.	5.385		575	575
26′ x 60′, 2′ deep	94.000	C.Y.	1.253		134	134
4′ deep	240.000	C.Y.	3.199		340	340
8′ deep	483.000	C.Y.	6.438		685	685
30′ x 66′, 2′ deep	105.000	C.Y.	1.400		149	149
4′ deep	268.000	C.Y.	3.572		380	380
8′ deep	539.000	C.Y.	7.185		765	765
Backfill, 24′ x 38′, 2″ lifts, no compaction	34.000	C.Y.	.227		35.50	35.50
Compaction, air tamped, add	34.000	C.Y.	6.800		315	315
4″ lifts, no compaction	87.000	C.Y.	.580		90.50	90.50
Compaction, air tamped, add	87.000	C.Y.	17.400		800	800
8″ lifts, no compaction	192.000	C.Y.	1.281		200	200
Compaction, air tamped, add	192.000	C.Y.	38.400		1750	1750
26′ x 46′, 2″ lifts, no compaction	40.000	C.Y.	.267		41.50	41.50
Compaction, air tamped, add	40.000	C.Y.	8.000		365	365
4″ lifts, no compaction	100.000	C.Y.	.667		104	104
Compaction, air tamped, add	100.000	C.Y.	20.000		920	920
8″ lifts, no compaction	202.000	C.Y.	1.347		211	211
Compaction, air tamped, add	202.000	C.Y.	40.400		1850	1850
26′ x 60′, 2″ lifts, no compaction	47.000	C.Y.	.313		49	49
Compaction, air tamped, add	47.000	C.Y.	9.400		435	435
4″ lifts, no compaction	120.000	C.Y.	.800		125	125
Compaction, air tamped, add	120.000	C.Y.	24.000		1100	1100
8″ lifts, no compaction	242.000	C.Y.	1.614		252	252
Compaction, air tamped, add	242.000	C.Y.	48.4000		2200	2200
30′ x 66′, 2″ lifts, no compaction	53.000	C.Y.	.354		55.50	55.50
Compaction, air tamped, add	53.000	C.Y.	10.600		485	485
4″ lifts, no compaction	134.000	C.Y.	.894		140	140
Compaction, air tamped, add	134.000	C.Y.	26.800		1225	1225
8″ lifts, no compaction	269.000	C.Y.	1.794		281	281
Compaction, air tamped, add	269.000	C.Y.	53.800		2450	2450
Rough grade, 30′ from building, 24′ x 38′	87.000	C.Y.	.580		90.50	90.50
26′ x 46′	100.000	C.Y.	.667		104	104
26′ x 60′	120.000	C.Y.	.800		125	125
30′ x 66′	134.000	C.Y.	.894		140	140

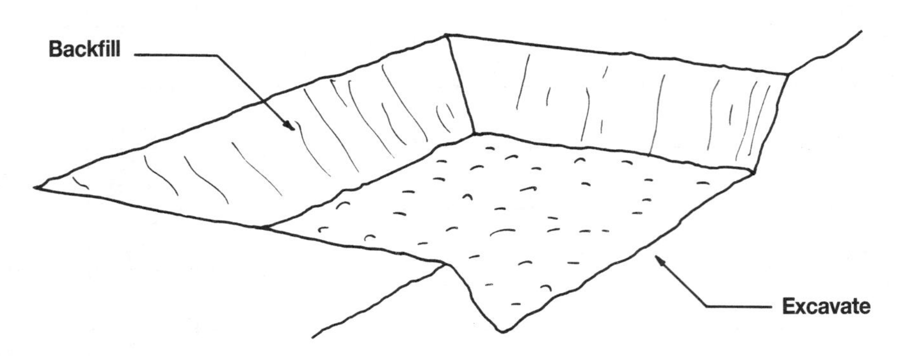

System Description	QUAN.	UNIT	LABOR HOURS	COST EACH MAT.	COST EACH INST.	COST EACH TOTAL
BUILDING, 24′ X 38′, 8′ DEEP						
Clear & grub, dozer, medium brush, 30′ from building	.190	Acre	2.027		188.10	188.10
Excavate, track loader, 1-1/2 C.Y. bucket	550.000	C.Y.	7.860		627	627
Backfill, dozer, 8″ lifts, no compaction	180.000	C.Y.	1.201		187.20	187.20
Rough grade, dozer, 30′ from building	280.000	C.Y.	1.868		291.20	291.20
TOTAL		Ea.	12.956		1,293.50	1,293.50
BUILDING, 26′ X 46′, 8′ DEEP						
Clear & grub, dozer, medium brush, 30′ from building	.210	Acre	2.240		207.90	207.90
Excavate, track loader, 1-1/2 C.Y. bucket	672.000	C.Y.	9.603		766.08	766.08
Backfill, dozer, 8″ lifts, no compaction	220.000	C.Y.	1.467		228.80	228.80
Rough grade, dozer, 30′ from building	340.000	C.Y.	2.268		353.60	353.60
TOTAL		Ea.	15.578		1,556.38	1,556.38
BUILDING, 26′ X 60′, 8′ DEEP						
Clear & grub, dozer, medium brush, 30′ from building	.240	Acre	2.560		237.60	237.60
Excavate, track loader, 1-1/2 C.Y. bucket	829.000	C.Y.	11.846		945.06	945.06
Backfill, dozer, 8″ lifts, no compaction	270.000	C.Y.	1.801		280.80	280.80
Rough grade, dozer, 30′ from building	420.000	C.Y.	2.801		436.80	436.80
TOTAL		Ea.	19.008		1,900.26	1,900.26
BUILDING, 30′ X 66′, 8′ DEEP						
Clear & grub, dozer, medium brush, 30′ from building	.260	Acre	2.773		257.40	257.40
Excavate, track loader, 1-1/2 C.Y. bucket	990.000	C.Y.	14.147		1,128.60	1,128.60
Backfill dozer, 8″ lifts, no compaction	320.000	C.Y.	2.134		332.80	332.80
Rough grade, dozer, 30′ from building	500.000	C.Y.	3.335		520	520
TOTAL		Ea.	22.389		2,238.80	2,238.80

The costs in this system are on a cost each basis.
Quantities are based on 1′-0″ clearance beyond footing projection.

Description	QUAN.	UNIT	LABOR HOURS	COST EACH MAT.	COST EACH INST.	COST EACH TOTAL

 Important: See the Reference Section for critical supporting data - Reference Nos., Crews & Location Factors

Foundation Excavation Price Sheet

Foundation Excavation Price Sheet	QUAN.	UNIT	LABOR HOURS	COST EACH		
				MAT.	INST.	TOTAL
Clear & grub, medium brush, 30′ from building, 24′ x 38′	.190	Acre	2.027		189	189
26′ x 46′	.210	Acre	2.240		208	208
26′ x 60′	.240	Acre	2.560		237	237
30′ x 66′	.260	Acre	2.773		258	258
Light trees, to 6″ dia. cut & chip, 24′ x 38′	.190	Acre	9.120		490	490
26′ x 46′	.210	Acre	10.080		540	540
26′ x 60′	.240	Acre	11.520		620	620
30′ x 66′	.260	Acre	12.480		670	670
Medium trees, to 10″ dia. cut & chip, 24′ x 38′	.190	Acre	13.029		700	700
26′ x 46′	.210	Acre	14.400		770	770
26′ x 60′	.240	Acre	16.457		880	880
30′ x 66′	.260	Acre	17.829		960	960
Excavation, basement, 24′ x 38′, 2′ deep	98.000	C.Y.	1.400		112	112
4′ deep	220.000	C.Y.	3.144		251	251
8′ deep	550.000	C.Y.	7.860		630	630
26′ x 46′, 2′ deep	123.000	C.Y.	1.758		140	140
4′ deep	274.000	C.Y.	3.915		315	315
8′ deep	672.000	C.Y.	9.603		765	765
26′ x 60′, 2′ deep	157.000	C.Y.	2.244		179	179
4′ deep	345.000	C.Y.	4.930		395	395
8′ deep	829.000	C.Y.	11.846		945	945
30′ x 66′, 2′ deep	192.000	C.Y.	2.744		219	219
4′ deep	419.000	C.Y.	5.988		475	475
8′ deep	990.000	C.Y.	14.147		1,125	1,125
Backfill, 24′ x 38′, 2″ lifts, no compaction	32.000	C.Y.	.213		33.50	33.50
Compaction, air tamped, add	32.000	C.Y.	6.400		294	294
4″ lifts, no compaction	72.000	C.Y.	.480		75	75
Compaction, air tamped, add	72.000	C.Y.	14.400		660	660
8″ lifts, no compaction	180.000	C.Y.	1.201		187	187
Compaction, air tamped, add	180.000	C.Y.	36.000		1650	1650
26′ x 46′, 2″ lifts, no compaction	40.000	C.Y.	.267		41.50	41.50
Compaction, air tamped, add	40.000	C.Y.	8.000		365	365
4″ lifts, no compaction	90.000	C.Y.	.600		93.50	93.50
Compaction, air tamped, add	90.000	C.Y.	18.000		825	825
8″ lifts, no compaction	220.000	C.Y.	1.467		229	229
Compacton, air tamped, add	220.000	C.Y.	44.000		2025	2025
26′ x 60′, 2″ lifts, no compaction	50.000	C.Y.	.334		52	52
Compaction, air tamped, add	50.000	C.Y.	10.000		460	460
4″ lifts, no compaction	110.000	C.Y.	.734		115	115
Compaction, air tamped, add	110.000	C.Y.	22.000		1000	1000
8″ lifts, no compaction	270.000	C.Y.	1.801		281	281
Compaction, air tamped, add	270.000	C.Y.	54.000		2475	2475
30′ x 66′, 2″ lifts, no compaction	60.000	C.Y.	.400		62.50	62.50
Compaction, air tamped, add	60.000	C.Y.	12.000		555	555
4″ lifts, no compaction	130.000	C.Y.	.867		136	136
Compaction, air tamped, add	130.000	C.Y.	26.000		1200	1200
8″ lifts, no compaction	320.000	C.Y.	2.134		335	335
Compaction, air tamped, add	320.000	C.Y.	64.000		2950	2950
Rough grade, 30′ from building, 24′ x 38′	280.000	C.Y.	1.868		291	291
26′ x 46′	340.000	C.Y.	2.268		355	355
26′ x 60′	420.000	C.Y.	2.801		435	435
30′ x 66′	500.000	C.Y.	3.335		520	520

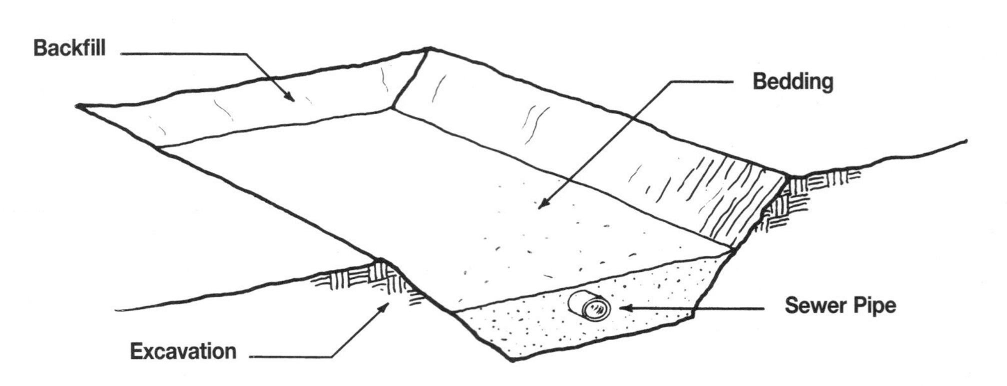

System Description	QUAN.	UNIT	LABOR HOURS	COST PER L.F.		
				MAT.	INST.	TOTAL
2′ DEEP						
Excavation, backhoe	.296	C.Y.	.032		1.51	1.51
Bedding, sand	.111	C.Y.	.044	1.38	1.47	2.85
Utility, sewer, 6″ cast iron	1.000	L.F.	.283	13.12	10.49	23.61
Backfill, incl. compaction	.185	C.Y.	.044		1.25	1.25
TOTAL		L.F.	.403	14.50	14.72	29.22
4′ DEEP						
Excavation, backhoe	.889	C.Y.	.095		4.52	4.52
Bedding, sand	.111	C.Y.	.044	1.38	1.47	2.85
Utility, sewer, 6″ cast iron	1.000	L.F.	.283	13.12	10.49	23.61
Backfill, incl. compaction	.778	C.Y.	.183		5.25	5.25
TOTAL		L.F.	.605	14.50	21.73	36.23
6′ DEEP						
Excavation, backhoe	1.770	C.Y.	.189		9.01	9.01
Bedding, sand	.111	C.Y.	.044	1.38	1.47	2.85
Utility, sewer, 6″ cast iron	1.000	L.F.	.283	13.12	10.49	23.61
Backfill, incl. compaction	1.660	C.Y.	.391		11.21	11.21
TOTAL		L.F.	.907	14.50	32.18	46.68
8′ DEEP						
Excavation, backhoe	2.960	C.Y.	.316		15.07	15.07
Bedding, sand	.111	C.Y.	.044	1.38	1.47	2.85
Utility, sewer, 6″ cast iron	1.000	L.F.	.283	13.12	10.49	23.61
Backfill, incl. compaction	2.850	C.Y.	.671		19.24	19.24
TOTAL		L.F.	1.314	14.50	46.27	60.77

The costs in this system are based on a cost per linear foot of trench, and based on 2′ wide at bottom of trench up to 6′ deep.

Description	QUAN.	UNIT	LABOR HOURS	COST PER L.F.		
				MAT.	INST.	TOTAL

Important: See the Reference Section for critical supporting data - Reference Nos., Crews & Location Factors

Utility Trenching Price Sheet

Utility Trenching Price Sheet	QUAN.	UNIT	LABOR HOURS	MAT.	INST.	TOTAL
Excavation, bottom of trench 2′ wide, 2′ deep	.296	C.Y.	.032		1.51	1.51
4′ deep	.889	C.Y.	.095		4.52	4.52
6′ deep	1.770	C.Y.	.142		7	7
8′ deep	2.960	C.Y.	.105		10.80	10.80
Bedding, sand, bottom of trench 2′ wide, no compaction, pipe, 2″ diameter	.070	C.Y.	.028	.87	.93	1.80
4″ diameter	.084	C.Y.	.034	1.04	1.11	2.15
6″ diameter	.105	C.Y.	.042	1.30	1.39	2.69
8″ diameter	.122	C.Y.	.049	1.51	1.61	3.12
Compacted, pipe, 2″ diameter	.074	C.Y.	.030	.92	.98	1.90
4″ diameter	.092	C.Y.	.037	1.14	1.22	2.36
6″ diameter	.111	C.Y.	.044	1.38	1.47	2.85
8″ diameter	.129	C.Y.	.052	1.60	1.71	3.31
3/4″ stone, bottom of trench 2′ wide, pipe, 4″ diameter	.082	C.Y.	.033	1.02	1.08	2.10
6″ diameter	.099	C.Y.	.040	1.23	1.31	2.54
3/8″ stone, bottom of trench 2′ wide, pipe, 4″ diameter	.084	C.Y.	.034	1.04	1.11	2.15
6″ diameter	.102	C.Y.	.041	1.26	1.35	2.61
Utilities, drainage & sewerage, corrugated plastic, 6″ diameter	1.000	L.F.	.069	3.44	2.05	5.49
8″ diameter	1.000	L.F.	.072	5.80	2.14	7.94
Bituminous fiber, 4″ diameter	1.000	L.F.	.064	1.93	1.91	3.84
6″ diameter	1.000	L.F.	.069	3.44	2.05	5.49
8″ diameter	1.000	L.F.	.072	5.80	2.14	7.94
Concrete, non-reinforced, 6″ diameter	1.000	L.F.	.181	4.46	6.40	10.86
8″ diameter	1.000	L.F.	.214	4.90	7.55	12.45
PVC, SDR 35, 4″ diameter	1.000	L.F.	.064	1.93	1.91	3.84
6″ diameter	1.000	L.F.	.069	3.44	2.05	5.49
8″ diameter	1.000	L.F.	.072	5.80	2.14	7.94
Vitrified clay, 4″ diameter	1.000	L.F.	.091	1.87	2.70	4.57
6″ diameter	1.000	L.F.	.120	3.12	3.58	6.70
8″ diameter	1.000	L.F.	.140	4.43	5.25	9.68
Gas & service, polyethylene, 1-1/4″ diameter	1.000	L.F.	.059	.94	2.03	2.97
Steel sched.40, 1″ diameter	1.000	L.F.	.107	2.88	4.63	7.51
2″ diameter	1.000	L.F.	.114	4.52	4.97	9.49
Sub-drainage, PVC, perforated, 3″ diameter	1.000	L.F.	.064	1.93	1.91	3.84
4″ diameter	1.000	L.F.	.064	1.93	1.91	3.84
5″ diameter	1.000	L.F.	.069	3.44	2.05	5.49
6″ diameter	1.000	L.F.	.069	3.44	2.05	5.49
Porous wall concrete, 4″ diameter	1.000	L.F.	.072	2.05	2.14	4.19
Vitrified clay, perforated, 4″ diameter	1.000	L.F.	.120	2.20	4.24	6.44
6″ diameter	1.000	L.F.	.152	3.64	5.40	9.04
Water service, copper, type K, 3/4″	1.000	L.F.	.083	1.49	3.57	5.06
1″ diameter	1.000	L.F.	.093	1.93	4	5.93
PVC, 3/4″	1.000	L.F.	.121	1.26	5.20	6.46
1″ diameter	1.000	L.F.	.134	1.41	5.75	7.16
Backfill, bottom of trench 2′ wide no compact, 2′ deep, pipe, 2″ diameter	.226	L.F.	.053		1.53	1.53
4″ diameter	.212	L.F.	.050		1.43	1.43
6″ diameter	.185	L.F.	.044		1.25	1.25
4′ deep, pipe, 2″ diameter	.819	C.Y.	.193		5.55	5.55
4″ diameter	.805	C.Y.	.189		5.45	5.45
6″ diameter	.778	C.Y.	.183		5.25	5.25
6′ deep, pipe, 2″ diameter	1.700	C.Y.	.400		11.50	11.50
4″ diameter	1.690	C.Y.	.398		11.40	11.40
6″ diameter	1.660	C.Y.	.391		11.20	11.20
8′ deep, pipe, 2″ diameter	2.890	C.Y.	.680		19.50	19.50
4″ diameter	2.870	C.Y.	.675		19.35	19.35
6″ diameter	2.850	C.Y.	.671		19.25	19.25

16 Sidewalk Systems

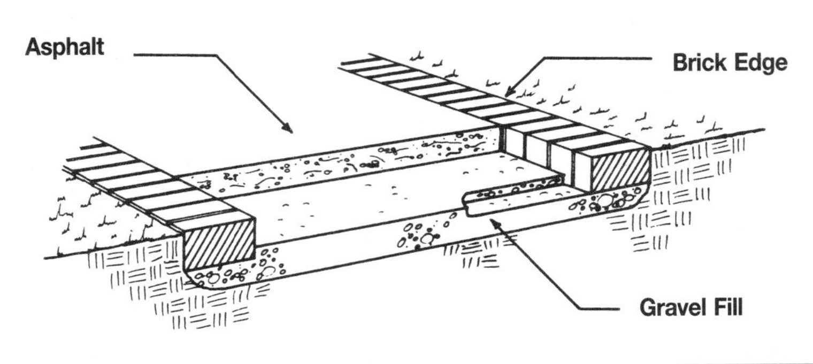

System Description	QUAN.	UNIT	LABOR HOURS	COST PER S.F.		
				MAT.	INST.	TOTAL
ASPHALT SIDEWALK SYSTEM, 3′ WIDE WALK						
Gravel fill, 4″ deep	1.000	S.F.	.001	.27	.03	.30
Compact fill	.012	C.Y.			.01	.01
Handgrade	1.000	S.F.	.004		.12	.12
Walking surface, bituminous paving, 2″ thick	1.000	S.F.	.007	.41	.25	.66
Edging, brick, laid on edge	.670	L.F.	.079	1.37	2.77	4.14
TOTAL		S.F.	.092	2.05	3.18	5.23
CONCRETE SIDEWALK SYSTEM, 3′ WIDE WALK						
Gravel fill, 4″ deep	1.000	S.F.	.001	.27	.03	.30
Compact fill	.012	C.Y.			.01	.01
Handgrade	1.000	S.F.	.004		.12	.12
Walking surface, concrete, 4″ thick	1.000	S.F.	.040	1.33	1.38	2.71
Edging, brick, laid on edge	.670	L.F.	.079	1.37	2.77	4.14
TOTAL		S.F.	.125	2.97	4.31	7.28
PAVERS, BRICK SIDEWALK SYSTEM, 3′ WIDE WALK						
Sand base fill, 4″ deep	1.000	S.F.	.001	.33	.06	.39
Compact fill	.012	C.Y.			.01	.01
Handgrade	1.000	S.F.	.004		.12	.12
Walking surface, brick pavers	1.000	S.F.	.160	2.62	5.55	8.17
Edging, redwood, untreated, 1″ x 4″	.670	L.F.	.032	2.04	1.27	3.31
TOTAL		S.F.	.198	4.99	7.01	12

The costs in this system are based on a cost per square foot of sidewalk area. Concrete used is 3000 p.s.i.

Description	QUAN.	UNIT	LABOR HOURS	COST PER S.F.		
				MAT.	INST.	TOTAL

Sidewalk Price Sheet

Sidewalk Price Sheet	QUAN.	UNIT	LABOR HOURS	COST PER S.F.		
				MAT.	INST.	TOTAL
Base, crushed stone, 3″ deep	1.000	S.F.	.001	.32	.07	.39
6″ deep	1.000	S.F.	.001	.64	.07	.71
9″ deep	1.000	S.F.	.002	.93	.10	1.03
12″ deep	1.000	S.F.	.002	1.46	.12	1.58
Bank run gravel, 6″ deep	1.000	S.F.	.001	.40	.05	.45
9″ deep	1.000	S.F.	.001	.59	.07	.66
12″ deep	1.000	S.F.	.001	.81	.09	.90
Compact base, 3″ deep	.009	C.Y.	.001		.01	.01
6″ deep	.019	C.Y.	.001		.02	.02
9″ deep	.028	C.Y.	.001		.03	.03
Handgrade	1.000	S.F.	.004		.12	.12
Surface, brick, pavers dry joints, laid flat, running bond	1.000	S.F.	.160	2.62	5.55	8.17
Basket weave	1.000	S.F.	.168	3.04	5.85	8.89
Herringbone	1.000	S.F.	.174	3.04	6.05	9.09
Laid on edge, running bond	1.000	S.F.	.229	2.37	7.95	10.32
Mortar jts. laid flat, running bond	1.000	S.F.	.192	3.14	6.65	9.79
Basket weave	1.000	S.F.	.202	3.65	7	10.65
Herringbone	1.000	S.F.	.209	3.65	7.25	10.90
Laid on edge, running bond	1.000	S.F.	.274	2.84	9.55	12.39
Bituminous paving, 1-1/2″ thick	1.000	S.F.	.006	.31	.18	.49
2″ thick	1.000	S.F.	.007	.41	.25	.66
2-1/2″ thick	1.000	S.F.	.008	.52	.27	.79
Sand finish, 3/4″ thick	1.000	S.F.	.001	.20	.08	.28
1″ thick	1.000	S.F.	.001	.24	.10	.34
Concrete, reinforced, broom finish, 4″ thick	1.000	S.F.	.040	1.33	1.38	2.71
5″ thick	1.000	S.F.	.044	1.77	1.52	3.29
6″ thick	1.000	S.F.	.047	2.07	1.62	3.69
Crushed stone, white marble, 3″ thick	1.000	S.F.	.009	.21	.27	.48
Bluestone, 3″ thick	1.000	S.F.	.009	.22	.27	.49
Flagging, bluestone, 1″	1.000	S.F.	.198	5.10	6.90	12
1-1/2″	1.000	S.F.	.188	9.25	6.55	15.80
Slate, natural cleft, 3/4″	1.000	S.F.	.174	5.70	6.05	11.75
Random rect., 1/2″	1.000	S.F.	.152	12.30	5.30	17.60
Granite blocks	1.000	S.F.	.174	6.90	6.05	12.95
Edging, corrugated aluminum, 4″, 3′ wide walk	.666	L.F.	.008	.23	.32	.55
4′ wide walk	.500	L.F.	.006	.18	.24	.42
6″, 3′ wide walk	.666	L.F.	.010	.29	.38	.67
4′ wide walk	.500	L.F.	.007	.22	.29	.51
Redwood-cedar-cypress, 1″ x 4″, 3′ wide walk	.666	L.F.	.021	1.03	.83	1.86
4′ wide walk	.500	L.F.	.016	.77	.63	1.40
2″ x 4″, 3′ wide walk	.666	L.F.	.032	2.04	1.27	3.31
4′ wide walk	.500	L.F.	.024	1.54	.95	2.49
Brick, dry joints, 3′ wide walk	.666	L.F.	.079	1.37	2.77	4.14
4′ wide walk	.500	L.F.	.059	1.03	2.07	3.10
Mortar joints, 3′ wide walk	.666	L.F.	.095	1.65	3.32	4.97
4′ wide walk	.500	L.F.	.071	1.23	2.48	3.71

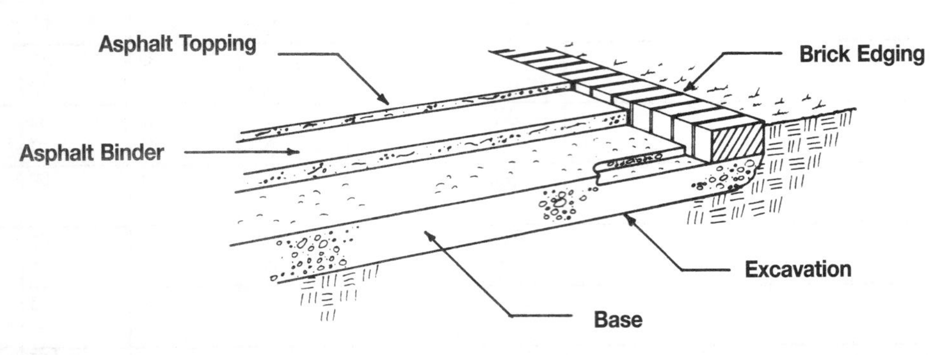

System Description	QUAN.	UNIT	LABOR HOURS	COST PER S.F.		
				MAT.	INST.	TOTAL
ASPHALT DRIVEWAY TO 10′ WIDE						
Excavation, driveway to 10′ wide, 6″ deep	.019	C.Y.			.03	.03
Base, 6″ crushed stone	1.000	S.F.	.001	.64	.07	.71
Handgrade base	1.000	S.F.	.004		.12	.12
2″ thick base	1.000	S.F.	.002	.41	.13	.54
1″ topping	1.000	S.F.	.001	.24	.10	.34
Edging, brick pavers	.200	L.F.	.024	.41	.83	1.24
TOTAL		S.F.	.033	1.70	1.28	2.98
CONCRETE DRIVEWAY TO 10′ WIDE						
Excavation, driveway to 10′ wide, 6″ deep	.019	C.Y.			.03	.03
Base, 6″ crushed stone	1.000	S.F.	.001	.64	.07	.71
Handgrade base	1.000	S.F.	.004		.12	.12
Surface, concrete, 4″ thick	1.000	S.F.	.040	1.33	1.38	2.71
Edging, brick pavers	.200	L.F.	.024	.41	.83	1.24
TOTAL		S.F.	.070	2.38	2.43	4.81
PAVERS, BRICK DRIVEWAY TO 10′ WIDE						
Excavation, driveway to 10′ wide, 6″ deep	.019	C.Y.			.03	.03
Base, 6″ sand	1.000	S.F.	.001	.52	.09	.61
Handgrade base	1.000	S.F.	.004		.12	.12
Surface, pavers, brick laid flat, running bond	1.000	S.F.	.160	2.62	5.55	8.17
Edging, redwood, untreated, 2″ x 4″	.200	L.F.	.010	.61	.38	.99
TOTAL		S.F.	.176	3.75	6.17	9.92

Description	QUAN.	UNIT	LABOR HOURS	COST PER S.F.		
				MAT.	INST.	TOTAL

 Important: See the Reference Section for critical supporting data - Reference Nos., Crews & Location Factors

Driveway Price Sheet	QUAN.	UNIT	LABOR HOURS	COST PER S.F.		
				MAT.	INST.	TOTAL
Excavation, by machine, 10′ wide, 6″ deep	.019	C.Y.	.001		.03	.03
12″ deep	.037	C.Y.	.001		.05	.05
18″ deep	.055	C.Y.	.001		.08	.08
20′ wide, 6″ deep	.019	C.Y.	.001		.03	.03
12″ deep	.037	C.Y.	.001		.05	.05
18″ deep	.055	C.Y.	.001		.08	.08
Base, crushed stone, 10′ wide, 3″ deep	1.000	S.F.	.001	.32	.04	.36
6″ deep	1.000	S.F.	.001	.64	.07	.71
9″ deep	1.000	S.F.	.002	.93	.10	1.03
20′ wide, 3″ deep	1.000	S.F.	.001	.32	.04	.36
6″ deep	1.000	S.F.	.001	.64	.07	.71
9″ deep	1.000	S.F.	.002	.93	.10	1.03
Bank run gravel, 10′ wide, 3″ deep	1.000	S.F.	.001	.20	.03	.23
6″ deep	1.000	S.F.	.001	.40	.05	.45
9″ deep	1.000	S.F.	.001	.59	.07	.66
20′ wide, 3″ deep	1.000	S.F.	.001	.20	.03	.23
6″ deep	1.000	S.F.	.001	.40	.05	.45
9″ deep	1.000	S.F.	.001	.59	.07	.66
Handgrade, 10′ wide	1.000	S.F.	.004		.12	.12
20′ wide	1.000	S.F.	.004		.12	.12
Surface, asphalt, 10′ wide, 3/4″ topping, 1″ base	1.000	S.F.	.002	.52	.17	.69
2″ base	1.000	S.F.	.003	.61	.21	.82
1″ topping, 1″ base	1.000	S.F.	.002	.56	.19	.75
2″ base	1.000	S.F.	.003	.65	.23	.88
20′ wide, 3/4″ topping, 1″ base	1.000	S.F.	.002	.52	.17	.69
2″ base	1.000	S.F.	.003	.61	.21	.82
1″ topping, 1″ base	1.000	S.F.	.002	.56	.19	.75
2″ base	1.000	S.F.	.003	.65	.23	.88
Concrete, 10′ wide, 4″ thick	1.000	S.F.	.040	1.33	1.38	2.71
6″ thick	1.000	S.F.	.047	2.07	1.62	3.69
20′ wide, 4″ thick	1.000	S.F.	.040	1.33	1.38	2.71
6″ thick	1.000	S.F.	.047	2.07	1.62	3.69
Paver, brick 10′ wide dry joints, running bond, laid flat	1.000	S.F.	.160	2.62	5.55	8.17
Laid on edge	1.000	S.F.	.229	2.37	7.95	10.32
Mortar joints, laid flat	1.000	S.F.	.192	3.14	6.65	9.79
Laid on edge	1.000	S.F.	.274	2.84	9.55	12.39
20′ wide, running bond, dry jts., laid flat	1.000	S.F.	.160	2.62	5.55	8.17
Laid on edge	1.000	S.F.	.229	2.37	7.95	10.32
Mortar joints, laid flat	1.000	S.F.	.192	3.14	6.65	9.79
Laid on edge	1.000	S.F.	.274	2.84	9.55	12.39
Crushed stone, 10′ wide, white marble, 3″	1.000	S.F.	.009	.21	.27	.48
Bluestone, 3″	1.000	S.F.	.009	.22	.27	.49
20′ wide, white marble, 3″	1.000	S.F.	.009	.21	.27	.48
Bluestone, 3″	1.000	S.F.	.009	.22	.27	.49
Soil cement, 10′ wide	1.000	S.F.	.007	.22	.65	.87
20′ wide	1.000	S.F.	.007	.22	.65	.87
Granite blocks, 10′ wide	1.000	S.F.	.174	6.90	6.05	12.95
20′ wide	1.000	S.F.	.174	6.90	6.05	12.95
Asphalt block, solid 1-1/4″ thick	1.000	S.F.	.119	4.42	4.13	8.55
Solid 3″ thick	1.000	S.F.	.123	6.20	4.29	10.49
Edging, brick, 10′ wide	.200	L.F.	.024	.41	.83	1.24
20′ wide	.100	L.F.	.012	.21	.41	.62
Redwood, untreated 2″ x 4″, 10′ wide	.200	L.F.	.010	.61	.38	.99
20′ wide	.100	L.F.	.005	.31	.19	.50
Granite, 4 1/2″ x 12″ straight, 10′ wide	.200	L.F.	.032	1.03	1.59	2.62
20′ wide	.100	L.F.	.016	.52	.80	1.32
Finishes, asphalt sealer, 10′ wide	1.000	S.F.	.023	.48	.68	1.16
20′ wide	1.000	S.F.	.023	.48	.68	1.16
Concrete, exposed aggregate 10′ wide	1.000	S.F.	.013	4.49	.46	4.95
20′ wide	1.000	S.F.	.013	4.49	.46	4.95

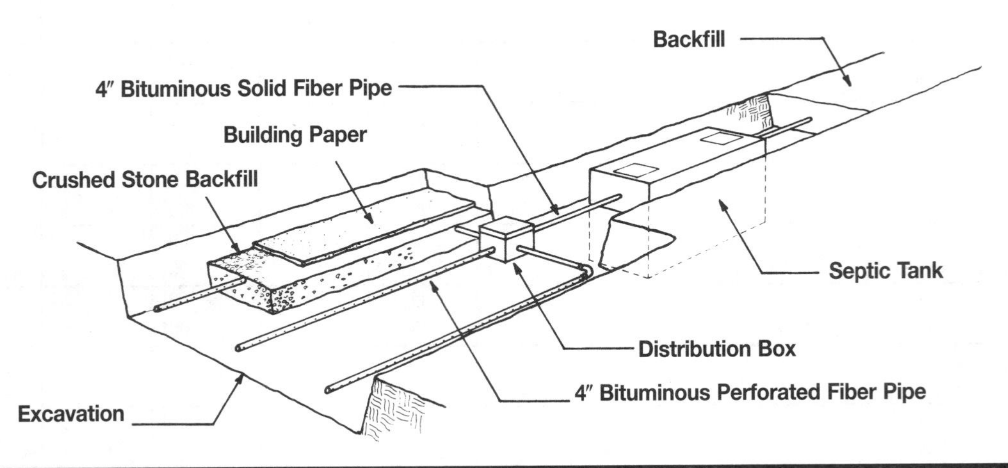

System Description	QUAN.	UNIT	LABOR HOURS	COST EACH		
				MAT.	INST.	TOTAL
SEPTIC SYSTEM WITH 1000 S.F. LEACHING FIELD, 1000 GALLON TANK						
Tank, 1000 gallon, concrete	1.000	Ea.	3.500	585	131	716
Distribution box, concrete	1.000	Ea.	1.000	109	28.50	137.50
4″ PVC pipe	25.000	L.F.	1.600	48.25	47.75	96
Tank and field excavation	119.000	C.Y.	6.565		673.54	673.54
Crushed stone backfill	76.000	C.Y.	12.160	1,862	495.52	2,357.52
Backfill with excavated material	36.000	C.Y.	.240		37.44	37.44
Building paper	125.000	S.Y.	2.430	33.75	90	123.75
4″ PVC perforated pipe	145.000	L.F.	9.280	279.85	276.95	556.80
4″ pipe fittings	2.000	Ea.	1.939	15.30	75	90.30
TOTAL		Ea.	38.714	2,933.15	1,855.70	4,788.85
SEPTIC SYSTEM WITH 2 LEACHING PITS, 1000 GALLON TANK						
Tank, 1000 gallon, concrete	1.000	Ea.	3.500	585	131	716
Distribution box, concrete	1.000	Ea.	1.000	109	28.50	137.50
4″ PVC pipe	75.000	L.F.	4.800	144.75	143.25	288
Excavation for tank only	20.000	C.Y.	1.103		113.20	113.20
Crushed stone backfill	10.000	C.Y.	1.600	245	65.20	310.20
Backfill with excavated material	55.000	C.Y.	.367		57.20	57.20
Pits, 6′ diameter, including excavation and stone backfill	2.000	Ea.		1,290		1,290
TOTAL		Ea.	12.370	2,373.75	538.35	2,912.10

The costs in this system include all necessary piping and excavation.

Description	QUAN.	UNIT	LABOR HOURS	COST EACH		
				MAT.	INST.	TOTAL

 Important: See the Reference Section for critical supporting data - Reference Nos., Crews & Location Factors

Septic Systems Price Sheet	QUAN.	UNIT	LABOR HOURS	COST EACH		
				MAT.	INST.	TOTAL
Tank, precast concrete, 1000 gallon	1.000	Ea.	3.500	585	131	716
2000 gallon	1.000	Ea.	5.600	1,150	210	1,360
Distribution box, concrete, 5 outlets	1.000	Ea.	1.000	109	28.50	137.50
12 outlets	1.000	Ea.	2.000	295	57.50	352.50
4″ pipe, PVC, solid	25.000	L.F.	1.600	48.50	48	96.50
Tank and field excavation, 1000 S.F. field	119.000	C.Y.	6.565		675	675
2000 S.F. field	190.000	C.Y.	10.482		1,075	1,075
Tank excavation only, 1000 gallon tank	20.000	C.Y.	1.103		113	113
2000 gallon tank	32.000	C.Y.	1.765		181	181
Backfill, crushed stone 1000 S.F. field	76.000	C.Y.	12.160	1,850	495	2,345
2000 S.F. field	140.000	C.Y.	22.400	3,425	915	4,340
Backfill with excavated material, 1000 S.F. field	36.000	C.Y.	.240		37.50	37.50
2000 S.F. field	60.000	C.Y.	.400		62.50	62.50
6′ diameter pits	55.000	C.Y.	.367		57.50	57.50
3′ diameter pits	42.000	C.Y.	.280		43.50	43.50
Building paper, 1000 S.F. field	125.000	S.Y.	2.376	33	88	121
2000 S.F. field	250.000	S.Y.	4.860	67.50	180	247.50
4″ pipe, PVC, perforated, 1000 S.F. field	145.000	L.F.	9.280	280	277	557
2000 S.F. field	265.000	L.F.	16.960	510	505	1,015
Pipe fittings, bituminous fiber, 1000 S.F. field	2.000	Ea.	1.939	15.30	75	90.30
2000 S.F. field	4.000	Ea.	3.879	30.50	150	180.50
Leaching pit, including excavation and stone backfill, 3′ diameter	1.000	Ea.		485		485
6′ diameter	1.000	Ea.		645		645

SITE WORK 1

60 Chain Link Fence Price Sheet

System Description	QUAN.	UNIT	LABOR HOURS	COST PER UNIT		
				MAT.	INST.	TOTAL
Chain link fence						
Galv.9ga. wire, 1-5/8"post 10'O.C., 1-3/8"top rail, 2"corner post, 3'hi	1.000	L.F.	.130	4.87	3.87	8.74
4' high	1.000	L.F.	.141	7.25	4.21	11.46
6' high	1.000	L.F.	.209	8.20	6.25	14.45
Add for gate 3' wide 1-3/8" frame 3' high	1.000	Ea.	2.000	43.50	59.50	103
4' high	1.000	Ea.	2.400	54	71.50	125.50
6' high	1.000	Ea.	2.400	97.50	71.50	169
Add for gate 4' wide 1-3/8" frame 3' high	1.000	Ea.	2.667	51	79.50	130.50
4' high	1.000	Ea.	2.667	67	79.50	146.50
6' high	1.000	Ea.	3.000	124	89.50	213.50
Alum.9ga. wire, 1-5/8"post, 10'O.C., 1-3/8"top rail, 2"corner post,3'hi	1.000	L.F.	.130	5.85	3.87	9.72
4' high	1.000	L.F.	.141	6.70	4.21	10.91
6' high	1.000	L.F.	.209	8.60	6.25	14.85
Add for gate 3' wide 1-3/8" frame 3' high	1.000	Ea.	2.000	57	59.50	116.50
4' high	1.000	Ea.	2.400	78	71.50	149.50
6' high	1.000	Ea.	2.400	117	71.50	188.50
Add for gate 4' wide 1-3/8" frame 3' high	1.000	Ea.	2.400	78	71.50	149.50
4' high	1.000	Ea.	2.667	104	79.50	183.50
6' high	1.000	Ea.	3.000	163	89.50	252.50
Vinyl 9ga. wire, 1-5/8"post 10'O.C., 1-3/8"top rail, 2"corner post,3'hi	1.000	L.F.	.130	5.20	3.87	9.07
4' high	1.000	L.F.	.141	8.55	4.21	12.76
6' high	1.000	L.F.	.209	9.75	6.25	16
Add for gate 3' wide 1-3/8" frame 3' high	1.000	Ea.	2.000	65	59.50	124.50
4' high	1.000	Ea.	2.400	84.50	71.50	156
6' high	1.000	Ea.	2.400	130	71.50	201.50
Add for gate 4' wide 1-3/8" frame 3' high	1.000	Ea.	2.400	88.50	71.50	160
4' high	1.000	Ea.	2.667	117	79.50	196.50
6' high	1.000	Ea.	3.000	169	89.50	258.50
Tennis court, chain link fence, 10' high						
Galv.11ga.wire, 2"post 10'O.C., 1-3/8"top rail, 2-1/2"corner post	1.000	L.F.	.253	13	7.55	20.55
Add for gate 3' wide 1-3/8" frame	1.000	Ea.	2.400	163	71.50	234.50
Alum.11ga.wire, 2"post 10'O.C., 1-3/8"top rail, 2-1/2"corner post	1.000	L.F.	.253	18.20	7.55	25.75
Add for gate 3' wide 1-3/8" frame	1.000	Ea.	2.400	208	71.50	279.50
Vinyl 11ga.wire,2"post 10' O.C.,1-3/8"top rail,2-1/2"corner post	1.000	L.F.	.253	15.60	7.55	23.15
Add for gate 3' wide 1-3/8" frame	1.000	Ea.	2.400	234	71.50	305.50
Railings, commercial						
Aluminum balcony rail, 1-1/2" posts with pickets	1.000	L.F.	.164	42	8.50	50.50
With expanded metal panels	1.000	L.F.	.164	54	8.50	62.50
With porcelain enamel panel inserts	1.000	L.F.	.164	48	8.50	56.50
Mild steel, ornamental rounded top rail	1.000	L.F.	.164	43	8.50	51.50
As above, but pitch down stairs	1.000	L.F.	.183	46.50	9.40	55.90
Steel pipe, welded, 1-1/2" round, painted	1.000	L.F.	.160	12.60	8.25	20.85
Galvanized	1.000	L.F.	.160	17.85	8.25	26.10
Residential, stock units, mild steel, deluxe	1.000	L.F.	.102	9.80	5.25	15.05
Economy	1.000	L.F.	.102	7.35	5.25	12.60

64 Wood Fence Price Sheet

System Description	QUAN.	UNIT	LABOR HOURS	COST PER UNIT		
				MAT.	INST.	TOTAL
Basketweave, 3/8"x4" boards, 2"x4" stringers on spreaders, 4"x4" posts						
No. 1 cedar, 6' high	1.000	L.F.	.150	8.60	4.47	13.07
Treated pine, 6' high	1.000	L.F.	.160	10.45	4.77	15.22
Board fence, 1"x4" boards, 2"x4" rails, 4"x4" posts						
Preservative treated, 2 rail, 3' high	1.000	L.F.	.166	6.40	4.94	11.34
4' high	1.000	L.F.	.178	7	5.30	12.30
3 rail, 5' high	1.000	L.F.	.185	7.90	5.50	13.40
6' high	1.000	L.F.	.192	9.05	5.75	14.80
Western cedar, No. 1, 2 rail, 3' high	1.000	L.F.	.166	6.95	4.94	11.89
3 rail, 4' high	1.000	L.F.	.178	8.25	5.30	13.55
5' high	1.000	L.F.	.185	9.50	5.50	15
6' high	1.000	L.F.	.192	10.40	5.75	16.15
No. 1 cedar, 2 rail, 3' high	1.000	L.F.	.166	10.45	4.94	15.39
4' high	1.000	L.F.	.178	11.85	5.30	17.15
3 rail, 5' high	1.000	L.F.	.185	13.75	5.50	19.25
6' high	1.000	L.F.	.192	15.30	5.75	21.05
Shadow box, 1"x6" boards, 2"x4" rails, 4"x4" posts						
Fir, pine or spruce, treated, 3 rail, 6' high	1.000	L.F.	.160	11.70	4.77	16.47
No. 1 cedar, 3 rail, 4' high	1.000	L.F.	.185	14.40	5.50	19.90
6' high	1.000	L.F.	.192	17.75	5.75	23.50
Open rail, split rails, No. 1 cedar, 2 rail, 3' high	1.000	L.F.	.150	5.80	4.47	10.27
3 rail, 4' high	1.000	L.F.	.160	7.80	4.77	12.57
No. 2 cedar, 2 rail, 3' high	1.000	L.F.	.150	4.50	4.47	8.97
3 rail, 4' high	1.000	L.F.	.160	5.10	4.77	9.87
Open rail, rustic rails, No. 1 cedar, 2 rail, 3' high	1.000	L.F.	.150	3.61	4.47	8.08
3 rail, 4' high	1.000	L.F.	.160	4.84	4.77	9.61
No. 2 cedar, 2 rail, 3' high	1.000	L.F.	.150	3.45	4.47	7.92
3 rail, 4' high	1.000	L.F.	.160	3.65	4.77	8.42
Rustic picket, molded pine pickets, 2 rail, 3' high	1.000	L.F.	.171	5.10	5.10	10.20
3 rail, 4' high	1.000	L.F.	.197	5.85	5.85	11.70
No. 1 cedar, 2 rail, 3' high	1.000	L.F.	.171	6.95	5.10	12.05
3 rail, 4' high	1.000	L.F.	.197	8	5.85	13.85
Picket fence, fir, pine or spruce, preserved, treated						
2 rail, 3' high	1.000	L.F.	.171	4.44	5.10	9.54
3 rail, 4' high	1.000	L.F.	.185	5.25	5.50	10.75
Western cedar, 2 rail, 3' high	1.000	L.F.	.171	5.55	5.10	10.65
3 rail, 4' high	1.000	L.F.	.185	5.70	5.50	11.20
No. 1 cedar, 2 rail, 3' high	1.000	L.F.	.171	11.10	5.10	16.20
3 rail, 4' high	1.000	L.F.	.185	12.95	5.50	18.45
Stockade, No. 1 cedar, 3-1/4" rails, 6' high	1.000	L.F.	.150	10.45	4.47	14.92
8' high	1.000	L.F.	.155	13.55	4.62	18.17
No. 2 cedar, treated rails, 6' high	1.000	L.F.	.150	10.45	4.47	14.92
Treated pine, treated rails, 6' high	1.000	L.F.	.150	10.25	4.47	14.72
Gates, No. 2 cedar, picket, 3'-6" wide 4' high	1.000	Ea.	2.667	55.50	79.50	135
No. 2 cedar, rustic round, 3' wide, 3' high	1.000	Ea.	2.667	70.50	79.50	150
No. 2 cedar, stockade screen, 3'-6" wide, 6' high	1.000	Ea.	3.000	61.50	89.50	151
General, wood, 3'-6" wide, 4' high	1.000	Ea.	2.400	53.50	71.50	125
6' high	1.000	Ea.	3.000	67	89.50	156.50

Division 2 Foundations

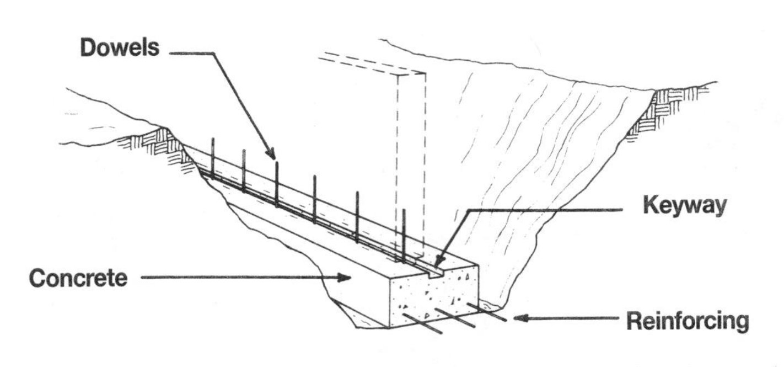

System Description	QUAN.	UNIT	LABOR HOURS	COST PER L.F.		
				MAT.	INST.	TOTAL
8" THICK BY 18" WIDE FOOTING						
Concrete, 3000 psi	.040	C.Y.		3.20		3.20
Place concrete, direct chute	.040	C.Y.	.016		.51	.51
Forms, footing, 4 uses	1.330	SFCA	.103	.76	3.48	4.24
Reinforcing, 1/2" diameter bars, 2 each	1.380	Lb.	.011	.46	.47	.93
Keyway, 2" x 4", beveled, 4 uses	1.000	L.F.	.015	.18	.59	.77
Dowels, 1/2" diameter bars, 2' long, 6' O.C.	.166	Ea.	.006	.08	.25	.33
TOTAL		L.F.	.151	4.68	5.30	9.98
12" THICK BY 24" WIDE FOOTING						
Concrete, 3000 psi	.070	C.Y.		5.60		5.60
Place concrete, direct chute	.070	C.Y.	.028		.88	.88
Forms, footing, 4 uses	2.000	SFCA	.155	1.14	5.24	6.38
Reinforcing, 1/2" diameter bars, 2 each	1.380	Lb.	.011	.46	.47	.93
Keyway, 2" x 4", beveled, 4 uses	1.000	L.F.	.015	.18	.59	.77
Dowels, 1/2" diameter bars, 2' long, 6' O.C.	.166	Ea.	.006	.08	.25	.33
TOTAL		L.F.	.215	7.46	7.43	14.89
12" THICK BY 36" WIDE FOOTING						
Concrete, 3000 psi	.110	C.Y.		8.80		8.80
Place concrete, direct chute	.110	C.Y.	.044		1.39	1.39
Forms, footing, 4 uses	2.000	SFCA	.155	1.14	5.24	6.38
Reinforcing, 1/2" diameter bars, 2 each	1.380	Lb.	.011	.46	.47	.93
Keyway, 2" x 4", beveled, 4 uses	1.000	L.F.	.015	.18	.59	.77
Dowels, 1/2" diameter bars, 2' long, 6' O.C.	.166	Ea.	.006	.08	.25	.33
TOTAL		L.F.	.231	10.66	7.94	18.60

The footing costs in this system are on a cost per linear foot basis

Description	QUAN.	UNIT	LABOR HOURS	COST PER S.F.		
				MAT.	INST.	TOTAL

Footing Price Sheet	QUAN.	UNIT	LABOR HOURS	COST PER L.F.		
				MAT.	INST.	TOTAL
Concrete, 8″ thick by 18″ wide footing						
2000 psi concrete	.040	C.Y.		3.06		3.06
2500 psi concrete	.040	C.Y.		3.12		3.12
3000 psi concrete	.040	C.Y.		3.20		3.20
3500 psi concrete	.040	C.Y.		3.26		3.26
4000 psi concrete	.040	C.Y.		3.36		3.36
12″ thick by 24″ wide footing						
2000 psi concrete	.070	C.Y.		5.35		5.35
2500 psi concrete	.070	C.Y.		5.45		5.45
3000 psi concrete	.070	C.Y.		5.60		5.60
3500 psi concrete	.070	C.Y.		5.70		5.70
4000 psi concrete	.070	C.Y.		5.90		5.90
12″ thick by 36″ wide footing						
2000 psi concrete	.110	C.Y.		8.40		8.40
2500 psi concrete	.110	C.Y.		8.60		8.60
3000 psi concrete	.110	C.Y.		8.80		8.80
3500 psi concrete	.110	C.Y.		8.95		8.95
4000 psi concrete	.110	C.Y.		9.25		9.25
Place concrete, 8″ thick by 18″ wide footing, direct chute	.040	C.Y.	.016		.51	.51
Pumped concrete	.040	C.Y.	.017		.76	.76
Crane & bucket	.040	C.Y.	.032		1.48	1.48
12″ thick by 24″ wide footing, direct chute	.070	C.Y.	.028		.88	.88
Pumped concrete	.070	C.Y.	.030		1.33	1.33
Crane & bucket	.070	C.Y.	.056		2.59	2.59
12″ thick by 36″ wide footing, direct chute	.110	C.Y.	.044		1.39	1.39
Pumped concrete	.110	C.Y.	.047		2.08	2.08
Crane & bucket	.110	C.Y.	.088		4.06	4.06
Forms, 8″ thick footing, 1 use	1.330	SFCA	.140	.21	4.72	4.93
4 uses	1.330	SFCA	.103	.76	3.48	4.24
12″ thick footing, 1 use	2.000	SFCA	.211	.32	7.10	7.42
4 uses	2.000	SFCA	.155	1.14	5.25	6.39
Reinforcing, 3/8″ diameter bar, 1 each	.400	Lb.	.003	.13	.14	.27
2 each	.800	Lb.	.006	.26	.27	.53
3 each	1.200	Lb.	.009	.40	.41	.81
1/2″ diameter bar, 1 each	.700	Lb.	.005	.23	.24	.47
2 each	1.380	Lb.	.011	.46	.47	.93
3 each	2.100	Lb.	.016	.69	.71	1.40
5/8″ diameter bar, 1 each	1.040	Lb.	.008	.34	.35	.69
2 each	2.080	Lb.	.016	.69	.71	1.40
Keyway, beveled, 2″ x 4″, 1 use	1.000	L.F.	.030	.36	1.18	1.54
2 uses	1.000	L.F.	.023	.27	.89	1.16
2″ x 6″, 1 use	1.000	L.F.	.032	.58	1.26	1.84
2 uses	1.000	L.F.	.024	.44	.95	1.39
Dowels, 2 feet long, 6′ O.C., 3/8″ bar	.166	Ea.	.005	.04	.23	.27
1/2″ bar	.166	Ea.	.006	.08	.25	.33
5/8″ bar	.166	Ea.	.006	.12	.27	.39
3/4″ bar	.166	Ea.	.006	.12	.27	.39

08 Block Wall Systems

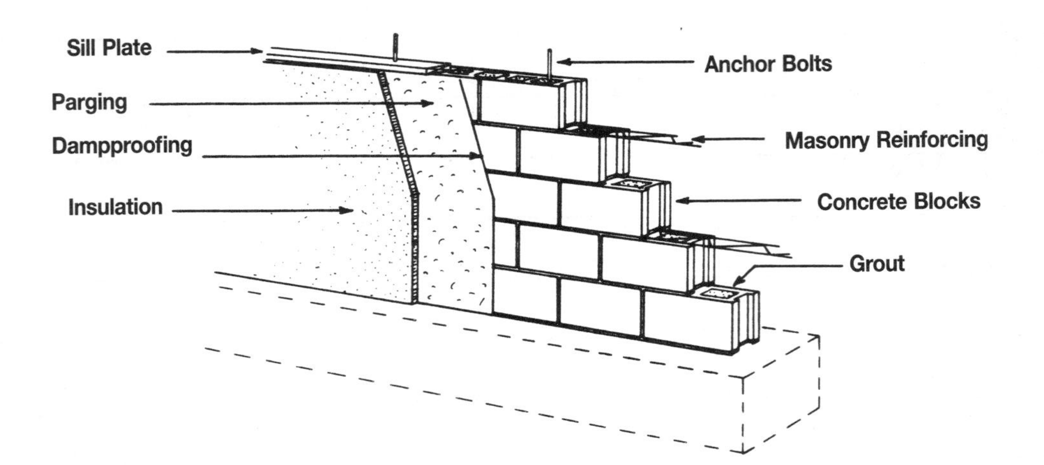

System Description	QUAN.	UNIT	LABOR HOURS	COST PER S.F.		
				MAT.	INST.	TOTAL
8″ WALL, GROUTED, FULL HEIGHT						
Concrete block, 8″ x 16″ x 8″	1.000	S.F.	.094	1.93	3.37	5.30
Masonry reinforcing, every second course	.750	L.F.	.002	.09	.08	.17
Parging, plastering with portland cement plaster, 1 coat	1.000	S.F.	.014	.23	.51	.74
Dampproofing, bituminous coating, 1 coat	1.000	S.F.	.012	.08	.43	.51
Insulation, 1″ rigid polystyrene	1.000	S.F.	.010	.40	.39	.79
Grout, solid, pumped	1.000	S.F.	.059	.94	2.08	3.02
Anchor bolts, 1/2″ diameter, 8″ long, 4′ O.C.	.060	Ea.	.002	.03	.10	.13
Sill plate, 2″ x 4″, treated	.250	L.F.	.007	.12	.29	.41
TOTAL		S.F.	.199	3.82	7.25	11.07
12″ WALL, GROUTED, FULL HEIGHT						
Concrete block, 8″ x 16″ x 12″	1.000	S.F.	.160	2.70	5.55	8.25
Masonry reinforcing, every second course	.750	L.F.	.003	.11	.12	.23
Parging, plastering with portland cement plaster, 1 coat	1.000	S.F.	.014	.23	.51	.74
Dampproofing, bituminous coating, 1 coat	1.000	S.F.	.012	.08	.43	.51
Insulation, 1″ rigid polystyrene	1.000	S.F.	.010	.40	.39	.79
Grout, solid, pumped	1.000	S.F.	.063	1.53	2.21	3.74
Anchor bolts, 1/2″ diameter, 8″ long, 4′ O.C.	.060	Ea.	.002	.03	.10	.13
Sill plate, 2″ x 4″, treated	.250	L.F.	.007	.12	.29	.41
TOTAL		S.F.	.233	5.20	9.60	14.80

The costs in this system are based on a square foot of wall. Do not subtract for window or door openings.

Description	QUAN.	UNIT	LABOR HOURS	COST PER S.F.		
				MAT.	INST.	TOTAL

Important: See the Reference Section for critical supporting data - Reference Nos., Crews & Location Factors

Block Wall Systems	QUAN.	UNIT	LABOR HOURS	COST PER S.F.		
				MAT.	INST.	TOTAL
Concrete, block, 8″ x 16″ x, 6″ thick	1.000	S.F.	.089	1.80	3.15	4.95
8″ thick	1.000	S.F.	.093	1.93	3.37	5.30
10″ thick	1.000	S.F.	.095	2.62	4.10	6.72
12″ thick	1.000	S.F.	.122	2.70	5.55	8.25
Solid block, 8″ x 16″ x, 6″ thick	1.000	S.F.	.091	1.95	3.26	5.21
8″ thick	1.000	S.F.	.096	2.78	3.45	6.23
10″ thick	1.000	S.F.	.096	2.78	3.45	6.23
12″ thick	1.000	S.F.	.126	4.04	4.78	8.82
Masonry reinforcing, wire strips, to 8″ wide, every course	1.500	L.F.	.004	.18	.17	.35
Every 2nd course	.750	L.F.	.002	.09	.08	.17
Every 3rd course	.500	L.F.	.001	.06	.06	.12
Every 4th course	.400	L.F.	.001	.05	.04	.09
Wire strips to 12″ wide, every course	1.500	L.F.	.006	.21	.24	.45
Every 2nd course	.750	L.F.	.003	.11	.12	.23
Every 3rd course	.500	L.F.	.002	.07	.08	.15
Every 4th course	.400	L.F.	.002	.06	.06	.12
Parging, plastering with portland cement plaster, 1 coat	1.000	S.F.	.014	.23	.51	.74
2 coats	1.000	S.F.	.022	.36	.79	1.15
Dampproofing, bituminous, brushed on, 1 coat	1.000	S.F.	.012	.08	.43	.51
2 coats	1.000	S.F.	.016	.11	.58	.69
Sprayed on, 1 coat	1.000	S.F.	.010	.08	.35	.43
2 coats	1.000	S.F.	.016	.16	.58	.74
Troweled on, 1/16″ thick	1.000	S.F.	.016	.18	.58	.76
1/8″ thick	1.000	S.F.	.020	.32	.72	1.04
1/2″ thick	1.000	S.F.	.023	1.03	.82	1.85
Insulation, rigid, fiberglass, 1.5#/C.F., unfaced						
1-1/2″ thick R 6.2	1.000	S.F.	.008	.45	.31	.76
2″ thick R 8.5	1.000	S.F.	.008	.51	.31	.82
3″ thick R 13	1.000	S.F.	.010	.63	.39	1.02
Foamglass, 1-1/2″ thick R 2.64	1.000	S.F.	.010	1.43	.39	1.82
2″ thick R 5.26	1.000	S.F.	.011	3.22	.43	3.65
Perlite, 1″ thick R 2.77	1.000	S.F.	.010	.29	.39	.68
2″ thick R 5.55	1.000	S.F.	.011	.55	.43	.98
Polystyrene, extruded, 1″ thick R 5.4	1.000	S.F.	.010	.40	.39	.79
2″ thick R 10.8	1.000	S.F.	.011	1.08	.43	1.51
Molded 1″ thick R 3.85	1.000	S.F.	.010	.18	.39	.57
2″ thick R 7.7	1.000	S.F.	.011	.57	.43	1
Grout, concrete block cores, 6″ thick	1.000	S.F.	.044	.71	1.57	2.28
8″ thick	1.000	S.F.	.059	.94	2.08	3.02
10″ thick	1.000	S.F.	.061	1.24	2.14	3.38
12″ thick	1.000	S.F.	.063	1.53	2.21	3.74
Anchor bolts, 2′ on center, 1/2″ diameter, 8″ long	.120	Ea.	.005	.07	.19	.26
12″ long	.120	Ea.	.005	.13	.20	.33
3/4″ diameter, 8″ long	.120	Ea.	.006	.15	.24	.39
12″ long	.120	Ea.	.006	.19	.26	.45
4′ on center, 1/2″ diameter, 8″ long	.060	Ea.	.002	.03	.10	.13
12″ long	.060	Ea.	.003	.06	.10	.16
3/4″ diameter, 8″ long	.060	Ea.	.003	.08	.12	.20
12″ long	.060	Ea.	.003	.10	.13	.23
Sill plates, treated, 2″ x 4″	.250	L.F.	.007	.12	.29	.41
4″ x 4″	.250	L.F.	.007	.35	.27	.62

12 Concrete Wall Systems

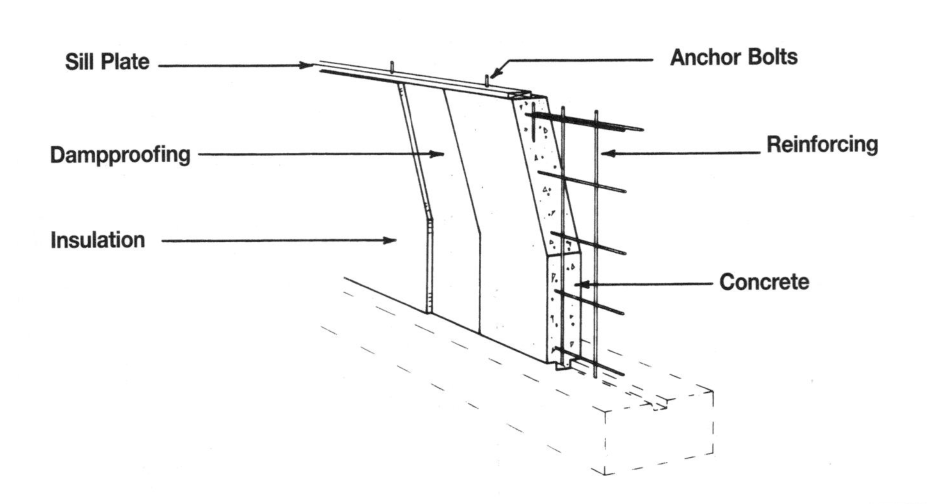

System Description	QUAN.	UNIT	LABOR HOURS	COST PER S.F.		
				MAT.	INST.	TOTAL
8″ THICK, POURED CONCRETE WALL						
Concrete, 8″ thick , 3000 psi	.025	C.Y.		2		2
Forms, prefabricated plywood, 4 uses per month	2.000	SFCA	.076	1.12	2.62	3.74
Reinforcing, light	.670	Lb.	.004	.22	.16	.38
Placing concrete, direct chute	.025	C.Y.	.013		.43	.43
Dampproofing, brushed on, 2 coats	1.000	S.F.	.016	.11	.58	.69
Rigid insulation, 1″ polystyrene	1.000	S.F.	.010	.40	.39	.79
Anchor bolts, 1/2″ diameter, 12″ long, 4′ O.C.	.060	Ea.	.003	.06	.10	.16
Sill plates, 2″ x 4″, treated	.250	L.F.	.007	.12	.29	.41
TOTAL		S.F.	.129	4.03	4.57	8.60
12″ THICK, POURED CONCRETE WALL						
Concrete, 12″ thick, 3000 psi	.040	C.Y.		3.20		3.20
Forms, prefabricated plywood, 4 uses per month	2.000	SFCA	.076	1.12	2.62	3.74
Reinforcing, light	1.000	Lb.	.005	.33	.24	.57
Placing concrete, direct chute	.040	C.Y.	.019		.60	.60
Dampproofing, brushed on, 2 coats	1.000	S.F.	.016	.11	.58	.69
Rigid insulation, 1″ polystyrene	1.000	S.F.	.010	.40	.39	.79
Anchor bolts, 1/2″ diameter, 12″ long, 4′ O.C.	.060	Ea.	.003	.06	.10	.16
Sill plates, 2″ x 4″ treated	.250	L.F.	.007	.12	.29	.41
TOTAL		S.F.	.136	5.34	4.82	10.16

The costs in this system are based on sq. ft. of wall. Do not subtract for window and door openings. The costs assume a 4′ high wall.

Description	QUAN.	UNIT	LABOR HOURS	COST PER S.F.		
				MAT.	INST.	TOTAL

 Important: See the Reference Section for critical supporting data - Reference Nos., Crews & Location Factors

Concrete Wall Price Sheet	QUAN.	UNIT	LABOR HOURS	COST PER S.F.		
				MAT.	INST.	TOTAL
Formwork, prefabricated plywood, 1 use per month	2.000	SFCA	.081	3.38	2.80	6.18
4 uses per month	2.000	SFCA	.076	1.12	2.62	3.74
Job built forms, 1 use per month	2.000	SFCA	.320	3.86	11	14.86
4 uses per month	2.000	SFCA	.221	1.48	7.60	9.08
Reinforcing, 8″ wall, light reinforcing	.670	Lb.	.004	.22	.16	.38
Heavy reinforcing	1.500	Lb.	.008	.50	.36	.86
10″ wall, light reinforcing	.850	Lb.	.005	.28	.20	.48
Heavy reinforcing	2.000	Lb.	.011	.66	.48	1.14
12″ wall light reinforcing	1.000	Lb.	.005	.33	.24	.57
Heavy reinforcing	2.250	Lb.	.012	.74	.54	1.28
Placing concrete, 8″ wall, direct chute	.025	C.Y.	.013		.43	.43
Pumped concrete	.025	C.Y.	.016		.71	.71
Crane & bucket	.025	C.Y.	.023		1.05	1.05
10″ wall, direct chute	.030	C.Y.	.016		.51	.51
Pumped concrete	.030	C.Y.	.019		.86	.86
Crane & bucket	.030	C.Y.	.027		1.26	1.26
12″ wall, direct chute	.040	C.Y.	.019		.60	.60
Pumped concrete	.040	C.Y.	.023		1.04	1.04
Crane & bucket	.040	C.Y.	.032		1.48	1.48
Dampproofing, bituminous, brushed on, 1 coat	1.000	S.F.	.012	.08	.43	.51
2 coats	1.000	S.F.	.016	.11	.58	.69
Sprayed on, 1 coat	1.000	S.F.	.010	.08	.35	.43
2 coats	1.000	S.F.	.016	.16	.58	.74
Troweled on, 1/16″ thick	1.000	S.F.	.016	.18	.58	.76
1/8″ thick	1.000	S.F.	.020	.32	.72	1.04
1/2″ thick	1.000	S.F.	.023	1.03	.82	1.85
Insulation rigid, fiberglass, 1.5#/C.F., unfaced						
1-1/2″ thick, R 6.2	1.000	S.F.	.008	.45	.31	.76
2″ thick, R 8.3	1.000	S.F.	.008	.51	.31	.82
3″ thick, R 12.4	1.000	S.F.	.010	.63	.39	1.02
Foamglass, 1-1/2″ thick R 2.64	1.000	S.F.	.010	1.43	.39	1.82
2″ thick R 5.26	1.000	S.F.	.011	3.22	.43	3.65
Perlite, 1″ thick R 2.77	1.000	S.F.	.010	.29	.39	.68
2″ thick R 5.55	1.000	S.F.	.011	.55	.43	.98
Polystyrene, extruded, 1″ thick R 5.40	1.000	S.F.	.010	.40	.39	.79
2″ thick R 10.8	1.000	S.F.	.011	1.08	.43	1.51
Molded, 1″ thick R 3.85	1.000	S.F.	.010	.18	.39	.57
2″ thick R 7.70	1.000	S.F.	.011	.57	.43	1
Anchor bolts, 2′ on center, 1/2″ diameter, 8″ long	.120	Ea.	.005	.07	.19	.26
12″ long	.120	Ea.	.005	.13	.20	.33
3/4″ diameter, 8″ long	.120	Ea.	.006	.15	.24	.39
12″ long	.120	Ea.	.006	.19	.26	.45
Sill plates, treated lumber, 2″ x 4″	.250	L.F.	.007	.12	.29	.41
4″ x 4″	.250	L.F.	.007	.35	.27	.62

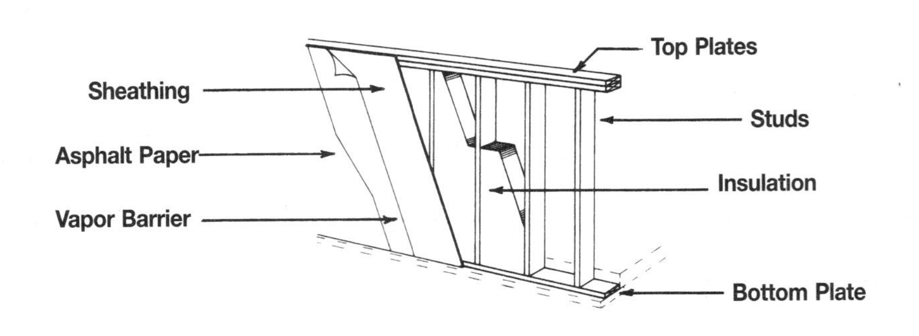

System Description	QUAN.	UNIT	LABOR HOURS	COST PER S.F.		
				MAT.	INST.	TOTAL
2″ X 4″ STUDS, 16″ O.C., WALL						
Studs, 2″ x 4″, 16″ O.C., treated	1.000	L.F.	.015	.46	.57	1.03
Plates, double top plate, single bottom plate, treated, 2″ x 4″	.750	L.F.	.011	.35	.43	.78
Sheathing, 1/2″, exterior grade, CDX, treated	1.000	S.F.	.014	.72	.56	1.28
Asphalt paper, 15# roll	1.100	S.F.	.002	.03	.09	.12
Vapor barrier, 4 mil polyethylene	1.000	S.F.	.002	.03	.08	.11
Insulation, batts, fiberglass, 3-1/2″ thick, R 11	1.000	S.F.	.005	.25	.20	.45
TOTAL		S.F.	.049	1.84	1.93	3.77
2″ X 6″ STUDS, 16″ O.C., WALL						
Studs, 2″ x 6″, 16″ O.C., treated	1.000	L.F.	.016	.84	.63	1.47
Plates, double top plate, single bottom plate, treated, 2″ x 6″	.750	L.F.	.012	.63	.47	1.10
Sheathing, 5/8″ exterior grade, CDX, treated	1.000	S.F.	.015	1.07	.60	1.67
Asphalt paper, 15# roll	1.100	S.F.	.002	.03	.09	.12
Vapor barrier, 4 mil polyethylene	1.000	S.F.	.002	.03	.08	.11
Insulation, batts, fiberglass, 6″ thick, R 19	1.000	S.F.	.006	.34	.23	.57
TOTAL		S.F.	.053	2.94	2.10	5.04
2″ X 8″ STUDS, 16″ O.C., WALL						
Studs, 2″ x 8″, 16″ O.C. treated	1.000	L.F.	.018	1.05	.70	1.75
Plates, double top plate, single bottom plate, treated, 2″ x 8″	.750	L.F.	.013	.79	.53	1.32
Sheathing, 3/4″ exterior grade, CDX, treated	1.000	S.F.	.016	1.07	.64	1.71
Asphalt paper, 15# roll	1.100	S.F.	.002	.03	.09	.12
Vapor barrier, 4 mil polyethylene	1.000	S.F.	.002	.03	.08	.11
Insulation, batts, fiberglass, 9″ thick, R 30	1.000	S.F.	.006	.66	.23	.89
TOTAL		S.F.	.057	3.63	2.27	5.90

The costs in this system are based on a sq. ft. of wall area. Do not Subtract for window or door openings. The costs assume a 4′ high wall.

Description	QUAN.	UNIT	LABOR HOURS	COST PER S.F.		
				MAT.	INST.	TOTAL

Important: See the Reference Section for critical supporting data - Reference Nos., Crews & Location Factors

Wood Wall Foundation Price Sheet	QUAN.	UNIT	LABOR HOURS	COST PER S.F.		
				MAT.	INST.	TOTAL
Studs, treated, 2″ x 4″, 12″ O.C.	1.250	L.F.	.018	.58	.71	1.29
16″ O.C.	1.000	L.F.	.015	.46	.57	1.03
2″ x 6″, 12″ O.C.	1.250	L.F.	.020	1.05	.79	1.84
16″ O.C.	1.000	L.F.	.016	.84	.63	1.47
2″ x 8″, 12″ O.C.	1.250	L.F.	.022	1.31	.88	2.19
16″ O.C.	1.000	L.F.	.018	1.05	.70	1.75
Plates, treated double top single bottom, 2″ x 4″	.750	L.F.	.011	.35	.43	.78
2″ x 6″	.750	L.F.	.012	.63	.47	1.10
2″ x 8″	.750	L.F.	.013	.79	.53	1.32
Sheathing, treated exterior grade CDX, 1/2″ thick	1.000	S.F.	.014	.72	.56	1.28
5/8″ thick	1.000	S.F.	.015	1.07	.60	1.67
3/4″ thick	1.000	S.F.	.016	1.07	.64	1.71
Asphalt paper, 15# roll	1.100	S.F.	.002	.03	.09	.12
Vapor barrier, polyethylene, 4 mil	1.000	S.F.	.002	.02	.08	.10
10 mil	1.000	S.F.	.002	.06	.08	.14
Insulation, rigid, fiberglass, 1.5#/C.F., unfaced	1.000	S.F.	.008	.32	.31	.63
1-1/2″ thick, R 6.2	1.000	S.F.	.008	.45	.31	.76
2″ thick, R 8.3	1.000	S.F.	.008	.51	.31	.82
3″ thick, R 12.4	1.000	S.F.	.010	.64	.40	1.04
Foamglass 1 1/2″ thick, R 2.64	1.000	S.F.	.010	1.43	.39	1.82
2″ thick, R 5.26	1.000	S.F.	.011	3.22	.43	3.65
Perlite 1″ thick, R 2.77	1.000	S.F.	.010	.29	.39	.68
2″ thick, R 5.55	1.000	S.F.	.011	.55	.43	.98
Polystyrene, extruded, 1″ thick, R 5.40	1.000	S.F.	.010	.40	.39	.79
2″ thick, R 10.8	1.000	S.F.	.011	1.08	.43	1.51
Molded 1″ thick, R 3.85	1.000	S.F.	.010	.18	.39	.57
2″ thick, R 7.7	1.000	S.F.	.011	.57	.43	1
Non rigid, batts, fiberglass, paper backed, 3-1/2″ thick roll, R 11	1.000	S.F.	.005	.25	.20	.45
6″, R 19	1.000	S.F.	.006	.34	.23	.57
9″, R 30	1.000	S.F.	.006	.66	.23	.89
12″, R 38	1.000	S.F.	.006	.84	.23	1.07
Mineral fiber, paper backed, 3-1/2″, R 13	1.000	S.F.	.005	.29	.20	.49
6″, R 19	1.000	S.F.	.005	.37	.20	.57
10″, R 30	1.000	S.F.	.006	.57	.23	.80

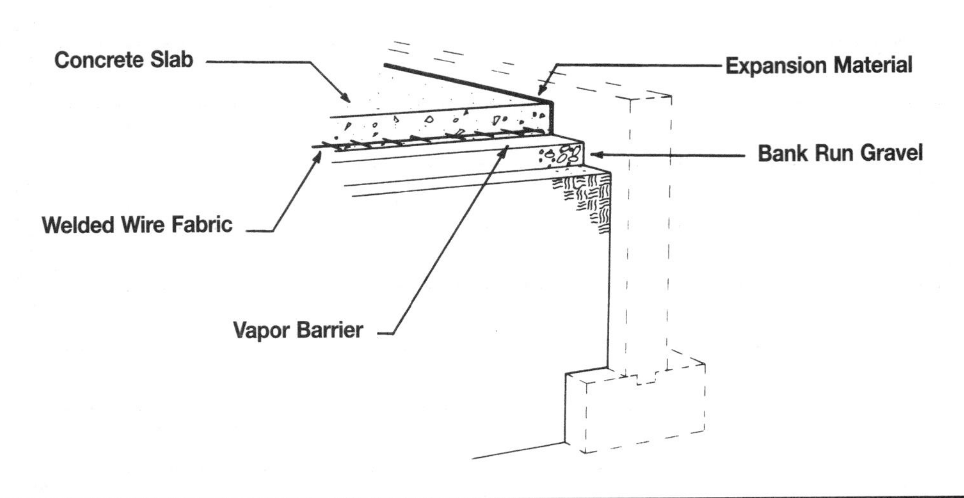

System Description	QUAN.	UNIT	LABOR HOURS	COST PER S.F.		
				MAT.	INST.	TOTAL
4″ THICK SLAB						
Concrete, 4″ thick, 3000 psi concrete	.012	C.Y.		.96		.96
Place concrete, direct chute	.012	C.Y.	.005		.17	.17
Bank run gravel, 4″ deep	1.000	S.F.	.001	.30	.04	.34
Polyethylene vapor barrier, .006″ thick	1.000	S.F.	.002	.03	.08	.11
Edge forms, expansion material	.100	L.F.	.005	.04	.18	.22
Welded wire fabric, 6 x 6, 10/10 (W1.4/W1.4)	1.100	S.F.	.005	.09	.22	.31
Steel trowel finish	1.000	S.F.	.015		.52	.52
TOTAL		S.F.	.033	1.42	1.21	2.63
6″ THICK SLAB						
Concrete, 6″ thick, 3000 psi concrete	.019	C.Y.		1.52		1.52
Place concrete, direct chute	.019	C.Y.	.008		.26	.26
Bank run gravel, 4″ deep	1.000	S.F.	.001	.30	.04	.34
Polyethylene vapor barrier, .006″ thick	1.000	S.F.	.002	.03	.08	.11
Edge forms, expansion material	.100	L.F.	.005	.04	.18	.22
Welded wire fabric, 6 x 6, 10/10 (W1.4/W1.4)	1.100	S.F.	.005	.09	.22	.31
Steel trowel finish	1.000	S.F.	.015		.52	.52
TOTAL		S.F.	.036	1.98	1.30	3.28

The slab costs in this section are based on a cost per square foot of floor area.

Description	QUAN.	UNIT	LABOR HOURS	COST PER S.F.		
				MAT.	INST.	TOTAL

Floor Slab Price Sheet	QUAN.	UNIT	LABOR HOURS	COST PER S.F.		
				MAT.	INST.	TOTAL
Concrete, 4″ thick slab, 2000 psi concrete	.012	C.Y.		.92		.92
2500 psi concrete	.012	C.Y.		.94		.94
3000 psi concrete	.012	C.Y.		.96		.96
3500 psi concrete	.012	C.Y.		.98		.98
4000 psi concrete	.012	C.Y.		1.01		1.01
4500 psi concrete	.012	C.Y.		1.03		1.03
5″ thick slab, 2000 psi concrete	.015	C.Y.		1.15		1.15
2500 psi concrete	.015	C.Y.		1.17		1.17
3000 psi concrete	.015	C.Y.		1.20		1.20
3500 psi concrete	.015	C.Y.		1.22		1.22
4000 psi concrete	.015	C.Y.		1.26		1.26
4500 psi concrete	.015	C.Y.		1.29		1.29
6″ thick slab, 2000 psi concrete	.019	C.Y.		1.45		1.45
2500 psi concrete	.019	C.Y.		1.48		1.48
3000 psi concrete	.019	C.Y.		1.52		1.52
3500 psi concrete	.019	C.Y.		1.55		1.55
4000 psi concrete	.019	C.Y.		1.60		1.60
4500 psi concrete	.019	C.Y.		1.63		1.63
Place concrete, 4″ slab, direct chute	.012	C.Y.	.005		.17	.17
Pumped concrete	.012	C.Y.	.006		.26	.26
Crane & bucket	.012	C.Y.	.008		.37	.37
5″ slab, direct chute	.015	C.Y.	.007		.21	.21
Pumped concrete	.015	C.Y.	.007		.33	.33
Crane & bucket	.015	C.Y.	.010		.46	.46
6″ slab, direct chute	.019	C.Y.	.008		.26	.26
Pumped concrete	.019	C.Y.	.009		.41	.41
Crane & bucket	.019	C.Y.	.012		.58	.58
Gravel, bank run, 4″ deep	1.000	S.F.	.001	.30	.04	.34
6″ deep	1.000	S.F.	.001	.40	.05	.45
9″ deep	1.000	S.F.	.001	.59	.07	.66
12″ deep	1.000	S.F.	.001	.81	.09	.90
3/4″ crushed stone, 3″ deep	1.000	S.F.	.001	.32	.04	.36
6″ deep	1.000	S.F.	.001	.64	.07	.71
9″ deep	1.000	S.F.	.002	.93	.10	1.03
12″ deep	1.000	S.F.	.002	1.46	.12	1.58
Vapor barrier polyethylene, .004″ thick	1.000	S.F.	.002	.02	.08	.10
.006″ thick	1.000	S.F.	.002	.03	.08	.11
Edge forms, expansion material, 4″ thick slab	.100	L.F.	.004	.03	.12	.15
6″ thick slab	.100	L.F.	.005	.04	.18	.22
Welded wire fabric 6 x 6, 10/10 (W1.4/W1.4)	1.100	S.F.	.005	.09	.22	.31
6 x 6, 6/6 (W2.9/W2.9)	1.100	S.F.	.006	.15	.28	.43
4 x 4, 10/10 (W1.4/W1.4)	1.100	S.F.	.006	.15	.25	.40
Finish concrete, screed finish	1.000	S.F.	.009		.32	.32
Float finish	1.000	S.F.	.011		.39	.39
Steel trowel, for resilient floor	1.000	S.F.	.013		.48	.48
For finished floor	1.000	S.F.	.015		.52	.52

Division 3
Framing

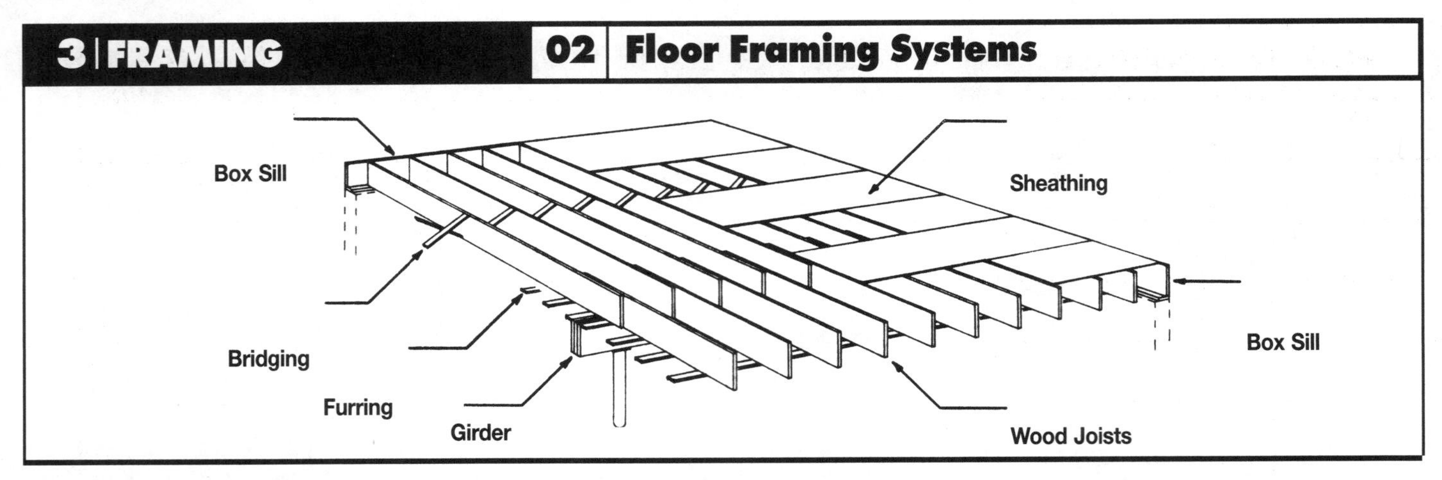

System Description	QUAN.	UNIT	LABOR HOURS	COST PER S.F. MAT.	COST PER S.F. INST.	COST PER S.F. TOTAL
2″ X 8″, 16″ O.C.						
Wood joists, 2″ x 8″, 16″ O.C.	1.000	L.F.	.015	.88	.57	1.45
Bridging, 1″ x 3″, 6′ O.C.	.080	Pr.	.005	.03	.19	.22
Box sills, 2″ x 8″	.150	L.F.	.002	.13	.09	.22
Concrete filled steel column, 4″ diameter	.125	L.F.	.002	.09	.09	.18
Girder, built up from three 2″ x 8″	.125	L.F.	.013	.33	.52	.85
Sheathing, plywood, subfloor, 5/8″ CDX	1.000	S.F.	.012	.50	.46	.96
Furring, 1″ x 3″, 16″ O.C.	1.000	L.F.	.023	.23	.90	1.13
TOTAL		S.F.	.072	2.19	2.82	5.01
2″ X 10″, 16″ O.C.						
Wood joists, 2″ x 10″, 16″ OC	1.000	L.F.	.018	1.19	.70	1.89
Bridging, 1″ x 3″, 6′ OC	.080	Pr.	.005	.03	.19	.22
Box sills, 2″ x 10″	.150	L.F.	.003	.18	.11	.29
Girder, built up from three 2″ x 10″	.125	L.F.	.002	.09	.09	.18
Sheathing, plywood, subfloor, 5/8″ CDX	1.000	S.F.	.012	.50	.46	.96
Furring, 1″ x 3″,16″ OC	1.000	L.F.	.023	.23	.90	1.13
TOTAL		S.F.	.077	2.66	3.01	5.67
2″ X 12″, 16″ O.C.						
Wood joists, 2″ x 12″, 16″ O.C.	1.000	L.F.	.018	1.62	.72	2.34
Bridging, 1″ x 3″, 6′ O.C.	.080	Pr.	.005	.03	.19	.22
Box sills, 2″ x 12″	.150	L.F.	.003	.24	.11	.35
Concrete filled steel column, 4″ diameter	.125	L.F.	.002	.09	.09	.18
Girder, built up from three 2″ x 12″	.125	L.F.	.015	.61	.59	1.20
Sheathing, plywood, subfloor, 5/8″ CDX	1.000	S.F.	.012	.50	.46	.96
Furring, 1″ x 3″, 16″ O.C.	1.000	L.F.	.023	.23	.90	1.13
TOTAL		S.F.	.078	3.32	3.06	6.38

Floor costs on this page are given on a cost per square foot basis.

Description	QUAN.	UNIT	LABOR HOURS	COST PER S.F. MAT.	COST PER S.F. INST.	COST PER S.F. TOTAL

Important: See the Reference Section for critical supporting data - Reference Nos., Crews & Location Factors

Floor Framing Price Sheet (Wood)

Floor Framing Price Sheet (Wood)	QUAN.	UNIT	LABOR HOURS	COST PER S.F.		
				MAT.	INST.	TOTAL
Joists, #2 or better, pine, 2″ x 4″, 12″ O.C.	1.250	L.F.	.016	.43	.63	1.06
16″ O.C.	1.000	L.F.	.013	.34	.50	.84
2″ x 6″, 12″ O.C.	1.250	L.F.	.016	.71	.63	1.34
16″ O.C.	1.000	L.F.	.013	.57	.50	1.07
2″ x 8″, 12″ O.C.	1.250	L.F.	.018	1.10	.71	1.81
16″ O.C.	1.000	L.F.	.015	.88	.57	1.45
2″ x 10″, 12″ O.C.	1.250	L.F.	.022	1.49	.88	2.37
16″ O.C.	1.000	L.F.	.018	1.19	.70	1.89
2″x 12″, 12″ O.C.	1.250	L.F.	.023	2.03	.90	2.93
16″ O.C.	1.000	L.F.	.018	1.62	.72	2.34
Bridging, wood 1″ x 3″, joists 12″ O.C.	.100	Pr.	.006	.03	.24	.27
16″ O.C.	.080	Pr.	.005	.03	.19	.22
Metal, galvanized, joists 12″ O.C.	.100	Pr.	.006	.10	.24	.34
16″ O.C.	.080	Pr.	.005	.08	.19	.27
Compression type, joists 12″ O.C.	.100	Pr.	.004	.12	.16	.28
16″ O.C.	.080	Pr.	.003	.10	.13	.23
Box sills, #2 or better pine, 2″ x 4″	.150	L.F.	.002	.05	.08	.13
2″ x 6″	.150	L.F.	.002	.09	.08	.17
2″ x 8″	.150	L.F.	.002	.13	.09	.22
2″ x 10″	.150	L.F.	.003	.18	.11	.29
2″ x 12″	.150	L.F.	.003	.24	.11	.35
Girders, including lally columns, 3 pieces spiked together, 2″ x 8″	.125	L.F.	.015	.42	.61	1.03
2″ x 10″	.125	L.F.	.016	.53	.65	1.18
2″ x 12″	.125	L.F.	.017	.70	.68	1.38
Solid girders, 3″ x 8″	.040	L.F.	.004	.19	.15	.34
3″ x 10″	.040	L.F.	.004	.21	.16	.37
3″ x 12″	.040	L.F.	.004	.24	.17	.41
4″ x 8″	.040	L.F.	.004	.24	.17	.41
4″ x 10″	.040	L.F.	.004	.28	.18	.46
4″ x 12″	.040	L.F.	.004	.32	.18	.50
Steel girders, bolted & including fabrication, wide flange shapes						
12″ deep, 14#/l.f.	.040	L.F.	.003	.50	.22	.72
10″ deep, 15#/l.f.	.040	L.F.	.003	.50	.22	.72
8″ deep, 10#/l.f.	.040	L.F.	.003	.34	.22	.56
6″ deep, 9#/l.f.	.040	L.F.	.003	.30	.22	.52
5″ deep, 16#/l.f.	.040	L.F.	.003	.50	.22	.72
Sheathing, plywood exterior grade CDX, 1/2″ thick	1.000	S.F.	.011	.43	.45	.88
5/8″ thick	1.000	S.F.	.012	.50	.46	.96
3/4″ thick	1.000	S.F.	.013	.63	.50	1.13
Boards, 1″ x 8″ laid regular	1.000	S.F.	.016	.88	.63	1.51
Laid diagonal	1.000	S.F.	.019	.88	.74	1.62
1″ x 10″ laid regular	1.000	S.F.	.015	1.16	.57	1.73
Laid diagonal	1.000	S.F.	.018	1.16	.70	1.86
Furring, 1″ x 3″, 12″ O.C.	1.250	L.F.	.029	.29	1.13	1.42
16″ O.C.	1.000	L.F.	.023	.23	.90	1.13
24″ O.C.	.750	L.F.	.017	.17	.68	.85

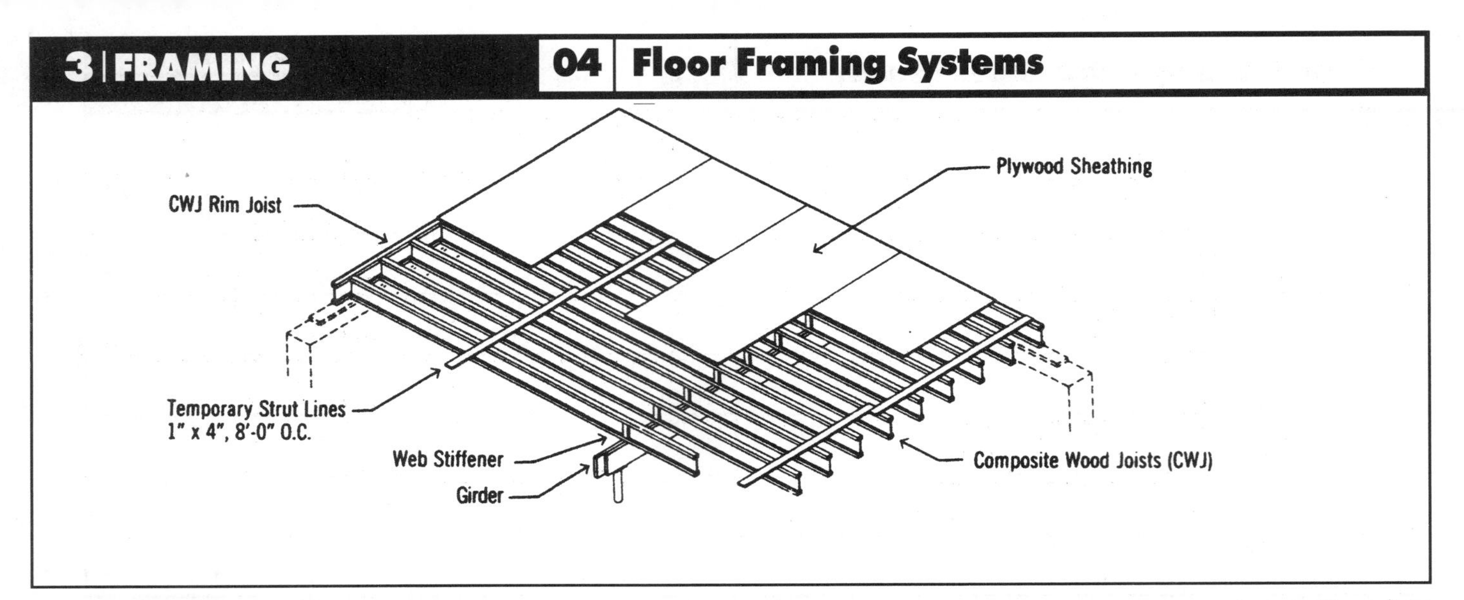

System Description	QUAN.	UNIT	LABOR HOURS	COST PER S.F.		
				MAT.	INST.	TOTAL
9-1/2" COMPOSITE WOOD JOISTS, 16" O.C.						
CWJ, 9-1/2", 16" O.C., 15' span	1.000	L.F.	.018	1.75	.70	2.45
Temp. strut line, 1" x 4", 8' O.C.	.160	L.F.	.003	.06	.12	.18
CWJ rim joist, 9-1/2"	.150	L.F.	.003	.26	.10	.36
Concrete filled steel column, 4" diameter	.125	L.F.	.002	.09	.09	.18
Girder, built up from three 2" x 8"	.125	L.F.	.013	.33	.52	.85
Sheathing, plywood, subfloor, 5/8" CDX	1.000	S.F.	.012	.50	.46	.96
TOTAL		S.F.	.051	2.99	1.99	4.98
11-1/2" COMPOSITE WOOD JOISTS, 16" O.C.						
CWJ, 11-1/2", 16" O.C., 18' span	1.000	L.F.	.018	1.88	.72	2.60
Temp. strut line, 1" x 4", 8' O.C.	.160	L.F.	.003	.06	.12	.18
CWJ rim joist, 11-1/2"	.150	L.F.	.003	.28	.11	.39
Concrete filled steel column, 4" diameter	.125	L.F.	.002	.09	.09	.18
Girder, built up from three 2" x 10"	.125	L.F.	.014	.44	.56	1
Sheathing, plywood, subfloor, 5/8" CDX	1.000	S.F.	.012	.50	.46	.96
TOTAL		S.F.	.052	3.25	2.06	5.31
14" COMPOSITE WOOD JOISTS, 16" O.C.						
CWJ, 14", 16" O.C., 22' span	1.000	L.F.	.020	2.05	.77	2.82
Temp. strut line, 1" x 4", 8' O.C.	.160	L.F.	.003	.06	.12	.18
CWJ rim joist, 14"	.150	L.F.	.003	.31	.11	.42
Concrete filled steel column, 4" diameter	.600	L.F.	.002	.09	.09	.18
Girder, built up from three 2" x 12"	.600	L.F.	.015	.61	.59	1.20
Sheathing, plywood, subfloor, 5/8" CDX	1.000	S.F.	.012	.50	.46	.96
TOTAL		S.F.	.055	3.62	2.14	5.76

Floor costs on this page are given on a cost per square foot basis.

Description	QUAN.	UNIT	LABOR HOURS	COST PER S.F.		
				MAT.	INST.	TOTAL

Important: See the Reference Section for critical supporting data - Reference Nos., Crews & Location Factors

Floor Framing Price Sheet (Wood)	QUAN.	UNIT	LABOR HOURS	COST PER S.F.		
				MAT.	INST.	TOTAL
Composite wood joist 9-1/2" deep, 12" O.C.	1.250	L.F.	.022	2.19	.87	3.06
16" O.C.	1.000	L.F.	.018	1.75	.70	2.45
11-1/2" deep, 12" O.C.	1.250	L.F.	.023	2.34	.89	3.23
16" O.C.	1.000	L.F.	.018	1.88	.72	2.60
14" deep, 12" O.C.	1.250	L.F.	.024	2.56	.96	3.52
16" O.C.	1.000	L.F.	.020	2.05	.77	2.82
16 " deep, 12" O.C.	1.250	L.F.	.026	3.50	1.01	4.51
16" O.C.	1.000	L.F.	.021	2.80	.81	3.61
CWJ rim joist, 9-1/2"	.150	L.F.	.003	.26	.10	.36
11-1/2"	.150	L.F.	.003	.28	.11	.39
14"	.150	L.F.	.003	.31	.11	.42
16"	.150	L.F.	.003	.42	.12	.54
Girders, including lally columns, 3 pieces spiked together, 2" x 8"	.125	L.F.	.015	.42	.61	1.03
2" x 10"	.125	L.F.	.016	.53	.65	1.18
2" x 12"	.125	L.F.	.017	.70	.68	1.38
Solid girders, 3" x 8"	.040	L.F.	.004	.19	.15	.34
3" x 10"	.040	L.F.	.004	.21	.16	.37
3" x 12"	.040	L.F.	.004	.24	.17	.41
4" x 8"	.040	L.F.	.004	.24	.17	.41
4" x 10"	.040	L.F.	.004	.28	.18	.46
4" x 12"	.040	L.F.	.004	.32	.18	.50
Steel girders, bolted & including fabrication, wide flange shapes						
12" deep, 14#/l.f.	.040	L.F.	.061	11.55	4.86	16.41
10" deep, 15#/l.f.	.040	L.F.	.067	12.60	5.30	17.90
8" deep, 10#/l.f.	.040	L.F.	.067	8.40	5.30	13.70
6" deep, 9#/l.f.	.040	L.F.	.067	7.55	5.30	12.85
5" deep, 16#/l.f.	.040	L.F.	.064	12.20	5.15	17.35
Sheathing, plywood exterior grade CDX, 1/2" thick	1.000	S.F.	.011	.43	.45	.88
5/8" thick	1.000	S.F.	.012	.50	.46	.96
3/4" thick	1.000	S.F.	.013	.63	.50	1.13
Boards, 1" x 8" laid regular	1.000	S.F.	.016	.88	.63	1.51
Laid diagonal	1.000	S.F.	.019	.88	.74	1.62
1" x 10" laid regular	1.000	S.F.	.015	1.16	.57	1.73
Laid diagonal	1.000	S.F.	.018	1.16	.70	1.86
Furring, 1" x 3", 12" O.C.	1.250	L.F.	.029	.29	1.13	1.42
16" O.C.	1.000	L.F.	.023	.23	.90	1.13
24" O.C.	.750	L.F.	.017	.17	.68	.85

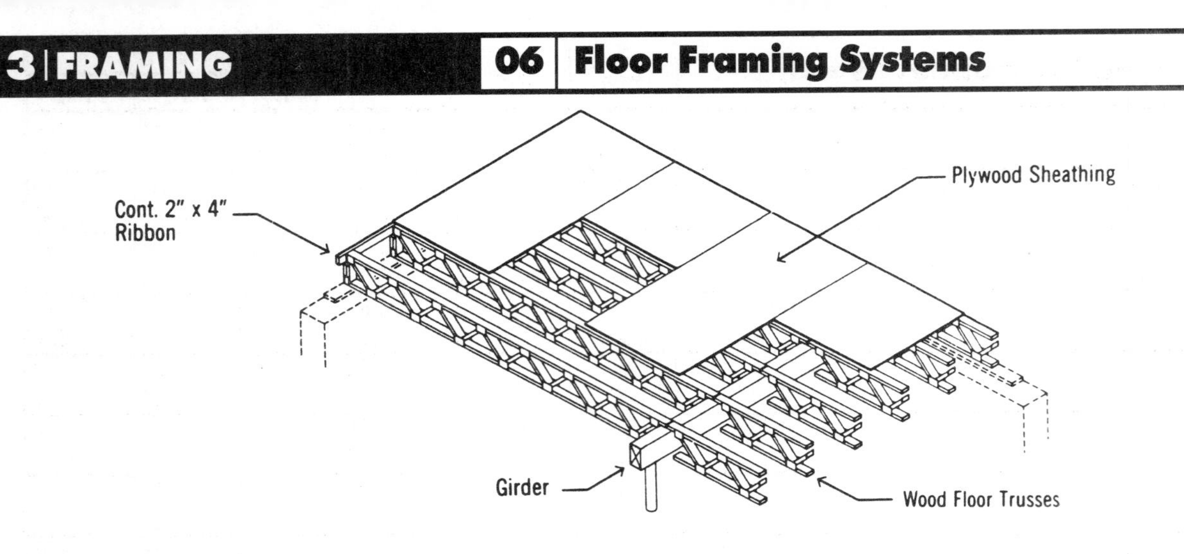

System Description	QUAN.	UNIT	LABOR HOURS	COST PER S.F.		
				MAT.	INST.	TOTAL
12" OPEN WEB JOISTS, 16" O.C.						
OWJ 12", 16" O.C., 21' span	1.000	L.F.	.018	1.85	.72	2.57
Continuous ribbing, 2" x 4"	.150	L.F.	.002	.05	.08	.13
Concrete filled steel column, 4" diameter	.125	L.F.	.002	.09	.09	.18
Girder, built up from three 2" x 8"	.125	L.F.	.013	.33	.52	.85
Sheathing, plywood, subfloor, 5/8" CDX	1.000	S.F.	.012	.50	.46	.96
Furring, 1" x 3", 16" O.C.	1.000	L.F.	.023	.23	.90	1.13
TOTAL		S.F.	.070	3.05	2.77	5.82
14" OPEN WEB WOOD JOISTS, 16" O.C.						
OWJ 14", 16" O.C., 22' span	1.000	L.F.	.020	2.15	.77	2.92
Continuous ribbing, 2" x 4"	.150	L.F.	.002	.05	.08	.13
Concrete filled steel column, 4" diameter	.125	L.F.	.002	.09	.09	.18
Girder, built up from three 2" x 10"	.125	L.F.	.014	.44	.56	1
Sheathing, plywood, subfloor, 5/8" CDX	1.000	S.F.	.012	.50	.46	.96
Furring, 1" x 3",16" O.C.	1.000	L.F.	.023	.23	.90	1.13
TOTAL		S.F.	.073	3.46	2.86	6.32
16" OPEN WEB WOOD JOISTS, 16" O.C.						
OWJ 16", 16" O.C., 24' span	1.000	L.F.	.021	2.23	.81	3.04
Continuous ribbing, 2" x 4"	.150	L.F.	.002	.05	.08	.13
Concrete filled steel column, 4" diameter	.125	L.F.	.002	.09	.09	.18
Girder, built up from three 2" x 12"	.125	L.F.	.015	.61	.59	1.20
Sheathing, plywood, subfloor, 5/8" CDX	1.000	S.F.	.012	.50	.46	.96
Furring, 1" x 3", 16" O.C.	1.000	L.F.	.023	.23	.90	1.13
TOTAL		S.F.	.075	3.71	2.93	6.64

Floor costs on this page are given on a cost per square foot basis.

Description	QUAN.	UNIT	LABOR HOURS	COST PER S.F.		
				MAT.	INST.	TOTAL

Floor Framing Price Sheet (Wood)	QUAN.	UNIT	LABOR HOURS	COST PER S.F.		
				MAT.	INST.	TOTAL
Open web joists, 12″ deep, 12″ O.C.	1.250	L.F.	.023	2.31	.89	3.20
16″ O.C.	1.000	L.F.	.018	1.85	.72	2.57
14″ deep, 12″ O.C.	1.250	L.F.	.024	2.69	.96	3.65
16″ O.C.	1.000	L.F.	.020	2.15	.77	2.92
16 ″ deep, 12″ O.C.	1.250	L.F.	.026	2.78	1.01	3.79
16″ O.C.	1.000	L.F.	.021	2.23	.81	3.04
18″ deep, 12″ O.C.	1.250	L.F.	.027	2.81	1.06	3.87
16″ O.C.	1.000	L.F.	.022	2.25	.85	3.10
Continuous ribbing, 2″ x 4″	.150	L.F.	.002	.05	.08	.13
2″ x 6″	.150	L.F.	.002	.09	.08	.17
2″ x 8″	.150	L.F.	.002	.13	.09	.22
2″ x 10″	.150	L.F.	.003	.18	.11	.29
2″ x 12″	.150	L.F.	.003	.24	.11	.35
Girders, including lally columns, 3 pieces spiked together, 2″ x 8″	.125	L.F.	.015	.42	.61	1.03
2″ x 10″	.125	L.F.	.016	.53	.65	1.18
2″ x 12″	.125	L.F.	.017	.70	.68	1.38
Solid girders, 3″ x 8″	.040	L.F.	.004	.19	.15	.34
3″ x 10″	.040	L.F.	.004	.21	.16	.37
3″ x 12″	.040	L.F.	.004	.24	.17	.41
4″ x 8″	.040	L.F.	.004	.24	.17	.41
4″ x 10″	.040	L.F.	.004	.28	.18	.46
4″ x 12″	.040	L.F.	.004	.32	.18	.50
Steel girders, bolted & including fabrication, wide flange shapes						
12″ deep, 14#/l.f.	.040	L.F.	.061	11.55	4.86	16.41
10″ deep, 15#/l.f.	.040	L.F.	.067	12.60	5.30	17.90
8″ deep, 10#/l.f.	.040	L.F.	.067	8.40	5.30	13.70
6″ deep, 9#/l.f.	.040	L.F.	.067	7.55	5.30	12.85
5″ deep, 16#/l.f.	.040	L.F.	.064	12.20	5.15	17.35
Sheathing, plywood exterior grade CDX, 1/2″ thick	1.000	S.F.	.011	.43	.45	.88
5/8″ thick	1.000	S.F.	.012	.50	.46	.96
3/4″ thick	1.000	S.F.	.013	.63	.50	1.13
Boards, 1″ x 8″ laid regular	1.000	S.F.	.016	.88	.63	1.51
Laid diagonal	1.000	S.F.	.019	.88	.74	1.62
1″ x 10″ laid regular	1.000	S.F.	.015	1.16	.57	1.73
Laid diagonal	1.000	S.F.	.018	1.16	.70	1.86
Furring, 1″ x 3″, 12″ O.C.	1.250	L.F.	.029	.29	1.13	1.42
16″ O.C.	1.000	L.F.	.023	.23	.90	1.13
24″ O.C.	.750	L.F.	.017	.17	.68	.85

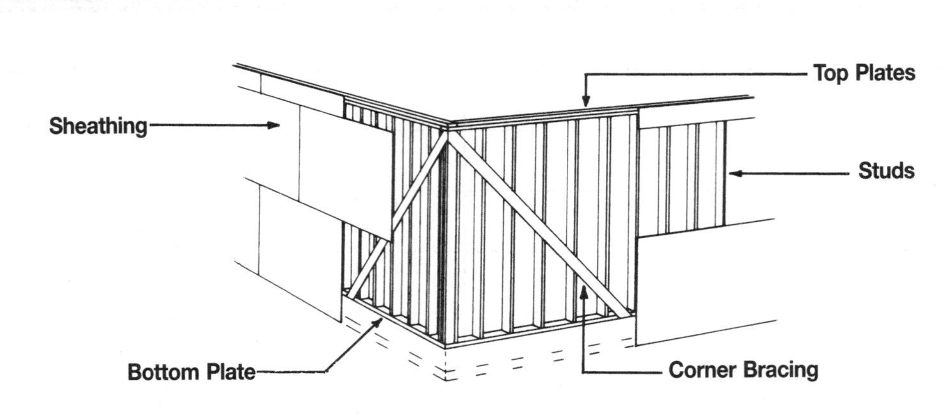

System Description	QUAN.	UNIT	LABOR HOURS	COST PER S.F.		
				MAT.	INST.	TOTAL
2″ X 4″, 16″ O.C.						
2″ x 4″ studs, 16″ O.C.	1.000	L.F.	.015	.34	.57	.91
Plates, 2″ x 4″, double top, single bottom	.375	L.F.	.005	.13	.21	.34
Corner bracing, let-in, 1″ x 6″	.063	L.F.	.003	.03	.13	.16
Sheathing, 1/2″ plywood, CDX	1.000	S.F.	.011	.43	.45	.88
TOTAL		S.F.	.034	.93	1.36	2.29
2″ X 4″, 24″ O.C.						
2″ x 4″ studs, 24″ O.C.	.750	L.F.	.011	.26	.43	.69
Plates, 2″ x 4″, double top, single bottom	.375	L.F.	.005	.13	.21	.34
Corner bracing, let-in, 1″ x 6″	.063	L.F.	.002	.03	.09	.12
Sheathing, 1/2″ plywood, CDX	1.000	S.F.	.011	.43	.45	.88
TOTAL		S.F.	.029	.85	1.18	2.03
2″ X 6″, 16″ O.C.						
2″ x 6″ studs, 16″ O.C.	1.000	L.F.	.016	.57	.63	1.20
Plates, 2″ x 6″, double top, single bottom	.375	L.F.	.006	.21	.24	.45
Corner bracing, let-in, 1″ x 6″	.063	L.F.	.003	.03	.13	.16
TOTAL		S.F.	.036	.81	1	1.81
2″ X 6″, 24″ O.C.						
2″ x 6″ studs, 24″ O.C.	.750	L.F.	.012	.43	.47	.90
Plates, 2″ x 6″, double top, single bottom	.375	L.F.	.006	.21	.24	.45
Corner bracing, let-in, 1″ x 6″	.063	L.F.	.002	.03	.09	.12
Sheathing, 1/2″ plywood, CDX	1.000	S.F.	.011	.43	.45	.88
TOTAL		S.F.	.031	1.10	1.25	2.35

The wall costs on this page are given in cost per square foot of wall.
For window and door openings see below.

Description	QUAN.	UNIT	LABOR HOURS	COST PER S.F.		
				MAT.	INST.	TOTAL

Important: See the Reference Section for critical supporting data - Reference Nos., Crews & Location Factors

Exterior Wall Framing Price Sheet

Exterior Wall Framing Price Sheet	QUAN.	UNIT	LABOR HOURS	COST PER S.F.		
				MAT.	INST.	TOTAL
Studs, #2 or better, 2″ x 4″, 12″ O.C.	1.250	L.F.	.018	.43	.71	1.14
16″ O.C.	1.000	L.F.	.015	.34	.57	.91
24″ O.C.	.750	L.F.	.011	.26	.43	.69
32″ O.C.	.600	L.F.	.009	.20	.34	.54
2″ x 6″, 12″ O.C.	1.250	L.F.	.020	.71	.79	1.50
16″ O.C.	1.000	L.F.	.016	.57	.63	1.20
24″ O.C.	.750	L.F.	.012	.43	.47	.90
32″ O.C.	.600	L.F.	.010	.34	.38	.72
2″ x 8″, 12″ O.C.	1.250	L.F.	.025	1.50	.98	2.48
16″ O.C.	1.000	L.F.	.020	1.20	.78	1.98
24″ O.C.	.750	L.F.	.015	.90	.59	1.49
32″ O.C.	.600	L.F.	.012	.72	.47	1.19
Plates, #2 or better, double top, single bottom, 2″ x 4″	.375	L.F.	.005	.13	.21	.34
2″ x 6″	.375	L.F.	.006	.21	.24	.45
2″ x 8″	.375	L.F.	.008	.45	.29	.74
Corner bracing, let-in 1″ x 6″ boards, studs, 12″ O.C.	.070	L.F.	.004	.03	.15	.18
16″ O.C.	.063	L.F.	.003	.03	.13	.16
24″ O.C.	.063	L.F.	.002	.03	.09	.12
32″ O.C.	.057	L.F.	.002	.02	.08	.10
Let-in steel ("T" shape), studs, 12″ O.C.	.070	L.F.	.001	.03	.04	.07
16″ O.C.	.063	L.F.	.001	.03	.03	.06
24″ O.C.	.063	L.F.	.001	.03	.03	.06
32″ O.C.	.057	L.F.	.001	.03	.03	.06
Sheathing, plywood CDX, 3/8″ thick	1.000	S.F.	.010	.48	.41	.89
1/2″ thick	1.000	S.F.	.011	.43	.45	.88
5/8″ thick	1.000	S.F.	.012	.50	.48	.98
3/4″ thick	1.000	S.F.	.013	.63	.52	1.15
Boards, 1″ x 6″, laid regular	1.000	S.F.	.025	.90	.97	1.87
Laid diagonal	1.000	S.F.	.027	.90	1.07	1.97
1″ x 8″, laid regular	1.000	S.F.	.021	.88	.82	1.70
Laid diagonal	1.000	S.F.	.025	.88	.97	1.85
Wood fiber, regular, no vapor barrier, 1/2″ thick	1.000	S.F.	.013	.58	.52	1.10
5/8″ thick	1.000	S.F.	.013	.78	.52	1.30
Asphalt impregnated 25/32″ thick	1.000	S.F.	.013	.37	.52	.89
1/2″ thick	1.000	S.F.	.013	.30	.52	.82
Polystyrene, regular, 3/4″ thick	1.000	S.F.	.010	.40	.39	.79
2″ thick	1.000	S.F.	.011	1.08	.43	1.51
Fiberglass, foil faced, 1″ thick	1.000	S.F.	.008	.88	.31	1.19
2″ thick	1.000	S.F.	.009	1.49	.35	1.84

Window & Door Openings

Window & Door Openings	QUAN.	UNIT	LABOR HOURS	COST EACH		
				MAT.	INST.	TOTAL
The following costs are to be added to the total costs of the wall for each opening. Do not subtract the area of the openings.						
Headers, 2″ x 6″ double, 2′ long	4.000	L.F.	.178	2.28	6.95	9.23
3′ long	6.000	L.F.	.267	3.42	10.45	13.87
4′ long	8.000	L.F.	.356	4.56	13.90	18.46
5′ long	10.000	L.F.	.444	5.70	17.40	23.10
2″ x 8″ double, 4′ long	8.000	L.F.	.376	7.05	14.70	21.75
5′ long	10.000	L.F.	.471	8.80	18.40	27.20
6′ long	12.000	L.F.	.565	10.55	22	32.55
8′ long	16.000	L.F.	.753	14.10	29.50	43.60
2″ x 10″ double, 4′ long	8.000	L.F.	.400	9.50	15.70	25.20
6′ long	12.000	L.F.	.600	14.30	23.50	37.80
8′ long	16.000	L.F.	.800	19.05	31.50	50.55
10′ long	20.000	L.F.	1.000	24	39	63
2″ x 12″ double, 8′ long	16.000	L.F.	.853	26	33.50	59.50
12′ long	24.000	L.F.	1.280	39	50	89

12 Gable End Roof Framing Systems

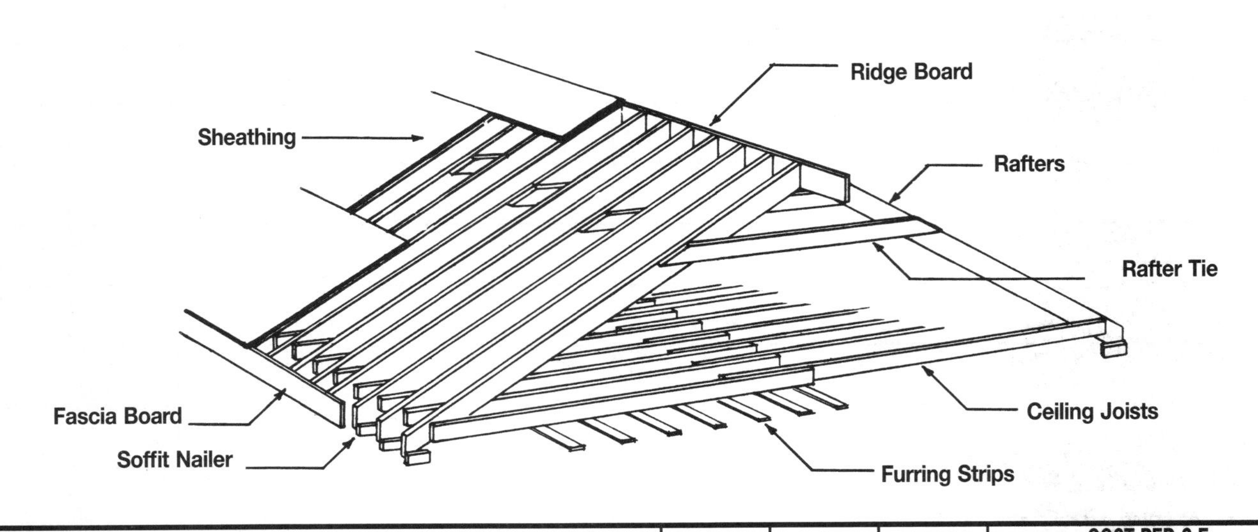

System Description	QUAN.	UNIT	LABOR HOURS	COST PER S.F.		
				MAT.	INST.	TOTAL
2″ X 6″ RAFTERS, 16″ O.C., 4/12 PITCH						
Rafters, 2″ x 6″, 16″ O.C., 4/12 pitch	1.170	L.F.	.019	.67	.74	1.41
Ceiling joists, 2″ x 4″, 16″ O.C.	1.000	L.F.	.013	.34	.50	.84
Ridge board, 2″ x 6″	.050	L.F.	.002	.03	.06	.09
Fascia board, 2″ x 6″	.100	L.F.	.005	.07	.21	.28
Rafter tie, 1″ x 4″, 4′ O.C.	.060	L.F.	.001	.02	.05	.07
Soffit nailer (outrigger), 2″ x 4″, 24″ O.C.	.170	L.F.	.004	.06	.17	.23
Sheathing, exterior, plywood, CDX, 1/2″ thick	1.170	S.F.	.013	.50	.53	1.03
Furring strips, 1″ x 3″, 16″ O.C.	1.000	L.F.	.023	.23	.90	1.13
TOTAL		S.F.	.080	1.92	3.16	5.08
2″ X 8″ RAFTERS, 16″ O.C., 4/12 PITCH						
Rafters, 2″ x 8″, 16″ O.C., 4/12 pitch	1.170	L.F.	.020	1.03	.77	1.80
Ceiling joists, 2″ x 6″, 16″ O.C.	1.000	L.F.	.013	.57	.50	1.07
Ridge board, 2″ x 8″	.050	L.F.	.002	.04	.07	.11
Fascia board, 2″ x 8″	.100	L.F.	.007	.09	.28	.37
Rafter tie, 1″ x 4″, 4′ O.C.	.060	L.F.	.001	.02	.05	.07
Soffit nailer (outrigger), 2″ x 4″, 24″ O.C.	.170	L.F.	.004	.06	.17	.23
Sheathing, exterior, plywood, CDX, 1/2″ thick	1.170	S.F.	.013	.50	.53	1.03
Furring strips, 1″ x 3″, 16″ O.C.	1.000	L.F.	.023	.23	.90	1.13
TOTAL		S.F.	.083	2.54	3.27	5.81

The cost of this system is based on the square foot of plan area.
All quantities have been adjusted accordingly.

Description	QUAN.	UNIT	LABOR HOURS	COST PER S.F.		
				MAT.	INST.	TOTAL

Important: See the Reference Section for critical supporting data - Reference Nos., Crews & Location Factors

Gable End Roof Framing Price Sheet	QUAN.	UNIT	LABOR HOURS	COST PER S.F.		
				MAT.	INST.	TOTAL
Rafters, #2 or better, 16″ O.C., 2″ x 6″, 4/12 pitch	1.170	L.F.	.019	.67	.74	1.41
8/12 pitch	1.330	L.F.	.027	.76	1.04	1.80
2″ x 8″, 4/12 pitch	1.170	L.F.	.020	1.03	.77	1.80
8/12 pitch	1.330	L.F.	.028	1.17	1.12	2.29
2″ x 10″, 4/12 pitch	1.170	L.F.	.030	1.39	1.17	2.56
8/12 pitch	1.330	L.F.	.043	1.58	1.69	3.27
24″ O.C., 2″ x 6″, 4/12 pitch	.940	L.F.	.015	.54	.59	1.13
8/12 pitch	1.060	L.F.	.021	.60	.83	1.43
2″ x 8″, 4/12 pitch	.940	L.F.	.016	.83	.62	1.45
8/12 pitch	1.060	L.F.	.023	.93	.89	1.82
2″ x 10″, 4/12 pitch	.940	L.F.	.024	1.12	.94	2.06
8/12 pitch	1.060	L.F.	.034	1.26	1.35	2.61
Ceiling joist, #2 or better, 2″ x 4″, 16″ O.C.	1.000	L.F.	.013	.34	.50	.84
24″ O.C.	.750	L.F.	.010	.26	.38	.64
2″ x 6″, 16″ O.C.	1.000	L.F.	.013	.57	.50	1.07
24″ O.C.	.750	L.F.	.010	.43	.38	.81
2″ x 8″, 16″ O.C.	1.000	L.F.	.015	.88	.57	1.45
24″ O.C.	.750	L.F.	.011	.66	.43	1.09
2″ x 10″, 16″ O.C.	1.000	L.F.	.018	1.19	.70	1.89
24″ O.C.	.750	L.F.	.013	.89	.53	1.42
Ridge board, #2 or better, 1″ x 6″	.050	L.F.	.001	.05	.05	.10
1″ x 8″	.050	L.F.	.001	.07	.06	.13
1″ x 10″	.050	L.F.	.002	.08	.06	.14
2″ x 6″	.050	L.F.	.002	.03	.06	.09
2″ x 8″	.050	L.F.	.002	.04	.07	.11
2″ x 10″	.050	L.F.	.002	.06	.08	.14
Fascia board, #2 or better, 1″ x 6″	.100	L.F.	.004	.05	.15	.20
1″ x 8″	.100	L.F.	.005	.06	.18	.24
1″ x 10″	.100	L.F.	.005	.06	.20	.26
2″ x 6″	.100	L.F.	.006	.07	.22	.29
2″ x 8″	.100	L.F.	.007	.09	.28	.37
2″ x 10″	.100	L.F.	.004	.24	.14	.38
Rafter tie, #2 or better, 4′ O.C., 1″ x 4″	.060	L.F.	.001	.02	.05	.07
1″ x 6″	.060	L.F.	.001	.02	.05	.07
2″ x 4″	.060	L.F.	.002	.03	.06	.09
2″ x 6″	.060	L.F.	.002	.04	.08	.12
Soffit nailer (outrigger), 2″ x 4″, 16″ O.C.	.220	L.F.	.006	.07	.22	.29
24″ O.C.	.170	L.F.	.004	.06	.17	.23
2″ x 6″, 16″ O.C.	.220	L.F.	.006	.09	.25	.34
24″ O.C.	.170	L.F.	.005	.07	.20	.27
Sheathing, plywood CDX, 4/12 pitch, 3/8″ thick.	1.170	S.F.	.012	.56	.48	1.04
1/2″ thick	1.170	S.F.	.013	.50	.53	1.03
5/8″ thick	1.170	S.F.	.014	.59	.56	1.15
8/12 pitch, 3/8″	1.330	S.F.	.014	.64	.55	1.19
1/2″ thick	1.330	S.F.	.015	.57	.60	1.17
5/8″ thick	1.330	S.F.	.016	.67	.64	1.31
Boards, 4/12 pitch roof, 1″ x 6″	1.170	S.F.	.026	1.05	1.02	2.07
1″ x 8″	1.170	S.F.	.021	1.03	.84	1.87
8/12 pitch roof, 1″ x 6″	1.330	S.F.	.029	1.20	1.16	2.36
1″ x 8″	1.330	S.F.	.024	1.17	.96	2.13
Furring, 1″ x 3″, 12″ O.C.	1.200	L.F.	.027	.28	1.08	1.36
16″ O.C.	1.000	L.F.	.023	.23	.90	1.13
24″ O.C.	.800	L.F.	.018	.18	.72	.90

3 FRAMING

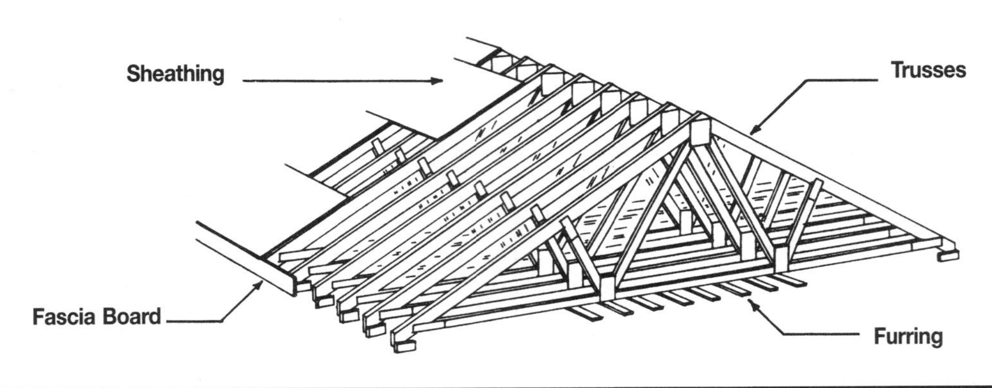

System Description	QUAN.	UNIT	LABOR HOURS	COST PER S.F.		
				MAT.	INST.	TOTAL
TRUSS, 16″ O.C., 4/12 PITCH, 1′ OVERHANG, 26′ SPAN						
Truss, 40# loading, 16″ O.C., 4/12 pitch, 26′ span	.030	Ea.	.021	2.15	1.09	3.24
Fascia board, 2″ x 6″	.100	L.F.	.005	.07	.21	.28
Sheathing, exterior, plywood, CDX, 1/2″ thick	1.170	S.F.	.013	.50	.53	1.03
Furring, 1″ x 3″, 16″ O.C.	1.000	L.F.	.023	.23	.90	1.13
TOTAL		S.F.	.062	2.95	2.73	5.68
TRUSS, 16″ O.C., 8/12 PITCH, 1′ OVERHANG, 26′ SPAN						
Truss, 40# loading, 16″ O.C., 8/12 pitch, 26′ span	.030	Ea.	.023	2.54	1.20	3.74
Fascia board, 2″ x 6″	.100	L.F.	.005	.07	.21	.28
Sheathing, exterior, plywood, CDX, 1/2″ thick	1.330	S.F.	.015	.57	.60	1.17
Furring, 1″ x 3″, 16″ O.C.	1.000	L.F.	.023	.23	.90	1.13
TOTAL		S.F.	.066	3.41	2.91	6.32
TRUSS, 24″ O.C., 4/12 PITCH, 1′ OVERHANG, 26′ SPAN						
Truss, 40# loading, 24″ O.C., 4/12 pitch, 26′ span	.020	Ea.	.014	1.43	.73	2.16
Fascia board, 2″ x 6″	.100	L.F.	.005	.07	.21	.28
Sheathing, exterior, plywood, CDX, 1/2″ thick	1.170	S.F.	.013	.50	.53	1.03
Furring, 1″ x 3″, 16″ O.C.	1.000	L.F.	.023	.23	.90	1.13
TOTAL		S.F.	.055	2.23	2.37	4.60
TRUSS, 24″ O.C., 8/12 PITCH, 1′ OVERHANG, 26′ SPAN						
Truss, 40# loading, 24″ O.C., 8/12 pitch, 26′ span	.020	Ea.	.015	1.69	.80	2.49
Fascia board, 2″ x 6″	.100	L.F.	.005	.07	.21	.28
Sheathing, exterior, plywood, CDX, 1/2″ thick	1.330	S.F.	.015	.57	.60	1.17
Furring, 1″ x 3″, 16″ O.C.	1.000	L.F.	.023	.23	.90	1.13
TOTAL		S.F.	.058	2.56	2.51	5.07

The cost of this system is based on the square foot of plan area.
A one foot overhang is included.

Description	QUAN.	UNIT	LABOR HOURS	COST PER S.F.		
				MAT.	INST.	TOTAL

Truss Roof Framing Price Sheet	QUAN.	UNIT	LABOR HOURS	COST PER S.F.		
				MAT.	INST.	TOTAL
Truss, 40# loading, including 1′ overhang, 4/12 pitch, 24′ span, 16″ O.C.	.033	Ea.	.022	1.68	1.15	2.83
24″ O.C.	.022	Ea.	.015	1.12	.77	1.89
26′ span, 16″ O.C.	.030	Ea.	.021	2.15	1.09	3.24
24″ O.C.	.020	Ea.	.014	1.43	.73	2.16
28′ span, 16″ O.C.	.027	Ea.	.020	1.67	1.06	2.73
24″ O.C.	.019	Ea.	.014	1.18	.74	1.92
32′ span, 16″ O.C.	.024	Ea.	.019	2.09	.99	3.08
24″ O.C.	.016	Ea.	.013	1.39	.67	2.06
36′ span, 16″ O.C.	.022	Ea.	.019	2.40	.99	3.39
24″ O.C.	.015	Ea.	.013	1.64	.68	2.32
8/12 pitch, 24′ span, 16″ O.C.	.033	Ea.	.024	2.57	1.24	3.81
24″ O.C.	.022	Ea.	.016	1.72	.83	2.55
26′ span, 16″ O.C.	.030	Ea.	.023	2.54	1.20	3.74
24″ O.C.	.020	Ea.	.015	1.69	.80	2.49
28′ span, 16″ O.C.	.027	Ea.	.022	2.46	1.14	3.60
24″ O.C.	.019	Ea.	.016	1.73	.80	2.53
32′ span, 16″ O.C.	.024	Ea.	.021	2.57	1.10	3.67
24″ O.C.	.016	Ea.	.014	1.71	.74	2.45
36′ span, 16″ O.C.	.022	Ea.	.021	2.79	1.11	3.90
24″ O.C.	.015	Ea.	.015	1.91	.76	2.67
Fascia board, #2 or better, 1″ x 6″	.100	L.F.	.004	.05	.15	.20
1″ x 8″	.100	L.F.	.005	.06	.18	.24
1″ x 10″	.100	L.F.	.005	.06	.20	.26
2″ x 6″	.100	L.F.	.006	.07	.22	.29
2″ x 8″	.100	L.F.	.007	.09	.28	.37
2″ x 10″	.100	L.F.	.009	.12	.35	.47
Sheathing, plywood CDX, 4/12 pitch, 3/8″ thick	1.170	S.F.	.012	.56	.48	1.04
1/2″ thick	1.170	S.F.	.013	.50	.53	1.03
5/8″ thick	1.170	S.F.	.014	.59	.56	1.15
8/12 pitch, 3/8″ thick	1.330	S.F.	.014	.64	.55	1.19
1/2″ thick	1.330	S.F.	.015	.57	.60	1.17
5/8″ thick	1.330	S.F.	.016	.67	.64	1.31
Boards, 4/12 pitch, 1″ x 6″	1.170	S.F.	.026	1.05	1.02	2.07
1″ x 8″	1.170	S.F.	.021	1.03	.84	1.87
8/12 pitch, 1″ x 6″	1.330	S.F.	.029	1.20	1.16	2.36
1″ x 8″	1.330	S.F.	.024	1.17	.96	2.13
Furring, 1″ x 3″, 12″ O.C.	1.200	L.F.	.027	.28	1.08	1.36
16″ O.C.	1.000	L.F.	.023	.23	.90	1.13
24″ O.C.	.800	L.F.	.018	.18	.72	.90

20 Hip Roof Framing Systems

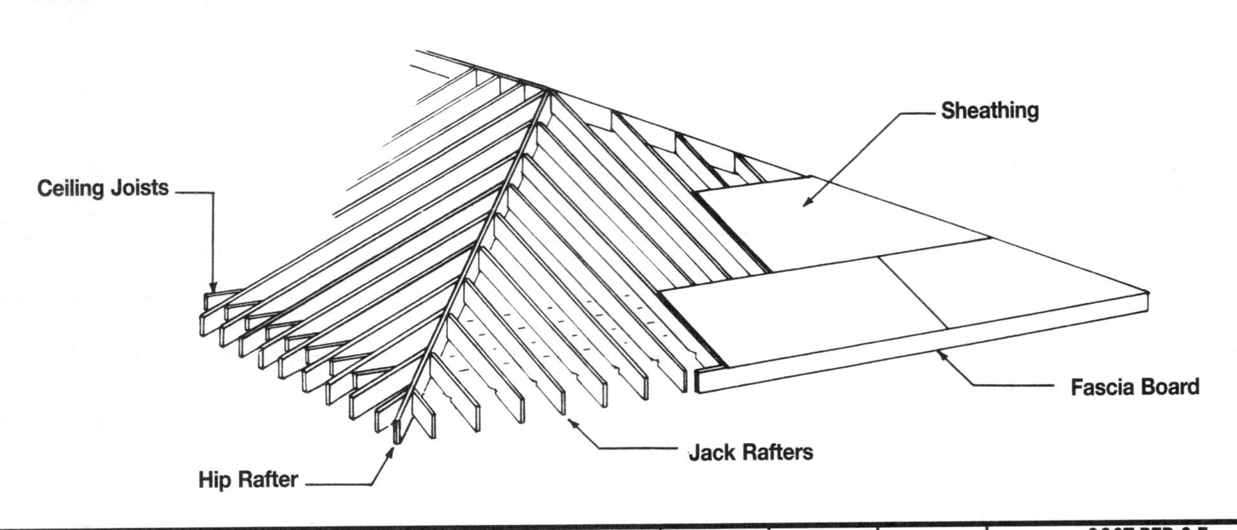

System Description	QUAN.	UNIT	LABOR HOURS	COST PER S.F.		
				MAT.	INST.	TOTAL
2" X 6", 16" O.C., 4/12 PITCH						
Hip and valley rafters, 2" x 6", ordinary	.160	L.F.	.003	.09	.13	.22
Jack rafters, 2" x 8", 16" O.C., 4/12 pitch	1.430	L.F.	.038	.82	1.50	2.32
Ceiling joists, 2" x 6", 16" O.C.	1.000	L.F.	.013	.34	.50	.84
Fascia board, 2" x 8"	.220	L.F.	.016	.19	.61	.80
Soffit nailer (outrigger), 2" x 4", 24" O.C.	.220	L.F.	.006	.07	.22	.29
Sheathing, 1/2" exterior plywood, CDX	1.570	S.F.	.018	.68	.71	1.39
Furring strips, 1" x 3", 16" O.C.	1.000	L.F.	.023	.23	.90	1.13
TOTAL		S.F.	.113	2.42	4.57	6.99
2" X 8", 16" O.C., 4/12 PITCH						
Hip rafters, 2" x 6", 4/12 pitch	.160	L.F.	.004	.14	.14	.28
Jack rafters, 2" x 6", 16" O.C., 4/12 pitch	1.430	L.F.	.047	1.26	1.83	3.09
Ceiling joists, 2" x 4", 16" O.C.	1.000	L.F.	.013	.57	.50	1.07
Fascia board, 2" x 6"	.220	L.F.	.012	.15	.47	.62
Soffit nailer (outrigger), 2" x 4", 24" O.C.	.220	L.F.	.006	.07	.22	.29
Sheathing, 1/2" exterior plywood, CDX	1.570	S.F.	.018	.68	.71	1.39
Furring strips, 1" x 3", 16" O.C.	1.000	L.F.	.023	.23	.90	1.13
TOTAL		S.F.	.127	3.10	4.77	7.87

The cost of this system is based on S.F. of plan area. Measurement is area under the hip roof only. See gable roof system for added costs.

Description	QUAN.	UNIT	LABOR HOURS	COST PER S.F.		
				MAT.	INST.	TOTAL

Important: See the Reference Section for critical supporting data - Reference Nos., Crews & Location Factors

Hip Roof Framing Price Sheet	QUAN.	UNIT	LABOR HOURS	COST PER S.F.		
				MAT.	INST.	TOTAL
Hip rafters, #2 or better, 2″ x 6″, 4/12 pitch	.160	L.F.	.003	.09	.13	.22
8/12 pitch	.210	L.F.	.006	.12	.22	.34
2″ x 8″, 4/12 pitch	.160	L.F.	.004	.14	.14	.28
8/12 pitch	.210	L.F.	.006	.18	.24	.42
2″ x 10″, 4/12 pitch	.160	L.F.	.004	.19	.18	.37
8/12 pitch roof	.210	L.F.	.008	.25	.30	.55
Jack rafters, #2 or better, 16″ O.C., 2″ x 6″, 4/12 pitch	1.430	L.F.	.038	.82	1.50	2.32
8/12 pitch	1.800	L.F.	.061	1.03	2.38	3.41
2″ x 8″, 4/12 pitch	1.430	L.F.	.047	1.26	1.83	3.09
8/12 pitch	1.800	L.F.	.075	1.58	2.93	4.51
2″ x 10″, 4/12 pitch	1.430	L.F.	.051	1.70	1.99	3.69
8/12 pitch	1.800	L.F.	.082	2.14	3.22	5.36
24″ O.C., 2″ x 6″, 4/12 pitch	1.150	L.F.	.031	.66	1.21	1.87
8/12 pitch	1.440	L.F.	.048	.82	1.90	2.72
2″ x 8″, 4/12 pitch	1.150	L.F.	.038	1.01	1.47	2.48
8/12 pitch	1.440	L.F.	.060	1.27	2.35	3.62
2″ x 10″, 4/12 pitch	1.150	L.F.	.041	1.37	1.60	2.97
8/12 pitch	1.440	L.F.	.066	1.71	2.58	4.29
Ceiling joists, #2 or better, 2″ x 4″, 16″ O.C.	1.000	L.F.	.013	.34	.50	.84
24″ O.C.	.750	L.F.	.010	.26	.38	.64
2″ x 6″, 16″ O.C.	1.000	L.F.	.013	.57	.50	1.07
24″ O.C.	.750	L.F.	.010	.43	.38	.81
2″ x 8″, 16″ O.C.	1.000	L.F.	.015	.88	.57	1.45
24″ O.C.	.750	L.F.	.011	.66	.43	1.09
2″ x 10″, 16″ O.C.	1.000	L.F.	.018	1.19	.70	1.89
24″ O.C.	.750	L.F.	.013	.89	.53	1.42
Fascia board, #2 or better, 1″ x 6″	.220	L.F.	.009	.11	.34	.45
1″ x 8″	.220	L.F.	.010	.12	.39	.51
1″ x 10″	.220	L.F.	.011	.14	.44	.58
2″ x 6″	.220	L.F.	.013	.15	.49	.64
2″ x 8″	.220	L.F.	.016	.19	.61	.80
2″ x 10″	.220	L.F.	.020	.26	.77	1.03
Soffit nailer (outrigger), 2″ x 4″, 16″ O.C.	.280	L.F.	.007	.10	.28	.38
24″ O.C.	.220	L.F.	.006	.07	.22	.29
2″ x 8″, 16″ O.C.	.280	L.F.	.007	.18	.26	.44
24″ O.C.	.220	L.F.	.005	.15	.21	.36
Sheathing, plywood CDX, 4/12 pitch, 3/8″ thick	1.570	S.F.	.016	.75	.64	1.39
1/2″ thick	1.570	S.F.	.018	.68	.71	1.39
5/8″ thick	1.570	S.F.	.019	.79	.75	1.54
8/12 pitch, 3/8″ thick	1.900	S.F.	.020	.91	.78	1.69
1/2″ thick	1.900	S.F.	.022	.82	.86	1.68
5/8″ thick	1.900	S.F.	.023	.95	.91	1.86
Boards, 4/12 pitch, 1″ x 6″ boards	1.450	S.F.	.032	1.31	1.26	2.57
1″ x 8″ boards	1.450	S.F.	.027	1.28	1.04	2.32
8/12 pitch, 1″ x 6″ boards	1.750	S.F.	.039	1.58	1.52	3.10
1″ x 8″ boards	1.750	S.F.	.032	1.54	1.26	2.80
Furring, 1″ x 3″, 12″ O.C.	1.200	L.F.	.027	.28	1.08	1.36
16″ O.C.	1.000	L.F.	.023	.23	.90	1.13
24″ O.C.	.800	L.F.	.018	.18	.72	.90

24 Gambrel Roof Framing Systems

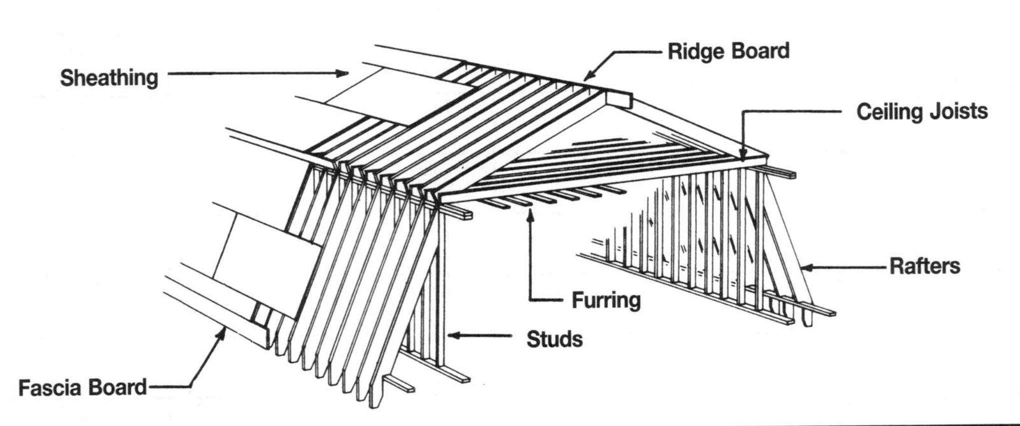

System Description	QUAN.	UNIT	LABOR HOURS	COST PER S.F.		
				MAT.	INST.	TOTAL
2" X 6" RAFTERS, 16" O.C.						
Roof rafters, 2" x 6", 16" O.C.	1.430	L.F.	.029	.82	1.12	1.94
Ceiling joists, 2" x 6", 16" O.C.	.710	L.F.	.009	.40	.36	.76
Stud wall, 2" x 4", 16" O.C., including plates	.790	L.F.	.012	.27	.49	.76
Furring strips, 1" x 3", 16" O.C.	.710	L.F.	.016	.16	.64	.80
Ridge board, 2" x 8"	.050	L.F.	.002	.04	.07	.11
Fascia board, 2" x 6"	.100	L.F.	.006	.07	.22	.29
Sheathing, exterior grade plywood, 1/2" thick	1.450	S.F.	.017	.62	.65	1.27
TOTAL		S.F.	.091	2.38	3.55	5.93
2" X 8" RAFTERS, 16" O.C.						
Roof rafters, 2" x 8", 16" O.C.	1.430	L.F.	.031	1.26	1.20	2.46
Ceiling joists, 2" x 6", 16" O.C.	.710	L.F.	.009	.40	.36	.76
Stud wall, 2" x 4", 16" O.C., including plates	.790	L.F.	.012	.27	.49	.76
Furring strips, 1" x 3", 16" O.C.	.710	L.F.	.016	.16	.64	.80
Ridge board, 2" x 8"	.050	L.F.	.002	.04	.07	.11
Fascia board, 2" x 8"	.100	L.F.	.007	.09	.28	.37
Sheathing, exterior grade plywood, 1/2" thick	1.450	S.F.	.017	.62	.65	1.27
TOTAL		S.F.	.094	2.84	3.69	6.53

The cost of this system is based on the square foot of plan area on the first floor.

Description	QUAN.	UNIT	LABOR HOURS	COST PER S.F.		
				MAT.	INST.	TOTAL

Important: See the Reference Section for critical supporting data - Reference Nos., Crews & Location Factors

Gambrel Roof Framing Price Sheet	QUAN.	UNIT	LABOR HOURS	COST PER S.F.		
				MAT.	INST.	TOTAL
Roof rafters, #2 or better, 2″ x 6″, 16″ O.C.	1.430	L.F.	.029	.82	1.12	1.94
24″ O.C.	1.140	L.F.	.023	.65	.89	1.54
2″ x 8″, 16″ O.C.	1.430	L.F.	.031	1.26	1.20	2.46
24″ O.C.	1.140	L.F.	.024	1	.96	1.96
2″ x 10″, 16″ O.C.	1.430	L.F.	.046	1.70	1.82	3.52
24″ O.C.	1.140	L.F.	.037	1.36	1.45	2.81
Ceiling joist, #2 or better, 2″ x 4″, 16″ O.C.	.710	L.F.	.009	.24	.36	.60
24″ O.C.	.570	L.F.	.007	.19	.29	.48
2″ x 6″, 16″ O.C.	.710	L.F.	.009	.40	.36	.76
24″ O.C.	.570	L.F.	.007	.32	.29	.61
2″ x 8″, 16″ O.C.	.710	L.F.	.010	.62	.40	1.02
24″ O.C.	.570	L.F.	.008	.50	.32	.82
Stud wall, #2 or better, 2″ x 4″, 16″ O.C.	.790	L.F.	.012	.27	.49	.76
24″ O.C.	.630	L.F.	.010	.21	.39	.60
2″ x 6″, 16″ O.C.	.790	L.F.	.014	.45	.55	1
24″ O.C.	.630	L.F.	.011	.36	.44	.80
Furring, 1″ x 3″, 16″ O.C.	.710	L.F.	.016	.16	.64	.80
24″ O.C.	.590	L.F.	.013	.14	.53	.67
Ridge board, #2 or better, 1″ x 6″	.050	L.F.	.001	.05	.05	.10
1″ x 8″	.050	L.F.	.001	.07	.06	.13
1″ x 10″	.050	L.F.	.002	.08	.06	.14
2″ x 6″	.050	L.F.	.002	.03	.06	.09
2″ x 8″	.050	L.F.	.002	.04	.07	.11
2″ x 10″	.050	L.F.	.002	.06	.08	.14
Fascia board, #2 or better, 1″ x 6″	.100	L.F.	.004	.05	.15	.20
1″ x 8″	.100	L.F.	.005	.06	.18	.24
1″ x 10″	.100	L.F.	.005	.06	.20	.26
2″ x 6″	.100	L.F.	.006	.07	.22	.29
2″ x 8″	.100	L.F.	.007	.09	.28	.37
2″ x 10″	.100	L.F.	.009	.12	.35	.47
Sheathing, plywood, exterior grade CDX, 3/8″ thick	1.450	S.F.	.015	.70	.59	1.29
1/2″ thick	1.450	S.F.	.017	.62	.65	1.27
5/8″ thick	1.450	S.F.	.018	.73	.70	1.43
3/4″ thick	1.450	S.F.	.019	.91	.75	1.66
Boards, 1″ x 6″, laid regular	1.450	S.F.	.032	1.31	1.26	2.57
Laid diagonal	1.450	S.F.	.036	1.31	1.41	2.72
1″ x 8″, laid regular	1.450	S.F.	.027	1.28	1.04	2.32
Laid diagonal	1.450	S.F.	.032	1.28	1.26	2.54

28 Mansard Roof Framing Systems

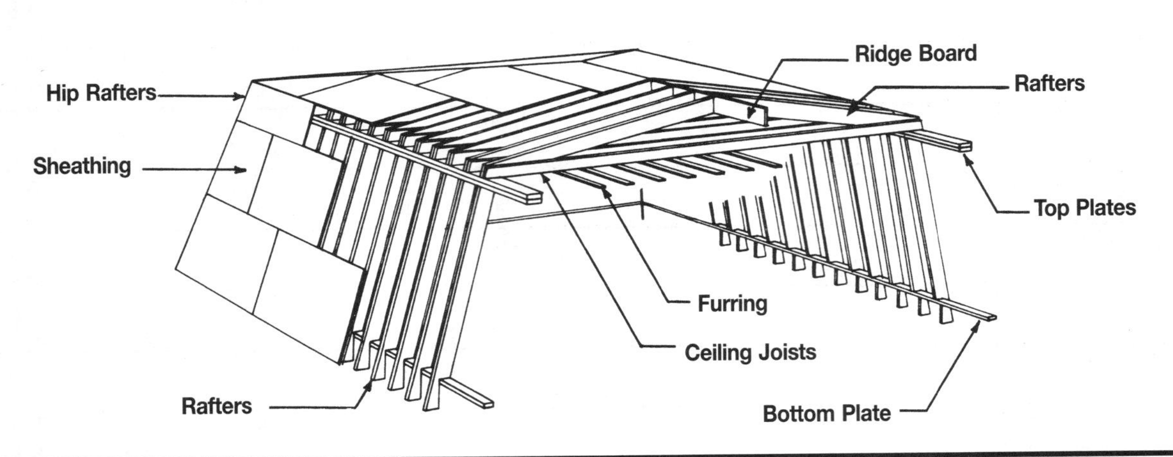

System Description	QUAN.	UNIT	LABOR HOURS	COST PER S.F.		
				MAT.	INST.	TOTAL
2″ X 6″ RAFTERS, 16″ O.C.						
Roof rafters, 2″ x 6″, 16″ O.C.	1.210	L.F.	.033	.69	1.28	1.97
Rafter plates, 2″ x 6″, double top, single bottom	.364	L.F.	.010	.21	.39	.60
Ceiling joists, 2″ x 4″, 16″ O.C.	.920	L.F.	.012	.31	.46	.77
Hip rafter, 2″ x 6″	.070	L.F.	.002	.04	.09	.13
Jack rafter, 2″ x 6″, 16″ O.C.	1.000	L.F.	.039	.57	1.53	2.10
Ridge board, 2″ x 6″	.018	L.F.	.001	.01	.02	.03
Sheathing, exterior grade plywood, 1/2″ thick	2.210	S.F.	.025	.95	.99	1.94
Furring strips, 1″ x 3″, 16″ O.C.	.920	L.F.	.021	.21	.83	1.04
TOTAL		S.F.	.143	2.99	5.59	8.58
2″ X 8″ RAFTERS, 16″ O.C.						
Roof rafters, 2″ x 8″, 16″ O.C.	1.210	L.F.	.036	1.06	1.40	2.46
Rafter plates, 2″ x 8″, double top, single bottom	.364	L.F.	.011	.32	.42	.74
Ceiling joists, 2″ x 6″, 16″ O.C.	.920	L.F.	.012	.52	.46	.98
Hip rafter, 2″ x 8″	.070	L.F.	.002	.06	.09	.15
Jack rafter, 2″ x 8″, 16″ O.C.	1.000	L.F.	.048	.88	1.87	2.75
Ridge board, 2″ x 8″	.018	L.F.	.001	.02	.03	.05
Sheathing, exterior grade plywood, 1/2″ thick	2.210	S.F.	.025	.95	.99	1.94
Furring strips, 1″ x 3″, 16″ O.C.	.920	L.F.	.021	.21	.83	1.04
TOTAL		S.F.	.156	4.02	6.09	10.11

The cost of this system is based on the square foot of plan area.

Description	QUAN.	UNIT	LABOR HOURS	COST PER S.F.		
				MAT.	INST.	TOTAL

Important: See the Reference Section for critical supporting data - Reference Nos., Crews & Location Factors

Mansard Roof Framing Price Sheet

	QUAN.	UNIT	LABOR HOURS	COST PER S.F. MAT.	INST.	TOTAL
Roof rafters, #2 or better, 2″ x 6″, 16″ O.C.	1.210	L.F.	.033	.69	1.28	1.97
24″ O.C.	.970	L.F.	.026	.55	1.03	1.58
2″ x 8″, 16″ O.C.	1.210	L.F.	.036	1.06	1.40	2.46
24″ O.C.	.970	L.F.	.029	.85	1.13	1.98
2″ x 10″, 16″ O.C.	1.210	L.F.	.046	1.44	1.79	3.23
24″ O.C.	.970	L.F.	.037	1.15	1.44	2.59
Rafter plates, #2 or better double top single bottom, 2″ x 6″	.364	L.F.	.010	.21	.39	.60
2″ x 8″	.364	L.F.	.011	.32	.42	.74
2″ x 10″	.364	L.F.	.014	.43	.54	.97
Ceiling joist, #2 or better, 2″ x 4″, 16″ O.C.	.920	L.F.	.012	.31	.46	.77
24″ O.C.	.740	L.F.	.009	.25	.37	.62
2″ x 6″, 16″ O.C.	.920	L.F.	.012	.52	.46	.98
24″ O.C.	.740	L.F.	.009	.42	.37	.79
2″ x 8″, 16″ O.C.	.920	L.F.	.013	.81	.52	1.33
24″ O.C.	.740	L.F.	.011	.65	.42	1.07
Hip rafter, #2 or better, 2″ x 6″	.070	L.F.	.002	.04	.09	.13
2″ x 8″	.070	L.F.	.002	.06	.09	.15
2″ x 10″	.070	L.F.	.003	.08	.12	.20
Jack rafter, #2 or better, 2″ x 6″, 16″ O.C.	1.000	L.F.	.039	.57	1.53	2.10
24″ O.C.	.800	L.F.	.031	.46	1.22	1.68
2″ x 8″, 16″ O.C.	1.000	L.F.	.048	.88	1.87	2.75
24″ O.C.	.800	L.F.	.038	.70	1.50	2.20
Ridge board, #2 or better, 1″ x 6″	.018	L.F.	.001	.02	.02	.04
1″ x 8″	.018	L.F.	.001	.02	.02	.04
1″ x 10″	.018	L.F.	.001	.03	.02	.05
2″ x 6″	.018	L.F.	.001	.01	.02	.03
2″ x 8″	.018	L.F.	.001	.02	.03	.05
2″ x 10″	.018	L.F.	.001	.02	.03	.05
Sheathing, plywood exterior grade CDX, 3/8″ thick	2.210	S.F.	.023	1.06	.91	1.97
1/2″ thick	2.210	S.F.	.025	.95	.99	1.94
5/8″ thick	2.210	S.F.	.027	1.11	1.06	2.17
3/4″ thick	2.210	S.F.	.029	1.39	1.15	2.54
Boards, 1″ x 6″, laid regular	2.210	S.F.	.049	1.99	1.92	3.91
Laid diagonal	2.210	S.F.	.054	1.99	2.14	4.13
1″ x 8″, laid regular	2.210	S.F.	.040	1.94	1.59	3.53
Laid diagonal	2.210	S.F.	.049	1.94	1.92	3.86
Furring, 1″ x 3″, 12″ O.C.	1.150	L.F.	.026	.26	1.04	1.30
24″ O.C.	.740	L.F.	.017	.17	.67	.84

32 Shed/Flat Roof Framing Systems

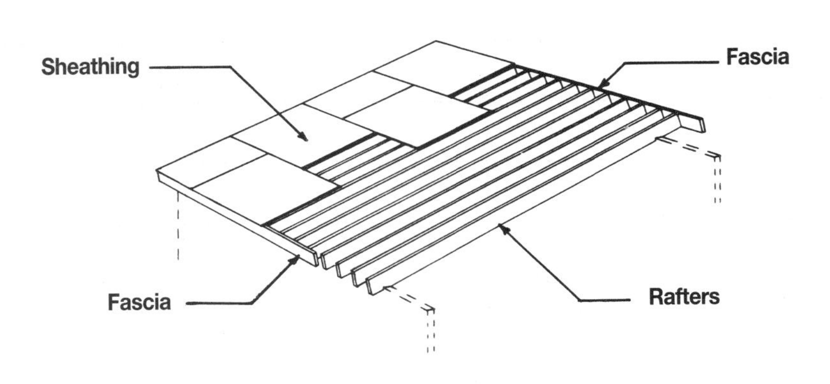

System Description	QUAN.	UNIT	LABOR HOURS	COST PER S.F. MAT.	COST PER S.F. INST.	COST PER S.F. TOTAL
2″ X 6″,16″ O.C., 4/12 PITCH						
Rafters, 2″ x 6″, 16″ O.C., 4/12 pitch	1.170	L.F.	.019	.67	.74	1.41
Fascia, 2″ x 6″	.100	L.F.	.006	.07	.22	.29
Bridging, 1″ x 3″, 6′ O.C.	.080	Pr.	.005	.03	.19	.22
Sheathing, exterior grade plywood, 1/2″ thick	1.230	S.F.	.014	.53	.55	1.08
TOTAL		S.F.	.044	1.30	1.70	3
2″ X 6″, 24″ O.C., 4/12 PITCH						
Rafters, 2″ x 6″, 24″ O.C., 4/12 pitch	.940	L.F.	.015	.54	.59	1.13
Fascia, 2″ x 6″	.100	L.F.	.006	.07	.22	.29
Bridging, 1″ x 3″, 6′ O.C.	.060	Pr.	.004	.02	.14	.16
Sheathing, exterior grade plywood, 1/2″ thick	1.230	S.F.	.014	.53	.55	1.08
TOTAL		S.F.	.039	1.16	1.50	2.66
2″ X 8″, 16″ O.C., 4/12 PITCH						
Rafters, 2″ x 8″, 16″ O.C., 4/12 pitch	1.170	L.F.	.020	1.03	.77	1.80
Fascia, 2″ x 8″	.100	L.F.	.007	.09	.28	.37
Bridging, 1″ x 3″, 6′ O.C.	.080	Pr.	.005	.03	.19	.22
Sheathing, exterior grade plywood, 1/2″ thick	1.230	S.F.	.014	.53	.55	1.08
TOTAL		S.F.	.046	1.68	1.79	3.47
2″ X 8″, 24″ O.C., 4/12 PITCH						
Rafters, 2″ x 8″, 24″ O.C., 4/12 pitch	.940	L.F.	.016	.83	.62	1.45
Fascia, 2″ x 8″	.100	L.F.	.007	.09	.28	.37
Bridging, 1″ x 3″, 6′ O.C.	.060	Pr.	.004	.02	.14	.16
Sheathing, exterior grade plywood, 1/2″ thick	1.230	S.F.	.014	.53	.55	1.08
TOTAL		S.F.	.041	1.47	1.59	3.06

The cost of this system is based on the square foot of plan area.
A 1′ overhang is assumed. No ceiling joists or furring are included.

Description	QUAN.	UNIT	LABOR HOURS	COST PER S.F. MAT.	COST PER S.F. INST.	COST PER S.F. TOTAL

Important: See the Reference Section for critical supporting data - Reference Nos., Crews & Location Factors

Shed/Flat Roof Framing Price Sheet

	QUAN.	UNIT	LABOR HOURS	COST PER S.F.		
				MAT.	INST.	TOTAL
Rafters, #2 or better, 16″ O.C., 2″ x 4″, 0 - 4/12 pitch	1.170	L.F.	.014	.50	.55	1.05
5/12 - 8/12 pitch	1.330	L.F.	.020	.57	.78	1.35
2″ x 6″, 0 - 4/12 pitch	1.170	L.F.	.019	.67	.74	1.41
5/12 - 8/12 pitch	1.330	L.F.	.027	.76	1.04	1.80
2″ x 8″, 0 - 4/12 pitch	1.170	L.F.	.020	1.03	.77	1.80
5/12 - 8/12 pitch	1.330	L.F.	.028	1.17	1.12	2.29
2″ x 10″, 0 - 4/12 pitch	1.170	L.F.	.030	1.39	1.17	2.56
5/12 - 8/12 pitch	1.330	L.F.	.043	1.58	1.69	3.27
24″ O.C., 2″ x 4″, 0 - 4/12 pitch	.940	L.F.	.011	.40	.45	.85
5/12 - 8/12 pitch	1.060	L.F.	.021	.60	.83	1.43
2″ x 6″, 0 - 4/12 pitch	.940	L.F.	.015	.54	.59	1.13
5/12 - 8/12 pitch	1.060	L.F.	.021	.60	.83	1.43
2″ x 8″, 0 - 4/12 pitch	.940	L.F.	.016	.83	.62	1.45
5/12 - 8/12 pitch	1.060	L.F.	.023	.93	.89	1.82
2″ x 10″, 0 - 4/12 pitch	.940	L.F.	.024	1.12	.94	2.06
5/12 - 8/12 pitch	1.060	L.F.	.034	1.26	1.35	2.61
Fascia, #2 or better,, 1″ x 4″	.100	L.F.	.003	.04	.11	.15
1″ x 6″	.100	L.F.	.004	.05	.15	.20
1″ x 8″	.100	L.F.	.005	.06	.18	.24
1″ x 10″	.100	L.F.	.005	.06	.20	.26
2″ x 4″	.100	L.F.	.005	.06	.19	.25
2″ x 6″	.100	L.F.	.006	.07	.22	.29
2″ x 8″	.100	L.F.	.007	.09	.28	.37
2″ x 10″	.100	L.F.	.009	.12	.35	.47
Bridging, wood 6′ O.C., 1″ x 3″, rafters, 16″ O.C.	.080	Pr.	.005	.03	.19	.22
24″ O.C.	.060	Pr.	.004	.02	.14	.16
Metal, galvanized, rafters, 16″ O.C.	.080	Pr.	.005	.08	.19	.27
24″ O.C.	.060	Pr.	.003	.11	.13	.24
Compression type, rafters, 16″ O.C.	.080	Pr.	.003	.10	.13	.23
24″ O.C.	.060	Pr.	.002	.07	.09	.16
Sheathing, plywood, exterior grade, 3/8″ thick, flat 0 - 4/12 pitch	1.230	S.F.	.013	.59	.50	1.09
5/12 - 8/12 pitch	1.330	S.F.	.014	.64	.55	1.19
1/2″ thick, flat 0 - 4/12 pitch	1.230	S.F.	.014	.53	.55	1.08
5/12 - 8/12 pitch	1.330	S.F.	.015	.57	.60	1.17
5/8″ thick, flat 0 - 4/12 pitch	1.230	S.F.	.015	.62	.59	1.21
5/12 - 8/12 pitch	1.330	S.F.	.016	.67	.64	1.31
3/4″ thick, flat 0 - 4/12 pitch	1.230	S.F.	.016	.77	.64	1.41
5/12 - 8/12 pitch	1.330	S.F.	.018	.84	.69	1.53
Boards, 1″ x 6″, laid regular, flat 0 - 4/12 pitch	1.230	S.F.	.027	1.11	1.07	2.18
5/12 - 8/12 pitch	1.330	S.F.	.041	1.20	1.61	2.81
Laid diagonal, flat 0 - 4/12 pitch	1.230	S.F.	.030	1.11	1.19	2.30
5/12 - 8/12 pitch	1.330	S.F.	.044	1.20	1.74	2.94
1″ x 8″, laid regular, flat 0 - 4/12 pitch	1.230	S.F.	.022	1.08	.89	1.97
5/12 - 8/12 pitch	1.330	S.F.	.034	1.17	1.32	2.49
Laid diagonal, flat 0 - 4/12 pitch	1.230	S.F.	.027	1.08	1.07	2.15
5/12 - 8/12 pitch	1.330	S.F.	.044	1.20	1.74	2.94

40 Gable Dormer Framing Systems

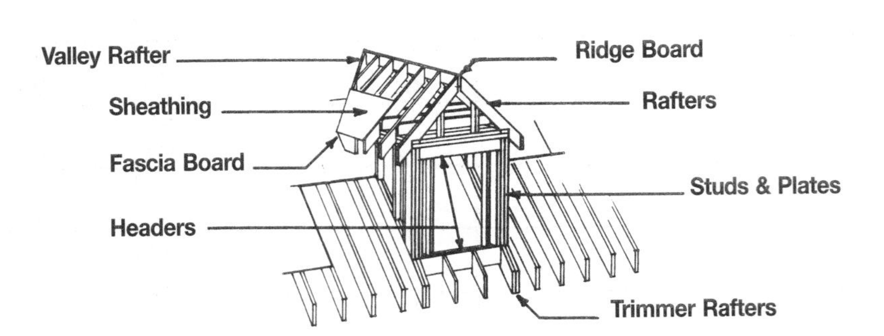

System Description	QUAN.	UNIT	LABOR HOURS	COST PER S.F.		
				MAT.	INST.	TOTAL
2" X 6", 16" O.C.						
Dormer rafter, 2" x 6", 16" O.C.	1.330	L.F.	.036	.76	1.41	2.17
Ridge board, 2" x 6"	.280	L.F.	.009	.16	.35	.51
Trimmer rafters, 2" x 6"	.880	L.F.	.014	.50	.55	1.05
Wall studs & plates, 2" x 4", 16" O.C.	3.160	L.F.	.056	1.07	2.18	3.25
Fascia, 2" x 6"	.220	L.F.	.012	.15	.47	.62
Valley rafter, 2" x 6", 16" O.C.	.280	L.F.	.009	.16	.34	.50
Cripple rafter, 2" x 6", 16" O.C.	.560	L.F.	.022	.32	.86	1.18
Headers, 2" x 6", doubled	.670	L.F.	.030	.38	1.17	1.55
Ceiling joist, 2" x 4", 16" O.C.	1.000	L.F.	.013	.34	.50	.84
Sheathing, exterior grade plywood, 1/2" thick	3.610	S.F.	.041	1.55	1.62	3.17
TOTAL		S.F.	.242	5.39	9.45	14.84
2" X 8", 16" O.C.						
Dormer rafter, 2" x 8", 16" O.C.	1.330	L.F.	.039	1.17	1.54	2.71
Ridge board, 2" x 8"	.280	L.F.	.010	.25	.39	.64
Trimmer rafter, 2" x 8"	.880	L.F.	.015	.77	.58	1.35
Wall studs & plates, 2" x 4", 16" O.C.	3.160	L.F.	.056	1.07	2.18	3.25
Fascia, 2" x 8"	.220	L.F.	.016	.19	.61	.80
Valley rafter, 2" x 8", 16" O.C.	.280	L.F.	.010	.25	.37	.62
Cripple rafter, 2" x 8", 16" O.C.	.560	L.F.	.027	.49	1.05	1.54
Headers, 2" x 8", doubled	.670	L.F.	.032	.59	1.23	1.82
Ceiling joist, 2" x 4", 16" O.C.	1.000	L.F.	.013	.34	.50	.84
Sheathing,, exterior grade plywood, 1/2" thick	3.610	S.F.	.041	1.55	1.62	3.17
TOTAL		S.F.	.259	6.67	10.07	16.74

The cost in this system is based on the square foot of plan area.
The measurement being the plan area of the dormer only.

Description	QUAN.	UNIT	LABOR HOURS	COST PER S.F.		
				MAT.	INST.	TOTAL

Important: See the Reference Section for critical supporting data - Reference Nos., Crews & Location Factors

Gable Dormer Framing Price Sheet	QUAN.	UNIT	LABOR HOURS	COST PER S.F.		
				MAT.	INST.	TOTAL
Dormer rafters, #2 or better, 2″ x 4″, 16″ O.C.	1.330	L.F.	.029	.61	1.13	1.74
24″ O.C.	1.060	L.F.	.023	.48	.90	1.38
2″ x 6″, 16″ O.C.	1.330	L.F.	.036	.76	1.41	2.17
24″ O.C.	1.060	L.F.	.029	.60	1.12	1.72
2″ x 8″, 16″ O.C.	1.330	L.F.	.039	1.17	1.54	2.71
24″ O.C.	1.060	L.F.	.031	.93	1.23	2.16
Ridge board, #2 or better, 1″ x 4″	.280	L.F.	.006	.22	.24	.46
1″ x 6″	.280	L.F.	.007	.28	.29	.57
1″ x 8″	.280	L.F.	.008	.38	.32	.70
2″ x 4″	.280	L.F.	.007	.13	.28	.41
2″ x 6″	.280	L.F.	.009	.16	.35	.51
2″ x 8″	.280	L.F.	.010	.25	.39	.64
Trimmer rafters, #2 or better, 2″ x 4″	.880	L.F.	.011	.40	.44	.84
2″ x 6″	.880	L.F.	.014	.50	.55	1.05
2″ x 8″	.880	L.F.	.015	.77	.58	1.35
2″ x 10″	.880	L.F.	.022	1.05	.88	1.93
Wall studs & plates, #2 or better, 2″ x 4″ studs, 16″ O.C.	3.160	L.F.	.056	1.07	2.18	3.25
24″ O.C.	2.800	L.F.	.050	.95	1.93	2.88
2″ x 6″ studs, 16″ O.C.	3.160	L.F.	.063	1.80	2.46	4.26
24″ O.C.	2.800	L.F.	.056	1.60	2.18	3.78
Fascia, #2 or better, 1″ x 4″	.220	L.F.	.006	.08	.25	.33
1″ x 6″	.220	L.F.	.008	.10	.31	.41
1″ x 8″	.220	L.F.	.009	.11	.36	.47
2″ x 4″	.220	L.F.	.011	.13	.42	.55
2″ x 6″	.220	L.F.	.014	.17	.53	.70
2″ x 8″	.220	L.F.	.016	.19	.61	.80
Valley rafter, #2 or better, 2″ x 4″	.280	L.F.	.007	.13	.28	.41
2″ x 6″	.280	L.F.	.009	.16	.34	.50
2″ x 8″	.280	L.F.	.010	.25	.37	.62
2″ x 10″	.280	L.F.	.012	.33	.46	.79
Cripple rafter, #2 or better, 2″ x 4″, 16″ O.C.	.560	L.F.	.018	.26	.69	.95
24″ O.C.	.450	L.F.	.014	.21	.55	.76
2″ x 6″, 16″ O.C.	.560	L.F.	.022	.32	.86	1.18
24″ O.C.	.450	L.F.	.018	.26	.69	.95
2″ x 8″, 16″ O.C.	.560	L.F.	.027	.49	1.05	1.54
24″ O.C.	.450	L.F.	.021	.40	.84	1.24
Headers, #2 or better double header, 2″ x 4″	.670	L.F.	.024	.31	.94	1.25
2″ x 6″	.670	L.F.	.030	.38	1.17	1.55
2″ x 8″	.670	L.F.	.032	.59	1.23	1.82
2″ x 10″	.670	L.F.	.034	.80	1.31	2.11
Ceiling joist, #2 or better, 2″ x 4″, 16″ O.C.	1.000	L.F.	.013	.34	.50	.84
24″ O.C.	.800	L.F.	.010	.27	.40	.67
2″ x 6″, 16″ O.C.	1.000	L.F.	.013	.57	.50	1.07
24″ O.C.	.800	L.F.	.010	.46	.40	.86
Sheathing, plywood exterior grade, 3/8″ thick	3.610	S.F.	.038	1.73	1.48	3.21
1/2″ thick	3.610	S.F.	.041	1.55	1.62	3.17
5/8″ thick	3.610	S.F.	.044	1.81	1.73	3.54
3/4″ thick	3.610	S.F.	.048	2.27	1.88	4.15
Boards, 1″ x 6″, laid regular	3.610	S.F.	.089	3.25	3.50	6.75
Laid diagonal	3.610	S.F.	.099	3.25	3.86	7.11
1″ x 8″, laid regular	3.610	S.F.	.076	3.18	2.96	6.14
Laid diagonal	3.610	S.F.	.089	3.18	3.50	6.68

3 FRAMING

44 Shed Dormer Framing Systems

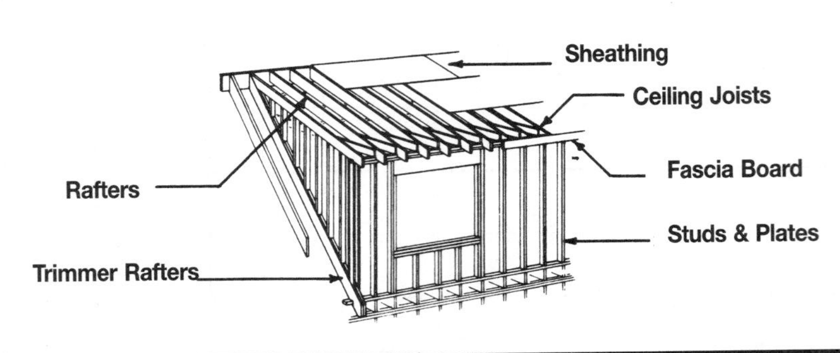

System Description	QUAN.	UNIT	LABOR HOURS	COST PER S.F.		
				MAT.	INST.	TOTAL
2" X 6" RAFTERS, 16" O.C.						
Dormer rafter, 2" x 6", 16" O.C.	1.080	L.F.	.029	.62	1.14	1.76
Trimmer rafter, 2" x 6"	.400	L.F.	.006	.23	.25	.48
Studs & plates, 2" x 4", 16" O.C.	2.750	L.F.	.049	.94	1.90	2.84
Fascia, 2" x 6"	.250	L.F.	.014	.17	.53	.70
Ceiling joist, 2" x 4", 16" O.C.	1.000	L.F.	.013	.34	.50	.84
Sheathing, exterior grade plywood, CDX, 1/2" thick	2.940	S.F.	.034	1.26	1.32	2.58
TOTAL		S.F.	.145	3.56	5.64	9.20
2" X 8" RAFTERS, 16" O.C.						
Dormer rafter, 2" x 8", 16" O.C.	1.080	L.F.	.032	.95	1.25	2.20
Trimmer rafter, 2" x 8"	.400	L.F.	.007	.35	.26	.61
Studs & plates, 2" x 4", 16" O.C.	2.750	L.F.	.049	.94	1.90	2.84
Fascia, 2" x 8"	.250	L.F.	.018	.22	.70	.92
Ceiling joist, 2" x 6", 16" O.C.	1.000	L.F.	.013	.57	.50	1.07
Sheathing, exterior grade plywood, CDX, 1/2" thick	2.940	S.F.	.034	1.26	1.32	2.58
TOTAL		S.F.	.153	4.29	5.93	10.22
2" X 10" RAFTERS, 16" O.C.						
Dormer rafter, 2" x 10", 16" O.C.	1.080	L.F.	.041	1.29	1.60	2.89
Trimmer rafter, 2" x 10"	.400	L.F.	.010	.48	.40	.88
Studs & plates, 2" x 4", 16" O.C.	2.750	L.F.	.049	.94	1.90	2.84
Fascia, 2" x 10"	.250	L.F.	.022	.30	.87	1.17
Ceiling joist, 2" x 6", 16" O.C.	1.000	L.F.	.013	.57	.50	1.07
Sheathing, exterior grade plywood, CDX, 1/2" thick	2.940	S.F.	.034	1.26	1.32	2.58
TOTAL		S.F.	.169	4.84	6.59	11.43

The cost in this system is based on the square foot of plan area.
The measurement is the plan area of the dormer only.

Description	QUAN.	UNIT	LABOR HOURS	COST PER S.F.		
				MAT.	INST.	TOTAL

Important: See the Reference Section for critical supporting data - Reference Nos., Crews & Location Factors

Shed Dormer Framing Price Sheet

	QUAN.	UNIT	LABOR HOURS	COST PER S.F.		
				MAT.	INST.	TOTAL
Dormer rafters, #2 or better, 2″ x 4″, 16″ O.C.	1.080	L.F.	.023	.49	.92	1.41
24″ O.C.	.860	L.F.	.019	.39	.73	1.12
2″ x 6″, 16″ O.C.	1.080	L.F.	.029	.62	1.14	1.76
24″ O.C.	.860	L.F.	.023	.49	.91	1.40
2″ x 8″, 16″ O.C.	1.080	L.F.	.032	.95	1.25	2.20
24″ O.C.	.860	L.F.	.025	.76	1	1.76
2″ x 10″, 16″ O.C.	1.080	L.F.	.041	1.29	1.60	2.89
24″ O.C.	.860	L.F.	.032	1.02	1.27	2.29
Trimmer rafter, #2 or better, 2″ x 4″	.400	L.F.	.005	.18	.20	.38
2″ x 6″	.400	L.F.	.006	.23	.25	.48
2″ x 8″	.400	L.F.	.007	.35	.26	.61
2″ x 10″	.400	L.F.	.010	.48	.40	.88
Studs & plates, #2 or better, 2″ x 4″, 16″ O.C.	2.750	L.F.	.049	.94	1.90	2.84
24″ O.C.	2.200	L.F.	.039	.75	1.52	2.27
2″ x 6″, 16″ O.C.	2.750	L.F.	.055	1.57	2.15	3.72
24″ O.C.	2.200	L.F.	.044	1.25	1.72	2.97
Fascia, #2 or better, 1″ x 4″	.250	L.F.	.006	.08	.25	.33
1″ x 6″	.250	L.F.	.008	.10	.31	.41
1″ x 8″	.250	L.F.	.009	.11	.36	.47
2″ x 4″	.250	L.F.	.011	.13	.42	.55
2″ x 6″	.250	L.F.	.014	.17	.53	.70
2″ x 8″	.250	L.F.	.018	.22	.70	.92
Ceiling joist, #2 or better, 2″ x 4″, 16″ O.C.	1.000	L.F.	.013	.34	.50	.84
24″ O.C.	.800	L.F.	.010	.27	.40	.67
2″ x 6″, 16″ O.C.	1.000	L.F.	.013	.57	.50	1.07
24″ O.C.	.800	L.F.	.010	.46	.40	.86
2″ x 8″, 16″ O.C.	1.000	L.F.	.015	.88	.57	1.45
24″ O.C.	.800	L.F.	.012	.70	.46	1.16
Sheathing, plywood exterior grade, 3/8″ thick	2.940	S.F.	.031	1.41	1.21	2.62
1/2″ thick	2.940	S.F.	.034	1.26	1.32	2.58
5/8″ thick	2.940	S.F.	.036	1.47	1.41	2.88
3/4″ thick	2.940	S.F.	.039	1.85	1.53	3.38
Boards, 1″ x 6″, laid regular	2.940	S.F.	.072	2.65	2.85	5.50
Laid diagonal	2.940	S.F.	.080	2.65	3.15	5.80
1″ x 8″, laid regular	2.940	S.F.	.062	2.59	2.41	5
Laid diagonal	2.940	S.F.	.072	2.59	2.85	5.44

Window Openings

	QUAN.	UNIT	LABOR HOURS	COST EACH		
				MAT.	INST.	TOTAL
The following are to be added to the total cost of the dormers for window openings. Do not subtract window area from the stud wall quantities.						
Headers, 2″ x 6″ doubled, 2′ long	4.000	L.F.	.178	2.28	6.95	9.23
3′ long	6.000	L.F.	.267	3.42	10.45	13.87
4′ long	8.000	L.F.	.356	4.56	13.90	18.46
5′ long	10.000	L.F.	.444	5.70	17.40	23.10
2″ x 8″ doubled, 4′ long	8.000	L.F.	.376	7.05	14.70	21.75
5′ long	10.000	L.F.	.471	8.80	18.40	27.20
6′ long	12.000	L.F.	.565	10.55	22	32.55
8′ long	16.000	L.F.	.753	14.10	29.50	43.60
2″ x 10″ doubled, 4′ long	8.000	L.F.	.400	9.50	15.70	25.20
6′ long	12.000	L.F.	.600	14.30	23.50	37.80
8′ long	16.000	L.F.	.800	19.05	31.50	50.55
10′ long	20.000	L.F.	1.000	24	39	63

48 Partition Framing Systems

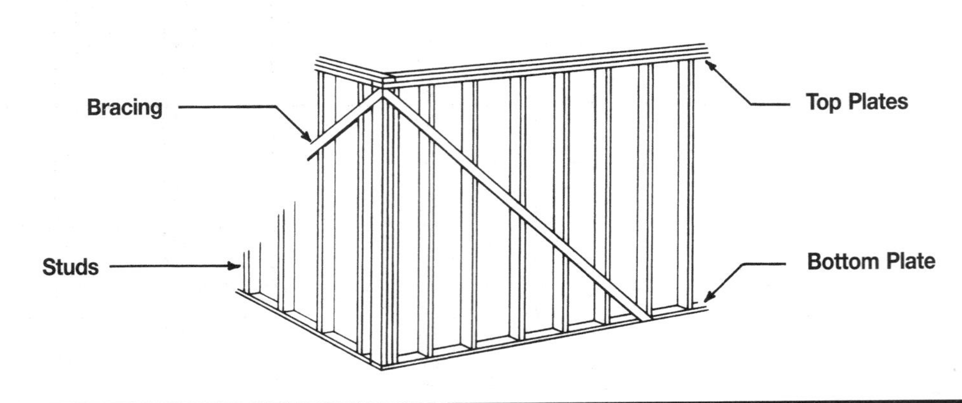

System Description	QUAN.	UNIT	LABOR HOURS	COST PER S.F. MAT.	COST PER S.F. INST.	COST PER S.F. TOTAL
2″ X 4″, 16″ O.C.						
2″ x 4″ studs, #2 or better, 16″ O.C.	1.000	L.F.	.015	.34	.57	.91
Plates, double top, single bottom	.375	L.F.	.005	.13	.21	.34
Cross bracing, let-in, 1″ x 6″	.080	L.F.	.004	.03	.17	.20
TOTAL		S.F.	.024	.50	.95	1.45
2″ X 4″, 24″ O.C.						
2″ x 4″ studs, #2 or better, 24″ O.C.	.800	L.F.	.012	.27	.46	.73
Plates, double top, single bottom	.375	L.F.	.005	.13	.21	.34
Cross bracing, let-in, 1″ x 6″	.080	L.F.	.003	.03	.11	.14
TOTAL		S.F.	.020	.43	.78	1.21
2″ X 6″, 16″ O.C.						
2″ x 6″ studs, #2 or better, 16″ O.C.	1.000	L.F.	.016	.57	.63	1.20
Plates, double top, single bottom	.375	L.F.	.006	.21	.24	.45
Cross bracing, let-in, 1″ x 6″	.080	L.F.	.004	.03	.17	.20
TOTAL		S.F.	.026	.81	1.04	1.85
2″ X 6″, 24″ O.C.						
2″ x 6″ studs, #2 or better, 24″ O.C.	.800	L.F.	.013	.46	.50	.96
Plates, double top, single bottom	.375	L.F.	.006	.21	.24	.45
Cross bracing, let-in, 1″ x 6″	.080	L.F.	.003	.03	.11	.14
TOTAL		S.F.	.022	.70	.85	1.55

The costs in this system are based on a square foot of wall area. Do not subtract for door or window openings.

Description	QUAN.	UNIT	LABOR HOURS	COST PER S.F. MAT.	COST PER S.F. INST.	COST PER S.F. TOTAL

Important: See the Reference Section for critical supporting data - Reference Nos., Crews & Location Factors

Partition Framing Price Sheet	QUAN.	UNIT	LABOR HOURS	COST PER S.F.		
				MAT.	INST.	TOTAL
Wood studs, #2 or better, 2″ x 4″, 12″ O.C.	1.250	L.F.	.018	.43	.71	1.14
16″ O.C.	1.000	L.F.	.015	.34	.57	.91
24″ O.C.	.800	L.F.	.012	.27	.46	.73
32″ O.C.	.650	L.F.	.009	.22	.37	.59
2″ x 6″, 12″ O.C.	1.250	L.F.	.020	.71	.79	1.50
16″ O.C.	1.000	L.F.	.016	.57	.63	1.20
24″ O.C.	.800	L.F.	.013	.46	.50	.96
32″ O.C.	.650	L.F.	.010	.37	.41	.78
Plates, #2 or better double top single bottom, 2″ x 4″	.375	L.F.	.005	.13	.21	.34
2″ x 6″	.375	L.F.	.006	.21	.24	.45
2″ x 8″	.375	L.F.	.005	.33	.21	.54
Cross bracing, let-in, 1″ x 6″ boards studs, 12″ O.C.	.080	L.F.	.005	.04	.21	.25
16″ O.C.	.080	L.F.	.004	.03	.17	.20
24″ O.C.	.080	L.F.	.003	.03	.11	.14
32″ O.C.	.080	L.F.	.002	.03	.09	.12
Let-in steel (T shaped) studs, 12″ O.C.	.080	L.F.	.001	.05	.05	.10
16″ O.C.	.080	L.F.	.001	.04	.04	.08
24″ O.C.	.080	L.F.	.001	.04	.04	.08
32″ O.C.	.080	L.F.	.001	.03	.03	.06
Steel straps studs, 12″ O.C.	.080	L.F.	.001	.06	.04	.10
16″ O.C.	.080	L.F.	.001	.06	.04	.10
24″ O.C.	.080	L.F.	.001	.06	.04	.10
32″ O.C.	.080	L.F.	.001	.05	.04	.09
Metal studs, load bearing 24″ O.C., 20 ga. galv., 2-1/2″ wide	1.000	S.F.	.015	.41	.59	1
3-5/8″ wide	1.000	S.F.	.015	.49	.60	1.09
4″ wide	1.000	S.F.	.016	.52	.61	1.13
6″ wide	1.000	S.F.	.016	.66	.62	1.28
16 ga., 2-1/2″ wide	1.000	S.F.	.017	.48	.67	1.15
3-5/8″ wide	1.000	S.F.	.017	.58	.68	1.26
4″ wide	1.000	S.F.	.018	.61	.70	1.31
6″ wide	1.000	S.F.	.018	.76	.72	1.48
Non-load bearing 24″ O.C., 25 ga. galv., 1-5/8″ wide	1.000	S.F.	.011	.15	.41	.56
2-1/2″ wide	1.000	S.F.	.011	.16	.42	.58
3-5/8″ wide	1.000	S.F.	.011	.17	.42	.59
4″ wide	1.000	S.F.	.011	.19	.42	.61
6″ wide	1.000	S.F.	.011	.27	.43	.70
20 ga., 2-1/2″ wide	1.000	S.F.	.013	.25	.52	.77
3-5/8″ wide	1.000	S.F.	.014	.30	.53	.83
4″ wide	1.000	S.F.	.014	.31	.53	.84
6″ wide	1.000	S.F.	.014	.41	.54	.95

Window & Door Openings	QUAN.	UNIT	LABOR HOURS	COST EACH		
				MAT.	INST.	TOTAL
The following costs are to be added to the total costs of the walls. Do not subtract openings from total wall area.						
Headers, 2″ x 6″ double, 2′ long	4.000	L.F.	.178	2.28	6.95	9.23
3′ long	6.000	L.F.	.267	3.42	10.45	13.87
4′ long	8.000	L.F.	.356	4.56	13.90	18.46
5′ long	10.000	L.F.	.444	5.70	17.40	23.10
2″ x 8″ double, 4′ long	8.000	L.F.	.376	7.05	14.70	21.75
5′ long	10.000	L.F.	.471	8.80	18.40	27.20
6′ long	12.000	L.F.	.565	10.55	22	32.55
8′ long	16.000	L.F.	.753	14.10	29.50	43.60
2″ x 10″ double, 4′ long	8.000	L.F.	.400	9.50	15.70	25.20
6′ long	12.000	L.F.	.600	14.30	23.50	37.80
8′ long	16.000	L.F.	.800	19.05	31.50	50.55
10′ long	20.000	L.F.	1.000	24	39	63
2″ x 12″ double, 8′ long	16.000	L.F.	.853	26	33.50	59.50
12′ long	24.000	L.F.	1.280	39	50	89

Division 4
Exterior Walls

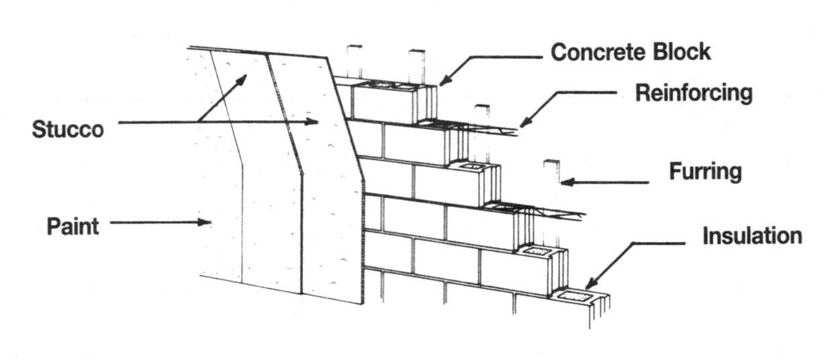

System Description	QUAN.	UNIT	LABOR HOURS	COST PER S.F. MAT.	COST PER S.F. INST.	COST PER S.F. TOTAL
6″ THICK CONCRETE BLOCK WALL						
6″ thick concrete block, 6″ x 8″ x 16″	1.000	S.F.	.100	1.52	3.58	5.10
Masonry reinforcing, truss strips every other course	.625	L.F.	.002	.08	.07	.15
Furring, 1″ x 3″, 16″ O.C.	1.000	L.F.	.016	.23	.63	.86
Masonry insulation, poured vermiculite	1.000	S.F.	.013	.61	.52	1.13
Stucco, 2 coats	1.000	S.F.	.069	.20	2.42	2.62
Masonry paint, 2 coats	1.000	S.F.	.016	.18	.55	.73
TOTAL		S.F.	.216	2.82	7.77	10.59
8″ THICK CONCRETE BLOCK WALL						
8″ thick concrete block, 8″ x 8″ x 16″	1.000	S.F.	.107	1.65	3.82	5.47
Masonry reinforcing, truss strips every other course	.625	L.F.	.002	.08	.07	.15
Furring, 1″ x 3″, 16″ O.C.	1.000	L.F.	.016	.23	.63	.86
Masonry insulation, poured vermiculite	1.000	S.F.	.018	.80	.69	1.49
Stucco, 2 coats	1.000	S.F.	.069	.20	2.42	2.62
Masonry paint, 2 coats	1.000	S.F.	.016	.18	.55	.73
TOTAL		S.F.	.228	3.14	8.18	11.32
12″ THICK CONCRETE BLOCK WALL						
12″ thick concrete block, 12″ x 8″ x 16″	1.000	S.F.	.141	2.40	4.92	7.32
Masonry reinforcing, truss strips every other course	.625	L.F.	.003	.09	.10	.19
Furring, 1″ x 3″, 16″ O.C.	1.000	L.F.	.016	.23	.63	.86
Masonry insulation, poured vermiculite	1.000	S.F.	.026	1.18	1.02	2.20
Stucco, 2 coats	1.000	S.F.	.069	.20	2.42	2.62
Masonry paint, 2 coats	1.000	S.F.	.016	.18	.55	.73
TOTAL		S.F.	.271	4.28	9.64	13.92

Costs for this system are based on a square foot of wall area. Do not subtract for window openings.

Description	QUAN.	UNIT	LABOR HOURS	COST PER S.F. MAT.	COST PER S.F. INST.	COST PER S.F. TOTAL

Masonry Block Price Sheet	QUAN.	UNIT	LABOR HOURS	COST PER S.F.		
				MAT.	INST.	TOTAL
Block concrete, 8″ x 16″ regular, 4″ thick	1.000	S.F.	.093	1.02	3.33	4.35
6″ thick	1.000	S.F.	.100	1.52	3.58	5.10
8″ thick	1.000	S.F.	.107	1.65	3.82	5.47
10″ thick	1.000	S.F.	.111	2.33	3.98	6.31
12″ thick	1.000	S.F.	.141	2.40	4.92	7.32
Solid block, 4″ thick	1.000	S.F.	.096	1.41	3.45	4.86
6″ thick	1.000	S.F.	.104	1.67	3.72	5.39
8″ thick	1.000	S.F.	.111	2.49	3.98	6.47
10″ thick	1.000	S.F.	.133	3.38	4.64	8.02
12″ thick	1.000	S.F.	.148	3.75	5.15	8.90
Lightweight, 4″ thick	1.000	S.F.	.093	1.02	3.33	4.35
6″ thick	1.000	S.F.	.100	1.52	3.58	5.10
8″ thick	1.000	S.F.	.107	1.65	3.82	5.47
10″ thick	1.000	S.F.	.111	2.33	3.98	6.31
12″ thick	1.000	S.F.	.141	2.40	4.92	7.32
Split rib profile, 4″ thick	1.000	S.F.	.116	2.30	4.16	6.46
6″ thick	1.000	S.F.	.123	2.65	4.41	7.06
8″ thick	1.000	S.F.	.131	3.05	4.78	7.83
10″ thick	1.000	S.F.	.157	3.24	5.50	8.74
12″ thick	1.000	S.F.	.175	3.60	6.10	9.70
Masonry reinforcing, wire truss strips, every course, 8″ block	1.375	L.F.	.004	.17	.15	.32
12″ block	1.375	L.F.	.006	.19	.22	.41
Every other course, 8″ block	.625	L.F.	.002	.08	.07	.15
12″ block	.625	L.F.	.003	.09	.10	.19
Furring, wood, 1″ x 3″, 12″ O.C.	1.250	L.F.	.020	.29	.79	1.08
16″ O.C.	1.000	L.F.	.016	.23	.63	.86
24″ O.C.	.800	L.F.	.013	.18	.50	.68
32″ O.C.	.640	L.F.	.010	.15	.40	.55
Steel, 3/4″ channels, 12″ O.C.	1.250	L.F.	.034	.18	1.18	1.36
16″ O.C.	1.000	L.F.	.030	.16	1.05	1.21
32″ O.C.	.640	L.F.	.018	.09	.63	.72
Masonry insulation, vermiculite or perlite poured 4″ thick	1.000	S.F.	.009	.39	.34	.73
6″ thick	1.000	S.F.	.013	.60	.52	1.12
8″ thick	1.000	S.F.	.018	.80	.69	1.49
10″ thick	1.000	S.F.	.021	.97	.84	1.81
12″ thick	1.000	S.F.	.026	1.18	1.02	2.20
Block inserts polystyrene, 6″ thick	1.000	S.F.		.94		.94
8″ thick	1.000	S.F.		.94		.94
10″ thick	1.000	S.F.		1.10		1.10
12″ thick	1.000	S.F.		1.16		1.16
Stucco, 1 coat	1.000	S.F.	.057	.16	2	2.16
2 coats	1.000	S.F.	.069	.20	2.42	2.62
3 coats	1.000	S.F.	.081	.23	2.85	3.08
Painting, 1 coat	1.000	S.F.	.011	.12	.38	.50
2 coats	1.000	S.F.	.016	.18	.55	.73
Primer & 1 coat	1.000	S.F.	.013	.20	.45	.65
2 coats	1.000	S.F.	.018	.26	.63	.89
Lath, metal lath expanded 2.5 lb/S.Y., painted	1.000	S.F.	.010	.17	.36	.53
Galvanized	1.000	S.F.	.012	.19	.40	.59

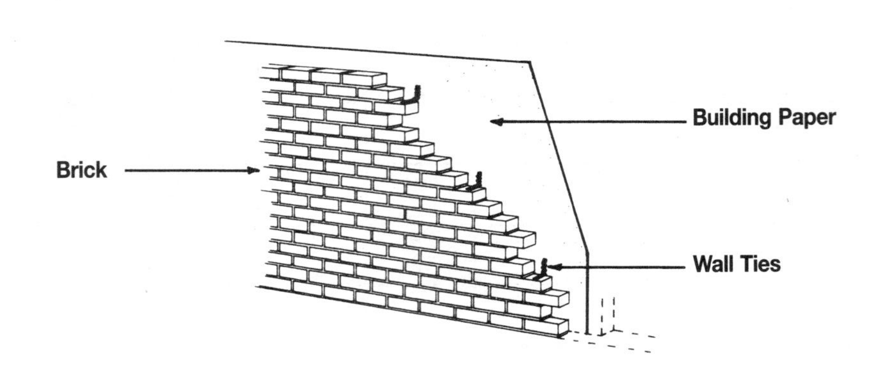

System Description	QUAN.	UNIT	LABOR HOURS	COST PER S.F.		
				MAT.	INST.	TOTAL
SELECT COMMON BRICK						
Brick, select common, running bond	1.000	S.F.	.174	3.03	6.25	9.28
Wall ties, 7/8" x 7", 22 gauge	1.000	Ea.	.008	.05	.31	.36
Building paper, #15 asphalt	1.100	S.F.	.002	.03	.09	.12
Trim, pine, painted	.125	L.F.	.004	.09	.16	.25
TOTAL		S.F.	.188	3.20	6.81	10.01
RED FACED COMMON BRICK						
Brick, common, red faced, running bond	1.000	S.F.	.182	3.03	6.50	9.53
Wall ties, 7/8" x 7", 22 gauge	1.000	Ea.	.008	.05	.31	.36
Building paper, #15 asphalt	1.100	S.F.	.002	.03	.09	.12
Trim, pine, painted	.125	L.F.	.004	.09	.16	.25
TOTAL		S.F.	.196	3.20	7.06	10.26
BUFF OR GREY FACE BRICK						
Brick, buff or grey	1.000	S.F.	.182	3.22	6.50	9.72
Wall ties, 7/8" x 7", 22 gauge	1.000	Ea.	.008	.05	.31	.36
Building paper, #15 asphalt	1.100	S.F.	.002	.03	.09	.12
Trim, pine, painted	.125	L.F.	.004	.09	.16	.25
TOTAL		S.F.	.196	3.39	7.06	10.45
STONE WORK, ROUGH STONE, AVERAGE						
Field stone veneer	1.000	S.F.	.223	4.82	8.01	12.83
Wall ties, 7/8" x 7", 22 gauge	1.000	Ea.	.008	.05	.31	.36
Building paper, #15 asphalt	1.000	S.F.	.002	.03	.09	.12
Trim, pine, painted	.125	L.F.	.004	.09	.16	.25
TOTAL		S.F.	.193	4.99	8.57	13.56

The costs in this system are based on a square foot of wall area. Do not subtract area for window & door openings.

Description	QUAN.	UNIT	LABOR HOURS	COST PER S.F.		
				MAT.	INST.	TOTAL

Brick/Stone Veneer Price Sheet	QUAN.	UNIT	LABOR HOURS	COST PER S.F.		
				MAT.	INST.	TOTAL
Brick						
Select common, running bond	1.000	S.F.	.174	3.03	6.25	9.28
Red faced, running bond	1.000	S.F.	.182	3.03	6.50	9.53
Buff or grey faced, running bond	1.000	S.F.	.182	3.22	6.50	9.72
Header every 6th course	1.000	S.F.	.216	3.53	7.75	11.28
English bond	1.000	S.F.	.286	4.53	10.25	14.78
Flemish bond	1.000	S.F.	.195	3.20	7	10.20
Common bond	1.000	S.F.	.267	4.03	9.55	13.58
Stack bond	1.000	S.F.	.182	3.22	6.50	9.72
Jumbo, running bond	1.000	S.F.	.092	3.97	3.30	7.27
Norman, running bond	1.000	S.F.	.125	4.22	4.48	8.70
Norwegian, running bond	1.000	S.F.	.107	3.24	3.82	7.06
Economy, running bond	1.000	S.F.	.129	3.73	4.62	8.35
Engineer, running bond	1.000	S.F.	.154	3.15	5.50	8.65
Roman, running bond	1.000	S.F.	.160	5.10	5.75	10.85
Utility, running bond	1.000	S.F.	.089	3.64	3.19	6.83
Glazed, running bond	1.000	S.F.	.190	8.75	6.85	15.60
Stone work, rough stone, average	1.000	S.F.	.179	4.82	8	12.82
Maximum	1.000	S.F.	.267	7.20	11.95	19.15
Wall ties, galvanized, corrugated 7/8" x 7", 22 gauge	1.000	Ea.	.008	.05	.31	.36
16 gauge	1.000	Ea.	.008	.15	.31	.46
Cavity wall, every 3rd course 6" long Z type, 1/4" diameter	1.330	L.F.	.010	.29	.40	.69
3/16" diameter	1.330	L.F.	.010	.20	.40	.60
8" long, Z type, 1/4" diameter	1.330	L.F.	.010	.34	.40	.74
3/16" diameter	1.330	L.F.	.010	.16	.40	.56
Building paper, aluminum and kraft laminated foil, 1 side	1.000	S.F.	.002	.04	.08	.12
2 sides	1.000	S.F.	.002	.07	.08	.15
#15 asphalt paper	1.100	S.F.	.002	.03	.09	.12
Polyethylene, .002" thick	1.000	S.F.	.002	.01	.08	.09
.004" thick	1.000	S.F.	.002	.02	.08	.10
.006" thick	1.000	S.F.	.002	.03	.08	.11
.010" thick	1.000	S.F.	.002	.06	.08	.14
Trim, 1" x 4", cedar	.125	L.F.	.005	.21	.20	.41
Fir	.125	L.F.	.005	.09	.20	.29
Redwood	.125	L.F.	.005	.21	.20	.41
White pine	.125	L.F.	.005	.09	.20	.29

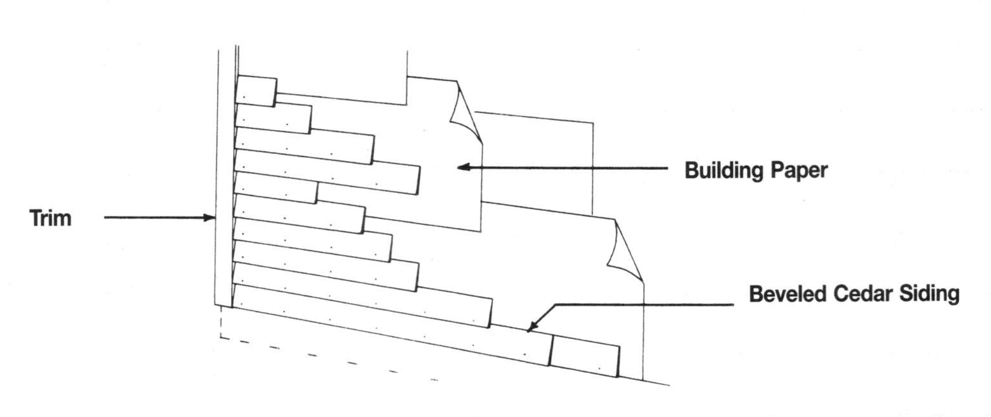

System Description	QUAN.	UNIT	LABOR HOURS	COST PER S.F.		
				MAT.	INST.	TOTAL
1/2" X 6" BEVELED CEDAR SIDING, "A" GRADE						
1/2" x 6" beveled cedar siding	1.000	S.F.	.032	2.10	1.25	3.35
#15 asphalt felt paper	1.100	S.F.	.002	.03	.09	.12
Trim, cedar	.125	L.F.	.005	.21	.20	.41
Paint, primer & 2 coats	1.000	S.F.	.017	.18	.59	.77
TOTAL		S.F.	.056	2.52	2.13	4.65
1/2" X 8" BEVELED CEDAR SIDING, "A" GRADE						
1/2" x 8" beveled cedar siding	1.000	S.F.	.029	2.71	1.14	3.85
#15 asphalt felt paper	1.100	S.F.	.002	.03	.09	.12
Trim, cedar	.125	L.F.	.005	.21	.20	.41
Paint, primer & 2 coats	1.000	S.F.	.017	.18	.59	.77
TOTAL		S.F.	.053	3.13	2.02	5.15
1" X 4" TONGUE & GROOVE, REDWOOD, VERTICAL GRAIN						
Redwood, clear, vertical grain, 1" x 10"	1.000	S.F.	.018	3.62	.72	4.34
#15 asphalt felt paper	1.100	S.F.	.002	.03	.09	.12
Trim, redwood	.125	L.F.	.005	.21	.20	.41
Sealer, 1 coat, stain, 1 coat	1.000	S.F.	.013	.12	.46	.58
TOTAL		S.F.	.053	3.98	1.47	5.45
1" X 6" TONGUE & GROOVE, REDWOOD, VERTICAL GRAIN						
Redwood, clear, vertical grain, 1" x 10"	1.000	S.F.	.019	3.73	.74	4.47
#15 asphalt felt paper	1.100	S.F.	.002	.03	.09	.12
Trim, redwood	.125	L.F.	.005	.21	.20	.41
Sealer, 1 coat, stain, 1 coat	1.000	S.F.	.013	.12	.46	.58
TOTAL		S.F.	.044	4.09	1.49	5.58

The costs in this system are based on a square foot of wall area.
Do not subtract area for door or window openings.

Description	QUAN.	UNIT	LABOR HOURS	COST PER S.F.		
				MAT.	INST.	TOTAL

Important: See the Reference Section for critical supporting data - Reference Nos., Crews & Location Factors

Wood Siding Price Sheet	QUAN.	UNIT	LABOR HOURS	COST PER S.F.		
				MAT.	INST.	TOTAL
Siding, beveled cedar, "A" grade, 1/2" x 6"	1.000	S.F.	.028	2.10	1.25	3.35
1/2" x 8"	1.000	S.F.	.023	2.71	1.14	3.85
"B" grade, 1/2" x 6"	1.000	S.F.	.032	2.33	1.39	3.72
1/2" x 8"	1.000	S.F.	.029	3.01	1.27	4.28
Clear grade, 1/2" x 6"	1.000	S.F.	.028	2.63	1.56	4.19
1/2" x 8"	1.000	S.F.	.023	3.39	1.43	4.82
Redwood, clear vertical grain, 1/2" x 6"	1.000	S.F.	.036	2.88	1.39	4.27
1/2" x 8"	1.000	S.F.	.032	2.33	1.25	3.58
Clear all heart vertical grain, 1/2" x 6"	1.000	S.F.	.028	3.20	1.54	4.74
1/2" x 8"	1.000	S.F.	.023	2.59	1.39	3.98
Siding board & batten, cedar, "B" grade, 1" x 10"	1.000	S.F.	.031	2.22	1.21	3.43
1" x 12"	1.000	S.F.	.031	2.22	1.21	3.43
Redwood, clear vertical grain, 1" x 6"	1.000	S.F.	.043	2.76	1.90	4.66
1" x 8"	1.000	S.F.	.018	2.52	1.67	4.19
White pine, #2 & better, 1" x 10"	1.000	S.F.	.029	.74	1.14	1.88
1" x 12"	1.000	S.F.	.029	.74	1.14	1.88
Siding vertical, tongue & groove, cedar "B" grade, 1" x 4"	1.000	S.F.	.033	2	.72	2.72
1" x 6"	1.000	S.F.	.024	2.06	.74	2.80
1" x 8"	1.000	S.F.	.024	2.12	.76	2.88
1" x 10"	1.000	S.F.	.021	2.18	.78	2.96
"A" grade, 1" x 4"	1.000	S.F.	.033	1.83	.66	2.49
1" x 6"	1.000	S.F.	.024	1.88	.67	2.55
1" x 8"	1.000	S.F.	.024	1.93	.69	2.62
1" x 10"	1.000	S.F.	.021	1.98	.71	2.69
Clear vertical grain, 1" x 4"	1.000	S.F.	.033	1.69	.60	2.29
1" x 6"	1.000	S.F.	.024	1.73	.62	2.35
1" x 8"	1.000	S.F.	.024	1.77	.63	2.40
1" x 10"	1.000	S.F.	.021	1.82	.65	2.47
Redwood, clear vertical grain, 1" x 4"	1.000	S.F.	.033	3.62	.72	4.34
1" x 6"	1.000	S.F.	.024	3.73	.74	4.47
1" x 8"	1.000	S.F.	.024	3.83	.76	4.59
1" x 10"	1.000	S.F.	.021	3.95	.78	4.73
Clear all heart vertical grain, 1" x 4"	1.000	S.F.	.033	3.32	.66	3.98
1" x 6"	1.000	S.F.	.024	3.41	.67	4.08
1" x 8"	1.000	S.F.	.024	3.50	.69	4.19
1" x 10"	1.000	S.F.	.021	3.59	.71	4.30
White pine, 1" x 10"	1.000	S.F.	.024	.77	.78	1.55
Siding plywood, texture 1-11 cedar, 3/8" thick	1.000	S.F.	.024	1.19	.93	2.12
5/8" thick	1.000	S.F.	.024	2.55	.93	3.48
Redwood, 3/8" thick	1.000	S.F.	.024	1.19	.93	2.12
5/8" thick	1.000	S.F.	.024	1.98	.93	2.91
Fir, 3/8" thick	1.000	S.F.	.024	.64	.93	1.57
5/8" thick	1.000	S.F.	.024	1.02	.93	1.95
Southern yellow pine, 3/8" thick	1.000	S.F.	.024	.64	.93	1.57
5/8" thick	1.000	S.F.	.024	.89	.93	1.82
Hard board, 7/16" thick primed, plain finish	1.000	S.F.	.025	1.14	.97	2.11
Board finish	1.000	S.F.	.023	.80	.90	1.70
Polyvinyl coated, 3/8" thick	1.000	S.F.	.021	.97	.84	1.81
5/8" thick	1.000	S.F.	.024	.89	.93	1.82
Paper, #15 asphalt felt	1.100	S.F.	.002	.03	.09	.12
Trim, cedar	.125	L.F.	.005	.21	.20	.41
Fir	.125	L.F.	.005	.09	.20	.29
Redwood	.125	L.F.	.005	.21	.20	.41
White pine	.125	L.F.	.005	.09	.20	.29
Painting, primer, & 1 coat	1.000	S.F.	.013	.12	.46	.58
2 coats	1.000	S.F.	.017	.18	.59	.77
Stain, sealer, & 1 coat	1.000	S.F.	.017	.09	.60	.69
2 coats	1.000	S.F.	.019	.15	.65	.80

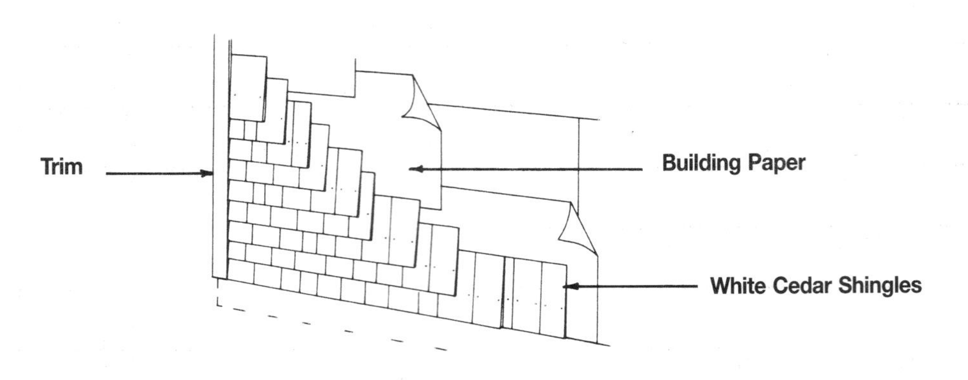

System Description	QUAN.	UNIT	LABOR HOURS	COST PER S.F.		
				MAT.	INST.	TOTAL
WHITE CEDAR SHINGLES, 5″ EXPOSURE						
White cedar shingles, 16″ long, grade "A", 5″ exposure	1.000	S.F.	.033	1.30	1.31	2.61
#15 asphalt felt paper	1.100	S.F.	.002	.03	.09	.12
Trim, cedar	.125	S.F.	.005	.21	.20	.41
Paint, primer & 1 coat	1.000	S.F.	.017	.09	.60	.69
TOTAL		S.F.	.057	1.63	2.20	3.83
NO. 1 PERFECTIONS, 5-1/2″ EXPOSURE						
No. 1 perfections, red cedar, 5-1/2″ exposure	1.000	S.F.	.029	1.76	1.14	2.90
#15 asphalt felt paper	1.100	S.F.	.002	.03	.09	.12
Trim, cedar	.125	S.F.	.005	.21	.20	.41
Stain, sealer & 1 coat	1.000	S.F.	.017	.09	.60	.69
TOTAL		S.F.	.053	2.09	2.03	4.12
RESQUARED & REBUTTED PERFECTIONS, 5-1/2″ EXPOSURE						
Resquared & rebutted perfections, 5-1/2″ exposure	1.000	S.F.	.027	2.19	1.05	3.24
#15 asphalt felt paper	1.100	S.F.	.002	.03	.09	.12
Trim, cedar	.125	S.F.	.005	.21	.20	.41
Stain, sealer & 1 coat	1.000	S.F.	.017	.09	.60	.69
TOTAL		S.F.	.051	2.52	1.94	4.46
HAND-SPLIT SHAKES, 8-1/2″ EXPOSURE						
Hand-split red cedar shakes, 18″ long, 8-1/2″ exposure	1.000	S.F.	.040	1.07	1.57	2.64
#15 asphalt felt paper	1.100	S.F.	.002	.03	.09	.12
Trim, cedar	.125	S.F.	.005	.21	.20	.41
Stain, sealer & 1 coat	1.000	S.F.	.017	.09	.60	.69
TOTAL		S.F.	.064	1.40	2.46	3.86

The costs in this system are based on a square foot of wall area.
Do not subtract area for door or window openings.

Description	QUAN.	UNIT	LABOR HOURS	COST PER S.F.		
				MAT.	INST.	TOTAL

Shingle Siding Price Sheet	QUAN.	UNIT	LABOR HOURS	COST PER S.F.		
				MAT.	INST.	TOTAL
Shingles wood, white cedar 16" long, "A" grade, 5" exposure	1.000	S.F.	.033	1.30	1.31	2.61
7" exposure	1.000	S.F.	.030	1.17	1.18	2.35
8-1/2" exposure	1.000	S.F.	.032	.74	1.26	2
10" exposure	1.000	S.F.	.028	.65	1.10	1.75
"B" grade, 5" exposure	1.000	S.F.	.040	1.21	1.57	2.78
7" exposure	1.000	S.F.	.028	.85	1.10	1.95
8-1/2" exposure	1.000	S.F.	.024	.73	.94	1.67
10" exposure	1.000	S.F.	.020	.61	.79	1.40
Fire retardant, "A" grade, 5" exposure	1.000	S.F.	.033	1.63	1.31	2.94
7" exposure	1.000	S.F.	.028	1.08	1.10	2.18
8-1/2" exposure	1.000	S.F.	.032	1.16	1.25	2.41
10" exposure	1.000	S.F.	.025	.90	.97	1.87
Fire retardant, 5" exposure	1.000	S.F.	.029	2.08	1.14	3.22
7" exposure	1.000	S.F.	.036	1.61	1.39	3
8-1/2" exposure	1.000	S.F.	.032	1.44	1.25	2.69
10" exposure	1.000	S.F.	.025	1.12	.97	2.09
Resquared & rebutted, 5-1/2" exposure	1.000	S.F.	.027	2.19	1.05	3.24
7" exposure	1.000	S.F.	.024	1.97	.95	2.92
8-1/2" exposure	1.000	S.F.	.021	1.75	.84	2.59
10" exposure	1.000	S.F.	.019	1.53	.74	2.27
Fire retardant, 5" exposure	1.000	S.F.	.027	2.51	1.05	3.56
7" exposure	1.000	S.F.	.024	2.25	.95	3.20
8-1/2" exposure	1.000	S.F.	.021	2	.84	2.84
10" exposure	1.000	S.F.	.023	1.35	.90	2.25
Hand-split, red cedar, 24" long, 7" exposure	1.000	S.F.	.045	2.13	1.75	3.88
8-1/2" exposure	1.000	S.F.	.038	1.82	1.50	3.32
10" exposure	1.000	S.F.	.032	1.52	1.25	2.77
12" exposure	1.000	S.F.	.026	1.22	1	2.22
Fire retardant, 7" exposure	1.000	S.F.	.045	2.59	1.75	4.34
8-1/2" exposure	1.000	S.F.	.038	2.22	1.50	3.72
10" exposure	1.000	S.F.	.032	1.85	1.25	3.10
12" exposure	1.000	S.F.	.026	1.48	1	2.48
18" long, 5" exposure	1.000	S.F.	.068	1.82	2.67	4.49
7" exposure	1.000	S.F.	.048	1.28	1.88	3.16
8-1/2" exposure	1.000	S.F.	.040	1.07	1.57	2.64
10" exposure	1.000	S.F.	.036	.96	1.41	2.37
Fire retardant, 5" exposure	1.000	S.F.	.068	2.38	2.67	5.05
7" exposure	1.000	S.F.	.048	1.68	1.88	3.56
8-1/2" exposure	1.000	S.F.	.040	1.40	1.57	2.97
10" exposure	1.000	S.F.	.036	1.26	1.41	2.67
Paper, #15 asphalt felt	1.100	S.F.	.002	.03	.08	.11
Trim, cedar	.125	S.F.	.005	.21	.20	.41
Fir	.125	S.F.	.005	.09	.20	.29
Redwood	.125	S.F.	.005	.21	.20	.41
White pine	.125	S.F.	.005	.09	.20	.29
Painting, primer, & 1 coat	1.000	S.F.	.013	.12	.46	.58
2 coats	1.000	S.F.	.017	.18	.59	.77
Staining, sealer, & 1 coat	1.000	S.F.	.017	.09	.60	.69
2 coats	1.000	S.F.	.019	.15	.65	.80

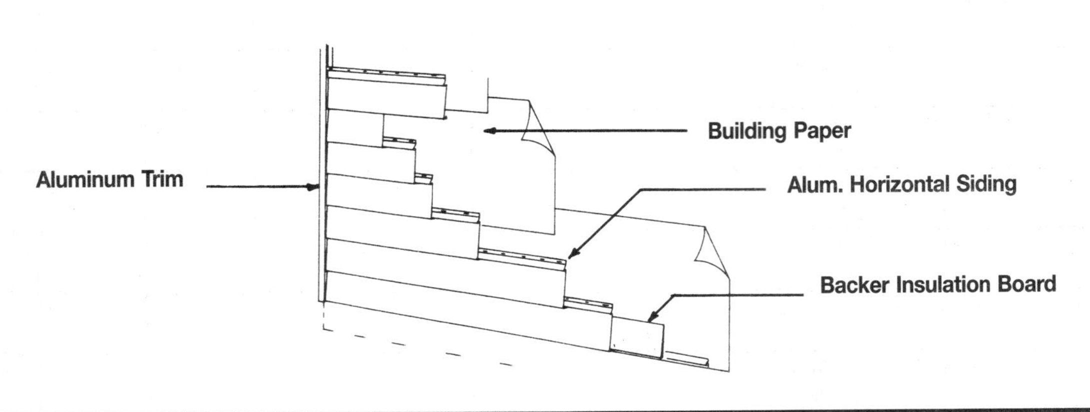

System Description	QUAN.	UNIT	LABOR HOURS	COST PER S.F.		
				MAT.	INST.	TOTAL
ALUMINUM CLAPBOARD SIDING, 8″ WIDE, WHITE						
Aluminum horizontal siding, 8″ clapboard	1.000	S.F.	.031	1.38	1.22	2.60
Backer, insulation board	1.000	S.F.	.008	.45	.31	.76
Trim, aluminum	.600	L.F.	.016	.62	.62	1.24
Paper, #15 asphalt felt	1.100	S.F.	.002	.03	.09	.12
TOTAL		S.F.	.057	2.48	2.24	4.72
ALUMINUM VERTICAL BOARD & BATTEN, WHITE						
Aluminum vertical board & batten	1.000	S.F.	.027	1.54	1.06	2.60
Backer insulation board	1.000	S.F.	.008	.45	.31	.76
Trim, aluminum	.600	L.F.	.016	.62	.62	1.24
Paper, #15 asphalt felt	1.100	S.F.	.002	.03	.09	.12
TOTAL		S.F.	.053	2.64	2.08	4.72
VINYL CLAPBOARD SIDING, 8″ WIDE, WHITE						
PVC vinyl horizontal siding, 8″ clapboard	1.000	S.F.	.032	.68	1.27	1.95
Backer, insulation board	1.000	S.F.	.008	.45	.31	.76
Trim, vinyl	.600	L.F.	.014	.45	.55	1
Paper, #15 asphalt felt	1.100	S.F.	.002	.03	.09	.12
TOTAL		S.F.	.056	1.61	2.22	3.83
VINYL VERTICAL BOARD & BATTEN, WHITE						
PVC vinyl vertical board & batten	1.000	S.F.	.029	1.51	1.14	2.65
Backer, insulation board	1.000	S.F.	.008	.45	.31	.76
Trim, vinyl	.600	L.F.	.014	.45	.55	1
Paper, #15 asphalt felt	1.100	S.F.	.002	.03	.09	.12
TOTAL		S.F.	.053	2.44	2.09	4.53

The costs in this system are on a square foot of wall basis.
subtract openings from wall area.

Description	QUAN.	UNIT	LABOR HOURS	COST PER S.F.		
				MAT.	INST.	TOTAL

 Important: See the Reference Section for critical supporting data - Reference Nos., Crews & Location Factors

Metal & Plastic Siding Price Sheet	QUAN.	UNIT	LABOR HOURS	COST PER S.F.		
				MAT.	INST.	TOTAL
Siding, aluminum, .024″ thick, smooth, 8″ wide, white	1.000	S.F.	.031	1.38	1.22	2.60
Color	1.000	S.F.	.031	1.47	1.22	2.69
Double 4″ pattern, 8″ wide, white	1.000	S.F.	.031	1.31	1.22	2.53
Color	1.000	S.F.	.031	1.40	1.22	2.62
Double 5″ pattern, 10″ wide, white	1.000	S.F.	.029	1.31	1.14	2.45
Color	1.000	S.F.	.029	1.40	1.14	2.54
Embossed, single, 8″ wide, white	1.000	S.F.	.031	1.63	1.22	2.85
Color	1.000	S.F.	.031	1.72	1.22	2.94
Double 4″ pattern, 8″ wide, white	1.000	S.F.	.031	1.49	1.22	2.71
Color	1.000	S.F.	.031	1.58	1.22	2.80
Double 5″ pattern, 10″ wide, white	1.000	S.F.	.029	1.49	1.14	2.63
Color	1.000	S.F.	.029	1.58	1.14	2.72
Alum siding with insulation board, smooth, 8″ wide, white	1.000	S.F.	.031	1.32	1.22	2.54
Color	1.000	S.F.	.031	1.41	1.22	2.63
Double 4″ pattern, 8″ wide, white	1.000	S.F.	.031	1.30	1.22	2.52
Color	1.000	S.F.	.031	1.39	1.22	2.61
Double 5″ pattern, 10″ wide, white	1.000	S.F.	.029	1.30	1.14	2.44
Color	1.000	S.F.	.029	1.39	1.14	2.53
Embossed, single, 8″ wide, white	1.000	S.F.	.031	1.52	1.22	2.74
Color	1.000	S.F.	.031	1.61	1.22	2.83
Double 4″ pattern, 8″ wide, white	1.000	S.F.	.031	1.54	1.22	2.76
Color	1.000	S.F.	.031	1.63	1.22	2.85
Double 5″ pattern, 10″ wide, white	1.000	S.F.	.029	1.54	1.14	2.68
Color	1.000	S.F.	.029	1.63	1.14	2.77
Aluminum, shake finish, 10″ wide, white	1.000	S.F.	.029	1.63	1.14	2.77
Color	1.000	S.F.	.029	1.72	1.14	2.86
Aluminum, vertical, 12″ wide, white	1.000	S.F.	.027	1.54	1.06	2.60
Color	1.000	S.F.	.027	1.63	1.06	2.69
Vinyl siding, 8″ wide, smooth, white	1.000	S.F.	.032	.68	1.27	1.95
Color	1.000	S.F.	.032	.77	1.27	2.04
10″ wide, Dutch lap, smooth, white	1.000	S.F.	.029	.70	1.14	1.84
Color	1.000	S.F.	.029	.79	1.14	1.93
Double 4″ pattern, 8″ wide, white	1.000	S.F.	.032	.61	1.27	1.88
Color	1.000	S.F.	.032	.70	1.27	1.97
Double 5″ pattern, 10″ wide, white	1.000	S.F.	.029	.62	1.14	1.76
Color	1.000	S.F.	.029	.71	1.14	1.85
Embossed, single, 8″ wide, white	1.000	S.F.	.032	.73	1.27	2
Color	1.000	S.F.	.032	.82	1.27	2.09
10″ wide, white	1.000	S.F.	.029	.74	1.14	1.88
Color	1.000	S.F.	.029	.83	1.14	1.97
Double 4″ pattern, 8″ wide, white	1.000	S.F.	.032	.65	1.27	1.92
Color	1.000	S.F.	.032	.74	1.27	2.01
Double 5″ pattern, 10″ wide, white	1.000	S.F.	.029	.67	1.14	1.81
Color	1.000	S.F.	.029	.76	1.14	1.90
Vinyl, shake finish, 10″ wide, white	1.000	S.F.	.029	2.09	1.14	3.23
Color	1.000	S.F.	.029	2.18	1.14	3.32
Vinyl, vertical, double 5″ pattern, 10″ wide, white	1.000	S.F.	.029	1.51	1.14	2.65
Color	1.000	S.F.	.029	1.60	1.14	2.74
Backer board, installed in siding panels 8″ or 10″ wide	1.000	S.F.	.008	.45	.31	.76
4′ x 8′ sheets, polystyrene, 3/4″ thick	1.000	S.F.	.010	.40	.39	.79
4′ x 8′ fiberboard, plain	1.000	S.F.	.008	.45	.31	.76
Trim, aluminum, white	.600	L.F.	.016	.62	.62	1.24
Color	.600	L.F.	.016	.67	.62	1.29
Vinyl, white	.600	L.F.	.014	.45	.55	1
Color	.600	L.F.	.014	.44	.55	.99
Paper, #15 asphalt felt	1.100	S.F.	.002	.03	.09	.12
Kraft paper, plain	1.100	S.F.	.002	.04	.09	.13
Foil backed	1.100	S.F.	.002	.08	.09	.17

Description	QUAN.	UNIT	LABOR HOURS	COST PER S.F. MAT.	INST.	TOTAL
Poured insulation, cellulose fiber, R3.8 per inch (1″ thick)	1.000	S.F.	.003	.04	.13	.17
Fiberglass , R4.0 per inch (1″ thick)	1.000	S.F.	.003	.03	.13	.16
Mineral wool, R3.0 per inch (1″ thick)	1.000	S.F.	.003	.03	.13	.16
Polystyrene, R4.0 per inch (1″ thick)	1.000	S.F.	.003	.20	.13	.33
Vermiculite, R2.7 per inch (1″ thick)	1.000	S.F.	.003	.15	.13	.28
Perlite, R2.7 per inch (1″ thick)	1.000	S.F.	.003	.15	.13	.28
Reflective insulation, aluminum foil reinforced with scrim	1.000	S.F.	.004	.15	.17	.32
Reinforced with woven polyolefin	1.000	S.F.	.004	.19	.17	.36
With single bubble air space, R8.8	1.000	S.F.	.005	.30	.21	.51
With double bubble air space, R9.8	1.000	S.F.	.005	.32	.21	.53
Rigid insulation, fiberglass, unfaced,						
1-1/2″ thick, R6.2	1.000	S.F.	.008	.45	.31	.76
2″ thick, R8.3	1.000	S.F.	.008	.51	.31	.82
2-1/2″ thick, R10.3	1.000	S.F.	.010	.63	.39	1.02
3″ thick, R12.4	1.000	S.F.	.010	.63	.39	1.02
Foil faced, 1″ thick, R4.3	1.000	S.F.	.008	.88	.31	1.19
1-1/2″ thick, R6.2	1.000	S.F.	.008	1.19	.31	1.50
2″ thick, R8.7	1.000	S.F.	.009	1.49	.35	1.84
2-1/2″ thick, R10.9	1.000	S.F.	.010	1.76	.39	2.15
3″ thick, R13.0	1.000	S.F.	.010	1.91	.39	2.30
Foam glass, 1-1/2″ thick R2.64	1.000	S.F.	.010	1.43	.39	1.82
2″ thick R5.26	1.000	S.F.	.011	3.22	.43	3.65
Perlite, 1″ thick R2.77	1.000	S.F.	.010	.29	.39	.68
2″ thick R5.55	1.000	S.F.	.011	.55	.43	.98
Polystyrene, extruded, blue, 2.2#/C.F., 3/4″ thick R4	1.000	S.F.	.010	.40	.39	.79
1-1/2″ thick R8.1	1.000	S.F.	.011	.78	.43	1.21
2″ thick R10.8	1.000	S.F.	.011	1.08	.43	1.51
Molded bead board, white, 1″ thick R3.85	1.000	S.F.	.010	.18	.39	.57
1-1/2″ thick, R5.6	1.000	S.F.	.011	.44	.43	.87
2″ thick, R7.7	1.000	S.F.	.011	.57	.43	1
Non-rigid insulation, batts						
Fiberglass, kraft faced, 3-1/2″ thick, R11, 11″ wide	1.000	S.F.	.005	.25	.20	.45
15″ wide	1.000	S.F.	.005	.25	.20	.45
23″ wide	1.000	S.F.	.005	.25	.20	.45
6″ thick, R19, 11″ wide	1.000	S.F.	.006	.34	.23	.57
15″ wide	1.000	S.F.	.006	.34	.23	.57
23″ wide	1.000	S.F.	.006	.34	.23	.57
9″ thick, R30, 15″ wide	1.000	S.F.	.006	.66	.23	.89
23″ wide	1.000	S.F.	.006	.66	.23	.89
12″ thick, R38, 15″ wide	1.000	S.F.	.006	.84	.23	1.07
23″ wide	1.000	S.F.	.006	.84	.23	1.07
Fiberglass, foil faced, 3-1/2″ thick, R11, 15″ wide	1.000	S.F.	.005	.37	.20	.57
23″ wide	1.000	S.F.	.005	.37	.20	.57
6″ thick, R19, 15″ thick	1.000	S.F.	.005	.45	.20	.65
23″ wide	1.000	S.F.	.005	.45	.20	.65
9″ thick, R30, 15″ wide	1.000	S.F.	.006	.78	.23	1.01
23″ wide	1.000	S.F.	.006	.78	.23	1.01

Important: See the Reference Section for critical supporting data - Reference Nos., Crews & Location Factors

Insulation Systems	QUAN.	UNIT	LABOR HOURS	COST PER S.F.		
				MAT.	INST.	TOTAL
Non-rigid insulation batts						
Fiberglass unfaced, 3-1/2" thick, R11, 15" wide	1.000	S.F.	.005	.23	.20	.43
23" wide	1.000	S.F.	.005	.23	.20	.43
6" thick, R19, 15" wide	1.000	S.F.	.006	.37	.23	.60
23" wide	1.000	S.F.	.006	.37	.23	.60
9" thick, R19, 15" wide	1.000	S.F.	.007	.66	.27	.93
23" wide	1.000	S.F.	.007	.66	.27	.93
12" thick, R38, 15" wide	1.000	S.F.	.007	.84	.27	1.11
23" wide	1.000	S.F.	.007	.84	.27	1.11
Mineral fiber batts, 3" thick, R11	1.000	S.F.	.005	.29	.20	.49
3-1/2" thick, R13	1.000	S.F.	.005	.29	.20	.49
6" thick, R19	1.000	S.F.	.005	.37	.20	.57
6-1/2" thick, R22	1.000	S.F.	.005	.37	.20	.57
10" thick, R30	1.000	S.F.	.006	.57	.23	.80

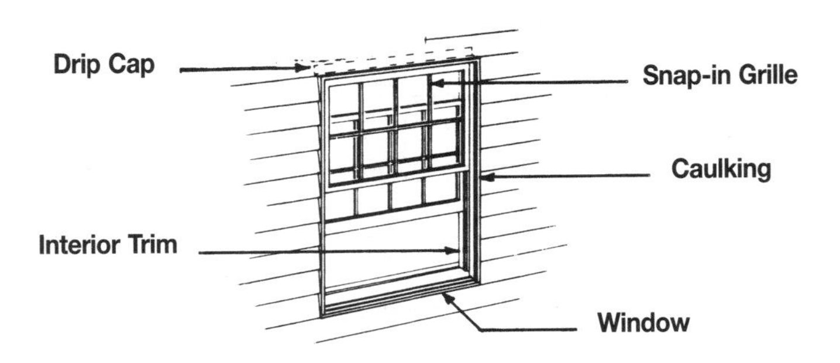

System Description	QUAN.	UNIT	LABOR HOURS	COST EACH		
				MAT.	INST.	TOTAL
BUILDER'S QUALITY WOOD WINDOW 2′ X 3′, DOUBLE HUNG						
Window, primed, builder's quality, 2′ x 3′, insulating glass	1.000	Ea.	.800	202	31.50	233.50
Trim, interior casing	11.000	L.F.	.367	10.34	14.41	24.75
Paint, interior & exterior, primer & 2 coats	2.000	Face	1.778	2.06	61	63.06
Caulking	10.000	L.F.	.323	1.60	12.90	14.50
Snap-in grille	1.000	Set	.333	44.50	13.05	57.55
Drip cap, metal	2.000	L.F.	.040	.56	1.56	2.12
TOTAL		Ea.	3.641	261.06	134.42	395.48
PLASTIC CLAD WOOD WINDOW 3′ X 4′, DOUBLE HUNG						
Window, plastic clad, premium, 3′ x 4′, insulating glass	1.000	Ea.	.889	280	35	315
Trim, interior casing	15.000	L.F.	.500	14.10	19.65	33.75
Paint, interior, primer & 2 coats	1.000	Face	.889	1.03	30.50	31.53
Caulking	14.000	L.F.	.452	2.24	18.06	20.30
Snap-in grille	1.000	Set	.333	44.50	13.05	57.55
TOTAL		Ea.	3.063	341.87	116.26	458.13
METAL CLAD WOOD WINDOW, 3′ X 5′, DOUBLE HUNG						
Window, metal clad, deluxe, 3′ x 5′, insulating glass	1.000	Ea.	1.000	297	39	336
Trim, interior casing	17.000	L.F.	.567	15.98	22.27	38.25
Paint, interior, primer & 2 coats	1.000	Face	.889	1.03	30.50	31.53
Caulking	16.000	L.F.	.516	2.56	20.64	23.20
Snap-in grille	1.000	Set	.235	124	9.20	133.20
Drip cap, metal	3.000	L.F.	.060	.84	2.34	3.18
TOTAL		Ea.	3.267	441.41	123.95	565.36

The cost of this system is on a cost per each window basis.

Description	QUAN.	UNIT	LABOR HOURS	COST EACH		
				MAT.	INST.	TOTAL

Important: See the Reference Section for critical supporting data - Reference Nos., Crews & Location Factors

Double Hung Window Price Sheet	QUAN.	UNIT	LABOR HOURS	COST EACH		
				MAT.	INST.	TOTAL
Windows, double-hung, builder's quality, 2′ x 3′, single glass	1.000	Ea.	.800	202	31.50	233.50
Insulating glass	1.000	Ea.	.800	202	31.50	233.50
3′ x 4′, single glass	1.000	Ea.	.889	269	35	304
Insulating glass	1.000	Ea.	.889	278	35	313
4′ x 4′-6″, single glass	1.000	Ea.	1.000	305	39	344
Insulating glass	1.000	Ea.	1.000	330	39	369
Plastic clad premium insulating glass, 2′-6″ x 3′	1.000	Ea.	.800	215	31.50	246.50
3′ x 3′-6″	1.000	Ea.	.800	253	31.50	284.50
3′ x 4′	1.000	Ea.	.889	280	35	315
3′ x 4′-6″	1.000	Ea.	.889	292	35	327
3′ x 5′	1.000	Ea.	1.000	305	39	344
3′-6″ x 6′	1.000	Ea.	1.000	355	39	394
Metal clad deluxe insulating glass, 2′-6″ x 3′	1.000	Ea.	.800	204	31.50	235.50
3′ x 3′-6″	1.000	Ea.	.800	242	31.50	273.50
3′ x 4′	1.000	Ea.	.889	257	35	292
3′ x 4′-6″	1.000	Ea.	.889	278	35	313
3′ x 5′	1.000	Ea.	1.000	297	39	336
3′-6″ x 6′	1.000	Ea.	1.000	360	39	399
Trim, interior casing, window 2′ x 3′	11.000	L.F.	.367	10.35	14.40	24.75
2′-6″ x 3′	12.000	L.F.	.400	11.30	15.70	27
3′ x 3′-6″	14.000	L.F.	.467	13.15	18.35	31.50
3′ x 4′	15.000	L.F.	.500	14.10	19.65	33.75
3′ x 4′-6″	16.000	L.F.	.533	15.05	21	36.05
3′ x 5′	17.000	L.F.	.567	16	22.50	38.50
3′-6″ x 6′	20.000	L.F.	.667	18.80	26	44.80
4′ x 4′-6″	18.000	L.F.	.600	16.90	23.50	40.40
Paint or stain, interior or exterior, 2′ x 3′ window, 1 coat	1.000	Face	.444	.38	15.35	15.73
2 coats	1.000	Face	.727	.76	25	25.76
Primer & 1 coat	1.000	Face	.727	.69	25	25.69
Primer & 2 coats	1.000	Face	.889	1.03	30.50	31.53
3′ x 4′ window, 1 coat	1.000	Face	.667	.78	23	23.78
2 coats	1.000	Face	.667	.87	23	23.87
Primer & 1 coat	1.000	Face	.727	1.05	25	26.05
Primer & 2 coats	1.000	Face	.889	1.03	30.50	31.53
4′ x 4′-6″ window, 1 coat	1.000	Face	.667	.78	23	23.78
2 coats	1.000	Face	.667	.87	23	23.87
Primer & 1 coat	1.000	Face	.727	1.05	25	26.05
Primer & 2 coats	1.000	Face	.889	1.03	30.50	31.53
Caulking, window, 2′ x 3′	10.000	L.F.	.323	1.60	12.90	14.50
2′-6″ x 3′	11.000	L.F.	.355	1.76	14.20	15.96
3′ x 3′-6″	13.000	L.F.	.419	2.08	16.75	18.83
3′ x 4′	14.000	L.F.	.452	2.24	18.05	20.29
3′ x 4′-6″	15.000	L.F.	.484	2.40	19.35	21.75
3′ x 5′	16.000	L.F.	.516	2.56	20.50	23.06
3′-6″ x 6′	19.000	L.F.	.613	3.04	24.50	27.54
4′ x 4′-6″	17.000	L.F.	.548	2.72	22	24.72
Grilles, glass size to, 16″ x 24″ per sash	1.000	Set	.333	44.50	13.05	57.55
32″ x 32″ per sash	1.000	Set	.235	124	9.20	133.20
Drip cap, aluminum, 2′ long	2.000	L.F.	.040	.56	1.56	2.12
3′ long	3.000	L.F.	.060	.84	2.34	3.18
4′ long	4.000	L.F.	.080	1.12	3.12	4.24
Wood, 2′ long	2.000	L.F.	.067	1.88	2.62	4.50
3′ long	3.000	L.F.	.100	2.82	3.93	6.75
4′ long	4.000	L.F.	.133	3.76	5.25	9.01

32 Casement Window Systems

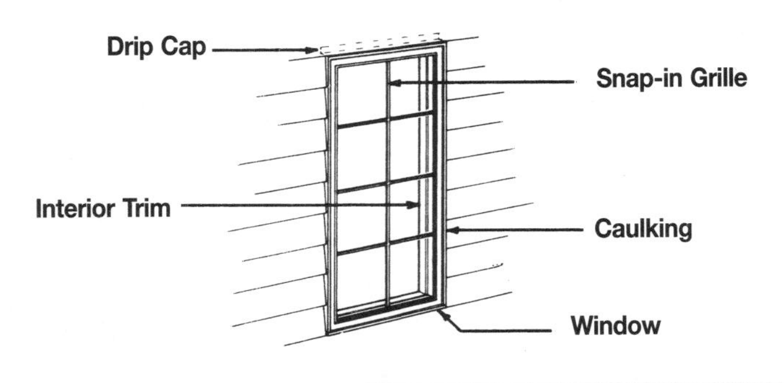

System Description	QUAN.	UNIT	LABOR HOURS	COST EACH		
				MAT.	INST.	TOTAL
BUILDER'S QUALITY WINDOW, WOOD, 2′ BY 3′, CASEMENT						
Window, primed, builder's quality, 2′ x 3′, insulating glass	1.000	Ea.	.800	257	31.50	288.50
Trim, interior casing	11.000	L.F.	.367	10.34	14.41	24.75
Paint, interior & exterior, primer & 2 coats	2.000	Face	1.778	2.06	61	63.06
Caulking	10.000	L.F.	.323	1.60	12.90	14.50
Snap-in grille	1.000	Ea.	.267	25	10.45	35.45
Drip cap, metal	2.000	L.F.	.040	.56	1.56	2.12
TOTAL		Ea.	3.575	296.56	131.82	428.38
PLASTIC CLAD WOOD WINDOW, 2′ X 4′, CASEMENT						
Window, plastic clad, premium, 2′ x 4′, insulating glass	1.000	Ea.	.889	305	35	340
Trim, interior casing	13.000	L.F.	.433	12.22	17.03	29.25
Paint, interior, primer & 2 coats	1.000	Ea.	.889	1.03	30.50	31.53
Caulking	12.000	L.F.	.387	1.92	15.48	17.40
Snap-in grille	1.000	Ea.	.267	25	10.45	35.45
TOTAL		Ea.	2.865	345.17	108.46	453.63
METAL CLAD WOOD WINDOW, 2′ X 5′, CASEMENT						
Window, metal clad, deluxe, 2′ x 5′, insulating glass	1.000	Ea.	1.000	263	39	302
Trim, interior casing	15.000	L.F.	.500	14.10	19.65	33.75
Paint, interior, primer & 2 coats	1.000	Ea.	.889	1.03	30.50	31.53
Caulking	14.000	L.F.	.452	2.24	18.06	20.30
Snap-in grille	1.000	Ea.	.250	36	9.80	45.80
Drip cap, metal	12.000	L.F.	.040	.56	1.56	2.12
TOTAL		Ea.	3.131	316.93	118.57	435.50

The cost of this system is on a cost per each window basis.

Description	QUAN.	UNIT	LABOR HOURS	COST EACH		
				MAT.	INST.	TOTAL

Important: See the Reference Section for critical supporting data - Reference Nos., Crews & Location Factors

Casement Window Price Sheet	QUAN.	UNIT	LABOR HOURS	COST EACH		
				MAT.	INST.	TOTAL
Window, casement, builders quality, 2′ x 3′, single glass	1.000	Ea.	.800	195	31.50	226.50
Insulating glass	1.000	Ea.	.800	257	31.50	288.50
2′ x 4′-6″, single glass	1.000	Ea.	.727	725	28.50	753.50
Insulating glass	1.000	Ea.	.727	680	28.50	708.50
2′ x 6′, single glass	1.000	Ea.	.889	1,050	35	1,085
Insulating glass	1.000	Ea.	.889	930	35	965
Plastic clad premium insulating glass, 2′ x 3′	1.000	Ea.	.800	187	31.50	218.50
2′ x 4′	1.000	Ea.	.889	210	35	245
2′ x 5′	1.000	Ea.	1.000	287	39	326
2′ x 6′	1.000	Ea.	1.000	360	39	399
Metal clad deluxe insulating glass, 2′ x 3′	1.000	Ea.	.800	193	31.50	224.50
2′ x 4′	1.000	Ea.	.889	232	35	267
2′ x 5′	1.000	Ea.	1.000	264	39	303
2′ x 6′	1.000	Ea.	1.000	305	39	344
Trim, interior casing, window 2′ x 3′	11.000	L.F.	.367	10.35	14.40	24.75
2′ x 4′	13.000	L.F.	.433	12.20	17.05	29.25
2′ x 4′-6″	14.000	L.F.	.467	13.15	18.35	31.50
2′ x 5′	15.000	L.F.	.500	14.10	19.65	33.75
2′ x 6′	17.000	L.F.	.567	16	22.50	38.50
Paint or stain, interior or exterior, 2′ x 3′ window, 1 coat	1.000	Face	.444	.38	15.35	15.73
2 coats	1.000	Face	.727	.76	25	25.76
Primer & 1 coat	1.000	Face	.727	.69	25	25.69
Primer & 2 coats	1.000	Face	.889	1.03	30.50	31.53
2′ x 4′ window, 1 coat	1.000	Face	.444	.38	15.35	15.73
2 coats	1.000	Face	.727	.76	25	25.76
Primer & 1 coat	1.000	Face	.727	.69	25	25.69
Primer & 2 coats	1.000	Face	.889	1.03	30.50	31.53
2′ x 6′ window, 1 coat	1.000	Face	.667	.78	23	23.78
2 coats	1.000	Face	.667	.87	23	23.87
Primer & 1 coat	1.000	Face	.727	1.05	25	26.05
Primer & 2 coats	1.000	Face	.889	1.03	30.50	31.53
Caulking, window, 2′ x 3′	10.000	L.F.	.323	1.60	12.90	14.50
2′ x 4′	12.000	L.F.	.387	1.92	15.50	17.42
2′ x 4′-6″	13.000	L.F.	.419	2.08	16.75	18.83
2′ x 5′	14.000	L.F.	.452	2.24	18.05	20.29
2′ x 6′	16.000	L.F.	.516	2.56	20.50	23.06
Grilles, glass size, to 20″ x 36″	1.000	Ea.	.267	25	10.45	35.45
To 20″ x 56″	1.000	Ea.	.250	36	9.80	45.80
Drip cap, metal, 2′ long	2.000	L.F.	.040	.56	1.56	2.12
Wood, 2′ long	2.000	L.F.	.067	1.88	2.62	4.50

36 Awning Window Systems

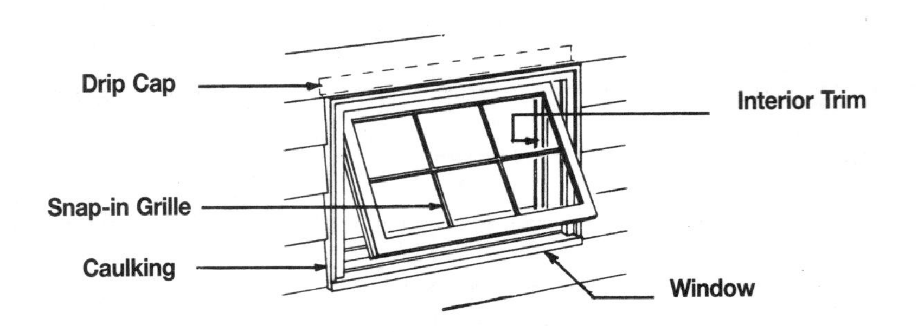

System Description	QUAN.	UNIT	LABOR HOURS	COST EACH		
				MAT.	INST.	TOTAL
BUILDER'S QUALITY WINDOW, WOOD, 34" X 22", AWNING						
Window, builder quality, 34" x 22", insulating glass	1.000	Ea.	.800	243	31.50	274.50
Trim, interior casing	10.500	L.F.	.350	9.87	13.76	23.63
Paint, interior & exterior, primer & 2 coats	2.000	Face	1.778	2.06	61	63.06
Caulking	9.500	L.F.	.306	1.52	12.26	13.78
Snap-in grille	1.000	Ea.	.267	20.50	10.45	30.95
Drip cap, metal	3.000	L.F.	.060	.84	2.34	3.18
TOTAL		Ea.	3.561	277.79	131.31	409.10
PLASTIC CLAD WOOD WINDOW, 40" X 28", AWNING						
Window, plastic clad, premium, 40" x 28", insulating glass	1.000	Ea.	.889	315	35	350
Trim interior casing	13.500	L.F.	.450	12.69	17.69	30.38
Paint, interior, primer & 2 coats	1.000	Face	.889	1.03	30.50	31.53
Caulking	12.500	L.F.	.403	2	16.13	18.13
Snap-in grille	1.000	Ea.	.267	20.50	10.45	30.95
TOTAL		Ea.	2.898	351.22	109.77	460.99
METAL CLAD WOOD WINDOW, 48" X 36", AWNING						
Window, metal clad, deluxe, 48" x 36", insulating glass	1.000	Ea.	1.000	340	39	379
Trim, interior casing	15.000	L.F.	.500	14.10	19.65	33.75
Paint, interior, primer & 2 coats	1.000	Face	.889	1.03	30.50	31.53
Caulking	14.000	L.F.	.452	2.24	18.06	20.30
Snap-in grille	1.000	Ea.	.250	29.50	9.80	39.30
Drip cap, metal	4.000	L.F.	.080	1.12	3.12	4.24
TOTAL		Ea.	3.171	387.99	120.13	508.12

The cost of this system is on a cost per each window basis.

Description	QUAN.	UNIT	LABOR HOURS	COST EACH		
				MAT.	INST.	TOTAL

Important: See the Reference Section for critical supporting data - Reference Nos., Crews & Location Factors

Awning Window Price Sheet	QUAN.	UNIT	LABOR HOURS	COST EACH		
				MAT.	INST.	TOTAL
Windows, awning, builder's quality, 34″ x 22″, insulated glass	1.000	Ea.	.800	213	31.50	244.50
Low E glass	1.000	Ea.	.800	243	31.50	274.50
40″ x 28″, insulated glass	1.000	Ea.	.889	290	35	325
Low E glass	1.000	Ea.	.889	310	35	345
48″ x 36″, insulated glass	1.000	Ea.	1.000	425	39	464
Low E glass	1.000	Ea.	1.000	445	39	484
Plastic clad premium insulating glass, 34″ x 22″	1.000	Ea.	.800	245	31.50	276.50
40″ x 22″	1.000	Ea.	.800	267	31.50	298.50
36″ x 28″	1.000	Ea.	.889	285	35	320
36″ x 36″	1.000	Ea.	.889	315	35	350
48″ x 28″	1.000	Ea.	1.000	340	39	379
60″ x 36″	1.000	Ea.	1.000	495	39	534
Metal clad deluxe insulating glass, 34″ x 22″	1.000	Ea.	.800	229	31.50	260.50
40″ x 22″	1.000	Ea.	.800	268	31.50	299.50
36″ x 25″	1.000	Ea.	.889	249	35	284
40″ x 30″	1.000	Ea.	.889	310	35	345
48″ x 28″	1.000	Ea.	1.000	320	39	359
60″ x 36″	1.000	Ea.	1.000	340	39	379
Trim, interior casing window, 34″ x 22″	10.500	L.F.	.350	9.85	13.75	23.60
40″ x 22″	11.500	L.F.	.383	10.80	15.05	25.85
36″ x 28″	12.500	L.F.	.417	11.75	16.40	28.15
40″ x 28″	13.500	L.F.	.450	12.70	17.70	30.40
48″ x 28″	14.500	L.F.	.483	13.65	19	32.65
48″ x 36″	15.000	L.F.	.500	14.10	19.65	33.75
Paint or stain, interior or exterior, 34″ x 22″, 1 coat	1.000	Face	.444	.38	15.35	15.73
2 coats	1.000	Face	.727	.76	25	25.76
Primer & 1 coat	1.000	Face	.727	.69	25	25.69
Primer & 2 coats	1.000	Face	.889	1.03	30.50	31.53
36″ x 28″, 1 coat	1.000	Face	.444	.38	15.35	15.73
2 coats	1.000	Face	.727	.76	25	25.76
Primer & 1 coat	1.000	Face	.727	.69	25	25.69
Primer & 2 coats	1.000	Face	.889	1.03	30.50	31.53
48″ x 36″, 1 coat	1.000	Face	.667	.78	23	23.78
2 coats	1.000	Face	.667	.87	23	23.87
Primer & 1 coat	1.000	Face	.727	1.05	25	26.05
Primer & 2 coats	1.000	Face	.889	1.03	30.50	31.53
Caulking, window, 34″ x 22″	9.500	L.F.	.306	1.52	12.25	13.77
40″ x 22″	10.500	L.F.	.339	1.68	13.55	15.23
36″ x 28″	11.500	L.F.	.371	1.84	14.85	16.69
40″ x 28″	12.500	L.F.	.403	2	16.15	18.15
48″ x 28″	13.500	L.F.	.436	2.16	17.40	19.56
48″ x 36″	14.000	L.F.	.452	2.24	18.05	20.29
Grilles, glass size, to 28″ by 16″	1.000	Ea.	.267	20.50	10.45	30.95
To 44″ by 24″	1.000	Ea.	.250	29.50	9.80	39.30
Drip cap, aluminum, 3′ long	3.000	L.F.	.060	.84	2.34	3.18
3′-6″ long	3.500	L.F.	.070	.98	2.73	3.71
4′ long	4.000	L.F.	.080	1.12	3.12	4.24
Wood, 3′ long	3.000	L.F.	.100	2.82	3.93	6.75
3′-6″ long	3.500	L.F.	.117	3.29	4.59	7.88
4′ long	4.000	L.F.	.133	3.76	5.25	9.01

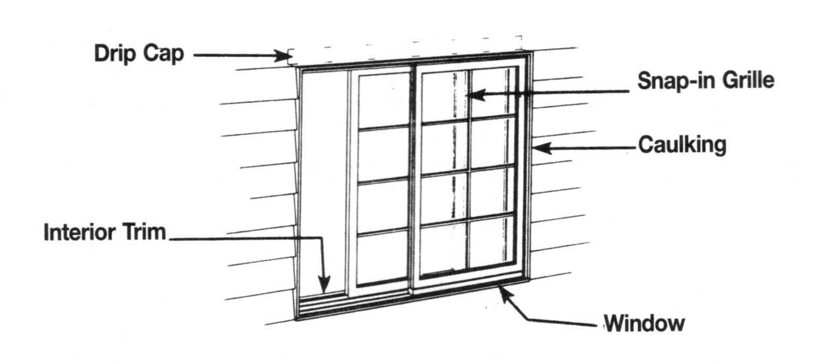

System Description	QUAN.	UNIT	LABOR HOURS	COST EACH		
				MAT.	INST.	TOTAL
BUILDER'S QUALITY WOOD WINDOW, 3′ X 2′, SLIDING						
Window, primed, builder's quality, 3′ x 2′, insul. glass	1.000	Ea.	.800	194	31.50	225.50
Trim, interior casing	11.000	L.F.	.367	10.34	14.41	24.75
Paint, interior & exterior, primer & 2 coats	2.000	Face	1.778	2.06	61	63.06
Caulking	10.000	L.F.	.323	1.60	12.90	14.50
Snap-in grille	1.000	Set	.333	25.50	13.05	38.55
Drip cap, metal	3.000	L.F.	.060	.84	2.34	3.18
TOTAL		Ea.	3.661	234.34	135.20	369.54
PLASTIC CLAD WOOD WINDOW, 4′ X 3′-6″, SLIDING						
Window, plastic clad, premium, 4′ x 3′-6″, insulating glass	1.000	Ea.	.889	705	35	740
Trim, interior casing	16.000	L.F.	.533	15.04	20.96	36
Paint, interior, primer & 2 coats	1.000	Face	.889	1.03	30.50	31.53
Caulking	17.000	L.F.	.548	2.72	21.93	24.65
Snap-in grille	1.000	Set	.333	25.50	13.05	38.55
TOTAL		Ea.	3.192	749.29	121.44	870.73
METAL CLAD WOOD WINDOW, 6′ X 5′, SLIDING						
Window, metal clad, deluxe, 6′ x 5′, insulating glass	1.000	Ea.	1.000	705	39	744
Trim, interior casing	23.000	L.F.	.767	21.62	30.13	51.75
Paint, interior, primer & 2 coats	1.000	Face	.889	1.03	30.50	31.53
Caulking	22.000	L.F.	.710	3.52	28.38	31.90
Snap-in grille	1.000	Set	.364	38	14.25	52.25
Drip cap, metal	6.000	L.F.	.120	1.68	4.68	6.36
TOTAL		Ea.	3.850	770.85	146.94	917.79

The cost of this system is on a cost per each window basis.

Description	QUAN.	UNIT	LABOR HOURS	COST EACH		
				MAT.	INST.	TOTAL

EXTERIOR WALLS 4

Important: See the Reference Section for critical supporting data - Reference Nos., Crews & Location Factors

Sliding Window Price Sheet	QUAN.	UNIT	LABOR HOURS	COST EACH		
				MAT.	INST.	TOTAL
Windows, sliding, builder's quality, 3′ x 3′, single glass	1.000	Ea.	.800	154	31.50	185.50
Insulating glass	1.000	Ea.	.800	194	31.50	225.50
4′ x 3′-6″, single glass	1.000	Ea.	.889	183	35	218
Insulating glass	1.000	Ea.	.889	229	35	264
6′ x 5′, single glass	1.000	Ea.	1.000	335	39	374
Insulating glass	1.000	Ea.	1.000	400	39	439
Plastic clad premium insulating glass, 3′ x 3′	1.000	Ea.	.800	565	31.50	596.50
4′ x 3′-6″	1.000	Ea.	.889	705	35	740
5′ x 4′	1.000	Ea.	.889	850	35	885
6′ x 5′	1.000	Ea.	1.000	1,025	39	1,064
Metal clad deluxe insulating glass, 3′ x 3′	1.000	Ea.	.800	310	31.50	341.50
4′ x 3′-6″	1.000	Ea.	.889	380	35	415
5′ x 4′	1.000	Ea.	.889	455	35	490
6′ x 5′	1.000	Ea.	1.000	705	39	744
Trim, interior casing, window 3′ x 2′	11.000	L.F.	.367	10.35	14.40	24.75
3′ x 3′	13.000	L.F.	.433	12.20	17.05	29.25
4′ x 3′-6″	16.000	L.F.	.533	15.05	21	36.05
5′ x 4′	19.000	L.F.	.633	17.85	25	42.85
6′ x 5′	23.000	L.F.	.767	21.50	30	51.50
Paint or stain, interior or exterior, 3′ x 2′ window, 1 coat	1.000	Face	.444	.38	15.35	15.73
2 coats	1.000	Face	.727	.76	25	25.76
Primer & 1 coat	1.000	Face	.727	.69	25	25.69
Primer & 2 coats	1.000	Face	.889	1.03	30.50	31.53
4′ x 3′-6″ window, 1 coat	1.000	Face	.667	.78	23	23.78
2 coats	1.000	Face	.667	.87	23	23.87
Primer & 1 coat	1.000	Face	.727	1.05	25	26.05
Primer & 2 coats	1.000	Face	.889	1.03	30.50	31.53
6′ x 5′ window, 1 coat	1.000	Face	.889	2.14	30.50	32.64
2 coats	1.000	Face	1.333	3.91	46	49.91
Primer & 1 coat	1.000	Face	1.333	3.69	46	49.69
Primer & 2 coats	1.000	Face	1.600	5.50	55	60.50
Caulking, window, 3′ x 2′	10.000	L.F.	.323	1.60	12.90	14.50
3′ x 3′	12.000	L.F.	.387	1.92	15.50	17.42
4′ x 3′-6″	15.000	L.F.	.484	2.40	19.35	21.75
5′ x 4′	18.000	L.F.	.581	2.88	23	25.88
6′ x 5′	22.000	L.F.	.710	3.52	28.50	32.02
Grilles, glass size, to 14″ x 36″	1.000	Set	.333	25.50	13.05	38.55
To 36″ x 36″	1.000	Set	.364	38	14.25	52.25
Drip cap, aluminum, 3′ long	3.000	L.F.	.060	.84	2.34	3.18
4′ long	4.000	L.F.	.080	1.12	3.12	4.24
5′ long	5.000	L.F.	.100	1.40	3.90	5.30
6′ long	6.000	L.F.	.120	1.68	4.68	6.36
Wood, 3′ long	3.000	L.F.	.100	2.82	3.93	6.75
4′ long	4.000	L.F.	.133	3.76	5.25	9.01
5′ long	5.000	L.F.	.167	4.70	6.55	11.25
6′ long	6.000	L.F.	.200	5.65	7.85	13.50

44 Bow/Bay Window Systems

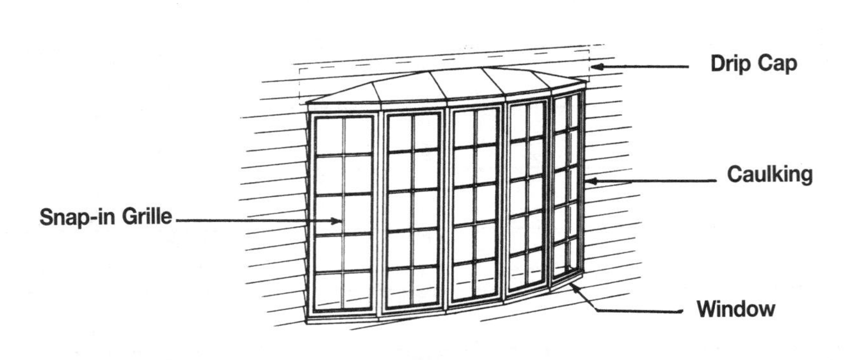

System Description	QUAN.	UNIT	LABOR HOURS	COST EACH		
				MAT.	INST.	TOTAL
AWNING TYPE BOW WINDOW, BUILDER'S QUALITY, 8′ X 5′						
Window, primed, builder's quality, 8′ x 5′, insulating glass	1.000	Ea.	1.600	1,275	62.50	1,337.50
Trim, interior casing	27.000	L.F.	.900	25.38	35.37	60.75
Paint, interior & exterior, primer & 1 coat	2.000	Face	3.200	11	110	121
Drip cap, vinyl	1.000	Ea.	.533	80	21	101
Caulking	26.000	L.F.	.839	4.16	33.54	37.70
Snap-in grilles	1.000	Set	1.067	100	41.80	141.80
TOTAL		Ea.	8.139	1,495.54	304.21	1,799.75
CASEMENT TYPE BOW WINDOW, PLASTIC CLAD, 10′ X 6′						
Window, plastic clad, premium, 10′ x 6′, insulating glass	1.000	Ea.	2.286	1,875	89.50	1,964.50
Trim, interior casing	33.000	L.F.	1.100	31.02	43.23	74.25
Paint, interior, primer & 1 coat	1.000	Face	1.778	2.06	61	63.06
Drip cap, vinyl	1.000	Ea.	.615	87	24	111
Caulking	32.000	L.F.	1.032	5.12	41.28	46.40
Snap-in grilles	1.000	Set	1.333	125	52.25	177.25
TOTAL		Ea.	8.144	2,125.20	311.26	2,436.46
DOUBLE HUNG TYPE, METAL CLAD, 9′ X 5′						
Window, metal clad, deluxe, 9′ x 5′, insulating glass	1.000	Ea.	2.667	1,200	105	1,305
Trim, interior casing	29.000	L.F.	.967	27.26	37.99	65.25
Paint, interior, primer & 1 coat	1.000	Face	1.778	2.06	61	63.06
Drip cap, vinyl	1.000	Set	.615	87	24	111
Caulking	28.000	L.F.	.903	4.48	36.12	40.60
Snap-in grilles	1.000	Set	1.067	100	41.80	141.80
TOTAL		Ea.	7.997	1,420.80	305.91	1,726.71

The cost of this system is on a cost per each window basis.

Description	QUAN.	UNIT	LABOR HOURS	COST EACH		
				MAT.	INST.	TOTAL

 Important: See the Reference Section for critical supporting data - Reference Nos., Crews & Location Factors

Bow/Bay Window Price Sheet

Bow/Bay Window Price Sheet	QUAN.	UNIT	LABOR HOURS	COST EACH MAT.	INST.	TOTAL
Windows, bow awning type, builder's quality, 8' x 5', insulating glass	1.000	Ea.	1.600	1,050	62.50	1,112.50
Low E glass	1.000	Ea.	1.600	1,275	62.50	1,337.50
12' x 6', insulating glass	1.000	Ea.	2.667	1,350	105	1,455
Low E glass	1.000	Ea.	2.667	1,425	105	1,530
Plastic clad premium insulating glass, 6' x 4'	1.000	Ea.	1.600	1,100	62.50	1,162.50
9' x 4'	1.000	Ea.	2.000	1,450	78.50	1,528.50
10' x 5'	1.000	Ea.	2.286	1,950	89.50	2,039.50
12' x 6'	1.000	Ea.	2.667	1,975	105	2,080
Metal clad deluxe insulating glass, 6' x 4'	1.000	Ea.	1.600	915	62.50	977.50
9' x 4'	1.000	Ea.	2.000	1,275	78.50	1,353.50
10' x 5'	1.000	Ea.	2.286	1,775	89.50	1,864.50
12' x 6'	1.000	Ea.	2.667	2,475	105	2,580
Bow casement type, builder's quality, 8' x 5', single glass	1.000	Ea.	1.600	1,550	62.50	1,612.50
Insulating glass	1.000	Ea.	1.600	1,875	62.50	1,937.50
12' x 6', single glass	1.000	Ea.	2.667	1,950	105	2,055
Insulating glass	1.000	Ea.	2.667	2,025	105	2,130
Plastic clad premium insulating glass, 8' x 5'	1.000	Ea.	1.600	1,250	62.50	1,312.50
10' x 5'	1.000	Ea.	2.000	1,775	78.50	1,853.50
10' x 6'	1.000	Ea.	2.286	1,875	89.50	1,964.50
12' x 6'	1.000	Ea.	2.667	2,225	105	2,330
Metal clad deluxe insulating glass, 8' x 5'	1.000	Ea.	1.600	1,375	62.50	1,437.50
10' x 5'	1.000	Ea.	2.000	1,475	78.50	1,553.50
10' x 6'	1.000	Ea.	2.286	1,750	89.50	1,839.50
12' x 6'	1.000	Ea.	2.667	2,425	105	2,530
Bow, double hung type, builder's quality, 8' x 4', single glass	1.000	Ea.	1.600	1,100	62.50	1,162.50
Insulating glass	1.000	Ea.	1.600	1,175	62.50	1,237.50
9' x 5', single glass	1.000	Ea.	2.667	1,175	105	1,280
Insulating glass	1.000	Ea.	2.667	1,250	105	1,355
Plastic clad premium insulating glass, 7' x 4'	1.000	Ea.	1.600	1,125	62.50	1,187.50
8' x 4'	1.000	Ea.	2.000	1,150	78.50	1,228.50
8' x 5'	1.000	Ea.	2.286	1,200	89.50	1,289.50
9' x 5'	1.000	Ea.	2.667	1,250	105	1,355
Metal clad deluxe insulating glass, 7' x 4'	1.000	Ea.	1.600	1,050	62.50	1,112.50
8' x 4'	1.000	Ea.	2.000	1,075	78.50	1,153.50
8' x 5'	1.000	Ea.	2.286	1,125	89.50	1,214.50
9' x 5'	1.000	Ea.	2.667	1,200	105	1,305
Trim, interior casing, window 7' x 4'	1.000	Ea.	.767	21.50	30	51.50
8' x 5'	1.000	Ea.	.900	25.50	35.50	61
10' x 6'	1.000	Ea.	1.100	31	43	74
12' x 6'	1.000	Ea.	1.233	35	48.50	83.50
Paint or stain, interior, or exterior, 7' x 4' window, 1 coat	1.000	Face	.889	2.14	30.50	32.64
Primer & 1 coat	1.000	Face	1.333	3.69	46	49.69
8' x 5' window, 1 coat	1.000	Face	.889	2.14	30.50	32.64
Primer & 1 coat	1.000	Face	1.333	3.69	46	49.69
10' x 6' window, 1 coat	1.000	Face	1.333	1.56	46	47.56
Primer & 1 coat	1.000	Face	1.778	2.06	61	63.06
12' x 6' window, 1 coat	1.000	Face	1.778	4.28	61	65.28
Primer & 1 coat	1.000	Face	2.667	7.40	92	99.40
Drip cap, vinyl moulded window, 7' long	1.000	Ea.	.533	80	21	101
8' long	1.000	Ea.	.533	80	21	101
10' long	1.000	Ea.	.615	87	24	111
12' long	1.000	Ea.	.615	87	24	111
Caulking, window, 7' x 4'	1.000	Ea.	.710	3.52	28.50	32.02
8' x 5'	1.000	Ea.	.839	4.16	33.50	37.66
10' x 6'	1.000	Ea.	1.032	5.10	41.50	46.60
12' x 6'	1.000	Ea.	1.161	5.75	46.50	52.25
Grilles, window, 7' x 4'	1.000	Set	.800	75	31.50	106.50
8' x 5'	1.000	Set	1.067	100	42	142
10' x 6'	1.000	Set	1.333	125	52.50	177.50
12' x 6'	1.000	Set	1.600	150	62.50	212.50

48 Fixed Window Systems

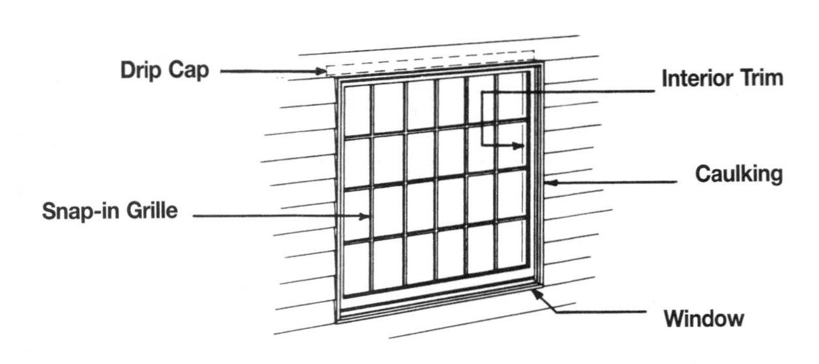

System Description	QUAN.	UNIT	LABOR HOURS	COST EACH		
				MAT.	INST.	TOTAL
BUILDER'S QUALITY PICTURE WINDOW, 4′ X 4′						
Window, primed, builder's quality, 4′ x 4′, insulating glass	1.000	Ea.	1.333	310	52.50	362.50
Trim, interior casing	17.000	L.F.	.567	15.98	22.27	38.25
Paint, interior & exterior, primer & 2 coats	2.000	Face	1.778	2.06	61	63.06
Caulking	16.000	L.F.	.516	2.56	20.64	23.20
Snap-in grille	1.000	Ea.	.267	142	10.45	152.45
Drip cap, metal	4.000	L.F.	.080	1.12	3.12	4.24
TOTAL		Ea.	4.541	473.72	169.98	643.70
PLASTIC CLAD WOOD WINDOW, 4′-6″ X 6′-6″						
Window, plastic clad, prem., 4′-6″ x 6′-6″, insul. glass	1.000	Ea.	1.455	710	57	767
Trim, interior casing	23.000	L.F.	.767	21.62	30.13	51.75
Paint, interior, primer & 2 coats	1.000	Face	.889	1.03	30.50	31.53
Caulking	22.000	L.F.	.710	3.52	28.38	31.90
Snap-in grille	1.000	Ea.	.267	142	10.45	152.45
TOTAL		Ea.	4.088	878.17	156.46	1,034.63
METAL CLAD WOOD WINDOW, 6′-6″ X 6′-6″						
Window, metal clad, deluxe, 6′-6″ x 6′-6″, insulating glass	1.000	Ea.	1.600	565	62.50	627.50
Trim interior casing	27.000	L.F.	.900	25.38	35.37	60.75
Paint, interior, primer & 2 coats	1.000	Face	1.600	5.50	55	60.50
Caulking	26.000	L.F.	.839	4.16	33.54	37.70
Snap-in grille	1.000	Ea.	.267	142	10.45	152.45
Drip cap, metal	6.500	L.F.	.130	1.82	5.07	6.89
TOTAL		Ea.	5.336	743.86	201.93	945.79

The cost of this system is on a cost per each window basis.

Description	QUAN.	UNIT	LABOR HOURS	COST EACH		
				MAT.	INST.	TOTAL

Fixed Window Price Sheet

Fixed Window Price Sheet	QUAN.	UNIT	LABOR HOURS	COST EACH		
				MAT.	INST.	TOTAL
Window-picture, builder's quality, 4′ x 4′, single glass	1.000	Ea.	1.333	286	52.50	338.50
Insulating glass	1.000	Ea.	1.333	310	52.50	362.50
4′ x 4′-6″, single glass	1.000	Ea.	1.455	295	57	352
Insulating glass	1.000	Ea.	1.455	325	57	382
5′ x 4′, single glass	1.000	Ea.	1.455	365	57	422
Insulating glass	1.000	Ea.	1.455	410	57	467
6′ x 4′-6″, single glass	1.000	Ea.	1.600	465	62.50	527.50
Insulating glass	1.000	Ea.	1.600	520	62.50	582.50
Plastic clad premium insulating glass, 4′ x 4′	1.000	Ea.	1.333	415	52.50	467.50
4′-6″ x 6′-6″	1.000	Ea.	1.455	710	57	767
5′-6″ x 6′-6″	1.000	Ea.	1.600	920	62.50	982.50
6′-6″ x 6′-6″	1.000	Ea.	1.600	940	62.50	1,002.50
Metal clad deluxe insulating glass, 4′ x 4′	1.000	Ea.	1.333	305	52.50	357.50
4′-6″ x 6′-6″	1.000	Ea.	1.455	445	57	502
5′-6″ x 6′-6″	1.000	Ea.	1.600	490	62.50	552.50
6′-6″ x 6′-6″	1.000	Ea.	1.600	565	62.50	627.50
Trim, interior casing, window 4′ x 4′	17.000	L.F.	.567	16	22.50	38.50
4′-6″ x 4′-6″	19.000	L.F.	.633	17.85	25	42.85
5′-0″ x 4′-0″	19.000	L.F.	.633	17.85	25	42.85
4′-6″ x 6′-6″	23.000	L.F.	.767	21.50	30	51.50
5′-6″ x 6′-6″	25.000	L.F.	.833	23.50	33	56.50
6′-6″ x 6′-6″	27.000	L.F.	.900	25.50	35.50	61
Paint or stain, interior or exterior, 4′ x 4′ window, 1 coat	1.000	Face	.667	.78	23	23.78
2 coats	1.000	Face	.667	.87	23	23.87
Primer & 1 coat	1.000	Face	.727	1.05	25	26.05
Primer & 2 coats	1.000	Face	.889	1.03	30.50	31.53
4′-6″ x 6′-6″ window, 1 coat	1.000	Face	.667	.78	23	23.78
2 coats	1.000	Face	.667	.87	23	23.87
Primer & 1 coat	1.000	Face	.727	1.05	25	26.05
Primer & 2 coats	1.000	Face	.889	1.03	30.50	31.53
6′-6″ x 6′-6″ window, 1 coat	1.000	Face	.889	2.14	30.50	32.64
2 coats	1.000	Face	1.333	3.91	46	49.91
Primer & 1 coat	1.000	Face	1.333	3.69	46	49.69
Primer & 2 coats	1.000	Face	1.600	5.50	55	60.50
Caulking, window, 4′ x 4′	1.000	Ea.	.516	2.56	20.50	23.06
4′-6″ x 4′-6″	1.000	Ea.	.581	2.88	23	25.88
5′-0″ x 4′-0″	1.000	Ea.	.581	2.88	23	25.88
4′-6″ x 6′-6″	1.000	Ea.	.710	3.52	28.50	32.02
5′-6″ x 6′-6″	1.000	Ea.	.774	3.84	31	34.84
6′-6″ x 6′-6″	1.000	Ea.	.839	4.16	33.50	37.66
Grilles, glass size, to 48″ x 48″	1.000	Ea.	.267	142	10.45	152.45
To 60″ x 68″	1.000	Ea.	.286	107	11.20	118.20
Drip cap, aluminum, 4′ long	4.000	L.F.	.080	1.12	3.12	4.24
4′-6″ long	4.500	L.F.	.090	1.26	3.51	4.77
5′ long	5.000	L.F.	.100	1.40	3.90	5.30
6′ long	6.000	L.F.	.120	1.68	4.68	6.36
Wood, 4′ long	4.000	L.F.	.133	3.76	5.25	9.01
4′-6″ long	4.500	L.F.	.150	4.23	5.90	10.13
5′ long	5.000	L.F.	.167	4.70	6.55	11.25
6′ long	6.000	L.F.	.200	5.65	7.85	13.50

52 Entrance Door Systems

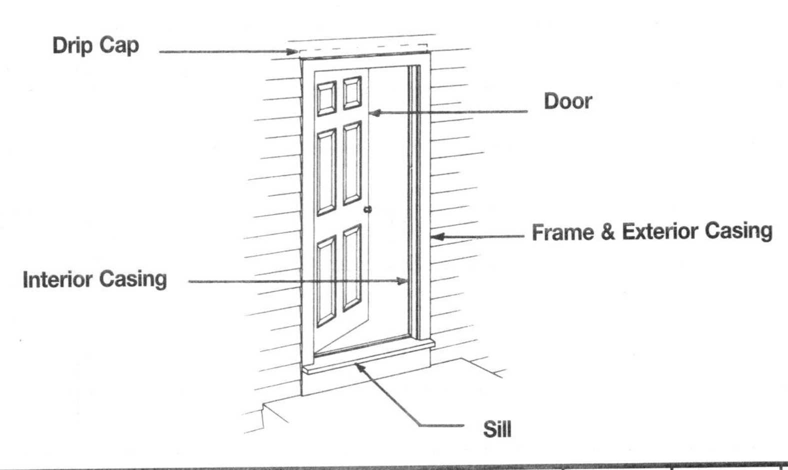

System Description	QUAN.	UNIT	LABOR HOURS	COST EACH		
				MAT.	INST.	TOTAL
COLONIAL, 6 PANEL, 3′ X 6′-8″, WOOD						
Door, 3′ x 6′-8″ x 1-3/4″ thick, pine, 6 panel colonial	1.000	Ea.	1.067	390	42	432
Frame, 5-13/16″ deep, incl. exterior casing & drip cap	17.000	L.F.	.725	137.70	28.39	166.09
Interior casing, 2-1/2″ wide	18.000	L.F.	.600	16.92	23.58	40.50
Sill, 8/4 x 8″ deep	3.000	L.F.	.480	43.95	18.75	62.70
Butt hinges, brass, 4-1/2″ x 4-1/2″	1.500	Pr.		19.58		19.58
Lockset	1.000	Ea.	.571	33	22.50	55.50
Weatherstripping, metal, spring type, bronze	1.000	Set	1.053	17.35	41.50	58.85
Paint, interior & exterior, primer & 2 coats	2.000	Face	1.778	10.90	61	71.90
TOTAL		Ea.	6.274	669.40	237.72	907.12
SOLID CORE BIRCH, FLUSH, 3′ X 6′-8″						
Door, 3′ x 6′-8″, 1-3/4″ thick, birch, flush solid core	1.000	Ea.	1.067	101	42	143
Frame, 5-13/16″ deep, incl. exterior casing & drip cap	17.000	L.F.	.725	137.70	28.39	166.09
Interior casing, 2-1/2″ wide	18.000	L.F.	.600	16.92	23.58	40.50
Sill, 8/4 x 8″ deep	3.000	L.F.	.480	43.95	18.75	62.70
Butt hinges, brass, 4-1/2″ x 4-1/2″	1.500	Pr.		19.58		19.58
Lockset	1.000	Ea.	.571	33	22.50	55.50
Weatherstripping, metal, spring type, bronze	1.000	Set	1.053	17.35	41.50	58.85
Paint, Interior & exterior, primer & 2 coats	2.000	Face	1.778	10.20	61	71.20
TOTAL		Ea.	6.274	379.70	237.72	617.42

These systems are on a cost per each door basis.

Description	QUAN.	UNIT	LABOR HOURS	COST EACH		
				MAT.	INST.	TOTAL

Entrance Door Price Sheet	QUAN.	UNIT	LABOR HOURS	COST EACH		
				MAT.	INST.	TOTAL
Door exterior wood 1-3/4" thick, pine, dutch door, 2'-8" x 6'-8" minimum	1.000	Ea.	1.333	655	52.50	707.50
Maximum	1.000	Ea.	1.600	690	62.50	752.50
3'-0" x 6'-8", minimum	1.000	Ea.	1.333	680	52.50	732.50
Maximum	1.000	Ea.	1.600	740	62.50	802.50
Colonial, 6 panel, 2'-8" x 6'-8"	1.000	Ea.	1.000	360	39	399
3'-0" x 6'-8"	1.000	Ea.	1.067	390	42	432
8 panel, 2'-6" x 6'-8"	1.000	Ea.	1.000	645	39	684
3'-0" x 6'-8"	1.000	Ea.	1.067	550	42	592
Flush, birch, solid core, 2'-8" x 6'-8"	1.000	Ea.	1.000	91.50	39	130.50
3'-0" x 6'-8"	1.000	Ea.	1.067	101	42	143
Porch door, 2'-8" x 6'-8"	1.000	Ea.	1.000	247	39	286
3'-0" x 6'-8"	1.000	Ea.	1.067	260	42	302
Hand carved mahogany, 2'-8" x 6'-8"	1.000	Ea.	1.067	490	42	532
3'-0" x 6'-8"	1.000	Ea.	1.067	525	42	567
Rosewood, 2'-8" x 6'-8"	1.000	Ea.	1.067	765	42	807
3'-0" x 6-8"	1.000	Ea.	1.067	795	42	837
Door, metal clad wood 1-3/8" thick raised panel, 2'-8" x 6'-8"	1.000	Ea.	1.067	273	42	315
3'-0" x 6'-8"	1.000	Ea.	1.067	271	42	313
Deluxe metal door, 2'-8" x 6'-8"	1.000	Ea.	1.231	420	48.50	468.50
3'-0" x 6'-8"	1.000	Ea.	1.231	415	48.50	463.50
Frame, pine, including exterior trim & drip cap, 5/4, x 4-9/16" deep	17.000	L.F.	.725	82.50	28.50	111
5-13/16" deep	17.000	L.F.	.725	138	28.50	166.50
6-9/16" deep	17.000	L.F.	.725	128	28.50	156.50
Safety glass lites, add	1.000	Ea.		27		27
Interior casing, 2'-8" x 6'-8" door	18.000	L.F.	.600	16.90	23.50	40.40
3'-0" x 6'-8" door	19.000	L.F.	.633	17.85	25	42.85
Sill, oak, 8/4 x 8" deep	3.000	L.F.	.480	44	18.75	62.75
8/4 x 10" deep	3.000	L.F.	.533	59.50	21	80.50
Butt hinges, steel plated, 4-1/2" x 4-1/2", plain	1.500	Pr.		19.60		19.60
Ball bearing	1.500	Pr.		43.50		43.50
Bronze, 4-1/2" x 4-1/2", plain	1.500	Pr.		22.50		22.50
Ball bearing	1.500	Pr.		46.50		46.50
Lockset, minimum	1.000	Ea.	.571	33	22.50	55.50
Maximum	1.000	Ea.	1.000	138	39	177
Weatherstripping, metal, interlocking, zinc	1.000	Set	2.667	13.60	105	118.60
Bronze	1.000	Set	2.667	21.50	105	126.50
Spring type, bronze	1.000	Set	1.053	17.35	41.50	58.85
Rubber, minimum	1.000	Set	1.053	4.73	41.50	46.23
Maximum	1.000	Set	1.143	5.40	45	50.40
Felt minimum	1.000	Set	.571	2.19	22.50	24.69
Maximum	1.000	Set	.615	2.37	24	26.37
Paint or stain, flush door, interior or exterior, 1 coat	2.000	Face	.941	3.70	32.50	36.20
2 coats	2.000	Face	1.455	7.40	50	57.40
Primer & 1 coat	2.000	Face	1.455	6.70	50	56.70
Primer & 2 coats	2.000	Face	1.778	10.20	61	71.20
Paneled door, interior & exterior, 1 coat	2.000	Face	1.143	3.96	39.50	43.46
2 coats	2.000	Face	2.000	7.90	69	76.90
Primer & 1 coat	2.000	Face	1.455	7.15	50	57.15
Primer & 2 coats	2.000	Face	1.778	10.90	61	71.90

53 Sliding Door Systems

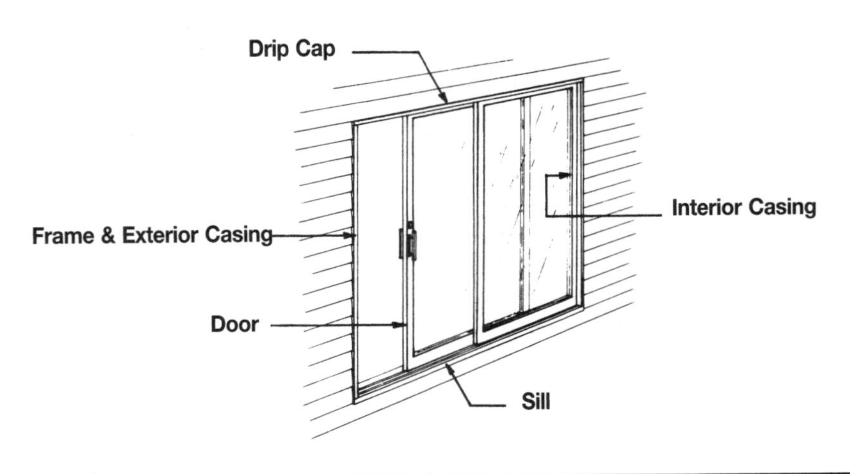

System Description	QUAN.	UNIT	LABOR HOURS	COST EACH		
				MAT.	INST.	TOTAL
WOOD SLIDING DOOR, 8′ WIDE, PREMIUM						
Wood, 5/8″ thick tempered insul. glass, 8′ wide, premium	1.000	Ea.	5.333	1,250	209	1,459
Interior casing	22.000	L.F.	.733	20.68	28.82	49.50
Exterior casing	22.000	L.F.	.733	20.68	28.82	49.50
Sill, oak, 8/4 x 8″ deep	8.000	L.F.	1.280	117.20	50	167.20
Drip cap	8.000	L.F.	.160	2.24	6.24	8.48
Paint, interior & exterior, primer & 2 coats	2.000	Face	2.816	15.84	96.80	112.64
TOTAL		Ea.	11.055	1,426.64	419.68	1,846.32
ALUMINUM SLIDING DOOR, 8′ WIDE, PREMIUM						
Aluminum, 5/8″ tempered insul. glass, 8′ wide, premium	1.000	Ea.	5.333	1,425	209	1,634
Interior casing	22.000	L.F.	.733	20.68	28.82	49.50
Exterior casing	22.000	L.F.	.733	20.68	28.82	49.50
Sill, oak, 8/4 x 8″ deep	8.000	L.F.	1.280	117.20	50	167.20
Drip cap	8.000	L.F.	.160	2.24	6.24	8.48
Paint, interior & exterior, primer & 2 coats	2.000	Face	2.816	15.84	96.80	112.64
TOTAL		Ea.	11.055	1,601.64	419.68	2,021.32

The cost of this system is on a cost per each door basis.

Description	QUAN.	UNIT	LABOR HOURS	COST EACH		
				MAT.	INST.	TOTAL

Important: See the Reference Section for critical supporting data - Reference Nos., Crews & Location Factors

Sliding Door Price Sheet

Sliding Door Price Sheet	QUAN.	UNIT	LABOR HOURS	MAT.	INST.	TOTAL
Sliding door, wood, 5/8″ thick, tempered insul. glass, 6′ wide, premium	1.000	Ea.	4.000	1,075	157	1,232
Economy	1.000	Ea.	4.000	765	157	922
8′wide, wood premium	1.000	Ea.	5.333	1,250	209	1,459
Economy	1.000	Ea.	5.333	885	209	1,094
12′ wide, wood premium	1.000	Ea.	6.400	2,975	251	3,226
Economy	1.000	Ea.	6.400	2,025	251	2,276
Aluminum, 5/8″ thick, tempered insul. glass, 6′wide, premium	1.000	Ea.	4.000	1,500	157	1,657
Economy	1.000	Ea.	4.000	780	157	937
8′wide, premium	1.000	Ea.	5.333	1,425	209	1,634
Economy	1.000	Ea.	5.333	1,200	209	1,409
12′ wide, premium	1.000	Ea.	6.400	2,375	251	2,626
Economy	1.000	Ea.	6.400	1,500	251	1,751
Interior casing, 6′ wide door	20.000	L.F.	.667	18.80	26	44.80
8′ wide door	22.000	L.F.	.733	20.50	29	49.50
12′ wide door	26.000	L.F.	.867	24.50	34	58.50
Exterior casing, 6′ wide door	20.000	L.F.	.667	18.80	26	44.80
8′ wide door	22.000	L.F.	.733	20.50	29	49.50
12′ wide door	26.000	L.F.	.867	24.50	34	58.50
Sill, oak, 8/4 x 8″ deep, 6′ wide door	6.000	L.F.	.960	88	37.50	125.50
8′ wide door	8.000	L.F.	1.280	117	50	167
12′ wide door	12.000	L.F.	1.920	176	75	251
8/4 x 10″ deep, 6′ wide door	6.000	L.F.	1.067	119	41.50	160.50
8′ wide door	8.000	L.F.	1.422	158	55.50	213.50
12′ wide door	12.000	L.F.	2.133	237	83.50	320.50
Drip cap, 6′ wide door	6.000	L.F.	.120	1.68	4.68	6.36
8′ wide door	8.000	L.F.	.160	2.24	6.25	8.49
12′ wide door	12.000	L.F.	.240	3.36	9.35	12.71
Paint or stain, interior & exterior, 6′ wide door, 1 coat	2.000	Face	1.600	4.80	55	59.80
2 coats	2.000	Face	1.600	4.80	55	59.80
Primer & 1 coat	2.000	Face	1.778	9.60	61.50	71.10
Primer & 2 coats	2.000	Face	2.560	14.40	88	102.40
8′ wide door, 1 coat	2.000	Face	1.760	5.30	60.50	65.80
2 coats	2.000	Face	1.760	5.30	60.50	65.80
Primer & 1 coat	2.000	Face	1.955	10.55	68	78.55
Primer & 2 coats	2.000	Face	2.816	15.85	97	112.85
12′ wide door, 1 coat	2.000	Face	2.080	6.25	72	78.25
2 coats	2.000	Face	2.080	6.25	72	78.25
Primer & 1 coat	2.000	Face	2.311	12.50	80	92.50
Primer & 2 coats	2.000	Face	3.328	18.70	114	132.70
Aluminum door, trim only, interior & exterior, 6′ door, 1 coat	2.000	Face	.800	2.40	27.50	29.90
2 coats	2.000	Face	.800	2.40	27.50	29.90
Primer & 1 coat	2.000	Face	.889	4.80	31	35.80
Primer & 2 coats	2.000	Face	1.280	7.20	44	51.20
8′ wide door, 1 coat	2.000	Face	.880	2.64	30.50	33.14
2 coats	2.000	Face	.880	2.64	30.50	33.14
Primer & 1 coat	2.000	Face	.978	5.30	34	39.30
Primer & 2 coats	2.000	Face	1.408	7.90	48.50	56.40
12′ wide door, 1 coat	2.000	Face	1.040	3.12	36	39.12
2 coats	2.000	Face	1.040	3.12	36	39.12
Primer & 1 coat	2.000	Face	1.155	6.25	40	46.25
Primer & 2 coats	2.000	Face	1.664	9.35	57	66.35

Drip Cap
Exterior Trim
Door
Weatherstripping
Jamb

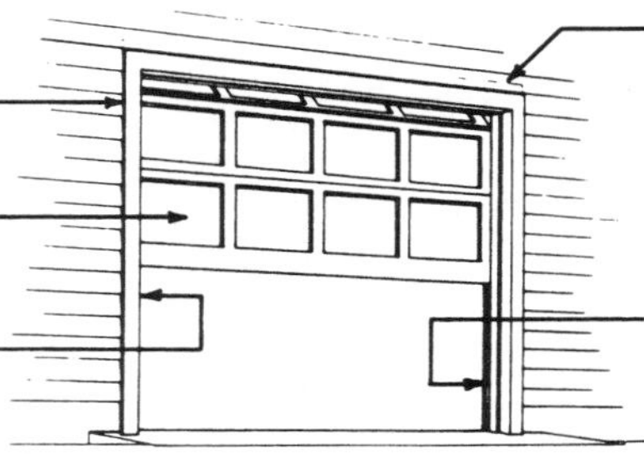

System Description	QUAN.	UNIT	LABOR HOURS	COST EACH MAT.	COST EACH INST.	COST EACH TOTAL
OVERHEAD, SECTIONAL GARAGE DOOR, 9′ X 7′						
Wood, overhead sectional door, std., incl. hardware, 9′ x 7′	1.000	Ea.	2.000	485	78.50	563.50
Jamb & header blocking, 2″ x 6″	25.000	L.F.	.901	14.25	35.25	49.50
Exterior trim	25.000	L.F.	.833	23.50	32.75	56.25
Paint, interior & exterior, primer & 2 coats	2.000	Face	3.556	21.80	122	143.80
Weatherstripping, molding type	1.000	Set	.767	21.62	30.13	51.75
Drip cap	9.000	L.F.	.180	2.52	7.02	9.54
TOTAL		Ea.	8.237	568.69	305.65	874.34
OVERHEAD, SECTIONAL GARAGE DOOR, 16′ X 7′						
Wood, overhead sectional, std., incl. hardware, 16′ x 7′	1.000	Ea.	2.667	980	105	1,085
Jamb & header blocking, 2″ x 6″	30.000	L.F.	1.081	17.10	42.30	59.40
Exterior trim	30.000	L.F.	1.000	28.20	39.30	67.50
Paint, interior & exterior, primer & 2 coats	2.000	Face	5.333	32.70	183	215.70
Weatherstripping, molding type	1.000	Set	1.000	28.20	39.30	67.50
Drip cap	16.000	L.F.	.320	4.48	12.48	16.96
TOTAL		Ea.	11.401	1,090.68	421.38	1,512.06
OVERHEAD, SWING-UP TYPE, GARAGE DOOR, 16′ X 7′						
Wood, overhead, swing-up, std., incl. hardware, 16′ x 7′	1.000	Ea.	2.667	670	105	775
Jamb & header blocking, 2″ x 6″	30.000	L.F.	1.081	17.10	42.30	59.40
Exterior trim	30.000	L.F.	1.000	28.20	39.30	67.50
Paint, interior & exterior, primer & 2 coats	2.000	Face	5.333	32.70	183	215.70
Weatherstripping, molding type	1.000	Set	1.000	28.20	39.30	67.50
Drip cap	16.000	L.F.	.320	4.48	12.48	16.96
TOTAL		Ea.	11.401	780.68	421.38	1,202.06

This system is on a cost per each door basis.

Description	QUAN.	UNIT	LABOR HOURS	COST EACH MAT.	COST EACH INST.	COST EACH TOTAL

 Important: See the Reference Section for critical supporting data - Reference Nos., Crews & Location Factors

Resi Garage Door Price Sheet	QUAN.	UNIT	LABOR HOURS	COST EACH		
				MAT.	INST.	TOTAL
Overhead, sectional, including hardware, fiberglass, 9′ x 7′, standard	1.000	Ea.	3.030	610	119	729
Deluxe	1.000	Ea.	3.030	770	119	889
16′ x 7′, standard	1.000	Ea.	2.667	1,050	105	1,155
Deluxe	1.000	Ea.	2.667	1,325	105	1,430
Hardboard, 9′ x 7′, standard	1.000	Ea.	2.000	410	78.50	488.50
Deluxe	1.000	Ea.	2.000	525	78.50	603.50
16′ x 7′, standard	1.000	Ea.	2.667	770	105	875
Deluxe	1.000	Ea.	2.667	900	105	1,005
Metal, 9′ x 7′, standard	1.000	Ea.	3.030	475	119	594
Deluxe	1.000	Ea.	2.000	640	78.50	718.50
16′ x 7′, standard	1.000	Ea.	5.333	605	209	814
Deluxe	1.000	Ea.	2.667	940	105	1,045
Wood, 9′ x 7′, standard	1.000	Ea.	2.000	485	78.50	563.50
Deluxe	1.000	Ea.	2.000	1,400	78.50	1,478.50
16′ x 7′, standard	1.000	Ea.	2.667	980	105	1,085
Deluxe	1.000	Ea.	2.667	2,050	105	2,155
Overhead swing-up type including hardware, fiberglass, 9′ x 7′, standard	1.000	Ea.	2.000	670	78.50	748.50
Deluxe	1.000	Ea.	2.000	740	78.50	818.50
16′ x 7′, standard	1.000	Ea.	2.667	845	105	950
Deluxe	1.000	Ea.	2.667	915	105	1,020
Hardboard, 9′ x 7′, standard	1.000	Ea.	2.000	320	78.50	398.50
Deluxe	1.000	Ea.	2.000	425	78.50	503.50
16′ x 7′, standard	1.000	Ea.	2.667	450	105	555
Deluxe	1.000	Ea.	2.667	670	105	775
Metal, 9′ x 7′, standard	1.000	Ea.	2.000	355	78.50	433.50
Deluxe	1.000	Ea.	2.000	605	78.50	683.50
16′ x 7′, standard	1.000	Ea.	2.667	555	105	660
Deluxe	1.000	Ea.	2.667	890	105	995
Wood, 9′ x 7′, standard	1.000	Ea.	2.000	385	78.50	463.50
Deluxe	1.000	Ea.	2.000	660	78.50	738.50
16′ x 7′, standard	1.000	Ea.	2.667	670	105	775
Deluxe	1.000	Ea.	2.667	945	105	1,050
Jamb & header blocking, 2″ x 6″, 9′ x 7′ door	25.000	L.F.	.901	14.25	35.50	49.75
16′ x 7′ door	30.000	L.F.	1.081	17.10	42.50	59.60
2″ x 8″, 9′ x 7′ door	25.000	L.F.	1.000	22	39.50	61.50
16′ x 7′ door	30.000	L.F.	1.200	26.50	47	73.50
Exterior trim, 9′ x 7′ door	25.000	L.F.	.833	23.50	33	56.50
16′ x 7′ door	30.000	L.F.	1.000	28	39.50	67.50
Paint or stain, interior & exterior, 9′ x 7′ door, 1 coat	1.000	Face	2.286	7.90	79	86.90
2 coats	1.000	Face	4.000	15.85	138	153.85
Primer & 1 coat	1.000	Face	2.909	14.30	100	114.30
Primer & 2 coats	1.000	Face	3.556	22	122	144
16′ x 7′ door, 1 coat	1.000	Face	3.429	11.90	118	129.90
2 coats	1.000	Face	6.000	24	207	231
Primer & 1 coat	1.000	Face	4.364	21.50	150	171.50
Primer & 2 coats	1.000	Face	5.333	32.50	183	215.50
Weatherstripping, molding type, 9′ x 7′ door	1.000	Set	.767	21.50	30	51.50
16′ x 7′ door	1.000	Set	1.000	28	39.50	67.50
Drip cap, 9′ door	9.000	L.F.	.180	2.52	7	9.52
16′ door	16.000	L.F.	.320	4.48	12.50	16.98
Garage door opener, economy	1.000	Ea.	1.000	279	39	318
Deluxe, including remote control	1.000	Ea.	1.000	410	39	449

58 Aluminum Window Systems

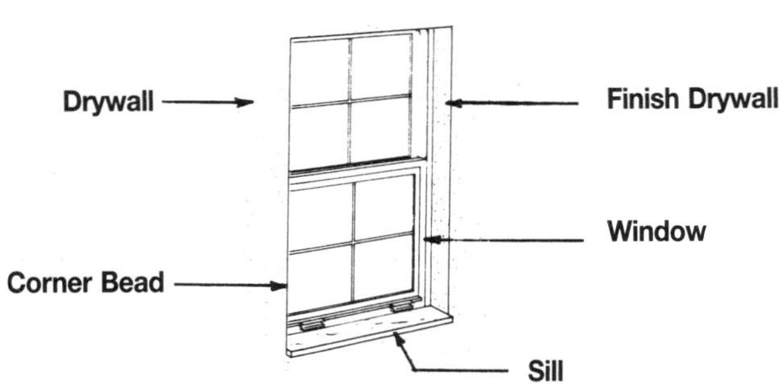

System Description	QUAN.	UNIT	LABOR HOURS	COST EACH MAT.	COST EACH INST.	COST EACH TOTAL
SINGLE HUNG, 2′ X 3′ OPENING						
Window, 2′ x 3′ opening, enameled, insulating glass	1.000	Ea.	1.600	182	77	259
Blocking, 1″ x 3″ furring strip nailers	10.000	L.F.	.146	2.30	5.70	8
Drywall, 1/2″ thick, standard	5.000	S.F.	.040	1.05	1.55	2.60
Corner bead, 1″ x 1″, galvanized steel	8.000	L.F.	.160	.88	6.24	7.12
Finish drywall, tape and finish corners inside and outside	16.000	L.F.	.269	1.44	10.56	12
Sill, slate	2.000	L.F.	.400	16.50	13.60	30.10
TOTAL		Ea.	2.579	204.17	114.65	318.82
SLIDING, 3′ X 2′ OPENING						
Window, 3′ x 2′ opening, enameled, insulating glass	1.000	Ea.	1.600	187	77	264
Blocking, 1″ x 3″ furring strip nailers	10.000	L.F.	.146	2.30	5.70	8
Drywall, 1/2″ thick, standard	5.000	S.F.	.040	1.05	1.55	2.60
Corner bead, 1″ x 1″, galvanized steel	7.000	L.F.	.140	.77	5.46	6.23
Finish drywall, tape and finish corners inside and outside	14.000	L.F.	.236	1.26	9.24	10.50
Sill, slate	3.000	L.F.	.600	24.75	20.40	45.15
TOTAL		Ea.	2.730	217.13	119.35	336.48
AWNING, 3′-1″ X 3′-2″						
Window, 3′-1″ x 3′-2″ opening, enameled, insul. glass	1.000	Ea.	1.600	207	77	284
Blocking, 1″ x 3″ furring strip, nailers	12.500	L.F.	.182	2.88	7.13	10.01
Drywall, 1/2″ thick, standard	4.500	S.F.	.036	.95	1.40	2.35
Corner bead, 1″ x 1″, galvanized steel	9.250	L.F.	.185	1.02	7.22	8.24
Finish drywall, tape and finish corners, inside and outside	18.500	L.F.	.312	1.67	12.21	13.88
Sill, slate	3.250	L.F.	.650	26.81	22.10	48.91
TOTAL		Ea.	2.922	240.33	127.06	367.39

Description	QUAN.	UNIT	LABOR HOURS	COST PER S.F. MAT.	COST PER S.F. INST.	COST PER S.F. TOTAL

Important: See the Reference Section for critical supporting data - Reference Nos., Crews & Location Factors

Aluminum Window Price Sheet	QUAN.	UNIT	LABOR HOURS	COST EACH		
				MAT.	INST.	TOTAL
Window, aluminum, awning, 3′-1″ x 3′-2″, standard glass	1.000	Ea.	1.600	223	77	300
Insulating glass	1.000	Ea.	1.600	207	77	284
4′-5″ x 5′-3″, standard glass	1.000	Ea.	2.000	315	96	411
Insulating glass	1.000	Ea.	2.000	360	96	456
Casement, 3′-1″ x 3′-2″, standard glass	1.000	Ea.	1.600	315	77	392
Insulating glass	1.000	Ea.	1.600	305	77	382
Single hung, 2′ x 3′, standard glass	1.000	Ea.	1.600	150	77	227
Insulating glass	1.000	Ea.	1.600	182	77	259
2′-8″ x 6′-8″, standard glass	1.000	Ea.	2.000	320	96	416
Insulating glass	1.000	Ea.	2.000	410	96	506
3′-4″ x 5′-0″, standard glass	1.000	Ea.	1.778	206	85.50	291.50
Insulating glass	1.000	Ea.	1.778	289	85.50	374.50
Sliding, 3′ x 2′, standard glass	1.000	Ea.	1.600	169	77	246
Insulating glass	1.000	Ea.	1.600	187	77	264
5′ x 3′, standard glass	1.000	Ea.	1.778	214	85.50	299.50
Insulating glass	1.000	Ea.	1.778	300	85.50	385.50
8′ x 4′, standard glass	1.000	Ea.	2.667	310	128	438
Insulating glass	1.000	Ea.	2.667	495	128	623
Blocking, 1″ x 3″ furring, opening 3′ x 2′	10.000	L.F.	.146	2.30	5.70	8
3′ x 3′	12.500	L.F.	.182	2.88	7.15	10.03
3′ x 5′	16.000	L.F.	.233	3.68	9.10	12.78
4′ x 4′	16.000	L.F.	.233	3.68	9.10	12.78
4′ x 5′	18.000	L.F.	.262	4.14	10.25	14.39
4′ x 6′	20.000	L.F.	.291	4.60	11.40	16
4′ x 8′	24.000	L.F.	.349	5.50	13.70	19.20
6′-8″ x 2′-8″	19.000	L.F.	.276	4.37	10.85	15.22
Drywall, 1/2″ thick, standard, opening 3′ x 2′	5.000	S.F.	.040	1.05	1.55	2.60
3′ x 3′	6.000	S.F.	.048	1.26	1.86	3.12
3′ x 5′	8.000	S.F.	.064	1.68	2.48	4.16
4′ x 4′	8.000	S.F.	.064	1.68	2.48	4.16
4′ x 5′	9.000	S.F.	.072	1.89	2.79	4.68
4′ x 6′	10.000	S.F.	.080	2.10	3.10	5.20
4′ x 8′	12.000	S.F.	.096	2.52	3.72	6.24
6′-8″ x 2′	9.500	S.F.	.076	2	2.95	4.95
Corner bead, 1″ x 1″, galvanized steel, opening 3′ x 2′	7.000	L.F.	.140	.77	5.45	6.22
3′ x 3′	9.000	L.F.	.180	.99	7	7.99
3′ x 5′	11.000	L.F.	.220	1.21	8.60	9.81
4′ x 4′	12.000	L.F.	.240	1.32	9.35	10.67
4′ x 5′	13.000	L.F.	.260	1.43	10.15	11.58
4′ x 6′	14.000	L.F.	.280	1.54	10.90	12.44
4′ x 8′	16.000	L.F.	.320	1.76	12.50	14.26
6′-8″ x 2′	15.000	L.F.	.300	1.65	11.70	13.35
Tape and finish corners, inside and outside, opening 3′ x 2′	14.000	L.F.	.204	1.26	9.25	10.51
3′ x 3′	18.000	L.F.	.262	1.62	11.90	13.52
3′ x 5′	22.000	L.F.	.320	1.98	14.50	16.48
4′ x 4′	24.000	L.F.	.349	2.16	15.85	18.01
4′ x 5′	26.000	L.F.	.378	2.34	17.15	19.49
4′ x 6′	28.000	L.F.	.407	2.52	18.50	21.02
4′ x 8′	32.000	L.F.	.466	2.88	21	23.88
6′-8″ x 2′	30.000	L.F.	.437	2.70	19.80	22.50
Sill, slate, 2′ long	2.000	L.F.	.400	16.50	13.60	30.10
3′ long	3.000	L.F.	.600	25	20.50	45.50
4′ long	4.000	L.F.	.800	33	27	60
Wood, 1-5/8″ x 6-1/4″, 2′ long	2.000	L.F.	.128	16.70	5	21.70
3′ long	3.000	L.F.	.192	25	7.55	32.55
4′ long	4.000	L.F.	.256	33.50	10.05	43.55

System Description	QUAN.	UNIT	LABOR HOURS	COST EACH		
				MAT.	INST.	TOTAL
Storm door, aluminum, combination, storm & screen, anodized, 2′-6″ x 6′-8″	1.000	Ea.	1.067	167	42	209
2′-8″ x 6′-8″	1.000	Ea.	1.143	191	45	236
3′-0″ x 6′-8″	1.000	Ea.	1.143	191	45	236
Mill finish, 2′-6″ x 6′-8″	1.000	Ea.	1.067	222	42	264
2′-8″ x 6′-8″	1.000	Ea.	1.143	222	45	267
3′-0″ x 6′-8″	1.000	Ea.	1.143	240	45	285
Painted, 2′-6″ x 6′-8″	1.000	Ea.	1.067	222	42	264
2′-8″ x 6′-8″	1.000	Ea.	1.143	225	45	270
3′-0″ x 6′-8″	1.000	Ea.	1.143	235	45	280
Wood, combination, storm & screen, crossbuck, 2′-6″ x 6′-9″	1.000	Ea.	1.455	288	57	345
2′-8″ x 6′-9″	1.000	Ea.	1.600	281	62.50	343.50
3′-0″ x 6′-9″	1.000	Ea.	1.778	287	69.50	356.50
Full lite, 2′-6″ x 6′-9″	1.000	Ea.	1.455	292	57	349
2′-8″ x 6′-9″	1.000	Ea.	1.600	292	62.50	354.50
3′-0″ x 6′-9″	1.000	Ea.	1.778	300	69.50	369.50
Windows, aluminum, combination storm & screen, basement, 1′-10″ x 1′-0″	1.000	Ea.	.533	31	21	52
2′-9″ x 1′-6″	1.000	Ea.	.533	33.50	21	54.50
3′-4″ x 2′-0″	1.000	Ea.	.533	40.50	21	61.50
Double hung, anodized, 2′-0″ x 3′-5″	1.000	Ea.	.533	80	21	101
2′-6″ x 5′-0″	1.000	Ea.	.571	107	22.50	129.50
4′-0″ x 6′-0″	1.000	Ea.	.640	227	25	252
Painted, 2′-0″ x 3′-5″	1.000	Ea.	.533	95.50	21	116.50
2′-6″ x 5′-0″	1.000	Ea.	.571	153	22.50	175.50
4′-0″ x 6′-0″	1.000	Ea.	.640	275	25	300
Fixed window, anodized, 4′-6″ x 4′-6″	1.000	Ea.	.640	122	25	147
5′-8″ x 4′-6″	1.000	Ea.	.800	139	31.50	170.50
Painted, 4′-6″ x 4′-6″	1.000	Ea.	.640	122	25	147
5′-8″ x 4′-6″	1.000	Ea.	.800	139	31.50	170.50

Important: See the Reference Section for critical supporting data - Reference Nos., Crews & Location Factors

64 Shutters/Blinds Systems

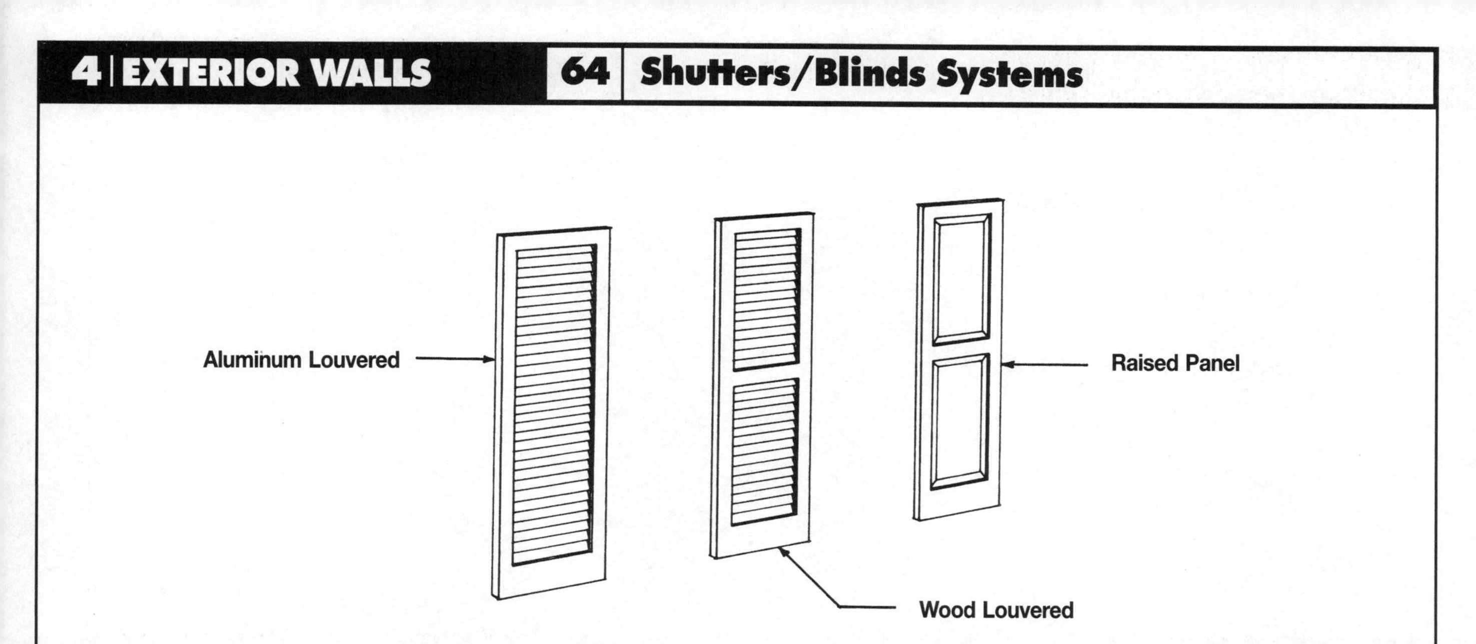

System Description	QUAN.	UNIT	LABOR HOURS	COST PER PAIR		
				MAT.	INST.	TOTAL
Shutters, exterior blinds, aluminum, louvered, 1′-4″ wide, 3″-0″ long	1.000	Set	.800	46	31.50	77.50
4′-0″ long	1.000	Set	.800	55	31.50	86.50
5′-4″ long	1.000	Set	.800	72.50	31.50	104
6′-8″ long	1.000	Set	.889	92.50	35	127.50
Wood, louvered, 1′-2″ wide, 3′-3″ long	1.000	Set	.800	87.50	31.50	119
4′-7″ long	1.000	Set	.800	118	31.50	149.50
5′-3″ long	1.000	Set	.800	134	31.50	165.50
1′-6″ wide, 3′-3″ long	1.000	Set	.800	93	31.50	124.50
4′-7″ long	1.000	Set	.800	130	31.50	161.50
Polystyrene, solid raised panel, 3′-3″ wide, 3′-0″ long	1.000	Set	.800	182	31.50	213.50
3′-11″ long	1.000	Set	.800	194	31.50	225.50
5′-3″ long	1.000	Set	.800	255	31.50	286.50
6′-8″ long	1.000	Set	.889	287	35	322
Polystyrene, louvered, 1′-2″ wide, 3′-3″ long	1.000	Set	.800	44	31.50	75.50
4′-7″ long	1.000	Set	.800	54.50	31.50	86
5′-3″ long	1.000	Set	.800	58	31.50	89.50
6′-8″ long	1.000	Set	.889	95.50	35	130.50
Vinyl, louvered, 1′-2″ wide, 4′-7″ long	1.000	Set	.720	51.50	28.50	80
1′-4″ x 6′-8″ long	1.000	Set	.889	87	35	122

Division 5
Roofing

04 Gable End Roofing Systems

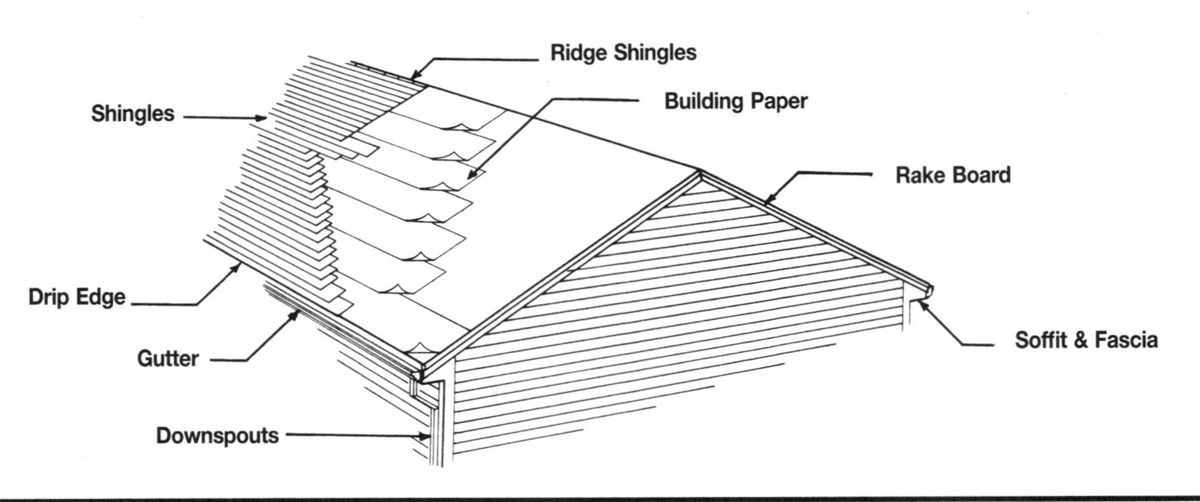

System Description	QUAN.	UNIT	LABOR HOURS	COST PER S.F. MAT.	COST PER S.F. INST.	COST PER S.F. TOTAL
ASPHALT, ROOF SHINGLES, CLASS A						
Shingles, inorganic class A, 210-235 lb./sq., 4/12 pitch	1.160	S.F.	.017	.41	.63	1.04
Drip edge, metal, 5″ wide	.150	L.F.	.003	.04	.12	.16
Building paper, #15 felt	1.300	S.F.	.002	.04	.06	.10
Ridge shingles, asphalt	.042	L.F.	.001	.03	.04	.07
Soffit & fascia, white painted aluminum, 1′ overhang	.083	L.F.	.012	.20	.47	.67
Rake trim, 1″ x 6″	.040	L.F.	.002	.04	.06	.10
Rake trim, prime and paint	.040	L.F.	.002	.01	.06	.07
Gutter, seamless, aluminum painted	.083	L.F.	.006	.10	.24	.34
Downspouts, aluminum painted	.035	L.F.	.002	.04	.07	.11
TOTAL		S.F.	.047	.91	1.75	2.66
WOOD, CEDAR SHINGLES NO. 1 PERFECTIONS, 18″ LONG						
Shingles, wood, cedar, No. 1 perfections, 4/12 pitch	1.160	S.F.	.035	2.11	1.37	3.48
Drip edge, metal, 5″ wide	.150	L.F.	.003	.04	.12	.16
Building paper, #15 felt	1.300	S.F.	.002	.04	.06	.10
Ridge shingles, cedar	.042	L.F.	.001	.11	.05	.16
Soffit & fascia, white painted aluminum, 1′ overhang	.083	L.F.	.012	.20	.47	.67
Rake trim, 1″ x 6″	.040	L.F.	.002	.04	.06	.10
Rake trim, prime and paint	.040	L.F.	.002	.01	.06	.07
Gutter, seamless, aluminum, painted	.083	L.F.	.006	.10	.24	.34
Downspouts, aluminum, painted	.035	L.F.	.002	.04	.07	.11
TOTAL		S.F.	.065	2.69	2.50	5.19

The prices in these systems are based on a square foot of plan area.
All quantities have been adjusted accordingly.

Description	QUAN.	UNIT	LABOR HOURS	COST PER S.F. MAT.	COST PER S.F. INST.	COST PER S.F. TOTAL

 Important: See the Reference Section for critical supporting data - Reference Nos., Crews & Location Factors

Gable End Roofing Price Sheet	QUAN.	UNIT	LABOR HOURS	COST PER S.F.		
				MAT.	INST.	TOTAL
Shingles, asphalt, inorganic, class A, 210-235 lb./sq., 4/12 pitch	1.160	S.F.	.017	.41	.63	1.04
8/12 pitch	1.330	S.F.	.019	.44	.68	1.12
Laminated, multi-layered, 240-260 lb./sq., 4/12 pitch	1.160	S.F.	.021	.56	.77	1.33
8/12 pitch	1.330	S.F.	.023	.60	.83	1.43
Premium laminated, multi-layered, 260-300 lb./sq., 4/12 pitch	1.160	S.F.	.027	.67	.99	1.66
8/12 pitch	1.330	S.F.	.030	.73	1.07	1.80
Clay tile, Spanish tile, red, 4/12 pitch	1.160	S.F.	.053	3.48	1.93	5.41
8/12 pitch	1.330	S.F.	.058	3.77	2.09	5.86
Mission tile, red, 4/12 pitch	1.160	S.F.	.083	8.35	3.01	11.36
8/12 pitch	1.330	S.F.	.090	9.05	3.26	12.31
French tile, red, 4/12 pitch	1.160	S.F.	.071	7.55	2.57	10.12
8/12 pitch	1.330	S.F.	.077	8.20	2.78	10.98
Slate, Buckingham, Virginia, black, 4/12 pitch	1.160	S.F.	.055	6.60	1.98	8.58
8/12 pitch	1.330	S.F.	.059	7.15	2.15	9.30
Vermont, black or grey, 4/12 pitch	1.160	S.F.	.055	4.32	1.98	6.30
8/12 pitch	1.330	S.F.	.059	4.68	2.15	6.83
Wood, No. 1 red cedar, 5X, 16″ long, 5″ exposure, 4/12 pitch	1.160	S.F.	.038	2.15	1.50	3.65
8/12 pitch	1.330	S.F.	.042	2.33	1.63	3.96
Fire retardant, 4/12 pitch	1.160	S.F.	.038	2.55	1.50	4.05
8/12 pitch	1.330	S.F.	.042	2.76	1.63	4.39
18″ long, No.1 perfections, 5″ exposure, 4/12 pitch	1.160	S.F.	.035	2.11	1.37	3.48
8/12 pitch	1.330	S.F.	.038	2.29	1.48	3.77
Fire retardant, 4/12 pitch	1.160	S.F.	.035	2.49	1.37	3.86
8/12 pitch	1.330	S.F.	.038	2.70	1.48	4.18
Resquared & rebutted, 18″ long, 6″ exposure, 4/12 pitch	1.160	S.F.	.032	2.63	1.26	3.89
8/12 pitch	1.330	S.F.	.035	2.85	1.37	4.22
Fire retardant, 4/12 pitch	1.160	S.F.	.032	3.01	1.26	4.27
8/12 pitch	1.330	S.F.	.035	3.26	1.37	4.63
Wood shakes hand split, 24″ long, 10″ exposure, 4/12 pitch	1.160	S.F.	.038	1.82	1.50	3.32
8/12 pitch	1.330	S.F.	.042	1.98	1.63	3.61
Fire retardant, 4/12 pitch	1.160	S.F.	.038	2.22	1.50	3.72
8/12 pitch	1.330	S.F.	.042	2.41	1.63	4.04
18″ long, 8″ exposure, 4/12 pitch	1.160	S.F.	.048	1.28	1.88	3.16
8/12 pitch	1.330	S.F.	.052	1.39	2.04	3.43
Fire retardant, 4/12 pitch	1.160	S.F.	.048	1.68	1.88	3.56
8/12 pitch	1.330	S.F.	.052	1.82	2.04	3.86
Drip edge, metal, 5″ wide	.150	L.F.	.003	.04	.12	.16
8″ wide	.150	L.F.	.003	.05	.12	.17
Building paper, #15 asphalt felt	1.300	S.F.	.002	.04	.06	.10
Ridge shingles, asphalt	.042	L.F.	.001	.03	.04	.07
Clay	.042	L.F.	.002	.40	.06	.46
Slate	.042	L.F.	.002	.39	.06	.45
Wood, shingles	.042	L.F.	.001	.11	.05	.16
Shakes	.042	L.F.	.001	.11	.05	.16
Soffit & fascia, aluminum, vented, 1′ overhang	.083	L.F.	.012	.20	.47	.67
2′ overhang	.083	L.F.	.013	.29	.52	.81
Vinyl, vented, 1′ overhang	.083	L.F.	.011	.14	.44	.58
2′ overhang	.083	L.F.	.012	.19	.47	.66
Wood, board fascia, plywood soffit, 1′ overhang	.083	L.F.	.004	.02	.13	.15
2′ overhang	.083	L.F.	.006	.03	.19	.22
Rake trim, painted, 1″ x 6″	.040	L.F.	.004	.05	.12	.17
1″ x 8″	.040	L.F.	.004	.08	.14	.22
Gutter, 5″ box, aluminum, seamless, painted	.083	L.F.	.006	.10	.24	.34
Vinyl	.083	L.F.	.006	.09	.24	.33
Downspout, 2″ x 3″, aluminum, one story house	.035	L.F.	.001	.04	.06	.10
Two story house	.060	L.F.	.003	.06	.11	.17
Vinyl, one story house	.035	L.F.	.002	.04	.07	.11
Two story house	.060	L.F.	.003	.06	.11	.17

08 Hip Roof Roofing Systems

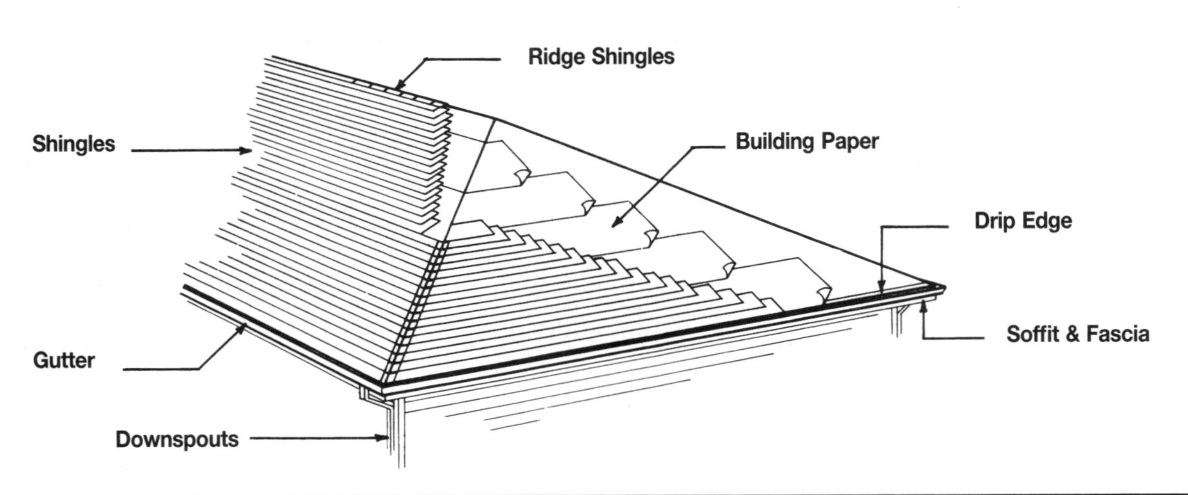

System Description	QUAN.	UNIT	LABOR HOURS	COST PER S.F.		
				MAT.	INST.	TOTAL
ASPHALT, ROOF SHINGLES, CLASS A						
Shingles, inorganic, class A, 210-235 lb./sq. 4/12 pitch	1.570	S.F.	.023	.54	.84	1.38
Drip edge, metal, 5″ wide	.122	L.F.	.002	.03	.10	.13
Building paper, #15 asphalt felt	1.800	S.F.	.002	.05	.09	.14
Ridge shingles, asphalt	.075	L.F.	.002	.06	.07	.13
Soffit & fascia, white painted aluminum, 1′ overhang	.120	L.F.	.017	.29	.68	.97
Gutter, seamless, aluminum, painted	.120	L.F.	.008	.14	.34	.48
Downspouts, aluminum, painted	.035	L.F.	.002	.04	.07	.11
TOTAL		S.F.	.056	1.15	2.19	3.34
WOOD, CEDAR SHINGLES, NO. 1 PERFECTIONS, 18″ LONG						
Shingles, red cedar, No. 1 perfections, 5″ exp., 4/12 pitch	1.570	S.F.	.047	2.82	1.82	4.64
Drip edge, metal, 5″ wide	.122	L.F.	.002	.03	.10	.13
Building paper, #15 asphalt felt	1.800	S.F.	.002	.05	.09	.14
Ridge shingles, wood, cedar	.075	L.F.	.002	.20	.08	.28
Soffit & fascia, white painted aluminum, 1′ overhang	.120	L.F.	.017	.29	.68	.97
Gutter, seamless, aluminum, painted	.120	L.F.	.008	.14	.34	.48
Downspouts, aluminum, painted	.035	L.F.	.002	.04	.07	.11
TOTAL		S.F.	.080	3.57	3.18	6.75

The prices in these systems are based on a square foot of plan area.
All quantities have been adjusted accordingly.

Description	QUAN.	UNIT	LABOR HOURS	COST PER S.F.		
				MAT.	INST.	TOTAL

Hip Roof - Roofing Price Sheet	QUAN.	UNIT	LABOR HOURS	COST PER S.F.		
				MAT.	INST.	TOTAL
Shingles, asphalt, inorganic, class A, 210-235 lb./sq., 4/12 pitch	1.570	S.F.	.023	.54	.84	1.38
8/12 pitch	1.850	S.F.	.028	.65	1	1.65
Laminated, multi-layered, 240-260 lb./sq., 4/12 pitch	1.570	S.F.	.028	.74	1.02	1.76
8/12 pitch	1.850	S.F.	.034	.88	1.22	2.10
Prem. laminated, multi-layered, 260-300 lb./sq., 4/12 pitch	1.570	S.F.	.037	.90	1.32	2.22
8/12 pitch	1.850	S.F.	.043	1.06	1.57	2.63
Clay tile, Spanish tile, red, 4/12 pitch	1.570	S.F.	.071	4.64	2.58	7.22
8/12 pitch	1.850	S.F.	.084	5.50	3.06	8.56
Mission tile, red, 4/12 pitch	1.570	S.F.	.111	11.10	4.02	15.12
8/12 pitch	1.850	S.F.	.132	13.20	4.77	17.97
French tile, red, 4/12 pitch	1.570	S.F.	.095	10.10	3.42	13.52
8/12 pitch	1.850	S.F.	.113	11.95	4.07	16.02
Slate, Buckingham, Virginia, black, 4/12 pitch	1.570	S.F.	.073	8.80	2.64	11.44
8/12 pitch	1.850	S.F.	.087	10.45	3.14	13.59
Vermont, black or grey, 4/12 pitch	1.570	S.F.	.073	5.75	2.64	8.39
8/12 pitch	1.850	S.F.	.087	6.85	3.14	9.99
Wood, red cedar, No.1 5X, 16″ long, 5″ exposure, 4/12 pitch	1.570	S.F.	.051	2.86	2	4.86
8/12 pitch	1.850	S.F.	.061	3.40	2.38	5.78
Fire retardant, 4/12 pitch	1.570	S.F.	.051	3.39	2	5.39
8/12 pitch	1.850	S.F.	.061	4.03	2.38	6.41
18″ long, No.1 perfections, 5″ exposure, 4/12 pitch	1.570	S.F.	.047	2.82	1.82	4.64
8/12 pitch	1.850	S.F.	.055	3.34	2.17	5.51
Fire retardant, 4/12 pitch	1.570	S.F.	.047	3.32	1.82	5.14
8/12 pitch	1.850	S.F.	.055	3.94	2.17	6.11
Resquared & rebutted, 18″ long, 6″ exposure, 4/12 pitch	1.570	S.F.	.043	3.50	1.68	5.18
8/12 pitch	1.850	S.F.	.051	4.16	2	6.16
Fire retardant, 4/12 pitch	1.570	S.F.	.043	4	1.68	5.68
8/12 pitch	1.850	S.F.	.051	4.76	2	6.76
Wood shakes hand split, 24″ long, 10″ exposure, 4/12 pitch	1.570	S.F.	.051	2.43	2	4.43
8/12 pitch	1.850	S.F.	.061	2.89	2.38	5.27
Fire retardant, 4/12 pitch	1.570	S.F.	.051	2.96	2	4.96
8/12 pitch	1.850	S.F.	.061	3.52	2.38	5.90
18″ long, 8″ exposure, 4/12 pitch	1.570	S.F.	.064	1.71	2.51	4.22
8/12 pitch	1.850	S.F.	.076	2.03	2.98	5.01
Fire retardant, 4/12 pitch	1.570	S.F.	.064	2.24	2.51	4.75
8/12 pitch	1.850	S.F.	.076	2.66	2.98	5.64
Drip edge, metal, 5″ wide	.122	L.F.	.002	.03	.10	.13
8″ wide	.122	L.F.	.002	.04	.10	.14
Building paper, #15 asphalt felt	1.800	S.F.	.002	.05	.09	.14
Ridge shingles, asphalt	.075	L.F.	.002	.06	.07	.13
Clay	.075	L.F.	.003	.71	.11	.82
Slate	.075	L.F.	.003	.70	.11	.81
Wood, shingles	.075	L.F.	.002	.20	.08	.28
Shakes	.075	L.F.	.002	.20	.08	.28
Soffit & fascia, aluminum, vented, 1′ overhang	.120	L.F.	.017	.29	.68	.97
2′ overhang	.120	L.F.	.019	.42	.75	1.17
Vinyl, vented, 1′ overhang	.120	L.F.	.016	.20	.63	.83
2′ overhang	.120	L.F.	.017	.28	.68	.96
Wood, board fascia, plywood soffit, 1′ overhang	.120	L.F.	.004	.02	.13	.15
2′ overhang	.120	L.F.	.006	.03	.19	.22
Gutter, 5″ box, aluminum, seamless, painted	.120	L.F.	.008	.14	.34	.48
Vinyl	.120	L.F.	.009	.13	.34	.47
Downspout, 2″ x 3″, aluminum, one story house	.035	L.F.	.002	.04	.07	.11
Two story house	.060	L.F.	.003	.06	.11	.17
Vinyl, one story house	.035	L.F.	.001	.04	.06	.10
Two story house	.060	L.F.	.003	.06	.11	.17

5 ROOFING

12 Gambrel Roofing Systems

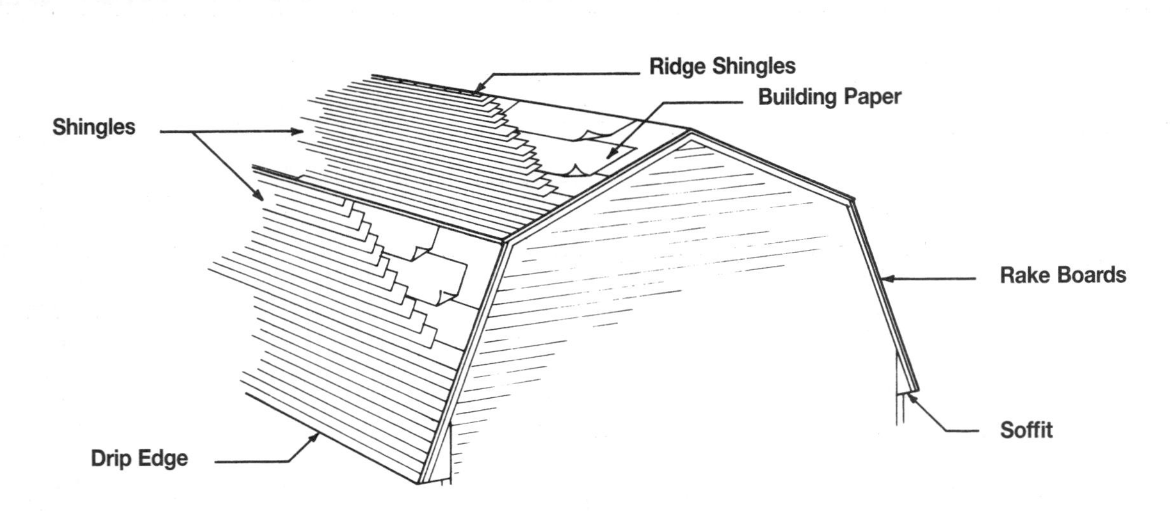

System Description	QUAN.	UNIT	LABOR HOURS	COST PER S.F.		
				MAT.	INST.	TOTAL
ASPHALT, ROOF SHINGLES, CLASS A						
Shingles, asphalt, inorganic, class A, 210-235 lb./sq.	1.450	S.F.	.022	.51	.79	1.30
Drip edge, metal, 5″ wide	.146	L.F.	.003	.04	.11	.15
Building paper, #15 asphalt felt	1.500	S.F.	.002	.04	.07	.11
Ridge shingles, asphalt	.042	L.F.	.001	.03	.04	.07
Soffit & fascia, painted aluminum, 1′ overhang	.083	L.F.	.012	.20	.47	.67
Rake trim, 1″ x 6″	.063	L.F.	.003	.07	.10	.17
Rake trim, prime and paint	.063	L.F.	.003	.02	.10	.12
Gutter, seamless, alumunum, painted	.083	L.F.	.006	.10	.24	.34
Downspouts, aluminum, painted	.042	L.F.	.002	.05	.08	.13
TOTAL		S.F.	.054	1.06	2	3.06
WOOD, CEDAR SHINGLES, NO. 1 PERFECTIONS, 18″ LONG						
Shingles, wood, red cedar, No. 1 perfections, 5″ exposure	1.450	S.F.	.044	2.64	1.71	4.35
Drip edge, metal, 5″ wide	.146	L.F.	.003	.04	.11	.15
Building paper, #15 asphalt felt	1.500	S.F.	.002	.04	.07	.11
Ridge shingles, wood	.042	L.F.	.001	.11	.05	.16
Soffit & fascia, white painted aluminum, 1′ overhang	.083	L.F.	.012	.20	.47	.67
Rake trim, 1″ x 6″	.063	L.F.	.003	.07	.10	.17
Rake trim, prime and paint	.063	L.F.	.001	.01	.04	.05
Gutter, seamless, aluminum, painted	.083	L.F.	.006	.10	.24	.34
Downspouts, aluminum, painted	.042	L.F.	.002	.05	.08	.13
TOTAL		S.F.	.074	3.26	2.87	6.13

The prices in this system are based on a square foot of plan area.
All quantities have been adjusted accordingly.

Description	QUAN.	UNIT	LABOR HOURS	COST PER S.F.		
				MAT.	INST.	TOTAL

 Important: See the Reference Section for critical supporting data - Reference Nos., Crews & Location Factors

Gambrel Roofing Price Sheet	QUAN.	UNIT	LABOR HOURS	COST PER S.F.		
				MAT.	INST.	TOTAL
Shingles, asphalt, standard, inorganic, class A, 210-235 lb./sq.	1.450	S.F.	.022	.51	.79	1.30
Laminated, multi-layered, 240-260 lb./sq.	1.450	S.F.	.027	.70	.96	1.66
Premium laminated, multi-layered, 260-300 lb./sq.	1.450	S.F.	.034	.84	1.24	2.08
Slate, Buckingham, Virginia, black	1.450	S.F.	.069	8.25	2.48	10.73
Vermont, black or grey	1.450	S.F.	.069	5.40	2.48	7.88
Wood, red cedar, No.1 5X, 16″ long, 5″ exposure, plain	1.450	S.F.	.048	2.69	1.88	4.57
Fire retardant	1.450	S.F.	.048	3.19	1.88	5.07
18″ long, No.1 perfections, 6″ exposure, plain	1.450	S.F.	.044	2.64	1.71	4.35
Fire retardant	1.450	S.F.	.044	3.11	1.71	4.82
Resquared & rebutted, 18″ long, 6″ exposure, plain	1.450	S.F.	.040	3.29	1.58	4.87
Fire retardant	1.450	S.F.	.040	3.75	1.58	5.33
Shakes, hand split, 24″ long, 10″ exposure, plain	1.450	S.F.	.048	2.28	1.88	4.16
Fire retardant	1.450	S.F.	.048	2.78	1.88	4.66
18″ long, 8″ exposure, plain	1.450	S.F.	.060	1.61	2.36	3.97
Fire retardant	1.450	S.F.	.060	2.11	2.36	4.47
Drip edge, metal, 5″ wide	.146	L.F.	.003	.04	.11	.15
8″ wide	.146	L.F.	.003	.05	.11	.16
Building paper, #15 asphalt felt	1.500	S.F.	.002	.04	.07	.11
Ridge shingles, asphalt	.042	L.F.	.001	.03	.04	.07
Slate	.042	L.F.	.002	.39	.06	.45
Wood, shingles	.042	L.F.	.001	.11	.05	.16
Shakes	.042	L.F.	.001	.11	.05	.16
Soffit & fascia, aluminum, vented, 1′ overhang	.083	L.F.	.012	.20	.47	.67
2′ overhang	.083	L.F.	.013	.29	.52	.81
Vinyl vented, 1′ overhang	.083	L.F.	.011	.14	.44	.58
2′ overhang	.083	L.F.	.012	.19	.47	.66
Wood board fascia, plywood soffit, 1′ overhang	.083	L.F.	.004	.02	.13	.15
2′ overhang	.083	L.F.	.006	.03	.19	.22
Rake trim, painted, 1″ x 6″	.063	L.F.	.006	.09	.20	.29
1″ x 8″	.063	L.F.	.007	.11	.26	.37
Gutter, 5″ box, aluminum, seamless, painted	.083	L.F.	.006	.10	.24	.34
Vinyl	.083	L.F.	.006	.09	.24	.33
Downspout 2″ x 3″, aluminum, one story house	.042	L.F.	.002	.04	.08	.12
Two story house	.070	L.F.	.003	.07	.13	.20
Vinyl, one story house	.042	L.F.	.002	.04	.08	.12
Two story house	.070	L.F.	.003	.07	.13	.20

16 Mansard Roofing Systems

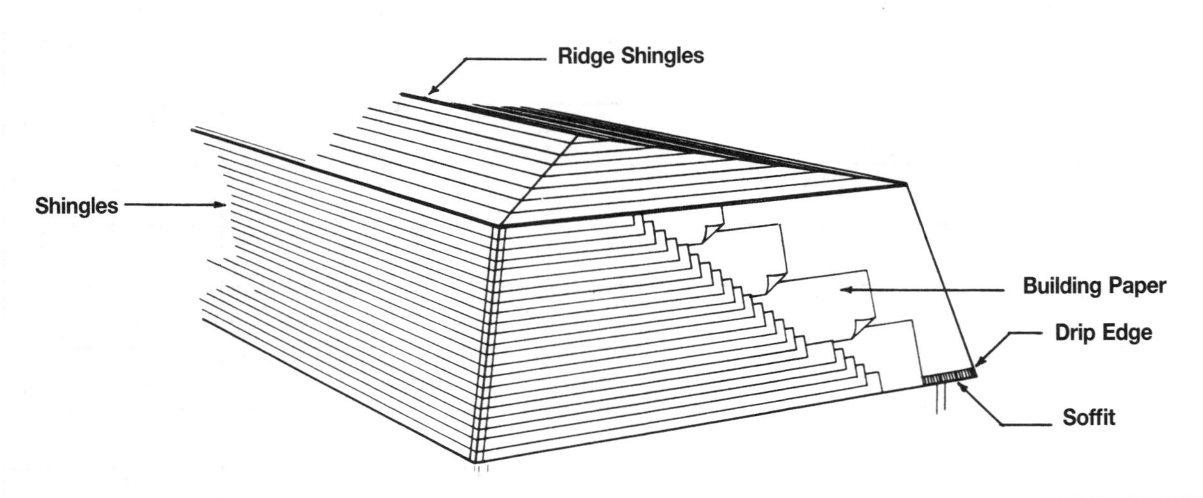

System Description	QUAN.	UNIT	LABOR HOURS	COST PER S.F.		
				MAT.	INST.	TOTAL
ASPHALT, ROOF SHINGLES, CLASS A						
Shingles, standard inorganic class A 210-235 lb./sq.	2.210	S.F.	.032	.75	1.16	1.91
Drip edge, metal, 5″ wide	.122	L.F.	.002	.03	.10	.13
Building paper, #15 asphalt felt	2.300	S.F.	.003	.07	.11	.18
Ridge shingles, asphalt	.090	L.F.	.002	.07	.08	.15
Soffit & fascia, white painted aluminum, 1′ overhang	.122	L.F.	.018	.29	.70	.99
Gutter, seamless, aluminum, painted	.122	L.F.	.008	.15	.35	.50
Downspouts, aluminum, painted	.042	L.F.	.002	.05	.08	.13
TOTAL		S.F.	.067	1.41	2.58	3.99
WOOD, CEDAR SHINGLES, NO. 1 PERFECTIONS, 18″ LONG						
Shingles, wood, red cedar, No. 1 perfections, 5″ exposure	2.210	S.F.	.064	3.87	2.51	6.38
Drip edge, metal, 5″ wide	.122	L.F.	.002	.03	.10	.13
Building paper, #15 asphalt felt	2.300	S.F.	.003	.07	.11	.18
Ridge shingles, wood	.090	L.F.	.003	.24	.10	.34
Soffit & fascia, white painted aluminum, 1′ overhang	.122	L.F.	.018	.29	.70	.99
Gutter, seamless, aluminum, painted	.122	L.F.	.008	.15	.35	.50
Downspouts, aluminum, painted	.042	L.F.	.002	.05	.08	.13
TOTAL		S.F.	.100	4.70	3.95	8.65

The prices in these systems are based on a square foot of plan area.
All quantities have been adjusted accordingly.

Description	QUAN.	UNIT	LABOR HOURS	COST PER S.F.		
				MAT.	INST.	TOTAL

 Important: See the Reference Section for critical supporting data - Reference Nos., Crews & Location Factors

Mansard Roofing Price Sheet

	QUAN.	UNIT	LABOR HOURS	COST PER S.F. MAT.	INST.	TOTAL
Shingles, asphalt, standard, inorganic, class A, 210-235 lb./sq.	2.210	S.F.	.032	.75	1.16	1.91
Laminated, multi-layered, 240-260 lb./sq.	2.210	S.F.	.039	1.02	1.41	2.43
Premium laminated, multi-layered, 260-300 lb./sq.	2.210	S.F.	.050	1.23	1.82	3.05
Slate Buckingham, Virginia, black	2.210	S.F.	.101	12.10	3.63	15.73
Vermont, black or grey	2.210	S.F.	.101	7.90	3.63	11.53
Wood, red cedar, No.1 5X, 16″ long, 5″ exposure, plain	2.210	S.F.	.070	3.94	2.75	6.69
Fire retardant	2.210	S.F.	.070	4.67	2.75	7.42
18″ long, No.1 perfections 6″ exposure, plain	2.210	S.F.	.064	3.87	2.51	6.38
Fire retardant	2.210	S.F.	.064	4.56	2.51	7.07
Resquared & rebutted, 18″ long, 6″ exposure, plain	2.210	S.F.	.059	4.82	2.31	7.13
Fire retardant	2.210	S.F.	.059	5.50	2.31	7.81
Shakes, hand split, 24″ long 10″ exposure, plain	2.210	S.F.	.070	3.34	2.75	6.09
Fire retardant	2.210	S.F.	.070	4.07	2.75	6.82
18″ long, 8″ exposure, plain	2.210	S.F.	.088	2.35	3.45	5.80
Fire retardant	2.210	S.F.	.088	3.08	3.45	6.53
Drip edge, metal, 5″ wide	.122	S.F.	.002	.03	.10	.13
8″ wide	.122	S.F.	.002	.04	.10	.14
Building paper, #15 asphalt felt	2.300	S.F.	.003	.07	.11	.18
Ridge shingles, asphalt	.090	L.F.	.002	.07	.08	.15
Slate	.090	L.F.	.004	.84	.13	.97
Wood, shingles	.090	L.F.	.003	.24	.10	.34
Shakes	.090	L.F.	.003	.24	.10	.34
Soffit & fascia, aluminum vented, 1′ overhang	.122	L.F.	.018	.29	.70	.99
2′ overhang	.122	L.F.	.020	.43	.76	1.19
Vinyl vented, 1′ overhang	.122	L.F.	.016	.20	.64	.84
2′ overhang	.122	L.F.	.018	.29	.70	.99
Wood board fascia, plywood soffit, 1′ overhang	.122	L.F.	.013	.28	.48	.76
2′ overhang	.122	L.F.	.019	.35	.72	1.07
Gutter, 5″ box, aluminum, seamless, painted	.122	L.F.	.008	.15	.35	.50
Vinyl	.122	L.F.	.009	.13	.35	.48
Downspout 2″ x 3″, aluminum, one story house	.042	L.F.	.002	.04	.08	.12
Two story house	.070	L.F.	.003	.07	.12	.19
Vinyl, one story house	.042	L.F.	.002	.04	.08	.12
Two story house	.070	L.F.	.003	.07	.12	.19

20 Shed Roofing Systems

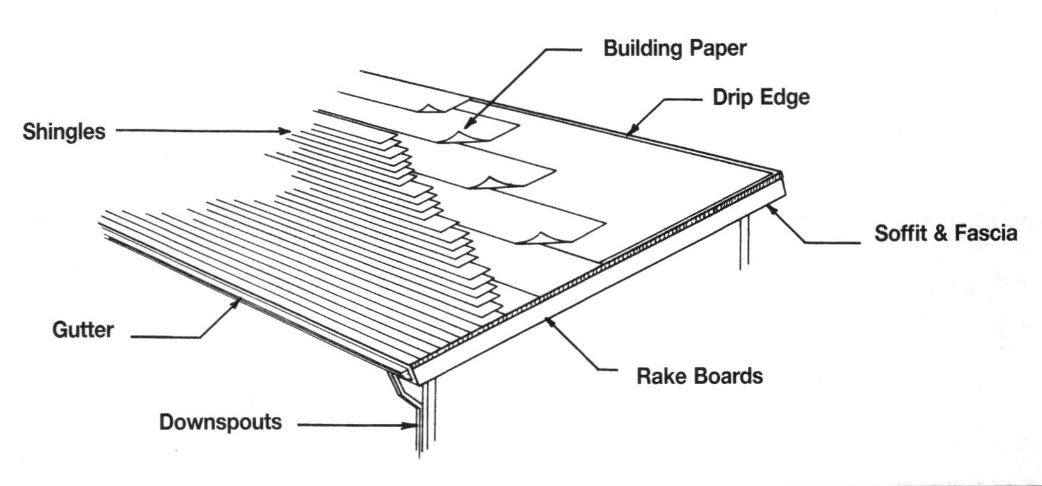

System Description	QUAN.	UNIT	LABOR HOURS	COST PER S.F.		
				MAT.	INST.	TOTAL
ASPHALT, ROOF SHINGLES, CLASS A						
Shingles, inorganic class A 210-235 lb./sq. 4/12 pitch	1.230	S.F.	.019	.44	.68	1.12
Drip edge, metal, 5″ wide	.100	L.F.	.002	.02	.08	.10
Building paper, #15 asphalt felt	1.300	S.F.	.002	.04	.06	.10
Soffit & fascia, white painted aluminum, 1′ overhang	.080	L.F.	.012	.19	.46	.65
Rake trim, 1″ x 6″	.043	L.F.	.002	.05	.07	.12
Rake trim, prime and paint	.043	L.F.	.002	.01	.07	.08
Gutter, seamless, aluminum, painted	.040	L.F.	.003	.05	.11	.16
Downspouts, painted aluminum	.020	L.F.	.001	.02	.04	.06
TOTAL		S.F.	.043	.82	1.57	2.39
WOOD, CEDAR SHINGLES, NO. 1 PERFECTIONS, 18″ LONG						
Shingles, red cedar, No. 1 perfections, 5″ exp., 4/12 pitch	1.230	S.F.	.035	2.11	1.37	3.48
Drip edge, metal, 5″ wide	.100	L.F.	.002	.02	.08	.10
Building paper, #15 asphalt felt	1.300	S.F.	.002	.04	.06	.10
Soffit & fascia, white painted aluminum, 1′ overhang	.080	L.F.	.012	.19	.46	.65
Rake trim, 1″ x 6″	.043	L.F.	.002	.05	.07	.12
Rake trim, prime and paint	.043	L.F.	.001	.01	.03	.04
Gutter, seamless, aluminum, painted	.040	L.F.	.003	.05	.11	.16
Downspouts, painted aluminum	.020	L.F.	.001	.02	.04	.06
TOTAL		S.F.	.058	2.49	2.22	4.71

The prices in these systems are based on a square foot of plan area.
All quantities have been adjusted accordingly.

Description	QUAN.	UNIT	LABOR HOURS	COST PER S.F.		
				MAT.	INST.	TOTAL

Important: See the Reference Section for critical supporting data - Reference Nos., Crews & Location Factors

Shed Roofing Price Sheet	QUAN.	UNIT	LABOR HOURS	COST PER S.F.		
				MAT.	INST.	TOTAL
Shingles, asphalt, inorganic, class A, 210-235 lb./sq., 4/12 pitch	1.230	S.F.	.017	.41	.63	1.04
8/12 pitch	1.330	S.F.	.019	.44	.68	1.12
Laminated, multi-layered, 240-260 lb./sq. 4/12 pitch	1.230	S.F.	.021	.56	.77	1.33
8/12 pitch	1.330	S.F.	.023	.60	.83	1.43
Premium laminated, multi-layered, 260-300 lb./sq. 4/12 pitch	1.230	S.F.	.027	.67	.99	1.66
8/12 pitch	1.330	S.F.	.030	.73	1.07	1.80
Clay tile, Spanish tile, red, 4/12 pitch	1.230	S.F.	.053	3.48	1.93	5.41
8/12 pitch	1.330	S.F.	.058	3.77	2.09	5.86
Mission tile, red, 4/12 pitch	1.230	S.F.	.083	8.35	3.01	11.36
8/12 pitch	1.330	S.F.	.090	9.05	3.26	12.31
French tile, red, 4/12 pitch	1.230	S.F.	.071	7.55	2.57	10.12
8/12 pitch	1.330	S.F.	.077	8.20	2.78	10.98
Slate, Buckingham, Virginia, black, 4/12 pitch	1.230	S.F.	.055	6.60	1.98	8.58
8/12 pitch	1.330	S.F.	.059	7.15	2.15	9.30
Vermont, black or grey, 4/12 pitch	1.230	S.F.	.055	4.32	1.98	6.30
8/12 pitch	1.330	S.F.	.059	4.68	2.15	6.83
Wood, red cedar, No.1 5X, 16″ long, 5″ exposure, 4/12 pitch	1.230	S.F.	.038	2.15	1.50	3.65
8/12 pitch	1.330	S.F.	.042	2.33	1.63	3.96
Fire retardant, 4/12 pitch	1.230	S.F.	.038	2.55	1.50	4.05
8/12 pitch	1.330	S.F.	.042	2.76	1.63	4.39
18″ long, 6″ exposure, 4/12 pitch	1.230	S.F.	.035	2.11	1.37	3.48
8/12 pitch	1.330	S.F.	.038	2.29	1.48	3.77
Fire retardant, 4/12 pitch	1.230	S.F.	.035	2.49	1.37	3.86
8/12 pitch	1.330	S.F.	.038	2.70	1.48	4.18
Resquared & rebutted, 18″ long, 6″ exposure, 4/12 pitch	1.230	S.F.	.032	2.63	1.26	3.89
8/12 pitch	1.330	S.F.	.035	2.85	1.37	4.22
Fire retardant, 4/12 pitch	1.230	S.F.	.032	3.01	1.26	4.27
8/12 pitch	1.330	S.F.	.035	3.26	1.37	4.63
Wood shakes, hand split, 24″ long, 10″ exposure, 4/12 pitch	1.230	S.F.	.038	1.82	1.50	3.32
8/12 pitch	1.330	S.F.	.042	1.98	1.63	3.61
Fire retardant, 4/12 pitch	1.230	S.F.	.038	2.22	1.50	3.72
8/12 pitch	1.330	S.F.	.042	2.41	1.63	4.04
18″ long, 8″ exposure, 4/12 pitch	1.230	S.F.	.048	1.28	1.88	3.16
8/12 pitch	1.330	S.F.	.052	1.39	2.04	3.43
Fire retardant, 4/12 pitch	1.230	S.F.	.048	1.68	1.88	3.56
8/12 pitch	1.330	S.F.	.052	1.82	2.04	3.86
Drip edge, metal, 5″ wide	.100	L.F.	.002	.02	.08	.10
8″ wide	.100	L.F.	.002	.04	.08	.12
Building paper, #15 asphalt felt	1.300	S.F.	.002	.04	.06	.10
Soffit & fascia, aluminum vented, 1′ overhang	.080	L.F.	.012	.19	.46	.65
2′ overhang	.080	L.F.	.013	.28	.50	.78
Vinyl vented, 1′ overhang	.080	L.F.	.011	.13	.42	.55
2′ overhang	.080	L.F.	.012	.19	.46	.65
Wood board fascia, plywood soffit, 1′ overhang	.080	L.F.	.010	.19	.35	.54
2′ overhang	.080	L.F.	.014	.24	.53	.77
Rake, trim, painted, 1″ x 6″	.043	L.F.	.004	.06	.14	.20
1″ x 8″	.043	L.F.	.004	.06	.14	.20
Gutter, 5″ box, aluminum, seamless, painted	.040	L.F.	.003	.05	.11	.16
Vinyl	.040	L.F.	.003	.04	.11	.15
Downspout 2″ x 3″, aluminum, one story house	.020	L.F.	.001	.02	.04	.06
Two story house	.020	L.F.	.001	.03	.06	.09
Vinyl, one story house	.020	L.F.	.001	.02	.04	.06
Two story house	.020	L.F.	.001	.03	.06	.09

24 Gable Dormer Roofing Systems

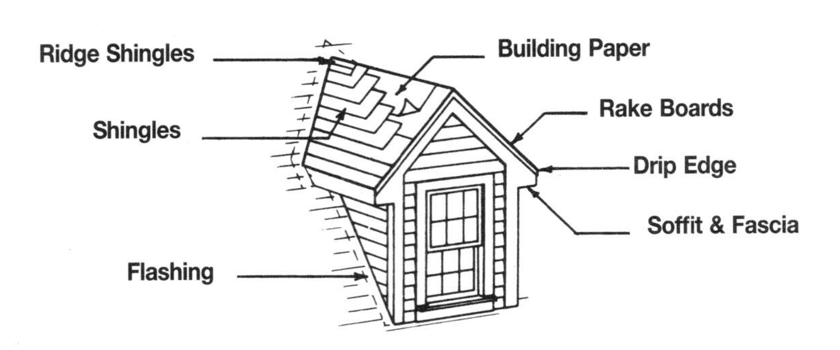

System Description	QUAN.	UNIT	LABOR HOURS	COST PER S.F.		
				MAT.	INST.	TOTAL
ASPHALT, ROOF SHINGLES, CLASS A						
Shingles, standard inorganic class A 210-235 lb./sq	1.400	S.F.	.020	.48	.74	1.22
Drip edge, metal, 5″ wide	.220	L.F.	.004	.05	.17	.22
Building paper, #15 asphalt felt	1.500	S.F.	.002	.04	.07	.11
Ridge shingles, asphalt	.280	L.F.	.007	.22	.24	.46
Soffit & fascia, aluminum, vented	.220	L.F.	.032	.52	1.25	1.77
Flashing, aluminum, mill finish, .013″ thick	1.500	S.F.	.083	.59	2.99	3.58
TOTAL		S.F.	.148	1.90	5.46	7.36
WOOD, CEDAR, NO. 1 PERFECTIONS						
Shingles, red cedar, No.1 perfections, 18″ long, 5″ exp.	1.400	S.F.	.041	2.46	1.60	4.06
Drip edge, metal, 5″ wide	.220	L.F.	.004	.05	.17	.22
Building paper, #15 asphalt felt	1.500	S.F.	.002	.04	.07	.11
Ridge shingles, wood	.280	L.F.	.008	.74	.31	1.05
Soffit & fascia, aluminum, vented	.220	L.F.	.032	.52	1.25	1.77
Flashing, aluminum, mill finish, .013″ thick	1.500	S.F.	.083	.59	2.99	3.58
TOTAL		S.F.	.170	4.40	6.39	10.79
SLATE, BUCKINGHAM, BLACK						
Shingles, Buckingham, Virginia, black	1.400	S.F.	.064	7.70	2.31	10.01
Drip edge, metal, 5″ wide	.220	L.F.	.004	.05	.17	.22
Building paper, #15 asphalt felt	1.500	S.F.	.002	.04	.07	.11
Ridge shingles, slate	.280	L.F.	.011	2.62	.41	3.03
Soffit & fascia, aluminum, vented	.220	L.F.	.032	.52	1.25	1.77
Flashing, copper, 16 oz.	1.500	S.F.	.104	3.68	3.75	7.43
TOTAL		S.F.	.217	14.61	7.96	22.57

The prices in these systems are based on a square foot of plan area under the dormer roof.

Description	QUAN.	UNIT	LABOR HOURS	COST PER S.F.		
				MAT.	INST.	TOTAL

Important: See the Reference Section for critical supporting data - Reference Nos., Crews & Location Factors

Gable Dormer Roofing Price Sheet	QUAN.	UNIT	LABOR HOURS	COST PER S.F.		
				MAT.	INST.	TOTAL
Shingles, asphalt, standard, inorganic, class A, 210-235 lb./sq.	1.400	S.F.	.020	.48	.74	1.22
Laminated, multi-layered, 240-260 lb./sq.	1.400	S.F.	.025	.65	.90	1.55
Premium laminated, multi-layered, 260-300 lb./sq.	1.400	S.F.	.032	.78	1.16	1.94
Clay tile, Spanish tile, red	1.400	S.F.	.062	4.06	2.25	6.31
Mission tile, red	1.400	S.F.	.097	9.75	3.51	13.26
French tile, red	1.400	S.F.	.083	8.80	3	11.80
Slate Buckingham, Virginia, black	1.400	S.F.	.064	7.70	2.31	10.01
Vermont, black or grey	1.400	S.F.	.064	5.05	2.31	7.36
Wood, red cedar, No.1 5X, 16″ long, 5″ exposure	1.400	S.F.	.045	2.51	1.75	4.26
Fire retardant	1.400	S.F.	.045	2.97	1.75	4.72
18″ long, No.1 perfections, 5″ exposure	1.400	S.F.	.041	2.46	1.60	4.06
Fire retardant	1.400	S.F.	.041	2.90	1.60	4.50
Resquared & rebutted, 18″ long, 5″ exposure	1.400	S.F.	.037	3.07	1.47	4.54
Fire retardant	1.400	S.F.	.037	3.51	1.47	4.98
Shakes hand split, 24″ long, 10″ exposure	1.400	S.F.	.045	2.13	1.75	3.88
Fire retardant	1.400	S.F.	.045	2.59	1.75	4.34
18″ long, 8″ exposure	1.400	S.F.	.056	1.50	2.20	3.70
Fire retardant	1.400	S.F.	.056	1.96	2.20	4.16
Drip edge, metal, 5″ wide	.220	L.F.	.004	.05	.17	.22
8″ wide	.220	L.F.	.004	.08	.17	.25
Building paper, #15 asphalt felt	1.500	S.F.	.002	.04	.07	.11
Ridge shingles, asphalt	.280	L.F.	.007	.22	.24	.46
Clay	.280	L.F.	.011	2.66	.41	3.07
Slate	.280	L.F.	.011	2.62	.41	3.03
Wood	.280	L.F.	.008	.74	.31	1.05
Soffit & fascia, aluminum, vented	.220	L.F.	.032	.52	1.25	1.77
Vinyl, vented	.220	L.F.	.029	.36	1.16	1.52
Wood, board fascia, plywood soffit	.220	L.F.	.026	.53	.98	1.51
Flashing, aluminum, .013″ thick	1.500	S.F.	.083	.59	2.99	3.58
.032″ thick	1.500	S.F.	.083	1.76	2.99	4.75
.040″ thick	1.500	S.F.	.083	2.30	2.99	5.29
.050″ thick	1.500	S.F.	.083	3.03	2.99	6.02
Copper, 16 oz.	1.500	S.F.	.104	3.68	3.75	7.43
20 oz.	1.500	S.F.	.109	6.45	3.93	10.38
24 oz.	1.500	S.F.	.114	7.75	4.11	11.86
32 oz.	1.500	S.F.	.120	10.30	4.32	14.62

28 Shed Dormer Roofing Systems

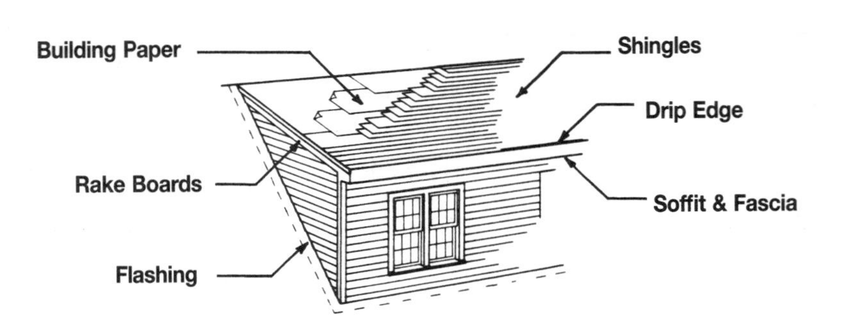

System Description	QUAN.	UNIT	LABOR HOURS	COST PER S.F.		
				MAT.	INST.	TOTAL
ASPHALT, ROOF SHINGLES, CLASS A						
Shingles, standard inorganic class A 210-235 lb./sq.	1.100	S.F.	.016	.37	.58	.95
Drip edge, aluminum, 5″ wide	.250	L.F.	.005	.07	.20	.27
Building paper, #15 asphalt felt	1.200	S.F.	.002	.04	.06	.10
Soffit & fascia, aluminum, vented, 1′ overhang	.250	L.F.	.036	.60	1.43	2.03
Flashing, aluminum, mill finish, 0.013″ thick	.800	L.F.	.044	.31	1.59	1.90
TOTAL		S.F.	.103	1.39	3.86	5.25
WOOD, CEDAR, NO. 1 PERFECTIONS, 18″ LONG						
Shingles, wood, red cedar, #1 perfections, 5″ exposure	1.100	S.F.	.032	1.94	1.25	3.19
Drip edge, aluminum, 5″ wide	.250	L.F.	.005	.07	.20	.27
Building paper, #15 asphalt felt	1.200	S.F.	.002	.04	.06	.10
Soffit & fascia, aluminum, vented, 1′ overhang	.250	L.F.	.036	.60	1.43	2.03
Flashing, aluminum, mill finish, 0.013″ thick	.800	L.F.	.044	.31	1.59	1.90
TOTAL		S.F.	.119	2.96	4.53	7.49
SLATE, BUCKINGHAM, BLACK						
Shingles, slate, Buckingham, black	1.100	S.F.	.050	6.05	1.82	7.87
Drip edge, aluminum, 5″ wide	.250	L.F.	.005	.07	.20	.27
Building paper, #15 asphalt felt	1.200	S.F.	.002	.04	.06	.10
Soffit & fascia, aluminum, vented, 1′ overhang	.250	L.F.	.036	.60	1.43	2.03
Flashing, copper, 16 oz.	.800	L.F.	.056	1.96	2	3.96
TOTAL		S.F.	.149	8.72	5.51	14.23

The prices in this system are based on a square foot of plan area under the dormer roof.

Description	QUAN.	UNIT	LABOR HOURS	COST PER S.F.		
				MAT.	INST.	TOTAL

Important: See the Reference Section for critical supporting data - Reference Nos., Crews & Location Factors

Shed Dormer Roofing Price Sheet	QUAN.	UNIT	LABOR HOURS	COST PER S.F.		
				MAT.	INST.	TOTAL
Shingles, asphalt, standard, inorganic, class A, 210-235 lb./sq.	1.100	S.F.	.016	.37	.58	.95
Laminated, multi-layered, 240-260 lb./sq.	1.100	S.F.	.020	.51	.70	1.21
Premium laminated, multi-layered, 260-300 lb./sq.	1.100	S.F.	.025	.62	.91	1.53
Clay tile, Spanish tile, red	1.100	S.F.	.049	3.19	1.77	4.96
Mission tile, red	1.100	S.F.	.077	7.65	2.76	10.41
French tile, red	1.100	S.F.	.065	6.95	2.35	9.30
Slate Buckingham, Virginia, black	1.100	S.F.	.050	6.05	1.82	7.87
Vermont, black or grey	1.100	S.F.	.050	3.96	1.82	5.78
Wood, red cedar, No. 1 5X, 16" long, 5" exposure	1.100	S.F.	.035	1.97	1.38	3.35
Fire retardant	1.100	S.F.	.035	2.33	1.38	3.71
18" long, No.1 perfections, 5" exposure	1.100	S.F.	.032	1.94	1.25	3.19
Fire retardant	1.100	S.F.	.032	2.29	1.25	3.54
Resquared & rebutted, 18" long, 5" exposure	1.100	S.F.	.029	2.41	1.16	3.57
Fire retardant	1.100	S.F.	.029	2.76	1.16	3.92
Shakes hand split, 24" long, 10" exposure	1.100	S.F.	.035	1.67	1.38	3.05
Fire retardant	1.100	S.F.	.035	2.03	1.38	3.41
18" long, 8" exposure	1.100	S.F.	.044	1.18	1.73	2.91
Fire retardant	1.100	S.F.	.044	1.54	1.73	3.27
Drip edge, metal, 5" wide	.250	L.F.	.005	.07	.20	.27
8" wide	.250	L.F.	.005	.09	.20	.29
Building paper, #15 asphalt felt	1.200	S.F.	.002	.04	.06	.10
Soffit & fascia, aluminum, vented	.250	L.F.	.036	.60	1.43	2.03
Vinyl, vented	.250	L.F.	.033	.41	1.31	1.72
Wood, board fascia, plywood soffit	.250	L.F.	.030	.61	1.11	1.72
Flashing, aluminum, .013" thick	.800	L.F.	.044	.31	1.59	1.90
.032" thick	.800	L.F.	.044	.94	1.59	2.53
.040" thick	.800	L.F.	.044	1.22	1.59	2.81
.050" thick	.800	L.F.	.044	1.62	1.59	3.21
Copper, 16 oz.	.800	L.F.	.056	1.96	2	3.96
20 oz.	.800	L.F.	.058	3.43	2.10	5.53
24 oz.	.800	L.F.	.061	4.12	2.19	6.31
32 oz.	.800	L.F.	.064	5.50	2.30	7.80

32 Skylight/Skywindow Systems

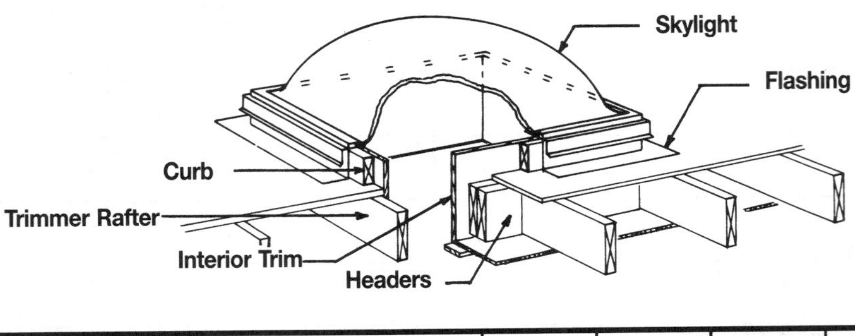

System Description	QUAN.	UNIT	LABOR HOURS	COST EACH		
				MAT.	INST.	TOTAL
SKYLIGHT, FIXED, 32" X 32"						
Skylight, fixed bubble, insulating, 32" x 32"	1.000	Ea.	1.422	104.89	50.84	155.73
Trimmer rafters, 2" x 6"	28.000	L.F.	.448	15.96	17.64	33.60
Headers, 2" x 6"	6.000	L.F.	.267	3.42	10.44	13.86
Curb, 2" x 4"	12.000	L.F.	.154	4.08	6	10.08
Flashing, aluminum, .013" thick	13.500	S.F.	.745	5.27	26.87	32.14
Trim, stock pine, 11/16" x 2-1/2"	12.000	L.F.	.400	11.28	15.72	27
Trim primer coat, oil base, brushwork	12.000	L.F.	.148	.36	5.04	5.40
Trim paint, 1 coat, brushwork	12.000	L.F.	.148	.36	5.04	5.40
TOTAL		Ea.	3.732	145.62	137.59	283.21
SKYLIGHT, FIXED, 48" X 48"						
Skylight, fixed bubble, insulating, 48" x 48"	1.000	Ea.	1.296	144.80	46.40	191.20
Trimmer rafters, 2" x 6"	28.000	L.F.	.448	15.96	17.64	33.60
Headers, 2" x 6"	8.000	L.F.	.356	4.56	13.92	18.48
Curb, 2" x 4"	16.000	L.F.	.205	5.44	8	13.44
Flashing, aluminum, .013" thick	16.000	S.F.	.883	6.24	31.84	38.08
Trim, stock pine, 11/16" x 2-1/2"	16.000	L.F.	.533	15.04	20.96	36
Trim primer coat, oil base, brushwork	16.000	L.F.	.197	.48	6.72	7.20
Trim paint, 1 coat, brushwork	16.000	L.F.	.197	.48	6.72	7.20
TOTAL		Ea.	4.115	193	152.20	345.20
SKYWINDOW, OPERATING, 24" X 48"						
Skywindow, operating, thermopane glass, 24" x 48"	1.000	Ea.	3.200	585	115	700
Trimmer rafters, 2" x 6"	28.000	L.F.	.448	15.96	17.64	33.60
Headers, 2" x 6"	8.000	L.F.	.267	3.42	10.44	13.86
Curb, 2" x 4"	14.000	L.F.	.179	4.76	7	11.76
Flashing, aluminum, .013" thick	14.000	S.F.	.772	5.46	27.86	33.32
Trim, stock pine, 11/16" x 2-1/2"	14.000	L.F.	.467	13.16	18.34	31.50
Trim primer coat, oil base, brushwork	14.000	L.F.	.172	.42	5.88	6.30
Trim paint, 1 coat, brushwork	14.000	L.F.	.172	.42	5.88	6.30
TOTAL		Ea.	5.677	628.60	208.04	836.64

The prices in these systems are on a cost each basis.

Description	QUAN.	UNIT	LABOR HOURS	COST EACH		
				MAT.	INST.	TOTAL

Skylight/Skywindow Price Sheet

	QUAN.	UNIT	LABOR HOURS	COST EACH		
				MAT.	INST.	TOTAL
Skylight, fixed bubble insulating, 24″ x 24″	1.000	Ea.	.800	59	28.50	87.50
32″ x 32″	1.000	Ea.	1.422	105	51	156
32″ x 48″	1.000	Ea.	.864	96.50	31	127.50
48″ x 48″	1.000	Ea.	1.296	145	46.50	191.50
Ventilating bubble insulating, 36″ x 36″	1.000	Ea.	2.667	405	95.50	500.50
52″ x 52″	1.000	Ea.	2.667	610	95.50	705.50
28″ x 52″	1.000	Ea.	3.200	475	115	590
36″ x 52″	1.000	Ea.	3.200	515	115	630
Skywindow, operating, thermopane glass, 24″ x 48″	1.000	Ea.	3.200	585	115	700
32″ x 48″	1.000	Ea.	3.556	610	127	737
Trimmer rafters, 2″ x 6″	28.000	L.F.	.448	15.95	17.65	33.60
2″ x 8″	28.000	L.F.	.472	24.50	18.50	43
2″ x 10″	28.000	L.F.	.711	33.50	28	61.50
Headers, 24″ window, 2″ x 6″	4.000	L.F.	.178	2.28	6.95	9.23
2″ x 8″	4.000	L.F.	.188	3.52	7.35	10.87
2″ x 10″	4.000	L.F.	.200	4.76	7.85	12.61
32″ window, 2″ x 6″	6.000	L.F.	.267	3.42	10.45	13.87
2″ x 8″	6.000	L.F.	.282	5.30	11.05	16.35
2″ x 10″	6.000	L.F.	.300	7.15	11.75	18.90
48″ window, 2″ x 6″	8.000	L.F.	.356	4.56	13.90	18.46
2″ x 8″	8.000	L.F.	.376	7.05	14.70	21.75
2″ x 10″	8.000	L.F.	.400	9.50	15.70	25.20
Curb, 2″ x 4″, skylight, 24″ x 24″	8.000	L.F.	.102	2.72	4	6.72
32″ x 32″	12.000	L.F.	.154	4.08	6	10.08
32″ x 48″	14.000	L.F.	.179	4.76	7	11.76
48″ x 48″	16.000	L.F.	.205	5.45	8	13.45
Flashing, aluminum .013″ thick, skylight, 24″ x 24″	9.000	S.F.	.497	3.51	17.90	21.41
32″ x 32″	13.500	S.F.	.745	5.25	27	32.25
32″ x 48″	14.000	S.F.	.772	5.45	28	33.45
48″ x 48″	16.000	S.F.	.883	6.25	32	38.25
Copper 16 oz., skylight, 24″ x 24″	9.000	S.F.	.626	22	22.50	44.50
32″ x 32″	13.500	S.F.	.939	33	34	67
32″ x 48″	14.000	S.F.	.974	34.50	35	69.50
48″ x 48″	16.000	S.F.	1.113	39	40	79
Trim, interior casing painted, 24″ x 24″	8.000	L.F.	.347	8.30	13.30	21.60
32″ x 32″	12.000	L.F.	.520	12.50	19.90	32.40
32″ x 48″	14.000	L.F.	.607	14.55	23	37.55
48″ x 48″	16.000	L.F.	.693	16.65	26.50	43.15

34 Built-up Roofing Systems

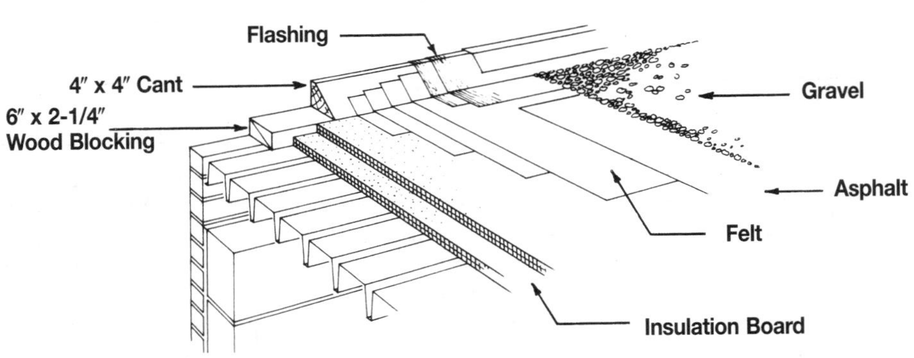

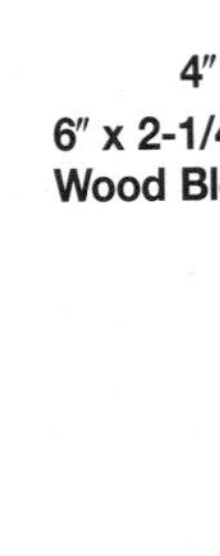

System Description	QUAN.	UNIT	LABOR HOURS	COST PER S.F.		
				MAT.	INST.	TOTAL
ASPHALT, ORGANIC, 4-PLY, INSULATED DECK						
Membrane, asphalt, 4-plies #15 felt, gravel surfacing	1.000	S.F.	.025	.68	1.01	1.69
Insulation board, 2-layers of 1-1/16″ glass fiber	2.000	S.F.	.016	1.68	.58	2.26
Wood blocking, 2″ x 6″	.040	L.F.	.004	.07	.17	.24
Treated 4″ x 4″ cant strip	.040	L.F.	.001	.05	.04	.09
Flashing, aluminum, 0.040″ thick	.050	S.F.	.003	.08	.10	.18
TOTAL		S.F.	.049	2.56	1.90	4.46
ASPHALT, INORGANIC, 3-PLY, INSULATED DECK						
Membrane, asphalt, 3-plies type IV glass felt, gravel surfacing	1.000	S.F.	.028	.66	1.12	1.78
Insulation board, 2-layers of 1-1/16″ glass fiber	2.000	S.F.	.016	1.68	.58	2.26
Wood blocking, 2″ x 6″	.040	L.F.	.004	.07	.17	.24
Treated 4″ x 4″ cant strip	.040	L.F.	.001	.05	.04	.09
Flashing, aluminum, 0.040″ thick	.050	S.F.	.003	.08	.10	.18
TOTAL		S.F.	.052	2.54	2.01	4.55
COAL TAR, ORGANIC, 4-PLY, INSULATED DECK						
Membrane, coal tar, 4-plies #15 felt, gravel surfacing	1.000	S.F.	.027	1.11	1.06	2.17
Insulation board, 2-layers of 1-1/16″ glass fiber	2.000	S.F.	.016	1.68	.58	2.26
Wood blocking, 2″ x 6″	.040	L.F.	.004	.07	.17	.24
Treated 4″ x 4″ cant strip	.040	L.F.	.001	.05	.04	.09
Flashing, aluminum, 0.040″ thick	.050	S.F.	.003	.08	.10	.18
TOTAL		S.F.	.051	2.99	1.95	4.94
COAL TAR, INORGANIC, 3-PLY, INSULATED DECK						
Membrane, coal tar, 3-plies type IV glass felt, gravel surfacing	1.000	S.F.	.029	.92	1.17	2.09
Insulation board, 2-layers of 1-1/16″ glass fiber	2.000	S.F.	.016	1.68	.58	2.26
Wood blocking, 2″ x 6″	.040	L.F.	.004	.07	.17	.24
Treated 4″ x 4″ cant strip	.040	L.F.	.001	.05	.04	.09
Flashing, aluminum, 0.040″ thick	.050	S.F.	.003	.08	.10	.18
TOTAL		S.F.	.053	2.80	2.06	4.86

Built-Up Roofing Price Sheet	QUAN.	UNIT	LABOR HOURS	COST PER S.F.		
				MAT.	INST.	TOTAL
Membrane, asphalt, 4-plies #15 organic felt, gravel surfacing	1.000	S.F.	.025	.68	1.01	1.69
Asphalt base sheet & 3-plies #15 asphalt felt	1.000	S.F.	.025	.52	1.01	1.53
3-plies type IV glass fiber felt	1.000	S.F.	.028	.66	1.12	1.78
4-plies type IV glass fiber felt	1.000	S.F.	.028	.81	1.12	1.93
Coal tar, 4-plies #15 organic felt, gravel surfacing	1.000	S.F.	.027			
4-plies tarred felt	1.000	S.F.	.027	1.11	1.06	2.17
3-plies type IV glass fiber felt	1.000	S.F.	.029	.92	1.17	2.09
4-plies type IV glass fiber felt	1.000	S.F.	.027	1.27	1.06	2.33
Roll, asphalt, 1-ply #15 organic felt, 2-plies mineral surfaced	1.000	S.F.	.021	.43	.82	1.25
3-plies type IV glass fiber, 1-ply mineral surfaced	1.000	S.F.	.022	.67	.89	1.56
Insulation boards, glass fiber, 1-1/16″ thick	1.000	S.F.	.008	.84	.29	1.13
2-1/16″ thick	1.000	S.F.	.010	1.23	.36	1.59
2-7/16″ thick	1.000	S.F.	.010	1.41	.36	1.77
Expanded perlite, 1″ thick	1.000	S.F.	.010	.33	.36	.69
1-1/2″ thick	1.000	S.F.	.010	.43	.36	.79
2″ thick	1.000	S.F.	.011	.67	.41	1.08
Fiberboard, 1″ thick	1.000	S.F.	.010	.37	.36	.73
1-1/2″ thick	1.000	S.F.	.010	.55	.36	.91
2″ thick	1.000	S.F.	.010	.75	.36	1.11
Extruded polystyrene, 15 PSI compressive strength, 2″ thick R10	1.000	S.F.	.006	.39	.23	.62
3″ thick R15	1.000	S.F.	.008	.75	.29	1.04
4″ thick R20	1.000	S.F.	.008	1.18	.29	1.47
Tapered for drainage	1.000	S.F.	.005	.39	.19	.58
40 PSI compressive strength, 1″ thick R5	1.000	S.F.	.005	.39	.19	.58
2″ thick R10	1.000	S.F.	.006	.76	.23	.99
3″ thick R15	1.000	S.F.	.008	1.11	.29	1.40
4″ thick R20	1.000	S.F.	.008	1.49	.29	1.78
Fiberboard high density, 1/2″ thick R1.3	1.000	S.F.	.008	.22	.29	.51
1″ thick R2.5	1.000	S.F.	.010	.40	.36	.76
1 1/2″ thick R3.8	1.000	S.F.	.010	.65	.36	1.01
Polyisocyanurate, 1 1/2″ thick R10.87	1.000	S.F.	.006	.41	.23	.64
2″ thick R14.29	1.000	S.F.	.007	.53	.26	.79
3 1/2″ thick R25	1.000	S.F.	.008	.81	.29	1.10
Tapered for drainage	1.000	S.F.	.006	.42	.21	.63
Expanded polystyrene, 1″ thick	1.000	S.F.	.005	.21	.19	.40
2″ thick R10	1.000	S.F.	.006	.41	.23	.64
3″ thick	1.000	S.F.	.006	.65	.23	.88
Wood blocking, treated, 6″ x 2″ & 4″ x 4″ cant	.040	L.F.	.002	.08	.10	.18
6″ x 4-1/2″ & 4″ x 4″ cant	.040	L.F.	.005	.14	.21	.35
6″ x 5″ & 4″ x 4″ cant	.040	L.F.	.007	.17	.27	.44
Flashing, aluminum, 0.019″ thick	.050	S.F.	.003	.04	.10	.14
0.032″ thick	.050	S.F.	.003	.06	.10	.16
0.040″ thick	.050	S.F.	.003	.08	.10	.18
Copper sheets, 16 oz., under 500 lbs.	.050	S.F.	.003	.12	.13	.25
Over 500 lbs.	.050	S.F.	.003	.16	.09	.25
20 oz., under 500 lbs.	.050	S.F.	.004	.21	.13	.34
Over 500 lbs.	.050	S.F.	.003	.20	.10	.30
Stainless steel, 32 gauge	.050	S.F.	.003	.12	.09	.21
28 gauge	.050	S.F.	.003	.15	.09	.24
26 gauge	.050	S.F.	.003	.18	.09	.27
24 gauge	.050	S.F.	.003	.23	.09	.32

Division 6
Interiors

04 Drywall & Thincoat Wall Systems

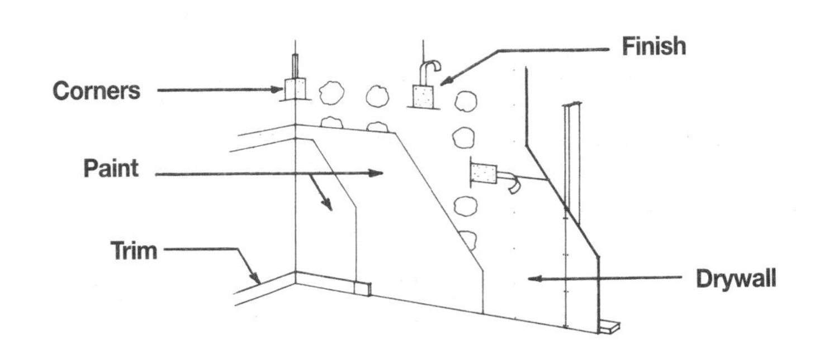

System Description	QUAN.	UNIT	LABOR HOURS	COST PER S.F.		
				MAT.	INST.	TOTAL
1/2" SHEETROCK, TAPED & FINISHED						
Drywall, 1/2" thick, standard	1.000	S.F.	.008	.21	.31	.52
Finish, taped & finished joints	1.000	S.F.	.008	.04	.31	.35
Corners, taped & finished, 32 L.F. per 12' x 12' room	.083	L.F.	.002	.01	.06	.07
Painting, primer & 2 coats	1.000	S.F.	.011	.15	.36	.51
Paint trim, to 6" wide, primer + 1 coat enamel	.125	L.F.	.001	.01	.04	.05
Trim, baseboard	.125	L.F.	.005	.21	.20	.41
TOTAL		S.F.	.034	.63	1.28	1.91
THINCOAT, SKIM-COAT, ON 1/2" BACKER DRYWALL						
Drywall, 1/2" thick, thincoat backer	1.000	S.F.	.008	.21	.31	.52
Thincoat plaster	1.000	S.F.	.011	.07	.39	.46
Corners, taped & finished, 32 L.F. per 12' x 12' room	.083	L.F.	.002	.01	.06	.07
Painting, primer & 2 coats	1.000	S.F.	.011	.15	.36	.51
Paint trim, to 6" wide, primer + 1 coat enamel	.125	L.F.	.001	.01	.04	.05
Trim, baseboard	.125	L.F.	.005	.21	.20	.41
TOTAL		S.F.	.037	.66	1.36	2.02
5/8" SHEETROCK, TAPED & FINISHED						
Drywall, 5/8" thick, standard	1.000	S.F.	.008	.24	.31	.55
Finish, taped & finished joints	1.000	S.F.	.008	.04	.31	.35
Corners, taped & finished, 32 L.F. per 12' x 12' room	.083	L.F.	.002	.01	.06	.07
Painting, primer & 2 coats	1.000	S.F.	.011	.15	.36	.51
Trim, baseboard	.125	L.F.	.005	.21	.20	.41
Paint trim, to 6" wide, primer + 1 coat enamel	.125	L.F.	.001	.01	.04	.05
TOTAL		S.F.	.034	.66	1.28	1.94

The costs in this system are based on a square foot of wall.
Do not deduct for openings.

Description	QUAN.	UNIT	LABOR HOURS	COST PER S.F.		
				MAT.	INST.	TOTAL

Drywall & Thincoat Wall Price Sheet	QUAN.	UNIT	LABOR HOURS	COST PER S.F.		
				MAT.	INST.	TOTAL
Drywall-sheetrock, 1/2" thick, standard	1.000	S.F.	.008	.21	.31	.52
Fire resistant	1.000	S.F.	.008	.22	.31	.53
Water resistant	1.000	S.F.	.008	.26	.31	.57
5/8" thick, standard	1.000	S.F.	.008	.24	.31	.55
Fire resistant	1.000	S.F.	.008	.25	.31	.56
Water resistant	1.000	S.F.	.008	.30	.31	.61
Drywall backer for thincoat system, 1/2" thick	1.000	S.F.	.008	.21	.31	.52
5/8" thick	1.000	S.F.	.008	.24	.31	.55
Finish drywall, taped & finished	1.000	S.F.	.008	.04	.31	.35
Texture spray	1.000	S.F.	.010	.04	.35	.39
Thincoat plaster, including tape	1.000	S.F.	.011	.07	.39	.46
Corners drywall, taped & finished, 32 L.F. per 4′ x 4′ room	.250	L.F.	.004	.02	.17	.19
6′ x 6′ room	.110	L.F.	.002	.01	.07	.08
10′ x 10′ room	.100	L.F.	.001	.01	.07	.08
12′ x 12′ room	.083	L.F.	.001	.01	.05	.06
16′ x 16′ room	.063	L.F.	.001	.01	.04	.05
Thincoat system, 32 L.F. per 4′ x 4′ room	.250	L.F.	.003	.02	.10	.12
6′ x 6′ room	.110	L.F.	.001	.01	.04	.05
10′ x 10′ room	.100	L.F.	.001	.01	.04	.05
12′ x 12′ room	.083	L.F.	.001	.01	.03	.04
16′ x 16′ room	.063	L.F.	.001		.02	.02
Painting, primer, & 1 coat	1.000	S.F.	.008	.10	.28	.38
& 2 coats	1.000	S.F.	.011	.15	.36	.51
Wallpaper, $7/double roll	1.000	S.F.	.013	.32	.43	.75
$17/double roll	1.000	S.F.	.015	.70	.51	1.21
$40/double roll	1.000	S.F.	.018	1.67	.63	2.30
Tile, ceramic adhesive thin set, 4 1/4" x 4 1/4" tiles	1.000	S.F.	.084	2.30	2.65	4.95
6" x 6" tiles	1.000	S.F.	.080	2.89	2.52	5.41
Pregrouted sheets	1.000	S.F.	.067	4.57	2.10	6.67
Trim, painted or stained, baseboard	.125	L.F.	.006	.22	.24	.46
Base shoe	.125	L.F.	.005	.15	.20	.35
Chair rail	.125	L.F.	.005	.13	.19	.32
Cornice molding	.125	L.F.	.004	.10	.16	.26
Cove base, vinyl	.125	L.F.	.003	.06	.11	.17
Paneling, not including furring or trim						
Plywood, prefinished, 1/4" thick, 4′ x 8′ sheets, vert. grooves						
Birch faced, minimum	1.000	S.F.	.032	.87	1.25	2.12
Average	1.000	S.F.	.038	1.33	1.49	2.82
Maximum	1.000	S.F.	.046	1.94	1.79	3.73
Mahogany, African	1.000	S.F.	.040	2.48	1.57	4.05
Philippine (lauan)	1.000	S.F.	.032	1.07	1.25	2.32
Oak or cherry, minimum	1.000	S.F.	.032	2.08	1.25	3.33
Maximum	1.000	S.F.	.040	3.19	1.57	4.76
Rosewood	1.000	S.F.	.050	4.52	1.96	6.48
Teak	1.000	S.F.	.040	3.19	1.57	4.76
Chestnut	1.000	S.F.	.043	4.71	1.67	6.38
Pecan	1.000	S.F.	.040	2.04	1.57	3.61
Walnut, minimum	1.000	S.F.	.032	2.72	1.25	3.97
Maximum	1.000	S.F.	.040	5.15	1.57	6.72

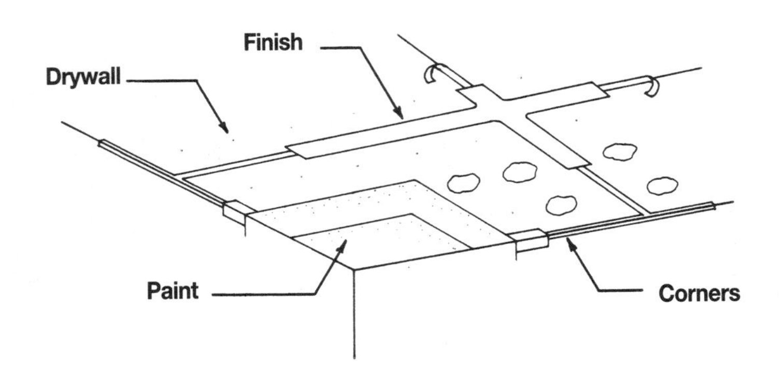

System Description	QUAN.	UNIT	LABOR HOURS	COST PER S.F.		
				MAT.	INST.	TOTAL
1/2″ SHEETROCK, TAPED & FINISHED						
Drywall, 1/2″ thick, standard	1.000	S.F.	.008	.21	.31	.52
Finish, taped & finished	1.000	S.F.	.008	.04	.31	.35
Corners, taped & finished, 12′ x 12′ room	.333	L.F.	.006	.03	.22	.25
Paint, primer & 2 coats	1.000	S.F.	.011	.15	.36	.51
TOTAL		S.F.	.032	.43	1.20	1.63
THINCOAT, SKIM COAT ON 1/2″ BACKER DRYWALL						
Drywall, 1/2″ thick, thincoat backer	1.000	S.F.	.008	.21	.31	.52
Thincoat plaster	1.000	S.F.	.011	.07	.39	.46
Corners, taped & finished, 12′ x 12′ room	.333	L.F.	.006	.03	.22	.25
Paint, primer & 2 coats	1.000	S.F.	.011	.15	.36	.51
TOTAL		S.F.	.035	.46	1.28	1.74
WATER-RESISTANT SHEETROCK, 1/2″ THICK, TAPED & FINISHED						
Drywall, 1/2″ thick, water-resistant	1.000	S.F.	.008	.26	.31	.57
Finish, taped & finished	1.000	S.F.	.008	.04	.31	.35
Corners, taped & finished, 12′ x 12′ room	.333	L.F.	.006	.03	.22	.25
Paint, primer & 2 coats	1.000	S.F.	.011	.15	.36	.51
TOTAL		S.F.	.032	.48	1.20	1.68
5/8″ SHEETROCK, TAPED & FINISHED						
Drywall, 5/8″ thick, standard	1.000	S.F.	.008	.24	.31	.55
Finish, taped & finished	1.000	S.F.	.008	.04	.31	.35
Corners, taped & finished, 12′ x 12′ room	.333	L.F.	.006	.03	.22	.25
Paint, primer & 2 coats	1.000	S.F.	.011	.15	.36	.51
TOTAL		S.F.	.032	.46	1.20	1.66

The costs in this system are based on a square foot of ceiling.

Description	QUAN.	UNIT	LABOR HOURS	COST PER S.F.		
				MAT.	INST.	TOTAL

Drywall & Thincoat Ceilings

	QUAN.	UNIT	LABOR HOURS	COST PER S.F.		
				MAT.	INST.	TOTAL
Drywall-sheetrock, 1/2" thick, standard	1.000	S.F.	.008	.21	.31	.52
Fire resistant	1.000	S.F.	.008	.22	.31	.53
Water resistant	1.000	S.F.	.008	.26	.31	.57
5/8" thick, standard	1.000	S.F.	.008	.24	.31	.55
Fire resistant	1.000	S.F.	.008	.25	.31	.56
Water resistant	1.000	S.F.	.008	.30	.31	.61
Drywall backer for thincoat system, 1/2" thick	1.000	S.F.	.016	.43	.62	1.05
5/8" thick	1.000	S.F.	.016	.46	.62	1.08
Finish drywall, taped & finished	1.000	S.F.	.008	.04	.31	.35
Texture spray	1.000	S.F.	.010	.04	.35	.39
Thincoat plaster	1.000	S.F.	.011	.07	.39	.46
Corners taped & finished, 4′ x 4′ room	1.000	L.F.	.015	.09	.66	.75
6′ x 6′ room	.667	L.F.	.010	.06	.44	.50
10′ x 10′ room	.400	L.F.	.006	.04	.26	.30
12′ x 12′ room	.333	L.F.	.005	.03	.22	.25
16′ x 16′ room	.250	L.F.	.003	.02	.12	.14
Thincoat system, 4′ x 4′ room	1.000	L.F.	.011	.07	.39	.46
6′ x 6′ room	.667	L.F.	.007	.05	.26	.31
10′ x 10′ room	.400	L.F.	.004	.03	.15	.18
12′ x 12′ room	.333	L.F.	.004	.02	.13	.15
16′ x 16′ room	.250	L.F.	.002	.01	.08	.09
Painting, primer & 1 coat	1.000	S.F.	.008	.10	.28	.38
& 2 coats	1.000	S.F.	.011	.15	.36	.51
Wallpaper, double roll, solid pattern, avg. workmanship	1.000	S.F.	.013	.32	.43	.75
Basic pattern, avg. workmanship	1.000	S.F.	.015	.70	.51	1.21
Basic pattern, quality workmanship	1.000	S.F.	.018	1.67	.63	2.30
Tile, ceramic adhesive thin set, 4 1/4" x 4 1/4" tiles	1.000	S.F.	.084	2.30	2.65	4.95
6" x 6" tiles	1.000	S.F.	.080	2.89	2.52	5.41
Pregrouted sheets	1.000	S.F.	.067	4.57	2.10	6.67

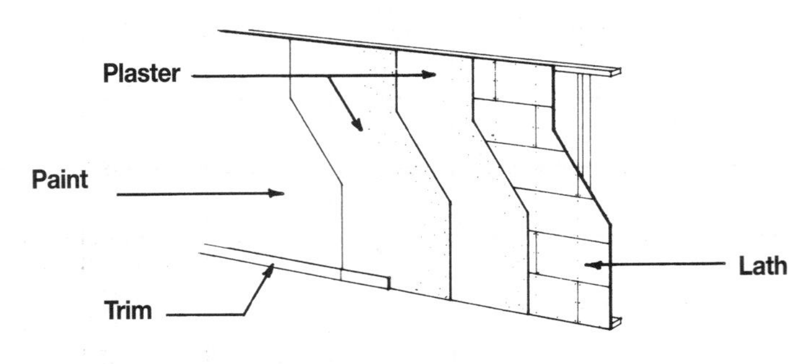

System Description	QUAN.	UNIT	LABOR HOURS	COST PER S.F.		
				MAT.	INST.	TOTAL
PLASTER ON GYPSUM LATH						
Plaster, gypsum or perlite, 2 coats	1.000	S.F.	.053	.38	1.88	2.26
Lath, 3/8″ gypsum	1.000	S.F.	.010	.42	.36	.78
Corners, expanded metal, 32 L.F. per 12′ x 12′ room	.083	L.F.	.002	.01	.06	.07
Painting, primer & 2 coats	1.000	S.F.	.011	.15	.36	.51
Paint trim, to 6″ wide, primer + 1 coat enamel	.125	L.F.	.001	.01	.04	.05
Trim, baseboard	.125	L.F.	.005	.21	.20	.41
TOTAL		S.F.	.082	1.18	2.90	4.08
PLASTER ON METAL LATH						
Plaster, gypsum or perlite, 2 coats	1.000	S.F.	.053	.38	1.88	2.26
Lath, 2.5 Lb. diamond, metal	1.000	S.F.	.010	.17	.36	.53
Corners, expanded metal, 32 L.F. per 12′ x 12′ room	.083	L.F.	.002	.01	.06	.07
Painting, primer & 2 coats	1.000	S.F.	.011	.15	.36	.51
Paint trim, to 6″ wide, primer + 1 coat enamel	.125	L.F.	.001	.01	.04	.05
Trim, baseboard	.125	L.F.	.005	.21	.20	.41
TOTAL		S.F.	.082	.93	2.90	3.83
STUCCO ON METAL LATH						
Stucco, 2 coats	1.000	S.F.	.041	.24	1.44	1.68
Lath, 2.5 Lb. diamond, metal	1.000	S.F.	.010	.17	.36	.53
Corners, expanded metal, 32 L.F. per 12′ x 12′ room	.083	L.F.	.002	.01	.06	.07
Painting, primer & 2 coats	1.000	S.F.	.011	.15	.36	.51
Paint trim, to 6″ wide, primer + 1 coat enamel	.125	L.F.	.001	.01	.04	.05
Trim, baseboard	.125	L.F.	.005	.21	.20	.41
TOTAL		S.F.	.070	.79	2.46	3.25

The costs in these systems are based on a per square foot of wall area.
Do not deduct for openings.

Description	QUAN.	UNIT	LABOR HOURS	COST PER S.F.		
				MAT.	INST.	TOTAL

Plaster & Stucco Wall Price Sheet

	QUAN.	UNIT	LABOR HOURS	COST PER S.F.		
				MAT.	INST.	TOTAL
Plaster, gypsum or perlite, 2 coats	1.000	S.F.	.053	.38	1.88	2.26
3 coats	1.000	S.F.	.065	.54	2.28	2.82
Lath, gypsum, standard, 3/8" thick	1.000	S.F.	.010	.42	.36	.78
1/2" thick	1.000	S.F.	.013	.43	.44	.87
Fire resistant, 3/8" thick	1.000	S.F.	.013	.43	.44	.87
1/2" thick	1.000	S.F.	.014	.48	.47	.95
Metal, diamond, 2.5 Lb.	1.000	S.F.	.010	.17	.36	.53
3.4 Lb.	1.000	S.F.	.012	.27	.41	.68
Rib, 2.75 Lb.	1.000	S.F.	.012	.25	.41	.66
3.4 Lb.	1.000	S.F.	.013	.36	.44	.80
Corners, expanded metal, 32 L.F. per 4' x 4' room	.250	L.F.	.005	.03	.20	.23
6' x 6' room	.110	L.F.	.002	.01	.09	.10
10' x 10' room	.100	L.F.	.002	.01	.08	.09
12' x 12' room	.083	L.F.	.002	.01	.06	.07
16' x 16' room	.063	L.F.	.001	.01	.05	.06
Painting, primer & 1 coats	1.000	S.F.	.008	.10	.28	.38
Primer & 2 coats	1.000	S.F.	.011	.15	.36	.51
Wallpaper, low price double roll	1.000	S.F.	.013	.32	.43	.75
Medium price double roll	1.000	S.F.	.015	.70	.51	1.21
High price double roll	1.000	S.F.	.018	1.67	.63	2.30
Tile, ceramic thin set, 4-1/4" x 4-1/4" tiles	1.000	S.F.	.084	2.30	2.65	4.95
6" x 6" tiles	1.000	S.F.	.080	2.89	2.52	5.41
Pregrouted sheets	1.000	S.F.	.067	4.57	2.10	6.67
Trim, painted or stained, baseboard	.125	L.F.	.006	.22	.24	.46
Base shoe	.125	L.F.	.005	.15	.20	.35
Chair rail	.125	L.F.	.005	.13	.19	.32
Cornice molding	.125	L.F.	.004	.10	.16	.26
Cove base, vinyl	.125	L.F.	.003	.06	.11	.17
Paneling not including furring or trim						
Plywood, prefinished, 1/4" thick, 4' x 8' sheets, vert. grooves						
Birch faced, minimum	1.000	S.F.	.032	.87	1.25	2.12
Average	1.000	S.F.	.038	1.33	1.49	2.82
Maximum	1.000	S.F.	.046	1.94	1.79	3.73
Mahogany, African	1.000	S.F.	.040	2.48	1.57	4.05
Philippine (lauan)	1.000	S.F.	.032	1.07	1.25	2.32
Oak or cherry, minimum	1.000	S.F.	.032	2.08	1.25	3.33
Maximum	1.000	S.F.	.040	3.19	1.57	4.76
Rosewood	1.000	S.F.	.050	4.52	1.96	6.48
Teak	1.000	S.F.	.040	3.19	1.57	4.76
Chestnut	1.000	S.F.	.043	4.71	1.67	6.38
Pecan	1.000	S.F.	.040	2.04	1.57	3.61
Walnut, minimum	1.000	S.F.	.032	2.72	1.25	3.97
Maximum	1.000	S.F.	.040	5.15	1.57	6.72

16 | Plaster & Stucco Ceiling Systems

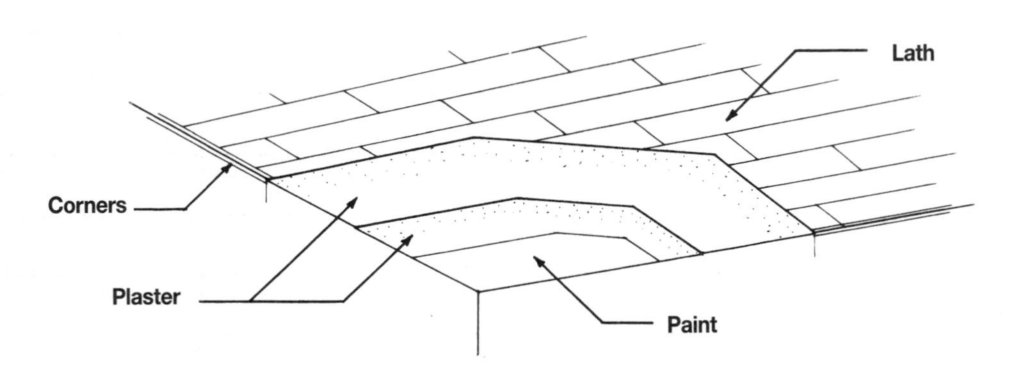

System Description	QUAN.	UNIT	LABOR HOURS	COST PER S.F.		
				MAT.	INST.	TOTAL
PLASTER ON GYPSUM LATH						
Plaster, gypsum or perlite, 2 coats	1.000	S.F.	.061	.38	2.14	2.52
Gypsum lath, plain or perforated, nailed, 3/8″ thick	1.000	S.F.	.010	.42	.36	.78
Gypsum lath, ceiling installation adder	1.000	S.F.	.004		.14	.14
Corners, expanded metal, 12′ x 12′ room	.330	L.F.	.007	.04	.26	.30
Painting, primer & 2 coats	1.000	S.F.	.011	.15	.36	.51
TOTAL		S.F.	.093	.99	3.26	4.25
PLASTER ON METAL LATH						
Plaster, gypsum or perlite, 2 coats	1.000	S.F.	.061	.38	2.14	2.52
Lath, 2.5 Lb. diamond, metal	1.000	S.F.	.012	.17	.41	.58
Corners, expanded metal, 12′ x 12′ room	.330	L.F.	.007	.04	.26	.30
Painting, primer & 2 coats	1.000	S.F.	.011	.15	.36	.51
TOTAL		S.F.	.091	.74	3.17	3.91
STUCCO ON GYPSUM LATH						
Stucco, 2 coats	1.000	S.F.	.041	.24	1.44	1.68
Gypsum lath, plain or perforated, nailed, 3/8″ thick	1.000	S.F.	.010	.42	.36	.78
Gypsum lath, ceiling installation adder	1.000	S.F.	.004		.14	.14
Corners, expanded metal, 12′ x 12′ room	.330	L.F.	.007	.04	.26	.30
Painting, primer & 2 coats	1.000	S.F.	.011	.15	.36	.51
TOTAL		S.F.	.073	.85	2.56	3.41
STUCCO ON METAL LATH						
Stucco, 2 coats	1.000	S.F.	.041	.24	1.44	1.68
Lath, 2.5 Lb. diamond, metal	1.000	S.F.	.012	.17	.41	.58
Corners, expanded metal, 12′ x 12′ room	.330	L.F.	.007	.04	.26	.30
Painting, primer & 2 coats	1.000	S.F.	.011	.15	.36	.51
TOTAL		S.F.	.071	.60	2.47	3.07

The costs in these systems are based on a square foot of ceiling area.

Description	QUAN.	UNIT	LABOR HOURS	COST PER S.F.		
				MAT.	INST.	TOTAL

 Important: See the Reference Section for critical supporting data - Reference Nos., Crews & Location Factors

Plaster & Stucco Ceiling Price Sheet

Plaster & Stucco Ceiling Price Sheet	QUAN.	UNIT	LABOR HOURS	COST PER S.F. MAT.	INST.	TOTAL
Plaster, gypsum or perlite, 2 coats	1.000	S.F.	.061	.38	2.14	2.52
3 coats	1.000	S.F.	.065	.54	2.28	2.82
Lath, gypsum, standard, 3/8″ thick	1.000	S.F.	.014	.42	.50	.92
1/2″ thick	1.000	S.F.	.015	.43	.53	.96
Fire resistant, 3/8″ thick	1.000	S.F.	.017	.43	.58	1.01
1/2″ thick	1.000	S.F.	.018	.48	.61	1.09
Metal, diamond, 2.5 Lb.	1.000	S.F.	.012	.17	.41	.58
3.4 Lb.	1.000	S.F.	.015	.27	.51	.78
Rib, 2.75 Lb.	1.000	S.F.	.012	.25	.41	.66
3.4 Lb.	1.000	S.F.	.013	.36	.44	.80
Corners expanded metal, 4′ x 4′ room	1.000	L.F.	.020	.11	.78	.89
6′ x 6′ room	.667	L.F.	.013	.07	.52	.59
10′ x 10′ room	.400	L.F.	.008	.04	.31	.35
12′ x 12′ room	.333	L.F.	.007	.04	.26	.30
16′ x 16′ room	.250	L.F.	.004	.02	.15	.17
Painting, primer & 1 coat	1.000	S.F.	.008	.10	.28	.38
Primer & 2 coats	1.000	S.F.	.011	.15	.36	.51

6 INTERIORS

18 Suspended Ceiling Systems

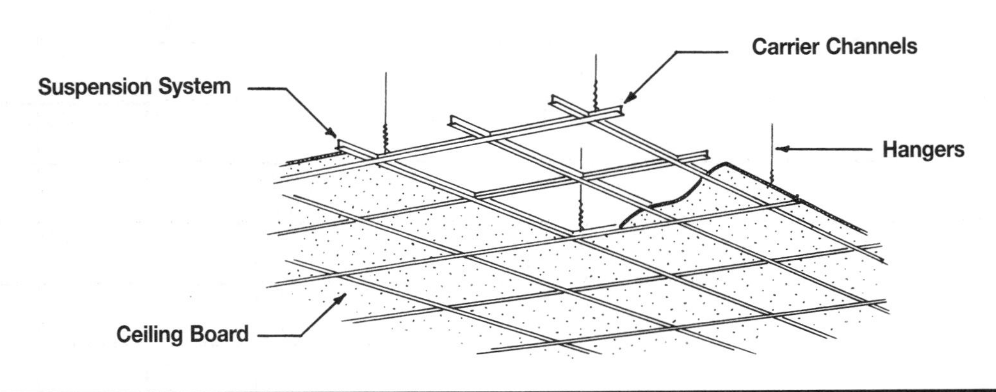

System Description	QUAN.	UNIT	LABOR HOURS	COST PER S.F.		
				MAT.	INST.	TOTAL
2′ X 2′ GRID, FILM FACED FIBERGLASS, 5/8″ THICK						
Suspension system, 2′ x 2′ grid, T bar	1.000	S.F.	.012	.56	.48	1.04
Ceiling board, film faced fiberglass, 5/8″ thick	1.000	S.F.	.013	.56	.50	1.06
Carrier channels, 1-1/2″ x 3/4″	1.000	S.F.	.017	.08	.67	.75
Hangers, #12 wire	1.000	S.F.	.002	.05	.07	.12
TOTAL		S.F.	.044	1.25	1.72	2.97
2′ X 4′ GRID, FILM FACED FIBERGLASS, 5/8″ THICK						
Suspension system, 2′ x 4′ grid, T bar	1.000	S.F.	.010	.45	.39	.84
Ceiling board, film faced fiberglass, 5/8″ thick	1.000	S.F.	.013	.56	.50	1.06
Carrier channels, 1-1/2″ x 3/4″	1.000	S.F.	.017	.08	.67	.75
Hangers, #12 wire	1.000	S.F.	.002	.05	.07	.12
TOTAL		S.F.	.042	1.14	1.63	2.77
2′ X 2′ GRID, MINERAL FIBER, REVEAL EDGE, 1″ THICK						
Suspension system, 2′ x 2′ grid, T bar	1.000	S.F.	.012	.56	.48	1.04
Ceiling board, mineral fiber, reveal edge, 1″ thick	1.000	S.F.	.013	1.32	.52	1.84
Carrier channels, 1-1/2″ x 3/4″	1.000	S.F.	.017	.08	.67	.75
Hangers, #12 wire	1.000	S.F.	.002	.05	.07	.12
TOTAL		S.F.	.044	2.01	1.74	3.75
2′ X 4′ GRID, MINERAL FIBER, REVEAL EDGE, 1″ THICK						
Suspension system, 2′ x 4′ grid, T bar	1.000	S.F.	.010	.45	.39	.84
Ceiling board, mineral fiber, reveal edge, 1″ thick	1.000	S.F.	.013	1.32	.52	1.84
Carrier channels, 1-1/2″ x 3/4″	1.000	S.F.	.017	.08	.67	.75
Hangers, #12 wire	1.000	S.F.	.002	.05	.07	.12
TOTAL		S.F.	.042	1.90	1.65	3.55

Description	QUAN.	UNIT	LABOR HOURS	COST PER S.F.		
				MAT.	INST.	TOTAL

 Important: See the Reference Section for critical supporting data - Reference Nos., Crews & Location Factors

Suspended Ceiling Price Sheet

	QUAN.	UNIT	LABOR HOURS	COST PER S.F.		
				MAT.	INST.	TOTAL
Suspension systems, T bar, 2′ x 2′ grid	1.000	S.F.	.012	.56	.48	1.04
2′ x 4′ grid	1.000	S.F.	.010	.45	.39	.84
Concealed Z bar, 12″ module	1.000	S.F.	.015	.41	.60	1.01
Ceiling boards, fiberglass, film faced, 2′ x 2′ or 2′ x 4′, 5/8″ thick	1.000	S.F.	.013	.56	.50	1.06
3/4″ thick	1.000	S.F.	.013	1.25	.52	1.77
3″ thick thermal R11	1.000	S.F.	.018	1.39	.70	2.09
Glass cloth faced, 3/4″ thick	1.000	S.F.	.016	1.78	.63	2.41
1″ thick	1.000	S.F.	.016	1.97	.65	2.62
1-1/2″ thick, nubby face	1.000	S.F.	.017	2.45	.66	3.11
Mineral fiber boards, 5/8″ thick, aluminum face 2′ x 2′	1.000	S.F.	.013	1.62	.52	2.14
2′ x 4′	1.000	S.F.	.012	1.09	.48	1.57
Standard faced, 2′ x 2′ or 2′ x 4′	1.000	S.F.	.012	.68	.46	1.14
Plastic coated face, 2′ x 2′ or 2′ x 4′	1.000	S.F.	.020	1.06	.78	1.84
Fire rated, 2 hour rating, 5/8″ thick	1.000	S.F.	.012	.92	.46	1.38
Tegular edge, 2′ x 2′ or 2′ x 4′, 5/8″ thick, fine textured	1.000	S.F.	.013	1.14	.67	1.81
Rough textured	1.000	S.F.	.015	1.49	.67	2.16
3/4″ thick, fine textured	1.000	S.F.	.016	1.62	.70	2.32
Rough textured	1.000	S.F.	.018	1.83	.70	2.53
Luminous panels, prismatic, acrylic	1.000	S.F.	.020	1.90	.78	2.68
Polystyrene	1.000	S.F.	.020	.97	.78	1.75
Flat or ribbed, acrylic	1.000	S.F.	.020	3.31	.78	4.09
Polystyrene	1.000	S.F.	.020	2.27	.78	3.05
Drop pan, white, acrylic	1.000	S.F.	.020	4.85	.78	5.63
Polystyrene	1.000	S.F.	.020	4.06	.78	4.84
Carrier channels, 4′-0″ on center, 3/4″ x 1-1/2″	1.000	S.F.	.017	.08	.67	.75
1-1/2″ x 3-1/2″	1.000	S.F.	.017	.20	.67	.87
Hangers, #12 wire	1.000	S.F.	.002	.05	.07	.12

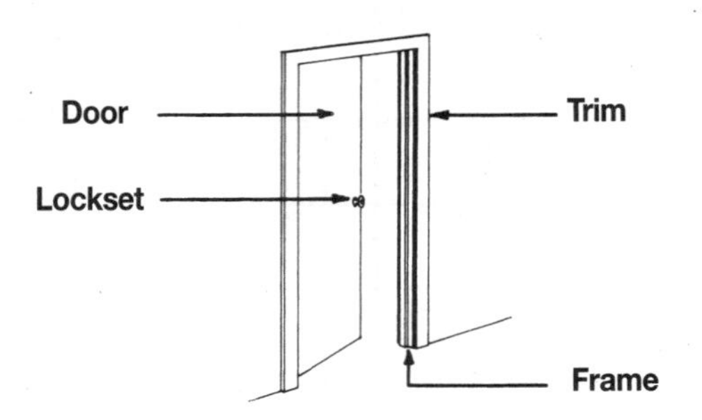

System Description	QUAN.	UNIT	LABOR HOURS	COST EACH		
				MAT.	INST.	TOTAL
LAUAN, FLUSH DOOR, HOLLOW CORE						
Door, flush, lauan, hollow core, 2′-8″ wide x 6′-8″ high	1.000	Ea.	.889	33	35	68
Frame, pine, 4-5/8″ jamb	17.000	L.F.	.725	102	28.39	130.39
Trim, stock pine, 11/16″ x 2-1/2″	34.000	L.F.	1.133	31.96	44.54	76.50
Paint trim, to 6″ wide, primer + 1 coat enamel	34.000	L.F.	.340	3.40	11.90	15.30
Butt hinges, chrome, 3-1/2″ x 3-1/2″	1.500	Pr.		35.25		35.25
Lockset, passage	1.000	Ea.	.500	14.70	19.60	34.30
Prime door & frame, oil, brushwork	2.000	Face	1.600	4.60	55	59.60
Paint door and frame, oil, 2 coats	2.000	Face	2.667	7.46	92	99.46
TOTAL		Ea.	7.473	232.37	286.43	518.80
BIRCH, FLUSH DOOR, HOLLOW CORE						
Door, flush, birch, hollow core, 2′-8″ wide x 6′-8″ high	1.000	Ea.	.889	47.50	35	82.50
Frame, pine, 4-5/8″ jamb	17.000	L.F.	.725	102	28.39	130.39
Trim, stock pine, 11/16″ x 2-1/2″	34.000	L.F.	1.133	31.96	44.54	76.50
Butt hinges, chrome, 3-1/2″ x 3-1/2″	1.500	Pr.		35.25		35.25
Lockset, passage	1.000	Ea.	.500	14.70	19.60	34.30
Prime door & frame, oil, brushwork	2.000	Face	1.600	4.60	55	59.60
Paint door and frame, oil, 2 coats	2.000	Face	2.667	7.46	92	99.46
TOTAL		Ea.	7.473	243.47	274.53	518
RAISED PANEL, SOLID, PINE DOOR						
Door, pine, raised panel, 2′-8″ wide x 6′-8″ high	1.000	Ea.	.889	156	35	191
Frame, pine, 4-5/8″ jamb	17.000	L.F.	.725	102	28.39	130.39
Detailed	34.000	L.F.	1.133	31.96	44.54	76.50
Butt hinges, bronze, 3-1/2″ x 3-1/2″	1.500	Pr.		39.75		39.75
Lockset, passage	1.000	Ea.	.500	14.70	19.60	34.30
Prime door & frame, oil, brushwork	2.000		1.600	4.60	55	59.60
Paint door and frame, oil, 2 coats	2.000		2.667	7.46	92	99.46
TOTAL		Ea.	11.587	356.47	274.53	631

The costs in these systems are based on a cost per each door.

Description	QUAN.	UNIT	LABOR HOURS	COST EACH		
				MAT.	INST.	TOTAL

Interior Door Price Sheet	QUAN.	UNIT	LABOR HOURS	COST EACH		
				MAT.	INST.	TOTAL
Door, hollow core, lauan 1-3/8″ thick, 6′-8″ high x 1′-6″ wide	1.000	Ea.	.889	26	35	61
2′-0″ wide	1.000	Ea.	.889	28	35	63
2′-6″ wide	1.000	Ea.	.889	31.50	35	66.50
2′-8″ wide	1.000	Ea.	.889	33	35	68
3′-0″ wide	1.000	Ea.	.941	35	37	72
Birch 1-3/8″ thick, 6′-8″ high x 1′-6″ wide	1.000	Ea.	.889	33.50	35	68.50
2′-0″ wide	1.000	Ea.	.889	41	35	76
2′-6″ wide	1.000	Ea.	.889	46	35	81
2′-8″ wide	1.000	Ea.	.889	47.50	35	82.50
3′-0″ wide	1.000	Ea.	.941	52	37	89
Louvered pine 1-3/8″ thick, 6′-8″ high x 1′-6″ wide	1.000	Ea.	.842	98.50	33	131.50
2′-0″ wide	1.000	Ea.	.889	124	35	159
2′-6″ wide	1.000	Ea.	.889	135	35	170
2′-8″ wide	1.000	Ea.	.889	143	35	178
3′-0″ wide	1.000	Ea.	.941	153	37	190
Paneled pine 1-3/8″ thick, 6′-8″ high x 1′-6″ wide	1.000	Ea.	.842	111	33	144
2′-0″ wide	1.000	Ea.	.889	128	35	163
2′-6″ wide	1.000	Ea.	.889	144	35	179
2′-8″ wide	1.000	Ea.	.889	156	35	191
3′-0″ wide	1.000	Ea.	.941	163	37	200
Frame, pine, 1′-6″ thru 2′-0″ wide door, 3-5/8″ deep	16.000	L.F.	.683	70	26.50	96.50
4-5/8″ deep	16.000	L.F.	.683	96	26.50	122.50
5-5/8″ deep	16.000	L.F.	.683	62	26.50	88.50
2′-6″ thru 3′0″ wide door, 3-5/8″ deep	17.000	L.F.	.725	74.50	28.50	103
4-5/8″ deep	17.000	L.F.	.725	102	28.50	130.50
5-5/8″ deep	17.000	L.F.	.725	66	28.50	94.50
Trim, casing, painted, both sides, 1′-6″ thru 2′-6″ wide door	32.000	L.F.	1.855	32	69	101
2′-6″ thru 3′-0″ wide door	34.000	L.F.	1.971	34	73	107
Butt hinges 3-1/2″ x 3-1/2″, steel plated, chrome	1.500	Pr.		35.50		35.50
Bronze	1.500	Pr.		40		40
Locksets, passage, minimum	1.000	Ea.	.500	14.70	19.60	34.30
Maximum	1.000	Ea.	.575	16.90	22.50	39.40
Privacy, miniumum	1.000	Ea.	.625	18.40	24.50	42.90
Maximum	1.000	Ea.	.675	19.85	26.50	46.35
Paint 2 sides, primer & 2 cts., flush door, 1′-6″ to 2′-0″ wide	2.000	Face	5.547	13.90	191	204.90
2′-6″ thru 3′-0″ wide	2.000	Face	6.933	17.40	239	256.40
Louvered door, 1′-6″ thru 2′-0″ wide	2.000	Face	6.400	13.40	221	234.40
2′-6″ thru 3′-0″ wide	2.000	Face	8.000	16.75	276	292.75
Paneled door, 1′-6″ thru 2′-0″ wide	2.000	Face	6.400	13.40	221	234.40
2′-6″ thru 3′-0″ wide	2.000	Face	8.000	16.75	276	292.75

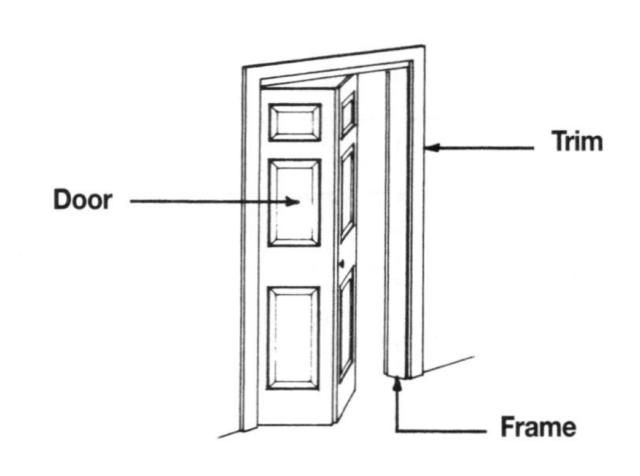

System Description	QUAN.	UNIT	LABOR HOURS	COST EACH MAT.	COST EACH INST.	COST EACH TOTAL
BI-PASSING, FLUSH, LAUAN, HOLLOW CORE, 4′-0″ X 6′-8″						
Door, flush, lauan, hollow core, 4′-0″ x 6′-8″ opening	1.000	Ea.	1.333	172	52.50	224.50
Frame, pine, 4-5/8″ jamb	18.000	L.F.	.768	108	30.06	138.06
Trim, stock pine, 11/16″ x 2-1/2″	36.000	L.F.	1.200	33.84	47.16	81
Prime door & frame, oil, brushwork	2.000	Face	1.600	4.60	55	59.60
Paint door and frame, oil, 2 coats	2.000	Face	2.667	7.46	92	99.46
TOTAL		Ea.	8.073	325.90	276.72	602.62
BI-PASSING, FLUSH, BIRCH, HOLLOW CORE, 6′-0″ X 6′-8″						
Door, flush, birch, hollow core, 6′-0″ x 6′-8″ opening	1.000	Ea.	1.600	248	62.50	310.50
Frame, pine, 4-5/8″ jamb	19.000	L.F.	.811	114	31.73	145.73
Trim, stock pine, 11/16″ x 2-1/2″	38.000	L.F.	1.267	35.72	49.78	85.50
Prime door & frame, oil, brushwork	2.000	Face	2.000	5.75	68.75	74.50
Paint door and frame, oil, 2 coats	2.000	Face	3.333	9.33	115	124.33
TOTAL		Ea.	9.471	412.80	327.76	740.56
BI-FOLD, PINE, PANELED, 3′-0″ X 6′-8″						
Door, pine, paneled, 3′-0″ x 6′-8″ opening	1.000	Ea.	1.231	162	48.50	210.50
Frame, pine, 4-5/8″ jamb	17.000	L.F.	.725	102	28.39	130.39
Trim, stock pine, 11/16″ x 2-1/2″	34.000	L.F.	1.133	31.96	44.54	76.50
Prime door & frame, oil, brushwork	2.000	Face	1.600	4.60	55	59.60
Paint door and frame, oil, 2 coats	2.000	Face	2.667	7.46	92	99.46
TOTAL		Ea.	11.927	308.02	268.43	576.45
BI-FOLD, PINE, LOUVERED, 6′-0″ X 6′-8″						
Door, pine, louvered, 6′-0″ x 6′-8″ opening	1.000	Ea.	1.600	225	62.50	287.50
Frame, pine, 4-5/8″ jamb	19.000	L.F.	.811	114	31.73	145.73
Trim, stock pine, 11/16″ x 2-1/2″	38.000	L.F.	1.267	35.72	49.78	85.50
Prime door & frame, oil, brushwork	2.500	Face	2.000	5.75	68.75	74.50
Paint door and frame, oil, 2 coats	2.500	Face	3.333	9.33	115	124.33
TOTAL		Ea.	14.614	389.80	327.76	717.56

The costs in this system are based on a cost per each door.

Description	QUAN.	UNIT	LABOR HOURS	COST EACH MAT.	COST EACH INST.	COST EACH TOTAL

Closet Door Price Sheet

Closet Door Price Sheet	QUAN.	UNIT	LABOR HOURS	COST EACH		
				MAT.	INST.	TOTAL
Doors, bi-passing, pine, louvered, 4′-0″ x 6′-8″ opening	1.000	Ea.	1.333	405	52.50	457.50
6′-0″ x 6′-8″ opening	1.000	Ea.	1.600	500	62.50	562.50
Paneled, 4′-0″ x 6′-8″ opening	1.000	Ea.	1.333	390	52.50	442.50
6′-0″ x 6′-8″ opening	1.000	Ea.	1.600	480	62.50	542.50
Flush, birch, hollow core, 4′-0″ x 6′-8″ opening	1.000	Ea.	1.333	207	52.50	259.50
6′-0″ x 6′-8″ opening	1.000	Ea.	1.600	248	62.50	310.50
Flush, lauan, hollow core, 4′-0″ x 6′-8″ opening	1.000	Ea.	1.333	172	52.50	224.50
6′-0″ x 6′-8″ opening	1.000	Ea.	1.600	201	62.50	263.50
Bi-fold, pine, louvered, 3′-0″ x 6′-8″ opening	1.000	Ea.	1.231	162	48.50	210.50
6′-0″ x 6′-8″ opening	1.000	Ea.	1.600	225	62.50	287.50
Paneled, 3′-0″ x 6′-8″ opening	1.000	Ea.	1.231	162	48.50	210.50
6′-0″ x 6′-8″ opening	1.000	Ea.	1.600	225	62.50	287.50
Flush, birch, hollow core, 3′-0″ x 6′-8″ opening	1.000	Ea.	1.231	54.50	48.50	103
6′-0″ x 6′-8″ opening	1.000	Ea.	1.600	108	62.50	170.50
Flush, lauan, hollow core, 3′-0″ x 6′8″ opening	1.000	Ea.	1.231	187	48.50	235.50
6′-0″ x 6′-8″ opening	1.000	Ea.	1.600	310	62.50	372.50
Frame pine, 3′-0″ door, 3-5/8″ deep	17.000	L.F.	.725	74.50	28.50	103
4-5/8″ deep	17.000	L.F.	.725	102	28.50	130.50
5-5/8″ deep	17.000	L.F.	.725	66	28.50	94.50
4′-0″ door, 3-5/8″ deep	18.000	L.F.	.768	79	30	109
4-5/8″ deep	18.000	L.F.	.768	108	30	138
5-5/8″ deep	18.000	L.F.	.768	70	30	100
6′-0″ door, 3-5/8″ deep	19.000	L.F.	.811	83.50	31.50	115
4-5/8″ deep	19.000	L.F.	.811	114	31.50	145.50
5-5/8″ deep	19.000	L.F.	.811	73.50	31.50	105
Trim both sides, painted 3′-0″ x 6′-8″ door	34.000	L.F.	1.971	34	73	107
4′-0″ x 6′-8″ door	36.000	L.F.	2.086	36	77.50	113.50
6′-0″ x 6′-8″ door	38.000	L.F.	2.203	38	81.50	119.50
Paint 2 sides, primer & 2 cts., flush door & frame, 3′ x 6′-8″ opng	2.000	Face	2.914	9.05	110	119.05
4′-0″ x 6′-8″ opening	2.000	Face	3.886	12.05	147	159.05
6′-0″ x 6′-8″ opening	2.000	Face	4.857	15.10	184	199.10
Paneled door & frame, 3′-0″ x 6′-8″ opening	2.000	Face	6.000	12.55	207	219.55
4′-0″ x 6′-8″ opening	2.000	Face	8.000	16.75	276	292.75
6′-0″ x 6′-8″ opening	2.000	Face	10.000	21	345	366
Louvered door & frame, 3′-0″ x 6′-8″ opening	2.000	Face	6.000	12.55	207	219.55
4′-0″ x 6′-8″ opening	2.000	Face	8.000	16.75	276	292.75
6′-0″ x 6′-8″ opening	2.000	Face	10.000	21	345	366

60 Carpet Systems

System Description	QUAN.	UNIT	LABOR HOURS	COST PER S.F.		
				MAT.	INST.	TOTAL
Carpet, direct glue-down, nylon, level loop, 26 oz.	1.000	S.F.	.018	1.83	.64	2.47
32 oz.	1.000	S.F.	.018	2.59	.64	3.23
40 oz.	1.000	S.F.	.018	3.84	.64	4.48
Nylon, plush, 20 oz.	1.000	S.F.	.018	1.21	.64	1.85
24 oz.	1.000	S.F.	.018	1.28	.64	1.92
30 oz.	1.000	S.F.	.018	1.91	.64	2.55
42 oz.	1.000	S.F.	.022	2.50	.77	3.27
48 oz.	1.000	S.F.	.022	3.32	.77	4.09
54 oz.	1.000	S.F.	.022	3.75	.77	4.52
Padding, sponge rubber cushion, minimum	1.000	S.F.	.006	.39	.21	.60
Maximum	1.000	S.F.	.006	1.01	.21	1.22
Felt, 32 oz. to 56 oz., minimum	1.000	S.F.	.006	.42	.21	.63
Maximum	1.000	S.F.	.006	.78	.21	.99
Bonded urethane, 3/8" thick, minimum	1.000	S.F.	.006	.46	.21	.67
Maximum	1.000	S.F.	.006	.79	.21	1
Prime urethane, 1/4" thick, minimum	1.000	S.F.	.006	.26	.21	.47
Maximum	1.000	S.F.	.006	.49	.21	.70
Stairs, for stairs, add to above carpet prices	1.000	Riser	.267		9.45	9.45
Underlayment plywood, 3/8" thick	1.000	S.F.	.011	.58	.42	1
1/2" thick	1.000	S.F.	.011	.68	.43	1.11
5/8" thick	1.000	S.F.	.011	.79	.45	1.24
3/4" thick	1.000	S.F.	.012	.91	.48	1.39
Particle board, 3/8" thick	1.000	S.F.	.011	.41	.42	.83
1/2" thick	1.000	S.F.	.011	.43	.43	.86
5/8" thick	1.000	S.F.	.011	.54	.45	.99
Hardboard, 4' x 4', 0.215" thick	1.000	S.F.	.011	.43	.42	.85

System Description	QUAN.	UNIT	LABOR HOURS	COST PER S.F.		
				MAT.	INST.	TOTAL
Resilient flooring, asphalt tile on concrete, 1/8″ thick						
Color group B	1.000	S.F.	.020	1.08	.71	1.79
Color group C & D	1.000	S.F.	.020	1.19	.71	1.90
Asphalt tile on wood subfloor, 1/8″ thick						
Color group B	1.000	S.F.	.020	1.28	.71	1.99
Color group C & D	1.000	S.F.	.020	1.39	.71	2.10
Vinyl composition tile, 12″ x 12″, 1/16″ thick	1.000	S.F.	.016	.83	.57	1.40
Embossed	1.000	S.F.	.016	1.05	.57	1.62
Marbleized	1.000	S.F.	.016	1.05	.57	1.62
Plain	1.000	S.F.	.016	1.17	.57	1.74
.080″ thick, embossed	1.000	S.F.	.016	1.06	.57	1.63
Marbleized	1.000	S.F.	.016	1.18	.57	1.75
Plain	1.000	S.F.	.016	1.72	.57	2.29
1/8″ thick, marbleized	1.000	S.F.	.016	1.12	.57	1.69
Plain	1.000	S.F.	.016	2.10	.57	2.67
Vinyl tile, 12″ x 12″, .050″ thick, minimum	1.000	S.F.	.016	1.85	.57	2.42
Maximum	1.000	S.F.	.016	3.62	.57	4.19
1/8″ thick, minimum	1.000	S.F.	.016	2.33	.57	2.90
Maximum	1.000	S.F.	.016	5.05	.57	5.62
1/8″ thick, solid colors	1.000	S.F.	.016	3.73	.57	4.30
Florentine pattern	1.000	S.F.	.016	4.29	.57	4.86
Marbleized or travertine pattern	1.000	S.F.	.016	8.80	.57	9.37
Vinyl sheet goods, backed, .070″ thick, minimum	1.000	S.F.	.032	2.17	1.14	3.31
Maximum	1.000	S.F.	.040	2.77	1.42	4.19
.093″ thick, minimum	1.000	S.F.	.035	2.33	1.23	3.56
Maximum	1.000	S.F.	.040	3.33	1.42	4.75
.125″ thick, minimum	1.000	S.F.	.035	2.63	1.23	3.86
Maximum	1.000	S.F.	.040	4.17	1.42	5.59
Wood, oak, finished in place, 25/32″ x 2-1/2″ clear	1.000	S.F.	.074	3.91	2.62	6.53
Select	1.000	S.F.	.074	4.72	2.62	7.34
No. 1 common	1.000	S.F.	.074	4.85	2.62	7.47
Prefinished, oak, 2-1/2″ wide	1.000	S.F.	.047	6.80	1.84	8.64
3-1/4″ wide	1.000	S.F.	.043	8.75	1.70	10.45
Ranch plank, oak, random width	1.000	S.F.	.055	8.45	2.16	10.61
Parquet, 5/16″ thick, finished in place, oak, minimum	1.000	S.F.	.077	3.90	2.74	6.64
Maximum	1.000	S.F.	.107	6.50	3.92	10.42
Teak, minimum	1.000	S.F.	.077	5.60	2.74	8.34
Maximum	1.000	S.F.	.107	9.20	3.92	13.12
Sleepers, treated, 16″ O.C., 1″ x 2″	1.000	S.F.	.007	.10	.27	.37
1″ x 3″	1.000	S.F.	.008	.19	.31	.50
2″ x 4″	1.000	S.F.	.011	.46	.42	.88
2″ x 6″	1.000	S.F.	.012	.84	.48	1.32
Subfloor, plywood, 1/2″ thick	1.000	S.F.	.011	.43	.42	.85
5/8″ thick	1.000	S.F.	.012	.50	.46	.96
3/4″ thick	1.000	S.F.	.013	.63	.50	1.13
Ceramic tile, color group 2, 1″ x 1″	1.000	S.F.	.087	4.61	2.75	7.36
2″ x 2″ or 2″ x 1″	1.000	S.F.	.084	4.41	2.65	7.06
Color group 1, 8″ x 8″		S.F.	.064	3.39	2.02	5.41
12″ x 12″		S.F.	.049	4.25	1.55	5.80
16″ x 16″		S.F.	.029	5.45	.92	6.37

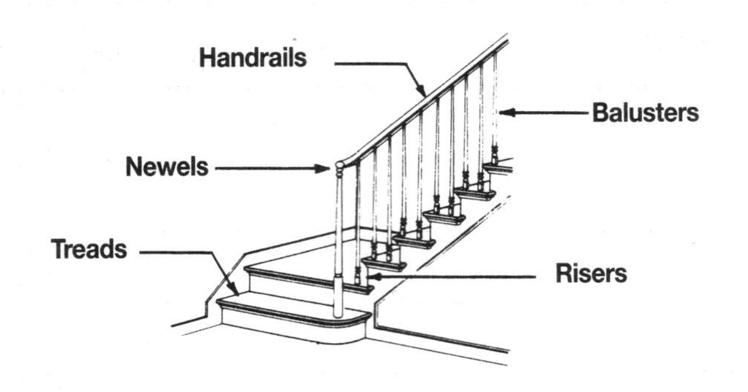

System Description	QUAN.	UNIT	LABOR HOURS	COST EACH MAT.	COST EACH INST.	COST EACH TOTAL
7 RISERS, OAK TREADS, BOX STAIRS						
Treads, oak, 1-1/4" x 10" wide, 3' long	6.000	Ea.	2.667	486	104.40	590.40
Risers, 3/4" thick, beech	7.000	Ea.	2.625	133.35	102.90	236.25
Balusters, birch, 30" high	12.000	Ea.	3.429	83.40	134.40	217.80
Newels, 3-1/4" wide	2.000	Ea.	2.286	82	90	172
Handrails, oak laminated	7.000	L.F.	.933	231	36.75	267.75
Stringers, 2" x 10", 3 each	21.000	L.F.	.306	7.14	11.97	19.11
TOTAL		Ea.	12.246	1,022.89	480.42	1,503.31
14 RISERS, OAK TREADS, BOX STAIRS						
Treads, oak, 1-1/4" x 10" wide, 3' long	13.000	Ea.	5.778	1,053	226.20	1,279.20
Risers, 3/4" thick, beech	14.000	Ea.	5.250	266.70	205.80	472.50
Balusters, birch, 30" high	26.000	Ea.	7.428	180.70	291.20	471.90
Newels, 3-1/4" wide	2.000	Ea.	2.286	82	90	172
Handrails, oak, laminated	14.000	L.F.	1.867	462	73.50	535.50
Stringers, 2" x 10", 3 each	42.000	L.F.	5.169	49.98	202.44	252.42
TOTAL		Ea.	27.778	2,094.38	1,089.14	3,183.52
14 RISERS, PINE TREADS, BOX STAIRS						
Treads, pine, 9-1/2" x 3/4" thick	13.000	Ea.	5.778	224.25	226.20	450.45
Risers, 3/4" thick, pine	14.000	Ea.	5.091	141.12	199.50	340.62
Balusters, pine, 30" high	26.000	Ea.	7.428	117.26	291.20	408.46
Newels, 3-1/4" wide	2.000	Ea.	2.286	82	90	172
Handrails, oak, laminated	14.000	L.F.	1.867	462	73.50	535.50
Stringers, 2" x 10", 3 each	42.000	L.F.	5.169	49.98	202.44	252.42
TOTAL		Ea.	27.619	1,076.61	1,082.84	2,159.45

Description	QUAN.	UNIT	LABOR HOURS	COST EACH MAT.	COST EACH INST.	COST EACH TOTAL

Important: See the Reference Section for critical supporting data - Reference Nos., Crews & Location Factors

Stairway Price Sheet	QUAN.	UNIT	LABOR HOURS	COST EACH		
				MAT.	INST.	TOTAL
Treads, oak, 1-1/16" x 9-1/2", 3' long, 7 riser stair	6.000	Ea.	2.667	485	104	589
14 riser stair	13.000	Ea.	5.778	1,050	226	1,276
1-1/16" x 11-1/2", 3' long, 7 riser stair	6.000	Ea.	2.667	171	104	275
14 riser stair	13.000	Ea.	5.778	370	226	596
Pine, 3/4" x 9-1/2", 3' long, 7 riser stair	6.000	Ea.	2.667	104	104	208
14 riser stair	13.000	Ea.	5.778	224	226	450
3/4" x 11-1/4", 3' long, 7 riser stair	6.000	Ea.	2.667	104	104	208
14 riser stair	13.000	Ea.	5.778	224	226	450
Risers, oak, 3/4" x 7-1/2" high, 7 riser stair	7.000	Ea.	2.625	139	103	242
14 riser stair	14.000	Ea.	5.250	277	206	483
Beech, 3/4" x 7-1/2" high, 7 riser stair	7.000	Ea.	2.625	133	103	236
14 riser stair	14.000	Ea.	5.250	267	206	473
Baluster, turned, 30" high, pine, 7 riser stair	12.000	Ea.	3.429	54	134	188
14 riser stair	26.000	Ea.	7.428	117	291	408
30" birch, 7 riser stair	12.000	Ea.	3.429	83.50	134	217.50
14 riser stair	26.000	Ea.	7.428	181	291	472
42" pine, 7 riser stair	12.000	Ea.	3.556	69.50	139	208.50
14 riser stair	26.000	Ea.	7.704	151	300	451
42" birch, 7 riser stair	12.000	Ea.	3.556	69.50	139	208.50
14 riser stair	26.000	Ea.	7.704	151	300	451
Newels, 3-1/4" wide, starting, 7 riser stair	2.000	Ea.	2.286	82	90	172
14 riser stair	2.000	Ea.	2.286	82	90	172
Landing, 7 riser stair	2.000	Ea.	3.200	216	125	341
14 riser stair	2.000	Ea.	3.200	216	125	341
Handrails, oak, laminated, 7 riser stair	7.000	L.F.	.933	231	37	268
14 riser stair	14.000	L.F.	1.867	460	73.50	533.50
Stringers, fir, 2" x 10" 7 riser stair	21.000	L.F.	2.585	25	101	126
14 riser stair	42.000	L.F.	5.169	50	202	252
2" x 12", 7 riser stair	21.000	L.F.	2.585	34	101	135
14 riser stair	42.000	L.F.	5.169	68	202	270

Special Stairways	QUAN.	UNIT	LABOR HOURS	COST EACH		
				MAT.	INST.	TOTAL
Basement stairs, prefabricated, open risers	1.000	Flight	4.000	50.50	12.05	62.55
Spiral stairs, oak, 4'-6" diameter, prefabricated, 9' high	1.000	Flight	10.667	4,850	420	5,270
Aluminum, 5'-0" diameter stock unit	1.000	Flight	9.956	3,100	515	3,615
Custom unit	1.000	Flight	9.956	5,875	515	6,390
Cast iron, 4'-0" diameter, minimum	1.000	Flight	9.956	2,775	515	3,290
Maximum	1.000	Flight	17.920	3,775	925	4,700
Steel, industrial, pre-erected, 3'-6" wide, bar rail	1.000	Flight	7.724	2,825	615	3,440
Picket rail	1.000	Flight	7.724	3,175	615	3,790

Division 7
Specialties

System Description	QUAN.	UNIT	LABOR HOURS	COST PER L.F.		
				MAT.	INST.	TOTAL
KITCHEN, ECONOMY GRADE						
Top cabinets, economy grade	1.000	L.F.	.171	31.04	6.72	37.76
Bottom cabinets, economy grade	1.000	L.F.	.256	46.56	10.08	56.64
Counter top, laminated plastic, post formed	1.000	L.F.	.267	9.70	10.45	20.15
Blocking, wood, 2″ x 4″	1.000	L.F.	.032	.34	1.25	1.59
Soffit, framing, wood, 2″ x 4″	4.000	L.F.	.071	1.36	2.80	4.16
Soffit drywall	2.000	S.F.	.047	.56	1.86	2.42
Drywall painting	2.000	S.F.	.013	.10	.52	.62
TOTAL		L.F.	.857	89.66	33.68	123.34
AVERAGE GRADE						
Top cabinets, average grade	1.000	L.F.	.213	38.80	8.40	47.20
Bottom cabinets, average grade	1.000	L.F.	.320	58.20	12.60	70.80
Counter top, laminated plastic, square edge, incl. backsplash	1.000	L.F.	.267	30.50	10.45	40.95
Blocking, wood, 2″ x 4″	1.000	L.F.	.032	.34	1.25	1.59
Soffit framing, wood, 2″ x 4″	4.000	L.F.	.071	1.36	2.80	4.16
Soffit drywall	2.000	S.F.	.047	.56	1.86	2.42
Drywall painting	2.000	S.F.	.013	.10	.52	.62
TOTAL		L.F.	.963	129.86	37.88	167.74
CUSTOM GRADE						
Top cabinets, custom grade	1.000	L.F.	.256	101.20	10	111.20
Bottom cabinets, custom grade	1.000	L.F.	.384	151.80	15	166.80
Counter top, laminated plastic, square edge, incl. backsplash	1.000	L.F.	.267	30.50	10.45	40.95
Blocking, wood, 2″ x 4″	1.000	L.F.	.032	.34	1.25	1.59
Soffit framing, wood, 2″ x 4″	4.000	L.F.	.071	1.36	2.80	4.16
Soffit drywall	2.000	S.F.	.047	.56	1.86	2.42
Drywall painting	2.000	S.F.	.013	.10	.52	.62
TOTAL		L.F.	1.070	285.86	41.88	327.74

Description	QUAN.	UNIT	LABOR HOURS	COST PER L.F.		
				MAT.	INST.	TOTAL

Important: See the Reference Section for critical supporting data - Reference Nos., Crews & Location Factors

Kitchen Price Sheet	QUAN.	UNIT	LABOR HOURS	COST PER L.F.		
				MAT.	INST.	TOTAL
Top cabinets, economy grade	1.000	L.F.	.171	31	6.70	37.70
Average grade	1.000	L.F.	.213	39	8.40	47.40
Custom grade	1.000	L.F.	.256	101	10	111
Bottom cabinets, economy grade	1.000	L.F.	.256	46.50	10.10	56.60
Average grade	1.000	L.F.	.320	58	12.60	70.60
Custom grade	1.000	L.F.	.384	152	15	167
Counter top, laminated plastic, 7/8″ thick, no splash	1.000	L.F.	.267	18.70	10.45	29.15
With backsplash	1.000	L.F.	.267	24.50	10.45	34.95
1-1/4″ thick, no splash	1.000	L.F.	.286	22	11.20	33.20
With backsplash	1.000	L.F.	.286	27.50	11.20	38.70
Post formed, laminated plastic	1.000	L.F.	.267	9.70	10.45	20.15
Marble, with backsplash, minimum	1.000	L.F.	.471	34	18.80	52.80
Maximum	1.000	L.F.	.615	86.50	24.50	111
Maple, solid laminated, no backsplash	1.000	L.F.	.286	57.50	11.20	68.70
With backsplash	1.000	L.F.	.286	68.50	11.20	79.70
Blocking, wood, 2″ x 4″	1.000	L.F.	.032	.34	1.25	1.59
2″ x 6″	1.000	L.F.	.036	.57	1.41	1.98
2″ x 8″	1.000	L.F.	.040	.88	1.57	2.45
Soffit framing, wood, 2″ x 3″	4.000	L.F.	.064	1.08	2.52	3.60
2″ x 4″	4.000	L.F.	.071	1.36	2.80	4.16
Soffit, drywall, painted	2.000	S.F.	.060	.66	2.38	3.04
Paneling, standard	2.000	S.F.	.064	1.74	2.50	4.24
Deluxe	2.000	S.F.	.091	3.88	3.58	7.46
Sinks, porcelain on cast iron, single bowl, 21″ x 24″	1.000	Ea.	10.334	299	400	699
21″ x 30″	1.000	Ea.	10.334	350	400	750
Double bowl, 20″ x 32″	1.000	Ea.	10.810	395	420	815
Stainless steel, single bowl, 16″ x 20″	1.000	Ea.	10.334	445	400	845
22″ x 25″	1.000	Ea.	10.334	485	400	885
Double bowl, 20″ x 32″	1.000	Ea.	10.810	238	420	658

Kitchen Price Sheet	QUAN.	UNIT	LABOR HOURS	COST PER L.F.		
				MAT.	INST.	TOTAL
Range, free standing, minimum	1.000	Ea.	3.600	330	132	462
Maximum	1.000	Ea.	6.000	1,650	201	1,851
Built-in, minimum	1.000	Ea.	3.333	555	144	699
Maximum	1.000	Ea.	10.000	1,525	400	1,925
Counter top range, 4-burner, minimum	1.000	Ea.	3.333	259	144	403
Maximum	1.000	Ea.	4.667	550	201	751
Compactor, built-in, minimum	1.000	Ea.	2.215	470	89	559
Maximum	1.000	Ea.	3.282	530	132	662
Dishwasher, built-in, minimum	1.000	Ea.	6.735	330	289	619
Maximum	1.000	Ea.	9.235	380	395	775
Garbage disposer, minimum	1.000	Ea.	2.810	70	121	191
Maximum	1.000	Ea.	2.810	183	121	304
Microwave oven, minimum	1.000	Ea.	2.615	96.50	113	209.50
Maximum	1.000	Ea.	4.615	445	199	644
Range hood, ducted, minimum	1.000	Ea.	4.658	72.50	188	260.50
Maximum	1.000	Ea.	5.991	685	241	926
Ductless, minimum	1.000	Ea.	2.615	71	107	178
Maximum	1.000	Ea.	3.948	685	160	845
Refrigerator, 16 cu.ft., minimum	1.000	Ea.	2.000	605	57.50	662.50
Maximum	1.000	Ea.	3.200	970	92	1,062
16 cu.ft. with icemaker, minimum	1.000	Ea.	4.210	750	147	897
Maximum	1.000	Ea.	5.410	1,125	181	1,306
19 cu.ft., minimum	1.000	Ea.	2.667	750	76.50	826.50
Maximum	1.000	Ea.	4.667	1,325	134	1,459
19 cu.ft. with icemaker, minimum	1.000	Ea.	5.143	960	173	1,133
Maximum	1.000	Ea.	7.143	1,525	231	1,756
Sinks, porcelain on cast iron single bowl, 21″ x 24″	1.000	Ea.	10.334	299	400	699
21″ x 30″	1.000	Ea.	10.334	350	400	750
Double bowl, 20″ x 32″	1.000	Ea.	10.810	395	420	815
Stainless steel, single bowl 16″ x 20″	1.000	Ea.	10.334	445	400	845
22″ x 25″	1.000	Ea.	10.334	485	400	885
Double bowl, 20″ x 32″	1.000	Ea.	10.810	238	420	658
Water heater, electric, 30 gallon	1.000	Ea.	3.636	310	156	466
40 gallon	1.000	Ea.	4.000	330	172	502
Gas, 30 gallon	1.000	Ea.	4.000	425	172	597
75 gallon	1.000	Ea.	5.333	820	229	1,049
Wall, packaged terminal heater/air conditioner cabinet, wall sleeve, louver, electric heat, thermostat, manual changeover, 208V						
6000 BTUH cooling, 8800 BTU heating	1.000	Ea.	2.667	1,050	103	1,153
9000 BTUH cooling, 13,900 BTU heating	1.000	Ea.	3.200	1,100	124	1,224
12,000 BTUH cooling, 13,900 BTU heating	1.000	Ea.	4.000	1,225	155	1,380
15,000 BTUH cooling, 13,900 BTU heating	1.000	Ea.	5.333	1,450	207	1,657

System Description	QUAN.	UNIT	LABOR HOURS	COST EACH		
				MAT.	INST.	TOTAL
Curtain rods, stainless, 1″ diameter, 3′ long	1.000	Ea.	.615	33.50	24	57.50
5′ long	1.000	Ea.	.615	33.50	24	57.50
Grab bar, 1″ diameter, 12″ long	1.000	Ea.	.283	19.55	11.10	30.65
36″ long	1.000	Ea.	.340	24.50	13.35	37.85
1-1/4″ diameter, 12″ long	1.000	Ea.	.333	23	13.05	36.05
36″ long	1.000	Ea.	.400	29	15.70	44.70
1-1/2″ diameter, 12″ long	1.000	Ea.	.383	26.50	15	41.50
36″ long	1.000	Ea.	.460	33.50	18.05	51.55
Mirror, 18″ x 24″	1.000	Ea.	.400	64.50	15.70	80.20
72″ x 24″	1.000	Ea.	1.333	216	52.50	268.50
Medicine chest with mirror, 18″ x 24″	1.000	Ea.	.400	99.50	15.70	115.20
36″ x 24″	1.000	Ea.	.600	149	23.50	172.50
Toilet tissue dispenser, surface mounted, minimum	1.000	Ea.	.267	12.30	10.45	22.75
Maximum	1.000	Ea.	.400	18.45	15.70	34.15
Flush mounted, minimum	1.000	Ea.	.293	13.55	11.50	25.05
Maximum	1.000	Ea.	.427	19.70	16.70	36.40
Towel bar, 18″ long, minimum	1.000	Ea.	.278	26.50	10.90	37.40
Maximum	1.000	Ea.	.348	33	13.65	46.65
24″ long, minimum	1.000	Ea.	.313	29.50	12.30	41.80
Maximum	1.000	Ea.	.383	36.50	15	51.50
36″ long, minimum	1.000	Ea.	.381	56	14.95	70.95
Maximum	1.000	Ea.	.419	61.50	16.45	77.95

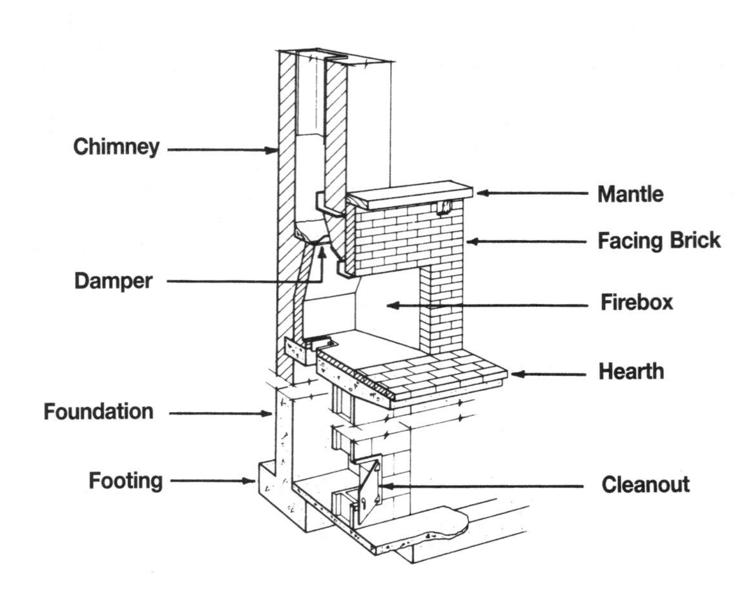

System Description	QUAN.	UNIT	LABOR HOURS	COST EACH		
				MAT.	INST.	TOTAL
MASONRY FIREPLACE						
Footing, 8″ thick, concrete, 4′ x 7′	.700	C.Y.	2.110	78.40	72.16	150.56
Foundation, concrete block, 32″ x 60″ x 4′ deep	1.000	Ea.	5.275	108	189	297
Fireplace, brick firebox, 30″ x 29″ opening	1.000	Ea.	40.000	420	1,400	1,820
Damper, cast iron, 30″ opening	1.000	Ea.	1.333	78	53	131
Facing brick, standard size brick, 6′ x 5′	30.000	S.F.	5.217	90.90	187.50	278.40
Hearth, standard size brick, 3′ x 6′	1.000	Ea.	8.000	154	279	433
Chimney, standard size brick, 8″ x 12″ flue, one story house	12.000	V.L.F.	12.000	312	420	732
Mantle, 4″ x 8″, wood	6.000	L.F.	1.333	33	52.20	85.20
Cleanout, cast iron, 8″ x 8″	1.000	Ea.	.667	32	26.50	58.50
TOTAL		Ea.	75.993	1,306.30	2,679.36	3,985.66

The costs in this system are on a cost each basis.

Description	QUAN.	UNIT	LABOR HOURS	COST EACH		
				MAT.	INST.	TOTAL

Masonry Fireplace Price Sheet	QUAN.	UNIT	LABOR HOURS	COST EACH		
				MAT.	INST.	TOTAL
Footing 8″ thick, 3′ x 6′	.440	C.Y.	1.326	49.50	45.50	95
4′ x 7′	.700	C.Y.	2.110	78.50	72.50	151
5′ x 8′	1.000	C.Y.	3.014	112	103	215
1′ thick, 3′ x 6′	.670	C.Y.	2.020	75	69	144
4′ x 7′	1.030	C.Y.	3.105	115	106	221
5′ x 8′	1.480	C.Y.	4.461	166	153	319
Foundation-concrete block, 24″ x 48″, 4′ deep	1.000	Ea.	4.267	86.50	151	237.50
8′ deep	1.000	Ea.	8.533	173	300	473
24″ x 60″, 4′ deep	1.000	Ea.	4.978	101	176	277
8′ deep	1.000	Ea.	9.956	202	355	557
32″ x 48″, 4′ deep	1.000	Ea.	4.711	95.50	167	262.50
8′ deep	1.000	Ea.	9.422	191	335	526
32″ x 60″, 4′ deep	1.000	Ea.	5.333	108	189	297
8′ deep	1.000	Ea.	10.845	220	385	605
32″ x 72″, 4′ deep	1.000	Ea.	6.133	124	217	341
8′ deep	1.000	Ea.	12.267	248	435	683
Fireplace, brick firebox 30″ x 29″ opening	1.000	Ea.	40.000	420	1,400	1,820
48″ x 30″ opening	1.000	Ea.	60.000	630	2,100	2,730
Steel fire box with registers, 25″ opening	1.000	Ea.	26.667	830	945	1,775
48″ opening	1.000	Ea.	44.000	1,250	1,550	2,800
Damper, cast iron, 30″ opening	1.000	Ea.	1.333	78	53	131
36″ opening	1.000	Ea.	1.556	91	62	153
Steel, 30″ opening	1.000	Ea.	1.333	70	53	123
36″ opening	1.000	Ea.	1.556	81.50	62	143.50
Facing for fireplace, standard size brick, 6′ x 5′	30.000	S.F.	5.217	91	188	279
7′ x 5′	35.000	S.F.	6.087	106	219	325
8′ x 6′	48.000	S.F.	8.348	145	300	445
Fieldstone, 6′ x 5′	30.000	S.F.	5.217	420	188	608
7′ x 5′	35.000	S.F.	6.087	490	219	709
8′ x 6′	48.000	S.F.	8.348	675	300	975
Sheetrock on metal, studs, 6′ x 5′	30.000	S.F.	.980	14.70	38	52.70
7′ x 5′	35.000	S.F.	1.143	17.15	44.50	61.65
8′ x 6′	48.000	S.F.	1.568	23.50	61	84.50
Hearth, standard size brick, 3′ x 6′	1.000	Ea.	8.000	154	279	433
3′ x 7′	1.000	Ea.	9.280	179	325	504
3′ x 8′	1.000	Ea.	10.640	205	370	575
Stone, 3′ x 6′	1.000	Ea.	8.000	166	279	445
3′ x 7′	1.000	Ea.	9.280	193	325	518
3′ x 8′	1.000	Ea.	10.640	221	370	591
Chimney, standard size brick , 8″ x 12″ flue, one story house	12.000	V.L.F.	12.000	310	420	730
Two story house	20.000	V.L.F.	20.000	520	700	1,220
Mantle wood, beams, 4″ x 8″	6.000	L.F.	1.333	33	52	85
4″ x 10″	6.000	L.F.	1.371	39.50	53.50	93
Ornate, prefabricated, 6′ x 3′-6″ opening, minimum	1.000	Ea.	1.600	149	62.50	211.50
Maximum	1.000	Ea.	1.600	185	62.50	247.50
Cleanout, door and frame, cast iron, 8″ x 8″	1.000	Ea.	.667	32	26.50	58.50
12″ x 12″	1.000	Ea.	.800	36.50	32	68.50

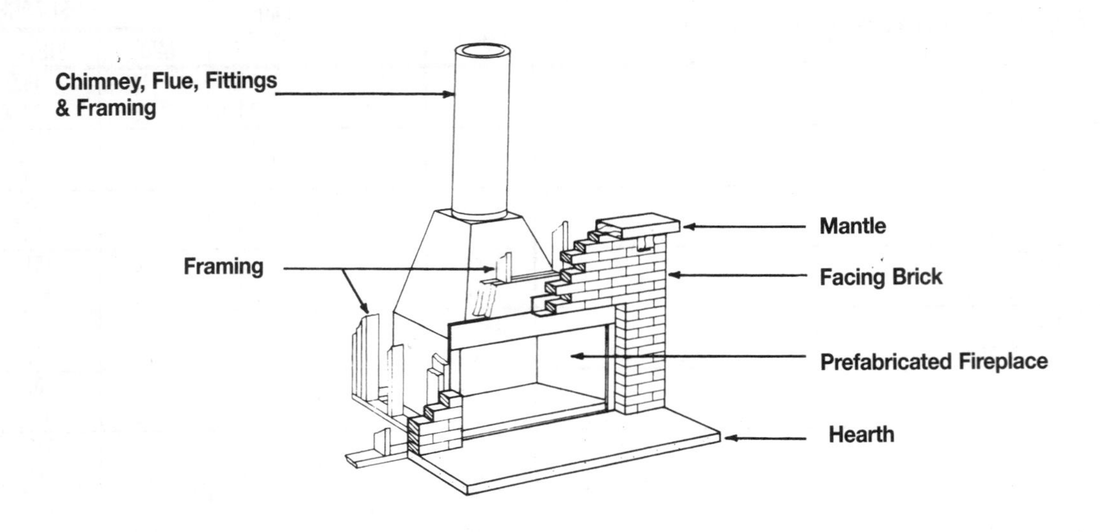

System Description	QUAN.	UNIT	LABOR HOURS	COST EACH		
				MAT.	INST.	TOTAL
PREFABRICATED FIREPLACE						
Prefabricated fireplace, metal, minimum	1.000	Ea.	6.154	1,150	241	1,391
Framing, 2″ x 4″ studs, 6′ x 5′	35.000	L.F.	.509	11.90	19.95	31.85
Fire resistant gypsum drywall, unfinished	40.000	S.F.	.320	8.80	12.40	21.20
Drywall finishing adder	40.000	S.F.	.320	1.60	12.40	14
Facing, brick, standard size brick, 6′ x 5′	30.000	S.F.	5.217	90.90	187.50	278.40
Hearth, standard size brick, 3′ x 6′	1.000	Ea.	8.000	154	279	433
Chimney, one story house, framing, 2″ x 4″ studs	80.000	L.F.	1.164	27.20	45.60	72.80
Sheathing, plywood, 5/8″ thick	32.000	S.F.	.758	81.60	29.76	111.36
Flue, 10″ metal, insulated pipe	12.000	V.L.F.	4.000	282	154.20	436.20
Fittings, ceiling support	1.000	Ea.	.667	127	25.50	152.50
Fittings, joist shield	1.000	Ea.	.667	71.50	25.50	97
Fittings, roof flashing	1.000	Ea.	.667	145	25.50	170.50
Mantle beam, wood, 4″ x 8″	6.000	L.F.	1.333	33	52.20	85.20
TOTAL		Ea.	29.776	2,184.50	1,110.51	3,295.01

The costs in this system are on a cost each basis.

Description	QUAN.	UNIT	LABOR HOURS	COST EACH		
				MAT.	INST.	TOTAL

 Important: See the Reference Section for critical supporting data - Reference Nos., Crews & Location Factors

Prefabricated Fireplace Price Sheet	QUAN.	UNIT	LABOR HOURS	COST EACH		
				MAT.	INST.	TOTAL
Prefabricated fireplace, minimum	1.000	Ea.	6.154	1,150	241	1,391
Average	1.000	Ea.	8.000	1,375	315	1,690
Maximum	1.000	Ea.	8.889	3,375	350	3,725
Framing, 2″ x 4″ studs, fireplace, 6′ x 5′	35.000	L.F.	.509	11.90	19.95	31.85
7′ x 5′	40.000	L.F.	.582	13.60	23	36.60
8′ x 6′	45.000	L.F.	.655	15.30	25.50	40.80
Sheetrock, 1/2″ thick, fireplace, 6′ x 5′	40.000	S.F.	.640	10.40	25	35.40
7′ x 5′	45.000	S.F.	.720	11.70	28	39.70
8′ x 6′	50.000	S.F.	.800	13	31	44
Facing for fireplace, brick, 6′ x 5′	30.000	S.F.	5.217	91	188	279
7′ x 5′	35.000	S.F.	6.087	106	219	325
8′ x 6′	48.000	S.F.	8.348	145	300	445
Fieldstone, 6′ x 5′	30.000	S.F.	5.217	455	188	643
7′ x 5′	35.000	S.F.	6.087	530	219	749
8′ x 6′	48.000	S.F.	8.348	725	300	1,025
Hearth, standard size brick, 3′ x 6′	1.000	Ea.	8.000	154	279	433
3′ x 7′	1.000	Ea.	9.280	179	325	504
3′ x 8′	1.000	Ea.	10.640	205	370	575
Stone, 3′ x 6′	1.000	Ea.	8.000	166	279	445
3′ x 7′	1.000	Ea.	9.280	193	325	518
3′ x 8′	1.000	Ea.	10.640	221	370	591
Chimney, framing, 2″ x 4″, one story house	80.000	L.F.	1.164	27	45.50	72.50
Two story house	120.000	L.F.	1.746	41	68.50	109.50
Sheathing, plywood, 5/8″ thick	32.000	S.F.	.758	81.50	30	111.50
Stucco on plywood	32.000	S.F.	1.125	37	43	80
Flue, 10″ metal pipe, insulated, one story house	12.000	V.L.F.	4.000	282	154	436
Two story house	20.000	V.L.F.	6.667	470	257	727
Fittings, ceiling support	1.000	Ea.	.667	127	25.50	152.50
Fittings joist sheild, one story house	1.000	Ea.	.667	71.50	25.50	97
Two story house	2.000	Ea.	1.333	143	51	194
Fittings roof flashing	1.000	Ea.	.667	145	25.50	170.50
Mantle, wood beam, 4″ x 8″	6.000	L.F.	1.333	33	52	85
4″ x 10″	6.000	L.F.	1.371	39.50	53.50	93
Ornate prefabricated, 6′ x 3′-6″ opening, minimum	1.000	Ea.	1.600	149	62.50	211.50
Maximum	1.000	Ea.	1.600	185	62.50	247.50

System Description	QUAN.	UNIT	LABOR HOURS	COST EACH		
				MAT.	INST.	TOTAL
Economy, lean to, shell only, not including 2′ stub wall, fndtn, flrs, heat 4′ x 16′	1.000	Ea.	26.212	2,000	1,025	3,025
4′ x 24′	1.000	Ea.	30.259	2,325	1,175	3,500
6′ x 10′	1.000	Ea.	16.552	1,675	650	2,325
6′ x 16′	1.000	Ea.	23.034	2,350	900	3,250
6′ x 24′	1.000	Ea.	29.793	3,025	1,175	4,200
8′ x 10′	1.000	Ea.	22.069	2,250	865	3,115
8′ x 16′	1.000	Ea.	38.400	3,900	1,500	5,400
8′ x 24′	1.000	Ea.	49.655	5,050	1,950	7,000
Free standing, 8′ x 8′	1.000	Ea.	17.356	2,625	680	3,305
8′ x 16′	1.000	Ea.	30.211	4,575	1,175	5,750
8′ x 24′	1.000	Ea.	39.051	5,900	1,525	7,425
10′ x 10′	1.000	Ea.	18.824	3,150	740	3,890
10′ x 16′	1.000	Ea.	24.095	4,025	945	4,970
10′ x 24′	1.000	Ea.	31.624	5,300	1,250	6,550
14′ x 10′	1.000	Ea.	20.741	3,925	810	4,735
14′ x 16′	1.000	Ea.	24.889	4,700	975	5,675
14′ x 24′	1.000	Ea.	33.349	6,300	1,300	7,600
Standard,lean to,shell only,not incl.2′stub wall, fndtn,flrs, heat 4′x10′	1.000	Ea.	28.235	2,150	1,100	3,250
4′ x 16′	1.000	Ea.	39.341	3,000	1,550	4,550
4′ x 24′	1.000	Ea.	45.412	3,475	1,775	5,250
6′ x 10′	1.000	Ea.	24.827	2,525	970	3,495
6′ x 16′	1.000	Ea.	34.538	3,500	1,350	4,850
6′ x 24′	1.000	Ea.	44.689	4,525	1,750	6,275
8′ x 10′	1.000	Ea.	33.103	3,350	1,300	4,650
8′ x 16′	1.000	Ea.	57.600	5,850	2,250	8,100
8′ x 24′	1.000	Ea.	74.482	7,550	2,925	10,475
Free standing, 8′ x 8′	1.000	Ea.	26.034	3,925	1,025	4,950
8′ x 16′	1.000	Ea.	45.316	6,850	1,775	8,625
8′ x 24′	1.000	Ea.	58.577	8,850	2,300	11,150
10′ x 10′	1.000	Ea.	28.236	4,725	1,100	5,825
10′ x 16′	1.000	Ea.	36.142	6,050	1,425	7,475
10′ x 24′	1.000	Ea.	47.436	7,950	1,875	9,825
14′ x 10′	1.000	Ea.	31.112	5,875	1,225	7,100
14′ x 16′	1.000	Ea.	37.334	7,050	1,450	8,500
14′ x 24′	1.000	Ea.	50.030	9,450	1,950	11,400
Deluxe,lean to,shell only,not incl.2′stub wall, fndtn, flrs or heat, 4′x10′	1.000	Ea.	20.645	3,750	800	4,550
4′ x 16′	1.000	Ea.	33.032	5,975	1,275	7,250
4′ x 24′	1.000	Ea.	49.548	8,975	1,925	10,900
6′ x 10′	1.000	Ea.	30.968	5,600	1,200	6,800
6′ x 16′	1.000	Ea.	49.548	8,975	1,925	10,900
6′ x 24′	1.000	Ea.	74.323	13,500	2,875	16,375
8′ x 10′	1.000	Ea.	41.290	7,475	1,600	9,075
8′ x 16′	1.000	Ea.	66.065	12,000	2,550	14,550
8′ x 24′	1.000	Ea.	99.097	18,000	3,850	21,850
Freestanding, 8′ x 8′	1.000	Ea.	18.618	5,150	730	5,880
8′ x 16′	1.000	Ea.	37.236	10,300	1,450	11,750
8′ x 24′	1.000	Ea.	55.855	15,500	2,200	17,700
10′ x 10′	1.000	Ea.	29.091	8,050	1,150	9,200
10′ x 16′	1.000	Ea.	46.546	12,900	1,825	14,725
10′ x 24′	1.000	Ea.	69.818	19,300	2,725	22,025
14′ x 10′	1.000	Ea.	40.727	11,300	1,600	12,900
14′ x 16′	1.000	Ea.	65.164	18,000	2,550	20,550
14′ x 24′	1.000	Ea.	97.746	27,000	3,825	30,825

System Description	QUAN.	UNIT	LABOR HOURS	COST EACH		
				MAT.	INST.	TOTAL
Swimming pools, vinyl lined, metal sides, sand bottom, 12′ x 28′	1.000	Ea.	50.177	3,125	2,000	5,125
12′ x 32′	1.000	Ea.	55.366	3,450	2,200	5,650
12′ x 36′	1.000	Ea.	60.061	3,750	2,400	6,150
16′ x 32′	1.000	Ea.	66.798	4,175	2,675	6,850
16′ x 36′	1.000	Ea.	71.190	4,450	2,825	7,275
16′ x 40′	1.000	Ea.	74.703	4,650	2,975	7,625
20′ x 36′	1.000	Ea.	77.860	4,850	3,100	7,950
20′ x 40′	1.000	Ea.	82.135	5,125	3,250	8,375
20′ x 44′	1.000	Ea.	90.348	5,650	3,600	9,250
24′ x 40′	1.000	Ea.	98.562	6,150	3,925	10,075
24′ x 44′	1.000	Ea.	108.418	6,775	4,325	11,100
24′ x 48′	1.000	Ea.	118.274	7,375	4,700	12,075
Vinyl lined, concrete sides, 12′ x 28′	1.000	Ea.	79.447	4,950	3,175	8,125
12′ x 32′	1.000	Ea.	88.818	5,550	3,525	9,075
12′ x 36′	1.000	Ea.	97.656	6,100	3,900	10,000
16′ x 32′	1.000	Ea.	111.393	6,950	4,425	11,375
16′ x 36′	1.000	Ea.	121.354	7,575	4,825	12,400
16′ x 40′	1.000	Ea.	130.445	8,150	5,200	13,350
28′ x 36′	1.000	Ea.	140.585	8,775	5,600	14,375
20′ x 40′	1.000	Ea.	149.336	9,325	5,950	15,275
20′ x 44′	1.000	Ea.	164.270	10,300	6,550	16,850
24′ x 40′	1.000	Ea.	179.203	11,200	7,125	18,325
24′ x 44′	1.000	Ea.	197.124	12,300	7,875	20,175
24′ x 48′	1.000	Ea.	215.044	13,400	8,575	21,975
Gunite, bottom and sides, 12′ x 28′	1.000	Ea.	129.767	6,500	5,200	11,700
12′ x 32′	1.000	Ea.	142.164	7,125	5,700	12,825
12′ x 36′	1.000	Ea.	153.028	7,675	6,125	13,800
16′ x 32′	1.000	Ea.	167.743	8,400	6,700	15,100
16′ x 36′	1.000	Ea.	176.421	8,850	7,050	15,900
16′ x 40′	1.000	Ea.	182.368	9,125	7,300	16,425
20′ x 36′	1.000	Ea.	187.949	9,425	7,500	16,925
20′ x 40′	1.000	Ea.	179.200	12,400	7,175	19,575
20′ x 44′	1.000	Ea.	197.120	13,700	7,900	21,600
24′ x 40′	1.000	Ea.	215.040	14,900	8,600	23,500
24′ x 44′	1.000	Ea.	273.244	13,700	10,900	24,600
24′ x 48′	1.000	Ea.	298.077	14,900	11,900	26,800

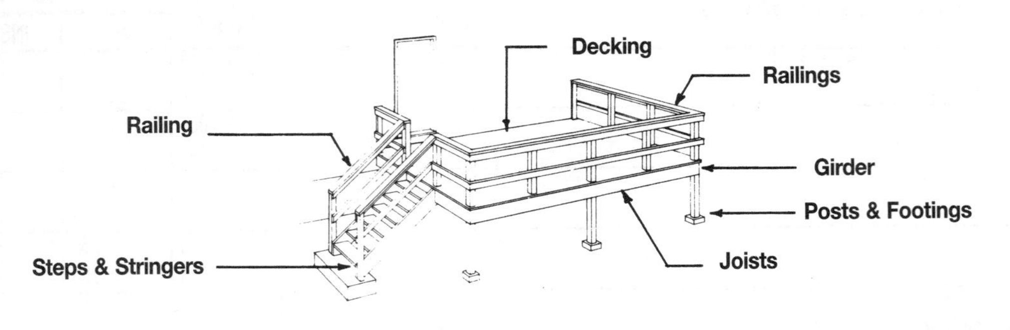

System Description	QUAN.	UNIT	LABOR HOURS	COST PER S.F.		
				MAT.	INST.	TOTAL
8′ X 12′ DECK, PRESSURE TREATED LUMBER, JOISTS 16″ O.C.						
Decking, 2″ x 6″ lumber	2.080	L.F.	.027	1.19	1.04	2.23
Lumber preservative	2.080	L.F.		.27		.27
Joists, 2″ x 8″, 16″ O.C.	1.000	L.F.	.015	.88	.57	1.45
Lumber preservative	1.000	L.F.		.17		.17
Girder, 2″ x 10″	.125	L.F.	.002	.15	.09	.24
Lumber preservative	.125	L.F.		.03		.03
Hand excavation for footings	.250	L.F.	.006		.17	.17
Concrete footings	.250	L.F.	.006	.22	.20	.42
4″ x 4″ Posts	.250	L.F.	.010	.32	.40	.72
Lumber preservative	.250	L.F.		.04		.04
Framing, pressure treated wood stairs, 3′ wide, 8 closed risers	1.000	Set	.080	1.43	3.15	4.58
Railings, 2″ x 4″	1.000	L.F.	.026	.34	1.01	1.35
Lumber preservative	1.000	L.F.		.09		.09
TOTAL		S.F.	.112	5.13	6.63	11.76
12′ X 16′ DECK, PRESSURE TREATED LUMBER, JOISTS 24″ O.C.						
Decking, 2″ x 6″	2.080	L.F.	.027	1.19	1.04	2.23
Lumber preservative	2.080	L.F.		.27		.27
Joists, 2″ x 10″, 24″ O.C.	.800	L.F.	.014	.95	.56	1.51
Lumber preservative	.800	L.F.		.17		.17
Girder, 2″ x 10″	.083	L.F.	.001	.10	.06	.16
Lumber preservative	.083	L.F.		.02		.02
Hand excavation for footings	.122	L.F.	.006		.17	.17
Concrete footings	.122	L.F.	.006	.22	.20	.42
4″ x 4″ Posts	.122	L.F.	.005	.16	.20	.36
Lumber preservative	.122	L.F.		.02		.02
Framing, pressure treated wood stairs, 3′ wide, 8 closed risers	1.000	Set	.040	.72	1.58	2.30
Railings, 2″ x 4″	.670	L.F.	.017	.23	.68	.91
Lumber preservative	.670	L.F.		.06		.06
TOTAL		S.F.	.088	4.11	4.49	8.60
12′ X 24′ DECK, REDWOOD OR CEDAR, JOISTS 16″ O.C.						
Decking, 2″ x 6″ redwood	2.080	L.F.	.027	4.85	1.04	5.89
Joists, 2″ x 10″, 16″ O.C.	1.000	L.F.	.018	4.02	.70	4.72
Girder, 2″ x 10″	.083	L.F.	.001	.33	.06	.39
Hand excavation for footings	.111	L.F.	.006		.17	.17
Concrete footings	.111	L.F.	.006	.22	.20	.42
Lumber preservative	.111	L.F.		.02		.02
Post, 4″ x 4″, including concrete footing	.111	L.F.	.009	.73	.35	1.08
Framing, redwood or cedar stairs, 3′ wide, 8 closed risers	1.000	Set	.028	1.56	1.10	2.66
Railings, 2″ x 4″	.540	L.F.	.005	.84	.18	1.02
TOTAL		S.F.	.084	12.57	3.80	16.37

The costs in this system are on a square foot basis.

Wood Deck Price Sheet	QUAN.	UNIT	LABOR HOURS	COST PER S.F. MAT.	INST.	TOTAL
Decking, treated lumber, 1″ x 4″	3.430	L.F.	.031	2.52	1.23	3.75
1″ x 6″	2.180	L.F.	.033	2.65	1.29	3.94
2″ x 4″	3.200	L.F.	.041	1.38	1.60	2.98
2″ x 6″	2.080	L.F.	.027	1.46	1.04	2.50
Redwood or cedar,, 1″ x 4″	3.430	L.F.	.035	2.95	1.35	4.30
1″ x 6″	2.180	L.F.	.036	3.08	1.41	4.49
2″ x 4″	3.200	L.F.	.028	5.15	1.10	6.25
2″ x 6″	2.080	L.F.	.027	4.85	1.04	5.89
Joists for deck, treated lumber, 2″ x 8″, 16″ O.C.	1.000	L.F.	.015	1.05	.57	1.62
24″ O.C.	.800	L.F.	.012	.84	.46	1.30
2″ x 10″, 16″ O.C.	1.000	L.F.	.018	1.41	.70	2.11
24″ O.C.	.800	L.F.	.014	1.12	.56	1.68
Redwood or cedar, 2″ x 8″, 16″ O.C.	1.000	L.F.	.015	3.28	.57	3.85
24″ O.C.	.800	L.F.	.012	2.62	.46	3.08
2″ x 10″, 16″ O.C.	1.000	L.F.	.018	4.02	.70	4.72
24″ O.C.	.800	L.F.	.014	3.22	.56	3.78
Girder for joists, treated lumber, 2″ x 10″, 8′ x 12′ deck	.125	L.F.	.002	.18	.09	.27
12′ x 16′ deck	.083	L.F.	.001	.12	.06	.18
12′ x 24′ deck	.083	L.F.	.001	.12	.06	.18
Redwood or cedar, 2″ x 10″, 8′ x 12′ deck	.125	L.F.	.002	.50	.09	.59
12′ x 16′ deck	.083	L.F.	.001	.33	.06	.39
12′ x 24′ deck	.083	L.F.	.001	.33	.06	.39
Posts, 4″ x 4″, including concrete footing, 8′ x 12′ deck	.250	S.F.	.022	.58	.77	1.35
12′ x 16′ deck	.122	L.F.	.017	.40	.57	.97
12′ x 24′ deck	.111	L.F.	.017	.38	.55	.93
Stairs 2″ x 10″ stringers, treated lumber, 8′ x 12′ deck	1.000	Set	.020	1.43	3.15	4.58
12′ x 16′ deck	1.000	Set	.012	.72	1.58	2.30
12′ x 24′ deck	1.000	Set	.008	.50	1.10	1.60
Redwood or cedar, 8′ x 12′ deck	1.000	Set	.040	4.45	3.15	7.60
12′ x 16′ deck	1.000	Set	.020	2.23	1.58	3.81
12′ x 24′ deck	1.000	Set	.012	1.56	1.10	2.66
Railings 2″ x 4″, treated lumber, 8′ x 12′ deck	1.000	L.F.	.026	.43	1.01	1.44
12′ x 16′ deck	.670	L.F.	.017	.29	.68	.97
12′ x 24′ deck	.540	L.F.	.014	.23	.55	.78
Redwood or cedar, 8′ x 12′ deck	1.000	L.F.	.009	1.56	.34	1.90
12′ x 16′ deck	.670	L.F.	.006	1.03	.22	1.25
12′ x 24′ deck	.540	L.F.	.005	.84	.18	1.02

Division 8
Mechanical

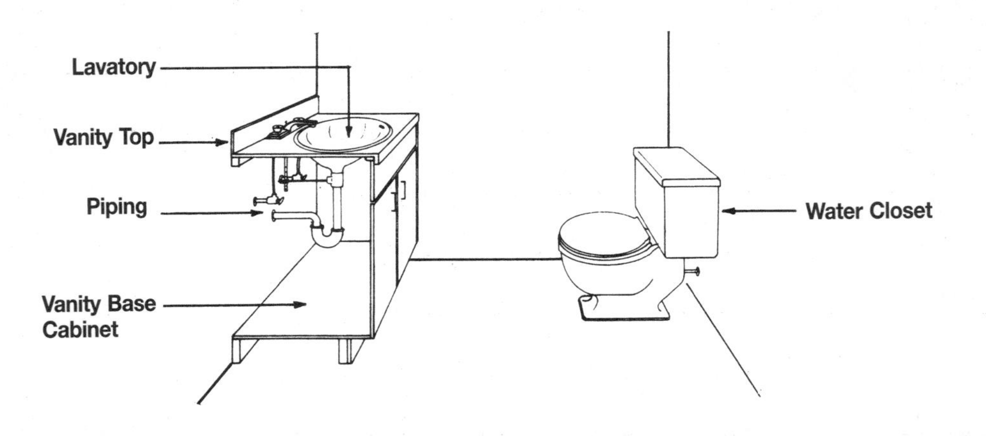

System Description	QUAN.	UNIT	LABOR HOURS	COST EACH		
				MAT.	INST.	TOTAL
LAVATORY INSTALLED WITH VANITY, PLUMBING IN 2 WALLS						
Water closet, floor mounted, 2 piece, close coupled, white	1.000	Ea.	3.019	176	117	293
Rough-in, vent, 2" diameter DWV piping	1.000	Ea.	.955	24.20	36.80	61
Waste, 4" diameter DWV piping	1.000	Ea.	.828	30.60	31.95	62.55
Supply, 1/2" diameter type "L" copper supply piping	1.000	Ea.	.593	6.66	25.44	32.10
Lavatory, 20" x 18", P.E. cast iron white	1.000	Ea.	2.500	227	96.50	323.50
Rough-in, vent, 1-1/2" diameter DWV piping	1.000	Ea.	.901	23.60	34.80	58.40
Waste, 2" diameter DWV piping	1.000	Ea.	.955	24.20	36.80	61
Supply, 1/2" diameter type "L" copper supply piping	1.000	Ea.	.988	11.10	42.40	53.50
Piping, supply, 1/2" diameter type "L" copper supply piping	10.000	L.F.	.988	11.10	42.40	53.50
Waste, 4" diameter DWV piping	7.000	L.F.	1.931	71.40	74.55	145.95
Vent, 2" diameter DWV piping	12.000	L.F.	2.866	72.60	110.40	183
Vanity base cabinet, 2 door, 30" wide	1.000	Ea.	1.000	214	39	253
Vanity top, plastic & laminated, square edge	2.670	L.F.	.712	81.44	27.90	109.34
TOTAL		Ea.	18.236	973.90	715.94	1,689.84
LAVATORY WITH WALL-HUNG LAVATORY, PLUMBING IN 2 WALLS						
Water closet, floor mounted, 2 piece close coupled, white	1.000	Ea.	3.019	176	117	293
Rough-in, vent, 2" diameter DWV piping	1.000	Ea.	.955	24.20	36.80	61
Waste, 4" diameter DWV piping	1.000	Ea.	.828	30.60	31.95	62.55
Supply, 1/2" diameter type "L" copper supply piping	1.000	Ea.	.593	6.66	25.44	32.10
Lavatory, 20" x 18", P.E. cast iron, wall hung, white	1.000	Ea.	2.000	261	77	338
Rough-in, vent, 1-1/2" diameter DWV piping	1.000	Ea.	.901	23.60	34.80	58.40
Waste, 2" diameter DWV piping	1.000	Ea.	.955	24.20	36.80	61
Supply, 1/2" diameter type "L" copper supply piping	1.000	Ea.	.988	11.10	42.40	53.50
Piping, supply, 1/2" diameter type "L" copper supply piping	10.000	L.F.	.988	11.10	42.40	53.50
Waste, 4" diameter DWV piping	7.000	L.F.	1.931	71.40	74.55	145.95
Vent, 2" diameter DWV piping	12.000	L.F.	2.866	72.60	110.40	183
Carrier, steel for studs, no arms	1.000	Ea.	1.143	29.50	49	78.50
TOTAL		Ea.	17.167	741.96	678.54	1,420.50

Description	QUAN.	UNIT	LABOR HOURS	COST EACH		
				MAT.	INST.	TOTAL

Important: See the Reference Section for critical supporting data - Reference Nos., Crews & Location Factors

Two Fixture Lavatory Price Sheet	QUAN.	UNIT	LABOR HOURS	COST EACH		
				MAT.	INST.	TOTAL
Water closet, close coupled standard 2 piece, white	1.000	Ea.	3.019	176	117	293
Color	1.000	Ea.	3.019	211	117	328
One piece elongated bowl, white	1.000	Ea.	3.019	535	117	652
Color	1.000	Ea.	3.019	665	117	782
Low profile, one piece elongated bowl, white	1.000	Ea.	3.019	730	117	847
Color	1.000	Ea.	3.019	945	117	1,062
Rough-in for water closet						
1/2" copper supply, 4" cast iron waste, 2" cast iron vent	1.000	Ea.	2.376	61.50	94	155.50
4" PVC waste, 2" PVC vent	1.000	Ea.	2.678	28.50	106	134.50
4" copper waste , 2" copper vent	1.000	Ea.	2.520	54.50	103	157.50
3" cast iron waste, 1-1/2" cast iron vent	1.000	Ea.	2.244	54.50	89	143.50
3" PVC waste, 1-1/2" PVC vent	1.000	Ea.	2.388	25	98.50	123.50
3" copper waste, 1-1/2" copper vent	1.000	Ea.	2.524	52	100	152
1/2" PVC supply, 4" PVC waste, 2" PVC vent	1.000	Ea.	2.974	39	118	157
3" PVC waste, 1-1/2" PVC vent	1.000	Ea.	2.684	35.50	111	146.50
1/2" steel supply, 4" cast iron waste, 2" cast iron vent	1.000	Ea.	2.545	64	101	165
4" cast iron waste, 2" steel vent	1.000	Ea.	2.590	57.50	103	160.50
4" PVC waste, 2" PVC vent	1.000	Ea.	2.847	30.50	113	143.50
Lavatory, vanity top mounted, P.E. on cast iron 20" x 18" white	1.000	Ea.	2.500	227	96.50	323.50
Color	1.000	Ea.	2.500	256	96.50	352.50
Steel, enameled 10" x 17" white	1.000	Ea.	2.759	142	106	248
Color	1.000	Ea.	2.500	148	96.50	244.50
Vitreous china 20" x 16", white	1.000	Ea.	2.963	244	114	358
Color	1.000	Ea.	2.963	244	114	358
Wall hung, P.E. on cast iron, 20" x 18", white	1.000	Ea.	2.000	261	77	338
Color	1.000	Ea.	2.000	297	77	374
Vitreous china 19" x 17", white	1.000	Ea.	2.286	183	88	271
Color	1.000	Ea.	2.286	204	88	292
Rough-in supply waste and vent for lavatory						
1/2" copper supply, 2" cast iron waste, 1-1/2" cast iron vent	1.000	Ea.	2.844	59	114	173
2" PVC waste, 1-1/2" PVC vent	1.000	Ea.	2.962	27	122	149
2" copper waste, 1-1/2" copper vent	1.000	Ea.	2.308	36.50	99	135.50
1-1/2" PVC waste, 1-1/4" PVC vent	1.000	Ea.	2.639	26	113	139
1-1/2" copper waste, 1-1/4" copper vent	1.000	Ea.	2.114	31.50	90.50	122
1/2" PVC supply, 2" PVC waste, 1-1/2" PVC vent	1.000	Ea.	3.456	44.50	144	188.50
1-1/2" PVC waste, 1-1/4" PVC vent	1.000	Ea.	3.133	43.50	134	177.50
1/2" steel supply, 2" cast iron waste, 1-1/2" cast iron vent	1.000	Ea.	3.126	63	126	189
2" cast iron waste, 2" steel vent	1.000	Ea.	3.225	57	130	187
2" PVC waste, 1-1/2" PVC vent	1.000	Ea.	3.244	31	135	166
1-1/2" PVC waste, 1-1/4" PVC vent	1.000	Ea.	2.921	30	125	155
Piping, supply, 1/2" copper, type "L"	10.000	L.F.	.988	11.10	42.50	53.60
1/2" steel	10.000	L.F.	1.270	15.10	54.50	69.60
1/2" PVC	10.000	L.F.	1.482	28.50	63.50	92
Waste, 4" cast iron	7.000	L.F.	1.931	71.50	74.50	146
4" copper	7.000	L.F.	2.800	77.50	108	185.50
4" PVC	7.000	L.F.	2.333	31	90	121
Vent, 2" cast iron	12.000	L.F.	2.866	72.50	110	182.50
2" copper	12.000	L.F.	2.182	43.50	93.50	137
2" PVC	12.000	L.F.	3.254	25	125	150
2" steel	12.000	Ea.	3.000	53.50	116	169.50
Vanity base cabinet, 2 door, 24" x 30"	1.000	Ea.	1.000	214	39	253
24" x 36"	1.000	Ea.	1.200	285	47	332
Vanity top, laminated plastic, square edge 25" x 32"	2.670	L.F.	.712	81.50	28	109.50
25" x 38"	3.170	L.F.	.845	96.50	33	129.50
Post formed, laminated plastic, 25" x 32"	2.670	L.F.	.712	26	28	54
25" x 38"	3.170	L.F.	.845	31	33	64
Cultured marble, 25" x 32" with bowl	1.000	Ea.	2.500	176	96.50	272.50
25" x 38" with bowl	1.000	Ea.	2.500	207	96.50	303.50
Carrier for lavatory, steel for studs	1.000	Ea.	1.143	29.50	49	78.50
Wood 2" x 8" blocking	1.330	L.F.	.053	1.17	2.09	3.26

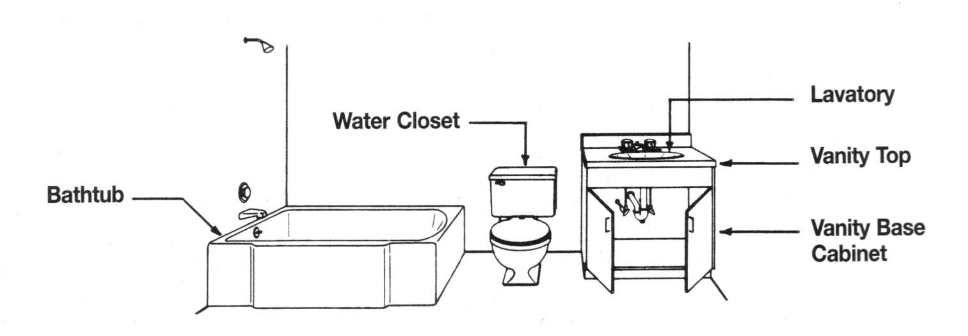

System Description	QUAN.	UNIT	LABOR HOURS	COST EACH		
				MAT.	INST.	TOTAL
BATHROOM INSTALLED WITH VANITY						
Water closet, floor mounted, 2 piece, close coupled, white	1.000	Ea.	3.019	176	117	293
Rough-in, waste, 4" diameter DWV piping	1.000	Ea.	.828	30.60	31.95	62.55
Vent, 2" diameter DWV piping	1.000	Ea.	.955	24.20	36.80	61
Supply, 1/2" diameter type "L" copper supply piping	1.000	Ea.	.593	6.66	25.44	32.10
Lavatory, 20" x 18", P.E. cast iron with accessories, white	1.000	Ea.	2.500	227	96.50	323.50
Rough-in, supply, 1/2" diameter type "L" copper supply piping	1.000	Ea.	.988	11.10	42.40	53.50
Waste, 1-1/2" diameter DWV piping	1.000	Ea.	1.803	47.20	69.60	116.80
Bathtub, P.E. cast iron, 5' long with accessories, white	1.000	Ea.	3.636	410	140	550
Rough-in, waste, 4" diameter DWV piping	1.000	Ea.	.828	30.60	31.95	62.55
Vent, 1-1/2" diameter DWV piping	1.000	Ea.	.593	11.12	25.40	36.52
Supply, 1/2" diameter type "L" copper supply piping	1.000	Ea.	.988	11.10	42.40	53.50
Piping, supply, 1/2" diameter type "L" copper supply piping	20.000	L.F.	1.975	22.20	84.80	107
Waste, 4" diameter DWV piping	9.000	L.F.	2.483	91.80	95.85	187.65
Vent, 2" diameter DWV piping	6.000	L.F.	1.500	26.82	57.90	84.72
Vanity base cabinet, 2 door, 30" wide	1.000	Ea.	1.000	214	39	253
Vanity top, plastic laminated square edge	2.670	L.F.	.712	62.75	27.90	90.65
TOTAL		Ea.	24.401	1,403.15	964.89	2,368.04
BATHROOM WITH WALL HUNG LAVATORY						
Water closet, floor mounted, 2 piece, close coupled, white	1.000	Ea.	3.019	176	117	293
Rough-in, vent, 2" diameter DWV piping	1.000	Ea.	.955	24.20	36.80	61
Waste, 4" diameter DWV piping	1.000	Ea.	.828	30.60	31.95	62.55
Supply, 1/2" diameter type "L" copper supply piping	1.000	Ea.	.593	6.66	25.44	32.10
Lavatory, 20" x 18" P.E. cast iron, wall hung, white	1.000	Ea.	2.000	261	77	338
Rough-in, waste, 1-1/2" diameter DWV piping	1.000	Ea.	1.803	47.20	69.60	116.80
Supply, 1/2" diameter type "L" copper supply piping	1.000	Ea.	.988	11.10	42.40	53.50
Bathtub, P.E. cast iron, 5' long with accessories, white	1.000	Ea.	3.636	410	140	550
Rough-in, waste, 4" diameter DWV piping	1.000	Ea.	.828	30.60	31.95	62.55
Supply, 1/2" diameter type "L" copper supply piping	1.000	Ea.	.988	11.10	42.40	53.50
Vent, 1-1/2" diameter DWV piping	1.000	Ea.	1.482	27.80	63.50	91.30
Piping, supply, 1/2" diameter type "L" copper supply piping	20.000	L.F.	1.975	22.20	84.80	107
Waste, 4" diameter DWV piping	9.000	L.F.	2.483	91.80	95.85	187.65
Vent, 2" diameter DWV piping	6.000	L.F.	1.500	26.82	57.90	84.72
Carrier, steel, for studs, no arms	1.000	Ea.	1.143	29.50	49	78.50
TOTAL		Ea.	24.221	1,206.58	965.59	2,172.17

The costs in this system are a cost each basis, all necessary piping is included.

Three Fixture Bathroom Price Sheet	QUAN.	UNIT	LABOR HOURS	COST EACH		
				MAT.	INST.	TOTAL
Water closet, close coupled standard 2 piece, white	1.000	Ea.	3.019	176	117	293
Color	1.000	Ea.	3.019	211	117	328
One piece, elongated bowl, white	1.000	Ea.	3.019	535	117	652
Color	1.000	Ea.	3.019	665	117	782
Low profile, one piece elongated bowl, white	1.000	Ea.	3.019	730	117	847
Color	1.000	Ea.	3.019	945	117	1,062
Rough-in, for water closet						
1/2″ copper supply, 4″ cast iron waste, 2″ cast iron vent	1.000	Ea.	2.376	61.50	94	155.50
4″ PVC/DWV waste, 2″ PVC vent	1.000	Ea.	2.678	28.50	106	134.50
4″ copper waste, 2″ copper vent	1.000	Ea.	2.520	54.50	103	157.50
3″ cast iron waste, 1-1/2″ cast iron vent	1.000	Ea.	2.244	54.50	89	143.50
3″ PVC waste, 1-1/2″ PVC vent	1.000	Ea.	2.388	25	98.50	123.50
3″ copper waste, 1-1/2″ copper vent	1.000	Ea.	2.014	37.50	83	120.50
1/2″ PVC supply, 4″ PVC waste, 2″ PVC vent	1.000	Ea.	2.974	39	118	157
3″ PVC waste, 1-1/2″ PVC supply	1.000	Ea.	2.684	35.50	111	146.50
1/2″ steel supply, 4″ cast iron waste, 2″ cast iron vent	1.000	Ea.	2.545	64	101	165
4″ cast iron waste, 2″ steel vent	1.000	Ea.	2.590	57.50	103	160.50
4″ PVC waste, 2″ PVC vent	1.000	Ea.	2.847	30.50	113	143.50
Lavatory, wall hung, P.E. cast iron 20″ x 18″, white	1.000	Ea.	2.000	261	77	338
Color	1.000	Ea.	2.000	297	77	374
Vitreous china 19″ x 17″, white	1.000	Ea.	2.286	183	88	271
Color	1.000	Ea.	2.286	204	88	292
Lavatory, for vanity top, P.E. cast iron 20″ x 18″″, white	1.000	Ea.	2.500	227	96.50	323.50
Color	1.000	Ea.	2.500	256	96.50	352.50
Steel, enameled 20″ x 17″, white	1.000	Ea.	2.759	142	106	248
Color	1.000	Ea.	2.500	148	96.50	244.50
Vitreous china 20″ x 16″, white	1.000	Ea.	2.963	244	114	358
Color	1.000	Ea.	2.963	244	114	358
Rough-in, for lavatory						
1/2″ copper supply, 1-1/2″ C.I. waste, 1-1/2″ C.I. vent	1.000	Ea.	2.791	58.50	112	170.50
1-1/2″ PVC waste, 1-1/4″ PVC vent	1.000	Ea.	2.639	26	113	139
1/2″ steel supply, 1-1/4″ cast iron waste, 1-1/4″ steel vent	1.000	Ea.	2.890	51	117	168
1-1/4″ PVC@ waste, 1-1/4″ PVC vent	1.000	Ea.	2.794	30	120	150
1/2″ PVC supply, 1-1/2″ PVC waste, 1-1/2″ PVC vent	1.000	Ea.	3.260	44	140	184
Bathtub, P.E. cast iron, 5′ long corner with fittings, white	1.000	Ea.	3.636	410	140	550
Color	1.000	Ea.	3.636	475	140	615
Rough-in, for bathtub						
1/2″ copper supply, 4″ cast iron waste, 1-1/2″ copper vent	1.000	Ea.	2.409	53	100	153
4″ PVC waste, 1-1/2″ PVC vent	1.000	Ea.	2.877	32	119	151
1/2″ steel supply, 4″ cast iron waste, 1-1/2″ steel vent	1.000	Ea.	2.898	59	117	176
4″ PVC waste, 1-1/2″ PVC vent	1.000	Ea.	3.159	36	131	167
1/2″ PVC supply, 4″ PVC waste, 1-1/2″ PVC vent	1.000	Ea.	3.371	49.50	140	189.50
Piping, supply 1/2″ copper	20.000	L.F.	1.975	22	85	107
1/2″ steel	20.000	L.F.	2.540	30	109	139
1/2″ PVC	20.000	L.F.	2.963	57	127	184
Piping, waste, 4″ cast iron no hub	9.000	L.F.	2.483	92	96	188
4″ PVC/DWV	9.000	L.F.	3.000	40	116	156
4″ copper/DWV	9.000	L.F.	3.600	100	139	239
Piping, vent 2″ cast iron no hub	6.000	L.F.	1.433	36.50	55	91.50
2″ copper/DWV	6.000	L.F.	1.091	21.50	47	68.50
2″ PVC/DWV	6.000	L.F.	1.627	12.55	62.50	75.05
2″ steel, galvanized	6.000	L.F.	1.500	27	58	85
Vanity base cabinet, 2 door, 24″ x 30″	1.000	Ea.	1.000	214	39	253
24″ x 36″	1.000	Ea.	1.200	285	47	332
Vanity top, laminated plastic square edge 25″ x 32″	2.670	L.F.	.712	63	28	91
25″ x 38″	3.160	L.F.	.843	74.50	33	107.50
Cultured marble, 25″ x 32″, with bowl	1.000	Ea.	2.500	176	96.50	272.50
25″ x 38″, with bowl	1.000	Ea.	2.500	207	96.50	303.50
Carrier, for lavatory, steel for studs, no arms	1.000	Ea.	1.143	29.50	49	78.50
Wood, 2″ x 8″ blocking	1.300	L.F.	.052	1.14	2.04	3.18

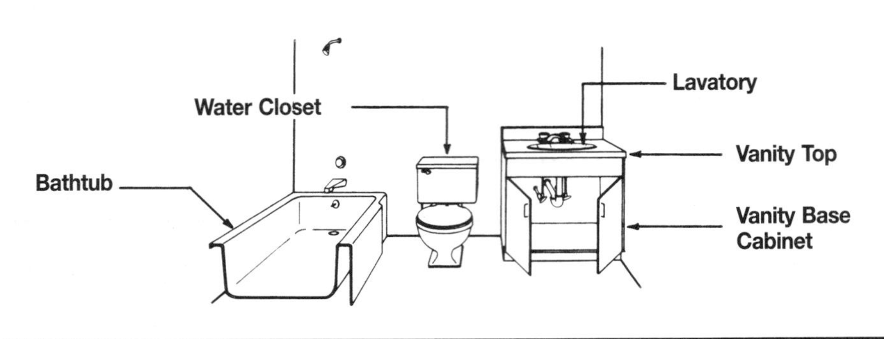

System Description	QUAN.	UNIT	LABOR HOURS	COST EACH		
				MAT.	INST.	TOTAL
BATHROOM WITH LAVATORY INSTALLED IN VANITY						
Water closet, floor mounted, 2 piece, close coupled, white	1.000	Ea.	3.019	176	117	293
Rough-in, waste, 4" diameter DWV piping	1.000	Ea.	.828	30.60	31.95	62.55
Vent, 2" diameter DWV piping	1.000	Ea.	.955	24.20	36.80	61
Supply, 1/2" diameter type "L" copper supply piping	1.000	Ea.	.593	6.66	25.44	32.10
Lavatory, 20" x 18", P.E. cast iron with accessories, white	1.000	Ea.	2.500	227	96.50	323.50
Rough-in, waste, 1-1/2" diameter DWV piping	1.000	Ea.	1.803	47.20	69.60	116.80
Supply, 1/2" diameter type "L" copper supply piping	1.000	Ea.	.988	11.10	42.40	53.50
Bathtub, P.E. cast iron 5' long with accessories, white	1.000	Ea.	3.636	410	140	550
Rough-in, waste, 4" diameter DWV piping	1.000	Ea.	.828	30.60	31.95	62.55
Vent, 1-1/2" diameter DWV piping	1.000	Ea.	.593	11.12	25.40	36.52
Supply, 1/2" diameter type "L" copper supply piping	1.000	Ea.	.988	11.10	42.40	53.50
Piping, supply, 1/2" diameter type "L" copper supply piping	10.000	L.F.	.988	11.10	42.40	53.50
Waste, 4" diameter DWV piping	6.000	L.F.	1.655	61.20	63.90	125.10
Vent, 2" diameter DWV piping	6.000	L.F.	1.500	26.82	57.90	84.72
Vanity base cabinet, 2 door, 30" wide	1.000	Ea.	1.000	214	39	253
Vanity top, plastic laminated square edge	2.670	L.F.	.712	62.75	27.90	90.65
TOTAL		Ea.	22.586	1,361.45	890.54	2,251.99
BATHROOM WITH WALL HUNG LAVATORY						
Water closet, floor mounted, 2 piece, close coupled, white	1.000	Ea.	3.019	176	117	293
Rough-in, vent, 2" diameter DWV piping	1.000	Ea.	.955	24.20	36.80	61
Waste, 4" diameter DWV piping	1.000	Ea.	.828	30.60	31.95	62.55
Supply, 1/2" diameter type "L" copper supply piping	1.000	Ea.	.593	6.66	25.44	32.10
Lavatory, 20" x 18" P.E. cast iron, wall hung, white	1.000	Ea.	2.000	261	77	338
Rough-in, waste, 1-1/2" diameter DWV piping	1.000	Ea.	1.803	47.20	69.60	116.80
Supply, 1/2" diameter type "L" copper supply piping	1.000	Ea.	.988	11.10	42.40	53.50
Bathtub, P.E. cast iron, 5' long with accessories, white	1.000	Ea.	3.636	410	140	550
Rough-in, waste, 4" diameter DWV piping	1.000	Ea.	.828	30.60	31.95	62.55
Supply, 1/2" diameter type "L" copper supply piping	1.000	Ea.	.988	11.10	42.40	53.50
Vent, 1-1/2" diameter DWV piping	1.000	Ea.	.593	11.12	25.40	36.52
Piping, supply, 1/2" diameter type "L" copper supply piping	10.000	L.F.	.988	11.10	42.40	53.50
Waste, 4" diameter DWV piping	6.000	L.F.	1.655	61.20	63.90	125.10
Vent, 2" diameter DWV piping	6.000	L.F.	1.500	26.82	57.90	84.72
Carrier, steel, for studs, no arms	1.000	Ea.	1.143	29.50	49	78.50
TOTAL		Ea.	21.517	1,148.20	853.14	2,001.34

The costs in this system are on a cost each basis. All necessary piping is included.

Three Fixture Bathroom Price Sheet	QUAN.	UNIT	LABOR HOURS	COST EACH		
				MAT.	INST.	TOTAL
Water closet, close coupled standard 2 piece, white	1.000	Ea.	3.019	176	117	293
Color	1.000	Ea.	3.019	211	117	328
One piece elongated bowl, white	1.000	Ea.	3.019	535	117	652
Color	1.000	Ea.	3.019	665	117	782
Low profile, one piece elongated bowl, white	1.000	Ea.	3.019	730	117	847
Color	1.000	Ea.	3.019	945	117	1,062
Rough-in for water closet						
1/2" copper supply, 4" cast iron waste, 2" cast iron vent	1.000	Ea.	2.376	61.50	94	155.50
4" PVC/DWV waste, 2" PVC vent	1.000	Ea.	2.678	28.50	106	134.50
4" carrier waste, 2" copper vent	1.000	Ea.	2.520	54.50	103	157.50
3" cast iron waste, 1-1/2" cast iron vent	1.000	Ea.	2.244	54.50	89	143.50
3" PVC waste, 1-1/2" PVC vent	1.000	Ea.	2.388	25	98.50	123.50
3" copper waste, 1-1/2" copper vent	1.000	Ea.	2.014	37.50	83	120.50
1/2" PVC supply, 4" PVC waste, 2" PVC vent	1.000	Ea.	2.974	39	118	157
3" PVC waste, 1-1/2" PVC supply	1.000	Ea.	2.684	35.50	111	146.50
1/2" steel supply, 4" cast iron waste, 2" cast iron vent	1.000	Ea.	2.545	64	101	165
4" cast iron waste, 2" steel vent	1.000	Ea.	2.590	57.50	103	160.50
4" PVC waste, 2" PVC vent	1.000	Ea.	2.847	30.50	113	143.50
Lavatory, wall hung, PE cast iron 20" x 18", white	1.000	Ea.	2.000	261	77	338
Color	1.000	Ea.	2.000	297	77	374
Vitreous china 19" x 17", white	1.000	Ea.	2.286	183	88	271
Color	1.000	Ea.	2.286	204	88	292
Lavatory, for vanity top, PE cast iron 20" x 18", white	1.000	Ea.	2.500	227	96.50	323.50
Color	1.000	Ea.	2.500	256	96.50	352.50
Steel enameled 20" x 17", white	1.000	Ea.	2.759	142	106	248
Color	1.000	Ea.	2.500	148	96.50	244.50
Vitreous china 20" x 16", white	1.000	Ea.	2.963	244	114	358
Color	1.000	Ea.	2.963	244	114	358
Rough-in for lavatory						
1/2" copper supply, 1-1/2" cast iron waste, 1-1/2" cast iron vent	1.000	Ea.	2.791	58.50	112	170.50
1-1/2" PVC waste, 1-1/4" PVC vent	1.000	Ea.	2.639	26	113	139
1/2" steel supply, 1-1/4" cast iron waste, 1-1/4" steel vent	1.000	Ea.	2.890	51	117	168
1-1/4" PVC waste, 1-1/4" PVC vent	1.000	Ea.	2.794	30	120	150
1/2" PVC supply, 1-1/2" PVC waste, 1-1/2" PVC vent	1.000	Ea.	3.260	44	140	184
Bathtub, PE cast iron, 5' long corner with fittings, white	1.000	Ea.	3.636	410	140	550
Color	1.000	Ea.	3.636	475	140	615
Rough-in for bathtub						
1/2" copper supply, 4" cast iron waste, 1-1/2" copper vent	1.000	Ea.	2.409	53	100	153
4" PVC waste, 1/2" PVC vent	1.000	Ea.	2.877	32	119	151
1/2" steel supply, 4" cast iron waste, 1-1/2" steel vent	1.000	Ea.	2.898	59	117	176
4" PVC waste, 1-1/2" PVC vent	1.000	Ea.	3.159	36	131	167
1/2" PVC supply, 4" PVC waste, 1-1/2" PVC vent	1.000	Ea.	3.371	49.50	140	189.50
Piping supply, 1/2" copper	10.000	L.F.	.988	11.10	42.50	53.60
1/2" steel	10.000	L.F.	1.270	15.10	54.50	69.60
1/2" PVC	10.000	L.F.	1.482	28.50	63.50	92
Piping waste, 4" cast iron no hub	6.000	L.F.	1.655	61	64	125
4" PVC/DWV	6.000	L.F.	2.000	26.50	77	103.50
4" copper/DWV	6.000	L.F.	2.400	66.50	92.50	159
Piping vent 2" cast iron no hub	6.000	L.F.	1.433	36.50	55	91.50
2" copper/DWV	6.000	L.F.	1.091	21.50	47	68.50
2" PVC/DWV	6.000	L.F.	1.627	12.55	62.50	75.05
2" steel, galvanized	6.000	L.F.	1.500	27	58	85
Vanity base cabinet, 2 door, 24" x 30"	1.000	Ea.	1.000	214	39	253
24" x 36"	1.000	Ea.	1.200	285	47	332
Vanity top, laminated plastic square edge 25" x 32"	2.670	L.F.	.712	63	28	91
25" x 38"	3.160	L.F.	.843	74.50	33	107.50
Cultured marble, 25" x 32", with bowl	1.000	Ea.	2.500	176	96.50	272.50
25" x 38", with bowl	1.000	Ea.	2.500	207	96.50	303.50
Carrier, for lavatory, steel for studs, no arms	1.000	Ea.	1.143	29.50	49	78.50
Wood, 2" x 8" blocking	1.300	L.F.	.052	1.14	2.04	3.18

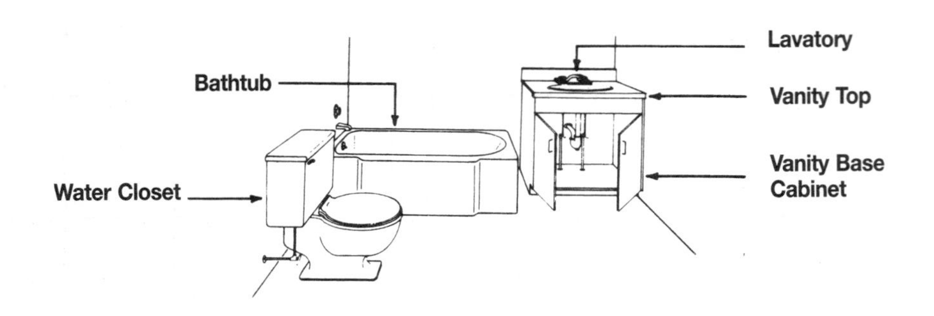

System Description	QUAN.	UNIT	LABOR HOURS	COST EACH		
				MAT.	INST.	TOTAL
BATHROOM WITH LAVATORY INSTALLED IN VANITY						
Water closet, floor mounted, 2 piece, close coupled, white	1.000	Ea.	3.019	176	117	293
Rough-in, vent, 2" diameter DWV piping	1.000	Ea.	.955	24.20	36.80	61
Waste, 4" diameter DWV piping	1.000	Ea.	.828	30.60	31.95	62.55
Supply, 1/2" diameter type "L" copper supply piping	1.000	Ea.	.593	6.66	25.44	32.10
Lavatory, 20" x 18", PE cast iron with accessories, white	1.000	Ea.	2.500	227	96.50	323.50
Rough-in, vent, 1-1/2" diameter DWV piping	1.000	Ea.	1.803	47.20	69.60	116.80
Supply, 1/2" diameter type "L" copper supply piping	1.000	Ea.	.988	11.10	42.40	53.50
Bathtub, P.E. cast iron, 5' long with accessories, white	1.000	Ea.	3.636	410	140	550
Rough-in, waste, 4" diameter DWV piping	1.000	Ea.	.828	30.60	31.95	62.55
Supply, 1/2" diameter type "L" copper supply piping	1.000	Ea.	.988	11.10	42.40	53.50
Vent, 1-1/2" diameter DWV piping	1.000	Ea.	.593	11.12	25.40	36.52
Piping, supply, 1/2" diameter type "L" copper supply piping	32.000	L.F.	3.161	35.52	135.68	171.20
Waste, 4" diameter DWV piping	12.000	L.F.	3.310	122.40	127.80	250.20
Vent, 2" diameter DWV piping	6.000	L.F.	1.500	26.82	57.90	84.72
Vanity base cabinet, 2 door, 30" wide	1.000	Ea.	1.000	214	39	253
Vanity top, plastic laminated square edge	2.670	L.F.	.712	62.75	27.90	90.65
TOTAL		Ea.	26.414	1,447.07	1,047.72	2,494.79
BATHROOM WITH WALL HUNG LAVATORY						
Water closet, floor mounted, 2 piece, close coupled, white	1.000	Ea.	3.019	176	117	293
Rough-in, vent, 2" diameter DWV piping	1.000	Ea.	.955	24.20	36.80	61
Waste, 4" diameter DWV piping	1.000	Ea.	.828	30.60	31.95	62.55
Supply, 1/2" diameter type "L" copper supply piping	1.000	Ea.	.593	6.66	25.44	32.10
Lavatory, 20" x 18" P.E. cast iron, wall hung, white	1.000	Ea.	2.000	261	77	338
Rough-in, waste, 1-1/2" diameter DWV piping	1.000	Ea.	1.803	47.20	69.60	116.80
Supply, 1/2" diameter type "L" copper supply piping	1.000	Ea.	.988	11.10	42.40	53.50
Bathtub, P.E. cast iron, 5' long with accessories, white	1.000	Ea.	3.636	410	140	550
Rough-in, waste, 4" diameter DWV piping	1.000	Ea.	.828	30.60	31.95	62.55
Supply, 1/2" diameter type "L" copper supply piping	1.000	Ea.	.988	11.10	42.40	53.50
Vent, 1-1/2" diameter DWV piping	1.000	Ea.	.593	11.12	25.40	36.52
Piping, supply, 1/2" diameter type "L" copper supply piping	32.000	L.F.	3.161	35.52	135.68	171.20
Waste, 4" diameter DWV piping	12.000	L.F.	3.310	122.40	127.80	250.20
Vent, 2" diameter DWV piping	6.000	L.F.	1.500	26.82	57.90	84.72
Carrier steel, for studs, no arms	1.000	Ea.	1.143	29.50	49	78.50
TOTAL		Ea.	25.345	1,233.82	1,010.32	2,244.14

The costs in this system are on a cost each basis. All necessary piping is included.

Three Fixture Bathroom Price Sheet

Three Fixture Bathroom Price Sheet	QUAN.	UNIT	LABOR HOURS	COST EACH		
				MAT.	INST.	TOTAL
Water closet, close coupled, standard 2 piece, white	1.000	Ea.	3.019	176	117	293
Color	1.000	Ea.	3.019	211	117	328
One piece, elongated bowl, white	1.000	Ea.	3.019	535	117	652
Color	1.000	Ea.	3.019	665	117	782
Low profile, one piece, elongated bowl, white	1.000	Ea.	3.019	730	117	847
Color	1.000	Ea.	3.019	945	117	1,062
Rough-in, for water closet						
1/2" copper supply, 4" cast iron waste, 2" cast iron vent	1.000	Ea.	2.376	61.50	94	155.50
4" PVC/DWV waste, 2" PVC vent	1.000	Ea.	2.678	28.50	106	134.50
4" copper waste, 2" copper vent	1.000	Ea.	2.520	54.50	103	157.50
3" cast iron waste, 1-1/2" cast iron vent	1.000	Ea.	2.244	54.50	89	143.50
3" PVC waste, 1-1/2" PVC vent	1.000	Ea.	2.388	25	98.50	123.50
3" copper waste, 1-1/2" copper vent	1.000	Ea.	2.014	37.50	83	120.50
1/2" PVC supply, 4" PVC waste, 2" PVC vent	1.000	Ea.	2.974	39	118	157
3" PVC waste, 1-1/2" PVC supply	1.000	Ea.	2.684	35.50	111	146.50
1/2" steel supply, 4" cast iron waste, 2" cast iron vent	1.000	Ea.	2.545	64	101	165
4" cast iron waste, 2" steel vent	1.000	Ea.	2.590	57.50	103	160.50
4" PVC waste, 2" PVC vent	1.000	Ea.	2.847	30.50	113	143.50
Lavatory wall hung, P.E. cast iron, 20" x 18", white	1.000	Ea.	2.000	261	77	338
Color	1.000	Ea.	2.000	297	77	374
Vitreous china, 19" x 17", white	1.000	Ea.	2.286	183	88	271
Color	1.000	Ea.	2.286	204	88	292
Lavatory, for vanity top, P.E., cast iron, 20" x 18", white	1.000	Ea.	2.500	227	96.50	323.50
Color	1.000	Ea.	2.500	256	96.50	352.50
Steel, enameled, 20" x 17", white	1.000	Ea.	2.759	142	106	248
Color	1.000	Ea.	2.500	148	96.50	244.50
Vitreous china, 20" x 16", white	1.000	Ea.	2.963	244	114	358
Color	1.000	Ea.	2.963	244	114	358
Rough-in, for lavatory						
1/2" copper supply, 1-1/2" C.I. waste, 1-1/2" C.I. vent	1.000	Ea.	2.791	58.50	112	170.50
1-1/2" PVC waste, 1-1/4" PVC vent	1.000	Ea.	2.639	26	113	139
1/2" steel supply, 1-1/4" cast iron waste, 1-1/4" steel vent	1.000	Ea.	2.890	51	117	168
1-1/4" PVC waste, 1-1/4" PVC vent	1.000	Ea.	2.794	30	120	150
1/2" PVC supply, 1-1/2" PVC waste, 1-1/2" PVC vent	1.000	Ea.	3.260	44	140	184
Bathtub, P.E. cast iron, 5' long corner with fittings, white	1.000	Ea.	3.636	410	140	550
Color	1.000	Ea.	3.636	475	140	615
Rough-in, for bathtub						
1/2" copper supply, 4" cast iron waste, 1-1/2" copper vent	1.000	Ea.	2.409	53	100	153
4" PVC waste, 1/2" PVC vent	1.000	Ea.	2.877	32	119	151
1/2" steel supply, 4" cast iron waste, 1-1/2" steel vent	1.000	Ea.	2.898	59	117	176
4" PVC waste, 1-1/2" PVC vent	1.000	Ea.	3.159	36	131	167
1/2" PVC supply, 4" PVC waste, 1-1/2" PVC vent	1.000	Ea.	3.371	49.50	140	189.50
Piping, supply, 1/2" copper	32.000	L.F.	3.161	35.50	136	171.50
1/2" steel	32.000	L.F.	4.063	48.50	174	222.50
1/2" PVC	32.000	L.F.	4.741	91.50	203	294.50
Piping, waste, 4" cast iron no hub	12.000	L.F.	3.310	122	128	250
4" PVC/DWV	12.000	L.F.	4.000	53.50	154	207.50
4" copper/DWV	12.000	L.F.	4.800	133	185	318
Piping, vent, 2" cast iron no hub	6.000	L.F.	1.433	36.50	55	91.50
2" copper/DWV	6.000	L.F.	1.091	21.50	47	68.50
2" PVC/DWV	6.000	L.F.	1.627	12.55	62.50	75.05
2" steel, galvanized	6.000	L.F.	1.500	27	58	85
Vanity base cabinet, 2 door, 24" x 30"	1.000	Ea.	1.000	214	39	253
24" x 36"	1.000	Ea.	1.200	285	47	332
Vanity top, laminated plastic square edge, 25" x 32"	2.670	L.F.	.712	63	28	91
25" x 38"	3.160	L.F.	.843	74.50	33	107.50
Cultured marble, 25" x 32", with bowl	1.000	Ea.	2.500	176	96.50	272.50
25" x 38", with bowl	1.000	Ea.	2.500	207	96.50	303.50
Carrier, for lavatory, steel for studs, no arms	1.000	Ea.	1.143	29.50	49	78.50
Wood, 2" x 8" blocking	1.300	L.F.	.052	1.14	2.04	3.18

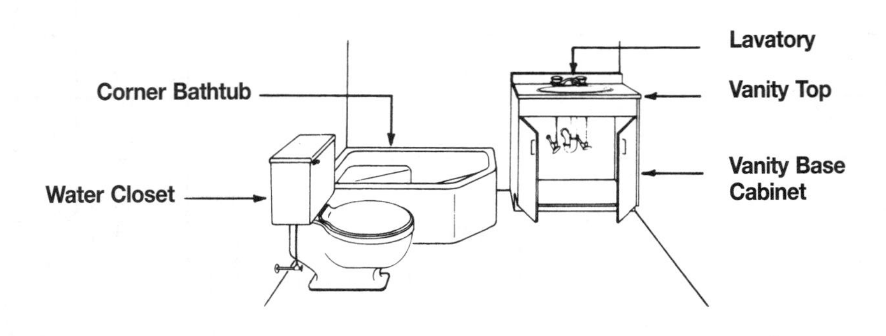

System Description	QUAN.	UNIT	LABOR HOURS	COST EACH		
				MAT.	INST.	TOTAL
BATHROOM WITH LAVATORY INSTALLED IN VANITY						
Water closet, floor mounted, 2 piece, close coupled, white	1.000	Ea.	3.019	176	117	293
Rough-in, vent, 2" diameter DWV piping	1.000	Ea.	.955	24.20	36.80	61
Waste, 4" diameter DWV piping	1.000	Ea.	.828	30.60	31.95	62.55
Supply, 1/2" diameter type "L" copper supply piping	1.000	Ea.	.593	6.66	25.44	32.10
Lavatory, 20" x 18", P.E. cast iron with fittings, white	1.000	Ea.	2.500	227	96.50	323.50
Rough-in, waste, 1-1/2" diameter DWV piping	1.000	Ea.	1.803	47.20	69.60	116.80
Supply, 1/2" diameter type "L" copper supply piping	1.000	Ea.	.988	11.10	42.40	53.50
Bathtub, P.E. cast iron, corner with fittings, white	1.000	Ea.	3.636	1,525	140	1,665
Rough-in, waste, 4" diameter DWV piping	1.000	Ea.	.828	30.60	31.95	62.55
Supply, 1/2" diameter type "L" copper supply piping	1.000	Ea.	.988	11.10	42.40	53.50
Vent, 1-1/2" diameter DWV piping	1.000	Ea.	.593	11.12	25.40	36.52
Piping, supply, 1/2" diameter type "L" copper supply piping	32.000	L.F.	3.161	35.52	135.68	171.20
Waste, 4" diameter DWV piping	12.000	L.F.	3.310	122.40	127.80	250.20
Vent, 2" diameter DWV piping	6.000	L.F.	1.500	26.82	57.90	84.72
Vanity base cabinet, 2 door, 30" wide	1.000	Ea.	1.000	214	39	253
Vanity top, plastic laminated, square edge	2.670	L.F.	.712	81.44	27.90	109.34
TOTAL		Ea.	26.414	2,580.76	1,047.72	3,628.48
BATHROOM WITH WALL HUNG LAVATORY						
Water closet, floor mounted, 2 piece, close coupled, white	1.000	Ea.	3.019	176	117	293
Rough-in, vent, 2" diameter DWV piping	1.000	Ea.	.955	24.20	36.80	61
Waste, 4" diameter DWV piping	1.000	Ea.	.828	30.60	31.95	62.55
Supply, 1/2" diameter type "L" copper supply piping	1.000	Ea.	.593	6.66	25.44	32.10
Lavatory, 20" x 18", P.E. cast iron, with fittings, white	1.000	Ea.	2.000	261	77	338
Rough-in, waste, 1-1/2" diameter DWV piping	1.000	Ea.	1.803	47.20	69.60	116.80
Supply, 1/2" diameter type "L" copper supply piping	1.000	Ea.	.988	11.10	42.40	53.50
Bathtub, P.E. cast iron, corner, with fittings, white	1.000	Ea.	3.636	1,525	140	1,665
Rough-in, waste, 4" diameter DWV piping	1.000	Ea.	.828	30.60	31.95	62.55
Supply, 1/2" diameter type "L" copper supply piping	1.000	Ea.	.988	11.10	42.40	53.50
Vent, 1-1/2" diameter DWV piping	1.000	Ea.	.593	11.12	25.40	36.52
Piping, supply, 1/2" diameter type "L" copper supply piping	32.000	L.F.	3.161	35.52	135.68	171.20
Waste, 4" diameter DWV piping	12.000	L.F.	3.310	122.40	127.80	250.20
Vent, 2" diameter DWV piping	6.000	L.F.	1.500	26.82	57.90	84.72
Carrier, steel, for studs, no arms	1.000	Ea.	1.143	29.50	49	78.50
TOTAL		Ea.	25.345	2,348.82	1,010.32	3,359.14

The costs in this system are on a cost each basis. All necessary piping is included.

Three Fixture Bathroom Price Sheet	QUAN.	UNIT	LABOR HOURS	COST EACH		
				MAT.	INST.	TOTAL
Water closet, close coupled, standard 2 piece, white	1.000	Ea.	3.019	176	117	293
Color	1.000	Ea.	3.019	211	117	328
One piece elongated bowl, white	1.000	Ea.	3.019	535	117	652
Color	1.000	Ea.	3.019	665	117	782
Low profile, one piece elongated bowl, white	1.000	Ea.	3.019	730	117	847
Color	1.000	Ea.	3.019	945	117	1,062
Rough-in, for water closet						
1/2" copper supply, 4" cast iron waste, 2" cast iron vent	1.000	Ea.	2.376	61.50	94	155.50
4" PVC/DWV waste, 2" PVC vent	1.000	Ea.	2.678	28.50	106	134.50
4" copper waste, 2" copper vent	1.000	Ea.	2.520	54.50	103	157.50
3" cast iron waste, 1-1/2" cast iron vent	1.000	Ea.	2.244	54.50	89	143.50
3" PVC waste, 1-1/2" PVC vent	1.000	Ea.	2.388	25	98.50	123.50
3" copper waste, 1-1/2" copper vent	1.000	Ea.	2.014	37.50	83	120.50
1/2" PVC supply, 4" PVC waste, 2" PVC vent	1.000	Ea.	2.974	39	118	157
3" PVC waste, 1-1/2" PVC supply	1.000	Ea.	2.684	35.50	111	146.50
1/2" steel supply, 4" cast iron waste, 2" cast iron vent	1.000	Ea.	2.545	64	101	165
4" cast iron waste, 2" steel vent	1.000	Ea.	2.590	57.50	103	160.50
4" PVC waste, 2" PVC vent	1.000	Ea.	2.847	30.50	113	143.50
Lavatory, wall hung P.E. cast iron 20" x 18", white	1.000	Ea.	2.000	261	77	338
Color	1.000	Ea.	2.000	297	77	374
Vitreous china 19" x 17", white	1.000	Ea.	2.286	183	88	271
Color	1.000	Ea.	2.286	204	88	292
Lavatory, for vanity top, P.E., cast iron, 20" x 18", white	1.000	Ea.	2.500	227	96.50	323.50
Color	1.000	Ea.	2.500	256	96.50	352.50
Steel enameled 20" x 17", white	1.000	Ea.	2.759	142	106	248
Color	1.000	Ea.	2.500	148	96.50	244.50
Vitreous china 20" x 16", white	1.000	Ea.	2.963	244	114	358
Color	1.000	Ea.	2.963	244	114	358
Rough-in, for lavatory						
1/2" copper supply, 1-1/2" cast iron waste, 1-1/2" cast iron vent	1.000	Ea.	2.791	58.50	112	170.50
1-1/2" PVC waste, 1-1/4" PVC vent	1.000	Ea.	2.639	26	113	139
1/2" steel supply, 1-1/4" cast iron waste, 1-1/4" steel vent	1.000	Ea.	2.890	51	117	168
1-1/4" PVC waste, 1-1/4" PVC vent	1.000	Ea.	2.794	30	120	150
1/2" PVC supply, 1-1/2" PVC waste, 1-1/2" PVC vent	1.000	Ea.	3.260	44	140	184
Bathtub, P.E. cast iron, corner with fittings, white	1.000	Ea.	3.636	1,525	140	1,665
Color	1.000	Ea.	4.000	1,700	154	1,854
Rough-in, for bathtub						
1/2" copper supply, 4" cast iron waste, 1-1/2" copper vent	1.000	Ea.	2.409	53	100	153
4" PVC waste, 1-1/2" PVC vent	1.000	Ea.	2.877	32	119	151
1/2" steel supply, 4" cast iron waste, 1-1/2" steel vent	1.000	Ea.	2.898	59	117	176
4" PVC waste, 1-1/2" PVC vent	1.000	Ea.	3.159	36	131	167
1/2" PVC supply, 4" PVC waste, 1-1/2" PVC vent	1.000	Ea.	3.371	49.50	140	189.50
Piping, supply, 1/2" copper	32.000	L.F.	3.161	35.50	136	171.50
1/2" steel	32.000	L.F.	4.063	48.50	174	222.50
1/2" PVC	32.000	L.F.	4.741	91.50	203	294.50
Piping, waste, 4" cast iron, no hub	12.000	L.F.	3.310	122	128	250
4" PVC/DWV	12.000	L.F.	4.000	53.50	154	207.50
4" copper/DWV	12.000	L.F.	4.800	133	185	318
Piping, vent 2" cast iron, no hub	6.000	L.F.	1.433	36.50	55	91.50
2" copper/DWV	6.000	L.F.	1.091	21.50	47	68.50
2" PVC/DWV	6.000	L.F.	1.627	12.55	62.50	75.05
2" steel, galvanized	6.000	L.F.	1.500	27	58	85
Vanity base cabinet, 2 door, 24" x 30"	1.000	Ea.	1.000	214	39	253
24" x 36"	1.000	Ea.	1.200	285	47	332
Vanity top, laminated plastic square edge 25" x 32"	2.670	L.F.	.712	81.50	28	109.50
25" x 38"	3.160	L.F.	.843	96.50	33	129.50
Cultured marble, 25" x 32", with bowl	1.000	Ea.	2.500	176	96.50	272.50
25" x 38", with bowl	1.000	Ea.	2.500	207	96.50	303.50
Carrier, for lavatory, steel for studs, no arms	1.000	Ea.	1.143	29.50	49	78.50
Wood, 2" x 8" blocking	1.300	L.F.	.053	1.17	2.09	3.26

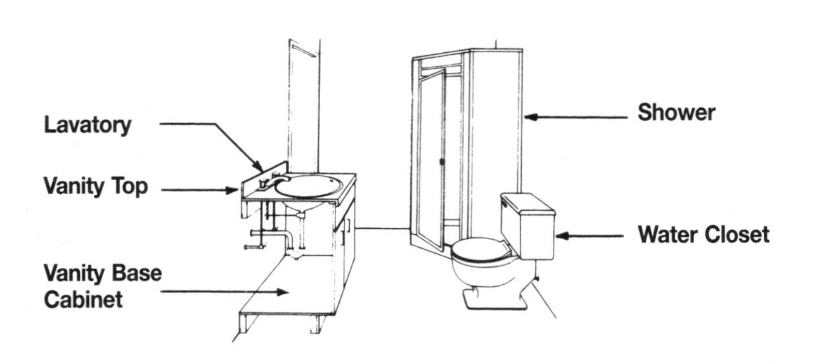

System Description	QUAN.	UNIT	LABOR HOURS	COST EACH MAT.	COST EACH INST.	COST EACH TOTAL
BATHROOM WITH SHOWER, LAVATORY INSTALLED IN VANITY						
Water closet, floor mounted, 2 piece, close coupled, white	1.000	Ea.	3.019	176	117	293
Rough-in, vent, 2" diameter DWV piping	1.000	Ea.	.955	24.20	36.80	61
Waste, 4" diameter DWV piping	1.000	Ea.	.828	30.60	31.95	62.55
Supply, 1/2" diameter type "L" copper supply piping	1.000	Ea.	.593	6.66	25.44	32.10
Lavatory, 20" x 18" P.E. cast iron with fittings, white	1.000	Ea.	2.500	227	96.50	323.50
Rough-in, waste, 1-1/2" diameter DWV piping	1.000	Ea.	1.803	47.20	69.60	116.80
Supply, 1/2" diameter type "L" copper supply piping	1.000	Ea.	.988	11.10	42.40	53.50
Shower, steel enameled, stone base, corner, white	1.000	Ea.	8.000	360	310	670
Rough-in, vent, 1-1/2" diameter DWV piping	1.000	Ea.	.225	5.90	8.70	14.60
Waste, 2" diameter DWV piping	1.000	Ea.	1.433	36.30	55.20	91.50
Supply, 1/2" diameter type "L" copper supply piping	1.000	Ea.	1.580	17.76	67.84	85.60
Piping, supply, 1/2" diameter type "L" copper supply piping	36.000	L.F.	4.148	46.62	178.08	224.70
Waste, 4" diameter DWV piping	7.000	L.F.	2.759	102	106.50	208.50
Vent, 2" diameter DWV piping	6.000	L.F.	2.250	40.23	86.85	127.08
Vanity base 2 door, 30" wide	1.000	Ea.	1.000	214	39	253
Vanity top, plastic laminated, square edge	2.170	L.F.	.712	65.42	27.90	93.32
TOTAL		Ea.	32.793	1,410.99	1,299.76	2,710.75
BATHROOM WITH SHOWER, WALL HUNG LAVATORY						
Water closet, floor mounted, close coupled	1.000	Ea.	3.019	176	117	293
Rough-in, vent, 2" diameter DWV piping	1.000	Ea.	.955	24.20	36.80	61
Waste, 4" diameter DWV piping	1.000	Ea.	.828	30.60	31.95	62.55
Supply, 1/2" diameter type "L" copper supply piping	1.000	Ea.	.593	6.66	25.44	32.10
Lavatory, 20" x 18" P.E. cast iron with fittings, white	1.000	Ea.	2.000	261	77	338
Rough-in, waste, 1-1/2" diameter DWV piping	1.000	Ea.	1.803	47.20	69.60	116.80
Supply, 1/2" diameter type "L" copper supply piping	1.000	Ea.	.988	11.10	42.40	53.50
Shower, steel enameled, stone base, white	1.000	Ea.	8.000	360	310	670
Rough-in, vent, 1-1/2" diameter DWV piping	1.000	Ea.	.225	5.90	8.70	14.60
Waste, 2" diameter DWV piping	1.000	Ea.	1.433	36.30	55.20	91.50
Supply, 1/2" diameter type "L" copper supply piping	1.000	Ea.	1.580	17.76	67.84	85.60
Piping, supply, 1/2" diameter type "L" copper supply piping	36.000	L.F.	4.148	46.62	178.08	224.70
Waste, 4" diameter DWV piping	7.000	L.F.	2.759	102	106.50	208.50
Vent, 2" diameter DWV piping	6.000	L.F.	2.250	40.23	86.85	127.08
Carrier, steel, for studs, no arms	1.000	Ea.	1.143	29.50	49	78.50
TOTAL		Ea.	31.724	1,195.07	1,262.36	2,457.43

The costs in this system are on a cost each basis. All necessary piping is included.

Three Fixture Bathroom Price Sheet	QUAN.	UNIT	LABOR HOURS	COST EACH		
				MAT.	INST.	TOTAL
Water closet, close coupled, standard 2 piece, white	1.000	Ea.	3.019	176	117	293
Color	1.000	Ea.	3.019	211	117	328
One piece elongated bowl, white	1.000	Ea.	3.019	535	117	652
Color	1.000	Ea.	3.019	665	117	782
Low profile, one piece elongated bowl, white	1.000	Ea.	3.019	730	117	847
Color	1.000	Ea.	3.019	945	117	1,062
Rough-in, for water closet						
1/2" copper supply, 4" cast iron waste, 2" cast iron vent	1.000	Ea.	2.376	61.50	94	155.50
4" PVC/DWV waste, 2" PVC vent	1.000	Ea.	2.678	28.50	106	134.50
4" copper waste, 2" copper vent	1.000	Ea.	2.520	54.50	103	157.50
3" cast iron waste, 1-1/2" cast iron vent	1.000	Ea.	2.244	54.50	89	143.50
3" PVC waste, 1-1/2" PVC vent	1.000	Ea.	2.388	25	98.50	123.50
3" copper waste, 1-1/2" copper vent	1.000	Ea.	2.014	37.50	83	120.50
1/2" PVC supply, 4" PVC waste, 2" PVC vent	1.000	Ea.	2.974	39	118	157
3" PVC waste, 1-1/2" PVC supply	1.000	Ea.	2.684	35.50	111	146.50
1/2" steel supply, 4" cast iron waste, 2" cast iron vent	1.000	Ea.	2.545	64	101	165
4" cast iron waste, 2" steel vent	1.000	Ea.	2.590	57.50	103	160.50
4" PVC waste, 2" PVC vent	1.000	Ea.	2.847	30.50	113	143.50
Lavatory, wall hung, P.E. cast iron 20" x 18", white	1.000	Ea.	2.000	261	77	338
Color	1.000	Ea.	2.000	297	77	374
Vitreous china 19" x 17", white	1.000	Ea.	2.286	183	88	271
Color	1.000	Ea.	2.286	204	88	292
Lavatory, for vanity top, P.E. cast iron 20" x 18", white	1.000	Ea.	2.500	227	96.50	323.50
Color	1.000	Ea.	2.500	256	96.50	352.50
Steel enameled 20" x 17", white	1.000	Ea.	2.759	142	106	248
Color	1.000	Ea.	2.500	148	96.50	244.50
Vitreous china 20" x 16", white	1.000	Ea.	2.963	244	114	358
Color	1.000	Ea.	2.963	244	114	358
Rough-in, for lavatory						
1/2" copper supply, 1-1/2" cast iron waste, 1-1/2" cast iron vent	1.000	Ea.	2.791	58.50	112	170.50
1-1/2" PVC waste, 1-1/2" PVC vent	1.000	Ea.	2.639	26	113	139
1/2" steel supply, 1-1/4" cast iron waste, 1-1/4" steel vent	1.000	Ea.	2.890	51	117	168
1-1/4" PVC waste, 1-1/4" PVC vent	1.000	Ea.	2.921	30	125	155
1/2" PVC supply, 1-1/2" PVC waste, 1-1/2" PVC vent	1.000	Ea.	3.260	44	140	184
Shower, steel enameled stone base, 32" x 32", white	1.000	Ea.	8.000	360	310	670
Color	1.000	Ea.	7.822	900	305	1,205
36" x 36" white	1.000	Ea.	8.889	910	345	1,255
Color	1.000	Ea.	8.889	1,025	345	1,370
Rough-in, for shower						
1/2" copper supply, 4" cast iron waste, 1-1/2" copper vent	1.000	Ea.	3.238	60	132	192
4" PVC waste, 1-1/2" PVC vent	1.000	Ea.	3.429	32	140	172
1/2" steel supply, 4" cast iron waste, 1-1/2" steel vent	1.000	Ea.	3.665	64	150	214
4" PVC waste, 1-1/2" PVC vent	1.000	Ea.	3.881	38.50	159	197.50
1/2" PVC supply, 4" PVC waste, 1-1/2" PVC vent	1.000	Ea.	4.219	60	174	234
Piping, supply, 1/2" copper	36.000	L.F.	4.148	46.50	178	224.50
1/2" steel	36.000	L.F.	5.333	63.50	229	292.50
1/2" PVC	36.000	L.F.	6.222	120	267	387
Piping, waste, 4" cast iron no hub	7.000	L.F.	2.759	102	107	209
4" PVC/DWV	7.000	L.F.	3.333	44.50	129	173.50
4" copper/DWV	7.000	L.F.	4.000	111	155	266
Piping, vent, 2" cast iron no hub	6.000	L.F.	2.149	54.50	83	137.50
2" copper/DWV	6.000	L.F.	1.636	32.50	70	102.50
2" PVC/DWV	6.000	L.F.	2.441	18.80	94	112.80
2" steel, galvanized	6.000	L.F.	2.250	40	87	127
Vanity base cabinet, 2 door, 24" x 30"	1.000	Ea.	1.000	214	39	253
24" x 36"	1.000	Ea.	1.200	285	47	332
Vanity top, laminated plastic square edge, 25" x 32"	2.170	L.F.	.712	65.50	28	93.50
25" x 38"	2.670	L.F.	.845	77.50	33	110.50
Carrier, for lavatory, steel for studs, no arms	1.000	Ea.	1.143	29.50	49	78.50
Wood, 2" x 8" blocking	1.300	L.F.	.052	1.14	2.04	3.18

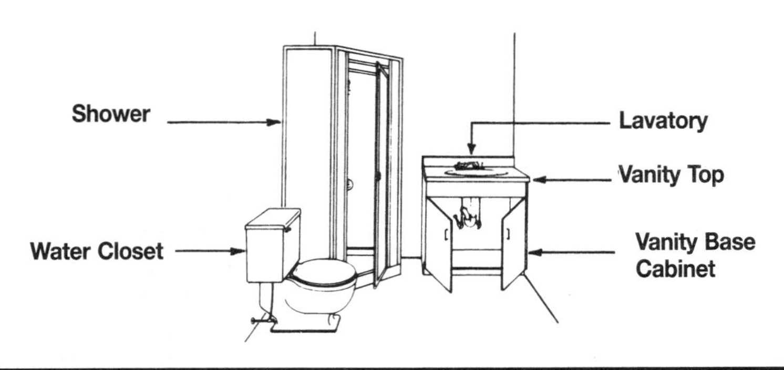

System Description	QUAN.	UNIT	LABOR HOURS	COST EACH		
				MAT.	INST.	TOTAL
BATHROOM WITH LAVATORY INSTALLED IN VANITY						
Water closet, floor mounted, 2 piece, close coupled, white	1.000	Ea.	3.019	176	117	293
Rough-in, vent, 2" diameter DWV piping	1.000	Ea.	.955	24.20	36.80	61
Waste, 4" diameter DWV piping	1.000	Ea.	.828	30.60	31.95	62.55
Supply, 1/2" diameter type "L" copper supply piping	1.000	Ea.	.593	6.66	25.44	32.10
Lavatory, 20" x 18", P.E. cast iron with fittings, white	1.000	Ea.	2.500	227	96.50	323.50
Rough-in, waste, 1-1/2" diameter DWV piping	1.000	Ea.	1.803	47.20	69.60	116.80
Supply, 1/2" diameter type "L" copper supply piping	1.000	Ea.	.988	11.10	42.40	53.50
Shower, steel enameled, stone base, corner, white	1.000	Ea.	8.000	360	310	670
Rough-in, vent, 1-1/2" diameter DWV piping	1.000	Ea.	.225	5.90	8.70	14.60
Waste, 2" diameter DWV piping	1.000	Ea.	1.433	36.30	55.20	91.50
Supply, 1/2" diameter type "L" copper supply piping	1.000	Ea.	1.580	17.76	67.84	85.60
Piping, supply, 1/2" diameter type "L" copper supply piping	36.000	L.F.	3.556	39.96	152.64	192.60
Waste, 4" diameter DWV piping	7.000	L.F.	1.931	71.40	74.55	145.95
Vent, 2" diameter DWV piping	6.000	L.F.	1.500	26.82	57.90	84.72
Vanity base, 2 door, 30" wide	1.000	Ea.	1.000	214	39	253
Vanity top, plastic laminated, square edge	2.670	L.F.	.712	62.75	27.90	90.65
TOTAL		Ea.	30.623	1,357.65	1,213.42	2,571.07
BATHROOM, WITH WALL HUNG LAVATORY						
Water closet, floor mounted, 2 piece, close coupled, white	1.000	Ea.	3.019	176	117	293
Rough-in, vent, 2" diameter DWV piping	1.000	Ea.	.955	24.20	36.80	61
Waste, 4" diameter DWV piping	1.000	Ea.	.828	30.60	31.95	62.55
Supply, 1/2" diameter type "L" copper supply piping	1.000	Ea.	.593	6.66	25.44	32.10
Lavatory, wall hung, 20" x 18" P.E. cast iron with fittings, white	1.000	Ea.	2.000	261	77	338
Rough-in, waste, 1-1/2" diameter DWV piping	1.000	Ea.	1.803	47.20	69.60	116.80
Supply, 1/2" diameter type "L" copper supply piping	1.000	Ea.	.988	11.10	42.40	53.50
Shower, steel enameled, stone base, corner, white	1.000	Ea.	8.000	360	310	670
Rough-in, waste, 1-1/2" diameter DWV piping	1.000	Ea.	.225	5.90	8.70	14.60
Waste, 2" diameter DWV piping	1.000	Ea.	1.433	36.30	55.20	91.50
Supply, 1/2" diameter type "L" copper supply piping	1.000	Ea.	1.580	17.76	67.84	85.60
Piping, supply, 1/2" diameter type "L" copper supply piping	36.000	L.F.	3.556	39.96	152.64	192.60
Waste, 4" diameter DWV piping	7.000	L.F.	1.931	71.40	74.55	145.95
Vent, 2" diameter DWV piping	6.000	L.F.	1.500	26.82	57.90	84.72
Carrier, steel, for studs, no arms	1.000	Ea.	1.143	29.50	49	78.50
TOTAL		Ea.	29.554	1,144.40	1,176.02	2,320.42

The costs in this system are on a cost each basis. All necessary piping is included.

Three Fixture Bathroom Price Sheet	QUAN.	UNIT	LABOR HOURS	COST EACH		
				MAT.	INST.	TOTAL
Water closet, close coupled, standard 2 piece, white	1.000	Ea.	3.019	176	117	293
Color	1.000	Ea.	3.019	211	117	328
One piece elongated bowl, white	1.000	Ea.	3.019	535	117	652
Color	1.000	Ea.	3.019	665	117	782
Low profile one piece elongated bowl, white	1.000	Ea.	3.019	730	117	847
Color	1.000	Ea.	3.623	1,125	140	1,265
Rough-in, for water closet						
1/2" copper supply, 4" cast iron waste, 2" cast iron vent	1.000	Ea.	2.376	61.50	94	155.50
4" P.V.C./DWV waste, 2" PVC vent	1.000	Ea.	2.678	28.50	106	134.50
4" copper waste, 2" copper vent	1.000	Ea.	2.520	54.50	103	157.50
3" cast iron waste, 1-1/2" cast iron vent	1.000	Ea.	2.244	54.50	89	143.50
3" PVC waste, 1-1/2" PVC vent	1.000	Ea.	2.388	25	98.50	123.50
3" copper waste, 1-1/2" copper vent	1.000	Ea.	2.014	37.50	83	120.50
1/2" P.V.C. supply, 4" P.V.C. waste, 2" P.V.C. vent	1.000	Ea.	2.974	39	118	157
3" P.V.C. waste, 1-1/2" P.V.C. vent	1.000	Ea.	2.684	35.50	111	146.50
1/2" steel supply, 4" cast iron waste, 2" cast iron vent	1.000	Ea.	2.545	64	101	165
4" cast iron waste, 2" steel vent	1.000	Ea.	2.590	57.50	103	160.50
4" P.V.C. waste, 2" P.V.C. vent	1.000	Ea.	2.847	30.50	113	143.50
Lavatory, wall hung P.E. cast iron 20" x 18", white	1.000	Ea.	2.000	261	77	338
Color	1.000	Ea.	2.000	297	77	374
Vitreous china 19" x 17", white	1.000	Ea.	2.286	183	88	271
Color	1.000	Ea.	2.286	204	88	292
Lavatory, for vanity top P.E. cast iron 20" x 18", white	1.000	Ea.	2.500	227	96.50	323.50
Color	1.000	Ea.	2.500	256	96.50	352.50
Steel enameled 20" x 17", white	1.000	Ea.	2.759	142	106	248
Color	1.000	Ea.	2.500	148	96.50	244.50
Vitreous china 20" x 16", white	1.000	Ea.	2.963	244	114	358
Color	1.000	Ea.	2.963	244	114	358
Rough-in, for lavatory						
1/2" copper supply, 1-1/2" cast iron waste, 1-1/2" cast iron vent	1.000	Ea.	2.791	58.50	112	170.50
1-1/2" P.V.C. waste, 1-1/2" P.V.C. vent	1.000	Ea.	2.639	26	113	139
1/2" steel supply, 1-1/2" cast iron waste, 1-1/4" steel vent	1.000	Ea.	2.890	51	117	168
1-1/2" P.V.C. waste, 1-1/4" P.V.C. vent	1.000	Ea.	2.921	30	125	155
1/2" P.V.C. supply, 1-1/2" P.V.C. waste, 1-1/2" P.V.C. vent	1.000	Ea.	3.260	44	140	184
Shower, steel enameled stone base, 32" x 32", white	1.000	Ea.	8.000	360	310	670
Color	1.000	Ea.	7.822	900	305	1,205
36" x 36", white	1.000	Ea.	8.889	910	345	1,255
Color	1.000	Ea.	8.889	1,025	345	1,370
Rough-in, for shower						
1/2" copper supply, 2" cast iron waste, 1-1/2" copper vent	1.000	Ea.	3.161	57	129	186
2" P.V.C. waste, 1-1/2" P.V.C. vent	1.000	Ea.	3.429	32	140	172
1/2" steel supply, 2" cast iron waste, 1-1/2" steel vent	1.000	Ea.	3.887	88.50	159	247.50
2" P.V.C. waste, 1-1/2" P.V.C. vent	1.000	Ea.	3.881	38.50	159	197.50
1/2" P.V.C. supply, 2" P.V.C. waste, 1-1/2" P.V.C. vent	1.000	Ea.	4.219	60	174	234
Piping, supply, 1/2" copper	36.000	L.F.	3.556	40	153	193
1/2" steel	36.000	L.F.	4.571	54.50	196	250.50
1/2" P.V.C.	36.000	L.F.	5.333	103	229	332
Waste, 4" cast iron, no hub	7.000	L.F.	1.931	71.50	74.50	146
4" P.V.C./DWV	7.000	L.F.	2.333	31	90	121
4" copper/DWV	7.000	L.F.	2.800	77.50	108	185.50
Vent, 2" cast iron, no hub	6.000	L.F.	1.091	21.50	47	68.50
2" copper/DWV	6.000	L.F.	1.091	21.50	47	68.50
2" P.V.C./DWV	6.000	L.F.	1.627	12.55	62.50	75.05
2" steel, galvanized	6.000	L.F.	1.500	27	58	85
Vanity base cabinet, 2 door, 24" x 30"	1.000	Ea.	1.000	214	39	253
24" x 36"	1.000	Ea.	1.200	285	47	332
Vanity top, laminated plastic square edge, 25" x 32"	2.670	L.F.	.712	63	28	91
25" x 38"	3.170	L.F.	.845	74.50	33	107.50
Carrier , for lavatory, steel, for studs, no arms	1.000	Ea.	1.143	29.50	49	78.50
Wood, 2" x 8" blocking	1.300	L.F.	.052	1.14	2.04	3.18

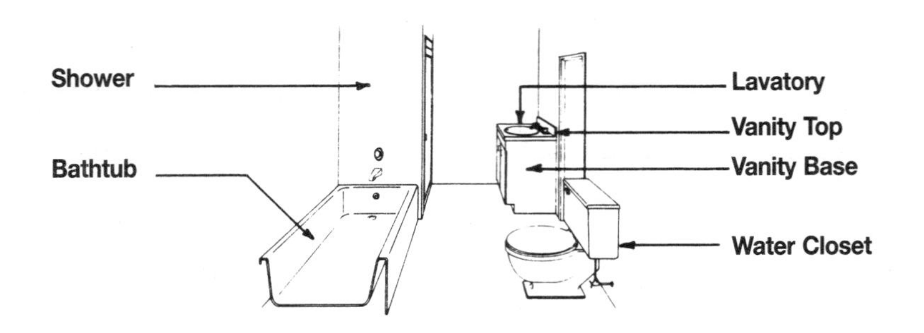

System Description	QUAN.	UNIT	LABOR HOURS	COST EACH		
				MAT.	INST.	TOTAL
BATHROOM WITH LAVATORY INSTALLED IN VANITY						
Water closet, floor mounted, 2 piece, close coupled, white	1.000	Ea.	3.019	176	117	293
Rough-in, vent, 2" diameter DWV piping	1.000	Ea.	.955	24.20	36.80	61
Waste, 4" diameter DWV piping	1.000	Ea.	.828	30.60	31.95	62.55
Supply, 1/2" diameter type "L" copper supply piping	1.000	Ea.	.593	6.66	25.44	32.10
Lavatory, 20" x 18" P.E. cast iron with fittings, white	1.000	Ea.	2.500	227	96.50	323.50
Shower, steel, enameled, stone base, corner, white	1.000	Ea.	8.889	830	345	1,175
Rough-in, waste, 1-1/2" diameter DWV piping	2.000	Ea.	4.507	118	174	292
Supply, 1/2" diameter type "L" copper supply piping	2.000	Ea.	3.161	35.52	135.68	171.20
Bathtub, P.E. cast iron, 5' long with fittings, white	1.000	Ea.	3.636	410	140	550
Rough-in, waste, 4" diameter DWV piping	1.000	Ea.	.828	30.60	31.95	62.55
Supply, 1/2" diameter type "L" copper supply piping	1.000	Ea.	.988	11.10	42.40	53.50
Vent, 1-1/2" diameter DWV piping	1.000	Ea.	.593	11.12	25.40	36.52
Piping, supply, 1/2" diameter type "L" copper supply piping	42.000	L.F.	4.148	46.62	178.08	224.70
Waste, 4" diameter DWV piping	10.000	L.F.	2.759	102	106.50	208.50
Vent, 2" diameter DWV piping	13.000	L.F.	3.250	58.11	125.45	183.56
Vanity base, 2 doors, 30" wide	1.000	Ea.	1.000	214	39	253
Vanity top, plastic laminated, square edge	2.670	L.F.	.712	62.75	27.90	90.65
TOTAL		Ea.	42.366	2,394.28	1,679.05	4,073.33
BATHROOM WITH WALL HUNG LAVATORY						
Water closet, floor mounted, 2 piece, close coupled, white	1.000	Ea.	3.019	176	117	293
Rough-in, vent, 2" diameter DWV piping	1.000	Ea.	.955	24.20	36.80	61
Waste, 4" diameter DWV piping	1.000	Ea.	.828	30.60	31.95	62.55
Supply, 1/2" diameter type "L" copper supply piping	1.000	Ea.	.593	6.66	25.44	32.10
Lavatory, 20" x 18" P.E. cast iron with fittings, white	1.000	Ea.	2.000	261	77	338
Shower, steel enameled, stone base, corner , white	1.000	Ea.	8.889	830	345	1,175
Rough-in, waste, 1-1/2" diameter DWV piping	2.000	Ea.	4.507	118	174	292
Supply, 1/2" diameter type "L" copper supply piping	2.000	Ea.	3.161	35.52	135.68	171.20
Bathtub, P.E. cast iron, 5' long with fittings, white	1.000	Ea.	3.636	410	140	550
Rough-in, waste, 4" diameter DWV piping	1.000	Ea.	.828	30.60	31.95	62.55
Supply, 1/2" diameter type "L" copper supply piping	1.000	Ea.	.988	11.10	42.40	53.50
Vent, 1-1/2" diameter copper DWV piping	1.000	Ea.	.593	11.12	25.40	36.52
Piping, supply, 1/2" diameter type "L" copper supply piping	42.000	L.F.	4.148	46.62	178.08	224.70
Waste, 4" diameter DWV piping	10.000	L.F.	2.759	102	106.50	208.50
Vent, 2" diameter DWV piping	13.000	L.F.	3.250	58.11	125.45	183.56
Carrier, steel, for studs, no arms	1.000	Ea.	1.143	29.50	49	78.50
TOTAL		Ea.	41.297	2,181.03	1,641.65	3,822.68

The costs in this system are on a cost each basis. All necessary piping is included.

Important: See the Reference Section for critical supporting data - Reference Nos., Crews & Location Factors

Four Fixture Bathroom Price Sheet	QUAN.	UNIT	LABOR HOURS	COST EACH		
				MAT.	INST.	TOTAL
Water closet, close coupled, standard 2 piece, white	1.000	Ea.	3.019	176	117	293
Color	1.000	Ea.	3.019	211	117	328
One piece elongated bowl, white	1.000	Ea.	3.019	535	117	652
Color	1.000	Ea.	3.019	665	117	782
Low profile, one piece elongated bowl, white	1.000	Ea.	3.019	730	117	847
Color	1.000	Ea.	3.019	945	117	1,062
1/2″ copper supply, 4″ cast iron waste, 2″ cast iron vent	1.000	Ea.	2.376	61.50	94	155.50
4″ PVC/DWV waste, 2″ PVC vent	1.000	Ea.	2.678	28.50	106	134.50
4″ copper waste, 2″ copper vent	1.000	Ea.	2.520	54.50	103	157.50
3″ cast iron waste, 1-1/2″ cast iron vent	1.000	Ea.	2.244	54.50	89	143.50
3″ P.V.C. waste, 1-1/2″ P.V.C. vent	1.000	Ea.	2.388	25	98.50	123.50
3″ copper waste, 1-1/2″ copper vent	1.000	Ea.	2.014	37.50	83	120.50
1/2″ P.V.C. supply, 4″ P.V.C. waste, 2″ P.V.C. vent	1.000	Ea.	2.974	39	118	157
3″ P.V.C. waste, 1-1/2″ P.V.C. vent	1.000	Ea.	2.684	35.50	111	146.50
1/2″ steel supply, 4″ cast iron waste, 2″ cast iron vent	1.000	Ea.	2.545	64	101	165
4″ cast iron waste, 2″ steel vent	1.000	Ea.	2.590	57.50	103	160.50
4″ P.V.C. waste, 2″ P.V.C. vent	1.000	Ea.	2.847	30.50	113	143.50
Lavatory, wall hung P.E. cast iron 20″ x 18″, white	1.000	Ea.	2.000	261	77	338
Color	1.000	Ea.	2.000	297	77	374
Vitreous china 19″ x 17″, white	1.000	Ea.	2.286	183	88	271
Color	1.000	Ea.	2.286	204	88	292
Lavatory for vanity top, P.E. cast iron 20″ x 18″, white	1.000	Ea.	2.500	227	96.50	323.50
Color	1.000	Ea.	2.500	256	96.50	352.50
Steel enameled, 20″ x 17″, white	1.000	Ea.	2.759	142	106	248
Color	1.000	Ea.	2.500	148	96.50	244.50
Vitreous china 20″ x 16″, white	1.000	Ea.	2.963	244	114	358
Color	1.000	Ea.	2.963	244	114	358
Shower, steel enameled stone base, 36″ square, white	1.000	Ea.	8.889	830	345	1,175
Color	1.000	Ea.	8.889	895	345	1,240
Rough-in, for lavatory or shower						
1/2″ copper supply, 1-1/2″ cast iron waste, 1-1/2″ cast iron vent	1.000	Ea.	3.834	77	155	232
1-1/2″ P.V.C. waste, 1-1/4″ P.V.C. vent	1.000	Ea.	3.675	36.50	158	194.50
1/2″ steel supply, 1-1/4″ cast iron waste, 1-1/4″ steel vent	1.000	Ea.	4.103	71.50	167	238.50
1-1/4″ P.V.C. waste, 1-1/4″ P.V.C. vent	1.000	Ea.	3.937	43	169	212
1/2″ P.V.C. supply, 1-1/2″ P.V.C. waste, 1-1/2″ P.V.C. vent	1.000	Ea.	4.592	65	197	262
Bathtub, P.E. cast iron, 5′ long with fittings, white	1.000	Ea.	3.636	410	140	550
Color	1.000	Ea.	3.636	475	140	615
Steel, enameled 5′ long with fittings, white	1.000	Ea.	2.909	325	112	437
Color	1.000	Ea.	2.909	325	112	437
Rough-in, for bathtub						
1/2″ copper supply, 4″ cast iron waste, 1-1/2″ copper vent	1.000	Ea.	2.409	53	100	153
4″ P.V.C. waste, 1-1/2″ P.V.C. vent	1.000	Ea.	2.877	32	119	151
1/2″ steel supply, 4″ cast iron waste, 1-1/2″ steel vent	1.000	Ea.	2.898	59	117	176
4″ P.V.C. waste, 1-1/2″ P.V.C. vent	1.000	Ea.	3.159	36	131	167
1/2″ P.V.C. supply, 4″ P.V.C. waste, 1-1/2″ P.V.C. vent	1.000	Ea.	3.371	49.50	140	189.50
Piping, supply, 1/2″ copper	42.000	L.F.	4.148	46.50	178	224.50
1/2″ steel	42.000	L.F.	5.333	63.50	229	292.50
1/2″ P.V.C.	42.000	L.F.	6.222	120	267	387
Waste, 4″ cast iron, no hub	10.000	L.F.	2.759	102	107	209
4″ P.V.C./DWV	10.000	L.F.	3.333	44.50	129	173.50
4″ copper/DWV	10.000	Ea.	4.000	111	155	266
Vent 2″ cast iron, no hub	13.000	L.F.	3.105	78.50	120	198.50
2″ copper/DWV	13.000	L.F.	2.364	47	101	148
2″ P.V.C./DWV	13.000	L.F.	3.525	27	136	163
2″ steel, galvanized	13.000	L.F.	3.250	58	125	183
Vanity base cabinet, 2 doors, 30″ wide	1.000	Ea.	1.000	214	39	253
Vanity top, plastic laminated, square edge	2.670	L.F.	.712	63	28	91
Carrier, steel for studs, no arms	1.000	Ea.	1.143	29.50	49	78.50
Wood, 2″ x 8″ blocking	1.300	L.F.	.052	1.14	2.04	3.18

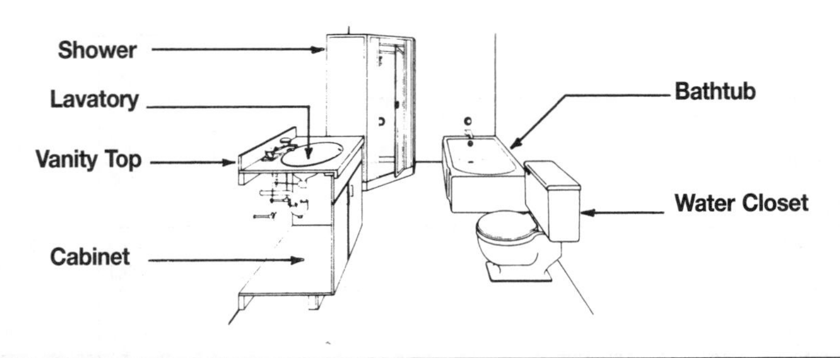

System Description	QUAN.	UNIT	LABOR HOURS	COST EACH		
				MAT.	INST.	TOTAL
BATHROOM WITH LAVATORY INSTALLED IN VANITY						
Water closet, floor mounted, 2 piece, close coupled, white	1.000	Ea.	3.019	176	117	293
Rough-in, vent, 2″ diameter DWV piping	1.000	Ea.	.955	24.20	36.80	61
Waste, 4″ diameter DWV piping	1.000	Ea.	.828	30.60	31.95	62.55
Supply, 1/2″ diameter type "L" copper supply piping	1.000	Ea.	.593	6.66	25.44	32.10
Lavatory, 20″ x 18″ P.E. cast iron with fittings, white	1.000	Ea.	2.500	227	96.50	323.50
Shower, steel, enameled, stone base, corner, white	1.000	Ea.	8.889	830	345	1,175
Rough-in, waste, 1-1/2″ diameter DWV piping	2.000	Ea.	4.507	118	174	292
Supply, 1/2″ diameter type "L" copper supply piping	2.000	Ea.	3.161	35.52	135.68	171.20
Bathtub, P.E. cast iron, 5′ long with fittings, white	1.000	Ea.	3.636	410	140	550
Rough-in, waste, 4″ diameter DWV piping	1.000	Ea.	.828	30.60	31.95	62.55
Supply, 1/2″ diameter type "L" copper supply piping	1.000	Ea.	.988	11.10	42.40	53.50
Vent, 1-1/2″ diameter DWV piping	1.000	Ea.	.593	11.12	25.40	36.52
Piping, supply, 1/2″ diameter type "L" copper supply piping	42.000	L.F.	4.939	55.50	212	267.50
Waste, 4″ diameter DWV piping	10.000	L.F.	4.138	153	159.75	312.75
Vent, 2″ diameter DWV piping	13.000	L.F.	4.500	80.46	173.70	254.16
Vanity base, 2 doors, 30″ wide	1.000	Ea.	1.000	214	39	253
Vanity top, plastic laminated, square edge	2.670	L.F.	.712	65.42	27.90	93.32
TOTAL		Ea.	45.786	2,479.18	1,814.47	4,293.65
BATHROOM WITH WALL HUNG LAVATORY						
Water closet, floor mounted, 2 piece, close coupled, white	1.000	Ea.	3.019	176	117	293
Rough-in, vent, 2″ diameter DWV piping	1.000	Ea.	.955	24.20	36.80	61
Waste, 4″ diameter DWV piping	1.000	Ea.	.828	30.60	31.95	62.55
Supply, 1/2″ diameter type "L" copper supply piping	1.000	Ea.	.593	6.66	25.44	32.10
Lavatory, 20″ x 18″ P.E. cast iron with fittings, white	1.000	Ea.	2.000	261	77	338
Shower, steel enameled, stone base, corner, white	1.000	Ea.	8.889	830	345	1,175
Rough-in, waste, 1-1/2″ diameter DWV piping	2.000	Ea.	4.507	118	174	292
Supply, 1/2″ diameter type "L" copper supply piping	2.000	Ea.	3.161	35.52	135.68	171.20
Bathtub, P.E. cast iron, 5″ long with fittings, white	1.000	Ea.	3.636	410	140	550
Rough-in, waste, 4″ diameter DWV piping	1.000	Ea.	.828	30.60	31.95	62.55
Supply, 1/2″ diameter type "L" copper supply piping	1.000	Ea.	.988	11.10	42.40	53.50
Vent, 1-1/2″ diameter DWV piping	1.000	Ea.	.593	11.12	25.40	36.52
Piping, supply, 1/2″ diameter type "L" copper supply piping	42.000	L.F.	4.939	55.50	212	267.50
Waste, 4″ diameter DWV piping	10.000	L.F.	4.138	153	159.75	312.75
Vent, 2″ diameter DWV piping	13.000	L.F.	4.500	80.46	173.70	254.16
Carrier, steel for studs, no arms	1.000	Ea.	1.143	29.50	49	78.50
TOTAL		Ea.	44.717	2,263.26	1,777.07	4,040.33

The costs in this system are on a cost each basis. All necessary piping is included.

Four Fixture Bathroom Price Sheet	QUAN.	UNIT	LABOR HOURS	COST EACH		
				MAT.	INST.	TOTAL
Water closet, close coupled, standard 2 piece, white	1.000	Ea.	3.019	176	117	293
Color	1.000	Ea.	3.019	211	117	328
One piece, elongated bowl, white	1.000	Ea.	3.019	535	117	652
Color	1.000	Ea.	3.019	665	117	782
Low profile, one piece elongated bowl, white	1.000	Ea.	3.019	730	117	847
Color	1.000	Ea.	3.019	945	117	1,062
Rough-in, for water closet						
1/2" copper supply, 4" cast iron waste, 2" cast iron vent	1.000	Ea.	2.376	61.50	94	155.50
4" PVC/DWV waste, 2" PVC vent	1.000	Ea.	2.678	28.50	106	134.50
4" copper waste, 2" copper vent	1.000	Ea.	2.520	54.50	103	157.50
3" cast iron waste, 1-1/2" cast iron vent	1.000	Ea.	2.244	54.50	89	143.50
3" PVC waste, 1-1/2" PVC vent	1.000	Ea.	2.388	25	98.50	123.50
3" PVC waste, 1-1/2" PVC vent	1.000	Ea.	2.014	37.50	83	120.50
1/2" PVC supply, 4" PVC waste, 2" PVC vent	1.000	Ea.	2.974	39	118	157
3" PVC waste, 1-1/2" PVC vent	1.000	Ea.	2.684	35.50	111	146.50
1/2" steel supply, 4" cast iron waste, 2" cast iron vent	1.000	Ea.	2.545	64	101	165
4" cast iron waste, 2" steel vent	1.000	Ea.	2.590	57.50	103	160.50
4" PVC waste, 2" PVC vent	1.000	Ea.	2.847	30.50	113	143.50
Lavatory wall hung, P.E. cast iron 20" x 18", white	1.000	Ea.	2.000	261	77	338
Color	1.000	Ea.	2.000	297	77	374
Vitreous china 19" x 17", white	1.000	Ea.	2.286	183	88	271
Color	1.000	Ea.	2.286	204	88	292
Lavatory for vanity top, P.E. cast iron, 20" x 18", white	1.000	Ea.	2.500	227	96.50	323.50
Color	1.000	Ea.	2.500	256	96.50	352.50
Steel, enameled 20" x 17", white	1.000	Ea.	2.759	142	106	248
Color	1.000	Ea.	2.500	148	96.50	244.50
Vitreous china 20" x 16", white	1.000	Ea.	2.963	244	114	358
Color	1.000	Ea.	2.963	244	114	358
Shower, steel enameled, stone base 36" square, white	1.000	Ea.	8.889	830	345	1,175
Color	1.000	Ea.	8.889	895	345	1,240
Rough-in, for lavatory and shower						
1/2" copper supply, 1-1/2" cast iron waste, 1-1/2" cast iron vent	1.000	Ea.	7.668	154	310	464
1-1/2" PVC waste, 1-1/4" PVC vent	1.000	Ea.	7.352	73.50	315	388.50
1/2" steel supply, 1-1/4" cast iron waste, 1-1/4" steel vent	1.000	Ea.	8.205	143	335	478
1-1/4" PVC waste, 1-1/4" PVC vent	1.000	Ea.	7.873	85.50	335	420.50
1/2" PVC supply, 1-1/2" PVC waste, 1-1/2" PVC vent	1.000	Ea.	9.185	130	395	525
Bathtub, P.E. cast iron, 5' long with fittings, white	1.000	Ea.	3.636	410	140	550
Color	1.000	Ea.	3.636	475	140	615
Steel enameled, 5' long with fittings, white	1.000	Ea.	2.909	325	112	437
Color	1.000	Ea.	2.909	325	112	437
Rough-in, for bathtub						
1/2" copper supply, 4" cast iron waste, 1-1/2" copper vent	1.000	Ea.	2.409	53	100	153
4" PVC waste, 1-1/2" PVC vent	1.000	Ea.	2.877	32	119	151
1/2" steel supply, 4" cast iron waste, 1-1/2" steel vent	1.000	Ea.	2.898	59	117	176
4" PVC waste, 1-1/2" PVC vent	1.000	Ea.	3.159	36	131	167
1/2" PVC supply, 4" PVC waste, 1-1/2" PVC vent	1.000	Ea.	3.371	49.50	140	189.50
Piping supply, 1/2" copper	42.000	L.F.	4.148	46.50	178	224.50
1/2" steel	42.000	L.F.	5.333	63.50	229	292.50
1/2" PVC	42.000	L.F.	6.222	120	267	387
Piping, waste, 4" cast iron, no hub	10.000	L.F.	3.586	133	138	271
4" PVC/DWV	10.000	L.F.	4.333	57.50	167	224.50
4" copper/DWV	10.000	L.F.	5.200	144	201	345
Piping, vent, 2" cast iron, no hub	13.000	L.F.	3.105	78.50	120	198.50
2" copper/DWV	13.000	L.F.	2.364	47	101	148
2" PVC/DWV	13.000	L.F.	3.525	27	136	163
2" steel, galvanized	13.000	L.F.	3.250	58	125	183
Vanity base cabinet, 2 doors, 30" wide	1.000	Ea.	1.000	214	39	253
Vanity top, plastic laminated, square edge	3.160	L.F.	.843	74.50	33	107.50
Carrier, steel, for studs, no arms	1.000	Ea.	1.143	29.50	49	78.50
Wood, 2" x 8" blocking	1.300	L.F.	.052	1.14	2.04	3.18

44 | Five Fixture Bathroom Systems

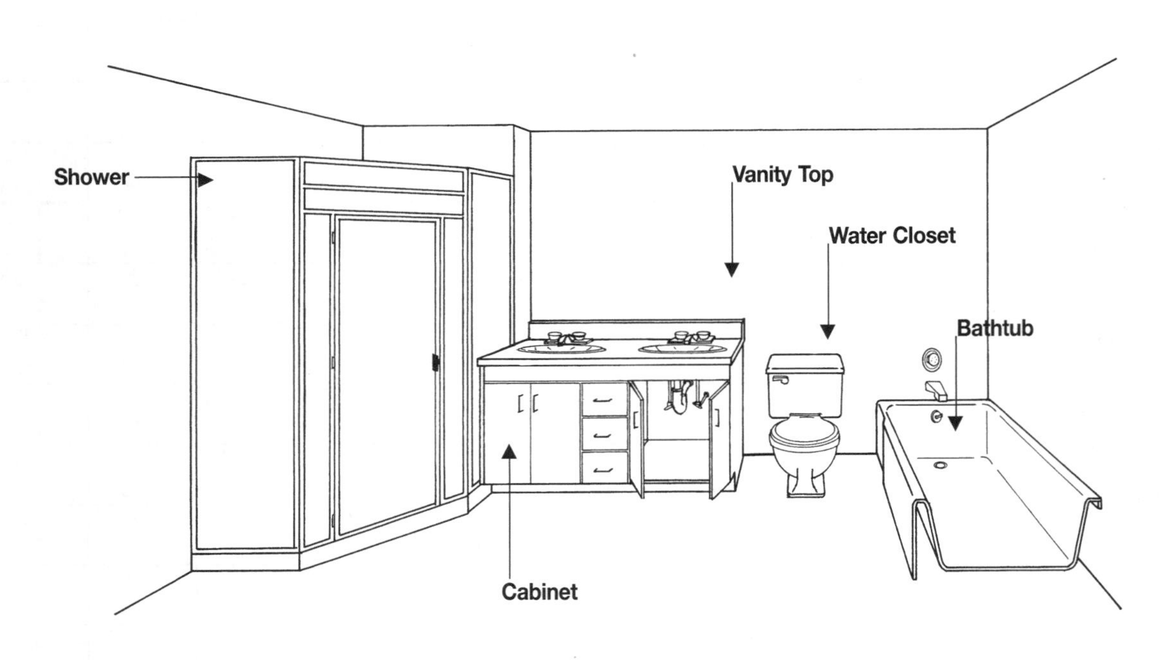

System Description	QUAN.	UNIT	LABOR HOURS	COST EACH		
				MAT.	INST.	TOTAL
BATHROOM WITH SHOWER, BATHTUB, LAVATORIES IN VANITY						
Water closet, floor mounted, 1 piece combination, white	1.000	Ea.	3.019	730	117	847
Rough-in, vent, 2" diameter DWV piping	1.000	Ea.	.955	24.20	36.80	61
Waste, 4" diameter DWV piping	1.000	Ea.	.828	30.60	31.95	62.55
Supply, 1/2" diameter type "L" copper supply piping	1.000	Ea.	.593	6.66	25.44	32.10
Lavatory, 20" x 16", vitreous china oval, with fittings, white	2.000	Ea.	5.926	488	228	716
Shower, steel enameled, stone base, corner, white	1.000	Ea.	8.889	830	345	1,175
Rough-in, waste, 1-1/2" diameter DWV piping	3.000	Ea.	5.408	141.60	208.80	350.40
Supply, 1/2" diameter type "L" copper supply piping	3.000	Ea.	2.963	33.30	127.20	160.50
Bathtub, P.E. cast iron, 5' long with fittings, white	1.000	Ea.	3.636	410	140	550
Rough-in, waste, 4" diameter DWV piping	1.000	Ea.	1.103	40.80	42.60	83.40
Supply, 1/2" diameter type "L" copper supply piping	1.000	Ea.	.988	11.10	42.40	53.50
Vent, 1-1/2" diameter copper DWV piping	1.000	Ea.	.593	11.12	25.40	36.52
Piping, supply, 1/2" diameter type "L" copper supply piping	42.000	L.F.	4.148	46.62	178.08	224.70
Waste, 4" diameter DWV piping	10.000	L.F.	2.759	102	106.50	208.50
Vent, 2" diameter DWV piping	13.000	L.F.	3.250	58.11	125.45	183.56
Vanity base, 2 door, 24" x 48"	1.000	Ea.	1.400	340	55	395
Vanity top, plastic laminated, square edge	4.170	L.F.	1.112	98	43.58	141.58
TOTAL		Ea.	47.570	3,402.11	1,879.20	5,281.31

The costs in this system are on a cost each basis. All necessary piping is included.

Description	QUAN.	UNIT	LABOR HOURS	COST EACH		
				MAT.	INST.	TOTAL

Five Fixture Bathroom Price Sheet	QUAN.	UNIT	LABOR HOURS	COST EACH		
				MAT.	INST.	TOTAL
Water closet, close coupled, standard 2 piece, white	1.000	Ea.	3.019	176	117	293
Color	1.000	Ea.	3.019	211	117	328
One piece elongated bowl, white	1.000	Ea.	3.019	535	117	652
Color	1.000	Ea.	3.019	665	117	782
Low profile, one piece elongated bowl, white	1.000	Ea.	3.019	730	117	847
Color	1.000	Ea.	3.019	945	117	1,062
Rough-in, supply, waste and vent for water closet						
1/2" copper supply, 4" cast iron waste, 2" cast iron vent	1.000	Ea.	2.376	61.50	94	155.50
4" P.V.C./DWV waste, 2" P.V.C. vent	1.000	Ea.	2.678	28.50	106	134.50
4" copper waste, 2" copper vent	1.000	Ea.	2.520	54.50	103	157.50
3" cast iron waste, 1-1/2" cast iron vent	1.000	Ea.	2.244	54.50	89	143.50
3" P.V.C. waste, 1-1/2" P.V.C. vent	1.000	Ea.	2.388	25	98.50	123.50
3" copper waste, 1-1/2" copper vent	1.000	Ea.	2.014	37.50	83	120.50
1/2" P.V.C. supply, 4" P.V.C. waste, 2" P.V.C. vent	1.000	Ea.	2.974	39	118	157
3" P.V.C. waste, 1-1/2" P.V.C. supply	1.000	Ea.	2.684	35.50	111	146.50
1/2" steel supply, 4" cast iron waste, 2" cast iron vent	1.000	Ea.	2.545	64	101	165
4" cast iron waste, 2" steel vent	1.000	Ea.	2.590	57.50	103	160.50
4" P.V.C. waste, 2" P.V.C. vent	1.000	Ea.	2.847	30.50	113	143.50
Lavatory, wall hung, P.E. cast iron 20" x 18", white	2.000	Ea.	4.000	520	154	674
Color	2.000	Ea.	4.000	595	154	749
Vitreous china, 19" x 17", white	2.000	Ea.	4.571	365	176	541
Color	2.000	Ea.	4.571	410	176	586
Lavatory, for vanity top, P.E. cast iron, 20" x 18", white	2.000	Ea.	5.000	455	193	648
Color	2.000	Ea.	5.000	510	193	703
Steel enameled 20" x 17", white	2.000	Ea.	5.517	284	212	496
Color	2.000	Ea.	5.000	296	193	489
Vitreous china 20" x 16", white	2.000	Ea.	5.926	490	228	718
Color	2.000	Ea.	5.926	490	228	718
Shower, steel enameled, stone base 36" square, white	1.000	Ea.	8.889	830	345	1,175
Color	1.000	Ea.	8.889	895	345	1,240
Rough-in, for lavatory or shower						
1/2" copper supply, 1-1/2" cast iron waste, 1-1/2" cast iron vent	3.000	Ea.	8.371	175	335	510
1-1/2" P.V.C. waste, 1-1/4" P.V.C. vent	3.000	Ea.	7.916	78.50	340	418.50
1/2" steel supply, 1-1/4" cast iron waste, 1-1/4" steel vent	3.000	Ea.	8.670	152	350	502
1-1/4" P.V.C. waste, 1-1/4" P.V.C. vent	3.000	Ea.	8.381	90	360	450
1/2" P.V.C. supply, 1-1/2" P.V.C. waste, 1-1/2" P.V.C. vent	3.000	Ea.	9.778	131	420	551
Bathtub, P.E. cast iron 5' long with fittings, white	1.000	Ea.	3.636	410	140	550
Color	1.000	Ea.	3.636	475	140	615
Steel, enameled 5' long with fittings, white	1.000	Ea.	2.909	325	112	437
Color	1.000	Ea.	2.909	325	112	437
Rough-in, for bathtub						
1/2" copper supply, 4" cast iron waste, 1-1/2" copper vent	1.000	Ea.	2.684	63	110	173
4" P.V.C. waste, 1-1/2" P.V.C. vent	1.000	Ea.	3.210	36.50	132	168.50
1/2" steel supply, 4" cast iron waste, 1-1/2" steel vent	1.000	Ea.	3.173	69.50	128	197.50
4" P.V.C. waste, 1-1/2" P.V.C. vent	1.000	Ea.	3.492	40.50	144	184.50
1/2" P.V.C. supply, 4" P.V.C. waste, 1-1/2" P.V.C. vent	1.000	Ea.	3.704	54	153	207
Piping, supply, 1/2" copper	42.000	L.F.	4.148	46.50	178	224.50
1/2" steel	42.000	L.F.	5.333	63.50	229	292.50
1/2" P.V.C.	42.000	L.F.	6.222	120	267	387
Piping, waste, 4" cast iron, no hub	10.000	L.F.	2.759	102	107	209
4" P.V.C./DWV	10.000	L.F.	3.333	44.50	129	173.50
4" copper/DWV	10.000	L.F.	4.000	111	155	266
Piping, vent, 2" cast iron, no hub	13.000	L.F.	3.105	78.50	120	198.50
2" copper/DWV	13.000	L.F.	2.364	47	101	148
2" P.V.C./DWV	13.000	L.F.	3.525	27	136	163
2" steel, galvanized	13.000	L.F.	3.250	58	125	183
Vanity base cabinet, 2 doors, 24" x 48"	1.000	Ea.	1.400	340	55	395
Vanity top, plastic laminated, square edge	4.170	L.F.	1.112	98	43.50	141.50
Carrier, steel, for studs, no arms	1.000	Ea.	1.143	29.50	49	78.50
Wood, 2" x 8" blocking	1.300	L.F.	.052	1.14	2.04	3.18

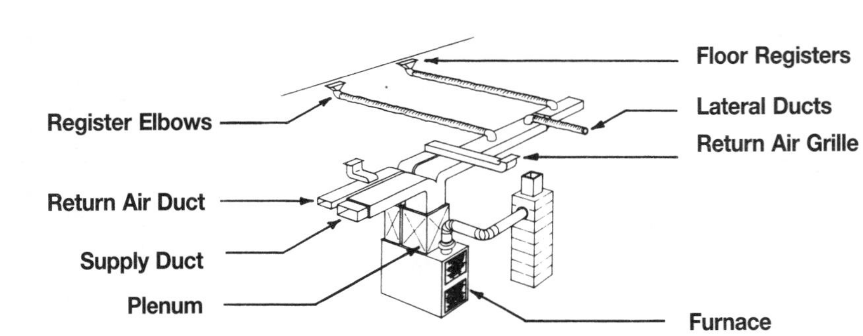

System Description	QUAN.	UNIT	LABOR HOURS	COST PER SYSTEM		
				MAT.	INST.	TOTAL
HEATING ONLY, GAS FIRED HOT AIR, ONE ZONE, 1200 S.F. BUILDING						
Furnace, gas, up flow	1.000	Ea.	5.000	720	193	913
Intermittent pilot	1.000	Ea.		145		145
Supply duct, rigid fiberglass	176.000	S.F.	12.068	109.12	482.24	591.36
Return duct, sheet metal, galvanized	158.000	Lb.	16.137	142.20	646.22	788.42
Lateral ducts, 6″ flexible fiberglass	144.000	L.F.	8.862	239.04	341.28	580.32
Register, elbows	12.000	Ea.	3.200	318	123.60	441.60
Floor registers, enameled steel	12.000	Ea.	3.000	227.40	128.40	355.80
Floor grille, return air	2.000	Ea.	.727	51	31.20	82.20
Thermostat	1.000	Ea.	1.000	30	43	73
Plenum	1.000	Ea.	1.000	67.50	38.50	106
TOTAL		System	50.994	2,049.26	2,027.44	4,076.70
HEATING/COOLING, GAS FIRED FORCED AIR, ONE ZONE, 1200 S.F. BUILDING						
Furnace, including plenum, compressor, coil	1.000	Ea.	14.720	3,588	565.80	4,153.80
Intermittent pilot	1.000	Ea.		145		145
Supply duct, rigid fiberglass	176.000	S.F.	12.068	109.12	482.24	591.36
Return duct, sheet metal, galvanized	158.000	Lb.	16.137	142.20	646.22	788.42
Lateral duct, 6″ flexible fiberglass	144.000	L.F.	8.862	239.04	341.28	580.32
Register elbows	12.000	Ea.	3.200	318	123.60	441.60
Floor registers, enameled steel	12.000	Ea.	3.000	227.40	128.40	355.80
Floor grille return air	2.000	Ea.	.727	51	31.20	82.20
Thermostat	1.000	Ea.	1.000	30	43	73
Refrigeration piping, 25 ft. (pre-charged)	1.000	Ea.		184		184
TOTAL		System	59.714	5,033.76	2,361.74	7,395.50

The costs in these systems are based on complete system basis. For larger buildings use the price sheet on the opposite page.

Description	QUAN.	UNIT	LABOR HOURS	COST PER SYSTEM		
				MAT.	INST.	TOTAL

Gas Heating/Cooling Price Sheet	QUAN.	UNIT	LABOR HOURS	COST EACH		
				MAT.	INST.	TOTAL
Furnace, heating only, 100 MBH, area to 1200 S.F.	1.000	Ea.	5.000	720	193	913
120 MBH, area to 1500 S.F.	1.000	Ea.	5.000	720	193	913
160 MBH, area to 2000 S.F.	1.000	Ea.	5.714	965	220	1,185
200 MBH, area to 2400 S.F.	1.000	Ea.	6.154	1,825	237	2,062
Heating/cooling, 100 MBH heat, 36 MBH cool, to 1200 S.F.	1.000	Ea.	16.000	3,900	615	4,515
120 MBH heat, 42 MBH cool, to 1500 S.F.	1.000	Ea.	18.462	4,175	740	4,915
144 MBH heat, 47 MBH cool, to 2000 S.F.	1.000	Ea.	20.000	4,800	800	5,600
200 MBH heat, 60 MBH cool, to 2400 S.F.	1.000	Ea.	34.286	5,050	1,375	6,425
Intermittent pilot, 100 MBH furnace	1.000	Ea.		145		145
200 MBH furnace	1.000	Ea.		145		145
Supply duct, rectangular, area to 1200 S.F., rigid fiberglass	176.000	S.F.	12.068	109	480	589
Sheet metal insulated	228.000	Lb.	31.331	298	1,225	1,523
Area to 1500 S.F., rigid fiberglass	176.000	S.F.	12.068	109	480	589
Sheet metal insulated	228.000	Lb.	31.331	298	1,225	1,523
Area to 2400 S.F., rigid fiberglass	205.000	S.F.	14.057	127	560	687
Sheet metal insulated	271.000	Lb.	37.048	355	1,450	1,805
Round flexible, insulated 6″ diameter, to 1200 S.F.	156.000	L.F.	9.600	259	370	629
To 1500 S.F.	184.000	L.F.	11.323	305	435	740
8″ diameter, to 2000 S.F.	269.000	L.F.	23.911	560	925	1,485
To 2400 S.F.	248.000	L.F.	22.045	520	850	1,370
Return duct, sheet metal galvanized, to 1500 S.F.	158.000	Lb.	16.137	142	645	787
To 2400 S.F.	191.000	Lb.	19.507	172	780	952
Lateral ducts, flexible round 6″ insulated, to 1200 S.F.	144.000	L.F.	8.862	239	340	579
To 1500 S.F.	172.000	L.F.	10.585	286	410	696
To 2000 S.F.	261.000	L.F.	16.062	435	620	1,055
To 2400 S.F.	300.000	L.F.	18.462	500	710	1,210
Spiral steel insulated, to 1200 S.F.	144.000	L.F.	20.067	330	750	1,080
To 1500 S.F.	172.000	L.F.	23.952	395	900	1,295
To 2000 S.F.	261.000	L.F.	36.352	600	1,350	1,950
To 2400 S.F.	300.000	L.F.	41.825	690	1,575	2,265
Rectangular sheet metal galvanized insulated, to 1200 S.F.	228.000	Lb.	39.056	390	1,500	1,890
To 1500 S.F.	344.000	Lb.	53.966	530	2,100	2,630
To 2000 S.F.	522.000	Lb.	81.926	800	3,175	3,975
To 2400 S.F.	600.000	Lb.	94.189	920	3,675	4,595
Register elbows, to 1500 S.F.	12.000	Ea.	3.200	320	124	444
To 2400 S.F.	14.000	Ea.	3.733	370	144	514
Floor registers, enameled steel w/damper, to 1500 S.F.	12.000	Ea.	3.000	227	128	355
To 2400 S.F.	14.000	Ea.	4.308	315	185	500
Return air grille, area to 1500 S.F. 12″ x 12″	2.000	Ea.	.727	51	31	82
Area to 2400 S.F. 8″ x 16″	2.000	Ea.	.444	46	19.05	65.05
Area to 2400 S.F. 8″ x 16″	2.000	Ea.	.727	51	31	82
16″ x 16″	1.000	Ea.	.364	37	15.60	52.60
Thermostat, manual, 1 set back	1.000	Ea.	1.000	30	43	73
Electric, timed, 1 set back	1.000	Ea.	1.000	89	43	132
2 set back	1.000	Ea.	1.000	196	43	239
Plenum, heating only, 100 M.B.H.	1.000	Ea.	1.000	67.50	38.50	106
120 MBH	1.000	Ea.	1.000	67.50	38.50	106
160 MBH	1.000	Ea.	1.000	67.50	38.50	106
200 MBH	1.000	Ea.	1.000	67.50	38.50	106
Refrigeration piping, 3/8″	25.000	L.F.		9.45		9.45
3/4″	25.000	L.F.		19		19
7/8″	25.000	L.F.		22		22
Refrigerant piping, 25 ft. (precharged)	1.000	Ea.		184		184
Diffusers, ceiling, 6″ diameter, to 1500 S.F.	10.000	Ea.	4.444	180	191	371
To 2400 S.F.	12.000	Ea.	6.000	232	258	490
Floor, aluminum, adjustable, 2-1/4″ x 12″ to 1500 S.F.	12.000	Ea.	3.000	164	128	292
To 2400 S.F.	14.000	Ea.	3.500	192	150	342
Side wall, aluminum, adjustable, 8″ x 4″, to 1500 S.F.	12.000	Ea.	3.000	360	128	488
5″ x 10″ to 2400 S.F.	12.000	Ea.	3.692	475	158	633

64 Oil Fired Heating/Cooling Systems

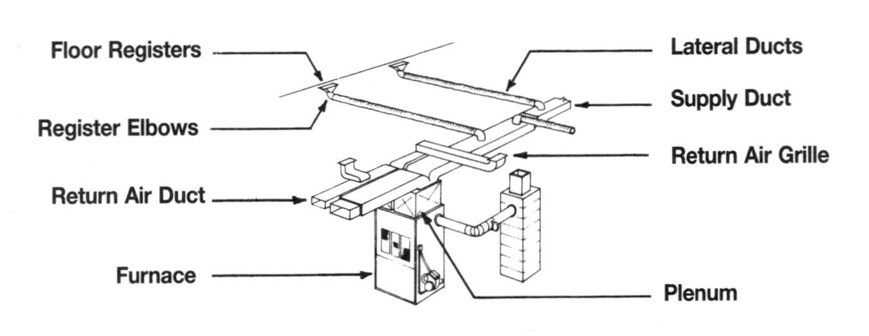

System Description	QUAN.	UNIT	LABOR HOURS	COST PER SYSTEM		
				MAT.	INST.	TOTAL
HEATING ONLY, OIL FIRED HOT AIR, ONE ZONE, 1200 S.F. BUILDING						
Furnace, oil fired, atomizing gun type burner	1.000	Ea.	4.571	850	176	1,026
3/8″ diameter copper supply pipe	1.000	Ea.	2.759	26.10	118.20	144.30
Shut off valve	1.000	Ea.	.333	7.40	14.30	21.70
Oil tank, 275 gallon, on legs	1.000	Ea.	3.200	276	124	400
Supply duct, rigid fiberglass	176.000	S.F.	12.068	109.12	482.24	591.36
Return duct, sheet metal, galvanized	158.000	Lb.	16.137	142.20	646.22	788.42
Lateral ducts, 6″ flexible fiberglass	144.000	L.F.	8.862	239.04	341.28	580.32
Register elbows	12.000	Ea.	3.200	318	123.60	441.60
Floor register, enameled steel	12.000	Ea.	3.000	227.40	128.40	355.80
Floor grille, return air	2.000	Ea.	.727	51	31.20	82.20
Thermostat	1.000	Ea.	1.000	30	43	73
TOTAL		System	55.946	2,276.26	2,228.44	4,504.70
HEATING/COOLING, OIL FIRED, FORCED AIR, ONE ZONE, 1200 S.F. BUILDING						
Furnace, including plenum, compressor, coil	1.000	Ea.	16.000	4,175	615	4,790
3/8″ diameter copper supply pipe	1.000	Ea.	2.759	26.10	118.20	144.30
Shut off valve	1.000	Ea.	.333	7.40	14.30	21.70
Oil tank, 275 gallon on legs	1.000	Ea.	3.200	276	124	400
Supply duct, rigid fiberglass	176.000	S.F.	12.068	109.12	482.24	591.36
Return duct, sheet metal, galvanized	158.000	Lb.	16.137	142.20	646.22	788.42
Lateral ducts, 6″ flexible fiberglass	144.000	L.F.	8.862	239.04	341.28	580.32
Register elbows	12.000	Ea.	3.200	318	123.60	441.60
Floor registers, enameled steel	12.000	Ea.	3.000	227.40	128.40	355.80
Floor grille, return air	2.000	Ea.	.727	51	31.20	82.20
Refrigeration piping (precharged)	25.000	L.F.		184		184
TOTAL		System	66.606	5,755.26	2,624.44	8,379.70

Description	QUAN.	UNIT	LABOR HOURS	COST EACH		
				MAT.	INST.	TOTAL

Important: See the Reference Section for critical supporting data - Reference Nos., Crews & Location Factors

Oil Fired Heating/Cooling	QUAN.	UNIT	LABOR HOURS	COST EACH		
				MAT.	INST.	TOTAL
Furnace, heating, 95.2 MBH,area to 1200 S.F.	1.000	Ea.	4.706	870	182	1,052
123.2 MBH, area to 1500 S.F.	1.000	Ea.	5.000	1,200	193	1,393
151.2 MBH, area to 2000 S.F.	1.000	Ea.	5.333	1,325	206	1,531
200 MBH, area to 2400 S.F.	1.000	Ea.	6.154	1,900	237	2,137
Heating/cooling, 95.2 MBH heat, 36 MBH cool, to 1200 S.F.	1.000	Ea.	16.000	4,175	615	4,790
112 MBH heat, 42 MBH cool, to 1500 S.F.	1.000	Ea.	24.000	6,275	925	7,200
151 MBH heat, 47 MBH cool, to 2000 S.F.	1.000	Ea.	20.800	5,425	800	6,225
184.8 MBH heat, 60 MBH cool, to 2400 S.F.	1.000	Ea.	24.000	5,725	960	6,685
Oil piping to furnace, 3/8″ dia., copper	1.000	Ea.	3.412	117	145	262
Oil tank, on legs above ground, 275 gallons	1.000	Ea.	3.200	276	124	400
550 gallons	1.000	Ea.	5.926	1,250	230	1,480
Below ground, 275 gallons	1.000	Ea.	3.200	276	124	400
550 gallons	1.000	Ea.	5.926	1,250	230	1,480
1000 gallons	1.000	Ea.	6.400	1,975	248	2,223
Supply duct, rectangular, area to 1200 S.F., rigid fiberglass	176.000	S.F.	12.068	109	480	589
Sheet metal, insulated	228.000	Lb.	31.331	298	1,225	1,523
Area to 1500 S.F., rigid fiberglass	176.000	S.F.	12.068	109	480	589
Sheet metal, insulated	228.000	Lb.	31.331	298	1,225	1,523
Area to 2400 S.F., rigid fiberglass	205.000	S.F.	14.057	127	560	687
Sheet metal, insulated	271.000	Lb.	37.048	355	1,450	1,805
Round flexible, insulated, 6″ diameter to 1200 S.F.	156.000	L.F.	9.600	259	370	629
To 1500 S.F.	184.000	L.F.	11.323	305	435	740
8″ diameter to 2000 S.F.	269.000	L.F.	23.911	560	925	1,485
To 2400 S.F.	269.000	L.F.	22.045	520	850	1,370
Return duct, sheet metal galvanized, to 1500 S.F.	158.000	Lb.	16.137	142	645	787
To 2400 S.F.	191.000	Lb.	19.507	172	780	952
Lateral ducts, flexible round, 6″, insulated to 1200 S.F.	144.000	L.F.	8.862	239	340	579
To 1500 S.F.	172.000	L.F.	10.585	286	410	696
To 2000 S.F.	261.000	L.F.	16.062	435	620	1,055
To 2400 S.F.	300.000	L.F.	18.462	500	710	1,210
Spiral steel, insulated to 1200 S.F.	144.000	L.F.	20.067	330	750	1,080
To 1500 S.F.	172.000	L.F.	23.952	395	900	1,295
To 2000 S.F.	261.000	L.F.	36.352	600	1,350	1,950
To 2400 S.F.	300.000	L.F.	41.825	690	1,575	2,265
Rectangular sheet metal galvanized insulated, to 1200 S.F.	288.000	Lb.	45.183	440	1,750	2,190
To 1500 S.F.	344.000	Lb.	53.966	530	2,100	2,630
To 2000 S.F.	522.000	Lb.	81.926	800	3,175	3,975
To 2400 S.F.	600.000	Lb.	94.189	920	3,675	4,595
Register elbows, to 1500 S.F.	12.000	Ea.	3.200	320	124	444
To 2400 S.F.	14.000	Ea.	3.733	370	144	514
Floor registers, enameled steel w/damper, to 1500 S.F.	12.000	Ea.	3.000	227	128	355
To 2400 S.F.	14.000	Ea.	4.308	315	185	500
Return air grille, area to 1500 S.F., 12″ x 12″	2.000	Ea.	.727	51	31	82
12″ x 24″	1.000	Ea.	.444	46	19.05	65.05
Area to 2400 S.F., 8″ x 16″	2.000	Ea.	.727	51	31	82
16″ x 16″	1.000	Ea.	.364	37	15.60	52.60
Thermostat, manual, 1 set back	1.000	Ea.	1.000	30	43	73
Electric, timed, 1 set back	1.000	Ea.	1.000	89	43	132
2 set back	1.000	Ea.	1.000	196	43	239
Refrigeration piping, 3/8″	25.000	L.F.		9.45		9.45
3/4″	25.000	L.F.		19		19
Diffusers, ceiling, 6″ diameter, to 1500 S.F.	10.000	Ea.	4.444	180	191	371
To 2400 S.F.	12.000	Ea.	6.000	232	258	490
Floor, aluminum, adjustable, 2-1/4″ x 12″ to 1500 S.F.	12.000	Ea.	3.000	164	128	292
To 2400 S.F.	14.000	Ea.	3.500	192	150	342
Side wall, aluminum, adjustable, 8″ x 4″, to 1500 S.F.	12.000	Ea.	3.000	360	128	488
5″ x 10″ to 2400 S.F.	12.000	Ea.	3.692	475	158	633

68 Hot Water Heating Systems

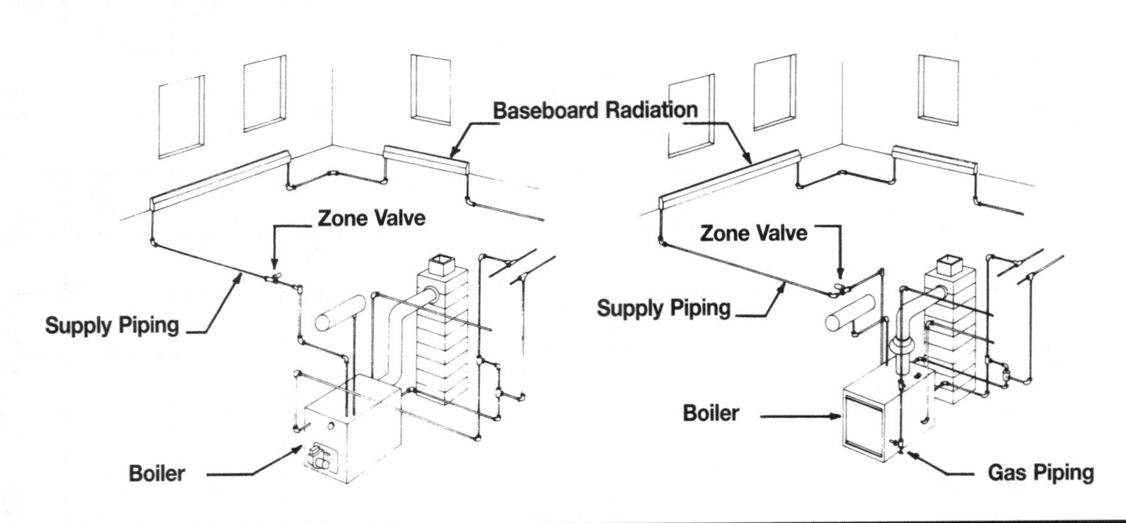

System Description	QUAN.	UNIT	LABOR HOURS	COST EACH		
				MAT.	INST.	TOTAL
OIL FIRED HOT WATER HEATING SYSTEM, AREA TO 1200 S.F.						
Boiler package, oil fired, 97 MBH, area to 1200 S.F. building	1.000	Ea.	15.000	1,300	560	1,860
3/8″ diameter copper supply pipe	1.000	Ea.	2.759	26.10	118.20	144.30
Shut off valve	1.000	Ea.	.333	7.40	14.30	21.70
Oil tank, 275 gallon, with black iron filler pipe	1.000	Ea.	3.200	276	124	400
Supply piping, 3/4″ copper tubing	176.000	L.F.	18.526	274.56	795.52	1,070.08
Supply fittings, copper 3/4″	36.000	Ea.	15.158	26.64	649.80	676.44
Supply valves, 3/4″	2.000	Ea.	.800	104	34.30	138.30
Baseboard radiation, 3/4″	106.000	L.F.	35.333	368.88	1,372.70	1,741.58
Zone valve	1.000	Ea.	.400	74.50	17.25	91.75
TOTAL		Ea.	91.659	2,458.08	3,686.07	6,144.15
OIL FIRED HOT WATER HEATING SYSTEM, AREA TO 2400 S.F.						
Boiler package, oil fired, 225 MBH, area to 2400 S.F. building	1.000	Ea.	19.704	3,225	735	3,960
3/8″ diameter copper supply pipe	1.000	Ea.	2.759	26.10	118.20	144.30
Shut off valve	1.000	Ea.	.333	7.40	14.30	21.70
Oil tank, 550 gallon, with black iron pipe filler pipe	1.000	Ea.	5.926	1,250	230	1,480
Supply piping, 3/4″ copper tubing	228.000	L.F.	23.999	355.68	1,030.56	1,386.24
Supply fittings, copper	46.000	Ea.	19.368	34.04	830.30	864.34
Supply valves	2.000	Ea.	.800	104	34.30	138.30
Baseboard radiation	212.000	L.F.	70.666	737.76	2,745.40	3,483.16
Zone valve	1.000	Ea.	.400	74.50	17.25	91.75
TOTAL		Ea.	144.152	5,814.48	5,755.31	11,569.79

The costs in this system are on a cost each basis. the costs represent total cost for the system based on a gross square foot of plan area.

Description	QUAN.	UNIT	LABOR HOURS	COST EACH		
				MAT.	INST.	TOTAL

Important: See the Reference Section for critical supporting data - Reference Nos., Crews & Location Factors

Hot Water Heating Price Sheet	QUAN.	UNIT	LABOR HOURS	COST EACH		
				MAT.	INST.	TOTAL
Boiler, oil fired, 97 MBH, area to 1200 S.F.	1.000	Ea.	15.000	1,300	560	1,860
118 MBH, area to 1500 S.F.	1.000	Ea.	16.506	2,500	615	3,115
161 MBH, area to 2000 S.F.	1.000	Ea.	18.405	3,125	685	3,810
215 MBH, area to 2400 S.F.	1.000	Ea.	19.704	3,225	735	3,960
Oil piping, (valve & filter), 3/8″ copper	1.000	Ea.	3.289	66	140	206
1/4″ copper	1.000	Ea.	3.242	46.50	138	184.50
Oil tank, filler pipe and cap on legs, 275 gallon	1.000	Ea.	3.200	276	124	400
550 gallon	1.000	Ea.	5.926	1,250	230	1,480
Buried underground, 275 gallon	1.000	Ea.	3.200	276	124	400
550 gallon	1.000	Ea.	5.926	1,250	230	1,480
1000 gallon	1.000	Ea.	6.400	1,975	248	2,223
Supply piping copper, area to 1200 S.F., 1/2″ tubing	176.000	L.F.	17.384	195	745	940
3/4″ tubing	176.000	L.F.	18.526	275	795	1,070
Area to 1500 S.F., 1/2″ tubing	186.000	L.F.	18.371	206	790	996
3/4″ tubing	186.000	L.F.	19.578	290	840	1,130
Area to 2000 S.F., 1/2″ tubing	204.000	L.F.	20.149	226	865	1,091
3/4″ tubing	204.000	L.F.	21.473	320	920	1,240
Area to 2400 S.F., 1/2″ tubing	228.000	L.F.	22.520	253	965	1,218
3/4″ tubing	228.000	L.F.	23.999	355	1,025	1,380
Supply pipe fittings copper, area to 1200 S.F., 1/2″	36.000	Ea.	14.400	11.90	615	626.90
3/4″	36.000	Ea.	15.158	26.50	650	676.50
Area to 1500 S.F., 1/2″	40.000	Ea.	16.000	13.20	685	698.20
3/4″	40.000	Ea.	16.842	29.50	720	749.50
Area to 2000 S.F., 1/2″	44.000	Ea.	17.600	14.50	755	769.50
3/4″	44.000	Ea.	18.526	32.50	795	827.50
Area to 2400, S.F., 1/2″	46.000	Ea.	18.400	15.20	790	805.20
3/4″	46.000	Ea.	19.368	34	830	864
Supply valves, 1/2″ pipe size	2.000	Ea.	.667	76	28.50	104.50
3/4″	2.000	Ea.	.800	104	34.50	138.50
Baseboard radiation, area to 1200 S.F., 1/2″ tubing	106.000	L.F.	28.267	625	1,100	1,725
3/4″ tubing	106.000	L.F.	35.333	370	1,375	1,745
Area to 1500 S.F., 1/2″ tubing	134.000	L.F.	35.734	790	1,375	2,165
3/4″ tubing	134.000	L.F.	44.666	465	1,725	2,190
Area to 2000 S.F., 1/2″ tubing	178.000	L.F.	47.467	1,050	1,850	2,900
3/4″ tubing	178.000	L.F.	59.333	620	2,300	2,920
Area to 2400 S.F., 1/2″ tubing	212.000	L.F.	56.534	1,250	2,200	3,450
3/4″ tubing	212.000	L.F.	70.666	740	2,750	3,490
Zone valves, 1/2″ tubing	1.000	Ea.	.400	74.50	17.25	91.75
3/4″ tubing	1.000	Ea.	.400	79.50	17.25	96.75

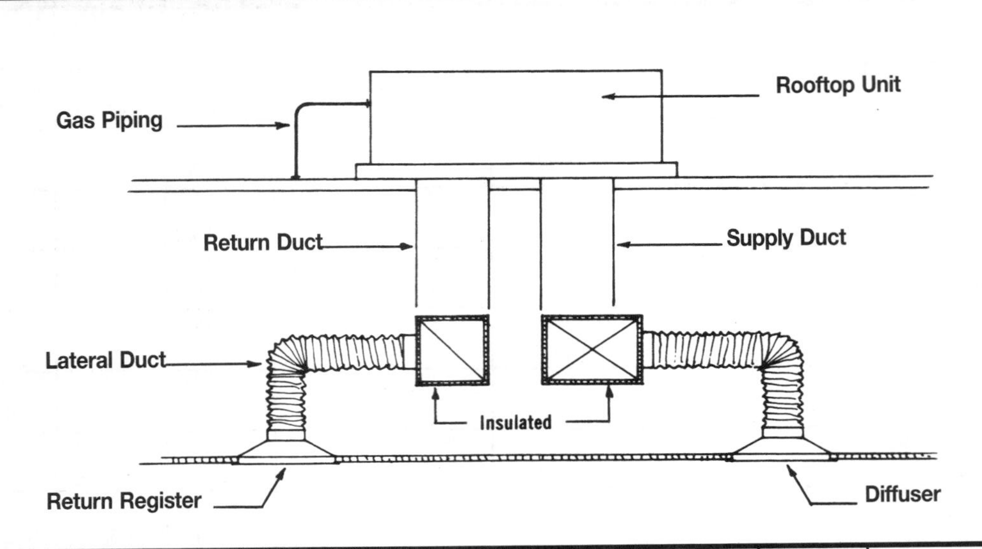

System Description	QUAN.	UNIT	LABOR HOURS	COST EACH		
				MAT.	INST.	TOTAL
ROOFTOP HEATING/COOLING UNIT, AREA TO 2000 S.F.						
Rooftop unit, single zone, electric cool, gas heat, to 2000 s.f.	1.000	Ea.	28.521	4,050	1,100	5,150
Gas piping	34.500	L.F.	5.207	77.28	224.25	301.53
Duct, supply and return, galvanized steel	38.000	Lb.	3.881	34.20	155.42	189.62
Insulation, ductwork	33.000	S.F.	1.508	17.49	55.44	72.93
Lateral duct, flexible duct 12″ diameter, insulated	72.000	L.F.	11.520	226.80	442.80	669.60
Diffusers	4.000	Ea.	4.571	1,240	196	1,436
Return registers	1.000	Ea.	.727	123	31	154
TOTAL		Ea.	55.935	5,768.77	2,204.91	7,973.68
ROOFTOP HEATING/COOLING UNIT, AREA TO 5000 S.F.						
Rooftop unit, single zone, electric cool, gas heat, to 5000 s.f.	1.000	Ea.	42.032	11,900	1,575	13,475
Gas piping	86.250	L.F.	13.019	193.20	560.63	753.83
Duct supply and return, galvanized steel	95.000	Lb.	9.702	85.50	388.55	474.05
Insulation, ductwork	82.000	S.F.	3.748	43.46	137.76	181.22
Lateral duct, flexible duct, 12″ diameter, insulated	180.000	L.F.	28.800	567	1,107	1,674
Diffusers	10.000	Ea.	11.429	3,100	490	3,590
Return registers	3.000	Ea.	2.182	369	93	462
TOTAL		Ea.	110.912	16,258.16	4,351.94	20,610.10

Description	QUAN.	UNIT	LABOR HOURS	COST EACH		
				MAT.	INST.	TOTAL

Important: See the Reference Section for critical supporting data - Reference Nos., Crews & Location Factors

Rooftop Price Sheet	QUAN.	UNIT	LABOR HOURS	COST EACH		
				MAT.	INST.	TOTAL
Rooftop unit, single zone, electric cool, gas heat to 2000 S.F.	1.000	Ea.	28.521	4,050	1,100	5,150
Area to 3000 S.F.	1.000	Ea.	35.982	7,600	1,350	8,950
Area to 5000 S.F.	1.000	Ea.	42.032	11,900	1,575	13,475
Area to 10000 S.F.	1.000	Ea.	68.376	26,300	2,650	28,950
Gas piping, area 2000 through 4000 S.F.	34.500	L.F.	5.207	77.50	224	301.50
Area 5000 to 10000 S.F.	86.250	L.F.	13.019	193	560	753
Duct, supply and return, galvanized steel, to 2000 S.F.	38.000	Lb.	3.881	34	155	189
Area to 3000 S.F.	57.000	Lb.	5.821	51.50	233	284.50
Area to 5000 S.F.	95.000	Lb.	9.702	85.50	390	475.50
Area to 10000 S.F.	190.000	Lb.	19.405	171	775	946
Rigid fiberglass, area to 2000 S.F.	33.000	S.F.	2.263	20.50	90.50	111
Area to 3000 S.F.	49.000	S.F.	3.360	30.50	134	164.50
Area to 5000 S.F.	82.000	S.F.	5.623	51	225	276
Area to 10000 S.F.	164.000	S.F.	11.245	102	450	552
Insulation, supply and return, blanket type, area to 2000 S.F.	33.000	S.F.	1.508	17.50	55.50	73
Area to 3000 S.F.	49.000	S.F.	2.240	26	82.50	108.50
Area to 5000 S.F.	82.000	S.F.	3.748	43.50	138	181.50
Area to 10000 S.F.	164.000	S.F.	7.496	87	276	363
Lateral ducts, flexible round, 12″ insulated, to 2000 S.F.	72.000	L.F.	11.520	227	445	672
Area to 3000 S.F.	108.000	L.F.	17.280	340	665	1,005
Area to 5000 S.F.	180.000	L.F.	28.800	565	1,100	1,665
Area to 10000 S.F.	360.000	L.F.	57.600	1,125	2,225	3,350
Rectangular, galvanized steel, to 2000 S.F.	239.000	Lb.	24.409	215	980	1,195
Area to 3000 S.F.	360.000	Lb.	36.767	325	1,475	1,800
Area to 5000 S.F.	599.000	Lb.	61.176	540	2,450	2,990
Area to 10000 S.F.	998.000	Lb.	101.926	900	4,075	4,975
Diffusers, ceiling, 1 to 4 way blow, 24″ x 24″, to 2000 S.F.	4.000	Ea.	4.571	1,250	196	1,446
Area to 3000 S.F.	6.000	Ea.	6.857	1,850	294	2,144
Area to 5000 S.F.	10.000	Ea.	11.429	3,100	490	3,590
Area to 10000 S.F.	20.000	Ea.	22.857	6,200	980	7,180
Return grilles, 24″ x 24″, to 2000 S.F.	1.000	Ea.	.727	123	31	154
Area to 3000 S.F.	2.000	Ea.	1.455	246	62	308
Area to 5000 S.F.	3.000	Ea.	2.182	370	93	463
Area to 10000 S.F.	5.000	Ea.	3.636	615	155	770

Division 9
Electrical

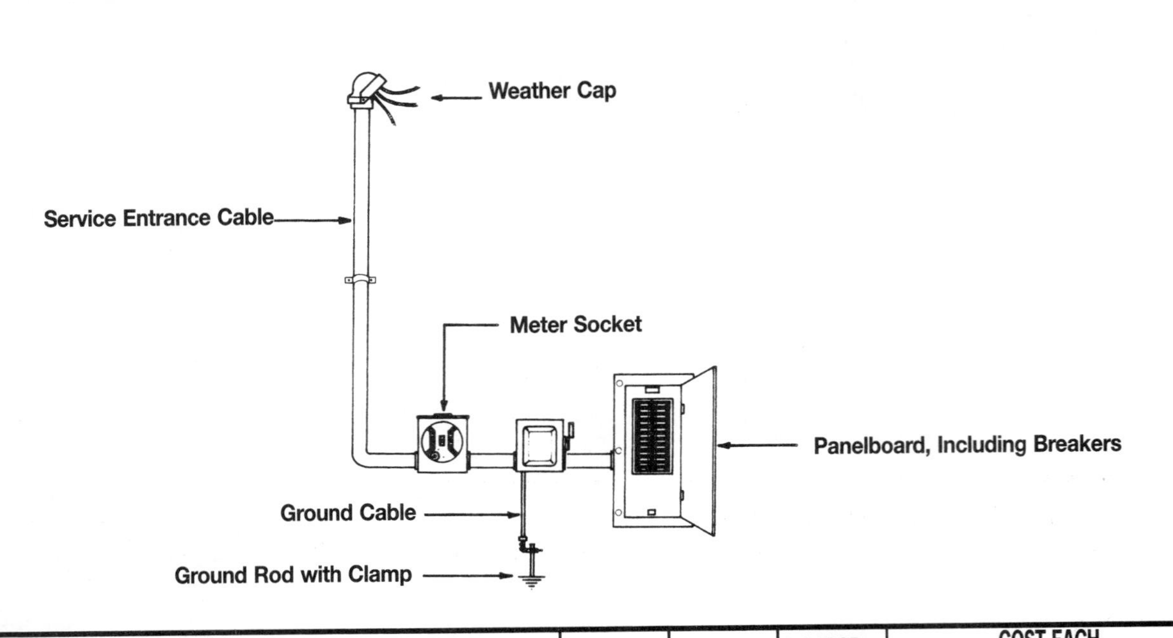

System Description	QUAN.	UNIT	LABOR HOURS	COST EACH MAT.	COST EACH INST.	COST EACH TOTAL
100 AMP SERVICE						
Weather cap	1.000	Ea.	.667	9.15	28.50	37.65
Service entrance cable	10.000	L.F.	.762	23.10	32.70	55.80
Meter socket	1.000	Ea.	2.500	35	107	142
Ground rod with clamp	1.000	Ea.	1.455	14.40	62.50	76.90
Ground cable	5.000	L.F.	.250	7.30	10.75	18.05
Panel board, 12 circuit	1.000	Ea.	6.667	182	286	468
TOTAL		Ea.	12.301	270.95	527.45	798.40
200 AMP SERVICE						
Weather cap	1.000	Ea.	1.000	20	43	63
Service entrance cable	10.000	L.F.	1.143	48.30	49.10	97.40
Meter socket	1.000	Ea.	4.211	52	181	233
Ground rod with clamp	1.000	Ea.	1.818	33	78	111
Ground cable	10.000	L.F.	.500	14.60	21.50	36.10
3/4" EMT	5.000	L.F.	.308	3.30	13.20	16.50
Panel board, 24 circuit	1.000	Ea.	12.308	490	440	930
TOTAL		Ea.	21.288	661.20	825.80	1,487
400 AMP SERVICE						
Weather cap	1.000	Ea.	2.963	365	127	492
Service entrance cable	180.000	L.F.	5.760	216	246.60	462.60
Meter socket	1.000	Ea.	4.211	52	181	233
Ground rod with clamp	1.000	Ea.	2.000	87	86	173
Ground cable	20.000	L.F.	.485	15.80	20.80	36.60
3/4" greenfield	20.000	L.F.	1.000	9.40	43	52.40
Current transformer cabinet	1.000	Ea.	6.154	140	264	404
Panel board, 42 circuit	1.000	Ea.	33.333	2,575	1,425	4,000
TOTAL		Ea.	55.906	3,460.20	2,393.40	5,853.60

Important: See the Reference Section for critical supporting data - Reference Nos., Crews & Location Factors

20 Electric Perimeter Heating Systems

Thermostat

Electric Baseboard

System Description	QUAN.	UNIT	LABOR HOURS	COST EACH		
				MAT.	INST.	TOTAL
4′ BASEBOARD HEATER						
Electric baseboard heater, 4′ long	1.000	Ea.	1.194	45	51.50	96.50
Thermostat, integral	1.000	Ea.	.500	19.60	21.50	41.10
Romex, 12-3 with ground	40.000	L.F.	1.600	13.60	68.80	82.40
Panel board breaker, 20 Amp	1.000	Ea.	.300	7.65	12.90	20.55
TOTAL		Ea.	3.594	85.85	154.70	240.55
6′ BASEBOARD HEATER						
Electric baseboard heater, 6′ long	1.000	Ea.	1.600	59.50	68.50	128
Thermostat, integral	1.000	Ea.	.500	19.60	21.50	41.10
Romex, 12-3 with ground	40.000	L.F.	1.600	13.60	68.80	82.40
Panel board breaker, 20 Amp	1.000	Ea.	.400	10.20	17.20	27.40
TOTAL		Ea.	4.100	102.90	176	278.90
8′ BASEBOARD HEATER						
Electric baseboard heater, 8′ long	1.000	Ea.	2.000	75	86	161
Thermostat, integral	1.000	Ea.	.500	19.60	21.50	41.10
Romex, 12-3 with ground	40.000	L.F.	1.600	13.60	68.80	82.40
Panel board breaker, 20 Amp	1.000	Ea.	.500	12.75	21.50	34.25
TOTAL		Ea.	4.600	120.95	197.80	318.75
10′ BASEBOARD HEATER						
Electric baseboard heater, 10′ long	1.000	Ea.	2.424	92.50	104	196.50
Thermostat, integral	1.000	Ea.	.500	19.60	21.50	41.10
Romex, 12-3 with ground	40.000	L.F.	1.600	13.60	68.80	82.40
Panel board breaker, 20 Amp	1.000	Ea.	.750	19.13	32.25	51.38
TOTAL		Ea.	5.274	144.83	226.55	371.38

The costs in this system are on a cost each basis and include all necessary conduit fittings.

Description	QUAN.	UNIT	LABOR HOURS	COST EACH		
				MAT.	INST.	TOTAL

System Description	QUAN.	UNIT	LABOR HOURS	COST EACH MAT.	COST EACH INST.	COST EACH TOTAL
Air conditioning receptacles						
Using non-metallic sheathed cable	1.000	Ea.	.800	13.40	34.50	47.90
Using BX cable	1.000	Ea.	.964	25	41.50	66.50
Using EMT conduit	1.000	Ea.	1.194	26	51.50	77.50
Disposal wiring						
Using non-metallic sheathed cable	1.000	Ea.	.889	10.55	38	48.55
Using BX cable	1.000	Ea.	1.067	21.50	46	67.50
Using EMT conduit	1.000	Ea.	1.333	23.50	57.50	81
Dryer circuit						
Using non-metallic sheathed cable	1.000	Ea.	1.455	31.50	62.50	94
Using BX cable	1.000	Ea.	1.739	46	74.50	120.50
Using EMT conduit	1.000	Ea.	2.162	37	93	130
Duplex receptacles						
Using non-metallic sheathed cable	1.000	Ea.	.615	13.40	26.50	39.90
Using BX cable	1.000	Ea.	.741	25	32	57
Using EMT conduit	1.000	Ea.	.920	26	39.50	65.50
Exhaust fan wiring						
Using non-metallic sheathed cable	1.000	Ea.	.800	13.10	34.50	47.60
Using BX cable	1.000	Ea.	.964	24.50	41.50	66
Using EMT conduit	1.000	Ea.	1.194	25.50	51.50	77
Furnace circuit & switch						
Using non-metallic sheathed cable	1.000	Ea.	1.333	19.80	57.50	77.30
Using BX cable	1.000	Ea.	1.600	32	68.50	100.50
Using EMT conduit	1.000	Ea.	2.000	32.50	86	118.50
Ground fault						
Using non-metallic sheathed cable	1.000	Ea.	1.000	42	43	85
Using BX cable	1.000	Ea.	1.212	54.50	52	106.50
Using EMT conduit	1.000	Ea.	1.481	66	63.50	129.50
Heater circuits						
Using non-metallic sheathed cable	1.000	Ea.	1.000	12.05	43	55.05
Using BX cable	1.000	Ea.	1.212	21	52	73
Using EMT conduit	1.000	Ea.	1.481	22.50	63.50	86
Lighting wiring						
Using non-metallic sheathed cable	1.000	Ea.	.500	12.75	21.50	34.25
Using BX cable	1.000	Ea.	.602	21.50	26	47.50
Using EMT conduit	1.000	Ea.	.748	21.50	32	53.50
Range circuits						
Using non-metallic sheathed cable	1.000	Ea.	2.000	65.50	86	151.50
Using BX cable	1.000	Ea.	2.424	94.50	104	198.50
Using EMT conduit	1.000	Ea.	2.963	64.50	127	191.50
Switches, single pole						
Using non-metallic sheathed cable	1.000	Ea.	.500	13.10	21.50	34.60
Using BX cable	1.000	Ea.	.602	24.50	26	50.50
Using EMT conduit	1.000	Ea.	.748	25.50	32	57.50
Switches, 3-way						
Using non-metallic sheathed cable	1.000	Ea.	.667	17.15	28.50	45.65
Using BX cable	1.000	Ea.	.800	28.50	34.50	63
Using EMT conduit	1.000	Ea.	1.333	36	57.50	93.50
Water heater						
Using non-metallic sheathed cable	1.000	Ea.	1.600	19.40	68.50	87.90
Using BX cable	1.000	Ea.	1.905	37	82	119
Using EMT conduit	1.000	Ea.	2.353	28.50	101	129.50
Weatherproof receptacle						
Using non-metallic sheathed cable	1.000	Ea.	1.333	104	57.50	161.50
Using BX cable	1.000	Ea.	1.600	113	68.50	181.50
Using EMT conduit	1.000	Ea.	2.000	114	86	200

DESCRIPTION	QUAN.	UNIT	LABOR HOURS	COST EACH		
				MAT.	INST.	TOTAL
Fluorescent strip, 4′ long, 1 light, average	1.000	Ea.	.941	29	40.50	69.50
Deluxe	1.000	Ea.	1.129	35	48.50	83.50
2 lights, average	1.000	Ea.	1.000	31	43	74
Deluxe	1.000	Ea.	1.200	37	51.50	88.50
8′ long, 1 light, average	1.000	Ea.	1.194	43.50	51.50	95
Deluxe	1.000	Ea.	1.433	52	62	114
2 lights, average	1.000	Ea.	1.290	52.50	55.50	108
Deluxe	1.000	Ea.	1.548	63	66.50	129.50
Surface mounted, 4′ x 1′, economy	1.000	Ea.	.914	61.50	39	100.50
Average	1.000	Ea.	1.143	77	49	126
Deluxe	1.000	Ea.	1.371	92.50	59	151.50
4′ x 2′, economy	1.000	Ea.	1.208	78.50	52	130.50
Average	1.000	Ea.	1.509	98	65	163
Deluxe	1.000	Ea.	1.811	118	78	196
Recessed, 4′x 1′, 2 lamps, economy	1.000	Ea.	1.123	39.50	48.50	88
Average	1.000	Ea.	1.404	49.50	60.50	110
Deluxe	1.000	Ea.	1.684	59.50	72.50	132
4′ x 2′, 4′ lamps, economy	1.000	Ea.	1.362	48	58.50	106.50
Average	1.000	Ea.	1.702	60	73	133
Deluxe	1.000	Ea.	2.043	72	87.50	159.50
Incandescent, exterior, 150W, single spot	1.000	Ea.	.500	18.60	21.50	40.10
Double spot	1.000	Ea.	1.167	75.50	50	125.50
Recessed, 100W, economy	1.000	Ea.	.800	49.50	34.50	84
Average	1.000	Ea.	1.000	62	43	105
Deluxe	1.000	Ea.	1.200	74.50	51.50	126
150W, economy	1.000	Ea.	.800	71	34.50	105.50
Average	1.000	Ea.	1.000	89	43	132
Deluxe	1.000	Ea.	1.200	107	51.50	158.50
Surface mounted, 60W, economy	1.000	Ea.	.800	37.50	34.50	72
Average	1.000	Ea.	1.000	42	43	85
Deluxe	1.000	Ea.	1.194	60.50	51.50	112
Metal halide, recessed 2′ x 2′ 250W	1.000	Ea.	2.500	294	107	401
2′ x 2′, 400W	1.000	Ea.	2.759	335	118	453
Surface mounted, 2′ x 2′, 250W	1.000	Ea.	2.963	294	127	421
2′ x 2′, 400W	1.000	Ea.	3.333	345	143	488
High bay, single, unit, 400W	1.000	Ea.	3.478	350	149	499
Twin unit, 400W	1.000	Ea.	5.000	705	215	920
Low bay, 250W	1.000	Ea.	2.500	340	107	447

Unit Price Section

Table of Contents

How to Use the Unit Price Pages

The following is a detailed explanation of a sample entry in the Unit Price Section. Next to each bold number below is the item being described with appropriate component of the sample entry following in parenthesis. Some prices are listed as bare costs, others as costs that include overhead and profit of the installing contractor. In most cases, if the work is to be subcontracted, the general contractor will need to add an additional markup (RSMeans suggests using 10%) to the figures in the column "Total Incl. O&P."

1 *Division Number/Title (03300/Cast-In-Place Concrete)*

Use the Unit Price Section Table of Contents to locate specific items. The sections are classified according to the CSI MasterFormat (1995 Edition).

2 *Line Numbers (03310 240 3900)*

Each unit price line item has been assigned a unique 12-digit code based on the CSI MasterFormat classification.

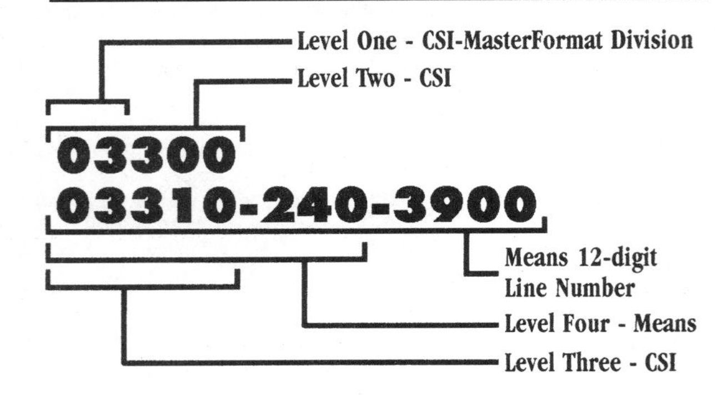

3 *Description (Concrete-In-Place, etc.)*

Each line item is described in detail. Sub-items and additional sizes are indented beneath the appropriate line items. The first line or two after the main item (in boldface) may contain descriptive information that pertains to all line items beneath this boldface listing.

4 *Reference Number Information*

R03310 -010 You'll see reference numbers shown in bold rectangles at the beginning of some sections. These refer to related items in the Reference Section, visually identified by a vertical gray bar on the edge of pages.

The relation may be: (1) an estimating procedure that should be read before estimating, (2) an alternate pricing method, or (3) technical information.

The "R" designates the Reference Section. The numbers refer to the MasterFormat classification system.

It is strongly recommended that you review all reference numbers that appear within the section in which you are working.

Note: Not all reference numbers appear in all Means publications.

03300 | Cast-In-Place Concrete 1

	03310 | Structural Concrete		CREW	DAILY OUTPUT	LABOR-HOURS	UNIT	2004 BARE COSTS MAT.	LABOR	EQUIP.	TOTAL	TOTAL INCL O&P	
240	0010	**CONCRETE IN PLACE** Including forms (4 us[illegible]s), reinforcing										240
	0050	steel and finishing unless [illegible]wise i[illegible]dicated 2 4										
	0500	Chimney foundations, industri[illegible]nimum	C-14[illegible]	32.22	3.476	C.Y.	145	75.50	.80	221.30	288	
	0510	Maximum		23.71	4.724		168	102	[illegible]9	271.09	360	
	3800	Footings, spread under 1 C.Y.		38.07	2.942		105	63.50	[illegible]68	169.18	226	
	3850	Over 5 C.Y.		81.04	1.382		96.50	30	.32	126.82	157	
	3900	Footings, strip, 18″ x 9″, unreinforced		40	2.800		91	60.50	.64	152.14	204	
	3920	18″ x 9″, reinforced		35	3.200		105	69.50	.74	175.24	234	
	3925	20″ x 10″, unreinforced 3		45	2.489		88	54	.57	142.57	190	
	3930	20″ x 10″, reinforced		40	2.800		99.50	60.50	.64	160.64	213	
	3935	24″ x 12″, unreinforced 5		5[illegible]	2.036		87	44	.47	131.47	[illegible]	
	3940	24″ x 12″, reinforced		48	2.333		98.50	50.50	.54	149.54	196	
	3945	36″ x 12″, unreinforced		70	1.600		83.50	34.50	.37	118.37	151	
	3950	36″ x 12″, reinforced		60	1.867		94	40.50	.43	134.93	172	

6 7 8 9

Crew (C-14C)

The "Crew" column designates the typical trade or crew used to install the item. If an installation can be accomplished by one trade and requires no power equipment, that trade and the number of workers are listed (for example, "2 Carpenters"). If an installation requires a composite crew, a crew code designation is listed (for example, "C-14C"). You'll find full details on all composite crews in the Crew Listings.

- For a complete list of all trades utilized in this book and their abbreviations, see the inside back cover.

Crews

Crew No.	Bare Costs		Incl. Subs O & P		Cost Per Labor-Hour	
Crew C-14C	Hr.	Daily	Hr.	Daily	Bare Costs	Incl. O&P
1 Carpenter Foreman (out)	$25.10	$200.80	$42.60	$340.80	$21.65	$36.95
6 Carpenters	23.10	1108.80	39.20	1881.60		
2 Rodmen (reinf.)	24.85	397.60	44.50	712.00		
4 Laborers	16.90	540.80	28.70	918.40		
1 Cement Finisher	22.10	176.80	35.65	285.20		
1 Gas Engine Vibrator		26.00		28.60	.23	.26
112 L.H., Daily Totals		$2450.80		$4166.60	$21.88	$37.21

Productivity: Daily Output (40)/Labor-Hours (2.800)

The "Daily Output" represents the typical number of units the designated crew will install in a normal 8-hour day. To find out the number of days the given crew would require to complete the installation, divide your quantity by the daily output. For example:

Quantity	÷	Daily Output	=	Duration
100 C.Y.	÷	40/ Crew Day	=	2.5 Crew Days

The "Labor-Hours" figure represents the number of labor-hours required to install one unit of work. To find out the number of labor-hours required for your particular task, multiply the quantity of the item times the number of labor-hours shown. For example:

Quantity	x	Productivity Rate	=	Duration
100 C.Y.	x	2.800 Labor-Hours/ C.Y.	=	280 Labor-Hours

Unit (C.Y.)

The abbreviated designation indicates the unit of measure upon which the price, production, and crew are based (C.Y. = Cubic Yard). For a complete listing of abbreviations refer to the Abbreviations Listing in the Reference Section of this book.

Bare Costs:

Mat. (Bare Material Cost) (91)

The unit material cost is the "bare" material cost with no overhead and profit included. *Costs shown reflect national average material prices for January of the current year and include delivery to the job site. No sales taxes are included.*

Labor (60.50)

The unit labor cost is derived by multiplying bare labor-hour costs for Crew C-14C by labor-hour units. The bare labor-hour cost is found in the Crew Section under C-14C. (If a trade is listed, the hourly labor cost—the wage rate—is found on the inside back cover.)

Labor-Hour Cost Crew C-14C	x	Labor-Hour Units	=	Labor
$21.65	x	2.800	=	$60.50

Equip. (Equipment) (.64)

Equipment costs for each crew are listed in the description of each crew. Tools or equipment whose value justifies purchase or ownership by a contractor are considered overhead as shown on the inside back cover. The unit equipment cost is derived by multiplying the bare equipment hourly cost by the labor-hour units.

Equipment Cost Crew C-14C	x	Labor-Hour Units	=	Equip.
.23	x	2.800	=	.64

Total (152.14)

The total of the bare costs is the arithmetic total of the three previous columns: mat., labor, and equip.

Material	+	Labor	+	Equip.	=	Total
$91	+	$60.50	+	$.64	=	$152.14

Total Costs Including O&P

This figure is the sum of the bare material cost plus 10% for profit; the bare labor cost plus total overhead and profit (per the inside back cover or, if a crew is listed, from the crew listings); and the bare equipment cost plus 10% for profit.

Material is Bare Material Cost + 10% = 91 + 9.10	=	$100.10
Labor for Crew C-14C = Labor-Hour Cost (36.95) x Labor-Hour Units (2.800)	=	$103.46
Equip. is Bare Equip. Cost + 10% = .64 + .06	=	$ .70
Total (Rounded)	=	$204

Division 1
General Requirements

Estimating Tips

The General Requirements of any contract are very important to both the bidder and the owner. These lay the ground rules under which the contract will be executed and have a significant influence on the cost of operations. Therefore, it is extremely important to thoroughly read and understand the General Requirements both before preparing an estimate and when the estimate is complete, to ascertain that nothing in the contract is overlooked. Caution should be exercised when applying items listed in Division 1 to an estimate. Many of the items are included in the unit prices listed in the other divisions such as mark-ups on labor and company overhead.

01200 Price & Payment Procedures

- When estimating historic preservation projects (depending on the condition of the existing structure and the owner's requirements), a 15-20% contingency or allowance is recommended, regardless of the stage of the drawings.

01300 Administrative Requirements

- Before determining a final cost estimate, it is a good practice to review all the items listed in subdivision 01300 to make final adjustments for items that may need customizing to specific job conditions.
- Historic preservation projects may require specialty labor and methods, as well as extra time to protect existing materials that must be preserved and/or restored. Some additional expenses may be incurred in architectural fees for facility surveys and other special inspections and analyses.

01330 Submittal Procedures

- Requirements for initial and periodic submittals can represent a significant cost to the General Requirements of a job. Thoroughly check the submittal specifications when estimating a project to determine any costs that should be included.

01400 Quality Requirements

- All projects will require some degree of Quality Control. This cost is not included in the unit cost of construction listed in each division. Depending upon the terms of the contract, the various costs of inspection and testing can be the responsibility of either the owner or the contractor. Be sure to include the required costs in your estimate.

01500 Temporary Facilities & Controls

- Barricades, access roads, safety nets, scaffolding, security and many more requirements for the execution of a safe project are elements of direct cost. These costs can easily be overlooked when preparing an estimate. When looking through the major classifications of this subdivision, determine which items apply to each division in your estimate.

01590 Equipment Rental

- This subdivision contains transportation, handling, storage, protection and product options and substitutions. Listed in this cost manual are average equipment rental rates for all types of equipment. This is useful information when estimating the time and materials requirement of any particular operation in order to establish a unit or total cost.
- A good rule of thumb is that weekly rental is 3 times daily rental and that monthly rental is 3 times weekly rental.
- The figures in the column for Crew Equipment Cost represent the rental rate used in determining the daily cost of equipment in a crew. It is calculated by dividing the weekly rate by 5 days and adding the hourly operating cost times 8 hours.

01770 Closeout Procedures

- When preparing an estimate, read the specifications to determine the requirements for Contract Closeout thoroughly. Final cleaning, record documentation, operation and maintenance data, warranties and bonds, and spare parts and maintenance materials can all be elements of cost for the completion of a contract. Do not overlook these in your estimate.

01830 Operations & Maintenance

- If maintenance and repair are included in your contract, they require special attention. To estimate the cost to remove and replace any unit usually requires a site visit to determine the accessibility and the specific difficulty at that location. Obstructions, dust control, safety, and often overtime hours must be considered when preparing your estimate.

Reference Numbers

Reference numbers are shown in bold squares at the beginning of some major classifications. These numbers refer to related items in the Reference Section. The reference information may be an estimating procedure, an alternate pricing method or technical information.

Note: Not all subdivisions listed here necessarily appear in this publication.

01100 | Summary

	01103 | Models & Renderings		CREW	DAILY OUTPUT	LABOR-HOURS	UNIT	2004 BARE COSTS MAT.	LABOR	EQUIP.	TOTAL	TOTAL INCL O&P	
500	0010	**RENDERINGS** Color, matted, 20" x 30", eye level,										500
	0050	Average				Ea.	2,500			2,500	2,750	

	01107 | Professional Consultant		CREW	DAILY OUTPUT	LABOR-HOURS	UNIT	2004 BARE COSTS MAT.	LABOR	EQUIP.	TOTAL	TOTAL INCL O&P	
100	0011	**ARCHITECTURAL FEES** R01107 -010										100
	0020	For new construction										
	0060	Minimum				Project					4.90%	
	0090	Maximum									16%	
	0100	For alteration work, to $500,000, add to fee									50%	
	0150	Over $500,000, add to fee									25%	
200	0011	**CONSTRUCTION MANAGEMENT FEES**										200
	0060	For work to $10,000				Project					10%	
	0070	To $25,000									9%	
	0090	To $100,000									6%	
700	0010	**SURVEYING** Conventional, topographical, minimum	A-7	3.30	7.273	Acre	16.30	177	18.05	211.35	335	700
	0100	Maximum	A-8	.60	53.333		49	1,275	99	1,423	2,325	
	0300	Lot location and lines, minimum, for large quantities	A-7	2	12		25.50	292	30	347.50	550	
	0320	Average	"	1.25	19.200		46	470	47.50	563.50	890	
	0400	Maximum, for small quantities	A-8	1	32		73.50	765	59.50	898	1,425	
	0600	Monuments, 3' long	A-7	10	2.400	Ea.	20	58.50	5.95	84.45	127	
	0800	Property lines, perimeter, cleared land	"	1,000	.024	L.F.	.03	.58	.06	.67	1.08	
	0900	Wooded land	A-8	875	.037	"	.05	.88	.07	1	1.61	
	1100	Crew for layout of building, trenching or pipe laying, 2 person crew	A-6	1	16	Day		365	59.50	424.50	680	
	1200	3 person crew	A-7	1	24	"		585	59.50	644.50	1,050	

01200 | Price & Payment Procedures

	01250 | Contract Modification Procedures		CREW	DAILY OUTPUT	LABOR-HOURS	UNIT	2004 BARE COSTS MAT.	LABOR	EQUIP.	TOTAL	TOTAL INCL O&P	
200	0010	**CONTINGENCIES** for estimate at conceptual stage				Project					20%	200
	0150	Final working drawing stage				"					3%	

	01290 | Payment Procedures		CREW	DAILY OUTPUT	LABOR-HOURS	UNIT	2004 BARE COSTS MAT.	LABOR	EQUIP.	TOTAL	TOTAL INCL O&P	
800	0010	**TAXES** Sales tax, State, average R01100 -090				%	4.65%					800
	0050	Maximum					7%					
	0200	Social Security, on first $87,000 of wages R01100 -100						7.65%				
	0300	Unemployment, MA, combined Federal and State, minimum						2.10%				
	0350	Average						6.20%				
	0400	Maximum						8%				

01300 | Administrative Requirements

	01310 | Project Management/Coordination		CREW	DAILY OUTPUT	LABOR-HOURS	UNIT	2004 BARE COSTS MAT.	LABOR	EQUIP.	TOTAL	TOTAL INCL O&P	
150	0010	**PERMITS** Rule of thumb, most cities, minimum				Job					.50%	150
	0100	Maximum				"					2%	

01300 | Administrative Requirements

		01310 Project Management/Coordination	CREW	DAILY OUTPUT	LABOR-HOURS	UNIT	2004 BARE COSTS MAT.	LABOR	EQUIP.	TOTAL	TOTAL INCL O&P	
350	0010	**INSURANCE** Builders risk, standard, minimum (R01100-040)				Job					.22%	350
	0050	Maximum									.59%	
	0200	All-risk type, minimum (R01100-060)									.25%	
	0250	Maximum				↓					.62%	
	0400	Contractor's equipment floater, minimum				Value					.50%	
	0450	Maximum				"					1.50%	
	0600	Public liability, average				Job					1.55%	
	0800	Workers' compensation & employer's liability, average										
	0850	by trade, carpentry, general				Payroll		18.51%				
	0900	Clerical						.60%				
	0950	Concrete						15.79%				
	1000	Electrical						6.40%				
	1050	Excavation						10.34%				
	1100	Glazing						13.82%				
	1150	Insulation						15.24%				
	1200	Lathing						10.71%				
	1250	Masonry						14.97%				
	1300	Painting & decorating						12.89%				
	1350	Pile driving						22.94%				
	1400	Plastering						14.62%				
	1450	Plumbing						7.78%				
	1500	Roofing						31.75%				
	1550	Sheet metal work (HVAC)						11.09%				
	1600	Steel erection, structural						38.86%				
	1650	Tile work, interior ceramic						9.63%				
	1700	Waterproofing, brush or hand caulking						7.27%				
	1800	Wrecking						40.51%				
	2000	Range of 35 trades in 50 states, excl. wrecking, min.						2.50%				
	2100	Average						16.20%				
	2200	Maximum				↓		110.10%				

01500 | Temporary Facilities & Controls

		01540 Construction Aids	CREW	DAILY OUTPUT	LABOR-HOURS	UNIT	2004 BARE COSTS MAT.	LABOR	EQUIP.	TOTAL	TOTAL INCL O&P	
550	0010	**PUMP STAGING**, Aluminum										550
	1300	System in place, 50′ working height, per use based on 50 uses	2 Carp	84.80	.189	C.S.F.	5.30	4.36		9.66	13.20	
	1400	100 uses (R01540-200)		84.80	.189		2.64	4.36		7	10.30	
	1500	150 uses	↓	84.80	.189	↓	1.77	4.36		6.13	9.35	
750	0010	**SCAFFOLDING** (R01540-100)										750
	0015	Steel tubular, reg, rent/mo, no plank, incl erect or dismantle										
	0090	Building exterior, wall face, 1 to 5 stories, 6′-4″ x 5′ frames	3 Carp	24	1	C.S.F.	24.50	23		47.50	65.50	
	0200	6 to 12 stories	4 Carp	21.20	1.509		24.50	35		59.50	85.50	
	0310	13 to 20 stories	5 Carp	20	2		24.50	46		70.50	105	
	0460	Building interior, wall face area, up to 16′ high	3 Carp	25	.960		24.50	22		46.50	64	
	0560	16′ to 40′ high		23	1.043	↓	24.50	24		48.50	67.50	
	0800	Building interior floor area, up to 30′ high	↓	312	.077	C.C.F.	2.57	1.78		4.35	5.85	
	0900	Over 30′ high	4 Carp	275	.116	"	2.57	2.69		5.26	7.40	
	0910	Steel tubular, heavy duty shoring, buy										

01540 | Construction Aids

	01540 Construction Aids		CREW	DAILY OUTPUT	LABOR-HOURS	UNIT	2004 BARE COSTS MAT.	LABOR	EQUIP.	TOTAL	TOTAL INCL O&P	
750	0920	Frames 5′ high 2′ wide R01540-100				Ea.	75			75	82.50	750
	0925	5′ high 4′ wide					85			85	93.50	
	0930	6′ high 2′ wide					86			86	94.50	
	0935	6′ high 4′ wide				↓	101			101	111	
	0940	Accessories										
	0945	Cross braces				Ea.	16			16	17.60	
	0950	U-head, 8″ x 8″					17.50			17.50	19.25	
	0955	J-head, 4″ x 8″					12.80			12.80	14.10	
	0960	Base plate, 8″ x 8″					14.20			14.20	15.60	
	0965	Leveling jack				↓	30.50			30.50	33.50	
	1000	Steel tubular, regular, buy										
	1100	Frames 3′ high 5′ wide				Ea.	58			58	64	
	1150	5′ high 5′ wide					67			67	73.50	
	1200	6′-4″ high 5′ wide					84			84	92.50	
	1350	7′-6″ high 6′ wide					145			145	160	
	1500	Accessories cross braces					15			15	16.50	
	1550	Guardrail post					15			15	16.50	
	1600	Guardrail 7′ section					7.25			7.25	8	
	1650	Screw jacks & plates					24			24	26.50	
	1700	Sidearm brackets					28			28	31	
	1750	8″ casters					33			33	36.50	
	1800	Plank 2″ x 10″ x 16′-0″					42.50			42.50	47	
	1900	Stairway section					245			245	270	
	1910	Stairway starter bar					29			29	32	
	1920	Stairway inside handrail					53			53	58.50	
	1930	Stairway outside handrail					73			73	80.50	
	1940	Walk-thru frame guardrail				↓	37			37	40.50	
	2000	Steel tubular, regular, rent/mo.										
	2100	Frames 3′ high 5′ wide				Ea.	3.75			3.75	4.13	
	2150	5′ high 5′ wide					3.75			3.75	4.13	
	2200	6′-4″ high 5′ wide					3.75			3.75	4.13	
	2250	7′-6″ high 6′ wide					7			7	7.70	
	2500	Accessories, cross braces					.60			.60	.66	
	2550	Guardrail post					1			1	1.10	
	2600	Guardrail 7′ section					.75			.75	.83	
	2650	Screw jacks & plates					1.50			1.50	1.65	
	2700	Sidearm brackets					1.50			1.50	1.65	
	2750	8″ casters					6			6	6.60	
	2800	Outrigger for rolling tower					3			3	3.30	
	2850	Plank 2″ x 10″ x 16′-0″					5			5	5.50	
	2900	Stairway section					10			10	11	
	2910	Stairway starter bar					.10			.10	.11	
	2920	Stairway inside handrail					5			5	5.50	
	2930	Stairway outside handrail					5			5	5.50	
	2940	Walk-thru frame guardrail				↓	2			2	2.20	
	3000	Steel tubular, heavy duty shoring, rent/mo.										
	3250	5′ high 2′ & 4′ wide				Ea.	5			5	5.50	
	3300	6′ high 2′ & 4′ wide					5			5	5.50	
	3500	Accessories, cross braces					1			1	1.10	
	3600	U - head, 8″ x 8″					1			1	1.10	
	3650	J - head, 4″ x 8″					1			1	1.10	
	3700	Base plate, 8″ x 8″					1			1	1.10	
	3750	Leveling jack					2			2	2.20	
	5700	Planks, 2x10x16′-0″, labor only, erect or remove to 50′ H	3 Carp	144	.167			3.85		3.85	6.55	
	5800	Over 50′ high	4 Carp	160	.200			4.62		4.62	7.85	
	6820	Erect or dismantle frames, 1st tier	4 Clab	45	.711	↓		12		12	20.50	

	01540 | Construction Aids		CREW	DAILY OUTPUT	LABOR-HOURS	UNIT	2004 BARE COSTS MAT.	LABOR	EQUIP.	TOTAL	TOTAL INCL O&P	
750	6830	2nd tier R01540-100	4 Clab	93	.344	Ea.		5.80		5.80	9.90	750
	6840	3rd tier		87	.368			6.20		6.20	10.55	
	6850	4th tier	↓	75	.427	↓		7.20		7.20	12.25	
760	0010	**STAGING AIDS** and fall protection equipment										760
	0100	Sidewall staging bracket, tubular, buy				Ea.	29			29	32	
	0110	Cost each per day, based on 250 days use				Day	.12			.12	.13	
	0200	Guard post, buy				Ea.	15			15	16.50	
	0210	Cost each per day, based on 250 days use				Day	.06			.06	.07	
	0300	End guard chains, buy per pair				Pair	25			25	27.50	
	0310	Cost per set per day, based on 250 days use				Day	.12			.12	.13	
	1010	Cost each per day, based on 250 days use				"	.03			.03	.03	
	1100	Wood bracket, buy				Ea.	13.15			13.15	14.45	
	1110	Cost each per day, based on 250 days use				Day	.05			.05	.06	
	2010	Cost per pair per day, based on 250 days use				"	.33			.33	.36	
	2100	Steel siderail jack, buy per pair				Pair	63			63	69.50	
	2110	Cost per pair per day, based on 250 days use				Day	.25			.25	.28	
	3010	Cost each per day, based on 250 days use				"	.17			.17	.19	
	3100	Aluminum scaffolding plank, 20" wide x 24' long, buy				Ea.	690			690	760	
	3110	Cost each per day, based on 250 days use				Day	2.76			2.76	3.04	
	4010	Cost each per day, based on 250 days use				"	.83			.83	.91	
	4100	Rope for safety line, 5/8" x 100' nylon, buy				Ea.	41			41	45	
	4110	Cost each per day, based on 250 days use				Day	.16			.16	.18	
	4200	Permanent U-Bolt roof anchor, buy				Ea.	31			31	34	
	4300	Temporary (one use) roof ridge anchor, buy				"	24			24	26.50	
	5000	Installation (setup and removal) of staging aids										
	5010	Sidewall staging bracket	2 Carp	64	.250	Ea.		5.80		5.80	9.80	
	5020	Guard post with 2 wood rails	"	64	.250			5.80		5.80	9.80	
	5030	End guard chains, set	1 Carp	64	.125			2.89		2.89	4.90	
	5100	Roof shingling bracket		96	.083			1.92		1.92	3.27	
	5200	Ladder jack	↓	64	.125			2.89		2.89	4.90	
	5300	Wood plank, 2x10x16'	2 Carp	80	.200			4.62		4.62	7.85	
	5310	Aluminum scaffold plank, 20" x 24'	"	40	.400			9.25		9.25	15.70	
	5410	Safety rope	1 Carp	40	.200			4.62		4.62	7.85	
	5420	Permanent U-Bolt roof anchor (install only)	2 Carp	40	.400			9.25		9.25	15.70	
	5430	Temporary roof ridge anchor (install only)	1 Carp	64	.125	↓		2.89		2.89	4.90	
800	0010	**TARPAULINS** Cotton duck, 10 oz. to 13.13 oz. per S.Y., minimum				S.F.	.48			.48	.53	800
	0050	Maximum					.57			.57	.63	
	0200	Reinforced polyethylene 3 mils thick, white					.11			.11	.12	
	0300	4 mils thick, white, clear or black					.14			.14	.15	
	0730	Polyester reinforced w/ integral fastening system 11 mils thick				↓	1.05			1.05	1.16	
820	0010	**SMALL TOOLS** As % of contractor's work, minimum				Total					.50%	820
	0100	Maximum	"			"					2%	

01590 | Equipment Rental

			UNIT	HOURLY OPER. COST	RENT PER DAY	RENT PER WEEK	RENT PER MONTH	CREW EQUIPMENT COST/DAY
100	0010	**CONCRETE EQUIPMENT RENTAL**						
	0100	without operators						
	0150	For batch plant, see div. 01590-500						
	0200	Bucket, concrete lightweight, 1/2 C.Y.	Ea.	.50	15.35	46	138	13.20
	0300	1 C.Y.		.55	18.65	56	168	15.60
	0400	1-1/2 C.Y.		.70	25.50	77	231	21
	0500	2 C.Y.		.80	30.50	92	276	24.80
	0580	8 C.Y.		4.25	202	605	1,825	155
	0600	Cart, concrete, self propelled, operator walking, 10 C.F.		1.85	56.50	170	510	48.80
	0700	Operator riding, 18 C.F.		2.85	86.50	260	780	74.80
	0800	Conveyer for concrete, portable, gas, 16" wide, 26' long		6.15	117	350	1,050	119.20
	0900	46' long		6.55	143	430	1,300	138.40
	1000	56' long		6.65	153	460	1,375	145.20
	1100	Core drill, electric, 2-1/2 H.P., 1" to 8" bit diameter		1.53	58.50	176	530	47.45
	1150	11 HP, 8" to 18" cores		6.73	87	260.80	780	106
	1200	Finisher, concrete floor, gas, riding trowel, 48" diameter		3.55	86.50	260	780	80.40
	1300	Gas, manual, 3 blade, 36" trowel		.80	25.50	76	228	21.60
	1400	4 blade, 48" trowel		1.15	30	90	270	27.20
	1500	Float, hand-operated (Bull float) 48" wide		.08	13.35	40	120	8.65
	1570	Curb builder, 14 H.P., gas, single screw		8.50	150	450	1,350	158
	1590	Double screw		9.25	190	570	1,700	188
	1600	Grinder, concrete and terrazzo, electric, floor		1.82	75	225	675	59.55
	1700	Wall grinder		.91	37.50	113	340	29.90
	1800	Mixer, powered, mortar and concrete, gas, 6 C.F., 18 H.P.		4.50	90	270	810	90
	1900	10 C.F., 25 H.P.		5.40	103	310	930	105.20
	2000	16 C.F.		5.70	127	380	1,150	121.60
	2100	Concrete, stationary, tilt drum, 2 C.Y.		5	202	605	1,825	161
	2120	Pump, concrete, truck mounted 4" line 80' boom		21.15	915	2,750	8,250	719.20
	2140	5" line, 110' boom		28.60	1,275	3,800	11,400	988.80
	2160	Mud jack, 50 C.F. per hr.		4.91	119	358	1,075	110.90
	2180	225 C.F. per hr.		7.70	159	476.80	1,425	156.95
	2190	Shotcrete pump rig, 12 CY/hr		10.80	235	705	2,125	227.40
	2600	Saw, concrete, manual, gas, 18 H.P.		2.85	33.50	100	300	42.80
	2650	Self-propelled, gas, 30 H.P.		5.40	81.50	245	735	92.20
	2700	Vibrators, concrete, electric, 60 cycle, 2 H.P.		.38	15	45	135	12.05
	2800	3 H.P.		.58	21.50	65	195	17.65
	2900	Gas engine, 5 H.P.		.75	24	72	216	20.40
	3000	8 H.P.		1.05	29.50	88	264	26
	3050	Vibrating screed, gas engine, 8HP		1.53	42.50	127	380	37.65
	3100	Concrete transit mixer, hydraulic drive						
	3120	6 x 4, 250 H.P., 8 C.Y., rear discharge		32.85	630	1,885	5,650	639.80
	3200	Front discharge		36.20	680	2,045	6,125	698.60
	3300	6 x 6, 285 H.P., 12 C.Y., rear discharge		35.10	640	1,915	5,750	663.80
	3400	Front discharge		36.90	690	2,070	6,200	709.20
200	0010	**EARTHWORK EQUIPMENT RENTAL** Without operators						
	0040	Aggregate spreader, push type 8' to 12' wide	Ea.	1.80	45	135	405	41.40
	0045	Tailgate type, 8' wide	"	1.75	31.50	95	285	33
	0050	Augers for truck or trailer mounting, vertical drilling						
	0055	Fence post auger, truck mounted	Ea.	8.30	485	1,455	4,375	357.40
	0060	4" to 36" diam., 54 H.P., gas, 10' spindle travel		28.35	645	1,940	5,825	614.80
	0070	14' spindle travel		32.25	805	2,410	7,225	740
	0075	Auger, truck mounted, vertical drilling, to 25' depth		138.75	2,950	8,835	26,500	2,877
	0080	Auger, horizontal boring machine, 12" to 36" diameter, 45 H.P.		15.45	193	580	1,750	239.60
	0090	12" to 48" diameter, 65 H.P.		21.70	345	1,040	3,125	381.60
	0100	Excavator, diesel hydraulic, crawler mounted, 1/2 C.Y. cap.		14.25	350	1,045	3,125	323
	0120	5/8 C.Y. capacity		17.50	465	1,390	4,175	418
	0140	3/4 C.Y. capacity		20.35	495	1,480	4,450	458.80
	0150	1 C.Y. capacity		22.30	515	1,545	4,625	487.40

01590 | Equipment Rental

200		UNIT	HOURLY OPER. COST	RENT PER DAY	RENT PER WEEK	RENT PER MONTH	CREW EQUIPMENT COST/DAY
0200	1-1/2 C.Y. capacity	Ea.	29.30	750	2,245	6,725	683.40
0300	2 C.Y. capacity		38.15	970	2,910	8,725	887.20
0320	2-1/2 C.Y. capacity		49.75	1,300	3,915	11,700	1,181
0340	3-1/2 C.Y. capacity		84.80	2,150	6,435	19,300	1,965
0341	Attachments						
0342	Bucket thumbs		2.40	208	625	1,875	144.20
0345	Grapples		1	220	658.80	1,975	139.75
0350	Gradall type, truck mounted, 3 ton @ 15' radius, 5/8 C.Y.		35.55	900	2,695	8,075	823.40
0370	1 C.Y. capacity		41.10	1,050	3,150	9,450	958.80
0400	Backhoe-loader, 40 to 45 H.P., 5/8 C.Y. capacity		7.55	180	540	1,625	168.40
0450	45 H.P. to 60 H.P., 3/4 C.Y. capacity		9.10	225	675	2,025	207.80
0460	80 H.P., 1-1/4 C.Y. capacity		11.10	237	710	2,125	230.80
0470	112 H.P., 1-1/2 C.Y. capacity		16.35	390	1,170	3,500	364.80
0480	Attachments						
0482	Compactor, 20,000 lb		4.25	117	350	1,050	104
0485	Hydraulic hammer, 750 ft-lbs		1.95	68.50	205	615	56.60
0486	Hydraulic hammer, 1200 ft-lbs		4	133	400	1,200	112
0500	Brush chipper, gas engine, 6" cutter head, 35 H.P.		5.90	93.50	280	840	103.20
0550	12" cutter head, 130 H.P.		9.35	150	450	1,350	164.80
0600	15" cutter head, 165 H.P.		13.40	158	475	1,425	202.20
0750	Bucket, clamshell, general purpose, 3/8 C.Y.		.95	33.50	100	300	27.60
0800	1/2 C.Y.		1.05	41.50	125	375	33.40
0850	3/4 C.Y.		1.20	51.50	155	465	40.60
0900	1 C.Y.		1.25	56.50	170	510	44
0950	1-1/2 C.Y.		1.95	75	225	675	60.60
1000	2 C.Y.		2.05	85	255	765	67.40
1010	Bucket, dragline, medium duty, 1/2 C.Y.		.55	22.50	67	201	17.80
1020	3/4 C.Y.		.55	23.50	71	213	18.60
1030	1 C.Y.		.60	25.50	77	231	20.20
1040	1-1/2 C.Y.		.90	38.50	115	345	30.20
1050	2 C.Y.		1	43.50	130	390	34
1070	3 C.Y.		1.45	60	180	540	47.60
1200	Compactor, roller, 2 drum, 2000 lb., operator walking		6.10	137	410	1,225	130.80
1250	Rammer compactor, gas, 1000 lb. blow		1.55	38.50	115	345	35.40
1300	Vibratory plate, gas, 18" plate, 3000 lb. blow		1.45	28.50	85	255	28.60
1350	21" plate, 5000 lb. blow		1.60	47.50	143	430	41.40
1370	Curb builder/extruder, 14 H.P., gas, single screw		8.50	152	456	1,375	159.20
1390	Double screw		9.25	191	572	1,725	188.40
1500	Disc harrow attachment, for tractor	↓	.35	59	177	530	38.20
1750	Extractor, piling, see lines 2500 to 2750						
1810	Feller buncher, shearing & accumulating trees, 100 H.P.	Ea.	19.80	455	1,370	4,100	432.40
1860	Grader, self-propelled, 25,000 lb.		16.25	410	1,235	3,700	377
1910	30,000 lb.		18.25	475	1,430	4,300	432
1920	40,000 lb.		28.20	735	2,205	6,625	666.60
1930	55,000 lb.		37.70	1,025	3,060	9,175	913.60
1950	Hammer, pavement demo., hyd., gas, self-prop., 1000 to 1250 lb.		20.15	390	1,170	3,500	395.20
2000	Diesel 1300 to 1500 lb.		27.80	590	1,765	5,300	575.40
2050	Pile driving hammer, steam or air, 4150 ft.-lb. @ 225 BPM		6.35	275	825	2,475	215.80
2100	8750 ft.-lb. @ 145 BPM		8.25	450	1,350	4,050	336
2150	15,000 ft.-lb. @ 60 BPM		8.60	485	1,455	4,375	359.80
2200	24,450 ft.-lb. @ 111 BPM	↓	11.35	535	1,605	4,825	411.80
2250	Leads, 15,000 ft.-lb. hammers	L.F.	.03	2.10	6.29	18.85	1.50
2300	24,450 ft.-lb. hammers and heavier	"	.05	3	9	27	2.20
2350	Diesel type hammer, 22,400 ft.-lb.	Ea.	24	620	1,860	5,575	564
2400	41,300 ft.-lb.		32.20	675	2,025	6,075	662.60
2450	141,000 ft.-lb.		60.90	1,475	4,430	13,300	1,373
2500	Vib. elec. hammer/extractor, 200 KW diesel generator, 34 H.P.		26.25	655	1,965	5,900	603
2550	80 H.P.	↓	44.80	960	2,885	8,650	935.40

01590 | Equipment Rental

		01590 Equipment Rental	UNIT	HOURLY OPER. COST	RENT PER DAY	RENT PER WEEK	RENT PER MONTH	CREW EQUIPMENT COST/DAY
200	2600	150 H.P.	Ea.	64	1,475	4,455	13,400	1,403
	2700	Extractor, steam or air, 700 ft.-lb.		14.50	345	1,040	3,125	324
	2750	1000 ft.-lb.		16.60	450	1,350	4,050	402.80
	2800	Log chipper, up to 22" diam, 600 H.P.		41.93	1,450	4,360	13,100	1,207
	2850	Logger, for skidding & stacking logs, 150 H.P.		33.80	790	2,375	7,125	745.40
	2900	Rake, spring tooth, with tractor		7.04	214	643	1,925	184.90
	3000	Roller, tandem, gas, 3 to 5 ton		4.90	115	345	1,025	108.20
	3050	Diesel, 8 to 12 ton		6.95	213	640	1,925	183.60
	3100	Towed type vibratory compactor, diesel, 50 HP, 72" smooth drum		31.15	600	1,805	5,425	610.20
	3150	Sheepsfoot, double 60" x 60"		2.35	98.50	295	885	77.80
	3170	Landfill compactor, 220 HP		43.05	1,175	3,505	10,500	1,045
	3200	Pneumatic tire diesel roller, 12 ton		7.95	300	905	2,725	244.60
	3250	21 to 25 ton		13	550	1,650	4,950	434
	3300	Sheepsfoot roller, self-propelled, 4 wheel, 130 H.P.		31.95	845	2,535	7,600	762.60
	3320	300 H.P.		43.50	1,200	3,590	10,800	1,066
	3350	Vibratory steel drum & pneumatic tire, diesel, 18,000 lb.		16.45	320	960	2,875	323.60
	3400	29,000 lb.		24.10	405	1,220	3,650	436.80
	3410	Rotary mower, brush, 60", with tractor		10.40	232	695	2,075	222.20
	3450	Scrapers, towed type, 9 to 12 C.Y. capacity		4.65	221	664	2,000	170
	3500	12 to 17 C.Y. capacity		1.51	295	885.20	2,650	189.10
	3550	Scrapers, self-propelled, 4 x 4 drive, 2 engine, 14 C.Y. capacity		76.50	1,425	4,280	12,800	1,468
	3600	2 engine, 24 C.Y. capacity		116.80	2,275	6,815	20,400	2,297
	3640	32 - 44 C.Y. capacity		139.15	2,650	7,935	23,800	2,700
	3650	Self-loading, 11 C.Y. capacity		39.80	810	2,435	7,300	805.40
	3700	22 C.Y. capacity		75.20	1,650	4,975	14,900	1,597
	3710	Screening plant 110 hp. w / 5' x 10'screen		20.35	375	1,125	3,375	387.80
	3720	5' x 16' screen	▼	22.40	475	1,425	4,275	464.20
	3850	Shovels, see Cranes division 01590-600						
	3860	Shovel/backhoe bucket, 1/2 C.Y.	Ea.	1.65	53.50	160	480	45.20
	3870	3/4 C.Y.		1.75	60	180	540	50
	3880	1 C.Y.		1.85	70	210	630	56.80
	3890	1-1/2 C.Y.		1.95	83.50	250	750	65.60
	3910	3 C.Y.		2.30	118	355	1,075	89.40
	3950	Stump chipper, 18" deep, 30 H.P.		4.65	83.50	250	750	87.20
	4110	Tractor, crawler, with bulldozer, torque converter, diesel 75 H.P.		14.40	310	925	2,775	300.20
	4150	105 H.P.		19.25	465	1,400	4,200	434
	4200	140 H.P.		23.90	585	1,755	5,275	542.20
	4260	200 H.P.		35.55	965	2,895	8,675	863.40
	4310	300 H.P.		46	1,225	3,655	11,000	1,099
	4360	410 H.P.		64.80	1,600	4,775	14,300	1,473
	4370	500 H.P.		83.45	2,075	6,205	18,600	1,909
	4380	700 H.P.		128	3,375	10,115	30,300	3,047
	4400	Loader, crawler, torque conv., diesel, 1-1/2 C.Y., 80 H.P.		13.45	320	965	2,900	300.60
	4450	1-1/2 to 1-3/4 C.Y., 95 H.P.		15.55	385	1,150	3,450	354.40
	4510	1-3/4 to 2-1/4 C.Y., 130 H.P.		21.25	615	1,840	5,525	538
	4530	2-1/2 to 3-1/4 C.Y., 190 H.P.		32	840	2,525	7,575	761
	4560	3-1/2 to 5 C.Y., 275 H.P.		42.85	1,200	3,590	10,800	1,061
	4610	Tractor loader, wheel, torque conv., 4 x 4, 1 to 1-1/4 C.Y., 65 H.P.		9	200	600	1,800	192
	4620	1-1/2 to 1-3/4 C.Y., 80 H.P.		11.15	245	735	2,200	236.20
	4650	1-3/4 to 2 C.Y., 100 H.P.		12.35	278	835	2,500	265.80
	4710	2-1/2 to 3-1/2 C.Y., 130 H.P.		13.45	315	940	2,825	295.60
	4730	3 to 4-1/2 C.Y., 170 H.P.		19.05	475	1,430	4,300	438.40
	4760	5-1/4 to 5-3/4 C.Y., 270 H.P.		32.10	695	2,090	6,275	674.80
	4810	7 to 8 C.Y., 375 H.P.		57.75	1,275	3,830	11,500	1,228
	4870	12-1/2 C.Y., 690 H.P.		82.45	2,025	6,085	18,300	1,877
	4880	Wheeled, skid steer, 10 C.F., 30 H.P. gas		7.40	140	420	1,250	143.20
	4890	1 C.Y., 78 H.P., diesel	▼	8.85	188	565	1,700	183.80
	4891	Attachments for all skid steer loaders						

01590 | Equipment Rental

			UNIT	HOURLY OPER. COST	RENT PER DAY	RENT PER WEEK	RENT PER MONTH	CREW EQUIPMENT COST/DAY	
200	4892	Auger	Ea.	.39	65	195	585	42.10	200
	4893	Backhoe		.65	108	325	975	70.20	
	4894	Broom		.63	106	317	950	68.45	
	4895	Forks		.22	36.50	109	325	23.55	
	4896	Grapple		.50	82.50	248	745	53.60	
	4897	Concrete hammer		.96	161	482	1,450	104.10	
	4898	Tree spade		.93	155	464	1,400	100.25	
	4899	Trencher		.68	113	340	1,025	73.45	
	4900	Trencher, chain, boom type, gas, operator walking, 12 H.P.		2.30	68.50	205	615	59.40	
	4910	Operator riding, 40 H.P.		7.70	245	735	2,200	208.60	
	5000	Wheel type, diesel, 4' deep, 12" wide		41.30	730	2,185	6,550	767.40	
	5100	Diesel, 6' deep, 20" wide		59.45	1,425	4,250	12,800	1,326	
	5150	Ladder type, diesel, 5' deep, 8" wide		21.05	650	1,950	5,850	558.40	
	5200	Diesel, 8' deep, 16" wide		52.05	1,500	4,470	13,400	1,310	
	5210	Tree spade, self-propelled		9.60	267	800	2,400	236.80	
	5250	Truck, dump, tandem, 12 ton payload		20.25	272	815	2,450	325	
	5300	Three axle dump, 16 ton payload		27.95	420	1,265	3,800	476.60	
	5350	Dump trailer only, rear dump, 16-1/2 C.Y.		4.15	117	350	1,050	103.20	
	5400	20 C.Y.		4.55	133	400	1,200	116.40	
	5450	Flatbed, single axle, 1-1/2 ton rating		11	56.50	170	510	122	
	5500	3 ton rating		14.05	80	240	720	160.40	
	5550	Off highway rear dump, 25 ton capacity		40.55	995	2,980	8,950	920.40	
	5600	35 ton capacity		41.45	1,025	3,075	9,225	946.60	
	5610	50 ton capacity		53.40	1,325	3,995	12,000	1,226	
	5620	65 ton capacity		57.20	1,425	4,260	12,800	1,310	
	5630	100 ton capacity		73.40	1,800	5,380	16,100	1,663	
	6000	Vibratory plow, 25 H.P., walking	↓	3.90	58.50	175	525	66.20	
400	0010	**GENERAL EQUIPMENT RENTAL** Without operators							400
	0150	Aerial lift, scissor type, to 15' high, 1000 lb. cap., electric	Ea.	2.20	43.50	130	390	43.60	
	0160	To 25' high, 2000 lb. capacity		2.60	63.50	190	570	58.80	
	0170	Telescoping boom to 40' high, 500 lb. capacity, gas		10.90	272	815	2,450	250.20	
	0180	To 45' high, 500 lb. capacity		11.70	315	940	2,825	281.60	
	0190	To 60' high, 600 lb. capacity		13.65	415	1,250	3,750	359.20	
	0195	Air compressor, portable, 6.5 CFM, electric		.31	17.65	53	159	13.10	
	0196	gasoline		.35	26.50	79	237	18.60	
	0200	Air compressor, portable, gas engine, 60 C.F.M.		4.10	32.50	98	294	52.40	
	0300	160 C.F.M.		7	43.50	130	390	82	
	0400	Diesel engine, rotary screw, 250 C.F.M.		7.95	107	320	960	127.60	
	0500	365 C.F.M.		9.95	130	390	1,175	157.60	
	0550	450 C.F.M.		13.40	158	475	1,425	202.20	
	0600	600 C.F.M.		19.50	217	650	1,950	286	
	0700	750 C.F.M.		21	228	685	2,050	305	
	0800	For silenced models, small sizes, add		3%	5%	5%	5%		
	0900	Large sizes, add	↓	5%	7%	7%	7%		
	0920	Air tools and accessories							
	0930	Breaker, pavement, 60 lb.	Ea.	.35	14	42	126	11.20	
	0940	80 lb.		.35	17	51	153	13	
	0950	Drills, hand (jackhammer) 65 lb.		.45	14.35	43	129	12.20	
	0960	Track or wagon, swing boom, 4" drifter		33.90	605	1,815	5,450	634.20	
	0970	5" drifter		46.05	765	2,295	6,875	827.40	
	0975	Track mounted quarry drill, 6" diameter drill		48.65	825	2,480	7,450	885.20	
	0980	Dust control per drill		.79	12.35	37	111	13.70	
	0990	Hammer, chipping, 12 lb.		.40	21.50	64	192	16	
	1000	Hose, air with couplings, 50' long, 3/4" diameter		.03	5.35	16	48	3.45	
	1100	1" diameter		.03	5.35	16	48	3.45	
	1200	1-1/2" diameter		.05	7.65	23	69	5	
	1300	2" diameter	↓	.10	16.65	50	150	10.80	

01590 | Equipment Rental

400		UNIT	HOURLY OPER. COST	RENT PER DAY	RENT PER WEEK	RENT PER MONTH	CREW EQUIPMENT COST/DAY
1400	2-1/2" diameter	Ea.	.12	19.65	59	177	12.75
1410	3" diameter		.16	27.50	82	246	17.70
1450	Drill, steel, 7/8" x 2'		.05	6.65	20	60	4.40
1460	7/8" x 6'		.06	8	24	72	5.30
1520	Moil points		.03	4.67	14	42	3.05
1525	Pneumatic nailer w/accessories		.41	27	81	243	19.50
1530	Sheeting driver for 60 lb. breaker		.10	6.95	20.80	62.50	4.95
1540	For 90 lb. breaker		.15	10	30	90	7.20
1550	Spade, 25 lb.		.35	6.65	20	60	6.80
1560	Tamper, single, 35 lb.		.49	32.50	98	294	23.50
1570	Triple, 140 lb.		.74	49	147	440	35.30
1580	Wrenches, impact, air powered, up to 3/4" bolt		.20	10	30	90	7.60
1590	Up to 1-1/4" bolt		.30	20	60	180	14.40
1600	Barricades, barrels, reflectorized, 1 to 50 barrels		.02	3.10	9.30	28	2
1610	100 to 200 barrels		.01	2.33	7	21	1.50
1620	Barrels with flashers, 1 to 50 barrels		.02	3.77	11.30	34	2.40
1630	100 to 200 barrels		.02	3	9	27	1.95
1640	Barrels with steady burn type C lights		.03	5	15	45	3.25
1650	Illuminated board, trailer mounted, with generator		.65	117	350	1,050	75.20
1670	Portable barricade, stock, with flashers, 1 to 6 units		.02	3.77	11.30	34	2.40
1680	25 to 50 units		.02	3.50	10.50	31.50	2.25
1690	Butt fusion machine, electric		21.70	435	1,300	3,900	433.60
1695	Electro fusion machine		8.60	173	520	1,550	172.80
1700	Carts, brick, hand powered, 1000 lb. capacity		.23	38.50	115	345	24.85
1800	Gas engine, 1500 lb., 7-1/2' lift		2.80	95	285	855	79.40
1822	Dehumidifier, medium, 6 Lb/Hr, 150 CFM		.68	41.50	124	370	30.25
1824	Large, 18 Lb/Hr, 600 CFM		1.36	82.50	248	745	60.50
1830	Distributor, asphalt, trailer mtd, 2000 gal., 38 H.P. diesel		7	253	760	2,275	208
1840	3000 gal., 38 H.P. diesel		8.30	290	870	2,600	240.40
1850	Drill, rotary hammer, electric, 1-1/2" diameter		.40	24.50	74	222	18
1860	Carbide bit for above		.03	5.35	16	48	3.45
1865	Rotary, crawler, 250 HP		81.10	1,675	5,055	15,200	1,660
1870	Emulsion sprayer, 65 gal., 5 H.P. gas engine		1.83	74	222	665	59.05
1880	200 gal., 5 H.P. engine	↓	4.75	123	370	1,100	112
1900	Fencing, see division 01560-250 & 02820-000						
1920	Floodlight, mercury vapor, or quartz, on tripod						
1930	1000 watt	Ea.	.30	11.65	35	105	9.40
1940	2000 watt		.52	21.50	65	195	17.15
1950	Floodlights, trailer mounted with generator, 1 - 300 watt light		2.45	65	195	585	58.60
1960	2 - 1000 watt lights		3.30	108	325	975	91.40
2000	4 - 300 watt lights		2.75	76.50	230	690	68
2020	Forklift, wheeled, for brick, 18', 3000 lb., 2 wheel drive, gas		13.15	188	565	1,700	218.20
2040	28', 4000 lb., 4 wheel drive, diesel		11.35	248	745	2,225	239.80
2050	For rough terrain, 8000 lb., 16' lift, 68 HP		15	370	1,110	3,325	342
2060	For plant, 4 T. capacity, 80 H.P., 2 wheel drive, gas		7.50	95	285	855	117
2080	10 T. capacity, 120 H.P., 2 wheel drive, diesel		11.25	172	515	1,550	193
2100	Generator, electric, gas engine, 1.5 KW to 3 KW		1.65	21.50	64	192	26
2200	5 KW		2.25	31.50	95	285	37
2300	10 KW		3.70	68.50	205	615	70.60
2400	25 KW		7.10	92.50	278	835	112.40
2500	Diesel engine, 20 KW		5.95	63.50	190	570	85.60
2600	50 KW		11.15	70	210	630	131.20
2700	100 KW		16.45	80	240	720	179.60
2800	250 KW		46.45	143	430	1,300	457.60
2850	Hammer, hydraulic, for mounting on boom, to 500 ft.-lb.		1.75	63.50	190	570	52
2860	1000 ft.-lb.		3.10	102	305	915	85.80
2900	Heaters, space, oil or electric, 50 MBH		.92	12.35	37	111	14.75
3000	100 MBH	↓	1.66	16.65	50	150	23.30

01590 | Equipment Rental

			UNIT	HOURLY OPER. COST	RENT PER DAY	RENT PER WEEK	RENT PER MONTH	CREW EQUIPMENT COST/DAY
400	3100	300 MBH	Ea.	5.33	35	105	315	63.65
	3150	500 MBH		10.70	50	150	450	115.60
	3200	Hose, water, suction with coupling, 20' long, 2" diameter		.02	5.65	17	51	3.55
	3210	3" diameter		.03	8.65	26	78	5.45
	3220	4" diameter		.04	11.65	35	105	7.30
	3230	6" diameter		.09	23	69	207	14.50
	3240	8" diameter		.26	43.50	130	390	28.10
	3250	Discharge hose with coupling, 50' long, 2" diameter		.02	5	15	45	3.15
	3260	3" diameter		.02	6	18	54	3.75
	3270	4" diameter		.03	8.35	25	75	5.25
	3280	6" diameter		.06	19	57	171	11.90
	3290	8" diameter		.36	59.50	178	535	38.50
	3295	Insulation blower		.11	7.35	22	66	5.30
	3300	Ladders, extension type, 16' to 36' long		.16	26	78	234	16.90
	3400	40' to 60' long		.19	31	93	279	20.10
	3405	Lance for cutting concrete		2.84	109	327	980	88.10
	3407	Lawn mower, rotary, 22", 5HP		.98	27.50	83	249	24.45
	3408	48" self propelled		2.44	85.50	257	770	70.90
	3410	Level, laser type, transit		1.24	82.50	248	745	59.50
	3430	For pipe laying, manual leveling		.71	47.50	142	425	34.10
	3440	Rotary beacon		.92	61	183	550	43.95
	3460	Builders level with tripod and rod		.09	14.65	44	132	9.50
	3500	Light towers, towable, with diesel generator, 2000 watt		2.75	76.50	230	690	68
	3600	4000 watt		3.30	108	325	975	91.40
	3700	Mixer, powered, plaster and mortar, 6 C.F., 7 H.P.		1.05	36.50	110	330	30.40
	3800	10 C.F., 9 H.P.		1.30	53.50	160	480	42.40
	3850	Nailer, pneumatic		.41	27	81	243	19.50
	3900	Paint sprayers complete, 8 CFM		.69	45.50	137	410	32.90
	4000	17 CFM		1.10	73.50	220	660	52.80
	4020	Pavers, bituminous, rubber tires, 8' wide, 52 H.P., gas		25.75	770	2,315	6,950	669
	4030	8' wide, 64 H.P., diesel		38.40	11.20	33.65	101	313.95
	4050	Crawler, 10' wide, 78 H.P., gas		44.95	1,325	3,940	11,800	1,148
	4060	10' wide, 87 H.P., diesel		58.05	1,650	4,965	14,900	1,457
	4070	Concrete paver, 12' to 24' wide, 250 H.P.		51.75	1,225	3,710	11,100	1,156
	4080	Placer-spreader-trimmer, 24' wide, 300 H.P.		64.05	1,775	5,340	16,000	1,580
	4100	Pump, centrifugal gas pump, 1-1/2", 4 MGPH		2.50	35	105	315	41
	4200	2", 8 MGPH		3.15	43.50	130	390	51.20
	4300	3", 15 MGPH		3.35	45	135	405	53.80
	4400	6", 90 MGPH		15.60	175	525	1,575	229.80
	4500	Submersible electric pump, 1-1/4", 55 GPM		.35	20.50	62	186	15.20
	4600	1-1/2", 83 GPM		.41	23.50	70	210	17.30
	4700	2", 120 GPM		.57	29.50	88	264	22.15
	4800	3", 300 GPM		.97	38.50	115	345	30.75
	4900	4", 560 GPM		6.24	124	372	1,125	124.30
	5000	6", 1590 GPM		9.05	185	555	1,675	183.40
	5100	Diaphragm pump, gas, single, 1-1/2" diameter		.74	36	108	325	27.50
	5200	2" diameter		2.40	45	135	405	46.20
	5300	3" diameter		2.45	48.50	145	435	48.60
	5400	Double, 4" diameter		4.30	88.50	265	795	87.40
	5500	Trash pump, self-priming, gas, 2" diameter		2.90	35	105	315	44.20
	5600	Diesel, 4" diameter		4.35	80	240	720	82.80
	5650	Diesel, 6" diameter		9	128	385	1,150	149
	5655	Grout Pump		3.90	38.50	116	350	54.40
	5660	Rollers, see division 01590-200						
	5700	Salamanders, L.P. gas fired, 100,000 B.T.U.	Ea.	1.66	10.65	32	96	19.70
	5705	50,000 BTU		1.25	7.65	23	69	14.60
	5720	Sandblaster, portable, open top, 3 C.F. capacity		.40	20.50	62	186	15.60
	5730	6 C.F. capacity		.65	30	90	270	23.20

01590 | Equipment Rental

			UNIT	HOURLY OPER. COST	RENT PER DAY	RENT PER WEEK	RENT PER MONTH	CREW EQUIPMENT COST/DAY	
400	5740	Accessories for above	Ea.	.11	18	54	162	11.70	400
	5750	Sander, floor		.70	16.35	49	147	15.40	
	5760	Edger		.60	20	60	180	16.80	
	5800	Saw, chain, gas engine, 18" long		1.15	16	48	144	18.80	
	5900	36" long		.55	48.50	145	435	33.40	
	5950	60" long		.55	50	150	450	34.40	
	6000	Masonry, table mounted, 14" diameter, 5 H.P.		1.30	56	168	505	44	
	6050	Portable cut-off, 8 H.P.		1.20	25.50	76	228	24.80	
	6100	Circular, hand held, electric, 7-1/4" diameter	Ea.	.20	10	30	90	7.60	
	6200	12" diameter		.27	14	42	126	10.55	
	6250	Wall saw, w/hydraulic power, 10 H.P		2.08	97.50	292.40	875	75.10	
	6275	Shot blaster, walk behind, 20" wide		1.04	445	1,330	4,000	274.30	
	6300	Steam cleaner, 100 gallons per hour		2.20	63.50	190	570	55.60	
	6310	200 gallons per hour		2.90	78.50	235	705	70.20	
	6340	Tar Kettle/Pot, 400 gallon		2.69	51.50	155	465	52.50	
	6350	Torch, cutting, acetylene-oxygen, 150' hose		1.50	13.35	40	120	20	
	6360	Hourly operating cost includes tips and gas		8.10				64.80	
	6410	Toilet, portable chemical		.10	17.35	52	156	11.20	
	6420	Recycle flush type		.13	21.50	64	192	13.85	
	6430	Toilet, fresh water flush, garden hose,		.14	24	72	216	15.50	
	6440	Hoisted, non-flush, for high rise		.13	21	63	189	13.65	
	6450	Toilet, trailers, minimum		.22	36	108	325	23.35	
	6460	Maximum		.65	108	324	970	70	
	6465	Tractor, farm with attachment		9.30	225	675	2,025	209.40	
	6470	Trailer, office, see division 01520-500							
	6500	Trailers, platform, flush deck, 2 axle, 25 ton capacity	Ea.	4.20	91.50	275	825	88.60	
	6600	40 ton capacity		5.45	128	385	1,150	120.60	
	6700	3 axle, 50 ton capacity		5.90	142	425	1,275	132.20	
	6800	75 ton capacity		7.40	185	555	1,675	170.20	
	6810	Trailer mounted cable reel for H.V. line work		4.38	209	626	1,875	160.25	
	6820	Trailer mounted cable tensioning rig		8.61	410	1,230	3,700	314.90	
	6830	Cable pulling rig		54.46	2,325	6,980	20,900	1,832	
	6850	Trailer, storage, see division 01520-500							
	6900	Water tank, engine driven discharge, 5000 gallons	Ea.	5.50	123	370	1,100	118	
	6925	10,000 gallons		7.65	175	525	1,575	166.20	
	6950	Water truck, off highway, 6000 gallons		49.05	715	2,150	6,450	822.40	
	7010	Tram car for H.V. line work, powered, 2 conductor		5.68	113	340	1,025	113.45	
	7020	Transit (builder's level) with tripod		.09	14.65	44	132	9.50	
	7030	Trench box, 3000 lbs. 6'x8'		.42	70	210	630	45.35	
	7040	7200 lbs. 6'x20'		.82	136	409	1,225	88.35	
	7050	8000 lbs., 8' x 16'		.87	145	436	1,300	94.15	
	7060	9500 lbs., 8'x20'		1.17	195	584	1,750	126.15	
	7065	11,000 lbs., 8'x24'		1.31	218	654	1,950	141.30	
	7070	12,000 lbs., 10' x 20'		1.63	272	817	2,450	176.45	
	7100	Truck, pickup, 3/4 ton, 2 wheel drive		5.35	55	165	495	75.80	
	7200	4 wheel drive		5.50	63.50	190	570	82	
	7250	Crew carrier, 9 passenger		5.53	108	324.40	975	109.10	
	7290	Tool van, 24,000 G.V.W.		9.09	94.50	283.20	850	129.35	
	7300	Tractor, 4 x 2, 30 ton capacity, 195 H.P.		13.90	173	520	1,550	215.20	
	7410	250 H.P.		18.90	272	815	2,450	314.20	
	7500	6 x 2, 40 ton capacity, 240 H.P.		17.50	272	815	2,450	303	
	7600	6 x 4, 45 ton capacity, 240 H.P.		21.45	267	800	2,400	331.60	
	7620	Vacuum truck, hazardous material, 2500 gallon		6.72	315	943	2,825	242.35	
	7625	5,000 gallon		9.09	420	1,256.80	3,775	324.10	
	7640	Tractor, with A frame, boom and winch, 225 H.P.		14.30	233	700	2,100	254.40	
	7650	Vacuum, H.E.P.A., 16 gal., wet/dry		.27	24	72	216	16.55	
	7655	55 gal, wet/dry		.60	36	108	325	26.40	
	7660	Water tank, portable		1	9.35	28	84	13.60	

01590 | Equipment Rental

			UNIT	HOURLY OPER. COST	RENT PER DAY	RENT PER WEEK	RENT PER MONTH	CREW EQUIPMENT COST/DAY	
400	7690	Large production vacuum loader, 3150 CFM	Ea.	15.63	630	1,890	5,675	503.05	400
	7700	Welder, electric, 200 amp		3.74	57	171	515	64.10	
	7800	300 amp		5.22	60	180	540	77.75	
	7900	Gas engine, 200 amp		5.30	35.50	106	320	63.60	
	8000	300 amp		6.20	42.50	128	385	75.20	
	8100	Wheelbarrow, any size		.06	10.35	31	93	6.70	
	8200	Wrecking ball, 4000 lb.	↓	1.85	68.50	205	615	55.80	
500	0010	**HIGHWAY EQUIPMENT RENTAL**							500
	0050	Asphalt batch plant, portable drum mixer, 100 ton/hr.	Ea.	55.10	1,350	4,020	12,100	1,245	
	0060	200 ton/hr.		61.10	1,400	4,215	12,600	1,332	
	0070	300 ton/hr.		71.20	1,675	4,990	15,000	1,568	
	0100	Backhoe attachment, long stick, up to 185 HP, 10.5' long		.30	19.65	59	177	14.20	
	0140	Up to 250 HP, 12' long		.32	21.50	64	192	15.35	
	0180	Over 250 HP, 15' long		.42	27.50	83	249	19.95	
	0200	Special dipper arm, up to 100 HP, 32' long		.86	57.50	172	515	41.30	
	0240	Over 100 HP, 33' long		1.08	72	216	650	51.85	
	0300	Concrete batch plant, portable, electric, 200 CY/Hr	↓	12.67	645	1,940	5,825	489.35	
	0500	Grader attachment, ripper/scarifier, rear mounted							
	0520	Up to 135 HP	Ea.	2.70	58.50	175	525	56.60	
	0540	Up to 180 HP		3.20	75	225	675	70.60	
	0580	Up to 250 HP		3.55	86.50	260	780	80.40	
	0700	Pvmt. removal bucket, for hyd. excavator, up to 90 HP		1.30	41.50	125	375	35.40	
	0740	Up to 200 HP		1.50	63.50	190	570	50	
	0780	Over 200 HP		1.65	76.50	230	690	59.20	
	0900	Aggregate spreader, self-propelled, 187 HP		37.15	790	2,370	7,100	771.20	
	1000	Chemical spreader, 3 C.Y.		2.20	80.50	242	725	66	
	1900	Hammermill, traveling, 250 HP		38.85	1,675	5,020	15,100	1,315	
	2000	Horizontal borer, 3" diam, 13 HP gas driven		3.85	53.50	160	480	62.80	
	2200	Hydromulchers, gas power, 3000 gal., for truck mounting		10.50	187	560	1,675	196	
	2400	Joint & crack cleaner, walk behind, 25 HP		2.05	46.50	140	420	44.40	
	2500	Filler, trailer mounted, 400 gal., 20 HP		6	182	545	1,625	157	
	3000	Paint striper, self propelled, double line, 30 HP		5.15	155	465	1,400	134.20	
	3200	Post drivers, 6" I-Beam frame, for truck mounting		8.15	415	1,250	3,750	315.20	
	3400	Road sweeper, self propelled, 8' wide, 90 HP		23.50	385	1,160	3,475	420	
	4000	Road mixer, self-propelled, 130 HP		28.35	590	1,770	5,300	580.80	
	4100	310 HP	↓	53.85	1,975	5,895	17,700	1,610	
	4200	Cold mix paver, incl pug mill and bitumen tank,							
	4220	165 HP	Ea.	65.05	1,925	5,770	17,300	1,674	
	4250	Paver, asphalt, wheel or crawler, 130 H.P., diesel		58.05	1,650	4,965	14,900	1,457	
	4300	Paver, road widener, gas 1' to 6', 67 HP		28.70	635	1,900	5,700	609.60	
	4400	Diesel, 2' to 14', 88 HP		38.80	995	2,980	8,950	906.40	
	4600	Slipform pavers, curb and gutter, 2 track, 75 HP		22.95	650	1,950	5,850	573.60	
	4700	4 track, 165 HP		31.05	730	2,195	6,575	687.40	
	4800	Median barrier, 215 HP		31.50	755	2,260	6,775	704	
	4901	Trailer, low bed, 75 ton capacity		7.85	182	545	1,625	171.80	
	5000	Road planer, walk behind, 10" cutting width, 10 HP		2	26	78	234	31.60	
	5100	Self propelled, 12" cutting width, 64 HP		5.30	288	865	2,600	215.40	
	5200	Pavement profiler, 4' to 6' wide, 450 HP		141.40	2,800	8,380	25,100	2,807	
	5300	8' to 10' wide, 750 HP		225.75	4,250	12,725	38,200	4,351	
	5400	Roadway plate, steel, 1"x8'x20'		.06	9.65	29	87	6.30	
	5600	Stabilizer,self-propelled, 150 HP		25.25	560	1,685	5,050	539	
	5700	310 HP		42.50	1,200	3,605	10,800	1,061	
	5800	Striper, thermal, truck mounted 120 gal. paint, 150H.P.		32.40	485	1,450	4,350	549.20	
	6000	Tar kettle, 330 gal., trailer mounted		2.37	36.50	110	330	40.95	
	7000	Tunnel locomotive, diesel, 8 to 12 ton		20.45	560	1,675	5,025	498.60	
	7005	Electric, 10 ton		20.45	635	1,905	5,725	544.60	
	7010	Muck cars, 1/2 C.Y. capacity	↓	1.50	20.50	62	186	24.40	

01590 | Equipment Rental

		01590 Equipment Rental	UNIT	HOURLY OPER. COST	RENT PER DAY	RENT PER WEEK	RENT PER MONTH	CREW EQUIPMENT COST/DAY
500	7020	1 C.Y. capacity	Ea.	1.70	29	87	261	31
	7030	2 C.Y. capacity		1.80	33.50	100	300	34.40
	7040	Side dump, 2 C.Y. capacity		2	41.50	125	375	41
	7050	3 C.Y. capacity		2.70	48.50	145	435	50.60
	7060	5 C.Y. capacity		3.80	61.50	185	555	67.40
	7100	Ventilating blower for tunnel, 7-1/2 H.P.		1.25	40	120	360	34
	7110	10 H.P.		1.44	41.50	125	375	36.50
	7120	20 H.P.		2.33	48.50	145	435	47.65
	7140	40 H.P.		4.07	73.50	220	660	76.55
	7160	60 H.P.		6.16	110	330	990	115.30
	7175	75 H.P.		7.89	148	445	1,325	152.10
	7180	200 H.P.		17.60	205	615	1,850	263.80
	7800	Windrow loader, elevating	↓	32.70	905	2,720	8,150	805.60
600	0010	**LIFTING AND HOISTING EQUIPMENT RENTAL**						
	0100	without operators						
	0120	Aerial lift truck, 2 person, to 80'	Ea.	17.85	600	1,795	5,375	501.80
	0140	Boom work platform, 40' snorkel		9.25	205	615	1,850	197
	0150	Crane, flatbed mntd, 3 ton cap.		11.60	182	545	1,625	201.80
	0200	Crane, climbing, 106' jib, 6000 lb. capacity, 410 FPM		41.95	1,325	3,990	12,000	1,134
	0300	101' jib, 10,250 lb. capacity, 270 FPM	↓	47.25	1,675	5,050	15,200	1,388
	0400	Tower, static, 130' high, 106' jib,						
	0500	6200 lb. capacity at 400 FPM	Ea.	45.05	1,525	4,610	13,800	1,282
	0600	Crawler mounted, lattice boom, 1/2 C.Y., 15 tons at 12' radius		20.11	485	1,450	4,350	450.90
	0700	3/4 C.Y., 20 tons at 12' radius		26.81	650	1,950	5,850	604.50
	0800	1 C.Y., 25 tons at 12' radius		35.75	845	2,530	7,600	792
	0900	1-1/2 C.Y., 40 tons at 12' radius		39.90	995	2,980	8,950	915.20
	1000	2 C.Y., 50 tons at 12' radius		49.90	1,425	4,305	12,900	1,260
	1100	3 C.Y., 75 tons at 12' radius		47.75	1,400	4,175	12,500	1,217
	1200	100 ton capacity, 60' boom		61.05	1,850	5,570	16,700	1,602
	1300	165 ton capacity, 60' boom		90.10	2,400	7,235	21,700	2,168
	1400	200 ton capacity, 70' boom		95.85	2,625	7,855	23,600	2,338
	1500	350 ton capacity, 80' boom		138	3,750	11,215	33,600	3,347
	1600	Truck mounted, lattice boom, 6 x 4, 20 tons at 10' radius		20.93	800	2,400	7,200	647.45
	1700	25 tons at 10' radius		22.32	855	2,560	7,675	690.55
	1800	8 x 4, 30 tons at 10' radius		28.67	905	2,720	8,150	773.35
	1900	40 tons at 12' radius		24.56	960	2,880	8,650	772.50
	2000	8 x 4, 60 tons at 15' radius		30.91	1,100	3,290	9,875	905.30
	2050	82 tons at 15' radius		45.40	1,600	4,770	14,300	1,317
	2100	90 tons at 15' radius		46.24	1,750	5,230	15,700	1,416
	2200	115 tons at 15' radius		51.10	1,950	5,815	17,400	1,572
	2300	150 tons at 18' radius		47.45	2,050	6,150	18,500	1,610
	2350	165 tons at 18' radius		73.35	2,400	7,235	21,700	2,034
	2400	Truck mounted, hydraulic, 12 ton capacity		31	605	1,820	5,450	612
	2500	25 ton capacity		31.15	625	1,875	5,625	624.20
	2550	33 ton capacity		32	655	1,965	5,900	649
	2560	40 ton capacity		30.40	645	1,940	5,825	631.20
	2600	55 ton capacity		44.75	920	2,765	8,300	911
	2700	80 ton capacity		54	980	2,940	8,825	1,020
	2720	100 ton capacity		78.90	2,600	7,770	23,300	2,185
	2740	120 ton capacity		82.60	2,825	8,460	25,400	2,353
	2760	150 ton capacity		100.45	3,550	10,625	31,900	2,929
	2800	Self-propelled, 4 x 4, with telescoping boom, 5 ton		14.60	335	1,005	3,025	317.80
	2900	12-1/2 ton capacity		21.90	510	1,535	4,600	482.20
	3000	15 ton capacity		23.85	600	1,800	5,400	550.80
	3050	20 ton capacity		24.80	625	1,880	5,650	574.40
	3100	25 ton capacity		25.90	615	1,840	5,525	575.20
	3150	40 ton capacity	↓	41.65	880	2,640	7,925	861.20

01590 | Equipment Rental

			UNIT	HOURLY OPER. COST	RENT PER DAY	RENT PER WEEK	RENT PER MONTH	CREW EQUIPMENT COST/DAY
600	3200	Derricks, guy, 20 ton capacity, 60' boom, 75' mast	Ea.	12.33	325	976	2,925	293.85
	3300	100' boom, 115' mast		20.01	560	1,680	5,050	496.10
	3400	Stiffleg, 20 ton capacity, 70' boom, 37' mast		14.25	415	1,250	3,750	364
	3500	100' boom, 47' mast		22.53	680	2,040	6,125	588.25
	3550	Helicopter, small, lift to 1250 lbs. maximum, w/pilot		66.23	2,625	7,890	23,700	2,108
	3600	Hoists, chain type, overhead, manual, 3/4 ton		.10	3	9	27	2.60
	3900	10 ton		.55	16	48	144	14
	4000	Hoist and tower, 5000 lb. cap., portable electric, 40' high		4.14	188	563	1,700	145.70
	4100	For each added 10' section, add		.09	14.65	44	132	9.50
	4200	Hoist and single tubular tower, 5000 lb. electric, 100' high		5.58	262	785	2,350	201.65
	4300	For each added 6'-6" section, add		.15	24.50	74	222	16
	4400	Hoist and double tubular tower, 5000 lb., 100' high		5.98	288	865	2,600	220.85
	4500	For each added 6'-6" section, add		.17	27.50	83	249	17.95
	4550	Hoist and tower, mast type, 6000 lb., 100' high		6.47	299	897	2,700	231.15
	4570	For each added 10' section, add		.11	18	54	162	11.70
	4600	Hoist and tower, personnel, electric, 2000 lb., 100' @ 125 FPM		13.27	795	2,390	7,175	584.15
	4700	3000 lb., 100' @ 200 FPM		15.15	900	2,700	8,100	661.20
	4800	3000 lb., 150' @ 300 FPM		16.80	1,000	3,030	9,100	740.40
	4900	4000 lb., 100' @ 300 FPM		17.43	1,025	3,090	9,275	757.45
	5000	6000 lb., 100' @ 275 FPM		18.84	1,075	3,240	9,725	798.70
	5100	For added heights up to 500', add	L.F.	.01	1.67	5	15	1.10
	5200	Jacks, hydraulic, 20 ton	Ea.	.05	8	24	72	5.20
	5500	100 ton	"	.30	23	69	207	16.20
	6000	Jacks, hydraulic, climbing with 50' jackrods						
	6010	and control consoles, minimum 3 mo. rental						
	6100	30 ton capacity	Ea.	1.62	108	323	970	77.55
	6150	For each added 10' jackrod section, add		.05	3.33	10	30	2.40
	6300	50 ton capacity		2.60	173	520	1,550	124.80
	6350	For each added 10' jackrod section, add		.06	4	12	36	2.90
	6500	125 ton capacity		6.80	455	1,360	4,075	326.40
	6550	For each added 10' jackrod section, add		.47	31	93	279	22.35
	6600	Cable jack, 10 ton capacity with 200' cable		1.35	90	270	810	64.80
	6650	For each added 50' of cable, add		.15	9.65	29	87	7
700	0010	**WELLPOINT EQUIPMENT RENTAL** See also division 02240						
	0020	Based on 2 months rental						
	0100	Combination jetting & wellpoint pump, 60 H.P. diesel	Ea.	8.91	267	801	2,400	231.50
	0200	High pressure gas jet pump, 200 H.P., 300 psi	"	15.79	228	684	2,050	263.10
	0300	Discharge pipe, 8" diameter	L.F.	.01	.43	1.29	3.87	.35
	0350	12" diameter		.01	.64	1.92	5.75	.45
	0400	Header pipe, flows up to 150 G.P.M., 4" diameter		.01	.39	1.18	3.54	.30
	0500	400 G.P.M., 6" diameter		.01	.46	1.39	4.17	.35
	0600	800 G.P.M., 8" diameter		.01	.64	1.92	5.75	.45
	0700	1500 G.P.M., 10" diameter		.01	.67	2.02	6.05	.50
	0800	2500 G.P.M., 12" diameter		.02	1.27	3.81	11.45	.90
	0900	4500 G.P.M., 16" diameter		.02	1.63	4.88	14.65	1.15
	0950	For quick coupling aluminum and plastic pipe, add		.03	1.68	5.05	15.15	1.25
	1100	Wellpoint, 25' long, with fittings & riser pipe, 1-1/2" or 2" diameter	Ea.	.05	3.36	10.08	30	2.40
	1200	Wellpoint pump, diesel powered, 4" diameter, 20 H.P.		4.33	154	462	1,375	127.05
	1300	6" diameter, 30 H.P.		5.66	191	573	1,725	159.90
	1400	8" suction, 40 H.P.		7.70	262	785	2,350	218.60
	1500	10" suction, 75 H.P.		10.55	305	918	2,750	268
	1600	12" suction, 100 H.P.		15.79	490	1,470	4,400	420.30
	1700	12" suction, 175 H.P.		20.97	540	1,620	4,850	491.75
800	0010	**MARINE EQUIPMENT RENTAL**						
	0200	Barge, 400 Ton, 30' wide x 90' long	Ea.	16.15	240	720	2,150	273.20
	0240	800 Ton, 45' wide x 90' long		26.65	345	1,030	3,100	419.20
	2000	Tugboat, diesel, 100 HP		16.95	168	505	1,525	236.60

01590 | Equipment Rental

			UNIT	HOURLY OPER. COST	RENT PER DAY	RENT PER WEEK	RENT PER MONTH	CREW EQUIPMENT COST/DAY	
800	2040	250 HP	Ea.	31.45	315	945	2,825	440.60	800
	2080	380 HP	↓	73.10	925	2,780	8,350	1,141	

01740 | Cleaning			CREW	DAILY OUTPUT	LABOR-HOURS	UNIT	2004 BARE COSTS MAT.	LABOR	EQUIP.	TOTAL	TOTAL INCL O&P	
500	0010	**CLEANING UP** After job completion, allow, minimum				Job					.30%	500
	0040	Maximum				"					1%	

Division Notes

		CREW	DAILY OUTPUT	LABOR-HOURS	UNIT	2004 BARE COSTS MAT.	LABOR	EQUIP.	TOTAL	TOTAL INCL O&P

Division 2 Site Construction

Estimating Tips

02200 Site Preparation

- If possible visit the site and take an inventory of the type, quantity and size of the trees. Certain trees may have a landscape resale value or firewood value. Stump disposal can be very expensive, particularly if they cannot be buried at the site. Consider using a bulldozer in lieu of hand cutting trees.
- Estimators should visit the site to determine the need for haul road, access, storage of materials, and security considerations. When estimating for access roads on unstable soil, consider using a geotextile stabilization fabric. It can greatly reduce the quantity of crushed stone or gravel. Sites of limited size and access can cause cost overruns due to lost productivity. Theft and damage is another consideration if the location is isolated. A temporary fence or security guards may be required. Investigate the site thoroughly.

02210 Subsurface Investigation

In preparing estimates on structures involving earthwork or foundations, all information concerning soil characteristics should be obtained. Look particularly for hazardous waste, evidence of prior dumping of debris, and previous stream beds.

02220 Selective Demolition

The costs shown for selective demolition do not include rubbish handling or disposal. These items should be estimated separately using Means data or other sources.

- Historic preservation often requires that the contractor remove materials from the existing structure, rehab them and replace them. The estimator must be aware of any related measures and precautions that must be taken when doing selective demolition, and cutting and patching. Requirements may include special handling and storage, as well as security.
- In addition to Section 02220, you can find selective demolition items in each division. Example: Roofing demolition is in division 7.

02300 Earthwork

- Estimating the actual cost of performing earthwork requires careful consideration of the variables involved. This includes items such as type of soil, whether or not water will be encountered, dewatering, whether or not banks need bracing, disposal of excavated earth, length of haul to fill or spoil sites, etc. If the project has large quantities of cut or fill, consider raising or lowering the site to reduce costs while paying close attention to the effect on site drainage and utilities if doing this.
- If the project has large quantities of fill, creating a borrow pit on the site can significantly lower the costs.
- It is very important to consider what time of year the project is scheduled for completion. Bad weather can create large cost overruns from dewatering, site repair and lost productivity from cold weather.

02500 Utility Services
02600 Drainage & Containment

- Never assume that the water, sewer and drainage lines will go in at the early stages of the project. Consider the site access needs before dividing the site in half with open trenches, loose pipe, and machinery obstructions. Always inspect the site to establish that the site drawings are complete. Check off all existing utilities on your drawing as you locate them. If you find any discrepancies, mark up the site plan for further research. Differing site conditions can be very costly if discovered later in the project.
- See also Section 02955 for restoration of pipe where removal/replacement may be undesirable.

02700 Bases, Ballasts, Pavements/Appurtenances

- When estimating paving, keep in mind the project schedule. If an asphaltic paving project is in a colder climate and runs through to the spring, consider placing the base course in the autumn, then topping it in the spring just prior to completion. This could save considerable costs in spring repair. Keep in mind that prices for asphalt and concrete are generally higher in the cold seasons.
- See also Sections 02960/02965.

02900 Planting

- The timing of planting and guarantee specifications often dictate the costs for establishing tree and shrub growth and a stand of grass or ground cover. Establish the work performance schedule to coincide with the local planting season. Maintenance and growth guarantees can add from 20% to 100% to the total landscaping cost. The cost to replace trees and shrubs can be as high as 5% of the total cost depending on the planting zone, soil conditions and time of year.

02960 & 02965 Flexible Pavement Surfacing Recovery

- Recycling of asphalt pavement is becoming very popular and is an alternative to removal and replacement of asphalt pavement. It can be a good value engineering proposal if removed pavement can be recycled either at the site or another site that is reasonably close to the project site.

Reference Numbers

Reference numbers are shown in bold squares at the beginning of some major classifications. These numbers refer to related items in the Reference Section. The reference information may be an estimating procedure, an alternate pricing method or technical information.

Note: Not all subdivisions listed here necessarily appear in this publication.

02050 | Basic Site Materials & Methods

02055 | Soils

		02055 Soils	CREW	DAILY OUTPUT	LABOR-HOURS	UNIT	2004 BARE COSTS MAT.	LABOR	EQUIP.	TOTAL	TOTAL INCL O&P	
150	0010	**BORROW**										150
	0020	And spread, with 200 H.P. dozer, no compaction										
	0200	Common borrow	B-15	600	.047	C.Y.	5.30	.92	3.03	9.25	10.65	

02060 | Aggregate

		02060 Aggregate	CREW	DAILY OUTPUT	LABOR-HOURS	UNIT	MAT.	LABOR	EQUIP.	TOTAL	TOTAL INCL O&P	
150	0010	**BORROW**										150
	0020	And spread, with 200 H.P. dozer, no compaction										
	0100	Bank run gravel	B-15	600	.047	C.Y.	15.40	.92	3.03	19.35	22	
	0300	Crushed stone, (1.40 tons per CY), 1-1/2"		600	.047		24.50	.92	3.03	28.45	32	
	0320	3/4"		600	.047		27	.92	3.03	30.95	34.50	
	0340	1/2"		600	.047		17.75	.92	3.03	21.70	24.50	
	0360	3/8"		600	.047		17.90	.92	3.03	21.85	24.50	
	0400	Sand, washed, concrete		600	.047		25	.92	3.03	28.95	32.50	
	0500	Dead or bank sand	↓	600	.047	↓	3.93	.92	3.03	7.88	9.20	

02080 | Utility Materials

		02080 Utility Materials	CREW	DAILY OUTPUT	LABOR-HOURS	UNIT	MAT.	LABOR	EQUIP.	TOTAL	TOTAL INCL O&P	
400	0010	**UTILITY BOXES** Precast concrete, 6" thick										400
	0050	5' x 10' x 6' high, I.D.	B-13	2	24	Ea.	1,450	440	310	2,200	2,700	
	0350	Hand hole, precast concrete, 1-1/2" thick										
	0400	1'-0" x 2'-0" x 1'-9", I.D., light duty	B-1	4	6	Ea.	248	105		353	450	
	0450	4'-6" x 3'-2" x 2'-0", O.D., heavy duty	B-6	3	8	"	760	149	69.50	978.50	1,150	

02200 | Site Preparation

02210 | Subsurface Investigation

		02210 Subsurface Investigation	CREW	DAILY OUTPUT	LABOR-HOURS	UNIT	2004 BARE COSTS MAT.	LABOR	EQUIP.	TOTAL	TOTAL INCL O&P	
120	0010	**BORING AND EXPLORATORY DRILLING**										120
	0020	Borings, initial field stake out & determination of elevations	A-6	1	16	Day		365	59.50	424.50	680	
	0100	Drawings showing boring details				Total		185		185	270	
	0200	Report and recommendations from P.E.						415		415	595	
	0300	Mobilization and demobilization, minimum	B-55	4	4	↓		70.50	194	264.50	330	
	0350	For over 100 miles, per added mile		450	.036	Mile		.63	1.72	2.35	2.95	
	0600	Auger holes in earth, no samples, 2-1/2" diameter		78.60	.204	L.F.		3.58	9.85	13.43	16.85	
	0800	Cased borings in earth, with samples, 2-1/2" diameter		55.50	.288	"	14.20	5.05	13.95	33.20	39.50	
	1400	Drill rig and crew with truck mounted auger	↓	1	16	Day		281	775	1,056	1,325	
	1500	For inner city borings add, minimum									10%	
	1510	Maximum									20%	

02220 | Site Demolition

		02220 Site Demolition	CREW	DAILY OUTPUT	LABOR-HOURS	UNIT	MAT.	LABOR	EQUIP.	TOTAL	TOTAL INCL O&P	
110	0010	**BUILDING DEMOLITION** Large urban projects, incl. 20 Mi. haul										110
	0500	Small bldgs, or single bldgs, no salvage included, steel	B-3	14,800	.003	C.F.		.06	.12	.18	.23	
	0600	Concrete	"	11,300	.004	"		.08	.15	.23	.30	
	0605	Concrete, plain	B-5	33	1.212	C.Y.		22.50	28	50.50	68.50	
	0610	Reinforced		25	1.600			29.50	37	66.50	90.50	
	0615	Concrete walls		34	1.176			22	27	49	67	
	0620	Elevated slabs	↓	26	1.538	↓		28.50	35.50	64	87	
	0650	Masonry	B-3	14,800	.003	C.F.		.06	.12	.18	.23	

02220 | Site Demolition

			CREW	DAILY OUTPUT	LABOR-HOURS	UNIT	2004 BARE COSTS				TOTAL INCL O&P
							MAT.	LABOR	EQUIP.	TOTAL	
110	0700	Wood	B-3	14,800	.003	C.F.		.06	.12	.18	.23
	1000	Single family, one story house, wood, minimum				Ea.				2,525	2,975
	1020	Maximum								4,400	5,275
	1200	Two family, two story house, wood, minimum								3,300	3,950
	1220	Maximum								6,375	7,700
	1300	Three family, three story house, wood, minimum								4,400	5,275
	1320	Maximum								7,700	9,250
	1400	Gutting building, see division 02225-400									
130	0010	**BLDG. FOOTINGS AND FOUNDATIONS DEMOLITION**									
	0200	Floors, concrete slab on grade,									
	0240	4″ thick, plain concrete	B-9C	500	.080	S.F.		1.38	.32	1.70	2.70
	0280	Reinforced, wire mesh		470	.085			1.47	.34	1.81	2.87
	0300	Rods		400	.100			1.73	.40	2.13	3.38
	0400	6″ thick, plain concrete		375	.107			1.85	.43	2.28	3.60
	0420	Reinforced, wire mesh		340	.118			2.04	.47	2.51	3.98
	0440	Rods		300	.133			2.31	.53	2.84	4.51
	1000	Footings, concrete, 1′ thick, 2′ wide	B-5	300	.133	L.F.		2.47	3.07	5.54	7.55
	1080	1′-6″ thick, 2′ wide		250	.160			2.97	3.68	6.65	9.05
	1120	3′ wide		200	.200			3.71	4.60	8.31	11.30
	1200	Average reinforcing, add								10%	10%
	2000	Walls, block, 4″ thick	1 Clab	180	.044	S.F.		.75		.75	1.28
	2040	6″ thick		170	.047			.80		.80	1.35
	2080	8″ thick		150	.053			.90		.90	1.53
	2100	12″ thick		150	.053			.90		.90	1.53
	2400	Concrete, plain concrete, 6″ thick	B-9	160	.250			4.33	1	5.33	8.45
	2420	8″ thick		140	.286			4.94	1.14	6.08	9.65
	2440	10″ thick		120	.333			5.75	1.33	7.08	11.25
	2500	12″ thick		100	.400			6.90	1.60	8.50	13.50
	2600	For average reinforcing, add								10%	10%
	4000	For congested sites or small quantities, add up to								200%	200%
	4200	Add for disposal, on site	B-11A	232	.069	C.Y.		1.38	3.72	5.10	6.40
	4250	To five miles	B-30	220	.109	″		2.21	7.45	9.66	11.85
220	0010	**FENCING DEMOLITION**									
	1600	Fencing, barbed wire, 3 strand	2 Clab	430	.037	L.F.		.63		.63	1.07
	1650	5 strand	″	280	.057			.97		.97	1.64
	1700	Chain link, posts & fabric, remove only, 8′ to 10′ high	B-6	445	.054			1.01	.47	1.48	2.20
	1750	Remove and reset	″	70	.343			6.40	2.97	9.37	14
240	0010	**MINOR SITE DEMOLITION**									
	0015	Minor site demolition, no hauling, abandon catch basin or manhole	B-6	7	3.429	Ea.		64	29.50	93.50	140
	0020	Remove existing catch basin or manhole, masonry		4	6			112	52	164	245
	0030	Catch basin or manhole frames and covers, stored		13	1.846			34.50	16	50.50	75.50
	0040	Remove and reset		7	3.429			64	29.50	93.50	140
	1000	Masonry walls, block or tile, solid, remove	B-5	1,800	.022	C.F.		.41	.51	.92	1.26
	1100	Cavity wall		2,200	.018			.34	.42	.76	1.03
	1200	Brick, solid		900	.044			.82	1.02	1.84	2.52
	1300	With block back-up		1,130	.035			.66	.81	1.47	2.01
	1400	Stone, with mortar		900	.044			.82	1.02	1.84	2.52
	1500	Dry set		1,500	.027			.50	.61	1.11	1.51
	2900	Pipe removal, sewer/water, no excavation, 12″ diameter	B-6	175	.137	L.F.		2.56	1.19	3.75	5.60
	2960	24″ diameter		120	.200	″		3.73	1.73	5.46	8.15
	4000	Sidewalk removal, bituminous, 2-1/2″ thick		325	.074	S.Y.		1.38	.64	2.02	3.01
	4050	Brick, set in mortar		185	.130			2.42	1.12	3.54	5.30
	4100	Concrete, plain, 4″		160	.150			2.80	1.30	4.10	6.15
	4200	Mesh reinforced		150	.160			2.99	1.39	4.38	6.50
	5000	Slab on grade removal, plain	B-5	45	.889	C.Y.		16.50	20.50	37	50.50

	02220 | Site Demolition		CREW	DAILY OUTPUT	LABOR-HOURS	UNIT	MAT.	LABOR	EQUIP.	TOTAL	TOTAL INCL O&P	
							2004 BARE COSTS					
240	5100	Mesh reinforced	B-5	33	1.212	C.Y.		22.50	28	50.50	68.50	240
	5200	Rod reinforced	↓	25	1.600			29.50	37	66.50	90.50	
	5500	For congested sites or small quantities, add up to								200%	200%	
	5550	For disposal on site, add	B-11A	232	.069			1.38	3.72	5.10	6.40	
	5600	To 5 miles, add	B-34D	76	.105	↓		1.98	5.50	7.48	9.35	
250	0010	**DEMOLISH, REMOVE PAVEMENT AND CURB**										250
	5010	Pavement removal, bituminous roads, 3" thick	B-38	690	.035	S.Y.		.65	.46	1.11	1.60	
	5050	4" to 6" thick		420	.057			1.07	.76	1.83	2.63	
	5100	Bituminous driveways		640	.037			.70	.50	1.20	1.72	
	5200	Concrete to 6" thick, hydraulic hammer, mesh reinforced		255	.094			1.76	1.25	3.01	4.33	
	5300	Rod reinforced	↓	200	.120	↓		2.24	1.60	3.84	5.50	
	5600	With hand held air equipment, bituminous, to 6" thick	B-39	1,900	.025	S.F.		.44	.08	.52	.83	
	5700	Concrete to 6" thick, no reinforcing		1,600	.030			.52	.10	.62	.99	
	5800	Mesh reinforced		1,400	.034			.59	.11	.70	1.13	
	5900	Rod reinforced	↓	765	.063	↓		1.08	.21	1.29	2.07	
	6000	Curbs, concrete, plain	B-6	360	.067	L.F.		1.24	.58	1.82	2.72	
	6100	Reinforced		275	.087			1.63	.76	2.39	3.56	
	6200	Granite		360	.067			1.24	.58	1.82	2.72	
	6300	Bituminous	↓	528	.045	↓		.85	.39	1.24	1.85	
310	0010	**SELECTIVE DEMOLITION, CUTOUT**										310
	0020	Concrete, elev. slab, light reinforcement, under 6 CF	B-9C	65	.615	C.F.		10.65	2.46	13.11	21	
	0050	Light reinforcing, over 6 C.F.	"	75	.533	"		9.25	2.13	11.38	18	
	0200	Slab on grade to 6" thick, not reinforced, under 8 S.F.	B-9	85	.471	S.F.		8.15	1.88	10.03	15.90	
	0250	Not reinforced, over 8 S.F.		175	.229	"		3.95	.91	4.86	7.70	
	0600	Walls, not reinforced, under 6 C.F.		60	.667	C.F.		11.55	2.67	14.22	22.50	
	0650	Not reinforced, over 6 C.F.	↓	65	.615			10.65	2.46	13.11	21	
	1000	Concrete, elevated slab, bar reinforced, under 6 C.F.	B-9C	45	.889			15.40	3.56	18.96	30	
	1050	Bar reinforced, over 6 C.F.	"	50	.800	↓		13.85	3.20	17.05	27	
	1200	Slab on grade to 6" thick, bar reinforced, under 8 S.F.	B-9	75	.533	S.F.		9.25	2.13	11.38	18	
	1250	Bar reinforced, over 8 S.F.	"	105	.381	"		6.60	1.52	8.12	12.90	
	1400	Walls, bar reinforced, under 6 C.F.	B-9C	50	.800	C.F.		13.85	3.20	17.05	27	
	1450	Bar reinforced, over 6 C.F.	"	55	.727	"		12.60	2.91	15.51	24.50	
	2000	Brick, to 4 S.F. opening, not including toothing										
	2040	4" thick	B-9C	30	1.333	Ea.		23	5.35	28.35	45	
	2060	8" thick		18	2.222			38.50	8.90	47.40	75.50	
	2080	12" thick		10	4			69	16	85	136	
	2400	Concrete block, to 4 S.F. opening, 2" thick		35	1.143			19.75	4.57	24.32	38.50	
	2420	4" thick		30	1.333			23	5.35	28.35	45	
	2440	8" thick		27	1.481			25.50	5.95	31.45	50	
	2460	12" thick	↓	24	1.667			29	6.65	35.65	56.50	
	2600	Gypsum block, to 4 S.F. opening, 2" thick	B-9	80	.500			8.65	2	10.65	16.90	
	2620	4" thick		70	.571			9.90	2.29	12.19	19.30	
	2640	8" thick		55	.727			12.60	2.91	15.51	24.50	
	2800	Terra cotta, to 4 S.F. opening, 4" thick		70	.571			9.90	2.29	12.19	19.30	
	2840	8" thick		65	.615			10.65	2.46	13.11	21	
	2880	12" thick	↓	50	.800	↓		13.85	3.20	17.05	27	
	3000	Toothing masonry cutouts, brick, soft old mortar	1 Brhe	40	.200	V.L.F.		3.58		3.58	5.95	
	3100	Hard mortar		30	.267			4.77		4.77	7.95	
	3200	Block, soft old mortar		70	.114			2.05		2.05	3.40	
	3400	Hard mortar	↓	50	.160	↓		2.86		2.86	4.76	
	6000	Walls, interior, not including re-framing,										
	6010	openings to 5 S.F.										
	6100	Drywall to 5/8" thick	1 Clab	24	.333	Ea.		5.65		5.65	9.55	
	6200	Paneling to 3/4" thick		20	.400			6.75		6.75	11.50	
	6300	Plaster, on gypsum lath	↓	20	.400	↓		6.75		6.75	11.50	

02220 | Site Demolition

			CREW	DAILY OUTPUT	LABOR-HOURS	UNIT	2004 BARE COSTS MAT.	LABOR	EQUIP.	TOTAL	TOTAL INCL O&P
310	6340	On wire lath	1 Clab	14	.571	Ea.		9.65		9.65	16.40
	7000	Wood frame, not including re-framing, openings to 5 S.F.									
	7200	Floors, sheathing and flooring to 2″ thick	1 Clab	5	1.600	Ea.		27		27	46
	7310	Roofs, sheathing to 1″ thick, not including roofing		6	1.333			22.50		22.50	38.50
	7410	Walls, sheathing to 1″ thick, not including siding		7	1.143			19.30		19.30	33
330	0010	**SELECTIVE DEMOLITION, DUMP CHARGES**									
	0020	Dump charges, typical urban city, tipping fees only									
	0100	Building construction materials				Ton					70
	0200	Trees, brush, lumber									50
	0300	Rubbish only									60
	0500	Reclamation station, usual charge									85
340	0010	**SELECTIVE DEMOLITION, GUTTING**									
	0020	Building interior, including disposal, dumpter fees not included									
	0500	Residential building									
	0560	Minimum	B-16	400	.080	SF Flr.		1.43	1.19	2.62	3.73
	0580	Maximum	"	360	.089	"		1.59	1.32	2.91	4.14
	0900	Commercial building									
	1000	Minimum	B-16	350	.091	SF Flr.		1.63	1.36	2.99	4.26
	1020	Maximum	"	250	.128	"		2.29	1.91	4.20	5.95
350	0010	**SELECTIVE DEMOLITION, RUBBISH HANDLING**									
	0020	The following are to be added to the demolition prices									
	0400	Chute, circular, prefabricated steel, 18″ diameter	B-1	40	.600	L.F.	28	10.55		38.55	49
	0440	30″ diameter	"	30	.800	"	37.50	14.05		51.55	65
	0700	10 C.Y. capacity (4 Tons)				Week					375
	0725	Dumpster, weekly rental, 1 dump/week, 20 C.Y. capacity (8 Tons)									440
	0800	30 C.Y. capacity (10 Tons)									665
	0840	40 C.Y. capacity (13 Tons)									805
	1000	Dust partition, 6 mil polyethylene, 1″ x 3″ frame	2 Carp	2,000	.008	S.F.	.16	.18		.34	.48
	1080	2″ x 4″ frame	"	2,000	.008	"	.26	.18		.44	.60
	2000	Load, haul, and dump, 50′ haul	2 Clab	24	.667	C.Y.		11.25		11.25	19.15
	2040	100′ haul		16.50	.970			16.40		16.40	28
	2080	Over 100′ haul, add per 100 L.F.		35.50	.451			7.60		7.60	12.95
	2120	In elevators, per 10 floors, add		140	.114			1.93		1.93	3.28
	3000	Loading & trucking, including 2 mile haul, chute loaded	B-16	45	.711			12.70	10.60	23.30	33
	3040	Hand loading truck, 50′ haul	"	48	.667			11.90	9.95	21.85	31
	3080	Machine loading truck	B-17	120	.267			4.99	4.44	9.43	13.25
	5000	Haul, per mile, up to 8 C.Y. truck	B-34B	1,165	.007			.13	.41	.54	.67
	5100	Over 8 C.Y. truck	"	1,550	.005			.10	.31	.41	.50

02230 | Site Clearing

			CREW	DAILY OUTPUT	LABOR-HOURS	UNIT	MAT.	LABOR	EQUIP.	TOTAL	TOTAL INCL O&P
100	0010	**CLEAR AND GRUB**									
	0020	Cut & chip light trees to 6″ diam.	B-7	1	48	Acre		875	995	1,870	2,575
	0150	Grub stumps and remove	B-30	2	12			243	820	1,063	1,300
	0200	Cut & chip medium, trees to 12″ diam.	B-7	.70	68.571			1,250	1,425	2,675	3,675
	0250	Grub stumps and remove	B-30	1	24			485	1,625	2,110	2,600
	0300	Cut & chip heavy, trees to 24″ diam.	B-7	.30	160			2,925	3,300	6,225	8,575
	0350	Grub stumps and remove	B-30	.50	48			975	3,275	4,250	5,225
	0400	If burning is allowed, reduce cut & chip									40%
200	0010	**SELECTIVE CLEARING**									
	0020	Clearing brush with brush saw	A-1C	.25	32	Acre		540	75	615	1,000
	0100	By hand	1 Clab	.12	66.667			1,125		1,125	1,925
	0300	With dozer, ball and chain, light clearing	B-11A	2	8			160	430	590	745
	0400	Medium clearing	"	1.50	10.667			214	575	789	990

02200 | Site Preparation

	02230 | Site Clearing		CREW	DAILY OUTPUT	LABOR-HOURS	UNIT	2004 BARE COSTS MAT.	LABOR	EQUIP.	TOTAL	TOTAL INCL O&P	
500	0010	**STRIPPING & STOCKPILING OF SOIL**										500
	1400	Loam or topsoil, remove and stockpile on site										
	1420	6″ deep, 200′ haul	B-10B	865	.009	C.Y.		.21	1	1.21	1.45	
	1430	300′ haul		520	.015			.36	1.66	2.02	2.42	
	1440	500′ haul		225	.036			.83	3.84	4.67	5.60	
	1450	Alternate method: 6″ deep, 200′ haul		5,090	.002	S.Y.		.04	.17	.21	.25	
	1460	500′ haul		1,325	.006	″		.14	.65	.79	.95	

02300 | Earthwork

	02305 | Equipment		CREW	DAILY OUTPUT	LABOR-HOURS	UNIT	2004 BARE COSTS MAT.	LABOR	EQUIP.	TOTAL	TOTAL INCL O&P	
250	0010	**MOBILIZATION OR DEMOB.** (One or the other, unless noted)										250
	0015	Up to 25 mi haul dist (50 mi round trip for mob/demob crew)										
	0020	Dozer, loader, backhoe, excav., grader, paver, roller, 70 to 150 H.P.	B-34N	4	2	Ea.		37.50	111	148.50	186	
	0900	Shovel or dragline, 3/4 C.Y.	B-34K	3.60	2.222			42	140	182	224	
	1100	Small equipment, placed in rear of, or towed by pickup truck	A-3A	8	1			18.25	10.25	28.50	42	
	1150	Equip up to 70 HP, on flatbed trailer behind pickup truck	A-3D	4	2			36.50	42.50	79	108	
	2000	Crane, truck-mounted, up to 75 ton (costs incl both mob & demob)	1 EQHV	3.60	2.222			53.50		53.50	88	
	2200	Crawler-mounted, up to 75 ton	A-3F	2	8			171	278	449	590	
	2500	For each additional 5 miles haul distance, add						10%	10%			
	3000	For large pieces of equipment, allow for assembly/knockdown										
	3100	For mob/demob of micro-tunneling equip, see section 02441-400										

	02310 | Grading		CREW	DAILY OUTPUT	LABOR-HOURS	UNIT	MAT.	LABOR	EQUIP.	TOTAL	TOTAL INCL O&P	
100	0010	**FINISH GRADING**										100
	0012	Finish grading area to be paved with grader, small area	B-11L	400	.040	S.Y.		.80	1.08	1.88	2.53	
	0100	Large area		2,000	.008			.16	.22	.38	.51	
	0200	Grade subgrade for base course, roadways		3,500	.005			.09	.12	.21	.29	
	1020	For large parking lots	B-32C	5,000	.010			.20	.30	.50	.66	
	1050	For small irregular areas	″	2,000	.024			.49	.74	1.23	1.63	
	1100	Fine grade for slab on grade, machine	B-11L	1,040	.015			.31	.42	.73	.97	
	1150	Hand grading	B-18	700	.034			.60	.06	.66	1.09	
	1200	Fine grade granular base for sidewalks and bikeways	B-62	1,200	.020			.37	.12	.49	.76	
	2550	Hand grade select gravel	2 Clab	60	.267	C.S.F.		4.51		4.51	7.65	
	3000	Hand grade select gravel, including compaction, 4″ deep	B-18	555	.043	S.Y.		.76	.07	.83	1.37	
	3100	6″ deep		400	.060			1.05	.10	1.15	1.90	
	3120	8″ deep		300	.080			1.41	.14	1.55	2.54	
	3300	Finishing grading slopes, gentle	B-11L	8,900	.002			.04	.05	.09	.11	
	3310	Steep slopes	″	7,100	.002			.05	.06	.11	.15	

	02315 | Excavation and Fill		CREW	DAILY OUTPUT	LABOR-HOURS	UNIT	MAT.	LABOR	EQUIP.	TOTAL	TOTAL INCL O&P	
110	0010	**BACKFILL, GENERAL**										110
	0015	By hand, no compaction, light soil	1 Clab	14	.571	C.Y.		9.65		9.65	16.40	
	0100	Heavy soil		11	.727	L.C.Y.		12.30		12.30	21	
	0300	Compaction in 6″ layers, hand tamp, add to above		20.60	.388	E.C.Y.		6.55		6.55	11.15	
	0500	Air tamp, add	B-9D	190	.211			3.64	.97	4.61	7.25	
	0600	Vibrating plate, add	A-1D	60	.133			2.25	.48	2.73	4.35	

02315 | Excavation and Fill

			CREW	DAILY OUTPUT	LABOR-HOURS	UNIT	MAT.	LABOR	EQUIP.	TOTAL	TOTAL INCL O&P	
110	0800	Compaction in 12" layers, hand tamp, add to above	1 Clab	34	.235	E.C.Y.		3.98		3.98	6.75	110
	1300	Dozer backfilling, bulk, up to 300' haul, no compaction	B-10B	1,200	.007	L.C.Y.		.15	.72	.87	1.04	
	1400	Air tamped	B-11B	240	.067	E.C.Y.		1.30	.80	2.10	3.06	
320	0010	**COMPACTION, STRUCTURAL**										320
	0020	Steel wheel tandem roller, 5 tons	B-10E	8	1	Hr.		23	13.55	36.55	53	
	0050	Air tamp, 6" to 8" lifts, common fill	B-9	250	.160	E.C.Y.		2.77	.64	3.41	5.40	
	0060	Select fill	"	300	.133			2.31	.53	2.84	4.51	
	0600	Vibratory plate, 8" lifts, common fill	A-1D	200	.040			.68	.14	.82	1.31	
	0700	Select fill	"	216	.037	↓		.63	.13	.76	1.21	
424	0010	**EXCAVATING, BULK BANK MEASURE** Common earth piled										424
	0020	For loading onto trucks, add								15%	15%	
	0200	Backhoe, hydraulic, crawler mtd., 1 C.Y. cap. = 75 C.Y./hr.	B-12A	600	.013	B.C.Y.		.32	.81	1.13	1.42	
	0310	Wheel mounted, 1/2 C.Y. cap. = 30 C.Y./hr.	B-12E	240	.033			.80	1.35	2.15	2.80	
	1200	Front end loader, track mtd., 1-1/2 C.Y. cap. = 70 C.Y./hr.	B-10N	560	.014			.33	.54	.87	1.14	
	1500	Wheel mounted, 3/4 C.Y. cap. = 45 C.Y./hr.	B-10R	360	.022	↓		.52	.53	1.05	1.44	
	8000	For hauling excavated material, see div. 02320-200										
462	0010	**EXCAVATION, STRUCTURAL**										462
	0015	Hand, pits to 6' deep, sandy soil	1 Clab	8	1	C.Y.		16.90		16.90	28.50	
	0100	Heavy soil or clay		4	2	B.C.Y.		34		34	57.50	
	1100	Hand loading trucks from stock pile, sandy soil		12	.667			11.25		11.25	19.15	
	1300	Heavy soil or clay	↓	8	1	↓		16.90		16.90	28.50	
	1500	For wet or muck hand excavation, add to above				%				50%	50%	
490	0010	**HAULING**, excavated or borrow, loose cubic yards										490
	0012	no loading included, highway haulers										
	0020	6 C.Y. dump truck, 1/4 mile round trip, 5.0 loads/hr.	B-34A	195	.041	L.C.Y.		.77	1.67	2.44	3.11	
	0200	4 mile round trip, 1.8 loads/hr.	"	70	.114			2.15	4.64	6.79	8.70	
	0310	12 C.Y. dump truck, 1/4 mile round trip 3.7 loads/hr.	B-34B	288	.028			.52	1.66	2.18	2.69	
	0500	4 mile round trip, 1.6 loads/hr.	"	125	.064	↓		1.20	3.81	5.01	6.20	
520	0010	**FILL**, spread dumped material, no compaction										520
	0020	By dozer, no compaction	B-10B	1,000	.008	C.Y.		.19	.86	1.05	1.26	
	0100	By hand	1 Clab	12	.667	L.C.Y.		11.25		11.25	19.15	
	0500	Gravel fill, compacted, under floor slabs, 4" deep	B-37	10,000	.005	S.F.	.15	.09	.01	.25	.33	
	0600	6" deep		8,600	.006		.23	.10	.01	.34	.43	
	0700	9" deep		7,200	.007		.38	.12	.02	.52	.63	
	0800	12" deep		6,000	.008	↓	.53	.15	.02	.70	.84	
	1000	Alternate pricing method, 4" deep		120	.400	E.C.Y.	11.25	7.25	.90	19.40	25.50	
	1100	6" deep		160	.300		11.25	5.45	.68	17.38	22.50	
	1200	9" deep		200	.240		11.25	4.35	.54	16.14	20.50	
	1300	12" deep	↓	220	.218	↓	11.25	3.95	.49	15.69	19.60	
610	0010	**EXCAVATING, TRENCH** or continuous footing, common earth										610
	0050	1' to 4' deep, 3/8 C.Y. tractor loader/backhoe	B-11C	150	.107	B.C.Y.		2.14	1.39	3.53	5.10	
	0060	1/2 C.Y. tractor loader/backhoe	B-11M	200	.080			1.60	1.15	2.75	3.95	
	0090	4' to 6' deep, 1/2 C.Y. tractor loader/backhoe	"	200	.080			1.60	1.15	2.75	3.95	
	0100	5/8 C.Y. hydraulic backhoe	B-12Q	250	.032			.77	1.67	2.44	3.10	
	0300	1/2 C.Y. hydraulic excavator, truck mounted	B-12J	200	.040			.96	4.12	5.08	6.10	
	1400	By hand with pick and shovel 2' to 6' deep, light soil	1 Clab	8	1			16.90		16.90	28.50	
	1500	Heavy soil	"	4	2	↓		34		34	57.50	
620	0010	**EXCAVATING, UTILITY TRENCH** Common earth										620
	0050	Trenching with chain trencher, 12 H.P., operator walking										
	0100	4" wide trench, 12" deep	B-53	800	.010	L.F.		.17	.07	.24	.37	
	1000	Backfill by hand including compaction, add										
	1050	4" wide trench, 12" deep	A-1G	800	.010	L.F.		.17	.04	.21	.34	

(MAT., LABOR, EQUIP. and TOTAL are grouped under "2004 BARE COSTS".)

02300 | Earthwork

		02315 | Excavation and Fill	CREW	DAILY OUTPUT	LABOR-HOURS	UNIT	2004 BARE COSTS MAT.	LABOR	EQUIP.	TOTAL	TOTAL INCL O&P	
640	0010	**UTILITY BEDDING** For pipe and conduit, not incl. compaction										640
	0050	Crushed or screened bank run gravel	B-6	150	.160	L.C.Y.	19.25	2.99	1.39	23.63	27.50	
	0100	Crushed stone 3/4" to 1/2"		150	.160		24.50	2.99	1.39	28.88	33.50	
	0200	Sand, dead or bank	↓	150	.160	↓	3.93	2.99	1.39	8.31	10.85	
	0500	Compacting bedding in trench	A-1D	90	.089	E.C.Y.		1.50	.32	1.82	2.90	

		02360 | Soil Treatment	CREW	DAILY OUTPUT	LABOR-HOURS	UNIT	MAT.	LABOR	EQUIP.	TOTAL	TOTAL INCL O&P	
200	0010	**TERMITE PRETREATMENT**										200
	0020	Slab and walls, residential	1 Skwk	1,200	.007	SF Flr.	.26	.15		.41	.55	
	0400	Insecticides for termite control, minimum		14.20	.563	Gal.	11.10	13		24.10	34.50	
	0500	Maximum	↓	11	.727	"	19	16.75		35.75	49.50	

		02370 | Erosion & Sedimentation Control	CREW	DAILY OUTPUT	LABOR-HOURS	UNIT	MAT.	LABOR	EQUIP.	TOTAL	TOTAL INCL O&P	
700	0010	**SYNTHETIC EROSION CONTROL**										700
	0020	Jute mesh, 100 SY per roll, 4' wide, stapled	B-80A	2,400	.010	S.Y.	.66	.17	.07	.90	1.09	
	0100	Plastic netting, stapled, 2" x 1" mesh, 20 mil	B-1	2,500	.010		.60	.17		.77	.95	
	0200	Polypropylene mesh, stapled, 6.5 oz./S.Y.		2,500	.010		1.24	.17		1.41	1.65	
	0300	Tobacco netting, or jute mesh #2, stapled	↓	2,500	.010	↓	.07	.17		.24	.37	
	1000	Silt fence, polypropylene, 3' high, ideal conditions	2 Clab	1,600	.010	L.F.	.30	.17		.47	.62	
	1100	Adverse conditions	"	950	.017	"	.30	.28		.58	.81	

02400 | Tunneling, Boring & Jacking

		02441 | Microtunneling	CREW	DAILY OUTPUT	LABOR-HOURS	UNIT	2004 BARE COSTS MAT.	LABOR	EQUIP.	TOTAL	TOTAL INCL O&P	
400	0010	**MICROTUNNELING** Not including excavation, backfill, shoring,										400
	0020	or dewatering, average 50'/day, slurry method										
	0100	24" to 48" outside diameter, minimum				L.F.					640	
	0110	Adverse conditions, add				%					50%	
	1000	Rent microtunneling machine, average monthly lease				Month					85,500	
	1010	Operating technician				Day					640	
	1100	Mobilization and demobilization, minimum				Job					42,800	
	1110	Maximum				"					430,000	

02500 | Utility Services

		02510 | Water Distribution	CREW	DAILY OUTPUT	LABOR-HOURS	UNIT	2004 BARE COSTS MAT.	LABOR	EQUIP.	TOTAL	TOTAL INCL O&P	
710	0010	**CROSSES AND SLEEVES**										710
	4700	10" main, 4" branch	B-21	2.70	10.370	Ea.		192	59	251	390	
	4750	6" branch	"	2.35	11.915	"		220	67.50	287.50	445	
730	0010	**WATER SUPPLY, DUCTILE IRON PIPE** cement lined										730
	0020	Not including excavation or backfill										

02510 | Water Distribution

			CREW	DAILY OUTPUT	LABOR-HOURS	UNIT	2004 BARE COSTS MAT.	LABOR	EQUIP.	TOTAL	TOTAL INCL O&P	
730	2000	Pipe, class 50 water piping, 18′ lengths										730
	2020	Mechanical joint, 4″ diameter	B-21A	200	.200	L.F.	9.60	4.27	2.41	16.28	20.50	
	2040	6″ diameter		160	.250		11.35	5.35	3.02	19.72	24.50	
	3000	Tyton, Push-on joint, 4″ diameter		400	.100		7.20	2.14	1.21	10.55	12.85	
	3020	6″ diameter	↓	333.33	.120	↓	8.20	2.56	1.45	12.21	14.90	
	4000	Drill and tap pressurized main (labor only)										
	4100	6″ main, 1″ to 2″ service	Q-1	3	5.333	Ea.		125		125	206	
	4150	8″ main, 1″ to 2″ service	″	2.75	5.818	″		137		137	225	
	4500	Tap and insert gate valve										
	4600	8″ main, 4″ branch	B-21	3.20	8.750	Ea.		162	49.50	211.50	330	
	4650	6″ branch		2.70	10.370			192	59	251	390	
	4651	Piping, drill, tap & insert gate valve, 8″ main, 6″ branch		2.70	10.370			192	59	251	390	
	4800	12″ main, 6″ branch	↓	2.35	11.915	↓		220	67.50	287.50	445	
	8000	Fittings, mechanical joint										
	8006	90° bend, 4″ diameter	B-20A	16	2	Ea.	127	41.50		168.50	208	
	8020	6″ diameter		12.80	2.500		171	52		223	275	
	8200	Wye or tee, 4″ diameter		10.67	2.999		150	62		212	268	
	8220	6″ diameter		8.53	3.751		188	77.50		265.50	335	
	8398	45° bends, 4″ diameter		16	2		132	41.50		173.50	215	
	8400	6″ diameter		12.80	2.500		146	52		198	248	
	8450	Decreaser, 6″ x 4″ diameter		14.22	2.250		102	46.50		148.50	191	
	8460	8″ x 6″ diameter	↓	11.64	2.749	↓	209	57		266	325	
	8550	Butterfly valves with boxes, cast iron										
	8560	4″ diameter	B-20	6	4	Ea.	360	70.50		430.50	515	
	9600	Steel sleeve and tap, 4″ diameter		3	8		400	141		541	675	
	9620	6″ diameter	↓	2	12	↓	470	211		681	875	
750	0010	**WATER SUPPLY, POLYVINYL CHLORIDE PIPE**										750
	2100	Class 160, S.D.R. 26, 1-1/2″ diameter	B-20	750	.032	L.F.	.44	.56		1	1.43	
	2120	2″ diameter		686	.035		1.07	.61		1.68	2.22	
	2140	2-1/2″ diameter		500	.048		1.59	.84		2.43	3.18	
	2160	3″ diameter		430	.056		2.27	.98		3.25	4.16	
	2180	4″ diameter	↓	375	.064	↓	3.70	1.12		4.82	6	

02520 | Wells

			CREW	DAILY OUTPUT	LABOR-HOURS	UNIT	MAT.	LABOR	EQUIP.	TOTAL	TOTAL INCL O&P	
510	0010	**WELLS & ACCESSORIES**, domestic										510
	0100	Drilled, 4″ to 6″ diameter	B-23	120	.333	L.F.		5.75	25.50	31.25	38	
	1500	Pumps, installed in wells to 100′ deep, 4″ submersible										
	1520	3/4 H.P.	Q-1	2.66	6.015	Ea.	375	142		517	645	
	1600	1 H.P.	″	2.29	6.987	″	395	164		559	705	

02530 | Sanitary Sewerage

			CREW	DAILY OUTPUT	LABOR-HOURS	UNIT	MAT.	LABOR	EQUIP.	TOTAL	TOTAL INCL O&P	
730	0010	**SEWAGE COLLECTION, CONCRETE PIPE**										730
	0020	Not including excavation or backfill										
	1020	8″ diameter	B-14	224	.214	L.F.	4.46	3.88	.93	9.27	12.45	
	1030	10″ diameter	″	216	.222	″	4.94	4.03	.96	9.93	13.30	
	3780	Concrete slotted pipe, class 4 mortar joint										
	3800	12″ diameter	B-21	168	.167	L.F.	12.90	3.08	.95	16.93	20.50	
	3840	18″ diameter	″	152	.184	″	19.95	3.41	1.05	24.41	29	
	3900	Class 4 O-ring										
	3940	12″ diameter	B-21	168	.167	L.F.	13.50	3.08	.95	17.53	21	
	3960	18″ diameter	″	152	.184	″	18.10	3.41	1.05	22.56	27	
780	0010	**SEWAGE COLLECTION, POLYVINYL CHLORIDE PIPE**										780
	0020	Not including excavation or backfill										
	2000	10′ lengths, S.D.R. 35, B&S, 4″ diameter	B-20	375	.064	L.F.	1.75	1.12		2.87	3.84	
	2040	6″ diameter	↓	350	.069	↓	3.13	1.20		4.33	5.50	

02500 | Utility Services

02530 | Sanitary Sewerage

		02530 Sanitary Sewerage	CREW	DAILY OUTPUT	LABOR-HOURS	UNIT	2004 BARE COSTS MAT.	LABOR	EQUIP.	TOTAL	TOTAL INCL O&P	
780	2080	8″ diameter	B-20	335	.072	L.F.	5.25	1.26		6.51	7.95	780
	2120	10″ diameter	B-21	330	.085		7.95	1.57	.48	10	11.95	
	4000	Piping, DWV PVC,no exc/bkfill, 10′ L, Sch 40, 4″ dia	B-20	375	.064		1.71	1.12		2.83	3.80	
	4010	6″ dia		350	.069		3.65	1.20		4.85	6.05	
	4020	8″ dia		335	.072		11.15	1.26		12.41	14.40	

02540 | Septic Tank Systems

		02540 Septic Tank Systems	CREW	DAILY OUTPUT	LABOR-HOURS	UNIT	MAT.	LABOR	EQUIP.	TOTAL	TOTAL INCL O&P	
400	0010	**SEPTIC TANKS**										400
	0015	Septic tanks, not incl exc or piping, precast, 1,000 gal	B-21	8	3.500	Ea.	530	64.50	19.90	614.40	715	
	0100	2,000 gallon		5	5.600		1,050	104	32	1,186	1,350	
	0600	High density polyethylene, 1,000 gallon		6	4.667		850	86.50	26.50	963	1,100	
	0700	1,500 gallon		4	7		1,100	129	40	1,269	1,450	
	1000	Distribution boxes, concrete, 7 outlets	2 Clab	16	1		99	16.90		115.90	138	
	1100	9 outlets	″	8	2		268	34		302	355	
	1150	Leaching field chambers, 13′ x 3′-7″ x 1′-4″, standard	B-13	16	3		535	55.50	39	629.50	725	
	1420	Leaching pit, 6′, dia, 3′ deep complete					590			590	645	
	2200	Excavation for septic tank, 3/4 C.Y. backhoe	B-12F	145	.055	C.Y.		1.32	3.16	4.48	5.65	
	2400	4′ trench for disposal field, 3/4 C.Y. backhoe	″	335	.024	L.F.		.57	1.37	1.94	2.45	
	2600	Gravel fill, run of bank	B-6	150	.160	C.Y.	15.40	2.99	1.39	19.78	23.50	
	2800	Crushed stone, 3/4″	″	150	.160	″	22.50	2.99	1.39	26.88	31	

02550 | Piped Energy Distribution

		02550 Piped Energy Distribution	CREW	DAILY OUTPUT	LABOR-HOURS	UNIT	MAT.	LABOR	EQUIP.	TOTAL	TOTAL INCL O&P	
464	0010	**PIPING, GAS SERVICE & DISTRIBUTION, POLYETHYLENE**										464
	0020	not including excavation or backfill										
	1000	60 psi coils, comp cplg @ 100′, 1/2″ diameter, SDR 9.3	B-20A	608	.053	L.F.	.47	1.09		1.56	2.34	
	1040	1-1/4″ diameter, SDR 11		544	.059		.86	1.22		2.08	2.97	
	1100	2″ diameter, SDR 11		488	.066		1.07	1.36		2.43	3.43	
	1160	3″ diameter, SDR 11		408	.078		2.24	1.62		3.86	5.15	
	1500	60 PSI 40′ joints with coupling, 3″ diameter, SDR 11	B-21A	408	.098		2.24	2.10	1.18	5.52	7.25	
	1540	4″ diameter, SDR 11		352	.114		5.10	2.43	1.37	8.90	11.20	
	1600	6″ diameter, SDR 11		328	.122		15.95	2.61	1.47	20.03	23.50	
	1640	8″ diameter, SDR 11		272	.147		21.50	3.14	1.77	26.41	31	
466	0010	**PIPING, GAS SERVICE & DISTRIBUTION, STEEL**										466
	0020	not including excavation or backfill, tar coated and wrapped										
	4000	Schedule 40, plain end										
	4040	1″ diameter	Q-4	300	.107	L.F.	2.62	2.65	.26	5.53	7.50	
	4080	2″ diameter	″	280	.114	″	4.11	2.84	.28	7.23	9.50	

02600 | Drainage & Containment

02620 | Subdrainage

		02620 Subdrainage	CREW	DAILY OUTPUT	LABOR-HOURS	UNIT	2004 BARE COSTS MAT.	LABOR	EQUIP.	TOTAL	TOTAL INCL O&P	
280	0010	**PIPING, SUBDRAINAGE, VITRIFIED CLAY**										280
	4000	Channel pipe, 4″ diameter	B-20	430	.056	L.F.	1.89	.98		2.87	3.74	
	4060	8″ diameter	″	295	.081	″	5.05	1.43		6.48	8.05	
610	0010	**PIPING, SUBDRAINAGE, CONCRETE**										610
	0021	Not including excavation and backfill										

02600 | Drainage & Containment

02620 | Subdrainage

		02620 Subdrainage	CREW	DAILY OUTPUT	LABOR-HOURS	UNIT	2004 BARE COSTS MAT.	LABOR	EQUIP.	TOTAL	TOTAL INCL O&P	
610	3000	Porous wall concrete underdrain, std. strength, 4" diameter	B-20	335	.072	L.F.	1.86	1.26		3.12	4.19	610
	3020	6" diameter	"	315	.076		2.42	1.34		3.76	4.93	
	3040	8" diameter	B-21	310	.090	↓	2.99	1.67	.51	5.17	6.65	
620	0010	**PIPING, SUBDRAINAGE, CORRUGATED METAL**										620
	0021	Not including excavation and backfill										
	2010	Aluminum, perforated										
	2020	6" diameter, 18 ga.	B-14	380	.126	L.F.	2.67	2.29	.55	5.51	7.40	
	2200	8" diameter, 16 ga.		370	.130		3.89	2.35	.56	6.80	8.85	
	2220	10" diameter, 16 ga.	↓	360	.133	↓	4.86	2.42	.58	7.86	10.05	
	3000	Uncoated galvanized, perforated										
	3020	6" diameter, 18 ga.	B-20	380	.063	L.F.	4.28	1.11		5.39	6.60	
	3200	8" diameter, 16 ga.	"	370	.065		5.90	1.14		7.04	8.45	
	3220	10" diameter, 16 ga.	B-21	360	.078		8.85	1.44	.44	10.73	12.60	
	3240	12" diameter, 16 ga.	"	285	.098	↓	9.25	1.82	.56	11.63	13.85	
	4000	Steel, perforated, asphalt coated										
	4020	6" diameter 18 ga.	B-20	380	.063	L.F.	3.42	1.11		4.53	5.65	
	4030	8" diameter 18 ga	"	370	.065		5.35	1.14		6.49	7.85	
	4040	10" diameter 16 ga	B-21	360	.078		6.15	1.44	.44	8.03	9.70	
	4050	12" diameter 16 ga		285	.098		7.05	1.82	.56	9.43	11.50	
	4060	18" diameter 16 ga	↓	205	.137	↓	9.65	2.53	.78	12.96	15.70	

02630 | Storm Drainage

		02630 Storm Drainage	CREW	DAILY OUTPUT	LABOR-HOURS	UNIT	MAT.	LABOR	EQUIP.	TOTAL	TOTAL INCL O&P	
110	0010	**CATCH BASIN GRATES AND FRAMES** not including footing, excavation										110
	1600	Frames & covers, C.I., 24" square, 500 lb.	B-6	7.80	3.077	Ea.	231	57.50	26.50	315	380	
400	0010	**STORM DRAINAGE MANHOLES, FRAMES & COVERS** not including										400
	0020	footing, excavation, backfill (See line items for frame & cover)										
	0050	Brick, 4' inside diameter, 4' deep	D-1	1	16	Ea.	290	335		625	875	
	1110	Precast, 4' I.D., 4' deep	B-22	4.10	7.317	"	450	138	58	646	790	

02700 | Bases, Ballasts, Pavements & Appurtenances

02710 | Bound Base Courses

		02710 Bound Base Courses	CREW	DAILY OUTPUT	LABOR-HOURS	UNIT	2004 BARE COSTS MAT.	LABOR	EQUIP.	TOTAL	TOTAL INCL O&P	
200	0010	**ASPHALT-TREATED PERMEABLE BASE COURSE** for roadways										200
	0020	and large paved areas										
	0700	Liquid application to gravel base, asphalt emulsion	B-45	6,000	.003	Gal.	2.78	.05	.04	2.87	3.18	
	0800	Prime and seal, cut back asphalt		6,000	.003	"	3.28	.05	.04	3.37	3.73	
	1000	Macadam penetration crushed stone, 2 gal. per S.Y., 4" thick		6,000	.003	S.Y.	5.55	.05	.04	5.64	6.20	
	1100	6" thick, 3 gal. per S.Y.		4,000	.004		8.35	.07	.06	8.48	9.35	
	1200	8" thick, 4 gal. per S.Y.	↓	3,000	.005	↓	11.10	.10	.08	11.28	12.50	
	8900	For small and irregular areas, add						50%	50%			

02720 | Unbound Base Courses & Ballasts

		02720 Unbound Base Courses & Ballasts	CREW	DAILY OUTPUT	LABOR-HOURS	UNIT	MAT.	LABOR	EQUIP.	TOTAL	TOTAL INCL O&P	
200	0010	**AGGREGATE BASE COURSE** For roadways and large paved areas										200
	0051	3/4" stone compacted to 3" deep	B-36	36,000	.001	S.F.	.29	.02	.03	.34	.39	
	0101	6" deep		35,100	.001		.58	.02	.03	.63	.71	
	0201	9" deep		25,875	.002		.84	.03	.04	.91	1.03	
	0305	12" deep		21,150	.002		1.32	.04	.05	1.41	1.58	
	0306	Crushed 1-1/2" stone base, compacted to 4" deep	↓	47,000	.001	↓	.05	.02	.02	.09	.11	

	02720 | Unbound Base Courses & Ballasts		Crew	Daily Output	Labor-Hours	Unit	2004 Bare Costs Mat.	Labor	Equip.	Total	Total Incl O&P	
200	0307	6″ deep	B-36	35,100	.001	S.F.	.66	.02	.03	.71	.80	200
	0308	8″ deep		27,000	.001		.88	.03	.04	.95	1.06	
	0309	12″ deep		16,200	.002		1.32	.05	.07	1.44	1.61	
	0350	Bank run gravel, spread and compacted										
	0371	6″ deep	B-32	54,000	.001	S.F.	.37	.01	.03	.41	.45	
	0391	9″ deep		39,600	.001		.53	.02	.04	.59	.66	
	0401	12″ deep		32,400	.001		.73	.02	.05	.80	.90	
	6900	For small and irregular areas, add						50%	50%			

	02740 | Flexible Pavement		Crew	Daily Output	Labor-Hours	Unit	Mat.	Labor	Equip.	Total	Total Incl O&P	
315	0010	**PAVING** Asphaltic concrete R02065-300										315
	0020	6″ stone base, 2″ binder course, 1″ topping	B-25C	9,000	.005	S.F.	1.30	.10	.18	1.58	1.80	
	0300	Binder course, 1-1/2″ thick		35,000	.001		.29	.03	.05	.37	.41	
	0400	2″ thick		25,000	.002		.37	.04	.07	.48	.54	
	0500	3″ thick		15,000	.003		.57	.06	.11	.74	.85	
	0600	4″ thick		10,800	.004		.75	.09	.15	.99	1.14	
	0800	Sand finish course, 3/4″ thick		41,000	.001		.18	.02	.04	.24	.28	
	0900	1″ thick		34,000	.001		.22	.03	.05	.30	.34	
	1000	Fill pot holes, hot mix, 2″ thick	B-16	4,200	.008		.40	.14	.11	.65	.79	
	1100	4″ thick		3,500	.009		.58	.16	.14	.88	1.07	
	1120	6″ thick		3,100	.010		.78	.18	.15	1.11	1.34	
	1140	Cold patch, 2″ thick	B-51	3,000	.016		.48	.28	.04	.80	1.03	
	1160	4″ thick		2,700	.018		.90	.31	.05	1.26	1.58	
	1180	6″ thick		1,900	.025		1.41	.44	.06	1.91	2.37	

	02750 | Rigid Pavement		Crew	Daily Output	Labor-Hours	Unit	Mat.	Labor	Equip.	Total	Total Incl O&P	
300	0010	**PLAIN CEMENT CONCRETE PAVEMENT**										300
	0015	Including joints, finishing and curing										
	0021	Fixed form, 12′ pass, unreinforced, 6″ thick	B-26	18,000	.005	S.F.	2.26	.10	.11	2.47	2.77	
	0101	8″ thick	″	13,500	.007		3.04	.13	.15	3.32	3.72	
	0701	Finishing, broom finish small areas	2 Cefi	1,215	.013			.29		.29	.47	

	02770 | Curbs and Gutters		Crew	Daily Output	Labor-Hours	Unit	Mat.	Labor	Equip.	Total	Total Incl O&P	
100	0010	**BITUMINOUS CONCRETE CURBS**										100
	0012	Curbs, asphaltic, machine formed, 8″ wide, 6″ high, 40 L.F./ton	B-27	1,000	.032	L.F.	.59	.56	.19	1.34	1.81	
	0100	8″ wide, 8″ high, 30 L.F. per ton		900	.036		.68	.62	.21	1.51	2.03	
	0150	Asphaltic berm, 12″ W, 3″-6″ H, 35 L.F./ton, before pavement		700	.046		.91	.80	.27	1.98	2.65	
	0200	12″ W, 1-1/2″ to 4″ H, 60 L.F. per ton, laid with pavement	B-2	1,050	.038		.55	.66		1.21	1.73	
300	0010	**CEMENT CONCRETE CURBS**										300
	0300	Concrete, wood forms, 6″ x 18″, straight	C-2A	500	.096	L.F.	2.10	2.13		4.23	5.90	
	0400	6″ x 18″, radius	″	200	.240		2.20	5.35		7.55	11.40	
	0550	Precast, 6″ x 18″, straight	B-29	700	.069		7	1.26	1.18	9.44	11.10	
	0600	6″ x 18″, radius	″	325	.148		8	2.72	2.53	13.25	16.20	
500	0010	**STONE CURBS**										500
	1000	Granite, split face, straight, 5″ x 16″	D-13	500	.096	L.F.	9.35	2.09	.96	12.40	14.80	
	1100	6″ x 18″	″	450	.107		12.25	2.33	1.07	15.65	18.55	
	1300	Radius curbing, 6″ x 18″, over 10′ radius	B-29	260	.185		15	3.40	3.17	21.57	25.50	
	1400	Corners, 2′ radius		80	.600	Ea.	50.50	11.05	10.30	71.85	85.50	
	1600	Edging, 4-1/2″ x 12″, straight		300	.160	L.F.	4.67	2.95	2.74	10.36	13.15	
	1800	Curb inlets, (guttermouth) straight		41	1.171	Ea.	112	21.50	20	153.50	182	
	2000	Indian granite (belgian block)										
	2100	Jumbo, 10-1/2″ x 7-1/2″ x 4″, grey	D-1	150	.107	L.F.	1.54	2.23		3.77	5.40	
	2150	Pink		150	.107		2.06	2.23		4.29	6	

02700 | Bases, Ballasts, Pavements & Appurtenances

02770 | Curbs and Gutters

			CREW	DAILY OUTPUT	LABOR-HOURS	UNIT	2004 BARE COSTS MAT.	LABOR	EQUIP.	TOTAL	TOTAL INCL O&P
500	2200	Regular, 9" x 4-1/2" x 4-1/2", grey	D-1	160	.100	L.F.	1.42	2.10		3.52	5.05
	2250	Pink		160	.100		2	2.10		4.10	5.70
	2300	Cubes, 4" x 4" x 4", grey		175	.091		1.40	1.92		3.32	4.72
	2350	Pink		175	.091		1.48	1.92		3.40	4.81
	2400	6" x 6" x 6", pink		155	.103		3.60	2.16		5.76	7.55
	2500	Alternate pricing method for indian granite									
	2550	Jumbo, 10-1/2" x 7-1/2" x 4" (30lb), grey				Ton	88			88	97
	2600	Pink					120			120	132
	2650	Regular, 9" x 4-1/2" x 4-1/2" (20lb), grey					100			100	110
	2700	Pink					140			140	154
	2750	Cubes, 4" x 4" x 4" (5lb), grey					170			170	187
	2800	Pink					190			190	209
	2850	6" x 6" x 6" (25lb), pink					140			140	154
	2900	For pallets, add					16.50			16.50	18.15

02775 | Sidewalks

			CREW	DAILY OUTPUT	LABOR-HOURS	UNIT	MAT.	LABOR	EQUIP.	TOTAL	TOTAL INCL O&P
275	0010	**SIDEWALKS, DRIVEWAYS, & PATIOS** No base									
	0021	Asphaltic concrete, 2" thick	B-37	6,480	.007	S.F.	.37	.13	.02	.52	.66
	0101	2-1/2" thick	"	5,950	.008	"	.47	.15	.02	.64	.79
	0300	Concrete, 3000 psi, CIP, 6 x 6 - W1.4 x W1.4 mesh,									
	0310	broomed finish, no base, 4" thick	B-24	600	.040	S.F.	1.21	.83		2.04	2.71
	0350	5" thick		545	.044		1.61	.91		2.52	3.29
	0400	6" thick		510	.047		1.88	.97		2.85	3.69
	0450	For bank run gravel base, 4" thick, add	B-18	2,500	.010		.39	.17	.02	.58	.74
	0520	8" thick, add	"	1,600	.015		.80	.26	.03	1.09	1.36
	1000	Crushed stone, 1" thick, white marble	2 Clab	1,700	.009		.19	.16		.35	.48
	1050	Bluestone	"	1,700	.009		.20	.16		.36	.49
	1700	Redwood, prefabricated, 4' x 4' sections	2 Carp	316	.051		8.10	1.17		9.27	10.90
	1750	Redwood planks, 1" thick, on sleepers	"	240	.067		5.65	1.54		7.19	8.85
	2250	Stone dust, 4" thick	B-62	900	.027	S.Y.	2.74	.50	.16	3.40	4.04

02780 | Unit Pavers

			CREW	DAILY OUTPUT	LABOR-HOURS	UNIT	MAT.	LABOR	EQUIP.	TOTAL	TOTAL INCL O&P
100	0010	**ASPHALT BLOCKS**									
	0020	Asphalt blocks, 6" x 12" x 1-1/4", W/bed & neopr. adhesive	D-1	135	.119	S.F.	4.02	2.48		6.50	8.55
	0100	3" thick		130	.123		5.65	2.58		8.23	10.50
	0300	Hexagonal tile, 8" wide, 1-1/4" thick		135	.119		4.02	2.48		6.50	8.55
	0400	2" thick		130	.123		5.65	2.58		8.23	10.50
	0500	Square, 8" x 8", 1-1/4" thick		135	.119		4.02	2.48		6.50	8.55
	0600	2" thick		130	.123		5.65	2.58		8.23	10.50
200	0010	**BRICK PAVING** 4" x 8" x 1-1/2", without joints (4.5 brick/S.F.)	D-1	110	.145	S.F.	2.20	3.05		5.25	7.50
	0100	Grouted, 3/8" joint (3.9 brick/S.F.)		90	.178		2.55	3.72		6.27	9
	0200	4" x 8" x 2-1/4", without joints (4.5 bricks/S.F.)		110	.145		2.82	3.05		5.87	8.15
	0300	Grouted, 3/8" joint (3.9 brick/S.F.)		90	.178		2.60	3.72		6.32	9.05
	0500	Bedding, asphalt, 3/4" thick	B-25	5,130	.017		.32	.32	.37	1.01	1.29
	0540	Course washed sand bed, 1" thick	B-18	5,000	.005		.16	.08	.01	.25	.33
	0580	Mortar, 1" thick	D-1	300	.053		.29	1.12		1.41	2.18
	0620	2" thick		200	.080		.58	1.68		2.26	3.43
	1500	Brick on 1" thick sand bed laid flat, 4.5 per S.F.		100	.160		2.38	3.35		5.73	8.15
	2000	Brick pavers, laid on edge, 7.2 per S.F.		70	.229		2.16	4.79		6.95	10.30
600	0010	**STONE PAVERS**									
	1100	Flagging, bluestone, irregular, 1" thick,	D-1	81	.198	S.F.	4.63	4.14		8.77	12
	1150	Snapped random rectangular, 1" thick		92	.174		7	3.64		10.64	13.75
	1200	1-1/2" thick		85	.188		8.45	3.94		12.39	15.80
	1250	2" thick		83	.193		9.80	4.04		13.84	17.50
	1300	Slate, natural cleft, irregular, 3/4" thick		92	.174		5.15	3.64		8.79	11.75

02700 | Bases, Ballasts, Pavements & Appurtenances

02780 | Unit Pavers

		02780 Unit Pavers	CREW	DAILY OUTPUT	LABOR-HOURS	UNIT	2004 BARE COSTS MAT.	LABOR	EQUIP.	TOTAL	TOTAL INCL O&P	
600	1310	1" thick	D-1	85	.188	S.F.	6	3.94		9.94	13.15	600
	1351	Random rectangular, gauged, 1/2" thick		105	.152		11.15	3.19		14.34	17.60	
	1400	Random rectangular, butt joint, gauged, 1/4" thick		150	.107		12	2.23		14.23	16.95	
	1450	For sand rubbed finish, add					5.60			5.60	6.15	
	1500	For interior setting, add								25%	25%	
	1550	Granite blocks, 3-1/2" x 3-1/2" x 3-1/2"	D-1	92	.174	S.F.	6.25	3.64		9.89	12.95	
700	0010	**STEPS** Incl. excav., borrow & concrete base, where applicable										700
	0100	Brick steps	B-24	35	.686	LF Riser	8.35	14.20		22.55	32.50	
	0200	Railroad ties	2 Clab	25	.640		2.83	10.80		13.63	21.50	
	0300	Bluestone treads, 12" x 2" or 12" x 1-1/2"	B-24	30	.800		21	16.55		37.55	51	
	0600	Precast concrete, see division 03480-800										
	4025	Steel edge strips, incl. stakes, 1/4" x 5"	B-1	390	.062	L.F.	2.91	1.08		3.99	5.05	
	4050	Edging, landscape timber or railroad ties, 6" x 8"	2 Carp	170	.094	"	2.44	2.17		4.61	6.35	

02785 | Flexible Pavement Coating

		02785 Flexible Pavement Coating	CREW	DAILY OUTPUT	LABOR-HOURS	UNIT	2004 BARE COSTS MAT.	LABOR	EQUIP.	TOTAL	TOTAL INCL O&P	
250	0010	**FOG SEAL**										250
	0012	Sealcoating, 2 coat coal tar pitch emulsion over 10,000 SY	B-45	5,000	.003	S.Y.	.44	.06	.05	.55	.63	
	0030	1000 to 10,000 S.Y.	"	3,000	.005		.44	.10	.08	.62	.73	
	0100	Under 1000 S.Y.	B-1	1,050	.023		.44	.40		.84	1.16	
	0300	Petroleum resistant, over 10,000 S.Y.	B-45	5,000	.003		.54	.06	.05	.65	.74	
	0320	1000 to 10,000 S.Y.	"	3,000	.005		.54	.10	.08	.72	.84	
	0400	Under 1000 S.Y.	B-1	1,050	.023		.54	.40		.94	1.27	

02800 | Site Improvements and Amenities

02810 | Irrigation System

		02810 Irrigation System	CREW	DAILY OUTPUT	LABOR-HOURS	UNIT	2004 BARE COSTS MAT.	LABOR	EQUIP.	TOTAL	TOTAL INCL O&P	
300	0010	**SPRINKLER IRRIGATION SYSTEM** For lawns										300
	0800	Residential system, custom, 1" supply	B-20	2,000	.012	S.F.	.26	.21		.47	.65	
	0900	1-1/2" supply	"	1,800	.013	"	.29	.23		.52	.72	

02820 | Fences & Gates

		02820 Fences & Gates	CREW	DAILY OUTPUT	LABOR-HOURS	UNIT	2004 BARE COSTS MAT.	LABOR	EQUIP.	TOTAL	TOTAL INCL O&P	
120	0010	**CHAIN LINK FENCE** 11 ga. wire										120
	0020	1-5/8" post 10'O.C.,1-3/8" top rail,2" corner post galv. stl., 3' high	B-1	185	.130	L.F.	4.43	2.28		6.71	8.75	
	0050	4' high		170	.141		6.60	2.48		9.08	11.45	
	0100	6' high		115	.209		7.45	3.67		11.12	14.45	
	0150	Add for gate 3' wide, 1-3/8" frame 3' high		12	2	Ea.	39.50	35		74.50	103	
	0170	4' high		10	2.400		49	42		91	126	
	0190	6' high		10	2.400		88.50	42		130.50	169	
	0200	Add for gate 4' wide, 1-3/8" frame 3' high		9	2.667		46.50	47		93.50	131	
	0220	4' high		9	2.667		61	47		108	147	
	0240	6' high		8	3		112	52.50		164.50	214	
	0350	Aluminized steel, 9 ga. wire, 3' high		185	.130	L.F.	5.35	2.28		7.63	9.70	
	0380	4' high		170	.141		6.10	2.48		8.58	10.90	
	0400	6' high		115	.209		7.80	3.67		11.47	14.85	
	0450	Add for gate 3' wide, 1-3/8" frame 3' high		12	2	Ea.	52	35		87	117	
	0470	4' high		10	2.400		71	42		113	150	
	0490	6' high		10	2.400		106	42		148	189	

02820 | Fences & Gates

			CREW	DAILY OUTPUT	LABOR-HOURS	UNIT	2004 BARE COSTS MAT.	LABOR	EQUIP.	TOTAL	TOTAL INCL O&P
120	0500	Add for gate 4′ wide, 1-3/8″ frame 3′ high	B-1	10	2.400	Ea.	71	42		113	150
	0520	4′ high		9	2.667		94.50	47		141.50	184
	0540	6′ high		8	3		148	52.50		200.50	253
	0620	Vinyl covered 9 ga. wire, 3′ high		185	.130	L.F.	4.72	2.28		7	9.05
	0640	4′ high		170	.141		7.75	2.48		10.23	12.75
	0660	6′ high		115	.209		8.85	3.67		12.52	16
	0720	Add for gate 3′ wide, 1-3/8″ frame 3′ high		12	2	Ea.	59	35		94	125
	0740	4′ high		10	2.400		77	42		119	156
	0760	6′ high		10	2.400		118	42		160	202
	0780	Add for gate 4′ wide, 1-3/8″ frame 3′ high		10	2.400		80.50	42		122.50	160
	0800	4′ high		9	2.667		106	47		153	197
	0820	6′ high		8	3		154	52.50		206.50	259
	0860	Tennis courts, 11 ga. wire, 2 1/2″ post 10′ O.C., 1-5/8″ top rail									
	0900	2-1/2″ corner post, 10′ high	B-1	95	.253	L.F.	11.80	4.44		16.24	20.50
	0920	12′ high		80	.300	″	14.15	5.25		19.40	24.50
	1000	Add for gate 3′ wide, 1-5/8″ frame 10′ high		10	2.400	Ea.	148	42		190	235
	1040	Aluminized, 11 ga. wire 10′ high		95	.253	L.F.	16.55	4.44		20.99	26
	1100	12′ high		80	.300	″	17.15	5.25		22.40	28
	1140	Add for gate 3′ wide, 1-5/8″ frame, 10′ high		10	2.400	Ea.	189	42		231	280
	1250	Vinyl covered 11 ga. wire, 10′ high		95	.253	L.F.	14.15	4.44		18.59	23
	1300	12′ high		80	.300	″	16.55	5.25		21.80	27
	1400	Add for gate 3′ wide, 1-3/8″ frame, 10′ high		10	2.400	Ea.	213	42		255	305
300	0010	**FENCE, VINYL,** white, steel reinforced, stainless steel fasteners									
	0020	Picket, 4″ x 4″ posts @ 6′ - 0″ OC, 3′ high	B-1	140	.171	L.F.	15.45	3.01		18.46	22
	0030	4′ high		130	.185		17.75	3.24		20.99	25
	0040	5′ high		120	.200		20.50	3.51		24.01	28.50
	0100	Board (semi-privacy), 5″ x 5″ posts @ 7′ - 6″ OC, 5′ high		130	.185		21	3.24		24.24	29
	0120	6′ high		125	.192		24	3.37		27.37	32.50
	0200	Basketweave, 5″ x 5″ posts @ 7′ - 6″ OC, 5′ high		160	.150		18.75	2.64		21.39	25
	0220	6′ high		150	.160		22	2.81		24.81	29.50
	0300	Privacy, 5″ x 5″ posts @ 7′ - 6″ OC, 5′ high		130	.185		23	3.24		26.24	30.50
	0320	6′ high		150	.160		26	2.81		28.81	33.50
	0350	Gate, 5′ high		9	2.667	Ea.	320	47		367	435
	0360	6′ high		9	2.667		330	47		377	445
	0400	For posts set in concrete, add		25	.960		6.80	16.85		23.65	36
410	0010	**FENCES, MISC. METAL**									
	0012	Chicken wire, posts @ 4′, 1″ mesh, 4′ high	B-80	410	.059	L.F.	1.25	1.03	1.26	3.54	4.52
	0100	2″ mesh, 6′ high		350	.069		1.13	1.20	1.48	3.81	4.92
	0200	Galv. steel, 12 ga., 2″ x 4″ mesh, posts 5′ O.C., 3′ high		300	.080		1.70	1.41	1.73	4.84	6.15
	0300	5′ high		300	.080		2.27	1.41	1.73	5.41	6.80
	0400	14 ga., 1″ x 2″ mesh, 3′ high		300	.080		1.81	1.41	1.73	4.95	6.30
	0500	5′ high		300	.080		2.50	1.41	1.73	5.64	7.05
	1000	Kennel fencing, 1-1/2″ mesh, 6′ long, 3′-6″ wide, 6′-2″ high	2 Clab	4	4	Ea.	284	67.50		351.50	425
	1050	12′ long		4	4		340	67.50		407.50	490
	1200	Top covers, 1-1/2″ mesh, 6′ long		15	1.067		57.50	18.05		75.55	94
	1250	12′ long		12	1.333		92.50	22.50		115	141
	1300	For kennel doors, see division 08344-350									
510	0010	**FENCE, WOOD** Basket weave, 3/8″ x 4″ boards, 2″ x 4″									
	0020	stringers on spreaders, 4″ x 4″ posts									
	0050	No. 1 cedar, 6′ high	B-1	160	.150	L.F.	7.80	2.64		10.44	13.05
	0070	Treated pine, 6′ high		150	.160		9.50	2.81		12.31	15.20
	0090	Vertical weave 6′ high		145	.166		11.45	2.91		14.36	17.50
	0200	Board fence, 1″ x 4″ boards, 2″ x 4″ rails, 4″ x 4″ post									
	0220	Preservative treated, 2 rail, 3′ high	B-1	145	.166	L.F.	5.80	2.91		8.71	11.35
	0240	4′ high		135	.178		6.35	3.12		9.47	12.30

02820 | Fences & Gates			CREW	DAILY OUTPUT	LABOR-HOURS	UNIT	2004 BARE COSTS MAT.	LABOR	EQUIP.	TOTAL	TOTAL INCL O&P	
510	0260	3 rail, 5′ high	B-1	130	.185	L.F.	7.20	3.24		10.44	13.40	510
	0300	6′ high		125	.192		8.25	3.37		11.62	14.80	
	0320	No. 2 grade western cedar, 2 rail, 3′ high		145	.166		6.35	2.91		9.26	11.90	
	0340	4′ high		135	.178		7.50	3.12		10.62	13.55	
	0360	3 rail, 5′ high		130	.185		8.65	3.24		11.89	15	
	0400	6′ high		125	.192		9.45	3.37		12.82	16.15	
	0420	No. 1 grade cedar, 2 rail, 3′ high		145	.166		9.50	2.91		12.41	15.40	
	0440	4′ high		135	.178		10.80	3.12		13.92	17.15	
	0460	3 rail, 5′ high		130	.185		12.50	3.24		15.74	19.25	
	0500	6′ high		125	.192		13.90	3.37		17.27	21	
	0540	Shadow box, 1″ x 6″ board, 2″ x 4″ rail, 4″ x 4″post										
	0560	Pine, pressure treated, 3 rail, 6′ high	B-1	150	.160	L.F.	10.65	2.81		13.46	16.45	
	0600	Gate, 3′-6″ wide		8	3	Ea.	61	52.50		113.50	157	
	0620	No. 1 cedar, 3 rail, 4′ high		130	.185	L.F.	13.10	3.24		16.34	19.90	
	0640	6′ high		125	.192		16.15	3.37		19.52	23.50	
	0860	Open rail fence, split rails, 2 rail 3′ high, no. 1 cedar		160	.150		5.25	2.64		7.89	10.25	
	0870	No. 2 cedar		160	.150		4.09	2.64		6.73	8.95	
	0880	3 rail, 4′ high, no. 1 cedar		150	.160		7.05	2.81		9.86	12.55	
	0890	No. 2 cedar		150	.160		4.66	2.81		7.47	9.85	
	0920	Rustic rails, 2 rail 3′ high, no. 1 cedar		160	.150		3.28	2.64		5.92	8.10	
	0930	No. 2 cedar		160	.150		3.14	2.64		5.78	7.90	
	0940	3 rail, 4′ high		150	.160		4.40	2.81		7.21	9.60	
	0950	No. 2 cedar		150	.160		3.32	2.81		6.13	8.40	
	0960	Picket fence, gothic, pressure treated pine										
	1000	2 rail, 3′ high	B-1	140	.171	L.F.	4.04	3.01		7.05	9.55	
	1020	3 rail, 4′ high		130	.185	″	4.76	3.24		8	10.75	
	1040	Gate, 3′-6″ wide		9	2.667	Ea.	42.50	47		89.50	127	
	1060	No. 2 cedar, 2 rail, 3′ high		140	.171	L.F.	5.05	3.01		8.06	10.65	
	1100	3 rail, 4′ high		130	.185	″	5.15	3.24		8.39	11.20	
	1120	Gate, 3′-6″ wide		9	2.667	Ea.	50.50	47		97.50	135	
	1140	No. 1 cedar, 2 rail 3′ high		140	.171	L.F.	10.10	3.01		13.11	16.20	
	1160	3 rail, 4′ high		130	.185		11.75	3.24		14.99	18.45	
	1200	Rustic picket, molded pine, 2 rail, 3′ high		140	.171		4.63	3.01		7.64	10.20	
	1220	No. 1 cedar, 2 rail, 3′ high		140	.171		6.30	3.01		9.31	12.05	
	1240	Stockade fence, no. 1 cedar, 3-1/4″ rails, 6′ high		160	.150		9.50	2.64		12.14	14.90	
	1260	8′ high		155	.155		12.30	2.72		15.02	18.15	
	1300	No. 2 cedar, treated wood rails, 6′ high		160	.150		9.50	2.64		12.14	14.90	
	1320	Gate, 3′-6″ wide		8	3	Ea.	56	52.50		108.50	151	
	1360	Treated pine, treated rails, 6′ high		160	.150	L.F.	9.30	2.64		11.94	14.70	
	1400	8′ high		150	.160	″	14	2.81		16.81	20	
520	0010	**FENCE, WOOD RAIL** Picket, No. 2 cedar, Gothic, 2 rail, 3′ high	B-1	160	.150	L.F.	5	2.64		7.64	9.95	520
	0050	Gate, 3′-6″ wide		9	2.667	Ea.	43	47		90	127	
	0400	3 rail, 4′ high		150	.160	L.F.	5.75	2.81		8.56	11.10	
	0500	Gate, 3′-6″ wide		9	2.667	Ea.	52	47		99	137	
	1200	Stockade, No. 2 cedar, treated wood rails, 6′ high		160	.150	L.F.	6.25	2.64		8.89	11.30	
	1250	Gate, 3′ wide		9	2.667	Ea.	51	47		98	136	
	1300	No. 1 cedar, 3-1/4″ cedar rails, 6′ high		160	.150	L.F.	15.65	2.64		18.29	21.50	
	1500	Gate, 3′ wide		9	2.667	Ea.	122	47		169	214	
	2700	Prefabricated redwood or cedar, 4′ high		160	.150	L.F.	11.85	2.64		14.49	17.50	
	2800	6′ high		150	.160		15.70	2.81		18.51	22	
	3300	Board, shadow box, 1″ x 6″, treated pine, 6′ high		160	.150		9.20	2.64		11.84	14.60	
	3400	No. 1 cedar, 6′ high		150	.160		18.20	2.81		21.01	25	
	3900	Basket weave, No. 1 cedar, 6′ high		160	.150		18	2.64		20.64	24.50	
	4200	Gate, 3′-6″ wide		9	2.667	Ea.	54	47		101	139	
	5000	Fence rail, redwood, 2″ x 4″, merch grade 8′		2,400	.010	L.F.	.98	.18		1.16	1.38	

02830 | Retaining Walls

			CREW	DAILY OUTPUT	LABOR-HOURS	UNIT	2004 BARE COSTS MAT.	LABOR	EQUIP.	TOTAL	TOTAL INCL O&P	
100	0010	**CAST IN PLACE RETAINING WALLS**										100
	1800	Concrete gravity wall with vertical face including excavation & backfill										
	1850	No reinforcing										
	1900	6′ high, level embankment	C-17C	36	2.306	L.F.	53	54	10.65	117.65	162	
	2000	33° slope embankment	"	32	2.594	"	48.50	61	11.95	121.45	169	
	2800	Reinforced concrete cantilever, incl. excavation, backfill & reinf.										
	2900	6′ high, 33° slope embankment	C-17C	35	2.371	L.F.	48.50	55.50	10.95	114.95	159	
200	0010	**INTERLOCKING SEGMENTAL RETAINING WALLS**										200
	7100	Segmental Retaining Wall system, incl pins, and void fill										
	7120	base not included										
	7140	Large unit, 8″ high x 18″ wide x 20″ deep, 3 plane split	B-62	300	.080	S.F.	8.55	1.49	.48	10.52	12.50	
	7150	straight split		300	.080		8.55	1.49	.48	10.52	12.50	
	7160	Medium, ltwt, 8″ high x 18″ wide x 12″ deep, 3 plane split		400	.060		7.55	1.12	.36	9.03	10.55	
	7170	straight split		400	.060		7.55	1.12	.36	9.03	10.55	
	7180	Small unit, 4″ x 18″ x 10″ deep, 3 plane split		400	.060		9.45	1.12	.36	10.93	12.65	
	7190	straight split		400	.060		9.45	1.12	.36	10.93	12.65	
	7200	Cap unit, 3 plane split		300	.080		10.60	1.49	.48	12.57	14.75	
	7210	Cap unit, st split	↓	300	.080		10.60	1.49	.48	12.57	14.75	
	7260	For reinforcing, add				↓				2.20	2.75	
	8000	For higher walls, add components as necessary										
600	0010	**METAL BIN RETAINING WALLS**, Aluminized steel bin, excavation										600
	0020	and backfill not included, 10′ wide										
	0100	4′ high, 5.5′ deep	B-13	650	.074	S.F.	13.05	1.36	.96	15.37	17.70	
	0200	8′ high, 5.5′ deep		615	.078		15	1.44	1.01	17.45	20	
	0300	10′ high, 7.7′ deep		580	.083		15.80	1.52	1.08	18.40	21	
	0400	12′ high, 7.7′ deep		530	.091		17.05	1.67	1.18	19.90	23	
	0500	16′ high, 7.7′ deep	↓	515	.093	↓	18	1.72	1.21	20.93	24	
800	0010	**STONE RETAINING WALLS**										800
	0015	Including excavation, concrete footing and										
	0020	stone 3′ below grade. Price is exposed face area.										
	0200	Decorative random stone, to 6′ high, 1′-6″ thick, dry set	D-1	35	.457	S.F.	8	9.60		17.60	24.50	
	0300	Mortar set		40	.400		9.70	8.40		18.10	24.50	
	0500	Cut stone, to 6′ high, 1′-6″ thick, dry set		35	.457		12.10	9.60		21.70	29	
	0600	Mortar set		40	.400		14.20	8.40		22.60	29.50	
	0800	Retaining wall, random stone, 6′ to 10′ high, 2′ thick, drv set		45	.356		10	7.45		17.45	23.50	
	0900	Mortar set		50	.320		12.10	6.70		18.80	24.50	
	1100	Cut stone, 6′ to 10′ high, 2′ thick, dry set		45	.356		15.25	7.45		22.70	29	
	1200	Mortar set	↓	50	.320	↓	16.55	6.70		23.25	29.50	

02870 | Site Furnishings

			CREW	DAILY OUTPUT	LABOR-HOURS	UNIT	MAT.	LABOR	EQUIP.	TOTAL	TOTAL INCL O&P	
310	0010	**BENCHES**										310
	0012	Seating, benches, park, precast conc, w/backs, wood rails, 4′ long	2 Clab	5	3.200	Ea.	370	54		424	495	
	0100	8′ long		4	4		645	67.50		712.50	825	
	0500	Steel barstock pedestals w/backs, 2″ x 3″ wood rails, 4′ long		10	1.600		770	27		797	895	
	0510	8′ long		7	2.286		910	38.50		948.50	1,075	
	0800	Cast iron pedestals, back & arms, wood slats, 4′ long		8	2		300	34		334	390	
	0820	8′ long		5	3.200		825	54		879	1,000	
	1700	Steel frame, fir seat, 10′ long	↓	10	1.600	↓	153	27		180	214	

For expanded coverage of these items see *Means Site Work and Landscape Cost Data 2004*

		02905 | Plants, Planting, Transplanting	Crew	Daily Output	Labor-Hours	Unit	2004 Bare Costs Mat.	Labor	Equip.	Total	Total Incl O&P
725	0010	**PLANTING**									
	0012	Moving shrubs on site, 12″ ball	B-62	28	.857	Ea.		16	5.10	21.10	32.50
	0100	24″ ball	″	22	1.091			20.50	6.50	27	41
	0300	Moving trees on site, 36″ ball	B-6	3.75	6.400			119	55.50	174.50	261
	0400	60″ ball	″	1	24	↓		450	208	658	980

		02910 | Plant Preparation	Crew	Daily Output	Labor-Hours	Unit	Mat.	Labor	Equip.	Total	Total Incl O&P
500	0010	**MULCHING**									
	0100	Aged barks, 3″ deep, hand spread	1 Clab	100	.080	S.Y.	1.63	1.35		2.98	4.09
	0150	Skid steer loader	B-63	13.50	2.963	M.S.F.	181	50	10.60	241.60	296
	0200	Hay, 1″ deep, hand spread	1 Clab	475	.017	S.Y.	.52	.28		.80	1.05
	0250	Power mulcher, small	B-64	180	.089	M.S.F.	58	1.56	1.25	60.81	67.50
	0350	Large	B-65	530	.030	″	58	.53	.61	59.14	65
	0400	Humus peat, 1″ deep, hand spread	1 Clab	700	.011	S.Y.	2.10	.19		2.29	2.64
	0450	Push spreader	″	2,500	.003	″	2.10	.05		2.15	2.40
	0550	Tractor spreader	B-66	700	.011	M.S.F.	233	.25	.24	233.49	258
	0600	Oat straw, 1″ deep, hand spread	1 Clab	475	.017	S.Y.	.30	.28		.58	.81
	0650	Power mulcher, small	B-64	180	.089	M.S.F.	33.50	1.56	1.25	36.31	40.50
	0700	Large	B-65	530	.030	″	33.50	.53	.61	34.64	38
	0750	Add for asphaltic emulsion	B-45	1,770	.009	Gal.	1.70	.16	.14	2	2.29
	0800	Peat moss, 1″ deep, hand spread	1 Clab	900	.009	S.Y.	1.67	.15		1.82	2.10
	0850	Push spreader	″	2,500	.003	″	1.67	.05		1.72	1.93
	0950	Tractor spreader	B-66	700	.011	M.S.F.	186	.25	.24	186.49	205
	1000	Polyethylene film, 6 mil.	2 Clab	2,000	.008	S.Y.	.15	.14		.29	.40
	1100	Redwood nuggets, 3″ deep, hand spread	1 Clab	150	.053	″	4.15	.90		5.05	6.10
	1150	Skid steer loader	B-63	13.50	2.963	M.S.F.	460	50	10.60	520.60	600
	1200	Stone mulch, hand spread, ceramic chips, economy	1 Clab	125	.064	S.Y.	6.10	1.08		7.18	8.55
	1250	Deluxe	″	95	.084	″	9.40	1.42		10.82	12.75
	1300	Granite chips	B-1	10	2.400	C.Y.	30.50	42		72.50	105
	1400	Marble chips		10	2.400		115	42		157	198
	1500	Onyx gemstone		10	2.400		335	42		377	440
	1600	Pea gravel		28	.857		56	15.05		71.05	87
	1700	Quartz	↓	10	2.400	↓	148	42		190	234
	1800	Tar paper, 15 Lb. felt	1 Clab	800	.010	S.Y.	.24	.17		.41	.56
	1900	Wood chips, 2″ deep, hand spread	″	220	.036	″	1.70	.61		2.31	2.91
	1950	Skid steer loader	B-63	20.30	1.970	M.S.F.	189	33.50	7.05	229.55	272
720	0010	**PLANT BED PREPARATION, SHRUB & TREE**									
	0100	Backfill planting pit, by hand, on site topsoil	2 Clab	18	.889	C.Y.		15		15	25.50
	0200	Prepared planting mix	″	24	.667			11.25		11.25	19.15
	0300	Skid steer loader, on site topsoil	B-62	340	.071			1.32	.42	1.74	2.67
	0400	Prepared planting mix	″	410	.059			1.09	.35	1.44	2.21
	1000	Excavate planting pit, by hand, sandy soil	2 Clab	16	1			16.90		16.90	28.50
	1100	Heavy soil or clay	″	8	2			34		34	57.50
	1200	1/2 C.Y. backhoe, sandy soil	B-11C	150	.107			2.14	1.39	3.53	5.10
	1300	Heavy soil or clay	″	115	.139			2.79	1.81	4.60	6.65
	2000	Mix planting soil, incl. loam, manure, peat, by hand	2 Clab	60	.267		35.50	4.51		40.01	46.50
	2100	Skid steer loader	B-62	150	.160	↓	35.50	2.99	.96	39.45	45
	3000	Pile sod, skid steer loader	″	2,800	.009	S.Y.		.16	.05	.21	.33
	3100	By hand	2 Clab	400	.040			.68		.68	1.15
	4000	Remove sod, F.E. loader	B-10S	2,000	.004			.09	.12	.21	.28
	4100	Sod cutter	B-12K	3,200	.002			.06	.30	.36	.43
	4200	By hand	2 Clab	240	.067	↓		1.13		1.13	1.91
810	0010	**LOAM & TOPSOIL**									
	0300	Fine grade, base course for paving, see div. 02720-200									

02910 | Plant Preparation

			CREW	DAILY OUTPUT	LABOR-HOURS	UNIT	2004 BARE COSTS MAT.	LABOR	EQUIP.	TOTAL	TOTAL INCL O&P	
810	0701	Furnish and place, truck dumped, unscreened, 4″ deep	B-10S	12,000	.001	S.F.	.34	.02	.02	.38	.42	810
	0801	6″ deep	″	7,400	.001	″	.36	.03	.03	.42	.48	
	0900	Fine grading and seeding, incl. lime, fertilizer & seed,										
	1001	With equipment	B-14	9,000	.005	S.F.	.05	.10	.02	.17	.25	

02915 | Shrub and Tree Transplanting

			CREW	DAILY OUTPUT	LABOR-HOURS	UNIT	MAT.	LABOR	EQUIP.	TOTAL	TOTAL INCL O&P	
200	0010	**GROUND COVER** Plants, pachysandra, in prepared beds	B-1	15	1.600	C	25	28		53	75	200
	0200	Vinca minor, 1 yr, bare root		12	2	″	26	35		61	88	
	0600	Stone chips, in 50 lb. bags, Georgia marble		520	.046	Bag	2.35	.81		3.16	3.97	
	0700	Onyx gemstone		260	.092		17	1.62		18.62	21.50	
	0800	Quartz		260	.092		6.35	1.62		7.97	9.75	
	0900	Pea gravel, truckload lots		28	.857	Ton	23.50	15.05		38.55	51	

02920 | Lawns & Grasses

			CREW	DAILY OUTPUT	LABOR-HOURS	UNIT	MAT.	LABOR	EQUIP.	TOTAL	TOTAL INCL O&P	
310	0010	**SEEDING, GENERAL** R02920-500										310
	0020	Mechanical seeding, 215 lb./acre	B-66	1.50	5.333	Acre	500	118	112	730	870	
	0101	$2.00/lb., 44 lb./M.S.Y.	1 Clab	13,950	.001	S.F.	.02	.01		.03	.04	
	0300	Fine grading and seeding incl. lime, fertilizer & seed,										
	0310	with equipment	B-14	1,000	.048	S.Y.	.16	.87	.21	1.24	1.88	
	0600	Limestone hand push spreader, 50 lbs. per M.S.F.	1 Clab	180	.044	M.S.F.	3.31	.75		4.06	4.92	
	0800	Grass seed hand push spreader, 4.5 lbs. per M.S.F.	″	180	.044	″	16.65	.75		17.40	19.65	
400	0010	**SODDING**										400
	0020	Sodding, 1″ deep, bluegrass sod, on level ground, over 8 MSF	B-63	22	1.818	M.S.F.	217	30.50	6.50	254	298	
	0200	4 M.S.F.		17	2.353		243	40	8.40	291.40	345	
	0300	1000 S.F.		13.50	2.963		265	50	10.60	325.60	390	
	0500	Sloped ground, over 8 M.S.F.		6	6.667		217	113	24	354	455	
	0600	4 M.S.F.		5	8		243	135	28.50	406.50	530	
	0700	1000 S.F.		4	10		265	169	36	470	620	
	1000	Bent grass sod, on level ground, over 6 M.S.F.		20	2		485	34	7.15	526.15	600	
	1100	3 M.S.F.		18	2.222		540	37.50	7.95	585.45	670	
	1200	Sodding 1000 S.F. or less		14	2.857		615	48.50	10.25	673.75	775	
	1500	Sloped ground, over 6 M.S.F.		15	2.667		485	45	9.55	539.55	620	
	1600	3 M.S.F.		13.50	2.963		540	50	10.60	600.60	690	
	1700	1000 S.F.		12	3.333		615	56.50	11.95	683.45	790	

02930 | Exterior Plants

			CREW	DAILY OUTPUT	LABOR-HOURS	UNIT	MAT.	LABOR	EQUIP.	TOTAL	TOTAL INCL O&P	
310	0010	**SHRUBS AND TREES** Evergreen, in prepared beds, B & B										310
	0100	Arborvitae pyramidal, 4′-5′	B-17	30	1.067	Ea.	42.50	19.95	17.75	80.20	100	
	0150	Globe, 12″-15″	B-1	96	.250		10.70	4.39		15.09	19.25	
	0300	Cedar, blue, 8′-10′	B-17	18	1.778		118	33	29.50	180.50	217	
	0500	Hemlock, canadian, 2-1/2′-3′	B-1	36	.667		17.40	11.70		29.10	39	
	0550	Holly, Savannah, 8′ - 10′ H		9.68	2.479		530	43.50		573.50	655	
	0600	Juniper, andorra, 18″-24″		80	.300		15.60	5.25		20.85	26	
	0620	Wiltoni, 15″-18″		80	.300		12	5.25		17.25	22	
	0640	Skyrocket, 4-1/2′-5′	B-17	55	.582		47	10.90	9.70	67.60	80.50	
	0660	Blue pfitzer, 2′-2-1/2′	B-1	44	.545		20.50	9.60		30.10	39.50	
	0680	Ketleerie, 2-1/2′-3′		50	.480		31	8.45		39.45	49	
	0700	Pine, black, 2-1/2′-3′		50	.480		33	8.45		41.45	50.50	
	0720	Mugo, 18″-24″		60	.400		33.50	7.05		40.55	49	
	0740	White, 4′-5′	B-17	75	.427		48	8	7.10	63.10	73.50	
	0800	Spruce, blue, 18″-24″	B-1	60	.400		32	7.05		39.05	47.50	
	0840	Norway, 4′-5′	B-17	75	.427		61	8	7.10	76.10	88	
	0900	Yew, denisforma, 12″-15″	B-1	60	.400		22	7.05		29.05	36	
	1000	Capitata, 18″-24″		30	.800		18.95	14.05		33	45	

	02930 | Exterior Plants		CREW	DAILY OUTPUT	LABOR-HOURS	UNIT	2004 BARE COSTS MAT.	LABOR	EQUIP.	TOTAL	TOTAL INCL O&P	
310	1100	Hicksi, 2'-2-1/2'	B-1	30	.800	Ea.	28	14.05		42.05	55	310
320	0010	**SHRUBS** Broadleaf evergreen, planted in prepared beds										320
	0100	Andromeda, 15"-18", container	B-1	96	.250	Ea.	17.15	4.39		21.54	26.50	
	0200	Azalea, 15" - 18", container		96	.250		22	4.39		26.39	32	
	0300	Barberry, 9"-12", container		130	.185		9.75	3.24		12.99	16.20	
	0400	Boxwood, 15"-18", B & B		96	.250		21	4.39		25.39	30.50	
	0500	Euonymus, emerald gaiety, 12" to 15", container		115	.209		14.30	3.67		17.97	22	
	0600	Holly, 15"-18", B & B		96	.250		15.35	4.39		19.74	24.50	
	0900	Mount laurel, 18" - 24", B & B		80	.300		48.50	5.25		53.75	62.50	
	1000	Paxistema, 9 - 12" high		130	.185		15.75	3.24		18.99	23	
	1100	Rhododendron, 18"-24", container		48	.500		27	8.80		35.80	44.50	
	1200	Rosemary, 1 gal container		600	.040		57.50	.70		58.20	64.50	
	2000	Deciduous, amelanchier, 2'-3', B & B		57	.421		78	7.40		85.40	98.50	
	2100	Azalea, 15"-18", B & B		96	.250		20.50	4.39		24.89	30	
	2300	Bayberry, 2'-3', B & B		57	.421		24	7.40		31.40	39	
	2600	Cotoneaster, 15"-18", B & B		80	.300		14.25	5.25		19.50	24.50	
	2800	Dogwood, 3'-4', B & B	B-17	40	.800		22.50	14.95	13.30	50.75	64.50	
	2900	Euonymus, alatus compacta, 15" to 18", container	B-1	80	.300		18.75	5.25		24	29.50	
	3200	Forsythia, 2'-3', container	"	60	.400		16.50	7.05		23.55	30	
	3300	Hibiscus, 3'-4', B & B	B-17	75	.427		12.90	8	7.10	28	35.50	
	3400	Honeysuckle, 3'-4', B & B	B-1	60	.400		17.10	7.05		24.15	31	
	3500	Hydrangea, 2'-3', B & B	"	57	.421		20.50	7.40		27.90	35	
	3600	Lilac, 3'-4', B & B	B-17	40	.800		25	14.95	13.30	53.25	67	
	3900	Privet, bare root, 18"-24"	B-1	80	.300		10.25	5.25		15.50	20	
	4100	Quince, 2'-3', B & B	"	57	.421		18.95	7.40		26.35	33.50	
	4200	Russian olive, 3'-4', B & B	B-17	75	.427		20.50	8	7.10	35.60	43.50	
	4400	Spirea, 3'-4', B & B	B-1	70	.343		27.50	6		33.50	40.50	
	4500	Viburnum, 3'-4', B & B	B-17	40	.800		19.95	14.95	13.30	48.20	61.50	
410	0010	**TREES** Deciduous, in prep. beds, balled & burlapped (B&B)										410
	0100	Ash, 2" caliper	B-17	8	4	Ea.	110	75	66.50	251.50	320	
	0200	Beech, 5'-6'		50	.640		215	11.95	10.65	237.60	269	
	0300	Birch, 6'-8', 3 stems		20	1.600		126	30	26.50	182.50	218	
	0500	Crabapple, 6'-8'		20	1.600		155	30	26.50	211.50	250	
	0600	Dogwood, 4'-5'		40	.800		66	14.95	13.30	94.25	112	
	0700	Eastern redbud 4'-5'		40	.800		120	14.95	13.30	148.25	172	
	0800	Elm, 8'-10'		20	1.600		113	30	26.50	169.50	204	
	0900	Ginkgo, 6'-7'		24	1.333		159	25	22	206	242	
	1000	Hawthorn, 8'-10', 1" caliper		20	1.600		128	30	26.50	184.50	221	
	1100	Honeylocust, 10'-12', 1-1/2" caliper		10	3.200		136	60	53.50	249.50	310	
	1300	Larch, 8'		32	1		90	18.70	16.65	125.35	149	
	1400	Linden, 8'-10', 1" caliper		20	1.600		114	30	26.50	170.50	205	
	1500	Magnolia, 4'-5'		20	1.600		61	30	26.50	117.50	147	
	1600	Maple, red, 8'-10', 1-1/2" caliper		10	3.200		152	60	53.50	265.50	325	
	1700	Mountain ash, 8'-10', 1" caliper		16	2		147	37.50	33.50	218	261	
	1800	Oak, 2-1/2"-3" caliper		6	5.333		229	99.50	89	417.50	515	
	2100	Planetree, 9'-11', 1-1/4" caliper		10	3.200		99.50	60	53.50	213	268	
	2200	Plum, 6'-8', 1" caliper		20	1.600		93.50	30	26.50	150	183	
	2300	Poplar, 9'-11', 1-1/4" caliper		10	3.200		46	60	53.50	159.50	209	
	2500	Sumac, 2'-3'		75	.427		23	8	7.10	38.10	46.50	
	2700	Tulip, 5'-6'		40	.800		47.50	14.95	13.30	75.75	91.50	
	2800	Willow, 6'-8', 1" caliper		20	1.600		59	30	26.50	115.50	145	

		02945 | Planting Accessories	CREW	DAILY OUTPUT	LABOR-HOURS	UNIT	2004 BARE COSTS MAT.	LABOR	EQUIP.	TOTAL	TOTAL INCL O&P	
120	0010	**EDGING**										120
	0050	Aluminum alloy, including stakes, 1/8" x 4", mill finish	B-1	390	.062	L.F.	1.92	1.08		3	3.95	
	0051	Black paint	↓	390	.062		2.23	1.08		3.31	4.29	
	0052	Black anodized		390	.062		2.57	1.08		3.65	4.67	
	0100	Brick, set horizontally, 1-1/2 bricks per L.F.	D-1	370	.043		.93	.91		1.84	2.53	
	0150	Set vertically, 3 bricks per L.F.	"	135	.119		2.64	2.48		5.12	7.05	
	0200	Corrugated aluminum, roll, 4" wide	1 Carp	650	.012		.32	.28		.60	.83	
	0250	6" wide	"	550	.015		.40	.34		.74	1.01	
	0600	Railroad ties, 6" x 8"	2 Carp	170	.094		2.44	2.17		4.61	6.35	
	0650	7" x 9"	"	136	.118	↓	2.71	2.72		5.43	7.60	
	0700	Redwood										
	0750	2" x 4"	2 Carp	330	.048	L.F.	2.79	1.12		3.91	4.97	
	0800	Steel edge strips, incl. stakes, 1/4" x 5"	B-1	390	.062		2.91	1.08		3.99	5.05	
	0850	3/16" x 4"	"	390	.062	↓	2.30	1.08		3.38	4.37	
300	0010	**PLANTERS** Concrete, sandblasted, precast, 48" diameter, 24" high	2 Clab	15	1.067	Ea.	530	18.05		548.05	615	300
	0300	Fiberglass, circular, 36" diameter, 24" high		15	1.067		365	18.05		383.05	435	
	1200	Wood, square, 48" side, 24" high		15	1.067		820	18.05		838.05	935	
	1300	Circular, 48" diameter, 30" high		10	1.600		675	27		702	790	
	1600	Planter/bench, 72"	↓	5	3.200	↓	2,650	54		2,704	3,000	
510	0010	**TREE GUYING**										510
	0015	Tree guying Including stakes, guy wire and wrap										
	0100	Less than 3" caliper, 2 stakes	2 Clab	35	.457	Ea.	15.10	7.75		22.85	29.50	
	0200	3" to 4" caliper, 3 stakes	"	21	.762	"	17.65	12.90		30.55	41.50	
	1000	Including arrowhead anchor, cable, turnbuckles and wrap										
	1100	Less than 3" caliper, 3" anchors	2 Clab	20	.800	Ea.	46.50	13.50		60	74	
	1200	3" to 6" caliper, 4" anchors		15	1.067		66.50	18.05		84.55	104	
	1300	6" caliper, 6" anchors		12	1.333		82.50	22.50		105	129	
	1400	8" caliper, 8" anchors	↓	9	1.778	↓	94.50	30		124.50	155	

Division Notes

		CREW	DAILY OUTPUT	LABOR-HOURS	UNIT	2004 BARE COSTS MAT.	LABOR	EQUIP.	TOTAL	TOTAL INCL O&P

Division 3 Concrete

Estimating Tips

General

- Carefully check all the plans and specifications. Concrete often appears on drawings other than structural drawings, including mechanical and electrical drawings for equipment pads. The cost of cutting and patching is often difficult to estimate. See Subdivision 02220 for demolition costs.
- Always obtain concrete prices from suppliers near the job site. A volume discount can often be negotiated depending upon competition in the area. Remember to add for waste, particularly for slabs and footings on grade.

03100 Concrete Forms & Accessories

- A primary cost for concrete construction is forming. Most jobs today are constructed with prefabricated forms. The selection of the forms best suited for the job and the total square feet of forms required for efficient concrete forming and placing are key elements in estimating concrete construction. Enough forms must be available for erection to make efficient use of the concrete placing equipment and crew.
- Concrete accessories for forming and placing depend upon the systems used. Study the plans and specifications to assure that all special accessory requirements have been included in the cost estimate such as anchor bolts, inserts and hangers.

03200 Concrete Reinforcement

- Ascertain that the reinforcing steel supplier has included all accessories, cutting, bending and an allowance for lapping, splicing and waste. A good rule of thumb is 10% for lapping, splicing and waste. Also, 10% waste should be allowed for welded wire fabric.

03300 Cast-in-Place Concrete

- When estimating structural concrete, pay particular attention to requirements for concrete additives, curing methods and surface treatments. Special consideration for climate, hot or cold, must be included in your estimate. Be sure to include requirements for concrete placing equipment and concrete finishing.

03400 Precast Concrete
03500 Cementitious Decks & Toppings

- The cost of hauling precast concrete structural members is often an important factor. For this reason, it is important to get a quote from the nearest supplier. It may become economically feasible to set up precasting beds on the site if the hauling costs are prohibitive.

Reference Numbers

Reference numbers are shown in bold squares at the beginning of some major classifications. These numbers refer to related items in the Reference Section. The reference information may be an estimating procedure, an alternate pricing method or technical information.

Note: Not all subdivisions listed here necessarily appear in this publication.

03050 | Basic Concrete Materials & Methods

03055 | Selective Demolition

			CREW	DAILY OUTPUT	LABOR-HOURS	UNIT	2004 BARE COSTS MAT.	LABOR	EQUIP.	TOTAL	TOTAL INCL O&P	
870	0010	**WINTER PROTECTION** For heated ready mix, add, minimum				C.Y.	4.25			4.25	4.68	870
	0050	Maximum				"	5.25			5.25	5.80	
	0100	Protecting concrete and temporary heat, add, minimum	2 Clab	6,000	.003	S.F.	.25	.05		.30	.36	
	0200	Temporary shelter for slab on grade, wood frame and polyethylene										
	0201	sheeting, minimum	2 Carp	10	1.600	M.S.F.	237	37		274	325	
	0210	Maximum	"	3	5.333	"	285	123		408	525	
	0300	See also Division 03390-200										

03100 | Concrete Forms & Accessories

03110 | Structural C.I.P. Forms

			CREW	DAILY OUTPUT	LABOR-HOURS	UNIT	2004 BARE COSTS MAT.	LABOR	EQUIP.	TOTAL	TOTAL INCL O&P	
410	0010	**FORMS IN PLACE, COLUMNS**										410
	1500	Round fiber tube, 1 use, 8″ diameter	C-1	155	.206	L.F.	1.46	4.13		5.59	8.60	
	1550	10″ diameter		155	.206		1.83	4.13		5.96	9	
	1600	12″ diameter		150	.213		2.10	4.27		6.37	9.55	
	1700	16″ diameter		140	.229		3.48	4.57		8.05	11.60	
	5000	Job-built plywood, 8″ x 8″ columns, 1 use		165	.194	SFCA	1.53	3.88		5.41	8.30	
	5500	12″ x 12″ columns, 1 use		180	.178		1.53	3.56		5.09	7.75	
	7500	Steel framed plywood, 4 use per mo., rent, 8″ x 8″		340	.094		2.85	1.88		4.73	6.35	
	7550	10″ x 10″		350	.091		2.40	1.83		4.23	5.75	
	7600	12″ x 12″		370	.086		2.58	1.73		4.31	5.75	
430	0010	**FORMS IN PLACE, FOOTINGS** Continuous wall, plywood, 1 use	C-1	375	.085	SFCA	2.20	1.71		3.91	5.30	430
	0150	4 use	"	485	.066	"	.72	1.32		2.04	3.03	
	1500	Keyway, 4 use, tapered wood, 2″ x 4″	1 Carp	530	.015	L.F.	.17	.35		.52	.77	
	1550	2″ x 6″	"	500	.016	"	.26	.37		.63	.92	
	5000	Spread footings, job-built lumber, 1 use	C-1	305	.105	SFCA	1.61	2.10		3.71	5.35	
	5150	4 use	"	414	.077	"	.52	1.55		2.07	3.19	
435	0010	**FORMS IN PLACE, GRADE BEAM** Job-built plywood, 1 use	C-2	530	.091	SFCA	1.50	1.84		3.34	4.77	435
	0150	4 use	"	605	.079	"	.49	1.61		2.10	3.28	
445	0010	**FORMS IN PLACE, SLAB ON GRADE**										445
	1000	Bulkhead forms with keyway, wood, 1 use, 2 piece	C-1	510	.063	L.F.	.93	1.26		2.19	3.15	
	1400	Bulkhead forms w/keyway, 1 piece expanded metal, left in place										
	1410	In lieu of 2 piece form	C-1	1,375	.023	L.F.	1.28	.47		1.75	2.20	
	1420	In lieu of 3 piece form		1,200	.027		1.28	.53		1.81	2.31	
	1430	In lieu of 4 piece form		1,050	.030		1.28	.61		1.89	2.44	
	2000	Curb forms, wood, 6″ to 12″ high, on grade, 1 use		215	.149	SFCA	1.63	2.98		4.61	6.85	
	2150	4 use		275	.116	"	.53	2.33		2.86	4.53	
	3000	Edge forms, wood, 4 use, on grade, to 6″ high		600	.053	L.F.	.40	1.07		1.47	2.25	
	3050	7″ to 12″ high		435	.074	SFCA	.85	1.47		2.32	3.44	
	4000	For slab blockouts, to 12″ high, 1 use		200	.160	L.F.	.67	3.20		3.87	6.20	
	4100	Plastic (extruded), to 6″ high, multiple use, on grade		800	.040	"	3.73	.80		4.53	5.45	
450	0010	**FORMS IN PLACE, STAIRS** (Slant length x width), 1 use	C-2	165	.291	S.F.	2.94	5.90		8.84	13.30	450
	0150	4 use		190	.253		.96	5.15		6.11	9.75	
	2000	Stairs, cast on sloping ground (length x width), 1 use		220	.218		2.81	4.44		7.25	10.60	
	2100	4 use		240	.200		.93	4.07		5	7.90	
455	0010	**FORMS IN PLACE, WALLS**										455
	0100	Box out for wall openings, to 16″ thick, to 10 S.F.	C-2	24	2	Ea.	17.35	40.50		57.85	88	
	0150	Over 10 S.F. (use perimeter)	"	280	.171	L.F.	1.55	3.49		5.04	7.60	
	0250	Brick shelf, 4″ w, add to wall forms, use wall area abv shelf										

03110 | Structural C.I.P. Forms

		03110 Structural C.I.P. Forms	CREW	DAILY OUTPUT	LABOR-HOURS	UNIT	2004 BARE COSTS MAT.	LABOR	EQUIP.	TOTAL	TOTAL INCL O&P	
455	0260	1 use	C-2	240	.200	SFCA	1.73	4.07		5.80	8.80	455
	0350	4 use		300	.160	"	.69	3.25		3.94	6.25	
	0500	Bulkhead, with keyway, 1 use, 2 piece		265	.181	L.F.	2.71	3.68		6.39	9.25	
	0550	3 piece		175	.274	"	3.41	5.60		9.01	13.20	
	0600	Bulkhead w/keyway, 1 piece expanded metal, left in place										
	0610	In lieu of 2 piece form	C-1	800	.040	L.F.	1.31	.80		2.11	2.80	
	2000	Wall, below grade, job-built plywood, to 8′ high, 1 use	C-2	300	.160	SFCA	1.76	3.25		5.01	7.45	
	2150	4 use		435	.110		.68	2.24		2.92	4.54	
	2400	Over 8′ to 16′ high, 1 use		280	.171		3.70	3.49		7.19	9.95	
	2420	2 use		345	.139		1.58	2.83		4.41	6.55	
	2430	3 use		375	.128		1.32	2.60		3.92	5.85	
	2440	4 use		395	.122		1.19	2.47		3.66	5.50	
	2445	Exterior wall, 8′ to 16′ high, 1 use		280	.171		1.66	3.49		5.15	7.75	
	2550	4 use		395	.122		.54	2.47		3.01	4.78	
	3000	For architectural finish, add		1,820	.026		.68	.54		1.22	1.66	
	3500	Polystyrene (expanded) wall forms										
	3510	To 8′ high, 1 use, left in place	1 Carp	295	.027	SFCA	1.77	.63		2.40	3.01	
	7800	Modular prefabricated plywood, to 8′ high, 1 use	C-2	1,180	.041		1.54	.83		2.37	3.09	
	7860	4 use		1,260	.038		.51	.77		1.28	1.87	
	8000	To 16′ high, 1 use		715	.067		1.99	1.36		3.35	4.50	
	8060	4 use		790	.061		.66	1.24		1.90	2.82	
	8100	Over 16′ high, 1 use		715	.067		2.39	1.36		3.75	4.94	
	8160	4 use		790	.061		.80	1.24		2.04	2.98	
800	0010	**SCAFFOLDING** See division 01540-750										800

03150 | Concrete Accessories

		03150 Concrete Accessories	CREW	DAILY OUTPUT	LABOR-HOURS	UNIT	2004 BARE COSTS MAT.	LABOR	EQUIP.	TOTAL	TOTAL INCL O&P	
080	0010	**ACCESSORIES, ANCHOR BOLTS** J-type, incl. nut and washer										080
	0020	1/2″ diameter, 6″ long	1 Carp	90	.089	Ea.	.78	2.05		2.83	4.34	
	0050	10″ long		85	.094		.89	2.17		3.06	4.67	
	0100	12″ long		85	.094		.98	2.17		3.15	4.77	
	0200	5/8″ diameter, 12″ long		80	.100		.99	2.31		3.30	5	
	0250	18″ long		70	.114		1.16	2.64		3.80	5.75	
	0300	24″ long		60	.133		1.33	3.08		4.41	6.70	
	0350	3/4″ diameter, 8″ long		80	.100		1.16	2.31		3.47	5.20	
	0400	12″ long		70	.114		1.45	2.64		4.09	6.10	
	0450	18″ long		60	.133		1.89	3.08		4.97	7.30	
	0500	24″ long		50	.160		2.47	3.70		6.17	8.95	
085	0012	**ANCHOR BOLTS** See division 04080-070										085
160	0010	**ACCESSORIES, CHAMFER STRIPS**										160
	5000	Wood, 1/2″ wide	1 Carp	535	.015	L.F.	.09	.35		.44	.69	
	5200	3/4″ wide		525	.015		.23	.35		.58	.85	
	5400	1″ wide		515	.016		.31	.36		.67	.95	
170	0010	**ACCESSORIES, COLUMN FORM**										170
	1000	Column clamps, adjustable to 24″ x 24″, buy				Set	80			80	88	
	1400	Rent per month				"	7.45			7.45	8.20	
250	0010	**EXPANSION JOINT** Keyed, cold, 24 ga, incl. stakes, 3-1/2″ high	1 Carp	200	.040	L.F.	.56	.92		1.48	2.19	250
	0050	4-1/2″ high		200	.040		.68	.92		1.60	2.32	
	0100	5-1/2″ high		195	.041		.76	.95		1.71	2.45	
	2000	Premolded, bituminous fiber, 1/2″ x 6″		375	.021		.38	.49		.87	1.26	
	2050	1″ x 12″		300	.027		1.34	.62		1.96	2.52	
	2500	Neoprene sponge, closed cell, 1/2″ x 6″		375	.021		1.28	.49		1.77	2.25	

03100 | Concrete Forms & Accessories

		03150 | Concrete Accessories	CREW	DAILY OUTPUT	LABOR-HOURS	UNIT	2004 BARE COSTS MAT.	LABOR	EQUIP.	TOTAL	TOTAL INCL O&P	
250	2550	1" x 12"	1 Carp	300	.027	L.F.	5.90	.62		6.52	7.55	250
	5000	For installation in walls, add						75%				
	5250	For installation in boxouts, add						25%				
400	0010	**ACCESSORIES, INSERTS**										400
	1000	All size nut insert, 5/8" & 3/4", incl. nut	1 Carp	84	.095	Ea.	3.39	2.20		5.59	7.45	
	2000	Continuous slotted, 1-5/8" x 1-3/8"										
	2100	12 ga., 3" long	1 Carp	65	.123	Ea.	3.08	2.84		5.92	8.20	
	2150	6" long		65	.123		3.98	2.84		6.82	9.20	
	2200	8 ga., 12" long		65	.123		9.55	2.84		12.39	15.30	
	2300	36" long	↓	60	.133	↓	22	3.08		25.08	29.50	
620	0010	**ACCESSORIES, SLEEVES AND CHASES**										620
	0100	Plastic, 1 use, 9" long, 2" diameter	1 Carp	100	.080	Ea.	.53	1.85		2.38	3.72	
	0150	4" diameter		90	.089		1.56	2.05		3.61	5.20	
	0200	6" diameter	↓	75	.107	↓	2.75	2.46		5.21	7.20	
640	0010	**ACCESSORIES, SNAP TIES, FLAT WASHER**, 4-3/4" L&W										640
	0100	3000 lb., to 8"				C	83.50			83.50	92	
	0250	16"					106			106	117	
	0300	18"					109			109	120	
	0500	With plastic cone, to 8"					77			77	85	
	0600	11" & 12"					91.50			91.50	101	
	0650	16"					97			97	106	
	0700	18"				↓	101			101	111	
850	0010	**ACCESSORIES, WALL AND FOUNDATION**										850
	0020	Coil tie system										
	0700	1-1/4", 36,000 lb., to 8"				C	720			720	795	
	1200	1-1/4" diameter x 3" long				"	1,400			1,400	1,550	
	4200	30" long				Ea.	3.90			3.90	4.29	
	4250	36" long				"	4.43			4.43	4.87	
860	0010	**WATERSTOP** PVC, ribbed 3/16" thick, 4" wide	1 Carp	155	.052	L.F.	.76	1.19		1.95	2.86	860
	0050	6" wide		145	.055		1.30	1.27		2.57	3.59	
	0500	Ribbed, PVC, with center bulb, 9" wide, 3/16" thick		135	.059		1.95	1.37		3.32	4.47	
	0550	3/8" thick	↓	130	.062	↓	2.95	1.42		4.37	5.65	

03200 | Concrete Reinforcement

		03210 | Reinforcing Steel	CREW	DAILY OUTPUT	LABOR-HOURS	UNIT	2004 BARE COSTS MAT.	LABOR	EQUIP.	TOTAL	TOTAL INCL O&P	
600	0010	**REINFORCING IN PLACE** A615 Grade 60, incl. reinf. access.										600
	0502	Footings, #4 to #7	4 Rodm	4,200	.008	Lb.	.30	.19		.49	.67	
	0550	#8 to #18		3.60	8.889	Ton	515	221		736	960	
	0702	Walls, #3 to #7		6,000	.005	Lb.	.30	.13		.43	.57	
	0750	#8 to #18	↓	4	8	Ton	540	199		739	950	
	2400	Dowels, 2 feet long, deformed, #3	2 Rodm	520	.031	Ea.	.24	.76		1	1.63	
	2410	#4		480	.033		.42	.83		1.25	1.94	
	2420	#5		435	.037		.66	.91		1.57	2.37	
	2430	#6	↓	360	.044	↓	.94	1.10		2.04	3.02	

03200 | Concrete Reinforcement

03220 | Welded Wire Fabric

			CREW	DAILY OUTPUT	LABOR-HOURS	UNIT	2004 BARE COSTS MAT.	LABOR	EQUIP.	TOTAL	TOTAL INCL O&P	
200	0011	**WELDED WIRE FABRIC** Sheets, 6 x 6 - W1.4 x W1.4 (10 x 10)	2 Rodm	3,500	.005	S.F.	.07	.11		.18	.28	200
	0301	6 x 6 - W2.9 x W2.9 (6 x 6) 42 lb. per C.S.F.		2,900	.006		.13	.14		.27	.39	
	0501	4 x 4 - W1.4 x W1.4 (10 x 10) 31 lb. per C.S.F.		3,100	.005		.12	.13		.25	.37	
	0750	Rolls										
	0901	2 x 2 - #12 galv. for gunite reinforcing	2 Rodm	650	.025	S.F.	.20	.61		.81	1.32	

03240 | Fibrous Reinforcing

			CREW	DAILY OUTPUT	LABOR-HOURS	UNIT	MAT.	LABOR	EQUIP.	TOTAL	TOTAL INCL O&P	
300	0010	**FIBROUS REINFORCING**										300
	0100	Synthetic fibers, add to concrete				Lb.	3.85			3.85	4.24	
	0110	1-1/2 lb. per C.Y.				C.Y.	5.95			5.95	6.55	
	0150	Steel fibers, add to concrete				Lb.	.44			.44	.48	
	0155	25 lb. per C.Y.				C.Y.	11			11	12.10	
	0160	50 lb. per C.Y.					22			22	24	
	0170	75 lb. per C.Y.					34			34	37.50	
	0180	100 lb. per C.Y.					44			44	48.50	

03300 | Cast-In-Place Concrete

03310 | Structural Concrete

			CREW	DAILY OUTPUT	LABOR-HOURS	UNIT	2004 BARE COSTS MAT.	LABOR	EQUIP.	TOTAL	TOTAL INCL O&P	
220	0010	**CONCRETE, READY MIX** Normal weight										220
	0020	2000 psi				C.Y.	69.50			69.50	76.50	
	0100	2500 psi					71			71	78	
	0150	3000 psi					72.50			72.50	80	
	0200	3500 psi					74			74	81.50	
	0300	4000 psi					76.50			76.50	84	
	0350	4500 psi					78.50			78.50	86	
	0400	5000 psi					81			81	89	
	0411	6000 psi					92.50			92.50	102	
	0412	8000 psi					151			151	166	
	0413	10,000 psi					214			214	235	
	0414	12,000 psi					258			258	284	
	1000	For high early strength cement, add					10%					
	2000	For all lightweight aggregate, add					45%					
240	0010	**CONCRETE IN PLACE** Including forms (4 uses), reinforcing										240
	0050	steel and finishing unless otherwise indicated										
	0500	Chimney foundations, industrial, minimum	C-14C	32.22	3.476	C.Y.	145	75.50	.80	221.30	288	
	0510	Maximum		23.71	4.724		168	102	1.09	271.09	360	
	3800	Footings, spread under 1 C.Y.		38.07	2.942		105	63.50	.68	169.18	226	
	3850	Over 5 C.Y.		81.04	1.382		96.50	30	.32	126.82	157	
	3900	Footings, strip, 18″ x 9″, unreinforced		40	2.800		91	60.50	.64	152.14	204	
	3920	18″ x 9″, reinforced		35	3.200		105	69.50	.74	175.24	234	
	3925	20″ x 10″, unreinforced		45	2.489		88	54	.57	142.57	190	
	3930	20″ x 10″, reinforced		40	2.800		99.50	60.50	.64	160.64	213	
	3935	24″ x 12″, unreinforced		55	2.036		87	44	.47	131.47	172	
	3940	24″ x 12″, reinforced		48	2.333		98.50	50.50	.54	149.54	196	
	3945	36″ x 12″, unreinforced		70	1.600		83.50	34.50	.37	118.37	151	
	3950	36″ x 12″, reinforced		60	1.867		94	40.50	.43	134.93	172	
	4000	Foundation mat, under 10 C.Y.		38.67	2.896		129	62.50	.67	192.17	250	
	4050	Over 20 C.Y.		56.40	1.986		115	43	.46	158.46	201	

03310 | Structural Concrete

		03310 Structural Concrete	Crew	Daily Output	Labor-Hours	Unit	2004 Bare Costs Mat.	Labor	Equip.	Total	Total Incl O&P	
240	4520	Handicap access ramp, railing both sides, 3′ wide	C-14H	14.58	3.292	L.F.	149	74	1.78	224.78	291	240
	4525	5′ wide		12.22	3.928		156	88.50	2.12	246.62	325	
	4530	With 6″ curb and rails both sides, 3′ wide		8.55	5.614		155	126	3.03	284.03	390	
	4535	5′ wide		7.31	6.566		159	148	3.55	310.55	430	
	4650	Slab on grade, not including finish, 4″ thick	C-14E	60.75	1.449	C.Y.	88.50	32	.43	120.93	153	
	4700	6″ thick	″	92	.957	″	85	21	.29	106.29	130	
	4751	Slab on grade, incl. troweled finish, not incl. forms										
	4760	or reinforcing, over 10,000 S.F., 4″ thick	C-14F	3,425	.021	S.F.	.96	.43	.01	1.40	1.78	
	4820	6″ thick	″	3,350	.021	″	1.41	.44	.01	1.86	2.28	
	5000	Slab on grade, incl. textured finish, not incl. forms										
	5001	or reinforcing, 4″ thick	C-14G	2,873	.019	S.F.	.94	.39	.01	1.34	1.70	
	5010	6″ thick		2,590	.022		1.47	.44	.01	1.92	2.35	
	5020	8″ thick		2,320	.024		1.92	.49	.01	2.42	2.92	
	6203	Retaining walls, gravity, 4′ high				C.Y.	88.50			88.50	97.50	
	6800	Stairs, not including safety treads, free standing, 3′-6″ wide	C-14H	83	.578	LF Nose	6.90	13.05	.31	20.26	30	
	6850	Cast on ground		125	.384	″	4.85	8.65	.21	13.71	20.50	
	7000	Stair landings, free standing		200	.240	S.F.	2.65	5.40	.13	8.18	12.25	
	7050	Cast on ground		475	.101	″	1.53	2.28	.05	3.86	5.60	
700	0010	**PLACING CONCRETE** and vibrating, including labor & equipment										700
	1900	Footings, continuous, shallow, direct chute	C-6	120	.400	C.Y.		7.25	.43	7.68	12.65	
	1950	Pumped	C-20	150	.427			7.95	5.15	13.10	18.95	
	2000	With crane and bucket	C-7	90	.800			15	10.85	25.85	37	
	2400	Footings, spread, under 1 C.Y., direct chute	C-6	55	.873			15.80	.94	16.74	27.50	
	2600	Over 5 C.Y., direct chute		120	.400			7.25	.43	7.68	12.65	
	2900	Foundation mats, over 20 C.Y., direct chute		350	.137			2.48	.15	2.63	4.33	
	4300	Slab on grade, 4″ thick, direct chute		110	.436			7.90	.47	8.37	13.80	
	4350	Pumped	C-20	130	.492			9.15	5.95	15.10	22	
	4400	With crane and bucket	C-7	110	.655			12.30	8.90	21.20	30.50	
	4900	Walls, 8″ thick, direct chute	C-6	90	.533			9.65	.58	10.23	16.90	
	4950	Pumped	C-20	100	.640			11.90	7.70	19.60	28.50	
	5000	With crane and bucket	C-7	80	.900			16.90	12.25	29.15	42	
	5050	12″ thick, direct chute	C-6	100	.480			8.70	.52	9.22	15.15	
	5100	Pumped	C-20	110	.582			10.80	7	17.80	26	
	5200	With crane and bucket	C-7	90	.800			15	10.85	25.85	37	
	5600	Wheeled concrete dumping, add to placing costs above										
	5610	Walking cart, 50′ haul, add	C-18	32	.281	C.Y.		4.82	1.52	6.34	9.90	
	5620	150′ haul, add		24	.375			6.40	2.03	8.43	13.15	
	5700	250′ haul, add		18	.500			8.55	2.71	11.26	17.55	
	5800	Riding cart, 50′ haul, add	C-19	80	.112			1.93	.93	2.86	4.30	
	5810	150′ haul, add		60	.150			2.57	1.25	3.82	5.75	
	5900	250′ haul, add		45	.200			3.42	1.66	5.08	7.65	

03350 | Concrete Finishing

		03350 Concrete Finishing	Crew	Daily Output	Labor-Hours	Unit	2004 Bare Costs Mat.	Labor	Equip.	Total	Total Incl O&P	
300	0010	**FINISHING FLOORS** Monolithic, screed finish	1 Cefi	900	.009	S.F.		.20		.20	.32	300
	0100	Screed and bull float (darby) finish		725	.011			.24		.24	.39	
	0150	Screed, float, and broom finish		630	.013			.28		.28	.45	
	0200	Screed, float, and hand trowel		600	.013			.29		.29	.48	
	0250	Machine trowel		550	.015			.32		.32	.52	
	1600	Exposed local aggregate finish, minimum		625	.013		4.08	.28		4.36	4.95	
	1650	Maximum		465	.017		12.90	.38		13.28	14.80	
325	0010	**CONTROL JOINT**, concrete floor slab										325
	0100	Sawcut in green concrete										
	0120	1″ depth	C-27	2,000	.008	L.F.		.18	.05	.23	.34	
	0140	1-1/2″ depth		1,800	.009			.20	.05	.25	.38	

		03350 Concrete Finishing	CREW	DAILY OUTPUT	LABOR-HOURS	UNIT	2004 BARE COSTS MAT.	LABOR	EQUIP.	TOTAL	TOTAL INCL O&P	
325	0160	2" depth	C-27	1,600	.010	L.F.		.22	.06	.28	.42	325
	0200	Clean out control joint of debris	C-28	6,000	.001	↓		.03		.03	.05	
	0300	Joint sealant										
	0320	Backer rod, polyethylene, 1/4" diameter	1 Cefi	460	.017	L.F.	.02	.38		.40	.64	
	0340	Sealant, polyurethane										
	0360	1/4" x 1/4" (308 LF/Gal)	1 Cefi	270	.030	L.F.	.15	.65		.80	1.23	
	0380	1/4" x 1/2" (154 LF/Gal)	"	255	.031	"	.31	.69		1	1.46	
350	0010	**FINISHING WALLS** Break ties and patch voids	1 Cefi	540	.015	S.F.	.03	.33		.36	.56	350
	0050	Burlap rub with grout	"	450	.018		.03	.39		.42	.66	
	0300	Bush hammer, green concrete	B-39	1,000	.048		.03	.83	.16	1.02	1.62	
	0350	Cured concrete	"	650	.074	↓	.03	1.27	.25	1.55	2.46	
600	0010	**SLAB TEXTURE STAMPING,** buy										600
	0020	Approx. 3 S.F.- 5 S.F. each, minimum				Ea.	44			44	48.50	
	0030	Average				"	48.50			48.50	53	
	0120	Per S.F. of tool, average				S.F.	53			53	58	
	0200	Commonly used chemicals for texture systems										
	0210	Hardeners w/colors average				S.F.	.44			.44	.48	
	0220	Release agents w/colors, average					.18			.18	.20	
	0225	Clear, average					.14			.14	.15	
	0230	Sealers, clear, average					.12			.12	.13	
	0240	Colors, average				↓	.15			.15	.17	

		03390 Concrete Curing	CREW	DAILY OUTPUT	LABOR-HOURS	UNIT	MAT.	LABOR	EQUIP.	TOTAL	TOTAL INCL O&P	
200	0011	**CURING** With burlap, 4 uses assumed, 7.5 oz.	2 Clab	5,500	.003	S.F.	.07	.05		.12	.16	200
	0101	10 oz.		5,500	.003		.11	.05		.16	.20	
	0201	With waterproof curing paper, 2 ply, reinforced		7,000	.002		.06	.04		.10	.14	
	0301	With sprayed membrane curing compound	↓	9,500	.002		.05	.03		.08	.11	
	0710	Electrically, heated pads, 15 watts/S.F., 20 uses, minimum					.16			.16	.18	
	0800	Maximum				↓	.27			.27	.30	

03400 | Precast Concrete

		03450 Plant-Precast Architectural Concrete	CREW	DAILY OUTPUT	LABOR-HOURS	UNIT	2004 BARE COSTS MAT.	LABOR	EQUIP.	TOTAL	TOTAL INCL O&P	
855	0010	**PRECAST WINDOW SILLS**										855
	0600	Precast concrete, 4" tapers to 3", 9" wide	D-1	70	.229	L.F.	9.70	4.79		14.49	18.60	
	0650	11" wide		60	.267		13.60	5.60		19.20	24.50	
	0700	13" wide, 3 1/2" tapers to 2 1/2", 12" wall	↓	50	.320	↓	13.50	6.70		20.20	26	

		03480 Precast Concrete Specialties	CREW	DAILY OUTPUT	LABOR-HOURS	UNIT	MAT.	LABOR	EQUIP.	TOTAL	TOTAL INCL O&P	
400	0010	**LINTELS**										400
	0800	Precast concrete, 4" wide, 8" high, to 5' long	D-10	28	1.429	Ea.	24	31	17.25	72.25	96.50	
	0850	5'-12' long		24	1.667		68	36	20	124	157	
	1000	6" wide, 8" high, to 5' long		26	1.538		29	33	18.55	80.55	108	
	1050	5'-12' long	↓	22	1.818	↓	84	39	22	145	182	
800	0010	**STAIRS,** Precast concrete treads on steel stringers, 3' wide	C-12	75	.640	Riser	56.50	14.45	8.15	79.10	96	800
	0300	Front entrance, 5' wide with 48" platform, 2 risers		16	3	Flight	289	67.50	38.50	395	475	
	0350	5 risers		12	4		330	90	51	471	575	
	0500	6' wide, 2 risers	↓	15	3.200	↓	330	72	41	443	530	

03400 | Precast Concrete

03480 | Precast Concrete Specialties

			CREW	DAILY OUTPUT	LABOR-HOURS	UNIT	2004 BARE COSTS MAT.	LABOR	EQUIP.	TOTAL	TOTAL INCL O&P	
800	1200	Basement entrance stairs, steel bulkhead doors, minimum	B-51	22	2.182	Flight	485	38	5.55	528.55	600	800
	1250	Maximum	"	11	4.364	↓	840	76	11.10	927.10	1,075	

3 CONCRETE

03900 | Concrete Restoration & Cleaning

03930 | Concrete Rehabilitation

			CREW	DAILY OUTPUT	LABOR-HOURS	UNIT	2004 BARE COSTS MAT.	LABOR	EQUIP.	TOTAL	TOTAL INCL O&P	
400	0012	**FLOOR PATCHING** 1/4" thick, small areas, regular	1 Cefi	170	.047	S.F.	2.90	1.04		3.94	4.87	400
	0100	Epoxy	"	100	.080	"	4.05	1.77		5.82	7.30	

Division 4 Masonry

Estimating Tips

04050 Basic Masonry Materials & Methods

- The terms *mortar* and *grout* are often used interchangeably, and incorrectly. Mortar is used to bed masonry units, seal the entry of air and moisture, provide architectural appearance, and allow for size variations in the units. Grout is used primarily in reinforced masonry construction and is used to bond the masonry to the reinforcing steel. Common mortar types are M(2500 psi), S(1800 psi), N(750 psi), and O(350 psi), and conform to ASTM C270. Grout is either fine or coarse, conforms to ASTM C476, and in-place strengths generally exceed 2500 psi. Mortar and grout are different components of masonry construction and are placed by entirely different methods. An estimator should be aware of their unique uses and costs.
- Waste, specifically the loss/droppings of mortar and the breakage of brick and block, is included in all masonry assemblies in this division. A factor of 25% is added for mortar and 3% for brick and concrete masonry units.
- Scaffolding or staging is not included in any of the Division 4 costs. Refer to section 01540 for scaffolding and staging costs.

04800 Masonry Assemblies

- The most common types of unit masonry are brick and concrete masonry. The major classifications of brick are building brick (ASTM C62), facing brick (ASTM C216) and glazed brick, fire brick and pavers. Many varieties of texture and appearance can exist within these classifications, and the estimator would be wise to check local custom and availability within the project area. On repair and remodeling jobs, matching the existing brick may be the most important criteria.
- Brick and concrete block are priced by the piece and then converted into a price per square foot of wall. Openings less than two square feet are generally ignored by the estimator because any savings in units used is offset by the cutting and trimming required.
- It is often difficult and expensive to find and purchase small lots of historic brick. Costs can vary widely. Many design issues affect costs, selection of mortar mix, and repairs or replacement of masonry materials. Cleaning techniques must be reflected in the estimate.
- All masonry walls, whether interior or exterior, require bracing. The cost of bracing walls during construction should be included by the estimator and this bracing must remain in place until permanent bracing is complete. Permanent bracing of masonry walls is accomplished by masonry itself, in the form of pilasters or abutting wall corners, or by anchoring the walls to the structural frame. Accessories in the form of anchors, anchor slots and ties are used, but their supply and installation can be by different trades. For instance, anchor slots on spandrel beams and columns are supplied and welded in place by the steel fabricator, but the ties from the slots into the masonry are installed by the bricklayer. Regardless of the installation method the estimator must be certain that these accessories are accounted for in pricing.

Reference Numbers

Reference numbers are shown in bold squares at the beginning of some major classifications. These numbers refer to related items in the Reference Section. The reference information may be an estimating procedure, an alternate pricing method or technical information.

Note: Not all subdivisions listed here necessarily appear in this publication.

		04055 | Selective Demolition	CREW	DAILY OUTPUT	LABOR-HOURS	UNIT	2004 BARE COSTS MAT.	LABOR	EQUIP.	TOTAL	TOTAL INCL O&P	
110	0010	**SELECTIVE DEMOLITION, MASONRY**										110
	1000	Chimney, 16″ x 16″, soft old mortar	1 Clab	24	.333	V.L.F.		5.65		5.65	9.55	
	1020	Hard mortar		18	.444			7.50		7.50	12.75	
	1080	20″ x 20″, soft old mortar		12	.667			11.25		11.25	19.15	
	1100	Hard mortar		10	.800			13.50		13.50	23	
	1140	20″ x 32″, soft old mortar		10	.800			13.50		13.50	23	
	1160	Hard mortar		8	1			16.90		16.90	28.50	
	1200	48″ x 48″, soft old mortar		5	1.600			27		27	46	
	1220	Hard mortar		4	2			34		34	57.50	
	2000	Columns, 8″ x 8″, soft old mortar		48	.167			2.82		2.82	4.78	
	2020	Hard mortar		40	.200			3.38		3.38	5.75	
	2060	16″ x 16″, soft old mortar		16	.500			8.45		8.45	14.35	
	2100	Hard mortar		14	.571			9.65		9.65	16.40	
	2140	24″ x 24″, soft old mortar		8	1			16.90		16.90	28.50	
	2160	Hard mortar		6	1.333			22.50		22.50	38.50	
	2200	36″ x 36″, soft old mortar		4	2			34		34	57.50	
	2220	Hard mortar		3	2.667			45		45	76.50	
	3000	Copings, precast or masonry, to 8″ wide										
	3020	Soft old mortar	1 Clab	180	.044	L.F.		.75		.75	1.28	
	3040	Hard mortar	″	160	.050	″		.85		.85	1.44	
	3100	To 12″ wide										
	3120	Soft old mortar	1 Clab	160	.050	L.F.		.85		.85	1.44	
	3140	Hard mortar	″	140	.057	″		.97		.97	1.64	
	4000	Fireplace, brick, 30″ x 24″ opening										
	4020	Soft old mortar	1 Clab	2	4	Ea.		67.50		67.50	115	
	4040	Hard mortar		1.25	6.400			108		108	184	
	4100	Stone, soft old mortar		1.50	5.333			90		90	153	
	4120	Hard mortar		1	8			135		135	230	
	5000	Veneers, brick, soft old mortar		140	.057	S.F.		.97		.97	1.64	
	5020	Hard mortar		125	.064			1.08		1.08	1.84	
	5100	Granite and marble, 2″ thick		180	.044			.75		.75	1.28	
	5120	4″ thick		170	.047			.80		.80	1.35	
	5140	Stone, 4″ thick		180	.044			.75		.75	1.28	
	5160	8″ thick		175	.046			.77		.77	1.31	
	5400	Alternate pricing method, stone, 4″ thick		60	.133	C.F.		2.25		2.25	3.83	
	5420	8″ thick		85	.094	″		1.59		1.59	2.70	
		04060 | Masonry Mortar										
200	0010	**CEMENT** Gypsum 80 lb. bag, T.L. lots (R04060-100)				Bag	11.75			11.75	12.95	200
	0050	L.T.L. lots					12.15			12.15	13.35	
	0100	Masonry, 70 lb. bag, T.L. lots					6.15			6.15	6.75	
	0150	L.T.L. lots					5.80			5.80	6.35	
	0200	White, 70 lb. bag, T.L. lots					16.35			16.35	18	
	0250	L.T.L. lots					17.25			17.25	19	
		04070 | Masonry Grout										
420	0010	**GROUTING** Bond bms. & lintels, 8″ dp., pumped, not incl. block										420
	0200	Concrete block cores, solid, 4″ thk., by hand, 0.067 C.F./S.F. of wall	D-8	1,100	.036	S.F.	.22	.78		1	1.54	
	0210	6″ thick, pumped, 0.175 C.F. per S.F.	D-4	720	.056		.58	1.05	.19	1.82	2.60	
	0250	8″ thick, pumped, 0.258 C.F. per S.F.		680	.059		.85	1.11	.20	2.16	3.02	
	0300	10″ thick, pumped, 0.340 C.F. per S.F.		660	.061		1.12	1.15	.21	2.48	3.38	
	0350	12″ thick, pumped, 0.422 C.F. per S.F.		640	.063		1.39	1.18	.22	2.79	3.74	
		04080 | Anchorage & Reinforcement										
070	0010	**ANCHOR BOLTS** Hooked type with nut and washer, 1/2″ diam., 8″ long	1 Bric	200	.040	Ea.	.52	.96		1.48	2.17	070
	0030	12″ long		190	.042		.98	1.01		1.99	2.76	

04050 | Basic Masonry Materials and Methods

04080 | Anchorage & Reinforcement

			CREW	DAILY OUTPUT	LABOR-HOURS	UNIT	2004 BARE COSTS MAT.	LABOR	EQUIP.	TOTAL	TOTAL INCL O&P	
070	0060	3/4" diameter, 8" long	1 Bric	160	.050	Ea.	1.16	1.20		2.36	3.28	070
	0070	12" long	↓	150	.053	↓	1.45	1.28		2.73	3.73	
200	0010	**REINFORCING** Steel bars A615, placed horiz., #3 & #4 bars R04080-500	1 Bric	450	.018	Lb.	.29	.43		.72	1.02	200
	0050	Placed vertical, #3 & #4 bars		350	.023		.29	.55		.84	1.22	
	0060	#5 & #6 bars		650	.012	↓	.29	.30		.59	.80	
	0200	Joint reinforcing, regular truss, to 6" wide, mill std galvanized		30	.267	C.L.F.	11.05	6.40		17.45	23	
	0250	12" wide		20	.400		11.75	9.60		21.35	29	
	0400	Cavity truss with drip section, to 6" wide		30	.267		9.70	6.40		16.10	21.50	
	0450	12" wide	↓	20	.400	↓	11.05	9.60		20.65	28	
650	0010	**WALL TIES** To brick veneer, galv., corrugated, 7/8" x 7", 22 Ga.	1 Bric	10.50	.762	C	4.47	18.30		22.77	35.50	650
	0100	24 Ga.		10.50	.762		4	18.30		22.30	35	
	0150	16 Ga.		10.50	.762		13.45	18.30		31.75	45.50	
	0200	Buck anchors, galv., corrugated, 16 gauge, 2" bend. 8" x 2"		10.50	.762		93	18.30		111.30	133	
	0250	8" x 3"		10.50	.762		99	18.30		117.30	140	
	0600	Cavity wall, Z type, galvanized, 6" long, 1/4" diameter		10.50	.762		20	18.30		38.30	52.50	
	0650	3/16" diameter		10.50	.762		13.60	18.30		31.90	45.50	
	0800	8" long, 1/4" diameter		10.50	.762		24	18.30		42.30	57	
	0850	3/16" diameter		10.50	.762		10.95	18.30		29.25	42.50	
	1000	Rectangular type, galvanized, 1/4" diameter, 2" x 6"		10.50	.762		25.50	18.30		43.80	59	
	1050	4" x 6"		10.50	.762		29	18.30		47.30	62	
	1100	3/16" diameter, 2" x 6"		10.50	.762		16.95	18.30		35.25	49	
	1150	4" x 6"		10.50	.762		19.40	18.30		37.70	52	
	1500	Rigid partition anchors, plain, 8" long, 1" x 1/8"		10.50	.762		48	18.30		66.30	83.50	
	1550	1" x 1/4"		10.50	.762		94	18.30		112.30	134	
	1580	1-1/2" x 1/8"		10.50	.762		66.50	18.30		84.80	104	
	1600	1-1/2" x 1/4"		10.50	.762		158	18.30		176.30	204	
	1650	2" x 1/8"		10.50	.762		83	18.30		101.30	122	
	1700	2" x 1/4"	↓	10.50	.762	↓	225	18.30		243.30	279	

04200 | Masonry Units

04210 | Clay Masonry Units

			CREW	DAILY OUTPUT	LABOR-HOURS	UNIT	2004 BARE COSTS MAT.	LABOR	EQUIP.	TOTAL	TOTAL INCL O&P	
100	0010	**COMMON BUILDING BRICK** C62, TL lots, material only R04210-120										100
	0020	Standard, minimum				M	270			270	297	
	0050	Average (select)				"	325			325	360	
300	0010	**FACE BRICK** C216, TL lots, material only R04210-120										300
	0300	Standard modular, 4" x 2-2/3" x 8", minimum				M	355			355	395	
	0350	Maximum					465			465	510	
	2170	For less than truck load lots, add					10			10	11	
	2180	For buff or gray brick, add				↓	15			15	16.50	

04500 | Refractories

04550 | Flue Liners

			CREW	DAILY OUTPUT	LABOR-HOURS	UNIT	2004 BARE COSTS MAT.	LABOR	EQUIP.	TOTAL	TOTAL INCL O&P	
250	0010	**FLUE LINING** Including mortar joints, 8″ x 8″	D-1	125	.128	V.L.F.	3.71	2.68		6.39	8.55	250
	0100	8″ x 12″		103	.155		5.25	3.25		8.50	11.15	
	0200	12″ x 12″		93	.172		6.15	3.60		9.75	12.80	
	0300	12″ x 18″		84	.190		11.35	3.99		15.34	19.15	
	0400	18″ x 18″		75	.213		16.50	4.47		20.97	25.50	
	0500	20″ x 20″		66	.242		29	5.10		34.10	40.50	
	0600	24″ x 24″		56	.286		40.50	6		46.50	54.50	
	1000	Round, 18″ diameter		66	.242		28	5.10		33.10	39.50	
	1100	24″ diameter	↓	47	.340	↓	37.50	7.15		44.65	53	

04800 | Masonry Assemblies

04810 | Unit Masonry Assemblies

				CREW	DAILY OUTPUT	LABOR-HOURS	UNIT	2004 BARE COSTS MAT.	LABOR	EQUIP.	TOTAL	TOTAL INCL O&P	
040	0010	**ADOBE BRICK** Semi-stabilized, with cement mortar											040
	0060	Brick, 10″ x 4″ x 14″ , 2.6/S.F.		D-8	560	.071	S.F.	2.06	1.54		3.60	4.82	
	0080	12″ x 4″ x 16″, 2.3/S.F.			580	.069		3.49	1.49		4.98	6.30	
	0100	10″ x 4″ x 16″, 2.3/S.F.			590	.068		3.17	1.46		4.63	5.90	
	0120	8″ x 4″ x 16″, 2.3/S.F.			560	.071		2.43	1.54		3.97	5.25	
	0140	4″ x 4″ x 16″, 2.3/S.F.			540	.074		2.55	1.60		4.15	5.45	
	0160	6″ x 4″ x 16″, 2.3/S.F.			540	.074		1.94	1.60		3.54	4.78	
	0180	4″ x 4″ x 12″, 3.0/S.F.			520	.077		2.42	1.66		4.08	5.40	
	0200	8″ x 4″ x 12″, 3.0/S.F.		↓	520	.077	↓	2.63	1.66		4.29	5.65	
050	0010	**AUTOCLAVED AERATED CONCRETE BLOCK** Scaffolding not incl											050
	0050	Solid, 4″ x 12″ x 24″, incl mortar		D-8	600	.067	S.F.	1.81	1.44		3.25	4.38	
	0060	6″ x 12″ x 24″			600	.067		2.42	1.44		3.86	5.05	
	0070	8″ x 8″ x 24″			575	.070		3.04	1.50		4.54	5.85	
	0080	10″ x 12″ x 24″			575	.070		3.96	1.50		5.46	6.85	
	0090	12″ x 12″ x 24″		↓	550	.073	↓	4.61	1.57		6.18	7.70	
100	0010	**BRICK VENEER** Scaffolding not included, truck load lots											100
	0015	Material costs incl. 3% brick and 25% mortar waste											
	2000	Standard, sel. common, 4″ x 2-2/3″ x 8″, (6.75/S.F.)	R04210 -100	D-8	230	.174	S.F.	2.76	3.75		6.51	9.30	
	2020	Standard, red, 4″ x 2-2/3″ x 8″, running bond (6.75/SF)			220	.182		2.76	3.92		6.68	9.55	
	2050	Full header every 6th course (7.88/S.F.)	R04210 -120		185	.216		3.21	4.66		7.87	11.30	
	2100	English, full header every 2nd course (10.13/S.F.)			140	.286		4.12	6.15		10.27	14.80	
	2150	Flemish, alternate header every course (9.00/S.F.)	R04210 -180		150	.267		3.67	5.75		9.42	13.60	
	2200	Flemish, alt. header every 6th course (7.13/S.F.)			205	.195		2.91	4.21		7.12	10.20	
	2250	Full headers throughout (13.50/S.F.)	R04210 -500		105	.381		5.50	8.20		13.70	19.70	
	2300	Rowlock course (13.50/S.F.)			100	.400		5.50	8.60		14.10	20.50	
	2350	Rowlock stretcher (4.50/S.F.)			310	.129		1.85	2.78		4.63	6.65	
	2400	Soldier course (6.75/S.F.)			200	.200		2.76	4.31		7.07	10.20	
	2450	Sailor course (4.50/S.F.)			290	.138		1.85	2.97		4.82	6.95	
	2600	Buff or gray face, running bond, (6.75/S.F.)			220	.182		2.93	3.92		6.85	9.70	
	2700	Glazed face brick, running bond			210	.190		7.95	4.11		12.06	15.60	
	2750	Full header every 6th course (7.88/S.F.)			170	.235		9.30	5.05		14.35	18.65	
	3000	Jumbo, 6″ x 4″ x 12″ running bond (3.00/S.F.)			435	.092		3.61	1.98		5.59	7.25	
	3050	Norman, 4″ x 2-2/3″ x 12″ running bond, (4.5/S.F.)			320	.125		3.83	2.70		6.53	8.70	
	3100	Norwegian, 4″ x 3-1/5″ x 12″ (3.75/S.F.)			375	.107		2.94	2.30		5.24	7.05	
	3150	Economy, 4″ x 4″ x 8″ (4.50/S.F.)	↓	↓	310	.129	↓	3.39	2.78		6.17	8.35	

Important: See the Reference Section for critical supporting data - Reference Nos., Crews, & Location Factors

04810 | Unit Masonry Assemblies

				DAILY	LABOR-		2004 BARE COSTS				TOTAL
			CREW	OUTPUT	HOURS	UNIT	MAT.	LABOR	EQUIP.	TOTAL	INCL O&P
100	3200	Engineer, 4" x 3-1/5" x 8" (5.63/S.F.) R04210-100	D-8	260	.154	S.F.	2.86	3.32		6.18	8.65
	3250	Roman, 4" x 2" x 12" (6.00/S.F.)		250	.160		4.62	3.45		8.07	10.85
	3300	SCR, 6" x 2-2/3" x 12" (4.50/S.F.) R04210-120		310	.129		4.23	2.78		7.01	9.30
	3350	Utility, 4" x 4" x 12" (3.00/S.F.)		450	.089		3.31	1.92		5.23	6.85
	3400	For cavity wall construction, add R04210-180						15%			
	3450	For stacked bond, add						10%			
	3500	For interior veneer construction, add R04210-500						15%			
	3550	For curved walls, add						30%			
160	0010	**CHIMNEY** See Div. 03310-240 for foundation, add to prices below									
	0100	Brick, 16" x 16", 8" flue, scaff. not incl.	D-1	18.20	.879	V.L.F.	14.90	18.40		33.30	47
	0150	16" x 20" with one 8" x 12" flue		16	1		23.50	21		44.50	61
	0200	16" x 24" with two 8" x 8" flues		14	1.143		34.50	24		58.50	78
	0250	20" x 20" with one 12" x 12" flue		13.70	1.168		26	24.50		50.50	69
	0300	20" x 24" with two 8" x 12" flues		12	1.333		38.50	28		66.50	89
	0350	20" x 32" with two 12" x 12" flues		10	1.600		45	33.50		78.50	105
170	0010	**COLUMNS** Face brick, includes mortar, scaffolding not included									
	0050	8" x 8", 9 brick per course	D-1	56	.286	V.L.F.	3.60	6		9.60	13.90
	0100	12" x 8", 13.5 brick		37	.432		5.40	9.05		14.45	21
	0200	12" x 12", 20 brick		25	.640		8	13.40		21.40	31.50
	0300	16" x 12", 27 brick		19	.842		10.80	17.65		28.45	41.50
	0400	16" x 16", 36 brick		14	1.143		14.40	24		38.40	56
	0500	20" x 16", 45 brick		11	1.455		18	30.50		48.50	70.50
	0600	20" x 20", 56 brick		9	1.778		22.50	37		59.50	86.50
172	0010	**CONCRETE BLOCK, BACK-UP**, C90, 2000 psi R04220-200									
	0020	Normal weight, 8" x 16" units, tooled joint 1 side									
	0050	Not-reinforced, 2000 psi, 2" thick	D-8	475	.084	S.F.	.82	1.82		2.64	3.92
	0200	4" thick		460	.087		.97	1.87		2.84	4.19
	0300	6" thick		440	.091		1.42	1.96		3.38	4.82
	0350	8" thick		400	.100		1.54	2.16		3.70	5.25
	0400	10" thick		330	.121		2.16	2.61		4.77	6.70
	0450	12" thick	D-9	310	.155		2.22	3.24		5.46	7.85
	1000	Reinforced, alternate courses, 4" thick	D-8	450	.089		1.05	1.92		2.97	4.34
	1100	6" thick		430	.093		1.50	2.01		3.51	4.98
	1150	8" thick		395	.101		1.62	2.18		3.80	5.40
	1200	10" thick		320	.125		2.24	2.70		4.94	6.95
	1250	12" thick	D-9	300	.160		2.31	3.35		5.66	8.10
175	0010	**CONCRETE BLOCK BOND BEAM** C90, 2000 psi									
	0020	Not including grout or reinforcing									
	0100	Regular block, 8" high, 8" thick	D-8	565	.071	L.F.	1.62	1.53		3.15	4.32
	0150	12" thick	D-9	510	.094	"	2.24	1.97		4.21	5.75
182	0010	**CONCRETE BLOCK, DECORATIVE** C90, 2000 psi									
	5000	Split rib profile units, 1" deep ribs, 8 ribs									
	5100	8" x 16" x 4" thick	D-8	345	.116	S.F.	2.09	2.50		4.59	6.45
	5150	6" thick		325	.123		2.41	2.65		5.06	7.05
	5200	8" thick		300	.133		2.77	2.87		5.64	7.85
	5250	12" thick	D-9	275	.175		3.27	3.66		6.93	9.70
	5400	For special deeper colors, 4" thick, add					.79			.79	.87
	5450	12" thick, add					.68			.68	.75
	5600	For white, 4" thick, add					.79			.79	.87
	5650	6" thick, add					.79			.79	.87
	5700	8" thick, add					.73			.73	.81
	5750	12" thick, add					.68			.68	.75
184	0010	**CONCRETE BLOCK, EXTERIOR** C90, 2000 psi									
	0020	Reinforced alt courses, tooled joints 2 sides									

MASONRY 4

04810 | Unit Masonry Assemblies

		04810 Unit Masonry Assemblies	CREW	DAILY OUTPUT	LABOR-HOURS	UNIT	2004 BARE COSTS MAT.	LABOR	EQUIP.	TOTAL	TOTAL INCL O&P	
184	0100	Normal weight, 8" x 16" x 6" thick	D-8	395	.101	S.F.	1.65	2.18		3.83	5.45	184
	0200	8" thick		360	.111		2.46	2.40		4.86	6.70	
	0250	10" thick		290	.138		2.93	2.97		5.90	8.15	
	0300	12" thick	D-9	250	.192		2.99	4.02		7.01	10	
186	0010	**CONCRETE BLOCK FOUNDATION WALL** C90/C145										186
	0050	Normal-weight, cut joints, horiz joint reinf, no vert reinf										
	0200	Hollow, 8" x 16" x 6" thick	D-8	455	.088	S.F.	1.63	1.90		3.53	4.95	
	0250	8" thick		425	.094		1.76	2.03		3.79	5.30	
	0300	10" thick		350	.114		2.38	2.46		4.84	6.70	
	0350	12" thick	D-9	300	.160		2.46	3.35		5.81	8.25	
	0500	Solid, 8" x 16" block, 6" thick	D-8	440	.091		1.77	1.96		3.73	5.20	
	0550	8" thick	"	415	.096		2.52	2.08		4.60	6.25	
	0600	12" thick	D-9	350	.137		3.68	2.87		6.55	8.80	
188	0010	**CONCRETE BLOCK INSULATION INSERTS**										188
	0100	Inserts, styrofoam, plant installed, add to block prices										
	0200	8" x 16" units, 6" thick				S.F.	.85			.85	.94	
	0250	8" thick					.85			.85	.94	
	0300	10" thick					1			1	1.10	
	0350	12" thick					1.05			1.05	1.16	
190	0010	**CONCRETE BLOCK, LINTELS** C90, normal weight										190
	0100	Including grout and horizontal reinforcing										
	0200	8" x 8" x 8", 1 #4 bar	D-4	300	.133	L.F.	3.57	2.52	.46	6.55	8.65	
	0250	2 #4 bars		295	.136		3.69	2.57	.47	6.73	8.85	
	1000	12" x 8" x 8", 1 #4 bar		275	.145		5.05	2.75	.50	8.30	10.70	
	1150	2 #5 bars		270	.148		5.30	2.80	.51	8.61	11.10	
210	0010	**CONCRETE BLOCK, PARTITIONS**, scaffolding not included										210
	1000	Lightweight block, tooled joints, 2 sides, hollow										
	1100	Not reinforced, 8" x 16" x 4" thick	D-8	440	.091	S.F.	1.11	1.96		3.07	4.48	
	1150	6" thick		410	.098		1.51	2.10		3.61	5.15	
	1200	8" thick		385	.104		1.86	2.24		4.10	5.75	
	1250	10" thick		370	.108		2.45	2.33		4.78	6.55	
	1300	12" thick	D-9	350	.137		2.87	2.87		5.74	7.95	
	4000	Regular block, tooled joints, 2 sides, hollow										
	4100	Not reinforced, 8" x 16" x 4" thick	D-8	430	.093	S.F.	.93	2.01		2.94	4.35	
	4150	6" thick		400	.100		1.38	2.16		3.54	5.10	
	4200	8" thick		375	.107		1.50	2.30		3.80	5.45	
	4250	10" thick		360	.111		2.12	2.40		4.52	6.30	
	4300	12" thick	D-9	340	.141		2.19	2.96		5.15	7.30	
250	0010	**COPING** Stock units										250
	0050	Precast concrete, 10" wide, 4" tapers to 3-1/2", 8" wall	D-1	75	.213	L.F.	17.55	4.47		22.02	27	
	0100	12" wide, 3-1/2" tapers to 3", 10" wall		70	.229		10.50	4.79		15.29	19.50	
	0150	16" wide, 4" tapers to 3-1/2", 14" wall		60	.267		14.50	5.60		20.10	25.50	
	0300	Limestone for 12" wall, 4" thick		90	.178		13.50	3.72		17.22	21	
	0350	6" thick		80	.200		15.75	4.19		19.94	24.50	
	0500	Marble, to 4" thick, no wash, 9" wide		90	.178		19.25	3.72		22.97	27	
	0550	12" wide		80	.200		29	4.19		33.19	39	
	0700	Terra cotta, 9" wide		90	.178		4.75	3.72		8.47	11.45	
	0800	Aluminum, for 12" wall		80	.200		11.25	4.19		15.44	19.35	
260	0010	**CORNICES** Brick, on existing building										260
	0110	Face bricks, 12 brick/S.F., minimum	D-1	30	.533	SF Face	5	11.15		16.15	24	
	0150	15 brick/S.F., maximum	"	23	.696	"	6	14.55		20.55	30.50	
325	0010	**GLASS BLOCK** R04060-200										325
	0150	8" x 8"	D-8	160	.250	S.F.	8.65	5.40		14.05	18.45	

04800 | Masonry Assemblies

04810 | Unit Masonry Assemblies

	Line	Description	Crew	Daily Output	Labor-Hours	Unit	2004 Bare Costs Mat.	Labor	Equip.	Total	Total Incl O&P
325	0160	end block R04060-200	D-8	160	.250	S.F.	26.50	5.40		31.90	38
	0170	90 deg corner		160	.250		27	5.40		32.40	38.50
	0180	45 deg corner		160	.250		12.95	5.40		18.35	23
	0200	12″ x 12″		175	.229		11.40	4.93		16.33	21
	0210	4″ x 8″		160	.250		7.65	5.40		13.05	17.40
	0220	6″ x 8″		160	.250		9.40	5.40		14.80	19.30
	0700	For solar reflective blocks, add					100%				
	1000	Thinline, plain, 3-1/8″ thick, under 1,000 S.F., 6″ x 6″	D-8	115	.348	S.F.	13.05	7.50		20.55	27
	1050	8″ x 8″		160	.250		7.70	5.40		13.10	17.40
	1400	For cleaning block after installation (both sides), add		1,000	.040		.10	.86		.96	1.54
650	0016	**WALLS** R04210-500									
	0800	4″ wall, face, 4″ x 2-2/3″ x 8″	D-8	215	.186	S.F.	2.72	4.01		6.73	9.65
	0850	4″ thick, as back up, 6.75 bricks per S.F.		240	.167		2.12	3.59		5.71	8.30
	0900	8″ thick wall, 13.50 brick per S.F.		135	.296		4.34	6.40		10.74	15.40
	1000	12″ thick wall, 20.25 bricks per S.F.		95	.421		6.55	9.10		15.65	22.50
	1050	16″ thick wall, 27.00 bricks per S.F.		75	.533		8.80	11.50		20.30	29
	1200	Reinforced, 4″ x 2-2/3″ x 8″ , 4″ wall		205	.195		2.13	4.21		6.34	9.35
	1250	8″ thick wall, 13.50 brick per S.F.		130	.308		4.35	6.65		11	15.85
	1300	12″ thick wall, 20.25 bricks per S.F.		90	.444		6.55	9.60		16.15	23
	1350	16″ thick wall, 27.00 bricks per S.F.		70	.571		8.80	12.30		21.10	30

04850 | Stone Assemblies

	Line	Description	Crew	Daily Output	Labor-Hours	Unit	2004 Bare Costs Mat.	Labor	Equip.	Total	Total Incl O&P
300	0010	**GRANITE** Cut to size									
	2450	For radius under 5′, add				L.F.	100%				
	2500	Steps, copings, etc., finished on more than one surface									
	2550	Minimum	D-10	50	.800	C.F.	75	17.25	9.65	101.90	122
	2600	Maximum	″	50	.800	″	120	17.25	9.65	146.90	171
	2800	Pavers, 4″ x 4″ x 4″ blocks, split face and joints									
	2850	Minimum	D-11	80	.300	S.F.	11	6.40		17.40	23
	2900	Maximum	″	80	.300	″	22	6.40		28.40	34.50
	3500	Curbing, city street type, See Division 02770-225									
400	0010	**LIMESTONE,** Cut to size									
	0020	Veneer facing panels									
	0500	Texture finish, light stick, 4-1/2″ thick, 5′x 12′	D-4	300	.133	S.F.	21	2.52	.46	23.98	27.50
	0750	5″ thick, 5′ x 14′ panels	D-10	275	.145		21.50	3.14	1.75	26.39	30.50
	1000	Sugarcube finish, 2″ Thick, 3′ x 5′ panels		275	.145		8.35	3.14	1.75	13.24	16.30
	1050	3″ Thick, 4′ x 9′ panels		275	.145		12.50	3.14	1.75	17.39	21
	1200	4″ Thick, 5′ x 11′ panels		275	.145		16.35	3.14	1.75	21.24	25
	1400	Sugarcube, textured finish, 4-1/2″ thick, 5′ x 12′		275	.145		21	3.14	1.75	25.89	30
	1450	5″ thick, 5′ x 14′ panels		275	.145		21.50	3.14	1.75	26.39	30.50
	2000	Coping, sugarcube finish, top & 2 sides		30	1.333	C.F.	37	29	16.10	82.10	106
	2100	Sills, lintels, jambs, trim, stops, sugarcube finish, average		20	2		55	43	24	122	159
	2150	Detailed		20	2		55	43	24	122	159
	2300	Steps, extra hard, 14″ wide, 6″ rise		50	.800	L.F.	19.50	17.25	9.65	46.40	60.50
	3000	Quoins, plain finish, 6″x12″x12″	D-12	25	1.280	Ea.	50	26		76	98.50
	3050	6″x16″x24″	″	25	1.280	″	66.50	26		92.50	117
500	0011	**MARBLE**, ashlar, split face, 4″ + or - thick, random									
	0040	lengths 1′ to 4′ & heights 2″ to 7-1/2″, average	D-8	175	.229	S.F.	14.15	4.93		19.08	24
	0100	Base, polished, 3/4″ or 7/8″ thick, polished, 6″ high	D-10	65	.615	L.F.	12.85	13.25	7.40	33.50	44.50
	1000	Facing, polished finish, cut to size, 3/4″ to 7/8″ thick									
	1050	Average	D-10	130	.308	S.F.	18.85	6.65	3.71	29.21	36
	1100	Maximum	″	130	.308	″	43.50	6.65	3.71	53.86	63
	2200	Window sills, 6″ x 3/4″ thick	D-1	85	.188	L.F.	7.10	3.94		11.04	14.35
	2500	Flooring, polished tiles, 12″ x 12″ x 3/8″ thick									

04850 | Stone Assemblies

	Line	Description	Crew	Daily Output	Labor-Hours	Unit	2004 Bare Costs Mat.	Labor	Equip.	Total	Total Incl O&P
500	2510	Thin set, average	D-11	90	.267	S.F.	8.75	5.70		14.45	19.05
	2600	Maximum		90	.267		31.50	5.70		37.20	44.50
	2700	Mortar bed, average		65	.369		8.90	7.85		16.75	23
	2740	Maximum		65	.369		29	7.85		36.85	45
	2780	Travertine, 3/8" thick, average	D-10	130	.308		12.05	6.65	3.71	22.41	28.50
	2790	Maximum	"	130	.308		29.50	6.65	3.71	39.86	47.50
	3500	Thresholds, 3' long, 7/8" thick, 4" to 5" wide, plain	D-12	24	1.333	Ea.	13.60	27.50		41.10	60.50
	3550	Beveled		24	1.333	"	15.85	27.50		43.35	63
	3700	Window stools, polished, 7/8" thick, 5" wide		85	.376	L.F.	12.75	7.70		20.45	27
600	0011	**ROUGH STONE WALL**, Dry									
	0100	Random fieldstone, under 18" thick	D-12	60	.533	C.F.	7.45	10.90		18.35	26.50
	0150	Over 18" thick	"	63	.508	"	8.90	10.40		19.30	27
700	0011	**SANDSTONE OR BROWNSTONE**									
	0100	Sawed face veneer, 2-1/2" thick, to 2' x 4' panels	D-10	130	.308	S.F.	16.65	6.65	3.71	27.01	33.50
	0150	4' thick, to 3'-6" x 8'panels		100	.400		16.65	8.60	4.82	30.07	38
	0300	Split face, random sizes		100	.400		9.65	8.60	4.82	23.07	30.50
	0350	Cut stone trim (limestone)									
	0360	Ribbon stone, 4" thick, 5' pieces	D-8	120	.333	Ea.	119	7.20		126.20	143
	0370	Cove stone, 4" thick, 5' pieces		105	.381		119	8.20		127.20	145
	0380	Cornice stone, 10" to 12" wide		90	.444		147	9.60		156.60	178
	0390	Band stone, 4" thick, 5' pieces		145	.276		76	5.95		81.95	93.50
	0410	Window and door trim, 3" to 4" wide		160	.250		64.50	5.40		69.90	80
	0420	Key stone, 18" long		60	.667		67.50	14.35		81.85	98.50
800	0010	**SLATE** Pennsylvania, blue gray to gray black; Vermont,									
	3500	Stair treads, sand finish, 1" thick x 12" wide									
	3600	3 L.F. to 6 L.F.	D-10	120	.333	L.F.	15.50	7.20	4.02	26.72	33.50
	3700	Ribbon, sand finish, 1" thick x 12" wide									
	3750	To 6 L.F.	D-10	120	.333	L.F.	10.25	7.20	4.02	21.47	27.50
900	0010	**WINDOW SILL** Bluestone, thermal top, 10" wide, 1-1/2" thick	D-1	85	.188	S.F.	13.50	3.94		17.44	21.50
	0050	2" thick		75	.213	"	15.75	4.47		20.22	25
	0100	Cut stone, 5" x 8" plain		48	.333	L.F.	10.20	7		17.20	23
	0200	Face brick on edge, brick, 8" wide		80	.200		2.15	4.19		6.34	9.30
	0400	Marble, 9" wide, 1" thick		85	.188		7.50	3.94		11.44	14.80
	0900	Slate, colored, unfading, honed, 12" wide, 1" thick		85	.188		15.25	3.94		19.19	23.50
	0950	2" thick		70	.229		21.50	4.79		26.29	31.50

04880 | Masonry Fireplaces

	Line	Description	Crew	Daily Output	Labor-Hours	Unit	2004 Bare Costs Mat.	Labor	Equip.	Total	Total Incl O&P
600	0010	**FIREPLACE** For prefabricated fireplace, see div. 10305-100									
	0100	Brick fireplace, not incl. foundations or chimneys									
	0110	30" x 29" opening, incl. chamber, plain brickwork	D-1	.40	40	Ea.	380	840		1,220	1,825
	0200	Fireplace box only (110 brick)	"	2	8	"	125	168		293	415
	0300	For elaborate brickwork and details, add					35%	35%			
	0400	For hearth, brick & stone, add	D-1	2	8	Ea.	140	168		308	435
	0410	For steel angle, damper, cleanouts, add		4	4		98	84		182	247
	0600	Plain brickwork, incl. metal circulator		.50	32		725	670		1,395	1,925
	0800	Face brick only, standard size, 8" x 2-2/3" x 4"		.30	53.333	M	380	1,125		1,505	2,275
	0900	Stone fireplace, fieldstone, add				SF Face	10			10	11
	1000	Cut stone, add				"	11			11	12.10

04910 | Unit Masonry Restoration

			CREW	DAILY OUTPUT	LABOR-HOURS	UNIT	2004 BARE COSTS MAT.	LABOR	EQUIP.	TOTAL	TOTAL INCL O&P	
720	0010	**POINTING MASONRY**										720
	0300	Cut and repoint brick, hard mortar, running bond	1 Bric	80	.100	S.F.	.27	2.40		2.67	4.28	
	0320	Common bond		77	.104		.27	2.49		2.76	4.44	
	0360	Flemish bond		70	.114		.28	2.74		3.02	4.87	
	0400	English bond		65	.123		.28	2.95		3.23	5.20	
	0600	Soft old mortar, running bond		100	.080		.27	1.92		2.19	3.49	
	0620	Common bond		96	.083		.27	2		2.27	3.61	
	0640	Flemish bond		90	.089		.28	2.13		2.41	3.86	
	0680	English bond		82	.098	↓	.28	2.34		2.62	4.20	
	0700	Stonework, hard mortar		140	.057	L.F.	.35	1.37		1.72	2.67	
	0720	Soft old mortar		160	.050	"	.35	1.20		1.55	2.39	
	1000	Repoint, mask and grout method, running bond		95	.084	S.F.	.35	2.02		2.37	3.75	
	1020	Common bond		90	.089		.35	2.13		2.48	3.94	
	1040	Flemish bond		86	.093		.35	2.23		2.58	4.10	
	1060	English bond		77	.104		.35	2.49		2.84	4.54	
	2000	Scrub coat, sand grout on walls, minimum		120	.067		2.80	1.60		4.40	5.75	
	2020	Maximum	↓	98	.082	↓	2.01	1.96		3.97	5.50	

04930 | Unit Masonry Cleaning

			CREW	DAILY OUTPUT	LABOR-HOURS	UNIT	MAT.	LABOR	EQUIP.	TOTAL	TOTAL INCL O&P	
900	0010	**BRICK WASHING** Acid, smooth brick (R04930 -100)	1 Bric	560	.014	S.F.	.02	.34		.36	.59	900
	0050	Rough brick		400	.020		.03	.48		.51	.83	
	0060	Stone, acid wash	↓	600	.013	↓	.03	.32		.35	.57	
	1000	Muriatic acid, price per gallon in 5 gallon lots				Gal.	4.12			4.12	4.53	

Division Notes

		CREW	DAILY OUTPUT	LABOR-HOURS	UNIT	2004 BARE COSTS MAT.	LABOR	EQUIP.	TOTAL	TOTAL INCL O&P

Division 5 Metals

Estimating Tips

05050 Basic Metal Materials & Methods

- Nuts, bolts, washers, connection angles and plates can add a significant amount to both the tonnage of a structural steel job as well as the estimated cost. As a rule of thumb add 10% to the total weight to account for these accessories.
- Type 2 steel construction, commonly referred to as "simple construction," consists generally of field bolted connections with lateral bracing supplied by other elements of the building, such as masonry walls or x-bracing. The estimator should be aware, however, that shop connections may be accomplished by welding or bolting. The method may be particular to the fabrication shop and may have an impact on the estimated cost.

05200 Metal Joists

- In any given project the total weight of open web steel joists is determined by the loads to be supported and the design. However, economies can be realized in minimizing the amount of labor used to place the joists. This is done by maximizing the joist spacing and therefore minimizing the number of joists required to be installed on the job. Certain spacings and locations may be required by the design, but in other cases maximizing the spacing and keeping it as uniform as possible will keep the costs down.

05300 Metal Deck

- The takeoff and estimating of metal deck involves more than simply the area of the floor or roof and the type of deck specified or shown on the drawings. Many different sizes and types of openings may exist. Small openings for individual pipes or conduits may be drilled after the floor/roof is installed, but larger openings may require special deck lengths as well as reinforcing or structural support. The estimator should determine who will be supplying this reinforcing. Additionally, some deck terminations are part of the deck package, such as screed angles and pour stops, and others will be part of the steel contract, such as angles attached to structural members and cast-in-place angles and plates. The estimator must ensure that all pieces are accounted for in the complete estimate.

05500 Metal Fabrications

- The most economical steel stairs are those that use common materials, standard details and most importantly, a uniform and relatively simple method of field assembly. Commonly available A36 channels and plates are very good choices for the main stringers of the stairs, as are angles and tees for the carrier members. Risers and treads are usually made by specialty shops, and it is most economical to use a typical detail in as many places as possible. The stairs should be pre-assembled and shipped directly to the site. The field connections should be simple and straightforward to be accomplished efficiently and with a minimum of equipment and labor.

Reference Numbers

Reference numbers are shown in bold squares at the beginning of some major classifications. These numbers refer to related items in the Reference Section. The reference information may be an estimating procedure, an alternate pricing method or technical information.

Note: Not all subdivisions listed here necessarily appear in this publication.

05090 | Metal Fastenings

	Line	05090 Metal Fastenings	Crew	Daily Output	Labor-Hours	Unit	2004 Bare Costs Mat.	Labor	Equip.	Total	Total Incl O&P
150	0010	**BOLTS & HEX NUTS** Steel, A307									
	0100	1/4" diameter, 1/2" long				Ea.	.05			.05	.06
	0200	1" long					.06			.06	.07
	0300	2" long					.08			.08	.09
	0400	3" long					.12			.12	.13
	0500	4" long					.13			.13	.14
	0600	3/8" diameter, 1" long					.09			.09	.09
	0700	2" long					.11			.11	.12
	0800	3" long					.15			.15	.17
	0900	4" long					.19			.19	.21
	1000	5" long					.24			.24	.26
	1100	1/2" diameter, 1-1/2" long					.17			.17	.19
	1200	2" long					.19			.19	.21
	1300	4" long					.30			.30	.33
	1400	6" long					.41			.41	.45
	1500	8" long					.53			.53	.59
	1600	5/8" diameter, 1-1/2" long					.35			.35	.39
	1700	2" long					.38			.38	.42
	1800	4" long					.53			.53	.59
	1900	6" long					.67			.67	.74
	2000	8" long					.97			.97	1.07
	2100	10" long					1.21			1.21	1.33
	2200	3/4" diameter, 2" long					.55			.55	.61
	2300	4" long					.77			.77	.85
	2400	6" long					.98			.98	1.08
	2500	8" long					1.45			1.45	1.60
	2600	10" long					1.89			1.89	2.07
	2700	12" long					2.20			2.20	2.42
	2800	1" diameter, 3" long					1.44			1.44	1.58
	2900	6" long					2.21			2.21	2.43
	3000	12" long					4.16			4.16	4.58
	3100	For galvanized, add					75%				
	3200	For stainless, add					350%				
300	0010	**CHEMICAL ANCHORS**, Includes layout & drilling									
	1430	Chemical anchor, w/rod & epoxy cartridge, 3/4" diam. x 9-1/2" long	B-89A	27	.593	Ea.	10.25	11.85	3.93	26.03	35.50
	1435	1" diameter x 11-3/4" long		24	.667		19.65	13.30	4.42	37.37	49
	1440	1-1/4" diameter x 14" long		21	.762		37.50	15.20	5.05	57.75	72.50
	1445	Concrete anchor, w/rod & epoxy cartridge, 1-3/4" diameter x 15" long		20	.800		70.50	16	5.30	91.80	110
	1450	18" long		17	.941		84.50	18.80	6.25	109.55	132
	1455	2" diameter x 18" long		16	1		108	20	6.65	134.65	160
	1460	24" long		15	1.067		141	21.50	7.05	169.55	199
340	0010	**DRILLING** For anchors, up to 4" deep, incl. bit and layout									
	0050	in concrete or brick walls and floors, no anchor									
	0100	Holes, 1/4" diameter	1 Carp	75	.107	Ea.	.08	2.46		2.54	4.26
	0150	For each additional inch of depth, add		430	.019		.02	.43		.45	.75
	0200	3/8" diameter		63	.127		.07	2.93		3	5.05
	0250	For each additional inch of depth, add		340	.024		.02	.54		.56	.94
	0300	1/2" diameter		50	.160		.07	3.70		3.77	6.35
	0350	For each additional inch of depth, add		250	.032		.02	.74		.76	1.27
	0400	5/8" diameter		48	.167		.13	3.85		3.98	6.70
	0450	For each additional inch of depth, add		240	.033		.03	.77		.80	1.35
	0500	3/4" diameter		45	.178		.16	4.11		4.27	7.15
	0550	For each additional inch of depth, add		220	.036		.04	.84		.88	1.47
	0600	7/8" diameter		43	.186		.19	4.30		4.49	7.50
	0650	For each additional inch of depth, add		210	.038		.05	.88		.93	1.54

05090 | Metal Fastenings

			CREW	DAILY OUTPUT	LABOR-HOURS	UNIT	MAT.	LABOR	EQUIP.	TOTAL	TOTAL INCL O&P
340	0700	1" diameter	1 Carp	40	.200	Ea.	.22	4.62		4.84	8.10
	0750	For each additional inch of depth, add		190	.042		.06	.97		1.03	1.71
	0800	1-1/4" diameter		38	.211		.31	4.86		5.17	8.60
	0850	For each additional inch of depth, add		180	.044		.08	1.03		1.11	1.83
	0900	1-1/2" diameter		35	.229		.48	5.30		5.78	9.50
	0950	For each additional inch of depth, add	↓	165	.048	↓	.12	1.12		1.24	2.03
	1000	For ceiling installations, add						40%			
	1100	Drilling & layout for drywall/plaster walls, up to 1" deep, no anchor									
	1200	Holes, 1/4" diameter	1 Carp	150	.053	Ea.	.01	1.23		1.24	2.10
	1300	3/8" diameter		140	.057		.01	1.32		1.33	2.25
	1400	1/2" diameter		130	.062		.01	1.42		1.43	2.42
	1500	3/4" diameter		120	.067		.02	1.54		1.56	2.63
	1600	1" diameter		110	.073		.03	1.68		1.71	2.88
	1700	1-1/4" diameter		100	.080		.04	1.85		1.89	3.18
	1800	1-1/2" diameter	↓	90	.089		.06	2.05		2.11	3.55
	1900	For ceiling installations, add				↓		40%			
	1910	Drilling & layout for steel, up to 1/4" deep, no anchor									
	1920	Holes, 1/4" diameter	1 Sswk	112	.071	Ea.	.10	1.78		1.88	3.55
	1925	For each additional 1/4" depth, add		336	.024		.10	.59		.69	1.26
	1930	3/8" diameter		104	.077		.12	1.92		2.04	3.83
	1935	For each additional 1/4" depth, add		312	.026		.12	.64		.76	1.36
	1940	1/2" diameter		96	.083		.13	2.07		2.20	4.15
	1945	For each additional 1/4" depth, add		288	.028		.13	.69		.82	1.48
	1950	5/8" diameter		88	.091		.21	2.26		2.47	4.60
	1955	For each additional 1/4" depth, add		264	.030		.21	.75		.96	1.69
	1960	3/4" diameter		80	.100		.24	2.49		2.73	5.05
	1965	For each additional 1/4" depth, add		240	.033		.24	.83		1.07	1.86
	1970	7/8" diameter		72	.111		.28	2.77		3.05	5.65
	1975	For each additional 1/4" depth, add		216	.037		.28	.92		1.20	2.09
	1980	1" diameter		64	.125		.32	3.11		3.43	6.35
	1985	For each additional 1/4" depth, add	↓	192	.042		.32	1.04		1.36	2.35
	1990	For drilling up, add				↓		40%			
380	0010	**EXPANSION ANCHORS** & shields									
	0100	Bolt anchors for concrete, brick or stone, no layout and drilling									
	0200	Expansion shields, zinc, 1/4" diameter, 1-5/16" long, single	1 Carp	90	.089	Ea.	.96	2.05		3.01	4.54
	0300	1-3/8" long, double		85	.094		1.06	2.17		3.23	4.86
	0500	2" long, double		80	.100		1.95	2.31		4.26	6.05
	0700	2-1/2" long, double		75	.107		2.52	2.46		4.98	6.95
	0900	2-3/4" long, double		70	.114		3.74	2.64		6.38	8.60
	1100	3-15/16" long, double	↓	65	.123	↓	7.40	2.84		10.24	12.95
	2100	Hollow wall anchors for gypsum wall board, plaster or tile									
	2500	3/16" diameter, short	1 Carp	150	.053	Ea.	.48	1.23		1.71	2.62
	3000	Toggle bolts, bright steel, 1/8" diameter, 2" long		85	.094		.21	2.17		2.38	3.92
	3100	4" long		80	.100		.32	2.31		2.63	4.27
	3200	3/16" diameter, 3" long		80	.100		.36	2.31		2.67	4.32
	3300	6" long		75	.107		.50	2.46		2.96	4.73
	3400	1/4" diameter, 3" long		75	.107		.41	2.46		2.87	4.63
	3500	6" long		70	.114		.57	2.64		3.21	5.10
	3600	3/8" diameter, 3" long		70	.114		.78	2.64		3.42	5.35
	3700	6" long		60	.133		1.37	3.08		4.45	6.75
	3800	1/2" diameter, 4" long		60	.133		2.03	3.08		5.11	7.50
	3900	6" long	↓	50	.160	↓	3.35	3.70		7.05	9.95
	4000	Nailing anchors									
	4100	Nylon nailing anchor, 1/4" diameter, 1" long	1 Carp	3.20	2.500	C	16.80	58		74.80	117
	4200	1-1/2" long		2.80	2.857		21.50	66		87.50	136
	4300	2" long	↓	2.40	3.333	↓	36	77		113	171

05090 | Metal Fastenings

		05090 Metal Fastenings	CREW	DAILY OUTPUT	LABOR-HOURS	UNIT	2004 BARE COSTS MAT.	LABOR	EQUIP.	TOTAL	TOTAL INCL O&P	
380	4400	Metal nailing anchor, 1/4" diameter, 1" long	1 Carp	3.20	2.500	C	25	58		83	126	380
	4500	1-1/2" long		2.80	2.857		34	66		100	150	
	4600	2" long		2.40	3.333		43.50	77		120.50	179	
	5000	Screw anchors for concrete, masonry,										
	5100	stone & tile, no layout or drilling included										
	5200	Jute fiber, #6, #8, & #10, 1" long	1 Carp	240	.033	Ea.	.23	.77		1	1.56	
	5300	#12, 1-1/2" long		200	.040		.33	.92		1.25	1.93	
	5400	#14, 2" long		160	.050		.52	1.16		1.68	2.53	
	5500	#16, 2" long		150	.053		.55	1.23		1.78	2.70	
	5600	#20, 2" long		140	.057		.87	1.32		2.19	3.20	
	5700	Lag screw shields, 1/4" diameter, short		90	.089		.38	2.05		2.43	3.90	
	5800	Long		85	.094		.45	2.17		2.62	4.19	
	5900	3/8" diameter, short		85	.094		.70	2.17		2.87	4.46	
	6000	Long		80	.100		.82	2.31		3.13	4.82	
	6100	1/2" diameter, short		80	.100		.97	2.31		3.28	4.99	
	6200	Long		75	.107		1.21	2.46		3.67	5.50	
	6300	3/4" diameter, short		70	.114		2.71	2.64		5.35	7.45	
	6400	Long		65	.123		3.28	2.84		6.12	8.45	
	6600	Lead, #6 & #8, 3/4" long		260	.031		.15	.71		.86	1.38	
	6700	#10 - #14, 1-1/2" long		200	.040		.22	.92		1.14	1.81	
	6800	#16 & #18, 1-1/2" long		160	.050		.30	1.16		1.46	2.29	
	6900	Plastic, #6 & #8, 3/4" long		260	.031		.09	.71		.80	1.31	
	7000	#8 & #10, 7/8" long		240	.033		.04	.77		.81	1.35	
	7100	#10 & #12, 1" long		220	.036		.12	.84		.96	1.56	
	7200	#14 & #16, 1-1/2" long		160	.050		.07	1.16		1.23	2.04	
460	0010	**LAG SCREWS**										460
	0020	Steel, 1/4" diameter, 2" long	1 Carp	200	.040	Ea.	.07	.92		.99	1.65	
	0100	3/8" diameter, 3" long		150	.053		.20	1.23		1.43	2.31	
	0200	1/2" diameter, 3" long		130	.062		.33	1.42		1.75	2.77	
	0300	5/8" diameter, 3" long		120	.067		.65	1.54		2.19	3.33	
580	0010	**POWDER ACTUATED** Tools & fasteners										580
	0020	Stud driver, .22 caliber, buy, minimum				Ea.	280			280	310	
	0100	Maximum				"	450			450	495	
	0300	Powder charges for above, low velocity				C	15.80			15.80	17.40	
	0400	Standard velocity					22.50			22.50	24.50	
	0600	Drive pins & studs, 1/4" & 3/8" diam., to 3" long, minimum					10.30			10.30	11.35	
	0700	Maximum					40.50			40.50	44.50	
600	0010	**RIVETS**										600
	0100	Aluminum rivet & mandrel, 1/2" grip length x 1/8" diameter				C	4.76			4.76	5.25	
	0200	3/16" diameter					7.35			7.35	8.05	
	0300	Aluminum rivet, steel mandrel, 1/8" diameter					6.35			6.35	7	
	0400	3/16" diameter					5.80			5.80	6.40	
	0500	Copper rivet, steel mandrel, 1/8" diameter					5.75			5.75	6.35	
	0600	Monel rivet, steel mandrel, 1/8" diameter					20.50			20.50	22.50	
	0700	3/16" diameter					59			59	65	
	0800	Stainless rivet & mandrel, 1/8" diameter					10.35			10.35	11.35	
	0900	3/16" diameter					19.35			19.35	21.50	
	1000	Stainless rivet, steel mandrel, 1/8" diameter					8.05			8.05	8.85	
	1100	3/16" diameter					14.55			14.55	16.05	
	1200	Steel rivet and mandrel, 1/8" diameter					4.99			4.99	5.50	
	1300	3/16" diameter					7.50			7.50	8.25	
	1400	Hand riveting tool, minimum				Ea.	101			101	111	
	1500	Maximum					193			193	212	

05050 | Basic Metal Materials & Methods

		05090 Metal Fastenings	CREW	DAILY OUTPUT	LABOR-HOURS	UNIT	2004 BARE COSTS MAT.	LABOR	EQUIP.	TOTAL	TOTAL INCL O&P	
600	1600	Power riveting tool, minimum				Ea.	730			730	800	600
	1700	Maximum				↓	1,825			1,825	2,025	

05100 | Structural Metal Framing

		05120 Structural Steel	CREW	DAILY OUTPUT	LABOR-HOURS	UNIT	2004 BARE COSTS MAT.	LABOR	EQUIP.	TOTAL	TOTAL INCL O&P	
220	0010	**CEILING SUPPORTS**										220
	1000	Entrance door/folding partition supports	E-4	60	.533	L.F.	12.75	13.55	1.25	27.55	41.50	
	1100	Linear accelerator door supports		14	2.286		58	58	5.35	121.35	182	
	1200	Lintels or shelf angles, hung, exterior hot dipped galv.		267	.120		8.70	3.04	.28	12.02	15.75	
	1250	Two coats primer paint instead of galv.		267	.120	↓	7.55	3.04	.28	10.87	14.50	
	1400	Monitor support, ceiling hung, expansion bolted		4	8	Ea.	202	203	18.80	423.80	635	
	1450	Hung from pre-set inserts		6	5.333		217	135	12.55	364.55	515	
	1600	Motor supports for overhead doors		4	8	↓	103	203	18.80	324.80	525	
	1700	Partition support for heavy folding partitions, without pocket		24	1.333	L.F.	29	34	3.13	66.13	101	
	1750	Supports at pocket only		12	2.667		58	67.50	6.25	131.75	202	
	2000	Rolling grilles & fire door supports		34	.941	↓	25	24	2.21	51.21	76	
	2100	Spider-leg light supports, expansion bolted to ceiling slab		8	4	Ea.	83	102	9.40	194.40	297	
	2150	Hung from pre-set inserts		12	2.667	"	89.50	67.50	6.25	163.25	236	
	2400	Toilet partition support		36	.889	L.F.	29	22.50	2.09	53.59	78	
	2500	X-ray travel gantry support	↓	12	2.667	"	99.50	67.50	6.25	173.25	247	
250	0010	**COLUMNS, LIGHTWEIGHT**										250
	1000	Lightweight units (lally), 3-1/2" diameter	E-2	780	.062	L.F.	2.36	1.54	1.82	5.72	7.50	
	1050	4" diameter	"	900	.053	"	3.47	1.34	1.57	6.38	8.05	
	8000	Lally columns, to 8', 3-1/2" diameter	2 Carp	24	.667	Ea.	18.90	15.40		34.30	47	
	8080	4" diameter	"	20	.800	"	28	18.50		46.50	62	
260	0010	**COLUMNS, STRUCTURAL**										260
	0020	Shop fab'd for 100-ton, 1-2 story project, bolted conn's.										
	0800	Steel, concrete filled, extra strong pipe, 3-1/2" diameter	E-2	660	.073	L.F.	21	1.82	2.15	24.97	29	
	0830	4" diameter		780	.062		23.50	1.54	1.82	26.86	31	
	0890	5" diameter		1,020	.047		28	1.18	1.39	30.57	34.50	
	0930	6" diameter	↓	1,200	.040	↓	37	1	1.18	39.18	43.50	
	1100	For galvanizing, add				Lb.	.20			.20	.22	
	1300	For web ties, angles, etc., add per added lb.	1 Sswk	945	.008		.64	.21		.85	1.11	
	1500	Steel pipe, extra strong, no concrete, 3" to 5" diameter	E-2	16,000	.003		.64	.08	.09	.81	.94	
	1600	6" to 12" diameter		14,000	.003		.64	.09	.10	.83	.97	
	2400	Structural tubing, rect, 5" to 6" wide, light section		11,200	.004		.64	.11	.13	.88	1.04	
	2700	12" x 8" x 1/2" thk wall		24,000	.002		.64	.05	.06	.75	.85	
	2800	Heavy section	↓	32,000	.002	↓	.64	.04	.04	.72	.82	
	8090	For projects 75 to 99 tons, add				All	10%					
	8092	50 to 74 tons, add					20%					
	8094	25 to 49 tons, add					30%	10%				
	8096	10 to 24 tons, add					50%	25%				
	8098	2 to 9 tons, add					75%	50%				
	8099	Less than 2 tons, add				↓	100%	100%				
480	0010	**LINTELS**										480
	0020	Plain steel angles, under 500 lb.	1 Bric	550	.015	Lb.	.49	.35		.84	1.12	

METALS 5

05100 | Structural Metal Framing

05120 Structural Steel			CREW	DAILY OUTPUT	LABOR-HOURS	UNIT	2004 BARE COSTS MAT.	LABOR	EQUIP.	TOTAL	TOTAL INCL O&P	
480	0100	500 to 1000 lb.	1 Bric	640	.013	Lb.	.48	.30		.78	1.03	480
	2000	Steel angles, 3-1/2" x 3", 1/4" thick, 2'-6" long		47	.170	Ea.	6.90	4.09		10.99	14.35	
	2100	4'-6" long		26	.308		12.40	7.40		19.80	26	
	2600	4" x 3-1/2", 1/4" thick, 5'-0" long		21	.381		15.80	9.15		24.95	32.50	
	2700	9'-0" long	↓	12	.667	↓	28.50	16		44.50	58	
	3500	For precast concrete lintels, see div. 03480-400										
720	0010	**STRUCTURAL STEEL** Bolted, incl. fabrication										720
	0050	Beams, W 6 x 9	E-2	720	.067	L.F.	6.90	1.67	1.97	10.54	12.85	
	0100	W 8 x 10		720	.067		7.65	1.67	1.97	11.29	13.70	
	0200	Columns, W 6 x 15		540	.089		12.45	2.23	2.62	17.30	20.50	
	0250	W 8 x 31	↓	540	.089	↓	25.50	2.23	2.62	30.35	35.50	
	7990	For projects 75 to 99 tons, add				All	10%					
	7992	50 to 75 tons, add					20%					
	7994	25 to 49 tons, add					30%	10%				
	7996	10 to 24 tons, add					50%	25%				
	7998	2 to 9 tons, add					75%	50%				
	7999	Less than 2 tons, add				↓	100%	100%				

5 METALS

05300 | Metal Deck

05310 Steel Deck			CREW	DAILY OUTPUT	LABOR-HOURS	UNIT	2004 BARE COSTS MAT.	LABOR	EQUIP.	TOTAL	TOTAL INCL O&P	
300	0010	**METAL DECKING** Steel decking										300
	1900	For multi-story or congested site, add						50%				
	2100	Open type, galv., 1-1/2" deep wide rib, 22 gauge, under 50 squares	E-4	4,500	.007	S.F.	.84	.18	.02	1.04	1.29	
	2600	20 gauge, under 50 squares		3,865	.008		.99	.21	.02	1.22	1.52	
	2900	18 gauge, under 50 squares		3,800	.008		1.27	.21	.02	1.50	1.83	
	3050	16 gauge, under 50 squares		3,700	.009		1.73	.22	.02	1.97	2.34	
	3700	4-1/2" deep, long span roof, over 50 squares, 20 gauge		2,700	.012		2.16	.30	.03	2.49	2.99	
	6100	Slab form, steel, 28 gauge, 9/16" deep, uncoated		4,000	.008		.55	.20	.02	.77	1.02	
	6200	Galvanized		4,000	.008		.49	.20	.02	.71	.95	
	6220	24 gauge, 1" deep, uncoated		3,900	.008		.61	.21	.02	.84	1.09	
	6240	Galvanized		3,900	.008		.72	.21	.02	.95	1.21	
	6300	24 gauge, 1-5/16" deep, uncoated		3,800	.008		.65	.21	.02	.88	1.15	
	6400	Galvanized		3,800	.008		.77	.21	.02	1	1.28	
	6500	22 gauge, 1-5/16" deep, uncoated		3,700	.009		.81	.22	.02	1.05	1.33	
	6600	Galvanized		3,700	.009		.83	.22	.02	1.07	1.35	
	6700	22 gauge, 2" deep uncoated		3,600	.009		1.08	.23	.02	1.33	1.65	
	6800	Galvanized	↓	3,600	.009	↓	1.06	.23	.02	1.31	1.63	

Important: See the Reference Section for critical supporting data - Reference Nos., Crews, & Location Factors

05410 | Load-Bearing Metal Studs

			Crew	Daily Output	Labor-Hours	Unit	2004 Bare Costs Mat.	Labor	Equip.	Total	Total Incl O&P
100	0010	**BRACING**, shear wall X-bracing, per 10′ x 10′ bay, one face									
	0120	Metal strap, 20 ga x 4″ wide	2 Carp	18	.889	Ea.	12.60	20.50		33.10	49
	0130	6″ wide		18	.889		19.80	20.50		40.30	57
	0160	18 ga x 4″ wide		16	1		18.45	23		41.45	59.50
	0170	6″ wide		16	1		27.50	23		50.50	69
	0410	Continuous strap bracing, per horizontal row on both faces									
	0420	Metal strap, 20 ga x 2″ wide, studs 12″ O.C.	1 Carp	7	1.143	C.L.F.	33	26.50		59.50	81.50
	0430	16″ O.C.		8	1		33	23		56	75.50
	0440	24″ O.C.		10	.800		33	18.50		51.50	68
	0450	18 ga x 2″ wide, studs 12″ O.C.		6	1.333		46	31		77	104
	0460	16″ O.C.		7	1.143		46	26.50		72.50	96
	0470	24″ O.C.		8	1		46	23		69	90
120	0010	**BRIDGING**, solid between studs w/ 1-1/4″ leg track, per stud bay									
	0200	Studs 12″ O.C., 18 ga x 2-1/2″ wide	1 Carp	125	.064	Ea.	.55	1.48		2.03	3.12
	0210	3-5/8″ wide		120	.067		.67	1.54		2.21	3.35
	0220	4″ wide		120	.067		.71	1.54		2.25	3.39
	0230	6″ wide		115	.070		.93	1.61		2.54	3.75
	0240	8″ wide		110	.073		1.17	1.68		2.85	4.14
	0300	16 ga x 2-1/2″ wide		115	.070		.69	1.61		2.30	3.49
	0310	3-5/8″ wide		110	.073		.84	1.68		2.52	3.77
	0320	4″ wide		110	.073		.90	1.68		2.58	3.84
	0330	6″ wide		105	.076		1.15	1.76		2.91	4.25
	0340	8″ wide		100	.080		1.47	1.85		3.32	4.75
	1200	Studs 16″ O.C., 18 ga x 2-1/2″ wide		125	.064		.71	1.48		2.19	3.29
	1210	3-5/8″ wide		120	.067		.86	1.54		2.40	3.55
	1220	4″ wide		120	.067		.91	1.54		2.45	3.61
	1230	6″ wide		115	.070		1.19	1.61		2.80	4.04
	1240	8″ wide		110	.073		1.50	1.68		3.18	4.50
	1300	16 ga x 2-1/2″ wide		115	.070		.89	1.61		2.50	3.71
	1310	3-5/8″ wide		110	.073		1.07	1.68		2.75	4.03
	1320	4″ wide		110	.073		1.16	1.68		2.84	4.12
	1330	6″ wide		105	.076		1.47	1.76		3.23	4.61
	1340	8″ wide		100	.080		1.88	1.85		3.73	5.20
	2200	Studs 24″ O.C., 18 ga x 2-1/2″ wide		125	.064		1.03	1.48		2.51	3.64
	2210	3-5/8″ wide		120	.067		1.24	1.54		2.78	3.98
	2220	4″ wide		120	.067		1.31	1.54		2.85	4.05
	2230	6″ wide		115	.070		1.72	1.61		3.33	4.62
	2240	8″ wide		110	.073		2.17	1.68		3.85	5.25
	2300	16 ga x 2-1/2″ wide		115	.070		1.29	1.61		2.90	4.15
	2310	3-5/8″ wide		110	.073		1.55	1.68		3.23	4.56
	2320	4″ wide		110	.073		1.67	1.68		3.35	4.69
	2330	6″ wide		105	.076		2.12	1.76		3.88	5.35
	2340	8″ wide		100	.080		2.72	1.85		4.57	6.15
	3000	Continuous bridging, per row									
	3100	16 ga x 1-1/2″ channel thru studs 12″ O.C.	1 Carp	6	1.333	C.L.F.	31	31		62	87
	3110	16″ O.C.		7	1.143		31	26.50		57.50	79.50
	3120	24″ O.C.		8.80	.909		31	21		52	70
	4100	2″ x 2″ angle x 18 ga, studs 12″ O.C.		7	1.143		45	26.50		71.50	94.50
	4110	16″ O.C.		9	.889		45	20.50		65.50	84.50
	4120	24″ O.C.		12	.667		45	15.40		60.40	75.50
	4200	16 ga, studs 12″ O.C.		5	1.600		57	37		94	126
	4210	16″ O.C.		7	1.143		57	26.50		83.50	108
	4220	24″ O.C.		10	.800		57	18.50		75.50	94.50
300	0010	**FRAMING**, boxed headers/beams									
	0200	Double, 18 ga x 6″ deep	2 Carp	220	.073	L.F.	3.20	1.68		4.88	6.35

	05410 | Load-Bearing Metal Studs	CREW	DAILY OUTPUT	LABOR-HOURS	UNIT	2004 BARE COSTS MAT.	LABOR	EQUIP.	TOTAL	TOTAL INCL O&P	
300	0210 8″ deep	2 Carp	210	.076	L.F.	3.56	1.76		5.32	6.90	300
	0220 10″ deep		200	.080		4.32	1.85		6.17	7.90	
	0230 12″ deep		190	.084		4.74	1.95		6.69	8.50	
	0300 16 ga x 8″ deep		180	.089		4.08	2.05		6.13	7.95	
	0310 10″ deep		170	.094		4.90	2.17		7.07	9.10	
	0320 12″ deep		160	.100		5.35	2.31		7.66	9.75	
	0400 14 ga x 10″ deep		140	.114		5.70	2.64		8.34	10.75	
	0410 12″ deep		130	.123		6.25	2.84		9.09	11.70	
	1210 Triple, 18 ga x 8″ deep		170	.094		5.15	2.17		7.32	9.40	
	1220 10″ deep		165	.097		6.20	2.24		8.44	10.60	
	1230 12″ deep		160	.100		6.80	2.31		9.11	11.40	
	1300 16 ga x 8″ deep		145	.110		5.95	2.55		8.50	10.90	
	1310 10″ deep		140	.114		7.05	2.64		9.69	12.25	
	1320 12″ deep		135	.119		7.70	2.74		10.44	13.15	
	1400 14 ga x 10″ deep		115	.139		7.80	3.21		11.01	14	
	1410 12″ deep	↓	110	.145	↓	8.60	3.36		11.96	15.20	
400	0010 **FRAMING, STUD WALLS** w/ top & bottom track, no openings,										400
	0020 headers, beams, bridging or bracing										
	4100 8′ high walls, 18 ga x 2-1/2″ wide, studs 12″ O.C.	2 Carp	54	.296	L.F.	5.35	6.85		12.20	17.50	
	4110 16″ O.C.		77	.208		4.29	4.80		9.09	12.85	
	4120 24″ O.C.		107	.150		3.21	3.45		6.66	9.40	
	4130 3-5/8″ wide, studs 12″ O.C.		53	.302		6.40	6.95		13.35	18.90	
	4140 16″ O.C.		76	.211		5.10	4.86		9.96	13.90	
	4150 24″ O.C.		105	.152		3.83	3.52		7.35	10.15	
	4160 4″ wide, studs 12″ O.C.		52	.308		6.70	7.10		13.80	19.45	
	4170 16″ O.C.		74	.216		5.35	4.99		10.34	14.40	
	4180 24″ O.C.		103	.155		4.03	3.59		7.62	10.55	
	4190 6″ wide, studs 12″ O.C.		51	.314		8.50	7.25		15.75	21.50	
	4200 16″ O.C.		73	.219		6.80	5.05		11.85	16.10	
	4210 24″ O.C.		101	.158		5.10	3.66		8.76	11.85	
	4220 8″ wide, studs 12″ O.C.		50	.320		10.40	7.40		17.80	24	
	4230 16″ O.C.		72	.222		8.35	5.15		13.50	17.90	
	4240 24″ O.C.		100	.160		6.30	3.70		10	13.20	
	4300 16 ga x 2-1/2″ wide, studs 12″ O.C.		47	.340		6.35	7.85		14.20	20.50	
	4310 16″ O.C.		68	.235		5	5.45		10.45	14.70	
	4320 24″ O.C.		94	.170		3.69	3.93		7.62	10.70	
	4330 3-5/8″ wide, studs 12″ O.C.		46	.348		7.60	8.05		15.65	22	
	4340 16″ O.C.		66	.242		6	5.60		11.60	16.10	
	4350 24″ O.C.		92	.174		4.43	4.02		8.45	11.70	
	4360 4″ wide, studs 12″ O.C.		45	.356		8	8.20		16.20	23	
	4370 16″ O.C.		65	.246		6.35	5.70		12.05	16.60	
	4380 24″ O.C.		90	.178		4.67	4.11		8.78	12.10	
	4390 6″ wide, studs 12″ O.C.		44	.364		10	8.40		18.40	25.50	
	4400 16″ O.C.		64	.250		7.95	5.80		13.75	18.55	
	4410 24″ O.C.		88	.182		5.90	4.20		10.10	13.60	
	4420 8″ wide, studs 12″ O.C.		43	.372		12.40	8.60		21	28.50	
	4430 16″ O.C.		63	.254		9.85	5.85		15.70	21	
	4440 24″ O.C.		86	.186		7.30	4.30		11.60	15.35	
	5100 10′ high walls, 18 ga x 2-1/2″ wide, studs 12″ O.C.		54	.296		6.45	6.85		13.30	18.70	
	5110 16″ O.C.		77	.208		5.10	4.80		9.90	13.75	
	5120 24″ O.C.		107	.150		3.75	3.45		7.20	10	
	5130 3-5/8″ wide, studs 12″ O.C.		53	.302		7.65	6.95		14.60	20.50	
	5140 16″ O.C.		76	.211		6.05	4.86		10.91	14.95	
	5150 24″ O.C.		105	.152		4.47	3.52		7.99	10.85	
	5160 4″ wide, studs 12″ O.C.		52	.308		8.05	7.10		15.15	21	
	5170 16″ O.C.	↓	74	.216	↓	6.35	4.99		11.34	15.50	

05410 | Load-Bearing Metal Studs

		Description	Crew	Daily Output	Labor-Hours	Unit	2004 Bare Costs Mat.	Labor	Equip.	Total	Total Incl O&P	
400	5180	24" O.C.	2 Carp	103	.155	L.F.	4.70	3.59		8.29	11.25	400
	5190	6" wide, studs 12" O.C.		51	.314		10.15	7.25		17.40	23.50	
	5200	16" O.C.		73	.219		8.05	5.05		13.10	17.45	
	5210	24" O.C.		101	.158		5.95	3.66		9.61	12.75	
	5220	8" wide, studs 12" O.C.		50	.320		12.45	7.40		19.85	26	
	5230	16" O.C.		72	.222		9.90	5.15		15.05	19.55	
	5240	24" O.C.		100	.160		7.35	3.70		11.05	14.30	
	5300	16 ga x 2-1/2" wide, studs 12" O.C.		47	.340		7.65	7.85		15.50	22	
	5310	16" O.C.		68	.235		6	5.45		11.45	15.80	
	5320	24" O.C.		94	.170		4.35	3.93		8.28	11.45	
	5330	3-5/8" wide, studs 12" O.C.		46	.348		9.15	8.05		17.20	24	
	5340	16" O.C.		66	.242		7.20	5.60		12.80	17.40	
	5350	24" O.C.		92	.174		5.20	4.02		9.22	12.55	
	5360	4" wide, studs 12" O.C.		45	.356		9.65	8.20		17.85	24.50	
	5370	16" O.C.		65	.246		7.55	5.70		13.25	18	
	5380	24" O.C.		90	.178		5.50	4.11		9.61	13	
	5390	6" wide, studs 12" O.C.		44	.364		12.05	8.40		20.45	27.50	
	5400	16" O.C.		64	.250		9.50	5.80		15.30	20.50	
	5410	24" O.C.		88	.182		6.90	4.20		11.10	14.75	
	5420	8" wide, studs 12" O.C.		43	.372		14.95	8.60		23.55	31	
	5430	16" O.C.		63	.254		11.75	5.85		17.60	23	
	5440	24" O.C.		86	.186		8.60	4.30		12.90	16.75	
	6190	12' high walls, 18 ga x 6" wide, studs 12" O.C.		41	.390		11.85	9		20.85	28.50	
	6200	16" O.C.		58	.276		9.30	6.35		15.65	21	
	6210	24" O.C.		81	.198		6.80	4.56		11.36	15.25	
	6220	8" wide, studs 12" O.C.		40	.400		14.45	9.25		23.70	31.50	
	6230	16" O.C.		57	.281		11.40	6.50		17.90	23.50	
	6240	24" O.C.		80	.200		8.35	4.62		12.97	17.05	
	6390	16 ga x 6" wide, studs 12" O.C.		35	.457		14.10	10.55		24.65	33.50	
	6400	16" O.C.		51	.314		11.05	7.25		18.30	24.50	
	6410	24" O.C.		70	.229		7.95	5.30		13.25	17.70	
	6420	8" wide, studs 12" O.C.		34	.471		17.45	10.85		28.30	37.50	
	6430	16" O.C.		50	.320		13.65	7.40		21.05	27.50	
	6440	24" O.C.		69	.232		9.85	5.35		15.20	19.95	
	6530	14 ga x 3-5/8" wide, studs 12" O.C.		34	.471		13.45	10.85		24.30	33.50	
	6540	16" O.C.		48	.333		10.50	7.70		18.20	24.50	
	6550	24" O.C.		65	.246		7.55	5.70		13.25	17.95	
	6560	4" wide, studs 12" O.C.		33	.485		14.20	11.20		25.40	34.50	
	6570	16" O.C.		47	.340		11.10	7.85		18.95	25.50	
	6580	24" O.C.		64	.250		7.95	5.80		13.75	18.55	
	6730	12 ga x 3-5/8" wide, studs 12" O.C.		31	.516		18.65	11.90		30.55	40.50	
	6740	16" O.C.		43	.372		14.35	8.60		22.95	30.50	
	6750	24" O.C.		59	.271		10.10	6.25		16.35	22	
	6760	4" wide, studs 12" O.C.		30	.533		19.95	12.30		32.25	43	
	6770	16" O.C.		42	.381		15.40	8.80		24.20	32	
	6780	24" O.C.		58	.276		10.85	6.35		17.20	22.50	
	7390	16' high walls, 16 ga x 6" wide, studs 12" O.C.		33	.485		18.25	11.20		29.45	39	
	7400	16" O.C.		48	.333		14.10	7.70		21.80	28.50	
	7410	24" O.C.		67	.239		10	5.50		15.50	20.50	
	7420	8" wide, studs 12" O.C.		32	.500		22.50	11.55		34.05	44.50	
	7430	16" O.C.		47	.340		17.45	7.85		25.30	32.50	
	7440	24" O.C.		66	.242		12.40	5.60		18	23	
	7560	14 ga x 4" wide, studs 12" O.C.		31	.516		18.35	11.90		30.25	40	
	7570	16" O.C.		45	.356		14.20	8.20		22.40	29.50	
	7580	24" O.C.		61	.262		10.05	6.05		16.10	21.50	
	7590	6" wide, studs 12" O.C.	↓	30	.533	↓	23	12.30		35.30	46.50	

05410 | Load-Bearing Metal Studs

			CREW	DAILY OUTPUT	LABOR-HOURS	UNIT	2004 BARE COSTS MAT.	LABOR	EQUIP.	TOTAL	TOTAL INCL O&P	
400	7600	16" O.C.	2 Carp	44	.364	L.F.	17.90	8.40		26.30	34	400
	7610	24" O.C.		60	.267		12.65	6.15		18.80	24.50	
	7760	12 ga x 4" wide, studs 12" O.C.		29	.552		26	12.75		38.75	50	
	7770	16" O.C.		40	.400		19.95	9.25		29.20	37.50	
	7780	24" O.C.		55	.291		13.90	6.70		20.60	26.50	
	7790	6" wide, studs 12" O.C.		28	.571		33	13.20		46.20	58.50	
	7800	16" O.C.		39	.410		25	9.50		34.50	43.50	
	7810	24" O.C.		54	.296		17.55	6.85		24.40	31	
	8590	20' high walls, 14 ga x 6" wide, studs 12" O.C.		29	.552		28.50	12.75		41.25	52.50	
	8600	16" O.C.		42	.381		22	8.80		30.80	39	
	8610	24" O.C.		57	.281		15.30	6.50		21.80	28	
	8620	8" wide, studs 12" O.C.		28	.571		35	13.20		48.20	61	
	8630	16" O.C.		41	.390		27	9		36	45	
	8640	24" O.C.		56	.286		18.80	6.60		25.40	31.50	
	8790	12 ga x 6" wide, studs 12" O.C.		27	.593		40.50	13.70		54.20	67.50	
	8800	16" O.C.		37	.432		31	10		41	51	
	8810	24" O.C.		51	.314		21.50	7.25		28.75	36	
	8820	8" wide, studs 12" O.C.		26	.615		49.50	14.20		63.70	78.50	
	8830	16" O.C.		36	.444		37.50	10.25		47.75	59	
	8840	24" O.C.	↓	50	.320	↓	26	7.40		33.40	41	

05420 | Cold-Formed Metal Joists

			CREW	DAILY OUTPUT	LABOR-HOURS	UNIT	MAT.	LABOR	EQUIP.	TOTAL	TOTAL INCL O&P	
100	0010	**BRACING**, continuous, per row, top & bottom										100
	0120	Flat strap, 20 ga x 2" wide, joists at 12" O.C.	1 Carp	4.67	1.713	C.L.F.	34.50	39.50		74	105	
	0130	16" O.C.		5.33	1.501		33.50	34.50		68	95.50	
	0140	24" O.C.		6.66	1.201		32	28		60	82.50	
	0150	18 ga x 2" wide, joists at 12" O.C.		4	2		46	46		92	129	
	0160	16" O.C.		4.67	1.713		45	39.50		84.50	117	
	0170	24" O.C.	↓	5.33	1.501	↓	44	34.50		78.50	108	
120	0010	**BRIDGING**, solid between joists w/ 1-1/4" leg track, per joist bay										120
	0230	Joists 12" O.C., 18 ga track x 6" wide	1 Carp	80	.100	Ea.	.93	2.31		3.24	4.94	
	0240	8" wide		75	.107		1.17	2.46		3.63	5.45	
	0250	10" wide		70	.114		1.44	2.64		4.08	6.05	
	0260	12" wide		65	.123		1.67	2.84		4.51	6.65	
	0330	16 ga track x 6" wide		70	.114		1.15	2.64		3.79	5.75	
	0340	8" wide		65	.123		1.47	2.84		4.31	6.45	
	0350	10" wide		60	.133		1.80	3.08		4.88	7.25	
	0360	12" wide		55	.145		2.08	3.36		5.44	8	
	0440	14 ga track x 8" wide		60	.133		1.85	3.08		4.93	7.30	
	0450	10" wide		55	.145		2.28	3.36		5.64	8.20	
	0460	12" wide		50	.160		2.63	3.70		6.33	9.15	
	0550	12 ga track x 10" wide		45	.178		3.35	4.11		7.46	10.65	
	0560	12" wide		40	.200		3.77	4.62		8.39	12	
	1230	16" O.C., 18 ga track x 6" wide		80	.100		1.19	2.31		3.50	5.25	
	1240	8" wide		75	.107		1.50	2.46		3.96	5.85	
	1250	10" wide		70	.114		1.85	2.64		4.49	6.50	
	1260	12" wide		65	.123		2.15	2.84		4.99	7.20	
	1330	16 ga track x 6" wide		70	.114		1.47	2.64		4.11	6.10	
	1340	8" wide		65	.123		1.88	2.84		4.72	6.90	
	1350	10" wide		60	.133		2.31	3.08		5.39	7.80	
	1360	12" wide		55	.145		2.67	3.36		6.03	8.65	
	1440	14 ga track x 8" wide		60	.133		2.38	3.08		5.46	7.85	
	1450	10" wide		55	.145		2.92	3.36		6.28	8.90	
	1460	12" wide		50	.160		3.37	3.70		7.07	9.95	
	1550	12 ga track x 10" wide	↓	45	.178	↓	4.29	4.11		8.40	11.65	

05420 | Cold-Formed Metal Joists

			CREW	DAILY OUTPUT	LABOR-HOURS	UNIT	2004 BARE COSTS MAT.	LABOR	EQUIP.	TOTAL	TOTAL INCL O&P
120	1560	12" wide	1 Carp	40	.200	Ea.	4.83	4.62		9.45	13.15
	2230	24" O.C., 18 ga track x 6" wide		80	.100		1.72	2.31		4.03	5.80
	2240	8" wide		75	.107		2.17	2.46		4.63	6.55
	2250	10" wide		70	.114		2.67	2.64		5.31	7.40
	2260	12" wide		65	.123		3.10	2.84		5.94	8.25
	2330	16 ga track x 6" wide		70	.114		2.12	2.64		4.76	6.80
	2340	8" wide		65	.123		2.72	2.84		5.56	7.80
	2350	10" wide		60	.133		3.34	3.08		6.42	8.95
	2360	12" wide		55	.145		3.87	3.36		7.23	9.95
	2440	14 ga track x 8" wide		60	.133		3.44	3.08		6.52	9.05
	2450	10" wide		55	.145		4.22	3.36		7.58	10.35
	2460	12" wide		50	.160		4.87	3.70		8.57	11.60
	2550	12 ga track x 10" wide		45	.178		6.20	4.11		10.31	13.80
	2560	12" wide	↓	40	.200	↓	7	4.62		11.62	15.55
200	0010	**FRAMING, BAND JOIST** (track) fastened to bearing wall									
	0220	18 ga track x 6" deep	2 Carp	1,000	.016	L.F.	.76	.37		1.13	1.46
	0230	8" deep		920	.017		.96	.40		1.36	1.73
	0240	10" deep		860	.019		1.18	.43		1.61	2.02
	0320	16 ga track x 6" deep		900	.018		.93	.41		1.34	1.73
	0330	8" deep		840	.019		1.20	.44		1.64	2.07
	0340	10" deep		780	.021		1.47	.47		1.94	2.42
	0350	12" deep		740	.022		1.70	.50		2.20	2.72
	0430	14 ga track x 8" deep		750	.021		1.51	.49		2	2.50
	0440	10" deep		720	.022		1.86	.51		2.37	2.91
	0450	12" deep		700	.023		2.14	.53		2.67	3.26
	0540	12 ga track x 10" deep		670	.024		2.73	.55		3.28	3.94
	0550	12" deep	↓	650	.025	↓	3.08	.57		3.65	4.35
300	0010	**FRAMING, BOXED HEADERS/BEAMS**									
	0200	Double, 18 ga x 6" deep	2 Carp	220	.073	L.F.	3.20	1.68		4.88	6.35
	0210	8" deep		210	.076		3.56	1.76		5.32	6.90
	0220	10" deep		200	.080		4.32	1.85		6.17	7.90
	0230	12" deep		190	.084		4.74	1.95		6.69	8.50
	0300	16 ga x 8" deep		180	.089		4.08	2.05		6.13	7.95
	0310	10" deep		170	.094		4.90	2.17		7.07	9.10
	0320	12" deep		160	.100		5.35	2.31		7.66	9.75
	0400	14 ga x 10" deep		140	.114		5.70	2.64		8.34	10.75
	0410	12" deep		130	.123		6.25	2.84		9.09	11.70
	0500	12 ga x 10" deep		110	.145		7.55	3.36		10.91	14
	0510	12" deep		100	.160		8.35	3.70		12.05	15.45
	1210	Triple, 18 ga x 8" deep		170	.094		5.15	2.17		7.32	9.40
	1220	10" deep		165	.097		6.20	2.24		8.44	10.60
	1230	12" deep		160	.100		6.80	2.31		9.11	11.40
	1300	16 ga x 8" deep		145	.110		5.95	2.55		8.50	10.90
	1310	10" deep		140	.114		7.05	2.64		9.69	12.25
	1320	12" deep		135	.119		7.70	2.74		10.44	13.15
	1400	14 ga x 10" deep		115	.139		8.25	3.21		11.46	14.55
	1410	12" deep		110	.145		9.10	3.36		12.46	15.70
	1500	12 ga x 10" deep		90	.178		11.05	4.11		15.16	19.10
	1510	12" deep	↓	85	.188	↓	12.25	4.35		16.60	21
410	0010	**FRAMING, JOISTS**, no band joists (track), web stiffeners, headers,									
	0020	beams, bridging or bracing									
	0030	Joists (2" flange) and fasteners, materials only									
	0220	18 ga x 6" deep				L.F.	.99			.99	1.09
	0230	8" deep					1.18			1.18	1.29
	0240	10" deep				↓	1.39			1.39	1.52

05420 | Cold-Formed Metal Joists

			CREW	DAILY OUTPUT	LABOR-HOURS	UNIT	2004 BARE COSTS MAT.	LABOR	EQUIP.	TOTAL	TOTAL INCL O&P	
410	0320	16 ga x 6" deep				L.F.	1.22			1.22	1.34	410
	0330	8" deep					1.45			1.45	1.59	
	0340	10" deep					1.69			1.69	1.86	
	0350	12" deep					1.92			1.92	2.11	
	0430	14 ga x 8" deep					1.84			1.84	2.02	
	0440	10" deep					2.11			2.11	2.32	
	0450	12" deep					2.40			2.40	2.64	
	0540	12 ga x 10" deep					3.09			3.09	3.40	
	0550	12" deep				↓	3.51			3.51	3.86	
	1010	Installation of joists to band joists, beams & headers, labor only										
	1220	18 ga x 6" deep	2 Carp	110	.145	Ea.		3.36		3.36	5.70	
	1230	8" deep		90	.178			4.11		4.11	6.95	
	1240	10" deep		80	.200			4.62		4.62	7.85	
	1320	16 ga x 6" deep		95	.168			3.89		3.89	6.60	
	1330	8" deep		70	.229			5.30		5.30	8.95	
	1340	10" deep		60	.267			6.15		6.15	10.45	
	1350	12" deep		55	.291			6.70		6.70	11.40	
	1430	14 ga x 8" deep		65	.246			5.70		5.70	9.65	
	1440	10" deep		45	.356			8.20		8.20	13.95	
	1450	12" deep		35	.457			10.55		10.55	17.90	
	1540	12 ga x 10" deep		40	.400			9.25		9.25	15.70	
	1550	12" deep	↓	30	.533	↓		12.30		12.30	21	
500	0010	**FRAMING, WEB STIFFENERS** at joist bearing, fabricated from										500
	0020	stud piece (1-5/8" flange) to stiffen joist (2" flange)										
	2120	For 6" deep joist, with 18 ga x 2-1/2" stud	1 Carp	120	.067	Ea.	1.19	1.54		2.73	3.92	
	2130	3-5/8" stud		110	.073		1.31	1.68		2.99	4.29	
	2140	4" stud		105	.076		1.27	1.76		3.03	4.39	
	2150	6" stud		100	.080		1.39	1.85		3.24	4.66	
	2160	8" stud		95	.084		1.43	1.95		3.38	4.87	
	2220	8" deep joist, with 2-1/2" stud		120	.067		1.30	1.54		2.84	4.04	
	2230	3-5/8" stud		110	.073		1.42	1.68		3.10	4.41	
	2240	4" stud		105	.076		1.39	1.76		3.15	4.52	
	2250	6" stud		100	.080		1.52	1.85		3.37	4.81	
	2260	8" stud		95	.084		1.64	1.95		3.59	5.10	
	2320	10" deep joist, with 2-1/2" stud		110	.073		1.84	1.68		3.52	4.87	
	2330	3-5/8" stud		100	.080		2.02	1.85		3.87	5.35	
	2340	4" stud		95	.084		2	1.95		3.95	5.50	
	2350	6" stud		90	.089		2.16	2.05		4.21	5.85	
	2360	8" stud		85	.094		2.20	2.17		4.37	6.10	
	2420	12" deep joist, with 2-1/2" stud		110	.073		1.94	1.68		3.62	4.99	
	2430	3-5/8" stud		100	.080		2.11	1.85		3.96	5.45	
	2440	4" stud		95	.084		2.08	1.95		4.03	5.60	
	2450	6" stud		90	.089		2.27	2.05		4.32	5.95	
	2460	8" stud		85	.094		2.45	2.17		4.62	6.40	
	3130	For 6" deep joist, with 16 ga x 3-5/8" stud		100	.080		1.38	1.85		3.23	4.66	
	3140	4" stud		95	.084		1.37	1.95		3.32	4.81	
	3150	6" stud		90	.089		1.49	2.05		3.54	5.10	
	3160	8" stud		85	.094		1.59	2.17		3.76	5.45	
	3230	8" deep joist, with 3-5/8" stud		100	.080		1.53	1.85		3.38	4.83	
	3240	4" stud		95	.084		1.50	1.95		3.45	4.95	
	3250	6" stud		90	.089		1.66	2.05		3.71	5.30	
	3260	8" stud		85	.094		1.79	2.17		3.96	5.65	
	3330	10" deep joist, with 3-5/8" stud		85	.094		2.10	2.17		4.27	6	
	3340	4" stud		80	.100		2.14	2.31		4.45	6.25	
	3350	6" stud		75	.107		2.31	2.46		4.77	6.70	
	3360	8" stud	↓	70	.114	↓	2.42	2.64		5.06	7.15	

05420 | Cold-Formed Metal Joists

			CREW	DAILY OUTPUT	LABOR-HOURS	UNIT	2004 BARE COSTS MAT.	LABOR	EQUIP.	TOTAL	TOTAL INCL O&P	
500	3430	12" deep joist, with 3-5/8" stud	1 Carp	85	.094	Ea.	2.29	2.17		4.46	6.20	500
	3440	4" stud		80	.100		2.24	2.31		4.55	6.40	
	3450	6" stud		75	.107		2.47	2.46		4.93	6.90	
	3460	8" stud		70	.114		2.67	2.64		5.31	7.40	
	4230	For 8" deep joist, with 14 ga x 3-5/8" stud		90	.089		1.99	2.05		4.04	5.65	
	4240	4" stud		85	.094		2.02	2.17		4.19	5.90	
	4250	6" stud		80	.100		2.19	2.31		4.50	6.35	
	4260	8" stud		75	.107		2.36	2.46		4.82	6.75	
	4330	10" deep joist, with 3-5/8" stud		75	.107		2.79	2.46		5.25	7.25	
	4340	4" stud		70	.114		2.76	2.64		5.40	7.50	
	4350	6" stud		65	.123		3.04	2.84		5.88	8.15	
	4360	8" stud		60	.133		3.19	3.08		6.27	8.75	
	4430	12" deep joist, with 3-5/8" stud		75	.107		2.97	2.46		5.43	7.45	
	4440	4" stud		70	.114		3.02	2.64		5.66	7.80	
	4450	6" stud		65	.123		3.28	2.84		6.12	8.40	
	4460	8" stud		60	.133		3.52	3.08		6.60	9.10	
	5330	For 10" deep joist, with 12 ga x 3-5/8" stud		65	.123		2.95	2.84		5.79	8.05	
	5340	4" stud		60	.133		3.03	3.08		6.11	8.60	
	5350	6" stud		55	.145		3.35	3.36		6.71	9.40	
	5360	8" stud		50	.160		3.67	3.70		7.37	10.30	
	5430	12" deep joist, with 3-5/8" stud		65	.123		3.27	2.84		6.11	8.40	
	5440	4" stud		60	.133		3.19	3.08		6.27	8.75	
	5450	6" stud		55	.145		3.65	3.36		7.01	9.70	
	5460	8" stud	↓	50	.160	↓	4.19	3.70		7.89	10.85	

05425 | Cold-Formed Roof Framing

			CREW	DAILY OUTPUT	LABOR-HOURS	UNIT	MAT.	LABOR	EQUIP.	TOTAL	TOTAL INCL O&P	
100	0010	**FRAMING, BRACING**										100
	0020	Continuous bracing, per row										
	0100	16 ga x 1-1/2" channel thru rafters/trusses @ 16" O.C.	1 Carp	4.50	1.778	C.L.F.	31	41		72	104	
	0120	24" O.C.		6	1.333		31	31		62	87	
	0300	2" x 2" angle x 18 ga, rafters/trusses @ 16" O.C.		6	1.333		45	31		76	102	
	0320	24" O.C.		8	1		45	23		68	88.50	
	0400	16 ga, rafters/trusses @ 16" O.C.		4.50	1.778		57	41		98	133	
	0420	24" O.C.	↓	6.50	1.231	↓	57	28.50		85.50	112	
200	0010	**FRAMING, BRIDGING**										200
	0020	Solid, between rafters w/ 1-1/4" leg track, per rafter bay										
	1200	Rafters 16" O.C., 18 ga x 4" deep	1 Carp	60	.133	Ea.	.91	3.08		3.99	6.25	
	1210	6" deep		57	.140		1.19	3.24		4.43	6.80	
	1220	8" deep		55	.145		1.50	3.36		4.86	7.35	
	1230	10" deep		52	.154		1.85	3.55		5.40	8.10	
	1240	12" deep		50	.160		2.15	3.70		5.85	8.60	
	2200	24" O.C., 18 ga x 4" deep		60	.133		1.31	3.08		4.39	6.70	
	2210	6" deep		57	.140		1.72	3.24		4.96	7.40	
	2220	8" deep		55	.145		2.17	3.36		5.53	8.10	
	2230	10" deep		52	.154		2.67	3.55		6.22	9	
	2240	12" deep	↓	50	.160	↓	3.10	3.70		6.80	9.65	
500	0010	**FRAMING, PARAPETS**										500
	0100	3' high installed on 1st story, 18 ga x 4" wide studs, 12" O.C.	2 Carp	100	.160	L.F.	3.36	3.70		7.06	9.95	
	0110	16" O.C.		150	.107		2.86	2.46		5.32	7.30	
	0120	24" O.C.		200	.080		2.35	1.85		4.20	5.75	
	0200	6" wide studs, 12" O.C.		100	.160		4.28	3.70		7.98	10.95	
	0210	16" O.C.		150	.107		3.65	2.46		6.11	8.20	
	0220	24" O.C.		200	.080		3.02	1.85		4.87	6.45	
	1100	Installed on 2nd story, 18 ga x 4" wide studs, 12" O.C.		95	.168		3.36	3.89		7.25	10.30	
	1110	16" O.C.		145	.110		2.86	2.55		5.41	7.45	
	1120	24" O.C.	↓	190	.084	↓	2.35	1.95		4.30	5.90	

	05425 Cold-Formed Roof Framing		CREW	DAILY OUTPUT	LABOR-HOURS	UNIT	MAT.	LABOR	EQUIP.	TOTAL	TOTAL INCL O&P	
500	1200	6" wide studs, 12" O.C.	2 Carp	95	.168	L.F.	4.28	3.89		8.17	11.30	500
	1210	16" O.C.		145	.110		3.65	2.55		6.20	8.35	
	1220	24" O.C.		190	.084		3.02	1.95		4.97	6.65	
	2100	Installed on gable, 18 ga x 4" wide studs, 12" O.C.		85	.188		3.36	4.35		7.71	11.10	
	2110	16" O.C.		130	.123		2.86	2.84		5.70	7.95	
	2120	24" O.C.		170	.094		2.35	2.17		4.52	6.30	
	2200	6" wide studs, 12" O.C.		85	.188		4.28	4.35		8.63	12.10	
	2210	16" O.C.		130	.123		3.65	2.84		6.49	8.85	
	2220	24" O.C.	↓	170	.094	↓	3.02	2.17		5.19	7	
550	0010	**FRAMING, ROOF RAFTERS**										550
	0100	Boxed ridge beam, double, 18 ga x 6" deep	2 Carp	160	.100	L.F.	3.20	2.31		5.51	7.45	
	0110	8" deep		150	.107		3.56	2.46		6.02	8.10	
	0120	10" deep		140	.114		4.32	2.64		6.96	9.25	
	0130	12" deep		130	.123		4.74	2.84		7.58	10	
	0200	16 ga x 6" deep		150	.107		3.64	2.46		6.10	8.20	
	0210	8" deep		140	.114		4.08	2.64		6.72	8.95	
	0220	10" deep		130	.123		4.90	2.84		7.74	10.20	
	0230	12" deep	↓	120	.133		5.35	3.08		8.43	11.10	
	1100	Rafters, 2" flange, material only, 18 ga x 6" deep					.99			.99	1.09	
	1110	8" deep					1.18			1.18	1.29	
	1120	10" deep					1.39			1.39	1.52	
	1130	12" deep					1.61			1.61	1.77	
	1200	16 ga x 6" deep					1.22			1.22	1.34	
	1210	8" deep					1.45			1.45	1.59	
	1220	10" deep					1.69			1.69	1.86	
	1230	12" deep				↓	1.92			1.92	2.11	
	2100	Installation only, ordinary rafter to 4:12 pitch, 18 ga x 6" deep	2 Carp	35	.457	Ea.		10.55		10.55	17.90	
	2110	8" deep		30	.533			12.30		12.30	21	
	2120	10" deep		25	.640			14.80		14.80	25	
	2130	12" deep		20	.800			18.50		18.50	31.50	
	2200	16 ga x 6" deep		30	.533			12.30		12.30	21	
	2210	8" deep		25	.640			14.80		14.80	25	
	2220	10" deep		20	.800			18.50		18.50	31.50	
	2230	12" deep	↓	15	1.067	↓		24.50		24.50	42	
	8100	Add to labor, ordinary rafters on steep roofs						25%				
	8110	Dormers & complex roofs						50%				
	8200	Hip & valley rafters to 4:12 pitch						25%				
	8210	Steep roofs						50%				
	8220	Dormers & complex roofs						75%				
	8300	Hip & valley jack rafters to 4:12 pitch						50%				
	8310	Steep roofs						75%				
	8320	Dormers & complex roofs						100%				
600	0010	**FRAMING, ROOF TRUSSES**										600
	0020	Fabrication of trusses on ground, Fink (W) or King Post, to 4:12 pitch										
	0120	18 ga x 4" chords, 16' span	2 Carp	12	1.333	Ea.	37.50	31		68.50	94	
	0130	20' span		11	1.455		47	33.50		80.50	109	
	0140	24' span		11	1.455		56.50	33.50		90	119	
	0150	28' span		10	1.600		65.50	37		102.50	135	
	0160	32' span		10	1.600		75	37		112	145	
	0250	6" chords, 28' span		9	1.778		82.50	41		123.50	160	
	0260	32' span		9	1.778		94	41		135	173	
	0270	36' span		8	2		106	46		152	195	
	0280	40' span		8	2		118	46		164	208	
	1120	5:12 to 8:12 pitch, 18 ga x 4" chords, 16' span	↓	10	1.600	↓	43	37		80	110	

Important: See the Reference Section for critical supporting data - Reference Nos., Crews, & Location Factors

05425 | Cold-Formed Roof Framing

			CREW	DAILY OUTPUT	LABOR-HOURS	UNIT	2004 BARE COSTS MAT.	LABOR	EQUIP.	TOTAL	TOTAL INCL O&P
600	1130	20′ span	2 Carp	9	1.778	Ea.	53.50	41		94.50	129
	1140	24′ span		9	1.778		64.50	41		105.50	141
	1150	28′ span		8	2		75	46		121	161
	1160	32′ span		8	2		86	46		132	173
	1250	6″ chords, 28′ span		7	2.286		94	53		147	193
	1260	32′ span		7	2.286		108	53		161	208
	1270	36′ span		6	2.667		121	61.50		182.50	238
	1280	40′ span		6	2.667		134	61.50		195.50	253
	2120	9:12 to 12:12 pitch, 18 ga x 4″ chords, 16′ span		8	2		53.50	46		99.50	138
	2130	20′ span		7	2.286		67	53		120	163
	2140	24′ span		7	2.286		80.50	53		133.50	178
	2150	28′ span		6	2.667		94	61.50		155.50	208
	2160	32′ span		6	2.667		107	61.50		168.50	223
	2250	6″ chords, 28′ span		5	3.200		118	74		192	254
	2260	32′ span		5	3.200		134	74		208	273
	2270	36′ span		4	4		151	92.50		243.50	325
	2280	40′ span		4	4		168	92.50		260.50	340
	5120	Erection only of roof trusses, to 4:12 pitch, 16′ span	F-6	48	.833			17.35	12.75	30.10	43
	5130	20′ span		46	.870			18.10	13.30	31.40	45
	5140	24′ span		44	.909			18.90	13.90	32.80	47.50
	5150	28′ span		42	.952			19.80	14.55	34.35	49.50
	5160	32′ span		40	1			21	15.30	36.30	52
	5170	36′ span		38	1.053			22	16.10	38.10	54.50
	5180	40′ span		36	1.111			23	17	40	57.50
	5220	5:12 to 8:12 pitch, 16′ span		42	.952			19.80	14.55	34.35	49.50
	5230	20′ span		40	1			21	15.30	36.30	52
	5240	24′ span		38	1.053			22	16.10	38.10	54.50
	5250	28′ span		36	1.111			23	17	40	57.50
	5260	32′ span		34	1.176			24.50	18	42.50	61.50
	5270	36′ span		32	1.250			26	19.15	45.15	65
	5280	40′ span		30	1.333			27.50	20.50	48	69.50
	5320	9:12 to 12:12 pitch, 16′ span		36	1.111			23	17	40	57.50
	5330	20′ span		34	1.176			24.50	18	42.50	61.50
	5340	24′ span		32	1.250			26	19.15	45.15	65
	5350	28′ span		30	1.333			27.50	20.50	48	69.50
	5360	32′ span		28	1.429			29.50	22	51.50	74
	5370	36′ span		26	1.538			32	23.50	55.50	80
	5380	40′ span		24	1.667			34.50	25.50	60	86.50
650	0010	**FRAMING, SOFFITS & CANOPIES**									
	0130	Continuous ledger track @ wall, studs @ 16″ O.C., 18 ga x 4″ wide	2 Carp	535	.030	L.F.	.61	.69		1.30	1.84
	0140	6″ wide		500	.032		.79	.74		1.53	2.12
	0150	8″ wide		465	.034		1	.79		1.79	2.45
	0160	10″ wide		430	.037		1.23	.86		2.09	2.82
	0230	Studs @ 24″ O.C., 18 ga x 4″ wide		800	.020		.58	.46		1.04	1.42
	0240	6″ wide		750	.021		.76	.49		1.25	1.67
	0250	8″ wide		700	.023		.96	.53		1.49	1.95
	0260	10″ wide		650	.025		1.18	.57		1.75	2.26
	1000	Horizontal soffit and canopy members, material only									
	1030	1-5/8″ flange studs, 18 ga x 4″ deep				L.F.	.80			.80	.88
	1040	6″ deep					1.01			1.01	1.11
	1050	8″ deep					1.22			1.22	1.35
	1140	2″ flange joists, 18 ga x 6″ deep					1.13			1.13	1.24
	1150	8″ deep					1.34			1.34	1.48
	1160	10″ deep					1.58			1.58	1.74
	4030	Installation only, 18 ga, 1-5/8″ flange x 4″ deep	2 Carp	130	.123	Ea.		2.84		2.84	4.82
	4040	6″ deep		110	.145			3.36		3.36	5.70

05400 | Cold Formed Metal Framing

	05425 | Cold-Formed Roof Framing		CREW	DAILY OUTPUT	LABOR-HOURS	UNIT	2004 BARE COSTS MAT.	LABOR	EQUIP.	TOTAL	TOTAL INCL O&P	
650	4050	8″ deep	2 Carp	90	.178	Ea.		4.11		4.11	6.95	650
	4140	2″ flange, 18 ga x 6″ deep		110	.145			3.36		3.36	5.70	
	4150	8″ deep		90	.178			4.11		4.11	6.95	
	4160	10″ deep	↓	80	.200			4.62		4.62	7.85	
	6010	Clips to attach facia to rafter tails, 2″ x 2″ x 18 ga angle	1 Carp	120	.067		.53	1.54		2.07	3.20	
	6020	16 ga angle	″	100	.080	↓	.68	1.85		2.53	3.88	

05500 | Metal Fabrications

	05517 | Metal Stairs		CREW	DAILY OUTPUT	LABOR-HOURS	UNIT	2004 BARE COSTS MAT.	LABOR	EQUIP.	TOTAL	TOTAL INCL O&P	
350	0010	**FIRE ESCAPE STAIRS**										350
	0020	One story, disappearing, stainless steel	2 Sswk	20	.800	V.L.F.	152	19.90		171.90	206	
	0100	Portable ladder				Ea.	50.50			50.50	55.50	
700	0010	**STAIR** Steel, safety nosing, steel stringers										700
	1700	Pre-erected, steel pan tread, 3′-6″ wide, 2 line pipe rail	E-2	87	.552	Riser	184	13.85	16.30	214.15	246	
	1800	With flat bar picket rail	″	87	.552		206	13.85	16.30	236.15	271	
	1810	Spiral aluminum, 5′-0″ diameter, stock units	E-4	45	.711		202	18.05	1.67	221.72	259	
	1820	Custom units		45	.711		380	18.05	1.67	399.72	455	
	1900	Spiral, cast iron, 4′-0″ diameter, ornamental, minimum		45	.711		180	18.05	1.67	199.72	235	
	1920	Maximum	↓	25	1.280	↓	246	32.50	3.01	281.51	335	

	05520 | Handrails & Railings		CREW	DAILY OUTPUT	LABOR-HOURS	UNIT	MAT.	LABOR	EQUIP.	TOTAL	TOTAL INCL O&P	
700	0010	**RAILING, PIPE**										700
	0020	Aluminum, 2 rail, satin finish, 1-1/4″ diameter	E-4	160	.200	L.F.	13.90	5.10	.47	19.47	25.50	
	0030	Clear anodized		160	.200		17.20	5.10	.47	22.77	29.50	
	0040	Dark anodized		160	.200		19.45	5.10	.47	25.02	32	
	0080	1-1/2″ diameter, satin finish		160	.200		16.65	5.10	.47	22.22	28.50	
	0090	Clear anodized		160	.200		18.60	5.10	.47	24.17	31	
	0100	Dark anodized		160	.200		20.50	5.10	.47	26.07	33	
	0140	Aluminum, 3 rail, 1-1/4″ diam., satin finish		137	.234		21.50	5.95	.55	28	35.50	
	0150	Clear anodized		137	.234		26.50	5.95	.55	33	41.50	
	0160	Dark anodized		137	.234		29.50	5.95	.55	36	44.50	
	0200	1-1/2″ diameter, satin finish		137	.234		25.50	5.95	.55	32	40	
	0210	Clear anodized		137	.234		29	5.95	.55	35.50	44	
	0220	Dark anodized		137	.234		32	5.95	.55	38.50	47	
	0500	Steel, 2 rail, on stairs, primed, 1-1/4″ diameter		160	.200		10.45	5.10	.47	16.02	22	
	0520	1-1/2″ diameter		160	.200		11.45	5.10	.47	17.02	23	
	0540	Galvanized, 1-1/4″ diameter		160	.200		14.45	5.10	.47	20.02	26	
	0560	1-1/2″ diameter		160	.200		16.20	5.10	.47	21.77	28	
	0580	Steel, 3 rail, primed, 1-1/4″ diameter		137	.234		15.55	5.95	.55	22.05	29	
	0600	1-1/2″ diameter		137	.234		16.50	5.95	.55	23	30	
	0620	Galvanized, 1-1/4″ diameter		137	.234		22	5.95	.55	28.50	36	
	0640	1-1/2″ diameter		137	.234		25.50	5.95	.55	32	40	
	0700	Stainless steel, 2 rail, 1-1/4″ diam. #4 finish		137	.234		37.50	5.95	.55	44	53.50	
	0720	High polish		137	.234		60.50	5.95	.55	67	78.50	
	0740	Mirror polish		137	.234		76	5.95	.55	82.50	95.50	
	0760	Stainless steel, 3 rail, 1-1/2″ diam., #4 finish		120	.267		56.50	6.75	.63	63.88	76.50	
	0770	High polish	↓	120	.267	↓	93.50	6.75	.63	100.88	117	

05520 | Handrails & Railings

			CREW	DAILY OUTPUT	LABOR-HOURS	UNIT	2004 BARE COSTS MAT.	LABOR	EQUIP.	TOTAL	TOTAL INCL O&P	
700	0780	Mirror finish	E-4	120	.267	L.F.	114	6.75	.63	121.38	140	700
	0900	Wall rail, alum. pipe, 1-1/4" diam., satin finish		213	.150		7.95	3.82	.35	12.12	16.50	
	0905	Clear anodized		213	.150		9.70	3.82	.35	13.87	18.45	
	0910	Dark anodized		213	.150		11.75	3.82	.35	15.92	20.50	
	0915	1-1/2" diameter, satin finish		213	.150		8.80	3.82	.35	12.97	17.45	
	0920	Clear anodized		213	.150		11.10	3.82	.35	15.27	19.95	
	0925	Dark anodized		213	.150		13.70	3.82	.35	17.87	23	
	0930	Steel pipe, 1-1/4" diameter, primed		213	.150		6.35	3.82	.35	10.52	14.70	
	0935	Galvanized		213	.150		9.15	3.82	.35	13.32	17.85	
	0940	1-1/2" diameter		176	.182		6.50	4.62	.43	11.55	16.50	
	0945	Galvanized		213	.150		9.20	3.82	.35	13.37	17.90	
	0955	Stainless steel pipe, 1-1/2" diam., #4 finish		107	.299		30	7.60	.70	38.30	48.50	
	0960	High polish		107	.299		61	7.60	.70	69.30	83	
	0965	Mirror polish	↓	107	.299	↓	72	7.60	.70	80.30	94.50	

05580 | Formed Metal Fabrications

			CREW	DAILY OUTPUT	LABOR-HOURS	UNIT	MAT.	LABOR	EQUIP.	TOTAL	TOTAL INCL O&P	
600	0010	**LAMP POSTS**										600
	0020	Aluminum, 7' high, stock units, post only	1 Carp	16	.500	Ea.	32	11.55		43.55	54.50	
	0100	Mild steel, plain	"	16	.500	"	28	11.55		39.55	50.50	

Division Notes

	CREW	DAILY OUTPUT	LABOR-HOURS	UNIT	2004 BARE COSTS				TOTAL INCL O&P
					MAT.	LABOR	EQUIP.	TOTAL	

Division 6 Wood & Plastics

Estimating Tips

06050 Basic Wood & Plastic Materials & Methods

- Common to any wood framed structure are the accessory connector items such as screws, nails, adhesives, hangers, connector plates, straps, angles and holdowns. For typical wood framed buildings, such as residential projects, the aggregate total for these items can be significant, especially in areas where seismic loading is a concern. For floor and wall framing, the material cost is based on 10 to 25 lbs. per MBF. Holdowns, hangers and other connectors should be taken off by the piece.

06100 Rough Carpentry

- Lumber is a traded commodity and therefore sensitive to supply and demand in the marketplace. Even in "budgetary" estimating of wood framed projects, it is advisable to call local suppliers for the latest market pricing.
- Common quantity units for wood framed projects are "thousand board feet" (MBF). A board foot is a volume of wood, 1″ x 1′ x 1′, or 144 cubic inches. Board foot quantities are generally calculated using nominal material dimensions—dressed sizes are ignored. Board foot per lineal foot of any stick of lumber can be calculated by dividing the nominal cross sectional area by 12. As an example, 2,000 lineal feet of 2 x 12 equates to 4 MBF by dividing the nominal area, 2 x 12, by 12, which equals 2, and multiplying by 2,000 to give 4,000 board feet. This simple rule applies to all nominal dimensioned lumber.
- Waste is an issue of concern at the quantity takeoff for any area of construction. Framing lumber is sold in even foot lengths, i.e., 10′, 12′, 14′, 16′, and depending on spans, wall heights and the grade of lumber, waste is inevitable. A rule of thumb for lumber waste is 5% to 10% depending on material quality and the complexity of the framing.
- Wood in various forms and shapes is used in many projects, even where the main structural framing is steel, concrete or masonry. Plywood as a back-up partition material and 2x boards used as blocking and cant strips around roof edges are two common examples. The estimator should ensure that the costs of all wood materials are included in the final estimate.

06200 Finish Carpentry

- It is necessary to consider the grade of workmanship when estimating labor costs for erecting millwork and interior finish. In practice, there are three grades: premium, custom and economy. The Means daily output for base and case moldings is in the range of 200 to 250 L.F. per carpenter per day. This is appropriate for most average custom grade projects. For premium projects an adjustment to productivity of 25% to 50% should be made depending on the complexity of the job.

Reference Numbers

Reference numbers are shown in bold squares at the beginning of some major classifications. These numbers refer to related items in the Reference Section. The reference information may be an estimating procedure, an alternate pricing method or technical information.

Note: Not all subdivisions listed here necessarily appear in this publication.

	06052 Selective Demolition		CREW	DAILY OUTPUT	LABOR-HOURS	UNIT	2004 BARE COSTS MAT.	LABOR	EQUIP.	TOTAL	TOTAL INCL O&P	
110	0010	SELECTIVE DEMOLITION, WOOD FRAMING										110
	3000	Wood framing, beams, 6″ x 8″	B-2	275	.145	L.F.		2.52		2.52	4.27	
	3040	6″ x 10″		220	.182			3.15		3.15	5.35	
	3080	6″ x 12″		185	.216			3.74		3.74	6.35	
	3120	8″ x 12″		140	.286			4.94		4.94	8.40	
	3160	10″ x 12″	↓	110	.364			6.30		6.30	10.70	
	3400	Fascia boards, 1″ x 6″	1 Clab	500	.016			.27		.27	.46	
	3440	1″ x 8″		450	.018			.30		.30	.51	
	3480	1″ x 10″		400	.020			.34		.34	.57	
	3800	Headers over openings, 2 @ 2″ x 6″		110	.073			1.23		1.23	2.09	
	3840	2 @ 2″ x 8″		100	.080			1.35		1.35	2.30	
	3880	2 @ 2″ x 10″	↓	90	.089			1.50		1.50	2.55	
	4230	Joists, 2″ x 6″	2 Clab	970	.016			.28		.28	.47	
	4240	2″ x 8″		940	.017			.29		.29	.49	
	4250	2″ x 10″		910	.018			.30		.30	.50	
	4280	2″ x 12″		880	.018			.31		.31	.52	
	5400	Posts, 4″ x 4″		800	.020			.34		.34	.57	
	5440	6″ x 6″		400	.040			.68		.68	1.15	
	5480	8″ x 8″		300	.053			.90		.90	1.53	
	5500	10″ x 10″		240	.067			1.13		1.13	1.91	
	5800	Rafters, ordinary, 2″ x 6″		850	.019			.32		.32	.54	
	5840	2″ x 8″		837	.019	↓		.32		.32	.55	
	6200	Stairs and stringers, minimum		40	.400	Riser		6.75		6.75	11.50	
	6240	Maximum		26	.615	″		10.40		10.40	17.65	
	6600	Studs, 2″ x 4″		2,000	.008	L.F.		.14		.14	.23	
	6640	2″ x 6″	↓	1,600	.010	″		.17		.17	.29	
	7000	Trusses, 2″ x 4″ flat wood construction										
	7050	12′ span	2 Clab	74	.216	Ea.		3.65		3.65	6.20	
	7150	24′ span		66	.242			4.10		4.10	6.95	
	7200	26′ span		64	.250			4.23		4.23	7.20	
	7250	28′ span		62	.258			4.36		4.36	7.40	
	7301	30′ span		56	.286			4.83		4.83	8.20	
	7350	32′ span		56	.286			4.83		4.83	8.20	
	7400	34′ span		54	.296			5		5	8.50	
	7450	36′ span	↓	52	.308	↓		5.20		5.20	8.85	
	9000	Minimum labor/equipment charge	1 Clab	4	2	Job		34		34	57.50	
	9500	See Div. 02220-350 for rubbish handling										
120	0010	SELECTIVE DEMOLITION, MILLWORK AND TRIM										120
	1000	Cabinets, wood, base cabinets	2 Clab	80	.200	L.F.		3.38		3.38	5.75	
	1020	Wall cabinets		80	.200			3.38		3.38	5.75	
	1100	Steel, painted, base cabinets		60	.267			4.51		4.51	7.65	
	1500	Counter top, minimum		200	.080			1.35		1.35	2.30	
	1510	Maximum		120	.133	↓		2.25		2.25	3.83	
	2000	Paneling, 4′ x 8′ sheets, 1/4″ thick		2,000	.008	S.F.		.14		.14	.23	
	2100	Boards, 1″ x 4″		700	.023			.39		.39	.66	
	2120	1″ x 6″		750	.021			.36		.36	.61	
	2140	1″ x 8″		800	.020	↓		.34		.34	.57	
	3000	Trim, baseboard, to 6″ wide		1,200	.013	L.F.		.23		.23	.38	
	3040	12″ wide		1,000	.016			.27		.27	.46	
	3100	Ceiling trim		1,000	.016			.27		.27	.46	
	3120	Chair rail		1,200	.013			.23		.23	.38	
	3140	Railings with balusters		240	.067	↓		1.13		1.13	1.91	
	3160	Wainscoting	↓	700	.023	S.F.		.39		.39	.66	
	4000	Curtain rod	1 Clab	80	.100	L.F.		1.69		1.69	2.87	
	9000	Minimum labor/equipment charge	″	4	2	Job		34		34	57.50	

06070 | Lumber Treatment

			CREW	DAILY OUTPUT	LABOR-HOURS	UNIT	2004 BARE COSTS MAT.	LABOR	EQUIP.	TOTAL	TOTAL INCL O&P	
400	0011	**LUMBER TREATMENT**										400
	0400	Fire retardant, wet				M.B.F.	291			291	320	
	0500	KDAT					277			277	305	
	0700	Salt treated, water borne, .40 lb. retention					133			133	146	
	0800	Oil borne, 8 lb. retention					156			156	171	
	1000	Kiln dried lumber, 1″ & 2″ thick, softwoods					88.50			88.50	97.50	
	1100	Hardwoods					94.50			94.50	104	
	1500	For small size 1″ stock, add				↓	11.95			11.95	13.15	
	1700	For full size rough lumber, add					20%					
600	0010	**PLYWOOD TREATMENT** Fire retardant, 1/4″ thick				M.S.F.	222			222	244	600
	0030	3/8″ thick					244			244	268	
	0050	1/2″ thick					261			261	287	
	0070	5/8″ thick					277			277	305	
	0100	3/4″ thick					305			305	335	
	0200	For KDAT, add					66.50			66.50	73	
	0500	Salt treated water borne, .25 lb., wet, 1/4″ thick					122			122	134	
	0530	3/8″ thick					127			127	140	
	0550	1/2″ thick					133			133	146	
	0570	5/8″ thick					145			145	159	
	0600	3/4″ thick					150			150	165	
	0800	For KDAT add					66.50			66.50	73	
	0900	For .40 lb., per C.F. retention, add					55.50			55.50	61	
	1000	For certification stamp, add				↓	32.50			32.50	36	

06090 | Wood & Plastic Fastenings

			CREW	DAILY OUTPUT	LABOR-HOURS	UNIT	MAT.	LABOR	EQUIP.	TOTAL	TOTAL INCL O&P	
600	0010	**NAILS** Prices of material only, based on 50# box purchase, copper, plain				Lb.	4.66			4.66	5.15	600
	0400	Stainless steel, plain					5.25			5.25	5.80	
	0500	Box, 3d to 20d, bright					1.19			1.19	1.31	
	0520	Galvanized					1.43			1.43	1.57	
	0600	Common, 3d to 60d, plain					.78			.78	.86	
	0700	Galvanized					.95			.95	1.05	
	0800	Aluminum					3.38			3.38	3.72	
	1000	Annular or spiral thread, 4d to 60d, plain					1.60			1.60	1.76	
	1200	Galvanized					1.70			1.70	1.87	
	1400	Drywall nails, plain					.70			.70	.77	
	1600	Galvanized					1.49			1.49	1.64	
	1800	Finish nails, 4d to 10d, plain					.81			.81	.89	
	2000	Galvanized					1.12			1.12	1.23	
	2100	Aluminum					4.10			4.10	4.51	
	2300	Flooring nails, hardened steel, 2d to 10d, plain					1.37			1.37	1.51	
	2400	Galvanized					2.21			2.21	2.43	
	2500	Gypsum lath nails, 1-1/8″, 13 ga. flathead, blued					1.36			1.36	1.50	
	2600	Masonry nails, hardened steel, 3/4″ to 3″ long, plain					1.41			1.41	1.55	
	2700	Galvanized					1.74			1.74	1.91	
	2900	Roofing nails, threaded, galvanized					1.32			1.32	1.45	
	3100	Aluminum					4.79			4.79	5.25	
	3300	Compressed lead head, threaded, galvanized					1.47			1.47	1.62	
	3600	Siding nails, plain shank, galvanized					1.36			1.36	1.50	
	3800	Aluminum					4.10			4.10	4.51	
	5000	Add to prices above for cement coating					.07			.07	.08	
	5200	Zinc or tin plating					.12			.12	.13	
	5500	Vinyl coated sinkers, 8d to 16d				↓	.56			.56	.62	
650	0010	**NAILS** mat. only, for pneumatic tools, framing, per carton of 5000, 2″				Ea.	37			37	41	650
	0100	2-3/8″				↓	42.50			42.50	46.50	

		06090 | Wood & Plastic Fastenings	CREW	DAILY OUTPUT	LABOR-HOURS	UNIT	2004 BARE COSTS MAT.	LABOR	EQUIP.	TOTAL	TOTAL INCL O&P	
650	0200	Per carton of 4000, 3"				Ea.	37.50			37.50	41.50	650
	0300	3-1/4"					40			40	44	
	0400	Per carton of 5000, 2-3/8", galv.					57.50			57.50	63.50	
	0500	Per carton of 4000, 3", galv.					65			65	71.50	
	0600	3-1/4", galv.					80.50			80.50	88.50	
	0700	Roofing, per carton of 7200, 1"					35			35	38.50	
	0800	1-1/4"					32.50			32.50	36	
	0900	1-1/2"					37.50			37.50	41.50	
	1000	1-3/4"				↓	45.50			45.50	50	
700	0010	**SHEET METAL SCREWS** Steel, standard, #8 x 3/4", plain				C	2.81			2.81	3.09	700
	0100	Galvanized					3.44			3.44	3.78	
	0300	#10 x 1", plain					3.76			3.76	4.14	
	0400	Galvanized					4.35			4.35	4.79	
	1500	Self-drilling, with washers, (pinch point) #8 x 3/4", plain					6.10			6.10	6.70	
	1600	Galvanized					6.10			6.10	6.70	
	1800	#10 x 3/4", plain					6.10			6.10	6.70	
	1900	Galvanized					6.10			6.10	6.70	
	3000	Stainless steel w/aluminum or neoprene washers, #14 x 1", plain					18.35			18.35	20	
	3100	#14 x 2", plain				↓	25			25	27.50	
750	0010	**WOOD SCREWS** #8, 1" long, steel				C	3.36			3.36	3.70	750
	0100	Brass					11.20			11.20	12.35	
	0200	#8, 2" long, steel					3.76			3.76	4.14	
	0300	Brass					11.70			11.70	12.90	
	0400	#10, 1" long, steel					4.30			4.30	4.73	
	0500	Brass					23			23	25	
	0600	#10, 2" long, steel					7.65			7.65	8.45	
	0700	Brass					40.50			40.50	44.50	
	0800	#10, 3" long, steel					11.95			11.95	13.15	
	1000	#12, 2" long, steel					4.89			4.89	5.40	
	1100	Brass					16.30			16.30	17.95	
	1500	#12, 3" long, steel					16.20			16.20	17.85	
	2000	#12, 4" long, steel				↓	29			29	32	
800	0010	**TIMBER CONNECTORS** Add up cost of each part for total										800
	0020	cost of connection										
	0100	Connector plates, steel, with bolts, straight	2 Carp	75	.213	Ea.	18.60	4.93		23.53	29	
	0110	Tee		50	.320		27.50	7.40		34.90	42.50	
	0120	T- Strap, 14 gauge12" x 8" x 2"		50	.320		27.50	7.40		34.90	42.50	
	0150	Anchor plate, 7 gauge, 9" x 7"	↓	75	.213		18.60	4.93		23.53	29	
	0200	Bolts, machine, sq. hd. with nut & washer, 1/2" diameter, 4" long	1 Carp	140	.057		.30	1.32		1.62	2.57	
	0300	7-1/2" long		130	.062		.53	1.42		1.95	2.99	
	0500	3/4" diameter, 7-1/2" long		130	.062		1.45	1.42		2.87	4.01	
	0610	Machine bolts, w/ nut, washer, 3/4" dia, 15" L, HD's & beam hangers		95	.084	↓	2.66	1.95		4.61	6.25	
	0720	Machine bolts, sq. hd. w/nut & wash		150	.053	Lb.	2.12	1.23		3.35	4.42	
	0800	Drilling bolt holes in timber, 1/2" diameter		450	.018	Inch		.41		.41	.70	
	0900	1" diameter		350	.023	"		.53		.53	.90	
	1100	Framing anchors, 2 or 3 dimensional, 10 gauge, no nails incl.		175	.046	Ea.	.46	1.06		1.52	2.30	
	1150	Framing anchors, 18 gauge, 4 1/2" x 2 3/4"		175	.046		.46	1.06		1.52	2.30	
	1160	Framing anchors, 18 gauge, 4 1/2" x 3"		175	.046		.46	1.06		1.52	2.30	
	1170	Clip anchors plates, 18 gauge, 12" x 1 1/8"		175	.046		.46	1.06		1.52	2.30	
	1250	Holdowns, 3 gauge base, 10 gauge body		8	1		15.45	23		38.45	56	
	1260	Holdowns, 7 gauge 11 1/16" x 3 1/4"		8	1		15.45	23		38.45	56	
	1270	Holdowns, 7 gauge 14 3/8" x 3 1/8"		8	1		15.45	23		38.45	56	
	1275	Holdowns, 12 gauge 8" x 2 1/2"		8	1		15.45	23		38.45	56	
	1300	Joist and beam hangers, 18 ga. galv., for 2" x 4" joist	↓	175	.046	↓	.57	1.06		1.63	2.42	

06090 | Wood & Plastic Fastenings

06090	Wood & Plastic Fastenings	CREW	DAILY OUTPUT	LABOR-HOURS	UNIT	2004 BARE COSTS MAT.	LABOR	EQUIP.	TOTAL	TOTAL INCL O&P
800 1400	2" x 6" to 2" x 10" joist	1 Carp	165	.048	Ea.	.49	1.12		1.61	2.44
1600	16 ga. galv., 3" x 6" to 3" x 10" joist		160	.050		2.61	1.16		3.77	4.83
1700	3" x 10" to 3" x 14" joist		160	.050		3.03	1.16		4.19	5.30
1800	4" x 6" to 4" x 10" joist		155	.052		2.30	1.19		3.49	4.55
1900	4" x 10" to 4" x 14" joist		155	.052		3.11	1.19		4.30	5.45
2000	Two-2" x 6" to two-2" x 10" joists		150	.053		2.50	1.23		3.73	4.84
2100	Two-2" x 10" to two-2" x 14" joists		150	.053		2.50	1.23		3.73	4.84
2300	3/16" thick, 6" x 8" joist		145	.055		5.45	1.27		6.72	8.15
2400	6" x 10" joist		140	.057		6.45	1.32		7.77	9.30
2500	6" x 12" joist		135	.059		7.75	1.37		9.12	10.80
2700	1/4" thick, 6" x 14" joist	↓	130	.062		9.60	1.42		11.02	12.95
2900	Plywood clips, extruded aluminum H clip, for 3/4" panels					.15			.15	.17
3000	Galvanized 18 ga. back-up clip					.14			.14	.15
3200	Post framing, 16 ga. galv. for 4" x 4" base, 2 piece	1 Carp	130	.062		5.40	1.42		6.82	8.35
3300	Cap		130	.062		2.64	1.42		4.06	5.30
3500	Rafter anchors, 18 ga. galv., 1-1/2" wide, 5-1/4" long		145	.055		.46	1.27		1.73	2.67
3600	10-3/4" long		145	.055		.90	1.27		2.17	3.15
3800	Shear plates, 2-5/8" diameter		120	.067		1.61	1.54		3.15	4.38
3900	4" diameter		115	.070		3.71	1.61		5.32	6.80
4000	Sill anchors, embedded in concrete or block, 18-5/8" long		115	.070		1.09	1.61		2.70	3.93
4100	Spike grids, 4" x 4", flat or curved		120	.067		.41	1.54		1.95	3.06
4400	Split rings, 2-1/2" diameter		120	.067		1.31	1.54		2.85	4.05
4500	4" diameter		110	.073		2.02	1.68		3.70	5.05
4550	Tie plate, 20 gauge, 7" x 3 1/8"		110	.073		2.02	1.68		3.70	5.05
4560	Tie plate, 20 gauge, 5" x 4 1/8"		110	.073		2.02	1.68		3.70	5.05
4575	Twist straps, 18 gauge, 12" x 1 1/4"		110	.073		2.02	1.68		3.70	5.05
4580	Twist straps, 18 gauge, 16" x 1 1/4"		110	.073		2.02	1.68		3.70	5.05
4600	Strap ties, 20 ga., 2 -1/16" wide, 12 13/16" long		180	.044		1.18	1.03		2.21	3.04
4700	Strap ties, 16 ga., 1-3/8" wide, 12" long		180	.044		1.18	1.03		2.21	3.04
4800	24" long		160	.050		1.62	1.16		2.78	3.74
5000	Toothed rings, 2-5/8" or 4" diameter		90	.089	↓	1.12	2.05		3.17	4.71
5200	Truss plates, nailed, 20 gauge, up to 32' span	↓	17	.471	Truss	8.05	10.85		18.90	27.50
5400	Washers, 2" x 2" x 1/8"				Ea.	.26			.26	.29
5500	3" x 3" x 3/16"				"	.66			.66	.73
6101	Beam hangers, polymer painted									
6102	Bolted, 3 ga., (W x H x L)									
6104	3-1/4" x 9" x 12" top flange	1 Carp	1	8	C	6,325	185		6,510	7,275
6106	5-1/4" x 9" x 12" top flange		1	8		6,575	185		6,760	7,550
6108	5-1/4" x 11" x 11-3/4" top flange		1	8		14,800	185		14,985	16,600
6110	6-7/8" x 9" x 12" top flange		1	8		6,800	185		6,985	7,800
6112	6-7/8" x 11" x 13-1/2" top flange		1	8		15,600	185		15,785	17,400
6114	8-7/8" x 11" x 15-1/2" top flange	↓	1	8	↓	16,600	185		16,785	18,600
6116	Nailed, 3 ga., (W x H x L)									
6118	3-1/4" x 10-1/2" x 10" top flange	1 Carp	1.80	4.444	C	4,375	103		4,478	4,975
6120	3-1/4" x 10-1/2" x 12" top flange		1.80	4.444		5,050	103		5,153	5,725
6122	5-1/4" x 9-1/2" x 10" top flange		1.80	4.444		4,700	103		4,803	5,350
6124	5-1/4" x 9-1/2" x 12" top flange		1.80	4.444		5,350	103		5,453	6,050
6126	5-1/2" x 9-1/2" x 12" top flange		1.80	4.444		4,700	103		4,803	5,350
6128	6-7/8" x 8-1/2" x 12" top flange		1.80	4.444		4,875	103		4,978	5,525
6130	7-1/2" x 8-1/2" x 12" top flange		1.80	4.444		5,025	103		5,128	5,700
6132	8-7/8" x 7-1/2" x 14" top flange	↓	1.80	4.444	↓	5,550	103		5,653	6,300
6201	Beam and purlin hangers, galvanized, 12 ga.									
6202	Purlin or joist size, 3" x 8"	1 Carp	1.70	4.706	C	800	109		909	1,075
6204	3" x 10"		1.70	4.706		895	109		1,004	1,175
6206	3" x 12"		1.65	4.848		1,050	112		1,162	1,375
6208	3" x 14"	↓	1.65	4.848	↓	1,250	112		1,362	1,550

06090 | Wood & Plastic Fastenings			CREW	DAILY OUTPUT	LABOR-HOURS	UNIT	2004 BARE COSTS MAT.	LABOR	EQUIP.	TOTAL	TOTAL INCL O&P	
800	6210	3" x 16"	1 Carp	1.65	4.848	C	1,425	112		1,537	1,775	800
	6212	4" x 8"		1.65	4.848		800	112		912	1,075	
	6214	4" x 10"		1.65	4.848		930	112		1,042	1,225	
	6216	4" x 12"		1.60	5		1,000	116		1,116	1,300	
	6218	4" x 14"		1.60	5		1,075	116		1,191	1,400	
	6220	4" x 16"		1.60	5		1,250	116		1,366	1,550	
	6224	6" x 10"		1.55	5.161		1,350	119		1,469	1,700	
	6226	6" x 12"		1.55	5.161		1,450	119		1,569	1,800	
	6228	6" x 14"		1.50	5.333		1,575	123		1,698	1,925	
	6230	6" x 16"	↓	1.50	5.333	↓	1,800	123		1,923	2,175	
	6300	Column bases										
	6302	4 x 4, 16 ga.	1 Carp	1.80	4.444	C	1,100	103		1,203	1,400	
	6306	7 ga.		1.80	4.444		2,050	103		2,153	2,450	
	6314	6 x 6, 16 ga.		1.75	4.571		1,425	106		1,531	1,725	
	6318	7 ga.		1.75	4.571		2,900	106		3,006	3,350	
	6326	8 x 8, 7 ga.		1.65	4.848		4,825	112		4,937	5,500	
	6330	8 x 10, 7 ga.	↓	1.65	4.848	↓	5,150	112		5,262	5,875	
	6590	Joist hangers, heavy duty 12 ga., galvanized										
	6592	2" x 4"	1 Carp	1.75	4.571	C	910	106		1,016	1,175	
	6594	2" x 6"		1.65	4.848		970	112		1,082	1,275	
	6595	2" x 6", 16 gauge		1.65	4.848		970	112		1,082	1,275	
	6596	2" x 8"		1.65	4.848		1,025	112		1,137	1,325	
	6597	2" x 8", 16 gauge		1.65	4.848		1,025	112		1,137	1,325	
	6598	2" x 10"		1.65	4.848		1,100	112		1,212	1,425	
	6600	2" x 12"		1.65	4.848		1,250	112		1,362	1,575	
	6622	(2) 2" x 6"		1.60	5		1,200	116		1,316	1,525	
	6624	(2) 2" x 8"		1.60	5		1,325	116		1,441	1,650	
	6626	(2) 2" x 10"		1.55	5.161		1,475	119		1,594	1,825	
	6628	(2) 2" x 12"	↓	1.55	5.161	↓	1,775	119		1,894	2,175	
	6890	Purlin hangers, painted										
	6892	12 ga., 2" x 6"	1 Carp	1.80	4.444	C	1,100	103		1,203	1,400	
	6894	2" x 8"		1.80	4.444		1,175	103		1,278	1,450	
	6896	2" x 10"		1.80	4.444		1,200	103		1,303	1,500	
	6898	2" x 12"		1.75	4.571		1,300	106		1,406	1,600	
	6934	(2) 2" x 6"		1.70	4.706		1,125	109		1,234	1,400	
	6936	(2) 2" x 8"		1.70	4.706		1,225	109		1,334	1,525	
	6938	(2) 2" x 10"		1.70	4.706		1,350	109		1,459	1,675	
	6940	(2) 2" x 12"	↓	1.65	4.848	↓	1,450	112		1,562	1,800	
825	0010	**ROUGH HARDWARE** Average % of carpentry material, minimum					.50%					825
	0200	Maximum					1.50%					
850	0010	**BRACING**										850
	0302	Let-in, "T" shaped, 22 ga. galv. steel, studs at 16" O.C.	1 Carp	580	.014	L.F.	.43	.32		.75	1.01	
	0402	Studs at 24" O.C.		600	.013		.43	.31		.74	.99	
	0502	16 ga. galv. steel straps, studs at 16" O.C.		600	.013		.64	.31		.95	1.22	
	0602	Studs at 24" O.C.	↓	620	.013	↓	.64	.30		.94	1.21	

	06110 | Wood Framing		CREW	DAILY OUTPUT	LABOR-HOURS	UNIT	2004 BARE COSTS MAT.	LABOR	EQUIP.	TOTAL	TOTAL INCL O&P	
100	0010	**BLOCKING** R06100-010										100
	1950	Miscellaneous, to wood construction										
	2000	2″ x 4″	1 Carp	250	.032	L.F.	.31	.74		1.05	1.59	
	2005	Pneumatic nailed		305	.026		.31	.61		.92	1.37	
	2050	2″ x 6″		222	.036		.52	.83		1.35	1.98	
	2055	Pneumatic nailed		271	.030		.52	.68		1.20	1.73	
	2100	2″ x 8″		200	.040		.80	.92		1.72	2.45	
	2105	Pneumatic nailed		244	.033		.80	.76		1.56	2.17	
	2150	2″ x 10″		178	.045		1.08	1.04		2.12	2.95	
	2155	Pneumatic nailed		217	.037		1.08	.85		1.93	2.64	
	2200	2″ x 12″		151	.053		1.47	1.22		2.69	3.70	
	2205	Pneumatic nailed		185	.043		1.47	1		2.47	3.32	
	2300	To steel construction										
	2320	2″ x 4″	1 Carp	208	.038	L.F.	.31	.89		1.20	1.85	
	2340	2″ x 6″		180	.044		.52	1.03		1.55	2.31	
	2360	2″ x 8″		158	.051		.80	1.17		1.97	2.86	
	2380	2″ x 10″		136	.059		1.08	1.36		2.44	3.50	
	2400	2″ x 12″		109	.073		1.47	1.70		3.17	4.50	
150	0012	**BRACING** Let-in, with 1″ x 6″ boards, studs @ 16″ O.C.	1 Carp	150	.053	L.F.	.38	1.23		1.61	2.50	150
	0202	Studs @ 24″ O.C.	″	230	.035	″	.38	.80		1.18	1.77	
200	0012	**BRIDGING** Wood, for joists 16″ O.C., 1″ x 3″	1 Carp	130	.062	Pr.	.31	1.42		1.73	2.75	200
	0017	Pneumatic nailed		170	.047		.31	1.09		1.40	2.18	
	0102	2″ x 3″ bridging		130	.062		.36	1.42		1.78	2.81	
	0107	Pneumatic nailed		170	.047		.36	1.09		1.45	2.24	
	0302	Steel, galvanized, 18 ga., for 2″ x 10″ joists at 12″ O.C.		130	.062		.87	1.42		2.29	3.37	
	0402	24″ O.C.		140	.057		1.59	1.32		2.91	3.99	
	0602	For 2″ x 14″ joists at 16″ O.C.		130	.062		1.05	1.42		2.47	3.57	
	0902	Compression type, 16″ O.C., 2″ x 8″ joists		200	.040		1.12	.92		2.04	2.80	
	1002	2″ x 12″ joists		200	.040		1.12	.92		2.04	2.80	
300	0010	**DECK, WOOD, PRESSURE TREATED LUMBER**										300
	0100	Railings and trim , 1″ x 4″	1 Carp	300	.027	L.F.	.57	.62		1.19	1.68	
	0150	2″ x 2″		300	.027		.29	.62		.91	1.37	
	0200	2″ x 4″		300	.027		.42	.62		1.04	1.51	
	0300	2″ x 6″		300	.027		.76	.62		1.38	1.89	
	0400	Decking, 1″ x 4″		275	.029	S.F.	1.89	.67		2.56	3.22	
	0500	2″ x 4″		300	.027		1.43	.62		2.05	2.63	
	0600	2″ x 6″		320	.025		1.67	.58		2.25	2.81	
	0650	5/4″ x 6″		320	.025		2.05	.58		2.63	3.24	
	0700	Redwood decking, 1″ x 4″		275	.029		5.05	.67		5.72	6.70	
	0800	2″ x 6″		340	.024		12.55	.54		13.09	14.70	
	0900	5/4″ x 6″		320	.025		8.10	.58		8.68	9.90	
505	0010	**FRAMING, BEAMS & GIRDERS** R06100-010										505
	1002	Single, 2″ x 6″ R06110-030	2 Carp	700	.023	L.F.	.52	.53		1.05	1.47	
	1007	Pneumatic nailed		812	.020		.52	.46		.98	1.34	
	1022	2″ x 8″		650	.025		.80	.57		1.37	1.85	
	1027	Pneumatic nailed		754	.021		.80	.49		1.29	1.71	
	1042	2″ x 10″		600	.027		1.08	.62		1.70	2.24	
	1047	Pneumatic nailed		696	.023		1.08	.53		1.61	2.09	
	1062	2″ x 12″		550	.029		1.47	.67		2.14	2.76	
	1067	Pneumatic nailed		638	.025		1.47	.58		2.05	2.60	
	1082	2″ x 14″		500	.032		1.77	.74		2.51	3.20	
	1087	Pneumatic nailed		580	.028		1.77	.64		2.41	3.03	
	1102	3″ x 8″		550	.029		2.19	.67		2.86	3.55	
	1122	3″ x 10″		500	.032		2.73	.74		3.47	4.26	
	1142	3″ x 12″		450	.036		3.28	.82		4.10	5	

	06110 | Wood Framing		CREW	DAILY OUTPUT	LABOR-HOURS	UNIT	2004 BARE COSTS MAT.	LABOR	EQUIP.	TOTAL	TOTAL INCL O&P	
505	1162	3" x 14" R06100-010	2 Carp	400	.040	L.F.	3.83	.92		4.75	5.80	505
	1170	4" x 6"	F-3	1,100	.036		2.51	.76	.56	3.83	4.64	
	1182	4" x 8" R06110-030		1,000	.040		3.34	.83	.61	4.78	5.75	
	1202	4" x 10"		950	.042		4.18	.88	.64	5.70	6.80	
	1222	4" x 12"		900	.044		5	.92	.68	6.60	7.80	
	1242	4" x 14"		850	.047		5.85	.98	.72	7.55	8.90	
	2002	Double, 2" x 6"	2 Carp	625	.026		1.04	.59		1.63	2.15	
	2007	Pneumatic nailed		725	.022		1.04	.51		1.55	2.02	
	2022	2" x 8"		575	.028		1.60	.64		2.24	2.85	
	2027	Pneumatic nailed		667	.024		1.60	.55		2.15	2.70	
	2042	2" x 10"		550	.029		2.16	.67		2.83	3.52	
	2047	Pneumatic nailed		638	.025		2.16	.58		2.74	3.36	
	2062	2" x 12"		525	.030		2.95	.70		3.65	4.43	
	2067	Pneumatic nailed		610	.026		2.95	.61		3.56	4.27	
	2082	2" x 14"		475	.034		3.54	.78		4.32	5.20	
	2087	Pneumatic nailed		551	.029		3.54	.67		4.21	5.05	
	3002	Triple, 2" x 6"		550	.029		1.56	.67		2.23	2.86	
	3007	Pneumatic nailed		638	.025		1.56	.58		2.14	2.70	
	3022	2" x 8"		525	.030		2.40	.70		3.10	3.83	
	3027	Pneumatic nailed		609	.026		2.40	.61		3.01	3.67	
	3042	2" x 10"		500	.032		3.24	.74		3.98	4.81	
	3047	Pneumatic nailed		580	.028		3.24	.64		3.88	4.64	
	3062	2" x 12"		475	.034		4.42	.78		5.20	6.20	
	3067	Pneumatic nailed		551	.029		4.42	.67		5.09	6	
	3082	2" x 14"		450	.036		5.30	.82		6.12	7.25	
	3087	Pneumatic nailed		522	.031		5.30	.71		6.01	7.05	
510	0010	**FRAMING, CEILINGS**										510
	6002	Suspended, 2" x 3"	2 Carp	1,000	.016	L.F.	.24	.37		.61	.90	
	6052	2" x 4"		900	.018		.31	.41		.72	1.04	
	6102	2" x 6"		800	.020		.52	.46		.98	1.35	
	6152	2" x 8"		650	.025		.80	.57		1.37	1.85	
515	0010	**FRAMING, COLUMNS**										515
	0101	4" x 4"	2 Carp	390	.041	L.F.	1.16	.95		2.11	2.89	
	0151	4" x 6"		275	.058		2.51	1.34		3.85	5.05	
	0201	4" x 8"		220	.073		3.34	1.68		5.02	6.55	
	0251	6" x 6"		215	.074		5.05	1.72		6.77	8.45	
	0301	6" x 8"		175	.091		7.10	2.11		9.21	11.40	
	0351	6" x 10"		150	.107		9.20	2.46		11.66	14.30	
520	0010	**FRAMING, HEAVY** Mill timber, beams, single 6" x 10"	2 Carp	1.10	14.545	M.B.F.	2,400	335		2,735	3,200	520
	0100	Single 8" x 16"		1.20	13.333	"	2,675	310		2,985	3,475	
	0202	Built from 2" lumber, multiple 2" x 14"		900	.018	B.F.	.76	.41		1.17	1.54	
	0212	Built from 3" lumber, multiple 3" x 6"		700	.023		1.09	.53		1.62	2.10	
	0222	Multiple 3" x 8"		800	.020		1.09	.46		1.55	1.98	
	0232	Multiple 3" x 10"		900	.018		1.09	.41		1.50	1.90	
	0242	Multiple 3" x 12"		1,000	.016		1.09	.37		1.46	1.83	
	0252	Built from 4" lumber, multiple 4" x 6"		800	.020		1.25	.46		1.71	2.16	
	0262	Multiple 4" x 8"		900	.018		1.25	.41		1.66	2.08	
	0272	Multiple 4" x 10"		1,000	.016		1.25	.37		1.62	2.01	
	0282	Multiple 4" x 12"		1,100	.015		1.25	.34		1.59	1.95	
	0292	Columns, structural grade, 1500f, 4" x 4"		450	.036	L.F.	2.23	.82		3.05	3.84	
	0302	6" x 6"		225	.071		6.50	1.64		8.14	9.95	
	0402	8" x 8"		240	.067		11.80	1.54		13.34	15.55	
	0502	10" x 10"		90	.178		19.30	4.11		23.41	28	
	0602	12" x 12"		70	.229		28.50	5.30		33.80	40	

06110 | Wood Framing

		06110 Wood Framing	CREW	DAILY OUTPUT	LABOR-HOURS	UNIT	2004 BARE COSTS MAT.	LABOR	EQUIP.	TOTAL	TOTAL INCL O&P	
520	0802	Floor planks, 2" thick, T & G, 2" x 6"	2 Carp	1,050	.015	B.F.	1.72	.35		2.07	2.49	520
	0902	2" x 10"		1,100	.015		1.73	.34		2.07	2.47	
	1102	3" thick, 3" x 6"		1,050	.015		1.24	.35		1.59	1.96	
	1202	3" x 10"		1,100	.015		1.24	.34		1.58	1.93	
	1402	Girders, structural grade, 12" x 12"		800	.020		1.78	.46		2.24	2.73	
	1502	10" x 16"		1,000	.016		1.73	.37		2.10	2.54	
	2050	Roof planks, see division 06150-600										
	2302	Roof purlins, 4" thick, structural grade	2 Carp	1,050	.015	B.F.	1.21	.35		1.56	1.93	
	2502	Roof trusses, add timber connectors, division 06090-800	"	450	.036	"	1.18	.82		2	2.69	
530	0010	**FRAMING, JOISTS**										530
	2002	Joists, 2" x 4"	2 Carp	1,250	.013	L.F.	.31	.30		.61	.84	
	2007	Pneumatic nailed		1,438	.011		.31	.26		.57	.78	
	2100	2" x 6"		1,250	.013		.52	.30		.82	1.07	
	2105	Pneumatic nailed		1,438	.011		.52	.26		.78	1.01	
	2152	2" x 8"		1,100	.015		.80	.34		1.14	1.45	
	2157	Pneumatic nailed		1,265	.013		.80	.29		1.09	1.38	
	2202	2" x 10"		900	.018		1.08	.41		1.49	1.89	
	2207	Pneumatic nailed		1,035	.015		1.08	.36		1.44	1.80	
	2252	2" x 12"		875	.018		1.47	.42		1.89	2.34	
	2257	Pneumatic nailed		1,006	.016		1.47	.37		1.84	2.24	
	2302	2" x 14"		770	.021		1.77	.48		2.25	2.76	
	2307	Pneumatic nailed		886	.018		1.77	.42		2.19	2.66	
	2352	3" x 6"		925	.017		1.64	.40		2.04	2.48	
	2402	3" x 10"		780	.021		2.73	.47		3.20	3.81	
	2452	3" x 12"		600	.027		3.28	.62		3.90	4.66	
	2502	4" x 6"		800	.020		2.51	.46		2.97	3.54	
	2552	4" x 10"		600	.027		4.18	.62		4.80	5.65	
	2602	4" x 12"		450	.036		5	.82		5.82	6.90	
	2607	Sister joist, 2" x 6"		800	.020		.52	.46		.98	1.35	
	2608	Pneumatic nailed		960	.017		.52	.39		.91	1.22	
	3000	Composite wood joist 9-1/2" deep		.90	17.778	M.L.F.	1,600	410		2,010	2,450	
	3010	11-1/2" deep		.88	18.182		1,700	420		2,120	2,600	
	3020	14" deep		.82	19.512		1,875	450		2,325	2,825	
	3030	16" deep		.78	20.513		2,550	475		3,025	3,600	
	4000	Open web joist 12" deep		.88	18.182		1,675	420		2,095	2,575	
	4010	14" deep		.82	19.512		1,950	450		2,400	2,925	
	4020	16" deep		.78	20.513		2,025	475		2,500	3,025	
	4030	18" deep		.74	21.622		2,050	500		2,550	3,100	
	6000	Composite rim joist, 1-1/4" x 9-1/2"		90	.178		1,975	4.11		1,979.11	2,150	
	6010	1-1/4" x 11-1/2"		.88	18.182		2,225	420		2,645	3,175	
	6020	1-1/4" x 14-1/2"		.82	19.512		2,650	450		3,100	3,700	
	6030	1-1/4" x 16-1/2"		.78	20.513		2,875	475		3,350	3,950	
545	0010	**FRAMING, MISCELLANEOUS**										545
	2002	Firestops, 2" x 4"	2 Carp	780	.021	L.F.	.31	.47		.78	1.14	
	2007	Pneumatic nailed		952	.017		.31	.39		.70	1	
	2102	2" x 6"		600	.027		.52	.62		1.14	1.62	
	2107	Pneumatic nailed		732	.022		.52	.51		1.03	1.43	
	5002	Nailers, treated, wood construction, 2" x 4"		800	.020		.42	.46		.88	1.24	
	5007	Pneumatic nailed		960	.017		.42	.39		.81	1.11	
	5102	2" x 6"		750	.021		.76	.49		1.25	1.68	
	5107	Pneumatic nailed		900	.018		.76	.41		1.17	1.54	
	5122	2" x 8"		700	.023		.95	.53		1.48	1.95	
	5127	Pneumatic nailed		840	.019		.95	.44		1.39	1.80	
	5202	Steel construction, 2" x 4"		750	.021		.42	.49		.91	1.30	

		06110 | Wood Framing	CREW	DAILY OUTPUT	LABOR-HOURS	UNIT	2004 BARE COSTS MAT.	LABOR	EQUIP.	TOTAL	TOTAL INCL O&P	
545	5222	2" x 6"	2 Carp	700	.023	L.F.	.76	.53		1.29	1.74	545
	5242	2" x 8"		650	.025		.95	.57		1.52	2.02	
	7002	Rough bucks, treated, for doors or windows, 2" x 6"		400	.040		.76	.92		1.68	2.41	
	7007	Pneumatic nailed		480	.033		.76	.77		1.53	2.15	
	7102	2" x 8"		380	.042		.95	.97		1.92	2.70	
	7107	Pneumatic nailed		456	.035		.95	.81		1.76	2.43	
	8001	Stair stringers, 2" x 10"		130	.123		1.08	2.84		3.92	6	
	8101	2" x 12"		130	.123		1.47	2.84		4.31	6.45	
	8151	3" x 10"		125	.128		2.73	2.96		5.69	8	
	8201	3" x 12"		125	.128		3.28	2.96		6.24	8.60	
	8870	Composite LSL, 1-1/4" x 11-1/2"		130	.123		2.23	2.84		5.07	7.25	
	8880	1-1/4" x 14-1/2"	↓	130	.123	↓	2.65	2.84		5.49	7.75	
550	0010	**PARTITIONS** Wood stud with single bottom plate and										550
	0020	double top plate, no waste, std. & better lumber										
	0182	2" x 4" studs, 8' high, studs 12" O.C.	2 Carp	80	.200	L.F.	3.74	4.62		8.36	11.95	
	0187	12" O.C., pneumatic nailed		96	.167		3.74	3.85		7.59	10.65	
	0202	16" O.C.		100	.160		3.06	3.70		6.76	9.60	
	0207	16" O.C., pneumatic nailed		120	.133		3.06	3.08		6.14	8.60	
	0302	24" O.C.		125	.128		2.38	2.96		5.34	7.60	
	0307	24" O.C., pneumatic nailed		150	.107		2.38	2.46		4.84	6.80	
	0382	10' high, studs 12" O.C.		80	.200		4.42	4.62		9.04	12.70	
	0387	12" O.C., pneumatic nailed		96	.167		4.42	3.85		8.27	11.40	
	0402	16" O.C.		100	.160		3.57	3.70		7.27	10.20	
	0407	16" O.C., pneumatic nailed		120	.133		3.57	3.08		6.65	9.20	
	0502	24" O.C.		125	.128		2.72	2.96		5.68	8	
	0507	24" O.C., pneumatic nailed		150	.107		2.72	2.46		5.18	7.15	
	0582	12' high, studs 12" O.C.		65	.246		5.10	5.70		10.80	15.25	
	0587	12" O.C., pneumatic nailed		78	.205		5.10	4.74		9.84	13.65	
	0602	16" O.C.		80	.200		4.08	4.62		8.70	12.35	
	0607	16" O.C., pneumatic nailed		96	.167		4.08	3.85		7.93	11.05	
	0700	24" O.C.		100	.160		2.78	3.70		6.48	9.30	
	0705	24" O.C., pneumatic nailed		120	.133		2.78	3.08		5.86	8.30	
	0782	2" x 6" studs, 8' high, studs 12" O.C.		70	.229		6.30	5.30		11.60	15.90	
	0787	12" O.C., pneumatic nailed		84	.190		6.30	4.40		10.70	14.40	
	0802	16" O.C.		90	.178		5.15	4.11		9.26	12.65	
	0807	16" O.C., pneumatic nailed		108	.148		5.15	3.42		8.57	11.50	
	0902	24" O.C.		115	.139		4.01	3.21		7.22	9.85	
	0907	24" O.C., pneumatic nailed		138	.116		4.01	2.68		6.69	8.95	
	0982	10' high, studs 12" O.C.		70	.229		7.45	5.30		12.75	17.15	
	0987	12" O.C., pneumatic nailed		84	.190		7.45	4.40		11.85	15.65	
	1002	16" O.C.		90	.178		6	4.11		10.11	13.55	
	1007	16" O.C., pneumatic nailed		108	.148		6	3.42		9.42	12.40	
	1102	24" O.C.		115	.139		4.59	3.21		7.80	10.50	
	1107	24" O.C., pneumatic nailed		138	.116		4.59	2.68		7.27	9.60	
	1182	12' high, studs 12" O.C.		55	.291		8.60	6.70		15.30	21	
	1187	12" O.C., pneumatic nailed		66	.242		8.60	5.60		14.20	18.95	
	1202	16" O.C.		70	.229		6.90	5.30		12.20	16.50	
	1207	16" O.C., pneumatic nailed		84	.190		6.90	4.40		11.30	15	
	1302	24" O.C.		90	.178		5.15	4.11		9.26	12.65	
	1307	24" O.C., pneumatic nailed		108	.148		5.15	3.42		8.57	11.50	
	1402	For horizontal blocking, 2" x 4", add		600	.027		.34	.62		.96	1.42	
	1502	2" x 6", add		600	.027		.57	.62		1.19	1.68	
	1600	For openings, add	↓	250	.064	↓		1.48		1.48	2.51	
	1702	Headers for above openings, material only, add				B.F.	.66			.66	.73	
555	0010	**FRAMING, ROOFS**										555
	2001	Fascia boards, 2" x 8"	2 Carp	225	.071	L.F.	.80	1.64		2.44	3.67	

6 WOOD & PLASTICS

06110 | Wood Framing

555		CREW	DAILY OUTPUT	LABOR-HOURS	UNIT	2004 BARE COSTS MAT.	LABOR	EQUIP.	TOTAL	TOTAL INCL O&P
2101	2" x 10"	2 Carp	180	.089	L.F.	1.08	2.05		3.13	4.67
5002	Rafters, to 4 in 12 pitch, 2" x 6", ordinary		1,000	.016		.52	.37		.89	1.20
5021	On steep roofs		800	.020		.52	.46		.98	1.35
5041	On dormers or complex roofs		590	.027		.52	.63		1.15	1.63
5062	2" x 8", ordinary		950	.017		.80	.39		1.19	1.54
5081	On steep roofs		750	.021		.80	.49		1.29	1.72
5101	On dormers or complex roofs		540	.030		.80	.68		1.48	2.04
5122	2" x 10", ordinary		630	.025		1.08	.59		1.67	2.19
5141	On steep roofs		495	.032		1.08	.75		1.83	2.46
5161	On dormers or complex roofs		425	.038		1.08	.87		1.95	2.67
5182	2" x 12", ordinary		575	.028		1.47	.64		2.11	2.71
5201	On steep roofs		455	.035		1.47	.81		2.28	3
5221	On dormers or complex roofs		395	.041		1.47	.94		2.41	3.21
5250	Composite rafter, 9-1/2" deep		575	.028		1.59	.64		2.23	2.84
5260	11-1/2" deep		575	.028		1.70	.64		2.34	2.96
5301	Hip and valley rafters, 2" x 6", ordinary		760	.021		.52	.49		1.01	1.40
5321	On steep roofs		585	.027		.52	.63		1.15	1.64
5341	On dormers or complex roofs		510	.031		.52	.72		1.24	1.80
5361	2" x 8", ordinary		720	.022		.80	.51		1.31	1.75
5381	On steep roofs		545	.029		.80	.68		1.48	2.03
5401	On dormers or complex roofs		470	.034		.80	.79		1.59	2.21
5421	2" x 10", ordinary		570	.028		1.08	.65		1.73	2.29
5441	On steep roofs		440	.036		1.08	.84		1.92	2.62
5461	On dormers or complex roofs	↓	380	.042	↓	1.08	.97		2.05	2.84
5470										
5481	Hip and valley rafters, 2" x 12", ordinary	2 Carp	525	.030	L.F.	1.47	.70		2.17	2.81
5501	On steep roofs		410	.039		1.47	.90		2.37	3.15
5521	On dormers or complex roofs		355	.045		1.47	1.04		2.51	3.39
5541	Hip and valley jacks, 2" x 6", ordinary		600	.027		.52	.62		1.14	1.62
5561	On steep roofs		475	.034		.52	.78		1.30	1.89
5581	On dormers or complex roofs		410	.039		.52	.90		1.42	2.10
5601	2" x 8", ordinary		490	.033		.80	.75		1.55	2.16
5621	On steep roofs		385	.042		.80	.96		1.76	2.51
5641	On dormers or complex roofs		335	.048		.80	1.10		1.90	2.75
5661	2" x 10", ordinary		450	.036		1.08	.82		1.90	2.58
5681	On steep roofs		350	.046		1.08	1.06		2.14	2.98
5701	On dormers or complex roofs		305	.052		1.08	1.21		2.29	3.25
5721	2" x 12", ordinary		375	.043		1.47	.99		2.46	3.29
5741	On steep roofs		295	.054		1.47	1.25		2.72	3.75
5762	On dormers or complex roofs		255	.063		1.47	1.45		2.92	4.08
5781	Rafter tie, 1" x 4", #3		800	.020		.34	.46		.80	1.15
5791	2" x 4", #3		800	.020		.31	.46		.77	1.12
5801	Ridge board, #2 or better, 1" x 6"		600	.027		.91	.62		1.53	2.05
5821	1" x 8"		550	.029		1.22	.67		1.89	2.48
5841	1" x 10"		500	.032		1.52	.74		2.26	2.92
5861	2" x 6"		500	.032		.52	.74		1.26	1.82
5881	2" x 8"		450	.036		.80	.82		1.62	2.27
5901	2" x 10"		400	.040		1.08	.92		2	2.76
5921	Roof cants, split, 4" x 4"		650	.025		1.16	.57		1.73	2.25
5941	6" x 6"		600	.027		5.05	.62		5.67	6.60
5961	Roof curbs, untreated, 2" x 6"		520	.031		.52	.71		1.23	1.78
5981	2" x 12"		400	.040		1.47	.92		2.39	3.19
6001	Sister rafters, 2" x 6"		800	.020		.52	.46		.98	1.35
6021	2" x 8"		640	.025		.80	.58		1.38	1.86
6041	2" x 10"		535	.030		1.08	.69		1.77	2.36
6061	2" x 12"	↓	455	.035	↓	1.47	.81		2.28	3

	06110 | Wood Framing	CREW	DAILY OUTPUT	LABOR-HOURS	UNIT	2004 BARE COSTS MAT.	LABOR	EQUIP.	TOTAL	TOTAL INCL O&P	
560	0010 **FRAMING, SILLS**										560
	2002 Ledgers, nailed, 2" x 4"	2 Carp	755	.021	L.F.	.31	.49		.80	1.17	
	2052 2" x 6"		600	.027		.52	.62		1.14	1.62	
	2102 Bolted, not including bolts, 3" x 6"		325	.049		1.64	1.14		2.78	3.73	
	2152 3" x 12"		233	.069		3.28	1.59		4.87	6.30	
	2602 Mud sills, redwood, construction grade, 2" x 4"		895	.018		2.79	.41		3.20	3.77	
	2622 2" x 6"		780	.021		4.18	.47		4.65	5.40	
	4002 Sills, 2" x 4"		600	.027		.31	.62		.93	1.39	
	4052 2" x 6"		550	.029		.52	.67		1.19	1.71	
	4082 2" x 8"		500	.032		.80	.74		1.54	2.13	
	4101 2" x 10"		450	.036		1.08	.82		1.90	2.58	
	4121 2" x 12"		400	.040		1.47	.92		2.39	3.19	
	4202 Treated, 2" x 4"		550	.029		.42	.67		1.09	1.60	
	4222 2" x 6"		500	.032		.76	.74		1.50	2.09	
	4242 2" x 8"		450	.036		.95	.82		1.77	2.44	
	4261 2" x 10"		400	.040		1.18	.92		2.10	2.87	
	4281 2" x 12"		350	.046		1.81	1.06		2.87	3.78	
	4402 4" x 4"		450	.036		1.29	.82		2.11	2.80	
	4422 4" x 6"		350	.046		2.07	1.06		3.13	4.06	
	4462 4" x 8"		300	.053		2.77	1.23		4	5.15	
	4480 4" x 10"	▼	260	.062	▼	3.18	1.42		4.60	5.90	
565	0010 **FRAMING, SLEEPERS**										565
	0100 On concrete, treated, 1" x 2"	2 Carp	2,350	.007	L.F.	.09	.16		.25	.37	
	0150 1" x 3"		2,000	.008		.17	.18		.35	.50	
	0200 2" x 4"		1,500	.011		.42	.25		.67	.88	
	0250 2" x 6"	▼	1,300	.012	▼	.76	.28		1.04	1.32	
570	0010 **FRAMING, SOFFITS & CANOPIES**										570
	1002 Canopy or soffit framing , 1" x 4"	2 Carp	900	.018	L.F.	.60	.41		1.01	1.36	
	1021 1" x 6"		850	.019		.91	.43		1.34	1.74	
	1042 1" x 8"		750	.021		1.22	.49		1.71	2.18	
	1102 2" x 4"		620	.026		.31	.60		.91	1.35	
	1121 2" x 6"		560	.029		.52	.66		1.18	1.69	
	1142 2" x 8"		500	.032		.80	.74		1.54	2.13	
	1202 3" x 4"		500	.032		1.09	.74		1.83	2.45	
	1221 3" x 6"		400	.040		1.64	.92		2.56	3.37	
	1242 3" x 10"	▼	300	.053	▼	2.73	1.23		3.96	5.10	
	1250										
575	0010 **FRAMING, TREATED LUMBER**										575
	0020 Water-borne salt, C.C.A., A.C.A., wet, .40 P.C.F. retention										
	0100 2" x 4"				M.B.F.	630			630	690	
	0110 2" x 6"					765			765	840	
	0120 2" x 8"					715			715	785	
	0130 2" x 10"					710			710	780	
	0140 2" x 12"					905			905	995	
	0200 4" x 4"					965			965	1,050	
	0210 4" x 6"					1,025			1,025	1,125	
	0220 4" x 8"				▼	1,050			1,050	1,150	
	0250 Add for .60 P.C.F. retention					40%					
	0260 Add for 2.5 P.C.F. retention					200%					
	0270 Add for K.D.A.T.					20%					
590	0010 **FRAMING, WALLS** R06100-010										590
	0100 Door buck, studs, header & access, 8' high 2" x 4" wall, 3'W	1 Carp	32	.250	Ea.	13.30	5.80		19.10	24.50	

06110 | Wood Framing

R06100-010

R06110-030

	06110 Wood Framing	CREW	DAILY OUTPUT	LABOR-HOURS	UNIT	2004 BARE COSTS MAT.	LABOR	EQUIP.	TOTAL	TOTAL INCL O&P
590										
0110	4' wide	1 Carp	32	.250	Ea.	14.35	5.80		20.15	25.50
0120	5' wide		32	.250		18.30	5.80		24.10	30
0130	6' wide		32	.250		19.90	5.80		25.70	32
0140	8' wide		30	.267		27.50	6.15		33.65	41
0150	10' wide		30	.267		40	6.15		46.15	54.50
0160	12' wide		30	.267		53.50	6.15		59.65	69
0170	2" x 6" wall, 3' wide		32	.250		20	5.80		25.80	32
0180	4' wide		32	.250		21	5.80		26.80	33
0190	5' wide		32	.250		25	5.80		30.80	37.50
0200	6' wide		32	.250		26.50	5.80		32.30	39.50
0210	8' wide		30	.267		34.50	6.15		40.65	48.50
0220	10' wide		30	.267		47	6.15		53.15	62
0230	12' wide		30	.267		60	6.15		66.15	76.50
0240	Window buck, studs, header & access, 8' high 2" x 4" wall, 2' wide		24	.333		13.75	7.70		21.45	28
0250	3' wide		24	.333		16.40	7.70		24.10	31
0260	4' wide		24	.333		18.05	7.70		25.75	33
0270	5' wide		24	.333		22	7.70		29.70	37
0280	6' wide		24	.333		24.50	7.70		32.20	40
0300	8' wide		22	.364		34	8.40		42.40	52
0310	10' wide		22	.364		47.50	8.40		55.90	67
0320	12' wide		22	.364		62.50	8.40		70.90	83.50
0330	2" x 6" wall, 2' wide		24	.333		22	7.70		29.70	37.50
0340	3' wide		24	.333		25.50	7.70		33.20	41
0350	4' wide		24	.333		27	7.70		34.70	43
0360	5' wide		24	.333		31.50	7.70		39.20	47.50
0370	6' wide		24	.333		34.50	7.70		42.20	51
0380	7' wide		24	.333		41.50	7.70		49.20	59
0390	8' wide		22	.364		45.50	8.40		53.90	64.50
0400	10' wide		22	.364		59.50	8.40		67.90	80
0410	12' wide	↓	22	.364	↓	75.50	8.40		83.90	98
2002	Headers over openings, 2" x 6"	2 Carp	360	.044	L.F.	.52	1.03		1.55	2.31
2007	2" x 6", pneumatic nailed		432	.037		.52	.86		1.38	2.02
2052	2" x 8"		340	.047		.80	1.09		1.89	2.72
2057	2" x 8", pneumatic nailed		408	.039		.80	.91		1.71	2.42
2101	2" x 10"		320	.050		1.08	1.16		2.24	3.15
2106	2" x 10", pneumatic nailed		384	.042		1.08	.96		2.04	2.82
2152	2" x 12"		300	.053		1.47	1.23		2.70	3.71
2157	2" x 12", pneumatic nailed		360	.044		1.47	1.03		2.50	3.36
2191	4" x 10"		240	.067		4.18	1.54		5.72	7.20
2196	4" x 10", pneumatic nailed		288	.056		4.18	1.28		5.46	6.80
2202	4" x 12"		190	.084		5	1.95		6.95	8.80
2207	4" x 12", pneumatic nailed		228	.070		5	1.62		6.62	8.25
2241	6" x 10"		165	.097		9.20	2.24		11.44	13.90
2246	6" x 10", pneumatic nailed		198	.081		9.20	1.87		11.07	13.25
2251	6" x 12"		140	.114		10.85	2.64		13.49	16.40
2256	6" x 12", pneumatic nailed		168	.095		10.85	2.20		13.05	15.65
3000	Radius, 2" x 6"		270	.059		.81	1.37		2.18	3.22
3010	2" x 6", pneumatic nailed		324	.049		.81	1.14		1.95	2.84
3020	2" x 8"		255	.063		1.25	1.45		2.70	3.84
3030	2" x 8", pneumatic nailed		296	.054		1.25	1.25		2.50	3.50
3040	2" x 10"		240	.067		1.69	1.54		3.23	4.47
3050	2" x 10", pneumatic nailed		285	.056		1.69	1.30		2.99	4.06
3060	2" x 12"		225	.071		2.30	1.64		3.94	5.30
3070	2" x 12", pneumatic nailed		270	.059		2.30	1.37		3.67	4.85
5002	Plates, untreated, 2" x 3"		850	.019		.24	.43		.67	1.01
5007	2" x 3", pneumatic nailed	↓	1,020	.016	↓	.24	.36		.60	.89

WOOD & PLASTICS 6

06110 | Wood Framing		CREW	DAILY OUTPUT	LABOR-HOURS	UNIT	2004 BARE COSTS MAT.	LABOR	EQUIP.	TOTAL	TOTAL INCL O&P
590 5022	2" x 4" R06100-010	2 Carp	800	.020	L.F.	.31	.46		.77	1.12
5027	2" x 4", pneumatic nailed		960	.017		.31	.39		.70	.99
5040	2" x 6" R06110-030		750	.021		.52	.49		1.01	1.41
5045	2" x 6", pneumatic nailed		900	.018		.52	.41		.93	1.27
5061	Treated, 2" x 3"		850	.019		.45	.43		.88	1.23
5066	2" x 3", treated, pneumatic nailed		1,020	.016		.45	.36		.81	1.11
5081	2" x 4"		800	.020		.42	.46		.88	1.24
5086	2" x 4", treated, pneumatic nailed		960	.017		.42	.39		.81	1.11
5101	2" x 6"		750	.021		.76	.49		1.25	1.68
5106	2" x 6", treated, pneumatic nailed		900	.018		.76	.41		1.17	1.54
5122	Studs, 8' high wall, 2" x 3"		1,200	.013		.24	.31		.55	.79
5127	2" x 3", pneumatic nailed		1,440	.011		.24	.26		.50	.71
5142	2" x 4"		1,100	.015		.31	.34		.65	.91
5147	2" x 4", pneumatic nailed		1,320	.012		.31	.28		.59	.82
5162	2" x 6"		1,000	.016		.52	.37		.89	1.20
5167	2" x 6", pneumatic nailed		1,200	.013		.52	.31		.83	1.09
5182	3" x 4"		800	.020		1.09	.46		1.55	1.98
5187	3" x 4", pneumatic nailed		960	.017		1.09	.39		1.48	1.85
5201	Installed on second story, 2" x 3"		1,170	.014		.24	.32		.56	.81
5206	2" x 3", pneumatic nailed		1,200	.013		.24	.31		.55	.79
5221	2" x 4"		1,015	.016		.31	.36		.67	.96
5226	2" x 4", pneumatic nailed		1,080	.015		.31	.34		.65	.92
5241	2" x 6"		890	.018		.52	.42		.94	1.27
5246	2" x 6", pneumatic nailed		1,020	.016		.52	.36		.88	1.19
5261	3" x 4"		800	.020		1.09	.46		1.55	1.98
5266	3" x 4", pneumatic nailed		960	.017		1.09	.39		1.48	1.85
5281	Installed on dormer or gable, 2" x 3"		1,045	.015		.24	.35		.59	.87
5286	2" x 3", pneumatic nailed		1,254	.013		.24	.29		.53	.77
5301	2" x 4"		905	.018		.31	.41		.72	1.03
5306	2" x 4", pneumatic nailed		1,086	.015		.31	.34		.65	.92
5321	2" x 6"		800	.020		.52	.46		.98	1.35
5326	2" x 6", pneumatic nailed		960	.017		.52	.39		.91	1.22
5341	3" x 4"		700	.023		1.09	.53		1.62	2.10
5346	3" x 4", pneumatic nailed		840	.019		1.09	.44		1.53	1.95
5361	6' high wall, 2" x 3"		970	.016		.24	.38		.62	.92
5366	2" x 3", pneumatic nailed		1,164	.014		.24	.32		.56	.81
5381	2" x 4"		850	.019		.31	.43		.74	1.08
5386	2" x 4", pneumatic nailed		1,020	.016		.31	.36		.67	.96
5401	2" x 6"		740	.022		.52	.50		1.02	1.42
5406	2" x 6", pneumatic nailed		888	.018		.52	.42		.94	1.28
5421	3" x 4"		600	.027		1.09	.62		1.71	2.25
5426	3" x 4", pneumatic nailed		720	.022		1.09	.51		1.60	2.07
5441	Installed on second story, 2" x 3"		950	.017		.24	.39		.63	.93
5446	2" x 3", pneumatic nailed		1,140	.014		.24	.32		.56	.82
5461	2" x 4"		810	.020		.31	.46		.77	1.11
5466	2" x 4", pneumatic nailed		972	.016		.31	.38		.69	.99
5481	2" x 6"		700	.023		.52	.53		1.05	1.47
5486	2" x 6", pneumatic nailed		840	.019		.52	.44		.96	1.32
5501	3" x 4"		550	.029		1.09	.67		1.76	2.34
5506	3" x 4", pneumatic nailed		660	.024		1.09	.56		1.65	2.15
5521	Installed on dormer or gable, 2" x 3"		850	.019		.24	.43		.67	1.01
5526	2" x 3", pneumatic nailed		1,020	.016		.24	.36		.60	.89
5541	2" x 4"		720	.022		.31	.51		.82	1.21
5546	2" x 4", pneumatic nailed		864	.019		.31	.43		.74	1.07
5561	2" x 6"		620	.026		.52	.60		1.12	1.58
5566	2" x 6", pneumatic nailed		744	.022		.52	.50		1.02	1.41

06110 | Wood Framing

		06110 Wood Framing	CREW	DAILY OUTPUT	LABOR-HOURS	UNIT	2004 BARE COSTS MAT.	LABOR	EQUIP.	TOTAL	TOTAL INCL O&P	
590	5581	3″ x 4″ (R06100-010)	2 Carp	480	.033	L.F.	1.09	.77		1.86	2.51	590
	5586	3″ x 4″, pneumatic nailed		576	.028		1.09	.64		1.73	2.29	
	5601	3′ high wall, 2″ x 3″ (R06110-030)		740	.022		.24	.50		.74	1.12	
	5606	2″ x 3″, pneumatic nailed		888	.018		.24	.42		.66	.98	
	5621	2″ x 4″		640	.025		.31	.58		.89	1.32	
	5626	2″ x 4″, pneumatic nailed		768	.021		.31	.48		.79	1.16	
	5641	2″ x 6″		550	.029		.52	.67		1.19	1.71	
	5646	2″ x 6″, pneumatic nailed		660	.024		.52	.56		1.08	1.52	
	5661	3″ x 4″		440	.036		1.09	.84		1.93	2.63	
	5666	3″ x 4″, pneumatic nailed		528	.030		1.09	.70		1.79	2.39	
	5681	Installed on second story, 2″ x 3″		700	.023		.24	.53		.77	1.17	
	5686	2″ x 3″, pneumatic nailed		840	.019		.24	.44		.68	1.02	
	5701	2″ x 4″		610	.026		.31	.61		.92	1.37	
	5706	2″ x 4″, pneumatic nailed		732	.022		.31	.51		.82	1.20	
	5721	2″ x 6″		520	.031		.52	.71		1.23	1.78	
	5726	2″ x 6″, pneumatic nailed		624	.026		.52	.59		1.11	1.58	
	5741	3″ x 4″		430	.037		1.09	.86		1.95	2.66	
	5746	3″ x 4″, pneumatic nailed		516	.031		1.09	.72		1.81	2.42	
	5761	Installed on dormer or gable, 2″ x 3″		625	.026		.24	.59		.83	1.27	
	5766	2″ x 3″, pneumatic nailed		750	.021		.24	.49		.73	1.11	
	5781	2″ x 4″		545	.029		.31	.68		.99	1.49	
	5786	2″ x 4″, pneumatic nailed		654	.024		.31	.57		.88	1.30	
	5801	2″ x 6″		465	.034		.52	.79		1.31	1.92	
	5806	2″ x 6″, pneumatic nailed		558	.029		.52	.66		1.18	1.69	
	5821	3″ x 4″		380	.042		1.09	.97		2.06	2.85	
	5826	3″ x 4″, pneumatic nailed		456	.035		1.09	.81		1.90	2.58	
	8250	For second story & above, add						5%				
	8300	For dormer & gable, add						15%				
600	0010	**FURRING** Wood strips, 1″ x 2″, on walls, on wood	1 Carp	550	.015	L.F.	.19	.34		.53	.78	600
	0017	Pneumatic nailed		710	.011		.19	.26		.45	.65	
	0302	On masonry		495	.016		.19	.37		.56	.84	
	0402	On concrete		260	.031		.19	.71		.90	1.42	
	0602	1″ x 3″, on walls, on wood		550	.015		.21	.34		.55	.80	
	0607	Pneumatic nailed		710	.011		.21	.26		.47	.67	
	0702	On masonry		495	.016		.21	.37		.58	.86	
	0802	On concrete		260	.031		.21	.71		.92	1.44	
	0852	On ceilings, on wood		350	.023		.21	.53		.74	1.13	
	0857	Pneumatic nailed		450	.018		.21	.41		.62	.93	
	0902	On masonry		320	.025		.21	.58		.79	1.21	
	0952	On concrete		210	.038		.21	.88		1.09	1.72	
700	0010	**GROUNDS** For casework, 1″ x 2″ wood strips, on wood	1 Carp	330	.024	L.F.	.19	.56		.75	1.16	700
	0102	On masonry		285	.028		.19	.65		.84	1.31	
	0202	On concrete		250	.032		.19	.74		.93	1.46	
	0402	For plaster, 3/4″ deep, on wood		450	.018		.19	.41		.60	.91	
	0502	On masonry		225	.036		.19	.82		1.01	1.60	
	0602	On concrete		175	.046		.19	1.06		1.25	2	
	0702	On metal lath		200	.040		.19	.92		1.11	1.78	

06150 | Wood Decking

		06150 Wood Decking	CREW	DAILY OUTPUT	LABOR-HOURS	UNIT	MAT.	LABOR	EQUIP.	TOTAL	TOTAL INCL O&P	
600	0010	**ROOF DECKS**										600
	0400	Cedar planks, 3″ thick	2 Carp	320	.050	S.F.	5.95	1.16		7.11	8.45	
	0500	4″ thick		250	.064		8	1.48		9.48	11.30	
	0702	Douglas fir, 3″ thick		320	.050		2.22	1.16		3.38	4.40	
	0802	4″ thick		250	.064		2.97	1.48		4.45	5.80	
	1002	Hemlock, 3″ thick		320	.050		2.22	1.16		3.38	4.40	

	06150 | Wood Decking		Crew	Daily Output	Labor-Hours	Unit	2004 Bare Costs Mat.	Labor	Equip.	Total	Total Incl O&P	
600	1102	4" thick	2 Carp	250	.064	S.F.	2.96	1.48		4.44	5.75	600
	1302	Western white spruce, 3" thick		320	.050		2.14	1.16		3.30	4.31	
	1402	4" thick		250	.064		2.85	1.48		4.33	5.65	
	06160 | Sheathing											
800	0010	**SHEATHING** Plywood on roof, CDX R06160-020										800
	0032	5/16" thick	2 Carp	1,600	.010	S.F.	.42	.23		.65	.85	
	0037	Pneumatic nailed		1,952	.008		.42	.19		.61	.78	
	0052	3/8" thick		1,525	.010		.44	.24		.68	.89	
	0057	Pneumatic nailed		1,860	.009		.44	.20		.64	.82	
	0102	1/2" thick		1,400	.011		.39	.26		.65	.88	
	0103	Pneumatic nailed		1,708	.009		.39	.22		.61	.80	
	0202	5/8" thick		1,300	.012		.45	.28		.73	.98	
	0207	Pneumatic nailed		1,586	.010		.45	.23		.68	.90	
	0302	3/4" thick		1,200	.013		.57	.31		.88	1.15	
	0307	Pneumatic nailed		1,464	.011		.57	.25		.82	1.06	
	0502	Plywood on walls with exterior CDX, 3/8" thick		1,200	.013		.44	.31		.75	1	
	0507	Pneumatic nailed		1,488	.011		.44	.25		.69	.90	
	0603	1/2" thick		1,125	.014		.39	.33		.72	.99	
	0608	Pneumatic nailed		1,395	.011		.39	.27		.66	.88	
	0702	5/8" thick		1,050	.015		.45	.35		.80	1.10	
	0707	Pneumatic nailed		1,302	.012		.45	.28		.73	.98	
	0803	3/4" thick		975	.016		.57	.38		.95	1.27	
	0808	Pneumatic nailed		1,209	.013		.57	.31		.88	1.15	
	1000	For shear wall construction, add						20%				
	1200	For structural 1 exterior plywood, add				S.F.	10%					
	1402	With boards, on roof 1" x 6" boards, laid horizontal	2 Carp	725	.022		.82	.51		1.33	1.77	
	1502	Laid diagonal		650	.025		.82	.57		1.39	1.87	
	1702	1" x 8" boards, laid horizontal		875	.018		.80	.42		1.22	1.60	
	1802	Laid diagonal		725	.022		.80	.51		1.31	1.75	
	2000	For steep roofs, add						40%				
	2200	For dormers, hips and valleys, add					5%	50%				
	2402	Boards on walls, 1" x 6" boards, laid regular	2 Carp	650	.025		.82	.57		1.39	1.87	
	2502	Laid diagonal		585	.027		.82	.63		1.45	1.97	
	2702	1" x 8" boards, laid regular		765	.021		.80	.48		1.28	1.70	
	2802	Laid diagonal		650	.025		.80	.57		1.37	1.85	
	2852	Gypsum, weatherproof, 1/2" thick		1,050	.015		.28	.35		.63	.91	
	2902	Sealed, 4/10" thick		1,100	.015		.44	.34		.78	1.05	
	3000	Wood fiber, regular, no vapor barrier, 1/2" thick		1,200	.013		.53	.31		.84	1.10	
	3100	5/8" thick		1,200	.013		.71	.31		1.02	1.30	
	3300	No vapor barrier, in colors, 1/2" thick		1,200	.013		.77	.31		1.08	1.37	
	3400	5/8" thick		1,200	.013		.95	.31		1.26	1.57	
	3600	With vapor barrier one side, white, 1/2" thick		1,200	.013		.54	.31		.85	1.11	
	3700	Vapor barrier 2 sides, 1/2" thick		1,200	.013		.75	.31		1.06	1.35	
	3800	Asphalt impregnated, 25/32" thick		1,200	.013		.34	.31		.65	.89	
	3850	Intermediate, 1/2" thick		1,200	.013		.27	.31		.58	.82	
	4000	Wafer board on roof, 1/2" thick		1,455	.011		.28	.25		.53	.74	
	4100	5/8" thick		1,330	.012		.42	.28		.70	.93	
850	0010	**SUBFLOOR** Plywood, CDX, 1/2" thick	2 Carp	1,500	.011	SF Flr.	.39	.25		.64	.85	850
	0017	Pneumatic nailed		1,860	.009		.39	.20		.59	.77	
	0102	5/8" thick		1,350	.012		.45	.27		.72	.96	
	0107	Pneumatic nailed		1,674	.010		.45	.22		.67	.87	
	0202	3/4" thick		1,250	.013		.57	.30		.87	1.13	
	0207	Pneumatic nailed		1,550	.010		.57	.24		.81	1.03	

06160 | Sheathing

			CREW	DAILY OUTPUT	LABOR-HOURS	UNIT	2004 BARE COSTS				TOTAL INCL O&P	
							MAT.	LABOR	EQUIP.	TOTAL		
850	0302	1-1/8" thick, 2-4-1 including underlayment	2 Carp	1,050	.015	SF Flr.	1.80	.35		2.15	2.58	850
	0502	With boards, 1" x 10" S4S, laid regular		1,100	.015		1.06	.34		1.40	1.73	
	0602	Laid diagonal		900	.018		1.06	.41		1.47	1.86	
	0802	1" x 8" S4S, laid regular		1,000	.016		.80	.37		1.17	1.51	
	0902	Laid diagonal		850	.019		.80	.43		1.23	1.62	
	1500	Wafer board, 5/8" thick		1,330	.012	S.F.	.49	.28		.77	1.01	
	1600	3/4" thick		1,230	.013	"	.54	.30		.84	1.10	
900	0011	**UNDERLAYMENT** Plywood, underlayment grade, 3/8" thick	2 Carp	1,500	.011	SF Flr.	.53	.25		.78	1	900
	0017	Pneumatic nailed		1,860	.009		.53	.20		.73	.92	
	0102	1/2" thick		1,450	.011		.62	.25		.87	1.11	
	0107	Pneumatic nailed		1,798	.009		.62	.21		.83	1.03	
	0202	5/8" thick		1,400	.011		.72	.26		.98	1.24	
	0207	Pneumatic nailed		1,736	.009		.72	.21		.93	1.15	
	0302	3/4" thick		1,300	.012		.83	.28		1.11	1.39	
	0306	Pneumatic nailed		1,612	.010		.83	.23		1.06	1.30	
	0502	Particle board, 3/8" thick		1,500	.011		.37	.25		.62	.83	
	0507	Pneumatic nailed		1,860	.009		.37	.20		.57	.75	
	0602	1/2" thick		1,450	.011		.39	.25		.64	.86	
	0607	Pneumatic nailed		1,798	.009		.39	.21		.60	.78	
	0802	5/8" thick		1,400	.011		.49	.26		.75	.99	
	0807	Pneumatic nailed		1,736	.009		.49	.21		.70	.90	
	0902	3/4" thick		1,300	.012		.54	.28		.82	1.07	
	0907	Pneumatic nailed		1,612	.010		.54	.23		.77	.98	
	1102	Hardboard, underlayment grade, 4' x 4', .215" thick		1,500	.011		.39	.25		.64	.85	

06170 | Prefabricated Structural Wood

			CREW	DAILY OUTPUT	LABOR-HOURS	UNIT	MAT.	LABOR	EQUIP.	TOTAL	TOTAL INCL O&P	
200	0010	**LAMINATED BEAMS** Fb 2800										200
	0050	3-1/2" x 18"	F-3	480	.083	L.F.	23.50	1.73	1.27	26.50	30.50	
	0100	5-1/4" x 11-7/8"		450	.089		23	1.85	1.36	26.21	30	
	0150	5-1/4" x 16"		360	.111		31.50	2.31	1.70	35.51	40.50	
	0200	5-1/4" x 18"		290	.138		35	2.87	2.11	39.98	45.50	
	0250	5-1/4" x 24"		220	.182		40	3.78	2.78	46.56	53.50	
	0300	7" x 11-7/8"		320	.125		31	2.60	1.91	35.51	40.50	
	0350	7" x 16"		260	.154		41.50	3.20	2.35	47.05	54	
	0400	7" x 18"		210	.190		47	3.96	2.91	53.87	61.50	
	0500	For premium appearance, add to S.F. prices					5%					
	0550	For industrial type, deduct					15%					
	0600	For stain and varnish, add					5%					
	0650	For 3/4" laminations, add					25%					
550	0010	**LAMINATED ROOF DECK** Pine or hemlock, 3" thick	2 Carp	425	.038	S.F.	2.80	.87		3.67	4.56	550
	0100	4" thick		325	.049		3.72	1.14		4.86	6	
	0300	Cedar, 3" thick		425	.038		3.42	.87		4.29	5.25	
	0400	4" thick		325	.049		4.35	1.14		5.49	6.70	
	0600	Fir, 3" thick		425	.038		2.85	.87		3.72	4.62	
	0700	4" thick		325	.049		3.57	1.14		4.71	5.85	
600	0010	**STRUCTURAL JOISTS** Fabricated "I" joists with wood flanges,										600
	0100	Plywood webs, incl. bridging & blocking, panels 24" O.C.										
	1200	15' to 24' span, 50 psf live load	F-5	2,400	.013	SF Flr.	1.81	.27		2.08	2.44	
	1300	55 psf live load		2,250	.014		1.94	.28		2.22	2.61	
	1400	24' to 30' span, 45 psf live load		2,600	.012		2.13	.25		2.38	2.76	
	1500	55 psf live load		2,400	.013		2.90	.27		3.17	3.64	
	1600	Tubular steel open webs, 45 psf, 24" O.C., 40' span	F-3	6,250	.006		1.80	.13	.10	2.03	2.31	
	1700	55' span		7,750	.005		1.75	.11	.08	1.94	2.20	

06170 | Prefabricated Structural Wood

		06170 Prefabricated Structural Wood	CREW	DAILY OUTPUT	LABOR-HOURS	UNIT	2004 BARE COSTS MAT.	LABOR	EQUIP.	TOTAL	TOTAL INCL O&P	
600	1800	70′ span	F-3	9,250	.004	SF Flr.	2.27	.09	.07	2.43	2.72	600
	1900	85 psf live load, 26′ span	↓	2,300	.017	↓	2.11	.36	.27	2.74	3.22	
980	0010	**ROOF TRUSSES**										980
	0020	For timber connectors, see div. 06090-800										
	5000	Common wood, 2″ x 4″ metal plate connected, 24″ O.C., 4/12 slope										
	5010	1′ overhang, 12′ span	F-5	55	.582	Ea.	24	11.65		35.65	46	
	5050	20′ span R06170-100	F-6	62	.645		39	13.40	9.85	62.25	76.50	
	5100	24′ span		60	.667		46.50	13.85	10.20	70.55	85.50	
	5150	26′ span		57	.702		65	14.60	10.75	90.35	108	
	5200	28′ span		53	.755		56.50	15.70	11.55	83.75	101	
	5240	30′ span		51	.784		76	16.30	12	104.30	124	
	5250	32′ span		50	.800		79.50	16.65	12.25	108.40	128	
	5280	34′ span		48	.833		97	17.35	12.75	127.10	150	
	5350	8/12 pitch, 1′ overhang, 20′ span		57	.702		60	14.60	10.75	85.35	102	
	5400	24′ span		55	.727		71	15.15	11.15	97.30	116	
	5450	26′ span		52	.769		77	16	11.75	104.75	124	
	5500	28′ span		49	.816		83	17	12.50	112.50	133	
	5550	32′ span		45	.889		97.50	18.50	13.60	129.60	153	
	5600	36′ span		41	.976		116	20.50	14.95	151.45	177	
	5650	38′ span		40	1		126	21	15.30	162.30	190	
	5700	40′ span	↓	40	1	↓	143	21	15.30	179.30	210	

06180 | Glued-Laminated Construction

		06180 Glued-Laminated Construction	CREW	DAILY OUTPUT	LABOR-HOURS	UNIT	2004 BARE COSTS MAT.	LABOR	EQUIP.	TOTAL	TOTAL INCL O&P	
400	0010	**LAMINATED FRAMING** Not including decking										400
	0020	30 lb., short term live load, 15 lb. dead load										
	0200	Straight roof beams, 20′ clear span, beams 8′ O.C.	F-3	2,560	.016	SF Flr.	1.58	.33	.24	2.15	2.55	
	0300	Beams 16′ O.C.		3,200	.013		1.14	.26	.19	1.59	1.90	
	0500	40′ clear span, beams 8′ O.C.		3,200	.013		3.03	.26	.19	3.48	3.98	
	0600	Beams 16′ O.C.	↓	3,840	.010		2.47	.22	.16	2.85	3.26	
	0800	60′ clear span, beams 8′ O.C.	F-4	2,880	.014		5.20	.29	.32	5.81	6.55	
	0900	Beams 16′ O.C.	″	3,840	.010		3.87	.22	.24	4.33	4.88	
	1100	Tudor arches, 30′ to 40′ clear span, frames 8′ O.C.	F-3	1,680	.024		6.80	.50	.36	7.66	8.70	
	1200	Frames 16′ O.C.	″	2,240	.018		5.30	.37	.27	5.94	6.80	
	1400	50′ to 60′ clear span, frames 8′ O.C.	F-4	2,200	.018		7.30	.38	.41	8.09	9.15	
	1500	Frames 16′ O.C.		2,640	.015		6.20	.32	.35	6.87	7.75	
	1700	Radial arches, 60′ clear span, frames 8′ O.C.		1,920	.021		6.85	.43	.47	7.75	8.75	
	1800	Frames 16′ O.C.		2,880	.014		5.25	.29	.32	5.86	6.65	
	2000	100′ clear span, frames 8′ O.C.		1,600	.025		7.10	.52	.57	8.19	9.30	
	2100	Frames 16′ O.C.		2,400	.017		6.20	.35	.38	6.93	7.85	
	2300	120′ clear span, frames 8′ O.C.		1,440	.028		9.40	.58	.63	10.61	12	
	2400	Frames 16′ O.C.	↓	1,920	.021		8.60	.43	.47	9.50	10.70	
	2600	Bowstring trusses, 20′ O.C., 40′ clear span	F-3	2,400	.017		4.24	.35	.26	4.85	5.50	
	2700	60′ clear span	F-4	3,600	.011		3.81	.23	.25	4.29	4.86	
	2800	100′ clear span		4,000	.010		5.40	.21	.23	5.84	6.55	
	2900	120′ clear span	↓	3,600	.011		5.80	.23	.25	6.28	7	
	3100	For premium appearance, add to S.F. prices					5%					
	3300	For industrial type, deduct					15%					
	3500	For stain and varnish, add					5%					
	3900	For 3/4″ laminations, add to straight					25%					
	4100	Add to curved				↓	15%					
	4300	Alternate pricing method: (use nominal footage of										
	4310	components). Straight beams, camber less than 6″	F-3	3.50	11.429	M.B.F.	2,350	238	175	2,763	3,175	
	4400	Columns, including hardware		2	20		2,525	415	305	3,245	3,800	
	4600	Curved members, radius over 32′		2.50	16		2,575	335	245	3,155	3,675	
	4700	Radius 10′ to 32′	↓	3	13.333	↓	2,550	277	204	3,031	3,525	

06100 | Rough Carpentry

		06180 Glued-Laminated Construction	CREW	DAILY OUTPUT	LABOR-HOURS	UNIT	2004 BARE COSTS MAT.	LABOR	EQUIP.	TOTAL	TOTAL INCL O&P	
400	4900	For complicated shapes, add maximum				M.B.F.	100%					400
	5100	For pressure treating, add to straight					35%					
	5200	Add to curved					45%					
	6000	Laminated veneer members, southern pine or western species										
	6050	1-3/4" wide x 5-1/2" deep	2 Carp	480	.033	L.F.	2.69	.77		3.46	4.27	
	6100	9-1/2" deep		480	.033		3.35	.77		4.12	5	
	6150	14" deep		450	.036		4.99	.82		5.81	6.90	
	6200	18" deep		450	.036		6.75	.82		7.57	8.85	
	6300	Parallel strand members, southern pine or western species										
	6350	1-3/4" wide x 9-1/4" deep	2 Carp	480	.033	L.F.	3.17	.77		3.94	4.80	
	6400	11-1/4" deep		450	.036		3.90	.82		4.72	5.70	
	6450	14" deep		400	.040		4.65	.92		5.57	6.65	
	6500	3-1/2" wide x 9-1/4" deep		480	.033		7.70	.77		8.47	9.80	
	6550	11-1/4" deep		450	.036		9.55	.82		10.37	11.90	
	6600	14" deep		400	.040		11.30	.92		12.22	14	
	6650	7" wide x 9-1/4" deep		450	.036		16.10	.82		16.92	19.10	
	6700	11-1/4" deep		420	.038		20	.88		20.88	23.50	
	6750	14" deep		400	.040		24	.92		24.92	28	

06200 | Finish Carpentry

		06220 Millwork	CREW	DAILY OUTPUT	LABOR-HOURS	UNIT	2004 BARE COSTS MAT.	LABOR	EQUIP.	TOTAL	TOTAL INCL O&P	
100	0010	**MILLWORK** Rule of thumb: Milled material cost										100
	0020	equals three times cost of lumber										
	1000	Typical finish hardwood milled material										
	1020	1" x 12", custom birch				L.F.	5.10			5.10	5.60	
	1040	Cedar					2.73			2.73	3	
	1060	Oak					6.25			6.25	6.90	
	1080	Redwood					5.90			5.90	6.50	
	1100	Southern yellow pine					1.90			1.90	2.09	
	1120	Sugar pine					2.51			2.51	2.76	
	1140	Teak					16.15			16.15	17.75	
	1160	Walnut					7.95			7.95	8.75	
	1180	White pine					3.09			3.09	3.40	
200	0010	**MOLDINGS, BASE**										200
	0500	Base, stock pine, 9/16" x 3-1/2"	1 Carp	240	.033	L.F.	1.22	.77		1.99	2.65	
	0501	Oak or birch, 9/16" x 3-1/2"		240	.033		3.10	.77		3.87	4.72	
	0550	9/16" x 4-1/2"		200	.040		1.54	.92		2.46	3.26	
	0561	Base shoe, oak, 3/4" x 1"		240	.033		1.04	.77		1.81	2.45	
	0570	Base, prefinished, 2-1/2" x 9/16"		242	.033		1.05	.76		1.81	2.46	
	0580	Shoe, prefinished, 3/8" x 5/8"		266	.030		.45	.69		1.14	1.68	
	0585	Flooring cant strip, 3/4" x 1/2"		500	.016		.46	.37		.83	1.14	
400	0010	**MOLDINGS, CASINGS**										400
	0090	Apron, stock pine, 5/8" x 2"	1 Carp	250	.032	L.F.	.99	.74		1.73	2.34	
	0110	5/8" x 3-1/2"		220	.036		1.44	.84		2.28	3.01	
	0300	Band, stock pine, 11/16" x 1-1/8"		270	.030		.54	.68		1.22	1.75	
	0350	11/16" x 1-3/4"		250	.032		.86	.74		1.60	2.20	
	0700	Casing, stock pine, 11/16" x 2-1/2"		240	.033		.85	.77		1.62	2.25	
	0701	Oak or birch		240	.033		2.25	.77		3.02	3.79	
	0750	11/16" x 3-1/2"		215	.037		1.33	.86		2.19	2.92	

	06220 | Millwork		CREW	DAILY OUTPUT	LABOR-HOURS	UNIT	2004 BARE COSTS MAT.	LABOR	EQUIP.	TOTAL	TOTAL INCL O&P	
400	0760	Door & window casing, exterior, 1-1/4" x 2"	1 Carp	200	.040	L.F.	1.64	.92		2.56	3.37	400
	0770	Finger jointed, 1-1/4" x 2"		200	.040		1.14	.92		2.06	2.82	
	4600	Mullion casing, stock pine, 5/16" x 2"		200	.040		.72	.92		1.64	2.36	
	4601	Oak or birch, 9/16" x 2-1/2"		200	.040		2.43	.92		3.35	4.24	
	4700	Teak, custom, nominal 1" x 1"		215	.037		1.15	.86		2.01	2.73	
	4800	Nominal 1" x 3"	↓	200	.040	↓	3.46	.92		4.38	5.40	
450	0010	**MOLDINGS, CEILINGS**										450
	0600	Bed, stock pine, 9/16" x 1-3/4"	1 Carp	270	.030	L.F.	.62	.68		1.30	1.84	
	0650	9/16" x 2"		240	.033		.76	.77		1.53	2.15	
	1200	Cornice molding, stock pine, 9/16" x 1-3/4"		330	.024		.67	.56		1.23	1.69	
	1300	9/16" x 2-1/4"		300	.027		.89	.62		1.51	2.03	
	2400	Cove scotia, stock pine, 9/16" x 1-3/4"		270	.030		.59	.68		1.27	1.81	
	2401	Oak or birch, 9/16" x 1-3/4"		270	.030		1.70	.68		2.38	3.03	
	2500	11/16" x 2-3/4"		255	.031		1.25	.72		1.97	2.61	
	2600	Crown, stock pine, 9/16" x 3-5/8"		250	.032		1.67	.74		2.41	3.09	
	2700	11/16" x 4-5/8"	↓	220	.036	↓	3.60	.84		4.44	5.40	
500	0010	**MOLDINGS, EXTERIOR**										500
	1500	Cornice, boards, pine, 1" x 2"	1 Carp	330	.024	L.F.	.32	.56		.88	1.30	
	1600	1" x 4"		250	.032		.66	.74		1.40	1.98	
	1700	1" x 6"		250	.032		1.15	.74		1.89	2.52	
	1800	1" x 8"		200	.040		1.42	.92		2.34	3.13	
	1900	1" x 10"		180	.044		1.57	1.03		2.60	3.47	
	2000	1" x 12"		180	.044		1.72	1.03		2.75	3.63	
	2200	Three piece, built-up, pine, minimum		80	.100		2.13	2.31		4.44	6.25	
	2300	Maximum		65	.123		4.71	2.84		7.55	10	
	3000	Corner board, sterling pine, 1" x 4"		200	.040		.66	.92		1.58	2.30	
	3100	1" x 6"		200	.040		.97	.92		1.89	2.64	
	3200	2" x 6"		165	.048		2.06	1.12		3.18	4.17	
	3300	2" x 8"		165	.048		2.79	1.12		3.91	4.97	
	3350	Fascia, sterling pine, 1" x 6"		250	.032		.97	.74		1.71	2.32	
	3370	1" x 8"		225	.036		1.51	.82		2.33	3.05	
	3372	2" x 6"		225	.036		2.06	.82		2.88	3.66	
	3374	2" x 8"		200	.040		2.79	.92		3.71	4.64	
	3376	2" x 10"		180	.044		3.39	1.03		4.42	5.45	
	3395	Grounds, 1" x 1" redwood		300	.027		.19	.62		.81	1.26	
	3400	Trim, exterior, sterling pine, back band		250	.032		.66	.74		1.40	1.98	
	3500	Casing		250	.032		1.75	.74		2.49	3.18	
	3600	Crown		250	.032		1.68	.74		2.42	3.10	
	3700	Porch rail with balusters		22	.364		13.70	8.40		22.10	29.50	
	3800	Screen		395	.020		1.06	.47		1.53	1.96	
	4100	Verge board, sterling pine, 1" x 4"		200	.040		.64	.92		1.56	2.27	
	4200	1" x 6"		200	.040		.96	.92		1.88	2.63	
	4300	2" x 6"		165	.048		1.56	1.12		2.68	3.62	
	4400	2" x 8"	↓	165	.048	↓	2.07	1.12		3.19	4.18	
	4700	For redwood trim, add					200%					
	5000	Casing/fascia, rough-sawn cedar										
	5100	1" x 2"	1 Carp	275	.029	L.F.	.26	.67		.93	1.43	
	5200	1" x 6"		250	.032		.78	.74		1.52	2.11	
	5300	1" x 8"		230	.035		1	.80		1.80	2.46	
	5400	2" x 4"		220	.036		1	.84		1.84	2.53	
	5500	2" x 6"		220	.036		1.52	.84		2.36	3.10	
	5600	2" x 8"		200	.040		2.02	.92		2.94	3.79	
	5700	2" x 10"		180	.044		2.53	1.03		3.56	4.52	
	5800	2" x 12"	↓	170	.047	↓	3.02	1.09		4.11	5.15	
700	0010	**MOLDINGS, TRIM**										700
	0200	Astragal, stock pine, 11/16" x 1-3/4"	1 Carp	255	.031	L.F.	.96	.72		1.68	2.29	

06220 Millwork			CREW	DAILY OUTPUT	LABOR-HOURS	UNIT	2004 BARE COSTS MAT.	LABOR	EQUIP.	TOTAL	TOTAL INCL O&P	
700	0250	1-5/16" x 2-3/16"	1 Carp	240	.033	L.F.	2.94	.77		3.71	4.54	700
	0800	Chair rail, stock pine, 5/8" x 2-1/2"		270	.030		.85	.68		1.53	2.10	
	0900	5/8" x 3-1/2"		240	.033		1.41	.77		2.18	2.86	
	1000	Closet pole, stock pine, 1-1/8" diameter		200	.040		.80	.92		1.72	2.45	
	1100	Fir, 1-5/8" diameter		200	.040		1.19	.92		2.11	2.88	
	1150	Corner, inside, 5/16" x 1"		225	.036		.32	.82		1.14	1.74	
	1160	Outside, 1-1/16" x 1-1/16"		240	.033		.88	.77		1.65	2.28	
	1161	1-5/16" x 1-5/16"		240	.033		1.03	.77		1.80	2.44	
	3300	Half round, stock pine, 1/4" x 1/2"		270	.030		.16	.68		.84	1.34	
	3350	1/2" x 1"	↓	255	.031	↓	.39	.72		1.11	1.66	
	3400	Handrail, fir, single piece, stock, hardware not included										
	3450	1-1/2" x 1-3/4"	1 Carp	80	.100	L.F.	1.22	2.31		3.53	5.25	
	3470	Pine, 1-1/2" x 1-3/4"		80	.100		1.07	2.31		3.38	5.10	
	3500	1-1/2" x 2-1/2"		76	.105		1.28	2.43		3.71	5.55	
	3600	Lattice, stock pine, 1/4" x 1-1/8"		270	.030		.23	.68		.91	1.41	
	3700	1/4" x 1-3/4"		250	.032		.35	.74		1.09	1.64	
	3800	Miscellaneous, custom, pine, 1" x 1"		270	.030		.31	.68		.99	1.50	
	3900	1" x 3"		240	.033		.63	.77		1.40	2	
	4100	Birch or oak, nominal 1" x 1"		240	.033		.49	.77		1.26	1.85	
	4200	Nominal 1" x 3"		215	.037		1.66	.86		2.52	3.29	
	4400	Walnut, nominal 1" x 1"		215	.037		.80	.86		1.66	2.34	
	4500	Nominal 1" x 3"		200	.040		2.39	.92		3.31	4.20	
	4700	Teak, nominal 1" x 1"		215	.037		1.13	.86		1.99	2.70	
	4800	Nominal 1" x 3"		200	.040		3.22	.92		4.14	5.10	
	4900	Quarter round, stock pine, 1/4" x 1/4"		275	.029		.15	.67		.82	1.31	
	4950	3/4" x 3/4"		255	.031	↓	.36	.72		1.08	1.63	
	5600	Wainscot moldings, 1-1/8" x 9/16", 2' high, minimum		76	.105	S.F.	8.75	2.43		11.18	13.75	
	5700	Maximum	↓	65	.123	"	18.05	2.84		20.89	24.50	
800	0010	**MOLDINGS, WINDOW AND DOOR**										800
	2800	Door moldings, stock, decorative, 1-1/8" wide, plain	1 Carp	17	.471	Set	29	10.85		39.85	50.50	
	2900	Detailed		17	.471	"	76.50	10.85		87.35	102	
	2960	Clear pine door jamb, no stops, 11/16" x 4-9/16"		240	.033	L.F.	2.53	.77		3.30	4.09	
	3150	Door trim set, 1 head and 2 sides, pine, 2-1/2 wide		5.90	1.356	Opng.	14.45	31.50		45.95	69	
	3170	3-1/2" wide		5.30	1.509	"	22.50	35		57.50	84	
	3250	Glass beads, stock pine, 3/8" x 1/2"		275	.029	L.F.	.36	.67		1.03	1.54	
	3270	3/8" x 7/8"		270	.030		.40	.68		1.08	1.60	
	4850	Parting bead, stock pine, 3/8" x 3/4"		275	.029		.26	.67		.93	1.43	
	4870	1/2" x 3/4"		255	.031		.31	.72		1.03	1.57	
	5000	Stool caps, stock pine, 11/16" x 3-1/2"		200	.040		1.53	.92		2.45	3.25	
	5100	1-1/16" x 3-1/4"		150	.053	↓	2.67	1.23		3.90	5.05	
	5300	Threshold, oak, 3' long, inside, 5/8" x 3-5/8"		32	.250	Ea.	7.60	5.80		13.40	18.15	
	5400	Outside, 1-1/2" x 7-5/8"	↓	16	.500	"	33	11.55		44.55	56	
	5900	Window trim sets, including casings, header, stops,										
	5910	stool and apron, 2-1/2" wide, minimum	1 Carp	13	.615	Opng.	19.90	14.20		34.10	46	
	5950	Average		10	.800		23.50	18.50		42	57	
	6000	Maximum	↓	6	1.333	↓	55	31		86	113	
900	0005	**SOFFITS**										900
	0200	Soffits, pine, 1" x 4"	2 Carp	420	.038	L.F.	.34	.88		1.22	1.86	
	0210	1" x 6"		420	.038		.38	.88		1.26	1.90	
	0220	1" x 8"		420	.038		.50	.88		1.38	2.04	
	0230	1" x 10"		400	.040		.84	.92		1.76	2.49	
	0240	1" x 12"		400	.040		.96	.92		1.88	2.62	
	0250	STK cedar, 1" x 4"		420	.038		.50	.88		1.38	2.04	
	0260	1" x 6"		420	.038		.81	.88		1.69	2.38	
	0270	1" x 8"		420	.038		1.06	.88		1.94	2.66	
	0280	1" x 10"	↓	400	.040	↓	1.27	.92		2.19	2.97	

	06220 | Millwork	CREW	DAILY OUTPUT	LABOR-HOURS	UNIT	2004 BARE COSTS MAT.	LABOR	EQUIP.	TOTAL	TOTAL INCL O&P	
900	0290 1″ x 12″	2 Carp	400	.040	L.F.	1.56	.92		2.48	3.29	900
	1000 Exterior AC plywood, 1/4″ thick		420	.038	S.F.	.45	.88		1.33	1.99	
	1050 3/8″ thick		420	.038		.53	.88		1.41	2.07	
	1100 1/2″ thick		420	.038		.62	.88		1.50	2.17	
	1150 Polyvinyl chloride, white, solid	1 Carp	230	.035		.66	.80		1.46	2.09	
	1160 Perforated	″	230	.035		.66	.80		1.46	2.09	
	1170 Accessories, "J" channel 5/8″	2 Carp	700	.023	L.F.	.26	.53		.79	1.19	

	06250 | Prefinished Paneling	CREW	DAILY OUTPUT	LABOR-HOURS	UNIT	MAT.	LABOR	EQUIP.	TOTAL	TOTAL INCL O&P	
200	0010 PANELING, HARDBOARD										200
	0050 Not incl. furring or trim, hardboard, tempered, 1/8″ thick	2 Carp	500	.032	S.F.	.30	.74		1.04	1.58	
	0100 1/4″ thick		500	.032		.38	.74		1.12	1.67	
	0300 Tempered pegboard, 1/8″ thick		500	.032		.37	.74		1.11	1.66	
	0400 1/4″ thick		500	.032		.42	.74		1.16	1.71	
	0600 Untempered hardboard, natural finish, 1/8″ thick		500	.032		.33	.74		1.07	1.61	
	0700 1/4″ thick		500	.032		.32	.74		1.06	1.60	
	0900 Untempered pegboard, 1/8″ thick		500	.032		.33	.74		1.07	1.61	
	1000 1/4″ thick		500	.032		.37	.74		1.11	1.66	
	1200 Plastic faced hardboard, 1/8″ thick		500	.032		.54	.74		1.28	1.84	
	1300 1/4″ thick		500	.032		.72	.74		1.46	2.04	
	1500 Plastic faced pegboard, 1/8″ thick		500	.032		.51	.74		1.25	1.81	
	1600 1/4″ thick		500	.032		.63	.74		1.37	1.94	
	1800 Wood grained, plain or grooved, 1/4″ thick, minimum		500	.032		.48	.74		1.22	1.78	
	1900 Maximum		425	.038		1.01	.87		1.88	2.59	
	2100 Moldings for hardboard, wood or aluminum, minimum		500	.032	L.F.	.33	.74		1.07	1.61	
	2200 Maximum		425	.038	″	.91	.87		1.78	2.48	
500	0010 PANELING, PLYWOOD										500
	2400 Plywood, prefinished, 1/4″ thick, 4′ x 8′ sheets										
	2410 with vertical grooves. Birch faced, minimum	2 Carp	500	.032	S.F.	.79	.74		1.53	2.12	
	2420 Average		420	.038		1.21	.88		2.09	2.82	
	2430 Maximum		350	.046		1.76	1.06		2.82	3.73	
	2600 Mahogany, African		400	.040		2.25	.92		3.17	4.05	
	2700 Philippine (Lauan)		500	.032		.97	.74		1.71	2.32	
	2900 Oak or Cherry, minimum		500	.032		1.89	.74		2.63	3.33	
	3000 Maximum		400	.040		2.90	.92		3.82	4.76	
	3200 Rosewood		320	.050		4.11	1.16		5.27	6.50	
	3400 Teak		400	.040		2.90	.92		3.82	4.76	
	3600 Chestnut		375	.043		4.28	.99		5.27	6.40	
	3800 Pecan		400	.040		1.85	.92		2.77	3.61	
	3900 Walnut, minimum		500	.032		2.47	.74		3.21	3.97	
	3950 Maximum		400	.040		4.68	.92		5.60	6.70	
	4000 Plywood, prefinished, 3/4″ thick, stock grades, minimum		320	.050		1.12	1.16		2.28	3.19	
	4100 Maximum		224	.071		4.83	1.65		6.48	8.10	
	4300 Architectural grade, minimum		224	.071		3.57	1.65		5.22	6.75	
	4400 Maximum		160	.100		5.45	2.31		7.76	9.90	
	4600 Plywood, "A" face, birch, V.C., 1/2″ thick, natural		450	.036		1.69	.82		2.51	3.25	
	4700 Select		450	.036		1.85	.82		2.67	3.43	
	4900 Veneer core, 3/4″ thick, natural		320	.050		1.78	1.16		2.94	3.92	
	5000 Select		320	.050		2.01	1.16		3.17	4.17	
	5200 Lumber core, 3/4″ thick, natural		320	.050		2.68	1.16		3.84	4.91	
	5500 Plywood, knotty pine, 1/4″ thick, A2 grade		450	.036		1.46	.82		2.28	3	
	5600 A3 grade		450	.036		1.85	.82		2.67	3.43	
	5800 3/4″ thick, veneer core, A2 grade		320	.050		1.90	1.16		3.06	4.05	
	5900 A3 grade		320	.050		2.13	1.16		3.29	4.30	

06200 | Finish Carpentry

		06250 | Prefinished Paneling	CREW	DAILY OUTPUT	LABOR-HOURS	UNIT	MAT.	LABOR	EQUIP.	TOTAL	TOTAL INCL O&P	
500	6100	Aromatic cedar, 1/4" thick, plywood	2 Carp	400	.040	S.F.	1.87	.92		2.79	3.63	500
	6200	1/4" thick, particle board	↓	400	.040	↓	.91	.92		1.83	2.57	

		06260 | Board Paneling	CREW	DAILY OUTPUT	LABOR-HOURS	UNIT	MAT.	LABOR	EQUIP.	TOTAL	TOTAL INCL O&P	
400	0010	**PANELING, BOARDS**										400
	6400	Wood board paneling, 3/4" thick, knotty pine	2 Carp	300	.053	S.F.	1.26	1.23		2.49	3.48	
	6500	Rough sawn cedar		300	.053		1.61	1.23		2.84	3.86	
	6700	Redwood, clear, 1" x 4" boards		300	.053		3.76	1.23		4.99	6.25	
	6900	Aromatic cedar, closet lining, boards	↓	275	.058	↓	2.90	1.34		4.24	5.45	

		06270 | Closet/Utility Wood Shelving	CREW	DAILY OUTPUT	LABOR-HOURS	UNIT	MAT.	LABOR	EQUIP.	TOTAL	TOTAL INCL O&P	
200	0010	**SHELVING** Pine, clear grade, no edge band, 1" x 8"	1 Carp	115	.070	L.F.	1.73	1.61		3.34	4.63	200
	0100	1" x 10"		110	.073		2.21	1.68		3.89	5.30	
	0200	1" x 12"		105	.076		2.97	1.76		4.73	6.25	
	0450	1" x 18"		95	.084		4.46	1.95		6.41	8.20	
	0460	1" x 24"		85	.094		5.95	2.17		8.12	10.25	
	0600	Plywood, 3/4" thick with lumber edge, 12" wide		75	.107		.31	2.46		2.77	4.52	
	0700	24" wide		70	.114	↓	1.97	2.64		4.61	6.65	
	0900	Bookcase, clear grade pine, shelves 12" O.C., 8" deep, /SF shelf		70	.114	S.F.	5.60	2.64		8.24	10.70	
	1000	12" deep shelves		65	.123	"	9.65	2.84		12.49	15.40	
	1200	Adjustable closet rod and shelf, 12" wide, 3' long		20	.400	Ea.	8.50	9.25		17.75	25	
	1300	8' long		15	.533	"	20.50	12.30		32.80	43.50	
	1500	Prefinished shelves with supports, stock, 8" wide		75	.107	L.F.	3.68	2.46		6.14	8.25	
	1600	10" wide	↓	70	.114	"	4.09	2.64		6.73	9	

06400 | Architectural Woodwork

		06410 | Custom Cabinets	CREW	DAILY OUTPUT	LABOR-HOURS	UNIT	MAT.	LABOR	EQUIP.	TOTAL	TOTAL INCL O&P	
100	0010	**CABINETS** Corner china cabinets, stock pine,										100
	0020	80" high, unfinished, minimum	2 Carp	6.60	2.424	Ea.	435	56		491	575	
	0100	Maximum	"	4.40	3.636	"	960	84		1,044	1,200	
	0700	Kitchen base cabinets, hardwood, not incl. counter tops,										
	0710	24" deep, 35" high, prefinished										
	0800	One top drawer, one door below, 12" wide	2 Carp	24.80	.645	Ea.	127	14.90		141.90	165	
	0820	15" wide		24	.667		173	15.40		188.40	216	
	0840	18" wide		23.30	.687		188	15.85		203.85	234	
	0860	21" wide		22.70	.705		196	16.30		212.30	244	
	0880	24" wide		22.30	.717		225	16.55		241.55	276	
	1000	Four drawers, 12" wide		24.80	.645		300	14.90		314.90	355	
	1020	15" wide		24	.667		231	15.40		246.40	280	
	1040	18" wide		23.30	.687		257	15.85		272.85	310	
	1060	24" wide		22.30	.717		280	16.55		296.55	340	
	1200	Two top drawers, two doors below, 27" wide		22	.727		250	16.80		266.80	305	
	1220	30" wide		21.40	.748		268	17.25		285.25	325	
	1240	33" wide		20.90	.766		277	17.70		294.70	335	
	1260	36" wide		20.30	.788		288	18.20		306.20	345	
	1280	42" wide		19.80	.808		310	18.65		328.65	370	
	1300	48" wide	↓	18.90	.847	↓	330	19.55		349.55	395	

06410 | Custom Cabinets

		06410 Custom Cabinets	CREW	DAILY OUTPUT	LABOR-HOURS	UNIT	2004 BARE COSTS MAT.	LABOR	EQUIP.	TOTAL	TOTAL INCL O&P	
100	1500	Range or sink base, two doors below, 30" wide	2 Carp	21.40	.748	Ea.	221	17.25		238.25	273	100
	1520	33" wide		20.90	.766		237	17.70		254.70	291	
	1540	36" wide		20.30	.788		248	18.20		266.20	305	
	1560	42" wide		19.80	.808		265	18.65		283.65	325	
	1580	48" wide	↓	18.90	.847		278	19.55		297.55	340	
	1800	For sink front units, deduct					48			48	53	
	2000	Corner base cabinets, 36" wide, standard	2 Carp	18	.889		365	20.50		385.50	435	
	2100	Lazy Susan with revolving door	"	16.50	.970	↓	355	22.50		377.50	430	
	4000	Kitchen wall cabinets, hardwood, 12" deep with two doors										
	4050	12" high, 30" wide	2 Carp	24.80	.645	Ea.	139	14.90		153.90	179	
	4100	36" wide		24	.667		162	15.40		177.40	204	
	4400	15" high, 30" wide		24	.667		146	15.40		161.40	186	
	4420	33" wide		23.30	.687		163	15.85		178.85	206	
	4440	36" wide		22.70	.705		165	16.30		181.30	209	
	4450	42" wide		22.70	.705		191	16.30		207.30	238	
	4700	24" high, 30" wide		23.30	.687		181	15.85		196.85	226	
	4720	36" wide		22.70	.705		200	16.30		216.30	248	
	4740	42" wide		22.30	.717		221	16.55		237.55	271	
	5000	30" high, one door, 12" wide		22	.727		122	16.80		138.80	163	
	5020	15" wide		21.40	.748		138	17.25		155.25	182	
	5040	18" wide		20.90	.766		151	17.70		168.70	196	
	5060	24" wide		20.30	.788		170	18.20		188.20	218	
	5300	Two doors, 27" wide		19.80	.808		211	18.65		229.65	264	
	5320	30" wide		19.30	.829		204	19.15		223.15	257	
	5340	36" wide		18.80	.851		233	19.65		252.65	290	
	5360	42" wide		18.50	.865		254	20		274	315	
	5380	48" wide		18.40	.870		286	20		306	350	
	6000	Corner wall, 30" high, 24" wide		18	.889		139	20.50		159.50	188	
	6050	30" wide		17.20	.930		166	21.50		187.50	220	
	6100	36" wide		16.50	.970		180	22.50		202.50	236	
	6500	Revolving Lazy Susan		15.20	1.053		274	24.50		298.50	340	
	7000	Broom cabinet, 84" high, 24" deep, 18" wide		10	1.600		375	37		412	475	
	7500	Oven cabinets, 84" high, 24" deep, 27" wide		8	2	↓	540	46		586	675	
	7750	Valance board trim	↓	396	.040	L.F.	7.75	.93		8.68	10.15	
	9000	For deluxe models of all cabinets, add					40%					
	9500	For custom built in place, add					25%	10%				
	9550	Rule of thumb, kitchen cabinets not including										
	9560	appliances & counter top, minimum	2 Carp	30	.533	L.F.	88	12.30		100.30	118	
	9600	Maximum	"	25	.640	"	230	14.80		244.80	278	
210	0010	**CASEWORK, FRAMES**										210
	0050	Base cabinets, counter storage, 36" high, one bay										
	0100	18" wide	1 Carp	2.70	2.963	Ea.	98.50	68.50		167	224	
	0400	Two bay, 36" wide		2.20	3.636		150	84		234	310	
	1100	Three bay, 54" wide		1.50	5.333		179	123		302	405	
	2800	Book cases, one bay, 7' high, 18" wide		2.40	3.333		116	77		193	258	
	3500	Two bay, 36" wide		1.60	5		168	116		284	380	
	4100	Three bay, 54" wide		1.20	6.667		278	154		432	565	
	6100	Wall mounted cabinet, one bay, 24" high, 18" wide		3.60	2.222		63.50	51.50		115	157	
	6800	Two bay, 36" wide		2.20	3.636		93	84		177	245	
	7400	Three bay, 54" wide		1.70	4.706		116	109		225	310	
	8400	30" high, one bay, 18" wide		3.60	2.222		69.50	51.50		121	163	
	9000	Two bay, 36" wide		2.15	3.721		92	86		178	247	
	9400	Three bay, 54" wide		1.60	5		115	116		231	320	
	9800	Wardrobe, 7' high, single, 24" wide		2.70	2.963		127	68.50		195.50	256	
	9950	Partition, adjustable shelves & drawers, 48" wide	↓	1.40	5.714	↓	243	132		375	490	

06410 | Custom Cabinets

			CREW	DAILY OUTPUT	LABOR-HOURS	UNIT	2004 BARE COSTS MAT.	LABOR	EQUIP.	TOTAL	TOTAL INCL O&P
220	0010	**CABINET DOORS**									
	2000	Glass panel, hardwood frame									
	2200	12" wide, 18" high	1 Carp	34	.235	Ea.	22	5.45		27.45	33.50
	2400	24" high		33	.242		24	5.60		29.60	35.50
	2600	30" high		32	.250		25.50	5.80		31.30	38
	2800	36" high		30	.267		29.50	6.15		35.65	43
	3000	48" high		23	.348		37.50	8.05		45.55	55
	3200	60" high		17	.471		46	10.85		56.85	69
	3400	72" high		15	.533		55	12.30		67.30	81.50
	3600	15" wide x 18" high		33	.242		23	5.60		28.60	34.50
	3800	24" high		32	.250		24.50	5.80		30.30	37
	4000	30" high		30	.267		26	6.15		32.15	39.50
	4250	36" high		28	.286		30.50	6.60		37.10	44.50
	4300	48" high		22	.364		38.50	8.40		46.90	57
	4350	60" high		16	.500		54.50	11.55		66.05	79.50
	4400	72" high		14	.571		63.50	13.20		76.70	92.50
	4450	18" wide, 18" high		32	.250		24	5.80		29.80	36
	4500	24" high		30	.267		24.50	6.15		30.65	37.50
	4550	30" high		29	.276		26.50	6.35		32.85	40
	4600	36" high		27	.296		31.50	6.85		38.35	46
	4650	48" high		21	.381		39.50	8.80		48.30	58
	4700	60" high		15	.533		55.50	12.30		67.80	82
	4750	72" high	▼	13	.615	▼	64.50	14.20		78.70	95
	5000	Hardwood, raised panel									
	5100	12" wide, 18" high	1 Carp	16	.500	Ea.	33.50	11.55		45.05	56.50
	5150	24" high		15.50	.516		36.50	11.90		48.40	60
	5200	30" high		15	.533		40	12.30		52.30	65
	5250	36" high		14	.571		43	13.20		56.20	70
	5300	48" high		11	.727		52.50	16.80		69.30	86.50
	5320	60" high		8	1		70.50	23		93.50	117
	5340	72" high		7	1.143		81	26.50		107.50	134
	5360	15" wide x 18" high		15.50	.516		35	11.90		46.90	58
	5380	24" high		15	.533		38.50	12.30		50.80	63.50
	5400	30" high		14.50	.552		42.50	12.75		55.25	68
	5420	36" high		13.50	.593		48.50	13.70		62.20	76.50
	5440	48" high		10.50	.762		59	17.60		76.60	94.50
	5460	60" high		7.50	1.067		86	24.50		110.50	137
	5480	72" high		6.50	1.231		97	28.50		125.50	156
	5500	18" wide, 18" high		15	.533		37	12.30		49.30	61.50
	5550	24" high		14.50	.552		41	12.75		53.75	66.50
	5600	30" high		14	.571		45	13.20		58.20	72
	5650	36" high		13	.615		51.50	14.20		65.70	80.50
	5700	48" high		10	.800		62	18.50		80.50	99.50
	5750	60" high		7	1.143		89.50	26.50		116	144
	5800	72" high	▼	6	1.333	▼	101	31		132	164
	6000	Plastic laminate on particle board									
	6100	12" wide, 18" high	1 Carp	25	.320	Ea.	15.20	7.40		22.60	29.50
	6120	24" high		24	.333		19.90	7.70		27.60	35
	6140	30" high		23	.348		24.50	8.05		32.55	40.50
	6160	36" high		21	.381		30.50	8.80		39.30	48.50
	6200	48" high		16	.500		40	11.55		51.55	63
	6250	60" high		13	.615		50.50	14.20		64.70	79.50
	6300	72" high		12	.667		61	15.40		76.40	93
	6320	15" wide x 18" high		24.50	.327		18.70	7.55		26.25	33.50
	6340	24" high		23.50	.340		25.50	7.85		33.35	42
	6360	30" high	▼	22.50	.356	▼	31.50	8.20		39.70	48.50

	06410 | Custom Cabinets	CREW	DAILY OUTPUT	LABOR-HOURS	UNIT	2004 BARE COSTS MAT.	LABOR	EQUIP.	TOTAL	TOTAL INCL O&P	
220	6380 36" high	1 Carp	20.50	.390	Ea.	37.50	9		46.50	56.50	220
	6400 48" high		15.50	.516		50.50	11.90		62.40	75.50	
	6450 60" high		12.50	.640		63	14.80		77.80	94.50	
	6480 72" high		11.50	.696		76	16.05		92.05	111	
	6500 18" wide, 18" high		24	.333		22	7.70		29.70	37.50	
	6550 24" high		23	.348		31.50	8.05		39.55	48	
	6600 30" high		22	.364		37.50	8.40		45.90	55.50	
	6650 36" high		20	.400		45.50	9.25		54.75	65.50	
	6700 48" high		15	.533		61	12.30		73.30	88	
	6750 60" high		12	.667		76	15.40		91.40	110	
	6800 72" high	↓	11	.727	↓	90	16.80		106.80	128	
	7000 Plywood, with edge band										
	7010 12" wide, 18" high	1 Carp	27	.296	Ea.	20.50	6.85		27.35	34	
	7100 24" high		26	.308		27.50	7.10		34.60	42.50	
	7120 30" high		25	.320		35	7.40		42.40	51	
	7140 36" high		23	.348		42	8.05		50.05	59.50	
	7180 48" high		18	.444		55	10.25		65.25	78	
	7200 60" high		15	.533		69.50	12.30		81.80	97.50	
	7250 72" high		14	.571		82	13.20		95.20	113	
	7300 15" wide x 18" high		26.50	.302		26.50	6.95		33.45	41	
	7350 24" high		25.50	.314		34	7.25		41.25	50	
	7400 30" high		24.50	.327		42	7.55		49.55	59.50	
	7450 36" high		22.50	.356		51	8.20		59.20	70	
	7500 48" high		17.50	.457		68	10.55		78.55	92.50	
	7550 60" high		14.50	.552		85.50	12.75		98.25	116	
	7600 72" high		13.50	.593		102	13.70		115.70	135	
	7650 18" wide, 18" high		26	.308		30	7.10		37.10	45	
	7700 24" high		25	.320		40.50	7.40		47.90	57	
	7750 30" high	↓	24	.333	↓	51.50	7.70		59.20	69.50	
230	0010 **CABINET HARDWARE**										230
	1000 Catches, minimum	1 Carp	235	.034	Ea.	.75	.79		1.54	2.16	
	1020 Average		119.40	.067		2.45	1.55		4	5.35	
	1040 Maximum	↓	80	.100	↓	4.65	2.31		6.96	9	
	2000 Door/drawer pulls, handles										
	2200 Handles and pulls, projecting, metal, minimum	1 Carp	160	.050	Ea.	3.48	1.16		4.64	5.80	
	2220 Average		95.24	.084		4.64	1.94		6.58	8.40	
	2240 Maximum		68	.118		6.40	2.72		9.12	11.60	
	2300 Wood, minimum		160	.050		3.48	1.16		4.64	5.80	
	2320 Average		95.24	.084		4.64	1.94		6.58	8.40	
	2340 Maximum		68	.118		6.40	2.72		9.12	11.60	
	2600 Flush, metal, minimum		160	.050		3.48	1.16		4.64	5.80	
	2620 Average		95.24	.084		4.64	1.94		6.58	8.40	
	2640 Maximum		68	.118	↓	6.40	2.72		9.12	11.60	
	3000 Drawer tracks/glides, minimum		48	.167	Pr.	5.90	3.85		9.75	13.05	
	3020 Average		32	.250		10	5.80		15.80	21	
	3040 Maximum		24	.333		17.15	7.70		24.85	32	
	4000 Cabinet hinges, minimum		160	.050		2	1.16		3.16	4.16	
	4020 Average		95.24	.084		2.90	1.94		4.84	6.50	
	4040 Maximum	↓	68	.118	↓	7	2.72		9.72	12.30	
240	0010 **DRAWERS**										240
	0100 Solid hardwood front										
	1000 4" high, 12" wide	1 Carp	17	.471	Ea.	17.30	10.85		28.15	37.50	
	1200 18" wide		16	.500		24	11.55		35.55	46	
	1400 24" wide		15	.533		30.50	12.30		42.80	54.50	
	1600 6" high, 12" wide	↓	16	.500	↓	19.50	11.55		31.05	41	

Important: See the Reference Section for critical supporting data - Reference Nos., Crews, & Location Factors

06410 | Custom Cabinets

			CREW	DAILY OUTPUT	LABOR-HOURS	UNIT	2004 BARE COSTS MAT.	LABOR	EQUIP.	TOTAL	TOTAL INCL O&P	
240	1800	18" wide	1 Carp	15	.533	Ea.	26	12.30		38.30	49.50	240
	2000	24" wide		14	.571		32.50	13.20		45.70	58	
	2200	9" high, 12" wide		15	.533		23	12.30		35.30	46	
	2400	18" wide		14	.571		29	13.20		42.20	54.50	
	2600	24" wide	↓	13	.615	↓	35.50	14.20		49.70	63.50	
	2800	Plastic laminate on particle board front										
	3000	4" high, 12" wide	1 Carp	17	.471	Ea.	22.50	10.85		33.35	43.50	
	3200	18" wide		16	.500		26.50	11.55		38.05	48.50	
	3600	24" wide		15	.533		31	12.30		43.30	55	
	3800	6" high, 12" wide		16	.500		26.50	11.55		38.05	48.50	
	4000	18" wide		15	.533		32	12.30		44.30	56.50	
	4500	24" wide		14	.571		39.50	13.20		52.70	66	
	4800	9" high, 12" wide		15	.533		30.50	12.30		42.80	54.50	
	5000	18" wide		14	.571		31	13.20		44.20	56.50	
	5200	24" wide	↓	13	.615	↓	49.50	14.20		63.70	78	
	5400	Plywood, flush panel front										
	6000	4" high, 12" wide	1 Carp	17	.471	Ea.	23.50	10.85		34.35	44.50	
	6200	18" wide	"	16	.500	"	28.50	11.55		40.05	50.50	
400	0010	**VANITIES**										400
	8000	Vanity bases, 2 doors, 30" high, 21" deep, 24" wide	2 Carp	20	.800	Ea.	169	18.50		187.50	218	
	8050	30" wide		16	1		194	23		217	253	
	8100	36" wide		13.33	1.200		259	27.50		286.50	330	
	8150	48" wide	↓	11.43	1.400		310	32.50		342.50	395	
	9000	For deluxe models of all vanities, add to above					40%					
	9500	For custom built in place, add to above				↓	25%	10%				

06415 | Countertops

			CREW	DAILY OUTPUT	LABOR-HOURS	UNIT	MAT.	LABOR	EQUIP.	TOTAL	TOTAL INCL O&P	
100	0010	**COUNTER TOP** Stock, plastic lam., 24" wide w/backsplash, min.	1 Carp	30	.267	L.F.	8.30	6.15		14.45	19.60	100
	0100	Maximum		25	.320		15.85	7.40		23.25	30	
	0300	Custom plastic, 7/8" thick, aluminum molding, no splash		30	.267		17	6.15		23.15	29	
	0400	Cove splash		30	.267		22	6.15		28.15	35	
	0600	1-1/4" thick, no splash		28	.286		20	6.60		26.60	33	
	0700	Square splash		28	.286		25	6.60		31.60	38.50	
	0900	Square edge, plastic face, 7/8" thick, no splash		30	.267		21.50	6.15		27.65	34	
	1000	With splash	↓	30	.267		28	6.15		34.15	41	
	1200	For stainless channel edge, 7/8" thick, add					2.27			2.27	2.50	
	1300	1-1/4" thick, add					2.66			2.66	2.93	
	1500	For solid color suede finish, add				↓	2.12			2.12	2.33	
	1700	For end splash, add				Ea.	13.35			13.35	14.70	
	1901	For cut outs, standard, add, minimum	1 Carp	32	.250			5.80		5.80	9.80	
	2000	Maximum		8	1		3.23	23		26.23	42.50	
	2010	Cut out in blacksplash for elec. wall outlet		38	.211			4.86		4.86	8.25	
	2020	Cut out for sink		20	.400			9.25		9.25	15.70	
	2030	Cut out for stove top		18	.444	↓		10.25		10.25	17.40	
	2100	Postformed, including backsplash and front edge		30	.267	L.F.	8.80	6.15		14.95	20	
	2110	Mitred, add		12	.667	Ea.		15.40		15.40	26	
	2200	Built-in place, 25" wide, plastic laminate		25	.320	L.F.	11.70	7.40		19.10	25.50	
	2300	Ceramic tile mosaic	↓	25	.320		25.50	7.40		32.90	40.50	
	2500	Marble, stock, with splash, 1/2" thick, minimum	1 Bric	17	.471		31	11.30		42.30	53	
	2700	3/4" thick, maximum	"	13	.615		78.50	14.75		93.25	111	
	2900	Maple, solid, laminated, 1-1/2" thick, no splash	1 Carp	28	.286		52.50	6.60		59.10	68.50	
	3000	With square splash		28	.286	↓	62	6.60		68.60	79.50	
	3200	Stainless steel		24	.333	S.F.	94	7.70		101.70	116	
	3400	Recessed cutting block with trim, 16" x 20" x 1"		8	1	Ea.	61	23		84	106	
	3411	Replace cutting block only	↓	16	.500	"	29.50	11.55		41.05	52	

06430 | Stairs & Railings

		06430 Stairs & Railings	CREW	DAILY OUTPUT	LABOR-HOURS	UNIT	2004 BARE COSTS MAT.	LABOR	EQUIP.	TOTAL	TOTAL INCL O&P	
500	0010	**RAILING** Custom design, architectural grade, hardwood, minimum	1 Carp	38	.211	L.F.	5.60	4.86		10.46	14.40	500
	0100	Maximum		30	.267		45.50	6.15		51.65	60.50	
	0300	Stock interior railing with spindles 6" O.C., 4' long		40	.200		29	4.62		33.62	39.50	
	0400	8' long	↓	48	.167	↓	27	3.85		30.85	36	
620	0011	**STAIRS, PREFABRICATED**										620
	0100	Box stairs, prefabricated, 3'-0" wide										
	0110	Oak treads, up to 14 risers	2 Carp	39	.410	Riser	71.50	9.50		81	94.50	
	0600	With pine treads for carpet, up to 14 risers	"	39	.410	"	46	9.50		55.50	66.50	
	1100	For 4' wide stairs, add				Flight	25%					
	1550	Stairs, prefabricated stair handrail with balusters	1 Carp	30	.267	L.F.	49	6.15		55.15	64	
	1700	Basement stairs, prefabricated, pine treads										
	1710	Pine risers, 3' wide, up to 14 risers	2 Carp	52	.308	Riser	46	7.10		53.10	62.50	
	4000	Residential, wood, oak treads, prefabricated		1.50	10.667	Flight	930	246		1,176	1,450	
	4200	Built in place	↓	.44	36.364	"	1,675	840		2,515	3,275	
	4400	Spiral, oak, 4'-6" diameter, unfinished, prefabricated,										
	4500	incl. railing, 9' high	2 Carp	1.50	10.667	Flight	4,400	246		4,646	5,275	
630	0010	**STAIR PARTS** Balusters, turned, 30" high, pine, minimum	1 Carp	28	.286	Ea.	4.10	6.60		10.70	15.70	630
	0100	Maximum		26	.308		19	7.10		26.10	33	
	0300	30" high birch balusters, minimum		28	.286		6.30	6.60		12.90	18.15	
	0400	Maximum		26	.308		25.50	7.10		32.60	40	
	0600	42" high, pine balusters, minimum		27	.296		5.30	6.85		12.15	17.40	
	0700	Maximum		25	.320		27.50	7.40		34.90	42.50	
	0900	42" high birch balusters, minimum		27	.296		5.30	6.85		12.15	17.40	
	1000	Maximum		25	.320	↓	37	7.40		44.40	53	
	1050	Baluster, stock pine, 1-1/4" x 1-1/4"		240	.033	L.F.	1.55	.77		2.32	3.02	
	1100	1-3/4" x 1-3/4"		220	.036	"	7.60	.84		8.44	9.80	
	1200	Newels, 3-1/4" wide, starting, minimum		7	1.143	Ea.	37	26.50		63.50	86	
	1300	Maximum		6	1.333		425	31		456	525	
	1500	Landing, minimum		5	1.600		98	37		135	171	
	1600	Maximum		4	2	↓	460	46		506	585	
	1800	Railings, oak, built-up, minimum		60	.133	L.F.	30	3.08		33.08	38.50	
	1900	Maximum		55	.145		43.50	3.36		46.86	53.50	
	2100	Add for sub rail		110	.073		5	1.68		6.68	8.35	
	2300	Risers, beech, 3/4" x 7-1/2" high		64	.125		5.80	2.89		8.69	11.25	
	2400	Fir, 3/4" x 7-1/2" high		64	.125		1.60	2.89		4.49	6.65	
	2600	Oak, 3/4" x 7-1/2" high		64	.125		6	2.89		8.89	11.50	
	2800	Pine, 3/4" x 7-1/2" high		66	.121		3.05	2.80		5.85	8.10	
	2850	Skirt board, pine, 1" x 10"		55	.145		2.90	3.36		6.26	8.90	
	2900	1" x 12"		52	.154	↓	3.45	3.55		7	9.85	
	3000	Treads, oak, 1-1/4" x 10" wide, 3' long		18	.444	Ea.	74	10.25		84.25	98.50	
	3100	4' long, oak		17	.471		98.50	10.85		109.35	126	
	3300	1-1/4" x 11-1/2" wide, 3' long, oak		18	.444		26	10.25		36.25	46	
	3400	6' long, oak	↓	14	.571		155	13.20		168.20	193	
	3600	Beech treads, add				↓	40%					
	3800	For mitered return nosings, add				L.F.	3.01			3.01	3.31	

06440 | Wood Ornaments

		06440 Wood Ornaments	CREW	DAILY OUTPUT	LABOR-HOURS	UNIT	MAT.	LABOR	EQUIP.	TOTAL	TOTAL INCL O&P	
150	0010	**BEAMS, DECORATIVE** Rough sawn cedar, non-load bearing, 4" x 4"	2 Carp	180	.089	L.F.	1.35	2.05		3.40	4.97	150
	0100	4" x 6"		170	.094		2.60	2.17		4.77	6.55	
	0200	4" x 8"		160	.100		3.34	2.31		5.65	7.60	
	0300	4" x 10"		150	.107		4.64	2.46		7.10	9.30	
	0400	4" x 12"		140	.114		5.60	2.64		8.24	10.70	
	0500	8" x 8"		130	.123		7.85	2.84		10.69	13.45	
	0600	Plastic beam, "hewn finish", 6" x 2"		240	.067		2.89	1.54		4.43	5.80	
	0601	6" x 4"	↓	220	.073	↓	3.37	1.68		5.05	6.55	

Important: See the Reference Section for critical supporting data - Reference Nos., Crews, & Location Factors

06440 | Wood Ornaments

			CREW	DAILY OUTPUT	LABOR-HOURS	UNIT	2004 BARE COSTS MAT.	LABOR	EQUIP.	TOTAL	TOTAL INCL O&P	
150	1100	Beam connector plates see div. 06090-800										150
350	0010	**GRILLES** and panels, hardwood, sanded										350
	0020	2′ x 4′ to 4′ x 8′, custom designs, unfinished, minimum	1 Carp	38	.211	S.F.	12.10	4.86		16.96	21.50	
	0050	Average		30	.267		26.50	6.15		32.65	39.50	
	0100	Maximum		19	.421		40.50	9.75		50.25	61	
	0300	As above, but prefinished, minimum		38	.211		12.10	4.86		16.96	21.50	
	0400	Maximum	↓	19	.421	↓	45.50	9.75		55.25	66.50	
400	0010	**LOUVERS** Redwood, 2′-0″ diameter, full circle	1 Carp	16	.500	Ea.	129	11.55		140.55	162	400
	0100	Half circle		16	.500		124	11.55		135.55	156	
	0200	Octagonal		16	.500		99	11.55		110.55	129	
	0300	Triangular, 5/12 pitch, 5′-0″ at base	↓	16	.500	↓	209	11.55		220.55	250	
500	0010	**FIREPLACE MANTELS** 6″ molding, 6′ x 3′-6″ opening, minimum	1 Carp	5	1.600	Opng.	135	37		172	212	500
	0100	Maximum		5	1.600		168	37		205	248	
	0300	Prefabricated pine, colonial type, stock, deluxe		2	4		2,125	92.50		2,217.50	2,475	
	0400	Economy	↓	3	2.667	↓	365	61.50		426.50	505	
550	0010	**FIREPLACE MANTEL BEAMS** Rough texture wood, 4″ x 8″	1 Carp	36	.222	L.F.	5	5.15		10.15	14.20	550
	0100	4″ x 10″		35	.229	″	6	5.30		11.30	15.55	
	0300	Laminated hardwood, 2-1/4″ x 10-1/2″ wide, 6′ long		5	1.600	Ea.	95.50	37		132.50	168	
	0400	8′ long		5	1.600	″	133	37		170	209	
	0600	Brackets for above, rough sawn		12	.667	Pr.	8.80	15.40		24.20	35.50	
	0700	Laminated	↓	12	.667	″	13.30	15.40		28.70	40.50	
700	0010	**COLUMNS**										700
	0050	Aluminum, round colonial, 6″ diameter	2 Carp	80	.200	V.L.F.	18.05	4.62		22.67	27.50	
	0100	8″ diameter		62.25	.257		24	5.95		29.95	36.50	
	0200	10″ diameter		55	.291		29.50	6.70		36.20	44	
	0250	Fir, stock units, hollow round, 6″ diameter		80	.200		16.20	4.62		20.82	25.50	
	0300	8″ diameter		80	.200		19	4.62		23.62	29	
	0350	10″ diameter		70	.229		24.50	5.30		29.80	36	
	0400	Solid turned, to 8′ high, 3-1/2″ diameter		80	.200		7.45	4.62		12.07	16	
	0500	4-1/2″ diameter		75	.213		10.60	4.93		15.53	20	
	0600	5-1/2″ diameter		70	.229		14.85	5.30		20.15	25.50	
	0800	Square columns, built-up, 5″ x 5″		65	.246		13.80	5.70		19.50	25	
	0900	Solid, 3-1/2″ x 3-1/2″		130	.123		6.35	2.84		9.19	11.80	
	1600	Hemlock, tapered, T & G, 12″ diam, 10′ high		100	.160		32	3.70		35.70	41.50	
	1700	16′ high		65	.246		56	5.70		61.70	71.50	
	1900	10′ high, 14″ diameter		100	.160		81.50	3.70		85.20	96	
	2000	18′ high		65	.246		77.50	5.70		83.20	94.50	
	2200	18″ diameter, 12′ high		65	.246		109	5.70		114.70	130	
	2300	20′ high		50	.320		105	7.40		112.40	129	
	2500	20″ diameter, 14′ high		40	.400		129	9.25		138.25	158	
	2600	20′ high	↓	35	.457		134	10.55		144.55	165	
	2800	For flat pilasters, deduct				↓	33%					
	3000	For splitting into halves, add				Ea.	63.50			63.50	70	
	4000	Rough sawn cedar posts, 4″ x 4″	2 Carp	250	.064	V.L.F.	2.58	1.48		4.06	5.35	
	4100	4″ x 6″		235	.068		3.84	1.57		5.41	6.90	
	4200	6″ x 6″		220	.073		5.80	1.68		7.48	9.20	
	4300	8″ x 8″	↓	200	.080	↓	5.95	1.85		7.80	9.70	

06445 | Simulated Wood Ornaments

			CREW	DAILY OUTPUT	LABOR-HOURS	UNIT	MAT.	LABOR	EQUIP.	TOTAL	TOTAL INCL O&P	
100	0010	**MILLWORK, HIGH DENSITY POLYMER**										100
	0100	Base, 9/16″ x 3-3/16″	1 Carp	230	.035	L.F.	1.31	.80		2.11	2.80	
	0200	Casing, fluted, 5/8″ x 3-1/4″		215	.037		3.66	.86		4.52	5.50	
	0300	Chair rail, 9/16″ x 2-1/4″		260	.031		1.91	.71		2.62	3.31	
	0600	Cove, 13/16″ x 3-3/4″		260	.031		3.83	.71		4.54	5.40	
	0700	Crown, 3/4″ x 3-13/16″	↓	260	.031	↓	5.40	.71		6.11	7.15	

		06445 | Simulated Wood Ornaments	CREW	DAILY OUTPUT	LABOR-HOURS	UNIT	2004 BARE COSTS MAT.	LABOR	EQUIP.	TOTAL	TOTAL INCL O&P	
100	0800	Half round, 15/16" x 2"	1 Carp	240	.033	L.F.	17.10	.77		17.87	20	100

		06470 | Screen, Blinds & Shutters	CREW	DAILY OUTPUT	LABOR-HOURS	UNIT	MAT.	LABOR	EQUIP.	TOTAL	TOTAL INCL O&P	
100	0010	**SHUTTERS, EXTERIOR** Aluminum, louvered, 1'-4" wide, 3'-0" long	1 Carp	10	.800	Pr.	42	18.50		60.50	77.50	100
	0200	4'-0" long		10	.800		50	18.50		68.50	86.50	
	0300	5'-4" long		10	.800		66	18.50		84.50	104	
	0400	6'-8" long		9	.889		84	20.50		104.50	128	
	1000	Pine, louvered, primed, each 1'-2" wide, 3'-3" long		10	.800		79.50	18.50		98	119	
	1100	4'-7" long		10	.800		107	18.50		125.50	150	
	1250	Each 1'-4" wide, 3'-0" long		10	.800		81	18.50		99.50	121	
	1350	5'-3" long		10	.800		122	18.50		140.50	166	
	1500	Each 1'-6" wide, 3'-3" long		10	.800		84.50	18.50		103	125	
	1600	4'-7" long		10	.800		119	18.50		137.50	162	
	1620	Hemlock, louvered, 1'-2" wide, 5'-7" long		10	.800		130	18.50		148.50	175	
	1630	Each 1'-4" wide, 2'-2" long		10	.800		81	18.50		99.50	121	
	1640	3'-0" long		10	.800		81	18.50		99.50	121	
	1650	3'-3" long		10	.800		87.50	18.50		106	128	
	1660	3'-11" long		10	.800		99.50	18.50		118	141	
	1670	4'-3" long		10	.800		97	18.50		115.50	139	
	1680	5'-3" long		10	.800		121	18.50		139.50	165	
	1690	5'-11" long		10	.800		137	18.50		155.50	183	
	1700	Door blinds, 6'-9" long, each 1'-3" wide		9	.889		138	20.50		158.50	186	
	1710	1'-6" wide		9	.889		148	20.50		168.50	198	
	1720	Hemlock, solid raised panel, each 1'-4" wide, 3'-3" long		10	.800		131	18.50		149.50	176	
	1730	3'-11" long		10	.800		154	18.50		172.50	201	
	1740	4'-3" long		10	.800		166	18.50		184.50	215	
	1750	4'-7" long		10	.800		179	18.50		197.50	229	
	1760	4'-11" long		10	.800		195	18.50		213.50	246	
	1770	5'-11" long		10	.800		222	18.50		240.50	276	
	1800	Door blinds, 6'-9" long, each 1'-3" wide		9	.889		250	20.50		270.50	310	
	1900	1'-6" wide		9	.889		272	20.50		292.50	335	
	2500	Polystyrene, solid raised panel, each 1'-4" wide, 3'-3" long		10	.800		55	18.50		73.50	92	
	2600	3'-11" long		10	.800		63	18.50		81.50	101	
	2700	4'-7" long		10	.800		66.50	18.50		85	105	
	2800	5'-3" long		10	.800		74.50	18.50		93	114	
	2900	6'-8" long		9	.889		103	20.50		123.50	149	
	3500	Polystyrene, solid raised panel, each 3'-3" wide, 3'-0" long		10	.800		166	18.50		184.50	214	
	3600	3'-11" long		10	.800		177	18.50		195.50	226	
	3700	4'-7" long		10	.800		209	18.50		227.50	262	
	3800	5'-3" long		10	.800		231	18.50		249.50	287	
	3900	6'-8" long		9	.889		261	20.50		281.50	320	
	4500	Polystyrene, louvered, each 1'-2" wide, 3'-3" long		10	.800		40	18.50		58.50	75.50	
	4600	4'-7" long		10	.800		49.50	18.50		68	86	
	4750	5'-3" long		10	.800		53	18.50		71.50	89.50	
	4850	6'-8" long		9	.889		87	20.50		107.50	131	
	6000	Vinyl, louvered, each 1'-2" x 4'-7" long		10	.800		52	18.50		70.50	88.50	
	6200	Each 1'-4" x 6'-8" long	↓	9	.889	↓	79	20.50		99.50	122	
	8000	PVC exterior rolling shutters										
	8100	including crank control	1 Carp	8	1	Ea.	400	23		423	480	
	8500	Insulative - 6' x 6'8" stock unit	"	8	1	"	570	23		593	670	
200	0010	**SHUTTERS, INTERIOR** Wood, louvered,				Set	105			105	116	200
	0200	Two panel, 27" wide, 36" high	1 Carp	5	1.600		105	37		142	179	
	0300	33" wide, 36" high		5	1.600		125	37		162	201	
	0500	47" wide, 36" high	↓	5	1.600	↓	182	37		219	263	

Important: See the Reference Section for critical supporting data - Reference Nos., Crews, & Location Factors

06470 | Screen, Blinds & Shutters

			CREW	DAILY OUTPUT	LABOR-HOURS	UNIT	2004 BARE COSTS MAT.	LABOR	EQUIP.	TOTAL	TOTAL INCL O&P	
200	1000	Four panel, 27" wide, 36" high	1 Carp	5	1.600	Set	241	37		278	330	200
	1100	33" wide, 36" high		5	1.600		241	37		278	330	
	1300	47" wide, 36" high	↓	5	1.600	↓	320	37		357	420	

06600 | Plastic Fabrications

06620 | Non-Structural Plastics

			CREW	DAILY OUTPUT	LABOR-HOURS	UNIT	2004 BARE COSTS MAT.	LABOR	EQUIP.	TOTAL	TOTAL INCL O&P	
810	0010	**SOLID SURFACE COUNTERTOPS**, Acrylic polymer										810
	2000	Pricing for order of 1 - 50 L.F.										
	2100	25" wide, solid colors	2 Carp	20	.800	L.F.	59	18.50		77.50	96.50	
	2200	Patterned colors		20	.800		75	18.50		93.50	114	
	2300	Premium patterned colors		20	.800		93.50	18.50		112	135	
	2400	With silicone attached 4" backsplash, solid colors		19	.842		65	19.45		84.45	105	
	2500	Patterned colors		19	.842		82	19.45		101.45	124	
	2600	Premium patterned colors		19	.842		102	19.45		121.45	145	
	2700	With hard seam attached 4" backsplash, solid colors		4	4		65	92.50		157.50	229	
	2800	Patterned colors		15	1.067		82	24.50		106.50	133	
	2900	Premium patterned colors	↓	15	1.067	↓	102	24.50		126.50	154	
	3800	Sinks, pricing for order of 1 - 50 units										
	3900	Single bowl, hard seamed, solid colors, 13" x 17"	1 Carp	2	4	Ea.	400	92.50		492.50	595	
	4000	10" x 15"		4.55	1.758		184	40.50		224.50	272	
	4100	Cutouts for sinks	↓	5.25	1.524	↓		35		35	59.50	
850	0010	**VANITY TOPS**										850
	0015	Solid surface, center bowl, 17" x 19"	1 Carp	12	.667	Ea.	168	15.40		183.40	211	
	0020	19" x 25"		12	.667		204	15.40		219.40	250	
	0030	19" x 31"		12	.667		247	15.40		262.40	298	
	0040	19" x 37"		12	.667		288	15.40		303.40	340	
	0050	22" x 25"		10	.800		229	18.50		247.50	284	
	0060	22" x 31"		10	.800		267	18.50		285.50	325	
	0070	22" x 37"		10	.800		310	18.50		328.50	370	
	0080	22" x 43"		10	.800		355	18.50		373.50	420	
	0090	22" x 49"		10	.800		395	18.50		413.50	460	
	0110	22" x 55"		8	1		445	23		468	530	
	0120	22" x 61"		8	1		510	23		533	600	
	0220	Double bowl, 22" x 61"		8	1		575	23		598	670	
	0230	Double bowl, 22" x 73"	↓	8	1	↓	625	23		648	730	
	0240	For aggregate colors, add					35%					
	0250	For faucets and fittings see 15410-300										

Division Notes

		CREW	DAILY OUTPUT	LABOR-HOURS	UNIT	2004 BARE COSTS MAT.	LABOR	EQUIP.	TOTAL	TOTAL INCL O&P

Division 7
Thermal & Moisture Protection

Estimating Tips

07100 Dampproofing & Waterproofing

- Be sure of the job specifications before pricing this subdivision. The difference in cost between waterproofing and dampproofing can be great. Waterproofing will hold back standing water. Dampproofing prevents the transmission of water vapor. Also included in this section are vapor retarding membranes.

07200 Thermal Protection

- Insulation and fireproofing products are measured by area, thickness, volume or R value. Specifications may only give what the specific R value should be in a certain situation. The estimator may need to choose the type of insulation to meet that R value.

07300 Shingles, Roof Tiles & Roof Coverings
07400 Roofing & Siding Panels

- Many roofing and siding products are bought and sold by the square. One square is equal to an area that measures 100 square feet.

This simple change in unit of measure could create a large error if the estimator is not observant. Accessories necessary for a complete installation must be figured into any calculations for both material and labor.

07500 Membrane Roofing
07600 Flashing & Sheet Metal
07700 Roof Specialties & Accessories

- The items in these subdivisions compose a roofing system. No one component completes the installation and all must be estimated. Built-up or single ply membrane roofing systems are made up of many products and installation trades. Wood blocking at roof perimeters or penetrations, parapet coverings, reglets, roof drains, gutters, downspouts, sheet metal flashing, skylights, smoke vents or roof hatches all need to be considered along with the roofing material. Several different installation trades will need to work together on the roofing system. Inherent difficulties in the scheduling and coordination of various trades must be accounted for when estimating labor costs.

07900 Joint Sealers

- To complete the weather-tight shell the sealants and caulkings must be estimated. Where different materials meet—at expansion joints, at flashing penetrations, and at hundreds of other locations throughout a construction project—they provide another line of defense against water penetration. Often, an entire system is based on the proper location and placement of caulking or sealants. The detail drawings that are included as part of a set of architectural plans, show typical locations for these materials. When caulking or sealants are shown at typical locations, this means the estimator must include them for all the locations where this detail is applicable. Be careful to keep different types of sealants separate, and remember to consider backer rods and primers if necessary.

Reference Numbers

Reference numbers are shown in bold squares at the beginning of some major classifications. These numbers refer to related items in the Reference Section. The reference information may be an estimating procedure, an alternate pricing method or technical information.

Note: Not all subdivisions listed here necessarily appear in this publication.

07050 | Basic Thermal & Moisture Protection Materials & Methods

07060 | Selective Demolition

			CREW	DAILY OUTPUT	LABOR-HOURS	UNIT	2004 BARE COSTS MAT.	LABOR	EQUIP.	TOTAL	TOTAL INCL O&P	
110	0010	**SELECTIVE DEMOLITION, ROOFING AND SIDING**										110
	1200	Wood, boards, tongue and groove, 2″ x 6″	2 Clab	960	.017	S.F.		.28		.28	.48	
	1220	2″ x 10″		1,040	.015			.26		.26	.44	
	1280	Standard planks, 1″ x 6″		1,080	.015			.25		.25	.43	
	1320	1″ x 8″		1,160	.014			.23		.23	.40	
	1340	1″ x 12″		1,200	.013			.23		.23	.38	
	1350	Plywood, to 1″ thick		2,000	.008			.14		.14	.23	
	1360	Flashing, aluminum	1 Clab	290	.028			.47		.47	.79	
	2000	Gutters, aluminum or wood, edge hung	″	240	.033	L.F.		.56		.56	.96	
	2010	Remove and reset, aluminum	1 Shee	125	.064			1.64		1.64	2.74	
	2020	Remove and reset, vinyl	1 Carp	125	.064			1.48		1.48	2.51	
	2100	Built-in	1 Clab	100	.080			1.35		1.35	2.30	
	2500	Roof accessories, plumbing vent flashing		14	.571	Ea.		9.65		9.65	16.40	
	2600	Adjustable metal chimney flashing		9	.889	″		15		15	25.50	
	2650	Coping, sheet metal, up to 12″ wide		240	.033	L.F.		.56		.56	.96	
	2660	Concrete, up to 12″ wide	2 Clab	160	.100	″		1.69		1.69	2.87	
	3000	Roofing, built-up, 5 ply roof, no gravel	B-2	1,600	.025	S.F.		.43		.43	.73	
	3100	Gravel removal, minimum		5,000	.008			.14		.14	.24	
	3120	Maximum		2,000	.020			.35		.35	.59	
	3400	Roof insulation board, up to 2″ thick		3,900	.010			.18		.18	.30	
	3450	Roll roofing, cold adhesive	1 Clab	12	.667	Sq.		11.25		11.25	19.15	
	4000	Shingles, asphalt strip, 1 layer	B-2	3,500	.011	S.F.		.20		.20	.34	
	4100	Slate		2,500	.016			.28		.28	.47	
	4300	Wood		2,200	.018			.31		.31	.53	
	4500	Skylight to 10 S.F.	1 Clab	8	1	Ea.		16.90		16.90	28.50	
	5000	Siding, metal, horizontal		444	.018	S.F.		.30		.30	.52	
	5020	Vertical		400	.020			.34		.34	.57	
	5200	Wood, boards, vertical		400	.020			.34		.34	.57	
	5220	Clapboards, horizontal		380	.021			.36		.36	.60	
	5240	Shingles		350	.023			.39		.39	.66	
	5260	Textured plywood		725	.011			.19		.19	.32	

07100 | Dampproofing and Waterproofing

07110 | Dampproofing

			CREW	DAILY OUTPUT	LABOR-HOURS	UNIT	2004 BARE COSTS MAT.	LABOR	EQUIP.	TOTAL	TOTAL INCL O&P	
100	0010	**BITUMINOUS ASPHALT COATING** For foundation										100
	0030	Brushed on, below grade, 1 coat	1 Rofc	665	.012	S.F.	.07	.24		.31	.51	
	0100	2 coat		500	.016		.10	.31		.41	.69	
	0300	Sprayed on, below grade, 1 coat, 25.6 S.F./gal.		830	.010		.07	.19		.26	.43	
	0400	2 coat, 20.5 S.F./gal.		500	.016		.14	.31		.45	.74	
	0600	Troweled on, asphalt with fibers, 1/16″ thick		500	.016		.16	.31		.47	.76	
	0700	1/8″ thick		400	.020		.29	.39		.68	1.04	
	1000	1/2″ thick		350	.023		.94	.45		1.39	1.85	
200	0010	**CEMENT PARGING** 2 coats, 1/2″ thick, regular P.C. R07110-010	D-1	250	.064	S.F.	.17	1.34		1.51	2.42	200
	0100	Waterproofed Portland cement	″	250	.064	″	.18	1.34		1.52	2.43	

07190 | Water Repellents

			CREW	DAILY OUTPUT	LABOR-HOURS	UNIT	2004 BARE COSTS MAT.	LABOR	EQUIP.	TOTAL	TOTAL INCL O&P	
700	0010	**RUBBER COATING** Water base liquid, roller applied	2 Rofc	7,000	.002	S.F.	.55	.05		.60	.69	700
	0200	Silicone or stearate, sprayed on CMU, 1 coat	1 Rofc	4,000	.002		.27	.04		.31	.37	

07100 | Dampproofing and Waterproofing

	07190 | Water Repellents		Crew	Daily Output	Labor-Hours	Unit	Mat. (2004 Bare Costs)	Labor (2004 Bare Costs)	Equip. (2004 Bare Costs)	Total (2004 Bare Costs)	Total Incl O&P
700	0300	2 coats	1 Rofc	3,000	.003	S.F.	.54	.05		.59	.70

07200 | Thermal Protection

	07210 | Building Insulation		Crew	Daily Output	Labor-Hours	Unit	Mat. (2004 Bare Costs)	Labor (2004 Bare Costs)	Equip. (2004 Bare Costs)	Total (2004 Bare Costs)	Total Incl O&P
150	0010	**BLOWN-IN INSULATION** Ceilings, with open access									
	0020	Cellulose, 3-1/2" thick, R13	G-4	5,000	.005	S.F.	.13	.08	.04	.25	.32
	0030	5-3/16" thick, R19		3,800	.006		.21	.11	.05	.37	.48
	0050	6-1/2" thick, R22		3,000	.008		.26	.14	.07	.47	.60
	1000	Fiberglass, 5" thick, R11		3,800	.006		.16	.11	.05	.32	.43
	1050	6" thick, R13		3,000	.008		.22	.14	.07	.43	.55
	1100	8-1/2" thick, R19		2,200	.011		.31	.19	.09	.59	.77
	1300	12" thick, R26		1,500	.016		.43	.28	.14	.85	1.10
	2000	Mineral wool, 4" thick, R12		3,500	.007		.18	.12	.06	.36	.46
	2050	6" thick, R17		2,500	.010		.20	.17	.08	.45	.60
	2100	9" thick, R23		1,750	.014		.29	.24	.12	.65	.86
	2500	Wall installation, incl. drilling & patching from outside, two 1"									
	2510	diam. holes @ 16" O.C., top & mid-point of wall, add to above									
	2700	For masonry	G-4	415	.058	S.F.	.06	1.02	.49	1.57	2.34
	2800	For wood siding		840	.029		.06	.50	.24	.80	1.19
	2900	For stucco/plaster		665	.036		.06	.63	.31	1	1.49
350	0010	**FLOOR INSULATION, NONRIGID** Including									
	0020	spring type wire fasteners									
	2000	Fiberglass, blankets or batts, paper or foil backing									
	2100	1 side, 3-1/2" thick, R11	1 Carp	700	.011	S.F.	.28	.26		.54	.76
	2150	6" thick, R19		600	.013		.36	.31		.67	.92
	2200	8-1/2" thick, R30		550	.015		.65	.34		.99	1.29
500	0010	**POURED INSULATION** Cellulose fiber, R3.8 per inch	1 Carp	200	.040	C.F.	.48	.92		1.40	2.10
	0080	Fiberglass wool, R4 per inch		200	.040		.37	.92		1.29	1.98
	0100	Mineral wool, R3 per inch		200	.040		.33	.92		1.25	1.93
	0300	Polystyrene, R4 per inch		200	.040		2.23	.92		3.15	4.02
	0400	Vermiculite or perlite, R2.7 per inch		200	.040		1.65	.92		2.57	3.39
550	0010	**MASONRY INSULATION** Vermiculite or perlite, poured									
	0100	In cores of concrete block, 4" thick wall, .115 CF/SF	D-1	4,800	.003	S.F.	.19	.07		.26	.33
	0700	Foamed in place, urethane in 2-5/8" cavity	G-2	1,035	.023		.40	.43	.11	.94	1.28
	0800	For each 1" added thickness, add	"	2,372	.010		.12	.19	.05	.36	.50
600	0010	**PERIMETER INSULATION**									
	0600	Polystyrene, expanded, 1" thick, R4	1 Carp	680	.012	S.F.	.19	.27		.46	.67
	0700	2" thick, R8	"	675	.012	"	.37	.27		.64	.87
700	0011	**REFLECTIVE INSULATION**, aluminum foil on reinforced scrim	1 Carp	1,900	.004	S.F.	.14	.10		.24	.32
	0101	Reinforced with woven polyolefin		1,900	.004		.17	.10		.27	.36
	0501	With single bubble air space, R8.8		1,500	.005		.27	.12		.39	.51
	0601	With double bubble air space, R9.8		1,500	.005		.29	.12		.41	.53
900	0010	**WALL INSULATION, RIGID**									
	0040	Fiberglass, 1.5#/CF, unfaced, 1" thick, R4.1	1 Carp	1,000	.008	S.F.	.29	.18		.47	.63
	0060	1-1/2" thick, R6.2		1,000	.008		.41	.18		.59	.76
	0080	2" thick, R8.3		1,000	.008		.46	.18		.64	.82

	07210	Building Insulation	CREW	DAILY OUTPUT	LABOR-HOURS	UNIT	2004 BARE COSTS MAT.	LABOR	EQUIP.	TOTAL	TOTAL INCL O&P	
900	0120	3" thick, R12.4	1 Carp	800	.010	S.F.	.57	.23		.80	1.02	900
	0370	3#/CF, unfaced, 1" thick, R4.3		1,000	.008		.36	.18		.54	.71	
	0390	1-1/2" thick, R6.5		1,000	.008		.69	.18		.87	1.07	
	0400	2" thick, R8.7		890	.009		.83	.21		1.04	1.26	
	0420	2-1/2" thick, R10.9		800	.010		1.02	.23		1.25	1.51	
	0440	3" thick, R13		800	.010		1.21	.23		1.44	1.72	
	0520	Foil faced, 1" thick, R4.3		1,000	.008		.80	.18		.98	1.19	
	0540	1-1/2" thick, R6.5		1,000	.008		1.08	.18		1.26	1.50	
	0560	2" thick, R8.7		890	.009		1.35	.21		1.56	1.84	
	0580	2-1/2" thick, R10.9		800	.010		1.60	.23		1.83	2.15	
	0600	3" thick, R13		800	.010		1.74	.23		1.97	2.30	
	0670	6#/CF, unfaced, 1" thick, R4.3		1,000	.008		.77	.18		.95	1.16	
	0690	1-1/2" thick, R6.5		890	.009		1.19	.21		1.40	1.66	
	0700	2" thick, R8.7		800	.010		1.68	.23		1.91	2.24	
	0721	2-1/2" thick, R10.9		800	.010		1.84	.23		2.07	2.41	
	0741	3" thick, R13		730	.011		2.20	.25		2.45	2.85	
	0821	Foil faced, 1" thick, R4.3		1,000	.008		1.09	.18		1.27	1.51	
	0840	1-1/2" thick, R6.5		890	.009		1.57	.21		1.78	2.08	
	0850	2" thick, R8.7		800	.010		2.05	.23		2.28	2.65	
	0880	2-1/2" thick, R10.9		800	.010		2.46	.23		2.69	3.10	
	0900	3" thick, R13		730	.011		2.94	.25		3.19	3.66	
	1500	Foamglass, 1-1/2" thick, R4.5		800	.010		1.30	.23		1.53	1.82	
	1550	3" thick, R9	↓	730	.011	↓	2.93	.25		3.18	3.65	
	1600	Isocyanurate, 4' x 8' sheet, foil faced, both sides										
	1610	1/2" thick, R3.9	1 Carp	800	.010	S.F.	.29	.23		.52	.71	
	1620	5/8" thick, R4.5		800	.010		.30	.23		.53	.72	
	1630	3/4" thick, R5.4		800	.010		.24	.23		.47	.65	
	1640	1" thick, R7.2		800	.010		.33	.23		.56	.75	
	1650	1-1/2" thick, R10.8		730	.011		.35	.25		.60	.82	
	1660	2" thick, R14.4		730	.011		.46	.25		.71	.94	
	1670	3" thick, R21.6		730	.011		1.11	.25		1.36	1.65	
	1680	4" thick, R28.8		730	.011		1.37	.25		1.62	1.94	
	1700	Perlite, 1" thick, R2.77		800	.010		.26	.23		.49	.68	
	1750	2" thick, R5.55		730	.011		.50	.25		.75	.98	
	1900	Extruded polystyrene, 25 PSI compressive strength, 1" thick, R5		800	.010		.36	.23		.59	.79	
	1940	2" thick R10		730	.011		.71	.25		.96	1.21	
	1960	3" thick, R15		730	.011		.98	.25		1.23	1.51	
	2100	Expanded polystyrene, 1" thick, R3.85		800	.010		.16	.23		.39	.57	
	2120	2" thick, R7.69		730	.011		.40	.25		.65	.87	
	2140	3" thick, R11.49	↓	730	.011	↓	.52	.25		.77	1	
950	0010	**WALL OR CEILING INSUL., NON-RIGID**										950
	0040	Fiberglass, kraft faced, batts or blankets										
	0061	3-1/2" thick, R11, 11" wide	1 Carp	1,600	.005	S.F.	.23	.12		.35	.45	
	0080	15" wide		1,600	.005		.23	.12		.35	.45	
	0141	6" thick, R19, 11" wide		1,350	.006		.31	.14		.45	.57	
	0201	9" thick, R30, 15" wide		1,350	.006		.60	.14		.74	.89	
	0241	12" thick, R38, 15" wide	↓	1,350	.006	↓	.76	.14		.90	1.07	
	0400	Fiberglass, foil faced, batts or blankets										
	0420	3-1/2" thick, R11, 15" wide	1 Carp	1,600	.005	S.F.	.34	.12		.46	.57	
	0461	6" thick, R19, 15" wide		1,600	.005		.41	.12		.53	.65	
	0501	9" thick, R30, 15" wide	↓	1,350	.006	↓	.71	.14		.85	1.01	
	0800	Fiberglass, unfaced, batts or blankets										
	0821	3-1/2" thick, R11, 15" wide	1 Carp	1,600	.005	S.F.	.21	.12		.33	.43	
	0861	6" thick, R19, 15" wide		1,350	.006		.34	.14		.48	.60	
	0901	9" thick, R30, 15" wide		1,150	.007		.60	.16		.76	.93	
	0941	12" thick, R38, 15" wide	↓	1,150	.007	↓	.76	.16		.92	1.11	

07210 | Building Insulation

			CREW	DAILY OUTPUT	LABOR-HOURS	UNIT	2004 BARE COSTS MAT.	LABOR	EQUIP.	TOTAL	TOTAL INCL O&P	
950	1300	Mineral fiber batts, kraft faced										950
	1320	3-1/2" thick, R12	1 Carp	1,600	.005	S.F.	.26	.12		.38	.49	
	1340	6" thick, R19		1,600	.005		.34	.12		.46	.57	
	1380	10" thick, R30		1,350	.006		.52	.14		.66	.80	
	1850	Friction fit wire insulation supports, 16" O.C.		960	.008	Ea.	.05	.19		.24	.39	
	1900	For foil backing, add				S.F.	.04			.04	.04	

07220 | Roof and Deck Insulation

			CREW	DAILY OUTPUT	LABOR-HOURS	UNIT	2004 BARE COSTS MAT.	LABOR	EQUIP.	TOTAL	TOTAL INCL O&P	
700	0010	ROOF DECK INSULATION										700
	0020	Fiberboard low density, 1/2" thick R1.39	1 Rofc	1,000	.008	S.F.	.19	.16		.35	.50	
	0030	1" thick R2.78		800	.010		.34	.20		.54	.73	
	0080	1 1/2" thick R4.17		800	.010		.50	.20		.70	.91	
	0100	2" thick R5.56		800	.010		.68	.20		.88	1.11	
	0110	Fiberboard high density, 1/2" thick R1.3		1,000	.008		.20	.16		.36	.51	
	0120	1" thick R2.5		800	.010		.36	.20		.56	.76	
	0130	1-1/2" thick R3.8		800	.010		.59	.20		.79	1.01	
	0200	Fiberglass, 3/4" thick R2.78		1,000	.008		.46	.16		.62	.80	
	0400	15/16" thick R3.70		1,000	.008		.61	.16		.77	.96	
	0460	1-1/16" thick R4.17		1,000	.008		.76	.16		.92	1.13	
	0600	1-5/16" thick R5.26		1,000	.008		1.05	.16		1.21	1.45	
	0650	2-1/16" thick R8.33		800	.010		1.12	.20		1.32	1.59	
	0700	2-7/16" thick R10		800	.010		1.28	.20		1.48	1.77	
	1650	Perlite, 1/2" thick R1.32		1,050	.008		.27	.15		.42	.57	
	1655	3/4" thick R2.08		800	.010		.29	.20		.49	.68	
	1660	1" thick R2.78		800	.010		.30	.20		.50	.69	
	1670	1-1/2" thick R4.17		800	.010		.39	.20		.59	.79	
	1680	2" thick R5.56		700	.011		.61	.22		.83	1.08	
	1685	2-1/2" thick R6.67		700	.011		.77	.22		.99	1.26	
	1700	Polyisocyanurate, 2#/CF density, 3/4" thick, R5.1		1,500	.005		.28	.10		.38	.50	
	1705	1" thick R7.14		1,400	.006		.34	.11		.45	.58	
	1715	1-1/2" thick R10.87		1,250	.006		.37	.13		.50	.64	
	1725	2" thick R14.29		1,100	.007		.48	.14		.62	.79	
	1735	2-1/2" thick R16.67		1,050	.008		.50	.15		.65	.82	
	1745	3" thick R21.74		1,000	.008		.72	.16		.88	1.08	
	1755	3-1/2" thick R25		1,000	.008		.74	.16		.90	1.10	
	1765	Tapered for drainage		1,400	.006	B.F.	.38	.11		.49	.63	
	1900	Extruded Polystyrene										
	1910	15 PSI compressive strength, 1" thick, R5	1 Rofc	1,500	.005	S.F.	.23	.10		.33	.44	
	1920	2" thick, R10		1,250	.006		.35	.13		.48	.62	
	1930	3" thick R15		1,000	.008		.68	.16		.84	1.04	
	1932	4" thick R20		1,000	.008		1.07	.16		1.23	1.47	
	1934	Tapered for drainage		1,500	.005	B.F.	.35	.10		.45	.58	
	1940	25 PSI compressive strength, 1" thick R5		1,500	.005	S.F.	.42	.10		.52	.65	
	1942	2" thick R10		1,250	.006		.82	.13		.95	1.13	
	1944	3" thick R15		1,000	.008		1.24	.16		1.40	1.65	
	1946	4" thick R20		1,000	.008		1.19	.16		1.35	1.60	
	1948	Tapered for drainage		1,500	.005	B.F.	.40	.10		.50	.63	
	1950	40 psi compressive strength, 1" thick R5		1,500	.005	S.F.	.35	.10		.45	.58	
	1952	2" thick R10		1,250	.006		.69	.13		.82	.99	
	1954	3" thick R15		1,000	.008		1.01	.16		1.17	1.40	
	1956	4" thick R20		1,000	.008		1.35	.16		1.51	1.78	
	1958	Tapered for drainage		1,400	.006	B.F.	.50	.11		.61	.76	
	1960	60 PSI compressive strength, 1" thick R5		1,450	.006	S.F.	.42	.11		.53	.66	
	1962	2" thick R10		1,200	.007		.75	.13		.88	1.07	
	1964	3" thick R15		975	.008		1.12	.16		1.28	1.53	
	1966	4" thick R20		950	.008		1.56	.17		1.73	2.02	

07220	Roof and Deck Insulation	CREW	DAILY OUTPUT	LABOR-HOURS	UNIT	2004 BARE COSTS MAT.	LABOR	EQUIP.	TOTAL	TOTAL INCL O&P
700 1968	Tapered for drainage	1 Rofc	1,400	.006	B.F.	.61	.11		.72	.88
2010	Expanded polystyrene, 1#/CF density, 3/4" thick R2.89		1,500	.005	S.F.	.19	.10		.29	.40
2020	1" thick R3.85		1,500	.005		.19	.10		.29	.40
2100	2" thick R7.69		1,250	.006		.37	.13		.50	.64
2110	3" thick R11.49		1,250	.006		.59	.13		.72	.88
2120	4" thick R15.38		1,200	.007		.57	.13		.70	.87
2130	5" thick R19.23		1,150	.007		.71	.14		.85	1.03
2140	6" thick R23.26		1,150	.007		.83	.14		.97	1.16
2150	Tapered for drainage		1,500	.005	B.F.	.34	.10		.44	.56
2400	Composites with 2" EPS									
2410	1" fiberboard	1 Rofc	950	.008	S.F.	.78	.17		.95	1.16
2420	7/16" oriented strand board		800	.010		.93	.20		1.13	1.38
2430	1/2" plywood		800	.010		1	.20		1.20	1.46
2440	1" perlite		800	.010		.82	.20		1.02	1.26
2450	Composites with 1 1/2" polyisocyanurate									
2460	1" fiberboard	1 Rofc	800	.010	S.F.	.84	.20		1.04	1.28
2470	1" perlite		850	.009		.88	.18		1.06	1.31
2480	7/16" oriented strand board		800	.010		1.01	.20		1.21	1.47

07240	Ext. Insulation Finish Systems (EIFS)	CREW	DAILY OUTPUT	LABOR-HOURS	UNIT	MAT.	LABOR	EQUIP.	TOTAL	TOTAL INCL O&P
100 0010	**EXTERIOR INSULATION FINISH SYSTEM**									
0095	Field applied, 1" EPS insulation	J-1	295	.136	S.F.	1.83	2.68	.31	4.82	6.80
0100	With 1/2" cement board sheathing		220	.182		2.59	3.60	.41	6.60	9.25
0105	2" EPS insulation		295	.136		2.07	2.68	.31	5.06	7.05
0110	With 1/2" cement board sheathing		220	.182		2.83	3.60	.41	6.84	9.50
0115	3" EPS insulation		295	.136		2.19	2.68	.31	5.18	7.20
0120	With 1/2" cement board sheathing		220	.182		2.95	3.60	.41	6.96	9.65
0125	4" EPS insulation		295	.136		2.47	2.68	.31	5.46	7.50
0130	With 1/2" cement board sheathing		220	.182		3.98	3.60	.41	7.99	10.80
0140	Premium finish add		1,265	.032		.28	.63	.07	.98	1.43
0150	Heavy duty reinforcement add		914	.044		1.68	.87	.10	2.65	3.40
0160	2.5#/S.Y. metal lath substrate add	1 Lath	75	.107	S.Y.	2.16	2.28		4.44	6.10
0170	3.4#/S.Y. metal lath substrate add	"	75	.107	"	2.25	2.28		4.53	6.20
0180	Color or texture change,	J-1	1,265	.032	S.F.	.73	.63	.07	1.43	1.92
0190	With substrate leveling base coat	1 Plas	530	.015		.73	.32		1.05	1.32
0210	With substrate sealing base coat	1 Pord	1,224	.007		.07	.14		.21	.31
0370	V groove shape in panel face				L.F.	.52			.52	.57
0380	U groove shape in panel face				"	.69			.69	.76

07260	Vapor Retarders	CREW	DAILY OUTPUT	LABOR-HOURS	UNIT	MAT.	LABOR	EQUIP.	TOTAL	TOTAL INCL O&P
100 0011	**BUILDING PAPER** Aluminum and kraft laminated, foil 1 side	1 Carp	3,700	.002	S.F.	.04	.05		.09	.12
0101	Foil 2 sides		3,700	.002		.06	.05		.11	.15
0301	Asphalt, two ply, 30#, for subfloors		1,900	.004		.14	.10		.24	.33
0401	Asphalt felt sheathing paper, 15#		3,700	.002		.03	.05		.08	.11
0450	Housewrap, exterior, spun bonded polypropylene									
0470	Small roll	1 Carp	3,800	.002	S.F.	.16	.05		.21	.26
0480	Large roll	"	4,000	.002	"	.10	.05		.15	.19
0500	Material only, 3' x 111.1' roll				Ea.	55			55	60.50
0520	9' x 111.1' roll				"	100			100	110
0601	Polyethylene vapor barrier, standard, .002" thick	1 Carp	3,700	.002	S.F.	.01	.05		.06	.09
0701	.004" thick		3,700	.002		.02	.05		.07	.10
0901	.006" thick		3,700	.002		.03	.05		.08	.11
1201	.010" thick		3,700	.002		.05	.05		.10	.14
1501	Red rosin paper, 5 sq rolls, 4 lb per square		3,700	.002		.02	.05		.07	.10
1601	5 lbs. per square		3,700	.002		.02	.05		.07	.10
1801	Reinf. waterproof, .002" polyethylene backing, 1 side		3,700	.002		.05	.05		.10	.14

07200 | Thermal Protection

07260 | Vapor Retarders

			CREW	DAILY OUTPUT	LABOR-HOURS	UNIT	2004 BARE COSTS MAT.	LABOR	EQUIP.	TOTAL	TOTAL INCL O&P	
100	1901	2 sides	1 Carp	3,700	.002	S.F.	.07	.05		.12	.15	100
	3000	Building wrap, spunbonded polyethylene	2 Carp	8,000	.002		.11	.05		.16	.20	

07300 | Shingles, Roof Tiles and Roof Coverings

07310 | Shingles

			CREW	DAILY OUTPUT	LABOR-HOURS	UNIT	2004 BARE COSTS MAT.	LABOR	EQUIP.	TOTAL	TOTAL INCL O&P	
100	0010	**ASPHALT SHINGLES**										100
	0100	Standard strip shingles										
	0150	Inorganic, class A, 210-235 lb/sq	1 Rofc	5.50	1.455	Sq.	31	28.50		59.50	86.50	
	0155	Pneumatic nailed		7	1.143		31	22.50		53.50	75	
	0200	Organic, class C, 235-240 lb/sq		5	1.600		38	31.50		69.50	99.50	
	0205	Pneumatic nailed		6.25	1.280		38	25		63	88	
	0250	Standard, laminated multi-layered shingles										
	0300	Class A, 240-260 lb/sq	1 Rofc	4.50	1.778	Sq.	42.50	35		77.50	111	
	0305	Pneumatic nailed		5.63	1.422		42.50	28		70.50	97.50	
	0350	Class C, 260-300 lb/square, 4 bundles/square		4	2		49	39.50		88.50	126	
	0355	Pneumatic nailed		5	1.600		49	31.50		80.50	112	
	0400	Premium, laminated multi-layered shingles										
	0450	Class A, 260-300 lb, 4 bundles/sq	1 Rofc	3.50	2.286	Sq.	50.50	45		95.50	139	
	0455	Pneumatic nailed		4.37	1.831		50.50	36		86.50	122	
	0500	Class C, 300-385 lb/square, 5 bundles/square		3	2.667		65.50	52.50		118	168	
	0505	Pneumatic nailed		3.75	2.133		65.50	42		107.50	149	
	0800	#15 felt underlayment		64	.125		2.68	2.46		5.14	7.45	
	0825	#30 felt underlayment		58	.138		7.10	2.71		9.81	12.80	
	0850	Self adhering polyethylene and rubberized asphalt underlayment		22	.364		38	7.15		45.15	55	
	0900	Ridge shingles		330	.024	L.F.	.72	.48		1.20	1.66	
	0905	Pneumatic nailed		412.50	.019	"	.72	.38		1.10	1.49	
	1000	For steep roofs (7 to 12 pitch or greater), add						50%				
500	0010	**FIBER CEMENT** shingles, 16" x 9.35", 500 lb per square	1 Rofc	2.20	3.636	Sq.	244	71.50		315.50	400	500
	0200	Shakes, 16" x 9.35", 550 lb per square		2.20	3.636	"	221	71.50		292.50	375	
	0301	Hip & ridge, 4.75 x 14"		100	.080	L.F.	6	1.57		7.57	9.50	
	0400	Hexagonal, 16" x 16"		3	2.667	Sq.	165	52.50		217.50	278	
	0500	Square, 16" x 16"		3	2.667		148	52.50		200.50	259	
	2000	For steep roofs (7/12 pitch or greater), add						50%				
800	0010	**SLATE**, Buckingham, Virginia, black (R07310-020)										800
	0100	3/16" - 1/4" thick	1 Rots	1.75	4.571	Sq.	500	90.50		590.50	715	
	0200	1/4" thick		1.75	4.571		570	90.50		660.50	790	
	0900	Pennsylvania black, Bangor, #1 clear		1.75	4.571		450	90.50		540.50	660	
	1200	Vermont, unfading, green, mottled green		1.75	4.571		345	90.50		435.50	540	
	1300	Semi-weathering green & gray		1.75	4.571		330	90.50		420.50	530	
	1400	Purple		1.75	4.571		370	90.50		460.50	570	
	1500	Black or gray		1.75	4.571		325	90.50		415.50	525	
	2700	Ridge shingles, slate		200	.040	L.F.	8.50	.79		9.29	10.80	
980	0010	**WOOD** 16" No. 1 red cedar shingles, 5" exposure, on roof	1 Carp	2.50	3.200	Sq.	163	74		237	305	980
	0015	Pneumatic nailed		3.25	2.462	"	163	57		220	276	
	0200	7-1/2" exposure, on walls		2.05	3.902	Sq.	108	90		198	272	
	0205	Pneumatic nailed		2.67	2.996		108	69		177	236	
	0300	18" No. 1 red cedar perfections, 5-1/2" exposure, on roof		2.75	2.909		160	67		227	290	
	0305	Pneumatic nailed		3.57	2.241		160	52		212	264	

07300 | Shingles, Roof Tiles and Roof Coverings

07310 | Shingles

		07310 Shingles	CREW	DAILY OUTPUT	LABOR-HOURS	UNIT	2004 BARE COSTS MAT.	LABOR	EQUIP.	TOTAL	TOTAL INCL O&P	
980	0500	7-1/2" exposure, on walls	1 Carp	2.25	3.556	Sq.	118	82		200	268	980
	0505	Pneumatic nailed		2.92	2.740		118	63.50		181.50	236	
	0600	Resquared, and rebutted, 5-1/2" exposure, on roof		3	2.667		199	61.50		260.50	325	
	0605	Pneumatic nailed		3.90	2.051		199	47.50		246.50	300	
	0900	7-1/2" exposure, on walls		2.45	3.265		146	75.50		221.50	289	
	0905	Pneumatic nailed		3.18	2.516		146	58		204	260	
	1000	Add to above for fire retardant shingles, 16" long					30			30	33	
	1050	18" long					28.50			28.50	31.50	
	1060	Preformed ridge shingles	1 Carp	400	.020	L.F.	1.65	.46		2.11	2.60	
	1100	Hand-split red cedar shakes, 1/2" thick x 24" long, 10" exp. on roof		2.50	3.200	Sq.	138	74		212	277	
	1105	Pneumatic nailed		3.25	2.462		138	57		195	249	
	1110	3/4" thick x 24" long, 10" exp. on roof		2.25	3.556		138	82		220	291	
	1115	Pneumatic nailed		2.92	2.740		138	63.50		201.50	259	
	1200	1/2" thick, 18" long, 8-1/2" exp. on roof		2	4		97.50	92.50		190	264	
	1205	Pneumatic nailed		2.60	3.077		97.50	71		168.50	228	
	1210	3/4" thick x 18" long, 8 1/2" exp. on roof		1.80	4.444		97.50	103		200.50	281	
	1215	Pneumatic nailed		2.34	3.419		97.50	79		176.50	241	
	1255	10" exp. on walls		2	4		110	92.50		202.50	278	
	1260	10" exposure on walls, pneumatic nailed		2.60	3.077		110	71		181	242	
	1700	Add to above for fire retardant shakes, 24" long					30			30	33	
	1800	18" long					30			30	33	
	1810	Ridge shakes	1 Carp	350	.023	L.F.	2.35	.53		2.88	3.49	
	2000	White cedar shingles, 16" long, extras, 5" exposure, on roof		2.40	3.333	Sq.	118	77		195	261	
	2005	Pneumatic nailed		3.12	2.564		118	59		177	231	
	2050	5" exposure on walls		2	4		118	92.50		210.50	287	
	2055	Pneumatic nailed		2.60	3.077		118	71		189	251	
	2100	7-1/2" exposure, on walls		2	4		84.50	92.50		177	250	
	2105	Pneumatic nailed		2.60	3.077		84.50	71		155.50	214	
	2150	"B" grade, 5" exposure on walls		2	4		110	92.50		202.50	278	
	2155	Pneumatic nailed		2.60	3.077		110	71		181	242	
	2300	For 15# organic felt underlayment on roof, 1 layer, add		64	.125		2.68	2.89		5.57	7.85	
	2400	2 layers, add		32	.250		5.35	5.80		11.15	15.70	
	2600	For steep roofs (7/12 pitch or greater), add to above						50%				
	3000	Ridge shakes or shingle wood	1 Carp	280	.029	L.F.	2.40	.66		3.06	3.76	

07320 | Roof Tiles

		07320 Roof Tiles	CREW	DAILY OUTPUT	LABOR-HOURS	UNIT	2004 BARE COSTS MAT.	LABOR	EQUIP.	TOTAL	TOTAL INCL O&P	
200	0010	**CLAY TILE** ASTM C1167, GR 1, severe weathering, acces. incl.										200
	0200	Lanai tile or Classic tile, 158 pc per sq	1 Rots	1.65	4.848	Sq.	395	96		491	610	
	0300	Americana, 158 pc per sq, most colors		1.65	4.848		485	96		581	705	
	0350	Green, gray or brown		1.65	4.848		490	96		586	715	
	0400	Blue		1.65	4.848		490	96		586	715	
	0600	Spanish tile, 171 pc per sq, red		1.80	4.444		263	88		351	450	
	0800	Blend		1.80	4.444		420	88		508	620	
	0900	Glazed white		1.80	4.444		500	88		588	710	
	1100	Mission tile, 192 pc per sq, machine scored finish, red		1.15	6.957		635	137		772	945	
	1700	French tile, 133 pc per sq, smooth finish, red		1.35	5.926		575	117		692	845	
	1750	Blue or green		1.35	5.926		685	117		802	970	
	1800	Norman black 317 pc per sq		1	8		755	158		913	1,125	
	2200	Williamsburg tile, 158 pc per sq, aged cedar		1.35	5.926		490	117		607	755	
	2250	Gray or green		1.35	5.926		490	117		607	755	
	2350	Ridge shingles, clay tile		200	.040	L.F.	8.65	.79		9.44	10.95	
	2510	One piece mission tile, natural red, 75 pc per square		1.65	4.848	Sq.	133	96		229	320	
	2530	Mission Tile, 134 pc per square		1.15	6.957		178	137		315	445	
	3000	For steep roofs (7/12 pitch or greater), add to above						50%				
300	0010	**CONCRETE TILE** Including installation of accessories										300
	0020	Corrugated, 13" x 16-1/2", 90 per sq, 950 lb per sq										

07300 | Shingles, Roof Tiles and Roof Coverings

07320 | Roof Tiles

			CREW	DAILY OUTPUT	LABOR-HOURS	UNIT	2004 BARE COSTS MAT.	LABOR	EQUIP.	TOTAL	TOTAL INCL O&P	
300	0050	Earthtone colors, nailed to wood deck	1 Rots	1.35	5.926	Sq.	90	117		207	315	300
	0150	Blues		1.35	5.926		102	117		219	325	
	0200	Greens		1.35	5.926		102	117		219	325	
	0250	Premium colors	↓	1.35	5.926	↓	153	117		270	380	
	0500	Shakes, 13" x 16-1/2", 90 per sq, 950 lb per sq										
	0600	All colors, nailed to wood deck	1 Rots	1.50	5.333	Sq.	185	105		290	395	
	1500	Accessory pieces, ridge & hip, 10" x 16-1/2", 8 lbs. each				Ea.	2.25			2.25	2.48	
	1700	Rake, 6-1/2" x 16-3/4", 9 lbs. each					2.25			2.25	2.48	
	1800	Mansard hip, 10" x 16-1/2", 9.2 lbs. each					2.25			2.25	2.48	
	1900	Hip starter, 10" x 16-1/2", 10.5 lbs. each					9.50			9.50	10.45	
	2000	3 or 4 way apex, 10" each side, 11.5 lbs. each				↓	10.25			10.25	11.30	

07400 | Roofing and Siding Panels

07410 | Metal Roof and Wall Panels

			CREW	DAILY OUTPUT	LABOR-HOURS	UNIT	2004 BARE COSTS MAT.	LABOR	EQUIP.	TOTAL	TOTAL INCL O&P	
100	0010	**ALUMINUM ROOFING** Corrugated or ribbed, .0155" thick, natural	G-3	1,200	.027	S.F.	.62	.57		1.19	1.63	100
	0300	Painted	"	1,200	.027	"	.89	.57		1.46	1.93	

07420 | Plastic Roof and Wall Panels

			CREW	DAILY OUTPUT	LABOR-HOURS	UNIT	2004 BARE COSTS MAT.	LABOR	EQUIP.	TOTAL	TOTAL INCL O&P	
770	0010	**FIBERGLASS** Corrugated panels, roofing, 8 oz per SF	G-3	1,000	.032	S.F.	2.24	.68		2.92	3.61	770
	0100	12 oz per SF		1,000	.032		3.23	.68		3.91	4.70	
	0300	Corrugated siding, 6 oz per SF		880	.036		1.94	.77		2.71	3.43	
	0400	8 oz per SF		880	.036		2.24	.77		3.01	3.76	
	0600	12 oz. siding, textured		880	.036		3.14	.77		3.91	4.75	
	0900	Flat panels, 6 oz per SF, clear or colors		880	.036		1.73	.77		2.50	3.20	
	1300	8 oz per SF, clear or colors	↓	880	.036	↓	2.24	.77		3.01	3.76	

07460 | Siding

			CREW	DAILY OUTPUT	LABOR-HOURS	UNIT	2004 BARE COSTS MAT.	LABOR	EQUIP.	TOTAL	TOTAL INCL O&P	
100	0011	**ALUMINUM SIDING**										100
	6040	.024 thick smooth white single 8" wide	2 Carp	515	.031	S.F.	1.25	.72		1.97	2.60	
	6060	Double 4" pattern		515	.031		1.19	.72		1.91	2.53	
	6080	Double 5" pattern		550	.029		1.19	.67		1.86	2.45	
	6120	Embossed white, 8" wide		515	.031		1.48	.72		2.20	2.85	
	6140	Double 4" pattern		515	.031		1.35	.72		2.07	2.71	
	6160	Double 5" pattern		550	.029		1.35	.67		2.02	2.63	
	6170	Vertical, embossed white, 12" wide		590	.027		1.35	.63		1.98	2.55	
	6320	.019 thick, insulated, smooth white, 8" wide		515	.031		1.20	.72		1.92	2.54	
	6340	Double 4" pattern		515	.031		1.18	.72		1.90	2.52	
	6360	Double 5" pattern		550	.029		1.18	.67		1.85	2.44	
	6400	Embossed white, 8" wide		515	.031		1.38	.72		2.10	2.74	
	6420	Double 4" pattern		515	.031		1.40	.72		2.12	2.76	
	6440	Double 5" pattern		550	.029		1.40	.67		2.07	2.68	
	6500	Shake finish 10" wide white		550	.029		1.48	.67		2.15	2.77	
	6600	Vertical pattern, 12" wide, white	↓	590	.027		1.40	.63		2.03	2.60	
	6640	For colors add				↓	.08			.08	.09	
	6700	Accessories, white										
	6720	Starter strip 2-1/8"	2 Carp	610	.026	L.F.	.19	.61		.80	1.24	
	6740	Sill trim	↓	450	.036	↓	.31	.82		1.13	1.73	

	07460 | Siding		CREW	DAILY OUTPUT	LABOR-HOURS	UNIT	2004 BARE COSTS MAT.	LABOR	EQUIP.	TOTAL	TOTAL INCL O&P	
100	6760	Inside corner	2 Carp	610	.026	L.F.	.94	.61		1.55	2.06	100
	6780	Outside corner post		610	.026		1.60	.61		2.21	2.79	
	6800	Door & window trim		440	.036		.30	.84		1.14	1.76	
	6820	For colors add					.08			.08	.09	
	6900	Soffit & fascia 1' overhang solid	2 Carp	110	.145		2.16	3.36		5.52	8.10	
	6920	Vented		110	.145		2.16	3.36		5.52	8.10	
	6940	2' overhang solid		100	.160		3.17	3.70		6.87	9.75	
	6960	Vented		100	.160		3.17	3.70		6.87	9.75	
300	0010	**FASCIA** Aluminum, reverse board and batten,										300
	0100	.032" thick, colored, no furring included	1 Shee	145	.055	S.F.	2.14	1.41		3.55	4.71	
	0200	Residential type, aluminum	1 Carp	200	.040	L.F.	1.15	.92		2.07	2.84	
	0300	Steel, galv and enameled, stock, no furring, long panels	1 Shee	145	.055	S.F.	2.23	1.41		3.64	4.81	
	0600	Short panels	"	115	.070	"	3.38	1.78		5.16	6.70	
500	0010	**FIBER CEMENT SIDING**										500
	0020	Lap siding, 5/16" thick, 6" wide, smooth texture	2 Carp	415	.039	S.F.	.98	.89		1.87	2.59	
	0025	Woodgrain texture		415	.039		.98	.89		1.87	2.59	
	0030	7-1/2" wide, smooth texture		425	.038		.88	.87		1.75	2.45	
	0035	Woodgrain texture		425	.038		.88	.87		1.75	2.45	
	0040	8" wide, smooth texture		425	.038		.87	.87		1.74	2.44	
	0045	Roughsawn texture		425	.038		.87	.87		1.74	2.44	
	0050	9-1/2" wide, smooth texture		440	.036		.84	.84		1.68	2.36	
	0055	Woodgrain texture		440	.036		.84	.84		1.68	2.36	
	0060	12" wide, smooth texture		455	.035		.81	.81		1.62	2.28	
	0065	Woodgrain texture		455	.035		.81	.81		1.62	2.28	
	0070	Panel siding, 5/16" thick, smooth texture		750	.021		.73	.49		1.22	1.64	
	0075	Stucco texture		750	.021		.73	.49		1.22	1.64	
	0080	Grooved woodgrain texture		750	.021		.73	.49		1.22	1.64	
	0085	V - grooved woodgrain texture		750	.021		.73	.49		1.22	1.64	
	0090	Wood starter strip		400	.040	L.F.	.17	.92		1.09	1.76	
600	0010	**VINYL SIDING** Solid PVC panels, 8" to 10" wide, plain	1 Carp	255	.031	S.F.	.69	.72		1.41	1.99	600
	2000	Smooth, white, single, 8" wide	2 Carp	495	.032		.62	.75		1.37	1.95	
	2020	Dutch lap, 10" wide		550	.029		.64	.67		1.31	1.84	
	2100	Double 4" pattern, 8" wide		495	.032		.55	.75		1.30	1.88	
	2120	Double 5" pattern, 10" wide		550	.029		.56	.67		1.23	1.76	
	2200	Embossed, white, single, 8" wide		495	.032		.66	.75		1.41	2	
	2220	10 " wide		550	.029		.67	.67		1.34	1.88	
	2300	Double 4" pattern, 8" wide		495	.032		.59	.75		1.34	1.92	
	2320	5" pattern, 10" wide		550	.029		.61	.67		1.28	1.81	
	2400	Shake finish, 10" wide, white		550	.029		1.90	.67		2.57	3.23	
	2600	Vertical pattern, double 5", 10" wide, white		550	.029		1.37	.67		2.04	2.65	
	2620											
	2700	For colors, add				S.F.	.08			.08	.09	
	2720	1/4" extruded polystyrene fan folded insulation	2 Carp	2,000	.008	"	.14	.18		.32	.46	
	3000	Accessories, starter strip		700	.023	L.F.	.24	.53		.77	1.16	
	3100	"J" channel, 1/2"		700	.023		.26	.53		.79	1.19	
	3120	5/8"		700	.023		.26	.53		.79	1.19	
	3140	3/4"		695	.023		.27	.53		.80	1.20	
	3160	1"		690	.023		.32	.54		.86	1.26	
	3180	1-1/8"		685	.023		.33	.54		.87	1.28	
	3190	1-1/4"		680	.024		.35	.54		.89	1.31	
	3200	Under sill trim		500	.032		.52	.74		1.26	1.82	
	3300	Outside corner post, 3" face, pocket 5/8"		700	.023		1.14	.53		1.67	2.15	
	3320	7/8"		690	.023		1.13	.54		1.67	2.15	
	3340	1-1/4"		680	.024		1.19	.54		1.73	2.23	
	3400	Inside corner post, pocket 5/8"		700	.023		.61	.53		1.14	1.57	

07460 | Siding

		07460 Siding	CREW	DAILY OUTPUT	LABOR-HOURS	UNIT	2004 BARE COSTS MAT.	LABOR	EQUIP.	TOTAL	TOTAL INCL O&P
600	3420	7/8″	2 Carp	690	.023	L.F.	.68	.54		1.22	1.66
	3440	1-1/4″		680	.024		.67	.54		1.21	1.66
	3500	Door & window trim, 2-1/2″ face, pocket 5/8″		510	.031		.58	.72		1.30	1.87
	3520	7/8″		500	.032		.62	.74		1.36	1.93
	3540	1-1/4″		490	.033		.62	.75		1.37	1.96
	3600	Soffit & fascia, 1′ overhang, solid		120	.133		1.44	3.08		4.52	6.85
	3620	Vented		120	.133		1.48	3.08		4.56	6.90
	3700	2′ overhang, solid		110	.145		2.13	3.36		5.49	8.05
	3720	Vented		110	.145		2.13	3.36		5.49	8.05
750	0010	**SOFFIT** Aluminum, residential, stock units, .020″ thick	1 Carp	210	.038	S.F.	1.01	.88		1.89	2.60
	0100	Baked enamel on steel, 16 or 18 gauge		105	.076		3.74	1.76		5.50	7.10
	0300	Polyvinyl chloride, white, solid		230	.035		.66	.80		1.46	2.09
	0400	Perforated		230	.035		.66	.80		1.46	2.09
	0500	For colors, add					.07			.07	.08
800	0010	**STEEL SIDING**, Beveled, vinyl coated, 8″ wide, including fasteners	1 Carp	265	.030	S.F.	1.15	.70		1.85	2.45
	0050	10″ wide	″	275	.029		1.23	.67		1.90	2.49
	0081	Galv., corrugated or ribbed, on steel frame, 30 gauge	G-3	775	.041		.77	.88		1.65	2.33
	0101	28 gauge		775	.041		.81	.88		1.69	2.37
	0301	26 gauge		775	.041		.78	.88		1.66	2.34
	0401	24 gauge		775	.041		1.14	.88		2.02	2.73
	0601	22 gauge		775	.041		1.31	.88		2.19	2.92
	0701	Colored, corrugated/ribbed, on steel frame, 10 yr fnsh, 28 ga.		775	.041		1.20	.88		2.08	2.80
	0901	26 gauge		775	.041		1.07	.88		1.95	2.66
	1001	24 gauge		775	.041		1.26	.88		2.14	2.87
900	0010	**WOOD SIDING, BOARDS**									
	2000	Board & batten, cedar, "B" grade, 1″ x 10″	1 Carp	400	.020	S.F.	1.98	.46		2.44	2.96
	2200	Redwood, clear, vertical grain, 1″ x 10″		400	.020		3.59	.46		4.05	4.73
	2400	White pine, #2 & better, 1″ x 10″		400	.020		.70	.46		1.16	1.55
	2410	Board & batten siding, white pine #2, 1″ x 12″		450	.018		.80	.41		1.21	1.58
	3200	Wood, cedar bevel, A grade, 1/2″ x 6″		250	.032		1.91	.74		2.65	3.35
	3300	1/2″ x 8″		275	.029		2.46	.67		3.13	3.85
	3500	3/4″ x 10″, clear grade		300	.027		3.54	.62		4.16	4.94
	3600	"B" grade		300	.027		2.80	.62		3.42	4.13
	3800	Cedar, rough sawn, 1″ x 4″, A grade, natural		240	.033		2.67	.77		3.44	4.25
	3900	Stained		240	.033		3.01	.77		3.78	4.62
	4100	1″ x 12″, board & batten, #3 & Btr., natural		260	.031		2.02	.71		2.73	3.43
	4200	Stained		260	.031		2.35	.71		3.06	3.80
	4400	1″ x 8″ channel siding, #3 & Btr., natural		250	.032		1.96	.74		2.70	3.41
	4500	Stained		250	.032		2.23	.74		2.97	3.70
	4700	Redwood, clear, beveled, vertical grain, 1/2″ x 4″		200	.040		3.12	.92		4.04	5
	4750	1/2″ x 6″		225	.036		2.62	.82		3.44	4.27
	4800	1/2″ x 8″		250	.032		2.12	.74		2.86	3.58
	5000	3/4″ x 10″		300	.027		3.46	.62		4.08	4.86
	5200	Channel siding, 1″ x 10″, B grade		285	.028		2.23	.65		2.88	3.55
	5250	Redwood, T&G boards, B grade, 1″ x 4″	2 Carp	300	.053		2.66	1.23		3.89	5
	5270	1″ x 8″	″	375	.043		2.29	.99		3.28	4.19
	5400	White pine, rough sawn, 1″ x 8″, natural	1 Carp	275	.029		.67	.67		1.34	1.88
	5500	Stained	″	275	.029		.99	.67		1.66	2.23
	5600	Tongue and groove, 1″ x 8″, horizontal	2 Carp	375	.043		.61	.99		1.60	2.34
950	0010	**WOOD PRODUCT SIDING**									
	0030	Lap siding, hardboard, 7/16″ x 8″, primed									
	0050	Wood grain texture finish	2 Carp	650	.025	S.F.	1.04	.57		1.61	2.11
	0100	Panels, 7/16″ thick, smooth, textured or grooved, primed		700	.023		.73	.53		1.26	1.70

07400 | Roofing and Siding Panels

07460 | Siding

		07460 Siding	CREW	DAILY OUTPUT	LABOR-HOURS	UNIT	2004 BARE COSTS MAT.	LABOR	EQUIP.	TOTAL	TOTAL INCL O&P	
950	0200	Stained	2 Carp	700	.023	S.F.	.88	.53		1.41	1.87	950
	0700	Particle board, overlaid, 3/8″ thick		750	.021		.63	.49		1.12	1.53	
	0900	Plywood, medium density overlaid, 3/8″ thick		750	.021		.84	.49		1.33	1.76	
	1000	1/2″ thick		700	.023		1.01	.53		1.54	2.01	
	1100	3/4″ thick		650	.025		1.39	.57		1.96	2.50	
	1600	Texture 1-11, cedar, 5/8″ thick, natural		675	.024		2.32	.55		2.87	3.48	
	1700	Factory stained		675	.024		2.21	.55		2.76	3.36	
	1900	Texture 1-11, fir, 5/8″ thick, natural		675	.024		.93	.55		1.48	1.95	
	2000	Factory stained		675	.024		1.08	.55		1.63	2.12	
	2050	Texture 1-11, S.Y.P., 5/8″ thick, natural		675	.024		.81	.55		1.36	1.82	
	2100	Factory stained		675	.024		.88	.55		1.43	1.90	
	2200	Rough sawn cedar, 3/8″ thick, natural		675	.024		1.08	.55		1.63	2.12	
	2300	Factory stained		675	.024		1.20	.55		1.75	2.25	
	2500	Rough sawn fir, 3/8″ thick, natural		675	.024		.58	.55		1.13	1.57	
	2600	Factory stained		675	.024		.65	.55		1.20	1.65	
	2800	Redwood, textured siding, 5/8″ thick		675	.024		1.80	.55		2.35	2.91	

07500 | Membrane Roofing

07510 | Built-Up Bituminous Roofing

		07510 Built-Up Bituminous Roofing	CREW	DAILY OUTPUT	LABOR-HOURS	UNIT	2004 BARE COSTS MAT.	LABOR	EQUIP.	TOTAL	TOTAL INCL O&P	
050	0010	**ASPHALT** Coated felt, #30, 2 sq per roll, not mopped	1 Rofc	58	.138	Sq.	7.10	2.71		9.81	12.80	050
	0200	#15, 4 sq per roll, plain or perforated, not mopped		58	.138		2.68	2.71		5.39	7.90	
	0250	Perforated		58	.138		2.68	2.71		5.39	7.90	
	0300	Roll roofing, smooth, #65		15	.533		6.35	10.50		16.85	26	
	0500	#90		15	.533		17.50	10.50		28	38.50	
	0520	Mineralized		15	.533		15.35	10.50		25.85	36	
	0540	D.C. (Double coverage), 19″ selvage edge		10	.800		29	15.70		44.70	61	
	0580	Adhesive (lap cement)				Gal.	3.68			3.68	4.05	
300	0010	**BUILT-UP ROOFING**										300
	0120	Asphalt flood coat with gravel/slag surfacing, not including										
	0140	Insulation, flashing or wood nailers										
	0200	Asphalt base sheet, 3 plies #15 asphalt felt, mopped	G-1	22	2.545	Sq.	46.50	47	13.65	107.15	153	
	0350	On nailable decks		21	2.667		51.50	49	14.30	114.80	162	
	0500	4 plies #15 asphalt felt, mopped		20	2.800		66.50	51.50	15	133	185	
	0550	On nailable decks		19	2.947		59.50	54.50	15.80	129.80	182	
	2000	Asphalt flood coat, smooth surface										
	2200	Asphalt base sheet & 3 plies #15 asphalt felt, mopped	G-1	24	2.333	Sq.	51	43	12.50	106.50	149	
	2400	On nailable decks		23	2.435		47.50	45	13.05	105.55	149	
	2600	4 plies #15 asphalt felt, mopped		24	2.333		59.50	43	12.50	115	158	
	2700	On nailable decks		23	2.435		56	45	13.05	114.05	158	
	4500	Coal tar pitch with gravel/slag surfacing										
	4600	4 plies #15 tarred felt, mopped	G-1	21	2.667	Sq.	101	49	14.30	164.30	217	
	4800	3 plies glass fiber felt (type IV), mopped	″	19	2.947	″	83	54.50	15.80	153.30	208	
400	0010	**CANTS** 4″ x 4″, treated timber, cut diagonally	1 Rofc	325	.025	L.F.	1.33	.48		1.81	2.35	400
	0100	Foamglass		325	.025		2.15	.48		2.63	3.26	
	0300	Mineral or fiber, trapezoidal, 1″x 4″ x 48″		325	.025		.17	.48		.65	1.08	
	0400	1-1/2″ x 5-5/8″ x 48″		325	.025		.29	.48		.77	1.21	

07500 | Membrane Roofing

07550 | Modified Bit. Membrane Roofing

		07550 Modified Bit. Membrane Roofing	CREW	DAILY OUTPUT	LABOR-HOURS	UNIT	2004 BARE COSTS MAT.	LABOR	EQUIP.	TOTAL	TOTAL INCL O&P	
500	0010	**MODIFIED BITUMEN ROOFING** R07550-030										500
	0020	Base sheet, #15 glass fiber felt, nailed to deck	1 Rofc	58	.138	Sq.	4.85	2.71		7.56	10.30	
	0030	Spot mopped to deck	G-1	295	.190		6.55	3.50	1.02	11.07	14.75	
	0040	Fully mopped to deck	"	192	.292		9.05	5.40	1.56	16.01	21.50	
	0050	#15 organic felt, nailed to deck	1 Rofc	58	.138		3.42	2.71		6.13	8.75	
	0060	Spot mopped to deck	G-1	295	.190		5.15	3.50	1.02	9.67	13.15	
	0070	Fully mopped to deck	"	192	.292	↓	7.60	5.40	1.56	14.56	19.90	
	0080	SBS modified, granule surf cap sheet, polyester rein., mopped										
	1500	Glass fiber reinforced, mopped, 160 mils	G-1	2,000	.028	S.F.	.42	.52	.15	1.09	1.57	
	1600	Smooth surface cap sheet, mopped, 145 mils		2,100	.027		.42	.49	.14	1.05	1.52	
	1700	Smooth surface flashing, 145 mils		1,260	.044		.42	.82	.24	1.48	2.22	
	1800	150 mils		1,260	.044		.41	.82	.24	1.47	2.21	
	1900	Granular surface flashing, 150 mils		1,260	.044		.45	.82	.24	1.51	2.26	
	2000	160 mils	↓	1,260	.044		.63	.82	.24	1.69	2.45	
	2100	APP mod., smooth surf. cap sheet, poly. reinf., torched, 160 mils	G-5	2,100	.019		.41	.34	.07	.82	1.15	
	2150	170 mils		2,100	.019		.46	.34	.07	.87	1.21	
	2200	Granule surface cap sheet, poly. reinf., torched, 180 mils		2,000	.020		.50	.36	.07	.93	1.29	
	2250	Smooth surface flashing, torched, 160 mils		1,260	.032		.41	.57	.11	1.09	1.61	
	2300	170 mils		1,260	.032		.46	.57	.11	1.14	1.67	
	2350	Granule surface flashing, torched, 180 mils	↓	1,260	.032		.50	.57	.11	1.18	1.71	
	2400	Fibrated aluminum coating ↓	1 Rofc	3,800	.002	↓	.10	.04		.14	.19	

07580 | Roll Roofing

		07580 Roll Roofing	CREW	DAILY OUTPUT	LABOR-HOURS	UNIT	MAT.	LABOR	EQUIP.	TOTAL	TOTAL INCL O&P	
200	0010	**ROLL ROOFING**										200
	0100	Asphalt, mineral surface										
	0200	1 ply #15 organic felt, 1 ply mineral surfaced										
	0300	Selvage roofing, lap 19", nailed & mopped	G-1	27	2.074	Sq.	39	38.50	11.10	88.60	125	
	0400	3 plies glass fiber felt (type IV), 1 ply mineral surfaced										
	0500	Selvage roofing, lapped 19", mopped	G-1	25	2.240	Sq.	61	41.50	12	114.50	156	
	0600	Coated glass fiber base sheet, 2 plies of glass fiber										
	0700	Felt (type IV), 1 ply mineral surfaced selvage										
	0800	Roofing, lapped 19", mopped	G-1	25	2.240	Sq.	66	41.50	12	119.50	161	
	0900	On nailable decks	"	24	2.333	"	61	43	12.50	116.50	160	
	1000	3 plies glass fiber felt (type III), 1 ply mineral surfaced										
	1100	Selvage roofing, lapped 19", mopped	G-1	25	2.240	Sq.	61	41.50	12	114.50	156	

07590 | Roof Maintenance and Repairs

		07590 Roof Maintenance and Repairs	CREW	DAILY OUTPUT	LABOR-HOURS	UNIT	MAT.	LABOR	EQUIP.	TOTAL	TOTAL INCL O&P	
300	0010	**ROOF COATINGS** Asphalt				Gal.	2.95			2.95	3.25	300
	0800	Glass fibered roof & patching cement, 5 gallon					4.43			4.43	4.87	
	1100	Roof patch & flashing cement, 5 gallon				↓	17.15			17.15	18.85	

07600 | Flashing and Sheet Metal

07610 | Sheet Metal Roofing

		07610 Sheet Metal Roofing	CREW	DAILY OUTPUT	LABOR-HOURS	UNIT	2004 BARE COSTS MAT.	LABOR	EQUIP.	TOTAL	TOTAL INCL O&P	
300	0010	**COPPER ROOFING** Batten seam, over 10 sq, 16 oz, 130 lb/sq	1 Shee	1.10	7.273	Sq.	405	186		591	755	300
	0200	18 oz, 145 lb per sq		1	8		450	205		655	840	
	0400	Standing seam, over 10 squares, 16 oz, 125 lb per sq		1.30	6.154		390	158		548	690	
	0600	18 oz, 140 lb per sq	↓	1.20	6.667	↓	435	171		606	765	

07600 | Flashing and Sheet Metal

07610 | Sheet Metal Roofing

	Line	Description	Crew	Daily Output	Labor-Hours	Unit	2004 Bare Costs Mat.	Labor	Equip.	Total	Total Incl O&P	
300	0900	Flat seam, over 10 squares, 16 oz, 115 lb per sq	1 Shee	1.20	6.667	Sq.	355	171		526	680	300
	1200	For abnormal conditions or small areas, add					25%	100%				
	1300	For lead-coated copper, add					25%					
900	0010	**ZINC** Copper alloy roofing, batten seam, .020″ thick	1 Shee	1.20	6.667	Sq.	520	171		691	855	900
	0100	.027″ thick		1.15	6.957		630	178		808	990	
	0300	.032″ thick		1.10	7.273		710	186		896	1,100	
	0400	.040″ thick		1.05	7.619		835	195		1,030	1,250	
	0600	For standing seam construction, deduct					2%					
	0700	For flat seam construction, deduct					3%					

07620 | Sheet Metal Flashing and Trim

	Line	Description	Crew	Daily Output	Labor-Hours	Unit	Mat.	Labor	Equip.	Total	Total Incl O&P	
100	0010	**SHEET METAL CLADDING**										100
	0100	Aluminum, up to 6 bends, .032″ thick, window casing	1 Carp	180	.044	S.F.	.58	1.03		1.61	2.38	
	0200	Window sill		72	.111	L.F.	.58	2.57		3.15	5	
	0300	Door casing		180	.044	S.F.	.58	1.03		1.61	2.38	
	0400	Fascia		250	.032		.58	.74		1.32	1.89	
	0500	Rake trim		225	.036		.58	.82		1.40	2.03	
	0700	.024″ thick, window casing		180	.044		.90	1.03		1.93	2.73	
	0800	Window sill		72	.111	L.F.	.90	2.57		3.47	5.35	
	0900	Door casing		180	.044	S.F.	.90	1.03		1.93	2.73	
	1000	Fascia		250	.032		.90	.74		1.64	2.24	
	1100	Rake trim		225	.036		.90	.82		1.72	2.38	
	1200	Vinyl coated aluminum, up to 6 bends, window casing		180	.044		.61	1.03		1.64	2.41	
	1300	Window sill		72	.111	L.F.	.61	2.57		3.18	5.05	
	1400	Door casing		180	.044	S.F.	.61	1.03		1.64	2.41	
	1500	Fascia		250	.032		.61	.74		1.35	1.92	
	1600	Rake trim		225	.036		.61	.82		1.43	2.06	

07650 | Flexible Flashing

	Line	Description	Crew	Daily Output	Labor-Hours	Unit	Mat.	Labor	Equip.	Total	Total Incl O&P	
600	0010	**FLASHING** Aluminum, mill finish, .013″ thick	1 Rofc	145	.055	S.F.	.35	1.08		1.43	2.38	600
	0030	.016″ thick		145	.055		.51	1.08		1.59	2.55	
	0060	.019″ thick		145	.055		.66	1.08		1.74	2.72	
	0100	.032″ thick		145	.055		1.06	1.08		2.14	3.16	
	0200	.040″ thick		145	.055		1.39	1.08		2.47	3.52	
	0300	.050″ thick		145	.055		1.84	1.08		2.92	4.01	
	0325	Mill finish 5″ x 7″ step flashing, .016″ thick		1,920	.004	Ea.	.10	.08		.18	.26	
	0350	Mill finish 12″ x 12″ step flashing, .016″ thick		1,600	.005	″	.40	.10		.50	.62	
	0400	Painted finish, add				S.F.	.24			.24	.26	
	1600	Copper, 16 oz, sheets, under 1000 lbs.	1 Rofc	115	.070		2.23	1.37		3.60	4.95	
	1900	20 oz sheets, under 1000 lbs.		110	.073		3.90	1.43		5.33	6.90	
	2200	24 oz sheets, under 1000 lbs.		105	.076		4.69	1.50		6.19	7.90	
	2500	32 oz sheets, under 1000 lbs.		100	.080		6.20	1.57		7.77	9.75	
	2700	W shape for valleys, 16 oz, 24″ wide		100	.080	L.F.	5.90	1.57		7.47	9.40	
	2800	Copper, paperbacked 1 side, 2 oz		330	.024	S.F.	.86	.48		1.34	1.82	
	2900	3 oz		330	.024		1.12	.48		1.60	2.10	
	3100	Paperbacked 2 sides, 2 oz		330	.024		.86	.48		1.34	1.82	
	3150	3 oz		330	.024		1.11	.48		1.59	2.09	
	3200	5 oz		330	.024		1.65	.48		2.13	2.69	
	5800	Lead, 2.5 lb. per SF, up to 12″ wide		135	.059		2.80	1.16		3.96	5.20	
	5900	Over 12″ wide		135	.059		3.25	1.16		4.41	5.70	
	6100	Lead-coated copper, fabric-backed, 2 oz		330	.024		1.41	.48		1.89	2.42	
	6200	5 oz		330	.024		1.62	.48		2.10	2.65	
	6400	Mastic-backed 2 sides, 2 oz		330	.024		1.10	.48		1.58	2.08	
	6500	5 oz		330	.024		1.36	.48		1.84	2.37	
	6700	Paperbacked 1 side, 2 oz		330	.024		.95	.48		1.43	1.92	

07600 | Flashing and Sheet Metal

07650 | Flexible Flashing

		07650 Flexible Flashing	CREW	DAILY OUTPUT	LABOR-HOURS	UNIT	2004 BARE COSTS MAT.	LABOR	EQUIP.	TOTAL	TOTAL INCL O&P
600	6800	3 oz	1 Rofc	330	.024	S.F.	1.12	.48		1.60	2.10
	7000	Paperbacked 2 sides, 2 oz		330	.024		.98	.48		1.46	1.95
	7100	5 oz		330	.024		1.59	.48		2.07	2.62
	8500	Shower pan, bituminous membrane, 7 oz		155	.052		1.08	1.01		2.09	3.05
	8550	3 ply copper and fabric, 3 oz		155	.052		1.63	1.01		2.64	3.65
	8600	7 oz		155	.052		3.37	1.01		4.38	5.55
	8650	Copper, 16 oz		100	.080		3.11	1.57		4.68	6.30
	8700	Lead on copper and fabric, 5 oz		155	.052		1.62	1.01		2.63	3.64
	8800	7 oz		155	.052		2.93	1.01		3.94	5.10
	8900	Stainless steel sheets, 32 ga, .010″ thick		155	.052		2.16	1.01		3.17	4.24
	9000	28 ga, .015″ thick		155	.052		2.68	1.01		3.69	4.81
	9100	26 ga, .018″ thick		155	.052		3.25	1.01		4.26	5.45
	9200	24 ga, .025″ thick		155	.052		4.22	1.01		5.23	6.50
	9290	For mechanically keyed flashing, add					40%				
	9300	Stainless steel, paperbacked 2 sides, .005″ thick	1 Rofc	330	.024	S.F.	1.90	.48		2.38	2.96
	9320	Steel sheets, galvanized, 20 gauge		130	.062		.73	1.21		1.94	3.02
	9340	30 gauge		160	.050		.31	.98		1.29	2.14
	9400	Terne coated stainless steel, .015″ thick, 28 ga		155	.052		4.05	1.01		5.06	6.30
	9500	.018″ thick, 26 ga		155	.052		4.57	1.01		5.58	6.90
	9600	Zinc and copper alloy (brass), .020″ thick		155	.052		3.26	1.01		4.27	5.45
	9700	.027″ thick		155	.052		4.37	1.01		5.38	6.65
	9800	.032″ thick		155	.052		5.10	1.01		6.11	7.45
	9900	.040″ thick		155	.052		6.20	1.01		7.21	8.70

07700 | Roof Specialties and Accessories

07710 | Manufactured Roof Specialties

		07710 Manufactured Roof Specialties	CREW	DAILY OUTPUT	LABOR-HOURS	UNIT	2004 BARE COSTS MAT.	LABOR	EQUIP.	TOTAL	TOTAL INCL O&P
400	0010	**DOWNSPOUTS** Aluminum 2″ x 3″, .020″ thick, embossed	1 Shee	190	.042	L.F.	.65	1.08		1.73	2.52
	0100	Enameled		190	.042		.95	1.08		2.03	2.85
	0300	Enameled, .024″ thick, 2″ x 3″		180	.044		1.12	1.14		2.26	3.13
	0400	3″ x 4″		140	.057		1.59	1.46		3.05	4.20
	0600	Round, corrugated aluminum, 3″ diameter, .020″ thick		190	.042		.98	1.08		2.06	2.88
	0700	4″ diameter, .025″ thick		140	.057		1.60	1.46		3.06	4.21
	0900	Wire strainer, round, 2″ diameter		155	.052	Ea.	1.75	1.32		3.07	4.14
	1000	4″ diameter		155	.052		1.82	1.32		3.14	4.21
	1200	Rectangular, perforated, 2″ x 3″		145	.055		2.15	1.41		3.56	4.73
	1300	3″ x 4″		145	.055		3.10	1.41		4.51	5.75
	1500	Copper, round, 16 oz., stock, 2″ diameter		190	.042	L.F.	4.53	1.08		5.61	6.80
	1600	3″ diameter		190	.042		4.51	1.08		5.59	6.75
	1800	4″ diameter		145	.055		4.65	1.41		6.06	7.45
	1900	5″ diameter		130	.062		6.40	1.58		7.98	9.70
	2100	Rectangular, corrugated copper, stock, 2″ x 3″		190	.042		3.71	1.08		4.79	5.90
	2200	3″ x 4″		145	.055		4.80	1.41		6.21	7.65
	2400	Rectangular, plain copper, stock, 2″ x 3″		190	.042		4.78	1.08		5.86	7.05
	2500	3″ x 4″		145	.055		6.30	1.41		7.71	9.25
	2700	Wire strainers, rectangular, 2″ x 3″		145	.055	Ea.	2.80	1.41		4.21	5.45
	2800	3″ x 4″		145	.055		4.44	1.41		5.85	7.25
	3000	Round, 2″ diameter		145	.055		2.63	1.41		4.04	5.25
	3100	3″ diameter		145	.055		3.67	1.41		5.08	6.40

	07710 | Manufactured Roof Specialties	CREW	DAILY OUTPUT	LABOR-HOURS	UNIT	2004 BARE COSTS MAT.	LABOR	EQUIP.	TOTAL	TOTAL INCL O&P	
400	3300 4" diameter	1 Shee	145	.055	Ea.	5.70	1.41		7.11	8.60	400
	3400 5" diameter		115	.070	↓	8.20	1.78		9.98	12.05	
	3600 Lead-coated copper, round, stock, 2" diameter		190	.042	L.F.	5.35	1.08		6.43	7.70	
	3700 3" diameter		190	.042		5.75	1.08		6.83	8.10	
	3900 4" diameter		145	.055		7.60	1.41		9.01	10.75	
	4300 Rectangular, corrugated, stock, 2" x 3"		190	.042		6.65	1.08		7.73	9.15	
	4500 Plain, stock, 2" x 3"		190	.042		7.35	1.08		8.43	9.90	
	4600 3" x 4"		145	.055		7.95	1.41		9.36	11.10	
	4800 Steel, galvanized, round, corrugated, 2" or 3" diam, 28 ga		190	.042		.74	1.08		1.82	2.61	
	4900 4" diameter, 28 gauge		145	.055		1.01	1.41		2.42	3.47	
	5700 Rectangular, corrugated, 28 gauge, 2" x 3"		190	.042		.54	1.08		1.62	2.39	
	5800 3" x 4"		145	.055		1.50	1.41		2.91	4.01	
	6000 Rectangular, plain, 28 gauge, galvanized, 2" x 3"		190	.042		.65	1.08		1.73	2.52	
	6100 3" x 4"		145	.055		1.23	1.41		2.64	3.71	
	6300 Epoxy painted, 24 gauge, corrugated, 2" x 3"		190	.042		1.03	1.08		2.11	2.93	
	6400 3" x 4"		145	.055	↓	1.78	1.41		3.19	4.32	
	6600 Wire strainers, rectangular, 2" x 3"		145	.055	Ea.	1.62	1.41		3.03	4.14	
	6700 3" x 4"		145	.055		2.61	1.41		4.02	5.25	
	6900 Round strainers, 2" or 3" diameter		145	.055		1.20	1.41		2.61	3.68	
	7000 4" diameter		145	.055		1.42	1.41		2.83	3.92	
	7200 5" diameter		145	.055		2.20	1.41		3.61	4.78	
	7300 6" diameter		115	.070	↓	2.63	1.78		4.41	5.85	
	8200 Vinyl, rectangular, 2" x 3"		210	.038	L.F.	.72	.98		1.70	2.42	
	8300 Round, 2-1/2"	↓	220	.036	"	.72	.93		1.65	2.35	
450	0010 **DRIP EDGE**, aluminum, .016" thick, 5" wide, mill finish	1 Carp	400	.020	L.F.	.25	.46		.71	1.06	450
	0100 White finish		400	.020		.22	.46		.68	1.02	
	0200 8" wide, mill finish		400	.020		.30	.46		.76	1.11	
	0300 Ice belt, 28" wide, mill finish		100	.080		3.42	1.85		5.27	6.90	
	0310 Vented, mill finish		400	.020		1.41	.46		1.87	2.33	
	0320 Painted finish		400	.020		1.54	.46		2	2.47	
	0400 Galvanized, 5" wide		400	.020		.22	.46		.68	1.02	
	0500 8" wide, mill finish		400	.020		.33	.46		.79	1.14	
	0510 Rake edge, aluminum, 1-1/2" x 1-1/2"		400	.020		.13	.46		.59	.92	
	0520 3-1/2" x 1-1/2"	↓	400	.020	↓	.19	.46		.65	.99	
500	0010 **ELBOWS** Aluminum, 2" x 3", embossed	1 Shee	100	.080	Ea.	.91	2.05		2.96	4.43	500
	0100 Enameled		100	.080		1.76	2.05		3.81	5.35	
	0200 3" x 4", .025" thick, embossed		100	.080		3.20	2.05		5.25	6.95	
	0300 Enameled		100	.080		3.20	2.05		5.25	6.95	
	0400 Round corrugated, 3", embossed, .020" thick		100	.080		1.98	2.05		4.03	5.60	
	0500 4", .025" thick		100	.080		2.99	2.05		5.04	6.70	
	0600 Copper, 16 oz. round, 2" diameter		100	.080		11.15	2.05		13.20	15.75	
	0700 3" diameter		100	.080		4.76	2.05		6.81	8.70	
	0800 4" diameter		100	.080		7.80	2.05		9.85	12.05	
	1000 2" x 3" corrugated		100	.080		5.10	2.05		7.15	9.10	
	1100 3" x 4" corrugated		100	.080		6.45	2.05		8.50	10.50	
	1300 Vinyl, 2-1/2" diameter, 45° or 75°		100	.080		2	2.05		4.05	5.65	
	1400 Tee Y junction	↓	75	.107	↓	8.50	2.73		11.23	13.90	
550	0010 **GRAVEL STOP** Aluminum, .050" thick, 4" face height, mill finish	1 Shee	145	.055	L.F.	3.45	1.41		4.86	6.15	550
	0080 Duranodic finish		145	.055		3.69	1.41		5.10	6.40	
	0100 Painted		145	.055		4.26	1.41		5.67	7.05	
	1350 Galv steel, 24 ga., 4" leg, plain, with continuous cleat, 4" face		145	.055		1.50	1.41		2.91	4.01	
	1500 Polyvinyl chloride, 6" face height		135	.059		3.28	1.52		4.80	6.15	
	1800 Stainless steel, 24 ga., 6" face height	↓	135	.059	↓	7.15	1.52		8.67	10.40	
650	0010 **GUTTERS** Aluminum, stock units, 5" box, .027" thick, plain	1 Shee	120	.067	L.F.	1.22	1.71		2.93	4.20	650
	0020 Inside corner	↓	25	.320	Ea.	5.25	8.20		13.45	19.50	

07700 | Roof Specialties and Accessories

07710 | Manufactured Roof Specialties

			CREW	DAILY OUTPUT	LABOR-HOURS	UNIT	2004 BARE COSTS MAT.	LABOR	EQUIP.	TOTAL	TOTAL INCL O&P	
650	0030	Outside corner	1 Shee	25	.320	Ea.	5.25	8.20		13.45	19.50	650
	0100	Enameled		120	.067	L.F.	1.09	1.71		2.80	4.06	
	0110	Inside corner		25	.320	Ea.	5.30	8.20		13.50	19.55	
	0120	Outside corner		25	.320	"	5.30	8.20		13.50	19.55	
	0300	5" box type, .032" thick, plain		120	.067	L.F.	1.22	1.71		2.93	4.20	
	0310	Inside corner		25	.320	Ea.	4.99	8.20		13.19	19.20	
	0320	Outside corner		25	.320	"	5.40	8.20		13.60	19.65	
	0400	Enameled		120	.067	L.F.	1.23	1.71		2.94	4.21	
	0410	Inside corner		25	.320	Ea.	5.60	8.20		13.80	19.85	
	0420	Outside corner		25	.320	"	5.60	8.20		13.80	19.85	
	0600	5" x 6" combination fascia & gutter, .032" thick, enameled		60	.133	L.F.	3.52	3.41		6.93	9.55	
	0700	Copper, half round, 16 oz, stock units, 4" wide		120	.067		4.01	1.71		5.72	7.25	
	0900	5" wide		120	.067		5.15	1.71		6.86	8.50	
	1000	6" wide		115	.070		6.20	1.78		7.98	9.80	
	1200	K type, 16 oz, stock, 4" wide		120	.067		4.74	1.71		6.45	8.05	
	1300	5" wide		120	.067		5.20	1.71		6.91	8.55	
	1500	Lead coated copper, half round, stock, 4" wide		120	.067		6.50	1.71		8.21	10	
	1600	6" wide		115	.070		9.95	1.78		11.73	13.95	
	1800	K type, stock, 4" wide		120	.067		7.65	1.71		9.36	11.25	
	1900	5" wide		120	.067		7.95	1.71		9.66	11.60	
	2100	Stainless steel, half round or box, stock, 4" wide		120	.067		4.65	1.71		6.36	7.95	
	2200	5" wide		120	.067		5	1.71		6.71	8.35	
	2400	Steel, galv, half round or box, 28 ga, 5" wide, plain		120	.067		.95	1.71		2.66	3.91	
	2500	Enameled		120	.067		1.08	1.71		2.79	4.05	
	2700	26 ga, stock, 5" wide		120	.067		.95	1.71		2.66	3.91	
	2800	6" wide		120	.067		1.43	1.71		3.14	4.43	
	3000	Vinyl, O.G., 4" wide	1 Carp	110	.073		.85	1.68		2.53	3.79	
	3100	5" wide		110	.073		1	1.68		2.68	3.95	
	3200	4" half round, stock units		110	.073		.68	1.68		2.36	3.60	
	3250	Joint connectors				Ea.	1.36			1.36	1.50	
	3300	Wood, clear treated cedar, fir or hemlock, 3" x 4"	1 Carp	100	.080	L.F.	6.30	1.85		8.15	10.05	
	3400	4" x 5"	"	100	.080	"	7.30	1.85		9.15	11.15	
700	0010	**GUTTER GUARD** 6" wide strip, aluminum mesh	1 Carp	500	.016	L.F.	.37	.37		.74	1.04	700
	0100	Vinyl mesh	"	500	.016	"	.40	.37		.77	1.07	

07720 | Roof Accessories

			CREW	DAILY OUTPUT	LABOR-HOURS	UNIT	MAT.	LABOR	EQUIP.	TOTAL	TOTAL INCL O&P	
550	0010	**RIDGE VENT**										550
	0100	Aluminum strips, mill finish	1 Rofc	160	.050	L.F.	1.20	.98		2.18	3.12	
	0150	Painted finish		160	.050	"	2.12	.98		3.10	4.14	
	0200	Connectors		48	.167	Ea.	1.88	3.28		5.16	8.05	
	0300	End caps		48	.167	"	.92	3.28		4.20	7	
	0400	Galvanized strips		160	.050	L.F.	2.07	.98		3.05	4.08	
	0430	Molded polyethylene, shingles not included		160	.050	"	2.55	.98		3.53	4.61	
	0440	End plugs		48	.167	Ea.	.92	3.28		4.20	7	
	0450	Flexible roll, shingles not included		160	.050	L.F.	1.99	.98		2.97	3.99	

07920 | Joint Sealants		CREW	DAILY OUTPUT	LABOR-HOURS	UNIT	2004 BARE COSTS MAT.	LABOR	EQUIP.	TOTAL	TOTAL INCL O&P
800 0010	CAULKING AND SEALANTS									
0020	Acoustical sealant, elastomeric, cartridges				Ea.	2.05			2.05	2.26
0100	Acrylic latex caulk, white									
0200	11 fl. oz cartridge				Ea.	1.82			1.82	2
0500	1/4" x 1/2"	1 Bric	248	.032	L.F.	.15	.77		.92	1.45
0600	1/2" x 1/2"		250	.032		.30	.77		1.07	1.61
0800	3/4" x 3/4"		230	.035		.67	.83		1.50	2.13
0900	3/4" x 1"		200	.040		.89	.96		1.85	2.58
1000	1" x 1"	↓	180	.044	↓	1.12	1.07		2.19	3
1400	Butyl based, bulk				Gal.	22			22	24.50
1500	Cartridges				"	27			27	29.50
1700	Bulk, in place 1/4" x 1/2", 154 L.F./gal.	1 Bric	230	.035	L.F.	.14	.83		.97	1.55
1800	1/2" x 1/2", 77 L.F./gal.	"	180	.044	"	.29	1.07		1.36	2.09
2000	Latex acrylic based, bulk				Gal.	23			23	25.50
2100	Cartridges				"	26			26	28.50
2200	Bulk in place, 1/4" x 1/2", 154 L.F./gal.	1 Bric	230	.035	L.F.	.15	.83		.98	1.56
2300	Polysulfide compounds, 1 component, bulk				Gal.	44			44	48
2400	Cartridges				"	46.50			46.50	51.50
2600	1 or 2 component, in place, 1/4" x 1/4", 308 L.F./gal.	1 Bric	145	.055	L.F.	.14	1.32		1.46	2.36
2700	1/2" x 1/4", 154 L.F./gal.		135	.059		.28	1.42		1.70	2.67
2900	3/4" x 3/8", 68 L.F./gal.		130	.062		.64	1.48		2.12	3.17
3000	1" x 1/2", 38 L.F./gal.	↓	130	.062	↓	1.15	1.48		2.63	3.73
3200	Polyurethane, 1 or 2 component				Gal.	47.50			47.50	52.50
3300	Cartridges				"	46.50			46.50	51.50
3500	Bulk, in place, 1/4" x 1/4"	1 Bric	150	.053	L.F.	.15	1.28		1.43	2.30
3600	1/2" x 1/4"		145	.055		.31	1.32		1.63	2.54
3800	3/4" x 3/8", 68 L.F./gal.		130	.062		.70	1.48		2.18	3.23
3900	1" x 1/2"	↓	110	.073	↓	1.24	1.75		2.99	4.26
4100	Silicone rubber, bulk				Gal.	34.50			34.50	38
4200	Cartridges				"	40.50			40.50	44.50

For information about Means Estimating Seminars, see yellow pages 12 and 13 in back of book

Division 8
Doors & Windows

Estimating Tips

08100 Metal Doors & Frames

- Most metal doors and frames look alike, but there may be significant differences among them. When estimating these items be sure to choose the line item that most closely compares to the specification or door schedule requirements regarding:
    - type of metal
    - metal gauge
    - door core material
    - fire rating
    - finish

08200 Wood & Plastic Doors

- Wood and plastic doors vary considerably in price. The primary determinant is the veneer material. Lauan, birch and oak are the most common veneers. Other variables include the following:
    - hollow or solid core
    - fire rating
    - flush or raised panel
    - finish
- If the specifications require compliance with AWI (Architectural Woodwork Institute) standards or acoustical standards, the cost of the door may increase substantially. All wood doors are priced pre-mortised for hinges and predrilled for cylindrical locksets.
- Frequently doors, frames, and windows are unique in old buildings. Specified replacement units could be stock, custom (similar to the original) or exact reproduction. The estimator should work closely with a window consultant to determine any extra costs that may be associated with the unusual installation requirements.

08300 Specialty Doors

- There are many varieties of special doors, and they are usually priced per each. Add frames, hardware or operators required for a complete installation.

08510 Steel Windows

- Most metal windows are delivered preglazed. However, some metal windows are priced without glass. Refer to 08800 Glazing for glass pricing. The grade C indicates commercial grade windows, usually ASTM C-35.

08550 Wood Windows

- All wood windows are priced preglazed. The two glazing options priced are single pane float glass and insulating glass 1/2″ thick. Add the cost of screens and grills if required.

08700 Hardware

- Hardware costs add considerably to the cost of a door. The most efficient method to determine the hardware requirements for a project is to review the door schedule. This schedule, in conjunction with the specifications, is all you should need to take off the door hardware.
- Door hinges are priced by the pair, with most doors requiring 1-1/2 pairs per door. The hinge prices do not include installation labor because it is included in door installation. Hinges are classified according to the frequency of use.

08800 Glazing

- Different openings require different types of glass. The three most common types are:
    - float
    - tempered
    - insulating
- Most exterior windows are glazed with insulating glass. Entrance doors and window walls, where the glass is less than 18″ from the floor, are generally glazed with tempered glass. Interior windows and some residential windows are glazed with float glass.

08900 Glazed Curtain Wall

- Glazed curtain walls consist of the metal tube framing and the glazing material. The cost data in this subdivision is presented for the metal tube framing alone or the composite wall. If your estimate requires a detailed takeoff of the framing, be sure to add the glazing cost.

Reference Numbers

Reference numbers are shown in bold squares at the beginning of some major classifications. These numbers refer to related items in the Reference Section. The reference information may be an estimating procedure, an alternate pricing method or technical information.

Note: Not all subdivisions listed here necessarily appear in this publication.

08050 | Basic Door and Window Materials and Methods

08060 Selective Demolition			CREW	DAILY OUTPUT	LABOR-HOURS	UNIT	2004 BARE COSTS MAT.	LABOR	EQUIP.	TOTAL	TOTAL INCL O&P	
110	0010	SELECTIVE DEMOLITION, DOORS										110
	0200	Doors, exterior, 1-3/4" thick, single, 3' x 7' high	1 Clab	16	.500	Ea.		8.45		8.45	14.35	
	0220	Double, 6' x 7' high		12	.667			11.25		11.25	19.15	
	0500	Interior, 1-3/8" thick, single, 3' x 7' high		20	.400			6.75		6.75	11.50	
	0520	Double, 6' x 7' high		16	.500			8.45		8.45	14.35	
	0700	Bi-folding, 3' x 6'-8" high		20	.400			6.75		6.75	11.50	
	0720	6' x 6'-8" high		18	.444			7.50		7.50	12.75	
	0900	Bi-passing, 3' x 6'-8" high		16	.500			8.45		8.45	14.35	
	0940	6' x 6'-8" high		14	.571			9.65		9.65	16.40	
	1500	Remove and reset, minimum	1 Carp	8	1			23		23	39	
	1520	Maximum		6	1.333			31		31	52.50	
	2000	Frames, including trim, metal		8	1			23		23	39	
	2200	Wood	2 Carp	32	.500			11.55		11.55	19.60	
	2201	Alternate pricing method	1 Carp	200	.040	L.F.		.92		.92	1.57	
	3000	Special doors, counter doors	2 Carp	6	2.667	Ea.		61.50		61.50	105	
	3300	Glass, sliding, including frames		12	1.333			31		31	52.50	
	3400	Overhead, commercial, 12' x 12' high		4	4			92.50		92.50	157	
	3500	Residential, 9' x 7' high		8	2			46		46	78.50	
	3540	16' x 7' high		7	2.286			53		53	89.50	
	3600	Remove and reset, minimum		4	4			92.50		92.50	157	
	3620	Maximum		2.50	6.400			148		148	251	
	3660	Remove and reset elec. garage door opener	1 Carp	8	1			23		23	39	
	4000	Residential lockset, exterior		30	.267			6.15		6.15	10.45	
	4200	Deadbolt lock		32	.250			5.80		5.80	9.80	
	9000	Minimum labor/equipment charge		4	2	Job		46		46	78.50	
120	0010	SELECTIVE DEMOLITION, WINDOWS										120
	0200	Aluminum, including trim, to 12 S.F.	1 Clab	16	.500	Ea.		8.45		8.45	14.35	
	0240	To 25 S.F.		11	.727			12.30		12.30	21	
	0280	To 50 S.F.		5	1.600			27		27	46	
	0320	Storm windows, to 12 S.F.		27	.296			5		5	8.50	
	0360	To 25 S.F.		21	.381			6.45		6.45	10.95	
	0400	To 50 S.F.		16	.500			8.45		8.45	14.35	
	0500	Screens, incl. aluminum frame, small		20	.400			6.75		6.75	11.50	
	0510	Large		16	.500			8.45		8.45	14.35	
	0600	Glass, minimum		200	.040	S.F.		.68		.68	1.15	
	0620	Maximum		150	.053	"		.90		.90	1.53	
	2000	Wood, including trim, to 12 S.F.		22	.364	Ea.		6.15		6.15	10.45	
	2020	To 25 S.F.		18	.444			7.50		7.50	12.75	
	2060	To 50 S.F.		13	.615			10.40		10.40	17.65	
	5020	Remove and reset window, minimum	1 Carp	6	1.333			31		31	52.50	
	5040	Average		4	2			46		46	78.50	
	5080	Maximum		2	4			92.50		92.50	157	
	9100	Window awning, residential	1 Clab	80	.100	L.F.		1.69		1.69	2.87	

08100 | Metal Doors and Frames

08110 Steel Doors and Frames			CREW	DAILY OUTPUT	LABOR-HOURS	UNIT	2004 BARE COSTS MAT.	LABOR	EQUIP.	TOTAL	TOTAL INCL O&P	
300	0010	FIRE DOOR										300
	0015	Steel, flush, "B" label, 90 minute										

08110 | Steel Doors and Frames

			CREW	DAILY OUTPUT	LABOR-HOURS	UNIT	2004 BARE COSTS MAT.	LABOR	EQUIP.	TOTAL	TOTAL INCL O&P	
300	0020	Full panel, 20 ga., 2′-0″ x 6′-8″	2 Carp	20	.800	Ea.	191	18.50		209.50	242	300
	0040	2′-8″ x 6′-8″		18	.889		198	20.50		218.50	253	
	0060	3′-0″ x 6′-8″		17	.941		198	21.50		219.50	255	
	0080	3′-0″ x 7′-0″		17	.941		206	21.50		227.50	264	
	0140	18 ga., 3′-0″ x 6′-8″		16	1		218	23		241	279	
	0160	2′-8″ x 7′-0″		17	.941		231	21.50		252.50	292	
	0180	3′-0″ x 7′-0″		16	1		226	23		249	287	
	0200	4′-0″ x 7′-0″	↓	15	1.067	↓	292	24.50		316.50	360	
	0210											
	0220	For "A" label, 3 hour, 18 ga., use same price as "B" label										
	0240	For vision lite, add				Ea.	47			47	51.50	
	0520	Flush, "B" label 90 min., composite, 20 ga., 2′-0″ x 6′-8″	2 Carp	18	.889		252	20.50		272.50	310	
	0540	2′-8″ x 6′-8″		17	.941		257	21.50		278.50	320	
	0560	3′-0″ x 6′-8″		16	1		259	23		282	325	
	0580	3′-0″ x 7′-0″		16	1		267	23		290	335	
	0640	Flush, "A" label 3 hour, composite, 18 ga., 3′-0″ x 6′-8″		15	1.067		218	24.50		242.50	282	
	0660	2′-8″ x 7′-0″		16	1		229	23		252	291	
	0680	3′-0″ x 7′-0″		15	1.067		226	24.50		250.50	290	
	0700	4′-0″ x 7′-0″	↓	14	1.143	↓	292	26.50		318.50	365	
600	0010	**RESIDENTIAL STEEL DOOR**										600
	0020	Prehung, insulated, exterior										
	0030	Embossed, full panel, 2′-8″ x 6′-8″	2 Carp	17	.941	Ea.	190	21.50		211.50	246	
	0040	3′-0″ x 6′-8″		15	1.067		190	24.50		214.50	251	
	0060	3′-0″ x 7′-0″		15	1.067		249	24.50		273.50	315	
	0070	5′-4″ x 6′-8″, double		8	2		375	46		421	495	
	0220	Half glass, 2′-8″ x 6′-8″		17	.941		230	21.50		251.50	290	
	0240	3′-0″ x 6′-8″		16	1		232	23		255	294	
	0260	3′-0″ x 7′-0″		16	1		282	23		305	350	
	0270	5′-4″ x 6′-8″, double		8	2		480	46		526	610	
	0720	Raised plastic face, full panel, 2′-8″ x 6′-8″		16	1		237	23		260	300	
	0740	3′-0″ x 6′-8″		15	1.067		239	24.50		263.50	305	
	0760	3′-0″ x 7′-0″		15	1.067		241	24.50		265.50	305	
	0780	5′-4″ x 6′-8″, double		8	2		445	46		491	570	
	0820	Half glass, 2′-8″ x 6′-8″		17	.941		265	21.50		286.50	330	
	0840	3′-0″ x 6′-8″		16	1		268	23		291	335	
	0860	3′-0″ x 7′-0″		16	1		295	23		318	365	
	0880	5′-4″ x 6′-8″, double		8	2		595	46		641	735	
	1320	Flush face, full panel, 2′-6″ x 6′-8″		16	1		192	23		215	250	
	1340	3′-0″ x 6′-8″		15	1.067		195	24.50		219.50	257	
	1360	3′-0″ x 7′-0″		15	1.067		247	24.50		271.50	315	
	1380	5′-4″ x 6′-8″, double		8	2		405	46		451	525	
	1420	Half glass, 2′-8″ x 6′-8″		17	.941		241	21.50		262.50	305	
	1440	3′-0″ x 6′-8″		16	1		245	23		268	310	
	1460	3′-0″ x 7′-0″		16	1		284	23		307	350	
	1480	5′-4″ x 6′-8″, double		8	2		475	46		521	600	
	2300	Interior, residential, closet, bi-fold, 6′-8″ x 2′-0″ wide		16	1		129	23		152	181	
	2330	3′-0″ wide		16	1		145	23		168	199	
	2360	4′-0″ wide		15	1.067		220	24.50		244.50	284	
	2400	5′-0″ wide		14	1.143		255	26.50		281.50	325	
	2420	6′-0″ wide	↓	13	1.231	↓	286	28.50		314.50	365	
	2510	Bi-passing closet, incl. hardware, no frame or trim incl.										
	2511	Mirrored, metal frame, 6′-8″ x 4′-0″ wide	2 Carp	10	1.600	Opng.	174	37		211	255	
	2512	5′-0″ wide		10	1.600		203	37		240	287	
	2513	6′-0″ wide		10	1.600		231	37		268	315	
	2514	7′-0″ wide	↓	9	1.778	↓	243	41		284	340	

	08110 | Steel Doors and Frames		CREW	DAILY OUTPUT	LABOR-HOURS	UNIT	2004 BARE COSTS MAT.	LABOR	EQUIP.	TOTAL	TOTAL INCL O&P	
600	2515	8'-0" wide	2 Carp	9	1.778	Opng.	480	41		521	600	600
	2611	Mirrored, metal, 8'-0" x 4'-0" wide		10	1.600		269	37		306	360	
	2612	5'-0" wide		10	1.600		294	37		331	390	
	2613	6'-0" wide		10	1.600		325	37		362	420	
	2614	7'-0" wide		9	1.778		350	41		391	455	
	2615	8'-0" wide	↓	9	1.778	↓	385	41		426	495	
820	0010	**STEEL FRAMES, KNOCK DOWN**										820
	0020	16 ga., up to 5-3/4" deep										
	0025	6'-8" high, 3'-0" wide, single	2 Carp	16	1	Ea.	75	23		98	122	
	0040	6'-0" wide, double		14	1.143		90.50	26.50		117	145	
	0100	7'-0" high, 3'-0" wide, single		16	1		77	23		100	124	
	0140	6'-0" wide, double		14	1.143		94.50	26.50		121	149	
	1000	16 ga., up to 4-7/8" deep, 7'-0" H, 3'-0" W, single		16	1		74	23		97	121	
	1140	6'-0" wide, double		14	1.143		91	26.50		117.50	145	
	2800	14 ga., up to 3-7/8" deep, 7'-0" high, 3'-0" wide, single		16	1		76	23		99	123	
	2840	6'-0" wide, double		14	1.143		93	26.50		119.50	147	
	3600	5-3/4" deep, 7'-0" high, 4'-0" wide, single		15	1.067		65	24.50		89.50	114	
	3640	8'-0" wide, double		12	1.333		86	31		117	147	
	3700	8'-0" high, 4'-0" wide, single		15	1.067		86	24.50		110.50	137	
	3740	8'-0" wide, double		12	1.333		107	31		138	170	
	4000	6-3/4" deep, 7'-0" high, 4'-0" wide, single		15	1.067		90	24.50		114.50	141	
	4020	6'-0" wide, double		12	1.333		103	31		134	167	
	4040	8'-0", wide double		12	1.333		120	31		151	185	
	4100	8'-0" high, 4'-0" wide, single		15	1.067		102	24.50		126.50	154	
	4140	8'-0" wide, double		12	1.333		119	31		150	184	
	4400	8-3/4" deep, 7'-0" high, 4'-0" wide, single		15	1.067		95.50	24.50		120	147	
	4440	8'-0" wide, double		12	1.333		126	31		157	192	
	4500	8'-0" high, 4'-0" wide, single		15	1.067		112	24.50		136.50	165	
	4540	8'-0" wide, double	↓	12	1.333		131	31		162	197	
	4900	For welded frames, add					28			28	31	
	5400	14 ga., "B" label, up to 5-3/4" deep, 7'-0" high, 4'-0" wide, single	2 Carp	15	1.067		99.50	24.50		124	151	
	5440	8'-0" wide, double		12	1.333		116	31		147	181	
	5800	6-3/4" deep, 7'-0" high, 4'-0" wide, single		15	1.067		88.50	24.50		113	139	
	5840	8'-0" wide, double		12	1.333		128	31		159	194	
	6200	8-3/4" deep, 7'-0" high, 4'-0" wide, single		15	1.067		108	24.50		132.50	160	
	6240	8'-0" wide, double	↓	12	1.333	↓	138	31		169	205	
	6300	For "A" label use same price as "B" label										
	6400	For baked enamel finish, add					30%	15%				
	6500	For galvanizing, add					15%					
	7900	Transom lite frames, fixed, add	2 Carp	155	.103	S.F.	27	2.38		29.38	33.50	
	8000	Movable, add	"	130	.123	"	32.50	2.84		35.34	40.50	
100	0010	**STORM DOORS & FRAMES** Aluminum, residential,										100
	0020	combination storm and screen										
	0400	Clear anodic coating, 6'-8" x 2'-6" wide	2 Carp	15	1.067	Ea.	152	24.50		176.50	209	
	0420	2'-8" wide		14	1.143		174	26.50		200.50	236	
	0440	3'-0" wide	↓	14	1.143	↓	174	26.50		200.50	236	
	0500	For 7' door height, add					5%					
	1000	Mill finish, 6'-8" x 2'-6" wide	2 Carp	15	1.067	Ea.	202	24.50		226.50	264	
	1020	2'-8" wide		14	1.143		202	26.50		228.50	267	
	1040	3'-0" wide	↓	14	1.143		218	26.50		244.50	285	
	1100	For 7'-0" door, add					5%					
	1500	White painted, 6'-8" x 2'-6" wide	2 Carp	15	1.067		202	24.50		226.50	264	
	1520	2'-8" wide		14	1.143		204	26.50		230.50	270	
	1540	3'-0" wide		14	1.143		214	26.50		240.50	280	
	1541	Storm door, painted, alum., insul., 6'-8" x 2'-6" wide	↓	14	1.143	↓	231	26.50		257.50	299	

08100 | Metal Doors and Frames

	08110 | Steel Doors and Frames	CREW	DAILY OUTPUT	LABOR-HOURS	UNIT	2004 BARE COSTS MAT.	LABOR	EQUIP.	TOTAL	TOTAL INCL O&P	
100	1545 2'-8" wide	2 Carp	14	1.143	Ea.	231	26.50		257.50	299	100
	1600 For 7'-0" door, add					5%					
	1800 Aluminum screen door, minimum, 6'-8" x 2'-8" wide	2 Carp	14	1.143		93.50	26.50		120	148	
	1810 3'-0" wide		14	1.143		213	26.50		239.50	280	
	1820 Average, 6'-8" x 2'-8" wide		14	1.143		156	26.50		182.50	217	
	1830 3'-0" wide		14	1.143		213	26.50		239.50	280	
	1840 Maximum, 6'-8" x 2'-8" wide		14	1.143		310	26.50		336.50	390	
	1850 3'-0" wide	↓	14	1.143	↓	231	26.50		257.50	299	
	2000 Wood door & screen, see division 08210-930										
	2020										

08200 | Wood and Plastic Doors

	08210 | Wood Doors	CREW	DAILY OUTPUT	LABOR-HOURS	UNIT	2004 BARE COSTS MAT.	LABOR	EQUIP.	TOTAL	TOTAL INCL O&P	
720	0010 **PRE-HUNG DOORS**										720
	0300 Exterior, wood, comb. storm & screen, 6'-9" x 2'-6" wide	2 Carp	15	1.067	Ea.	266	24.50		290.50	335	
	0320 2'-8" wide		15	1.067		266	24.50		290.50	335	
	0340 3'-0" wide	↓	15	1.067		273	24.50		297.50	340	
	0360 For 7'-0" high door, add				↓	23			23	25.50	
	0370 For aluminum storm doors, see division 08280-800										
	1600 Entrance door, flush, birch, solid core										
	1620 4-5/8" solid jamb, 1-3/4" x 6'-8" x 2'-8" wide	2 Carp	16	1	Ea.	272	23		295	340	
	1640 3'-0" wide	"	16	1		279	23		302	345	
	1680 For 7'-0" high door, add				↓	12.95			12.95	14.25	
	2000 Entrance door, colonial, 6 panel pine										
	2020 4-5/8" solid jamb, 1-3/4" x 6'-8" x 2'-8" wide	2 Carp	16	1	Ea.	445	23		468	530	
	2040 3'-0" wide	"	16	1		470	23		493	560	
	2060 For 7'-0" high door, add					35			35	38	
	2200 For 5-5/8" solid jamb, add				↓	22.50			22.50	24.50	
	2250 French door, 6'-8" x 4'-0" wide, 1/2" insul. glass and grille	2 Carp	7	2.286	Pr.	1,000	53		1,053	1,200	
	2260 5'-0" wide	"	7	2.286	"	1,100	53		1,153	1,325	
	2500 Exterior, metal face, insulated, incl. jamb, brickmold and										
	2520 threshold, flush, 2'-8" x 6'-8"	2 Carp	16	1	Ea.	175	23		198	232	
	2550 3'-0" x 6'-8"	"	16	1	"	178	23		201	235	
	2990										
	3500 Embossed, 6 panel, 2'-8" x 6'-8"	2 Carp	16	1	Ea.	159	23		182	213	
	3550 3'-0" x 6'-8"		16	1		184	23		207	241	
	3600 2 narrow lites, 2'-8" x 6'-8"		16	1		220	23		243	281	
	3650 3'-0" x 6'-8"		16	1		225	23		248	287	
	3700 Half glass, 2'-8" x 6'-8"		16	1		210	23		233	270	
	3750 3'-0" x 6'-8"		16	1		212	23		235	272	
	3800 2 top lites, 2'-8" x 6'-8"		16	1		204	23		227	264	
	3850 3'-0" x 6'-8"	↓	16	1	↓	206	23		229	266	
	4000 Interior, passage door, 4-5/8" solid jamb										
	4400 Lauan, flush, solid core, 1-3/8" x 6'-8" x 2'-6" wide	2 Carp	20	.800	Ea.	167	18.50		185.50	216	
	4420 2'-8" wide		20	.800		147	18.50		165.50	193	
	4440 3'-0" wide		19	.842		151	19.45		170.45	200	
	4600 Hollow core, 1-3/8" x 6'-8" x 2'-6" wide		20	.800		109	18.50		127.50	152	
	4620 2'-8" wide		20	.800		110	18.50		128.50	154	
	4640 3'-0" wide	↓	19	.842	↓	112	19.45		131.45	156	
100											

	08210 | Wood Doors	CREW	DAILY OUTPUT	LABOR-HOURS	UNIT	2004 BARE COSTS MAT.	LABOR	EQUIP.	TOTAL	TOTAL INCL O&P	
720	4700 For 7'-0" high door, add				Ea.	21.50			21.50	24	720
	5000 Birch, flush, solid core, 1-3/8" x 6'-8" x 2'-6" wide	2 Carp	20	.800		154	18.50		172.50	201	
	5020 2'-8" wide		20	.800		174	18.50		192.50	223	
	5040 3'-0" wide		19	.842		186	19.45		205.45	237	
	5200 Hollow core, 1-3/8" x 6'-8" x 2'-6" wide		20	.800		127	18.50		145.50	172	
	5220 2'-8" wide		20	.800		133	18.50		151.50	179	
	5240 3'-0" wide		19	.842		133	19.45		152.45	179	
	5280 For 7'-0" high door, add					15.45			15.45	17	
	5500 Hardboard paneled, 1-3/8" x 6'-8" x 2'-6" wide	2 Carp	20	.800		128	18.50		146.50	173	
	5520 2'-8" wide		20	.800		134	18.50		152.50	180	
	5540 3'-0" wide		19	.842		132	19.45		151.45	179	
	6000 Pine paneled, 1-3/8" x 6'-8" x 2'-6" wide		20	.800		220	18.50		238.50	274	
	6020 2'-8" wide		20	.800		237	18.50		255.50	293	
	6040 3'-0" wide		19	.842		246	19.45		265.45	305	
	6500 For 5-5/8" solid jamb, add					10.80			10.80	11.85	
	6520 For split jamb, deduct					13.10			13.10	14.40	
910	0010 **WOOD DOORS, DECORATOR**										910
	3000 Solid wood, 1-3/4" thick stile and rail										
	3020 Mahogany, 3'-0" x 7'-0", minimum	2 Carp	14	1.143	Ea.	855	26.50		881.50	985	
	3030 Maximum		10	1.600		1,075	37		1,112	1,275	
	3040 3'-6" x 8'-0", minimum		10	1.600		870	37		907	1,025	
	3050 Maximum		8	2		1,000	46		1,046	1,175	
	3100 Pine, 3'-0" x 7'-0", minimum		14	1.143		305	26.50		331.50	380	
	3110 Maximum		10	1.600		550	37		587	670	
	3120 3'-6" x 8'-0", minimum		10	1.600		605	37		642	735	
	3130 Maximum		8	2		1,000	46		1,046	1,175	
	3200 Red oak, 3'-0" x 7'-0", minimum		14	1.143		830	26.50		856.50	955	
	3210 Maximum		10	1.600		1,375	37		1,412	1,600	
	3220 3'-6" x 8'-0", minimum		10	1.600		555	37		592	675	
	3230 Maximum		8	2		1,600	46		1,646	1,825	
	4000 Hand carved door, mahogany										
	4020 3'-0" x 7'-0", minimum	2 Carp	14	1.143	Ea.	1,375	26.50		1,401.50	1,550	
	4030 Maximum		11	1.455		2,150	33.50		2,183.50	2,425	
	4040 3'-6" x 8'-0", minimum		10	1.600		1,375	37		1,412	1,600	
	4050 Maximum		8	2		2,450	46		2,496	2,750	
	4200 Red oak, 3'-0" x 7'-0", minimum		14	1.143		4,000	26.50		4,026.50	4,450	
	4210 Maximum		11	1.455		11,000	33.50		11,033.50	12,200	
	4220 3'-6" x 8'-0", minimum		10	1.600		4,500	37		4,537	5,025	
	4280 For 6'-8" high door, deduct from 7'-0" door					30			30	33	
	4400 For custom finish, add					310			310	340	
	4600 Side light, mahogany, 7'-0" x 1'-6" wide, minimum	2 Carp	18	.889		745	20.50		765.50	850	
	4610 Maximum		14	1.143		2,150	26.50		2,176.50	2,400	
	4620 8'-0" x 1'-6" wide, minimum		14	1.143		1,450	26.50		1,476.50	1,650	
	4630 Maximum		10	1.600		1,650	37		1,687	1,875	
	4640 Side light, oak, 7'-0" x 1'-6" wide, minimum		18	.889		840	20.50		860.50	960	
	4650 Maximum		14	1.143		1,575	26.50		1,601.50	1,800	
	4660 8'-0" x 1-6" wide, minimum		14	1.143		735	26.50		761.50	850	
	4670 Maximum		10	1.600		1,575	37		1,612	1,825	
	6520 Interior cafe doors, 2'-6" opening, stock, panel pine		16	1		168	23		191	224	
	6540 3'-0" opening		16	1		175	23		198	232	
	6550 Louvered pine										
	6560 2'-6" opening	2 Carp	16	1	Ea.	147	23		170	200	
	8000 3'-0" opening		16	1		157	23		180	212	
	8010 2'-6" opening, hardwood		16	1		260	23		283	325	
	8020 3'-0" opening		16	1		286	23		309	355	
	8800 Pre-hung doors, see division 08210-720										

08210 | Wood Doors

		08210 Wood Doors	CREW	DAILY OUTPUT	LABOR-HOURS	UNIT	2004 BARE COSTS MAT.	LABOR	EQUIP.	TOTAL	TOTAL INCL O&P	
920	0010	**WOOD DOORS, PANELED**										920
	0020	Interior, six panel, hollow core, 1-3/8" thick										
	0040	Molded hardboard, 2'-0" x 6'-8"	2 Carp	17	.941	Ea.	44.50	21.50		66	85.50	
	0060	2'-6" x 6'-8"		17	.941		48	21.50		69.50	89.50	
	0070	2'-8" x 6'-8"		17	.941		50	21.50		71.50	92	
	0080	3'-0" x 6'-8"		17	.941		52.50	21.50		74	95	
	0140	Embossed print, molded hardboard, 2'-0" x 6'-8"		17	.941		48	21.50		69.50	89.50	
	0160	2'-6" x 6'-8"		17	.941		48	21.50		69.50	89.50	
	0180	3'-0" x 6'-8"		17	.941		52.50	21.50		74	95	
	0540	Six panel, solid, 1-3/8" thick, pine, 2'-0" x 6'-8"		15	1.067		116	24.50		140.50	169	
	0560	2'-6" x 6'-8"		14	1.143		130	26.50		156.50	188	
	0580	3'-0" x 6'-8"		13	1.231		150	28.50		178.50	213	
	1020	Two panel, bored rail, solid, 1-3/8" thick, pine, 1'-6" x 6'-8"		16	1		211	23		234	271	
	1040	2'-0" x 6'-8"		15	1.067		277	24.50		301.50	345	
	1060	2'-6" x 6'-8"		14	1.143		315	26.50		341.50	395	
	1340	Two panel, solid, 1-3/8" thick, fir, 2'-0" x 6'-8"		15	1.067		116	24.50		140.50	169	
	1360	2'-6" x 6'-8"		14	1.143		130	26.50		156.50	188	
	1380	3'-0" x 6'-8"		13	1.231		315	28.50		343.50	400	
	1740	Five panel, solid, 1-3/8" thick, fir, 2'-0" x 6'-8"		15	1.067		207	24.50		231.50	270	
	1760	2'-6" x 6'-8"		14	1.143		335	26.50		361.50	410	
	1780	3'-0" x 6'-8"	↓	13	1.231	↓	335	28.50		363.50	415	
930	0010	**WOOD DOORS, RESIDENTIAL**										930
	0200	Exterior, combination storm & screen, pine										
	0260	2'-8" wide	2 Carp	10	1.600	Ea.	256	37		293	345	
	0280	3'-0" wide		9	1.778		261	41		302	355	
	0300	7'-1" x 3'-0" wide		9	1.778		279	41		320	375	
	0400	Full lite, 6'-9" x 2'-6" wide		11	1.455		266	33.50		299.50	350	
	0420	2'-8" wide		10	1.600		266	37		303	355	
	0440	3'-0" wide		9	1.778		273	41		314	370	
	0500	7'-1" x 3'-0" wide		9	1.778		295	41		336	395	
	0700	Dutch door, pine, 1-3/4" x 6'-8" x 2'-8" wide, minimum		12	1.333		595	31		626	710	
	0720	Maximum		10	1.600		630	37		667	755	
	0800	3'-0" wide, minimum		12	1.333		620	31		651	735	
	0820	Maximum		10	1.600		670	37		707	805	
	1000	Entrance door, colonial, 1-3/4" x 6'-8" x 2'-8" wide		16	1		330	23		353	400	
	1020	6 panel pine, 3'-0" wide		15	1.067		355	24.50		379.50	430	
	1100	8 panel pine, 2'-8" wide		16	1		585	23		608	685	
	1120	3'-0" wide	↓	15	1.067		500	24.50		524.50	590	
	1200	For tempered safety glass lites, add					24.50			24.50	27	
	1300	Flush, birch, solid core, 1-3/4" x 6'-8" x 2'-8" wide	2 Carp	16	1		83.50	23		106.50	131	
	1320	3'-0" wide		15	1.067		91.50	24.50		116	143	
	1350	7'-0" x 2'-8" wide		16	1		87	23		110	135	
	1360	3'-0" wide		15	1.067		99.50	24.50		124	151	
	1740	Mahogany, 2'-8" x 6'-8"		15	1.067		695	24.50		719.50	805	
	1760	3'-0"x 6'-8"	↓	15	1.067	↓	720	24.50		744.50	835	
	2700	Interior, closet, bi-fold, w/hardware, no frame or trim incl.										
	2720	Flush, birch, 6'-6" or 6'-8" x 2'-6" wide	2 Carp	13	1.231	Ea.	45.50	28.50		74	98.50	
	2740	3'-0" wide		13	1.231		49.50	28.50		78	103	
	2760	4'-0" wide		12	1.333		90	31		121	152	
	2780	5'-0" wide		11	1.455		91	33.50		124.50	157	
	2800	6'-0" wide		10	1.600		98	37		135	171	
	3000	Raised panel pine, 6'-6" or 6'-8" x 2'-6" wide		13	1.231		151	28.50		179.50	215	
	3020	3'-0" wide		13	1.231		170	28.50		198.50	236	
	3040	4'-0" wide		12	1.333		218	31		249	293	
	3060	5'-0" wide	↓	11	1.455	↓	252	33.50		285.50	335	

	08210 | Wood Doors	CREW	DAILY OUTPUT	LABOR-HOURS	UNIT	2004 BARE COSTS MAT.	LABOR	EQUIP.	TOTAL	TOTAL INCL O&P	
930	3080 6'-0" wide	2 Carp	10	1.600	Ea.	283	37		320	375	930
	3200 Louvered, pine 6'-6" or 6'-8" x 2'-6" wide		13	1.231		93.50	28.50		122	152	
	3220 3'-0" wide		13	1.231		147	28.50		175.50	211	
	3240 4'-0" wide		12	1.333		164	31		195	233	
	3260 5'-0" wide		11	1.455		185	33.50		218.50	260	
	3280 6'-0" wide		10	1.600		204	37		241	288	
	4400 Bi-passing closet, incl. hardware and frame, no trim incl.										
	4420 Flush, lauan, 6'-8" x 4'-0" wide	2 Carp	12	1.333	Opng.	156	31		187	225	
	4440 5'-0" wide		11	1.455		171	33.50		204.50	245	
	4460 6'-0" wide		10	1.600		183	37		220	264	
	4600 Flush, birch, 6'-8" x 4'-0" wide		12	1.333		188	31		219	260	
	4620 5'-0" wide		11	1.455		193	33.50		226.50	269	
	4640 6'-0" wide		10	1.600		226	37		263	310	
	4800 Louvered, pine, 6'-8" x 4'-0" wide		12	1.333		370	31		401	460	
	4820 5'-0" wide		11	1.455		360	33.50		393.50	450	
	4840 6'-0" wide		10	1.600		455	37		492	565	
	4900 Mirrored, 6'-8" x 4'-0" wide		12	1.333	Ea.	232	31		263	310	
	5000 Paneled, pine, 6'-8" x 4'-0" wide		12	1.333	Opng.	355	31		386	445	
	5020 5'-0" wide		11	1.455		375	33.50		408.50	465	
	5040 6'-0" wide		10	1.600		440	37		477	545	
	5061 Hardboard, 6'-8" x 4'-0" wide		10	1.600		185	37		222	267	
	5062 5'-0" wide		10	1.600		185	37		222	267	
	5063 6'-0" wide		10	1.600		218	37		255	305	
	6100 Folding accordion, closet, including track and frame										
	6121 Vinyl, 2 layer, stock	2 Carp	400	.040	S.F.	2.93	.92		3.85	4.79	
	6140 Woven mahogany and vinyl, stock		400	.040		1.57	.92		2.49	3.30	
	6160 Wood slats with vinyl overlay, stock		400	.040		9.45	.92		10.37	11.95	
	6180 Economy vinyl, stock		400	.040		1.61	.92		2.53	3.34	
	6200 Rigid PVC		400	.040		4.40	.92		5.32	6.40	
	6220 For custom partition, add					25%	10%				
	7310 Passage doors, flush, no frame included										
	7320 Hardboard, hollow core, 1-3/8" x 6'-8" x 1'-6" wide	2 Carp	18	.889	Ea.	37.50	20.50		58	76.50	
	7330 2'-0" wide		18	.889		38	20.50		58.50	76.50	
	7340 2'-6" wide		18	.889		42	20.50		62.50	81	
	7350 2'-8" wide		18	.889		44	20.50		64.50	83.50	
	7360 3'-0" wide		17	.941		46.50	21.50		68	88	
	7420 Lauan, hollow core, 1-3/8" x 6'-8" x 1'-6" wide		18	.889		24	20.50		44.50	61	
	7440 2'-0" wide		18	.889		25.50	20.50		46	63	
	7450 2'-4" wide		18	.889		28.50	20.50		49	66.50	
	7460 2'-6" wide		18	.889		28.50	20.50		49	66.50	
	7480 2'-8" wide		18	.889		30	20.50		50.50	68	
	7500 3'-0" wide		17	.941		31.50	21.50		53	72	
	7700 Birch, hollow core, 1-3/8" x 6'-8" x 1'-6" wide		18	.889		30.50	20.50		51	68.50	
	7720 2'-0" wide		18	.889		37.50	20.50		58	76	
	7740 2'-6" wide		18	.889		41.50	20.50		62	81	
	7760 2'-8" wide		18	.889		43.50	20.50		64	82.50	
	7780 3'-0" wide		17	.941		47	21.50		68.50	89	
	8000 Pine louvered, 1-3/8" x 6'-8" x 1'-6" wide		19	.842		89.50	19.45		108.95	132	
	8020 2'-0" wide		18	.889		113	20.50		133.50	159	
	8040 2'-6" wide		18	.889		123	20.50		143.50	170	
	8060 2'-8" wide		18	.889		130	20.50		150.50	178	
	8080 3'-0" wide		17	.941		139	21.50		160.50	190	
	8300 Pine paneled, 1-3/8" x 6'-8" x 1'-6" wide		19	.842		101	19.45		120.45	144	
	8320 2'-0" wide		18	.889		117	20.50		137.50	163	
	8330 2'-4" wide		18	.889		128	20.50		148.50	176	
	8340 2'-6" wide		18	.889		131	20.50		151.50	179	

		08210 | Wood Doors	CREW	DAILY OUTPUT	LABOR-HOURS	UNIT	2004 BARE COSTS MAT.	LABOR	EQUIP.	TOTAL	TOTAL INCL O&P	
930	8360	2'-8" wide	2 Carp	18	.889	Ea.	142	20.50		162.50	191	930
	8380	3'-0" wide		17	.941		148	21.50		169.50	200	
	8450	French door, pine, 15 lites, 1-3/8"x6'-8"x2'-6" wide		18	.889		209	20.50		229.50	265	
	8470	2'-8" wide		18	.889		252	20.50		272.50	310	
	8490	3'-0" wide	↓	17	.941	↓	266	21.50		287.50	330	
	8550	For over 20 doors, deduct					15%					
960	0010	WOOD FRAMES										960
	0400	Exterior frame, incl. ext. trim, pine, 5/4 x 4-9/16" deep	2 Carp	375	.043	L.F.	4.41	.99		5.40	6.50	
	0420	5-3/16" deep		375	.043		7.35	.99		8.34	9.75	
	0440	6-9/16" deep		375	.043		6.85	.99		7.84	9.15	
	0600	Oak, 5/4 x 4-9/16" deep		350	.046		8.45	1.06		9.51	11.05	
	0620	5-3/16" deep		350	.046		9.50	1.06		10.56	12.25	
	0640	6-9/16" deep		350	.046		10.55	1.06		11.61	13.40	
	0800	Walnut, 5/4 x 4-9/16" deep		350	.046		9.90	1.06		10.96	12.70	
	0820	5-3/16" deep		350	.046		14.40	1.06		15.46	17.60	
	0840	6-9/16" deep		350	.046		16.95	1.06		18.01	20.50	
	1000	Sills, 8/4 x 8" deep, oak, no horns		100	.160		11.95	3.70		15.65	19.40	
	1020	2" horns		100	.160		13.30	3.70		17	21	
	1040	3" horns		100	.160		15.35	3.70		19.05	23	
	1100	8/4 x 10" deep, oak, no horns		90	.178		16.10	4.11		20.21	24.50	
	1120	2" horns		90	.178		17.95	4.11		22.06	26.50	
	1140	3" horns		90	.178	↓	19.55	4.11		23.66	28.50	
	2000	Exterior, colonial, frame & trim, 3' opng., in-swing, minimum		22	.727	Ea.	279	16.80		295.80	335	
	2010	Average		21	.762		415	17.60		432.60	485	
	2020	Maximum		20	.800		940	18.50		958.50	1,050	
	2100	5'-4" opening, in-swing, minimum		17	.941		315	21.50		336.50	380	
	2120	Maximum		15	1.067		940	24.50		964.50	1,075	
	2140	Out-swing, minimum		17	.941		325	21.50		346.50	390	
	2160	Maximum		15	1.067		975	24.50		999.50	1,125	
	2400	6'-0" opening, in-swing, minimum		16	1		300	23		323	375	
	2420	Maximum		10	1.600		975	37		1,012	1,150	
	2460	Out-swing, minimum		16	1		325	23		348	395	
	2480	Maximum		10	1.600	↓	1,200	37		1,237	1,400	
	2600	For two sidelights, add, minimum		30	.533	Opng.	310	12.30		322.30	365	
	2620	Maximum		20	.800	"	995	18.50		1,013.50	1,125	
	2700	Custom birch frame, 3'-0" opening		16	1	Ea.	179	23		202	236	
	2750	6'-0" opening		16	1		270	23		293	335	
	2900	Exterior, modern, plain trim, 3' opng., in-swing, minimum		26	.615		30	14.20		44.20	57.50	
	2920	Average		24	.667		36	15.40		51.40	65.50	
	2940	Maximum		22	.727	↓	44	16.80		60.80	76.50	
	3000	Interior frame, pine, 11/16" x 3-5/8" deep		375	.043	L.F.	3.99	.99		4.98	6.05	
	3020	4-9/16" deep		375	.043		5.45	.99		6.44	7.65	
	3040	5-3/16" deep		375	.043		3.53	.99		4.52	5.55	
	3200	Oak, 11/16" x 3-5/8" deep		350	.046		3.59	1.06		4.65	5.75	
	3220	4-9/16" deep		350	.046		3.87	1.06		4.93	6.05	
	3240	5-3/16" deep		350	.046		3.98	1.06		5.04	6.15	
	3400	Walnut, 11/16" x 3-5/8" deep		350	.046		6.05	1.06		7.11	8.45	
	3420	4-9/16" deep		350	.046		6.40	1.06		7.46	8.80	
	3440	5-3/16" deep		350	.046		6.60	1.06		7.66	9.10	
	3800	Threshold, oak, 5/8" x 3-5/8" deep		200	.080		2.33	1.85		4.18	5.70	
	3820	4-5/8" deep		190	.084		2.96	1.95		4.91	6.55	
	3840	5-5/8" deep	↓	180	.089	↓	5	2.05		7.05	9	
	4000	For casing see division 06220-400 & 06220-800										

08200 | Wood and Plastic Doors

08260 | Sliding Wood and Plastic Doors

			CREW	DAILY OUTPUT	LABOR-HOURS	UNIT	2004 BARE COSTS MAT.	LABOR	EQUIP.	TOTAL	TOTAL INCL O&P
700	0010	**GLASS, SLIDING**									
	0012	Vinyl clad, 1" insul. glass, 6'-0" x 6'-10" high	2 Carp	4	4	Opng.	1,150	92.50		1,242.50	1,400
	0030	6'-0" x 8'-0" high		4	4	Ea.	1,725	92.50		1,817.50	2,050
	0100	8'-0" x 6'-10" high		4	4	Opng.	1,775	92.50		1,867.50	2,100
	0500	3 leaf, 9'-0" x 6'-10" high		3	5.333		1,600	123		1,723	1,975
	0600	12'-0" x 6'-10" high	↓	3	5.333	↓	2,000	123		2,123	2,400
900	0010	**GLASS, SLIDING**									
	0020	Wood, 5/8" tempered insul. glass, 6' wide, premium	2 Carp	4	4	Ea.	985	92.50		1,077.50	1,225
	0100	Economy		4	4		695	92.50		787.50	920
	0150	8' wide, wood, premium		3	5.333		1,150	123		1,273	1,450
	0200	Economy		3	5.333		805	123		928	1,100
	0250	12' wide, wood, premium		2.50	6.400		2,700	148		2,848	3,225
	0300	Economy	↓	2.50	6.400	↓	1,825	148		1,973	2,275
	0350	Aluminum, 5/8" tempered insulated glass, 6' wide									
	0400	Premium	2 Carp	4	4	Ea.	1,375	92.50		1,467.50	1,650
	0450	Economy		4	4		710	92.50		802.50	935
	0500	8' wide, premium		3	5.333		1,300	123		1,423	1,625
	0550	Economy		3	5.333		1,100	123		1,223	1,400
	0600	12' wide, premium		2.50	6.400		2,150	148		2,298	2,625
	0650	Economy	↓	2.50	6.400	↓	1,375	148		1,523	1,750
	1000	Replacement doors, wood									
	1050	6' wide, premium	2 Carp	4	4	Ea.	635	92.50		727.50	850

08300 | Specialty Doors

08310 | Access Doors and Panels

			CREW	DAILY OUTPUT	LABOR-HOURS	UNIT	2004 BARE COSTS MAT.	LABOR	EQUIP.	TOTAL	TOTAL INCL O&P
150	0010	**BULKHEAD CELLAR DOORS**									
	0020	Steel, not incl. sides, 44" x 62"	1 Carp	5.50	1.455	Ea.	204	33.50		237.50	282
	0100	52" x 73"		5.10	1.569		227	36		263	310
	0500	With sides and foundation plates, 57" x 45" x 24"		4.70	1.702		266	39.50		305.50	360
	0600	42" x 49" x 51"	↓	4.30	1.860	↓	320	43		363	430

08360 | Overhead Doors

			CREW	DAILY OUTPUT	LABOR-HOURS	UNIT	2004 BARE COSTS MAT.	LABOR	EQUIP.	TOTAL	TOTAL INCL O&P
600	0010	**RESIDENTIAL GARAGE DOORS** Including hardware, no frame									
	0050	Hinged, wood, custom, double door, 9' x 7'	2 Carp	4	4	Ea.	345	92.50		437.50	535
	0070	16' x 7'		3	5.333		585	123		708	855
	0200	Overhead, sectional, incl. hardware, fiberglass, 9' x 7', standard		5.28	3.030		555	70		625	730
	0220	Deluxe		5.28	3.030		700	70		770	890
	0300	16' x 7', standard		6	2.667		965	61.50		1,026.50	1,150
	0320	Deluxe		6	2.667		1,200	61.50		1,261.50	1,425
	0500	Hardboard, 9' x 7', standard		8	2		370	46		416	490
	0520	Deluxe		8	2		480	46		526	605
	0600	16' x 7', standard		6	2.667		700	61.50		761.50	875
	0620	Deluxe		6	2.667		815	61.50		876.50	1,000
	0700	Metal, 9' x 7', standard		5.28	3.030		435	70		505	595
	0720	Deluxe		8	2		585	46		631	720
	0800	16' x 7', standard		3	5.333		550	123		673	815
	0820	Deluxe		6	2.667		855	61.50		916.50	1,050
	0900	Wood, 9' x 7', standard	↓	8	2	↓	440	46		486	565

08300 | Specialty Doors

08360	Overhead Doors	CREW	DAILY OUTPUT	LABOR-HOURS	UNIT	2004 BARE COSTS MAT.	LABOR	EQUIP.	TOTAL	TOTAL INCL O&P
600 0920	Deluxe	2 Carp	8	2	Ea.	1,275	46		1,321	1,475
1000	16′ x 7′, standard		6	2.667		890	61.50		951.50	1,075
1020	Deluxe	↓	6	2.667		1,850	61.50		1,911.50	2,150
1800	Door hardware, sectional	1 Carp	4	2		195	46		241	293
1810	Door tracks only		4	2		93	46		139	181
1820	One side only	↓	7	1.143		63	26.50		89.50	115
3000	Swing-up, including hardware, fiberglass, 9′ x 7′, standard	2 Carp	8	2		610	46		656	750
3020	Deluxe		8	2		670	46		716	820
3100	16′ x 7′, standard		6	2.667		770	61.50		831.50	950
3120	Deluxe		6	2.667		830	61.50		891.50	1,025
3200	Hardboard, 9′ x 7′, standard		8	2		293	46		339	400
3220	Deluxe		8	2		390	46		436	505
3300	16′ x 7′, standard		6	2.667		410	61.50		471.50	555
3320	Deluxe		6	2.667		610	61.50		671.50	775
3400	Metal, 9′ x 7′, standard		8	2		320	46		366	435
3420	Deluxe		8	2		550	46		596	685
3500	16′ x 7′, standard		6	2.667		505	61.50		566.50	660
3520	Deluxe		6	2.667		810	61.50		871.50	995
3600	Wood, 9′ x 7′, standard		8	2		350	46		396	465
3620	Deluxe		8	2		600	46		646	740
3700	16′ x 7′, standard		6	2.667		610	61.50		671.50	775
3720	Deluxe	↓	6	2.667		860	61.50		921.50	1,050
3900	Door hardware only, swing up	1 Carp	4	2		98.50	46		144.50	187
3920	One side only		7	1.143		57	26.50		83.50	108
4000	For electric operator, economy, add		8	1		254	23		277	320
4100	Deluxe, including remote control	↓	8	1	↓	370	23		393	450
4500	For transmitter/receiver control , add to operator				Total	80.50			80.50	88.50
4600	Transmitters, additional				"	28			28	31
6000	Replace section, on sectional door, fiberglass, 9′ x 7′	1 Carp	4	2	Ea.	161	46		207	256
6020	16′ x 7′		3.50	2.286		244	53		297	360
6200	Hardboard, 9′ x 7′		4	2		85.50	46		131.50	173
6220	16′ x 7′		3.50	2.286		164	53		217	270
6300	Metal, 9′ x 7′		4	2		139	46		185	232
6320	16′ x 7′		3.50	2.286		228	53		281	340
6500	Wood, 9′ x 7′		4	2		84	46		130	171
6520	16′ x 7′	↓	3.50	2.286	↓	163	53		216	270

08500 | Windows

08510	Steel Windows	CREW	DAILY OUTPUT	LABOR-HOURS	UNIT	2004 BARE COSTS MAT.	LABOR	EQUIP.	TOTAL	TOTAL INCL O&P
700 0010	SCREENS									
0020	For metal sash, aluminum or bronze mesh, flat screen	2 Sswk	1,200	.013	S.F.	3.30	.33		3.63	4.27
0500	Wicket screen, inside window	"	1,000	.016	"	5.05	.40		5.45	6.30
0600	Residential, aluminum mesh and frame, 2′ x 3′	2 Carp	32	.500	Ea.	11.85	11.55		23.40	32.50
0610	Rescreen		50	.320		9.10	7.40		16.50	22.50
0620	3′ x 5′		32	.500		26	11.55		37.55	48
0630	Rescreen		45	.356		23.50	8.20		31.70	40
0640	4′ x 8′		25	.640		49.50	14.80		64.30	79
0650	Rescreen		40	.400		38	9.25		47.25	57.50
0660	Patio door	↓	25	.640	↓	135	14.80		149.80	173

DOORS & WINDOWS 8

		08510 | Steel Windows	CREW	DAILY OUTPUT	LABOR-HOURS	UNIT	2004 BARE COSTS MAT.	LABOR	EQUIP.	TOTAL	TOTAL INCL O&P	
700	0680	Rescreening	2 Carp	1,600	.010	S.F.	1.13	.23		1.36	1.63	700
	1000	For solar louvers, add	2 Sswk	160	.100	"	18.45	2.49		20.94	25.50	

		08520 | Aluminum Windows	CREW	DAILY OUTPUT	LABOR-HOURS	UNIT	MAT.	LABOR	EQUIP.	TOTAL	TOTAL INCL O&P	
120	0010	**ALUMINUM WINDOWS** Incl. frame and glazing, Commercial grade										120
	1000	Stock units, casement, 3'-1" x 3'-2" opening	2 Sswk	10	1.600	Ea.	284	40		324	390	
	1040	Insulating glass	"	10	1.600		279	40		319	380	
	1050	Add for storms					57			57	63	
	1600	Projected, with screen, 3'-1" x 3'-2" opening	2 Sswk	10	1.600		203	40		243	300	
	1650	Insulating glass	"	10	1.600		188	40		228	284	
	1700	Add for storms					54			54	59.50	
	2000	4'-5" x 5'-3" opening	2 Sswk	8	2		286	50		336	410	
	2050	Insulating glass	"	8	2		325	50		375	455	
	2100	Add for storms					74.50			74.50	82	
	2500	Enamel finish windows, 3'-1" x 3'-2"	2 Sswk	10	1.600		182	40		222	278	
	2550	Insulating glass		10	1.600		196	40		236	293	
	2600	4'-5" x 5'-3"		8	2		273	50		323	395	
	2700	Insulating glass		8	2		355	50		405	485	
	3000	Single hung, 2' x 3' opening, enameled, standard glazed		10	1.600		136	40		176	227	
	3100	Insulating glass		10	1.600		165	40		205	259	
	3300	2'-8" x 6'-8" opening, standard glazed		8	2		290	50		340	415	
	3400	Insulating glass		8	2		375	50		425	505	
	3700	3'-4" x 5'-0" opening, standard glazed		9	1.778		187	44.50		231.50	292	
	3800	Insulating glass		9	1.778		263	44.50		307.50	375	
	4000	Sliding aluminum, 3' x 2' opening, standard glazed		10	1.600		154	40		194	246	
	4100	Insulating glass		10	1.600		170	40		210	264	
	4300	5' x 3' opening, standard glazed		9	1.778		195	44.50		239.50	300	
	4400	Insulating glass		9	1.778		273	44.50		317.50	385	
	4600	8' x 4' opening, standard glazed		6	2.667		281	66.50		347.50	440	
	4700	Insulating glass		6	2.667		450	66.50		516.50	625	
	5000	9' x 5' opening, standard glazed		4	4		425	99.50		524.50	655	
	5100	Insulating glass		4	4		680	99.50		779.50	940	
	5500	Sliding, with thermal barrier and screen, 6' x 4', 2 track		8	2		580	50		630	730	
	5700	4 track	↓	8	2		705	50		755	870	
	6000	For above units with bronze finish, add					12%					
	6200	For installation in concrete openings, add				↓	5%					
500	0010	**JALOUSIES**										500
	0020	Aluminum incl. glazing & screens, stock, 1'-7" x 3'-2"	2 Sswk	10	1.600	Ea.	125	40		165	214	
	0100	2'-3" x 4'-0"		10	1.600		178	40		218	273	
	0200	3'-1" x 2'-0"		10	1.600		140	40		180	231	
	0300	3'-1" x 5'-3"		10	1.600		254	40		294	355	
	1000	Mullions for above, 2'-0" long		80	.200		9.60	4.98		14.58	20	
	1100	5'-3" long	↓	80	.200	↓	16.40	4.98		21.38	27.50	

		08550 | Wood Windows	CREW	DAILY OUTPUT	LABOR-HOURS	UNIT	MAT.	LABOR	EQUIP.	TOTAL	TOTAL INCL O&P	
100	0010	**AWNING WINDOW** Including frame, screens and grills										100
	0100	Average quality, builders model, 34" x 22", double insulated glass	1 Carp	10	.800	Ea.	193	18.50		211.50	245	
	0200	Low E glass		10	.800		221	18.50		239.50	275	
	0300	40" x 28", double insulated glass		9	.889		264	20.50		284.50	325	
	0400	Low E Glass		9	.889		280	20.50		300.50	345	
	0500	48" x 36", double insulated glass		8	1		385	23		408	465	
	0600	Low E glass		8	1		405	23		428	485	
	0800	Vinyl clad, premium, double insulated glass, 24" x 17"		12	.667		178	15.40		193.40	222	
	0840	24" x 28"		11	.727		213	16.80		229.80	263	
	0860	36" x 17"	↓	11	.727	↓	215	16.80		231.80	266	

		08550 | Wood Windows	CREW	DAILY OUTPUT	LABOR-HOURS	UNIT	2004 BARE COSTS MAT.	LABOR	EQUIP.	TOTAL	TOTAL INCL O&P	
100	0900	36" x 40"	1 Carp	10	.800	Ea.	330	18.50		348.50	395	100
	1100	40" x 22"		10	.800		243	18.50		261.50	299	
	1200	36" x 28"		9	.889		259	20.50		279.50	320	
	1300	36" x 36"		9	.889		288	20.50		308.50	350	
	1400	48" x 28"		8	1		310	23		333	380	
	1500	60" x 36"		8	1		450	23		473	535	
	2000	Metal clad, deluxe, double insulated glass, 34" x 22"		10	.800		208	18.50		226.50	261	
	2100	40" x 22"		10	.800		244	18.50		262.50	300	
	2200	36" x 25"		9	.889		226	20.50		246.50	284	
	2300	40" x 30"		9	.889		283	20.50		303.50	345	
	2400	48" x 28"		8	1		289	23		312	360	
	2500	60" x 36"	↓	8	1	↓	310	23		333	380	
150	0010	**BOW-BAY WINDOW** Including frame, screens and grills,										150
	0020	end panels operable										
	1000	Bow type, casement, wood, bldrs mdl, 8' x 5' dbl insltd glass, 4 panel	2 Carp	10	1.600	Ea.	960	37		997	1,125	
	1050	Low E glass		10	1.600		1,175	37		1,212	1,350	
	1100	10'-0" x 5'-0" , double insulated glass, 6 panels		6	2.667		1,225	61.50		1,286.50	1,450	
	1200	Low E glass, 6 panels		6	2.667		1,300	61.50		1,361.50	1,525	
	1300	Vinyl clad, bldrs model, double insulated glass, 6'-0" x 4'-0", 3 panel		10	1.600		1,000	37		1,037	1,175	
	1340	9'-0" x 4'-0", 4 panel		8	2		1,325	46		1,371	1,525	
	1380	10'-0" x 6'-0", 5 panels		7	2.286		1,775	53		1,828	2,050	
	1420	12'-0" x 6'-0", 6 panels		6	2.667		1,800	61.50		1,861.50	2,075	
	1600	Metal clad, casement, bldrs mdl, 6'-0" x 4'-0", dbl insltd gls, 3 panels		10	1.600		830	37		867	980	
	1640	9'-0" x 4'-0" , 4 panels		8	2		1,175	46		1,221	1,350	
	1680	10'-0" x 5'-0", 5 panels		7	2.286		1,600	53		1,653	1,875	
	1720	12'-0" x 6'-0", 6 panels		6	2.667		2,250	61.50		2,311.50	2,575	
	2000	Bay window, casement, builders model, 8' x 5' dbl insul glass, 4 panels		10	1.600		1,425	37		1,462	1,625	
	2050	Low E glass,		10	1.600		1,725	37		1,762	1,950	
	2100	12'-0" x 6'-0" , double insulated glass, 6 panels		6	2.667		1,775	61.50		1,836.50	2,050	
	2200	Low E glass		6	2.667		1,825	61.50		1,886.50	2,125	
	2280	6'-0" x 4'-0"		11	1.455		975	33.50		1,008.50	1,125	
	2300	Vinyl clad, premium, double insulated glass, 8'-0" x 5'-0"		10	1.600		1,150	37		1,187	1,325	
	2340	10'-0" x 5'-0"		8	2		1,625	46		1,671	1,850	
	2380	10'-0" x 6'-0"		7	2.286		1,700	53		1,753	1,975	
	2420	12'-0" x 6'-0"		6	2.667		2,025	61.50		2,086.50	2,325	
	2430	14'-0" x 3'-0"		7	2.286		1,400	53		1,453	1,625	
	2440	14'-0" x 6'-0"		5	3.200		2,200	74		2,274	2,550	
	2600	Metal clad, deluxe, dbl insul. glass, 8'-0" x 5'-0" high, 4 panels		10	1.600		1,250	37		1,287	1,450	
	2640	10'-0" x 5'-0" high, 5 panels		8	2		1,350	46		1,396	1,550	
	2680	10'-0" x 6'-0" high, 5 panels		7	2.286		1,600	53		1,653	1,850	
	2720	12'-0" x 6'-0" high, 6 panels		6	2.667		2,200	61.50		2,261.50	2,525	
	3000	Double hung, bldrs. model, bay, 8' x 4' high,dbl insulated glass		10	1.600		995	37		1,032	1,175	
	3050	Low E glass		10	1.600		1,075	37		1,112	1,250	
	3100	9'-0" x 5'-0" high, doublel insulated glass		6	2.667		1,075	61.50		1,136.50	1,275	
	3200	Low E glass		6	2.667		1,125	61.50		1,186.50	1,350	
	3300	Vinyl clad, premium, double insulated glass, 7'-0" x 4'-6"		10	1.600		1,025	37		1,062	1,200	
	3340	8'-0" x 4'-6"		8	2		1,050	46		1,096	1,225	
	3380	8'-0" x 5'-0"		7	2.286		1,100	53		1,153	1,300	
	3420	9'-0" x 5'-0"		6	2.667		1,125	61.50		1,186.50	1,350	
	3600	Metal clad, deluxe, dbl insul. glass, 7'-0" x 4'-0" high		10	1.600		955	37		992	1,125	
	3640	8'-0" x 4'-0" high		8	2		990	46		1,036	1,150	
	3680	8'-0" x 5'-0" high		7	2.286		1,025	53		1,078	1,225	
	3720	9'-0" x 5'-0" high		6	2.667		1,075	61.50		1,136.50	1,300	
	7000	Drip cap, premolded vinyl, 8' long		30	.533		73	12.30		85.30	101	
	7040	12' long	↓	26	.615	↓	79.50	14.20		93.70	111	

08550 | Wood Windows		CREW	DAILY OUTPUT	LABOR-HOURS	UNIT	2004 BARE COSTS MAT.	LABOR	EQUIP.	TOTAL	TOTAL INCL O&P
200 0010	**CASEMENT WINDOW** Including frame, screen, and grills R08550-010									
0100	Avg. quality, bldrs. model, 2'-0" x 3'-0" H, dbl. insulated glass	1 Carp	10	.800	Ea.	177	18.50		195.50	227
0150	Low E glass		10	.800		234	18.50		252.50	289
0200	2'-0" x 4'-6" high, double insulated glass		9	.889		230	20.50		250.50	288
0250	Low E glass		9	.889		320	20.50		340.50	385
0300	2'-3" x 6'-0" high, double insulated glass		8	1		269	23		292	335
0350	Low E glass		8	1		325	23		348	400
0522	Vinyl clad, premium, double insulated glass, 2'-0" x 3'-0"		10	.800		237	18.50		255.50	293
0524	2'-0" x 4'-0"		9	.889		277	20.50		297.50	340
0525	2'-0" x 5'-0"		8	1		315	23		338	390
0528	2'-0" x 6'-0"	↓	8	1	↓	360	23		383	435
3020	Vinyl clad, premium, double insulated glass, multiple leaf units									
3080	Single unit, 1'-6" x 5'-0"	2 Carp	20	.800	Ea.	272	18.50		290.50	330
3100	2'-0" x 2'-0"		20	.800		182	18.50		200.50	232
3140	2'-0" x 2'-6"		20	.800		237	18.50		255.50	293
3220	2'-0" x 3'-6"		20	.800		235	18.50		253.50	290
3260	2'-0" x 4'-0"		19	.842		277	19.45		296.45	340
3300	2'-0" x 4'-6"		19	.842		272	19.45		291.45	330
3340	2'-0" x 5'-0"		18	.889		315	20.50		335.50	385
3460	2'-4" x 3'-0"		20	.800		237	18.50		255.50	293
3500	2'-4" x 4'-0"		19	.842		294	19.45		313.45	360
3540	2'-4" x 5'-0"		18	.889		350	20.50		370.50	420
3700	Double unit, 2'-8" x 5'-0"		18	.889		490	20.50		510.50	575
3740	2'-8" x 6'-0"		17	.941		570	21.50		591.50	665
3840	3'-0" x 4'-6"		18	.889		445	20.50		465.50	525
3860	3'-0" x 5'-0"		17	.941		580	21.50		601.50	675
3880	3'-0" x 6'-0"		17	.941		620	21.50		641.50	715
3980	3'-4" x 2'-6"		19	.842		375	19.45		394.45	450
4000	3'-4" x 3'-0"		12	1.333		375	31		406	470
4030	3'-4" x 4'-0"		18	.889		460	20.50		480.50	540
4050	3'-4" x 5'-0"		12	1.333		580	31		611	695
4100	3'-4" x 6'-0"		11	1.455		620	33.50		653.50	735
4200	3'-6" x 3'-0"		18	.889		375	20.50		395.50	450
4340	4'-0" x 3'-0"		18	.889		425	20.50		445.50	500
4380	4'-0" x 3'-6"		17	.941		455	21.50		476.50	535
4420	4'-0" x 4'-0"		16	1		545	23		568	640
4460	4'-0" x 4'-4"		16	1		535	23		558	630
4540	4'-0" x 5'-0"		16	1		585	23		608	680
4580	4'-0" x 6'-0"		15	1.067		665	24.50		689.50	770
4740	4'-8" x 3'-0"		18	.889		475	20.50		495.50	560
4780	4'-8" x 3'-6"		17	.941		510	21.50		531.50	595
4820	4'-8" x 4'-0"		16	1		605	23		628	705
4860	4'-8" x 5'-0"		15	1.067		690	24.50		714.50	800
4900	4'-8" x 6'-0"		15	1.067		775	24.50		799.50	895
5060	Triple unit, 5'-0" x 5'-0"		15	1.067		880	24.50		904.50	1,000
5100	5'-6" x 3'-0"		17	.941		600	21.50		621.50	695
5140	5'-6" x 3'-6"		16	1		635	23		658	740
5180	5'-6" x 4'-6"		15	1.067		715	24.50		739.50	830
5220	5'-6" x 5'-6"		15	1.067		960	24.50		984.50	1,100
5300	6'-0" x 4'-6"		15	1.067		715	24.50		739.50	830
5850	5'-0" x 3'-0"		12	1.333		635	31		666	755
5900	5'-0" x 4'-0"		11	1.455		795	33.50		828.50	930
6000	5'-0" x 5'-0"		10	1.600		915	37		952	1,075
6100	5'-0" x 5'-6"		10	1.600		925	37		962	1,100
6150	5'-0" x 6'-0"		10	1.600		1,025	37		1,062	1,225
6200	6'-0" x 3'-0"	↓	12	1.333	↓	995	31		1,026	1,150

		08550 | Wood Windows	CREW	DAILY OUTPUT	LABOR-HOURS	UNIT	2004 BARE COSTS MAT.	LABOR	EQUIP.	TOTAL	TOTAL INCL O&P	
200	6250	6'-0" x 3'-4" R08550-010	2 Carp	12	1.333	Ea.	635	31		666	755	200
	6300	6'-0" x 4'-0"		11	1.455		685	33.50		718.50	810	
	6350	6'-0" x 5'-0"		10	1.600		765	37		802	905	
	6400	6'-0" x 6'-0"		10	1.600		990	37		1,027	1,150	
	6500	Quadruple unit, 7'-0" x 4'-0"		9	1.778		955	41		996	1,125	
	6700	8'-0" x 4'-6"		9	1.778		1,175	41		1,216	1,350	
	6950	6'-8" x 4'-0"		10	1.600		915	37		952	1,075	
	7000	6'-8" x 6'-0"	↓	10	1.600		1,225	37		1,262	1,425	
	8100	Metal clad, deluxe, dbl. insul. glass, 2'-0" x 3'-0" high	1 Carp	10	.800		175	18.50		193.50	225	
	8120	2'-0" x 4'-0" high		9	.889		211	20.50		231.50	267	
	8140	2'-0" x 5'-0" high		8	1		240	23		263	300	
	8160	2'-0" x 6'-0" high	↓	8	1	↓	275	23		298	345	
	8200	For multiple leaf units, deduct for stationary sash										
	8220	2' high				Ea.	18.20			18.20	20	
	8240	4'-6" high					21			21	23	
	8260	6' high					28			28	31	
	8300	For installation, add per leaf				↓		15%				
250	0010	**DOUBLE HUNG** Including frame, screens, and grills										250
	0100	Avg. quality, bldrs. model, 2'-0" x 3'-0" high, dbl insul. glass	1 Carp	10	.800	Ea.	183	18.50		201.50	234	
	0150	Low E glass		10	.800		183	18.50		201.50	234	
	0200	3'-0" x 4'-0" high, double insulated glass		9	.889		244	20.50		264.50	305	
	0250	Low E glass		9	.889		252	20.50		272.50	315	
	0300	4'-0" x 4'-6" high, double insulated glass		8	1		277	23		300	345	
	0350	Low E glass		8	1		300	23		323	370	
	1000	Vinyl clad, premium, double insulated glass, 2'-6" x 3'-0"		10	.800		195	18.50		213.50	247	
	1100	3'-0" x 3'-6"		10	.800		230	18.50		248.50	285	
	1200	3'-0" x 4'-0"		9	.889		255	20.50		275.50	315	
	1300	3'-0" x 4'-6"		9	.889		266	20.50		286.50	325	
	1400	3'-0" x 5'-0"		8	1		278	23		301	345	
	1500	3'-6" x 6'-0"		8	1		325	23		348	395	
	2000	Metal clad, deluxe, dbl. insul. glass, 2'-6" x 3'-0" high		10	.800		186	18.50		204.50	236	
	2100	3'-0" x 3'-6" high		10	.800		220	18.50		238.50	274	
	2200	3'-0" x 4'-0" high		9	.889		234	20.50		254.50	292	
	2300	3'-0" x 4'-6" high		9	.889		253	20.50		273.50	315	
	2400	3'-0" x 5'-0" high		8	1		270	23		293	335	
	2500	3'-6" x 6'-0" high	↓	8	1	↓	330	23		353	400	
260	0010	**HALF ROUND WINDOW**, Vinyl clad, double insulated glass, including grill										260
	0800	14" height x 24" base	2 Carp	9	1.778	Ea.	325	41		366	430	
	1040	15" height x 25" base		8	2		330	46		376	440	
	1060	16" height x 28" base		7	2.286		360	53		413	485	
	1080	17" height x 29" base	↓	7	2.286		375	53		428	500	
	2000	19" height x 33" base	1 Carp	6	1.333		400	31		431	495	
	2100	20" height x 35" base		6	1.333		445	31		476	545	
	2200	21" height x 37" base	↓	6	1.333		425	31		456	525	
	2250	23" height x 41" base	2 Carp	6	2.667		465	61.50		526.50	615	
	2300	26" height x 48" base		6	2.667		480	61.50		541.50	635	
	2350	30" height x 56" base	↓	6	2.667		565	61.50		626.50	725	
	3000	36" height x 67"base	1 Carp	4	2		975	46		1,021	1,150	
	3040	38" height x 71" base	2 Carp	5	3.200		900	74		974	1,125	
	3050	40" height x 75" base	"	5	3.200		1,175	74		1,249	1,425	
	5000	Elliptical, 71" x 16"	1 Carp	11	.727		740	16.80		756.80	845	
	5100	Elliptical, 95" x 21"	"	10	.800	↓	1,050	18.50		1,068.50	1,175	
650	0010	**PALLADIAN WINDOWS**										650
	0020	Aluminum clad, double insulated glass, including frame and grills										

08550 | Wood Windows			CREW	DAILY OUTPUT	LABOR-HOURS	UNIT	2004 BARE COSTS MAT.	LABOR	EQUIP.	TOTAL	TOTAL INCL O&P	
650	0040	3′-2″ x 2′-0″ high	2 Carp	11	1.455	Ea.	1,175	33.50		1,208.50	1,325	650
	0060	3′-2″ x 4′-10″		11	1.455		1,325	33.50		1,358.50	1,500	
	0080	3′-2″ x 6′-4″		10	1.600		1,575	37		1,612	1,800	
	0100	4′-0″ x 4′-0″	↓	10	1.600		1,275	37		1,312	1,500	
	0120	4′-0″ x 5′-4″	3 Carp	10	2.400		1,500	55.50		1,555.50	1,750	
	0140	4′-0″ x 6′-0″		9	2.667		1,550	61.50		1,611.50	1,800	
	0160	4′-0″ x 7′-4″		9	2.667		1,700	61.50		1,761.50	1,975	
	0180	5′-5″ x 4′-10″		9	2.667		1,700	61.50		1,761.50	1,975	
	0200	5′-5″ x 6′-10″		9	2.667		1,925	61.50		1,986.50	2,225	
	0220	5′-5″ x 7′-9″		9	2.667		2,100	61.50		2,161.50	2,425	
	0240	6′-0″ x 7′-11″		8	3		2,800	69.50		2,869.50	3,200	
	0260	8′-0″ x 6′-0″	↓	8	3	↓	2,400	69.50		2,469.50	2,750	
670	0010	**PICTURE WINDOW** Including frame and grills										670
	0100	Average quality, bldrs. model, 3′-6″ x 4′-0″ high, dbl insulated glass	2 Carp	12	1.333	Ea.	260	31		291	340	
	0150	Low E glass		12	1.333		284	31		315	365	
	0200	4′-0″ x 4′-6″ high, double insulated glass		11	1.455		268	33.50		301.50	350	
	0250	Low E glass		11	1.455		294	33.50		327.50	380	
	0300	5′-0″ x 4′-0″ high, double insulated glass		11	1.455		330	33.50		363.50	420	
	0350	Low E glass		11	1.455		370	33.50		403.50	465	
	0400	6′-0″ x 4′-6″ high, double insulated glass		10	1.600		420	37		457	530	
	0450	Low E glass		10	1.600		475	37		512	585	
	1000	Vinyl clad, premium, dbl. insul. glass, 4′-0″ x 4′-0″		12	1.333		375	31		406	470	
	1100	4′-0″ x 6′-0″		11	1.455		645	33.50		678.50	765	
	1200	5′-0″ x 6′-0″		10	1.600		840	37		877	985	
	1300	6′-0″ x 6′-0″		10	1.600		855	37		892	1,000	
	2000	Metal clad, deluxe, dbl. insul. glass, 4′-0″ x 4′-0″ high		12	1.333		275	31		306	360	
	2100	4′-0″ x 6′-0″ high		11	1.455		405	33.50		438.50	500	
	2200	5′-0″ x 6′-0″ high		10	1.600		445	37		482	555	
	2300	6′-0″ x 6′-0″ high	↓	10	1.600	↓	515	37		552	630	
750	0010	**SLIDING WINDOW** Including frame, screen, and grills										750
	0100	Average quality, bldrs. model, 3′-0″ x 3′-0″ high, double insulated	1 Carp	10	.800	Ea.	140	18.50		158.50	186	
	0120	Low E glass		10	.800		177	18.50		195.50	226	
	0200	4′-0″ x 3′-6″ high, double insulated		9	.889		166	20.50		186.50	218	
	0220	Low E glass		9	.889		208	20.50		228.50	264	
	0300	6′-0″ x 5′-0″ high, double insulated		8	1		305	23		328	375	
	0320	Low E glass		8	1		365	23		388	440	
	1000	Vinyl clad, premium, dbl. insulated glass, 3′-0″ x 3′-0″		10	.800		515	18.50		533.50	595	
	1020	4′-0″ x 1′-11″		11	.727		505	16.80		521.80	585	
	1040	4′-0″ x 3′-0″		10	.800		600	18.50		618.50	690	
	1050	4′-0″ x 3′-6″		9	.889		640	20.50		660.50	740	
	1090	4′-0″ x 5′-0″		9	.889		770	20.50		790.50	885	
	1100	5′-0″ x 4′-0″		9	.889		770	20.50		790.50	885	
	1120	5′-0″ x 5′-0″		8	1		855	23		878	985	
	1140	6′-0″ x 4′-0″		8	1		855	23		878	985	
	1150	6′-0″ x 5′-0″		8	1		945	23		968	1,075	
	2000	Metal clad, deluxe, double insulated glass, 3′-0″ x 3′-0″ high		10	.800		282	18.50		300.50	340	
	2050	4′-0″ x 3′-6″ high		9	.889		345	20.50		365.50	415	
	2100	5′-0″ x 4′-0″ high		9	.889		415	20.50		435.50	490	
	2150	6′-0″ x 5′-0″ high	↓	8	1	↓	640	23		663	745	
760	0010	**TRANSOM WINDOWS**										760
	0050	Vinyl clad, premium, double insulated glass, 32″ x 8″	1 Carp	16	.500	Ea.	144	11.55		155.55	178	
	0100	36″ x 8″		16	.500		154	11.55		165.55	189	
	0110	36″ x 12″		16	.500		173	11.55		184.55	210	
	0150	36″ x 48″	↓	12	.667	↓	203	15.40		218.40	250	

Important: See the Reference Section for critical supporting data - Reference Nos., Crews, & Location Factors

		08550 | Wood Windows	CREW	DAILY OUTPUT	LABOR-HOURS	UNIT	2004 BARE COSTS MAT.	LABOR	EQUIP.	TOTAL	TOTAL INCL O&P	
770	0010	**WEATHERSTRIPPING** See division 08720										770
780	0010	**TRAPEZOID WINDOWS**										780
	0100	Vinyl clad, including frame and exterior trim										
	0900	20" base x 44" leg x 53" leg	2 Carp	13	1.231	Ea.	350	28.50		378.50	430	
	1000	24" base x 90" leg x 102" leg		8	2		580	46		626	715	
	3000	36" base x 0" leg x 22" leg		12	1.333		370	31		401	460	
	3010	36" base x 4" leg x 25" leg		13	1.231		390	28.50		418.50	480	
	3050	36" base x 26" leg x 48" leg		9	1.778		405	41		446	515	
	3100	36" base x 42" legs, 50" peak		9	1.778		470	41		511	590	
	3200	36" base x 60" leg x 81" leg		11	1.455		615	33.50		648.50	735	
	4320	44" base x 23" leg x 56" leg		11	1.455		485	33.50		518.50	590	
	4350	44" base x 59" leg x 92" leg		10	1.600		730	37		767	865	
	4500	46" base x 15" leg x 46" leg		8	2		365	46		411	485	
	4550	46" base x 16" leg x 48" leg		8	2		390	46		436	510	
	4600	46" base x 50" leg x 80" leg		7	2.286		585	53		638	735	
	6600	66" base x 12" leg x 42" leg		8	2		495	46		541	620	
	6650	66" base x 12" legs, 28" peak		9	1.778		400	41		441	510	
	6700	68" base x 3" legs, 31" peak	↓	8	2	↓	500	46		546	630	
800	0010	**WINDOW GRILLE OR MUNTIN** Snap-in type										800
	0020	Standard pattern interior grills										
	2000	Wood, awning window, glass size 28" x 16" high	1 Carp	30	.267	Ea.	18.55	6.15		24.70	31	
	2060	44" x 24" high		32	.250		27	5.80		32.80	39.50	
	2100	Casement, glass size, 20" x 36" high		30	.267		22.50	6.15		28.65	35.50	
	2180	20" x 56" high		32	.250	↓	33	5.80		38.80	46	
	2200	Double hung, glass size, 16" x 24" high		24	.333	Set	40.50	7.70		48.20	57.50	
	2280	32" x 32" high		34	.235	"	113	5.45		118.45	133	
	2500	Picture, glass size, 48" x 48" high		30	.267	Ea.	129	6.15		135.15	152	
	2580	60" x 68" high		28	.286	"	97.50	6.60		104.10	118	
	2600	Sliding, glass size, 14" x 36" high		24	.333	Set	23	7.70		30.70	38.50	
	2680	36" x 36" high	↓	22	.364	"	35	8.40		43.40	52.50	
820	0010	**WOOD SASH** Including glazing but not including trim										820
	0050	Custom, 5'-0" x 4'-0", 1" dbl. glazed, 3/16" thick lites	2 Carp	3.20	5	Ea.	147	116		263	360	
	0100	1/4" thick lites		5	3.200		151	74		225	291	
	0200	1" thick, triple glazed		5	3.200		345	74		419	505	
	0300	7'-0" x 4'-6" high, 1" double glazed, 3/16" thick lites		4.30	3.721		350	86		436	530	
	0400	1/4" thick lites		4.30	3.721		395	86		481	580	
	0500	1" thick, triple glazed		4.30	3.721		455	86		541	645	
	0600	8'-6" x 5'-0" high, 1" double glazed, 3/16" thick lites		3.50	4.571		475	106		581	705	
	0700	1/4" thick lites		3.50	4.571		520	106		626	750	
	0800	1" thick, triple glazed	↓	3.50	4.571	↓	525	106		631	755	
	0900	Window frames only, based on perimeter length				L.F.	2.90			2.90	3.19	
	3000	Replacement sash, double hung, double glazing, to 12 S.F.	1 Carp	64	.125	S.F.	15.80	2.89		18.69	22.50	
	3100	12 S.F. to 20 S.F.		94	.085		16	1.97		17.97	21	
	3200	20 S.F. and over	↓	106	.075		13.95	1.74		15.69	18.30	
	3800	Triple glazing for above, add				↓	2.13			2.13	2.34	
	7000	Sash, single lite, 2'-0" x 2'-0" high	1 Carp	20	.400	Ea.	41	9.25		50.25	60.50	
	7050	2'-6" x 2'-0" high		19	.421		44	9.75		53.75	65	
	7100	2'-6" x 2'-6" high		18	.444		46.50	10.25		56.75	69	
	7150	3'-0" x 2'-0" high	↓	17	.471	↓	58.50	10.85		69.35	83	
840	0010	**WOOD SCREENS**										840
	0020	Over 3 S.F., 3/4" frames	2 Carp	375	.043	S.F.	3.34	.99		4.33	5.35	
	0100	1-1/8" frames	"	375	.043	"	5.95	.99		6.94	8.20	

		08560 | Plastic Windows	CREW	DAILY OUTPUT	LABOR-HOURS	UNIT	2004 BARE COSTS MAT.	LABOR	EQUIP.	TOTAL	TOTAL INCL O&P	
100	0010	VINYL SINGLE HUNG WINDOWS										100
	0100	Grids, low E, J fin, ext. jambs, 21" x 53"	2 Carp	18	.889	Ea.	130	20.50		150.50	178	
	0110	21" x 57"		17	.941		133	21.50		154.50	184	
	0120	21" x 65"		16	1		139	23		162	192	
	0130	25" x 41"		20	.800		123	18.50		141.50	167	
	0140	25" x 49"		18	.889		136	20.50		156.50	184	
	0150	25" x 57"		17	.941		139	21.50		160.50	190	
	0160	25" x 65"		16	1		145	23		168	198	
	0170	29" x 41"		18	.889		131	20.50		151.50	179	
	0180	29" x 53"		18	.889		140	20.50		160.50	189	
	0190	29" x 57"		17	.941		143	21.50		164.50	195	
	0200	29" x 65"		16	1		149	23		172	203	
	0210	33" x 41"		20	.800		135	18.50		153.50	181	
	0220	33" x 53"		18	.889		146	20.50		166.50	195	
	0230	33" x 57"		17	.941		149	21.50		170.50	201	
	0240	33" x 65"		16	1		155	23		178	209	
	0250	37" x 41"		20	.800		143	18.50		161.50	189	
	0260	37" x 53"		18	.889		153	20.50		173.50	203	
	0270	37" x 57"		17	.941		156	21.50		177.50	209	
	0280	37" x 65"	↓	16	1	↓	163	23		186	218	
200	0010	VINYL DOUBLE HUNG WINDOWS										200
	0100	Grids, low E, J fin, ext. jambs, 21" x 53"	2 Carp	18	.889	Ea.	149	20.50		169.50	199	
	0102	21" x 37"		18	.889		133	20.50		153.50	181	
	0104	21" x 41"		18	.889		136	20.50		156.50	185	
	0106	21" x 49"		18	.889		143	20.50		163.50	192	
	0110	21" x 57"		17	.941		152	21.50		173.50	204	
	0120	21" x 65"		16	1		158	23		181	213	
	0128	25" x 37"		20	.800		140	18.50		158.50	186	
	0130	25" x 41"		20	.800		143	18.50		161.50	190	
	0140	25" x 49"		18	.889		148	20.50		168.50	197	
	0145	25" x 53"		18	.889		154	20.50		174.50	204	
	0150	25" x 57"		17	.941		154	21.50		175.50	207	
	0160	25" x 65"		16	1		164	23		187	219	
	0162	25" x 69"		16	1		170	23		193	226	
	0164	25" x 77"		16	1		180	23		203	237	
	0168	29" x 37"		18	.889		145	20.50		165.50	194	
	0170	29" x 41"		18	.889		148	20.50		168.50	197	
	0172	29" x 49"		18	.889		155	20.50		175.50	205	
	0180	29" x 53"		18	.889		158	20.50		178.50	209	
	0190	29" x 57"		17	.941		162	21.50		183.50	215	
	0200	29" x 65"		16	1		168	23		191	224	
	0202	29" x 69"		16	1		174	23		197	231	
	0205	29" x 77"		16	1		184	23		207	242	
	0208	33" x 37"		20	.800		149	18.50		167.50	195	
	0210	33" x 41"		20	.800		152	18.50		170.50	199	
	0215	33" x 49"		20	.800		160	18.50		178.50	208	
	0220	33" x 53"		18	.889		163	20.50		183.50	215	
	0230	33" x 57"		17	.941		167	21.50		188.50	221	
	0240	33" x 65"		16	1		171	23		194	227	
	0242	33" x 69"		16	1		181	23		204	239	
	0246	33" x 77"		16	1		191	23		214	249	
	0250	37" x 41"		20	.800		156	18.50		174.50	203	
	0255	37" x 49"		20	.800		163	18.50		181.50	212	
	0260	37" x 53"		18	.889		170	20.50		190.50	222	
	0270	37" x 57"		17	.941		174	21.50		195.50	228	
	0280	37" x 65"	↓	16	1	↓	178	23		201	235	

08560 | Plastic Windows

			CREW	DAILY OUTPUT	LABOR-HOURS	UNIT	2004 BARE COSTS MAT.	LABOR	EQUIP.	TOTAL	TOTAL INCL O&P	
200	0282	37" x 69"	2 Carp	16	1	Ea.	233	23		256	296	200
	0286	37" x 77"		16	1		244	23		267	305	
	0300	Solid vinyl, average quality, double insulated glass, 2'-0" x 3'-0"	1 Carp	10	.800		127	18.50		145.50	171	
	0310	3'-0" x 4'-0"		9	.889		156	20.50		176.50	206	
	0320	4'-0" x 4'-6"		8	1		187	23		210	245	
	0330	Premium, double insulated glass, 2'-6" x 3'-0"		10	.800		148	18.50		166.50	195	
	0340	3'-0" x 3'-6"		9	.889		172	20.50		192.50	224	
	0350	3'-0" x 4"-0"		9	.889		183	20.50		203.50	236	
	0360	3'-0" x 4'-6"		9	.889		187	20.50		207.50	241	
	0370	3'-0" x 5'-0"		8	1		192	23		215	250	
	0380	3'-6" x 6"-0"		8	1		211	23		234	271	
300	0010	VINYL CASEMENT WINDOWS										300
	0100	Grids, low E, J fin, ext. jambs, 1 lt, 21" x 41"	2 Carp	20	.800	Ea.	193	18.50		211.50	244	
	0110	21" x 47"		20	.800		210	18.50		228.50	263	
	0120	21" x 53"		20	.800		227	18.50		245.50	282	
	0128	24" x 35"		19	.842		185	19.45		204.45	237	
	0130	24" x 41"		19	.842		201	19.45		220.45	254	
	0140	24" x 47"		19	.842		218	19.45		237.45	273	
	0150	24" x 53"		19	.842		235	19.45		254.45	291	
	0158	28" x 35"		19	.842		197	19.45		216.45	250	
	0160	28" x 41"		19	.842		213	19.45		232.45	267	
	0170	28" x 47"		19	.842		230	19.45		249.45	286	
	0180	28" x 53"		19	.842		254	19.45		273.45	310	
	0184	28" x 59"		19	.842		258	19.45		277.45	315	
	0188	Two lites, 33" x 35"		18	.889		315	20.50		335.50	385	
	0190	33" x 41"		18	.889		340	20.50		360.50	410	
	0200	33" x 47"		18	.889		365	20.50		385.50	435	
	0210	33" x 53"		18	.889		390	20.50		410.50	465	
	0212	33" x 59"		18	.889		415	20.50		435.50	490	
	0215	33" x 72"		18	.889		430	20.50		450.50	505	
	0220	41" x 41"		18	.889		370	20.50		390.50	445	
	0230	41" x 47"		18	.889		395	20.50		415.50	470	
	0240	41" x 53"		17	.941		420	21.50		441.50	500	
	0242	41" x 59"		17	.941		445	21.50		466.50	525	
	0246	41" x 72"		17	.941		465	21.50		486.50	545	
	0250	47" x 41"		17	.941		375	21.50		396.50	445	
	0260	47" x 47"		17	.941		400	21.50		421.50	475	
	0270	47" x 53"		17	.941		425	21.50		446.50	500	
	0272	47" x 59"		17	.941		465	21.50		486.50	545	
	0280	56" x 41"		15	1.067		400	24.50		424.50	480	
	0290	56" x 47"		15	1.067		425	24.50		449.50	505	
	0300	56" x 53"		15	1.067		460	24.50		484.50	550	
	0302	56" x 59"		15	1.067		480	24.50		504.50	570	
	0310	56" x 72"		15	1.067		525	24.50		549.50	615	
	0340	Solid vinyl, premium, double insulated glass, 2'-0" x 3'-0" high	1 Carp	10	.800		170	18.50		188.50	219	
	0360	2'-0" x 4'-0" high		9	.889		209	20.50		229.50	265	
	0380	2'-0" x 5'-0" high		8	1		244	23		267	310	
400	0010	VINYL PICTURE WINDOWS										400
	0100	Grids, low E, J fin, ext. jambs, 33" x 47"	2 Carp	12	1.333	Ea.	194	31		225	266	
	0110	35" x 71"		12	1.333		205	31		236	278	
	0120	41" x 47"		12	1.333		225	31		256	300	
	0130	41" x 71"		12	1.333		244	31		275	320	
	0140	47" x 47"		12	1.333		254	31		285	335	
	0150	47" x 71"		11	1.455		267	33.50		300.50	350	
	0160	53" x 47"		11	1.455		250	33.50		283.50	330	

08560 | Plastic Windows

			CREW	DAILY OUTPUT	LABOR-HOURS	UNIT	2004 BARE COSTS MAT.	LABOR	EQUIP.	TOTAL	TOTAL INCL O&P	
400	0170	53″ x 71″	2 Carp	11	1.455	Ea.	261	33.50		294.50	345	400
	0180	59″ x 47″		11	1.455		283	33.50		316.50	365	
	0190	59″ x 71″		11	1.455		305	33.50		338.50	390	
	0200	71″ x 47″		10	1.600		315	37		352	410	
	0210	71″ x 71″	↓	10	1.600	↓	330	37		367	430	
500	0010	**VINYL HALF ROUND WINDOW,** including grill, j fin, low E, ext. jambs										500
	0100	10″ height x 20″ base	2 Carp	9	1.778	Ea.	237	41		278	330	
	0110	15″ height x 30″ base		8	2		305	46		351	415	
	0120	17″ height x 34″ base		7	2.286		355	53		408	480	
	0130	19″ height x 38″ base	↓	7	2.286		405	53		458	535	
	0140	20″ height x 40″ base	1 Carp	6	1.333		345	31		376	430	
	0150	24″ height x 48″ base		6	1.333		415	31		446	510	
	0160	25″ height x 50″ base	↓	6	1.333		450	31		481	550	
	0170	30″ height x 60″ base	2 Carp	6	2.667	↓	555	61.50		616.50	715	

08580 | Special Function Windows

			CREW	DAILY OUTPUT	LABOR-HOURS	UNIT	MAT.	LABOR	EQUIP.	TOTAL	TOTAL INCL O&P	
900	0010	**STORM WINDOWS** Aluminum, residential										900
	0300	Basement, mill finish, incl. fiberglass screen										
	0320	1′-10″ x 1′-0″ high	2 Carp	30	.533	Ea.	28	12.30		40.30	52	
	0340	2′-9″ x 1′-6″ high		30	.533		30.50	12.30		42.80	54.50	
	0360	3′-4″ x 2′-0″ high	↓	30	.533	↓	37	12.30		49.30	61.50	
	1600	Double-hung, combination, storm & screen										
	1700	Custom, clear anodic coating, 2′-0″ x 3′-5″ high	2 Carp	30	.533	Ea.	73	12.30		85.30	101	
	1720	2′-6″ x 5′-0″ high		28	.571		97.50	13.20		110.70	130	
	1740	4′-0″ x 6′-0″ high		25	.640		206	14.80		220.80	252	
	1800	White painted, 2′-0″ x 3′-5″ high		30	.533		86.50	12.30		98.80	117	
	1820	2′-6″ x 5′-0″ high		28	.571		139	13.20		152.20	176	
	1840	4′-0″ x 6′-0″ high		25	.640		250	14.80		264.80	300	
	2000	Average quality, clear anodic coating, 2′-0″ x 3′-5″ high		30	.533		74	12.30		86.30	103	
	2020	2′-6″ x 5′-0″ high		28	.571		89.50	13.20		102.70	121	
	2040	4′-0″ x 6′-0″ high		25	.640		110	14.80		124.80	146	
	2400	White painted, 2′-0″ x 3′-5″ high		30	.533		72.50	12.30		84.80	101	
	2420	2′-6″ x 5′-0″ high		28	.571		80.50	13.20		93.70	111	
	2440	4′-0″ x 6′-0″ high		25	.640		88	14.80		102.80	122	
	2600	Mill finish, 2′-0″ x 3′-5″ high		30	.533		66.50	12.30		78.80	94	
	2620	2′-6″ x 5′-0″ high		28	.571		74	13.20		87.20	104	
	2640	4′-0″ x 6-8″ high	↓	25	.640	↓	83	14.80		97.80	116	
	4000	Picture window, storm, 1 lite, white or bronze finish										
	4020	4′-6″ x 4′-6″ high	2 Carp	25	.640	Ea.	111	14.80		125.80	147	
	4040	5′-8″ x 4′-6″ high		20	.800		126	18.50		144.50	171	
	4400	Mill finish, 4′-6″ x 4′-6″ high		25	.640		111	14.80		125.80	147	
	4420	5′-8″ x 4′-6″ high	↓	20	.800	↓	126	18.50		144.50	171	
	4600	3 lite, white or bronze finish										
	4620	4′-6″ x 4′-6″ high	2 Carp	25	.640	Ea.	135	14.80		149.80	174	
	4640	5′-8″ x 4′-6″ high		20	.800		150	18.50		168.50	198	
	4800	Mill finish, 4′-6″ x 4′-6″ high		25	.640		119	14.80		133.80	156	
	4820	5′-8″ x 4′-6″ high	↓	20	.800		126	18.50		144.50	171	
	5000	Sliding glass door, storm 6′ x 6′-8″, standard	1 Glaz	2	4		680	91		771	900	
	5100	Economy	″	2	4	↓	297	91		388	475	
	6000	Sliding window, storm, 2 lite, white or bronze finish										
	6020	3′-4″ x 2′-7″ high	2 Carp	28	.571	Ea.	92.50	13.20		105.70	125	
	6040	4′-4″ x 3′-3″ high		25	.640		126	14.80		140.80	164	
	6060	5′-4″ x 6′-0″ high	↓	20	.800	↓	202	18.50		220.50	255	
	6400	3 lite, white or bronze finish										

	08580 | Special Function Windows	CREW	DAILY OUTPUT	LABOR-HOURS	UNIT	2004 BARE COSTS MAT.	LABOR	EQUIP.	TOTAL	TOTAL INCL O&P	
900	6420 4'-4" x 3'-3" high	2 Carp	25	.640	Ea.	146	14.80		160.80	186	900
	6440 5'-4" x 6'-0" high		20	.800		264	18.50		282.50	320	
	6460 6'-0" x 6'-0" high		18	.889		265	20.50		285.50	325	
	6800 Mill finish, 4'-4" x 3'-3" high		25	.640		126	14.80		140.80	164	
	6820 5'-4" x 6'-0" high		20	.800		265	18.50		283.50	325	
	6840 6'-0" x 6-0" high	↓	18	.889	↓	273	20.50		293.50	335	
	9000 Magnetic interior storm window										
	9100 3/16" plate glass	1 Glaz	107	.075	S.F.	3.98	1.70		5.68	7.20	

	08590 | Window Restoration & Replace	CREW	DAILY OUTPUT	LABOR-HOURS	UNIT	MAT.	LABOR	EQUIP.	TOTAL	TOTAL INCL O&P	
600	0010 SOLID VINYL REPLACEMENT WINDOWS R08550-200										600
	0020 Double hung, insulated glass, up to 83 united inches	2 Carp	8	2	Ea.	196	46		242	295	
	0040 84 to 93		8	2		219	46		265	320	
	0060 94 to 101		6	2.667		249	61.50		310.50	380	
	0080 102 to 111		6	2.667		276	61.50		337.50	410	
	0100 112 to 120		6	2.667	↓	310	61.50		371.50	445	
	0120 For each united inch over 120 , add		800	.020	Inch	3.28	.46		3.74	4.39	
	0140 Casement windows, one operating sash , 42 to 60 united inches		8	2	Ea.	170	46		216	266	
	0160 61 to 70		8	2		193	46		239	291	
	0180 71 to 80		8	2		210	46		256	310	
	0200 81 to 96		8	2		223	46		269	325	
	0220 Two operating sash, 58 to 78 united inches		8	2		340	46		386	455	
	0240 79 to 88		8	2		365	46		411	480	
	0260 89 to 98		8	2		395	46		441	515	
	0280 99 to 108		6	2.667		415	61.50		476.50	560	
	0300 109 to 121		6	2.667		445	61.50		506.50	595	
	0320 Three operating sash, 73 to 108 united inches		8	2		535	46		581	670	
	0340 109 to 118		8	2		565	46		611	700	
	0360 119 to 128		6	2.667		580	61.50		641.50	745	
	0380 129 to 138		6	2.667		620	61.50		681.50	785	
	0400 139 to 156		6	2.667		655	61.50		716.50	825	
	0420 Four operating sash, 98 to 118 united inches		8	2		770	46		816	925	
	0440 119 to 128		8	2		825	46		871	990	
	0460 129 to 138		6	2.667		875	61.50		936.50	1,075	
	0480 139 to 148		6	2.667		920	61.50		981.50	1,100	
	0500 149 to 168		6	2.667		980	61.50		1,041.50	1,175	
	0520 169 to 178	↓	6	2.667		1,050	61.50		1,111.50	1,275	
	0540 For venting unit to fixed unit, deduct					15.50			15.50	17.05	
	0560 Fixed picture window, up to 63 united inches	2 Carp	8	2		123	46		169	214	
	0580 64 to 83		8	2		146	46		192	239	
	0600 84 to 101		8	2	↓	187	46		233	285	
	0620 For each united inch over 101, add	↓	900	.018	Inch	2.25	.41		2.66	3.18	
	0640 Picture window opt., low E glazing, up to 101 united inches				Ea.	17.50			17.50	19.25	
	0660 102 to 124					22.50			22.50	25	
	0680 124 and over					34			34	37.50	
	0700 Options, low E glazing, up to 101 united inches					8.75			8.75	9.65	
	0720 102 to 124					11.25			11.25	12.40	
	0740 124 and over					17			17	18.70	
	0760 Muntins, between glazing, square, per lite					1.65			1.65	1.82	
	0780 Diamond shape, per full or partial diamond				↓	2.60			2.60	2.86	
	0800 Celluose fiber insulation, poured into sash balance cavity	1 Carp	36	.222	C.F.	.48	5.15		5.63	9.25	
	0820 Silicone caulking at perimeter	"	800	.010	L.F.	.13	.23		.36	.53	

08610 | Roof Windows		CREW	DAILY OUTPUT	LABOR-HOURS	UNIT	2004 BARE COSTS MAT.	LABOR	EQUIP.	TOTAL	TOTAL INCL O&P	
600	0010 METAL ROOF WINDOW Fixed, high perf tmpd glazing, 46" x 21-1/2"	1 Carp	8	1	Ea.	216	23		239	277	600
	0100 46" x 28"		8	1		250	23		273	315	
	0125 57" x 44"		6	1.333		320	31		351	410	
	0130 72" x 28"		7	1.143		320	26.50		346.50	400	
	0150 Venting, high performance tempered glazing, 46" x 21-1/2"		8	1		310	23		333	380	
	0175 46" x 28"		8	1		298	23		321	370	
	0200 57" x 44"		6	1.333		430	31		461	530	
	0500 Flashing set for shingled roof, 46" x 21-1/2"		5	1.600		31	37		68	96.50	
	0525 46" x 28"		5	1.600		32.50	37		69.50	98	
	0550 57" x 44"		5	1.600		38	37		75	104	
	0560 72" x 28"		6	1.333		38	31		69	94	
	0575 Flashing set for low pitched roof, 46" x 21-1/2"		5	1.600		136	37		173	213	
	0600 46" x 28"		5	1.600		139	37		176	216	
	0625 57" x 44"		5	1.600		158	37		195	237	
	0650 Flashing set for tile roof 46" x 21-1/2"		5	1.600		80.50	37		117.50	151	
	0675 46" x 28"		5	1.600		81	37		118	152	
	0700 57" x 44"	↓	5	1.600	↓	92.50	37		129.50	165	

08620 | Unit Skylights		CREW	DAILY OUTPUT	LABOR-HOURS	UNIT	MAT.	LABOR	EQUIP.	TOTAL	TOTAL INCL O&P	
800	0010 SKYLIGHT Plastic domes, flush or curb mounted, ten or										800
	0100 more units, curb not included										
	0300 Nominal size under 10 S.F., double	G-3	130	.246	S.F.	18.30	5.25		23.55	29	
	0400 Single		160	.200		13.40	4.25		17.65	22	
	0600 10 S.F. to 20 S.F., double		315	.102		16.20	2.16		18.36	21.50	
	0700 Single		395	.081		8.25	1.72		9.97	11.95	
	0900 20 S.F. to 30 S.F., double		395	.081		14.70	1.72		16.42	19.05	
	1000 Single		465	.069		10.35	1.46		11.81	13.85	
	1200 30 S.F. to 65 S.F., double		465	.069		11.10	1.46		12.56	14.65	
	1300 Single	↓	610	.052		14.10	1.11		15.21	17.40	
	1500 For insulated 4" curbs, double, add					25%					
	1600 Single, add				↓	30%					
	1800 For integral insulated 9" curbs, double, add					30%					
	1900 Single, add					40%					
	2120 Ventilating insulated plexiglass dome with										
	2130 curb mounting, 36" x 36"	G-3	12	2.667	Ea.	370	56.50		426.50	500	
	2150 52" x 52"		12	2.667		555	56.50		611.50	705	
	2160 28" x 52"		10	3.200		430	68		498	590	
	2170 36" x 52"	↓	10	3.200		465	68		533	630	
	2180 For electric opening system, add					277			277	305	
	2210 Operating skylight, with thermopane glass, 24" x 48"	G-3	10	3.200		530	68		598	700	
	2220 32" x 48"	"	9	3.556	↓	555	75.50		630.50	735	
	2310 Non venting insulated plexiglass dome skylight with										
	2320 Flush mount 22" x 46"	G-3	15.23	2.101	Ea.	300	44.50		344.50	405	
	2330 30" x 30"		16	2		277	42.50		319.50	375	
	2340 46" x 46"		13.91	2.300		510	49		559	645	
	2350 Curb mount 22" x 46"		15.23	2.101		264	44.50		308.50	365	
	2360 30" x 30"		16	2		252	42.50		294.50	350	
	2370 46" x 46"		13.91	2.300		475	49		524	605	
	2381 Non-insulated flush mount 22" x 46"		15.23	2.101		203	44.50		247.50	298	
	2382 30" x 30"		16	2		184	42.50		226.50	275	
	2383 46" x 46"		13.91	2.300		345	49		394	465	
	2384 Curb mount 22" x 46"		15.23	2.101		172	44.50		216.50	264	
	2385 30" x 30"	↓	16	2	↓	166	42.50		208.50	255	

		08710 | Door Hardware	Crew	Daily Output	Labor-Hours	Unit	2004 Bare Costs Mat.	Labor	Equip.	Total	Total Incl O&P
150	0010	**AVERAGE** Percentage for hardware, total job cost, minimum									.75%
	0050	Maximum									3.50%
	0500	Total hardware for building, average distribution					85%	15%			
	1000	Door hardware, apartment, interior				Door	115			115	126
	2100	Pocket door				Ea.	115			115	126
	4000	Door knocker, bright brass	1 Carp	32	.250		39	5.80		44.80	53
	4100	Mail slot, bright brass, 2" x 11"	"	25	.320		51.50	7.40		58.90	69
	4200	Peep hole, add to price of door					14.25			14.25	15.70
340	0010	**DOORSTOPS** Holder and bumper, floor or wall	1 Carp	32	.250	Ea.	28	5.80		33.80	41
	1300	Wall bumper, 4" diameter, with rubber pad, aluminum		32	.250		8.50	5.80		14.30	19.15
	1600	Door bumper, floor type, aluminum		32	.250		4.44	5.80		10.24	14.70
	1900	Plunger type, door mounted		32	.250		24	5.80		29.80	36
400	0010	**ENTRANCE LOCKS** Cylinder, grip handle, deadlocking latch	1 Carp	9	.889	Ea.	108	20.50		128.50	154
	0020	Deadbolt		8	1		131	23		154	183
	0100	Push and pull plate, dead bolt		8	1		125	23		148	176
	0900	For handicapped lever, add					136			136	150
520	0010	**HINGES** Full mortise, avg. freq., steel base, 4-1/2" x 4-1/2", USP (R08700-100)				Pr.	19.35			19.35	21.50
	0100	5" x 5", USP					31.50			31.50	34.50
	0200	6" x 6", USP					67			67	74
	0400	Brass base, 4-1/2" x 4-1/2", US10					40			40	44
	0500	5" x 5", US10					56.50			56.50	62.50
	0600	6" x 6", US10					96.50			96.50	106
	0800	Stainless steel base, 4-1/2" x 4-1/2", US32					61.50			61.50	67.50
	0900	For non removable pin, add				Ea.	2.23			2.23	2.45
	0910	For floating pin, driven tips, add					2.50			2.50	2.75
	0930	For hospital type tip on pin, add					10.80			10.80	11.90
	0940	For steeple type tip on pin, add					9.45			9.45	10.40
	0950	Full mortise, high frequency, steel base, 3-1/2" x 3-1/2", US26D				Pr.	17.90			17.90	19.70
	1000	4-1/2" x 4-1/2", USP					40.50			40.50	44.50
	1100	5" x 5", USP					44.50			44.50	49
	1200	6" x 6", USP					107			107	117
	1400	Brass base, 3-1/2" x 3-1/2", US4					37			37	40.50
	1430	4-1/2" x 4-1/2", US10					63.50			63.50	69.50
	1500	5" x 5", US10					94			94	103
	1600	6" x 6", US10					136			136	150
	1800	Stainless steel base, 4-1/2" x 4-1/2", US32					100			100	110
	1930	For hospital type tip on pin, add				Ea.	6.70			6.70	7.35
	1950	Full mortise, low frequency, steel base, 3-1/2" x 3-1/2", US26D				Pr.	9.45			9.45	10.40
	2000	4-1/2" x 4-1/2", USP					8.85			8.85	9.70
	2100	5" x 5", USP					23.50			23.50	26
	2200	6" x 6", USP					47.50			47.50	52
	2300	4-1/2" x 4-1/2", US3					13.95			13.95	15.35
	2310	5" x 5", US3					34.50			34.50	37.50
	2400	Brass bass, 4-1/2" x 4-1/2", US10					33.50			33.50	36.50
	2500	5" x 5", US10					50.50			50.50	55.50
	2800	Stainless steel base, 4-1/2" x 4-1/2", US32					57.50			57.50	63
550	0010	**KICK PLATE** 6" high, for 3' door, stainless steel	1 Carp	15	.533	Ea.	25.50	12.30		37.80	49
	0500	Bronze	"	15	.533	"	31.50	12.30		43.80	55.50
650	0010	**LOCKSET** Standard duty, cylindrical, with sectional trim									
	0020	Non-keyed, passage	1 Carp	12	.667	Ea.	40	15.40		55.40	69.50
	0100	Privacy		12	.667		48	15.40		63.40	79
	0400	Keyed, single cylinder function		10	.800		68.50	18.50		87	107
	0500	Lever handled, keyed, single cylinder function		10	.800		121	18.50		139.50	165
	1700	Residential, interior door, minimum		16	.500		13.40	11.55		24.95	34.50
	1720	Maximum		8	1		35.50	23		58.50	78
	1800	Exterior, minimum		14	.571		30	13.20		43.20	55.50

08700 | Hardware

		08710 | Door Hardware	CREW	DAILY OUTPUT	LABOR-HOURS	UNIT	2004 BARE COSTS MAT.	LABOR	EQUIP.	TOTAL	TOTAL INCL O&P	
650	1810	Average	1 Carp	8	1	Ea.	63	23		86	109	650
	1820	Maximum	↓	8	1	↓	125	23		148	177	

		08720 | Weatherstripping & Seals	CREW	DAILY OUTPUT	LABOR-HOURS	UNIT	MAT.	LABOR	EQUIP.	TOTAL	TOTAL INCL O&P	
300	0010	**WEATHERSTRIPPING** Window, double hung, 3′ x 5′, zinc	1 Carp	7.20	1.111	Opng.	10.70	25.50		36.20	55.50	300
	0100	Bronze		7.20	1.111		21.50	25.50		47	67	
	0200	Vinyl V strip		7	1.143		3.65	26.50		30.15	49	
	0500	As above but heavy duty, zinc		4.60	1.739		13.80	40		53.80	83	
	0600	Bronze		4.60	1.739		24	40		64	94.50	
	1000	Doors, wood frame, interlocking, for 3′ x 7′ door, zinc		3	2.667		12.35	61.50		73.85	119	
	1100	Bronze		3	2.667		19.45	61.50		80.95	127	
	1300	6′ x 7′ opening, zinc		2	4		13.50	92.50		106	172	
	1400	Bronze		2	4	↓	25.50	92.50		118	185	
	1500	Vinyl V strip	↓	6.40	1.250	Ea.	6.95	29		35.95	56.50	
	1700	Wood frame, spring type, bronze										
	1800	3′ x 7′ door	1 Carp	7.60	1.053	Opng.	15.80	24.50		40.30	59	
	1900	6′ x 7′ door		7	1.143		16.60	26.50		43.10	63.50	
	1920	Felt, 3′ x 7′ door		14	.571		1.99	13.20		15.19	24.50	
	1930	6′ x 7′ door		13	.615		2.15	14.20		16.35	26.50	
	1950	Rubber, 3′ x 7′ door		7.60	1.053		4.30	24.50		28.80	46	
	1960	6′ x 7′ door	↓	7	1.143	↓	4.91	26.50		31.41	50.50	
	2200	Metal frame, spring type, bronze										
	2300	3′ x 7′ door	1 Carp	3	2.667	Opng.	26.50	61.50		88	134	
	2400	6′ x 7′ door	"	2.50	3.200	"	37	74		111	166	
	2500	For stainless steel, spring type, add					133%					
	2700	Metal frame, extruded sections, 3′ x 7′ door, aluminum	1 Carp	2	4	Opng.	36	92.50		128.50	197	
	2800	Bronze		2	4		90.50	92.50		183	257	
	3100	6′ x 7′ door, aluminum		1.20	6.667		45.50	154		199.50	310	
	3200	Bronze	↓	1.20	6.667	↓	107	154		261	380	
	3500	Threshold weatherstripping										
	3650	Door sweep, flush mounted, aluminum	1 Carp	25	.320	Ea.	10.65	7.40		18.05	24.50	
	3700	Vinyl		25	.320	"	12.55	7.40		19.95	26.50	
	4000	Astragal for double doors, aluminum		4	2	Opng.	17.65	46		63.65	98	
	4100	Bronze		4	2	"	28.50	46		74.50	110	
	5000	Garage door bottom weatherstrip, 12′ aluminum, clear		14	.571	Ea.	17	13.20		30.20	41	
	5010	Bronze		14	.571		65	13.20		78.20	94	
	5050	Bottom protection, 12′ aluminum, clear		14	.571		20	13.20		33.20	44.50	
	5100	Bronze	↓	14	.571	↓	80.50	13.20		93.70	111	
800	0010	**THRESHOLD** 3′ long door saddles, aluminum	1 Carp	48	.167	L.F.	3.43	3.85		7.28	10.30	800
	0100	Aluminum, 8″ wide, 1/2″ thick		12	.667	Ea.	29	15.40		44.40	58	
	0500	Bronze		60	.133	L.F.	27	3.08		30.08	35.50	
	0600	Bronze, panic threshold, 5″ wide, 1/2″ thick		12	.667	Ea.	57	15.40		72.40	88.50	
	0700	Rubber, 1/2″ thick, 5-1/2″ wide		20	.400		30	9.25		39.25	48.50	
	0800	2-3/4″ wide	↓	20	.400	↓	13.65	9.25		22.90	31	

		08750 | Window Hardware	CREW	DAILY OUTPUT	LABOR-HOURS	UNIT	MAT.	LABOR	EQUIP.	TOTAL	TOTAL INCL O&P	
400	0010	**WINDOW HARDWARE**										400
	1000	Handles, surface mounted, aluminum	1 Carp	24	.333	Ea.	1.84	7.70		9.54	15.05	
	1020	Brass		24	.333		2.14	7.70		9.84	15.40	
	1040	Chrome		24	.333		1.96	7.70		9.66	15.20	
	1500	Recessed, aluminum		12	.667		1.07	15.40		16.47	27	
	1520	Brass		12	.667		1.18	15.40		16.58	27.50	
	1540	Chrome		12	.667		1.12	15.40		16.52	27	
	2000	Latches, aluminum		20	.400		1.53	9.25		10.78	17.40	
	2020	Brass		20	.400		1.84	9.25		11.09	17.70	
	2040	Chrome	↓	20	.400	↓	1.73	9.25		10.98	17.60	

08700 | Hardware

08770 | Door/Window Accessories

		08770 Door/Window Accessories	Crew	Daily Output	Labor-Hours	Unit	2004 Bare Costs Mat.	Labor	Equip.	Total	Total Incl O&P
550	0010	**DETECTION SYSTEMS** See division 13851									
560	0010	**DOOR ACCESSORIES**									
	1000	Knockers, brass, standard	1 Carp	16	.500	Ea.	36	11.55		47.55	59
	1100	Deluxe		10	.800		111	18.50		129.50	154
	4000	Security chain, standard		18	.444		6.20	10.25		16.45	24
	4100	Deluxe		18	.444		37	10.25		47.25	58.50

08800 | Glazing

08810 | Glass

		08810 Glass	Crew	Daily Output	Labor-Hours	Unit	2004 Bare Costs Mat.	Labor	Equip.	Total	Total Incl O&P
260	0010	**FLOAT GLASS** 3/16" thick, clear, plain	2 Glaz	130	.123	S.F.	4.02	2.80		6.82	9.05
	0200	Tempered, clear		130	.123		4.77	2.80		7.57	9.85
	0300	Tinted		130	.123		6	2.80		8.80	11.20
	0600	1/4" thick, clear, plain		120	.133		4.83	3.03		7.86	10.30
	0700	Tinted		120	.133		4.76	3.03		7.79	10.25
	0800	Tempered, clear		120	.133		5.95	3.03		8.98	11.50
	0900	Tinted		120	.133		8.20	3.03		11.23	14
	1600	3/8" thick, clear, plain		75	.213		7.75	4.85		12.60	16.50
	1700	Tinted		75	.213		9.50	4.85		14.35	18.45
	1800	Tempered, clear		75	.213		11.95	4.85		16.80	21
	1900	Tinted		75	.213		14.70	4.85		19.55	24
	2200	1/2" thick, clear, plain		55	.291		15.35	6.60		21.95	28
	2300	Tinted		55	.291		16.65	6.60		23.25	29.50
	2400	Tempered, clear		55	.291		17.90	6.60		24.50	30.50
	2500	Tinted		55	.291		22	6.60		28.60	35.50
	2800	5/8" thick, clear, plain		45	.356		16.65	8.10		24.75	31.50
	2900	Tempered, clear		45	.356		19.05	8.10		27.15	34.50
	8900	For low emissivity coating for 3/16" and 1/4" only, add to above					15%				
300	0010	**GLAZING VARIABLES**									
	0600	For glass replacement, add				S.F.		100%			
	0700	For gasket settings, add				L.F.	3.57			3.57	3.93
	0900	For sloped glazing, add				S.F.		25%			
	2000	Fabrication, polished edges, 1/4" thick				Inch	.31			.31	.34
	2100	1/2" thick					.78			.78	.86
	2500	Mitered edges, 1/4" thick					.78			.78	.86
	2600	1/2" thick					1.26			1.26	1.39
460	0010	**INSULATING GLASS** 2 lites 1/8" float, 1/2" thk, under 15 S.F.									
	0100	Tinted	2 Glaz	95	.168	S.F.	10.15	3.83		13.98	17.45
	0280	Double glazed, 5/8" thk unit, 3/16" float, 15-30 S.F., clear		90	.178		8.35	4.04		12.39	15.85
	0400	1" thk, dbl. glazed, 1/4" float, 30-70 S.F., clear		75	.213		11.70	4.85		16.55	21
	0500	Tinted		75	.213		14.30	4.85		19.15	23.50
	2000	Both lites, light & heat reflective		85	.188		19.05	4.28		23.33	28
	2500	Heat reflective, film inside, 1" thick unit, clear		85	.188		16.65	4.28		20.93	25.50
	2600	Tinted		85	.188		17.95	4.28		22.23	27
	3000	Film on weatherside, clear, 1/2" thick unit		95	.168		11.90	3.83		15.73	19.40
	3100	5/8" thick unit		90	.178		14.65	4.04		18.69	23
	3200	1" thick unit		85	.188		16.40	4.28		20.68	25

08810 | Glass		CREW	DAILY OUTPUT	LABOR-HOURS	UNIT	2004 BARE COSTS MAT.	LABOR	EQUIP.	TOTAL	TOTAL INCL O&P	
850	0010 **WINDOW GLASS** Clear float, stops, putty bed, 1/8" thick	2 Glaz	480	.033	S.F.	3.31	.76		4.07	4.89	850
	0500 3/16" thick, clear		480	.033		4.07	.76		4.83	5.75	
	0600 Tinted		480	.033		4.59	.76		5.35	6.30	
	0700 Tempered	↓	480	.033	↓	5.55	.76		6.31	7.35	

08830 | Mirrors		CREW	DAILY OUTPUT	LABOR-HOURS	UNIT	MAT.	LABOR	EQUIP.	TOTAL	TOTAL INCL O&P	
100	0010 **MIRRORS** No frames, wall type, 1/4" plate glass, polished edge										100
	0100 Up to 5 S.F.	2 Glaz	125	.128	S.F.	6.40	2.91		9.31	11.85	
	0200 Over 5 S.F.		160	.100		6.20	2.28		8.48	10.60	
	0500 Door type, 1/4" plate glass, up to 12 S.F.		160	.100		6.70	2.28		8.98	11.10	
	1000 Float glass, up to 10 S.F., 1/8" thick		160	.100		3.86	2.28		6.14	8	
	1100 3/16" thick		150	.107		4.48	2.43		6.91	8.95	
	1500 12" x 12" wall tiles, square edge, clear		195	.082		1.59	1.87		3.46	4.83	
	1600 Veined		195	.082		4.17	1.87		6.04	7.65	
	2010 Bathroom, unframed, laminated	↓	160	.100	↓	10.80	2.28		13.08	15.65	

Division 9 Finishes

Estimating Tips

General

- Room Finish Schedule: A complete set of plans should contain a room finish schedule. If one is not available, it would be well worth the time and effort to put one together. A room finish schedule should contain the room number, room name (for clarity), floor materials, base materials, wainscot materials, wainscot height, wall materials (for each wall), ceiling materials, ceiling height and special instructions.
- Surplus Finishes: Review the specifications to determine if there is any requirement to provide certain amounts of extra materials for the owner's maintenance department. In some cases the owner may require a substantial amount of materials, especially when it is a special order item or long lead time item.

09200 Plaster & Gypsum Board

- Lath is estimated by the square yard for both gypsum and metal lath, plus usually 5% allowance for waste. Furring, channels and accessories are measured by the linear foot. An extra foot should be allowed for each accessory miter or stop.
- Plaster is also estimated by the square yard. Deductions for openings vary by preference, from zero deduction to 50% of all openings over 2 feet in width. Some estimators deduct a percentage of the total yardage for openings. The estimator should allow one extra square foot for each linear foot of horizontal interior or exterior angle located below the ceiling level. Also, double the areas of small radius work.
- Each room should be measured, perimeter times maximum wall height. Floors and ceiling areas are equal to length times width.
- Drywall accessories, studs, track, and acoustical caulking are all measured by the linear foot. Drywall taping is figured by the square foot. Gypsum wallboard is estimated by the square foot. No material deductions should be made for door or window openings under 32 S.F. Coreboard can be obtained in a 1″ thickness for solid wall and shaft work. Additions should be made to price out the inside or outside corners.
- Different types of partition construction should be listed separately on the quantity sheets. There may be walls with studs of various widths, double studded, and similar or dissimilar surface materials. Shaft work is usually different construction from surrounding partitions requiring separate quantities and pricing of the work.

09300 Tile
09400 Terrazzo

- Tile and terrazzo areas are taken off on a square foot basis. Trim and base materials are measured by the linear foot. Accent tiles are listed per each. Two basic methods of installation are used. Mud set is approximately 30% more expensive than the thin set. In terrazzo work, be sure to include the linear footage of embedded decorative strips, grounds, machine rubbing and power cleanup.

09600 Flooring

- Wood flooring is available in strip, parquet, or block configuration. The latter two types are set in adhesives with quantities estimated by the square foot. The laying pattern will influence labor costs and material waste. In addition to the material and labor for laying wood floors, the estimator must make allowances for sanding and finishing these areas unless the flooring is prefinished.
- Most of the various types of flooring are all measured on a square foot basis. Base is measured by the linear foot. If adhesive materials are to be quantified, they are estimated at a specified coverage rate by the gallon depending upon the specified type and the manufacturer's recommendations.
- Sheet flooring is measured by the square yard. Roll widths vary, so consideration should be given to use the most economical width, as waste must be figured into the total quantity. Consider also the installation methods available, direct glue down or stretched.

09700 Wall Finishes

- Wall coverings are estimated by the square foot. The area to be covered is measured, length by height of wall above baseboards, to calculate the square footage of each wall. This figure is divided by the number of square feet in the single roll which is being used. Deduct, in full, the areas of openings such as doors and windows. Where a pattern match is required allow 25%-30% waste. One gallon of paste should be sufficient to hang 12 single rolls of light to medium weight paper.

09800 Acoustical Treatment

- Acoustical systems fall into several categories. The takeoff of these materials should be by the square foot of area with a 5% allowance for waste. Do not forget about scaffolding, if applicable, when estimating these systems.

09900 Paints & Coatings

- A major portion of the work in painting involves surface preparation. Be sure to include cleaning, sanding, filling and masking costs in the estimate.
- Painting is one area where bids vary to a greater extent than almost any other section of a project. This arises from the many methods of measuring surfaces to be painted. The estimator should check the plans and specifications carefully to be sure of the required number of coats.
- Protection of adjacent surfaces is not included in painting costs. When considering the method of paint application, an important factor is the amount of protection and masking required. These must be estimated separately and may be the determining factor in choosing the method of application.

Reference Numbers

Reference numbers are shown in bold squares at the beginning of some major classifications. These numbers refer to related items in the Reference Section. The reference information may be an estimating procedure, an alternate pricing method or technical information.

Note: Not all subdivisions listed here necessarily appear in this publication.

09050 | Basic Material Finishes and Methods

09060 | Selective Demolition

			CREW	DAILY OUTPUT	LABOR-HOURS	UNIT	2004 BARE COSTS MAT.	LABOR	EQUIP.	TOTAL	TOTAL INCL O&P
110	0010	**SELECTIVE DEMOLITION, CEILINGS**									
	0200	Drywall, furred and nailed	2 Clab	800	.020	S.F.		.34		.34	.57
	1000	Plaster, lime and horse hair, on wood lath, incl. lath		700	.023			.39		.39	.66
	1200	Suspended ceiling, mineral fiber, 2'x2' or 2'x4'		1,500	.011			.18		.18	.31
	1250	On suspension system, incl. system		1,200	.013			.23		.23	.38
	1500	Tile, wood fiber, 12" x 12", glued		900	.018			.30		.30	.51
	1540	Stapled		1,500	.011			.18		.18	.31
	2000	Wood, tongue and groove, 1" x 4"		1,000	.016			.27		.27	.46
	2040	1" x 8"		1,100	.015			.25		.25	.42
	2400	Plywood or wood fiberboard, 4' x 8' sheets		1,200	.013			.23		.23	.38
120	0010	**SELECTIVE DEMOLITION, FLOORING**									
	0200	Brick with mortar	2 Clab	475	.034	S.F.		.57		.57	.97
	0400	Carpet, bonded, including surface scraping		2,000	.008			.14		.14	.23
	0480	Tackless		9,000	.002			.03		.03	.05
	0800	Resilient, sheet goods		1,400	.011			.19		.19	.33
	0900	Vinyl composition tile, 12" x 12"		1,000	.016			.27		.27	.46
	2000	Tile, ceramic, thin set		675	.024			.40		.40	.68
	2020	Mud set		625	.026			.43		.43	.73
	3000	Wood, block, on end	1 Carp	400	.020			.46		.46	.78
	3200	Parquet		450	.018			.41		.41	.70
	3400	Strip flooring, interior, 2-1/4" x 25/32" thick		325	.025			.57		.57	.97
	3500	Exterior, porch flooring, 1" x 4"		220	.036			.84		.84	1.43
	3800	Subfloor, tongue and groove, 1" x 6"		325	.025			.57		.57	.97
	3820	1" x 8"		430	.019			.43		.43	.73
	3840	1" x 10"		520	.015			.36		.36	.60
	4000	Plywood, nailed		600	.013			.31		.31	.52
	4100	Glued and nailed		400	.020			.46		.46	.78
130	0010	**SELECTIVE DEMOLITION, WALLS AND PARTITIONS**									
	1000	Drywall, nailed	1 Clab	1,000	.008	S.F.		.14		.14	.23
	1500	Fiberboard, nailed	"	900	.009			.15		.15	.26
	2200	Metal or wood studs, finish 2 sides, fiberboard	B-1	520	.046			.81		.81	1.38
	2250	Lath and plaster		260	.092			1.62		1.62	2.75
	2300	Plasterboard (drywall)		520	.046			.81		.81	1.38
	2350	Plywood		450	.053			.94		.94	1.59
	3000	Plaster, lime and horsehair, on wood lath	1 Clab	400	.020			.34		.34	.57
	3020	On metal lath	"	335	.024			.40		.40	.69

09100 | Metal Support Assemblies

09110 | Non-Load Bearing Wall Framing

			CREW	DAILY OUTPUT	LABOR-HOURS	UNIT	2004 BARE COSTS MAT.	LABOR	EQUIP.	TOTAL	TOTAL INCL O&P
100	0010	**METAL STUDS AND TRACK**									
	1600	Non-load bearing, galv, 8' high, 25 ga. 1-5/8" wide, 16" O.C.	1 Carp	619	.013	S.F.	.19	.30		.49	.72
	1610	24" O.C.		950	.008		.14	.19		.33	.49
	1620	2-1/2" wide, 16" O.C.		613	.013		.21	.30		.51	.74
	1630	24" O.C.		938	.009		.16	.20		.36	.50
	1640	3-5/8" wide, 16" O.C.		600	.013		.22	.31		.53	.77
	1650	24" O.C.		925	.009		.17	.20		.37	.53
	1660	4" wide, 16" O.C.		594	.013		.25	.31		.56	.81

09110 | Non-Load Bearing Wall Framing

			CREW	DAILY OUTPUT	LABOR-HOURS	UNIT	2004 BARE COSTS MAT.	LABOR	EQUIP.	TOTAL	TOTAL INCL O&P
100	1670	24" O.C.	1 Carp	925	.009	S.F.	.19	.20		.39	.55
	1680	6" wide, 16" O.C.		588	.014		.36	.31		.67	.92
	1690	24" O.C.		906	.009		.27	.20		.47	.64
	1700	20 ga. studs, 1-5/8" wide, 16" O.C.		494	.016		.29	.37		.66	.95
	1710	24" O.C.		763	.010		.22	.24		.46	.65
	1720	2-1/2" wide, 16" O.C.		488	.016		.32	.38		.70	.99
	1730	24" O.C.		750	.011		.24	.25		.49	.69
	1740	3-5/8" wide, 16" O.C.		481	.017		.38	.38		.76	1.07
	1750	24" O.C.		738	.011		.29	.25		.54	.74
	1760	4" wide, 16" O.C.		475	.017		.40	.39		.79	1.10
	1770	24" O.C.		738	.011		.30	.25		.55	.75
	1780	6" wide, 16" O.C.		469	.017		.53	.39		.92	1.25
	1790	24" O.C.		725	.011		.40	.25		.65	.87
	2000	Non-load bearing, galv, 10' high, 25 ga. 1-5/8" wide, 16" O.C.		495	.016		.18	.37		.55	.83
	2100	24" O.C.		760	.011		.13	.24		.37	.56
	2200	2-1/2" wide, 16" O.C.		490	.016		.20	.38		.58	.86
	2250	24" O.C.		750	.011		.15	.25		.40	.58
	2300	3-5/8" wide, 16" O.C.		480	.017		.21	.39		.60	.88
	2350	24" O.C.		740	.011		.16	.25		.41	.59
	2400	4" wide, 16" O.C.		475	.017		.24	.39		.63	.92
	2450	24" O.C.		740	.011		.18	.25		.43	.61
	2500	6" wide, 16" O.C.		470	.017		.34	.39		.73	1.04
	2550	24" O.C.		725	.011		.25	.25		.50	.70
	2600	20 ga. studs, 1-5/8" wide, 16" O.C.		395	.020		.28	.47		.75	1.09
	2650	24" O.C.		610	.013		.20	.30		.50	.74
	2700	2-1/2" wide, 16" O.C.		390	.021		.30	.47		.77	1.13
	2750	24" O.C.		600	.013		.23	.31		.54	.77
	2800	3-5/8" wide, 16" O.C.		385	.021		.36	.48		.84	1.21
	2850	24" O.C.		590	.014		.27	.31		.58	.83
	2900	4" wide, 16" O.C.		380	.021		.38	.49		.87	1.25
	2950	24" O.C.		590	.014		.28	.31		.59	.84
	3000	6" wide, 16" O.C.		375	.021		.50	.49		.99	1.39
	3050	24" O.C.		580	.014		.37	.32		.69	.95
	3060	Non-load bearing, galv, 12' high, 25 ga. 1-5/8" wide, 16" O.C.		413	.019		.17	.45		.62	.95
	3070	24" O.C.		633	.013		.13	.29		.42	.64
	3080	2-1/2" wide, 16" O.C.		408	.020		.19	.45		.64	.98
	3090	24" O.C.		625	.013		.14	.30		.44	.65
	3100	3-5/8" wide, 16" O.C.		400	.020		.20	.46		.66	1
	3110	24" O.C.		617	.013		.15	.30		.45	.67
	3120	4" wide, 16" O.C.		396	.020		.23	.47		.70	1.04
	3130	24" O.C.		617	.013		.17	.30		.47	.69
	3140	6" wide, 16" O.C.		392	.020		.33	.47		.80	1.16
	3150	24" O.C.		604	.013		.24	.31		.55	.78
	3160	20 ga. studs, 1-5/8" wide, 16" O.C.		329	.024		.26	.56		.82	1.24
	3170	24" O.C.		508	.016		.19	.36		.55	.83
	3180	2-1/2" wide, 16" O.C.		325	.025		.29	.57		.86	1.29
	3190	24" O.C.		500	.016		.21	.37		.58	.86
	3200	3-5/8" wide, 16" O.C.		321	.025		.35	.58		.93	1.36
	3210	24" O.C.		492	.016		.25	.38		.63	.92
	3220	4" wide, 16" O.C.		317	.025		.37	.58		.95	1.39
	3230	24" O.C.		492	.016		.27	.38		.65	.93
	3240	6" wide, 16" O.C.		313	.026		.48	.59		1.07	1.53
	3250	24" O.C.	↓	483	.017	↓	.35	.38		.73	1.03
	5000	Load bearing studs, see division 05410-400									

FINISHES 9

09100 | Metal Support Assemblies

		09120 | Ceiling Suspension	CREW	DAILY OUTPUT	LABOR-HOURS	UNIT	2004 BARE COSTS MAT.	LABOR	EQUIP.	TOTAL	TOTAL INCL O&P	
100	0010	**CEILING SUSPENSION SYSTEMS** For gypsum board or plaster										100
	8000	Suspended ceilings, including carriers										
	8200	1-1/2" carriers, 24" O.C. with:										
	8300	7/8" channels, 16" O.C.	1 Lath	165	.048	S.F.	.28	1.04		1.32	1.99	
	8320	24" O.C.		200	.040		.23	.86		1.09	1.64	
	8400	1-5/8" channels, 16" O.C.		155	.052		.35	1.10		1.45	2.17	
	8420	24" O.C.	↓	190	.042	↓	.27	.90		1.17	1.76	
	8600	2" carriers, 24" O.C. with:										
	8700	7/8" channels, 16" O.C.	1 Lath	155	.052	S.F.	.34	1.10		1.44	2.16	
	8720	24" O.C.		190	.042		.29	.90		1.19	1.78	
	8800	1-5/8" channels, 16" O.C.		145	.055		.40	1.18		1.58	2.35	
	8820	24" O.C.	↓	180	.044	↓	.33	.95		1.28	1.90	

		09130 | Acoustical Suspension	CREW	DAILY OUTPUT	LABOR-HOURS	UNIT	MAT.	LABOR	EQUIP.	TOTAL	TOTAL INCL O&P	
100	0010	**CEILING SUSPENSION SYSTEMS** For boards and tile										100
	0050	Class A suspension system, 15/16" T bar, 2' x 4' grid	1 Carp	800	.010	S.F.	.41	.23		.64	.84	
	0300	2' x 2' grid	"	650	.012		.51	.28		.79	1.04	
	0350	For 9/16" grid, add					.13			.13	.14	
	0360	For fire rated grid, add					.07			.07	.08	
	0370	For colored grid, add					.15			.15	.17	
	0400	Concealed Z bar suspension system, 12" module	1 Carp	520	.015		.37	.36		.73	1.01	
	0600	1-1/2" carrier channels, 4' O.C., add		470	.017		.07	.39		.46	.75	
	0650	1-1/2" x 3-1/2" channels	↓	470	.017	↓	.18	.39		.57	.87	
	0700	Carrier channels for ceilings with										
	0900	recessed lighting fixtures, add	1 Carp	460	.017	S.F.	.13	.40		.53	.83	
	5000	Wire hangers, #12 wire	"	300	.027	Ea.	.77	.62		1.39	1.90	

9 FINISHES

09200 | Plaster & Gypsum Board

		09205 | Furring & Lathing	CREW	DAILY OUTPUT	LABOR-HOURS	UNIT	2004 BARE COSTS MAT.	LABOR	EQUIP.	TOTAL	TOTAL INCL O&P	
530	0010	**FURRING** Beams & columns, 7/8" galvanized channels,										530
	0030	12" O.C.	1 Lath	155	.052	S.F.	.18	1.10		1.28	1.99	
	0050	16" O.C.		170	.047		.15	1.01		1.16	1.79	
	0070	24" O.C.		185	.043		.10	.93		1.03	1.61	
	0100	Ceilings, on steel, 7/8" channels, galvanized, 12" O.C.		210	.038		.16	.82		.98	1.50	
	0300	16" O.C.		290	.028		.15	.59		.74	1.12	
	0400	24" O.C.		420	.019		.10	.41		.51	.77	
	0600	1-5/8" channels, galvanized, 12" O.C.		190	.042		.24	.90		1.14	1.72	
	0700	16" O.C.		260	.031		.21	.66		.87	1.30	
	0900	24" O.C.		390	.021		.14	.44		.58	.87	
	1000	Walls, 7/8" channels, galvanized, 12" O.C.		235	.034		.16	.73		.89	1.36	
	1200	16" O.C.		265	.030		.15	.65		.80	1.21	
	1300	24" O.C.		350	.023		.10	.49		.59	.90	
	1500	1-5/8" channels, galvanized, 12" O.C.		210	.038		.24	.82		1.06	1.58	
	1600	16" O.C.		240	.033		.21	.71		.92	1.38	
	1800	24" O.C.	↓	305	.026	↓	.14	.56		.70	1.07	
540	0011	**GYPSUM LATH** Plain or perforated, nailed, 3/8" thick	1 Lath	765	.010	S.F.	.38	.22		.60	.78	540
	0101	1/2" thick, nailed		720	.011		.39	.24		.63	.82	
	0301	Clipped to steel studs, 3/8" thick		675	.012		.38	.25		.63	.83	
	0401	1/2" thick	↓	630	.013	↓	.39	.27		.66	.87	

09200 | Plaster & Gypsum Board

		09205 | Furring & Lathing	CREW	DAILY OUTPUT	LABOR-HOURS	UNIT	2004 BARE COSTS MAT.	LABOR	EQUIP.	TOTAL	TOTAL INCL O&P	
540	0601	Firestop gypsum base, to steel studs, 3/8" thick	1 Lath	630	.013	S.F.	.39	.27		.66	.87	540
	0701	1/2" thick		585	.014		.44	.29		.73	.95	
	0901	Foil back, to steel studs, 3/8" thick		675	.012		.40	.25		.65	.85	
	1001	1/2" thick		630	.013		.44	.27		.71	.92	
	1501	For ceiling installations, add		1,950	.004			.09		.09	.14	
	1601	For columns and beams, add		1,550	.005			.11		.11	.18	
560	0011	**METAL LATH**										560
	3601	2.5 lb. diamond painted, on wood framing, on walls	1 Lath	765	.010	S.F.	.15	.22		.37	.53	
	3701	On ceilings		675	.012		.15	.25		.40	.58	
	4201	3.4 lb. diamond painted, wired to steel framing, on walls		675	.012		.24	.25		.49	.68	
	4301	On ceilings		540	.015		.24	.32		.56	.78	
	5101	Rib lath, painted, wired to steel, on walls, 2.75 lb.		675	.012		.23	.25		.48	.66	
	5201	3.4 lb.		630	.013		.33	.27		.60	.80	
	5701	Suspended ceiling system, incl. 3.4 lb. diamond lath, painted		135	.059		.98	1.27		2.25	3.13	
	5801	Galvanized		135	.059		1.01	1.27		2.28	3.16	
700	0010	**ACCESSORIES, PLASTER** Casing bead, expanded flange, galvanized	1 Lath	2.70	2.963	C.L.F.	28	63.50		91.50	134	700
	0900	Channels, cold rolled, 16 ga., 3/4" deep, galvanized					16.45			16.45	18.10	
	1620	Corner bead, expanded bullnose, 3/4" radius, #10, galvanized	1 Lath	2.60	3.077		21	66		87	130	
	1650	#1, galvanized		2.55	3.137		35	67		102	148	
	1670	Expanded wing, 2-3/4" wide, galv. #1		2.65	3.019		22	64.50		86.50	130	
	1700	Inside corner, (corner rite) 3" x 3", painted		2.60	3.077		18.80	66		84.80	128	
	1750	Strip-ex, 4" wide, painted		2.55	3.137		16.45	67		83.45	127	
	1800	Expansion joint, 3/4" grounds, limited expansion, galv., 1 piece		2.70	2.963		68.50	63.50		132	178	
	2100	Extreme expansion, galvanized, 2 piece		2.60	3.077		120	66		186	239	
		09210 | Gypsum Plaster										
100	0010	**GYPSUM PLASTER** 80# bag, less than 1 ton				Bag	14.05			14.05	15.45	100
	0302	2 coats, no lath included, on walls	J-1	750	.053	S.F.	.35	1.06	.12	1.53	2.26	
	0402	On ceilings		660	.061		.35	1.20	.14	1.69	2.52	
	0903	3 coats, no lath included, on walls		620	.065		.49	1.28	.15	1.92	2.82	
	1002	On ceilings		560	.071		.66	1.41	.16	2.23	3.26	
	1600	For irregular or curved surfaces, add						30%				
	1800	For columns & beams, add						50%				
500	0010	**PERLITE OR VERMICULITE PLASTER** 100 lb. bags Under 200 bags				Bag	13			13	14.30	500
	0301	2 coats, no lath included, on walls	J-1	830	.048	S.F.	.36	.95	.11	1.42	2.10	
	0401	On ceilings		710	.056		.36	1.12	.13	1.61	2.39	
	0901	3 coats, no lath included, on walls		665	.060		.60	1.19	.14	1.93	2.78	
	1001	On ceilings		565	.071		.60	1.40	.16	2.16	3.16	
	1700	For irregular or curved surfaces, add to above				S.Y.		30%				
	1800	For columns and beams, add to above						50%				
	1900	For soffits, add to ceiling prices						40%				
900	0010	**THIN COAT** Plaster, 1 coat veneer, not incl. lath	J-1	3,600	.011	S.F.	.07	.22	.03	.32	.46	900
	1000	In 50 lb. bags				Bag	8.95			8.95	9.85	
		09220 | Portland Cement Plaster										
200	0011	**STUCCO** 3 coats 1" thick, float finish, with mesh, on wood frame	J-2	470	.102	S.F.	.40	2.05	.19	2.64	4.02	200
	0101	On masonry construction	J-1	495	.081		.21	1.60	.18	1.99	3.08	
	0151	2 coats, 3/4" thick, float finish, no lath incl.	"	980	.041		.22	.81	.09	1.12	1.68	
	0301	For trowel finish, add	1 Plas	1,530	.005			.11		.11	.18	
	0600	For coloring and special finish, add, minimum	J-1	685	.058	S.Y.	.36	1.16	.13	1.65	2.46	
	0700	Maximum		200	.200	"	1.26	3.96	.45	5.67	8.45	
	1001	Exterior stucco, with bonding agent, 3 coats, on walls		1,800	.022	S.F.	.33	.44	.05	.82	1.15	
	1201	Ceilings		1,620	.025		.33	.49	.06	.88	1.23	

		09220 | Portland Cement Plaster	CREW	DAILY OUTPUT	LABOR-HOURS	UNIT	2004 BARE COSTS MAT.	LABOR	EQUIP.	TOTAL	TOTAL INCL O&P	
200	1301	Beams	J-1	720	.056	S.F.	.33	1.10	.13	1.56	2.32	200
	1501	Columns	↓	900	.044		.33	.88	.10	1.31	1.93	
	1601	Mesh, painted, nailed to wood, 1.8 lb.	1 Lath	540	.015		.34	.32		.66	.89	
	1801	3.6 lb.		495	.016		.21	.35		.56	.79	
	1901	Wired to steel, painted, 1.8 lb.		477	.017		.34	.36		.70	.96	
	2101	3.6 lb.	↓	450	.018	↓	.21	.38		.59	.85	

		09250 | Gypsum Board	CREW	DAILY OUTPUT	LABOR-HOURS	UNIT	MAT.	LABOR	EQUIP.	TOTAL	TOTAL INCL O&P	
200	0010	**CEMENTITIOUS BACKERBOARD**										200
	0070	Cementitious backerboard, on floor, 3′ x 4′x 1/2″ sheets	2 Carp	525	.030	S.F.	1.04	.70		1.74	2.34	
	0080	3′ x 5′ x 1/2″ sheets		525	.030		1.04	.70		1.74	2.34	
	0090	3′ x 6′ x 1/2″ sheets		525	.030		.76	.70		1.46	2.02	
	0100	3′ x 4′x 5/8″ sheets		525	.030		1.08	.70		1.78	2.38	
	0110	3′ x 5′ x 5/8″ sheets		525	.030		1.06	.70		1.76	2.36	
	0120	3′ x 6′ x 5/8″ sheets		525	.030		1.07	.70		1.77	2.36	
	0150	On wall, 3′ x 4′x 1/2″ sheets		350	.046		1.04	1.06		2.10	2.94	
	0160	3′ x 5′ x 1/2″ sheets		350	.046		1.04	1.06		2.10	2.94	
	0170	3′ x 6′ x 1/2″ sheets		350	.046		.76	1.06		1.82	2.62	
	0180	3′ x 4′x 5/8″ sheets		350	.046		1.08	1.06		2.14	2.98	
	0190	3′ x 5′ x 5/8″ sheets		350	.046		1.06	1.06		2.12	2.96	
	0200	3′ x 6′ x 5/8″ sheets		350	.046		1.07	1.06		2.13	2.96	
	0250	On counter, 3′ x 4′x 1/2″ sheets		180	.089		1.04	2.05		3.09	4.63	
	0260	3′ x 5′ x 1/2″ sheets		180	.089		1.04	2.05		3.09	4.63	
	0270	3′ x 6′ x 1/2″ sheets		180	.089		.76	2.05		2.81	4.31	
	0300	3′ x 4′x 5/8″ sheets		180	.089		1.08	2.05		3.13	4.67	
	0310	3′ x 5′ x 5/8″ sheets		180	.089		1.06	2.05		3.11	4.65	
	0320	3′ x 6′ x 5/8″ sheets	↓	180	.089	↓	1.07	2.05		3.12	4.65	
300	0010	**BLUEBOARD** For use with thin coat										300
	0100	plaster application (see division 09210-900)										
	1000	3/8″ thick, on walls or ceilings, standard, no finish included	2 Carp	1,900	.008	S.F.	.18	.19		.37	.53	
	1100	With thin coat plaster finish		875	.018		.25	.42		.67	.99	
	1400	On beams, columns, or soffits, standard, no finish included		675	.024		.21	.55		.76	1.16	
	1450	With thin coat plaster finish		475	.034		.27	.78		1.05	1.62	
	3000	1/2″ thick, on walls or ceilings, standard, no finish included		1,900	.008		.20	.19		.39	.55	
	3100	With thin coat plaster finish		875	.018		.27	.42		.69	1.01	
	3300	Fire resistant, no finish included		1,900	.008		.20	.19		.39	.55	
	3400	With thin coat plaster finish		875	.018		.27	.42		.69	1.01	
	3450	On beams, columns, or soffits, standard, no finish included		675	.024		.23	.55		.78	1.18	
	3500	With thin coat plaster finish		475	.034		.30	.78		1.08	1.65	
	3700	Fire resistant, no finish included		675	.024		.23	.55		.78	1.18	
	3800	With thin coat plaster finish		475	.034		.30	.78		1.08	1.65	
	5000	5/8″ thick, on walls or ceilings, fire resistant, no finish included		1,900	.008		.21	.19		.40	.56	
	5100	With thin coat plaster finish		875	.018		.28	.42		.70	1.02	
	5500	On beams, columns, or soffits, no finish included		675	.024		.24	.55		.79	1.20	
	5600	With thin coat plaster finish		475	.034		.31	.78		1.09	1.66	
	6000	For high ceilings, over 8′ high, add		3,060	.005			.12		.12	.21	
	6500	For over 3 stories high, add per story	↓	6,100	.003	↓		.06		.06	.10	
700	0010	**DRYWALL** Gypsum plasterboard, nailed or screwed (R09250 -100)										700
	0100	to studs unless otherwise noted										
	0150	3/8″ thick, on walls, standard, no finish included	2 Carp	2,000	.008	S.F.	.20	.18		.38	.53	
	0200	On ceilings, standard, no finish included		1,800	.009		.20	.21		.41	.57	
	0250	On beams, columns, or soffits, no finish included		675	.024		.20	.55		.75	1.15	
	0300	1/2″ thick, on walls, standard, no finish included		2,000	.008		.19	.18		.37	.52	
	0350	Taped and finished (level 4 finish)		965	.017		.23	.38		.61	.90	
	0390	With compound skim coat (level 5 finish)	↓	775	.021	↓	.27	.48		.75	1.11	

9 FINISHES

09250 | Gypsum Board

			CREW	DAILY OUTPUT	LABOR-HOURS	UNIT	2004 BARE COSTS MAT.	LABOR	EQUIP.	TOTAL	TOTAL INCL O&P	
700	0400	Fire resistant, no finish included (R09250-100)	2 Carp	2,000	.008	S.F.	.20	.18		.38	.53	700
	0450	Taped and finished (level 4 finish)		965	.017		.24	.38		.62	.91	
	0490	With compound skim coat (level 5 finish)		775	.021		.28	.48		.76	1.12	
	0500	Water resistant, no finish included		2,000	.008		.24	.18		.42	.57	
	0550	Taped and finished (level 4 finish)		965	.017		.28	.38		.66	.96	
	0590	With compound skim coat (level 5 finish)		775	.021		.32	.48		.80	1.16	
	0600	Prefinished, vinyl, clipped to studs		900	.018		.56	.41		.97	1.32	
	1000	On ceilings, standard, no finish included		1,800	.009		.19	.21		.40	.56	
	1050	Taped and finished (level 4 finish)		765	.021		.23	.48		.71	1.07	
	1090	With compound skim coat (level 5 finish)		610	.026		.27	.61		.88	1.33	
	1100	Fire resistant, no finish included		1,800	.009		.20	.21		.41	.57	
	1150	Taped and finished (level 4 finish)		765	.021		.24	.48		.72	1.08	
	1195	With compound skim coat (level 5 finish)		610	.026		.28	.61		.89	1.34	
	1200	Water resistant, no finish included		1,800	.009		.24	.21		.45	.61	
	1250	Taped and finished (level 4 finish)		765	.021		.28	.48		.76	1.13	
	1290	With compound skim coat (level 5 finish)		610	.026		.32	.61		.93	1.38	
	1500	On beams, columns, or soffits, standard, no finish included		675	.024		.22	.55		.77	1.17	
	1550	Taped and finished (level 4 finish)		475	.034		.23	.78		1.01	1.57	
	1590	With compound skim coat (level 5 finish)		540	.030		.27	.68		.95	1.46	
	1600	Fire resistant, no finish included		675	.024		.23	.55		.78	1.18	
	1650	Taped and finished (level 4 finish)		475	.034		.24	.78		1.02	1.58	
	1690	With compound skim coat (level 5 finish)		540	.030		.28	.68		.96	1.47	
	1700	Water resistant, no finish included		675	.024		.28	.55		.83	1.23	
	1750	Taped and finished (level 4 finish)		475	.034		.28	.78		1.06	1.63	
	1790	With compound skim coat (level 5 finish)		540	.030		.32	.68		1	1.51	
	2000	5/8" thick, on walls, standard, no finish included		2,000	.008		.22	.18		.40	.55	
	2050	Taped and finished (level 4 finish)		965	.017		.26	.38		.64	.93	
	2090	With compound skim coat (level 5 finish)		775	.021		.30	.48		.78	1.14	
	2100	Fire resistant, no finish included		2,000	.008		.23	.18		.41	.56	
	2150	Taped and finished (level 4 finish)		965	.017		.27	.38		.65	.94	
	2195	With compound skim coat (level 5 finish)		775	.021		.31	.48		.79	1.15	
	2200	Water resistant, no finish included		2,000	.008		.27	.18		.45	.61	
	2250	Taped and finished (level 4 finish)		965	.017		.31	.38		.69	.99	
	2290	With compound skim coat (level 5 finish)		775	.021		.35	.48		.83	1.20	
	2300	Prefinished, vinyl, clipped to studs		900	.018		.65	.41		1.06	1.42	
	3000	On ceilings, standard, no finish included		1,800	.009		.22	.21		.43	.59	
	3050	Taped and finished (level 4 finish)		765	.021		.26	.48		.74	1.10	
	3090	With compound skim coat (level 5 finish)		615	.026		.30	.60		.90	1.35	
	3100	Fire resistant, no finish included		1,800	.009		.23	.21		.44	.60	
	3150	Taped and finished (level 4 finish)		765	.021		.27	.48		.75	1.11	
	3190	With compound skim coat (level 5 finish)		615	.026		.31	.60		.91	1.36	
	3200	Water resistant, no finish included		1,800	.009		.27	.21		.48	.65	
	3250	Taped and finished (level 4 finish)		765	.021		.31	.48		.79	1.16	
	3290	With compound skim coat (level 5 finish)		615	.026		.35	.60		.95	1.41	
	3500	On beams, columns, or soffits, no finish included		675	.024		.25	.55		.80	1.21	
	3550	Taped and finished (level 4 finish)		475	.034		.30	.78		1.08	1.65	
	3590	With compound skim coat (level 5 finish)		380	.042		.35	.97		1.32	2.03	
	3600	Fire resistant, no finish included		675	.024		.26	.55		.81	1.22	
	3650	Taped and finished (level 4 finish)		475	.034		.31	.78		1.09	1.66	
	3690	With compound skim coat (level 5 finish)		380	.042		.31	.97		1.28	1.99	
	3700	Water resistant, no finish included		675	.024		.31	.55		.86	1.27	
	3750	Taped and finished (level 4 finish)		475	.034		.35	.78		1.13	1.71	
	3790	With compound skim coat (level 5 finish)		380	.042		.35	.97		1.32	2.04	
	4000	Fireproofing, beams or columns, 2 layers, 1/2" thick, incl finish		330	.048		.44	1.12		1.56	2.38	
	4050	5/8" thick		300	.053		.54	1.23		1.77	2.68	
	4100	3 layers, 1/2" thick		225	.071		.64	1.64		2.28	3.49	

09200 | Plaster & Gypsum Board

	09250 | Gypsum Board	CREW	DAILY OUTPUT	LABOR-HOURS	UNIT	2004 BARE COSTS MAT.	LABOR	EQUIP.	TOTAL	TOTAL INCL O&P	
700	4150 5/8" thick R09250-100	2 Carp	210	.076	S.F.	.80	1.76		2.56	3.87	700
	5200 For work over 8' high, add		3,060	.005			.12		.12	.21	
	5270 For textured spray, add	2 Lath	1,600	.010		.04	.21		.25	.39	
	5350 For finishing inner corners, add	2 Carp	950	.017	L.F.	.08	.39		.47	.75	
	5355 For finishing outer corners, add	"	1,250	.013		.17	.30		.47	.68	
	5500 For acoustical sealant, add per bead	1 Carp	500	.016		.03	.37		.40	.66	
	5550 Sealant, 1 quart tube				Ea.	4.89			4.89	5.40	
	5600 Sound deadening board, 1/4" gypsum	2 Carp	1,800	.009	S.F.	.19	.21		.40	.56	
	5650 1/2" wood fiber	"	1,800	.009	"	.34	.21		.55	.72	

	09270 | Drywall Accessories	CREW	DAILY OUTPUT	LABOR-HOURS	UNIT	MAT.	LABOR	EQUIP.	TOTAL	TOTAL INCL O&P	
100	0011 **ACCESSORIES, DRYWALL** Casing bead, galvanized steel	1 Carp	290	.028	L.F.	.17	.64		.81	1.26	100
	0101 Vinyl		290	.028		.16	.64		.80	1.25	
	0401 Corner bead, galvanized steel, 1-1/4" x 1-1/4"		350	.023		.10	.53		.63	1	
	0411 1-1/4" x 1-1/4", 10' long		35	.229	Ea.	.95	5.30		6.25	10	
	0601 Vinyl corner bead		400	.020	L.F.	.21	.46		.67	1.01	
	0901 Furring channel, galv. steel, 7/8" deep, standard		260	.031		.17	.71		.88	1.40	
	1001 Resilient		260	.031		.18	.71		.89	1.41	
	1101 J trim, galvanized steel, 1/2" wide		300	.027			.62		.62	1.05	
	1121 5/8" wide		300	.027		.15	.62		.77	1.21	
	1160 Screws #6 x 1" A				M	6.80			6.80	7.50	
	1170 #6 x 1-5/8" A				"	8.90			8.90	9.80	
	1501 Z stud, galvanized steel, 1-1/2" wide	1 Carp	260	.031	L.F.	.31	.71		1.02	1.56	

	09280 | Gypsum Wallboard Repairs	CREW	DAILY OUTPUT	LABOR-HOURS	UNIT	MAT.	LABOR	EQUIP.	TOTAL	TOTAL INCL O&P	
100	0010 **GYPSUM WALLBOARD REPAIRS**										100
	0100 Fill and sand, pin / nail holes	1 Carp	960	.008	Ea.		.19		.19	.33	
	0110 Screw head pops		480	.017			.39		.39	.65	
	0120 Dents, up to 2" square		48	.167		.01	3.85		3.86	6.55	
	0130 2" to 4" square		24	.333		.03	7.70		7.73	13.10	
	0140 Cut square, patch, sand and finish, holes, up to 2" square		12	.667		.02	15.40		15.42	26	
	0150 2" to 4" square		11	.727		.06	16.80		16.86	28.50	
	0160 4" to 8" square		10	.800		.14	18.50		18.64	31.50	
	0170 8" to 12" square		8	1		.28	23		23.28	39.50	

09300 | Tile

	09310 | Ceramic Tile	CREW	DAILY OUTPUT	LABOR-HOURS	UNIT	2004 BARE COSTS MAT.	LABOR	EQUIP.	TOTAL	TOTAL INCL O&P	
100	0010 **CERAMIC TILE**										100
	0050 Base, using 1' x 4" high pc. with 1" x 1" tiles, mud set	D-7	82	.195	L.F.	3.99	3.82		7.81	10.55	
	0100 Thin set	"	128	.125		3.79	2.45		6.24	8.10	
	0300 For 6" high base, 1" x 1" tile face, add					.63			.63	.69	
	0400 For 2" x 2" tile face, add to above					.33			.33	.36	
	0600 Cove base, 4-1/4" x 4-1/4" high, mud set	D-7	91	.176		3.07	3.44		6.51	8.95	
	0700 Thin set		128	.125		3.09	2.45		5.54	7.35	
	0900 6" x 4-1/4" high, mud set		100	.160		2.77	3.13		5.90	8.10	
	1000 Thin set		137	.117		2.89	2.29		5.18	6.85	
	1200 Sanitary cove base, 6" x 4-1/4" high, mud set		93	.172		3.56	3.37		6.93	9.30	

Important: See the Reference Section for critical supporting data - Reference Nos., Crews, & Location Factors

09310 | Ceramic Tile

			CREW	DAILY OUTPUT	LABOR-HOURS	UNIT	2004 BARE COSTS MAT.	LABOR	EQUIP.	TOTAL	TOTAL INCL O&P	
100	1300	Thin set	D-7	124	.129	L.F.	3.62	2.53		6.15	8.05	100
	1500	6" x 6" high, mud set		84	.190		3.28	3.73		7.01	9.60	
	1600	Thin set		117	.137		3.32	2.68		6	7.95	
	1800	Bathroom accessories, average		82	.195	Ea.	9.50	3.82		13.32	16.60	
	1900	Bathtub, 5', rec. 4-1/4" x 4-1/4" tile wainscot, adhesive set 6' high		2.90	5.517		140	108		248	330	
	2100	7' high wainscot		2.50	6.400		160	125		285	380	
	2200	8' high wainscot		2.20	7.273		170	142		312	415	
	2400	Bullnose trim, 4-1/4" x 4-1/4", mud set		82	.195	L.F.	2.66	3.82		6.48	9.10	
	2500	Thin set		128	.125		2.60	2.45		5.05	6.80	
	2700	6" x 4-1/4" bullnose trim, mud set		84	.190		2.02	3.73		5.75	8.20	
	2800	Thin set		124	.129		2.02	2.53		4.55	6.30	
	3000	Floors, natural clay, random or uniform, thin set, color group 1		183	.087	S.F.	3.62	1.71		5.33	6.75	
	3100	Color group 2		183	.087		3.90	1.71		5.61	7.05	
	3260	Floors, glazed, thin set, 8" x 8", color group 1		250	.064		3.08	1.25		4.33	5.40	
	3270	12" x 12" tile		325	.049		3.86	.96		4.82	5.80	
	3280	16" x 16" tile		550	.029		4.95	.57		5.52	6.35	
	3300	Porcelain type, 1 color, color group 2, 1" x 1"		183	.087		4.19	1.71		5.90	7.35	
	3310	2" x 2" or 2" x 1", thin set		190	.084		4.01	1.65		5.66	7.05	
	3350	For random blend, 2 colors, add					.77			.77	.85	
	3360	4 colors, add					1.10			1.10	1.21	
	4300	Specialty tile, 4-1/4" x 4-1/4" x 1/2", decorator finish	D-7	183	.087		8.95	1.71		10.66	12.60	
	4500	Add for epoxy grout, 1/16" joint, 1" x 1" tile		800	.020		.54	.39		.93	1.22	
	4600	2" x 2" tile		820	.020		.50	.38		.88	1.16	
	4800	Pregrouted sheets, walls, 4-1/4" x 4-1/4", 6" x 4-1/4"										
	4810	and 8-1/2" x 4-1/4", 4 S.F. sheets, silicone grout	D-7	240	.067	S.F.	4.15	1.31		5.46	6.65	
	5100	Floors, unglazed, 2 S.F. sheets,										
	5110	urethane adhesive	D-7	180	.089	S.F.	4.13	1.74		5.87	7.35	
	5400	Walls, interior, thin set, 4-1/4" x 4-1/4" tile		190	.084		2.09	1.65		3.74	4.95	
	5500	6" x 4-1/4" tile		190	.084		2.30	1.65		3.95	5.20	
	5700	8-1/2" x 4-1/4" tile		190	.084		3.25	1.65		4.90	6.25	
	5800	6" x 6" tile		200	.080		2.63	1.57		4.20	5.40	
	5810	8" x 8" tile		225	.071		3.16	1.39		4.55	5.70	
	5820	12" x 12" tile		300	.053		2.98	1.04		4.02	4.96	
	5830	16" x 16" tile		500	.032		3.23	.63		3.86	4.56	
	6000	Decorated wall tile, 4-1/4" x 4-1/4", minimum		270	.059		3.50	1.16		4.66	5.70	
	6100	Maximum		180	.089		39	1.74		40.74	45.50	
	6600	Crystalline glazed, 4-1/4" x 4-1/4", mud set, plain		100	.160		3.25	3.13		6.38	8.65	
	6700	4-1/4" x 4-1/4", scored tile		100	.160		4.05	3.13		7.18	9.50	
	6900	6"x 6" plain		93	.172		3.25	3.37		6.62	9	
	7000	For epoxy grout, 1/16" joints, 4-1/4" tile, add		800	.020		.33	.39		.72	.99	
	7200	For tile set in dry mortar, add		1,735	.009			.18		.18	.29	
	7300	For tile set in portland cement mortar, add		290	.055			1.08		1.08	1.74	

09330 | Quarry Tile

			CREW	DAILY OUTPUT	LABOR-HOURS	UNIT	MAT.	LABOR	EQUIP.	TOTAL	TOTAL INCL O&P	
100	0010	**QUARRY TILE** Base, cove or sanitary, 2" or 5" high, mud set										100
	0100	1/2" thick	D-7	110	.145	L.F.	3.85	2.85		6.70	8.80	
	0300	Bullnose trim, red, mud set, 6" x 6" x 1/2" thick		120	.133		3.81	2.61		6.42	8.40	
	0400	4" x 4" x 1/2" thick		110	.145		4.30	2.85		7.15	9.30	
	0600	4" x 8" x 1/2" thick, using 8" as edge		130	.123		3.78	2.41		6.19	8.05	
	0700	Floors, mud set, 1,000 S.F. lots, red, 4" x 4" x 1/2" thick		120	.133	S.F.	3.81	2.61		6.42	8.40	
	0900	6" x 6" x 1/2" thick		140	.114		2.96	2.24		5.20	6.85	
	1000	4" x 8" x 1/2" thick		130	.123		3.81	2.41		6.22	8.05	
	1300	For waxed coating, add					.61			.61	.67	
	1500	For colors other than green, add					.36			.36	.40	
	1600	For abrasive surface, add					.43			.43	.47	
	1800	Brown tile, imported, 6" x 6" x 3/4"	D-7	120	.133		4.53	2.61		7.14	9.20	

09300 | Tile

		09330 Quarry Tile	CREW	DAILY OUTPUT	LABOR-HOURS	UNIT	2004 BARE COSTS MAT.	LABOR	EQUIP.	TOTAL	TOTAL INCL O&P	
100	1900	8″ x 8″ x 1″	D-7	110	.145	S.F.	5.15	2.85		8	10.25	100
	2100	For thin set mortar application, deduct		700	.023			.45		.45	.72	
	2700	Stair tread, 6″ x 6″ x 3/4″, plain		50	.320		4.20	6.25		10.45	14.70	
	2800	Abrasive		47	.340		4.68	6.65		11.33	15.85	
	3000	Wainscot, 6″ x 6″ x 1/2″, thin set, red		105	.152		3.49	2.98		6.47	8.65	
	3100	Colors other than green		105	.152	↓	3.89	2.98		6.87	9.10	
	3300	Window sill, 6″ wide, 3/4″ thick		90	.178	L.F.	4.47	3.48		7.95	10.50	
	3400	Corners	↓	80	.200	Ea.	4.91	3.92		8.83	11.70	

		09370 Metal Tile	CREW	DAILY OUTPUT	LABOR-HOURS	UNIT	MAT.	LABOR	EQUIP.	TOTAL	TOTAL INCL O&P	
100	0010	**METAL TILE** 4′ x 4′ sheet, 24 ga., tile pattern, nailed										100
	0200	Stainless steel	2 Carp	512	.031	S.F.	22	.72		22.72	25.50	
	0400	Aluminized steel	″	512	.031	″	11.90	.72		12.62	14.35	

09400 | Terrazzo

		09420 Prcast Terrazzo	CREW	DAILY OUTPUT	LABOR-HOURS	UNIT	2004 BARE COSTS MAT.	LABOR	EQUIP.	TOTAL	TOTAL INCL O&P	
900	0010	**TERRAZZO, PRECAST** Base, 6″ high, straight	1 Mstz	35	.229	L.F.	9.10	5.05		14.15	18.15	900
	0100	Cove		30	.267		10.15	5.90		16.05	20.50	
	0300	8″ high base, straight		30	.267		9.10	5.90		15	19.50	
	0400	Cove	↓	25	.320		13.40	7.05		20.45	26	
	0600	For white cement, add					.37			.37	.41	
	0700	For 16 ga. zinc toe strip, add					1.34			1.34	1.47	
	0900	Curbs, 4″ x 4″ high	1 Mstz	19	.421		25.50	9.30		34.80	43.50	
	1000	8″ x 8″ high	″	15	.533	↓	30	11.80		41.80	52	
	1200	Floor tiles, non-slip, 1″ thick, 12″ x 12″	D-1	29	.552	S.F.	15.40	11.55		26.95	36	
	1300	1-1/4″ thick, 12″ x 12″		29	.552		17.05	11.55		28.60	38	
	1500	16″ x 16″		23	.696		18.55	14.55		33.10	44.50	
	1600	1-1/2″ thick, 16″ x 16″	↓	21	.762		16.95	15.95		32.90	45	
	4800	Wainscot, 12″ x 12″ x 1″ tiles	1 Mstz	12	.667		5.45	14.75		20.20	29.50	
	4900	16″ x 16″ x 1-1/2″ tiles	″	8	1	↓	11.95	22		33.95	48.50	

09500 | Ceilings

		09510 Acoustical Ceilings	CREW	DAILY OUTPUT	LABOR-HOURS	UNIT	2004 BARE COSTS MAT.	LABOR	EQUIP.	TOTAL	TOTAL INCL O&P	
700	0010	**SUSPENDED ACOUSTIC CEILING TILES**, Not including										700
	0100	suspension system										
	0300	Fiberglass boards, film faced, 2′ x 2′ or 2′ x 4′, 5/8″ thick	1 Carp	625	.013	S.F.	.51	.30		.81	1.06	
	0400	3/4″ thick	↓	600	.013	↓	1.14	.31		1.45	1.77	

		09510 | Acoustical Ceilings	Crew	Daily Output	Labor-Hours	Unit	2004 Bare Costs Mat.	Labor	Equip.	Total	Total Incl O&P	
700	0500	3" thick, thermal, R11	1 Carp	450	.018	S.F.	1.26	.41		1.67	2.09	700
	0600	Glass cloth faced fiberglass, 3/4" thick		500	.016		1.62	.37		1.99	2.41	
	0700	1" thick		485	.016		1.79	.38		2.17	2.62	
	0820	1-1/2" thick, nubby face		475	.017		2.23	.39		2.62	3.11	
	1110	Mineral fiber tile, lay-in, 2' x 2' or 2' x 4', 5/8" thick, fine texture		625	.013		.43	.30		.73	.97	
	1115	Rough textured		625	.013		1.06	.30		1.36	1.67	
	1125	3/4" thick, fine textured		600	.013		1.17	.31		1.48	1.81	
	1130	Rough textured		600	.013		1.47	.31		1.78	2.14	
	1135	Fissured		600	.013		1.74	.31		2.05	2.43	
	1150	Tegular, 5/8" thick, fine textured		470	.017		1.04	.39		1.43	1.81	
	1155	Rough textured		470	.017		1.35	.39		1.74	2.16	
	1165	3/4" thick, fine textured		450	.018		1.47	.41		1.88	2.32	
	1170	Rough textured		450	.018		1.66	.41		2.07	2.53	
	1175	Fissured		450	.018		2.59	.41		3	3.55	
	1180	For aluminum face, add					4.72			4.72	5.20	
	1185	For plastic film face, add					.70			.70	.77	
	1190	For fire rating, add					.35			.35	.39	
	1300	Mirror faced panels, 15/16" thick, 2' x 2'	1 Carp	500	.016		9.80	.37		10.17	11.45	
	1900	Eggcrate, acrylic, 1/2" x 1/2" x 1/2" cubes		500	.016		1.42	.37		1.79	2.19	
	2100	Polystyrene eggcrate, 3/8" x 3/8" x 1/2" cubes		510	.016		1.19	.36		1.55	1.93	
	2200	1/2" x 1/2" x 1/2" cubes		500	.016		1.60	.37		1.97	2.39	
	2400	Luminous panels, prismatic, acrylic		400	.020		1.73	.46		2.19	2.68	
	2500	Polystyrene		400	.020		.88	.46		1.34	1.75	
	2700	Flat white acrylic		400	.020		3.01	.46		3.47	4.09	
	2800	Polystyrene		400	.020		2.06	.46		2.52	3.05	
	3000	Drop pan, white, acrylic		400	.020		4.41	.46		4.87	5.65	
	3100	Polystyrene		400	.020		3.69	.46		4.15	4.84	
	3600	Perforated aluminum sheets, .024" thick, corrugated, painted		490	.016		1.76	.38		2.14	2.58	
	3700	Plain		500	.016		3.07	.37		3.44	4.01	
	3750	Wood fiber in cementitious binder, 2' x 2' or 4', painted, 1" thick		600	.013		1.20	.31		1.51	1.84	
	3760	2" thick		550	.015		1.98	.34		2.32	2.75	
	3770	2-1/2" thick		500	.016		2.71	.37		3.08	3.61	
	3780	3" thick		450	.018		3.04	.41		3.45	4.04	
760	0010	**SUSPENDED CEILINGS, COMPLETE** Including standard										760
	0100	suspension system but not incl. 1-1/2" carrier channels										
	0600	Fiberglass ceiling board, 2' x 4' x 5/8", plain faced,	1 Carp	500	.016	S.F.	.92	.37		1.29	1.64	
	0700	Offices, 2' x 4' x 3/4"		380	.021		1.55	.49		2.04	2.53	
	1800	Tile, Z bar suspension, 5/8" mineral fiber tile		150	.053		1.55	1.23		2.78	3.80	
	1900	3/4" mineral fiber tile		150	.053		1.65	1.23		2.88	3.91	
900	0010	**CEILING TILE**, Stapled or cemented										900
	0100	12" x 12" or 12" x 24", not including furring										
	0600	Mineral fiber, vinyl coated, 5/8" thick	1 Carp	1,000	.008	S.F.	.89	.18		1.07	1.29	
	0700	3/4" thick		1,000	.008		1.30	.18		1.48	1.74	
	0900	Fire rated, 3/4" thick, plain faced		1,000	.008		.85	.18		1.03	1.25	
	1000	Plastic coated face		1,000	.008		1.04	.18		1.22	1.45	
	1200	Aluminum faced, 5/8" thick, plain		1,000	.008		1.12	.18		1.30	1.54	
	3300	For flameproofing, add					.09			.09	.10	
	3400	For sculptured 3 dimensional, add					.25			.25	.28	
	3900	For ceiling primer, add					.12			.12	.13	
	4000	For ceiling cement, add					.33			.33	.36	

09600 | Flooring

09620 | Specialty Flooring

			CREW	DAILY OUTPUT	LABOR-HOURS	UNIT	2004 BARE COSTS MAT.	LABOR	EQUIP.	TOTAL	TOTAL INCL O&P	
100	0010	**ATHLETIC FLOORING**										100
	3700	Polyethylene, in rolls, no base incl., landscape surfaces	1 Tilf	275	.029	S.F.	2.45	.64		3.09	3.73	
	3800	Nylon action surface, 1/8" thick		275	.029		2.63	.64		3.27	3.92	
	3900	1/4" thick		275	.029		3.79	.64		4.43	5.20	
	4000	3/8" thick	↓	275	.029	↓	4.76	.64		5.40	6.30	

09631 | Brick Flooring

			CREW	DAILY OUTPUT	LABOR-HOURS	UNIT	MAT.	LABOR	EQUIP.	TOTAL	TOTAL INCL O&P	
100	0010	**BRICK FLOORING**										100
	0020	Acid proof shales, red, 8" x 3-3/4" x 1-1/4" thick	D-7	.43	37.209	M	695	730		1,425	1,950	
	0050	2-1/4" thick	D-1	.40	40		755	840		1,595	2,225	
	0200	Acid proof clay brick, 8" x 3-3/4" x 2-1/4" thick	"	.40	40	↓	755	840		1,595	2,225	
	0260	Cast ceramic, pressed, 4" x 8" x 1/2", unglazed	D-7	100	.160	S.F.	4.94	3.13		8.07	10.50	
	0270	Glazed		100	.160		6.60	3.13		9.73	12.30	
	0280	Hand molded flooring, 4" x 8" x 3/4", unglazed		95	.168		6.55	3.30		9.85	12.50	
	0290	Glazed		95	.168		8.20	3.30		11.50	14.30	
	0300	8" hexagonal, 3/4" thick, unglazed		85	.188		7.15	3.69		10.84	13.85	
	0310	Glazed	↓	85	.188		12.95	3.69		16.64	20	
	0450	Acid proof joints, 1/4" wide	D-1	65	.246		1.13	5.15		6.28	9.80	
	0500	Pavers, 8" x 4", 1" to 1-1/4" thick, red	D-7	95	.168		2.88	3.30		6.18	8.45	
	0510	Ironspot	"	95	.168		4.07	3.30		7.37	9.80	
	0540	1-3/8" to 1-3/4" thick, red	D-1	95	.168		2.78	3.53		6.31	8.90	
	0560	Ironspot		95	.168		4.02	3.53		7.55	10.25	
	0580	2-1/4" thick, red		90	.178		2.83	3.72		6.55	9.30	
	0590	Ironspot	↓	90	.178	↓	4.38	3.72		8.10	11	
	0800	For sidewalks and patios with pavers, see division 02780-200										
	0870	For epoxy joints, add	D-1	600	.027	S.F.	2.15	.56		2.71	3.30	
	0880	For Furan underlayment, add	"	600	.027		1.78	.56		2.34	2.89	
	0890	For waxed surface, steam cleaned, add	D-5	1,000	.008	↓	.15	.14		.29	.41	

09635 | Marble Flooring

			CREW	DAILY OUTPUT	LABOR-HOURS	UNIT	MAT.	LABOR	EQUIP.	TOTAL	TOTAL INCL O&P	
100	0010	**MARBLE** Thin gauge tile, 12" x 6", 3/8", White Carara	D-7	60	.267	S.F.	8.75	5.20		13.95	18.05	100
	0100	Travertine		60	.267		9.65	5.20		14.85	19	
	0200	12" x 12" x 3/8", thin set, floors		60	.267		6.70	5.20		11.90	15.75	
	0300	On walls	↓	52	.308	↓	8.90	6		14.90	19.50	

09637 | Stone Flooring

			CREW	DAILY OUTPUT	LABOR-HOURS	UNIT	MAT.	LABOR	EQUIP.	TOTAL	TOTAL INCL O&P	
100	0010	**SLATE TILE** Vermont, 6" x 6" x 1/4" thick, thin set	D-7	180	.089	S.F.	4.14	1.74		5.88	7.35	100
200	0010	**SLATE & STONE FLOORS** See division 02780-600										200

09647 | Wood Parquet Flooring

			CREW	DAILY OUTPUT	LABOR-HOURS	UNIT	MAT.	LABOR	EQUIP.	TOTAL	TOTAL INCL O&P	
100	0010	**WOOD PARQUET** flooring										100
	5200	Parquetry, standard, 5/16" thick, not incl. finish, oak, minimum	1 Carp	160	.050	S.F.	2.85	1.16		4.01	5.10	
	5300	Maximum		100	.080		5.20	1.85		7.05	8.90	
	5500	Teak, minimum		160	.050		4.40	1.16		5.56	6.80	
	5600	Maximum		100	.080		7.70	1.85		9.55	11.60	
	5650	13/16" thick, select grade oak, minimum		160	.050		8.50	1.16		9.66	11.30	
	5700	Maximum		100	.080		12.90	1.85		14.75	17.30	
	5800	Custom parquetry, including finish, minimum		100	.080		14.20	1.85		16.05	18.75	
	5900	Maximum		50	.160		18.80	3.70		22.50	27	
	6700	Parquetry, prefinished white oak, 5/16" thick, minimum		160	.050		3.50	1.16		4.66	5.80	
	6800	Maximum		100	.080		7.15	1.85		9	11.05	
	7000	Walnut or teak, parquetry, minimum	↓	160	.050	↓	4.77	1.16		5.93	7.20	

09647 | Wood Parquet Flooring

			CREW	DAILY OUTPUT	LABOR-HOURS	UNIT	2004 BARE COSTS MAT.	LABOR	EQUIP.	TOTAL	TOTAL INCL O&P	
100	7100	Maximum	1 Carp	100	.080	S.F.	8.30	1.85		10.15	12.30	100
	7200	Acrylic wood parquet blocks, 12" x 12" x 5/16",										
	7210	irradiated, set in epoxy	1 Carp	160	.050	S.F.	6.95	1.16		8.11	9.60	

09648 | Wood Strip Flooring

			CREW	DAILY OUTPUT	LABOR-HOURS	UNIT	MAT.	LABOR	EQUIP.	TOTAL	TOTAL INCL O&P	
100	0010	**WOOD** Fir, vertical grain, 1" x 4", not incl. finish, B & better	1 Carp	255	.031	S.F.	2.44	.72		3.16	3.91	100
	0100	C grade & better		255	.031		2.29	.72		3.01	3.75	
	0300	Flat grain, 1" x 4", not incl. finish, B & better		255	.031		2.79	.72		3.51	4.30	
	0400	C & better		255	.031		2.68	.72		3.40	4.18	
	4000	Maple, strip, 25/32" x 2-1/4", not incl. finish, select		170	.047		4.65	1.09		5.74	6.95	
	4100	#2 & better		170	.047		2.81	1.09		3.90	4.93	
	4300	33/32" x 3-1/4", not incl. finish, #1 grade		170	.047		3.45	1.09		4.54	5.65	
	4400	#2 & better	↓	170	.047	↓	3.07	1.09		4.16	5.20	
	4600	Oak, white or red, 25/32" x 2-1/4", not incl. finish										
	4700	#1 common	1 Carp	170	.047	S.F.	2.86	1.09		3.95	4.99	
	4900	Select quartered, 2-1/4" wide		170	.047		3.60	1.09		4.69	5.80	
	5000	Clear		170	.047		3.72	1.09		4.81	5.95	
	6100	Prefinished, white oak, prime grade, 2-1/4" wide		170	.047		6.20	1.09		7.29	8.65	
	6200	3-1/4" wide		185	.043		7.95	1		8.95	10.45	
	6400	Ranch plank		145	.055		7.70	1.27		8.97	10.60	
	6500	Hardwood blocks, 9" x 9", 25/32" thick		160	.050		5.15	1.16		6.31	7.60	
	7400	Yellow pine, 3/4" x 3-1/8", T & G, C & better, not incl. finish	↓	200	.040		2.28	.92		3.20	4.08	
	7500	Refinish wood floor, sand, 2 cts poly, wax, soft wood, min.	1 Clab	400	.020		.69	.34		1.03	1.33	
	7600	Hard wood, max		130	.062		1.04	1.04		2.08	2.91	
	7800	Sanding and finishing, 2 coats polyurethane	↓	295	.027	↓	.69	.46		1.15	1.54	
	7900	Subfloor and underlayment, see division 06160										
	8015	Transition molding, 2 1/4" wide, 5' long	1 Carp	19.20	.417	Ea.	12.65	9.65		22.30	30.50	
	8300	Floating floor, wood composition strip, complete.	1 Clab	133	.060	S.F.	3.36	1.02		4.38	5.45	
	8310	Floating floor components, T & G wood composite strips					3.02			3.02	3.33	
	8320	Film					.14			.14	.15	
	8330	Foam					.17			.17	.18	
	8340	Adhesive					.07			.07	.08	
	8350	Installation kit				↓	.16			.16	.18	
	8360	Trim, 2" wide x 3' long				L.F.	2.05			2.05	2.26	
	8370	Reducer moulding				"	4.15			4.15	4.57	
200	0010	**RESILIENT BASE**										200
	0800	Base, cove, rubber or vinyl, .080" thick										
	1100	Standard colors, 2-1/2" high	1 Tilf	315	.025	L.F.	.44	.56		1	1.38	
	1150	4" high		315	.025		.49	.56		1.05	1.44	
	1200	6" high		315	.025		.80	.56		1.36	1.78	
	1450	1/8" thick, standard colors, 2-1/2" high		315	.025		.51	.56		1.07	1.46	
	1500	4" high		315	.025		.67	.56		1.23	1.64	
	1550	6" high		315	.025	↓	.88	.56		1.44	1.87	
	1600	Corners, 2-1/2" high		315	.025	Ea.	1.11	.56		1.67	2.12	
	1630	4" high		315	.025		1.16	.56		1.72	2.18	
	1660	6" high	↓	315	.025	↓	1.51	.56		2.07	2.56	

09653 | Resilient Sheet Flooring

			CREW	DAILY OUTPUT	LABOR-HOURS	UNIT	MAT.	LABOR	EQUIP.	TOTAL	TOTAL INCL O&P	
100	0010	**RESILIENT SHEET FLOORING**										100
	5900	Rubber, sheet goods, 36" wide, 1/8" thick	1 Tilf	120	.067	S.F.	3.37	1.47		4.84	6.10	
	5950	3/16" thick		100	.080		4.79	1.76		6.55	8.10	
	6000	1/4" thick		90	.089		5.55	1.96		7.51	9.25	
	8000	Vinyl sheet goods, backed, .065" thick, minimum		250	.032		1.97	.71		2.68	3.31	
	8050	Maximum	↓	200	.040	↓	2.52	.88		3.40	4.19	

		09653 Resilient Sheet Flooring	CREW	DAILY OUTPUT	LABOR-HOURS	UNIT	2004 BARE COSTS MAT.	LABOR	EQUIP.	TOTAL	TOTAL INCL O&P	
100	8100	.080" thick, minimum	1 Tilf	230	.035	S.F.	2.12	.77		2.89	3.56	100
	8150	Maximum		200	.040		3.03	.88		3.91	4.75	
	8200	.125" thick, minimum		230	.035		2.39	.77		3.16	3.86	
	8250	Maximum	↓	200	.040	↓	3.79	.88		4.67	5.60	
	8700	Adhesive cement, 1 gallon does 200 to 300 S.F.				Gal.	15.55			15.55	17.10	
	8800	Asphalt primer, 1 gallon per 300 S.F.					9.20			9.20	10.10	
	8900	Emulsion, 1 gallon per 140 S.F.					11.65			11.65	12.85	
	8950	Latex underlayment, liquid, fortified				↓	31.50			31.50	34.50	

		09658 Resilient Tile Flooring	CREW	DAILY OUTPUT	LABOR-HOURS	UNIT	MAT.	LABOR	EQUIP.	TOTAL	TOTAL INCL O&P	
100	0010	**RESILIENT TILE FLOORING**										100
	2200	Cork tile, standard finish, 1/8" thick	1 Tilf	315	.025	S.F.	3.57	.56		4.13	4.83	
	2250	3/16" thick		315	.025		3.56	.56		4.12	4.82	
	2300	5/16" thick		315	.025		4.78	.56		5.34	6.15	
	2350	1/2" thick		315	.025		5.50	.56		6.06	6.95	
	2500	Urethane finish, 1/8" thick		315	.025		4.50	.56		5.06	5.85	
	2550	3/16" thick		315	.025		4.88	.56		5.44	6.25	
	2600	5/16" thick		315	.025		6.05	.56		6.61	7.55	
	2650	1/2" thick		315	.025		8.50	.56		9.06	10.25	
	6050	Tile, marbleized colors, 12" x 12", 1/8" thick		400	.020		3.85	.44		4.29	4.95	
	6100	3/16" thick		400	.020		5.20	.44		5.64	6.40	
	6300	Special tile, plain colors, 1/8" thick		400	.020		4.18	.44		4.62	5.30	
	6350	3/16" thick		400	.020		5.65	.44		6.09	6.90	
	7000	Vinyl composition tile, 12" x 12", 1/16" thick		500	.016		.75	.35		1.10	1.40	
	7050	Embossed		500	.016		.95	.35		1.30	1.62	
	7100	Marbleized		500	.016		.95	.35		1.30	1.62	
	7150	Solid		500	.016		1.06	.35		1.41	1.74	
	7200	3/32" thick, embossed		500	.016		.96	.35		1.31	1.63	
	7250	Marbleized		500	.016		1.07	.35		1.42	1.75	
	7300	Solid		500	.016		1.56	.35		1.91	2.29	
	7350	1/8" thick, marbleized		500	.016		1.02	.35		1.37	1.69	
	7400	Solid		500	.016		1.91	.35		2.26	2.67	
	7450	Conductive		500	.016		3.58	.35		3.93	4.51	
	7500	Vinyl tile, 12" x 12", .050" thick, minimum		500	.016		1.68	.35		2.03	2.42	
	7550	Maximum		500	.016		3.29	.35		3.64	4.19	
	7600	1/8" thick, minimum		500	.016		2.12	.35		2.47	2.90	
	7650	Solid colors		500	.016		4.57	.35		4.92	5.60	
	7700	Marbleized or Travertine pattern		500	.016		3.39	.35		3.74	4.30	
	7750	Florentine pattern		500	.016		3.90	.35		4.25	4.86	
	7800	Maximum	↓	500	.016	↓	8	.35		8.35	9.35	

		09662 Static Control Flooring	CREW	DAILY OUTPUT	LABOR-HOURS	UNIT	MAT.	LABOR	EQUIP.	TOTAL	TOTAL INCL O&P	
100	0010	**CONDUCTIVE RESILIENT FLOORING**										100
	1700	Conductive flooring, rubber tile, 1/8" thick	1 Tilf	315	.025	S.F.	2.74	.56		3.30	3.91	
	1800	Homogeneous vinyl tile, 1/8" thick	"	315	.025	"	3.73	.56		4.29	5	

		09680 Carpet	CREW	DAILY OUTPUT	LABOR-HOURS	UNIT	MAT.	LABOR	EQUIP.	TOTAL	TOTAL INCL O&P	
600	0010	**CARPET PAD**, commercial grade										600
	9001	Sponge rubber pad, minimum	1 Tilf	1,350	.006	S.F.	.35	.13		.48	.60	
	9101	Maximum		1,350	.006		.92	.13		1.05	1.22	
	9201	Felt pad, minimum		1,350	.006		.39	.13		.52	.63	
	9301	Maximum		1,350	.006		.71	.13		.84	.99	
	9401	Bonded urethane pad, minimum		1,350	.006		.42	.13		.55	.67	
	9501	Maximum		1,350	.006		.72	.13		.85	1	
	9601	Prime urethane pad, minimum	↓	1,350	.006	↓	.24	.13		.37	.47	

09600 | Flooring

09680 | Carpet

		09680 Carpet	CREW	DAILY OUTPUT	LABOR-HOURS	UNIT	2004 BARE COSTS MAT.	LABOR	EQUIP.	TOTAL	TOTAL INCL O&P	
600	9701	Maximum	1 Tilf	1,350	.006	S.F.	.44	.13		.57	.70	600
800	0010	**CARPET** Commercial grades, direct cement										800
	0701	Nylon, level loop, 26 oz., light to medium traffic	1 Tilf	445	.018	S.F.	1.66	.40		2.06	2.47	
	0901	32 oz., medium traffic		445	.018		2.35	.40		2.75	3.23	
	1101	40 oz., medium to heavy traffic		445	.018		3.49	.40		3.89	4.48	
	2101	Nylon, plush, 20 oz., light traffic		445	.018		1.10	.40		1.50	1.85	
	2801	24 oz., light to medium traffic		445	.018		1.17	.40		1.57	1.92	
	2901	30 oz., medium traffic		445	.018		1.74	.40		2.14	2.55	
	3001	36 oz., medium traffic		445	.018		2.19	.40		2.59	3.05	
	3101	42 oz., medium to heavy traffic		370	.022		2.28	.48		2.76	3.27	
	3201	46 oz., medium to heavy traffic		370	.022		3.02	.48		3.50	4.09	
	3301	54 oz., heavy traffic		370	.022		3.41	.48		3.89	4.52	
	3501	Olefin, 15 oz., light traffic		445	.018		.59	.40		.99	1.29	
	3651	22 oz., light traffic		445	.018		.70	.40		1.10	1.41	
	4501	50 oz., medium to heavy traffic, level loop		445	.018		7.70	.40		8.10	9.15	
	4701	32 oz., medium to heavy traffic, patterned		400	.020		7.60	.44		8.04	9.05	
	4901	48 oz., heavy traffic, patterned	↓	400	.020	↓	7.75	.44		8.19	9.20	
	5000	For less than full roll, add					25%					
	5100	For small rooms, less than 12′ wide, add						25%				
	5200	For large open areas (no cuts), deduct						25%				
	5600	For bound carpet baseboard, add	1 Tilf	300	.027	L.F.	1.09	.59		1.68	2.15	
	5610	For stairs, not incl. price of carpet, add	"	30	.267	Riser		5.90		5.90	9.45	
	8950	For tackless, stretched installation, add padding to above										
	9850	For "branded" fiber, add				S.Y.	25%					
900	0010	**CARPET TILE**										900
	0100	Tufted nylon, 18" x 18", hard back, 20 oz.	1 Tilf	150	.053	S.Y.	19.65	1.18		20.83	23.50	
	0110	26 oz.		150	.053		33.50	1.18		34.68	39	
	0200	Cushion back, 20 oz.		150	.053		24.50	1.18		25.68	29	
	0210	26 oz.	↓	150	.053	↓	38.50	1.18		39.68	44.50	

09700 | Wall Finishes

09720 | Wall Coverings

		09720 Wall Coverings	CREW	DAILY OUTPUT	LABOR-HOURS	UNIT	2004 BARE COSTS MAT.	LABOR	EQUIP.	TOTAL	TOTAL INCL O&P	
100	0010	**WALL COVERING** Including sizing, add 10%-30% waste at takeoff R09700 -700										100
	0050	Aluminum foil	1 Pape	275	.029	S.F.	.83	.61		1.44	1.91	
	0100	Copper sheets, .025" thick, vinyl backing		240	.033		4.44	.69		5.13	6	
	0300	Phenolic backing		240	.033		5.75	.69		6.44	7.50	
	0600	Cork tiles, light or dark, 12" x 12" x 3/16"		240	.033		2.60	.69		3.29	4	
	0700	5/16" thick		235	.034		2.95	.71		3.66	4.42	
	0900	1/4" basketweave		240	.033		4.53	.69		5.22	6.10	
	1000	1/2" natural, non-directional pattern		240	.033		6.05	.69		6.74	7.80	
	1100	3/4" natural, non-directional pattern		240	.033		9.85	.69		10.54	12	
	1200	Granular surface, 12" x 36", 1/2" thick		385	.021		.98	.43		1.41	1.79	
	1300	1" thick		370	.022		1.27	.45		1.72	2.14	
	1500	Polyurethane coated, 12" x 12" x 3/16" thick		240	.033		3.06	.69		3.75	4.51	
	1600	5/16" thick		235	.034		4.35	.71		5.06	5.95	
	1800	Cork wallpaper, paperbacked, natural	↓	480	.017	↓	1.74	.35		2.09	2.48	

09700 | Wall Finishes

09720 | Wall Coverings

			Crew	Daily Output	Labor-Hours	Unit	2004 Bare Costs Mat.	Labor	Equip.	Total	Total Incl O&P	
100	1900	Colors (R09700-700)	1 Pape	480	.017	S.F.	2.15	.35		2.50	2.94	100
	2100	Flexible wood veneer, 1/32" thick, plain woods		100	.080		1.83	1.67		3.50	4.75	
	2200	Exotic woods	↓	95	.084	↓	2.78	1.76		4.54	5.95	
	2400	Gypsum-based, fabric-backed, fire										
	2500	resistant for masonry walls, minimum, 21 oz./S.Y.	1 Pape	800	.010	S.F.	.68	.21		.89	1.09	
	2600	Average		720	.011		1	.23		1.23	1.48	
	2700	Maximum, (small quantities)	↓	640	.013		1.11	.26		1.37	1.65	
	2750	Acrylic, modified, semi-rigid PVC, .028" thick	2 Carp	330	.048		.92	1.12		2.04	2.91	
	2800	.040" thick	"	320	.050		1.21	1.16		2.37	3.29	
	3000	Vinyl wall covering, fabric-backed, lightweight, (12-15 oz./S.Y.)	1 Pape	640	.013		.58	.26		.84	1.07	
	3300	Medium weight, type 2, (20-24 oz./S.Y.)		480	.017		.72	.35		1.07	1.36	
	3400	Heavy weight, type 3, (28 oz./S.Y.)	↓	435	.018	↓	1.16	.38		1.54	1.91	
	3600	Adhesive, 5 gal. lots, (18SY/Gal.)				Gal.	8.95			8.95	9.80	
	3700	Wallpaper, average workmanship, solid pattern, low cost paper	1 Pape	640	.013	S.F.	.29	.26		.55	.75	
	3900	basic patterns (matching required), avg. cost paper		535	.015		.64	.31		.95	1.21	
	4000	Paper at $85per double roll, quality workmanship	↓	435	.018	↓	1.52	.38		1.90	2.30	
	4100	Linen wall covering, paper backed										
	4150	Flame treatment, minimum				S.F.	.68			.68	.75	
	4180	Maximum					1.26			1.26	1.39	
	4200	Grass cloths with lining paper, minimum	1 Pape	400	.020		.63	.42		1.05	1.38	
	4300	Maximum	"	350	.023	↓	2.03	.48		2.51	3.01	

09770 | Special Wall Surfaces

			Crew	Daily Output	Labor-Hours	Unit	Mat.	Labor	Equip.	Total	Total Incl O&P	
700	0010	**PANEL SYSTEM**										700
	0100	Raised panel, eng. wood core w/ wood veneer, std., paint grade	2 Carp	300	.053	S.F.	6.90	1.23		8.13	9.70	
	0110	Oak veneer		300	.053		10.60	1.23		11.83	13.80	
	0120	Maple veneer		300	.053		17.90	1.23		19.13	22	
	0130	Cherry veneer		300	.053		21.50	1.23		22.73	26	
	0300	Class I fire rated, paint grade		300	.053		16.55	1.23		17.78	20.50	
	0310	Oak veneer		300	.053		28	1.23		29.23	33	
	0320	Maple veneer		300	.053		46.50	1.23		47.73	53	
	0330	Cherry veneer	↓	300	.053	↓	54	1.23		55.23	61.50	
	5000	For prefinished paneling, see division 06250-500 & 06250-200										

09900 | Paints & Coatings

09910 | Paints

			Crew	Daily Output	Labor-Hours	Unit	2004 Bare Costs Mat.	Labor	Equip.	Total	Total Incl O&P	
100	0010	**CABINETS AND CASEWORK**										100
	1000	Primer coat, oil base, brushwork	1 Pord	650	.012	S.F.	.05	.26		.31	.47	
	2000	Paint, oil base, brushwork, 1 coat		650	.012		.06	.26		.32	.49	
	2500	2 coats		400	.020		.12	.42		.54	.82	
	3000	Stain, brushwork, wipe off		650	.012		.05	.26		.31	.47	
	4000	Shellac, 1 coat, brushwork		650	.012		.06	.26		.32	.48	
	4500	Varnish, 3 coats, brushwork, sand after 1st coat	↓	325	.025		.17	.52		.69	1.04	
	5000	For latex paint, deduct				↓	10%					
300	0010	**DOORS AND WINDOWS, EXTERIOR** (R09910-220)										300
	0100	Door frames & trim, only										
	0110	Brushwork, primer	1 Pord	512	.016	L.F.	.05	.33		.38	.60	
	0120	Finish coat, exterior latex	↓	512	.016	↓	.06	.33		.39	.60	

9 FINISHES

09910 | Paints

	09910 Paints		CREW	DAILY OUTPUT	LABOR-HOURS	UNIT	2004 BARE COSTS MAT.	LABOR	EQUIP.	TOTAL	TOTAL INCL O&P	
300	0130	Primer & 1 coat, exterior latex (R09910-220)	1 Pord	300	.027	L.F.	.11	.56		.67	1.04	300
	0140	Primer & 2 coats, exterior latex		265	.030		.16	.63		.79	1.22	
	0150	Doors, flush, both sides, incl. frame & trim										
	0160	Roll & brush, primer	1 Pord	10	.800	Ea.	3.82	16.80		20.62	31.50	
	0170	Finish coat, exterior latex		10	.800		4.35	16.80		21.15	32.50	
	0180	Primer & 1 coat, exterior latex		7	1.143		8.15	24		32.15	48.50	
	0190	Primer & 2 coats, exterior latex		5	1.600		12.50	33.50		46	69	
	0200	Brushwork, stain, sealer & 2 coats polyurethane		4	2		15.55	42		57.55	86	
	0210	Doors, French, both sides, 10-15 lite, incl. frame & trim										
	0220	Brushwork, primer	1 Pord	6	1.333	Ea.	1.91	28		29.91	48	
	0230	Finish coat, exterior latex		6	1.333		2.17	28		30.17	48.50	
	0240	Primer & 1 coat, exterior latex		3	2.667		4.09	56		60.09	96.50	
	0250	Primer & 2 coats, exterior latex		2	4		6.15	84		90.15	145	
	0260	Brushwork, stain, sealer & 2 coats polyurethane		2.50	3.200		5.60	67		72.60	116	
	0270	Doors, louvered, both sides, incl. frame & trim										
	0280	Brushwork, primer	1 Pord	7	1.143	Ea.	3.82	24		27.82	43.50	
	0290	Finish coat, exterior latex		7	1.143		4.35	24		28.35	44.50	
	0300	Primer & 1 coat, exterior latex		4	2		8.15	42		50.15	78	
	0310	Primer & 2 coats, exterior latex		3	2.667		12.25	56		68.25	106	
	0320	Brushwork, stain, sealer & 2 coats polyurethane		4.50	1.778		15.55	37.50		53.05	78.50	
	0330	Doors, panel, both sides, incl. frame & trim										
	0340	Roll & brush, primer	1 Pord	6	1.333	Ea.	3.82	28		31.82	50	
	0350	Finish coat, exterior latex		6	1.333		4.35	28		32.35	51	
	0360	Primer & 1 coat, exterior latex		3	2.667		8.15	56		64.15	101	
	0370	Primer & 2 coats, exterior latex		2.50	3.200		12.25	67		79.25	124	
	0380	Brushwork, stain, sealer & 2 coats polyurethane		3	2.667		15.55	56		71.55	109	
	0400	Windows, per ext. side, based on 15 SF										
	0410	1 to 6 lite										
	0420	Brushwork, primer	1 Pord	13	.615	Ea.	.75	12.90		13.65	22	
	0430	Finish coat, exterior latex		13	.615		.86	12.90		13.76	22	
	0440	Primer & 1 coat, exterior latex		8	1		1.61	21		22.61	36.50	
	0450	Primer & 2 coats, exterior latex		6	1.333		2.42	28		30.42	48.50	
	0460	Stain, sealer & 1 coat varnish		7	1.143		2.20	24		26.20	42	
	0470	7 to 10 lite										
	0480	Brushwork, primer	1 Pord	11	.727	Ea.	.75	15.25		16	26	
	0490	Finish coat, exterior latex		11	.727		.86	15.25		16.11	26	
	0500	Primer & 1 coat, exterior latex		7	1.143		1.61	24		25.61	41.50	
	0510	Primer & 2 coats, exterior latex		5	1.600		2.42	33.50		35.92	57.50	
	0520	Stain, sealer & 1 coat varnish		6	1.333		2.20	28		30.20	48.50	
	0530	12 lite										
	0540	Brushwork, primer	1 Pord	10	.800	Ea.	.75	16.80		17.55	28.50	
	0550	Finish coat, exterior latex		10	.800		.86	16.80		17.66	28.50	
	0560	Primer & 1 coat, exterior latex		6	1.333		1.61	28		29.61	48	
	0570	Primer & 2 coats, exterior latex		5	1.600		2.42	33.50		35.92	57.50	
	0580	Stain, sealer & 1 coat varnish		6	1.333		2.20	28		30.20	48.50	
	0590	For oil base paint, add					10%					
310	0010	**DOORS & WINDOWS, INTERIOR LATEX** (R09910-220)										310
	0100	Doors flush, both sides, incl. frame & trim										
	0110	Roll & brush, primer	1 Pord	10	.800	Ea.	3.33	16.80		20.13	31	
	0120	Finish coat, latex		10	.800		3.60	16.80		20.40	31.50	
	0130	Primer & 1 coat latex		7	1.143		6.90	24		30.90	47	
	0140	Primer & 2 coats latex		5	1.600		10.30	33.50		43.80	66.50	
	0160	Spray, both sides, primer		20	.400		3.50	8.40		11.90	17.65	
	0170	Finish coat, latex		20	.400		3.77	8.40		12.17	17.95	
	0180	Primer & 1 coat latex		11	.727		7.30	15.25		22.55	33	
	0190	Primer & 2 coats latex		8	1		10.90	21		31.90	46.50	

FINISHES 9

09910 | Paints		CREW	DAILY OUTPUT	LABOR-HOURS	UNIT	2004 BARE COSTS MAT.	LABOR	EQUIP.	TOTAL	TOTAL INCL O&P
310	0200 Doors, French, both sides, 10-15 lite, incl. frame & trim (R09910-220)									
	0210 Roll & brush, primer	1 Pord	6	1.333	Ea.	1.66	28		29.66	48
	0220 Finish coat, latex		6	1.333		1.80	28		29.80	48
	0230 Primer & 1 coat latex		3	2.667		3.46	56		59.46	96
	0240 Primer & 2 coats latex		2	4		5.15	84		89.15	144
	0260 Doors, louvered, both sides, incl. frame & trim									
	0270 Roll & brush, primer	1 Pord	7	1.143	Ea.	3.33	24		27.33	43
	0280 Finish coat, latex		7	1.143		3.60	24		27.60	43.50
	0290 Primer & 1 coat, latex		4	2		6.75	42		48.75	76.50
	0300 Primer & 2 coats, latex		3	2.667		10.50	56		66.50	104
	0320 Spray, both sides, primer		20	.400		3.50	8.40		11.90	17.65
	0330 Finish coat, latex		20	.400		3.77	8.40		12.17	17.95
	0340 Primer & 1 coat, latex		11	.727		7.30	15.25		22.55	33
	0350 Primer & 2 coats, latex		8	1		11.15	21		32.15	47
	0360 Doors, panel, both sides, incl. frame & trim									
	0370 Roll & brush, primer	1 Pord	6	1.333	Ea.	3.50	28		31.50	50
	0380 Finish coat, latex		6	1.333		3.60	28		31.60	50
	0390 Primer & 1 coat, latex		3	2.667		6.90	56		62.90	99.50
	0400 Primer & 2 coats, latex		2.50	3.200		10.50	67		77.50	122
	0420 Spray, both sides, primer		10	.800		3.50	16.80		20.30	31.50
	0430 Finish coat, latex		10	.800		3.77	16.80		20.57	31.50
	0440 Primer & 1 coat, latex		5	1.600		7.30	33.50		40.80	63
	0450 Primer & 2 coats, latex		4	2		11.15	42		53.15	81.50
	0460 Windows, per interior side, based on 15 SF									
	0470 1 to 6 lite									
	0480 Brushwork, primer	1 Pord	13	.615	Ea.	.66	12.90		13.56	21.50
	0490 Finish coat, enamel		13	.615		.71	12.90		13.61	22
	0500 Primer & 1 coat enamel		8	1		1.37	21		22.37	36
	0510 Primer & 2 coats enamel		6	1.333		2.08	28		30.08	48.50
	0530 7 to 10 lite									
	0540 Brushwork, primer	1 Pord	11	.727	Ea.	.66	15.25		15.91	25.50
	0550 Finish coat, enamel		11	.727		.71	15.25		15.96	26
	0560 Primer & 1 coat enamel		7	1.143		1.37	24		25.37	41
	0570 Primer & 2 coats enamel		5	1.600		2.08	33.50		35.58	57.50
	0590 12 lite									
	0600 Brushwork, primer	1 Pord	10	.800	Ea.	.66	16.80		17.46	28
	0610 Finish coat, enamel		10	.800		.71	16.80		17.51	28.50
	0620 Primer & 1 coat enamel		6	1.333		1.37	28		29.37	47.50
	0630 Primer & 2 coats enamel		5	1.600		2.08	33.50		35.58	57.50
	0650 For oil base paint, add					10%				
320	0010 **DOORS AND WINDOWS, INTERIOR ALKYD (OIL BASE)**									
	0500 Flush door & frame, 3' x 7', oil, primer, brushwork	1 Pord	10	.800	Ea.	2.09	16.80		18.89	30
	1000 Paint, 1 coat		10	.800		2	16.80		18.80	29.50
	1200 2 coats		6	1.333		3.39	28		31.39	49.50
	1400 Stain, brushwork, wipe off		18	.444		1	9.35		10.35	16.45
	1600 Shellac, 1 coat, brushwork		25	.320		1.16	6.70		7.86	12.35
	1800 Varnish, 3 coats, brushwork, sand after 1st coat		9	.889		3.53	18.65		22.18	34.50
	2000 Panel door & frame, 3' x 7', oil, primer, brushwork		6	1.333		1.80	28		29.80	48
	2200 Paint, 1 coat		6	1.333		2	28		30	48
	2400 2 coats		3	2.667		5.80	56		61.80	98.50
	2600 Stain, brushwork, panel door, 3' x 7', not incl. frame		16	.500		1	10.50		11.50	18.35
	2800 Shellac, 1 coat, brushwork		22	.364		1.16	7.65		8.81	13.85
	3000 Varnish, 3 coats, brushwork, sand after 1st coat		7.50	1.067		3.53	22.50		26.03	41
	3020 French door, incl. 3' x 7', 6 lites, frame & trim									
	3022 Paint, 1 coat, over existing paint	1 Pord	5	1.600	Ea.	4.01	33.50		37.51	59.50
	3024 2 coats, over existing paint		5	1.600		7.80	33.50		41.30	63.50

	09910 | Paints		CREW	DAILY OUTPUT	LABOR-HOURS	UNIT	2004 BARE COSTS MAT.	LABOR	EQUIP.	TOTAL	TOTAL INCL O&P	
320	3026	Primer & 1 coat	1 Pord	3.50	2.286	Ea.	7.60	48		55.60	87.50	320
	3028	Primer & 2 coats		3	2.667		11.60	56		67.60	105	
	3032	Varnish or polyurethane, 1 coat		5	1.600		4.39	33.50		37.89	60	
	3034	2 coats, sanding between		3	2.667		8.80	56		64.80	102	
	4400	Windows, including frame and trim, per side										
	4600	Colonial type, 6/6 lites, 2′ x 3′, oil, primer, brushwork	1 Pord	14	.571	Ea.	.28	12		12.28	20	
	5800	Paint, 1 coat		14	.571		.32	12		12.32	20	
	6000	2 coats		9	.889		.61	18.65		19.26	31	
	6200	3′ x 5′ opening, 6/6 lites, primer coat, brushwork		12	.667		.71	14		14.71	24	
	6400	Paint, 1 coat		12	.667		.79	14		14.79	24	
	6600	2 coats		7	1.143		1.54	24		25.54	41	
	6800	4′ x 8′ opening, 6/6 lites, primer coat, brushwork		8	1		1.52	21		22.52	36	
	7000	Paint, 1 coat		8	1		1.69	21		22.69	36.50	
	7200	2 coats		5	1.600		3.28	33.50		36.78	58.50	
	8000	Single lite type, 2′ x 3′, oil base, primer coat, brushwork		33	.242		.28	5.10		5.38	8.65	
	8200	Paint, 1 coat		33	.242		.32	5.10		5.42	8.70	
	8400	2 coats		20	.400		.61	8.40		9.01	14.50	
	8600	3′ x 5′ opening, primer coat, brushwork		20	.400		.71	8.40		9.11	14.60	
	8800	Paint, 1 coat		20	.400		.79	8.40		9.19	14.65	
	9000	2 coats		13	.615		1.54	12.90		14.44	22.50	
	9200	4′ x 8′ opening, primer coat, brushwork		14	.571		1.52	12		13.52	21.50	
	9400	Paint, 1 coat		14	.571		1.69	12		13.69	21.50	
	9600	2 coats		8	1		3.28	21		24.28	38	
400	0010	**FENCES**										400
	0100	Chain link or wire metal, one side, water base										
	0110	Roll & brush, first coat R09910-220	1 Pord	960	.008	S.F.	.06	.17		.23	.35	
	0120	Second coat		1,280	.006		.05	.13		.18	.28	
	0130	Spray, first coat		2,275	.004		.06	.07		.13	.18	
	0140	Second coat		2,600	.003		.06	.06		.12	.17	
	0150	Picket, water base										
	0160	Roll & brush, first coat	1 Pord	865	.009	S.F.	.06	.19		.25	.39	
	0170	Second coat		1,050	.008		.06	.16		.22	.33	
	0180	Spray, first coat		2,275	.004		.06	.07		.13	.19	
	0190	Second coat		2,600	.003		.06	.06		.12	.18	
	0200	Stockade, water base										
	0210	Roll & brush, first coat	1 Pord	1,040	.008	S.F.	.06	.16		.22	.34	
	0220	Second coat		1,200	.007		.06	.14		.20	.30	
	0230	Spray, first coat		2,275	.004		.06	.07		.13	.19	
	0240	Second coat		2,600	.003		.06	.06		.12	.18	
500	0010	**FLOORS, INTERIOR**										500
	0100	Concrete										
	0120	1st coat	1 Pord	975	.008	S.F.	.11	.17		.28	.40	
	0130	2nd coat		1,150	.007		.07	.15		.22	.32	
	0140	3rd coat		1,300	.006		.06	.13		.19	.27	
	0150	Roll, latex, block filler										
	0160	1st coat	1 Pord	2,600	.003	S.F.	.15	.06		.21	.27	
	0170	2nd coat		3,250	.002		.09	.05		.14	.18	
	0180	3rd coat		3,900	.002		.06	.04		.10	.14	
	0190	Spray, latex, block filler										
	0200	1st coat	1 Pord	2,600	.003	S.F.	.13	.06		.19	.25	
	0210	2nd coat		3,250	.002		.07	.05		.12	.16	
	0220	3rd coat		3,900	.002		.05	.04		.09	.13	
620	0010	**MISCELLANEOUS, EXTERIOR** R09910-220										620
	0100	Railing, ext., decorative wood, incl. cap & baluster										

09910 | Paints

		09910 Paints	CREW	DAILY OUTPUT	LABOR-HOURS	UNIT	2004 BARE COSTS MAT.	LABOR	EQUIP.	TOTAL	TOTAL INCL O&P
620	0110	newels & spindles @ 12″ O.C. (R09910-220)									
	0120	Brushwork, stain, sand, seal & varnish									
	0130	First coat	1 Pord	90	.089	L.F.	.44	1.87		2.31	3.55
	0140	Second coat	″	120	.067	″	.44	1.40		1.84	2.78
	0150	Rough sawn wood, 42″ high, 2″x2″ verticals, 6″ O.C.									
	0160	Brushwork, stain, each coat	1 Pord	90	.089	L.F.	.14	1.87		2.01	3.23
	0170	Wrought iron, 1″ rail, 1/2″ sq. verticals									
	0180	Brushwork, zinc chromate, 60″ high, bars 6″ O.C.									
	0190	Primer	1 Pord	130	.062	L.F.	.49	1.29		1.78	2.66
	0200	Finish coat		130	.062		.15	1.29		1.44	2.29
	0210	Additional coat		190	.042		.18	.88		1.06	1.65
	0220	Shutters or blinds, single panel, 2′x4′, paint all sides									
	0230	Brushwork, primer	1 Pord	20	.400	Ea.	.55	8.40		8.95	14.40
	0240	Finish coat, exterior latex		20	.400		.46	8.40		8.86	14.30
	0250	Primer & 1 coat, exterior latex		13	.615		.89	12.90		13.79	22
	0260	Spray, primer		35	.229		.80	4.80		5.60	8.80
	0270	Finish coat, exterior latex		35	.229		.97	4.80		5.77	8.95
	0280	Primer & 1 coat, exterior latex		20	.400		.87	8.40		9.27	14.75
	0290	For louvered shutters, add				S.F.	10%				
	0300	Stair stringers, exterior, metal									
	0310	Roll & brush, zinc chromate, to 14″, each coat	1 Pord	320	.025	L.F.	.05	.53		.58	.92
	0320	Rough sawn wood, 4″ x 12″									
	0330	Roll & brush, exterior latex, each coat	1 Pord	215	.037	L.F.	.07	.78		.85	1.35
	0340	Trellis/lattice, 2″x2″ @ 3″ O.C. with 2″x8″ supports									
	0350	Spray, latex, per side, each coat	1 Pord	475	.017	S.F.	.07	.35		.42	.65
	0450	Decking, Ext., sealer, alkyd, brushwork, sealer coat		1,140	.007		.05	.15		.20	.30
	0460	1st coat		1,140	.007		.05	.15		.20	.30
	0470	2nd coat		1,300	.006		.04	.13		.17	.25
	0500	Paint, alkyd, brushwork, primer coat		1,140	.007		.07	.15		.22	.32
	0510	1st coat		1,140	.007		.07	.15		.22	.31
	0520	2nd coat		1,300	.006		.05	.13		.18	.26
	0600	Sand paint, alkyd, brushwork, 1 coat		150	.053		.09	1.12		1.21	1.93
630	0010	**MISCELLANEOUS, INTERIOR**									
	2400	Floors, conc./wood, oil base, primer/sealer coat, brushwork	2 Pord	1,950	.008	S.F.	.06	.17		.23	.34
	2450	Roller		5,200	.003		.06	.06		.12	.18
	2600	Spray		6,000	.003		.06	.06		.12	.16
	2650	Paint 1 coat, brushwork		1,950	.008		.05	.17		.22	.34
	2800	Roller		5,200	.003		.05	.06		.11	.17
	2850	Spray		6,000	.003		.06	.06		.12	.15
	3000	Stain, wood floor, brushwork, 1 coat		4,550	.004		.05	.07		.12	.17
	3200	Roller		5,200	.003		.05	.06		.11	.17
	3250	Spray		6,000	.003		.05	.06		.11	.14
	3400	Varnish, wood floor, brushwork		4,550	.004		.06	.07		.13	.18
	3450	Roller		5,200	.003		.06	.06		.12	.17
	3600	Spray		6,000	.003		.06	.06		.12	.16
	3800	Grilles, per side, oil base, primer coat, brushwork	1 Pord	520	.015		.09	.32		.41	.63
	3850	Spray		1,140	.007		.10	.15		.25	.35
	3880	Paint 1 coat, brushwork		520	.015		.11	.32		.43	.65
	3900	Spray		1,140	.007		.12	.15		.27	.37
	3920	Paint 2 coats, brushwork		325	.025		.20	.52		.72	1.08
	3940	Spray		650	.012		.23	.26		.49	.68
	4250	Paint 1 coat, brushwork	2 Pord	1,300	.012	L.F.	.11	.26		.37	.54
	4500	Louvers, one side, primer, brushwork	1 Pord	524	.015	S.F.	.06	.32		.38	.59
	4520	Paint one coat, brushwork		520	.015		.05	.32		.37	.59
	4530	Spray		1,140	.007		.06	.15		.21	.31
	4540	Paint two coats, brushwork		325	.025		.11	.52		.63	.97

	09910 | Paints		CREW	DAILY OUTPUT	LABOR-HOURS	UNIT	2004 BARE COSTS MAT.	LABOR	EQUIP.	TOTAL	TOTAL INCL O&P	
630	4550	Spray	1 Pord	650	.012	S.F.	.12	.26		.38	.55	630
	4560	Paint three coats, brushwork		270	.030		.16	.62		.78	1.19	
	4570	Spray	↓	500	.016	↓	.18	.34		.52	.74	
	5000	Pipe, to 4" diameter, primer or sealer coat, oil base, brushwork	2 Pord	1,250	.013	L.F.	.06	.27		.33	.51	
	5100	Spray		2,165	.007		.06	.16		.22	.32	
	5200	Paint 1 coat, brushwork		1,250	.013		.06	.27		.33	.51	
	5300	Spray		2,165	.007		.06	.16		.22	.32	
	5350	Paint 2 coats, brushwork		775	.021		.11	.43		.54	.83	
	5400	Spray		1,240	.013		.12	.27		.39	.59	
	5450	To 8" diameter, primer or sealer coat, brushwork		620	.026		.12	.54		.66	1.02	
	5500	Spray		1,085	.015		.20	.31		.51	.72	
	5550	Paint 1 coat, brushwork		620	.026		.17	.54		.71	1.08	
	5600	Spray		1,085	.015		.19	.31		.50	.72	
	5650	Paint 2 coats, brushwork		385	.042		.22	.87		1.09	1.68	
	5700	Spray	↓	620	.026	↓	.25	.54		.79	1.16	
	6600	Radiators, per side, primer, brushwork	1 Pord	520	.015	S.F.	.06	.32		.38	.59	
	6620	Paint one coat, brushwork		520	.015		.05	.32		.37	.59	
	6640	Paint two coats, brushwork		340	.024		.11	.49		.60	.93	
	6660	Paint three coats, brushwork	↓	283	.028	↓	.16	.59		.75	1.15	
	7000	Trim, wood, incl. puttying, under 6" wide										
	7200	Primer coat, oil base, brushwork	1 Pord	650	.012	L.F.	.02	.26		.28	.45	
	7250	Paint, 1 coat, brushwork		650	.012		.03	.26		.29	.45	
	7400	2 coats		400	.020		.05	.42		.47	.75	
	7450	3 coats		325	.025		.08	.52		.60	.93	
	7500	Over 6" wide, primer coat, brushwork		650	.012		.05	.26		.31	.47	
	7550	Paint, 1 coat, brushwork		650	.012		.05	.26		.31	.48	
	7600	2 coats		400	.020		.10	.42		.52	.80	
	7650	3 coats		325	.025	↓	.15	.52		.67	1.02	
	8000	Cornice, simple design, primer coat, oil base, brushwork		650	.012	S.F.	.05	.26		.31	.47	
	8250	Paint, 1 coat		650	.012		.05	.26		.31	.48	
	8300	2 coats		400	.020		.10	.42		.52	.80	
	8350	Ornate design, primer coat		350	.023		.05	.48		.53	.84	
	8400	Paint, 1 coat		350	.023		.05	.48		.53	.85	
	8450	2 coats		400	.020		.10	.42		.52	.80	
	8600	Balustrades, primer coat, oil base, brushwork		520	.015		.05	.32		.37	.58	
	8650	Paint, 1 coat		520	.015		.05	.32		.37	.59	
	8700	2 coats		325	.025		.10	.52		.62	.96	
	8900	Trusses and wood frames, primer coat, oil base, brushwork		800	.010		.05	.21		.26	.40	
	8950	Spray		1,200	.007		.05	.14		.19	.28	
	9000	Paint 1 coat, brushwork		750	.011		.05	.22		.27	.43	
	9200	Spray		1,200	.007		.06	.14		.20	.29	
	9220	Paint 2 coats, brushwork		500	.016		.10	.34		.44	.66	
	9240	Spray		600	.013		.11	.28		.39	.59	
	9260	Stain, brushwork, wipe off		600	.013		.05	.28		.33	.51	
	9280	Varnish, 3 coats, brushwork	↓	275	.029		.17	.61		.78	1.19	
	9350	For latex paint, deduct				↓	10%					
700	0010	**SIDING EXTERIOR**, Alkyd (oil base)										700
	0450	Steel siding, oil base, paint 1 coat, brushwork	2 Pord	2,015	.008	S.F.	.05	.17		.22	.33	
	0500	Spray		4,550	.004		.08	.07		.15	.21	
	0800	Paint 2 coats, brushwork		1,300	.012		.11	.26		.37	.54	
	1000	Spray		4,550	.004		.14	.07		.21	.27	
	1200	Stucco, rough, oil base, paint 2 coats, brushwork		1,300	.012		.11	.26		.37	.54	
	1400	Roller		1,625	.010		.11	.21		.32	.47	
	1600	Spray		2,925	.005		.12	.11		.23	.32	
	1800	Texture 1-11 or clapboard, oil base, primer coat, brushwork		1,300	.012		.09	.26		.35	.52	
	2000	Spray	↓	4,550	.004	↓	.09	.07		.16	.22	

		09910 | Paints	CREW	DAILY OUTPUT	LABOR-HOURS	UNIT	2004 BARE COSTS MAT.	LABOR	EQUIP.	TOTAL	TOTAL INCL O&P	
700	2100	Paint 1 coat, brushwork	2 Pord	1,300	.012	S.F.	.08	.26		.34	.51	700
	2200	Spray		4,550	.004		.08	.07		.15	.21	
	2400	Paint 2 coats, brushwork		810	.020		.16	.41		.57	.85	
	2600	Spray		2,600	.006		.18	.13		.31	.40	
	3000	Stain 1 coat, brushwork		1,520	.011		.05	.22		.27	.41	
	3200	Spray		5,320	.003		.05	.06		.11	.16	
	3400	Stain 2 coats, brushwork		950	.017		.10	.35		.45	.68	
	4000	Spray		3,050	.005		.11	.11		.22	.30	
	4200	Wood shingles, oil base primer coat, brushwork		1,300	.012		.08	.26		.34	.51	
	4400	Spray		3,900	.004		.08	.09		.17	.23	
	4600	Paint 1 coat, brushwork		1,300	.012		.07	.26		.33	.49	
	4800	Spray		3,900	.004		.08	.09		.17	.23	
	5000	Paint 2 coats, brushwork		810	.020		.13	.41		.54	.82	
	5200	Spray		2,275	.007		.12	.15		.27	.38	
	5800	Stain 1 coat, brushwork		1,500	.011		.05	.22		.27	.42	
	6000	Spray		3,900	.004		.05	.09		.14	.19	
	6500	Stain 2 coats, brushwork		950	.017		.10	.35		.45	.68	
	7000	Spray	↓	2,660	.006		.13	.13		.26	.35	
	8000	For latex paint, deduct					10%					
	8100	For work over 12′ H, from pipe scaffolding, add						15%				
	8200	For work over 12′ H, from extension ladder, add						25%				
	8300	For work over 12′ H, from swing staging, add				↓		35%				
710	0010	**SIDING, MISC.** R09910-220										710
	0100	Aluminum siding										
	0110	Brushwork, primer	2 Pord	2,275	.007	S.F.	.05	.15		.20	.30	
	0120	Finish coat, exterior latex		2,275	.007		.04	.15		.19	.28	
	0130	Primer & 1 coat exterior latex		1,300	.012		.10	.26		.36	.53	
	0140	Primer & 2 coats exterior latex	↓	975	.016	↓	.13	.34		.47	.72	
	0150	Mineral Fiber shingles										
	0160	Brushwork, primer	2 Pord	1,495	.011	S.F.	.09	.22		.31	.47	
	0170	Finish coat, industrial enamel		1,495	.011		.10	.22		.32	.48	
	0180	Primer & 1 coat enamel		810	.020		.19	.41		.60	.89	
	0190	Primer & 2 coats enamel		540	.030		.29	.62		.91	1.34	
	0200	Roll, primer		1,625	.010		.10	.21		.31	.45	
	0210	Finish coat, industrial enamel		1,625	.010		.11	.21		.32	.46	
	0220	Primer & 1 coat enamel		975	.016		.21	.34		.55	.80	
	0230	Primer & 2 coats enamel		650	.025		.32	.52		.84	1.20	
	0240	Spray, primer		3,900	.004		.08	.09		.17	.23	
	0250	Finish coat, industrial enamel		3,900	.004		.09	.09		.18	.24	
	0260	Primer & 1 coat enamel		2,275	.007		.17	.15		.32	.42	
	0270	Primer & 2 coats enamel		1,625	.010		.26	.21		.47	.62	
	0280	Waterproof sealer, first coat		4,485	.004		.06	.08		.14	.19	
	0290	Second coat	↓	5,235	.003	↓	.06	.06		.12	.18	
	0300	Rough wood incl. shingles, shakes or rough sawn siding										
	0310	Brushwork, primer	2 Pord	1,280	.013	S.F.	.11	.26		.37	.55	
	0320	Finish coat, exterior latex		1,280	.013		.07	.26		.33	.51	
	0330	Primer & 1 coat exterior latex		960	.017		.18	.35		.53	.78	
	0340	Primer & 2 coats exterior latex		700	.023		.25	.48		.73	1.06	
	0350	Roll, primer		2,925	.005		.15	.11		.26	.35	
	0360	Finish coat, exterior latex		2,925	.005		.08	.11		.19	.28	
	0370	Primer & 1 coat exterior latex		1,790	.009		.23	.19		.42	.56	
	0380	Primer & 2 coats exterior latex		1,300	.012		.31	.26		.57	.76	
	0390	Spray, primer		3,900	.004		.13	.09		.22	.28	
	0400	Finish coat, exterior latex		3,900	.004		.06	.09		.15	.21	
	0410	Primer & 1 coat exterior latex		2,600	.006		.19	.13		.32	.42	
	0420	Primer & 2 coats exterior latex	↓	2,080	.008	↓	.25	.16		.41	.55	

		09910 | Paints		CREW	DAILY OUTPUT	LABOR-HOURS	UNIT	2004 BARE COSTS MAT.	LABOR	EQUIP.	TOTAL	TOTAL INCL O&P	
710	0430	Waterproof sealer, first coat	R09910-220	2 Pord	4,485	.004	S.F.	.12	.08		.20	.25	710
	0440	Second coat		↓	4,485	.004	↓	.06	.08		.14	.19	
	0450	Smooth wood incl. butt, T&G, beveled, drop or B&B siding											
	0460	Brushwork, primer		2 Pord	2,325	.007	S.F.	.08	.14		.22	.33	
	0470	Finish coat, exterior latex			1,280	.013		.07	.26		.33	.51	
	0480	Primer & 1 coat exterior latex			800	.020		.15	.42		.57	.85	
	0490	Primer & 2 coats exterior latex			630	.025		.22	.53		.75	1.12	
	0500	Roll, primer			2,275	.007		.09	.15		.24	.34	
	0510	Finish coat, exterior latex			2,275	.007		.07	.15		.22	.32	
	0520	Primer & 1 coat exterior latex			1,300	.012		.16	.26		.42	.60	
	0530	Primer & 2 coats exterior latex			975	.016		.24	.34		.58	.83	
	0540	Spray, primer			4,550	.004		.07	.07		.14	.20	
	0550	Finish coat, exterior latex			4,550	.004		.06	.07		.13	.19	
	0560	Primer & 1 coat exterior latex			2,600	.006		.13	.13		.26	.36	
	0570	Primer & 2 coats exterior latex			1,950	.008		.20	.17		.37	.50	
	0580	Waterproof sealer, first coat			5,230	.003		.06	.06		.12	.18	
	0590	Second coat		↓	5,980	.003		.06	.06		.12	.16	
	0600	For oil base paint, add	↓				↓	10%					
800	0010	**TRIM, EXTERIOR**	R09910-220										800
	0100	Door frames & trim (see Doors, interior or exterior)											
	0110	Fascia, latex paint, one coat coverage											
	0120	1" x 4", brushwork		1 Pord	640	.013	L.F.	.02	.26		.28	.45	
	0130	Roll			1,280	.006		.02	.13		.15	.24	
	0140	Spray			2,080	.004		.01	.08		.09	.15	
	0150	1" x 6" to 1" x 10", brushwork			640	.013		.06	.26		.32	.50	
	0160	Roll			1,230	.007		.06	.14		.20	.29	
	0170	Spray			2,100	.004		.05	.08		.13	.18	
	0180	1" x 12", brushwork			640	.013		.06	.26		.32	.50	
	0190	Roll			1,050	.008		.06	.16		.22	.33	
	0200	Spray		↓	2,200	.004	↓	.05	.08		.13	.18	
	0210	Gutters & downspouts, metal, zinc chromate paint											
	0220	Brushwork, gutters, 5", first coat		1 Pord	640	.013	L.F.	.05	.26		.31	.49	
	0230	Second coat			960	.008		.05	.17		.22	.35	
	0240	Third coat			1,280	.006		.04	.13		.17	.27	
	0250	Downspouts, 4", first coat			640	.013		.05	.26		.31	.49	
	0260	Second coat			960	.008		.05	.17		.22	.35	
	0270	Third coat		↓	1,280	.006	↓	.04	.13		.17	.27	
	0280	Gutters & downspouts, wood											
	0290	Brushwork, gutters, 5", primer		1 Pord	640	.013	L.F.	.05	.26		.31	.49	
	0300	Finish coat, exterior latex			640	.013		.05	.26		.31	.49	
	0310	Primer & 1 coat exterior latex			400	.020		.11	.42		.53	.81	
	0320	Primer & 2 coats exterior latex			325	.025		.16	.52		.68	1.03	
	0330	Downspouts, 4", primer			640	.013		.05	.26		.31	.49	
	0340	Finish coat, exterior latex			640	.013		.05	.26		.31	.49	
	0350	Primer & 1 coat exterior latex			400	.020		.11	.42		.53	.81	
	0360	Primer & 2 coats exterior latex		↓	325	.025	↓	.08	.52		.60	.94	
	0370	Molding, exterior, up to 14" wide											
	0380	Brushwork, primer		1 Pord	640	.013	L.F.	.06	.26		.32	.50	
	0390	Finish coat, exterior latex			640	.013		.06	.26		.32	.50	
	0400	Primer & 1 coat exterior latex			400	.020		.13	.42		.55	.83	
	0410	Primer & 2 coats exterior latex			315	.025		.13	.53		.66	1.02	
	0420	Stain & fill			1,050	.008		.06	.16		.22	.32	
	0430	Shellac			1,850	.004		.07	.09		.16	.22	
	0440	Varnish	↓	↓	1,275	.006	↓	.07	.13		.20	.29	
910	0350	**WALLS, MASONRY (CMU), EXTERIOR**											910
	0360	Concrete masonry units (CMU), smooth surface											

09910 Paints			CREW	DAILY OUTPUT	LABOR-HOURS	UNIT	2004 BARE COSTS MAT.	LABOR	EQUIP.	TOTAL	TOTAL INCL O&P	
910	0370	Brushwork, latex, first coat	1 Pord	640	.013	S.F.	.03	.26		.29	.47	910
	0380	Second coat		960	.008		.03	.17		.20	.32	
	0390	Waterproof sealer, first coat		736	.011		.07	.23		.30	.46	
	0400	Second coat		1,104	.007		.05	.15		.20	.30	
	0410	Roll, latex, paint, first coat		1,465	.005		.04	.11		.15	.24	
	0420	Second coat		1,790	.004		.03	.09		.12	.18	
	0430	Waterproof sealer, first coat		1,680	.005		.07	.10		.17	.24	
	0440	Second coat		2,060	.004		.05	.08		.13	.18	
	0450	Spray, latex, paint, first coat		1,950	.004		.03	.09		.12	.18	
	0460	Second coat		2,600	.003		.03	.06		.09	.14	
	0470	Waterproof sealer, first coat		2,245	.004		.09	.07		.16	.21	
	0480	Second coat	↓	2,990	.003	↓	.03	.06		.09	.13	
	0490	Concrete masonry unit (CMU), porous										
	0500	Brushwork, latex, first coat	1 Pord	640	.013	S.F.	.07	.26		.33	.51	
	0510	Second coat		960	.008		.03	.17		.20	.33	
	0520	Waterproof sealer, first coat		736	.011		.09	.23		.32	.47	
	0530	Second coat		1,104	.007		.04	.15		.19	.30	
	0540	Roll latex, first coat		1,465	.005		.05	.11		.16	.25	
	0550	Second coat		1,790	.004		.03	.09		.12	.19	
	0560	Waterproof sealer, first coat		1,680	.005		.09	.10		.19	.25	
	0570	Second coat		2,060	.004		.05	.08		.13	.18	
	0580	Spray latex, first coat		1,950	.004		.04	.09		.13	.18	
	0590	Second coat		2,600	.003		.03	.06		.09	.14	
	0600	Waterproof sealer, first coat		2,245	.004		.07	.07		.14	.20	
	0610	Second coat	↓	2,990	.003	↓	.04	.06		.10	.13	
920	0010	**WALLS AND CEILINGS**, Interior										920
	0100	Concrete, dry wall or plaster, oil base, primer or sealer coat										
	0200	Smooth finish, brushwork	1 Pord	1,150	.007	S.F.	.05	.15		.20	.29	
	0240	Roller		2,040	.004		.04	.08		.12	.19	
	0300	Sand finish, brushwork		975	.008		.04	.17		.21	.33	
	0340	Roller		1,150	.007		.05	.15		.20	.29	
	0380	Spray		2,275	.004		.04	.07		.11	.16	
	0400	Paint 1 coat, smooth finish, brushwork		1,200	.007		.05	.14		.19	.28	
	0440	Roller		1,300	.006		.05	.13		.18	.26	
	0480	Spray		2,275	.004		.04	.07		.11	.16	
	0500	Sand finish, brushwork		1,050	.008		.04	.16		.20	.31	
	0540	Roller		1,600	.005		.05	.11		.16	.22	
	0580	Spray		2,100	.004		.04	.08		.12	.17	
	0800	Paint 2 coats, smooth finish, brushwork		680	.012		.09	.25		.34	.51	
	0840	Roller		800	.010		.09	.21		.30	.45	
	0880	Spray		1,625	.005		.08	.10		.18	.26	
	0900	Sand finish, brushwork		605	.013		.09	.28		.37	.56	
	0940	Roller		1,020	.008		.09	.16		.25	.37	
	0980	Spray		1,700	.005		.08	.10		.18	.25	
	1200	Paint 3 coats, smooth finish, brushwork		510	.016		.13	.33		.46	.69	
	1240	Roller		650	.012		.14	.26		.40	.58	
	1280	Spray		1,625	.005		.12	.10		.22	.30	
	1300	Sand finish, brushwork		454	.018		.13	.37		.50	.76	
	1340	Roller		680	.012		.14	.25		.39	.57	
	1380	Spray		1,133	.007		.12	.15		.27	.37	
	1600	Glaze coating, 5 coats, spray, clear		900	.009		.60	.19		.79	.97	
	1640	Multicolor	↓	900	.009		.83	.19		1.02	1.22	
	1700	For latex paint, deduct					10%					
	1800	For ceiling installations, add				↓		25%				
	2000	Masonry or concrete block, oil base, primer or sealer coat										

9 FINISHES

	09910 | Paints		CREW	DAILY OUTPUT	LABOR-HOURS	UNIT	2004 BARE COSTS MAT.	LABOR	EQUIP.	TOTAL	TOTAL INCL O&P	
920	2100	Smooth finish, brushwork	1 Pord	1,224	.007	S.F.	.05	.14		.19	.28	920
	2180	Spray		2,400	.003		.07	.07		.14	.18	
	2200	Sand finish, brushwork		1,089	.007		.07	.15		.22	.33	
	2280	Spray		2,400	.003		.07	.07		.14	.18	
	2400	Paint 1 coat, smooth finish, brushwork		1,100	.007		.07	.15		.22	.33	
	2480	Spray		2,400	.003		.07	.07		.14	.18	
	2500	Sand finish, brushwork		979	.008		.07	.17		.24	.36	
	2580	Spray		2,400	.003		.07	.07		.14	.18	
	2800	Paint 2 coats, smooth finish, brushwork		756	.011		.15	.22		.37	.53	
	2880	Spray		1,360	.006		.14	.12		.26	.35	
	2900	Sand finish, brushwork		672	.012		.15	.25		.40	.57	
	2980	Spray		1,360	.006		.14	.12		.26	.35	
	3200	Paint 3 coats, smooth finish, brushwork		560	.014		.22	.30		.52	.73	
	3280	Spray		1,088	.007		.20	.15		.35	.47	
	3300	Sand finish, brushwork		498	.016		.22	.34		.56	.79	
	3380	Spray		1,088	.007		.20	.15		.35	.47	
	3600	Glaze coating, 5 coats, spray, clear		900	.009		.60	.19		.79	.97	
	3620	Multicolor		900	.009		.83	.19		1.02	1.22	
	4000	Block filler, 1 coat, brushwork		425	.019		.12	.40		.52	.78	
	4100	Silicone, water repellent, 2 coats, spray		2,000	.004		.25	.08		.33	.41	
	4120	For latex paint, deduct					10%					
	8200	For work 8 - 15′ H, add						10%				
	8300	For work over 15′ H, add						20%				
940	0010	**DRY FALL PAINTING** R09910-220										940
	0100	Walls										
	0200	Wallboard and smooth plaster, one coat, brush	1 Pord	910	.009	S.F.	.04	.18		.22	.35	
	0210	Roll		1,560	.005		.04	.11		.15	.23	
	0220	Spray		2,600	.003		.04	.06		.10	.16	
	0230	Two coats, brush		520	.015		.08	.32		.40	.62	
	0240	Roll		877	.009		.08	.19		.27	.40	
	0250	Spray		1,560	.005		.08	.11		.19	.27	
	0260	Concrete or textured plaster, one coat, brush		747	.011		.04	.22		.26	.42	
	0270	Roll		1,300	.006		.04	.13		.17	.26	
	0280	Spray		1,560	.005		.04	.11		.15	.23	
	0290	Two coats, brush		422	.019		.08	.40		.48	.74	
	0300	Roll		747	.011		.08	.22		.30	.46	
	0310	Spray		1,300	.006		.08	.13		.21	.30	
	0320	Concrete block, one coat, brush		747	.011		.04	.22		.26	.42	
	0330	Roll		1,300	.006		.04	.13		.17	.26	
	0340	Spray		1,560	.005		.04	.11		.15	.23	
	0350	Two coats, brush		422	.019		.08	.40		.48	.74	
	0360	Roll		747	.011		.08	.22		.30	.46	
	0370	Spray		1,300	.006		.08	.13		.21	.30	
	0380	Wood, one coat, brush		747	.011		.04	.22		.26	.42	
	0390	Roll		1,300	.006		.04	.13		.17	.26	
	0400	Spray		877	.009		.04	.19		.23	.36	
	0410	Two coats, brush		487	.016		.08	.35		.43	.66	
	0420	Roll		747	.011		.08	.22		.30	.46	
	0430	Spray		650	.012		.08	.26		.34	.51	
	0440	Ceilings										
	0450	Wallboard and smooth plaster, one coat, brush	1 Pord	600	.013	S.F.	.04	.28		.32	.51	
	0460	Roll		1,040	.008		.04	.16		.20	.32	
	0470	Spray		1,560	.005		.04	.11		.15	.23	
	0480	Two coats, brush		341	.023		.08	.49		.57	.90	
	0490	Roll		650	.012		.08	.26		.34	.51	

09900 | Paints & Coatings

09910 | Paints

			CREW	DAILY OUTPUT	LABOR-HOURS	UNIT	2004 BARE COSTS MAT.	LABOR	EQUIP.	TOTAL	TOTAL INCL O&P	
940	0500	Spray (R09910-220)	1 Pord	1,300	.006	S.F.	.08	.13		.21	.30	940
	0510	Concrete or textured plaster, one coat, brush		487	.016		.04	.35		.39	.62	
	0520	Roll		877	.009		.04	.19		.23	.36	
	0530	Spray		1,560	.005		.04	.11		.15	.23	
	0540	Two coats, brush		276	.029		.08	.61		.69	1.09	
	0550	Roll		520	.015		.08	.32		.40	.62	
	0560	Spray		1,300	.006		.08	.13		.21	.30	
	0570	Structural steel, bar joists or metal deck, one coat, spray		1,560	.005		.04	.11		.15	.23	
	0580	Two coats, spray	↓	1,040	.008	↓	.08	.16		.24	.36	

09930 | Stains/Transp. Finishes

			CREW	DAILY OUTPUT	LABOR-HOURS	UNIT	MAT.	LABOR	EQUIP.	TOTAL	TOTAL INCL O&P	
100	0010	**VARNISH** 1 coat + sealer, on wood trim, no sanding included	1 Pord	400	.020	S.F.	.06	.42		.48	.76	100
	0100	Hardwood floors, 2 coats, no sanding included, roller	"	1,890	.004	"	.12	.09		.21	.28	

09963 | Glazed Coatings

			CREW	DAILY OUTPUT	LABOR-HOURS	UNIT	MAT.	LABOR	EQUIP.	TOTAL	TOTAL INCL O&P	
200	0010	**WALL COATINGS**										200
	0100	Acrylic glazed coatings, minimum	1 Pord	525	.015	S.F.	.25	.32		.57	.81	
	0200	Maximum		305	.026		.52	.55		1.07	1.47	
	0300	Epoxy coatings, minimum		525	.015		.32	.32		.64	.88	
	0400	Maximum		170	.047		.99	.99		1.98	2.71	
	0600	Exposed aggregate, troweled on, 1/16" to 1/4", minimum		235	.034		.49	.71		1.20	1.71	
	0700	Maximum (epoxy or polyacrylate)		130	.062		1.06	1.29		2.35	3.29	
	0900	1/2" to 5/8" aggregate, minimum		130	.062		.98	1.29		2.27	3.20	
	1000	Maximum		80	.100		1.67	2.10		3.77	5.30	
	1200	1" aggregate size, minimum		90	.089		1.70	1.87		3.57	4.94	
	1300	Maximum		55	.145		2.60	3.05		5.65	7.85	
	1500	Exposed aggregate, sprayed on, 1/8" aggregate, minimum		295	.027		.46	.57		1.03	1.45	
	1600	Maximum	↓	145	.055	↓	.84	1.16		2	2.82	

09990 | Paint Restoration

			CREW	DAILY OUTPUT	LABOR-HOURS	UNIT	MAT.	LABOR	EQUIP.	TOTAL	TOTAL INCL O&P	
500	0010	**SCRAPE AFTER FIRE DAMAGE**										500
	0050	Boards, 1" x 4"	1 Pord	336	.024	L.F.		.50		.50	.82	
	0060	1" x 6"		260	.031			.65		.65	1.06	
	0070	1" x 8"		207	.039			.81		.81	1.33	
	0080	1" x 10"		174	.046			.97		.97	1.59	
	0500	Framing, 2" x 4"		265	.030			.63		.63	1.04	
	0510	2" x 6"		221	.036			.76		.76	1.25	
	0520	2" x 8"		190	.042			.88		.88	1.45	
	0530	2" x 10"		165	.048			1.02		1.02	1.67	
	0540	2" x 12"		144	.056			1.17		1.17	1.92	
	1000	Heavy framing, 3" x 4"		226	.035			.74		.74	1.22	
	1010	4" x 4"		210	.038			.80		.80	1.31	
	1020	4" x 6"		191	.042			.88		.88	1.44	
	1030	4" x 8"		165	.048			1.02		1.02	1.67	
	1040	4" x 10"		144	.056			1.17		1.17	1.92	
	1060	4" x 12"		131	.061	↓		1.28		1.28	2.11	
	2900	For sealing, minimum		825	.010	S.F.	.12	.20		.32	.46	
	2920	Maximum	↓	460	.017	"	.25	.37		.62	.88	
800	0010	**SANDING** and puttying interior trim, compared to										800
	0100	Painting 1 coat, on quality work				L.F.		100%				
	0300	Medium work						50%				
	0400	Industrial grade				↓		25%				
	0500	Surface protection, placement and removal										
	0510	Basic drop cloths	1 Pord	6,400	.001	S.F.		.03		.03	.04	

9 FINISHES

09990 | Paint Restoration

		09990 Paint Restoration	Crew	Daily Output	Labor-Hours	Unit	2004 Bare Costs Mat.	Labor	Equip.	Total	Total Incl O&P
800	0520	Masking with paper	1 Pord	800	.010	S.F.	.03	.21		.24	.38
	0530	Volume cover up (using plastic sheathing, or building paper)		16,000	.001			.01		.01	.02
900	0010	**SURFACE PREPARATION, EXTERIOR**									
	0015	Doors, per side, not incl. frames or trim									
	0020	Scrape & sand									
	0030	Wood, flush	1 Pord	616	.013	S.F.		.27		.27	.45
	0040	Wood, detail		496	.016			.34		.34	.56
	0050	Wood, louvered		280	.029			.60		.60	.99
	0060	Wood, overhead		616	.013			.27		.27	.45
	0070	Wire brush									
	0080	Metal, flush	1 Pord	640	.013	S.F.		.26		.26	.43
	0090	Metal, detail		520	.015			.32		.32	.53
	0100	Metal, louvered		360	.022			.47		.47	.77
	0110	Metal or fibr., overhead		640	.013			.26		.26	.43
	0120	Metal, roll up		560	.014			.30		.30	.49
	0130	Metal, bulkhead		640	.013			.26		.26	.43
	0140	Power wash, based on 2500 lb. operating pressure									
	0150	Metal, flush	B-9	2,240	.018	S.F.		.31	.07	.38	.60
	0160	Metal, detail		2,120	.019			.33	.08	.41	.63
	0170	Metal, louvered		2,000	.020			.35	.08	.43	.68
	0180	Metal or fibr., overhead		2,400	.017			.29	.07	.36	.56
	0190	Metal, roll up		2,400	.017			.29	.07	.36	.56
	0200	Metal, bulkhead		2,200	.018			.31	.07	.38	.61
	0400	Windows, per side, not incl. trim									
	0410	Scrape & sand									
	0420	Wood, 1-2 lite	1 Pord	320	.025	S.F.		.53		.53	.86
	0430	Wood, 3-6 lite		280	.029			.60		.60	.99
	0440	Wood, 7-10 lite		240	.033			.70		.70	1.15
	0450	Wood, 12 lite		200	.040			.84		.84	1.38
	0460	Wood, Bay / Bow		320	.025			.53		.53	.86
	0470	Wire brush									
	0480	Metal, 1-2 lite	1 Pord	480	.017	S.F.		.35		.35	.58
	0490	Metal, 3-6 lite		400	.020			.42		.42	.69
	0500	Metal, Bay / Bow		480	.017			.35		.35	.58
	0510	Power wash, based on 2500 lb. operating pressure									
	0520	1-2 lite	B-9	4,400	.009	S.F.		.16	.04	.20	.31
	0530	3-6 lite		4,320	.009			.16	.04	.20	.31
	0540	7-10 lite		4,240	.009			.16	.04	.20	.32
	0550	12 lite		4,160	.010			.17	.04	.21	.32
	0560	Bay / Bow		4,400	.009			.16	.04	.20	.31
	0600	Siding, scrape and sand, light=10-30%, med.=30-70%									
	0610	Heavy=70-100%, % of surface to sand									
	0650	Texture 1-11, light	1 Pord	480	.017	S.F.		.35		.35	.58
	0660	Med.		440	.018			.38		.38	.63
	0670	Heavy		360	.022			.47		.47	.77
	0680	Wood shingles, shakes, light		440	.018			.38		.38	.63
	0690	Med.		360	.022			.47		.47	.77
	0700	Heavy		280	.029			.60		.60	.99
	0710	Clapboard, light		520	.015			.32		.32	.53
	0720	Med.		480	.017			.35		.35	.58
	0730	Heavy		400	.020			.42		.42	.69
	0740	Wire brush									
	0750	Aluminum, light	1 Pord	600	.013	S.F.		.28		.28	.46
	0760	Med.		520	.015			.32		.32	.53
	0770	Heavy		440	.018			.38		.38	.63
	0780	Pressure wash, based on 2500 lb.. operating pressure									

	09990 | Paint Restoration		CREW	DAILY OUTPUT	LABOR-HOURS	UNIT	2004 BARE COSTS MAT.	LABOR	EQUIP.	TOTAL	TOTAL INCL O&P
900	0790	Stucco	B-9	3,080	.013	S.F.		.22	.05	.27	.44
	0800	Aluminum or vinyl		3,200	.013			.22	.05	.27	.43
	0810	Siding, masonry, brick & block	↓	2,400	.017	↓		.29	.07	.36	.56
	1300	Miscellaneous, wire brush									
	1310	Metal, pedestrian gate	1 Pord	100	.080	S.F.		1.68		1.68	2.76
910	0010	**SURFACE PREPARATION, INTERIOR**									
	0020	Doors									
	0030	Scrape & sand									
	0040	Wood, flush	1 Pord	616	.013	S.F.		.27		.27	.45
	0050	Wood, detail		496	.016			.34		.34	.56
	0060	Wood, louvered	↓	280	.029	↓		.60		.60	.99
	0070	Wire brush									
	0080	Metal, flush	1 Pord	640	.013	S.F.		.26		.26	.43
	0090	Metal, detail		520	.015			.32		.32	.53
	0100	Metal, louvered	↓	360	.022	↓		.47		.47	.77
	0110	Hand wash									
	0120	Wood, flush	1 Pord	2,160	.004	S.F.		.08		.08	.13
	0130	Wood, detailed		2,000	.004			.08		.08	.14
	0140	Wood, louvered		1,360	.006			.12		.12	.20
	0150	Metal, flush		2,160	.004			.08		.08	.13
	0160	Metal, detail		2,000	.004			.08		.08	.14
	0170	Metal, louvered	↓	1,360	.006	↓		.12		.12	.20
	0400	Windows, per side, not incl. trim									
	0410	Scrape & sand									
	0420	Wood, 1-2 lite	1 Pord	360	.022	S.F.		.47		.47	.77
	0430	Wood, 3-6 lite		320	.025			.53		.53	.86
	0440	Wood, 7-10 lite		280	.029			.60		.60	.99
	0450	Wood, 12 lite		240	.033			.70		.70	1.15
	0460	Wood, Bay / Bow	↓	360	.022	↓		.47		.47	.77
	0470	Wire brush									
	0480	Metal, 1-2 lite	1 Pord	520	.015	S.F.		.32		.32	.53
	0490	Metal, 3-6 lite		440	.018			.38		.38	.63
	0500	Metal, Bay / Bow	↓	520	.015	↓		.32		.32	.53
	0600	Walls, sanding, light=10-30%									
	0610	Med.=30-70%, heavy=70-100%, % of surface to sand									
	0650	Walls, sand									
	0660	Drywall, gypsum, plaster, light	1 Pord	3,077	.003	S.F.		.05		.05	.09
	0670	Drywall, gypsum, plaster, med.		2,160	.004			.08		.08	.13
	0680	Drywall, gypsum, plaster, heavy		923	.009			.18		.18	.30
	0690	Wood, T&G, light		2,400	.003			.07		.07	.11
	0700	Wood, T&G, med.		1,600	.005			.11		.11	.17
	0710	Wood, T&G, heavy	↓	800	.010	↓		.21		.21	.35
	0720	Walls, wash									
	0730	Drywall, gypsum, plaster	1 Pord	3,200	.002	S.F.		.05		.05	.09
	0740	Wood, T&G		3,200	.002			.05		.05	.09
	0750	Masonry, brick & block, smooth		2,800	.003			.06		.06	.10
	0760	Masonry, brick & block, coarse	↓	2,000	.004	↓		.08		.08	.14
	8000	For Chemical Washing, see Division 04930									

For information about Means Estimating Seminars, see yellow pages 12 and 13 in back of book

Division 10 Specialties

Estimating Tips

General

- The items in this division are usually priced per square foot or each.
- Many items in Division 10 require some type of support system or special anchors that are not usually furnished with the item. The required anchors must be added to the estimate in the appropriate division.
- Some items in Division 10, such as lockers, may require assembly before installation. Verify the amount of assembly required. Assembly can often exceed installation time.

10150 Compartments & Cubicles

- Support angles and blocking are not included in the installation of toilet compartments, shower/dressing compartments or cubicles. Appropriate line items from Divisions 5 or 6 may need to be added to support the installations.
- Toilet partitions are priced by the stall. A stall consists of a side wall, pilaster and door with hardware. Toilet tissue holders and grab bars are extra.

10600 Partitions

- The required acoustical rating of a folding partition can have a significant impact on costs. Verify the sound transmission coefficient rating of the panel priced to the specification requirements.

10800 Toilet/Bath/Laundry Accessories

- Grab bar installation does not include supplemental blocking or backing to support the required load. When grab bars are installed at an existing facility provisions must be made to attach the grab bars to solid structure.

Reference Numbers

Reference numbers are shown in bold squares at the beginning of some major classifications. These numbers refer to related items in the Reference Section. The reference information may be an estimating procedure, an alternate pricing method or technical information.

Note: Not all subdivisions listed here necessarily appear in this publication.

10150 | Compartments & Cubicles

10185 | Shower/Dressing Compartments

		10185 Shower/Dressing Compartments	CREW	DAILY OUTPUT	LABOR-HOURS	UNIT	2004 BARE COSTS MAT.	LABOR	EQUIP.	TOTAL	TOTAL INCL O&P
100	0010	**PARTITIONS, SHOWER** Floor mounted, no plumbing									
	0100	Cabinet, incl. base, no door, painted steel, 1″ thick walls	2 Shee	5	3.200	Ea.	690	82		772	890
	0300	With door, fiberglass		4.50	3.556		560	91		651	770
	0600	Galvanized and painted steel, 1″ thick walls		5	3.200		730	82		812	940
	0800	Stall, 1″ thick wall, no base, enameled steel		5	3.200		795	82		877	1,000
	1500	Circular fiberglass, cabinet 36″ diameter,		4	4		575	102		677	805
	1700	One piece, 36″ diameter, less door		4	4		485	102		587	705
	1800	With door		3.50	4.571		800	117		917	1,075
	2400	Glass stalls, with doors, no receptors, chrome on brass		3	5.333		1,150	137		1,287	1,475
	2700	Anodized aluminum		4	4		790	102		892	1,050
	3200	Receptors, precast terrazzo, 32″ x 32″	2 Marb	14	1.143		214	25.50		239.50	279
	3300	48″ x 34″		9.50	1.684		365	37.50		402.50	470
	3500	Plastic, simulated terrazzo receptor, 32″ x 32″		14	1.143		90.50	25.50		116	143
	3600	32″ x 48″		12	1.333		134	30		164	197
	3800	Precast concrete, colors, 32″ x 32″		14	1.143		179	25.50		204.50	240
	3900	48″ x 48″		8	2		191	44.50		235.50	285
	4100	Shower doors, economy plastic, 24″ wide	1 Shee	9	.889		98.50	23		121.50	146
	4200	Tempered glass door, economy		8	1		164	25.50		189.50	223
	4400	Folding, tempered glass, aluminum frame		6	1.333		315	34		349	400
	4700	Deluxe, tempered glass, chrome on brass frame, minimum		8	1		244	25.50		269.50	310
	4800	Maximum		1	8		665	205		870	1,075
	4850	On anodized aluminum frame, minimum		2	4		116	102		218	299
	4900	Maximum		1	8		390	205		595	775
	5100	Shower enclosure, tempered glass, anodized alum. frame									
	5120	2 panel & door, corner unit, 32″ x 32″	1 Shee	2	4	Ea.	390	102		492	600
	5140	Neo-angle corner unit, 16″ x 24″ x 16″	″	2	4		720	102		822	960
	5200	Shower surround, 3 wall, polypropylene, 32″ x 32″	1 Carp	4	2		217	46		263	315
	5220	PVC, 32″ x 32″		4	2		247	46		293	350
	5240	Fiberglass		4	2		285	46		331	395
	5250	2 wall, polypropylene, 32″ x 32″		4	2		204	46		250	305
	5270	PVC		4	2		254	46		300	360
	5290	Fiberglass		4	2		283	46		329	390
	5300	Tub doors, tempered glass & frame, minimum	1 Shee	8	1		167	25.50		192.50	227
	5400	Maximum		6	1.333		385	34		419	480
	5600	Chrome plated, brass frame, minimum		8	1		218	25.50		243.50	282
	5700	Maximum		6	1.333		430	34		464	530
	5900	Tub/shower enclosure, temp. glass, alum. frame, minimum		2	4		295	102		397	495
	6200	Maximum		1.50	5.333		575	137		712	860
	6500	On chrome-plated brass frame, minimum		2	4		410	102		512	620
	6600	Maximum		1.50	5.333		840	137		977	1,150
	6800	Tub surround, 3 wall, polypropylene	1 Carp	4	2		165	46		211	261
	6900	PVC		4	2		251	46		297	355
	7000	Fiberglass, minimum		4	2		280	46		326	390
	7100	Maximum		3	2.667		480	61.50		541.50	635

10200 | Louvers & Vents

10210 | Wall Louvers

		10210 Wall Louvers	CREW	DAILY OUTPUT	LABOR-HOURS	UNIT	2004 BARE COSTS MAT.	LABOR	EQUIP.	TOTAL	TOTAL INCL O&P
800	0010	**LOUVERS** Aluminum with screen, residential, 8″ x 8″	1 Carp	38	.211	Ea.	7.65	4.86		12.51	16.65
	0100	12″ x 12″		38	.211		8.45	4.86		13.31	17.55

10200 | Louvers & Vents

10210 | Wall Louvers

			CREW	DAILY OUTPUT	LABOR-HOURS	UNIT	2004 BARE COSTS MAT.	LABOR	EQUIP.	TOTAL	TOTAL INCL O&P	
800	0200	12" x 18"	1 Carp	35	.229	Ea.	12.30	5.30		17.60	22.50	800
	0250	14" x 24"		30	.267		15.95	6.15		22.10	28	
	0300	18" x 24"		27	.296		18.55	6.85		25.40	32	
	0500	24" x 30"		24	.333		24	7.70		31.70	39	
	0700	Triangle, adjustable, small		20	.400		21	9.25		30.25	38.50	
	0800	Large		15	.533		40	12.30		52.30	65	
	2100	Midget, aluminum, 3/4" deep, 1" diameter		85	.094		.67	2.17		2.84	4.43	
	2150	3" diameter		60	.133		1.39	3.08		4.47	6.80	
	2200	4" diameter		50	.160		2.16	3.70		5.86	8.65	
	2250	6" diameter	↓	30	.267	↓	2.58	6.15		8.73	13.30	
	2300	Ridge vent strip, mill finish	1 Shee	155	.052	L.F.	2.30	1.32		3.62	4.74	
	2400	Under eaves vent, aluminum, mill finish, 16" x 4"	1 Carp	48	.167	Ea.	1.61	3.85		5.46	8.30	
	2500	16" x 8"		48	.167		1.79	3.85		5.64	8.50	
	7000	Vinyl gable vent, 8" x 8"		38	.211		8.95	4.86		13.81	18.10	
	7020	12" x 12"		38	.211		18.40	4.86		23.26	28.50	
	7080	12" x 18"		35	.229		23.50	5.30		28.80	35	
	7200	18" x 24"	↓	30	.267	↓	28	6.15		34.15	41	

10300 | Fireplaces & Stoves

10305 | Manufactured Fireplaces

			CREW	DAILY OUTPUT	LABOR-HOURS	UNIT	2004 BARE COSTS MAT.	LABOR	EQUIP.	TOTAL	TOTAL INCL O&P	
100	0010	**FIREPLACE, PREFABRICATED** Free standing or wall hung										100
	0100	with hood & screen, minimum	1 Carp	1.30	6.154	Ea.	1,050	142		1,192	1,400	
	0150	Average		1	8		1,250	185		1,435	1,700	
	0200	Maximum		.90	8.889	↓	3,075	205		3,280	3,725	
	0500	Chimney dbl. wall, all stainless, over 8'-6", 7" diam., add		33	.242	V.L.F.	47.50	5.60		53.10	62	
	0600	10" diameter, add		32	.250		50.50	5.80		56.30	65.50	
	0700	12" diameter, add		31	.258		66	5.95		71.95	82.50	
	0800	14" diameter, add		30	.267	↓	83.50	6.15		89.65	102	
	1000	Simulated brick chimney top, 4' high, 16" x 16"		10	.800	Ea.	180	18.50		198.50	230	
	1100	24" x 24"		7	1.143	"	335	26.50		361.50	415	
	1500	Simulated logs, gas fired, 40,000 BTU, 2' long, minimum		7	1.143	Set	440	26.50		466.50	530	
	1600	Maximum		6	1.333		615	31		646	730	
	1700	Electric, 1,500 BTU, 1'-6" long, minimum		7	1.143		124	26.50		150.50	182	
	1800	11,500 BTU, maximum		6	1.333	↓	269	31		300	350	
	2000	Fireplace, built-in, 36" hearth, radiant		1.30	6.154	Ea.	540	142		682	830	
	2100	Recirculating, small fan		1	8		770	185		955	1,175	
	2150	Large fan		.90	8.889		1,425	205		1,630	1,925	
	2200	42" hearth, radiant		1.20	6.667		685	154		839	1,025	
	2300	Recirculating, small fan		.90	8.889		900	205		1,105	1,350	
	2350	Large fan		.80	10		1,725	231		1,956	2,300	
	2400	48" hearth, radiant		1.10	7.273		1,275	168		1,443	1,675	
	2500	Recirculating, small fan		.80	10		1,575	231		1,806	2,150	
	2550	Large fan		.70	11.429		2,450	264		2,714	3,150	
	3000	See through, including doors		.80	10		2,025	231		2,256	2,625	
	3200	Corner (2 wall)	↓	1	8	↓	1,000	185		1,185	1,425	

10300 | Fireplaces & Stoves

	10310 Fireplace Specialties & Accessories	CREW	DAILY OUTPUT	LABOR-HOURS	UNIT	2004 BARE COSTS MAT.	LABOR	EQUIP.	TOTAL	TOTAL INCL O&P	
100	0010 FIREPLACE ACCESSORIES Chimney screens, galv., 13" x 13" flue	1 Bric	8	1	Ea.	33.50	24		57.50	77	100
	0050 Galv., 24" x 24" flue		5	1.600		100	38.50		138.50	174	
	0200 Stainless steel, 13" x 13" flue		8	1		264	24		288	330	
	0250 20" x 20" flue		5	1.600		360	38.50		398.50	460	
	0400 Cleanout doors and frames, cast iron, 8" x 8"		12	.667		29	16		45	58.50	
	0450 12" x 12"		10	.800		33.50	19.20		52.70	68.50	
	0500 18" x 24"		8	1		105	24		129	155	
	0550 Cast iron frame, steel door, 24" x 30"		5	1.600		227	38.50		265.50	315	
	0800 Damper, rotary control, steel, 30" opening		6	1.333		64	32		96	123	
	0850 Cast iron, 30" opening		6	1.333		71	32		103	131	
	1200 Steel plate, poker control, 60" opening		8	1		225	24		249	288	
	1250 84" opening, special opening		5	1.600		410	38.50		448.50	515	
	1400 "Universal" type, chain operated, 32" x 20" opening		8	1		156	24		180	211	
	1450 48" x 24" opening		5	1.600		261	38.50		299.50	350	
	1600 Dutch Oven door and frame, cast iron, 12" x 15" opening		13	.615		92	14.75		106.75	126	
	1650 Copper plated, 12" x 15" opening		13	.615		177	14.75		191.75	219	
	1800 Fireplace forms, no accessories, 32" opening		3	2.667		525	64		589	685	
	1900 36" opening		2.50	3.200		635	77		712	830	
	2000 40" opening		2	4		765	96		861	1,000	
	2100 78" opening		1.50	5.333		1,100	128		1,228	1,425	
	2400 Squirrel and bird screens, galvanized, 8" x 8" flue		16	.500		36	12		48	60	
	2450 13" x 13" flue	↓	12	.667	↓	41.50	16		57.50	72	
	10320 Stoves										
100	0010 WOODBURNING STOVES Cast iron, minimum	2 Carp	1.30	12.308	Ea.	775	284		1,059	1,325	100
	0020 Average		1	16		1,150	370		1,520	1,900	
	0030 Maximum	↓	.80	20		1,875	460		2,335	2,850	
	0050 For gas log lighter, add				↓	39			39	43	

10340 | Manufactured Exterior Specialties

	10342 Cupolas	CREW	DAILY OUTPUT	LABOR-HOURS	UNIT	2004 BARE COSTS MAT.	LABOR	EQUIP.	TOTAL	TOTAL INCL O&P	
100	0010 CUPOLA Stock units, pine, painted, 18" sq., 28" high, alum. roof	1 Carp	4.10	1.951	Ea.	141	45		186	232	100
	0100 Copper roof		3.80	2.105		143	48.50		191.50	241	
	0300 23" square, 33" high, aluminum roof		3.70	2.162		236	50		286	345	
	0400 Copper roof		3.30	2.424		238	56		294	355	
	0600 30" square, 37" high, aluminum roof		3.70	2.162		360	50		410	480	
	0700 Copper roof		3.30	2.424		370	56		426	505	
	0900 Hexagonal, 31" wide, 46" high, copper roof		4	2		540	46		586	675	
	1000 36" wide, 50" high, copper roof	↓	3.50	2.286		575	53		628	720	
	1200 For deluxe stock units, add to above					25%					
	1400 For custom built units, add to above				↓	50%	50%				
	10344 Weathervanes										
800	0010 WEATHERVANES										800
	0020 Residential types, minimum	1 Carp	8	1	Ea.	40	23		63	83	
	0100 Maximum	"	2	4	"	800	92.50		892.50	1,025	

10350 | Flagpoles

		10355 | Flagpoles	CREW	DAILY OUTPUT	LABOR-HOURS	UNIT	2004 BARE COSTS MAT.	LABOR	EQUIP.	TOTAL	TOTAL INCL O&P	
400	0010	**FLAGPOLE**, Ground set										400
	0050	Not including base or foundation										
	0100	Aluminum, tapered, ground set 20′ high	K-1	2	8	Ea.	675	165	80	920	1,100	
	0200	25′ high		1.70	9.412		885	195	94.50	1,174.50	1,400	
	0300	30′ high		1.50	10.667		885	221	107	1,213	1,450	
	0500	40′ high	↓	1.20	13.333	↓	1,975	276	134	2,385	2,775	

10520 | Fire Protection Specialties

		10525 | Fire Prot. Specialties	CREW	DAILY OUTPUT	LABOR-HOURS	UNIT	2004 BARE COSTS MAT.	LABOR	EQUIP.	TOTAL	TOTAL INCL O&P	
300	0010	**FIRE EXTINGUISHERS**										300
	0120	CO2, portable with swivel horn, 5 lb.				Ea.	101			101	111	
	0140	With hose and "H" horn, 10 lb.				"	150			150	165	
	1000	Dry chemical, pressurized										
	1040	Standard type, portable, painted, 2-1/2 lb.				Ea.	27.50			27.50	30.50	
	1080	10 lb.					67			67	73.50	
	1100	20 lb.					90			90	99	
	1120	30 lb.					157			157	173	
	2000	ABC all purpose type, portable, 2-1/2 lb.					27.50			27.50	30.50	
	2080	9-1/2 lb.				↓	60			60	66	

10530 | Protective Covers

		10535 | Awnings & Canopies	CREW	DAILY OUTPUT	LABOR-HOURS	UNIT	2004 BARE COSTS MAT.	LABOR	EQUIP.	TOTAL	TOTAL INCL O&P	
100	0010	**CANOPIES, RESIDENTIAL** Prefabricated										100
	0500	Carport, free standing, baked enamel, alum., .032″, 40 psf										
	0520	16′ x 8′, 4 posts	2 Carp	3	5.333	Ea.	2,775	123		2,898	3,250	
	0600	20′ x 10′, 6 posts	"	2	8		2,900	185		3,085	3,525	
	1000	Door canopies, extruded alum., .032″, 42″ projection, 4′ wide	1 Carp	8	1		335	23		358	410	
	1020	6′ wide	"	6	1.333		410	31		441	505	
	1040	8′ wide	2 Carp	9	1.778		525	41		566	645	
	1060	10′ wide		7	2.286		615	53		668	765	
	1080	12′ wide	↓	5	3.200		725	74		799	925	
	1200	54″ projection, 4′ wide	1 Carp	8	1		430	23		453	515	
	1220	6′ wide	"	6	1.333		550	31		581	660	
	1240	8′ wide	2 Carp	9	1.778		725	41		766	865	
	1260	10′ wide		7	2.286		820	53		873	990	
	1280	12′ wide	↓	5	3.200		930	74		1,004	1,150	
	1300	Painted, add					20%					
	1310	Bronze anodized, add					50%					
	3000	Window awnings, aluminum, window 3′ high, 4′ wide	1 Carp	10	.800		224	18.50		242.50	278	
	3020	6′ wide	"	8	1		261	23		284	325	
	3040	9′ wide	2 Carp	9	1.778		425	41		466	535	
	3060	12′ wide	"	5	3.200	↓	580	74		654	760	

10530 | Protective Covers

10535 Awnings & Canopies		CREW	DAILY OUTPUT	LABOR-HOURS	UNIT	2004 BARE COSTS MAT.	LABOR	EQUIP.	TOTAL	TOTAL INCL O&P	
100	3100 Window, 4' high, 4' wide	1 Carp	10	.800	Ea.	274	18.50		292.50	330	100
	3120 6' wide	"	8	1		370	23		393	445	
	3140 9' wide	2 Carp	9	1.778		500	41		541	620	
	3160 12' wide	"	5	3.200		645	74		719	835	
	3200 Window, 6' high, 4' wide	1 Carp	10	.800		415	18.50		433.50	485	
	3220 6' wide	"	8	1		570	23		593	670	
	3240 9' wide	2 Carp	9	1.778		785	41		826	930	
	3260 12' wide	"	5	3.200		1,075	74		1,149	1,300	
	3400 Roll-up aluminum, 2'-6" wide	1 Carp	14	.571		91.50	13.20		104.70	124	
	3420 3' wide		12	.667		110	15.40		125.40	147	
	3440 4' wide		10	.800		141	18.50		159.50	187	
	3460 6' wide	↓	8	1		174	23		197	231	
	3480 9' wide	2 Carp	9	1.778		253	41		294	350	
	3500 12' wide	"	5	3.200	↓	310	74		384	465	
	3600 Window awnings, canvas, 24" drop, 3' wide	1 Carp	30	.267	L.F.	36	6.15		42.15	50	
	3620 4' wide		40	.200		33	4.62		37.62	44.50	
	3700 30" drop, 3' wide		30	.267		51.50	6.15		57.65	67	
	3720 4' wide		40	.200		44.50	4.62		49.12	56.50	
	3740 5' wide		45	.178		40	4.11		44.11	51	
	3760 6' wide		48	.167		37	3.85		40.85	47	
	3780 8' wide		48	.167		30.50	3.85		34.35	40	
	3800 10' wide	↓	50	.160	↓	28	3.70		31.70	37.50	

10550 | Postal Specialties

10555 Mail Delivery Systems		CREW	DAILY OUTPUT	LABOR-HOURS	UNIT	2004 BARE COSTS MAT.	LABOR	EQUIP.	TOTAL	TOTAL INCL O&P	
600	0011 MAIL BOXES										600
	1900 Letter slot, residential	1 Carp	20	.400	Ea.	56.50	9.25		65.75	77.50	
	2400 Residential, galv. steel, small 20" x 7" x 9"	1 Clab	16	.500		102	8.45		110.45	126	
	2410 With galv. steel post, 54" long		6	1.333		168	22.50		190.50	224	
	2420 Large, 24" x 12" x 15"		16	.500		102	8.45		110.45	126	
	2430 With galv. steel post, 54" long		6	1.333		199	22.50		221.50	258	
	2440 Decorative, polyethylene, 22" x 10" x 10"		16	.500		33	8.45		41.45	51	
	2450 With alum. post, decorative, 54" long	↓	6	1.333	↓	52	22.50		74.50	96	

10670 | Storage Shelving

10674 Storage Shelving		CREW	DAILY OUTPUT	LABOR-HOURS	UNIT	2004 BARE COSTS MAT.	LABOR	EQUIP.	TOTAL	TOTAL INCL O&P	
500	0010 SHELVING Metal, industrial, cross-braced, 3' wide, 12" deep	1 Sswk	175	.046	SF Shlf	5.90	1.14		7.04	8.70	500
	0100 24" deep		330	.024		4.50	.60		5.10	6.10	
	2200 Wide span, 1600 lb. capacity per shelf, 6' wide, 24" deep		380	.021		7.35	.52		7.87	9.10	
	2400 36" deep	↓	440	.018	↓	6.45	.45		6.90	7.90	
	3000 Residential, vinyl covered wire, wardrobe, 12" deep	1 Carp	195	.041	L.F.	2.86	.95		3.81	4.76	
	3100 16" deep	↓	195	.041	↓	2.99	.95		3.94	4.90	

Important: See the Reference Section for critical supporting data - Reference Nos., Crews, & Location Factors

10670 | Storage Shelving

		10674 Storage Shelving	CREW	DAILY OUTPUT	LABOR-HOURS	UNIT	2004 BARE COSTS MAT.	LABOR	EQUIP.	TOTAL	TOTAL INCL O&P	
500	3200	Standard, 6" deep	1 Carp	195	.041	L.F.	2.65	.95		3.60	4.53	500
	3300	9" deep		195	.041		2.65	.95		3.60	4.53	
	3400	12" deep		195	.041		2.65	.95		3.60	4.53	
	3500	16" deep		195	.041		2.71	.95		3.66	4.59	
	3600	20" deep		195	.041	↓	2.76	.95		3.71	4.65	
	3700	Support bracket	↓	80	.100	Ea.	1.39	2.31		3.70	5.45	

10800 | Toilet/Bath/Laundry Accessories

		10810 Toilet Accessories	CREW	DAILY OUTPUT	LABOR-HOURS	UNIT	2004 BARE COSTS MAT.	LABOR	EQUIP.	TOTAL	TOTAL INCL O&P	
100	0010	**COMMERCIAL TOILET ACCESSORIES**										100
	0200	Curtain rod, stainless steel, 5' long, 1" diameter	1 Carp	13	.615	Ea.	30.50	14.20		44.70	57.50	
	0300	1-1/4" diameter		13	.615		29	14.20		43.20	55.50	
	0800	Grab bar, straight, 1-1/4" diameter, stainless steel, 18" long		24	.333		21	7.70		28.70	36	
	1100	36" long		20	.400		26.50	9.25		35.75	44.50	
	3000	Mirror, with stainless steel 3/4" square frame, 18" x 24"		20	.400		58.50	9.25		67.75	80	
	3100	36" x 24"		15	.533		101	12.30		113.30	132	
	3300	72" x 24"		6	1.333		196	31		227	269	
	4300	Robe hook, single, regular		36	.222		4.76	5.15		9.91	13.95	
	4400	Heavy duty, concealed mounting		36	.222		10.80	5.15		15.95	20.50	
	6400	Towel bar, stainless steel, 18" long		23	.348		30	8.05		38.05	46.50	
	6500	30" long		21	.381		51	8.80		59.80	71	
	7400	Tumbler holder, tumbler only		30	.267		25	6.15		31.15	38	
	7500	Soap, tumbler & toothbrush	↓	30	.267	↓	22.50	6.15		28.65	35.50	
		10820 Bath Accessories										
400	0010	**MEDICINE CABINETS** With mirror, st. st. frame, 16" x 22", unlighted	1 Carp	14	.571	Ea.	69.50	13.20		82.70	99	400
	0100	Wood frame		14	.571		96.50	13.20		109.70	129	
	0300	Sliding mirror doors, 20" x 16" x 4-3/4", unlighted		7	1.143		86.50	26.50		113	140	
	0400	24" x 19" x 8-1/2", lighted		5	1.600		136	37		173	213	
	0600	Triple door, 30" x 32", unlighted, plywood body		7	1.143		214	26.50		240.50	281	
	0700	Steel body		7	1.143		282	26.50		308.50	355	
	0900	Oak door, wood body, beveled mirror, single door		7	1.143		125	26.50		151.50	183	
	1000	Double door	↓	6	1.333	↓	320	31		351	405	

For information about Means Estimating Seminars, see yellow pages 12 and 13 in back of book

Division Notes

	CREW	DAILY OUTPUT	LABOR-HOURS	UNIT	2004 BARE COSTS				TOTAL INCL O&P
					MAT.	LABOR	EQUIP.	TOTAL	

Division 11 Equipment

Estimating Tips

General

- The items in this division are usually priced per square foot or each. Many of these items are purchased by the owner for installation by the contractor. Check the specifications for responsibilities, and include time for receiving, storage, installation and mechanical and electrical hook-ups in the appropriate divisions.
- Many items in Division 11 require some type of support system that is not usually furnished with the item. Examples of these systems include blocking for the attachment of casework and support angles for ceiling hung projection screens. The required blocking or supports must be added to the estimate in the appropriate division.
- Some items in Division 11 may require assembly or electrical hook-ups. Verify the amount of assembly required or the need for a hard electrical connection and add the appropriate costs.

Reference Numbers

Reference numbers are shown in bold squares at the beginning of some major classifications. These numbers refer to related items in the Reference Section. The reference information may be an estimating procedure, an alternate pricing method or technical information.

Note: Not all subdivisions listed here necessarily appear in this publication.

11010 | Maintenance Equipment

11013 Floor/Wall Cleaning Equipment			CREW	DAILY OUTPUT	LABOR-HOURS	UNIT	2004 BARE COSTS MAT.	LABOR	EQUIP.	TOTAL	TOTAL INCL O&P	
800	0010	**VACUUM CLEANING**										800
	0020	Central, 3 inlet, residential	1 Skwk	.90	8.889	Total	590	205		795	995	
	0400	5 inlet system, residential		.50	16		895	370		1,265	1,600	
	0600	7 inlet system, commercial		.40	20		1,000	460		1,460	1,875	
	0800	9 inlet system, residential	↓	.30	26.667		1,275	615		1,890	2,450	
	4010	Rule of thumb: First 1200 S.F., installed				↓					1,125	
	4020	For each additional S.F., add				S.F.					.18	

11400 | Food Service Equipment

11405 Food Storage Equipment			CREW	DAILY OUTPUT	LABOR-HOURS	UNIT	2004 BARE COSTS MAT.	LABOR	EQUIP.	TOTAL	TOTAL INCL O&P	
800	0010	**WINE CELLAR**, refrigerated, Redwood interior, carpeted, walk-in type										800
	0020	6'-8" high, including racks										
	0200	80 "W x 48"D for 900 bottles	2 Carp	1.50	10.667	Ea.	2,650	246		2,896	3,350	
	0250	80" W x 72" D for 1300 bottles		1.33	12.030		3,500	278		3,778	4,325	
	0300	80" W x 94" D for 1900 bottles	↓	1.17	13.675	↓	4,550	315		4,865	5,525	

11450 | Residential Equipment

11454 Residential Appliances			CREW	DAILY OUTPUT	LABOR-HOURS	UNIT	2004 BARE COSTS MAT.	LABOR	EQUIP.	TOTAL	TOTAL INCL O&P	
500	0010	**RESIDENTIAL APPLIANCES**										500
	0020	Cooking range, 30" free standing, 1 oven, minimum	2 Clab	10	1.600	Ea.	241	27		268	310	
	0050	Maximum		4	4		1,450	67.50		1,517.50	1,700	
	0150	2 oven, minimum		10	1.600		1,475	27		1,502	1,675	
	0200	Maximum	↓	10	1.600		1,450	27		1,477	1,650	
	0350	Built-in, 30" wide, 1 oven, minimum	1 Elec	6	1.333		445	35		480	550	
	0400	Maximum	2 Carp	2	8		1,325	185		1,510	1,775	
	0500	2 oven, conventional, minimum		4	4		930	92.50		1,022.50	1,175	
	0550	1 conventional, 1 microwave, maximum	↓	2	8		1,450	185		1,635	1,925	
	0700	Free-standing, 1 oven, 21" wide range, minimum	2 Clab	10	1.600		247	27		274	320	
	0750	21" wide, maximum	"	4	4		273	67.50		340.50	415	
	0900	Counter top cook tops, 4 burner, standard, minimum	1 Elec	6	1.333		176	35		211	251	
	0950	Maximum		3	2.667		440	70.50		510.50	600	
	1050	As above, but with grille and griddle attachment, minimum		6	1.333		435	35		470	535	
	1100	Maximum		3	2.667		640	70.50		710.50	820	
	1250	Microwave oven, minimum		4	2		75.50	53		128.50	169	
	1300	Maximum	↓	2	4		390	106		496	600	
	1750	Compactor, residential size, 4 to 1 compaction, minimum	1 Carp	5	1.600		415	37		452	520	
	1800	Maximum	"	3	2.667		470	61.50		531.50	620	
	2000	Deep freeze, 15 to 23 C.F., minimum	2 Clab	10	1.600		390	27		417	475	
	2050	Maximum		5	3.200		515	54		569	655	
	2200	30 C.F., minimum	↓	8	2	↓	750	34		784	885	

11454 | Residential Appliances		CREW	DAILY OUTPUT	LABOR-HOURS	UNIT	2004 BARE COSTS MAT.	LABOR	EQUIP.	TOTAL	TOTAL INCL O&P
500	2250 Maximum	2 Clab	3	5.333	Ea.	850	90		940	1,100
	2450 Dehumidifier, portable, automatic, 15 pint					149			149	164
	2550 40 pint					167			167	183
	2750 Dishwasher, built-in, 2 cycles, minimum	L-1	4	2.500		245	65.50		310.50	375
	2800 Maximum		2	5		289	131		420	535
	2950 4 or more cycles, minimum		4	2.500		262	65.50		327.50	395
	2960 Average		4	2.500		350	65.50		415.50	490
	3000 Maximum		2	5		535	131		666	805
	3200 Dryer, automatic, minimum	L-2	3	5.333		269	107		376	475
	3250 Maximum	"	2	8		765	160		925	1,100
	3300 Garbage disposer, sink type, minimum	L-1	10	1		42.50	26		68.50	89.50
	3350 Maximum	"	10	1		145	26		171	203
	3550 Heater, electric, built-in, 1250 watt, ceiling type, minimum	1 Elec	4	2		70	53		123	163
	3600 Maximum		3	2.667		114	70.50		184.50	241
	3700 Wall type, minimum		4	2		99.50	53		152.50	195
	3750 Maximum		3	2.667		132	70.50		202.50	260
	3900 1500 watt wall type, with blower		4	2		123	53		176	221
	3950 3000 watt		3	2.667		251	70.50		321.50	390
	4150 Hood for range, 2 speed, vented, 30" wide, minimum	L-3	5	2		37.50	47.50		85	121
	4200 Maximum		3	3.333		595	79		674	790
	4300 42" wide, minimum		5	2		225	47.50		272.50	325
	4330 Custom		5	2		625	47.50		672.50	765
	4350 Maximum		3	3.333		760	79		839	970
	4500 For ventless hood, 2 speed, add					15			15	16.50
	4650 For vented 1 speed, deduct from maximum					39			39	43
	4850 Humidifier, portable, 8 gallons per day					149			149	164
	5000 15 gallons per day					179			179	197
	5200 Icemaker, automatic, 20 lb. per day	1 Plum	7	1.143		360	30		390	445
	5350 51 lb. per day	"	2	4		1,000	105		1,105	1,275
	5380 Oven, built in, standard	1 Elec	4	2		385	53		438	505
	5390 Deluxe	"	2	4		1,625	106		1,731	1,950
	5500 Refrigerator, no frost, 10 C.F. to 12 C.F. minimum	2 Clab	10	1.600		440	27		467	530
	5600 Maximum		6	2.667		680	45		725	825
	5750 14 C.F. to 16 C.F., minimum		9	1.778		445	30		475	540
	5800 Maximum		5	3.200		480	54		534	620
	5950 18 C.F. to 20 C.F., minimum		8	2		510	34		544	620
	6000 Maximum		4	4		840	67.50		907.50	1,050
	6150 21 C.F. to 29 C.F., minimum		7	2.286		645	38.50		683.50	775
	6200 Maximum		3	5.333		2,075	90		2,165	2,425
	6400 Sump pump cellar drainer, pedestal, 1/3 H.P., molded PVC base	1 Plum	3	2.667		87	69.50		156.50	210
	6450 Solid brass	"	2	4		179	105		284	370
	6460 Sump pump, see also division 15440-940									
	6650 Washing machine, automatic, minimum	1 Plum	3	2.667	Ea.	273	69.50		342.50	415
	6700 Maximum	"	1	8		1,025	209		1,234	1,475
	6900 Water heater, electric, glass lined, 30 gallon, minimum	L-1	5	2		252	52.50		304.50	365
	6950 Maximum		3	3.333		350	87.50		437.50	530
	7100 80 gallon, minimum		2	5		485	131		616	745
	7150 Maximum		1	10		670	262		932	1,175
	7180 Water heater, gas, glass lined, 30 gallon, minimum	2 Plum	5	3.200		345	83.50		428.50	515
	7220 Maximum		3	5.333		480	139		619	760
	7260 50 gallon, minimum		2.50	6.400		450	167		617	770
	7300 Maximum		1.50	10.667		625	279		904	1,150
	7310 Water heater, see also division 15480-200									
	7350 Water softener, automatic, to 30 grains per gallon	2 Plum	5	3.200	Ea.	430	83.50		513.50	610
	7400 To 100 grains per gallon	"	4	4		600	105		705	830
	7450 Vent kits for dryers	1 Carp	10	.800		12.65	18.50		31.15	45.50

11450 | Residential Equipment

	11454 | Residential Appliances		CREW	DAILY OUTPUT	LABOR-HOURS	UNIT	2004 BARE COSTS MAT.	LABOR	EQUIP.	TOTAL	TOTAL INCL O&P	
550	0010	**DISAPPEARING STAIRWAY** No trim included										550
	0020	One piece, yellow pine, 8'-0" ceiling	2 Carp	4	4	Ea.	890	92.50		982.50	1,125	
	0030	9'-0" ceiling		4	4		900	92.50		992.50	1,150	
	0040	10'-0" ceiling		3	5.333		950	123		1,073	1,250	
	0050	11'-0" ceiling		3	5.333		1,125	123		1,248	1,425	
	0060	12'-0" ceiling	↓	3	5.333		1,150	123		1,273	1,475	
	0100	Custom grade, pine, 8'-6" ceiling, minimum	1 Carp	4	2		99.50	46		145.50	188	
	0150	Average		3.50	2.286		100	53		153	200	
	0200	Maximum		3	2.667		165	61.50		226.50	287	
	0500	Heavy duty, pivoted, from 7'-7" to 12'-10" floor to floor		3	2.667		325	61.50		386.50	465	
	0600	16'-0" ceiling		2	4		1,075	92.50		1,167.50	1,350	
	0800	Economy folding, pine, 8'-6" ceiling		4	2		90.50	46		136.50	179	
	0900	9'-6" ceiling	↓	4	2		98.50	46		144.50	187	
	1000	Fire escape, galvanized steel, 8'-0" to 10'-4" ceiling	2 Carp	1	16		1,150	370		1,520	1,900	
	1010	10'-6" to 13'-6" ceiling		1	16		1,450	370		1,820	2,225	
	1100	Automatic electric, aluminum, floor to floor height, 8' to 9'	↓	1	16	↓	5,675	370		6,045	6,875	
	11460 | Unit Kitchens											
100	0010	**UNIT KITCHENS**										100
	1500	Combination range, refrigerator and sink, 30" wide, minimum	L-1	2	5	Ea.	730	131		861	1,025	
	1550	Maximum		1	10		1,450	262		1,712	2,025	
	1570	60" wide, average		1.40	7.143		2,400	187		2,587	2,950	
	1590	72" wide, average	↓	1.20	8.333	↓	2,725	218		2,943	3,350	

Division 12 Furnishings

Estimating Tips

General

- The items in this division are usually priced per square foot or each. Most of these items are purchased by the owner and placed by the supplier. Do not assume the items in Division 12 will be purchased and installed by the supplier. Check the specifications for responsibilities and include receiving, storage, installation and mechanical and electrical hook-ups in the appropriate divisions.
- Some items in this division require some type of support system that is not usually furnished with the item. Examples of these systems include blocking for the attachment of casework and heavy drapery rods. The required blocking must be added to the estimate in the appropriate division.

Reference Numbers

Reference numbers are shown in bold squares at the beginning of some major classifications. These numbers refer to related items in the Reference Section. The reference information may be an estimating procedure, an alternate pricing method or technical information.

Note: Not all subdivisions listed here necessarily appear in this publication.

12300 | Manufactured Casework

12310 | Metal Casework

			CREW	DAILY OUTPUT	LABOR-HOURS	UNIT	2004 BARE COSTS MAT.	LABOR	EQUIP.	TOTAL	TOTAL INCL O&P	
560	0010	**IRONING CENTER**										560
	0020	Including cabinet, board & light, minimum	1 Carp	2	4	Ea.	261	92.50		353.50	445	

12400 | Furnishings & Accessories

12492 | Blinds and Shades

			CREW	DAILY OUTPUT	LABOR-HOURS	UNIT	2004 BARE COSTS MAT.	LABOR	EQUIP.	TOTAL	TOTAL INCL O&P	
100	0010	**BLINDS, INTERIOR**										100
	0020	Horizontal, 1" aluminum slats, solid color, stock	1 Carp	590	.014	S.F.	2.82	.31		3.13	3.63	
	0090	Custom, minimum		590	.014		2.57	.31		2.88	3.36	
	0100	Maximum		440	.018		6.40	.42		6.82	7.75	
	0450	Stock, minimum		590	.014		4.21	.31		4.52	5.15	
	0500	Maximum		440	.018		6.85	.42		7.27	8.20	
	3000	Wood folding panels with movable louvers, 7" x 20" each		17	.471	Pr.	42	10.85		52.85	65	
	3300	8" x 28" each		17	.471		61	10.85		71.85	85.50	
	3450	9" x 36" each		17	.471		72.50	10.85		83.35	98	
	3600	10" x 40" each		17	.471		82	10.85		92.85	108	
	4000	Fixed louver type, stock units, 8" x 20" each		17	.471		63	10.85		73.85	88	
	4150	10" x 28" each		17	.471		86.50	10.85		97.35	113	
	4300	12" x 36" each		17	.471		110	10.85		120.85	139	
	4450	18" x 40" each		17	.471		131	10.85		141.85	162	
	5000	Insert panel type, stock, 7" x 20" each		17	.471		14.85	10.85		25.70	35	
	5150	8" x 28" each		17	.471		27	10.85		37.85	48.50	
	5300	9" x 36" each		17	.471		34.50	10.85		45.35	56.50	
	5450	10" x 40" each		17	.471		37	10.85		47.85	59	
	5600	Raised panel type, stock, 10" x 24" each		17	.471		109	10.85		119.85	138	
	5650	12" x 26" each		17	.471		126	10.85		136.85	157	
	5700	14" x 30" each		17	.471		143	10.85		153.85	175	
	5750	16" x 36" each		17	.471		160	10.85		170.85	194	
	6000	For custom built pine, add					22%					
	6500	For custom built hardwood blinds, add					42%					
600	0011	**SHADES** Basswood roll-up, stain finish, 3/8" slats	1 Carp	300	.027	S.F.	9.85	.62		10.47	11.85	600
	5011	Insulative shades		125	.064		8.25	1.48		9.73	11.60	
	6011	Solar screening, fiberglass		85	.094		3.83	2.17		6	7.90	
	8011	Interior insulative shutter										
	8111	Stock unit, 15" x 60"	1 Carp	17	.471	Pr.	8.25	10.85		19.10	27.50	

12493 | Curtains and Drapes

			CREW	DAILY OUTPUT	LABOR-HOURS	UNIT	2004 BARE COSTS MAT.	LABOR	EQUIP.	TOTAL	TOTAL INCL O&P	
200	0010	**DRAPERY HARDWARE**										200
	0030	Standard traverse, per foot, minimum	1 Carp	59	.136	L.F.	1.90	3.13		5.03	7.40	
	0100	Maximum		51	.157	"	10.30	3.62		13.92	17.45	
	0200	Decorative traverse, 28"-48", minimum		22	.364	Ea.	13.95	8.40		22.35	29.50	
	0220	Maximum		21	.381		32.50	8.80		41.30	51	
	0300	48"-84", minimum		20	.400		18.60	9.25		27.85	36	
	0320	Maximum		19	.421		53	9.75		62.75	75	
	0400	66"-120", minimum		18	.444		21	10.25		31.25	40.50	
	0420	Maximum		17	.471		79	10.85		89.85	105	
	0500	84"-156", minimum		16	.500		23.50	11.55		35.05	45	
	0520	Maximum		15	.533		87.50	12.30		99.80	118	
	0600	130"-240", minimum		14	.571		28	13.20		41.20	53	

	12493 | Curtains and Drapes	CREW	DAILY OUTPUT	LABOR-HOURS	UNIT	2004 BARE COSTS MAT.	LABOR	EQUIP.	TOTAL	TOTAL INCL O&P	
200	0620 Maximum	1 Carp	13	.615	Ea.	123	14.20		137.20	159	200
	0700 Slide rings, each, minimum					.64			.64	.70	
	0720 Maximum					2.09			2.09	2.30	
	3000 Ripplefold, snap-a-pleat system, 3′ or less, minimum	1 Carp	15	.533		44	12.30		56.30	69.50	
	3020 Maximum	"	14	.571		64.50	13.20		77.70	93.50	
	3200 Each additional foot, add, minimum				L.F.	2.09			2.09	2.30	
	3220 Maximum				"	6.15			6.15	6.75	
	4000 Traverse rods, adjustable, 28″ to 48″	1 Carp	22	.364	Ea.	16.45	8.40		24.85	32.50	
	4020 48″ to 84″		20	.400		24	9.25		33.25	41.50	
	4040 66″ to 120″		18	.444		27	10.25		37.25	47.50	
	4060 84″ to 156″		16	.500		30	11.55		41.55	52.50	
	4080 100″ to 180″		14	.571		34.50	13.20		47.70	60.50	
	4100 228″ to 312″		13	.615		53	14.20		67.20	82.50	
	4500 Curtain rod, 28″ to 48″, single		22	.364		4.55	8.40		12.95	19.25	
	4510 Double		22	.364		7.75	8.40		16.15	23	
	4520 48″ to 86″, single		20	.400		7.80	9.25		17.05	24.50	
	4530 Double		20	.400		13	9.25		22.25	30	
	4540 66″ to 120″, single		18	.444		13.05	10.25		23.30	32	
	4550 Double		18	.444		20.50	10.25		30.75	40	
	4600 Valance, pinch pleated fabric, 12″ deep, up to 54″ long, minimum					32.50			32.50	36	
	4610 Maximum					81.50			81.50	89.50	
	4620 Up to 77″ long, minimum					50			50	55	
	4630 Maximum					132			132	145	
	5000 Stationary rods, first 2 feet					8.40			8.40	9.25	

For information about Means Estimating Seminars, see yellow pages 12 and 13 in back of book

Division Notes

		CREW	DAILY OUTPUT	LABOR-HOURS	UNIT	2004 BARE COSTS MAT.	LABOR	EQUIP.	TOTAL	TOTAL INCL O&P

Division 13 Special Construction

Estimating Tips

General

- The items and systems in this division are usually estimated, purchased, supplied and installed as a unit by one or more subcontractors. The estimator must ensure that all parties are operating from the same set of specifications and assumptions and that all necessary items are estimated and will be provided. Many times the complex items and systems are covered but the more common ones such as excavation or a crane are overlooked for the very reason that everyone assumes nobody could miss them. The estimator should be the central focus and be able to ensure that all systems are complete.
- Another area where problems can develop in this division is at the interface between systems. The estimator must ensure, for instance, that anchor bolts, nuts and washers are estimated and included for the air-supported structures and pre-engineered buildings to be bolted to their foundations. Utility supply is a common area where essential items or pieces of equipment can be missed or overlooked due to the fact that each subcontractor may feel it is the others' responsibility. The estimator should also be aware of certain items which may be supplied as part of a package but installed by others, and ensure that the installing contractor's estimate includes the cost of installation. Conversely, the estimator must also ensure that items are not costed by two different subcontractors, resulting in an inflated overall estimate.

13120 Pre-Engineered Structures

- The foundations and floor slab, as well as rough mechanical and electrical, should be estimated, as this work is required for the assembly and erection of the structure. Generally, as noted in the book, the pre-engineered building comes as a shell and additional features must be included by the estimator. Here again, the estimator must have a clear understanding of the scope of each portion of the work and all the necessary interfaces.

13200 Storage Tanks

- The prices in this subdivision for above and below ground storage tanks do not include foundations or hold-down slabs. The estimator should refer to Divisions 2 and 3 for foundation system pricing. In addition to the foundations, required tank accessories such as tank gauges, leak detection devices, and additional manholes and piping must be added to the tank prices.

Reference Numbers

Reference numbers are shown in bold squares at the beginning of some major classifications. These numbers refer to related items in the Reference Section. The reference information may be an estimating procedure, an alternate pricing method or technical information.

Note: Not all subdivisions listed here necessarily appear in this publication.

13030 | Special Purpose Rooms

	13035 | Special Purpose Rooms	CREW	DAILY OUTPUT	LABOR-HOURS	UNIT	2004 BARE COSTS MAT.	LABOR	EQUIP.	TOTAL	TOTAL INCL O&P	
800	0010 **SAUNA** Prefabricated, incl. heater & controls, 7′ high, 6′ x 4′, C/C	L-7	2.20	11.818	Ea.	3,450	231		3,681	4,200	800
	0050 6′ x 4′, C/P		2	13		3,225	254		3,479	3,950	
	0400 6′ x 5′, C/C		2	13		3,875	254		4,129	4,675	
	0450 6′ x 5′, C/P		2	13		3,625	254		3,879	4,400	
	0600 6′ x 6′, C/C		1.80	14.444		4,125	282		4,407	5,000	
	0650 6′ x 6′, C/P		1.80	14.444		3,850	282		4,132	4,725	
	0800 6′ x 9′, C/C		1.60	16.250		5,175	320		5,495	6,200	
	0850 6′ x 9′, C/P		1.60	16.250		4,900	320		5,220	5,925	
	1000 8′ x 12′, C/C		1.10	23.636		8,000	460		8,460	9,575	
	1050 8′ x 12′, C/P		1.10	23.636		7,350	460		7,810	8,875	
	1400 8′ x 10′, C/C		1.20	21.667		6,775	425		7,200	8,175	
	1450 8′ x 10′, C/P		1.20	21.667		6,325	425		6,750	7,675	
	1600 10′ x 12′, C/C		1	26		8,475	510		8,985	10,200	
	1650 10′ x 12′, C/P	↓	1	26		7,675	510		8,185	9,300	
	1700 Door only, cedar, 2′x6′, with tempered insulated glass window	2 Carp	3.40	4.706		455	109		564	690	
	1800 Prehung, incl. jambs, pulls & hardware	"	12	1.333		465	31		496	570	
	2500 Heaters only (incl. above), wall mounted, to 200 C.F.					450			450	495	
	2750 To 300 C.F.					545			545	600	
	3000 Floor standing, to 720 C.F., 10,000 watts, w/controls	1 Elec	3	2.667		1,400	70.50		1,470.50	1,675	
	3250 To 1,000 C.F., 16,000 watts	"	3	2.667	↓	1,450	70.50		1,520.50	1,700	
940	0010 **STEAM BATH** Heater, timer & head, single, to 140 C.F.	1 Plum	1.20	6.667	Ea.	930	174		1,104	1,300	940
	0500 To 300 C.F.	"	1.10	7.273		1,050	190		1,240	1,450	
	2700 Conversion unit for residential tub, including door				↓	2,925			2,925	3,225	

13100 | Lightning Protection

	13101 | Lightning Protection	CREW	DAILY OUTPUT	LABOR-HOURS	UNIT	2004 BARE COSTS MAT.	LABOR	EQUIP.	TOTAL	TOTAL INCL O&P	
055	0010 LIGHTNING PROTECTION										055
	0200 Air terminals & base, copper										
	0400 3/8″ diameter x 10″ (to 75′ high)	1 Elec	8	1	Ea.	23.50	26.50		50	68.50	
	1000 Aluminum, 1/2″ diameter x 12″ (to 75′ high)		8	1	"	18	26.50		44.50	63	
	2000 Cable, copper, 220 lb. per thousand ft. (to 75′ high)		320	.025	L.F.	1.07	.66		1.73	2.25	
	2500 Aluminum, 101 lb. per thousand ft. (to 75′ high)		280	.029	"	.58	.75		1.33	1.87	
	3000 Arrester, 175 volt AC to ground	↓	8	1	Ea.	36	26.50		62.50	82.50	

13120 | Pre-Engineered Structures

	13128 | Pre-Engineered Structures	CREW	DAILY OUTPUT	LABOR-HOURS	UNIT	2004 BARE COSTS MAT.	LABOR	EQUIP.	TOTAL	TOTAL INCL O&P	
540	0010 **GREENHOUSE** Shell only, stock units, not incl. 2′ stub walls,										540
	0020 foundation, floors, heat or compartments										
	0300 Residential type, free standing, 8′-6″ long x 7′-6″ wide	2 Carp	59	.271	SF Flr.	37	6.25		43.25	51.50	
	0400 10′-6″ wide	↓	85	.188	↓	28.50	4.35		32.85	39	

13120 | Pre-Engineered Structures

13128 | Pre-Engineered Structures			CREW	DAILY OUTPUT	LABOR-HOURS	UNIT	2004 BARE COSTS MAT.	LABOR	EQUIP.	TOTAL	TOTAL INCL O&P	
540	0600	13'-6" wide	2 Carp	108	.148	SF Flr.	25.50	3.42		28.92	34	540
	0700	17'-0" wide		160	.100		28.50	2.31		30.81	35.50	
	0900	Lean-to type, 3'-10" wide		34	.471		33	10.85		43.85	54.50	
	1000	6'-10" wide	↓	58	.276	↓	25.50	6.35		31.85	39	
	1100	Wall mounted, to existing window, 3' x 3'	1 Carp	4	2	Ea.	355	46		401	470	
	1120	4' x 5'	"	3	2.667	"	530	61.50		591.50	690	
	1200	Deluxe quality, free standing, 7'-6" wide	2 Carp	55	.291	SF Flr.	73	6.70		79.70	92	
	1220	10'-6" wide		81	.198		68	4.56		72.56	82.50	
	1240	13'-6" wide		104	.154		63.50	3.55		67.05	76	
	1260	17'-0" wide		150	.107		54	2.46		56.46	63.50	
	1400	Lean-to type, 3'-10" wide		31	.516		85	11.90		96.90	114	
	1420	6'-10" wide		55	.291		79.50	6.70		86.20	99	
	1440	8'-0" wide	↓	97	.165	↓	74	3.81		77.81	88	
880	0010	**SWIMMING POOL ENCLOSURE** Translucent, free standing,										880
	0020	not including foundations, heat or light										
	0200	Economy, minimum	2 Carp	200	.080	SF Hor.	11.35	1.85		13.20	15.60	
	0300	Maximum		100	.160		30	3.70		33.70	39.50	
	0400	Deluxe, minimum		100	.160		34	3.70		37.70	44	
	0600	Maximum	↓	70	.229	↓	163	5.30		168.30	188	

13150 | Swimming Pools

13151 | Swimming Pools			CREW	DAILY OUTPUT	LABOR-HOURS	UNIT	2004 BARE COSTS MAT.	LABOR	EQUIP.	TOTAL	TOTAL INCL O&P	
200	0010	**SWIMMING POOLS** Residential in-ground, vinyl lined, concrete sides										200
	0020	Sides including equipment, sand bottom	B-52	300	.187	SF Surf	10.60	3.56	1.27	15.43	19.10	
	0100	Metal or polystyrene sides R13128-520	B-14	410	.117		8.85	2.12	.51	11.48	13.90	
	0200	Add for vermiculite bottom				↓	.68			.68	.75	
	0500	Gunite bottom and sides, white plaster finish										
	0600	12' x 30' pool	B-52	145	.386	SF Surf	17.55	7.35	2.62	27.52	35	
	0720	16' x 32' pool		155	.361		15.85	6.90	2.45	25.20	32	
	0750	20' x 40' pool	↓	250	.224	↓	14.15	4.28	1.52	19.95	24.50	
	0810	Concrete bottom and sides, tile finish										
	0820	12' x 30' pool	B-52	80	.700	SF Surf	17.75	13.35	4.75	35.85	47.50	
	0830	16' x 32' pool		95	.589		14.65	11.25	4	29.90	39.50	
	0840	20' x 40' pool	↓	130	.431	↓	11.65	8.20	2.92	22.77	30	
	1600	For water heating system, see division 15510-880										
	1700	Filtration and deck equipment only, as % of total				Total				20%	20%	
	1800	Deck equipment, rule of thumb, 20' x 40' pool				SF Pool					1.30	
	3000	Painting pools, preparation + 3 coats, 20' x 40' pool, epoxy	2 Pord	.33	48.485	Total	605	1,025		1,630	2,350	
	3100	Rubber base paint, 18 gallons	"	.33	48.485	"	460	1,025		1,485	2,175	
700	0010	**SWIMMING POOL EQUIPMENT** Diving stand, stainless steel, 3 meter	2 Carp	.40	40	Ea.	4,825	925		5,750	6,875	700
	0600	Diving boards, 16' long, aluminum		2.70	5.926		2,375	137		2,512	2,850	
	0700	Fiberglass	↓	2.70	5.926	↓	1,925	137		2,062	2,325	
	0900	Filter system, sand or diatomite type, incl. pump, 6,000 gal./hr.	2 Plum	1.80	8.889	Total	1,075	232		1,307	1,550	
	1020	Add for chlorination system, 800 S.F. pool	"	3	5.333	Ea.	203	139		342	450	
	1200	Ladders, heavy duty, stainless steel, 2 tread	2 Carp	7	2.286		470	53		523	605	
	1500	4 tread	"	6	2.667		580	61.50		641.50	745	
	2100	Lights, underwater, 12 volt, with transformer, 300 watt	1 Elec	.40	20	↓	136	530		666	1,000	

13150 | Swimming Pools

	13151 Swimming Pools	CREW	DAILY OUTPUT	LABOR-HOURS	UNIT	2004 BARE COSTS MAT.	LABOR	EQUIP.	TOTAL	TOTAL INCL O&P	
700	2200 110 volt, 500 watt, standard	1 Elec	.40	20	Ea.	127	530		657	1,000	700
	3000 Pool covers, reinforced vinyl	3 Clab	1,800	.013	S.F.	.29	.23		.52	.70	
	3100 Vinyl water tube		3,200	.007		.22	.13		.35	.46	
	3200 Maximum	↓	3,000	.008	↓	.46	.14		.60	.74	
	3300 Slides, tubular, fiberglass, aluminum handrails & ladder, 5′-0″, straight	2 Carp	1.60	10	Ea.	2,200	231		2,431	2,825	
	3320 8′-0″, curved	″	3	5.333	″	5,875	123		5,998	6,650	

13200 | Storage Tanks

	13201 Storage Tanks	CREW	DAILY OUTPUT	LABOR-HOURS	UNIT	2004 BARE COSTS MAT.	LABOR	EQUIP.	TOTAL	TOTAL INCL O&P	
300	3001 **STEEL,** storage, above ground, including supports, coating										300
	3020 fittings, not including fdn, pumps or piping										
	3040 Single wall, interior, 275 gallon	Q-5	5	3.200	Ea.	251	75.50		326.50	400	
	3060 550 gallon	″	2.70	5.926		1,125	140		1,265	1,475	
	3080 1,000 gallon	Q-7	5	6.400		1,800	151		1,951	2,225	
	3320 Double wall, 500 gallon capacity	Q-5	2.40	6.667		2,000	158		2,158	2,450	
	3330 2000 gallon capacity	Q-7	4.15	7.711		4,575	182		4,757	5,325	
	3340 4000 gallon capacity		3.60	8.889		8,125	210		8,335	9,300	
	3350 6000 gallon capacity		2.40	13.333		9,600	315		9,915	11,100	
	3360 8000 gallon capacity		2	16		12,300	380		12,680	14,100	
	3370 10000 gallon capacity		1.80	17.778		13,600	420		14,020	15,600	
	3380 15000 gallon capacity		1.50	21.333		20,600	505		21,105	23,500	
	3390 20000 gallon capacity		1.30	24.615		23,500	580		24,080	26,800	
	3400 25000 gallon capacity		1.15	27.826		28,500	660		29,160	32,500	
	3410 30000 gallon capacity	↓	1	32	↓	31,300	755		32,055	35,700	
800	0010 **UNDERGROUND STORAGE TANKS**										800
	0210 Fiberglass, underground, single wall, U.L. listed, not including										
	0220 manway or hold-down strap										
	0230 1,000 gallon capacity	Q-5	2.46	6.504	Ea.	1,850	154		2,004	2,275	
	0240 2,000 gallon capacity	Q-7	4.57	7.002		2,425	165		2,590	2,950	
	0500 For manway, fittings and hold-downs, add				↓	20%	15%				
	2210 Fiberglass, underground, single wall, U.L. listed, including										
	2220 hold-down straps, no manways										
	2230 1,000 gallon capacity	Q-5	1.88	8.511	Ea.	2,050	201		2,251	2,575	
	2240 2,000 gallon capacity	Q-7	3.55	9.014	″	2,625	213		2,838	3,225	
	5000 Steel underground, sti-P3, set in place, not incl. hold-down bars.										
	5500 Excavation, pad, pumps and piping not included										
	5510 Single wall, 500 gallon capacity, 7 gauge shell	Q-5	2.70	5.926	Ea.	820	140		960	1,125	
	5520 1,000 gallon capacity, 7 gauge shell	″	2.50	6.400		1,225	151		1,376	1,600	
	5530 2,000 gallon capacity, 1/4″ thick shell	Q-7	4.60	6.957		2,475	164		2,639	3,000	
	5535 2,500 gallon capacity, 7 gauge shell	Q-5	3	5.333		2,625	126		2,751	3,100	
	5610 25,000 gallon capacity, 3/8″ thick shell	Q-7	1.30	24.615		17,600	580		18,180	20,400	
	5630 40,000 gallon capacity, 3/8″ thick shell		.90	35.556		30,700	840		31,540	35,100	
	5640 50,000 gallon capacity, 3/8″ thick shell	↓	.80	40	↓	38,300	945		39,245	43,800	

Important: See the Reference Section for critical supporting data - Reference Nos., Crews, & Location Factors

	13281	Hazardous Material Remediation	CREW	DAILY OUTPUT	LABOR-HOURS	UNIT	2004 BARE COSTS MAT.	LABOR	EQUIP.	TOTAL	TOTAL INCL O&P	
440	0010	**REMOVAL** Existing lead paint, by chemicals, per application										440
	0020	See also, Div. 13280, Haz. Mat'l. Abatement										
	0050	Baseboard, to 6" wide	1 Pord	64	.125	L.F.	1.49	2.63		4.12	5.95	
	0070	To 12" wide		32	.250	"	2.94	5.25		8.19	11.90	
	0200	Balustrades, one side		28	.286	S.F.	3.33	6		9.33	13.50	
	1400	Cabinets, simple design		32	.250		2.92	5.25		8.17	11.85	
	1420	Ornate design		25	.320		3.75	6.70		10.45	15.20	
	1600	Cornice, simple design		60	.133		1.57	2.80		4.37	6.35	
	1620	Ornate design		20	.400		4.62	8.40		13.02	18.90	
	2800	Doors, one side, flush		84	.095		1.13	2		3.13	4.53	
	2820	Two panel		80	.100		1.17	2.10		3.27	4.74	
	2840	Four panel		45	.178		2.07	3.73		5.80	8.45	
	2880	For trim, one side, add		64	.125	L.F.	1.49	2.63		4.12	5.95	
	3000	Fence, picket, one side		30	.267	S.F.	3.13	5.60		8.73	12.65	
	3200	Grilles, one side, simple design		30	.267		3.13	5.60		8.73	12.65	
	3220	Ornate design		25	.320		3.75	6.70		10.45	15.20	
	4400	Pipes, to 4" diameter		90	.089	L.F.	1.07	1.87		2.94	4.25	
	4420	To 8" diameter		50	.160		1.86	3.36		5.22	7.55	
	4440	To 12" diameter		36	.222		2.61	4.67		7.28	10.50	
	4460	To 16" diameter		20	.400		4.65	8.40		13.05	18.90	
	4500	For hangers, add		40	.200	Ea.	2.33	4.20		6.53	9.45	
	4800	Siding		90	.089	S.F.	1.07	1.87		2.94	4.25	
	5000	Trusses, open		55	.145	SF Face	1.71	3.05		4.76	6.90	
	6200	Windows, one side only, double hung, 1/1 light, 24" x 48" high		4	2	Ea.	23.50	42		65.50	95	
	6220	30" x 60" high		3	2.667		31.50	56		87.50	127	
	6240	36" x 72" high		2.50	3.200		37.50	67		104.50	152	
	6280	40" x 80" high		2	4		47	84		131	190	
	6400	Colonial window, 6/6 light, 24" x 48" high		2	4		47	84		131	190	
	6420	30" x 60" high		1.50	5.333		62.50	112		174.50	253	
	6440	36" x 72" high		1	8		94	168		262	380	
	6480	40" x 80" high		1	8		94	168		262	380	
	6600	8/8 light, 24" x 48" high		2	4		47	84		131	190	
	6620	40" x 80" high		1	8		94	168		262	380	
	6800	12/12 light, 24" x 48" high		1	8		94	168		262	380	
	6820	40" x 80" high		.75	10.667		125	224		349	510	
	6840	Window frame & trim items, included in pricing above										
460	0010	**LEAD PAINT ENCAPSULATION**, water based polymer coating,14 mil DFT										460
	0020	Interior, brushwork, trim, under 6"	1 Pord	240	.033	L.F.	2.20	.70		2.90	3.57	
	0030	6" to 12" wide		180	.044		2.93	.93		3.86	4.75	
	0040	Balustrades		300	.027		1.77	.56		2.33	2.87	
	0050	Pipe to 4" diameter		500	.016		1.06	.34		1.40	1.72	
	0060	To 8" diameter		375	.021		1.40	.45		1.85	2.28	
	0070	To 12" diameter		250	.032		2.11	.67		2.78	3.42	
	0080	To 16" diameter		170	.047		3.10	.99		4.09	5.05	
	0090	Cabinets, ornate design		200	.040	S.F.	2.65	.84		3.49	4.30	
	0100	Simple design		250	.032	"	2.11	.67		2.78	3.42	
	0110	Doors, 3'x 7', both sides, incl. frame & trim										
	0120	Flush	1 Pord	6	1.333	Ea.	27	28		55	75.50	
	0130	French, 10-15 lite		3	2.667		5.40	56		61.40	98	
	0140	Panel		4	2		32.50	42		74.50	105	
	0150	Louvered		2.75	2.909		29.50	61		90.50	133	
	0160	Windows, per interior side, per 15 S.F.										
	0170	1 to 6 lite	1 Pord	14	.571	Ea.	18.65	12		30.65	40	
	0180	7 to 10 lite		7.50	1.067		20.50	22.50		43	59.50	
	0190	12 lite		5.75	1.391		27.50	29		56.50	78.50	
	0200	Radiators		8	1		66	21		87	107	

13281 | Hazardous Material Remediation

			CREW	DAILY OUTPUT	LABOR-HOURS	UNIT	2004 BARE COSTS MAT.	LABOR	EQUIP.	TOTAL	TOTAL INCL O&P	
460	0210	Grilles, vents	1 Pord	275	.029	S.F.	1.92	.61		2.53	3.11	460
	0220	Walls, roller, drywall or plaster		1,000	.008		.53	.17		.70	.86	
	0230	With spunbonded reinforcing fabric		720	.011		.60	.23		.83	1.04	
	0240	Wood		800	.010		.66	.21		.87	1.08	
	0250	Ceilings, roller, drywall or plaster		900	.009		.60	.19		.79	.97	
	0260	Wood		700	.011		.75	.24		.99	1.22	
	0270	Exterior, brushwork, gutters and downspouts		300	.027	L.F.	1.77	.56		2.33	2.87	
	0280	Columns		400	.020	S.F.	1.31	.42		1.73	2.13	
	0290	Spray, siding		600	.013	"	.88	.28		1.16	1.43	
	0300	Miscellaneous										
	0310	Electrical conduit, brushwork, to 2" diameter	1 Pord	500	.016	L.F.	1.06	.34		1.40	1.72	
	0320	Brick, block or concrete, spray		500	.016	S.F.	1.06	.34		1.40	1.72	
	0330	Steel, flat surfaces and tanks to 12"		500	.016		1.06	.34		1.40	1.72	
	0340	Beams, brushwork		400	.020		1.31	.42		1.73	2.13	
	0350	Trusses		400	.020		1.31	.42		1.73	2.13	

13720 | Detection & Alarm

			CREW	DAILY OUTPUT	LABOR-HOURS	UNIT	MAT.	LABOR	EQUIP.	TOTAL	TOTAL INCL O&P	
065	0010	**DETECTION SYSTEMS**, not including wires & conduits										065
	0100	Burglar alarm, battery operated, mechanical trigger	1 Elec	4	2	Ea.	249	53		302	360	
	0200	Electrical trigger		4	2		297	53		350	410	
	0400	For outside key control, add		8	1		70.50	26.50		97	121	
	0600	For remote signaling circuitry, add		8	1		112	26.50		138.50	166	
	0800	Card reader, flush type, standard		2.70	2.963		835	78		913	1,050	
	1000	Multi-code		2.70	2.963		1,075	78		1,153	1,300	
	1200	Door switches, hinge switch		5.30	1.509		52.50	40		92.50	123	
	1400	Magnetic switch		5.30	1.509		62	40		102	133	
	2800	Ultrasonic motion detector, 12 volt		2.30	3.478		206	92		298	375	
	3000	Infrared photoelectric detector		2.30	3.478		170	92		262	335	
	3200	Passive infrared detector		2.30	3.478		254	92		346	430	
	3420	Switchmats, 30" x 5'		5.30	1.509		76	40		116	149	
	3440	30" x 25'		4	2		182	53		235	286	
	3460	Police connect panel		4	2		219	53		272	325	
	3480	Telephone dialer		5.30	1.509		345	40		385	445	
	3500	Alarm bell		4	2		69.50	53		122.50	163	
	3520	Siren		4	2		131	53		184	230	
	5200	Smoke detector, ceiling type		6.20	1.290		75	34		109	138	
	5600	Strobe and horn		5.30	1.509		95	40		135	170	
	5800	Fire alarm horn		6.70	1.194		36.50	31.50		68	91.50	
	6600	Drill switch		8	1		86.50	26.50		113	138	
	6800	Master box		2.70	2.963		3,100	78		3,178	3,525	
	7800	Remote annunciator, 8 zone lamp		1.80	4.444		175	117		292	385	
	8000	12 zone lamp	2 Elec	2.60	6.154		300	162		462	595	
	8200	16 zone lamp	"	2.20	7.273		300	192		492	640	
	8400	Standpipe or sprinkler alarm, alarm device	1 Elec	8	1		125	26.50		151.50	181	
	8600	Actuating device	"	8	1		290	26.50		316.50	365	

13838 | Pneumatic/Electric Controls

			CREW	DAILY OUTPUT	LABOR-HOURS	UNIT	2004 BARE COSTS MAT.	LABOR	EQUIP.	TOTAL	TOTAL INCL O&P	
200	0010	CONTROL COMPONENTS										200
	5000	Thermostats										
	5030	Manual	1 Shee	8	1	Ea.	27	25.50		52.50	73	
	5040	1 set back, electric, timed	↓	8	1	↓	81	25.50		106.50	132	
	5050	2 set back, electric, timed	↓	8	1	↓	178	25.50		203.50	239	

For information about Means Estimating Seminars, see yellow pages 12 and 13 in back of book

Division Notes

I	CREW	DAILY OUTPUT	LABOR-HOURS	UNIT	2004 BARE COSTS				TOTAL INCL O&P
					MAT.	LABOR	EQUIP.	TOTAL	

Division 14
Conveying Systems

Estimating Tips

General

- Many products in Division 14 will require some type of support or blocking for installation not included with the item itself. Examples are supports for conveyors or tube systems, attachment points for lifts, and footings for hoists or cranes. Add these supports in the appropriate division.

14100 Dumbwaiters
14200 Elevators

- Dumbwaiters and elevators are estimated and purchased in a method similar to buying a car. The manufacturer has a base unit with standard features. Added to this base unit price will be whatever options the owner or specifications require. Increased load capacity, additional vertical travel, additional stops, higher speed, and cab finish options are items to be considered. When developing an estimate for dumbwaiters and elevators, remember that some items needed by the installers may have to be included as part of the general contract.

  Examples are:
  - shaftway
  - rail support brackets
  - machine room
  - electrical supply
  - sill angles
  - electrical connections
  - pits
  - roof penthouses
  - pit ladders

  Check the job specifications and drawings before pricing.
- Installation of elevators and handicapped lifts in historic structures can require significant additional costs. The associated structural requirements may involve cutting into and repairing finishes, mouldings, flooring, etc. The estimator must account for these special conditions.

14300 Escalators & Moving Walks

- Escalators and moving walks are specialty items installed by specialty contractors. There are numerous options associated with these items. For specific options contact a manufacturer or contractor. In a method similar to estimating dumbwaiters and elevators, you should verify the extent of general contract work and add items as necessary.

14400 Lifts
14500 Material Handling
14600 Hoists & Cranes

- Products such as correspondence lifts, conveyors, chutes, pneumatic tube systems, material handling cranes and hoists as well as other items specified in this subdivision may require trained installers. The general contractor might not have any choice as to who will perform the installation or when it will be performed. Long lead times are often required for these products, making early decisions in scheduling necessary.

Reference Numbers

Reference numbers are shown in bold squares at the beginning of some major classifications. These numbers refer to related items in the Reference Section. The reference information may be an estimating procedure, an alternate pricing method or technical information.

Note: Not all subdivisions listed here necessarily appear in this publication.

14200 | Elevators

14210 | Electric Traction Elevators

			CREW	DAILY OUTPUT	LABOR-HOURS	UNIT	2004 BARE COSTS MAT.	LABOR	EQUIP.	TOTAL	TOTAL INCL O&P	
100	0012	ELEVATOR SYSTEMS										100
	7000	Residential, cab type, 1 floor, 2 stop, minimum	2 Elev	.20	80	Ea.	8,075	2,225		10,300	12,600	
	7100	Maximum		.10	160		13,700	4,450		18,150	22,300	
	7200	2 floor, 3 stop, minimum		.12	133		12,000	3,725		15,725	19,300	
	7300	Maximum	↓	.06	266	↓	19,600	7,425		27,025	33,600	

14420 | Wheelchair Lifts

			CREW	DAILY OUTPUT	LABOR-HOURS	UNIT	MAT.	LABOR	EQUIP.	TOTAL	TOTAL INCL O&P	
100	0010	CHAIR / WHEELCHAIR LIFT										100
	7700	Stair climber (chair lift), single seat, minimum	2 Elev	1	16	Ea.	3,925	445		4,370	5,025	
	7800	Maximum	"	.20	80	"	5,375	2,225		7,600	9,575	

Division 15 Mechanical

Estimating Tips

15100 Building Services Piping

This subdivision is primarily basic pipe and related materials. The pipe may be used by any of the mechanical disciplines, i.e., plumbing, fire protection, heating, and air conditioning.

- The piping section lists the add to labor for elevated pipe installation. These adds apply to all elevated pipe, fittings, valves, insulation, etc., that are placed above 10′ high. CAUTION: the correct percentage may vary for the same pipe. For example, the percentage add for the basic pipe installation should be based on the maximum height that the craftsman must install for that particular section. If the pipe is to be located 14′ above the floor but it is suspended on threaded rod from beams, the bottom flange of which is 18′ high (4′ rods), then the height is actually 18′ and the add is 20%. The pipe coverer, however, does not have to go above the 14′ and so his add should be 10%.
- Most pipe is priced first as straight pipe with a joint (coupling, weld, etc.) every 10′ and a hanger usually every 10′. There are exceptions with hanger spacing such as: for cast iron pipe (5′) and plastic pipe (3 per 10′). Following each type of pipe there are several lines listing sizes and the amount to be subtracted to delete couplings and hangers. This is for pipe that is to be buried or supported together on trapeze hangers. The reason that the couplings are deleted is that these runs are usually long and frequently longer lengths of pipe are used. By deleting the couplings the estimator is expected to look up and add back the correct reduced number of couplings.
- When preparing an estimate it may be necessary to approximate the fittings. Fittings usually run between 25% and 50% of the cost of the pipe. The lower percentage is for simpler runs, and the higher number is for complex areas like mechanical rooms.
- For historic restoration projects, the systems must be as invisible as possible, and pathways must be sought for pipes, conduits, and ductwork. While installations in accessible spaces (such as basements and attics) are relatively straightforward to estimate, labor costs may be more difficult to determine when delivery systems must be concealed.

15400 Plumbing Fixtures & Equipment

- Plumbing fixture costs usually require two lines, the fixture itself and its "rough-in, supply and waste".
- In the Assemblies Section (Plumbing D2010) for the desired fixture, the System Components Group in the center of the page shows the fixture itself on the first line while the rest of the list (fittings, pipe, tubing, etc.) will total up to what we refer to in the Unit Price section as "Rough-in, supply, waste and vent". Note that for most fixtures we allow a nominal 5′ of tubing to reach from the fixture to a main or riser.
- Remember that gas and oil fired units need venting.

15500 Heat Generation Equipment

- When estimating the cost of an HVAC system, check to see who is responsible for providing and installing the temperature control system. It is possible to overlook controls, assuming that they would be included in the electrical estimate.
- When looking up a boiler be careful on specified capacity. Some manufacturers rate their products on output while others use input.
- Include HVAC insulation for pipe, boiler and duct (wrap and liner).
- Be careful when looking up mechanical items to get the correct pressure rating and connection type (thread, weld, flange).

15700 Heating/Ventilation/ Air Conditioning Equipment

- Combination heating and cooling units are sized by the air conditioning requirements. (See Reference No. R15710-020 for preliminary sizing guide.)
- A ton of air conditioning is nominally 400 CFM.
- Rectangular duct is taken off by the linear foot for each size, but its cost is usually estimated by the pound. Remember that SMACNA standards now base duct on internal pressure.
- Prefabricated duct is estimated and purchased like pipe: straight sections and fittings.
- Note that cranes or other lifting equipment are not included on any lines in Division 15. For example, if a crane is required to lift a heavy piece of pipe into place high above a gym floor, or to put a rooftop unit on the roof of a four-story building, etc., it must be added. Due to the potential for extreme variation—from nothing additional required, to a major crane or helicopter—we feel that including a nominal amount for "lifting contingency" would be useless and detract from the accuracy of the estimate. When using equipment rental from Means do not forget to include the cost of the operator(s).

Reference Numbers

Reference numbers are shown in bold squares at the beginning of some major classifications. These numbers refer to related items in the Reference Section. The reference information may be an estimating procedure, an alternate pricing method or technical information.

Note: Not all subdivisions listed here necessarily appear in this publication.

15055 | Selective Mech Demolition

			CREW	DAILY OUTPUT	LABOR-HOURS	UNIT	2004 BARE COSTS MAT.	LABOR	EQUIP.	TOTAL	TOTAL INCL O&P	
300	0010	HVAC DEMOLITION										300
	0100	Air conditioner, split unit, 3 ton	Q-5	2	8	Ea.		189		189	310	
	0150	Package unit, 3 ton	Q-6	3	8	"		182		182	299	
	0260	Baseboard, hydronic fin tube, 1/2"	Q-5	117	.137	L.F.		3.23		3.23	5.30	
	0300	Boiler, electric	Q-19	2	12	Ea.		295		295	480	
	0340	Gas or oil, steel, under 150 MBH	Q-6	3	8	"		182		182	299	
	1000	Ductwork, 4" high, 8" wide	1 Clab	200	.040	L.F.		.68		.68	1.15	
	1100	6" high, 8" wide		165	.048			.82		.82	1.39	
	1200	10" high, 12" wide		125	.064			1.08		1.08	1.84	
	1300	12"-14" high, 16"-18" wide		85	.094			1.59		1.59	2.70	
	1500	30" high, 36" wide	↓	56	.143	↓		2.41		2.41	4.10	
	2200	Furnace, electric	Q-20	2	10	Ea.		237		237	395	
	2300	Gas or oil, under 120 MBH	Q-9	4	4			92		92	154	
	2340	Over 120 MBH	"	3	5.333			123		123	206	
	2800	Heat pump, package unit, 3 ton	Q-5	2.40	6.667			158		158	259	
	2840	Split unit, 3 ton		2	8			189		189	310	
	2950	Tank, steel, oil, 275 gal., above ground		10	1.600			38		38	62	
	2960	Remove and reset	↓	3	5.333	↓		126		126	207	
	9000	Minimum labor/equipment charge	Q-6	3	8	Job		182		182	299	
600	0010	PLUMBING DEMOLITION										600
	1020	Fixtures, including 10' piping										
	1100	Bath tubs, cast iron	1 Plum	4	2	Ea.		52.50		52.50	86	
	1120	Fiberglass		6	1.333			35		35	57	
	1140	Steel		5	1.600			42		42	68.50	
	1200	Lavatory, wall hung		10	.800			21		21	34.50	
	1220	Counter top		8	1			26		26	43	
	1300	Sink, steel or cast iron, single		8	1			26		26	43	
	1320	Double		7	1.143			30		30	49	
	1400	Water closet, floor mounted		8	1			26		26	43	
	1420	Wall mounted		7	1.143	↓		30		30	49	
	2000	Piping, metal, to 1-1/2" diameter		200	.040	L.F.		1.05		1.05	1.72	
	2050	2" to 3-1/2" diameter	↓	150	.053			1.39		1.39	2.29	
	2100	4" to 6" diameter	2 Plum	100	.160	↓		4.18		4.18	6.85	
	2250	Water heater, 40 gal.	1 Plum	6	1.333	Ea.		35		35	57	
	3000	Submersible sump pump		24	.333			8.70		8.70	14.30	
	6000	Remove and reset fixtures, minimum		6	1.333			35		35	57	
	6100	Maximum		4	2	↓		52.50		52.50	86	
	9000	Minimum labor/equipment charge	↓	2	4	Job		105		105	172	

15080 | Mechanical Insulation

			CREW	DAILY OUTPUT	LABOR-HOURS	UNIT	MAT.	LABOR	EQUIP.	TOTAL	TOTAL INCL O&P	
200	0010	DUCT INSULATION										200
	3000	Ductwork										
	3020	Blanket type, fiberglass, flexible										
	3030	Fire resistant liner, black coating one side										
	3050	1/2" thick, 2 lb. density	Q-14	380	.042	S.F.	.36	.90		1.26	1.94	
	3060	1" thick, 1-1/2 lb. density	"	350	.046	"	.48	.98		1.46	2.21	
	3140	FRK vapor barrier wrap, .75 lb. density										
	3160	1" thick	Q-14	350	.046	S.F.	.34	.98		1.32	2.05	
	3170	1-1/2" thick	"	320	.050	"	.35	1.07		1.42	2.22	
	3490	Board type, fiberglass liner, 3 lb. density										
	3500	Fire resistant, black pigmented, 1 side										
	3520	1" thick	Q-14	150	.107	S.F.	1.44	2.28		3.72	5.50	
	3540	1-1/2" thick	"	130	.123	"	1.76	2.63		4.39	6.45	
	9600	Minimum labor/equipment charge	1 Stpi	4	2	Job		52.50		52.50	86	

15050 | Basic Materials & Methods

15080 | Mechanical Insulation

			CREW	DAILY OUTPUT	LABOR-HOURS	UNIT	2004 BARE COSTS MAT.	LABOR	EQUIP.	TOTAL	TOTAL INCL O&P	
400	0010	**EQUIPMENT INSULATION**										400
	2900	Domestic water heater wrap kit										
	2920	1-1/2″ with vinyl jacket, 20-60 gal.	1 Plum	8	1	Ea.	16.60	26		42.60	61.50	
600	0010	**PIPING INSULATION**										600
	2930	Insulated protectors, (ADA)										
	2935	For exposed piping under sinks or lavatories.										
	2940	Vinyl coated foam, velcro tabs										
	2945	P Trap, 1-1/4″ or 1-1/2″	1 Plum	32	.250	Ea.	17.15	6.55		23.70	29.50	
	2960	Valve and supply cover										
	2965	1/2″, 3/8″, and 7/16″ pipe size	1 Plum	32	.250	Ea.	17.15	6.55		23.70	29.50	
	2970	Extension drain cover										
	2975	1-1/4″, or 1-1/2″ pipe size	1 Plum	32	.250	Ea.	17.55	6.55		24.10	30	
	2985	1-1/4″ pipe size	″	32	.250	″	20	6.55		26.55	33	
	4000	Pipe covering (price copper tube one size less than IPS)										
	6600	Fiberglass, with all service jacket										
	6840	1″ wall, 1/2″ iron pipe size	Q-14	240	.067	L.F.	.69	1.43		2.12	3.21	
	6860	3/4″ iron pipe size		230	.070		.79	1.49		2.28	3.42	
	6870	1″ iron pipe size		220	.073		.80	1.56		2.36	3.55	
	6900	2″ iron pipe size	↓	200	.080	↓	1.08	1.71		2.79	4.12	
	7879	Rubber tubing, flexible closed cell foam										
	8100	1/2″ wall, 1/4″ iron pipe size	1 Asbe	90	.089	L.F.	.34	2.11		2.45	3.99	
	8130	1/2″ iron pipe size		89	.090		.54	2.13		2.67	4.25	
	8140	3/4″ iron pipe size		89	.090		.56	2.13		2.69	4.28	
	8150	1″ iron pipe size		88	.091		.66	2.16		2.82	4.43	
	8170	1-1/2″ iron pipe size		87	.092		.98	2.18		3.16	4.83	
	8180	2″ iron pipe size		86	.093		1.61	2.21		3.82	5.55	
	8300	3/4″ wall, 1/4″ iron pipe size		90	.089		.67	2.11		2.78	4.36	
	8330	1/2″ iron pipe size		89	.090		.81	2.13		2.94	4.55	
	8340	3/4″ iron pipe size		89	.090		.97	2.13		3.10	4.73	
	8350	1″ iron pipe size		88	.091		1.18	2.16		3.34	5	
	8380	2″ iron pipe size		86	.093		2.43	2.21		4.64	6.45	
	8444	1″ wall, 1/2″ iron pipe size		86	.093		1.53	2.21		3.74	5.45	
	8445	3/4″ iron pipe size		84	.095		1.90	2.26		4.16	5.95	
	8446	1″ iron pipe size		84	.095		2.30	2.26		4.56	6.40	
	8447	1-1/4″ iron pipe size		82	.098		2.68	2.32		5	6.95	
	8448	1-1/2″ iron pipe size		82	.098		3.02	2.32		5.34	7.30	
	8449	2″ iron pipe size		80	.100		4.70	2.38		7.08	9.25	
	8450	2-1/2″ iron pipe size	↓	80	.100	↓	6.15	2.38		8.53	10.85	
	8456	Rubber insulation tape, 1/8″ x 2″ x 30′				Ea.	10.80			10.80	11.90	

15100 | Building Services Piping

15107 | Metal Pipe & Fittings

			CREW	DAILY OUTPUT	LABOR-HOURS	UNIT	2004 BARE COSTS MAT.	LABOR	EQUIP.	TOTAL	TOTAL INCL O&P	
320	0010	**PIPE, CAST IRON** Soil, on hangers 5′ O.C. R15100-050										320
	0020	Single hub, service wt., lead & oakum joints 10′ O.C.										
	2120	2″ diameter	Q-1	63	.254	L.F.	4.69	6		10.69	14.95	
	2140	3″ diameter		60	.267		6.50	6.25		12.75	17.45	
	2160	4″ diameter	↓	55	.291	↓	8.30	6.85		15.15	20.50	
	4000	No hub, couplings 10′ O.C.										

MECHANICAL 15

	15107 | Metal Pipe & Fittings		CREW	DAILY OUTPUT	LABOR-HOURS	UNIT	2004 BARE COSTS MAT.	LABOR	EQUIP.	TOTAL	TOTAL INCL O&P	
320	4100	1-1/2" diameter (R15100-050)	Q-1	71	.225	L.F.	5.35	5.30		10.65	14.60	320
	4120	2" diameter		67	.239		5.50	5.60		11.10	15.25	
	4140	3" diameter		64	.250		7.35	5.90		13.25	17.70	
	4160	4" diameter		58	.276		9.30	6.50		15.80	21	
360	0010	**PIPE, CAST IRON, FITTINGS** Soil										360
	0040	Hub and spigot, service weight, lead & oakum joints										
	0080	1/4 bend, 2"	Q-1	16	1	Ea.	9.05	23.50		32.55	48.50	
	0120	3"		14	1.143		12	27		39	57	
	0140	4"		13	1.231		18.80	29		47.80	68	
	0340	1/8 bend, 2"		16	1		6.40	23.50		29.90	45.50	
	0350	3"		14	1.143		10.10	27		37.10	55	
	0360	4"		13	1.231		14.70	29		43.70	63.50	
	0500	Sanitary tee, 2"		10	1.600		11.10	37.50		48.60	74	
	0540	3"		9	1.778		20.50	42		62.50	91	
	0620	4"		8	2		25	47		72	105	
	5990	No hub										
	6000	Cplg. & labor required at joints not incl. in fitting										
	6010	price. Add 1 coupling per joint for installed price										
	6020	1/4 Bend, 1-1/2"				Ea.	5.10			5.10	5.60	
	6060	2"					5.50			5.50	6.05	
	6080	3"					7.65			7.65	8.45	
	6120	4"					11			11	12.10	
	6184	1/4 Bend, long sweep, 1-1/2"					11.90			11.90	13.05	
	6186	2"					11.90			11.90	13.05	
	6188	3"					14.15			14.15	15.60	
	6189	4"					22.50			22.50	25	
	6190	5"					41.50			41.50	45.50	
	6191	6"					50.50			50.50	56	
	6192	8"					123			123	136	
	6193	10"					221			221	243	
	6200	1/8 Bend, 1-1/2"					4.22			4.22	4.64	
	6210	2"					4.69			4.69	5.15	
	6212	3"					6.35			6.35	6.95	
	6214	4"					8.05			8.05	8.85	
	6380	Sanitary Tee, tapped, 1-1/2"					9.30			9.30	10.25	
	6382	2" x 1-1/2"					8.45			8.45	9.30	
	6384	2"					9.35			9.35	10.30	
	6386	3" x 2"					13.05			13.05	14.40	
	6388	3"					24			24	26.50	
	6390	4" x 1-1/2"					11.70			11.70	12.85	
	6392	4" x 2"					13.15			13.15	14.45	
	6394	6" x 1-1/2"					27			27	30	
	6396	6" x 2"					27.50			27.50	30.50	
	6459	Sanitary Tee, 1-1/2"					7			7	7.70	
	6460	2"					7.65			7.65	8.45	
	6470	3"					9.30			9.30	10.25	
	6472	4"					14.40			14.40	15.85	
	8000	Coupling, standard (by CISPI Mfrs.)										
	8020	1-1/2"	Q-1	48	.333	Ea.	4.84	7.85		12.69	18.15	
	8040	2"		44	.364		4.84	8.55		13.39	19.35	
	8080	3"		38	.421		5.75	9.90		15.65	22.50	
	8120	4"		33	.485		6.80	11.40		18.20	26	
420	0010	**PIPE, COPPER** Solder joints										420
	1000	Type K tubing, couplings & clevis hangers 10' O.C.										
	1180	3/4" diameter	1 Plum	74	.108	L.F.	1.76	2.83		4.59	6.55	
	1200	1" diameter	"	66	.121	"	2.27	3.17		5.44	7.70	

15107 | Metal Pipe & Fittings

	Line	Description	Crew	Daily Output	Labor-Hours	Unit	2004 Bare Costs Mat.	Labor	Equip.	Total	Total Incl O&P
420	2000	Type L tubing, couplings & hangers 10′ O.C.									
	2140	1/2″ diameter	1 Plum	81	.099	L.F.	1.01	2.58		3.59	5.35
	2160	5/8″ diameter		79	.101		1.35	2.65		4	5.85
	2180	3/4″ diameter		76	.105		1.42	2.75		4.17	6.10
	2200	1″ diameter		68	.118		1.87	3.08		4.95	7.10
	2220	1-1/4″ diameter		58	.138		2.44	3.61		6.05	8.60
	3000	Type M tubing, couplings & hangers 10′ O.C.									
	3140	1/2″ diameter	1 Plum	84	.095	L.F.	.86	2.49		3.35	5.05
	3180	3/4″ diameter		78	.103		1.22	2.68		3.90	5.75
	3200	1″ diameter		70	.114		1.96	2.99		4.95	7.05
	3220	1-1/4″ diameter		60	.133		2.05	3.49		5.54	7.95
	3240	1-1/2″ diameter		54	.148		2.68	3.87		6.55	9.30
	3260	2″ diameter		44	.182		4.01	4.75		8.76	12.20
	4000	Type DWV tubing, couplings & hangers 10′ O.C.									
	4100	1-1/4″ diameter	1 Plum	60	.133	L.F.	2.08	3.49		5.57	8
	4120	1-1/2″ diameter		54	.148		2.53	3.87		6.40	9.15
	4140	2″ diameter		44	.182		3.29	4.75		8.04	11.40
	4160	3″ diameter	Q-1	58	.276		5.90	6.50		12.40	17.15
	4180	4″ diameter	″	40	.400		10.10	9.40		19.50	26.50
460	0010	**PIPE, COPPER, FITTINGS** Wrought unless otherwise noted									
	0040	Solder joints, copper x copper									
	0100	1/2″	1 Plum	20	.400	Ea.	.30	10.45		10.75	17.50
	0120	3/4″		19	.421		.67	11		11.67	18.80
	0250	45° elbow, 1/4″		22	.364		1.86	9.50		11.36	17.65
	0280	1/2″		20	.400		.55	10.45		11	17.75
	0290	5/8″		19	.421		2.89	11		13.89	21
	0300	3/4″		19	.421		.96	11		11.96	19.10
	0310	1″		16	.500		3.45	13.10		16.55	25.50
	0320	1-1/4″		15	.533		3.45	13.95		17.40	27
	0450	Tee, 1/4″		14	.571		2.10	14.95		17.05	27
	0480	1/2″		13	.615		.51	16.10		16.61	27
	0490	5/8″		12	.667		3.48	17.45		20.93	32.50
	0500	3/4″		12	.667		1.24	17.45		18.69	30
	0510	1″		10	.800		3.81	21		24.81	38.50
	0520	1-1/4″		9	.889		6.10	23		29.10	44.50
	0612	Tee, reducing on the outlet, 1/4″		15	.533		3.37	13.95		17.32	26.50
	0613	3/8″		15	.533		3.17	13.95		17.12	26.50
	0614	1/2″		14	.571		2.78	14.95		17.73	27.50
	0615	5/8″		13	.615		5.60	16.10		21.70	32.50
	0616	3/4″		12	.667		1.20	17.45		18.65	30
	0617	1″		11	.727		4.03	19		23.03	35.50
	0618	1-1/4″		10	.800		5.95	21		26.95	41
	0619	1-1/2″		9	.889		6.30	23		29.30	45
	0620	2″		8	1		10.30	26		36.30	54.50
	0621	2-1/2″	Q-1	9	1.778		24.50	42		66.50	95.50
	0622	3″		8	2		34	47		81	115
	0623	4″		6	2.667		66.50	63		129.50	176
	0624	5″		5	3.200		305	75.50		380.50	460
	0625	6″	Q-2	7	3.429		420	77.50		497.50	585
	0626	8″	″	6	4		1,700	90.50		1,790.50	2,025
	0630	Tee, reducing on the run, 1/4″	1 Plum	15	.533		4.22	13.95		18.17	27.50
	0631	3/8″		15	.533		5.65	13.95		19.60	29
	0632	1/2″		14	.571		5.05	14.95		20	30
	0633	5/8″		13	.615		5.65	16.10		21.75	32.50
	0634	3/4″		12	.667		2.88	17.45		20.33	31.50

15107 | Metal Pipe & Fittings

	Line	Description	Crew	Daily Output	Labor-Hours	Unit	2004 Bare Costs Mat.	Labor	Equip.	Total	Total Incl O&P
460	0635	1"	1 Plum	11	.727	Ea.	4.80	19		23.80	36.50
	0636	1-1/4"		10	.800		7.60	21		28.60	43
	0637	1-1/2"		9	.889		13.85	23		36.85	53.50
	0638	2"		8	1		18.45	26		44.45	63.50
	0639	2-1/2"	Q-1	9	1.778		44.50	42		86.50	117
	0640	3"		8	2		62.50	47		109.50	146
	0641	4"		6	2.667		132	63		195	248
	0642	5"		5	3.200		305	75.50		380.50	460
	0643	6"	Q-2	7	3.429		420	77.50		497.50	585
	0644	8"	"	6	4		1,700	90.50		1,790.50	2,025
	0650	Coupling, 1/4"	1 Plum	24	.333		.23	8.70		8.93	14.55
	0680	1/2"		22	.364		.23	9.50		9.73	15.85
	0690	5/8"		21	.381		.71	9.95		10.66	17.15
	0700	3/4"		21	.381		.45	9.95		10.40	16.85
	0710	1"		18	.444		.91	11.60		12.51	20
	0715	1-1/4"		17	.471		1.71	12.30		14.01	22
	2000	DWV, solder joints, copper x copper									
	2030	90° Elbow, 1-1/4"	1 Plum	13	.615	Ea.	3.06	16.10		19.16	30
	2050	1-1/2"		12	.667		4.12	17.45		21.57	33
	2070	2"		10	.800		5.95	21		26.95	41
	2090	3"	Q-1	10	1.600		14.75	37.50		52.25	78.50
	2100	4"	"	9	1.778		71	42		113	147
	2250	Tee, Sanitary, 1-1/4"	1 Plum	9	.889		6.05	23		29.05	44.50
	2270	1-1/2"		8	1		7.50	26		33.50	51.50
	2290	2"		7	1.143		8.75	30		38.75	58.50
	2310	3"	Q-1	7	2.286		32	54		86	123
	2330	4"	"	6	2.667		81	63		144	192
	2400	Coupling, 1-1/4"	1 Plum	14	.571		1.43	14.95		16.38	26
	2420	1-1/2"		13	.615		1.78	16.10		17.88	28.50
	2440	2"		11	.727		2.46	19		21.46	33.50
	2460	3"	Q-1	11	1.455		4.77	34		38.77	61.50
	2480	4"	"	10	1.600		15.20	37.50		52.70	78.50
620	0010	**PIPE, STEEL**									
	0050	Schedule 40, threaded, with couplings, and clevis type									
	0060	hangers sized for covering, 10' O.C.									
	0540	Black, 1/4" diameter	1 Plum	66	.121	L.F.	1.28	3.17		4.45	6.60
	0570	3/4" diameter		61	.131		1.42	3.43		4.85	7.20
	0580	1" diameter		53	.151		2.04	3.95		5.99	8.75
	0590	1-1/4" diameter	Q-1	89	.180		2.48	4.23		6.71	9.70
	0600	1-1/2" diameter		80	.200		2.83	4.71		7.54	10.80
	0610	2" diameter		64	.250		3.64	5.90		9.54	13.65
640	0010	**PIPE, STEEL, FITTINGS** Threaded									
	5000	Malleable iron, 150 lb.									
	5020	Black									
	5040	90° elbow, straight									
	5090	3/4"	1 Plum	14	.571	Ea.	1.57	14.95		16.52	26
	5100	1"	"	13	.615		2.73	16.10		18.83	29.50
	5120	1-1/2"	Q-1	20	.800		5.90	18.80		24.70	37.50
	5130	2"	"	18	.889		10.20	21		31.20	45.50
	5450	Tee, straight									
	5500	3/4"	1 Plum	9	.889	Ea.	2.50	23		25.50	41
	5510	1"	"	8	1		3.82	26		29.82	47
	5520	1-1/4"	Q-1	14	1.143		6.90	27		33.90	51.50
	5530	1-1/2"		13	1.231		8.60	29		37.60	57
	5540	2"		11	1.455		14.65	34		48.65	72

15107 | Metal Pipe & Fittings

			CREW	DAILY OUTPUT	LABOR-HOURS	UNIT	2004 BARE COSTS MAT.	LABOR	EQUIP.	TOTAL	TOTAL INCL O&P	
640	5650	Coupling										640
	5700	3/4"	1 Plum	18	.444	Ea.	2	11.60		13.60	21.50	
	5710	1"	"	15	.533		3.15	13.95		17.10	26.50	
	5730	1-1/2"	Q-1	24	.667		5.50	15.70		21.20	31.50	
	5740	2"	"	21	.762	↓	8.15	17.95		26.10	38.50	

15108 | Plastic Pipe & Fittings

			CREW	DAILY OUTPUT	LABOR-HOURS	UNIT	MAT.	LABOR	EQUIP.	TOTAL	TOTAL INCL O&P	
520	0010	**PIPE, PLASTIC**										520
	1800	PVC, couplings 10′ O.C., hangers 3 per 10′										
	1820	Schedule 40										
	1860	1/2" diameter	1 Plum	54	.148	L.F.	1.41	3.87		5.28	7.90	
	1870	3/4" diameter		51	.157		1.49	4.10		5.59	8.40	
	1880	1" diameter		46	.174		1.66	4.55		6.21	9.30	
	1890	1-1/4" diameter		42	.190		1.86	4.98		6.84	10.20	
	1900	1-1/2" diameter	↓	36	.222		1.97	5.80		7.77	11.70	
	1910	2" diameter	Q-1	59	.271		2.30	6.40		8.70	13	
	1920	2-1/2" diameter		56	.286		3	6.70		9.70	14.35	
	1930	3" diameter		53	.302		3.77	7.10		10.87	15.80	
	1940	4" diameter	↓	48	.333	↓	4.76	7.85		12.61	18.10	
	4100	DWV type, schedule 40, couplings 10′ O.C., hangers 3 per 10′										
	4120	ABS										
	4140	1-1/4" diameter	1 Plum	42	.190	L.F.	1.57	4.98		6.55	9.90	
	4150	1-1/2" diameter	"	36	.222		1.60	5.80		7.40	11.30	
	4160	2" diameter	Q-1	59	.271	↓	1.71	6.40		8.11	12.35	
	4400	PVC										
	4410	1-1/4" diameter	1 Plum	42	.190	L.F.	1.70	4.98		6.68	10	
	4420	1-1/2" diameter	"	36	.222		1.73	5.80		7.53	11.45	
	4460	2" diameter	Q-1	59	.271		1.90	6.40		8.30	12.55	
	4470	3" diameter		53	.302		3.19	7.10		10.29	15.15	
	4480	4" diameter	↓	48	.333	↓	4.04	7.85		11.89	17.30	
	5360	CPVC, couplings 10′ O.C., hangers 3 per 10′										
	5380	Schedule 40										
	5460	1/2" diameter	1 Plum	54	.148	L.F.	2.60	3.87		6.47	9.20	
	5470	3/4" diameter		51	.157		3.29	4.10		7.39	10.35	
	5480	1" diameter		46	.174		3.97	4.55		8.52	11.80	
	5490	1-1/4" diameter		42	.190		4.56	4.98		9.54	13.15	
	5500	1-1/2" diameter	↓	36	.222		5.05	5.80		10.85	15.10	
	5510	2" diameter	Q-1	59	.271	↓	6.10	6.40		12.50	17.20	
	6500	Residential installation, plastic pipe										
	6510	Couplings 10′ O.C., strap hangers 3 per 10′										
	6520	PVC, Schedule 40										
	6530	1/2" diameter	1 Plum	138	.058	L.F.	.87	1.52		2.39	3.44	
	6540	3/4" diameter		128	.063		.93	1.63		2.56	3.70	
	6550	1" diameter		119	.067		1.04	1.76		2.80	4.02	
	6560	1-1/4" diameter		111	.072		1.23	1.88		3.11	4.44	
	6570	1-1/2" diameter	↓	104	.077		1.52	2.01		3.53	4.97	
	6580	2" diameter	Q-1	197	.081		1.71	1.91		3.62	5	
	6590	2-1/2" diameter		162	.099		2.95	2.32		5.27	7.05	
	6600	4" diameter	↓	123	.130	↓	4.78	3.06		7.84	10.25	
	6700	PVC, DWV, Schedule 40										
	6720	1-1/4" diameter	1 Plum	100	.080	L.F.	1.35	2.09		3.44	4.92	
	6730	1-1/2" diameter	"	94	.085		1.49	2.23		3.72	5.30	
	6740	2" diameter	Q-1	178	.090		1.66	2.12		3.78	5.30	
	6760	4" diameter	"	110	.145	↓	4.51	3.42		7.93	10.55	

15108 | Plastic Pipe & Fittings		CREW	DAILY OUTPUT	LABOR-HOURS	UNIT	2004 BARE COSTS MAT.	LABOR	EQUIP.	TOTAL	TOTAL INCL O&P
560 0010	PIPE, PLASTIC, FITTINGS									
2700	PVC (white), schedule 40, socket joints									
2760	90° elbow, 1/2"	1 Plum	33.30	.240	Ea.	.29	6.30		6.59	10.60
2770	3/4"		28.60	.280		.32	7.30		7.62	12.35
2780	1"		25	.320		.57	8.35		8.92	14.40
2790	1-1/4"		22.20	.360		1	9.40		10.40	16.55
2800	1-1/2"	↓	20	.400		1.07	10.45		11.52	18.35
2810	2"	Q-1	36.40	.440		1.68	10.35		12.03	18.80
2820	2-1/2"		26.70	.599		5.10	14.10		19.20	28.50
2830	3"		22.90	.699		6.10	16.45		22.55	33.50
2840	4"	↓	18.20	.879		10.95	20.50		31.45	46
3180	Tee, 1/2"	1 Plum	22.20	.360		.35	9.40		9.75	15.85
3190	3/4"		19	.421		.40	11		11.40	18.50
3200	1"		16.70	.479		.75	12.55		13.30	21.50
3210	1-1/4"		14.80	.541		1.18	14.15		15.33	24.50
3220	1-1/2"	↓	13.30	.601		1.43	15.75		17.18	27.50
3230	2"	Q-1	24.20	.661		2.07	15.55		17.62	28
3240	2-1/2"		17.80	.899		6.85	21		27.85	42
3250	3"		15.20	1.053		13.55	25		38.55	55.50
3260	4"	↓	12.10	1.322		16.25	31		47.25	69
3380	Coupling, 1/2"	1 Plum	33.30	.240		.19	6.30		6.49	10.50
3390	3/4"		28.60	.280		.26	7.30		7.56	12.30
3400	1"		25	.320		.44	8.35		8.79	14.25
3410	1-1/4"		22.20	.360		.61	9.40		10.01	16.10
3420	1-1/2"	↓	20	.400		.65	10.45		11.10	17.85
3430	2"	Q-1	36.40	.440		1.01	10.35		11.36	18.05
3440	2-1/2"		26.70	.599		2.22	14.10		16.32	25.50
3450	3"		22.90	.699		3.49	16.45		19.94	31
3460	4"	↓	18.20	.879	↓	5	20.50		25.50	39.50
4500	DWV, ABS, non pressure, socket joints									
4540	1/4 Bend, 1-1/4"	1 Plum	20.20	.396	Ea.	2.94	10.35		13.29	20
4560	1-1/2"	"	18.20	.440		1.03	11.50		12.53	20
4570	2"	Q-1	33.10	.483	↓	1.55	11.35		12.90	20.50
4800	Tee, sanitary									
4820	1-1/4"	1 Plum	13.50	.593	Ea.	1.85	15.50		17.35	27.50
4830	1-1/2"	"	12.10	.661		1.56	17.30		18.86	30
4840	2"	Q-1	20	.800	↓	2.49	18.80		21.29	33.50
5000	DWV, PVC, schedule 40, socket joints									
5040	1/4 bend, 1-1/4"	1 Plum	20.20	.396	Ea.	1.46	10.35		11.81	18.60
5060	1-1/2"	"	18.20	.440		1.03	11.50		12.53	20
5070	2"	Q-1	33.10	.483		1.78	11.35		13.13	20.50
5080	3"		20.80	.769		4.37	18.10		22.47	34.50
5090	4"	↓	16.50	.970		6.95	23		29.95	45
5110	1/4 bend, long sweep, 1-1/2"	1 Plum	18.20	.440		1.94	11.50		13.44	21
5112	2"	Q-1	33.10	.483		1.76	11.35		13.11	20.50
5114	3"		20.80	.769		4.35	18.10		22.45	34.50
5116	4"	↓	16.50	.970		8.10	23		31.10	46.50
5250	Tee, sanitary 1-1/4"	1 Plum	13.50	.593		2.13	15.50		17.63	28
5254	1-1/2"	"	12.10	.661		1.17	17.30		18.47	30
5255	2"	Q-1	20	.800		1.90	18.80		20.70	33
5256	3"		13.90	1.151		4	27		31	49
5257	4"		11	1.455		7.35	34		41.35	64
5259	6"	↓	6.70	2.388		38	56		94	134
5261	8"	Q-2	6.20	3.871		114	87.50		201.50	269
5264	2" x 1-1/2"	Q-1	22	.727		4.21	17.10		21.31	32.50
5266	3" x 1-1/2"	↓	15.50	1.032	↓	2.67	24.50		27.17	43

	15108 | Plastic Pipe & Fittings		CREW	DAILY OUTPUT	LABOR-HOURS	UNIT	2004 BARE COSTS MAT.	LABOR	EQUIP.	TOTAL	TOTAL INCL O&P	
560	5268	4″ x 3″	Q-1	12.10	1.322	Ea.	12.05	31		43.05	64.50	560
	5271	6″ x 4″		6.90	2.319		58	54.50		112.50	153	
	5314	Combination Y & 1/8 bend, 1-1/2″	1 Plum	12.10	.661		2.91	17.30		20.21	31.50	
	5315	2″	Q-1	20	.800		3.76	18.80		22.56	35	
	5317	3″		13.90	1.151		6.65	27		33.65	52	
	5318	4″		11	1.455		12.90	34		46.90	70	
	5324	Combination Y & 1/8 bend, reducing										
	5325	2″ x 2″ x 1-1/2″	Q-1	22	.727	Ea.	4.17	17.10		21.27	32.50	
	5327	3″ x 3″ x 1-1/2″		15.50	1.032		7.25	24.50		31.75	48	
	5328	3″ x 3″ x 2″		15.30	1.046		5.05	24.50		29.55	46	
	5329	4″ x 4″ x 2″		12.20	1.311		10.10	31		41.10	61.50	
	5331	Wye, 1-1/4″	1 Plum	13.50	.593		3.21	15.50		18.71	29	
	5332	1-1/2″	″	12.10	.661		1.96	17.30		19.26	30.50	
	5333	2″	Q-1	20	.800		2.30	18.80		21.10	33.50	
	5334	3″		13.90	1.151		5.90	27		32.90	51	
	5335	4″		11	1.455		9.05	34		43.05	66	
	5336	6″		6.70	2.388		35.50	56		91.50	131	
	5337	8″	Q-2	6.20	3.871		64	87.50		151.50	215	
	5341	2″ x 1-1/2″	Q-1	22	.727		3.21	17.10		20.31	31.50	
	5342	3″ x 1-1/2″		15.50	1.032		4.25	24.50		28.75	44.50	
	5343	4″ x 3″		12.10	1.322		7.20	31		38.20	59	
	5344	6″ x 4″		6.90	2.319		28.50	54.50		83	121	
	5345	8″ x 6″	Q-2	6.40	3.750		48.50	85		133.50	193	
	5347	Double wye, 1-1/2″	1 Plum	9.10	.879		3.87	23		26.87	42	
	5348	2″	Q-1	16.60	.964		4.97	22.50		27.47	42.50	
	5349	3″		10.40	1.538		12.85	36		48.85	73.50	
	5350	4″		8.25	1.939		26	45.50		71.50	104	
	5354	2″ x 1-1/2″		16.80	.952		4.54	22.50		27.04	42	
	5355	3″ x 2″		10.60	1.509		9.55	35.50		45.05	69	
	5356	4″ x 3″		8.45	1.893		20.50	44.50		65	95.50	
	5357	6″ x 4″		7.25	2.207		43	52		95	132	
	5410	Reducer bushing, 2″ x 1-1/4″		36.50	.438		.69	10.30		10.99	17.65	
	5412	3″ x 1-1/2″		27.30	.586		3.02	13.80		16.82	26	
	5414	4″ x 2″		18.20	.879		6.35	20.50		26.85	41	
	5416	6″ x 4″		11.10	1.441		18.20	34		52.20	75.50	
	5418	8″ x 6″	Q-2	10.20	2.353		36.50	53.50		90	128	
	5500	CPVC, Schedule 80, threaded joints										
	5540	90° Elbow, 1/4″	1 Plum	32	.250	Ea.	7.05	6.55		13.60	18.50	
	5560	1/2″		30.30	.264		4.10	6.90		11	15.85	
	5570	3/4″		26	.308		6.15	8.05		14.20	19.95	
	5580	1″		22.70	.352		8.60	9.20		17.80	24.50	
	5590	1-1/4″		20.20	.396		16.55	10.35		26.90	35.50	
	5600	1-1/2″		18.20	.440		17.85	11.50		29.35	38.50	
	5610	2″	Q-1	33.10	.483		24	11.35		35.35	45	
	6000	Coupling, 1/4″	1 Plum	32	.250		9	6.55		15.55	20.50	
	6020	1/2″		30.30	.264		7.40	6.90		14.30	19.50	
	6030	3/4″		26	.308		11.95	8.05		20	26.50	
	6040	1″		22.70	.352		13.55	9.20		22.75	30	
	6050	1-1/4″		20.20	.396		14.40	10.35		24.75	33	
	6060	1-1/2″		18.20	.440		15.45	11.50		26.95	36	
	6070	2″	Q-1	33.10	.483		18.25	11.35		29.60	38.50	

	15110 | Valves											
160	0010	**VALVES, BRONZE**										160
	1750	Check, swing, class 150, regrinding disc, threaded										

15110 | Valves

			Crew	Daily Output	Labor-Hours	Unit	2004 Bare Costs Mat.	Labor	Equip.	Total	Total Incl O&P	
160	1860	3/4″	1 Plum	20	.400	Ea.	33	10.45		43.45	53.50	160
	1870	1″	″	19	.421	″	49.50	11		60.50	72.50	
	2850	Gate, N.R.S., soldered, 125 psi										
	2940	3/4″	1 Plum	20	.400	Ea.	18.30	10.45		28.75	37	
	2950	1″	″	19	.421	″	26	11		37	46.50	
	5600	Relief, pressure & temperature, self-closing, ASME, threaded										
	5650	1″	1 Plum	24	.333	Ea.	107	8.70		115.70	132	
	5660	1-1/4″	″	20	.400	″	215	10.45		225.45	253	
	6400	Pressure, water, ASME, threaded										
	6440	3/4″	1 Plum	28	.286	Ea.	44.50	7.45		51.95	61.50	
	6450	1″	″	24	.333	″	93.50	8.70		102.20	117	
	6900	Reducing, water pressure										
	6940	1/2″	1 Plum	24	.333	Ea.	125	8.70		133.70	152	
	6960	1″	″	19	.421	″	194	11		205	231	
	8350	Tempering, water, sweat connections										
	8400	1/2″	1 Plum	24	.333	Ea.	45.50	8.70		54.20	64.50	
	8440	3/4″	″	20	.400	″	55.50	10.45		65.95	78.50	
	8650	Threaded connections										
	8700	1/2″	1 Plum	24	.333	Ea.	55.50	8.70		64.20	76	
	8740	3/4″	″	20	.400	″	213	10.45		223.45	251	

15120 | Piping Specialties

			Crew	Daily Output	Labor-Hours	Unit	Mat.	Labor	Equip.	Total	Total Incl O&P	
320	0010	**EXPANSION TANKS**										320
	1505	Fiberglass and steel single / double wall storage, see Div 13201										
	2000	Steel, liquid expansion, ASME, painted, 15 gallon capacity	Q-5	17	.941	Ea.	365	22		387	435	
	2040	30 gallon capacity		12	1.333		405	31.50		436.50	495	
	3000	Steel ASME expansion, rubber diaphragm, 19 gal. cap. accept.		12	1.333		1,425	31.50		1,456.50	1,625	
	3020	31 gallon capacity		8	2		1,600	47.50		1,647.50	1,825	
940	0010	**WATER SUPPLY METERS**										940
	2000	Domestic/commercial, bronze										
	2020	Threaded										
	2060	5/8″ diameter, to 20 GPM	1 Plum	16	.500	Ea.	40	13.10		53.10	65.50	
	2080	3/4″ diameter, to 30 GPM		14	.571		67.50	14.95		82.45	99	
	2100	1″ diameter, to 50 GPM		12	.667		94	17.45		111.45	132	

15140 | Domestic Water Piping

			Crew	Daily Output	Labor-Hours	Unit	Mat.	Labor	Equip.	Total	Total Incl O&P	
100	0010	**BACKFLOW PREVENTER** Includes valves										100
	0020	and four test cocks, corrosion resistant, automatic operation										
	4100	Threaded, valves are ball										
	4120	3/4″ pipe size	1 Plum	16	.500	Ea.	715	13.10		728.10	805	
600	0010	**VACUUM BREAKERS** Hot or cold water										600
	1030	Anti-siphon, brass										
	1060	1/2″ size	1 Plum	24	.333	Ea.	16.35	8.70		25.05	32.50	
	1080	3/4″ size		20	.400		28	10.45		38.45	47.50	
	1100	1″ size		19	.421		43.50	11		54.50	66	
800	0010	**WATER HAMMER ARRESTORS / SHOCK ABSORBERS**										800
	0490	Copper										
	0500	3/4″ male I.P.S. For 1 to 11 fixtures	1 Plum	12	.667	Ea.	14.50	17.45		31.95	44.50	

15150 | Sanitary Waste and Vent Piping

			Crew	Daily Output	Labor-Hours	Unit	Mat.	Labor	Equip.	Total	Total Incl O&P	
200	0010	**CLEANOUTS**										200
	0080	Round or square, scoriated nickel bronze top										
	0100	2″ pipe size	1 Plum	10	.800	Ea.	75.50	21		96.50	118	
	0140	4″ pipe size	″	6	1.333	″	112	35		147	181	

15150 | Sanitary Waste and Vent Piping

			CREW	DAILY OUTPUT	LABOR-HOURS	UNIT	2004 BARE COSTS MAT.	LABOR	EQUIP.	TOTAL	TOTAL INCL O&P	
250	0010	**CLEANOUT TEE**										250
	0100	Cast iron, B&S, with countersunk plug										
	0220	3″ pipe size	1 Plum	3.60	2.222	Ea.	112	58		170	220	
	0240	4″ pipe size	″	3.30	2.424		140	63.50		203.50	258	
	0500	For round smooth access cover, same price				↓						
	4000	Plastic, tees and adapters. Add plugs										
	4010	ABS, DWV										
	4020	Cleanout tee, 1-1/2″ pipe size	1 Plum	15	.533	Ea.	2.94	13.95		16.89	26	
300	0010	**FLOOR AND AREA DRAINS**										300
	2000	Floor, medium duty, C.I., deep flange, 7″ dia top										
	2040	2″ and 3″ pipe size	Q-1	12	1.333	Ea.	77.50	31.50		109	137	
	2080	For galvanized body, add					32.50			32.50	35.50	
	2120	For polished bronze top, add				↓	39			39	43	
800	0010	**TRAPS**										800
	0030	Cast iron, service weight										
	0050	Running P trap, without vent										
	1100	2″	Q-1	16	1	Ea.	19.70	23.50		43.20	60	
	1150	4″	″	13	1.231		57	29		86	111	
	1160	6″	Q-2	17	1.412		264	32		296	345	
	3000	P trap, B&S, 2″ pipe size	Q-1	16	1		14.30	23.50		37.80	54.50	
	3040	3″ pipe size	″	14	1.143	↓	21.50	27		48.50	67.50	
	4700	Copper, drainage, drum trap										
	4840	3″ x 6″ swivel, 1-1/2″ pipe size	1 Plum	16	.500	Ea.	36	13.10		49.10	61	
	5100	P trap, standard pattern										
	5200	1-1/4″ pipe size	1 Plum	18	.444	Ea.	17.05	11.60		28.65	38	
	5240	1-1/2″ pipe size		17	.471		17.05	12.30		29.35	39	
	5260	2″ pipe size		15	.533		27	13.95		40.95	52.50	
	5280	3″ pipe size	↓	11	.727	↓	64.50	19		83.50	102	
	6710	ABS DWV P trap, solvent weld joint										
	6720	1-1/2″ pipe size	1 Plum	18	.444	Ea.	3.38	11.60		14.98	23	
	6722	2″ pipe size		17	.471		4.43	12.30		16.73	25	
	6724	3″ pipe size		15	.533		17.50	13.95		31.45	42.50	
	6726	4″ pipe size		14	.571		36	14.95		50.95	64	
	6860	PVC DWV hub x hub, basin trap, 1-1/4″ pipe size		18	.444		4.77	11.60		16.37	24.50	
	6870	Sink P trap, 1-1/2″ pipe size		18	.444		4.77	11.60		16.37	24.50	
	6880	Tubular S trap, 1-1/2″ pipe size	↓	17	.471	↓	9.80	12.30		22.10	31	
	6890	PVC sch. 40 DWV, drum trap										
	6900	1-1/2″ pipe size	1 Plum	16	.500	Ea.	11.05	13.10		24.15	33.50	
	6910	P trap, 1-1/2″ pipe size		18	.444		2.65	11.60		14.25	22	
	6920	2″ pipe size		17	.471		4.05	12.30		16.35	24.50	
	6930	3″ pipe size		15	.533		14.45	13.95		28.40	39	
	6940	4″ pipe size		14	.571		35	14.95		49.95	63	
	6950	P trap w/clean out, 1-1/2″ pipe size		18	.444		5.05	11.60		16.65	24.50	
	6960	2″ pipe size	↓	17	.471	↓	8.60	12.30		20.90	29.50	
900	0010	**VENT FLASHING, CAPS**										900
	0120	Vent caps										
	0140	Cast iron										
	0160	1-1/4″ - 1-1/2″ pipe	1 Plum	23	.348	Ea.	25	9.10		34.10	42.50	
	0170	2″ - 2-1/8″ pipe		22	.364		29	9.50		38.50	47.50	
	0180	2-1/2″ - 3-5/8″ pipe		21	.381		33	9.95		42.95	53	
	0190	4″ - 4-1/8″ pipe		19	.421		39.50	11		50.50	61.50	
	0200	5″ - 6″ pipe	↓	17	.471	↓	59	12.30		71.30	85	
	0300	PVC										
	0320	1-1/4″ - 1-1/2″ pipe	1 Plum	24	.333	Ea.	18.85	8.70		27.55	35	

15100 | Building Services Piping

	15150 Sanitary Waste and Vent Piping		CREW	DAILY OUTPUT	LABOR-HOURS	UNIT	2004 BARE COSTS MAT.	LABOR	EQUIP.	TOTAL	TOTAL INCL O&P	
900	0330	2" - 2-1/8" pipe	1 Plum	23	.348	Ea.	22	9.10		31.10	39	900
	0340	2-1/2" - 3-5/8" pipe		22	.364		25	9.50		34.50	43	
	0350	4" - 4-1/8" pipe		20	.400		30	10.45		40.45	50	
	0360	5" - 6" pipe	↓	18	.444	↓	44.50	11.60		56.10	67.50	

	15160 Storm Drainage Piping		CREW	DAILY OUTPUT	LABOR-HOURS	UNIT	MAT.	LABOR	EQUIP.	TOTAL	TOTAL INCL O&P	
500	0010	**STORM AREA DRAINS**										500
	3860	Roof, flat metal deck, C.I. body, 12" C.I. dome										
	3890	3" pipe size	Q-1	14	1.143	Ea.	151	27		178	210	

	15180 Heating and Cooling Piping		CREW	DAILY OUTPUT	LABOR-HOURS	UNIT	MAT.	LABOR	EQUIP.	TOTAL	TOTAL INCL O&P	
200	0010	**PUMPS, CIRCULATING** Heated or chilled water application										200
	0600	Bronze, sweat connections, 1/40 HP, in line										
	0640	3/4" size	Q-1	16	1	Ea.	120	23.50		143.50	171	
	1000	Flange connection, 3/4" to 1-1/2" size										
	1040	1/12 HP	Q-1	6	2.667	Ea.	325	63		388	465	
	1060	1/8 HP	"	6	2.667	"	560	63		623	720	

15400 | Plumbing Fixtures & Equipment

	15410 Plumbing Fixtures		CREW	DAILY OUTPUT	LABOR-HOURS	UNIT	2004 BARE COSTS MAT.	LABOR	EQUIP.	TOTAL	TOTAL INCL O&P	
200	0010	**CARRIERS/SUPPORTS** For plumbing fixtures										200
	0600	Plate type with studs, top back plate	1 Plum	7	1.143	Ea.	27	30		57	78.50	
	3000	Lavatory, concealed arm										
	3050	Floor mounted, single										
	3100	High back fixture	1 Plum	6	1.333	Ea.	169	35		204	243	
	3200	Flat slab fixture	"	6	1.333	"	197	35		232	274	
	8200	Water closet, residential										
	8220	Vertical centerline, floor mount										
	8240	Single, 3" caulk, 2" or 3" vent	1 Plum	6	1.333	Ea.	195	35		230	271	
	8260	4" caulk, 2" or 4" vent	"	6	1.333	"	251	35		286	335	
300	0010	**FAUCETS/FITTINGS**										300
	0150	Bath, faucets, diverter spout combination, sweat	1 Plum	8	1	Ea.	69.50	26		95.50	120	
	0200	For integral stops, IPS unions, add					73			73	80.50	
	0420	Bath, press-bal mix valve w/diverter, spout, shower hd, arm/flange	1 Plum	8	1		109	26		135	163	
	0500	Drain, central lift, 1-1/2" IPS male		20	.400		38	10.45		48.45	59	
	0600	Trip lever, 1-1/2" IPS male		20	.400		38.50	10.45		48.95	59.50	
	1000	Kitchen sink faucets, top mount, cast spout		10	.800		51.50	21		72.50	91	
	1100	For spray, add		24	.333		9.50	8.70		18.20	25	
	2000	Laundry faucets, shelf type, IPS or copper unions	↓	12	.667	↓	42.50	17.45		59.95	75	
	2020											
	2100	Lavatory faucet, centerset, without drain	1 Plum	10	.800	Ea.	37.50	21		58.50	75.50	
	2200	With pop-up drain		16	.500		52	13.10		65.10	79	
	2800	Self-closing, center set		10	.800		112	21		133	158	
	3000	Service sink faucet, cast spout, pail hook, hose end		14	.571		72	14.95		86.95	104	
	4000	Shower by-pass valve with union		18	.444		50.50	11.60		62.10	74.50	
	4200	Shower thermostatic mixing valve, concealed	↓	8	1		233	26		259	299	
	4300	For inlet strainer, check, and stops, add					31			31	34	
	5000	Sillcock, compact, brass, IPS or copper to hose	1 Plum	24	.333	↓	4.74	8.70		13.44	19.50	

15418 | Resi/Comm/Industrial Fixtures

			CREW	DAILY OUTPUT	LABOR-HOURS	UNIT	2004 BARE COSTS MAT.	LABOR	EQUIP.	TOTAL	TOTAL INCL O&P
100	0010	**BATHS** R15100-420									
	0100	Tubs, recessed porcelain enamel on cast iron, with trim									
	0180	48″ x 42″	Q-1	4	4	Ea.	1,450	94		1,544	1,750
	0220	72″ x 36″		3	5.333		1,350	125		1,475	1,675
	0300	Mat bottom, 4′ long		5.50	2.909		990	68.50		1,058.50	1,200
	0380	5′ long		4.40	3.636		375	85.50		460.50	550
	0480	Above floor drain, 5′ long		4	4		645	94		739	860
	0560	Corner 48″ x 44″		4.40	3.636		1,400	85.50		1,485.50	1,675
	2000	Enameled formed steel, 4′-6″ long		5.80	2.759		295	65		360	430
	2200	5′ long		5.50	2.909		286	68.50		354.50	425
	4600	Module tub & showerwall surround, molded fiberglass									
	4610	5′ long x 34″ wide x 76″ high	Q-1	4	4	Ea.	490	94		584	695
	6000	Whirlpool, bath with vented overflow, molded fiberglass									
	6100	66″ x 48″ x 24″	Q-1	1	16	Ea.	2,225	375		2,600	3,075
	6400	72″ x 36″ x 24″		1	16		2,175	375		2,550	3,025
	6500	60″ x 30″ x 21″		1	16		1,900	375		2,275	2,700
	6600	72″ x 42″ x 22″		1	16		3,075	375		3,450	4,025
	6700	83″ x 65″		.30	53.333		4,150	1,250		5,400	6,625
	7000	Redwood hot tub system									
	7050	4′ diameter x 4′ deep	Q-1	1	16	Ea.	1,300	375		1,675	2,050
	7150	6′ diameter x 4′ deep		.80	20		2,000	470		2,470	2,975
	7200	8′ diameter x 4′ deep		.80	20		3,025	470		3,495	4,100
	9600	Rough-in, supply, waste and vent, for all above tubs, add		2.07	7.729		115	182		297	425
400	0010	**LAUNDRY SINKS** With trim									
	0020	Porcelain enamel on cast iron, black iron frame									
	0050	24″ x 20″, single compartment	Q-1	6	2.667	Ea.	247	63		310	375
	0100	24″ x 23″, single compartment	″	6	2.667	″	269	63		332	400
	3000	Plastic, on wall hanger or legs									
	3020	18″ x 23″, single compartment	Q-1	6.50	2.462	Ea.	87.50	58		145.50	192
	3100	20″ x 24″, single compartment		6.50	2.462		114	58		172	220
	3200	36″ x 23″, double compartment		5.50	2.909		137	68.50		205.50	263
	3300	40″ x 24″, double compartment		5.50	2.909		198	68.50		266.50	330
	5000	Stainless steel, counter top, 22″ x 17″ single compartment		6	2.667		325	63		388	460
	5100	22″ x 22″, single compartment		6	2.667		410	63		473	555
	5200	33″ x 22″, double compartment		5	3.200		405	75.50		480.50	570
	9600	Rough-in, supply, waste and vent, for all laundry sinks		2.14	7.477		71	176		247	365
450	0010	**LAVATORIES** With trim, white unless noted otherwise									
	0500	Vanity top, porcelain enamel on cast iron									
	0600	20″ x 18″	Q-1	6.40	2.500	Ea.	206	59		265	325
	0640	33″ x 19″ oval		6.40	2.500		405	59		464	540
	0720	19″ round		6.40	2.500		214	59		273	330
	0860	For color, add					25%				
	1000	Cultured marble, 19″ x 17″, single bowl	Q-1	6.40	2.500	Ea.	160	59		219	273
	1120	25″ x 22″, single bowl		6.40	2.500		160	59		219	273
	1160	37″ x 22″, single bowl		6.40	2.500		188	59		247	305
	1560										
	1900	Stainless steel, self-rimming, 25″ x 22″, single bowl, ledge	Q-1	6.40	2.500	Ea.	224	59		283	345
	1960	17″ x 22″, single bowl		6.40	2.500		222	59		281	340
	2600	Steel, enameled, 20″ x 17″, single bowl		5.80	2.759		129	65		194	248
	2900	Vitreous china, 20″ x 16″, single bowl		5.40	2.963		222	69.50		291.50	360
	3200	22″ x 13″, single bowl		5.40	2.963		261	69.50		330.50	400
	3580	Rough-in, supply, waste and vent for all above lavatories		2.30	6.957		63	164		227	340
	4000	Wall hung									
	4040	Porcelain enamel on cast iron, 16″ x 14″, single bowl	Q-1	8	2	Ea.	315	47		362	420

15418 | Resi/Comm/Industrial Fixtures

			CREW	DAILY OUTPUT	LABOR-HOURS	UNIT	2004 BARE COSTS MAT.	LABOR	EQUIP.	TOTAL	TOTAL INCL O&P	
450	4180	20″ x 18″, single bowl	Q-1	8	2	Ea.	237	47		284	340	450
	4580	For color, add					30%					
	6000	Vitreous china, 18″ x 15″, single bowl with backsplash	Q-1	7	2.286	Ea.	184	54		238	290	
	6060	19″ x 17″, single bowl		7	2.286		167	54		221	271	
	6960	Rough-in, supply, waste and vent for above lavatories		1.66	9.639		220	227		447	610	
500	0010	**SHOWERS**										500
	1500	Stall, with drain only. Add for valve and door/curtain										
	1520	32″ square	Q-1	2	8	Ea.	325	188		513	670	
	1530	36″ square		2	8		410	188		598	760	
	1540	Terrazzo receptor, 32″ square		2	8		690	188		878	1,075	
	1560	36″ square		1.80	8.889		825	209		1,034	1,250	
	1580	36″ corner angle		1.80	8.889		755	209		964	1,175	
	3000	Fiberglass, one piece, with 3 walls, 32″ x 32″ square		2.40	6.667		350	157		507	645	
	3100	36″ x 36″ square		2.40	6.667		395	157		552	690	
	4200	Rough-in, supply, waste and vent for above showers		2.05	7.805		62.50	184		246.50	370	
600	0010	**SINKS** With faucets and drain										600
	2000	Kitchen, counter top style, P.E. on C.I., 24″ x 21″ single bowl	Q-1	5.60	2.857	Ea.	201	67		268	330	
	2100	31″ x 22″ single bowl		5.60	2.857		249	67		316	385	
	2200	32″ x 21″ double bowl		4.80	3.333		288	78.50		366.50	445	
	3000	Stainless steel, self rimming, 19″ x 18″ single bowl		5.60	2.857		335	67		402	475	
	3100	25″ x 22″ single bowl		5.60	2.857		370	67		437	515	
	3200	33″ x 22″ double bowl		4.80	3.333		530	78.50		608.50	715	
	3300	43″ x 22″ double bowl		4.80	3.333		615	78.50		693.50	805	
	4000	Steel, enameled, with ledge, 24″ x 21″ single bowl		5.60	2.857		107	67		174	227	
	4100	32″ x 21″ double bowl		4.80	3.333		146	78.50		224.50	289	
	4960	For color sinks except stainless steel, add					10%					
	4980	For rough-in, supply, waste and vent, counter top sinks	Q-1	2.14	7.477		71	176		247	365	
	5000	Kitchen, raised deck, P.E. on C.I.										
	5100	32″ x 21″, dual level, double bowl	Q-1	2.60	6.154	Ea.	258	145		403	520	
	5790	For rough-in, supply, waste & vent, sinks		1.85	8.649		71	204		275	415	
	6650	Service, floor, corner, P.E. on C.I., 28″ x 28″		4.40	3.636		490	85.50		575.50	680	
	6750	Vinyl coated rim guard, add					63.50			63.50	70	
	6760	Mop sink, molded stone, 24″ x 36″	1 Plum	3.33	2.402		194	63		257	315	
	6770	Mop sink, molded stone, 24″ x 36″, w/rim 3 sides	″	3.33	2.402		215	63		278	340	
	6790	For rough-in, supply, waste & vent, floor service sinks	Q-1	1.64	9.756		161	230		391	550	
900	0010	**WATER CLOSETS**										900
	0150	Tank type, vitreous china, incl. seat, supply pipe w/stop										
	0200	Wall hung, one piece	Q-1	5.30	3.019	Ea.	390	71		461	545	
	0400	Two piece, close coupled		5.30	3.019		470	71		541	630	
	0960	For rough-in, supply, waste, vent and carrier		2.73	5.861		264	138		402	515	
	1000	Floor mounted, one piece		5.30	3.019		485	71		556	650	
	1020	One piece, low profile		5.30	3.019		390	71		461	545	
	1100	Two piece, close coupled, water saver		5.30	3.019		160	71		231	293	
	1960	For color, add					30%					
	1980	For rough-in, supply, waste and vent	Q-1	3.05	5.246	Ea.	130	123		253	345	
	3000	Bowl only, with flush valve, seat										
	3100	Wall hung	Q-1	5.80	2.759	Ea.	340	65		405	480	
	3200	For rough-in, supply, waste and vent, single WC		2.56	6.250		274	147		421	540	
	3300	Floor mounted		5.80	2.759		305	65		370	440	
	3400	For rough-in, supply, waste and vent, single WC		2.84	5.634		139	133		272	370	

15440 | Plumbing Pumps

			CREW	DAILY OUTPUT	LABOR-HOURS	UNIT	MAT.	LABOR	EQUIP.	TOTAL	TOTAL INCL O&P	
940	0010	**PUMPS, SUBMERSIBLE** Sump										940
	7000	Sump pump, automatic										

15400 | Plumbing Fixtures & Equipment

15440 | Plumbing Pumps

			CREW	DAILY OUTPUT	LABOR-HOURS	UNIT	2004 BARE COSTS MAT.	LABOR	EQUIP.	TOTAL	TOTAL INCL O&P	
940	7100	Plastic, 1-1/4" discharge, 1/4 HP	1 Plum	6	1.333	Ea.	105	35		140	173	940
	7500	Cast iron, 1-1/4" discharge, 1/4 HP	"	6	1.333	"	125	35		160	194	

15480 | Domestic Water Heaters

			CREW	DAILY OUTPUT	LABOR-HOURS	UNIT	MAT.	LABOR	EQUIP.	TOTAL	TOTAL INCL O&P	
200	0010	**WATER HEATERS**										200
	1000	Residential, electric, glass lined tank, 5 yr, 10 gal., single element	1 Plum	2.30	3.478	Ea.	210	91		301	380	
	1060	30 gallon, double element		2.20	3.636		280	95		375	465	
	1080	40 gallon, double element		2	4		300	105		405	500	
	1100	52 gallon, double element		2	4		345	105		450	550	
	1120	66 gallon, double element		1.80	4.444		475	116		591	715	
	1140	80 gallon, double element	↓	1.60	5	↓	535	131		666	805	
	2000	Gas fired, foam lined tank, 10 yr, vent not incl.,										
	2040	30 gallon	1 Plum	2	4	Ea.	385	105		490	595	
	2100	75 gallon		1.50	5.333		745	139		884	1,050	
	3000	Oil fired, glass lined tank, 5 yr, vent not included, 30 gallon		2	4		785	105		890	1,025	
	3040	50 gallon	↓	1.80	4.444	↓	1,375	116		1,491	1,700	

15500 | Heat Generation Equipment

15510 | Heating Boilers and Accessories

			CREW	DAILY OUTPUT	LABOR-HOURS	UNIT	2004 BARE COSTS MAT.	LABOR	EQUIP.	TOTAL	TOTAL INCL O&P	
120	0010	**BURNERS**										120
	0990	Residential, conversion, gas fired, LP or natural										
	1000	Gun type, atmospheric input 72 to 200 MBH	Q-1	2.50	6.400	Ea.	655	151		806	965	
	1020	120 to 360 MBH		2	8		725	188		913	1,100	
	1040	280 to 800 MBH	↓	1.70	9.412	↓	1,400	221		1,621	1,925	
300	0010	**BOILERS, ELECTRIC, ASME** Standard controls and trim										300
	1000	Steam, 6 KW, 20.5 MBH	Q-19	1.20	20	Ea.	2,625	490		3,115	3,675	
	1160	60 KW, 205 MBH		1	24		4,575	590		5,165	6,025	
	2000	Hot water, 6 KW, 20.5 MBH		1.30	18.462		2,450	455		2,905	3,450	
	2040	24 KW, 82 MBH		1.20	20		2,925	490		3,415	4,025	
	2060	30 KW, 103 MBH	↓	1.20	20	↓	3,200	490		3,690	4,300	
400	0010	**BOILERS, GAS FIRED** Natural or propane, standard controls										400
	1000	Cast iron, with insulated jacket										
	3000	Hot water, gross output, 80 MBH	Q-7	1.46	21.918	Ea.	1,125	520		1,645	2,100	
	3020	100 MBH	"	1.35	23.704	"	1,325	560		1,885	2,375	
	4000	Steel, insulating jacket										
	6000	Hot water, including burner & one zone valve, gross output										
	6010	51.2 MBH	Q-6	2	12	Ea.	1,650	273		1,923	2,275	
	6020	72 MBH		2	12		1,850	273		2,123	2,475	
	6040	89 MBH		1.90	12.632		1,900	287		2,187	2,550	
	6060	105 MBH		1.80	13.333		2,125	305		2,430	2,825	
	6080	132 MBH		1.70	14.118		2,425	320		2,745	3,200	
	6100	155 MBH	↓	1.50	16	↓	2,800	365		3,165	3,675	
	7000	For tankless water heater on smaller gas units, add					10%					
	7050	For additional zone valves up to 312 MBH add				Ea.	107			107	117	
460	0010	**BOILERS, GAS/OIL** Combination with burners and controls										460
	1000	Cast iron with insulated jacket										
	2000	Steam, gross output, 720 MBH	Q-7	.43	74.074	Ea.	6,725	1,750		8,475	10,300	
	2900	Hot water, gross output										

	15510 | Heating Boilers and Accessories		CREW	DAILY OUTPUT	LABOR-HOURS	UNIT	2004 BARE COSTS MAT.	LABOR	EQUIP.	TOTAL	TOTAL INCL O&P	
460	2910	200 MBH	Q-6	.61	39.024	Ea.	5,400	890		6,290	7,400	460
	2920	300 MBH		.49	49.080		5,400	1,125		6,525	7,775	
	2930	400 MBH		.41	57.971		6,325	1,325		7,650	9,125	
	2940	500 MBH		.36	67.039		6,800	1,525		8,325	9,975	
	3000	584 MBH	Q-7	.44	72.072		6,000	1,700		7,700	9,400	
	4000	Steel, insulated jacket, skid base, tubeless										
	4500	Steam, 150 psi gross output, 335 MBH, 10 BHP	Q-6	.54	44.037	Ea.	10,900	1,000		11,900	13,700	
500	0010	**BOILERS, OIL FIRED** Standard controls, flame retention burner										500
	1000	Cast iron, with insulated flush jacket										
	2000	Steam, gross output, 109 MBH	Q-7	1.20	26.667	Ea.	1,500	630		2,130	2,650	
	2060	207 MBH	"	.90	35.556	"	2,025	840		2,865	3,625	
	3000	Hot water, same price as steam										
	7000	Hot water, gross output, 103 MBH	Q-6	1.60	15	Ea.	1,200	340		1,540	1,850	
	7020	122 MBH		1.45	16.506		2,275	375		2,650	3,125	
	7060	168 MBH		1.30	18.405		2,850	420		3,270	3,800	
	7080	225 MBH		1.22	19.704		2,925	450		3,375	3,950	
880	0010	**SWIMMING POOL HEATERS** Not including wiring, external										880
	0020	piping, base or pad,										
	0060	Gas fired, input, 115 MBH	Q-6	3	8	Ea.	1,700	182		1,882	2,150	
	0100	135 MBH		2	12		1,900	273		2,173	2,550	
	0160	155 MBH		1.50	16		2,000	365		2,365	2,800	
	0200	190 MBH		1	24		2,625	545		3,170	3,775	
	0280	500 MBH		.40	60		6,325	1,375		7,700	9,200	
	2000	Electric, 12 KW, 4,800 gallon pool	Q-19	3	8		1,550	196		1,746	2,050	
	2020	15 KW, 7,200 gallon pool		2.80	8.571		1,550	210		1,760	2,075	
	2040	24 KW, 9,600 gallon pool		2.40	10		2,100	246		2,346	2,725	
	2100	55 KW, 24,000 gallon pool		1.20	20		3,000	490		3,490	4,100	

15530 | Furnaces

			CREW	DAILY OUTPUT	LABOR-HOURS	UNIT	MAT.	LABOR	EQUIP.	TOTAL	TOTAL INCL O&P	
200	0010	**FURNACE COMPONENTS AND COMBINATIONS**										200
	0080	Coils, A/C evaporator, for gas or oil furnaces										
	0090	Add-on, with holding charge										
	0100	Upflow										
	0120	1-1/2 ton cooling	Q-5	4	4	Ea.	127	94.50		221.50	295	
	0130	2 ton cooling		3.70	4.324		153	102		255	335	
	0140	3 ton cooling		3.30	4.848		191	115		306	400	
	0150	4 ton cooling		3	5.333		270	126		396	505	
	0160	5 ton cooling		2.70	5.926		345	140		485	610	
	0300	Downflow										
	0330	2-1/2 ton cooling	Q-5	3	5.333	Ea.	178	126		304	400	
	0340	3-1/2 ton cooling		2.60	6.154		239	145		384	500	
	0350	5 ton cooling		2.20	7.273		345	172		517	660	
	0600	Horizontal										
	0630	2 ton cooling	Q-5	3.90	4.103	Ea.	180	97		277	355	
	0640	3 ton cooling		3.50	4.571		206	108		314	405	
	0650	4 ton cooling		3.20	5		259	118		377	480	
	0660	5 ton cooling		2.90	5.517		345	130		475	595	
	2000	Cased evaporator coils for air handlers										
	2100	1-1/2 ton cooling	Q-5	4.40	3.636	Ea.	191	86		277	350	
	2110	2 ton cooling		4.10	3.902		194	92		286	365	
	2120	2-1/2 ton cooling		3.90	4.103		218	97		315	400	
	2130	3 ton cooling		3.70	4.324		256	102		358	450	
	2140	3-1/2 ton cooling		3.50	4.571		240	108		348	440	

15530 | Furnaces

		15530 Furnaces	Crew	Daily Output	Labor-Hours	Unit	2004 Bare Costs Mat.	Labor	Equip.	Total	Total Incl O&P
200	2150	4 ton cooling	Q-5	3.20	5	Ea.	288	118		406	510
	2160	5 ton cooling		2.90	5.517		330	130		460	580
	3010	Air handler, modular									
	3100	With cased evaporator cooling coil									
	3120	1-1/2 ton cooling	Q-5	3.80	4.211	Ea.	545	99.50		644.50	765
	3130	2 ton cooling		3.50	4.571		570	108		678	800
	3140	2-1/2 ton cooling		3.30	4.848		620	115		735	870
	3150	3 ton cooling		3.10	5.161		675	122		797	945
	3160	3-1/2 ton cooling		2.90	5.517		810	130		940	1,100
	3170	4 ton cooling		2.50	6.400		920	151		1,071	1,275
	3180	5 ton cooling		2.10	7.619		975	180		1,155	1,375
	3500	With no cooling coil									
	3520	1-1/2 ton coil size	Q-5	12	1.333	Ea.	350	31.50		381.50	430
	3530	2 ton coil size		10	1.600		375	38		413	470
	3540	2-1/2 ton coil size		10	1.600		415	38		453	515
	3554	3 ton coil size		9	1.778		460	42		502	575
	3560	3-1/2 ton coil size		9	1.778		515	42		557	635
	3570	4 ton coil size		8.50	1.882		660	44.50		704.50	800
	3580	5 ton coil size		8	2		710	47.50		757.50	860
	4000	With heater									
	4120	5 kW, 17.1 MBH	Q-5	16	1	Ea.	315	23.50		338.50	385
	4130	7.5 kW, 25.6 MBH		15.60	1.026		335	24		359	410
	4140	10 kW, 34.2 MBH		15.20	1.053		390	25		415	470
	4150	12.5 KW, 42.7 MBH		14.80	1.081		440	25.50		465.50	520
	4160	15 KW, 51.2 MBH		14.40	1.111		525	26.50		551.50	625
	4170	25 KW, 85.4 MBH		14	1.143		595	27		622	695
	4180	30 KW, 102 MBH		13	1.231		695	29		724	815
400	0010	**FURNACES** Hot air heating, blowers, standard controls (R13600-610)									
	0020	not including gas, oil or flue piping									
	1000	Electric, UL listed									
	1020	10.2 MBH	Q-20	5	4	Ea.	315	95		410	505
	1100	34.1 MBH	″	4.40	4.545	″	405	108		513	625
	3000	Gas, AGA certified, upflow, direct drive models									
	3020	45 MBH input	Q-9	4	4	Ea.	435	92		527	635
	3040	60 MBH input		3.80	4.211		580	97		677	800
	3060	75 MBH input		3.60	4.444		615	102		717	850
	3100	100 MBH input		3.20	5		655	115		770	915
	3120	125 MBH input		3	5.333		755	123		878	1,025
	3130	150 MBH input		2.80	5.714		875	132		1,007	1,175
	3140	200 MBH input		2.60	6.154		1,675	142		1,817	2,050
	4000	For starter plenum, add		16	1		61.50	23		84.50	106
	6000	Oil, UL listed, atomizing gun type burner									
	6020	56 MBH output	Q-9	3.60	4.444	Ea.	745	102		847	990
	6030	84 MBH output		3.50	4.571		775	105		880	1,025
	6040	95 MBH output		3.40	4.706		790	108		898	1,050
	6060	134 MBH output		3.20	5		1,100	115		1,215	1,400
	6080	151 MBH output		3	5.333		1,225	123		1,348	1,525
	6100	200 MBH input		2.60	6.154		1,725	142		1,867	2,125
440	0010	**FURNACES, COMBINATION SYSTEMS** Heating, cooling,									
	0020	electric air cleaner, humidification, dehumidification.									
	2000	Gas fired, 80 MBH heat output, 24 MBH cooling	Q-9	1.20	13.333	Ea.	3,025	305		3,330	3,850
	2020	80 MBH heat output, 36 MBH cooling		1.20	13.333		3,300	305		3,605	4,150
	2040	100 MBH heat output, 29 MBH cooling		1	16		3,375	370		3,745	4,350
	2060	100 MBH heat output, 36 MBH cooling		1	16		3,550	370		3,920	4,525

15500 | Heat Generation Equipment

	15530	Furnaces	CREW	DAILY OUTPUT	LABOR-HOURS	UNIT	2004 BARE COSTS MAT.	LABOR	EQUIP.	TOTAL	TOTAL INCL O&P	
440	2080	100 MBH heat output, 47 MBH cooling	Q-9	.90	17.778	Ea.	4,200	410		4,610	5,300	440
	2100	120 MBH heat output, 29 MBH cooling	Q-10	1.30	18.462		3,575	440		4,015	4,675	
	2120	120 MBH heat output, 42 MBH cooling		1.30	18.462		3,775	440		4,215	4,925	
	2140	120 MBH heat output, 47 MBH cooling		1.20	20		4,225	480		4,705	5,450	
	2160	120 MBH heat output, 55 MBH cooling		1.10	21.818		4,550	520		5,070	5,875	
	2180	144 MBH heat output, 42 MBH cooling		1.20	20		3,950	480		4,430	5,125	
	2200	144 MBH heat output, 47 MBH cooling		1.20	20		4,375	480		4,855	5,600	
	2220	144 MBH heat output, 58 MBH cooling		1	24		4,600	575		5,175	6,000	
	2250	144 MBH heat output, 60 MBH cool		.70	34.286		4,600	820		5,420	6,425	
	3000	Oil fired, 84 MBH heat output, 24 MBH cooling	Q-9	1.20	13.333		3,375	305		3,680	4,250	
	3020	84 MBH heat output, 36 MBH cooling		1.20	13.333		3,725	305		4,030	4,625	
	3040	95.2 MBH heat output, 29 MBH cooling		1	16		3,625	370		3,995	4,625	
	3060	95.2 MBH heat output, 36 MBH cooling		1	16		3,775	370		4,145	4,800	
	3280	184.8 MBH heat, 60 MBH cooling	Q-10	1	24		5,200	575		5,775	6,675	
	3500	For precharged tubing with connection, add										
	3520	15 feet				Ea.	126			126	138	
	3540	25 feet					168			168	184	
	3560	35 feet					201			201	221	

	15550	Breechings, Chimneys & Stacks	CREW	DAILY OUTPUT	LABOR-HOURS	UNIT	MAT.	LABOR	EQUIP.	TOTAL	TOTAL INCL O&P	
440	0010	**VENT CHIMNEY** Prefab metal, U.L. listed										440
	0020	Gas, double wall, galvanized steel										
	0080	3" diameter	Q-9	72	.222	V.L.F.	3.68	5.10		8.78	12.60	
	0100	4" diameter		68	.235	"	4.57	5.40		9.97	14.15	
	5000	Vent damper bi-metal 6" flue		16	1	Ea.	105	23		128	155	
	5100	Gas, auto., electric		8	2	"	169	46		215	263	

15 MECHANICAL

15700 | Heating/Ventilating/Air Conditioning Equipment

	15730	Unitary Air Conditioning Equip	CREW	DAILY OUTPUT	LABOR-HOURS	UNIT	2004 BARE COSTS MAT.	LABOR	EQUIP.	TOTAL	TOTAL INCL O&P	
500	0010	**PACKAGED TERMINAL AIR CONDITIONER** Cabinet, wall sleeve,										500
	0100	louver, electric heat, thermostat, manual changeover, 208 V										
	0200	6,000 BTUH cooling, 8800 BTU heat	Q-5	6	2.667	Ea.	955	63		1,018	1,150	
	0220	9,000 BTUH cooling, 13,900 BTU heat		5	3.200		1,000	75.50		1,075.50	1,225	
	0240	12,000 BTUH cooling, 13,900 BTU heat		4	4		1,100	94.50		1,194.50	1,375	
	0260	15,000 BTUH cooling, 13,900 BTU heat		3	5.333		1,300	126		1,426	1,650	
600	0010	**ROOF TOP AIR CONDITIONERS** Standard controls, curb, economizer										600
	1000	Single zone, electric cool, gas heat										
	1140	5 ton cooling, 112 MBH heating	Q-5	.56	28.520	Ea.	3,675	675		4,350	5,150	
	1160	10 ton cooling, 200 MBH heating	Q-6	.67	35.982	"	6,900	820		7,720	8,950	
800	0010	**WINDOW UNIT AIR CONDITIONERS**										800
	4000	Portable/window, 15 amp 125V grounded receptacle required										
	4060	5000 BTUH	1 Carp	8	1	Ea.	279	23		302	345	
	4340	6000 BTUH		8	1		355	23		378	430	
	4480	8000 BTUH		6	1.333		410	31		441	505	
	4500	10,000 BTUH		6	1.333		485	31		516	585	
	4520	12,000 BTUH	L-2	8	2		525	40		565	650	
	4600	Window/thru-the-wall, 15 amp 230V grounded receptacle required										
	4780	17,000 BTUH	L-2	6	2.667	Ea.	755	53.50		808.50	925	
	4940	25,000 BTUH		4	4		1,075	80		1,155	1,325	

Important: See the Reference Section for critical supporting data - Reference Nos., Crews, & Location Factors

	15730 Unitary Air Conditioning Equip		CREW	DAILY OUTPUT	LABOR-HOURS	UNIT	2004 BARE COSTS MAT.	LABOR	EQUIP.	TOTAL	TOTAL INCL O&P	
800	4960	29,000 BTUH	L-2	4	4	Ea.	1,275	80		1,355	1,525	800
840	0010	**SELF-CONTAINED SINGLE PACKAGE**										840
	0100	Air cooled, for free blow or duct, not incl. remote condenser										
	0200	3 ton cooling	Q-5	1	16	Ea.	2,425	380		2,805	3,275	
	0210	4 ton cooling	"	.80	20	"	2,625	475		3,100	3,675	
	1000	Water cooled for free blow or duct, not including tower										
	1100	3 ton cooling	Q-6	1	24	Ea.	2,350	545		2,895	3,475	
900	0010	**SPLIT DUCTLESS SYSTEM**										900
	0100	Cooling only, single zone										
	0110	Wall mount										
	0120	3/4 ton cooling	Q-5	2	8	Ea.	1,125	189		1,314	1,525	
	0130	1 ton cooling		1.80	8.889		1,225	210		1,435	1,700	
	0140	1-1/2 ton cooling		1.60	10		1,575	236		1,811	2,150	
	0150	2 ton cooling	↓	1.40	11.429	↓	2,275	270		2,545	2,950	
	1000	Ceiling mount										
	1020	2 ton cooling	Q-5	1.40	11.429	Ea.	1,100	270		1,370	1,675	
	1030	3 ton cooling	"	1.20	13.333	"	3,225	315		3,540	4,075	
	2000	T-Bar mount										
	2010	2 ton cooling	Q-5	1.40	11.429	Ea.	2,400	270		2,670	3,100	
	2020	3 ton cooling		1.20	13.333		2,875	315		3,190	3,700	
	2030	3-1/2 ton cooling	↓	1.10	14.545	↓	3,475	345		3,820	4,400	
	3000	Multizone										
	3010	Wall mount										
	3020	2 @ 3/4 ton cooling	Q-5	1.80	8.889	Ea.	1,075	210		1,285	1,550	
	5000	Cooling / Heating										
	5110	1 ton cooling	Q-5	1.70	9.412	Ea.	795	222		1,017	1,250	
	5120	1-1/2 ton cooling	"	1.50	10.667	"	1,275	252		1,527	1,825	
	5300	Ceiling mount										
	5310	3 ton cooling	Q-5	1	16	Ea.	3,750	380		4,130	4,725	
	7000	Accessories for all split ductless systems										
	7010	Add for ambient frost control	Q-5	8	2	Ea.	118	47.50		165.50	208	
	7020	Add for tube / wiring kit										
	7030	15′ kit	Q-5	32	.500	Ea.	28.50	11.80		40.30	50.50	
	7040	35′ kit	"	24	.667	"	91.50	15.75		107.25	126	

	15740 Heat Pumps		CREW	DAILY OUTPUT	LABOR-HOURS	UNIT	MAT.	LABOR	EQUIP.	TOTAL	TOTAL INCL O&P	
100	0010	**AIR-SOURCE HEAT PUMPS** (Not including interconnecting tubing)										100
	1000	Air to air, split system, not including curbs, pads, or ductwork										
	1020	2 ton cooling, 8.5 MBH heat @ 0°F	Q-5	1.20	13.333	Ea.	985	315		1,300	1,600	
	1054	4 ton cooling, 24 MBH heat @ 0°F	"	.60	26.667	"	1,475	630		2,105	2,650	
	1500	Single package, not including curbs, pads, or plenums										
	1520	2 ton cooling, 6.5 MBH heat @ 0°F	Q-5	1.50	10.667	Ea.	1,950	252		2,202	2,550	
	1580	4 ton cooling, 13 MBH heat @ 0°F	"	.96	16.667	"	2,675	395		3,070	3,600	
800	0010	**WATER-SOURCE HEAT PUMPS** (Not including interconnecting tubing)										800
	2000	Water source to air, single package										
	2100	1 ton cooling, 13 MBH heat @ 75°F	Q-5	2	8	Ea.	895	189		1,084	1,300	
	2200	4 ton cooling, 31 MBH heat @ 75°F	"	1.20	13.333	"	1,425	315		1,740	2,100	
250	0010	**ELECTRIC HEATING**, not incl. conduit or feed wiring										250
	1100	Rule of thumb: Baseboard units, including control	1 Elec	4.40	1.818	kW	77	48		125	163	
	1300	Baseboard heaters, 2′ long, 375 watt		8	1	Ea.	29	26.50		55.50	75	
	1400	3′ long, 500 watt		8	1		34.50	26.50		61	81	
	1600	4′ long, 750 watt		6.70	1.194		41	31.50		72.50	96.50	
	1800	5′ long, 935 watt	↓	5.70	1.404	↓	48.50	37		85.50	114	

	15740 Heat Pumps	CREW	DAILY OUTPUT	LABOR-HOURS	UNIT	2004 BARE COSTS MAT.	LABOR	EQUIP.	TOTAL	TOTAL INCL O&P	
250	2000 6′ long, 1125 watt	1 Elec	5	1.600	Ea.	54	42		96	128	250
	2400 8′ long, 1500 watt		4	2		68	53		121	161	
	2800 10′ long, 1875 watt	↓	3.30	2.424	↓	84.50	64		148.50	197	
	2950 Wall heaters with fan, 120 to 277 volt										
	3600 Thermostats, integral	1 Elec	16	.500	Ea.	17.80	13.20		31	41	
	3800 Line voltage, 1 pole		8	1		21	26.50		47.50	66.50	
	5000 Radiant heating ceiling panels, 2′ x 4′, 500 watt		16	.500		195	13.20		208.20	236	
	5050 750 watt		16	.500		215	13.20		228.20	258	
	5300 Infrared quartz heaters, 120 volts, 1000 watts		6.70	1.194		119	31.50		150.50	183	
	5350 1500 watt		5	1.600		119	42		161	200	
	5400 240 volts, 1500 watt		5	1.600		119	42		161	200	
	5450 2000 watt		4	2		119	53		172	217	
	5500 3000 watt	↓	3	2.667	↓	139	70.50		209.50	267	
600	0010 **HYDRONIC HEATING** Terminal units, not incl. main supply pipe										600
	1000 Radiation										
	1310 Baseboard, pkgd, 1/2″ copper tube, alum. fin, 7″ high	Q-5	60	.267	L.F.	5.35	6.30		11.65	16.25	
	1320 3/4″ copper tube, alum. fin, 7″ high		58	.276		5.65	6.50		12.15	16.95	
	1340 1″ copper tube, alum. fin, 8-7/8″ high		56	.286		11.70	6.75		18.45	24	
	1360 1-1/4″ copper tube, alum. fin, 8-7/8″ high	↓	54	.296	↓	17.30	7		24.30	30.50	
	3000 Radiators, cast iron										
	3100 Free standing or wall hung, 6 tube, 25″ high	Q-5	96	.167	Section	21.50	3.94		25.44	30	

15 MECHANICAL

15800 | Air Distribution

	15810 Ducts	CREW	DAILY OUTPUT	LABOR-HOURS	UNIT	2004 BARE COSTS MAT.	LABOR	EQUIP.	TOTAL	TOTAL INCL O&P	
100	0010 **METAL DUCTWORK**										100
	0020 Fabricated rectangular, includes fittings, joints, supports,										
	0030 allowance for flexible connections, no insulation										
	0031 NOTE: Fabrication and installation are combined										
	0040 as LABOR cost. Approx. 25% fittings assumed.										
	0100 Aluminum, alloy 3003-H14, under 100 lb.	Q-10	75	.320	Lb.	2.45	7.65		10.10	15.50	
	0110 100 to 500 lb.		80	.300		1.95	7.15		9.10	14.15	
	0120 500 to 1,000 lb.		95	.253		1.65	6.05		7.70	11.90	
	0140 1,000 to 2,000 lb.		120	.200		1.45	4.78		6.23	9.60	
	0500 Galvanized steel, under 200 lb.		235	.102		.82	2.44		3.26	4.99	
	0520 200 to 500 lb.		245	.098		.62	2.34		2.96	4.60	
	0540 500 to 1,000 lb.	↓	255	.094	↓	.52	2.25		2.77	4.33	
500	0010 **FLEXIBLE DUCTS**										500
	1300 Flexible, coated fiberglass fabric on corr. resist. metal helix										
	1400 pressure to 12″ (WG) UL-181										
	1500 Non-insulated, 3″ diameter	Q-9	400	.040	L.F.	1.03	.92		1.95	2.67	
	1540 5″ diameter		320	.050		1.29	1.15		2.44	3.35	
	1560 6″ diameter		280	.057		1.52	1.32		2.84	3.87	
	1580 7″ diameter		240	.067		1.81	1.54		3.35	4.56	
	1900 Insulated, 1″ thick, PE jacket, 3″ diameter		380	.042		1.11	.97		2.08	2.84	
	1910 4″ diameter		340	.047		1.19	1.08		2.27	3.13	
	1920 5″ diameter	↓	300	.053	↓	1.31	1.23		2.54	3.50	

Important: See the Reference Section for critical supporting data - Reference Nos., Crews, & Location Factors

	15810 Ducts		CREW	DAILY OUTPUT	LABOR-HOURS	UNIT	2004 BARE COSTS MAT.	LABOR	EQUIP.	TOTAL	TOTAL INCL O&P	
500	1940	6″ diameter	Q-9	260	.062	L.F.	1.51	1.42		2.93	4.03	500
	1960	7″ diameter		220	.073		1.71	1.68		3.39	4.69	
	1980	8″ diameter		180	.089		1.90	2.05		3.95	5.50	
	2040	12″ diameter	↓	100	.160	↓	2.86	3.69		6.55	9.30	
	15820 Duct Accessories											
300	0010	**DUCT ACCESSORIES**										300
	0050	Air extractors, 12″ x 4″	1 Shee	24	.333	Ea.	15.55	8.55		24.10	31.50	
	0100	8″ x 6″		22	.364		15.55	9.30		24.85	33	
	3000	Fire damper, curtain type, 1-1/2 hr rated, vertical, 6″ x 6″		24	.333		13.40	8.55		21.95	29	
	3020	8″ x 6″		22	.364		21	9.30		30.30	38.50	
	6000	12″ x 12″	↓	21	.381	↓	26.50	9.75		36.25	45.50	
	8000	Multi-blade dampers, parallel blade										
	8100	8″ x 8″	1 Shee	24	.333	Ea.	53.50	8.55		62.05	73.50	
	15830 Fans											
100	0010	**FANS**										100
	8000	Ventilation, residential										
	8020	Attic, roof type										
	8030	Aluminum dome, damper & curb										
	8040	6″ diameter, 300 CFM	1 Elec	16	.500	Ea.	238	13.20		251.20	283	
	8050	7″ diameter, 450 CFM		15	.533		259	14.10		273.10	310	
	8060	9″ diameter, 900 CFM		14	.571		415	15.10		430.10	485	
	8080	12″ diameter, 1000 CFM (gravity)		10	.800		268	21		289	330	
	8090	16″ diameter, 1500 CFM (gravity)		9	.889		325	23.50		348.50	395	
	8100	20″ diameter, 2500 CFM (gravity)	↓	8	1	↓	395	26.50		421.50	480	
	8160	Plastic, ABS dome										
	8180	1050 CFM	1 Elec	14	.571	Ea.	79	15.10		94.10	111	
	8200	1600 CFM	″	12	.667	″	118	17.60		135.60	159	
	8240	Attic, wall type, with shutter, one speed										
	8250	12″ diameter, 1000 CFM	1 Elec	14	.571	Ea.	187	15.10		202.10	231	
	8260	14″ diameter, 1500 CFM		12	.667		203	17.60		220.60	252	
	8270	16″ diameter, 2000 CFM	↓	9	.889	↓	229	23.50		252.50	290	
	8290	Whole house, wall type, with shutter, one speed										
	8300	30″ diameter, 4800 CFM	1 Elec	7	1.143	Ea.	490	30		520	590	
	8310	36″ diameter, 7000 CFM		6	1.333		535	35		570	650	
	8320	42″ diameter, 10,000 CFM		5	1.600		600	42		642	730	
	8330	48″ diameter, 16,000 CFM	↓	4	2		745	53		798	905	
	8340	For two speed, add				↓	44.50			44.50	49	
	8350	Whole house, lay-down type, with shutter, one speed										
	8360	30″ diameter, 4500 CFM	1 Elec	8	1	Ea.	525	26.50		551.50	620	
	8370	36″ diameter, 6500 CFM		7	1.143		560	30		590	670	
	8380	42″ diameter, 9000 CFM		6	1.333		620	35		655	740	
	8390	48″ diameter, 12,000 CFM	↓	5	1.600		700	42		742	840	
	8440	For two speed, add					33.50			33.50	36.50	
	8450	For 12 hour timer switch, add	1 Elec	32	.250	↓	33.50	6.60		40.10	47.50	
	15850 Air Outlets & Inlets											
300	0010	**DIFFUSERS** Aluminum, opposed blade damper unless noted										300
	0100	Ceiling, linear, also for sidewall										
	0120	2″ wide	1 Shee	32	.250	L.F.	27.50	6.40		33.90	40.50	
	0160	4″ wide		26	.308	″	36	7.90		43.90	52.50	
	0500	Perforated, 24″ x 24″ lay-in panel size, 6″ x 6″		16	.500	Ea.	74	12.80		86.80	103	
	0520	8″ x 8″		15	.533		76	13.65		89.65	107	
	0530	9″ x 9″		14	.571		78	14.65		92.65	110	
	0590	16″ x 16″	↓	11	.727	↓	94.50	18.60		113.10	135	

15850 | Air Outlets & Inlets

			CREW	DAILY OUTPUT	LABOR-HOURS	UNIT	2004 BARE COSTS MAT.	LABOR	EQUIP.	TOTAL	TOTAL INCL O&P	
300	1000	Rectangular, 1 to 4 way blow, 6" x 6"	1 Shee	16	.500	Ea.	39.50	12.80		52.30	65	300
	1010	8" x 8"		15	.533		47	13.65		60.65	75	
	1014	9" x 9"		15	.533		50.50	13.65		64.15	78.50	
	1016	10" x 10"		15	.533		59.50	13.65		73.15	88.50	
	1020	12" x 6"		15	.533		67	13.65		80.65	96.50	
	1040	12" x 9"		14	.571		73	14.65		87.65	105	
	1060	12" x 12"		12	.667		69.50	17.05		86.55	105	
	1070	14" x 6"		13	.615		72	15.75		87.75	106	
	1074	14" x 14"		12	.667		101	17.05		118.05	140	
	1150	18" x 18"		9	.889		134	23		157	186	
	1170	24" x 12"		10	.800		121	20.50		141.50	168	
	1180	24" x 24"		7	1.143		281	29.50		310.50	360	
	1500	Round, butterfly damper, 6" diameter		18	.444		16.35	11.40		27.75	37	
	1520	8" diameter		16	.500		17.60	12.80		30.40	41	
	2000	T bar mounting, 24" x 24" lay-in frame, 6" x 6"		16	.500		82	12.80		94.80	112	
	2020	9" x 9"		14	.571		90.50	14.65		105.15	124	
	2040	12" x 12"		12	.667		117	17.05		134.05	158	
	2060	15" x 15"		11	.727		150	18.60		168.60	196	
	2080	18" x 18"	↓	10	.800		157	20.50		177.50	207	
	6000	For steel diffusers instead of aluminum, deduct				↓	10%					
500	0010	**GRILLES**										500
	0020	Aluminum										
	1000	Air return, 6" x 6"	1 Shee	26	.308	Ea.	13	7.90		20.90	27.50	
	1020	10" x 6"		24	.333		15.60	8.55		24.15	31.50	
	1080	16" x 8"		22	.364		23.50	9.30		32.80	41	
	1100	12" x 12"		22	.364		23.50	9.30		32.80	41	
	1120	24" x 12"		18	.444		41.50	11.40		52.90	65	
	1180	16" x 16"	↓	22	.364	↓	34	9.30		43.30	52.50	
700	0010	**REGISTERS**										700
	0980	Air supply										
	3000	Baseboard, hand adj. damper, enameled steel										
	3012	8" x 6"	1 Shee	26	.308	Ea.	12.05	7.90		19.95	26.50	
	3020	10" x 6"		24	.333		13.60	8.55		22.15	29.50	
	3040	12" x 5"		23	.348		15.55	8.90		24.45	32	
	3060	12" x 6"	↓	23	.348	↓	14.35	8.90		23.25	30.50	
	4000	Floor, toe operated damper, enameled steel										
	4020	4" x 8"	1 Shee	32	.250	Ea.	17.25	6.40		23.65	29.50	
	4040	4" x 12"	"	26	.308	"	20.50	7.90		28.40	35.50	

15860 | Air Cleaning Devices

			CREW	DAILY OUTPUT	LABOR-HOURS	UNIT	MAT.	LABOR	EQUIP.	TOTAL	TOTAL INCL O&P	
100	0010	**AIR FILTERS**										100
	0050	Activated charcoal type, full flow				MCFM	600			600	660	
	0060	Activated charcoal type, full flow, impregnated media 12" deep					175			175	193	
	0070	Activated charcoal type, HEPA filter & frame for field erection					175			175	193	
	0080	Activated charcoal type, HEPA filter-diffuser, ceiling install.				↓	165			165	182	
	2000	Electronic air cleaner, duct mounted										
	2150	400 - 1000 CFM	1 Shee	2.30	3.478	Ea.	675	89		764	890	
	2200	1000 - 1400 CFM		2.20	3.636		705	93		798	930	
	2250	1400 - 2000 CFM	↓	2.10	3.810	↓	780	97.50		877.50	1,025	
	2950	Mechanical media filtration units										
	3000	High efficiency type, with frame, non-supported				MCFM	45			45	49.50	
	3100	Supported type					55			55	60.50	
	4000	Medium efficiency, extended surface					5			5	5.50	
	4500	Permanent washable					20			20	22	
	5000	Renewable disposable roll				↓	120			120	132	
	5500	Throwaway glass or paper media type				Ea.	4.60			4.60	5.05	

Division 16 Electrical

Estimating Tips

16060 Grounding & Bonding

- When taking off grounding system, identify separately the type and size of wire and list each unique type of ground connection.

16100 Wiring Methods

- Conduit should be taken off in three main categories: power distribution, branch power, and branch lighting, so the estimator can concentrate on systems and components, therefore making it easier to ensure all items have been accounted for.
- For cost modifications for elevated conduit installation, add the percentages to labor according to the height of installation and only the quantities exceeding the different height levels, not to the total conduit quantities.
- Remember that aluminum wiring of equal ampacity is larger in diameter than copper and may require larger conduit.
- If more than three wires at a time are being pulled, deduct percentages from the labor hours of that grouping of wires.
- The estimator should take the weights of materials into consideration when completing a takeoff. Topics to consider include: How will the materials be supported? What methods of support are available? How high will the support structure have to reach? Will the final support structure be able to withstand the total burden? Is the support material included or separate from the fixture, equipment and material specified?

16200 Electrical Power

- Do not overlook the costs for equipment used in the installation. If scaffolding or highlifts are available in the field, contractors may use them in lieu of the proposed ladders and rolling staging.

16400 Low-Voltage Distribution

- Supports and concrete pads may be shown on drawings for the larger equipment, or the support system may be just a piece of plywood for the back of a panelboard. In either case, it must be included in the costs.

16500 Lighting

- Fixtures should be taken off room by room, using the fixture schedule, specifications, and the ceiling plan. For large concentrations of lighting fixtures in the same area deduct the percentages from labor hours.

16700 Communications
16800 Sound & Video

- When estimating material costs for special systems, it is always prudent to obtain manufacturers' quotations for equipment prices and special installation requirements which will affect the total costs.

Reference Numbers

Reference numbers are shown in bold squares at the beginning of some major classifications. These numbers refer to related items in the Reference Section. The reference information may be an estimating procedure, an alternate pricing method or technical information.

Note: Not all subdivisions listed here necessarily appear in this publication.

16055	Selective Demolition	CREW	DAILY OUTPUT	LABOR-HOURS	UNIT	2004 BARE COSTS MAT.	LABOR	EQUIP.	TOTAL	TOTAL INCL O&P
300 0010	ELECTRICAL DEMOLITION									
0020	Conduit to 15′ high, including fittings & hangers									
0100	Rigid galvanized steel, 1/2″ to 1″ diameter	1 Elec	242	.033	L.F.		.87		.87	1.42
0120	1-1/4″ to 2″	″	200	.040	″		1.06		1.06	1.72
0270	Armored cable, (BX) avg. 50′ runs									
0290	#14, 3 wire	1 Elec	571	.014	L.F.		.37		.37	.60
0300	#12, 2 wire		605	.013			.35		.35	.57
0310	#12, 3 wire		514	.016			.41		.41	.67
0320	#10, 2 wire		514	.016			.41		.41	.67
0330	#10, 3 wire		425	.019			.50		.50	.81
0340	#8, 3 wire		342	.023			.62		.62	1
0350	Non metallic sheathed cable (Romex)									
0360	#14, 2 wire	1 Elec	720	.011	L.F.		.29		.29	.48
0370	#14, 3 wire		657	.012			.32		.32	.52
0380	#12, 2 wire		629	.013			.34		.34	.55
0390	#10, 3 wire		450	.018			.47		.47	.76
0400	Wiremold raceway, including fittings & hangers									
0420	No. 3000	1 Elec	250	.032	L.F.		.84		.84	1.37
0440	No. 4000		217	.037			.97		.97	1.58
0460	No. 6000		166	.048			1.27		1.27	2.07
0500	Channels, steel, including fittings & hangers									
0520	3/4″ x 1-1/2″	1 Elec	308	.026	L.F.		.69		.69	1.12
0540	1-1/2″ x 1-1/2″		269	.030			.79		.79	1.28
0560	1-1/2″ x 1-7/8″		229	.035			.92		.92	1.50
1180	400 amp	2 Elec	6.80	2.353	Ea.		62		62	101
1210	Panel boards, incl. removal of all breakers,									
1220	conduit terminations & wire connections									
1230	3 wire, 120/240V, 100A, to 20 circuits	1 Elec	2.60	3.077	Ea.		81		81	132
1240	200 amps, to 42 circuits		1.30	6.154			162		162	264
1260	4 wire, 120/208V, 125A, to 20 circuits		2.40	3.333			88		88	143
1270	200 amps, to 42 circuits		1.20	6.667			176		176	286
1720	Junction boxes, 4″ sq. & oct.		80	.100			2.64		2.64	4.30
1760	Switch box		107	.075			1.97		1.97	3.21
1780	Receptacle & switch plates		257	.031			.82		.82	1.34
1800	Wire, THW-THWN-THHN, removed from									
1810	in place conduit, to 15′ high									
1830	#14	1 Elec	65	.123	C.L.F.		3.25		3.25	5.30
1840	#12		55	.145			3.84		3.84	6.25
1850	#10		45.50	.176			4.64		4.64	7.55
2000	Interior fluorescent fixtures, incl. supports									
2010	& whips, to 15′ high									
2100	Recessed drop-in 2′ x 2′, 2 lamp	2 Elec	35	.457	Ea.		12.05		12.05	19.65
2140	2′ x 4′, 4 lamp	″	30	.533	″		14.10		14.10	23
2180	Surface mount, acrylic lens & hinged frame									
2220	2′ x 2′, 2 lamp	2 Elec	44	.364	Ea.		9.60		9.60	15.60
2260	2′ x 4′, 4 lamp	″	33	.485	″		12.80		12.80	21
2300	Strip fixtures, surface mount									
2320	4′ long, 1 lamp	2 Elec	53	.302	Ea.		7.95		7.95	12.95
2380	8′ long, 2 lamp	″	40	.400	″		10.55		10.55	17.20
2460	Interior incandescent, surface, ceiling									
2470	or wall mount, to 12′ high									
2480	Metal cylinder type, 75 Watt	2 Elec	62	.258	Ea.		6.80		6.80	11.10
2600	Exterior fixtures, incandescent, wall mount									
2620	100 Watt	2 Elec	50	.320	Ea.		8.45		8.45	13.75
3000	Ceiling fan, tear out and remove	1 Elec	18	.444	″		11.75		11.75	19.10
9000	Minimum labor/equipment charge	″	4	2	Job		53		53	86

16050 | Basic Electrical Materials & Methods

16060 | Grounding & Bonding

		16060 Grounding & Bonding	CREW	DAILY OUTPUT	LABOR-HOURS	UNIT	2004 BARE COSTS MAT.	LABOR	EQUIP.	TOTAL	TOTAL INCL O&P	
800	0010	**GROUNDING**										800
	0030	Rod, copper clad, 8' long, 1/2" diameter	1 Elec	5.50	1.455	Ea.	13.10	38.50		51.60	77	
	0050	3/4" diameter		5.30	1.509		27	40		67	94.50	
	0080	10' long, 1/2" diameter		4.80	1.667		16.90	44		60.90	90	
	0100	3/4" diameter		4.40	1.818		30	48		78	111	
	0261	Wire, ground bare armored, #8-1 conductor		200	.040	L.F.	.72	1.06		1.78	2.51	
	0271	#6-1 conductor		180	.044	"	.91	1.17		2.08	2.91	
	0390	Bare copper wire, #8 stranded		11	.727	C.L.F.	9.80	19.20		29	42	
	0401	Bare copper, #6 wire		1,000	.008	L.F.	.17	.21		.38	.52	
	0601	#2 stranded	2 Elec	1,000	.016	"	.39	.42		.81	1.11	
	1800	Water pipe ground clamps, heavy duty										
	2000	Bronze, 1/2" to 1" diameter	1 Elec	8	1	Ea.	13.05	26.50		39.55	57.50	

16100 | Wiring Methods

16120 | Conductors & Cables

		16120 Conductors & Cables	CREW	DAILY OUTPUT	LABOR-HOURS	UNIT	2004 BARE COSTS MAT.	LABOR	EQUIP.	TOTAL	TOTAL INCL O&P	
120	0010	**ARMORED CABLE**										120
	0051	600 volt, copper (BX), #14, 2 conductor, solid	1 Elec	240	.033	L.F.	.52	.88		1.40	2	
	0101	3 conductor, solid		200	.040		.69	1.06		1.75	2.48	
	0151	#12, 2 conductor, solid		210	.038		.53	1.01		1.54	2.23	
	0201	3 conductor, solid		180	.044		.83	1.17		2	2.83	
	0251	#10, 2 conductor, solid		180	.044		.97	1.17		2.14	2.97	
	0301	3 conductor, solid		150	.053		1.30	1.41		2.71	3.72	
	0351	#8, 3 conductor, solid		120	.067		2.03	1.76		3.79	5.10	
550	0010	**NON-METALLIC SHEATHED CABLE** 600 volt										550
	0100	Copper with ground wire, (Romex)										
	0151	#14, 2 wire	1 Elec	250	.032	L.F.	.12	.84		.96	1.50	
	0201	3 wire		230	.035		.21	.92		1.13	1.73	
	0251	#12, 2 wire		220	.036		.17	.96		1.13	1.75	
	0301	3 wire		200	.040		.31	1.06		1.37	2.06	
	0351	#10, 2 wire		200	.040		.34	1.06		1.40	2.09	
	0401	3 wire		140	.057		.47	1.51		1.98	2.97	
	0451	#8, 3 conductor		130	.062		.91	1.62		2.53	3.64	
	0501	#6, 3 wire		120	.067		1.45	1.76		3.21	4.46	
	0550	SE type SER aluminum cable, 3 RHW and										
	0601	1 bare neutral, 3 #8 & 1 #8	1 Elec	150	.053	L.F.	.92	1.41		2.33	3.31	
	0651	3 #6 & 1 #6	"	130	.062		1.05	1.62		2.67	3.79	
	0701	3 #4 & 1 #6	2 Elec	220	.073		1.17	1.92		3.09	4.41	
	0751	3 #2 & 1 #4		200	.080		1.73	2.11		3.84	5.35	
	0801	3 #1/0 & 1 #2		180	.089		2.61	2.35		4.96	6.70	
	0851	3 #2/0 & 1 #1		160	.100		3.08	2.64		5.72	7.70	
	0901	3 #4/0 & 1 #2/0		140	.114		4.39	3.02		7.41	9.75	
	2401	SEU service entrance cable, copper 2 conductors, #8 + #8 neut.	1 Elec	150	.053		.78	1.41		2.19	3.15	
	2601	#6 + #8 neutral		130	.062		1.16	1.62		2.78	3.92	
	2801	#6 + #6 neutral		130	.062		1.27	1.62		2.89	4.04	
	3001	#4 + #6 neutral	2 Elec	220	.073		1.80	1.92		3.72	5.10	
	3201	#4 + #4 neutral		220	.073		1.96	1.92		3.88	5.25	
	3401	#3 + #5 neutral		210	.076		2.10	2.01		4.11	5.60	
	6500	Service entrance cap for copper SEU										
	6600	100 amp	1 Elec	12	.667	Ea.	8.35	17.60		25.95	37.50	

16120 | Conductors & Cables

			CREW	DAILY OUTPUT	LABOR-HOURS	UNIT	2004 BARE COSTS MAT.	LABOR	EQUIP.	TOTAL	TOTAL INCL O&P	
550	6700	150 amp	1 Elec	10	.800	Ea.	12.10	21		33.10	48	550
	6800	200 amp	↓	8	1	↓	18.30	26.50		44.80	63	
900	0010	**WIRE**										900
	0021	600 volt type THW, copper solid, #14	1 Elec	1,300	.006	L.F.	.03	.16		.19	.29	
	0031	#12		1,100	.007		.04	.19		.23	.36	
	0041	#10		1,000	.008		.07	.21		.28	.41	
	0161	#6	↓	650	.012		.20	.33		.53	.75	
	0181	#4	2 Elec	1,060	.015		.31	.40		.71	.99	
	0201	#3		1,000	.016		.38	.42		.80	1.10	
	0221	#2		900	.018		.47	.47		.94	1.28	
	0241	#1		800	.020		.60	.53		1.13	1.51	
	0261	1/0		660	.024		.72	.64		1.36	1.83	
	0281	2/0		580	.028		.88	.73		1.61	2.16	
	0301	3/0		500	.032		1.09	.84		1.93	2.57	
	0351	4/0	↓	440	.036	↓	1.38	.96		2.34	3.08	

16132 | Conduit & Tubing

			CREW	DAILY OUTPUT	LABOR-HOURS	UNIT	2004 BARE COSTS MAT.	LABOR	EQUIP.	TOTAL	TOTAL INCL O&P	
205	0010	**CONDUIT** To 15′ high, includes 2 terminations, 2 elbows and										205
	0020	11 beam clamps per 100 L.F.										
	1750	Rigid galvanized steel, 1/2″ diameter	1 Elec	90	.089	L.F.	1.71	2.35		4.06	5.70	
	1770	3/4″ diameter		80	.100		1.99	2.64		4.63	6.50	
	1800	1″ diameter		65	.123		2.80	3.25		6.05	8.40	
	1830	1-1/4″ diameter		60	.133		3.81	3.52		7.33	9.95	
	1850	1-1/2″ diameter		55	.145		4.40	3.84		8.24	11.10	
	1870	2″ diameter		45	.178		5.90	4.69		10.59	14.10	
	5000	Electric metallic tubing (EMT), 1/2″ diameter		170	.047		.40	1.24		1.64	2.45	
	5020	3/4″ diameter		130	.062		.60	1.62		2.22	3.30	
	5040	1″ diameter		115	.070		1.02	1.84		2.86	4.11	
	5060	1-1/4″ diameter		100	.080		1.53	2.11		3.64	5.15	
	5080	1-1/2″ diameter		90	.089		1.95	2.35		4.30	5.95	
	9100	PVC, #40, 1/2″ diameter		190	.042		.71	1.11		1.82	2.59	
	9110	3/4″ diameter		145	.055		.83	1.46		2.29	3.28	
	9120	1″ diameter		125	.064		1.21	1.69		2.90	4.08	
	9130	1-1/4″ diameter		110	.073		1.59	1.92		3.51	4.87	
	9140	1-1/2″ diameter		100	.080		1.88	2.11		3.99	5.50	
	9150	2″ diameter	↓	90	.089	↓	2.45	2.35		4.80	6.50	
230	0010	**CONDUIT IN CONCRETE SLAB** Including terminations,										230
	0020	fittings and supports										
	3230	PVC, schedule 40, 1/2″ diameter	1 Elec	270	.030	L.F.	.43	.78		1.21	1.75	
	3250	3/4″ diameter		230	.035		.52	.92		1.44	2.06	
	3270	1″ diameter		200	.040		.69	1.06		1.75	2.48	
	3300	1-1/4″ diameter		170	.047		.97	1.24		2.21	3.09	
	3330	1-1/2″ diameter		140	.057		1.20	1.51		2.71	3.76	
	3350	2″ diameter		120	.067		1.51	1.76		3.27	4.52	
	4350	Rigid galvanized steel, 1/2″ diameter		200	.040		1.43	1.06		2.49	3.30	
	4400	3/4″ diameter		170	.047		1.72	1.24		2.96	3.91	
	4450	1″ diameter		130	.062		2.53	1.62		4.15	5.40	
	4500	1-1/4″ diameter		110	.073		3.33	1.92		5.25	6.80	
	4600	1-1/2″ diameter		100	.080		3.93	2.11		6.04	7.75	
	4800	2″ diameter	↓	90	.089	↓	5.25	2.35		7.60	9.55	
240	0010	**CONDUIT IN TRENCH** Includes terminations and fittings										240
	0200	Rigid galvanized steel, 2″ diameter	1 Elec	150	.053	L.F.	5.05	1.41		6.46	7.85	
	0400	2-1/2″ diameter	″	100	.080		9.55	2.11		11.66	13.95	
	0600	3″ diameter	2 Elec	160	.100	↓	11.95	2.64		14.59	17.45	

16100 | Wiring Methods

16132 | Conduit & Tubing

		16132 Conduit & Tubing	CREW	DAILY OUTPUT	LABOR-HOURS	UNIT	2004 BARE COSTS MAT.	LABOR	EQUIP.	TOTAL	TOTAL INCL O&P	
240	0800	3-1/2″ diameter	2 Elec	140	.114	L.F.	14.90	3.02		17.92	21.50	240
250	0010	CONDUIT FITTINGS FOR RIGID GALVANIZED STEEL										250
	2280	LB, LR or LL fittings & covers, 1/2″ diameter	1 Elec	16	.500	Ea.	8.70	13.20		21.90	31	
	2290	3/4″ diameter		13	.615		10.50	16.25		26.75	38	
	2300	1″ diameter		11	.727		15.50	19.20		34.70	48	
	2330	1-1/4″ diameter		8	1		23	26.50		49.50	68.50	
	2350	1-1/2″ diameter		6	1.333		29	35		64	89.50	
	2370	2″ diameter		5	1.600		48	42		90	122	
	5280	Service entrance cap, 1/2″ diameter		16	.500		6.80	13.20		20	29	
	5300	3/4″ diameter		13	.615		7.85	16.25		24.10	35	
	5320	1″ diameter		10	.800		7.90	21		28.90	43	
	5340	1-1/4″ diameter		8	1		13.70	26.50		40.20	58	
	5360	1-1/2″ diameter		6.50	1.231		19.20	32.50		51.70	74	
	5380	2″ diameter		5.50	1.455		37.50	38.50		76	104	
320	0010	FLEXIBLE METALLIC CONDUIT										320
	0050	Steel, 3/8″ diameter	1 Elec	200	.040	L.F.	.25	1.06		1.31	2	
	0100	1/2″ diameter		200	.040		.31	1.06		1.37	2.06	
	0200	3/4″ diameter		160	.050		.43	1.32		1.75	2.62	
	0250	1″ diameter		100	.080		.84	2.11		2.95	4.36	
	0300	1-1/4″ diameter		70	.114		.92	3.02		3.94	5.90	
	0350	1-1/2″ diameter		50	.160		1.59	4.22		5.81	8.60	
	0370	2″ diameter		40	.200		1.94	5.30		7.24	10.75	

16133 | Multi-outlet Assemblies

		16133 Multi-outlet Assemblies	CREW	DAILY OUTPUT	LABOR-HOURS	UNIT	MAT.	LABOR	EQUIP.	TOTAL	TOTAL INCL O&P	
800	0010	SURFACE RACEWAY										800
	0100	No. 500	1 Elec	100	.080	L.F.	.73	2.11		2.84	4.24	
	0110	No. 700		100	.080		.82	2.11		2.93	4.34	
	0200	No. 1000		90	.089		1.40	2.35		3.75	5.35	
	0400	No. 1500, small pancake		90	.089		1.50	2.35		3.85	5.45	
	0600	No. 2000, base & cover, blank		90	.089		1.47	2.35		3.82	5.45	
	0800	No. 3000, base & cover, blank		75	.107		2.95	2.82		5.77	7.85	
	2400	Fittings, elbows, No. 500		40	.200	Ea.	1.33	5.30		6.63	10.05	
	2800	Elbow cover, No. 2000		40	.200		2.55	5.30		7.85	11.40	
	2880	Tee, No. 500		42	.190		2.56	5.05		7.61	11	
	2900	No. 2000		27	.296		8.05	7.80		15.85	21.50	
	3000	Switch box, No. 500		16	.500		8.40	13.20		21.60	31	
	3400	Telephone outlet, No. 1500		16	.500		9.65	13.20		22.85	32	
	3600	Junction box, No. 1500		16	.500		6.70	13.20		19.90	29	
	3800	Plugmold wired sections, No. 2000										
	4000	1 circuit, 6 outlets, 3 ft. long	1 Elec	8	1	Ea.	24.50	26.50		51	70	
	4100	2 circuits, 8 outlets, 6 ft. long	″	5.30	1.509	″	41	40		81	110	

16136 | Boxes

		16136 Boxes	CREW	DAILY OUTPUT	LABOR-HOURS	UNIT	MAT.	LABOR	EQUIP.	TOTAL	TOTAL INCL O&P	
600	0010	OUTLET BOXES										600
	0021	Pressed steel, octagon, 4″	1 Elec	18	.444	Ea.	1.48	11.75		13.23	20.50	
	0060	Covers, blank		64	.125		.62	3.30		3.92	6.05	
	0100	Extension rings		40	.200		2.39	5.30		7.69	11.25	
	0151	Square 4″		18	.444		2.10	11.75		13.85	21.50	
	0200	Extension rings		40	.200		2.48	5.30		7.78	11.35	
	0250	Covers, blank		64	.125		.70	3.30		4	6.10	
	0300	Plaster rings		64	.125		1.14	3.30		4.44	6.60	
	0651	Switchbox		24	.333		2.31	8.80		11.11	16.85	
	1101	Concrete, floor, 1 gang		4.80	1.667		61.50	44		105.50	140	

ELECTRICAL 16

16136 | Boxes

		16136 Boxes	CREW	DAILY OUTPUT	LABOR-HOURS	UNIT	2004 BARE COSTS MAT.	LABOR	EQUIP.	TOTAL	TOTAL INCL O&P	
620	0010	OUTLET BOXES, PLASTIC										620
	0051	4" diameter, round, with 2 mounting nails	1 Elec	23	.348	Ea.	1.86	9.20		11.06	17	
	0101	Bar hanger mounted		23	.348		3.28	9.20		12.48	18.55	
	0201	Square with 2 mounting nails		23	.348		3	9.20		12.20	18.25	
	0300	Plaster ring		64	.125		1.03	3.30		4.33	6.50	
	0401	Switch box with 2 mounting nails, 1 gang		27	.296		1.36	7.80		9.16	14.25	
	0501	2 gang		23	.348		2.28	9.20		11.48	17.45	
	0601	3 gang	↓	18	.444	↓	3.60	11.75		15.35	23	
700	0010	PULL BOXES & CABINETS										700
	0100	Sheet metal, pull box, NEMA 1, type SC, 6" W x 6" H x 4" D	1 Elec	8	1	Ea.	9.65	26.50		36.15	53.50	
	0200	8" W x 8" H x 4" D		8	1		13.20	26.50		39.70	57.50	
	0300	10" W x 12" H x 6" D	↓	5.30	1.509	↓	23.50	40		63.50	90.50	

16139 | Residential Wiring

		16139 Residential Wiring	CREW	DAILY OUTPUT	LABOR-HOURS	UNIT	MAT.	LABOR	EQUIP.	TOTAL	TOTAL INCL O&P	
700	0010	RESIDENTIAL WIRING										700
	0020	20' avg. runs and #14/2 wiring incl. unless otherwise noted										
	1000	Service & panel, includes 24' SE-AL cable, service eye, meter,										
	1010	Socket, panel board, main bkr., ground rod, 15 or 20 amp										
	1020	1-pole circuit breakers, and misc. hardware										
	1100	100 amp, with 10 branch breakers	1 Elec	1.19	6.723	Ea.	410	177		587	745	
	1110	With PVC conduit and wire		.92	8.696		445	230		675	865	
	1120	With RGS conduit and wire		.73	10.959		575	289		864	1,100	
	1150	150 amp, with 14 branch breakers		1.03	7.767		645	205		850	1,050	
	1170	With PVC conduit and wire		.82	9.756		715	258		973	1,200	
	1180	With RGS conduit and wire	↓	.67	11.940		955	315		1,270	1,575	
	1200	200 amp, with 18 branch breakers	2 Elec	1.80	8.889		840	235		1,075	1,300	
	1220	With PVC conduit and wire		1.46	10.959		915	289		1,204	1,475	
	1230	With RGS conduit and wire	↓	1.24	12.903		1,225	340		1,565	1,900	
	1800	Lightning surge suppressor for above services, add	1 Elec	32	.250	↓	40	6.60		46.60	55	
	2000	Switch devices										
	2100	Single pole, 15 amp, Ivory, with a 1-gang box, cover plate,										
	2110	Type NM (Romex) cable	1 Elec	17.10	.468	Ea.	6.75	12.35		19.10	27.50	
	2120	Type MC (BX) cable		14.30	.559		18.05	14.75		32.80	44	
	2130	EMT & wire		5.71	1.401		17.85	37		54.85	79.50	
	2150	3-way, #14/3, type NM cable		14.55	.550		10.10	14.50		24.60	34.50	
	2170	Type MC cable		12.31	.650		23	17.15		40.15	53.50	
	2180	EMT & wire		5	1.600		19.95	42		61.95	90.50	
	2200	4-way, #14/3, type NM cable		14.55	.550		23	14.50		37.50	49	
	2220	Type MC cable		12.31	.650		36	17.15		53.15	67.50	
	2230	EMT & wire		5	1.600		33	42		75	105	
	2250	S.P., 20 amp, #12/2, type NM cable		13.33	.600		11.90	15.85		27.75	39	
	2270	Type MC cable		11.43	.700		22.50	18.50		41	54.50	
	2280	EMT & wire		4.85	1.649		23.50	43.50		67	96.50	
	2290	S.P. rotary dimmer, 600W, no wiring		17	.471		16	12.40		28.40	37.50	
	2300	S.P. rotary dimmer, 600W, type NM cable		14.55	.550		18.45	14.50		32.95	44	
	2320	Type MC cable		12.31	.650		29.50	17.15		46.65	60.50	
	2330	EMT & wire		5	1.600		30	42		72	102	
	2350	3-way rotary dimmer, type NM cable		13.33	.600		16.15	15.85		32	44	
	2370	Type MC cable		11.43	.700		27.50	18.50		46	60	
	2380	EMT & wire	↓	4.85	1.649	↓	28	43.50		71.50	102	
	2400	Interval timer wall switch, 20 amp, 1-30 min., #12/2										
	2410	Type NM cable	1 Elec	14.55	.550	Ea.	29	14.50		43.50	55.50	
	2420	Type MC cable		12.31	.650		37.50	17.15		54.65	69	
	2430	EMT & wire	↓	5	1.600	↓	40.50	42		82.50	113	
	2500	Decorator style										
	2510	S.P., 15 amp, type NM cable	1 Elec	17.10	.468	Ea.	10.30	12.35		22.65	31.50	

16139 | Residential Wiring		CREW	DAILY OUTPUT	LABOR-HOURS	UNIT	2004 BARE COSTS MAT.	LABOR	EQUIP.	TOTAL	TOTAL INCL O&P
700	2520 Type MC cable	1 Elec	14.30	.559	Ea.	21.50	14.75		36.25	48
	2530 EMT & wire		5.71	1.401		21.50	37		58.50	83.50
	2550 3-way, #14/3, type NM cable		14.55	.550		13.65	14.50		28.15	38.50
	2570 Type MC cable		12.31	.650		26.50	17.15		43.65	57
	2580 EMT & wire		5	1.600		23.50	42		65.50	94.50
	2600 4-way, #14/3, type NM cable		14.55	.550		26.50	14.50		41	52.50
	2620 Type MC cable		12.31	.650		39.50	17.15		56.65	71.50
	2630 EMT & wire		5	1.600		36.50	42		78.50	109
	2650 S.P., 20 amp, #12/2, type NM cable		13.33	.600		15.45	15.85		31.30	43
	2670 Type MC cable		11.43	.700		26	18.50		44.50	58.50
	2680 EMT & wire		4.85	1.649		27	43.50		70.50	101
	2700 S.P., slide dimmer, type NM cable		17.10	.468		25.50	12.35		37.85	48
	2720 Type MC cable		14.30	.559		37	14.75		51.75	64.50
	2730 EMT & wire		5.71	1.401		37	37		74	101
	2750 S.P., touch dimmer, type NM cable		17.10	.468		22	12.35		34.35	44
	2770 Type MC cable		14.30	.559		33	14.75		47.75	60.50
	2780 EMT & wire		5.71	1.401		33.50	37		70.50	97
	2800 3-way touch dimmer, type NM cable		13.33	.600		40	15.85		55.85	70
	2820 Type MC cable		11.43	.700		51	18.50		69.50	86.50
	2830 EMT & wire	↓	4.85	1.649	↓	51.50	43.50		95	128
	3000 Combination devices									
	3100 S.P. switch/15 amp recpt., Ivory, 1-gang box, plate									
	3110 Type NM cable	1 Elec	11.43	.700	Ea.	14.80	18.50		33.30	46.50
	3120 Type MC cable		10	.800		26	21		47	63
	3130 EMT & wire		4.40	1.818		26.50	48		74.50	107
	3150 S.P. switch/pilot light, type NM cable		11.43	.700		15.45	18.50		33.95	47
	3170 Type MC cable		10	.800		26.50	21		47.50	64
	3180 EMT & wire		4.43	1.806		27	47.50		74.50	108
	3190 2-S.P. switches, 2-#14/2, no wiring		14	.571		5.70	15.10		20.80	31
	3200 2-S.P. switches, 2-#14/2, type NM cables		10	.800		16.55	21		37.55	52.50
	3220 Type MC cable		8.89	.900		36	24		60	78
	3230 EMT & wire		4.10	1.951		27.50	51.50		79	115
	3250 3-way switch/15 amp recpt., #14/3, type NM cable		10	.800		21	21		42	57.50
	3270 Type MC cable		8.89	.900		34	24		58	76
	3280 EMT & wire		4.10	1.951		31	51.50		82.50	118
	3300 2-3 way switches, 2-#14/3, type NM cables		8.89	.900		27.50	24		51.50	68.50
	3320 Type MC cable		8	1		50	26.50		76.50	98
	3330 EMT & wire		4	2		34.50	53		87.50	124
	3350 S.P. switch/20 amp recpt., #12/2, type NM cable		10	.800		25	21		46	62
	3370 Type MC cable		8.89	.900		33.50	24		57.50	75
	3380 EMT & wire	↓	4.10	1.951	↓	36.50	51.50		88	125
	3400 Decorator style									
	3410 S.P. switch/15 amp recpt., type NM cable	1 Elec	11.43	.700	Ea.	18.35	18.50		36.85	50
	3420 Type MC cable		10	.800		29.50	21		50.50	67
	3430 EMT & wire		4.40	1.818		30	48		78	111
	3450 S.P. switch/pilot light, type NM cable		11.43	.700		19	18.50		37.50	51
	3470 Type MC cable		10	.800		30.50	21		51.50	68
	3480 EMT & wire		4.40	1.818		30.50	48		78.50	112
	3500 2-S.P. switches, 2-#14/2, type NM cables		10	.800		20	21		41	56.50
	3520 Type MC cable		8.89	.900		39.50	24		63.50	82
	3530 EMT & wire		4.10	1.951		31	51.50		82.50	118
	3550 3-way/15 amp recpt., #14/3, type NM cable		10	.800		24.50	21		45.50	61.50
	3570 Type MC cable		8.89	.900		37.50	24		61.50	79.50
	3580 EMT & wire		4.10	1.951		34.50	51.50		86	122
	3650 2-3 way switches, 2-#14/3, type NM cables		8.89	.900		31	24		55	72.50
	3670 Type MC cable	↓	8	1	↓	53.50	26.50		80	102

16139 | Residential Wiring		CREW	DAILY OUTPUT	LABOR-HOURS	UNIT	2004 BARE COSTS MAT.	LABOR	EQUIP.	TOTAL	TOTAL INCL O&P
700	3680 EMT & wire	1 Elec	4	2	Ea.	38.50	53		91.50	128
	3700 S.P. switch/20 amp recpt., #12/2, type NM cable		10	.800		29	21		50	66
	3720 Type MC cable		8.89	.900		37	24		61	79
	3730 EMT & wire	↓	4.10	1.951	↓	40	51.50		91.50	128
	4000 Receptacle devices									
	4010 Duplex outlet, 15 amp recpt., Ivory, 1-gang box, plate									
	4015 Type NM cable	1 Elec	14.55	.550	Ea.	5.25	14.50		19.75	29.50
	4020 Type MC cable		12.31	.650		16.55	17.15		33.70	46
	4030 EMT & wire		5.33	1.501		16.35	39.50		55.85	82.50
	4050 With #12/2, type NM cable		12.31	.650		6.25	17.15		23.40	35
	4070 Type MC cable		10.67	.750		16.75	19.80		36.55	50.50
	4080 EMT & wire		4.71	1.699		17.65	45		62.65	92.50
	4100 20 amp recpt., #12/2, type NM cable		12.31	.650		12.15	17.15		29.30	41.50
	4120 Type MC cable		10.67	.750		22.50	19.80		42.30	57
	4130 EMT & wire	↓	4.71	1.699	↓	23.50	45		68.50	99
	4140 For GFI see line 4300 below									
	4150 Decorator style, 15 amp recpt., type NM cable	1 Elec	14.55	.550	Ea.	8.80	14.50		23.30	33
	4170 Type MC cable		12.31	.650		20	17.15		37.15	50
	4180 EMT & wire		5.33	1.501		19.90	39.50		59.40	86.50
	4200 With #12/2, type NM cable		12.31	.650		9.80	17.15		26.95	39
	4220 Type MC cable		10.67	.750		20.50	19.80		40.30	54.50
	4230 EMT & wire		4.71	1.699		21	45		66	96.50
	4250 20 amp recpt. #12/2, type NM cable		12.31	.650		15.70	17.15		32.85	45.50
	4270 Type MC cable		10.67	.750		26	19.80		45.80	61
	4280 EMT & wire		4.71	1.699		27	45		72	103
	4300 GFI, 15 amp recpt., type NM cable		12.31	.650		33	17.15		50.15	64
	4320 Type MC cable		10.67	.750		44	19.80		63.80	80.50
	4330 EMT & wire		4.71	1.699		44	45		89	122
	4350 GFI with #12/2, type NM cable		10.67	.750		34	19.80		53.80	69.50
	4370 Type MC cable		9.20	.870		44.50	23		67.50	86.50
	4380 EMT & wire		4.21	1.900		45.50	50		95.50	132
	4400 20 amp recpt., #12/2 type NM cable		10.67	.750		35.50	19.80		55.30	71
	4420 Type MC cable		9.20	.870		46	23		69	88
	4430 EMT & wire		4.21	1.900		47	50		97	133
	4500 Weather-proof cover for above receptacles, add	↓	32	.250	↓	4.40	6.60		11	15.60
	4550 Air conditioner outlet, 20 amp-240 volt recpt.									
	4560 30' of #12/2, 2 pole circuit breaker									
	4570 Type NM cable	1 Elec	10	.800	Ea.	40.50	21		61.50	79
	4580 Type MC cable		9	.889		54.50	23.50		78	98
	4590 EMT & wire		4	2		51.50	53		104.50	143
	4600 Decorator style, type NM cable		10	.800		44.50	21		65.50	83.50
	4620 Type MC cable		9	.889		58.50	23.50		82	103
	4630 EMT & wire	↓	4	2	↓	55.50	53		108.50	147
	4650 Dryer outlet, 30 amp-240 volt recpt., 20' of #10/3									
	4660 2 pole circuit breaker									
	4670 Type NM cable	1 Elec	6.41	1.248	Ea.	47	33		80	106
	4680 Type MC cable		5.71	1.401		60.50	37		97.50	127
	4690 EMT & wire	↓	3.48	2.299	↓	52.50	60.50		113	156
	4700 Range outlet, 50 amp-240 volt recpt., 30' of #8/3									
	4710 Type NM cable	1 Elec	4.21	1.900	Ea.	68.50	50		118.50	157
	4720 Type MC cable		4	2		106	53		159	203
	4730 EMT & wire		2.96	2.703		67	71.50		138.50	190
	4750 Central vacuum outlet, Type NM cable		6.40	1.250		42	33		75	100
	4770 Type MC cable		5.71	1.401		64	37		101	131
	4780 EMT & wire	↓	3.48	2.299	↓	53.50	60.50		114	158
	4800 30 amp-110 volt locking recpt., #10/2 circ. bkr.									

16139 | Residential Wiring

700		Crew	Daily Output	Labor-Hours	Unit	2004 Bare Costs Mat.	2004 Bare Costs Labor	2004 Bare Costs Equip.	2004 Bare Costs Total	Total Incl O&P
4810	Type NM cable	1 Elec	6.20	1.290	Ea.	50	34		84	111
4820	Type MC cable		5.40	1.481		76	39		115	147
4830	EMT & wire		3.20	2.500		61	66		127	174
4900	Low voltage outlets									
4910	Telephone recpt., 20′ of 4/C phone wire	1 Elec	26	.308	Ea.	6.70	8.10		14.80	20.50
4920	TV recpt., 20′ of RG59U coax wire, F type connector	"	16	.500	"	11.15	13.20		24.35	34
4950	Door bell chime, transformer, 2 buttons, 60′ of bellwire									
4970	Economy model	1 Elec	11.50	.696	Ea.	51.50	18.35		69.85	87
4980	Custom model		11.50	.696		82.50	18.35		100.85	121
4990	Luxury model, 3 buttons		9.50	.842		224	22		246	282
6000	Lighting outlets									
6050	Wire only (for fixture), type NM cable	1 Elec	32	.250	Ea.	3.31	6.60		9.91	14.40
6070	Type MC cable		24	.333		12.65	8.80		21.45	28.50
6080	EMT & wire		10	.800		11.45	21		32.45	47
6100	Box (4″), and wire (for fixture), type NM cable		25	.320		7.35	8.45		15.80	22
6120	Type MC cable		20	.400		16.70	10.55		27.25	35.50
6130	EMT & wire		11	.727		15.45	19.20		34.65	48
6200	Fixtures (use with lines 6050 or 6100 above)									
6210	Canopy style, economy grade	1 Elec	40	.200	Ea.	25.50	5.30		30.80	36.50
6220	Custom grade		40	.200		46	5.30		51.30	59
6250	Dining room chandelier, economy grade		19	.421		76	11.10		87.10	102
6260	Custom grade		19	.421		225	11.10		236.10	266
6270	Luxury grade		15	.533		495	14.10		509.10	570
6310	Kitchen fixture (fluorescent), economy grade		30	.267		51.50	7.05		58.55	68
6320	Custom grade		25	.320		161	8.45		169.45	191
6350	Outdoor, wall mounted, economy grade		30	.267		27	7.05		34.05	41
6360	Custom grade		30	.267		101	7.05		108.05	122
6370	Luxury grade		25	.320		227	8.45		235.45	264
6410	Outdoor PAR floodlights, 1 lamp, 150 watt		20	.400		21	10.55		31.55	40
6420	2 lamp, 150 watt each		20	.400		35	10.55		45.55	55.50
6430	For infrared security sensor, add		32	.250		87	6.60		93.60	107
6450	Outdoor, quartz-halogen, 300 watt flood		20	.400		38	10.55		48.55	59
6600	Recessed downlight, round, pre-wired, 50 or 75 watt trim		30	.267		35	7.05		42.05	50
6610	With shower light trim		30	.267		43	7.05		50.05	59
6620	With wall washer trim		28	.286		52.50	7.55		60.05	70.50
6630	With eye-ball trim		28	.286		52.50	7.55		60.05	70.50
6640	For direct contact with insulation, add					1.60			1.60	1.76
6700	Porcelain lamp holder	1 Elec	40	.200		3.50	5.30		8.80	12.45
6710	With pull switch		40	.200		3.84	5.30		9.14	12.80
6750	Fluorescent strip, 1-20 watt tube, wrap around diffuser, 24″		24	.333		51.50	8.80		60.30	71
6760	1-40 watt tube, 48″		24	.333		65	8.80		73.80	86
6770	2-40 watt tubes, 48″		20	.400		79	10.55		89.55	104
6780	With residential ballast		20	.400		89.50	10.55		100.05	116
6800	Bathroom heat lamp, 1-250 watt		28	.286		34.50	7.55		42.05	50.50
6810	2-250 watt lamps		28	.286		55	7.55		62.55	73
6820	For timer switch, see line 2400									
6900	Outdoor post lamp, incl. post, fixture, 35′ of #14/2									
6910	Type NMC cable	1 Elec	3.50	2.286	Ea.	181	60.50		241.50	297
6920	Photo-eye, add		27	.296		29	7.80		36.80	45
6950	Clock dial time switch, 24 hr., w/enclosure, type NM cable		11.43	.700		52.50	18.50		71	87.50
6970	Type MC cable		11	.727		63.50	19.20		82.70	101
6980	EMT & wire		4.85	1.649		63.50	43.50		107	141
7000	Alarm systems									
7050	Smoke detectors, box, #14/3, type NM cable	1 Elec	14.55	.550	Ea.	28	14.50		42.50	54
7070	Type MC cable		12.31	.650		38.50	17.15		55.65	70
7080	EMT & wire		5	1.600		35.50	42		77.50	108

16139 | Residential Wiring

			CREW	DAILY OUTPUT	LABOR-HOURS	UNIT	2004 BARE COSTS MAT.	LABOR	EQUIP.	TOTAL	TOTAL INCL O&P	
700	7090	For relay output to security system, add				Ea.	11.75			11.75	12.95	700
	8000	Residential equipment										
	8050	Disposal hook-up, incl. switch, outlet box, 3′ of flex										
	8060	20 amp-1 pole circ. bkr., and 25′ of #12/2										
	8070	Type NM cable	1 Elec	10	.800	Ea.	19.60	21		40.60	56	
	8080	Type MC cable		8	1		32	26.50		58.50	78	
	8090	EMT & wire	↓	5	1.600	↓	32.50	42		74.50	104	
	8100	Trash compactor or dishwasher hook-up, incl. outlet box,										
	8110	3′ of flex, 15 amp-1 pole circ. bkr., and 25′ of #14/2										
	8130	Type MC cable	1 Elec	8	1	Ea.	27.50	26.50		54	73	
	8140	EMT & wire	"	5	1.600	"	27	42		69	98	
	8150	Hot water sink dispensor hook-up, use line 8100										
	8200	Vent/exhaust fan hook-up, type NM cable	1 Elec	32	.250	Ea.	3.31	6.60		9.91	14.40	
	8220	Type MC cable		24	.333		12.65	8.80		21.45	28.50	
	8230	EMT & wire	↓	10	.800	↓	11.45	21		32.45	47	
	8250	Bathroom vent fan, 50 CFM (use with above hook-up)										
	8260	Economy model	1 Elec	15	.533	Ea.	21.50	14.10		35.60	46.50	
	8270	Low noise model		15	.533		30	14.10		44.10	56	
	8280	Custom model	↓	12	.667	↓	111	17.60		128.60	151	
	8300	Bathroom or kitchen vent fan, 110 CFM										
	8310	Economy model	1 Elec	15	.533	Ea.	56.50	14.10		70.60	85	
	8320	Low noise model	"	15	.533	"	75	14.10		89.10	106	
	8350	Paddle fan, variable speed (w/o lights)										
	8360	Economy model (AC motor)	1 Elec	10	.800	Ea.	100	21		121	145	
	8370	Custom model (AC motor)		10	.800		173	21		194	225	
	8380	Luxury model (DC motor)		8	1		340	26.50		366.50	420	
	8390	Remote speed switch for above, add	↓	12	.667	↓	24.50	17.60		42.10	55.50	
	8500	Whole house exhaust fan, ceiling mount, 36″, variable speed										
	8510	Remote switch, incl. shutters, 20 amp-1 pole circ. bkr.										
	8520	30′ of #12/2, type NM cable	1 Elec	4	2	Ea.	610	53		663	755	
	8530	Type MC cable		3.50	2.286		625	60.50		685.50	785	
	8540	EMT & wire	↓	3	2.667	↓	625	70.50		695.50	805	
	8600	Whirlpool tub hook-up, incl. timer switch, outlet box										
	8610	3′ of flex, 20 amp-1 pole GFI circ. bkr.										
	8620	30′ of #12/2, type NM cable	1 Elec	5	1.600	Ea.	79.50	42		121.50	156	
	8630	Type MC cable		4.20	1.905		88.50	50.50		139	179	
	8640	EMT & wire	↓	3.40	2.353	↓	89.50	62		151.50	199	
	8650	Hot water heater hook-up, incl. 1-2 pole circ. bkr., box;										
	8660	3′ of flex, 20′ of #10/2, type NM cable	1 Elec	5	1.600	Ea.	20.50	42		62.50	91	
	8670	Type MC cable		4.20	1.905		36.50	50.50		87	122	
	8680	EMT & wire	↓	3.40	2.353	↓	28.50	62		90.50	133	
	9000	Heating/air conditioning										
	9050	Furnace/boiler hook-up, incl. firestat, local on-off switch										
	9060	Emergency switch, and 40′ of type NM cable	1 Elec	4	2	Ea.	42.50	53		95.50	133	
	9070	Type MC cable		3.50	2.286		62	60.50		122.50	166	
	9080	EMT & wire	↓	1.50	5.333	↓	61	141		202	297	
	9100	Air conditioner hook-up, incl. local 60 amp disc. switch										
	9110	3′ sealtite, 40 amp, 2 pole circuit breaker										
	9130	40′ of #8/2, type NM cable	1 Elec	3.50	2.286	Ea.	142	60.50		202.50	254	
	9140	Type MC cable		3	2.667		199	70.50		269.50	335	
	9150	EMT & wire	↓	1.30	6.154	↓	152	162		314	430	
	9200	Heat pump hook-up, 1-40 & 1-100 amp 2 pole circ. bkr.										
	9210	Local disconnect switch, 3′ sealtite										
	9220	40′ of #8/2 & 30′ of #3/2										
	9230	Type NM cable	1 Elec	1.30	6.154	Ea.	325	162		487	620	
	9240	Type MC cable	↓	1.08	7.407	↓	500	196		696	870	

16100 | Wiring Methods

16139 | Residential Wiring

			Crew	Daily Output	Labor-Hours	Unit	2004 Bare Costs Mat.	Labor	Equip.	Total	Total Incl O&P
700	9250	EMT & wire	1 Elec	.94	8.511	Ea.	335	225		560	735
	9500	Thermostat hook-up, using low voltage wire									
	9520	Heating only	1 Elec	24	.333	Ea.	4.74	8.80		13.54	19.50
	9530	Heating/cooling	″	20	.400	″	5.70	10.55		16.25	23.50

16140 | Wiring Devices

			Crew	Daily Output	Labor-Hours	Unit	2004 Bare Costs Mat.	Labor	Equip.	Total	Total Incl O&P
500	0010	LOW VOLTAGE SWITCHING									
	3600	Relays, 120 V or 277 V standard	1 Elec	12	.667	Ea.	26	17.60		43.60	57
	3800	Flush switch, standard		40	.200		9.05	5.30		14.35	18.55
	4000	Interchangeable		40	.200		11.80	5.30		17.10	21.50
	4100	Surface switch, standard		40	.200		6.60	5.30		11.90	15.90
	4200	Transformer 115 V to 25 V		12	.667		93	17.60		110.60	131
	4400	Master control, 12 circuit, manual		4	2		94	53		147	189
	4500	25 circuit, motorized		4	2		102	53		155	198
	4600	Rectifier, silicon		12	.667		30.50	17.60		48.10	62
	4800	Switchplates, 1 gang, 1, 2 or 3 switch, plastic		80	.100		3	2.64		5.64	7.60
	5000	Stainless steel		80	.100		8.10	2.64		10.74	13.20
	5400	2 gang, 3 switch, stainless steel		53	.151		15.65	3.98		19.63	23.50
	5500	4 switch, plastic		53	.151		6.70	3.98		10.68	13.85
	5600	2 gang, 4 switch, stainless steel		53	.151		16.35	3.98		20.33	24.50
	5700	6 switch, stainless steel		53	.151		36	3.98		39.98	46
	5800	3 gang, 9 switch, stainless steel		32	.250		50	6.60		56.60	66
910	0010	WIRING DEVICES									
	0200	Toggle switch, quiet type, single pole, 15 amp	1 Elec	40	.200	Ea.	4.69	5.30		9.99	13.75
	0600	3 way, 15 amp		23	.348		6.70	9.20		15.90	22.50
	0900	4 way, 15 amp		15	.533		20	14.10		34.10	45
	1650	Dimmer switch, 120 volt, incandescent, 600 watt, 1 pole		16	.500		10.80	13.20		24	33.50
	2460	Receptacle, duplex, 120 volt, grounded, 15 amp		40	.200		1.14	5.30		6.44	9.85
	2470	20 amp		27	.296		7.05	7.80		14.85	20.50
	2490	Dryer, 30 amp		15	.533		10.35	14.10		24.45	34.50
	2500	Range, 50 amp		11	.727		10.75	19.20		29.95	43
	2600	Wall plates, stainless steel, 1 gang		80	.100		1.80	2.64		4.44	6.30
	2800	2 gang		53	.151		4.10	3.98		8.08	11
	3200	Lampholder, keyless		26	.308		9.20	8.10		17.30	23.50
	3400	Pullchain with receptacle		22	.364		8.90	9.60		18.50	25.50

16150 | Wiring Connections

			Crew	Daily Output	Labor-Hours	Unit	2004 Bare Costs Mat.	Labor	Equip.	Total	Total Incl O&P
275	0010	MOTOR CONNECTIONS									
	0020	Flexible conduit and fittings, 115 volt, 1 phase, up to 1 HP motor	1 Elec	8	1	Ea.	4.34	26.50		30.84	48

16200 | Electrical Power

16210 | Electrical Utility Services

			Crew	Daily Output	Labor-Hours	Unit	2004 Bare Costs Mat.	Labor	Equip.	Total	Total Incl O&P
600	0010	METER CENTERS AND SOCKETS									
	0100	Sockets, single position, 4 terminal, 100 amp	1 Elec	3.20	2.500	Ea.	31.50	66		97.50	142
	0200	150 amp		2.30	3.478		35.50	92		127.50	188
	0300	200 amp		1.90	4.211		47.50	111		158.50	233
	0500	Double position, 4 terminal, 100 amp		2.80	2.857		124	75.50		199.50	260
	0600	150 amp		2.10	3.810		141	101		242	320

16200 | Electrical Power

	16210 | Electrical Utility Services		CREW	DAILY OUTPUT	LABOR-HOURS	UNIT	2004 BARE COSTS MAT.	LABOR	EQUIP.	TOTAL	TOTAL INCL O&P	
600	0700	200 amp	1 Elec	1.70	4.706	Ea.	300	124		424	535	600
	2590	Basic meter device										
	2600	1P 3W 120/240V 4 jaw 125A sockets, 3 meter	2 Elec	1	16	Ea.	440	420		860	1,175	
	2620	5 meter		.80	20		660	530		1,190	1,575	
	2640	7 meter		.56	28.571		965	755		1,720	2,300	
	2660	10 meter		.48	33.333		1,325	880		2,205	2,875	
	2680	Rainproof 1P 3W 120/240V 4 jaw 125A sockets										
	2690	3 meter	2 Elec	1	16	Ea.	440	420		860	1,175	
	2710	6 meter		.60	26.667		760	705		1,465	1,975	
	2730	8 meter		.52	30.769		1,050	810		1,860	2,475	
	2750	1P 3W 120/240V 4 jaw sockets										
	2760	with 125A circuit breaker, 3 meter	2 Elec	1	16	Ea.	825	420		1,245	1,600	
	2780	5 meter		.80	20		1,300	530		1,830	2,275	
	2800	7 meter		.56	28.571		1,850	755		2,605	3,275	
	2820	10 meter		.48	33.333		2,600	880		3,480	4,275	
	2830	Rainproof 1P 3W 120/240V 4 jaw sockets										
	2840	with 125A circuit breaker, 3 meter	2 Elec	1	16	Ea.	825	420		1,245	1,600	
	2870	6 meter		.60	26.667		1,525	705		2,230	2,825	
	2890	8 meter		.52	30.769		2,075	810		2,885	3,600	
	3250	1P 3W 120/240V 4 jaw sockets										
	3260	with 200A circuit breaker, 3 meter	2 Elec	1	16	Ea.	1,225	420		1,645	2,025	
	3290	6 meter		.60	26.667		2,475	705		3,180	3,875	
	3310	8 meter		.56	28.571		3,325	755		4,080	4,900	
	3330	Rainproof 1P 3W 120/240V 4 jaw sockets										
	3350	with 200A circuit breaker, 3 meter	2 Elec	1	16	Ea.	1,225	420		1,645	2,025	
	3380	6 meter		.60	26.667		2,475	705		3,180	3,875	
	3400	8 meter		.52	30.769		3,325	810		4,135	5,000	

	16230 | Generator Assemblies		CREW	DAILY OUTPUT	LABOR-HOURS	UNIT	MAT.	LABOR	EQUIP.	TOTAL	TOTAL INCL O&P	
450	0010	**GENERATOR SET**										450
	0020	Gas or gasoline operated, includes battery,										
	0050	charger, muffler & transfer switch										
	0200	3 phase 4 wire, 277/480 volt, 7.5 kW	R-3	.83	24.096	Ea.	6,000	630	192	6,822	7,825	
	0300	11.5 kW		.71	28.169		8,500	735	224	9,459	10,800	
	0400	20 kW		.63	31.746		10,000	830	252	11,082	12,600	

16400 | Low-Voltage Distribution

	16410 | Encl Switches & Circuit Breakers		CREW	DAILY OUTPUT	LABOR-HOURS	UNIT	2004 BARE COSTS MAT.	LABOR	EQUIP.	TOTAL	TOTAL INCL O&P	
200	0010	**CIRCUIT BREAKERS** (in enclosure)										200
	0100	Enclosed (NEMA 1), 600 volt, 3 pole, 30 amp	1 Elec	3.20	2.500	Ea.	410	66		476	560	
	0200	60 amp		2.80	2.857		410	75.50		485.50	580	
	0400	100 amp		2.30	3.478		470	92		562	670	
800	0010	**SAFETY SWITCHES**										800
	0100	General duty 240 volt, 3 pole NEMA 1, fusible, 30 amp	1 Elec	3.20	2.500	Ea.	74.50	66		140.50	189	
	0200	60 amp		2.30	3.478		126	92		218	288	
	0300	100 amp		1.90	4.211		217	111		328	420	
	0400	200 amp		1.30	6.154		465	162		627	780	
	0500	400 amp	2 Elec	1.80	8.889		1,175	235		1,410	1,675	

16400 | Low-Voltage Distribution

		16410 | Encl Switches & Circuit Breakers	CREW	DAILY OUTPUT	LABOR-HOURS	UNIT	2004 BARE COSTS MAT.	LABOR	EQUIP.	TOTAL	TOTAL INCL O&P	
800	9010	Disc. switch, 600V 3 pole fusible, 30 amp, to 10 HP motor	1 Elec	3.20	2.500	Ea.	217	66		283	345	800
	9050	60 amp, to 30 HP motor		2.30	3.478		500	92		592	700	
	9070	100 amp, to 60 HP motor	↓	1.90	4.211	↓	500	111		611	730	
840	0010	TIME SWITCHES										840
	0100	Single pole, single throw, 24 hour dial	1 Elec	4	2	Ea.	82	53		135	176	
	0200	24 hour dial with reserve power		3.60	2.222		350	58.50		408.50	480	
	0300	Astronomic dial		3.60	2.222		141	58.50		199.50	251	
	0400	Astronomic dial with reserve power		3.30	2.424		455	64		519	605	
	0500	7 day calendar dial		3.30	2.424		126	64		190	243	
	0600	7 day calendar dial with reserve power		3.20	2.500		385	66		451	525	
	0700	Photo cell 2000 watt	↓	8	1	↓	15.15	26.50		41.65	59.50	

		16415 | Transfer Switches	CREW	DAILY OUTPUT	LABOR-HOURS	UNIT	MAT.	LABOR	EQUIP.	TOTAL	TOTAL INCL O&P	
600	0010	AUTOMATIC TRANSFER SWITCHES										600
	0100	Switches, enclosed 480 volt, 3 pole, 30 amp	1 Elec	2.30	3.478	Ea.	2,900	92		2,992	3,350	
	0200	60 amp	"	1.90	4.211	"	2,900	111		3,011	3,375	

		16440 | Swbds, Panels & Control Centers	CREW	DAILY OUTPUT	LABOR-HOURS	UNIT	MAT.	LABOR	EQUIP.	TOTAL	TOTAL INCL O&P	
500	0010	LOAD CENTERS (residential type)										500
	0100	3 wire, 120/240V, 1 phase, including 1 pole plug-in breakers										
	0200	100 amp main lugs, indoor, 8 circuits	1 Elec	1.40	5.714	Ea.	118	151		269	375	
	0300	12 circuits		1.20	6.667		165	176		341	470	
	0400	Rainproof, 8 circuits		1.40	5.714		142	151		293	400	
	0500	12 circuits	↓	1.20	6.667		200	176		376	505	
	0600	200 amp main lugs, indoor, 16 circuits	R-1A	1.80	8.889		272	192		464	620	
	0700	20 circuits		1.50	10.667		340	231		571	755	
	0800	24 circuits		1.30	12.308		450	266		716	930	
	1200	Rainproof, 16 circuits		1.80	8.889		325	192		517	675	
	1300	20 circuits		1.50	10.667		390	231		621	805	
	1400	24 circuits	↓	1.30	12.308	↓	580	266		846	1,075	

16500 | Lighting

		16510 | Interior Luminaires	CREW	DAILY OUTPUT	LABOR-HOURS	UNIT	2004 BARE COSTS MAT.	LABOR	EQUIP.	TOTAL	TOTAL INCL O&P	
440	0010	INTERIOR LIGHTING FIXTURES Including lamps, mounting										440
	0030	hardware and connections										
	0100	Fluorescent, C.W. lamps, troffer, recess mounted in grid, RS										
	0130	grid ceiling mount										
	0200	Acrylic lens, 1'W x 4'L, two 40 watt	1 Elec	5.70	1.404	Ea.	45	37		82	110	
	0300	2'W x 2'L, two U40 watt		5.70	1.404		48	37		85	114	
	0600	2'W x 4'L, four 40 watt	↓	4.70	1.702	↓	54.50	45		99.50	133	
	1000	Surface mounted, RS										
	1030	Acrylic lens with hinged & latched door frame										
	1100	1'W x 4'L, two 40 watt	1 Elec	7	1.143	Ea.	70	30		100	126	
	1200	2'W x 2'L, two U40 watt		7	1.143		75	30		105	132	
	1500	2'W x 4'L, four 40 watt	↓	5.30	1.509	↓	89	40		129	163	
	2100	Strip fixture										
	2200	4' long, one 40 watt RS	1 Elec	8.50	.941	Ea.	26.50	25		51.50	69.50	

16510 | Interior Luminaires

			CREW	DAILY OUTPUT	LABOR-HOURS	UNIT	2004 BARE COSTS MAT.	LABOR	EQUIP.	TOTAL	TOTAL INCL O&P	
440	2300	4′ long, two 40 watt RS	1 Elec	8	1	Ea.	28.50	26.50		55	74	440
	2600	8′ long, one 75 watt, SL	2 Elec	13.40	1.194		39.50	31.50		71	95	
	2700	8′ long, two 75 watt, SL	″	12.40	1.290		47.50	34		81.50	108	
	4450	Incandescent, high hat can, round alzak reflector, prewired										
	4470	100 watt	1 Elec	8	1	Ea.	56.50	26.50		83	105	
	4480	150 watt	″	8	1	″	81	26.50		107.50	132	
	5200	Ceiling, surface mounted, opal glass drum										
	5300	8″, one 60 watt lamp	1 Elec	10	.800	Ea.	34	21		55	72	
	5400	10″, two 60 watt lamps		8	1		38	26.50		64.50	85	
	5500	12″, four 60 watt lamps		6.70	1.194		55	31.50		86.50	112	
	6900	Mirror light, fluorescent, RS, acrylic enclosure, two 40 watt		8	1		84.50	26.50		111	136	
	6910	One 40 watt		8	1		66	26.50		92.50	116	
	6920	One 20 watt		12	.667		52	17.60		69.60	86	
800	0010	**RESIDENTIAL FIXTURES**										800
	0400	Fluorescent, interior, surface, circline, 32 watt & 40 watt	1 Elec	20	.400	Ea.	76.50	10.55		87.05	101	
	0500	2′ x 2′, two U 40 watt		8	1		93	26.50		119.50	145	
	0700	Shallow under cabinet, two 20 watt		16	.500		40.50	13.20		53.70	66.50	
	0900	Wall mounted, 4′L, one 40 watt, with baffle		10	.800		99	21		120	144	
	2000	Incandescent, exterior lantern, wall mounted, 60 watt		16	.500		31	13.20		44.20	55.50	
	2100	Post light, 150W, with 7′ post		4	2		110	53		163	207	
	2500	Lamp holder, weatherproof with 150W PAR		16	.500		16.90	13.20		30.10	40	
	2550	With reflector and guard		12	.667		52	17.60		69.60	85.50	
	2600	Interior pendent, globe with shade, 150 watt		20	.400		117	10.55		127.55	146	

16520 | Exterior Luminaires

			CREW	DAILY OUTPUT	LABOR-HOURS	UNIT	MAT.	LABOR	EQUIP.	TOTAL	TOTAL INCL O&P	
300	0010	**EXTERIOR FIXTURES** With lamps										300
	0400	Quartz, 500 watt	1 Elec	5.30	1.509	Ea.	53.50	40		93.50	124	
	1100	Wall pack, low pressure sodium, 35 watt		4	2		214	53		267	320	
	1150	55 watt		4	2		255	53		308	365	
	6420	Wood pole, 4-1/2″ x 5-1/8″, 8′ high		6	1.333		230	35		265	310	
	6440	12′ high		5.70	1.404		330	37		367	425	
	6460	20′ high		4	2		465	53		518	595	
	6500	Bollard light, lamp & ballast, 42″ high with polycarbonate lens										
	7200	Incandescent, 150 watt	1 Elec	3	2.667	Ea.	430	70.50		500.50	590	
	7380	Landscape recessed uplight, incl. housing, ballast, transformer										
	7390	& reflector										
	7420	Incandescent, 250 watt	1 Elec	5	1.600	Ea.	420	42		462	530	
	7440	Quartz, 250 watt	″	5	1.600	″	400	42		442	510	

16550 | Special Purpose Lighting

			CREW	DAILY OUTPUT	LABOR-HOURS	UNIT	MAT.	LABOR	EQUIP.	TOTAL	TOTAL INCL O&P	
820	0010	**TRACK LIGHTING**										820
	0100	8′ section	1 Elec	5.30	1.509	Ea.	62	40		102	133	
	0300	3 circuits, 4′ section		6.70	1.194		48	31.50		79.50	104	
	0400	8′ section		5.30	1.509		74	40		114	147	
	0500	12′ section		4.40	1.818		148	48		196	241	
	1000	Feed kit, surface mounting		16	.500		9.10	13.20		22.30	31.50	
	1100	End cover		24	.333		3.20	8.80		12	17.80	
	1200	Feed kit, stem mounting, 1 circuit		16	.500		25	13.20		38.20	49	
	1300	3 circuit		16	.500		25	13.20		38.20	49	
	2000	Electrical joiner, for continuous runs, 1 circuit		32	.250		11.90	6.60		18.50	24	
	2100	3 circuit		32	.250		28	6.60		34.60	41.50	
	2200	Fixtures, spotlight, 75W PAR halogen		16	.500		87	13.20		100.20	117	
	2210	50W MR16 halogen		16	.500		106	13.20		119.20	139	
	3000	Wall washer, 250 watt tungsten halogen		16	.500		101	13.20		114.20	133	

Important: See the Reference Section for critical supporting data - Reference Nos., Crews, & Location Factors

16500 | Lighting

16550 | Special Purpose Lighting

		16550 Special Purpose Lighting	CREW	DAILY OUTPUT	LABOR-HOURS	UNIT	2004 BARE COSTS MAT.	LABOR	EQUIP.	TOTAL	TOTAL INCL O&P	
820	3100	Low voltage, 25/50 watt, 1 circuit	1 Elec	16	.500	Ea.	102	13.20		115.20	134	820
	3120	3 circuit		16	.500		105	13.20		118.20	138	

16585 | Lamps

		16585 Lamps	CREW	DAILY OUTPUT	LABOR-HOURS	UNIT	MAT.	LABOR	EQUIP.	TOTAL	TOTAL INCL O&P	
600	0010	**LAMPS**										600
	0081	Fluorescent, rapid start, cool white, 2' long, 20 watt	1 Elec	100	.080	Ea.	3.01	2.11		5.12	6.75	
	0101	4' long, 40 watt		90	.089		2.73	2.35		5.08	6.80	
	1351	High pressure sodium, 70 watt		30	.267		46.50	7.05		53.55	62.50	
	1371	150 watt		30	.267		49.50	7.05		56.55	66	

16800 | Sound & Video

16820 | Sound Reinforcement

		16820 Sound Reinforcement	CREW	DAILY OUTPUT	LABOR-HOURS	UNIT	2004 BARE COSTS MAT.	LABOR	EQUIP.	TOTAL	TOTAL INCL O&P	
300	0010	**DOORBELL SYSTEM** Incl. transformer, button & signal										300
	1000	Door chimes, 2 notes, minimum	1 Elec	16	.500	Ea.	22	13.20		35.20	45.50	
	1020	Maximum		12	.667		115	17.60		132.60	156	
	1100	Tube type, 3 tube system		12	.667		163	17.60		180.60	208	
	1180	4 tube system		10	.800		261	21		282	320	
	1900	For transformer & button, minimum add		5	1.600		12.40	42		54.40	82	
	1960	Maximum, add		4.50	1.778		37	47		84	117	
	3000	For push button only, minimum		24	.333		2.42	8.80		11.22	16.95	
	3100	Maximum		20	.400		19.45	10.55		30	38.50	

16850 | Television Equipment

		16850 Television Equipment	CREW	DAILY OUTPUT	LABOR-HOURS	UNIT	MAT.	LABOR	EQUIP.	TOTAL	TOTAL INCL O&P	
600	0010	**T.V. SYSTEMS** not including rough-in wires, cables & conduits										600
	0100	Master TV antenna system										
	0200	VHF reception & distribution, 12 outlets	1 Elec	6	1.333	Outlet	155	35		190	229	
	0800	VHF & UHF reception & distribution, 12 outlets		6	1.333	"	154	35		189	227	
	5000	T.V. Antenna only, minimum		6	1.333	Ea.	34	35		69	95	
	5100	Maximum		4	2	"	143	53		196	243	

Division Notes

	CREW	DAILY OUTPUT	LABOR-HOURS	UNIT	2004 BARE COSTS MAT.	LABOR	EQUIP.	TOTAL	TOTAL INCL O&P

Reference Section

All the reference information is in one section making it easy to find what you need to know . . . and easy to use the book on a daily basis. This section is visually identified by a vertical gray bar on the edge of pages.

In the reference number information that follows, you'll see the background that relates to the "reference numbers" that appeared in the Unit Price Sections. You'll find reference tables, explanations and estimating information that support how we arrived at the unit price data. Also included are alternate pricing methods, technical data and estimating procedures along with information on design and economy in construction.

Also in this Reference Section, we've included Change Orders, information on pricing changes to contract documents; Crew Listings, a full listing of all the crews, equipment and their costs; Historical Cost Indexes for cost comparisons over time; City Cost Indexes and Location Factors for adjusting costs to the region you are in; and an explanation of all Abbreviations used in the book.

Table of Contents

Reference Numbers

Reference Numbers (cont.)

R01100-005 Tips for Accurate Estimating

1. Use pre-printed or columnar forms for orderly sequence of dimensions and locations and for recording telephone quotations.
2. Use only the front side of each paper or form except for certain pre-printed summary forms.
3. Be consistent in listing dimensions: For example, length x width x height. This helps in rechecking to ensure that, the total length of partitions is appropriate for the building area.
4. Use printed (rather than measured) dimensions where given.
5. Add up multiple printed dimensions for a single entry where possible.
6. Measure all other dimensions carefully.
7. Use each set of dimensions to calculate multiple related quantities.
8. Convert foot and inch measurements to decimal feet when listing. Memorize decimal equivalents to .01 parts of a foot (1/8″ equals approximately .01′).
9. Do not "round off" quantities until the final summary.
10. Mark drawings with different colors as items are taken off.
11. Keep similar items together, different items separate.
12. Identify location and drawing numbers to aid in future checking for completeness.
13. Measure or list everything on the drawings or mentioned in the specifications.
14. It may be necessary to list items not called for to make the job complete.
15. Be alert for: Notes on plans such as N.T.S. (not to scale); changes in scale throughout the drawings; reduced size drawings; discrepancies between the specifications and the drawings.
16. Develop a consistent pattern of performing an estimate. For example:
    a. Start the quantity takeoff at the lower floor and move to the next higher floor.
    b. Proceed from the main section of the building to the wings.
    c. Proceed from south to north or vice versa, clockwise or counterclockwise.
    d. Take off floor plan quantities first, elevations next, then detail drawings.
17. List all gross dimensions that can be either used again for different quantities, or used as a rough check of other quantities for verification (exterior perimeter, gross floor area, individual floor areas, etc.).
18. Utilize design symmetry or repetition (repetitive floors, repetitive wings, symmetrical design around a center line, similar room layouts, etc.). Note: Extreme caution is needed here so as not to omit or duplicate an area.
19. Do not convert units until the final total is obtained. For instance, when estimating concrete work, keep all units to the nearest cubic foot, then summarize and convert to cubic yards.
20. When figuring alternatives, it is best to total all items involved in the basic system, then total all items involved in the alternates. Therefore you work with positive numbers in all cases. When adds and deducts are used, it is often confusing whether to add or subtract a portion of an item; especially on a complicated or involved alternate.

R01100-040 Builder's Risk Insurance

Builder's Risk Insurance is insurance on a building during construction. Premiums are paid by the owner or the contractor. Blasting, collapse and underground insurance would raise total insurance costs above those listed. Floater policy for materials delivered to the job runs $.75 to $1.25 per $100 value. Contractor equipment insurance runs $.50 to $1.50 per $100 value. Insurance for miscellaneous tools to $1,500 value runs from $3.00 to $7.50 per $100 value.

Tabulated below are New England Builder's Risk insurance rates in dollars per $100 value for $1,000 deductible. For $25,000 deductible, rates can be reduced 13% to 34%. On contracts over $1,000,000, rates may be lower than those tabulated. Policies are written annually for the total completed value in place. For "all risk" insurance (excluding flood, earthquake and certain other perils) add $.025 to total rates below.

Coverage	Frame Construction (Class 1)		Brick Construction (Class 4)		Fire Resistive (Class 6)	
	Range	Average	Range	Average	Range	Average
Fire Insurance	$.350 to $ .850	$.600	$.158 to $.189	$.174	$.052 to $.080	$.070
Extended Coverage	.115 to .200	.158	.080 to .105	.101	.081 to .105	.100
Vandalism	.012 to .016	.014	.008 to .011	.011	.008 to .011	.010
Total Annual Rate	$.477 to $1.066	$.772	$.246 to $.305	$.286	$.141 to $.196	$.180

R01100-050 General Contractor's Overhead

There are two distinct types of overhead on a construction project: Project Overhead and Main Office Overhead. Project Overhead includes those costs at a construction site not directly associated with the installation of construction materials. Examples of Project Overhead costs include the following:

1. Superintendent
2. Construction office and storage trailers
3. Temporary sanitary facilities
4. Temporary utilities
5. Security fencing
6. Photographs
7. Clean up
8. Performance and payment bonds

The above Project Overhead items are also referred to as General Requirements and therefore are estimated in Division 1. Division 1 is the first division listed in the CSI MasterFormat but it is usually the last division estimated. The sum of the costs in Divisions 1 through 16 is referred to as the sum of the direct costs.

All construction projects also include indirect costs. The primary components of indirect costs are the contractor's Main Office Overhead and profit. The amount of the Main Office Overhead expense varies depending on the the following:

1. Owner's compensation
2. Project managers and estimator's wages
3. Clerical support wages
4. Office rent and utilities
5. Corporate legal and accounting costs
6. Advertising
7. Automobile expenses
8. Association dues
9. Travel and entertainment expenses

These costs are usually calculated as a percentage of annual sales volume. This percentage can range from 35% for a small contractor doing less than $500,000 to 5% for a large contractor with sales in excess of $100 million.

R01100-060 Workers' Compensation Insurance Rates by Trade

The table below tabulates the national averages for Workers' Compensation insurance rates by trade and type of building. The average "Insurance Rate" is multiplied by the "% of Building Cost" for each trade. This produces the "Workers' Compensation Cost" by % of total labor cost, to be added for each trade by building type to determine the weighted average Workers' Compensation rate for the building types analyzed.

Trade	Insurance Rate (% Labor Cost)		% of Building Cost			Workers' Compensation		
	Range	Average	Office Bldgs.	Schools & Apts.	Mfg.	Office Bldgs.	Schools & Apts.	Mfg.
Excavation, Grading, etc.	3.5 % to 18.5%	10.3%	4.8%	4.9%	4.5%	.49%	.50%	.46%
Piles & Foundations	7.3 to 76.7	22.9	7.1	5.2	8.7	1.63	1.19	1.99
Concrete	5.7 to 35.2	15.8	5.0	14.8	3.7	.79	2.34	.58
Masonry	5.1 to 31.0	15.0	6.9	7.5	1.9	1.04	1.13	.29
Structural Steel	7.2 to 112.0	38.9	10.7	3.9	17.6	4.16	1.52	6.85
Miscellaneous & Ornamental Metals	5.2 to 25.4	12.6	2.8	4.0	3.6	.35	.50	.45
Carpentry & Millwork	6.7 to 53.2	18.5	3.7	4.0	0.5	.68	.74	.09
Metal or Composition Siding	5.3 to 35.5	16.1	2.3	0.3	4.3	.37	.05	.69
Roofing	7.3 to 77.1	31.8	2.3	2.6	3.1	.73	.83	.99
Doors & Hardware	4.5 to 25.3	10.9	0.9	1.4	0.4	.10	.15	.04
Sash & Glazing	4.9 to 38.0	13.8	3.5	4.0	1.0	.48	.55	.14
Lath & Plaster	4.0 to 45.5	14.6	3.3	6.9	0.8	.48	1.01	.12
Tile, Marble & Floors	3.7 to 23.3	9.6	2.6	3.0	0.5	.25	.29	.05
Acoustical Ceilings	2.5 to 24.5	10.7	2.4	0.2	0.3	.26	.02	.03
Painting	4.3 to 29.6	12.9	1.5	1.6	1.6	.19	.21	.21
Interior Partitions	6.7 to 53.2	18.5	3.9	4.3	4.4	.72	.80	.81
Miscellaneous Items	2.5 to 110.1	17.3	5.2	3.7	9.7	.90	.64	1.68
Elevators	2.5 to 15.3	7.2	2.1	1.1	2.2	.15	.08	.16
Sprinklers	2.8 to 23.1	9.0	0.5	—	2.0	.05	—	.18
Plumbing	3.0 to 12.5	7.8	4.9	7.2	5.2	.38	.56	.41
Heat., Vent., Air Conditioning	4.0 to 28.1	11.1	13.5	11.0	12.9	1.50	1.22	1.43
Electrical	2.7 to 12.5	6.4	10.1	8.4	11.1	.65	.54	.71
Total	2.5 % to 110.1%	—	100.0%	100.0%	100.0%	16.35%	14.87%	18.36%
		Overall Weighted Average	16.53%					

Workers' Compensation Insurance Rates by States

The table below lists the weighted average Workers' Compensation base rate for each state with a factor comparing this with the national average of 16.2%.

State	Weighted Average	Factor	State	Weighted Average	Factor	State	Weighted Average	Factor
Alabama	28.0%	173	Kentucky	16.6%	102	North Dakota	12.9%	80
Alaska	17.9	110	Louisiana	28.2	174	Ohio	12.7	78
Arizona	7.1	44	Maine	20.8	128	Oklahoma	22.2	137
Arkansas	14.7	91	Maryland	12.0	74	Oregon	14.3	88
California	18.8	116	Massachusetts	14.6	90	Pennsylvania	16.0	99
Colorado	14.0	86	Michigan	19.0	117	Rhode Island	21.2	131
Connecticut	24.9	154	Minnesota	27.7	171	South Carolina	13.9	86
Delaware	11.9	73	Mississippi	17.0	105	South Dakota	14.5	90
District of Columbia	19.1	118	Missouri	20.7	128	Tennessee	16.4	101
Florida	31.4	194	Montana	20.6	127	Texas	14.6	90
Georgia	23.0	142	Nebraska	20.1	124	Utah	12.3	76
Hawaii	16.9	104	Nevada	16.2	100	Vermont	20.0	123
Idaho	10.4	64	New Hampshire	22.6	140	Virginia	12.5	77
Illinois	18.2	112	New Jersey	10.6	65	Washington	10.6	65
Indiana	6.2	38	New Mexico	15.2	94	West Virginia	12.8	79
Iowa	12.0	74	New York	13.5	83	Wisconsin	15.9	98
Kansas	8.8	54	North Carolina	13.7	85	Wyoming	8.0	49
			Weighted Average for U.S. is	16.5% of payroll = 100%				

Rates in the following table are the base or manual costs per $100 of payroll for Workers' Compensation in each state. Rates are usually applied to straight time wages only and not to premium time wages and bonuses.

The weighted average skilled worker rate for 35 trades is 16.2%. For bidding purposes, apply the full value of Workers' Compensation directly to total labor costs, or if labor is 38%, materials 42% and overhead and profit 20% of total cost, carry 38/80 x 16.2% =7.7% of cost (before overhead and profit) into overhead. Rates vary not only from state to state but also with the experience rating of the contractor.

Rates are the most current available at the time of publication.

R01100-060 Workers' Compensation Insurance Rates by Trade and State (cont.)

State	Carpentry — 3 stories or less 5651	Carpentry — interior cab. work 5437	Carpentry — general 5403	Concrete Work — NOC 5213	Concrete Work — flat (flr., sdwk.) 5221	Electrical Wiring — inside 5190	Excavation — earth NOC 6217	Excavation — rock 6217	Glaziers 5462	Insulation Work 5479	Lathing 5443	Masonry 5022	Painting & Decorating 5474	Pile Driving 6003	Plastering 5480	Plumbing 5183	Roofing 5551	Sheet Metal Work (HVAC) 5538	Steel Erection — door & sash 5102	Steel Erection — inter., ornam. 5102	Steel Erection — structure 5040	Steel Erection — NOC 5057	Tile Work — (interior ceramic) 5348	Waterproofing 9014	Wrecking 5701
AL	23.03	13.25	33.49	13.23	10.32	9.17	15.14	15.14	38.02	19.34	13.00	30.97	28.83	29.60	45.45	11.88	68.79	28.14	25.40	25.40	44.99	44.16	15.68	7.34	44.99
AK	12.28	13.50	11.97	11.89	9.38	9.65	18.53	18.53	16.81	21.09	8.51	16.75	13.92	51.57	13.48	8.75	36.97	9.15	11.46	11.46	29.37	18.17	8.49	8.15	29.37
AZ	6.40	4.59	13.14	6.28	3.94	3.56	5.17	5.17	6.37	9.71	4.46	6.18	4.25	8.37	5.21	3.64	11.20	4.75	7.99	7.99	12.75	6.86	3.76	2.71	50.31
AR	12.83	10.62	14.66	14.60	6.75	6.24	9.07	9.07	15.62	26.86	10.49	10.75	9.85	13.56	13.96	6.36	22.13	10.19	8.32	8.32	37.69	30.03	7.13	4.00	37.69
CA	28.28	9.07	28.28	13.18	13.18	9.93	7.88	7.88	16.37	23.34	10.90	14.55	19.45	21.18	18.12	11.59	39.79	15.33	13.55	13.55	23.93	21.91	7.74	19.45	21.91
CO	17.74	9.16	12.15	13.17	8.11	5.71	11.90	11.90	10.18	15.27	5.87	14.10	11.24	19.75	11.07	8.33	25.02	13.10	8.13	8.13	34.92	15.33	8.02	6.34	34.92
CT	21.93	17.69	29.90	28.54	17.17	9.55	12.88	12.88	18.03	32.51	15.16	28.12	18.82	29.36	25.85	11.47	52.53	14.89	16.46	16.46	70.71	21.51	13.29	7.07	50.86
DE	11.48	11.48	9.88	9.99	7.61	5.06	7.82	7.82	9.69	9.88	10.52	9.25	12.70	16.84	10.52	6.05	21.69	8.41	10.11	10.11	23.77	10.11	8.23	9.25	23.37
DC	12.49	11.98	16.79	21.64	21.97	7.25	8.71	8.71	31.70	12.33	8.71	19.06	9.35	23.54	15.51	11.65	25.17	8.78	17.99	17.99	54.08	22.86	23.27	3.99	54.08
FL	35.52	25.27	36.16	35.18	17.49	12.49	16.03	16.03	27.03	25.63	15.21	27.00	24.86	68.52	42.18	12.45	53.69	21.04	17.19	17.19	67.47	46.35	12.40	10.76	67.47
GA	32.80	16.96	24.66	16.82	12.15	9.13	14.76	14.76	18.12	19.52	19.99	21.28	18.47	30.93	20.75	11.27	40.25	15.09	14.49	14.49	38.29	55.91	10.34	9.92	38.29
HI	17.38	11.62	28.20	13.74	12.27	7.10	7.73	7.73	20.55	21.19	10.94	17.87	11.33	20.15	15.61	6.06	33.24	8.02	11.11	11.11	31.71	21.70	9.78	11.54	31.71
ID	9.86	5.74	13.99	9.13	6.98	4.97	5.36	5.36	9.28	7.65	4.97	8.01	7.10	11.84	8.68	3.92	27.36	8.02	9.87	9.87	26.44	12.60	4.78	4.98	26.44
IL	15.59	13.15	18.76	27.06	10.81	8.23	9.43	9.43	15.78	13.70	9.70	16.83	10.09	25.35	13.11	9.87	29.38	14.82	15.55	15.55	50.68	23.78	15.66	4.69	50.68
IN	5.29	4.52	6.96	5.65	3.48	2.67	3.50	3.50	4.94	3.74	2.67	5.12	4.84	7.38	4.04	2.97	11.69	4.41	5.18	5.18	21.38	10.19	3.69	2.46	21.38
IA	10.40	5.30	11.50	12.21	6.38	3.96	6.30	6.30	13.38	8.84	5.29	7.81	8.26	9.58	8.44	4.85	17.87	6.26	10.24	10.24	48.08	29.00	6.20	4.52	29.58
KS	10.94	7.55	10.15	7.52	5.93	3.67	5.22	5.22	6.91	8.78	4.57	7.23	6.46	8.35	9.08	4.91	18.79	6.05	7.00	7.00	20.07	12.71	4.25	3.02	20.07
KY	17.33	12.77	19.21	16.77	6.05	6.85	17.98	17.98	17.33	19.31	11.00	10.81	10.91	19.93	17.05	5.48	17.33	13.46	13.29	13.29	36.48	24.51	12.79	3.95	36.48
LA	23.14	24.93	53.17	26.43	15.61	9.88	17.49	17.49	20.14	21.43	24.51	28.33	29.58	31.25	22.37	8.64	77.12	21.20	18.98	18.98	51.76	24.21	13.77	13.78	66.41
ME	13.43	10.42	43.59	24.29	11.12	5.03	12.19	12.19	13.35	15.39	14.03	17.26	16.07	31.44	17.95	7.32	32.07	8.82	15.08	15.08	37.84	61.81	11.66	6.06	37.84
MD	10.55	5.95	10.55	11.35	5.05	5.15	9.25	9.25	13.20	13.05	6.45	11.35	6.75	27.45	6.65	5.55	22.80	7.00	9.15	9.15	26.80	18.50	6.35	3.50	26.80
MA	10.62	6.96	16.60	17.84	9.34	3.69	6.41	6.41	8.63	14.02	6.80	14.44	8.26	14.80	5.69	5.09	33.29	8.79	14.31	14.31	40.78	35.81	11.47	3.86	38.14
MI	21.68	13.59	19.80	22.12	9.35	5.71	11.69	11.69	13.87	12.38	13.59	19.33	14.93	39.06	15.59	8.24	41.33	10.41	11.36	11.36	39.06	29.15	11.31	6.40	39.06
MN	20.92	21.58	36.74	16.39	13.64	6.79	15.72	15.72	18.94	22.75	23.83	25.13	18.89	29.83	23.83	10.56	65.54	10.92	16.66	16.66	112.03	36.34	17.83	6.43	132.92
MS	15.88	14.44	19.71	11.45	9.13	7.19	10.97	10.97	11.49	12.48	7.98	13.65	12.76	23.19	17.21	7.92	31.71	19.53	10.85	10.85	42.19	33.21	9.10	7.01	42.19
MO	18.67	12.07	17.22	17.39	12.13	7.88	11.53	11.53	13.12	21.70	14.67	18.39	15.05	25.80	16.18	9.66	36.85	13.49	14.36	14.36	69.65	43.83	6.53	8.15	69.65
MT	22.03	10.06	20.00	12.80	10.83	5.54	17.20	17.20	11.51	16.19	19.02	14.58	12.03	76.71	13.82	8.93	59.20	9.44	9.73	9.73	41.23	17.93	6.54	5.33	41.23
NE	21.22	10.97	20.95	28.60	9.65	8.32	12.20	12.20	15.50	24.47	11.30	18.50	12.32	23.35	14.75	11.17	36.72	13.05	13.15	13.15	52.27	38.35	9.97	6.15	50.40
NV	17.85	9.06	13.89	10.87	10.44	7.67	11.47	11.47	14.46	17.06	7.20	12.16	11.24	13.14	12.28	10.71	23.68	20.06	16.04	16.04	36.06	33.10	9.57	7.25	44.29
NH	16.93	11.48	21.49	32.83	11.47	6.81	17.55	17.55	12.44	28.41	8.80	21.86	15.11	17.52	21.89	11.37	61.69	12.49	14.85	14.85	59.93	32.72	14.21	6.54	59.93
NJ	11.19	7.77	11.19	9.15	6.78	3.85	6.97	6.97	7.19	10.59	9.00	12.02	8.97	13.42	9.00	5.75	28.36	6.67	9.38	9.38	17.22	10.28	4.46	4.54	23.36
NM	22.18	6.84	14.66	12.96	8.24	6.41	8.27	8.27	12.43	11.53	6.24	12.92	10.22	16.44	9.38	7.70	28.79	10.65	19.01	19.01	43.96	21.54	5.89	6.30	43.96
NY	13.46	6.24	14.87	17.22	11.81	5.86	8.25	8.25	10.46	8.73	15.59	17.70	12.38	15.54	8.98	7.46	28.94	14.33	8.80	8.80	14.52	20.59	8.83	6.21	29.99
NC	13.68	10.74	18.04	13.14	6.63	7.71	8.44	8.44	10.19	11.90	7.68	9.44	9.58	15.60	15.16	8.08	25.69	10.77	8.00	8.00	41.99	18.79	7.12	4.14	41.99
ND	9.60	9.60	19.60	5.81	5.81	4.00	5.39	5.39	9.60	9.60	9.04	8.48	6.30	20.12	9.04	5.35	21.53	5.35	20.12	20.12	20.12	20.12	9.60	21.53	14.48
OH	6.07	12.00	9.68	12.78	9.99	5.55	8.44	8.44	9.99	12.41	2.48	11.45	15.04	25.09	4.63	6.40	21.67	8.11	10.30	10.30	26.21	16.39	8.70	15.16	26.21
OK	26.16	11.79	19.33	15.51	11.50	7.41	18.45	18.45	13.55	18.00	12.05	16.80	17.10	36.26	18.29	8.72	42.26	12.45	11.17	11.17	69.45	46.76	11.42	8.20	69.45
OR	17.06	9.31	16.98	13.61	8.81	5.00	10.55	10.55	16.83	8.87	7.87	15.13	14.15	16.07	11.45	6.07	22.99	11.17	10.25	10.25	35.39	16.67	10.80	5.00	35.39
PA	8.69	8.69	12.27	13.48	8.94	6.46	7.91	7.91	15.69	12.63	18.94	11.65	13.04	16.76	10.78	7.18	25.87	12.58	13.96	13.96	59.05	13.96	8.22	18.96	81.34
RI	19.53	11.65	18.07	18.23	16.24	4.43	10.38	10.38	12.85	22.78	11.97	25.11	24.13	37.66	17.25	8.52	33.92	10.29	14.07	14.07	59.49	37.50	14.36	7.70	78.79
SC	19.10	13.03	19.24	12.88	6.19	7.02	8.32	8.32	12.22	10.69	7.27	9.55	11.19	16.64	18.89	6.88	28.98	11.42	9.49	9.49	21.80	23.66	6.23	4.65	21.80
SD	15.40	7.40	15.04	18.64	5.48	5.64	11.42	11.42	9.88	13.70	7.72	8.28	13.08	25.78	11.65	8.56	20.66	8.33	11.93	11.93	45.92	17.38	6.33	4.17	45.92
TN	26.71	12.31	20.38	16.92	8.67	7.33	12.26	12.26	10.17	12.94	10.97	14.30	12.72	15.20	15.23	9.52	33.46	12.19	9.28	9.28	39.46	21.96	7.22	5.73	39.46
TX	16.89	11.33	13.25	12.10	8.85	7.36	10.49	10.49	11.61	15.68	9.53	13.68	9.89	13.95	20.99	8.06	22.80	14.58	11.23	11.23	33.39	14.78	7.29	8.09	17.86
UT	9.45	6.23	10.52	7.68	7.99	7.05	6.38	6.38	9.55	10.95	12.32	13.75	15.29	17.24	10.60	6.41	26.27	6.30	8.99	8.99	23.51	23.51	6.06	5.83	26.53
VT	19.90	11.34	19.46	34.13	12.01	6.10	10.61	10.61	21.25	25.54	10.85	19.53	10.41	23.91	18.23	10.10	28.73	12.62	13.94	13.94	47.05	37.23	8.52	9.59	47.05
VA	11.10	8.05	11.35	12.89	6.35	4.15	7.42	7.42	7.80	7.18	12.99	8.00	10.15	16.73	7.63	6.04	22.40	7.96	13.93	13.93	35.15	20.77	9.28	3.45	35.15
WA	7.57	7.57	8.66	6.22	7.51	2.85	8.31	8.31	13.01	7.88	8.66	12.82	9.07	18.93	10.38	4.68	19.63	4.02	12.96	12.96	9.25	9.25	9.89	10.23	9.25
WV	12.00	12.00	12.00	23.80	23.80	5.13	7.74	7.74	6.08	6.08	16.88	11.57	12.20	12.89	12.20	4.77	13.44	6.08	14.98	14.98	9.84	13.15	16.02	3.88	13.27
WI	11.29	9.54	18.46	11.96	10.19	4.78	7.16	7.16	14.43	14.33	10.61	17.44	11.44	19.25	12.41	6.40	40.73	7.27	16.38	16.38	39.18	19.60	13.98	5.38	39.18
WY	7.29	7.29	7.29	7.29	7.29	7.29	7.29	7.29	7.29	7.29	7.29	7.29	7.29	7.29	7.29	7.29	7.29	7.29	7.29	7.29	7.29	7.29	7.29	7.29	7.29
AVG.	16.06	10.91	18.51	15.79	9.94	6.40	10.34	10.34	13.82	15.24	10.71	14.97	12.89	22.94	14.62	7.78	31.75	11.09	12.61	12.61	38.86	24.78	9.63	7.27	40.51

R01100-060 Workers' Compensation (cont.) (Canada in Canadian dollars)

Province		Alberta	British Columbia	Manitoba	Ontario	New Brunswick	Newfndld. & Labrador	Northwest Territories	Nova Scotia	Prince Edward Island	Quebec	Saskat-chewan	Yukon
Carpentry—3 stories or less	Rate	2.72	4.63	3.80	5.00	4.40	7.20	4.62	8.06	7.25	12.69	5.01	3.25
	Code	25401	721028	40102	723	422	403	4-41	4226	401	80110	B12-02	202
Carpentry—interior cab. work	Rate	1.17	4.85	3.80	5.00	3.48	7.20	4.62	5.69	3.49	12.69	3.25	3.25
	Code	42133	721021	40102	723	427	403	4-41	4274	402	80110	B11-27	202
CARPENTRY—general	Rate	2.72	4.63	3.80	5.00	4.40	7.20	4.62	8.06	7.25	12.69	5.01	3.25
	Code	25401	721028	40102	723	422	403	4-41	4226	401	80110	B12-02	202
CONCRETE WORK—NOC	Rate	7.65	6.48	5.76	17.18	4.40	7.20	4.62	4.97	7.25	13.89	6.50	3.25
	Code	42104	721010	40110	748	422	403	4-41	4224	401	80100	B13-14	203
CONCRETE WORK—flat (flr. sidewalk)	Rate	7.65	6.48	5.76	17.18	4.40	7.20	4.62	4.97	7.25	13.89	6.50	3.25
	Code	42104	721010	40110	748	422	403	4-41	4224	401	80100	B13-14	203
ELECTRICAL Wiring—inside	Rate	2.98	2.90	2.03	3.03	1.74	4.83	3.46	2.61	3.49	5.69	3.25	2.35
	Code	42124	721019	40203	704	426	400	4-46	4261	402	80170	B11-05	206
EXCAVATION—earth NOC	Rate	2.57	3.40	3.80	4.21	2.93	7.20	3.46	4.25	3.61	7.71	3.70	3.25
	Code	40604	721031	40706	711	421	403	4-43	4214	404	80030	R11-06	207
EXCAVATION—rock	Rate	2.57	3.40	3.80	4.21	2.93	7.20	3.46	4.25	3.61	7.71	3.70	3.25
	Code	40604	721031	40706	711	421	403	4-43	4214	404	80030	R11-06	207
GLAZIERS	Rate	2.79	3.21	3.77	8.42	6.01	5.12	4.62	8.06	3.49	14.40	6.50	2.35
	Code	42121	715020	40109	751	423	402	4-41	4233	402	80150	B13-04	212
INSULATION WORK	Rate	2.89	5.80	3.80	8.42	6.01	5.12	4.62	8.06	7.25	14.15	5.01	3.25
	Code	42184	721029	40102	751	423	402	4-41	4234	401	80120	B12-07	202
LATHING	Rate	8.79	8.29	3.80	5.00	3.48	5.12	4.62	5.69	3.49	14.15	6.50	3.25
	Code	42135	721033	40102	723	427	402	4-41	4271	402	80120	B13-16	202
MASONRY	Rate	7.65	8.29	3.80	12.36	6.01	7.20	4.62	8.06	7.25	13.89	6.50	3.25
	Code	42102	721037	40102	741	423	403	4-41	4231	401	80100	B13-18	202
PAINTING & DECORATING	Rate	4.07	5.80	4.20	7.09	3.28	5.12	4.62	5.69	3.49	12.69	5.01	3.25
	Code	42111	721041	40105	719	427	402	4-41	4275	402	80110	B12-01	202
PILE DRIVING	Rate	7.65	13.55	3.80	5.84	4.40	10.89	3.46	4.97	7.25	7.71	6.50	3.25
	Code	42159	722004	40706	732	422	404	4-43	4221	401	80030	B13-10	202
PLASTERING	Rate	8.79	8.29	4.46	7.09	3.28	5.12	4.62	5.69	3.49	12.69	5.01	3.25
	Code	42135	721042	40108	719	427	402	4-41	4271	402	80110	B12-21	202
PLUMBING	Rate	1.99	4.30	2.37	3.96	2.30	4.60	3.46	2.94	3.49	7.13	3.25	1.40
	Code	42122	721043	40204	707	424	401	4-46	4241	402	80160	B11-01	214
ROOFING	Rate	9.70	8.74	5.42	12.36	8.05	7.20	4.62	8.06	7.25	22.17	6.50	3.25
	Code	42118	721036	40403	728	430	403	4-41	4236	401	80130	B13-20	202
SHEET METAL WORK (HVAC)	Rate	1.99	4.30	5.14	3.96	2.30	4.60	3.46	2.94	3.49	7.13	3.25	2.35
	Code	42117	721043	40402	707	424	401	4-46	4244	402	80160	B11-07	208
STEEL ERECTION—door & sash	Rate	3.84	13.55	6.56	17.18	4.40	10.89	4.62	8.06	7.25	32.47	6.50	3.25
	Code	42106	722005	40502	748	422	404	4-41	4227	401	80080	B13-22	202
STEEL ERECTION—inter., ornam.	Rate	3.84	13.55	6.56	17.18	4.40	7.20	4.62	8.06	7.25	32.47	6.50	3.25
	Code	42106	722005	40502	748	422	403	4-41	4227	401	80080	B13-22	202
STEEL ERECTION—structure	Rate	3.84	13.55	6.56	17.18	4.40	10.89	4.62	8.06	7.25	32.47	6.50	3.25
	Code	42106	722005	40502	748	422	404	4-41	4227	401	80080	B13-22	202
STEEL ERECTION—NOC	Rate	3.84	13.55	6.56	17.18	4.40	10.89	4.62	8.06	7.25	32.47	6.50	3.25
	Code	42106	722005	40502	748	422	404	4-41	4227	401	80080	B13-22	202
TILE WORK—inter. (ceramic)	Rate	2.94	3.27	1.88	7.09	3.28	5.12	4.62	5.69	3.49	12.69	6.50	3.25
	Code	42113	721054	40103	719	427	402	4-41	4276	402	80110	B13-01	202
WATERPROOFING	Rate	4.07	5.80	3.80	5.00	6.01	7.20	4.62	8.06	3.49	22.17	5.01	3.25
	Code	42139	721016	40102	723	423	403	4-41	4239	402	80130	B12-17	202
WRECKING	Rate	2.57	6.08	5.15	17.18	2.93	7.20	3.46	4.25	7.25	12.69	6.50	3.25
	Code	40604	721005	40106	748	421	403	4-43	4211	401	80110	B13-09	202

R01100-070 Contractor's Overhead & Profit

Below are the **average** installing contractor's percentage mark-ups applied to base labor rates to arrive at typical billing rates.

Column A: Labor rates are based on average open shop wages for 7 major U.S. regions. Base rates including fringe benefits are listed hourly and daily. These figures are the sum of the wage rate and employer-paid fringe benefits such as vacation pay, and employer-paid health costs.

Column B: Workers' Compensation rates are the national average of state rates established for each trade.

Column C: Column C lists average fixed overhead figures for all trades. Included are Federal and State Unemployment costs set at 6.2%; Social Security Taxes (FICA) set at 7.65%; Builder's Risk Insurance costs set at 0.44%; and Public Liability costs set at 2.02%. All the percentages except those for Social Security Taxes vary from state to state as well as from company to company.

Columns D and E: Percentages in Columns D and E are based on the presumption that the installing contractor has annual billing of $2,000,000 and up. Overhead percentages may increase with smaller annual billing. The overhead percentages for any given contractor may vary greatly and depend on a number of factors, such as the contractor's annual volume, engineering and logistical support costs, and staff requirements. The figures for overhead and profit will also vary depending on the type of job, the job location, and the prevailing economic conditions. All factors should be examined very carefully for each job.

Column F: Column F lists the total of Columns B, C, D, and E.

Column G: Column G is Column A (hourly base labor rate) multiplied by the percentage in Column F (O&P percentage).

Column H: Column H is the total of Column A (hourly base labor rate) plus Column G (Total O&P).

Column I: Column I is Column H multiplied by eight hours.

		A		B	C	D	E	F	G	H	I
		Base Rate Incl. Fringes		Workers' Comp. Ins.	Average Fixed Overhead	Overhead	Profit	Total Overhead & Profit		Rate with O & P	
Abbr.	Trade	Hourly	Daily					%	Amount	Hourly	Daily
Skwk	Skilled Workers Average (35 trades)	$23.05	$184.40	16.2%	16.3%	27.0%	10.0%	69.5%	$16.00	$39.05	$312.40
	Helpers Average (5 trades)	16.90	135.20	17.9		25.0		69.2	$11.70	28.60	228.80
	Foreman Average, Inside ($0.50 over trade)	23.55	188.40	16.2		27.0		69.5	16.35	39.90	319.20
	Foreman Average, Outside ($2.00 over trade)	25.05	200.40	16.2		27.0		69.5	17.40	42.45	339.60
Clab	Common Building Laborers	16.90	135.20	18.5		25.0		69.8	11.80	28.70	229.60
Asbe	Asbestos Workers	23.75	190.00	15.2		30.0		71.5	17.00	40.75	326.00
Boil	Boilermakers	27.25	218.00	13.1		30.0		69.4	18.90	46.15	369.20
Bric	Bricklayers	24.00	192.00	15.0		25.0		66.3	15.90	39.90	319.20
Brhe	Bricklayer Helpers	17.90	143.20	15.0		25.0		66.3	11.85	29.75	238.00
Carp	Carpenters	23.10	184.80	18.5		25.0		69.8	16.10	39.20	313.60
Cefi	Cement Finishers	22.10	176.80	9.9		25.0		61.2	13.55	35.65	285.20
Elec	Electricians	26.40	211.20	6.4		30.0		62.7	16.55	42.95	343.60
Elev	Elevator Constructors	27.85	222.80	7.2		30.0		63.5	17.70	45.55	364.40
Eqhv	Equipment Operators, Crane or Shovel	24.00	192.00	10.3		28.0		64.6	15.50	39.50	316.00
Eqmd	Equipment Operators, Medium Equipment	23.20	185.60	10.3		28.0		64.6	15.00	38.20	305.60
Eqlt	Equipment Operators, Light Equipment	22.20	177.60	10.3		28.0		64.6	14.35	36.55	292.40
Eqol	Equipment Operators, Oilers	20.15	161.20	10.3		28.0		64.6	13.00	33.15	265.20
Eqmm	Equipment Operators, Master Mechanics	24.30	194.40	10.3		28.0		64.6	15.70	40.00	320.00
Glaz	Glaziers	22.75	182.00	13.8		25.0		65.1	14.80	37.55	300.40
Lath	Lathers	21.40	171.20	10.7		25.0		62.0	13.25	34.65	277.20
Marb	Marble Setters	22.35	178.80	15.0		25.0		66.3	14.80	37.15	297.20
Mill	Millwrights	24.05	192.40	10.3		25.0		61.6	14.80	38.85	310.80
Mstz	Mosaic and Terrazzo Workers	22.10	176.80	9.6		25.0		60.9	13.45	35.55	284.40
Pord	Painters, Ordinary	21.00	168.00	12.9		25.0		64.2	13.50	34.50	276.00
Psst	Painters, Structural Steel	21.35	170.80	50.0		25.0		101.3	21.65	43.00	344.00
Pape	Paper Hangers	20.85	166.80	12.9		25.0		64.2	13.40	34.25	274.00
Pile	Pile Drivers	22.45	179.60	22.9		30.0		79.2	17.80	40.25	322.00
Plas	Plasterers	20.95	167.60	14.6		25.0		65.9	13.80	34.75	278.00
Plah	Plasterer Helpers	18.05	144.40	14.6		25.0		65.9	11.90	29.95	239.60
Plum	Plumbers	26.15	209.20	7.8		30.0		64.1	16.75	42.90	343.20
Rodm	Rodmen (Reinforcing)	24.85	198.80	24.8		28.0		79.1	19.65	44.50	356.00
Rofc	Roofers, Composition	19.65	157.20	31.8		25.0		83.1	16.35	36.00	288.00
Rots	Roofers, Tile and Slate	19.75	158.00	31.8		25.0		83.1	16.40	36.15	289.20
Rohe	Roofer Helpers (Composition)	14.45	115.60	31.8		25.0		83.1	12.00	26.45	211.60
Shee	Sheet Metal Workers	25.60	204.80	11.1		30.0		67.4	17.25	42.85	342.80
Spri	Sprinkler Installers	25.90	207.20	9.0		30.0		65.3	16.90	42.80	342.40
Stpi	Steamfitters or Pipefitters	26.25	210.00	7.8		30.0		64.1	16.85	43.10	344.80
Ston	Stone Masons	23.00	184.00	15.0		25.0		66.3	15.25	38.25	306.00
Sswk	Structural Steel Workers	24.90	199.20	38.9		28.0		93.2	23.20	48.10	384.80
Tilf	Tile Layers (Floor)	22.05	176.40	9.6		25.0		60.9	13.45	35.50	284.00
Tilh	Tile Layer Helpers	17.10	136.80	9.6		25.0		60.9	10.40	27.50	220.00
Trlt	Truck Drivers, Light	18.25	146.00	15.1		25.0		66.4	12.10	30.35	242.80
Trhv	Truck Drivers, Heavy	18.80	150.40	15.1		25.0		66.4	12.50	31.30	250.40
Sswl	Welders, Structural Steel	24.90	199.20	38.9		28.0		93.2	23.20	48.10	384.80
Wrck	*Wrecking	17.40	139.20	40.5		25.0		91.8	15.95	33.35	266.80

*Not included in Averages.

R01100-090 Sales Tax by State

State sales tax on materials is tabulated below (5 states have no sales tax). Many states allow local jurisdictions, such as a county or city, to levy additional sales tax.

Some projects may be sales tax exempt, particularly those constructed with public funds.

State	Tax (%)	State	Tax (%)	State	Tax (%)	State	Tax (%)
Alabama	4	Illinois	6.25	Montana	0	Rhode Island	7
Alaska	0	Indiana	5	Nebraska	5	South Carolina	5
Arizona	5	Iowa	5	Nevada	6.5	South Dakota	4
Arkansas	4.625	Kansas	4.9	New Hampshire	0	Tennessee	6
California	6	Kentucky	6	New Jersey	6	Texas	6.25
Colorado	3	Louisiana	4	New Mexico	5	Utah	4.75
Connecticut	6	Maine	5	New York	4	Vermont	5
Delaware	0	Maryland	5	North Carolina	4	Virginia	3.5
District of Columbia	5.75	Massachusetts	5	North Dakota	5	Washington	6.5
Florida	6	Michigan	6	Ohio	5	West Virginia	6
Georgia	4	Minnesota	6.5	Oklahoma	4.5	Wisconsin	5
Hawaii	4	Mississippi	7	Oregon	0	Wyoming	4
Idaho	5	Missouri	4.225	Pennsylvania	6	Average	4.65 %

Sales Tax by Province (Canada)

GST - a value-added tax, which the government imposes on most goods and services provided in or imported into Canada.
PST - a retail sales tax, which five of the provinces impose on the price of most goods and some services.

QST - a value-added tax, similar to the federal GST, which Quebec imposes.
HST - Three provinces have combined their retail sales tax with the federal GST into one harmonized tax.

Province	PST (%)	QST (%)	GST(%)	HST(%)
Alberta	0	0	7	0
British Columbia	7.5	0	7	0
Manitoba	7	0	7	0
New Brunswick	0	0	0	15
Newfoundland	0	0	0	15
Northwest Territories	0	0	7	0
Nova Scotia	0	0	0	15
Ontario	8	0	7	0
Prince Edward Island	10	0	7	0
Quebec	0	7.5	7	0
Saskatchewan	6	0	7	0
Yukon	0	0	7	0

R01100-100 Unemployment Taxes and Social Security Taxes

Mass. State Unemployment tax ranges from 1.325% to 7.225% plus an experience rating assessment the following year, on the first $10,800 of wages. Federal Unemployment tax is 6.2% of the first $7,000 of wages. This is reduced by a credit for payment to the state. The minimum Federal Unemployment tax is 0.8% after all credits.

Combined rates in Mass. thus vary from 2.125% to 8.025% of the first $10,800 of wages. Combined average U.S. rate is about 6.2% of the first $7,000. Contractors with permanent workers will pay less since the average annual wages for skilled workers is $23.05 x 2,000 hours or about $46,100 per year. The average combined rate for U.S. would thus be 6.2% x $7,000 ÷ $46,100 = 0.9% of total wages for permanent employees.

Rates vary not only from state to state but also with the experience rating of the contractor.

Social Security (FICA) for 2004 is estimated at time of publication to be 7.65% of wages up to $87,000.

R01107-010 Architectural Fees

Tabulated below are typical percentage fees by project size, for good professional architectural service. Fees may vary from those listed depending upon degree of design difficulty and economic conditions in any particular area.

Rates can be interpolated horizontally and vertically. Various portions of the same project requiring different rates should be adjusted proportionately. For alterations, add 50% to the fee for the first $500,000 of project cost and add 25% to the fee for project cost over $500,000.

Architectural fees tabulated below include Structural, Mechanical and Electrical Engineering Fees. They do not include the fees for special consultants such as kitchen planning, security, acoustical, interior design, etc.

Civil Engineering fees are included in the Architectural fee for project sites requiring minimal design such as city sites. However, separate Civil Engineering fees must be added when utility connections require design, drainage calculations are needed, stepped foundations are required, or provisions are required to protect adjacent wetlands.

Building Types	Total Project Size in Thousands of Dollars						
	100	250	500	1,000	5,000	10,000	50,000
Factories, garages, warehouses, repetitive housing	9.0%	8.0%	7.0%	6.2%	5.3%	4.9%	4.5%
Apartments, banks, schools, libraries, offices, municipal buildings	12.2	12.3	9.2	8.0	7.0	6.6	6.2
Churches, hospitals, homes, laboratories, museums, research	15.0	13.6	12.7	11.9	9.5	8.8	8.0
Memorials, monumental work, decorative furnishings	—	16.0	14.5	13.1	10.0	9.0	8.3

General Requirements | R012 | Special Project Procedures

R01250-010 Repair and Remodeling

Cost figures are based on new construction utilizing the most cost-effective combination of labor, equipment and material with the work scheduled in proper sequence to allow the various trades to accomplish their work in an efficient manner.

The costs for repair and remodeling work must be modified due to the following factors that may be present in any given repair and remodeling project.

1. Equipment usage curtailment due to the physical limitations of the project, with only hand-operated equipment being used.
2. Increased requirement for shoring and bracing to hold up the building while structural changes are being made and to allow for temporary storage of construction materials on above-grade floors.
3. Material handling becomes more costly due to having to move within the confines of an enclosed building. For multi-story construction, low capacity elevators and stairwells may be the only access to the upper floors.
4. Large amount of cutting and patching and attempting to match the existing construction is required. It is often more economical to remove entire walls rather than create many new door and window openings. This sort of trade-off has to be carefully analyzed.
5. Cost of protection of completed work is increased since the usual sequence of construction usually cannot be accomplished.
6. Economies of scale usually associated with new construction may not be present. If small quantities of components must be custom fabricated due to job requirements, unit costs will naturally increase. Also, if only small work areas are available at a given time, job scheduling between trades becomes difficult and subcontractor quotations may reflect the excessive start-up and shut-down phases of the job.
7. Work may have to be done on other than normal shifts and may have to be done around an existing production facility which has to stay in production during the course of the repair and remodeling.
8. Dust and noise protection of adjoining non-construction areas can involve substantial special protection and alter usual construction methods.
9. Job may be delayed due to unexpected conditions discovered during demolition or removal. These delays ultimately increase construction costs.
10. Piping and ductwork runs may not be as simple as for new construction. Wiring may have to be snaked through walls and floors.
11. Matching "existing construction" may be impossible because materials may no longer be manufactured. Substitutions may be expensive.
12. Weather protection of existing structure requires additional temporary structures to protect building at openings.
13. On small projects, because of local conditions, it may be necessary to pay a tradesman for a minimum of four hours for a task that is completed in one hour.

All of the above areas can contribute to increased costs for a repair and remodeling project. Each of the above factors should be considered in the planning, bidding and construction stage in order to minimize the increased costs associated with repair and remodeling jobs.

R01540-100 Steel Tubular Scaffolding

On new construction, tubular scaffolding is efficient up to 60′ high or five stories. Above this it is usually better to use a hung scaffolding if construction permits. Swing scaffolding operations may interfere with tenants. In this case, the tubular is more practical at all heights.

In repairing or cleaning the front of an existing building the cost of tubular scaffolding per S.F. of building front increases as the height increases above the first tier. The first tier cost is relatively high due to leveling and alignment.

The minimum efficient crew for erection is three workers. For heights over 50′, a crew of four is more efficient. Use two or more on top and two at the bottom for handing up or hoisting. Four workers can erect and dismantle about nine frames per hour up to five stories. From five to eight stories they will average six frames per hour. With 7′ horizontal spacing this will run about 400 S.F. and 265 S.F. of wall surface, respectively. Time for placing planks must be added to the above. On heights above 50′, five planks can be placed per labor-hour.

The table below shows the number of pieces required to erect tubular steel scaffolding for 1000 S.F. of building frontage. This area is made up of a scaffolding system that is 12 frames (11 bays) long by 2 frames high.

For jobs under twenty-five frames, add 50% to rental cost. Rental rates will be lower for jobs over three months duration. Large quantities for long periods can reduce rental rates by 20%.

Description of Component	CSI Line Item	Number of Pieces for 1000 S.F. of Building Front	Unit
5′ Wide Standard Frame, 6′-4″ High	01540-750-2200	24	Ea.
Leveling Jack & Plate	01540-750-2650	24	↓
Cross Brace	01540-750-2500	44	↓
Side Arm Bracket, 21″	01540-750-2700	12	↓
Guardrail Post	01540-750-2550	12	↓
Guardrail, 7′ section	01540-750-2600	22	↓
Stairway Section	01540-750-2900	2	↓
Stairway Starter Bar	01540-750-2910	1	↓
Stairway Inside Handrail	01540-750-2920	2	↓
Stairway Outside Handrail	01540-750-2930	2	↓
Walk-Thru Frame Guardrail	01540-750-2940	2	↓

Scaffolding is often used as falsework over 15′ high during construction of cast-in-place concrete beams and slabs. Two foot wide scaffolding is generally used for heavy beam construction. The span between frames depends upon the load to be carried with a maximum span of 5′.

Heavy duty shoring frames with a capacity of 10,000#/leg can be spaced up to 10′ O.C. depending upon form support design and loading.

Scaffolding used as horizontal shoring requires less than half the material required with conventional shoring.

On new construction, erection is done by carpenters.

Rolling towers supporting horizontal shores can reduce labor and speed the job. For maintenance work, catwalks with spans up to 70′ can be supported by the rolling towers.

R01540-200 Pump Staging

Pump staging is generally not available for rent. The table below shows the number of pieces required to erect pump staging for 2400 S.F. of building frontage. This area is made up of a pump jack system that is 3 poles (2 bays) wide by 2 poles high.

Item	CSI Line Item	Number of Pieces for 2400 S.F. of Building Front	Unit
Aluminum pole section, 24′ long	01540-550-0200	6	Ea.
Aluminum splice joint, 6′ long	01540-550-0600	3	↓
Aluminum foldable brace	01540-550-0900	3	↓
Aluminum pump jack	01540-550-0700	3	↓
Aluminum support for workbench/back safety rail	01540-550-1000	3	↓
Aluminum scaffold plank/workbench, 14″ wide x 24′ long	01540-550-1100	4	↓
Safety net, 22′ long	01540-550-1250	2	↓
Aluminum plank end safety rail	01540-550-1200	2	↓

The cost in place for this 2400 S.F. will depend on how many uses are realized during the life of the equipment. Several options are given in Division 01540-550.

R02065-300 Bituminous Paving

City	Sidewalk Mix Bituminous Asphalt per Ton*	Sidewalks (2") 9.2 S.Y./ton				Pavement (3") 6.13 S.Y./ton			
		Cost per S.Y.			Per Ton	Cost per S.Y.			Per Ton
		Material*	Installation	Total	Total	Material*	Installation	Total	Total
Atlanta	$30.00	$3.26	$1.31	$4.57	$42.04	$4.89	$.77	$5.66	$34.70
Baltimore	34.73	3.78	1.27	5.05	46.46	5.67	.74	6.41	39.29
Boston	39.50	4.29	1.35	5.64	51.89	6.44	.79	7.23	44.32
Buffalo	34.60	3.76	1.28	5.04	46.37	5.64	.75	6.39	39.17
Chicago	31.75	3.45	1.40	4.85	44.62	5.18	.82	6.00	36.78
Cincinnati	41.00	4.46	1.48	5.94	54.65	6.69	.87	7.56	46.34
Cleveland	29.50	3.21	1.46	4.67	42.96	4.81	.86	5.67	34.76
Columbus	30.50	3.32	1.39	4.71	43.33	4.98	.82	5.80	35.55
Dallas	29.25	3.18	1.18	4.36	40.11	4.77	.69	5.46	33.47
Denver	30.00	3.26	1.46	4.72	43.42	4.89	.85	5.74	35.19
Detroit	32.00	3.48	1.33	4.81	44.25	5.22	.78	6.00	36.78
Houston	34.50	3.75	1.21	4.96	45.63	5.63	.71	6.34	38.86
Indianapolis	29.75	3.23	1.37	4.60	42.32	4.85	.81	5.66	34.70
Kansas City	29.50	3.21	1.25	4.46	41.03	4.81	.73	5.54	33.96
Los Angeles	36.00	3.91	1.47	5.38	49.50	5.87	.86	6.73	41.25
Memphis	38.75	4.21	1.25	5.46	50.23	6.32	.74	7.06	43.28
Milwaukee	32.50	3.53	1.27	4.80	44.16	5.30	.75	6.05	37.09
Minneapolis	31.38	3.41	1.49	4.90	45.08	5.12	.87	5.99	36.72
Nashville	27.75	3.02	1.36	4.38	40.30	4.53	.80	5.33	32.67
New Orleans	33.75	3.67	1.20	4.87	44.80	5.51	.71	6.22	38.13
New York City	49.50	5.38	1.57	6.95	63.94	8.08	.92	9.00	55.17
Philadelphia	32.50	3.53	1.29	4.82	44.34	5.30	.76	6.06	37.15
Phoenix	30.00	3.26	1.43	4.69	43.15	4.89	.84	5.73	35.12
Pittsburgh	35.00	3.80	1.49	5.29	48.67	5.71	.88	6.59	40.40
St. Louis	32.00	3.48	1.32	4.80	44.16	5.22	.78	6.00	36.78
San Antonio	33.25	3.61	1.21	4.82	44.34	5.42	.71	6.13	37.58
San Diego	37.00	4.02	1.38	5.40	49.68	6.04	.81	6.85	41.99
San Francisco	40.25	4.38	1.52	5.90	54.28	6.57	.89	7.46	45.73
Seattle	39.50	4.29	1.55	5.84	53.73	6.44	.91	7.35	45.06
Washington, D.C.	35.41	3.85	1.22	5.07	46.64	5.78	.72	6.50	39.85
Average	$34.00	$3.70	$1.36	$5.06	$46.55	$5.55	$.80	$6.35	$38.93

Assumed density is 145 lb. per C.F.
*Includes delivery within 20 miles

Table below shows quantities and bare costs for 1000 S.Y. of Bituminous Paving.

Item	Sidewalks, 2" Thick		Roads and Parking Areas, 3" Thick	
	Quantities	Cost	Quantities	Cost
Bituminous asphalt	108.7 tons @ $34.00 per ton	$3,696.00	163.1 tons@ $34.00 per ton	$5,546.00
Installation using	Crew B-37 @ $977.80 /720 SY/day x 1000	1,358.06	Crew B-25B @ $3,908.80 /4900SY/ day x 1000	797.71
Total per 1000 S.Y.		$5,054.06		$6,343.71
Total per S.Y.		$ 5.06		$ 6.35
Total per Ton		$ 46.55		$ 38.93

R02510-800 Piping Designations

There are several systems currently in use to describe pipe and fittings. The following paragraphs will help to identify and clarify classifications of piping systems used for water distribution.

Piping may be classified by schedule. Piping schedules include 5S, 10S, 10, 20, 30, Standard, 40, 60, Extra Strong, 80, 100, 120, 140, 160 and Double Extra Strong. These schedules are dependent upon the pipe wall thickness. The wall thickness of a particular schedule may vary with pipe size.

Ductile iron pipe for water distribution is classified by Pressure Classes such as Class 150, 200, 250, 300 and 350. These classes are actually the rated water working pressure of the pipe in pounds per square inch (psi). The pipe in these pressure classes is designed to withstand the rated water working pressure plus a surge allowance of 100 psi.

The American Water Works Association (AWWA) provides standards for various types of **plastic pipe.** C-900 is the specification for polyvinyl chloride (PVC) piping used for water distribution in sizes ranging from 4″ through 12″. C-901 is the specification for polyethylene (PE) pressure pipe, tubing and fittings used for water distribution in sizes ranging from 1/2″ through 3″. C-905 is the specification for PVC piping sizes 14″ and greater.

PVC pressure-rated pipe is identified using the standard dimensional ratio (SDR) method. This method is defined by the American Society for Testing and Materials (ASTM) Standard D 2241. This pipe is available in SDR numbers 64, 41, 32.5, 26, 21, 17, and 13.5. Pipe with an SDR of 64 will have the thinnest wall while pipe with an SDR of 13.5 will have the thickest wall. When the pressure rating (PR) of a pipe is given in psi, it is based on a line supplying water at 73 degrees F.

The National Sanitation Foundation (NSF) seal of approval is applied to products that can be used with potable water. These products have been tested to ANSI/NSF Standard 14.

Valves and strainers are classified by American National Standards Institute (ANSI) Classes. These Classes are 125, 150, 200, 250, 300, 400, 600, 900, 1500 and 2500. Within each class there is an operating pressure range dependent upon temperature. Design parameters should be compared to the appropriate material dependent, pressure-temperature rating chart for accurate valve selection.

R02920-500 Seeding

The type of grass is determined by light, shade and moisture content of soil plus intended use. Fertilizer should be disked 4″ before seeding. For steep slopes disk five tons of mulch and lay two tons of hay or straw on surface per acre after seeding. Surface mulch can be staked, lightly disked or tar emulsion sprayed. Material for mulch can be wood chips, peat moss, partially rotted hay or straw, wood fibers and sprayed emulsions. Hemp seed blankets with fertilizer are also available. For spring seeding, watering is necessary. Late fall seeding may have to be reseeded in the spring. Hydraulic seeding, power mulching, and aerial seeding can be used on large areas.

R02930-900 Cost of Trees: Based on Pin Oak (Quercus palustris)

Tree Diameter	Normal Height	Catalog List Price of Tree	Guying Material	Bare Equipment Charge	Bare Installation Labor	Bare Total
2 to 3 inch	14 feet	$ 146	$15.35	$ 53.28	$ 59.84	$ 274.47
3 to 4 inch	16 feet	323	18.35	$ 88.80	$ 99.73	529.88
4 to 5 inch	18 feet	373	75.50	$106.56	$119.68	674.74
6 to 7 inch	22 feet	813	80.00	$133.20	$119.68	1,145.88
8 to 9 inch	26 feet	1,023	90.00	$177.60	$199.47	1,490.07

Installation Time & Cost for Planting Trees, Bare Costs														
Ball Size Diam. X Depth	Soil in Ball	Weight of Ball	Hole Diam. Req'd	Hole Excavation	Amount of Soil Displ.	Topsoil Handled	Time Required in Labor-Hours						Cost	
							Dig & Lace	Handle Ball	Dig Hole	Plant & Prune	Water & Guy	Total L.H.	Crew	Bare Total per Tree
Inches	C.F.	Lbs.	Feet	C.F.	C.F.	C.F.								
12 x 12	0.70	56.00	2.00	4.00	3.00	11.00	0.25	0.17	0.33	0.25	0.07	1.10	1 Clab	$ 18.59
18 x 16	2.00	160.00	2.50	8.00	6.00	21.00	0.50	0.33	0.47	0.35	0.08	1.70	2 Clab	28.73
24 x 18	4.00	320.00	3.00	13.00	9.00	38.00	1.00	0.67	1.08	0.82	0.20	3.80	3 Clab	64.22
30 x 21	7.50	600.00	4.00	27.00	19.50	76.00	0.82	0.71	0.79	1.22	0.26	3.80		103.85
36 x 24	12.50	980.00	4.50	38.00	25.50	114.00	1.08	0.95	1.11	1.32	0.30	4.76		130.09
42 x 27	19.00	1,520.00	5.50	64.00	45.00	185.00	1.90	1.27	1.87	1.43	0.34	6.80		185.84
48 x 30	28.00	2,040.00	6.00	85.00	57.00	254.00	2.41	1.60	2.06	1.55	0.39	8.00		218.64
54 x 33	38.50	3,060.00	7.00	127.00	88.50	370.00	2.86	1.90	2.39	1.76	0.45	9.40	B-6	256.90
60 x 36	52.00	4,160.00	7.50	159.00	107.00	474.00	3.26	2.17	2.73	2.00	0.51	10.70		292.43
66 x 39	68.00	5,440.00	8.00	196.00	128.00	596.00	3.61	2.41	3.07	2.26	0.58	11.90		325.23
72 x 42	87.00	7,160.00	9.00	267.00	180.00	785.00	3.90	2.60	3.71	2.78	0.70	13.70		374.42

R04060-100 Cement Mortar (material only)

Type N - 1:1:6 mix by volume. Use everywhere above grade except as noted below.

- 1:3 mix using conventional masonry cement which saves handling two separate bagged materials.

Type M - 1:1/4:3 mix by volume, or 1 part cement, 1/4 (10% by wt.) lime, 3 parts sand. Use for heavy loads and where earthquakes or hurricanes may occur. Also for reinforced brick, sewers, manholes and everywhere below grade.

Cost and Mix Proportions of Various Types of Mortar

Components	Type Mortar and Mix Proportions by Volume										
	M		S		N		O		K	PM	PL
	1:1:6	1:1/4:3	1/2:1:4	1:1/2:4	1:3	1:1:6	1:3	1:2:9	1:3:12	1:1:6	1:1/2:4
Portland cement	$ 8.10	$ 8.10	$ 4.05	$ 8.10	—	$ 8.10	—	$ 8.10	$ 8.10	$ 8.10	$ 8.10
Masonry cement	6.15	—	6.15	—	$6.15	—	$6.15	—	—	6.15	—
Lime	—	1.40	—	2.80	—	5.60	—	11.20	16.80	—	2.80
Masonry sand*	4.39	2.19	2.93	2.93	2.19	4.39	2.19	6.58	8.78	4.39	2.93
Mixing machine incl. fuel**	3.16	1.58	2.10	2.10	1.58	3.16	1.58	4.73	6.31	3.16	2.10
Total for Materials	$21.80	$13.27	$15.23	$15.93	$9.92	$21.25	$9.92	$30.61	$39.99	$21.80	$15.93
Total C.F.	6	3	4	4	3	6	3	9	12	6	4
Approximate Cost per C.F.	$ 3.63	$ 4.42	$ 3.81	$ 3.98	$3.31	$ 3.54	$3.31	$ 3.40	$ 3.33	$ 3.63	$ 3.98

*Includes 10 mile haul
**Based on a daily rental, 10 C.F., 25 H.P. mixer, mix 200 C.F./Day

Mix Proportions by Volume, Compressive Strength and Cost of Mortar

Where Used	Allowable Proportions by Volume					Compressive Strength @ 28 days	Cost per Cubic Foot
	Mortar Type	Portland Cement	Masonry Cement	Hydrated Lime	Masonry Sand		
Plain Masonry	M	1	1	—	6	2500 psi	$3.63
		1	—	1/4	3		4.42
	S	1/2	1	—	4	1800 psi	3.81
		1	—	1/4 to 1/2	4		3.98
	N	—	1	—	3	750 psi	3.31
		1	—	1/2 to 1-1/4	6		3.54
	O	—	1	—	3	350 psi	3.31
		1	—	1-1/4 to 2-1/2	9		3.40
	K	1	—	2-1/2 to 4	12	75 psi	3.33
Reinforced Masonry	PM	1	1	—	6	2500 psi	3.63
	PL	1	—	1/4 to 1/2	4	2500 psi	3.98

Note: The total aggregate should be between 2.25 to 3 times the sum of the cement and lime used.

The labor cost to mix the mortar is included in the labor cost on brickwork.

Machine mixing is usually specified on jobs of any size. There is a large price saving over hand mixing and mortar is more uniform.

There are two types of mortar color used. Prices in Section 04060-540 are for the inert additive type with about 100 lbs. per M brick as the typical quantity required. These colors are also available in smaller batch size bags (1 lb. to 15 lb.) which can be placed directly into the mixer without measuring. The other type is premixed and replaces the masonry cement. Dark green color has the highest cost.

R04060-200 Miscellaneous Mortar (material only)

Quantities	Glass Block Mortar		Gypsum Cement Mortar	
White Portland cement, 94 Lb bag	7 bags	$133.00		
Gypsum cement, 80 Lb bag			11.25 bags	$132.19
Lime, 50 Lb bag	280 lbs.	31.36		
Sand*	1 C.Y.	22.45	1 C.Y.	22.45
Mixing machine and fuel**		14.22		14.22
Total per C.Y.		$201.03		$168.86
Approximate Total per C.F.		$ 7.45		$ 6.25

* Includes 10 mile haul

** Based on a daily rental, 10 C.F., 25 HP mixer, mix 200 C.F./Day = 7.4 C.Y./Day

R04080-500 Masonry Reinforcing

Horizontal joint reinforcing helps prevent wall cracks where wall movement may occur and in many locations is required by code. Horizontal joint reinforcing is generally not considered to be structural reinforcing and an unreinforced wall may still contain joint reinforcing.

Reinforcing strips come in 10′ and 12′ lengths and in truss and ladder shapes, with and without drips. Field labor runs between 2.7 to 5.3 hours per 1000 L.F. for wall thicknesses up to 12″.

The wire meets ASTM A82 for cold drawn steel wire and the typical size is 9 ga. sides and ties with 3/16″ diameter also available. Typical finish is mill galvanized with zinc coating at .10 oz. per S.F. Class I (.40 oz. per S.F.) and Class III (.80 oz per S.F.) are also available, as is hot dipped galvanizing at 1.50 oz. per S.F.

R04210-100 Economy in Bricklaying

Have adequate supervision. Be sure bricklayers are always supplied with materials so there is no waiting. Place best bricklayers at corners and openings.

Use only screened sand for mortar. Otherwise, labor time will be wasted picking out pebbles. Use seamless metal tubs for mortar as they do not leak or catch the trowel. Locate stack and mortar for easy wheeling.

Have brick delivered for stacking. This makes for faster handling, reduces chipping and breakage, and requires less storage space. Many dealers will deliver select common in 2′ x 3′ x 4′ pallets or face brick packaged. This affords quick handling with a crane or forklift and easy tonging in units of ten, which reduces waste.

Use wider bricks for one wythe wall construction. Keep scaffolding away from wall to allow mortar to fall clear and not stain wall.

On large jobs develop specialized crews for each type of masonry unit.

Consider designing for prefabricated panel construction on high rise projects.

Avoid excessive corners or openings. Each opening adds about 50% to labor cost for area of opening.

Bolting stone panels and using window frames as stops reduces labor costs and speeds up erection.

R04210-120 Common and Face Brick Prices

Prices are based on truckload lot purchases for Common Building Brick and Facing Brick. Prices are per M, (thousand), brick.

City	Material			Installation				Total			
	Brick per M Delivered		Mortar	Common in 8″ Wall		Face Brick, 4″ Veneer		Common in 8″ Wall		Face Brick, 4″ Veneer	
	Common	Face	3/8″ Joint	Bare Costs	Incl. O & P	Bare Costs	Incl. O & P	Bare Costs	Incl. O & P	Bare Costs	Incl. O & P
Atlanta	$185	$259	$44.00	$330	$ 549	$396	$ 659	$ 565	$ 784	$ 699	$ 962
Baltimore	315	443	for 8″ Wall	386	641	463	769	754	1,009	955	1,261
Boston	410	575	and	690	1,148	828	1,377	1,156	1,614	1,456	2,005
Buffalo	380	532	$36.00	541	899	649	1,079	976	1,334	1,233	1,663
Chicago	245	343	for 4″ Wall	624	1,038	749	1,245	920	1,334	1,138	1,634
Cincinnati	180	254		468	778	561	933	697	1,007	859	1,231
Cleveland	225	319		524	871	628	1,045	800	1,147	993	1,410
Columbus	180	255		457	761	549	913	686	990	848	1,212
Dallas	290	408		316	526	379	631	659	869	835	1,087
Denver	240	339		382	635	459	763	673	926	844	1,148
Detroit	315	443		580	965	696	1,158	948	1,333	1,188	1,650
Houston	340	480		324	539	389	647	718	933	919	1,177
Indianapolis	215	300		456	759	548	911	721	1,024	893	1,256
Kansas City	355	503		498	829	598	994	908	1,239	1,152	1,548
Los Angeles	230	325		547	910	657	1,093	828	1,191	1,028	1,464
Memphis	180	251		395	656	474	788	624	885	769	1,083
Milwaukee	300	425		550	915	660	1,098	903	1,268	1,134	1,572
Minneapolis	340	475		607	1,010	729	1,212	1,001	1,404	1,254	1,737
Nashville	190	265		329	548	395	658	569	788	704	967
New Orleans	230	325		287	478	345	573	568	759	716	944
New York City	265	375		755	1,255	906	1,506	1,072	1,572	1,328	1,928
Philadelphia	290	405		611	1,016	733	1,219	954	1,359	1,186	1,672
Phoenix	230	325		344	573	413	687	625	854	784	1,058
Pittsburgh	295	418		476	792	572	951	824	1,140	1,039	1,418
St. Louis	245	346		294	488	352	586	590	784	744	978
San Antonio	225	315		490	816	588	979	766	1,092	948	1,339
San Diego	275	390		617	1,026	740	1,231	944	1,353	1,178	1,669
San Francisco	270	380		539	897	647	1,076	861	1,219	1,074	1,503
Seattle	405	567		548	911	658	1,094	1,009	1,372	1,278	1,714
Washington, D.C.	265	375		398	662	478	795	715	979	900	1,217
Average	$270	$380		$480	$ 795	$575	$ 955	$ 800	$1,120	$1,005	$1,385

Common building brick manufactured according to ASTM C62 and facing brick manufactured according to ASTM C216 are the two standard bricks available for general building use.

Building brick is made in three grades; SW, where high resistance to damage caused by cyclic freezing is required; MW, where moderate resistance to cyclic freezing is needed; and NW, where little resistance to cyclic freezing is needed. Facing brick is made in only the two grades SW and MW. Additionally, facing brick is available in three types; FBS, for general use; FBX, for general use where a higher degree of precision and lower permissible variation in size than FBS is needed; and FBA, for general use to produce characteristic architectural effects resulting from non-uniformity in size and texture of the units.

In figuring above installation costs, a D-8 Crew (with a daily output of 1.5 M) was used for the 4″ veneer. A D-8 Crew (with a daily output of 1.8 M) was used for the 8″ solid wall.

In figuring the total cost including overhead and profit, an allowance of 10% was added to the sum of the cost of the brick and mortar. Also, 3% breakage was included for both the bare costs and the costs with overhead and profit. If bricks are delivered palletized with 280 to 300 per pallet, or packaged, allow only 1-1/2% for breakage. Then add $10 per M to the cost of brick and deduct two hours helper time. The net result is a savings of $30 to $40 per M in place. Packaged or palletized delivery is practical when a job is big enough to have a crane or other equipment available to handle a package of brick. This is so on all industrial work but not always true on small commercial buildings.

The prices above are for bricks used in commercial, apartment house or industrial construction. If it is possible to obtain the price of the actual brick to be used, it should be done and substituted in the table. The use of buff and gray face is increasing, and there is a continuing trend to the Norman, Roman, Jumbo and SCR brick.

See R04210-500 for brick quantities per S.F. and mortar quantities per M brick. (Average prices for the various sizes are listed in Division 4.)

Common red clay brick for backup is not used that often. Concrete block is the most usual backup material with occasional use of sand lime or cement brick. Sand lime cost about $15 per M less than red clay and cement brick are about $5 per M less than red clay. These figures may be substituted in the common brick breakdown for the cost of these items in place, as labor is about the same. Building brick is commonly used in solid walls for strength and as a fire stop.

Brick panels built on the ground and then crane erected to the upper floors have proven to be economical. This allows the work to be done under cover and without scaffolding.

R04210-180 Brick in Place

Table below is for common bond with 3/8″ concave joints and includes 3% waste for brick and 25% waste for mortar.
Crew costs are bare costs.

Item	8″ Common Brick Wall 8″ x 2-2/3″ x 4″		Select Common Face 8″ x 2-2/3″ x 4″		Red Face Brick 8″ x 2-2/3″ x 4″	
1030 brick delivered		$278.10		$334.75		$ 391.40
Type N mortar	12.5 C.F.	44.25	10.3 C.F.	36.46	10.3 C.F.	36.46
Installation using indicated crew	Crew D-8 @ 0.556 Days	479.49	Crew D-8 @ 0.667 Days	575.22	Crew D-8 @ 0.667 Days	575.22
Total per M in place		$801.84		$946.43		$1,003.08
Total per S.F. of wall	13.5 bricks/S.F.	$ 10.82	6.75 bricks/S.F.	$ 6.39	6.75 bricks/S.F.	$ 6.77

R04210-500 Brick, Block & Mortar Quantities

Running Bond						For Other Bonds Standard Size Add to S.F. Quantities in Table to Left		
Number of Brick per S.F. of Wall - Single Wythe with 3/8″ Joints				C.F. of Mortar per M Bricks, Waste Included				
Type Brick	Nominal Size (incl. mortar) L H W	Modular Coursing	Number of Brick per S.F.	3/8″ Joint	1/2″ Joint	Bond Type	Description	Factor
Standard	8 x 2-2/3 x 4	3C=8″	6.75	10.3	12.9	Common	full header every fifth course	+20%
Economy	8 x 4 x 4	1C=4″	4.50	11.4	14.6		full header every sixth course	+16.7%
Engineer	8 x 3-1/5 x 4	5C=16″	5.63	10.6	13.6	English	full header every second course	+50%
Fire	9 x 2-1/2 x 4-1/2	2C=5″	6.40	550 # Fireclay	—	Flemish	alternate headers every course	+33.3%
Jumbo	12 x 4 x 6 or 8	1C=4″	3.00	23.8	30.8		every sixth course	+5.6%
Norman	12 x 2-2/3 x 4	3C=8″	4.50	14.0	17.9	Header = W x H exposed		+100%
Norwegian	12 x 3-1/5 x 4	5C=16″	3.75	14.6	18.6	Rowlock = H x W exposed		+100%
Roman	12 x 2 x 4	2C=4″	6.00	13.4	17.0	Rowlock stretcher = L x W exposed		+33.3%
SCR	12 x 2-2/3 x 6	3C=8″	4.50	21.8	28.0	Soldier = H x L exposed		—
Utility	12 x 4 x 4	1C=4″	3.00	15.4	19.6	Sailor = W x L exposed		-33.3%

Concrete Blocks Nominal Size	Approximate Weight per S.F.		Blocks per 100 S.F.	Mortar per M block, waste included	
	Standard	Lightweight		Partitions	Back up
2″ x 8″ x 16″	20 PSF	15 PSF	113	27 C.F.	36 C.F.
4″	30	20	↓	41	51
6″	42	30	↓	56	66
8″	55	38	↓	72	82
10″	70	47	↓	87	97
12″ ↓	85 ↓	55 ↓	↓	102 ↓	112 ↓

R04220-200 Concrete Block

Concrete masonry units, 8″ high x 16″ long, sand aggregate, 3/8 joints for partitions, tooled joints two sides, 113 blocks per 100 S.F., bare costs.

City	Material				Mortar 3/8″ Joint	Bare Installation		Bare Total Per 100 S.F.	
	Per Block, Delivered		113 Block, Delivered						
	4″ Thick	8″ Thick	4″ Thick	8″ Thick		4″ Thick	8″ Thick	4″ Thick	8″ Thick
Atlanta	$ .81	$1.38	$ 91.53	$155.94	$15.23 for 4″	$138.44	$158.64	$245.20	$341.39
Baltimore	.63	1.07	71.19	120.91	$26.81 for 8″	161.70	185.30	248.12	333.02
Boston	.65	1.11	73.45	125.43		289.48	331.73	378.16	483.97
Buffalo	.64	1.08	72.32	122.04		226.80	259.90	314.35	408.75
Chicago	.51	.87	57.63	98.31		261.74	299.94	334.60	425.06
Cincinnati	.44	.75	49.72	84.75		196.18	224.80	261.13	336.36
Cleveland	.63	1.07	71.19	120.91		219.65	251.70	306.07	399.42
Columbus	.41	.70	46.33	79.10		191.85	219.84	253.41	325.75
Dallas	.67	1.13	75.71	127.69		132.58	151.93	223.52	306.43
Denver	.86	1.45	97.18	163.85		160.26	183.64	272.67	374.30
Detroit	.65	1.10	73.45	124.30		243.41	278.93	332.09	430.04
Houston	1.01	1.71	114.13	193.23		135.96	155.80	265.32	375.84
Indianapolis	.78	1.33	88.14	150.29		191.36	219.29	294.73	396.39
Kansas City	.71	1.20	80.23	135.60		208.96	239.46	304.42	401.87
Los Angeles	.65	1.10	73.45	124.30		229.61	263.11	318.29	414.22
Memphis	.67	1.13	75.71	127.69		165.53	189.68	256.47	344.18
Milwaukee	.71	1.20	80.23	135.60		230.82	264.50	326.28	426.91
Minneapolis	.94	1.60	106.22	180.80		254.63	291.79	376.08	499.40
Nashville	.57	.96	64.41	108.48		138.19	158.36	217.83	293.65
New Orleans	.68	1.15	76.84	129.95		120.47	138.05	212.54	294.81
New York City	.63	1.07	71.19	120.91		316.51	362.69	402.93	510.41
Philadelphia	.52	.88	58.76	99.44		256.11	293.48	330.10	419.73
Phoenix	.51	.87	57.63	98.31		144.43	165.50	217.29	290.62
Pittsburgh	.60	1.01	67.80	114.13		199.82	228.98	282.85	369.92
St. Louis	.91	1.55	102.83	175.15		123.19	141.16	241.25	343.12
San Antonio	.66	1.12	74.58	126.56		205.68	235.69	295.49	389.06
San Diego	.51	.87	57.63	98.31		258.69	296.44	331.55	421.56
San Francisco	1.06	1.79	119.78	202.27		226.24	259.25	361.25	488.33
Seattle	.85	1.44	96.05	162.72		229.83	263.37	341.11	452.90
Washington D.C.	.74	1.26	83.62	142.38		167.01	191.39	265.86	360.58
Average	$ .69	$1.17	$ 77.97	$132.21		$200.84	$230.14	$294.04	$389.16

Cost for 100 S.F. of 8″ x 16″ Concrete Block Partitions to Four Stories High, Tooled Joints Two Sides						
8″ x 16″ Sand Aggregate	4″ Thick Block		8″ Thick Block		12″ Thick Block	
113 block delivered		$ 77.97		$132.21		$180.80
Mortar Type N, 1:3	4.6 C.F.	15.23	8.1 C.F.	26.81	11.5 C.F.	38.07
Installation Crew	D-8 @ 0.233 days	200.84	D-8 @ .267 days	230.14	D-9 @ .294 days	295.53
Total bare cost per 100 S.F.		$294.04		$389.16		$514.40
Add for filling cores solid	6.7 C.F.	$100.54	25.8 C.F.	$216.89	42.2 C.F.	$279.32

Cost for 100 S.F. of 8″ x 16″ Concrete Block Backup, Tooled Joints One Side						
8″ x 16″ Sand Aggregate	4″ Thick Block		8″ Thick Block		12″ Thick Block	
113 block delivered		$ 77.97		$132.21		$180.80
Mortar Type N, 1:3	5.8 C.F.	19.20	9.3 C.F.	30.78	12.7 C.F.	42.04
Installation crew	D-8 @ .217 days	187.05	D-8 @ .250 days	215.49	D-9 @ .323 days	324.68
Total bare cost per 100 S.F.		$284.22		$378.48		$547.52

Special block: corner, jamb and head block are same price as ordinary block of same size. Tabulated on the next page are national average prices per block. Labor on specials is about the same as equal sized regular block. Bond beam and 16″ high lintel blocks cost 30% more than regular units of equal size. Lintel blocks are 8″ long and 8″ or 16″ high. Costs in individual cities may be factored from the table above.

Use of motorized mortar spreader box will speed construction of continuous walls. Hollow non-load bearing units are made according to ASTM C129 and hollow load bearing units according to ASTM C90.

R04930-100 Cleaning Face Brick

On smooth brick a person can clean 70 S.F. an hour; on rough brick 50 S.F. per hour. Use one gallon muriatic acid to 20 gallons of water for 1000 S.F. Do not use acid solution until wall is at least seven days old, but a mild soap solution may be used after two days. Commercial cleaners cost from $9 to $12 per gallon.

Time has been allowed for clean-up in brick prices.

R06100-010 Thirty City Lumber Prices

Prices for boards are for #2 or better or sterling, whichever is in best supply. Dimension lumber is "Standard or Better" either Southern Yellow Pine (S.Y.P.), Spruce-Pine-Fir (S.P.F.), Hem-Fir (H.F.) or Douglas Fir (D.F.). The species of lumber used in a geographic area is listed by city. Plyform is 3/4″ BB oil sealed fir or S.Y.P. whichever prevails locally, 3/4″ CDX is S.Y.P. or Fir.

These are prices at the time of publication and should be checked against the current market price. Relative differences between cities will stay approximately constant.

		Contractor Purchases per M.B.F. S4S								Contractor Purchases per M.S.F.	
		Dimensions						Boards			
City	Species	2″x4″	2″x6″	2″x8″	2″x10″	2″x12″	4″x4″	1″x6″	1″x12″	3/4″ Ext. Plyform	3/4″ Thick CDX T&G
Atlanta	S.Y.P.	$461	$488	$562	$644	$732	$851	$1,065	$1,445	$ 928	$809
Baltimore	S.P.F.	432	457	527	603	686	798	998	1354	869	758
Boston	S.P.F.	417	441	508	582	662	770	963	1307	839	731
Buffalo	S.P.F.	411	434	501	573	652	758	949	1287	826	720
Chicago	H.F.	495	523	603	691	785	913	1143	1551	995	868
Cincinnati	S.P.F.	443	468	540	618	703	818	1023	1388	891	777
Cleveland	S.P.F.	429	454	523	599	681	792	991	1345	863	752
Columbus	S.P.F.	438	464	534	612	696	809	1012	1374	882	768
Dallas	S.Y.P.	455	481	555	635	722	840	1051	1426	916	798
Denver	H.F.	493	522	601	689	783	911	1139	1546	992	865
Detroit	H.F.	513	543	626	716	815	947	1185	1608	1032	900
Houston	S.Y.P.	450	476	549	629	715	832	1040	1412	906	790
Indianapolis	S.P.F.	499	529	609	698	793	923	1154	1566	1005	876
Kansas City	D.F.	493	522	601	689	783	911	1139	1546	992	865
Los Angeles	D.F.	485	514	592	678	771	896	1121	1522	977	851
Memphis	S.Y.P.	461	488	562	644	732	851	1065	1445	928	809
Milwaukee	H.F.	495	523	603	691	785	913	1143	1551	995	868
Minneapolis	H.F.	493	522	601	689	783	911	1139	1546	992	865
Nashville	S.Y.P.	461	488	562	644	732	851	1065	1445	928	809
New Orleans	S.Y.P.	464	491	566	648	737	857	1072	1455	934	814
New York City	S.P.F.	429	454	523	599	681	792	991	1345	863	752
Philadelphia	S.P.F.	432	457	527	603	686	798	998	1354	869	758
Phoenix	D.F.	508	538	620	710	807	939	1174	1594	1023	892
Pittsburgh	S.P.F.	438	464	534	612	696	809	1012	1374	882	768
St. Louis	H.F.	499	529	609	698	793	923	1154	1566	1005	876
San Antonio	S.Y.P.	459	486	560	642	730	848	1061	1441	925	806
San Diego	D.F.	490	519	598	685	779	905	1133	1537	987	860
San Francisco	D.F.	464	491	566	648	737	857	1072	1455	934	814
Seattle	D.F.	459	486	560	642	730	848	1061	1440	925	806
Washington, DC	S.P.F.	436	462	532	609	693	806	1008	1368	878	765
Average		$464	$491	$566	$648	$737	$857	$1,072	$1,455	$ 934	$814

To convert square feet of surface to board feet, 4% waste included.

S4S Size	Multiply S.F. by	T & G Size	Multiply S.F. by	Flooring Size	Multiply S.F. by
				25/32″ x 2-1/4″	1.37
1 x 4	1.18	1 x 4	1.27	25/32″ x 3-1/4″	1.29
1 x 6	1.13	1 x 6	1.18	15/32″ x 1-1/2″	1.54
1 x 8	1.11	1 x 8	1.14	1″ x 3″	1.28
1 x 10	1.09	2 x 6	2.36	1″ x 4″	1.24

R06110-030 Lumber Product Material Prices

The price of forest products fluctuates widely from location to location and from season to season depending upon economic conditions. The table below indicates National Average material prices in effect Jan. 1 of this book year. The table shows relative differences between various sizes, grades and species. These percentage differentials remain fairly constant even though lumber prices in general may change significantly during the year.

Availability of certain items depends upon geographic location and must be checked prior to firm price bidding.

	National Average Contractor Price, Quantity Purchase Dimension Lumber, S4S, #2 & Better, KD						Heavy Timbers, Fir	
	Species	2″x4″	2″x6″	2″x8″	2″x10″	2″x12″		
Framing Lumber per MBF	Douglas Fir	$ 392	$ 385	$ 412	$ 446	$ 432	3″ x 4″ thru 3″ x 12″	$527
	Spruce	311	293	302	359	367	4″ x 4″ thru 4″ x 12″	527
	Southern Yellow Pine	383	362	369	452	531	6″ x 6″ thru 6″ x 12″	770
	Hem-Fir	324	301	359	373	351	8″ x 8″ thru 8″ x 12″	1,060
							10″ x 10″ and 10″ x 12″	1,094

	S4S "D" Quality or Clear, KD						S4S # 2 & Better or Sterling, KD					
	Species	1″x4″	1″x6″	1″x8″	1″x10″	1″x12″	Species	1″x4″	1″x6″	1″x8″	1″x10″	1″x12″
Boards per MBF *See also Cedar Siding	Sugar Pine	$1,337	$1,553	$1,553	$2,261	$3,038	Sugar Pine	$560	$581	$601	$608	$864
	Idaho Pine	803	770	857	878	878	Idaho Pine	803	770	857	878	837
	Engleman Spruce	844	1,080	1,073	1,080	1,600	Engleman Spruce	601	628	614	540	810
	So. Yellow Pine	830	959	938	857	1,087	So. Yellow Pine	533	513	533	513	668
	Ponderosa Pine	770	1,094	945	1,195	1,890	Ponderosa Pine	594	601	628	493	736

R06160-020 Plywood

There are two types of plywood used in construction: interior, which is moisture resistant but not waterproofed, and exterior, which is waterproofed.

The grade of the exterior surface of the plywood sheets is designated by the first letter: A, for smooth surface with patches allowed; B, for solid surface with patches and plugs allowed; C, which may be surface plugged or may have knot holes up to 1″ wide; and D, which is used only for interior type plywood and may have knot holes up to 2-1/2″ wide. "Structural Grade" is specifically designed for engineered applications such as box beams. All CC & DD grades have roof and floor spans marked on them.

Underlayment grade plywood runs from 1/4″ to 1-1/4″ thick. Thicknesses 5/8″ and over have optional tongue and groove joints which eliminates the need for blocking the edges. Underlayment 19/32″ and over may be referred to as Sturd-i-Floor.

The price of plywood can fluctuate widely due to geographic and economic conditions. When one or two local prices are known, the relative prices for other types and sizes may be found by direct factoring of the prices in the table below.

Typical uses for various plywood grades are as follows:

AA-AD Interior — cupboards, shelving, paneling, furniture
BB Plyform — concrete form plywood
CDX — wall and roof sheathing
Structural — box beams, girders, stressed skin panels
AA-AC Exterior — fences, signs, siding, soffits, etc.
Underlayment — base for resilient floor coverings
Overlaid HDO — high density for concrete forms & highway signs
Overlaid MDO — medium density for painting, siding, soffits & signs
303 Siding — exterior siding, textured, striated, embossed, etc.

Grade	National Average Price in Lots of 10 MSF, per MSF-January				
	Type	4′x8′	Type	4′x8′	4′x10′
Sanded Grade	1/4″ Interior AD	$ 448	1/4″ Exterior AC	$ 450	$ 448
	3/8″	530	3/8″	530	530
	1/2″	619	1/2″	620	619
	5/8″	724	5/8″	724	724
	3/4″	800	3/4″	800	800
	1″	1,221	1″	1,342	1,342
	1-1/4″	1,464	Exterior AA, add	150	155
	Interior AA, add	150	Exterior AB, add	125	130
Unsanded Grade 4′ x 8′ Sheets			CD Structural 1		Underlayment
	5/16″ CDX	$ 420			
	3/8″	435	3/8″ 4′x8′ sheets	$340	3/8″ 4′x8′ sheets $ 530
	1/2″	392	1/2″	455	1/2″ 619
	5/8″	450	3/4″	655	5/8″T&G 540
	3/4″	574			3/4″T&G 839
	3/4″ T&G	839			1-1/8″ 2-4-1T&G 1,026
Form Plywood	5/8″ Exterior, oiled BB, plyform	$ 880	5/8″ HDO (overlay 2 sides)		$2,148
	3/4″ Exterior, oiled BB, plyform	934	3/4″ HDO (overlay 2 sides)		2,214
Overlaid 4′x8′ Sheets	Overlay 2 Sides MDO		Overlay 1 Side MDO		
	3/8″ thick	$1,345	3/8″ thick		$1,142
	1/2″	1,576	1/2″		1,316
	5/8″	1,776	5/8″		1,478
	3/4″	2,034	3/4″		1,734
303 Siding	Fir, rough sawn, natural finish, 3/8″ thick	$ 518	Texture 1-11 5/8″ thick, Fir		$ 945
	Redwood	1,796	Redwood		1,890
	Cedar	1,701	Cedar		1,537
	Southern Yellow Pine	418	Southern Yellow		763
Waferboard/O.S.B.	1/4″ sheathing	$ 250	19/32″ T&G		$ 424
	7/16″ sheathing	281	23/32″ T&G		510

For 2 MSF to 10 MSF, add 10%. For less than 2 MSF, add 15%.

R06170-100 Wood Roof Trusses

Loading figures represent live load. An additional load of 10 psf on the top chord and 10 psf on the bottom chord is included in the truss design. Spacing is 24" O.C.

Span in Feet	Cost per Truss for Different Live Loads and Roof Pitches					
	Flat	4 in 12 Pitch		5 in 12 Pitch		8 in 12 Pitch
	40 psf	30 psf	40 psf	30 psf	40 psf	30 psf
20	$ 64	$ 49	$ 51	$ 59	$ 54	$ 60
22	70	53	55	61	57	60
24	77	45	47	52	50	55
26	80	46	48	52	51	65
28	88	58	60	61	63	66
30	111	59	61	63	64	78
32	119	71	74	70	130	115
34	126	104	106	113	113	124
36	133	110	112	115	115	130
38	141	118	120	121	122	140
40	148	120	122	121	122	146

Thermal & Moist. Protection | R071 | Waterproofing & Dampproofing

R07110-010 1/2" Pargetting (rough dampproofing plaster)

1:2-1/2 Mix, 4.5 C.F. Covers 100 S.F., 2 Coats, Waste Included	Regular Portland Cement		Waterproofed Portland Cement	
1.7 Lbs. integral waterproofing admixture			$.83 per lb.	$ 1.41
1.7 Bags Portland cement	$8.10 per bag	$ 13.77	8.10 per bag	13.77
4.25 C.F. sand	18.80 per C.Y.	2.96	18.80 per C.Y.	2.96
.40 Days of labor to mix and apply (crew D-1)	335.20 per day	134.08	335.20 per day	134.08
Total Bare Cost per 100 S.F.		$150.81		$152.22

Thermal & Moist. Protection | R073 | Shingles & Roofing Tiles

R07310-020 Roof Slate

16", 18" and 20" are standard lengths, and slate usually comes in random widths. For standard 3/16" thickness use 1-1/2" copper nails. Allow for 3% breakage.

Quantities per Square	Unfading Vermont Colored	Weathering Sea Green	Buckingham, Virginia Black
Slate delivered (incl. punching)	$449.00	$325.00	$545.00
# 30 Felt, 2.5 lbs. copper nails	19.10	19.10	19.10
Slate roofer 4.6 hrs. @ $19.75 per hr.	90.85	90.85	90.85
Total Bare Cost per Square	$558.95	$434.95	$654.95

R07550-030 Modified Bitumen Roofing

The cost of modified bitumen roofing is highly dependent on the type of installation that is planned. Installation is based on the type of modifier used in the bitumen. The two most popular modifiers are atactic polypropylene (APP) and styrene butadiene styrene (SBS). The modifiers are added to heated bitumen during the manufacturing process to change its characteristics. A polyethylene, polyester or fiberglass reinforcing sheet is then sandwiched between layers of this bitumen. When completed, the result is a pre-assembled, built-up roof that has increased elasticity and weatherablility. Some manufacturers include a surfacing material such as ceramic or mineral granules, metal particles or sand.

The preferred method of adhering SBS-modified bitumen roofing to the substrate is with hot-mopped asphalt (much the same as built-up roofing). This installation method requires a tar kettle/pot to heat the asphalt, as well as the labor, tools and equipment necessary to distribute and spread the hot asphalt.

The alternative method for applying APP and SBS modified bitumen is as follows. A skilled installer uses a torch to melt a small pool of bitumen off the membrane. This pool must form across the entire roll for proper adhesion. The installer must unroll the roofing at a pace slow enough to melt the bitumen, but fast enough to prevent damage to the rest of the membrane.

Modified bitumen roofing provides the advantages of both built-up and single-ply roofing. Labor costs are reduced over those of built-up roofing because only a single ply is necessary. The elasticity of single-ply roofing is attained with the reinforcing sheet and polymer modifiers. Modifieds have some self-healing characteristics and because of their multi-layer construction, they offer the reliability and safety of built-up roofing.

Doors & Windows | R085 | Windows

R08550-010 Double Hung Windows - Tilt Wash

Ponderosa pine and vinyl clad sash, exterior primed with double insulated, low E glass.

Description	2′-0″ x 3′-0″				3′-0″ x 4′-0″			
	Wood		Vinyl Clad		Wood		Vinyl Clad	
Window, 2 lights w/ screens & grilles		$215.00		$244.90		$315.70		$338.70
Interior trim set		15.75		15.75		27.30		27.30
Carpenter @ $23.10 per hr.	1.7 hr.	39.27	1.7 hr.	39.27	2 hr.	46.20	2 hr.	46.20
Complete bare cost in place		$270.02		$299.92		$389.20		$412.20

R08550-200 Replacement Windows

Replacement windows are typically measured per United Inch.

United Inches are calculated by rounding the width and height of the window opening up to the nearest inch, then adding the two figures.

The labor cost for replacement windows includes removal of sash, existing sash balance or weights, parting bead where necessary and installation of new window.

Debris hauling and dump fees are not included.

Doors & Windows | R087 | Hardware

R08700-100 Hinges

All closer equipped doors should have ball bearing hinges. Lead lined or extremely heavy doors require special strength hinges. Usually 1-1/2 pair of hinges are used per door up to 7′-6″ high openings. Table below shows typical hinge requirements.

Use Frequency	Type Hinge Required	Type of Opening	Type of Structure
High	Heavy weight	Entrances	Banks, Office buildings, Schools, Stores & Theaters
	ball bearing	Toilet Rooms	Office buildings and Schools
Average	Standard	Entrances	Dwellings
	weight	Corridors	Office buildings and Schools
	ball bearing	Toilet Rooms	Stores
Low	Plain bearing	Interior	Dwellings

Door Thickness	Weight of Doors in Pounds per Square Foot				
	White Pine	Oak	Hollow Core	Solid Core	Hollow Metal
1-3/8″	3 psf	6 psf	1-1/2 psf	3-1/2 — 4 psf	6-1/2 psf
1-3/4″	3-1/2	7	2-1/2	4-1/2 — 5-1/4	6-1/2
2-1/4″	4-1/2	9	—	5-1/2 — 6-3/4	6-1/2

R09250-050 Lath, Plaster and Gypsum Board

Gypsum board lath is available in 3/8″ thick x 16″ wide x 4′ long sheets as a base material for multi-layer plaster applications. It is also available as a base for either multi-layer or veneer plaster applications in 1/2″ and 5/8″ thick–4′ wide x 8′, 10′ or 12′ long sheets. Fasteners are screws or blued ring shank nails for wood framing and screws for metal framing.

Metal lath is available in diamond mesh pattern with flat or self-furring profiles. Paper backing is available for applications where excessive plaster waste needs to be avoided. A slotted mesh ribbed lath should be used in areas where the span between structural supports is greater than normal. Most metal lath comes in 27″ x 96″ sheets. Diamond mesh weighs 1.75, 2.5 or 3.4 pounds per square yard, slotted mesh lath weighs 2.75 or 3.4 pounds per square yard. Metal lath can be nailed, screwed or tied in place.

Many **accessories** are available. Corner beads, flat reinforcing strips, casing beads, control and expansion joints, furring brackets and channels are some examples. Note that accessories are not included in plaster or stucco line items.

Plaster is defined as a material or combination of materials that when mixed with a suitable amount of water, forms a plastic mass or paste. When applied to a surface, the paste adheres to it and subsequently hardens, preserving in a rigid state the form or texture imposed during the period of elasticity.

Gypsum plaster is made from ground calcined gypsum. It is mixed with aggregates and water for use as a base coat plaster.

Vermiculite plaster is a fire-retardant plaster covering used on steel beams, concrete slabs and other heavy construction materials. Vermiculite is a group name for certain clay minerals, hydrous silicates or aluminum, magnesium and iron that have been expanded by heat.

Perlite plaster is a plaster using perlite as an aggregate instead of sand. Perlite is a volcanic glass that has been expanded by heat.

Gauging plaster is a mix of gypsum plaster and lime putty that when applied produces a quick drying finish coat.

Veneer plaster is a one or two component gypsum plaster used as a thin finish coat over special gypsum board.

Keenes cement is a white cementitious material manufactured from gypsum that has been burned at a high temperature and ground to a fine powder. Alum is added to accelerate the set. The resulting plaster is hard and strong and accepts and maintains a high polish, hence it is used as a finishing plaster.

Stucco is a Portland cement based plaster used primarily as an exterior finish.

Plaster is used on both interior and exterior surfaces. Generally it is applied in multiple-coat systems. A three-coat system uses the terms scratch, brown and finish to identify each coat. A two-coat system uses base and finish to describe each coat. Each type of plaster and application system has attributes that are chosen by the designer to best fit the intended use.

Gypsum Plaster Quantities for 100 S.Y.	2 Coat, 5/8″ Thick		3 Coat, 3/4″ Thick		
	Base 1:3 Mix	Finish 2:1 Mix	Scratch 1:2 Mix	Brown 1:3 Mix	Finish 2:1 Mix
Gypsum plaster	1,300 lb.		1,350 lb.	650 lb.	
Sand	1.75 C.Y.		1.85 C.Y.	1.35 C.Y.	
Finish hydrated lime		340 lb.			340 lb.
Gauging plaster		170 lb.			170 lb.

Vermiculite or Perlite Plaster Quantities for 100 S.Y.	2 Coat, 5/8″ Thick		3 Coat, 3/4″ Thick		
	Base	Finish	Scratch	Brown	Finish
Gypsum plaster	1,250 lb.		1,450 lb.	800 lb.	
Vermiculite or perlite	7.8 bags		8.0 bags	3.3 bags	
Finish hydrated lime		340 lb.			340 lb.
Gauging plaster		170 lb.			170 lb.

Stucco-Three-Coat System Quantities for 100 S.Y.	On Wood Frame	On Masonry
Portland cement	29 bags	21 bags
Sand	2.6 C.Y.	2.0 C.Y.
Hydrated lime	180 lb.	120 lb.

R09250-100 Levels of Gypsum Drywall Finish

In the past, contract documents often used phrases such as "industry standard" and "workmanlike finish" to specify the expected quality of gypsum board wall and ceiling installations. The vagueness of these descriptions led to unacceptable work and disputes.

In order to resolve this problem, four major trade associations concerned with the manufacture, erection, finish and decoration of gypsum board wall and ceiling systems have developed an industry-wide *Recommended Levels of Gypsum Board Finish.*

The finish of gypsum board walls and ceilings for specific final decoration is dependent on a number of factors. A primary consideration is the location of the surface and the degree of decorative treatment desired. Painted and unpainted surfaces in warehouses and other areas where appearance is normally not critical may simply require the taping of wallboard joints and 'spotting' of fastener heads. Blemish-free, smooth, monolithic surfaces often intended for painted and decorated walls and ceilings in habitated structures, ranging from single-family dwellings through monumental buildings, require additional finishing prior to the application of the final decoration.

Other factors to be considered in determining the level of finish of the gypsum board surface are (1) the type of angle of surface illumination (both natural and artificial lighting), and (2) the paint and method of application or the type and finish of wallcovering specified as the final decoration. Critical lighting conditions, gloss paints, and thin wallcoverings require a higher level of gypsum board finish than do heavily textured surfaces which are subsequently painted or surfaces which are to be decorated with heavy grade wallcoverings.

The following descriptions were developed jointly by the Association of the Wall and Ceiling Industries-International (AWCI), Ceiling & Interior Systems Construction Association (CISCA), Gypsum Association (GA), and Painting and Decorating Contractors of America (PDCA) as a guide.

Level 0: No taping, finishing, or accessories required. This level of finish may be useful in temporary construction or whenever the final decoration has not been determined.

Level 1: All joints and interior angles shall have tape set in joint compound. Surface shall be free of excess joint compound. Tool marks and ridges are acceptable. Frequently specified in plenum areas above ceilings, in attics, in areas where the assembly would generally be concealed or in building service corridors, and other areas not normally open to public view.

Level 2: All joints and interior angles shall have tape embedded in joint compound and wiped with a joint knife leaving a thin coating of joint compound over all joints and interior angles. Fastener heads and accessories shall be covered with a coat of joint compound. Surface shall be free of excess joint compound. Tool marks and ridges are acceptable. Joint compound applied over the body of the tape at the time of tape embedment shall be considered a separate coat of joint compound and shall satisfy the conditions of this level. Specified where water-resistant gypsum backing board is used as a substrate for tile; may be specified in garages, warehouse storage, or other similar areas where surface appearance is not of primary concern.

Level 3: All joints and interior angles shall have tape embedded in joint compound and one additional coat of joint compound applied over all joints and interior angles. Fastener heads and accessories shall be covered with two separate coats of joint compound. All joint compound shall be smooth and free of tool marks and ridges. Typically specified in appearance areas which are to receive heavy- or medium-texture (spray or hand applied) finishes before final painting, or where heavy-grade wallcoverings are to be applied as the final decoration. This level of finish is not recommended where smooth painted surfaces or light to medium wallcoverings are specified.

Level 4: All joints and interior angles shall have tape embedded in joint compound and two separate coats of joint compound applied over all flat joints and one separate coat of joint compound applied over interior angles. Fastener heads and accessories shall be covered with three separate coats of joint compound. All joint compound shall be smooth and free of tool marks and ridges. This level should be specified where flat paints, light textures, or wallcoverings are to be applied. In critical lighting areas, flat paints applied over light textures tend to reduce joint photographing. Gloss, semi-gloss, and enamel paints are not recommended over this level of finish. The weight, texture, and sheen level of wallcoverings applied over this level of finish should be carefully evaluated. Joints and fasteners must be adequately concealed if the wallcovering material is lightweight, contains limited pattern, has a gloss finish, or any combination of these finishes is present. Unbacked vinyl wallcoverings are not recommended over this level of finish.

Level 5: All joints and interior angles shall have tape embedded in joint compound and two separate coats of joint compound applied over all flat joints and one separate coat of joint compound applied over interior angles. Fastener heads and accessories shall be covered with three separate coats of joint compound. A thin skim coat of joint compound or a material manufactured especially for this purpose, shall be applied to the entire surface. The surface shall be smooth and free of tool marks and ridges. This level of finish is highly recommended where gloss, semi-gloss, enamel, or nontextured flat paints are specified or where severe lighting conditions occur. This highest quality finish is the most effective method to provide a uniform surface and minimize the possibility of joint photographing and of fasteners showing through the final decoration.

R09700-700 Wall Covering

Quantities for 100 S.F.				Medium Price Paper			Expensive Paper		
				Quantities	Bare Cost	Incl. O & P	Quantities	Bare Cost	Incl. O & P
Paper @	$30.00 and	$57.00	per double roll	1.6 dbl. rolls	$48.00	$ 52.80	1.6 dbl. rolls	$ 91.20	$100.32
Wall sizing @	$15.20 per gallon			.25 gallon	3.80	4.18	.25 gallon	3.80	4.18
Vinyl wall paste @	$ 9.00 per gallon			0.6 gallon	5.40	5.94	0.6 gallon	5.40	5.94
Apply sizing @	$20.85 and	$34.25	per hour	0.3 hour	6.26	10.28	0.3 hour	6.26	10.28
Apply paper @	$20.85 and	$34.25	per hour	1.2 hours	25.02	41.10	1.5 hours	31.28	51.38
Total cost in place per 100 S.F.					$88.48	$114.30		$137.94	$172.10
Total cost in place per double roll					$55.30	$ 71.44		$ 86.21	$107.56

Most wallpapers now come in double rolls only.
To remove old paper, allow 1.3 hours per 100 S.F.

R09910-220 Painting

Item	Coat	One Gallon Covers			In 8 Hours a Laborer Covers			Labor-Hours per 100 S.F.		
		Brush	Roller	Spray	Brush	Roller	Spray	Brush	Roller	Spray
Paint wood siding	prime	250 S.F.	225 S.F.	290 S.F.	1150 S.F.	1300 S.F.	2275 S.F.	.695	.615	.351
	others	270	250	290	1300	1625	2600	.615	.492	.307
Paint exterior trim	prime	400	—	—	650	—	—	1.230	—	—
	1st	475	—	—	800	—	—	1.000	—	—
	2nd	520	—	—	975	—	—	.820	—	—
Paint shingle siding	prime	270	255	300	650	975	1950	1.230	.820	.410
	others	360	340	380	800	1150	2275	1.000	.695	.351
Stain shingle siding	1st	180	170	200	750	1125	2250	1.068	.711	.355
	2nd	270	250	290	900	1325	2600	.888	.603	.307
Paint brick masonry	prime	180	135	160	750	800	1800	1.066	1.000	.444
	1st	270	225	290	815	975	2275	.981	.820	.351
	2nd	340	305	360	815	1150	2925	.981	.695	.273
Paint interior plaster or drywall	prime	400	380	495	1150	2000	3250	.695	.400	.246
	others	450	425	495	1300	2300	4000	.615	.347	.200
Paint interior doors and windows	prime	400	—	—	650	—	—	1.230	—	—
	1st	425	—	—	800	—	—	1.000	—	—
	2nd	450	—	—	975	—	—	.820	—	—

R13128-520 Swimming Pools

Pool prices given per square foot of surface area include pool structure, filter and chlorination equipment, pumps, related piping, ladders/steps, maintenance kit, skimmer and vacuum system. Decks and electrical service to equipment are not included.

Residential in-ground pool construction can be divided into two categories: vinyl lined and gunite. Vinyl lined pool walls are constructed of different materials including wood, concrete, plastic or metal. The bottom is often graded with sand over which the vinyl liner is installed. Vermiculite or soil cement bottoms may be substituted for an added cost.

Gunite pool construction is used both in residential and municipal installations. These structures are steel reinforced for strength and finished with a white cement limestone plaster.

Municipal pools will have a higher cost because plumbing codes require more expensive materials, chlorination equipment and higher filtration rates.

Municipal pools greater than 1,800 S.F. require gutter systems to control waves. This gutter may be formed into the concrete wall. Often a vinyl/stainless steel gutter or gutter/wall system is specified, which will raise the pool cost.

Competition pools usually require tile bottoms and sides with contrasting lane striping, which will also raise the pool cost.

Special Construction | R136 | Solar & Wind Energy Equip.

R13600-610 Solar Heating (Space and Hot Water)

Collectors should face as close to due South as possible, however, variations of up to 20 degrees on either side of true South are acceptable. Local climate and collector type may influence the choice between east or west deviations. Obviously they should be located so they are not shaded from the sun's rays. Incline collectors at a slope of latitude minus 5 degrees for domestic hot water and latitude plus 15 degrees for space heating.

Flat plate collectors consist of a number of components as follows: Insulation to reduce heat loss through the bottom and sides of the collector. The enclosure which contains all the components in this assembly is usually weatherproof and prevents dust, wind and water from coming in contact with the absorber plate. The cover plate usually consists of one or more layers of a variety of glass or plastic and reduces the reradiation by creating an air space which traps the heat between the cover and the absorber plates.

The absorber plate must have a good thermal bond with the fluid passages. The absorber plate is usually metallic and treated with a surface coating which improves absorptivity. Black or dark paints or selective coatings are used for this purpose, and the design of this passage and plate combination helps determine a solar system's effectiveness.

Heat transfer fluid passage tubes are attached above and below or integral with an absorber plate for the purpose of transferring thermal energy from the absorber plate to a heat transfer medium. The heat exchanger is a device for transferring thermal energy from one fluid to another.

Piping and storage tanks should be well insulated to minimize heat losses.

Size domestic water heating storage tanks to hold 20 gallons of water per user, minimum, plus 10 gallons per dishwasher or washing machine. For domestic water heating an optimum collector size is approximately 3/4 square foot of area per gallon of water storage. For space heating of residences and small commercial applications the collector is commonly sized between 30% and 50% of the internal floor area. For space heating of large commercial applications, collector areas less than 30% of the internal floor area can still provide significant heat reductions.

A supplementary heat source is recommended for Northern states for December through February.

The solar energy transmission per square foot of collector surface varies greatly with the material used. Initial cost, heat transmittance and useful life are obviously interrelated.

R15100-050 Pipe Material Considerations

1. Malleable fittings should be used for gas service.
2. Malleable fittings are used where there are stresses/strains due to expansion and vibration.
3. Cast fittings may be broken as an aid to disassembling of heating lines frozen by long use, temperature and minerals.
4. Cast iron pipe is extensively used for underground and submerged service.
5. Type M (light wall) copper tubing is available in hard temper only and is used for nonpressure and less severe applications than K and L.
6. Type L (medium wall) copper tubing, available hard or soft for interior service.
7. Type K (heavy wall) copper tubing, available in hard or soft temper for use where conditions are severe. For underground and interior service.
8. Hard drawn tubing requires fewer hangers or supports but should not be bent. Silver brazed fittings are recommended, however soft solder is normally used.
9. Type DMV (very light wall) copper tubing designed for drainage, waste and vent plus other non-critical pressure services.

Domestic/Imported Pipe and Fittings Cost

The prices shown in this publication for steel/cast iron pipe and steel, cast iron, malleable iron fittings are based on domestic production sold at the normal trade discounts. The above listed items of foreign manufacture may be available at prices of 1/3 to 1/2 those shown. Some imported items after minor machining or finishing operations are being sold as domestic to further complicate the system.

Caution: Most pipe prices in this book also include a coupling and pipe hangers which for the larger sizes can add significantly to the per foot cost and should be taken into account when comparing "book cost" with quoted supplier's cost.

R15100-420 Plumbing Fixture Installation Time

Item	Rough-In	Set	Total Hours	Item	Rough-In	Set	Total Hours
Bathtub	5	5	10	Shower head only	2	1	3
Bathtub and shower, cast iron	6	6	12	Shower drain	3	1	4
Fire hose reel and cabinet	4	2	6	Shower stall, slate		15	15
Floor drain to 4 inch diameter	3	1	4	Slop sink	5	3	8
Grease trap, single, cast iron	5	3	8	Test 6 fixtures			14
Kitchen gas range		4	4	Urinal, wall	6	2	8
Kitchen sink, single	4	4	8	Urinal, pedestal or floor	6	4	10
Kitchen sink, double	6	6	12	Water closet and tank	4	3	7
Laundry tubs	4	2	6	Water closet and tank, wall hung	5	3	8
Lavatory wall hung	5	3	8	Water heater, 45 gals. gas, automatic	5	2	7
Lavatory pedestal	5	3	8	Water heaters, 65 gals. gas, automatic	5	2	7
Shower and stall	6	4	10	Water heaters, electric, plumbing only	4	2	6

Fixture prices in front of book are based on the cost per fixture set in place. The rough-in cost, which must be added for each fixture, includes carrier, if required, some supply, waste and vent pipe connecting fittings and stops. The lengths of rough-in pipe are nominal runs which would connect to the larger runs and stacks. The supply runs and DWV runs and stacks must be accounted for in separate entries. In the eastern half of the United States it is common for the plumber to carry these to a point 5′ outside the building.

Crew No.	Bare Costs		Incl. Subs O & P		Cost Per Labor-Hour	

Crew A-1	Hr.	Daily	Hr.	Daily	Bare Costs	Incl. O&P
1 Building Laborer	$16.90	$135.20	$28.70	$229.60	$16.90	$28.70
1 Concrete saw, gas manual		42.80		47.10	5.35	5.89
8 L.H., Daily Totals		$178.00		$276.70	$22.25	$34.59

Crew A-1A	Hr.	Daily	Hr.	Daily	Bare Costs	Incl. O&P
1 Skilled Worker	$23.05	$184.40	$39.05	$312.40	$23.05	$39.05
1 Shot Blaster, 20″		274.30		301.75	34.29	37.72
8 L.H., Daily Totals		$458.70		$614.15	$57.34	$76.77

Crew A-1B	Hr.	Daily	Hr.	Daily	Bare Costs	Incl. O&P
1 Laborers, (Semi-Skilled)	$16.90	$135.20	$28.70	$229.60	$16.90	$28.70
1 Concr. saw, gas, self-prop.		92.20		101.40	11.53	12.68
8 L.H., Daily Totals		$227.40		$331.00	$28.43	$41.38

Crew A-1C	Hr.	Daily	Hr.	Daily	Bare Costs	Incl. O&P
1 Building Laborer	$16.90	$135.20	$28.70	$229.60	$16.90	$28.70
1 Brush saw		18.80		20.70	2.35	2.59
8 L.H., Daily Totals		$154.00		$250.30	$19.25	$31.29

Crew A-1D	Hr.	Daily	Hr.	Daily	Bare Costs	Incl. O&P
1 Building Laborer	$16.90	$135.20	$28.70	$229.60	$16.90	$28.70
1 Vibrating plate, gas, 18″		28.60		31.45	3.58	3.93
8 L.H., Daily Totals		$163.80		$261.05	$20.48	$32.63

Crew A-1E	Hr.	Daily	Hr.	Daily	Bare Costs	Incl. O&P
1 Building Laborer	$16.90	$135.20	$28.70	$229.60	$16.90	$28.70
1 Vibrating plate, gas, 21″		41.40		45.55	5.18	5.69
8 L.H., Daily Totals		$176.60		$275.15	$22.08	$34.39

Crew A-1F	Hr.	Daily	Hr.	Daily	Bare Costs	Incl. O&P
1 Building Laborer	$16.90	$135.20	$28.70	$229.60	$16.90	$28.70
1 Rammer/tamper, gas, 6″ - 11″		35.40		38.95	4.43	4.87
8 L.H., Daily Totals		$170.60		$268.55	$21.33	$33.57

Crew A-1G	Hr.	Daily	Hr.	Daily	Bare Costs	Incl. O&P
1 Building Laborer	$16.90	$135.20	$28.70	$229.60	$16.90	$28.70
1 Rammer/tamper, gas, 13″ - 18″		35.40		38.95	4.43	4.87
8 L.H., Daily Totals		$170.60		$268.55	$21.33	$33.57

Crew A-1H	Hr.	Daily	Hr.	Daily	Bare Costs	Incl. O&P
1 Building Laborer	$16.90	$135.20	$28.70	$229.60	$16.90	$28.70
1 Pressure washer		55.60		61.15	6.95	7.65
8 L.H., Daily Totals		$190.80		$290.75	$23.85	$36.35

Crew A-1J	Hr.	Daily	Hr.	Daily	Bare Costs	Incl. O&P
1 Building Laborer	$16.90	$135.20	$28.70	$229.60	$16.90	$28.70
1 Rototiller		70.90		78.00	8.86	9.75
8 L.H., Daily Totals		$206.10		$307.60	$25.76	$38.45

Crew A-1K	Hr.	Daily	Hr.	Daily	Bare Costs	Incl. O&P
1 Building Laborer	$16.90	$135.20	$28.70	$229.60	$16.90	$28.70
1 Lawn aerator		24.45		26.90	3.06	3.36
8 L.H., Daily Totals		$159.65		$256.50	$19.96	$32.06

Crew A-1L	Hr.	Daily	Hr.	Daily	Bare Costs	Incl. O&P
1 Building Laborer	$16.90	$135.20	$28.70	$229.60	$16.90	$28.70
1 Powwr blower/vacuum		24.45		26.90	3.06	3.36
8 L.H., Daily Totals		$159.65		$256.50	$19.96	$32.06

Crew A-1M	Hr.	Daily	Hr.	Daily	Bare Costs	Incl. O&P
1 Building Laborer	$16.90	$135.20	$28.70	$229.60	$16.90	$28.70
1 Snow blower		70.90		78.00	8.86	9.75
8 L.H., Daily Totals		$206.10		$307.60	$25.76	$38.45

Crew A-2	Hr.	Daily	Hr.	Daily	Bare Costs	Incl. O&P
2 Laborers	$16.90	$270.40	$28.70	$459.20	$17.35	$29.25
1 Truck Driver (light)	18.25	146.00	30.35	242.80		
1 Light Truck, 1.5 Ton		122.00		134.20	5.08	5.59
24 L.H., Daily Totals		$538.40		$836.20	$22.43	$34.84

Crew A-2A	Hr.	Daily	Hr.	Daily	Bare Costs	Incl. O&P
2 Laborers	$16.90	$270.40	$28.70	$459.20	$17.35	$29.25
1 Truck Driver (light)	18.25	146.00	30.35	242.80		
1 Light Truck, 1.5 Ton		122.00		134.20		
1 Concrete Saw		92.20		101.40	8.93	9.82
24 L.H., Daily Totals		$630.60		$937.60	$26.28	$39.07

Crew A-3	Hr.	Daily	Hr.	Daily	Bare Costs	Incl. O&P
1 Truck Driver (heavy)	$18.80	$150.40	$31.30	$250.40	$18.80	$31.30
1 Dump Truck, 12 Ton		325.00		357.50	40.63	44.69
8 L.H., Daily Totals		$475.40		$607.90	$59.43	$75.99

Crew A-3A	Hr.	Daily	Hr.	Daily	Bare Costs	Incl. O&P
1 Truck Driver (light)	$18.25	$146.00	$30.35	$242.80	$18.25	$30.35
1 Pickup Truck (4x4)		82.00		90.20	10.25	11.28
8 L.H., Daily Totals		$228.00		$333.00	$28.50	$41.63

Crew A-3B	Hr.	Daily	Hr.	Daily	Bare Costs	Incl. O&P
1 Equip. Oper. (medium)	$23.20	$185.60	$38.20	$305.60	$21.00	$34.75
1 Truck Driver (heavy)	18.80	150.40	31.30	250.40		
1 Dump Truck, 16 Ton		476.60		524.25		
1 F.E. Loader, 3 C.Y.		295.60		325.15	48.26	53.09
16 L.H., Daily Totals		$1108.20		$1405.40	$69.26	$87.84

Crew A-3C	Hr.	Daily	Hr.	Daily	Bare Costs	Incl. O&P
1 Equip. Oper. (light)	$22.20	$177.60	$36.55	$292.40	$22.20	$36.55
1 Wheeled Skid Steer Loader		183.80		202.20	22.98	25.27
8 L.H., Daily Totals		$361.40		$494.60	$45.18	$61.82

Crew A-3D	Hr.	Daily	Hr.	Daily	Bare Costs	Incl. O&P
1 Truck Driver, Light	$18.25	$146.00	$30.35	$242.80	$18.25	$30.35
1 Pickup Truck (4x4)		82.00		90.20		
1 Flatbed Trailer, 25 Ton		88.60		97.45	21.33	23.46
8 L.H., Daily Totals		$316.60		$430.45	$39.58	$53.81

Crew A-3E	Hr.	Daily	Hr.	Daily	Bare Costs	Incl. O&P
1 Equip. Oper. (crane)	$24.00	$192.00	$39.50	$316.00	$21.40	$35.40
1 Truck Driver (heavy)	18.80	150.40	31.30	250.40		
1 Pickup Truck (4x4)		82.00		90.20	5.13	5.64
16 L.H., Daily Totals		$424.40		$656.60	$26.53	$41.04

Crew A-3F	Hr.	Daily	Hr.	Daily	Bare Costs	Incl. O&P
1 Equip. Oper. (crane)	$24.00	$192.00	$39.50	$316.00	$21.40	$35.40
1 Truck Driver (heavy)	18.80	150.40	31.30	250.40		
1 Pickup Truck (4x4)		82.00		90.20		
1 Tractor, 6x2, 40 Ton Cap.		303.00		333.30		
1 Lowbed Trailer, 75 Ton		171.80		189.00	34.80	38.28
16 L.H., Daily Totals		$899.20		$1178.90	$56.20	$73.68

Crew No.	Bare Costs		Incl. Subs O & P		Cost Per Labor-Hour	
Crew A-3G	**Hr.**	**Daily**	**Hr.**	**Daily**	**Bare Costs**	**Incl. O&P**
1 Equip. Oper. (crane)	$24.00	$192.00	$39.50	$316.00	$21.40	$35.40
1 Truck Driver (heavy)	18.80	150.40	31.30	250.40		
1 Pickup Truck (4x4)		82.00		90.20		
1 Tractor, 6x4, 45 Ton Cap.		331.60		364.75		
1 Lowbed Trailer, 75 Ton		171.80		189.00	36.59	40.25
16 L.H., Daily Totals		$927.80		$1210.35	$57.99	$75.65

Crew A-4	Hr.	Daily	Hr.	Daily	Bare Costs	Incl. O&P
2 Carpenters	$23.10	$369.60	$39.20	$627.20	$22.40	$37.63
1 Painter, Ordinary	21.00	168.00	34.50	276.00		
24 L.H., Daily Totals		$537.60		$903.20	$22.40	$37.63

Crew A-5	Hr.	Daily	Hr.	Daily	Bare Costs	Incl. O&P
2 Laborers	$16.90	$270.40	$28.70	$459.20	$17.05	$28.88
.25 Truck Driver (light)	18.25	36.50	30.35	60.70		
.25 Light Truck, 1.5 Ton		30.50		33.55	1.69	1.86
18 L.H., Daily Totals		$337.40		$553.45	$18.74	$30.74

Crew A-6	Hr.	Daily	Hr.	Daily	Bare Costs	Incl. O&P
1 Instrument Man	$23.05	$184.40	$39.05	$312.40	$22.90	$38.30
1 Rodman/Chainman	22.75	182.00	37.55	300.40		
1 Laser Transit/Level		59.50		65.45	3.72	4.09
16 L.H., Daily Totals		$425.90		$678.25	$26.62	$42.39

Crew A-7	Hr.	Daily	Hr.	Daily	Bare Costs	Incl. O&P
1 Chief Of Party	$27.25	$218.00	$46.15	$369.20	$24.35	$40.92
1 Instrument Man	23.05	184.40	39.05	312.40		
1 Rodman/Chainman	22.75	182.00	37.55	300.40		
1 Laser Transit/Level		59.50		65.45	2.48	2.73
24 L.H., Daily Totals		$643.90		$1047.45	$26.83	$43.65

Crew A-8	Hr.	Daily	Hr.	Daily	Bare Costs	Incl. O&P
1 Chief Of Party	$27.25	$218.00	$46.15	$369.20	$23.95	$40.08
1 Instrument Man	23.05	184.40	39.05	312.40		
2 Rodmen/Chainmen	22.75	364.00	37.55	600.80		
1 Laser Transit/Level		59.50		65.45	1.86	2.05
32 L.H., Daily Totals		$825.90		$1347.85	$25.81	$42.13

Crew A-9	Hr.	Daily	Hr.	Daily	Bare Costs	Incl. O&P
1 Asbestos Foreman	$24.25	$194.00	$41.60	$332.80	$23.81	$40.86
7 Asbestos Workers	23.75	1330.00	40.75	2282.00		
64 L.H., Daily Totals		$1524.00		$2614.80	$23.81	$40.86

Crew A-10	Hr.	Daily	Hr.	Daily	Bare Costs	Incl. O&P
1 Asbestos Foreman	$24.25	$194.00	$41.60	$332.80	$23.81	$40.86
7 Asbestos Workers	23.75	1330.00	40.75	2282.00		
64 L.H., Daily Totals		$1524.00		$2614.80	$23.81	$40.86

Crew A-10A	Hr.	Daily	Hr.	Daily	Bare Costs	Incl. O&P
1 Asbestos Foreman	$24.25	$194.00	$41.60	$332.80	$23.92	$41.03
2 Asbestos Workers	23.75	380.00	40.75	652.00		
24 L.H., Daily Totals		$574.00		$984.80	$23.92	$41.03

Crew A-10B	Hr.	Daily	Hr.	Daily	Bare Costs	Incl. O&P
1 Asbestos Foreman	$24.25	$194.00	$41.60	$332.80	$23.88	$40.96
3 Asbestos Workers	23.75	570.00	40.75	978.00		
32 L.H., Daily Totals		$764.00		$1310.80	$23.88	$40.96

Crew A-10C	Hr.	Daily	Hr.	Daily	Bare Costs	Incl. O&P
3 Asbestos Workers	$23.75	$570.00	$40.75	$978.00	$23.75	$40.75
1 Flatbed Truck		122.00		134.20	5.08	5.59
24 L.H., Daily Totals		$692.00		$1112.20	$28.83	$46.34

Crew A-10D	Hr.	Daily	Hr.	Daily	Bare Costs	Incl. O&P
2 Asbestos Workers	$23.75	$380.00	$40.75	$652.00	$22.91	$38.54
1 Equip. Oper. (crane)	24.00	192.00	39.50	316.00		
1 Equip. Oper. Oiler	20.15	161.20	33.15	265.20		
1 Hydraulic Crane, 33 Ton		649.00		713.90	20.28	22.31
32 L.H., Daily Totals		$1382.20		$1947.10	$43.19	$60.85

Crew A-11	Hr.	Daily	Hr.	Daily	Bare Costs	Incl. O&P
1 Asbestos Foreman	$24.25	$194.00	$41.60	$332.80	$23.81	$40.86
7 Asbestos Workers	23.75	1330.00	40.75	2282.00		
2 Chipping Hammers		32.00		35.20	.50	.55
64 L.H., Daily Totals		$1556.00		$2650.00	$24.31	$41.41

Crew A-12	Hr.	Daily	Hr.	Daily	Bare Costs	Incl. O&P
1 Asbestos Foreman	$24.25	$194.00	$41.60	$332.80	$23.81	$40.86
7 Asbestos Workers	23.75	1330.00	40.75	2282.00		
1 Large Prod. Vac. Loader		503.05		553.35	7.86	8.65
64 L.H., Daily Totals		$2027.05		$3168.15	$31.67	$49.51

Crew A-13	Hr.	Daily	Hr.	Daily	Bare Costs	Incl. O&P
1 Equip. Oper. (light)	$22.20	$177.60	$36.55	$292.40	$22.20	$36.55
1 Large Prod. Vac. Loader		503.05		553.35	62.88	69.17
8 L.H., Daily Totals		$680.65		$845.75	$85.08	$105.72

Crew B-1	Hr.	Daily	Hr.	Daily	Bare Costs	Incl. O&P
1 Labor Foreman (outside)	$18.90	$151.20	$32.10	$256.80	$17.57	$29.83
2 Laborers	16.90	270.40	28.70	459.20		
24 L.H., Daily Totals		$421.60		$716.00	$17.57	$29.83

Crew B-2	Hr.	Daily	Hr.	Daily	Bare Costs	Incl. O&P
1 Labor Foreman (outside)	$18.90	$151.20	$32.10	$256.80	$17.30	$29.38
4 Laborers	16.90	540.80	28.70	918.40		
40 L.H., Daily Totals		$692.00		$1175.20	$17.30	$29.38

Crew B-3	Hr.	Daily	Hr.	Daily	Bare Costs	Incl. O&P
1 Labor Foreman (outside)	$18.90	$151.20	$32.10	$256.80	$18.92	$31.72
2 Laborers	16.90	270.40	28.70	459.20		
1 Equip. Oper. (med.)	23.20	185.60	38.20	305.60		
2 Truck Drivers (heavy)	18.80	300.80	31.30	500.80		
1 F.E. Loader, T.M., 2.5 C.Y.		761.00		837.10		
2 Dump Trucks, 16 Ton		953.20		1048.50	35.71	39.28
48 L.H., Daily Totals		$2622.20		$3408.00	$54.63	$71.00

Crew B-3A	Hr.	Daily	Hr.	Daily	Bare Costs	Incl. O&P
4 Laborers	$16.90	$540.80	$28.70	$918.40	$18.16	$30.60
1 Equip. Oper. (med.)	23.20	185.60	38.20	305.60		
1 Hyd. Excavator, 1.5 C.Y.		683.40		751.75	17.09	18.79
40 L.H., Daily Totals		$1409.80		$1975.75	$35.25	$49.39

Crew B-3B	Hr.	Daily	Hr.	Daily	Bare Costs	Incl. O&P
2 Laborers	$16.90	$270.40	$28.70	$459.20	$18.95	$31.73
1 Equip. Oper. (med.)	23.20	185.60	38.20	305.60		
1 Truck Driver (heavy)	18.80	150.40	31.30	250.40		
1 Backhoe Loader, 80 H.P.		230.80		253.90		
1 Dump Truck, 16 Ton		476.60		524.25	22.11	24.32
32 L.H., Daily Totals		$1313.80		$1793.35	$41.06	$56.05

Crew No.	Bare Costs		Incl. Subs O & P		Cost Per Labor-Hour	
Crew B-3C	**Hr.**	**Daily**	**Hr.**	**Daily**	**Bare Costs**	**Incl. O&P**
3 Laborers	$16.90	$405.60	$28.70	$688.80	$18.48	$31.08
1 Equip. Oper. (med.)	23.20	185.60	38.20	305.60		
1 F.E. Crawler Ldr, 4 C.Y.		1061.00		1167.10	33.16	36.47
32 L.H., Daily Totals		$1652.20		$2161.50	$51.64	$67.55

Crew B-4	Hr.	Daily	Hr.	Daily	Bare Costs	Incl. O&P
1 Labor Foreman (outside)	$18.90	$151.20	$32.10	$256.80	$17.55	$29.70
4 Laborers	16.90	540.80	28.70	918.40		
1 Truck Driver (heavy)	18.80	150.40	31.30	250.40		
1 Tractor, 4 x 2, 195 H.P.		215.20		236.70		
1 Platform Trailer		120.60		132.65	7.00	7.70
48 L.H., Daily Totals		$1178.20		$1794.95	$24.55	$37.40

Crew B-5	Hr.	Daily	Hr.	Daily	Bare Costs	Incl. O&P
1 Labor Foreman (outside)	$18.90	$151.20	$32.10	$256.80	$18.56	$31.28
3 Laborers	16.90	405.60	28.70	688.80		
1 Equip. Oper. (med.)	23.20	185.60	38.20	305.60		
1 Air Compr., 250 C.F.M.		127.60		140.35		
2 Air Tools & Accessories		22.40		24.65		
2-50 Ft. Air Hoses, 1.5″ Dia.		10.00		11.00		
1 F.E. Loader, T.M., 2.5 C.Y.		761.00		837.10	23.02	25.32
40 L.H., Daily Totals		$1663.40		$2264.30	$41.58	$56.60

Crew B-5A	Hr.	Daily	Hr.	Daily	Bare Costs	Incl. O&P
1 Foreman	$18.90	$151.20	$32.10	$256.80	$18.88	$31.65
6 Laborers	16.90	811.20	28.70	1377.60		
2 Equip. Oper. (med.)	23.20	371.20	38.20	611.20		
1 Equip. Oper. (light)	22.20	177.60	36.55	292.40		
2 Truck Drivers (heavy)	18.80	300.80	31.30	500.80		
1 Air Compr. 365 C.F.M.		157.60		173.35		
2 Pavement Breakers		22.40		24.65		
8 Air Hoses w/Coup.,1″		27.60		30.35		
2 Dump Trucks, 12 Ton		650.00		715.00	8.93	9.83
96 L.H., Daily Totals		$2669.60		$3982.15	$27.81	$41.48

Crew B-5B	Hr.	Daily	Hr.	Daily	Bare Costs	Incl. O&P
1 Powderman	$23.05	$184.40	$39.05	$312.40	$20.98	$34.89
2 Equip. Oper. (med.)	23.20	371.20	38.20	611.20		
3 Truck Drivers (heavy)	18.80	451.20	31.30	751.20		
1 F.E. Ldr. 2-1/2 CY		295.60		325.15		
3 Dump Trucks, 16 Ton		1429.80		1572.80		
1 Air Compr. 365 C.F.M.		157.60		173.35	39.23	43.15
48 L.H., Daily Totals		$2889.80		$3746.10	$60.21	$78.04

Crew B-5C	Hr.	Daily	Hr.	Daily	Bare Costs	Incl. O&P
3 Laborers	$16.90	$405.60	$28.70	$688.80	$19.46	$32.44
1 Equip. Oper. (medium)	23.20	185.60	38.20	305.60		
2 Truck Drivers (heavy)	18.80	300.80	31.30	500.80		
1 Equip. Oper. (crane)	24.00	192.00	39.50	316.00		
1 Equip. Oper. Oiler	20.15	161.20	33.15	265.20		
2 Dump Trucks, 16 Ton		953.20		1048.50		
1 F.E. Crawler Ldr, 4 C.Y.		1061.00		1167.10		
1 Hyd. Crane, 25 Ton		575.20		632.70	40.46	44.51
64 L.H., Daily Totals		$3834.60		$4924.70	$59.92	$76.95

Crew B-6	Hr.	Daily	Hr.	Daily	Bare Costs	Incl. O&P
2 Laborers	$16.90	$270.40	$28.70	$459.20	$18.67	$31.32
1 Equip. Oper. (light)	22.20	177.60	36.55	292.40		
1 Backhoe Loader, 48 H.P.		207.80		228.60	8.66	9.52
24 L.H., Daily Totals		$655.80		$980.20	$27.33	$40.84

Crew No.	Bare Costs		Incl. Subs O & P		Cost Per Labor-Hour	
Crew B-6B	**Hr.**	**Daily**	**Hr.**	**Daily**	**Bare Costs**	**Incl. O&P**
2 Labor Foremen (out)	$18.90	$302.40	$32.10	$513.60	$17.57	$29.83
4 Laborers	16.90	540.80	28.70	918.40		
1 Winch Truck		317.80		349.60		
1 Flatbed Truck		122.00		134.20		
1 Butt Fusion Machine		433.60		476.95	18.20	20.02
48 L.H., Daily Totals		$1716.60		$2392.75	$35.77	$49.85

Crew B-7	Hr.	Daily	Hr.	Daily	Bare Costs	Incl. O&P
1 Labor Foreman (outside)	$18.90	$151.20	$32.10	$256.80	$18.28	$30.85
4 Laborers	16.90	540.80	28.70	918.40		
1 Equip. Oper. (med.)	23.20	185.60	38.20	305.60		
1 Chipping Machine		164.80		181.30		
1 F.E. Loader, T.M., 2.5 C.Y.		761.00		837.10		
2 Chain Saws, 36″		66.80		73.50	20.68	22.75
48 L.H., Daily Totals		$1870.20		$2572.70	$38.96	$53.60

Crew B-7A	Hr.	Daily	Hr.	Daily	Bare Costs	Incl. O&P
2 Laborers	$16.90	$270.40	$28.70	$459.20	$18.67	$31.32
1 Equip. Oper. (light)	22.20	177.60	36.55	292.40		
1 Rake w/Tractor		184.90		203.40		
2 Chain Saws, 18″		37.60		41.35	9.27	10.20
24 L.H., Daily Totals		$670.50		$996.35	$27.94	$41.52

Crew B-8	Hr.	Daily	Hr.	Daily	Bare Costs	Incl. O&P
1 Labor Foreman (outside)	$18.90	$151.20	$32.10	$256.80	$19.53	$32.64
2 Laborers	16.90	270.40	28.70	459.20		
2 Equip. Oper. (med.)	23.20	371.20	38.20	611.20		
2 Truck Drivers (heavy)	18.80	300.80	31.30	500.80		
1 Hyd. Crane, 25 Ton		624.20		686.60		
1 F.E. Loader, T.M., 2.5 C.Y.		761.00		837.10		
2 Dump Trucks, 16 Ton		953.20		1048.50	41.76	45.93
56 L.H., Daily Totals		$3432.00		$4400.20	$61.29	$78.57

Crew B-9	Hr.	Daily	Hr.	Daily	Bare Costs	Incl. O&P
1 Labor Foreman (outside)	$18.90	$151.20	$32.10	$256.80	$17.30	$29.38
4 Laborers	16.90	540.80	28.70	918.40		
1 Air Compr., 250 C.F.M.		127.60		140.35		
2 Air Tools & Accessories		22.40		24.65		
2-50 Ft. Air Hoses, 1.5″ Dia.		10.00		11.00	4.00	4.40
40 L.H., Daily Totals		$852.00		$1351.20	$21.30	$33.78

Crew B-9A	Hr.	Daily	Hr.	Daily	Bare Costs	Incl. O&P
2 Laborers	$16.90	$270.40	$28.70	$459.20	$17.53	$29.57
1 Truck Driver (heavy)	18.80	150.40	31.30	250.40		
1 Water Tanker		118.00		129.80		
1 Tractor		215.20		236.70		
2-50 Ft. Disch. Hoses		7.50		8.25	14.20	15.62
24 L.H., Daily Totals		$761.50		$1084.35	$31.73	$45.19

Crew B-9B	Hr.	Daily	Hr.	Daily	Bare Costs	Incl. O&P
2 Laborers	$16.90	$270.40	$28.70	$459.20	$17.53	$29.57
1 Truck Driver (heavy)	18.80	150.40	31.30	250.40		
2-50 Ft. Disch. Hoses		7.50		8.25		
1 Water Tanker		118.00		129.80		
1 Tractor		215.20		236.70		
1 Pressure Washer		46.40		51.05	16.13	17.74
24 L.H., Daily Totals		$807.90		$1135.40	$33.66	$47.31

Crew No.	Bare Costs		Incl. Subs O & P		Cost Per Labor-Hour	

Crew B-9C	Hr.	Daily	Hr.	Daily	Bare Costs	Incl. O&P
1 Labor Foreman (outside)	$18.90	$151.20	$32.10	$256.80	$17.30	$29.38
4 Laborers	16.90	540.80	28.70	918.40		
1 Air Compr., 250 C.F.M.		127.60		140.35		
2-50 Ft. Air Hoses, 1.5″ Dia.		10.00		11.00		
2 Breaker, Pavement, 60 lb.		22.40		24.65	4.00	4.40
40 L.H., Daily Totals		$852.00		$1351.20	$21.30	$33.78

Crew B-9D	Hr.	Daily	Hr.	Daily	Bare Costs	Incl. O&P
1 Labor Foreman (Outside)	$18.90	$151.20	$32.10	$256.80	$17.30	$29.38
4 Common Laborers	16.90	540.80	28.70	918.40		
1 Air Compressor, 250 CFM		127.60		140.35		
2 Air hoses, 1.5″ x 50′		10.00		11.00		
2 Air tamper		47.00		51.70	4.61	5.07
40 L.H., Daily Totals		$876.60		$1378.25	$21.91	$34.45

Crew B-10	Hr.	Daily	Hr.	Daily	Bare Costs	Incl. O&P
1 Equip. Oper. (med.)	$23.20	$185.60	$38.20	$305.60	$23.20	$38.20
8 L.H., Daily Totals		$185.60		$305.60	$23.20	$38.20

Crew B-10A	Hr.	Daily	Hr.	Daily	Bare Costs	Incl. O&P
1 Equip. Oper. (med.)	$23.20	$185.60	$38.20	$305.60	$23.20	$38.20
1 Roll. Compact., 2K Lbs.		130.80		143.90	16.35	17.99
8 L.H., Daily Totals		$316.40		$449.50	$39.55	$56.19

Crew B-10B	Hr.	Daily	Hr.	Daily	Bare Costs	Incl. O&P
1 Equip. Oper. (med.)	$23.20	$185.60	$38.20	$305.60	$23.20	$38.20
1 Dozer, 200 H.P.		863.40		949.75	107.93	118.72
8 L.H., Daily Totals		$1049.00		$1255.35	$131.13	$156.92

Crew B-10C	Hr.	Daily	Hr.	Daily	Bare Costs	Incl. O&P
1 Equip. Oper. (med.)	$23.20	$185.60	$38.20	$305.60	$23.20	$38.20
1 Dozer, 200 H.P.		863.40		949.75		
1 Vibratory Roller, Towed		610.20		671.20	184.20	202.62
8 L.H., Daily Totals		$1659.20		$1926.55	$207.40	$240.82

Crew B-10D	Hr.	Daily	Hr.	Daily	Bare Costs	Incl. O&P
1 Equip. Oper. (med.)	$23.20	$185.60	$38.20	$305.60	$23.20	$38.20
1 Dozer, 200 H.P.		863.40		949.75		
1 Sheepsft. Roller, Towed		77.80		85.60	117.65	129.42
8 L.H., Daily Totals		$1126.80		$1340.95	$140.85	$167.62

Crew B-10E	Hr.	Daily	Hr.	Daily	Bare Costs	Incl. O&P
1 Equip. Oper. (med.)	$23.20	$185.60	$38.20	$305.60	$23.20	$38.20
1 Tandem Roller, 5 Ton		108.20		119.00	13.53	14.88
8 L.H., Daily Totals		$293.80		$424.60	$36.73	$53.08

Crew B-10F	Hr.	Daily	Hr.	Daily	Bare Costs	Incl. O&P
1 Equip. Oper. (med.)	$23.20	$185.60	$38.20	$305.60	$23.20	$38.20
1 Tandem Roller, 10 Ton		183.60		201.95	22.95	25.25
8 L.H., Daily Totals		$369.20		$507.55	$46.15	$63.45

Crew B-10G	Hr.	Daily	Hr.	Daily	Bare Costs	Incl. O&P
1 Equip. Oper. (med.)	$23.20	$185.60	$38.20	$305.60	$23.20	$38.20
1 Sheepsft. Roll., 130 H.P.		762.60		838.85	95.33	104.86
8 L.H., Daily Totals		$948.20		$1144.45	$118.53	$143.06

Crew No.	Bare Costs		Incl. Subs O & P		Cost Per Labor-Hour	

Crew B-10H	Hr.	Daily	Hr.	Daily	Bare Costs	Incl. O&P
1 Equip. Oper. (med.)	$23.20	$185.60	$38.20	$305.60	$23.20	$38.20
1 Diaphr. Water Pump, 2″		46.20		50.80		
1-20 Ft. Suction Hose, 2″		3.55		3.90		
2-50 Ft. Disch. Hoses, 2″		6.30		6.95	7.01	7.71
8 L.H., Daily Totals		$241.65		$367.25	$30.21	$45.91

Crew B-10I	Hr.	Daily	Hr.	Daily	Bare Costs	Incl. O&P
1 Equip. Oper. (med.)	$23.20	$185.60	$38.20	$305.60	$23.20	$38.20
1 Diaphr. Water Pump, 4″		87.40		96.15		
1-20 Ft. Suction Hose, 4″		7.30		8.05		
2-50 Ft. Disch. Hoses, 4″		10.50		11.55	13.15	14.47
8 L.H., Daily Totals		$290.80		$421.35	$36.35	$52.67

Crew B-10J	Hr.	Daily	Hr.	Daily	Bare Costs	Incl. O&P
1 Equip. Oper. (med.)	$23.20	$185.60	$38.20	$305.60	$23.20	$38.20
1 Centr. Water Pump, 3″		53.80		59.20		
1-20 Ft. Suction Hose, 3″		5.45		6.00		
2-50 Ft. Disch. Hoses, 3″		7.50		8.25	8.34	9.18
8 L.H., Daily Totals		$252.35		$379.05	$31.54	$47.38

Crew B-10K	Hr.	Daily	Hr.	Daily	Bare Costs	Incl. O&P
1 Equip. Oper. (med.)	$23.20	$185.60	$38.20	$305.60	$23.20	$38.20
1 Centr. Water Pump, 6″		229.80		252.80		
1-20 Ft. Suction Hose, 6″		14.50		15.95		
2-50 Ft. Disch. Hoses, 6″		23.80		26.20	33.51	36.86
8 L.H., Daily Totals		$453.70		$600.55	$56.71	$75.06

Crew B-10L	Hr.	Daily	Hr.	Daily	Bare Costs	Incl. O&P
1 Equip. Oper. (med.)	$23.20	$185.60	$38.20	$305.60	$23.20	$38.20
1 Dozer, 80 H.P.		300.20		330.20	37.53	41.28
8 L.H., Daily Totals		$485.80		$635.80	$60.73	$79.48

Crew B-10M	Hr.	Daily	Hr.	Daily	Bare Costs	Incl. O&P
1 Equip. Oper. (med.)	$23.20	$185.60	$38.20	$305.60	$23.20	$38.20
1 Dozer, 300 H.P.		1099.00		1208.90	137.38	151.11
8 L.H., Daily Totals		$1284.60		$1514.50	$160.58	$189.31

Crew B-10N	Hr.	Daily	Hr.	Daily	Bare Costs	Incl. O&P
1 Equip. Oper. (med.)	$23.20	$185.60	$38.20	$305.60	$23.20	$38.20
1 F.E. Loader, T.M., 1.5 C.Y		300.60		330.65	37.58	41.33
8 L.H., Daily Totals		$486.20		$636.25	$60.78	$79.53

Crew B-10O	Hr.	Daily	Hr.	Daily	Bare Costs	Incl. O&P
1 Equip. Oper. (med.)	$23.20	$185.60	$38.20	$305.60	$23.20	$38.20
1 F.E. Loader, T.M., 2.25 C.Y.		538.00		591.80	67.25	73.98
8 L.H., Daily Totals		$723.60		$897.40	$90.45	$112.18

Crew B-10P	Hr.	Daily	Hr.	Daily	Bare Costs	Incl. O&P
1 Equip. Oper. (med.)	$23.20	$185.60	$38.20	$305.60	$23.20	$38.20
1 F.E. Loader, T.M., 2.5 C.Y.		761.00		837.10	95.13	104.64
8 L.H., Daily Totals		$946.60		$1142.70	$118.33	$142.84

Crew B-10Q	Hr.	Daily	Hr.	Daily	Bare Costs	Incl. O&P
1 Equip. Oper. (med.)	$23.20	$185.60	$38.20	$305.60	$23.20	$38.20
1 F.E. Loader, T.M., 5 C.Y.		1061.00		1167.10	132.63	145.89
8 L.H., Daily Totals		$1246.60		$1472.70	$155.83	$184.09

Crew No.	Bare Costs		Incl. Subs O & P		Cost Per Labor-Hour	
Crew B-10R	**Hr.**	**Daily**	**Hr.**	**Daily**	**Bare Costs**	**Incl. O&P**
1 Equip. Oper. (med.)	$23.20	$185.60	$38.20	$305.60	$23.20	$38.20
1 F.E. Loader, W.M., 1 C.Y.		192.00		211.20	24.00	26.40
8 L.H., Daily Totals		$377.60		$516.80	$47.20	$64.60
Crew B-10S	**Hr.**	**Daily**	**Hr.**	**Daily**	**Bare Costs**	**Incl. O&P**
1 Equip. Oper. (med.)	$23.20	$185.60	$38.20	$305.60	$23.20	$38.20
1 F.E. Loader, W.M., 1.5 C.Y.		236.20		259.80	29.53	32.48
8 L.H., Daily Totals		$421.80		$565.40	$52.73	$70.68
Crew B-10T	**Hr.**	**Daily**	**Hr.**	**Daily**	**Bare Costs**	**Incl. O&P**
1 Equip. Oper. (med.)	$23.20	$185.60	$38.20	$305.60	$23.20	$38.20
1 F.E. Loader, W.M.,2.5 C.Y.		295.60		325.15	36.95	40.65
8 L.H., Daily Totals		$481.20		$630.75	$60.15	$78.85
Crew B-10U	**Hr.**	**Daily**	**Hr.**	**Daily**	**Bare Costs**	**Incl. O&P**
1 Equip. Oper. (med.)	$23.20	$185.60	$38.20	$305.60	$23.20	$38.20
1 F.E. Loader, W.M., 5.5 C.Y.		674.80		742.30	84.35	92.79
8 L.H., Daily Totals		$860.40		$1047.90	$107.55	$130.99
Crew B-10V	**Hr.**	**Daily**	**Hr.**	**Daily**	**Bare Costs**	**Incl. O&P**
1 Equip. Oper. (med.)	$23.20	$185.60	$38.20	$305.60	$23.20	$38.20
1 Dozer, 700 H.P.		3047.00		3351.70	380.88	418.96
8 L.H., Daily Totals		$3232.60		$3657.30	$404.08	$457.16
Crew B-10W	**Hr.**	**Daily**	**Hr.**	**Daily**	**Bare Costs**	**Incl. O&P**
1 Equip. Oper. (med.)	$23.20	$185.60	$38.20	$305.60	$23.20	$38.20
1 Dozer, 105 H.P.		434.00		477.40	54.25	59.68
8 L.H., Daily Totals		$619.60		$783.00	$77.45	$97.88
Crew B-10X	**Hr.**	**Daily**	**Hr.**	**Daily**	**Bare Costs**	**Incl. O&P**
1 Equip. Oper. (med.)	$23.20	$185.60	$38.20	$305.60	$23.20	$38.20
1 Dozer, 410 H.P.		1473.00		1620.30	184.13	202.54
8 L.H., Daily Totals		$1658.60		$1925.90	$207.33	$240.74
Crew B-10Y	**Hr.**	**Daily**	**Hr.**	**Daily**	**Bare Costs**	**Incl. O&P**
1 Equip. Oper. (med.)	$23.20	$185.60	$38.20	$305.60	$23.20	$38.20
1 Vibratory Drum Roller		323.60		355.95	40.45	44.50
8 L.H., Daily Totals		$509.20		$661.55	$63.65	$82.70
Crew B-11A	**Hr.**	**Daily**	**Hr.**	**Daily**	**Bare Costs**	**Incl. O&P**
1 Equipment Oper. (med.)	$23.20	$185.60	$38.20	$305.60	$20.05	$33.45
1 Laborer	16.90	135.20	28.70	229.60		
1 Dozer, 200 H.P.		863.40		949.75	53.96	59.36
16 L.H., Daily Totals		$1184.20		$1484.95	$74.01	$92.81
Crew B-11B	**Hr.**	**Daily**	**Hr.**	**Daily**	**Bare Costs**	**Incl. O&P**
1 Equipment Oper. (light)	$22.20	$177.60	$36.55	$292.40	$19.55	$32.63
1 Laborer	16.90	135.20	28.70	229.60		
1 Air Powered Tamper		23.50		25.85		
1 Air Compr. 365 C.F.M.		157.60		173.35		
2-50 Ft. Air Hoses, 1.5″ Dia.		10.00		11.00	11.94	13.13
16 L.H., Daily Totals		$503.90		$732.20	$31.49	$45.76
Crew B-11C	**Hr.**	**Daily**	**Hr.**	**Daily**	**Bare Costs**	**Incl. O&P**
1 Equipment Oper. (med.)	$23.20	$185.60	$38.20	$305.60	$20.05	$33.45
1 Laborer	16.90	135.20	28.70	229.60		
1 Backhoe Loader, 48 H.P.		207.80		228.60	12.99	14.29
16 L.H., Daily Totals		$528.60		$763.80	$33.04	$47.74

Crew No.	Bare Costs		Incl. Subs O & P		Cost Per Labor-Hour	
Crew B-11K	**Hr.**	**Daily**	**Hr.**	**Daily**	**Bare Costs**	**Incl. O&P**
1 Equipment Oper. (med.)	$23.20	$185.60	$38.20	$305.60	$20.05	$33.45
1 Laborer	16.90	135.20	28.70	229.60		
1 Trencher, 8′ D., 16″ W.		1310.00		1441.00	81.88	90.06
16 L.H., Daily Totals		$1630.80		$1976.20	$101.93	$123.51
Crew B-11L	**Hr.**	**Daily**	**Hr.**	**Daily**	**Bare Costs**	**Incl. O&P**
1 Equipment Oper. (med.)	$23.20	$185.60	$38.20	$305.60	$20.05	$33.45
1 Laborer	16.90	135.20	28.70	229.60		
1 Grader, 30,000 Lbs.		432.00		475.20	27.00	29.70
16 L.H., Daily Totals		$752.80		$1010.40	$47.05	$63.15
Crew B-11M	**Hr.**	**Daily**	**Hr.**	**Daily**	**Bare Costs**	**Incl. O&P**
1 Equipment Oper. (med.)	$23.20	$185.60	$38.20	$305.60	$20.05	$33.45
1 Laborer	16.90	135.20	28.70	229.60		
1 Backhoe Loader, 80 H.P.		230.80		253.90	14.43	15.87
16 L.H., Daily Totals		$551.60		$789.10	$34.48	$49.32
Crew B-11W	**Hr.**	**Daily**	**Hr.**	**Daily**	**Bare Costs**	**Incl. O&P**
1 Equipment Operator (med.)	$23.20	$185.60	$38.20	$305.60	$19.01	$31.66
1 Common Laborer	16.90	135.20	28.70	229.60		
10 Truck Drivers, Heavy	18.80	1504.00	31.30	2504.00		
1 Dozer, 200 H.P.		863.40		949.75		
1 Vib. roller, smth, towed, 23 Ton		610.20		671.20		
10 Dump Truck, 10 Ton		3250.00		3575.00	49.20	54.12
96 L.H., Daily Totals		$6548.40		$8235.15	$68.21	$85.78
Crew B-11Y	**Hr.**	**Daily**	**Hr.**	**Daily**	**Bare Costs**	**Incl. O&P**
1 Labor Foreman (Outside)	$18.90	$151.20	$32.10	$256.80	$19.22	$32.24
5 Common Laborers	16.90	676.00	28.70	1148.00		
3 Equipment Operator (med.)	23.20	556.80	38.20	916.80		
1 Dozer, 80 H.P.		300.20		330.20		
2 Wlk-Beh. Comp., 2-Drum, 1 Ton		261.60		287.75		
4 Vibratory plate, gas, 21″		165.60		182.15	10.10	11.11
72 L.H., Daily Totals		$2111.40		$3121.70	$29.32	$43.35
Crew B-12A	**Hr.**	**Daily**	**Hr.**	**Daily**	**Bare Costs**	**Incl. O&P**
1 Equip. Oper. (crane)	$24.00	$192.00	$39.50	$316.00	$24.00	$39.50
1 Hyd. Excavator, 1 C.Y.		487.40		536.15	60.93	67.02
8 L.H., Daily Totals		$679.40		$852.15	$84.93	$106.52
Crew B-12B	**Hr.**	**Daily**	**Hr.**	**Daily**	**Bare Costs**	**Incl. O&P**
1 Equip. Oper. (crane)	$24.00	$192.00	$39.50	$316.00	$24.00	$39.50
1 Hyd. Excavator, 1.5 C.Y.		683.40		751.75	85.43	93.97
8 L.H., Daily Totals		$875.40		$1067.75	$109.43	$133.47
Crew B-12C	**Hr.**	**Daily**	**Hr.**	**Daily**	**Bare Costs**	**Incl. O&P**
1 Equip. Oper. (crane)	$24.00	$192.00	$39.50	$316.00	$24.00	$39.50
1 Hyd. Excavator, 2 C.Y.		887.20		975.90	110.90	121.99
8 L.H., Daily Totals		$1079.20		$1291.90	$134.90	$161.49
Crew B-12D	**Hr.**	**Daily**	**Hr.**	**Daily**	**Bare Costs**	**Incl. O&P**
1 Equip. Oper. (crane)	$24.00	$192.00	$39.50	$316.00	$24.00	$39.50
1 Hyd. Excavator, 3.5 C.Y.		1965.00		2161.50	245.63	270.19
8 L.H., Daily Totals		$2157.00		$2477.50	$269.63	$309.69
Crew B-12E	**Hr.**	**Daily**	**Hr.**	**Daily**	**Bare Costs**	**Incl. O&P**
1 Equip. Oper. (crane)	$24.00	$192.00	$39.50	$316.00	$24.00	$39.50
1 Hyd. Excavator, .5 C.Y.		323.00		355.30	40.38	44.41
8 L.H., Daily Totals		$515.00		$671.30	$64.38	$83.91

Crew No.	Bare Costs		Incl. Subs O & P		Cost Per Labor-Hour	
Crew B-12F	**Hr.**	**Daily**	**Hr.**	**Daily**	**Bare Costs**	**Incl. O&P**
1 Equip. Oper. (crane)	$24.00	$192.00	$39.50	$316.00	$24.00	$39.50
1 Hyd. Excavator, .75 C.Y.		458.80		504.70	57.35	63.09
8 L.H., Daily Totals		$650.80		$820.70	$81.35	$102.59

Crew B-12G	Hr.	Daily	Hr.	Daily	Bare Costs	Incl. O&P
1 Equip. Oper. (crane)	$24.00	$192.00	$39.50	$316.00	$24.00	$39.50
1 Power Shovel, .5 C.Y.		450.90		496.00		
1 Clamshell Bucket, .5 C.Y.		33.40		36.75	60.54	66.59
8 L.H., Daily Totals		$676.30		$848.75	$84.54	$106.09

Crew B-12H	Hr.	Daily	Hr.	Daily	Bare Costs	Incl. O&P
1 Equip. Oper. (crane)	$24.00	$192.00	$39.50	$316.00	$24.00	$39.50
1 Power Shovel, 1 C.Y.		792.00		871.20		
1 Clamshell Bucket, 1 C.Y.		44.00		48.40	104.50	114.95
8 L.H., Daily Totals		$1028.00		$1235.60	$128.50	$154.45

Crew B-12I	Hr.	Daily	Hr.	Daily	Bare Costs	Incl. O&P
1 Equip. Oper. (crane)	$24.00	$192.00	$39.50	$316.00	$24.00	$39.50
1 Power Shovel, .75 C.Y.		604.50		664.95		
1 Dragline Bucket, .75 C.Y.		18.60		20.45	77.89	85.68
8 L.H., Daily Totals		$815.10		$1001.40	$101.89	$125.18

Crew B-12J	Hr.	Daily	Hr.	Daily	Bare Costs	Incl. O&P
1 Equip. Oper. (crane)	$24.00	$192.00	$39.50	$316.00	$24.00	$39.50
1 Gradall, 3 Ton, .5 C.Y.		823.40		905.75	102.93	113.22
8 L.H., Daily Totals		$1015.40		$1221.75	$126.93	$152.72

Crew B-12K	Hr.	Daily	Hr.	Daily	Bare Costs	Incl. O&P
1 Equip. Oper. (crane)	$24.00	$192.00	$39.50	$316.00	$24.00	$39.50
1 Gradall, 3 Ton, 1 C.Y.		958.80		1054.70	119.85	131.84
8 L.H., Daily Totals		$1150.80		$1370.70	$143.85	$171.34

Crew B-12L	Hr.	Daily	Hr.	Daily	Bare Costs	Incl. O&P
1 Equip. Oper. (crane)	$24.00	$192.00	$39.50	$316.00	$24.00	$39.50
1 Power Shovel, .5 C.Y.		450.90		496.00		
1 F.E. Attachment, .5 C.Y.		45.20		49.70	62.01	68.21
8 L.H., Daily Totals		$688.10		$861.70	$86.01	$107.71

Crew B-12M	Hr.	Daily	Hr.	Daily	Bare Costs	Incl. O&P
1 Equip. Oper. (crane)	$24.00	$192.00	$39.50	$316.00	$24.00	$39.50
1 Power Shovel, .75 C.Y.		604.50		664.95		
1 F.E. Attachment, .75 C.Y.		50.00		55.00	81.81	89.99
8 L.H., Daily Totals		$846.50		$1035.95	$105.81	$129.49

Crew B-12N	Hr.	Daily	Hr.	Daily	Bare Costs	Incl. O&P
1 Equip. Oper. (crane)	$24.00	$192.00	$39.50	$316.00	$24.00	$39.50
1 Power Shovel, 1 C.Y.		792.00		871.20		
1 F.E. Attachment, 1 C.Y.		56.80		62.50	106.10	116.71
8 L.H., Daily Totals		$1040.80		$1249.70	$130.10	$156.21

Crew B-12O	Hr.	Daily	Hr.	Daily	Bare Costs	Incl. O&P
1 Equip. Oper. (crane)	$24.00	$192.00	$39.50	$316.00	$24.00	$39.50
1 Power Shovel, 1.5 C.Y.		915.20		1006.70		
1 F.E. Attachment, 1.5 C.Y.		65.60		72.15	122.60	134.86
8 L.H., Daily Totals		$1172.80		$1394.85	$146.60	$174.36

Crew B-12P	Hr.	Daily	Hr.	Daily	Bare Costs	Incl. O&P
1 Equip. Oper. (crane)	$24.00	$192.00	$39.50	$316.00	$24.00	$39.50
1 Crawler Crane, 40 Ton		915.20		1006.70		
1 Dragline Bucket, 1.5 C.Y.		30.20		33.20	118.18	129.99
8 L.H., Daily Totals		$1137.40		$1355.90	$142.18	$169.49

Crew B-12Q	Hr.	Daily	Hr.	Daily	Bare Costs	Incl. O&P
1 Equip. Oper. (crane)	$24.00	$192.00	$39.50	$316.00	$24.00	$39.50
1 Hyd. Excavator, 5/8 C.Y.		418.00		459.80	52.25	57.48
8 L.H., Daily Totals		$610.00		$775.80	$76.25	$96.98

Crew B-12R	Hr.	Daily	Hr.	Daily	Bare Costs	Incl. O&P
1 Equip. Oper. (crane)	$24.00	$192.00	$39.50	$316.00	$24.00	$39.50
1 Hyd. Excavator, 1.5 C.Y.		683.40		751.75	85.43	93.97
8 L.H., Daily Totals		$875.40		$1067.75	$109.43	$133.47

Crew B-12S	Hr.	Daily	Hr.	Daily	Bare Costs	Incl. O&P
1 Equip. Oper. (crane)	$24.00	$192.00	$39.50	$316.00	$24.00	$39.50
1 Hyd. Excavator, 2.5 C.Y.		1181.00		1299.10	147.63	162.39
8 L.H., Daily Totals		$1373.00		$1615.10	$171.63	$201.89

Crew B-12T	Hr.	Daily	Hr.	Daily	Bare Costs	Incl. O&P
1 Equip. Oper. (crane)	$24.00	$192.00	$39.50	$316.00	$24.00	$39.50
1 Crawler Crane, 75 Ton		1217.00		1338.70		
1 F.E. Attachment, 3 C.Y.		89.40		98.35	163.30	179.63
8 L.H., Daily Totals		$1498.40		$1753.05	$187.30	$219.13

Crew B-12V	Hr.	Daily	Hr.	Daily	Bare Costs	Incl. O&P
1 Equip. Oper. (crane)	$24.00	$192.00	$39.50	$316.00	$24.00	$39.50
1 Crawler Crane, 75 Ton		1217.00		1338.70		
1 Dragline Bucket, 3 C.Y.		47.60		52.35	158.08	173.88
8 L.H., Daily Totals		$1456.60		$1707.05	$182.08	$213.38

Crew B-13	Hr.	Daily	Hr.	Daily	Bare Costs	Incl. O&P
1 Labor Foreman (outside)	$18.90	$151.20	$32.10	$256.80	$18.42	$31.07
4 Laborers	16.90	540.80	28.70	918.40		
1 Equip. Oper. (crane)	24.00	192.00	39.50	316.00		
1 Hyd. Crane, 25 Ton		624.20		686.60	13.00	14.30
48 L.H., Daily Totals		$1508.20		$2177.80	$31.42	$45.37

Crew B-13A	Hr.	Daily	Hr.	Daily	Bare Costs	Incl. O&P
1 Foreman	$18.90	$151.20	$32.10	$256.80	$19.53	$32.64
2 Laborers	16.90	270.40	28.70	459.20		
2 Equipment Operators	23.20	371.20	38.20	611.20		
2 Truck Drivers (heavy)	18.80	300.80	31.30	500.80		
1 Crane, 75 Ton		1217.00		1338.70		
1 F.E. Lder, 3.75 C.Y.		1061.00		1167.10		
2 Dump Trucks, 12 Ton		650.00		715.00	52.29	57.51
56 L.H., Daily Totals		$4021.60		$5048.80	$71.82	$90.15

Crew B-13B	Hr.	Daily	Hr.	Daily	Bare Costs	Incl. O&P
1 Labor Foreman (outside)	$18.90	$151.20	$32.10	$256.80	$18.66	$31.36
4 Laborers	16.90	540.80	28.70	918.40		
1 Equip. Oper. (crane)	24.00	192.00	39.50	316.00		
1 Equip. Oper. Oiler	20.15	161.20	33.15	265.20		
1 Hyd. Crane, 55 Ton		911.00		1002.10	16.27	17.89
56 L.H., Daily Totals		$1956.20		$2758.50	$34.93	$49.25

Crew No.	Bare Costs		Incl. Subs O & P		Cost Per Labor-Hour	
Crew B-13C	**Hr.**	**Daily**	**Hr.**	**Daily**	**Bare Costs**	**Incl. O&P**
1 Labor Foreman (outside)	$18.90	$151.20	$32.10	$256.80	$18.66	$31.36
4 Laborers	16.90	540.80	28.70	918.40		
1 Equip. Oper. (crane)	24.00	192.00	39.50	316.00		
1 Equip. Oper. Oiler	20.15	161.20	33.15	265.20		
1 Crawler Crane, 100 Ton		1602.00		1762.20	28.61	31.47
56 L.H., Daily Totals		$2647.20		$3518.60	$47.27	$62.83
Crew B-14	**Hr.**	**Daily**	**Hr.**	**Daily**	**Bare Costs**	**Incl. O&P**
1 Labor Foreman (outside)	$18.90	$151.20	$32.10	$256.80	$18.12	$30.58
4 Laborers	16.90	540.80	28.70	918.40		
1 Equip. Oper. (light)	22.20	177.60	36.55	292.40		
1 Backhoe Loader, 48 H.P.		207.80		228.60	4.33	4.76
48 L.H., Daily Totals		$1077.40		$1696.20	$22.45	$35.34
Crew B-15	**Hr.**	**Daily**	**Hr.**	**Daily**	**Bare Costs**	**Incl. O&P**
1 Equipment Oper. (med)	$23.20	$185.60	$38.20	$305.60	$19.79	$32.90
.5 Laborer	16.90	67.60	28.70	114.80		
2 Truck Drivers (heavy)	18.80	300.80	31.30	500.80		
2 Dump Trucks, 16 Ton		953.20		1048.50		
1 Dozer, 200 H.P.		863.40		949.75	64.88	71.37
28 L.H., Daily Totals		$2370.60		$2919.45	$84.67	$104.27
Crew B-16	**Hr.**	**Daily**	**Hr.**	**Daily**	**Bare Costs**	**Incl. O&P**
1 Labor Foreman (outside)	$18.90	$151.20	$32.10	$256.80	$17.88	$30.20
2 Laborers	16.90	270.40	28.70	459.20		
1 Truck Driver (heavy)	18.80	150.40	31.30	250.40		
1 Dump Truck, 16 Ton		476.60		524.25	14.89	16.38
32 L.H., Daily Totals		$1048.60		$1490.65	$32.77	$46.58
Crew B-17	**Hr.**	**Daily**	**Hr.**	**Daily**	**Bare Costs**	**Incl. O&P**
2 Laborers	$16.90	$270.40	$28.70	$459.20	$18.70	$31.31
1 Equip. Oper. (light)	22.20	177.60	36.55	292.40		
1 Truck Driver (heavy)	18.80	150.40	31.30	250.40		
1 Backhoe Loader, 48 H.P.		207.80		228.60		
1 Dump Truck, 12 Ton		325.00		357.50	16.65	18.32
32 L.H., Daily Totals		$1131.20		$1588.10	$35.35	$49.63
Crew B-18	**Hr.**	**Daily**	**Hr.**	**Daily**	**Bare Costs**	**Incl. O&P**
1 Labor Foreman (outside)	$18.90	$151.20	$32.10	$256.80	$17.57	$29.83
2 Laborers	16.90	270.40	28.70	459.20		
1 Vibrating Compactor		41.40		45.55	1.73	1.90
24 L.H., Daily Totals		$463.00		$761.55	$19.30	$31.73
Crew B-19	**Hr.**	**Daily**	**Hr.**	**Daily**	**Bare Costs**	**Incl. O&P**
1 Pile Driver Foreman	$24.45	$195.60	$43.80	$350.40	$22.16	$39.00
4 Pile Drivers	22.45	718.40	40.25	1288.00		
1 Equip. Oper. (crane)	24.00	192.00	39.50	316.00		
1 Building Laborer	16.90	135.20	28.70	229.60		
1 Crane, 40 Ton & Access.		915.20		1006.70		
60 L.F. Pile Leads		90.00		99.00		
1 Hammer, Diesel, 22k Ft-Lb		564.00		620.40	28.02	30.82
56 L.H., Daily Totals		$2810.40		$3910.10	$50.18	$69.82

Crew No.	Bare Costs		Incl. Subs O & P		Cost Per Labor-Hour	
Crew B-19A	**Hr.**	**Daily**	**Hr.**	**Daily**	**Bare Costs**	**Incl. O&P**
1 Pile Driver Foreman	$24.45	$195.60	$43.80	$350.40	$22.80	$39.62
4 Pile Drivers	22.45	718.40	40.25	1288.00		
2 Equip. Oper. (crane)	24.00	384.00	39.50	632.00		
1 Equip. Oper. Oiler	20.15	161.20	33.15	265.20		
1 Crawler Crane, 75 Ton		1217.00		1338.70		
60 L.F. Leads, 25K Ft. Lbs.		132.00		145.20		
1 Hammer, Diesel, 41k Ft-Lb		662.60		728.85	31.43	34.57
64 L.H., Daily Totals		$3470.80		$4748.35	$54.23	$74.19
Crew B-20	**Hr.**	**Daily**	**Hr.**	**Daily**	**Bare Costs**	**Incl. O&P**
1 Labor Foreman (out)	$18.90	$151.20	$32.10	$256.80	$17.57	$29.83
2 Laborer	16.90	270.40	28.70	459.20		
24 L.H., Daily Totals		$421.60		$716.00	$17.57	$29.83
Crew B-20A	**Hr.**	**Daily**	**Hr.**	**Daily**	**Bare Costs**	**Incl. O&P**
1 Labor Foreman	$18.90	$151.20	$32.10	$256.80	$20.71	$34.50
1 Laborer	16.90	135.20	28.70	229.60		
1 Plumber	26.15	209.20	42.90	343.20		
1 Plumber Apprentice	20.90	167.20	34.30	274.40		
32 L.H., Daily Totals		$662.80		$1104.00	$20.71	$34.50
Crew B-21	**Hr.**	**Daily**	**Hr.**	**Daily**	**Bare Costs**	**Incl. O&P**
1 Labor Foreman (out)	$18.90	$151.20	$32.10	$256.80	$18.49	$31.21
2 Laborer	16.90	270.40	28.70	459.20		
.5 Equip. Oper. (crane)	24.00	96.00	39.50	158.00		
.5 S.P. Crane, 5 Ton		158.90		174.80	5.68	6.24
28 L.H., Daily Totals		$676.50		$1048.80	$24.17	$37.45
Crew B-21A	**Hr.**	**Daily**	**Hr.**	**Daily**	**Bare Costs**	**Incl. O&P**
1 Labor Foreman	$18.90	$151.20	$32.10	$256.80	$21.37	$35.50
1 Laborer	16.90	135.20	28.70	229.60		
1 Plumber	26.15	209.20	42.90	343.20		
1 Plumber Apprentice	20.90	167.20	34.30	274.40		
1 Equip. Oper. (crane)	24.00	192.00	39.50	316.00		
1 S.P. Crane, 12 Ton		482.20		530.40	12.06	13.26
40 L.H., Daily Totals		$1337.00		$1950.40	$33.43	$48.76
Crew B-22	**Hr.**	**Daily**	**Hr.**	**Daily**	**Bare Costs**	**Incl. O&P**
1 Labor Foreman (out)	$18.90	$151.20	$32.10	$256.80	$18.85	$31.77
2 Laborer	16.90	270.40	28.70	459.20		
.75 Equip. Oper. (crane)	24.00	144.00	39.50	237.00		
.75 S.P. Crane, 5 Ton		238.35		262.20	7.95	8.74
30 L.H., Daily Totals		$803.95		$1215.20	$26.80	$40.51
Crew B-22A	**Hr.**	**Daily**	**Hr.**	**Daily**	**Bare Costs**	**Incl. O&P**
1 Labor Foreman (out)	$18.90	$151.20	$32.10	$256.80	$19.74	$33.30
1 Skilled Worker	23.05	184.40	39.05	312.40		
2 Laborers	16.90	270.40	28.70	459.20		
.75 Equipment Oper. (crane)	24.00	144.00	39.50	237.00		
.75 Crane, 5 Ton		238.35		262.20		
1 Generator, 5 KW		37.00		40.70		
1 Butt Fusion Machine		433.60		476.95	18.66	20.52
38 L.H., Daily Totals		$1458.95		$2045.25	$38.40	$53.82
Crew B-22B	**Hr.**	**Daily**	**Hr.**	**Daily**	**Bare Costs**	**Incl. O&P**
1 Skilled Worker	$23.05	$184.40	$39.05	$312.40	$19.98	$33.88
1 Laborer	16.90	135.20	28.70	229.60		
1 Electro Fusion Machine		172.80		190.10	10.80	11.88
16 L.H., Daily Totals		$492.40		$732.10	$30.78	$45.76

Crews

Crew B-23	Bare Costs Hr.	Bare Costs Daily	Incl. Subs O & P Hr.	Incl. Subs O & P Daily	Cost Per Labor-Hour Bare Costs	Cost Per Labor-Hour Incl. O&P
1 Labor Foreman (outside)	$18.90	$151.20	$32.10	$256.80	$17.30	$29.38
4 Laborers	16.90	540.80	28.70	918.40		
1 Drill Rig, Wells		2877.00		3164.70		
1 Light Truck, 3 Ton		160.40		176.45	75.94	83.53
40 L.H., Daily Totals		$3729.40		$4516.35	$93.24	$112.91

Crew B-23A	Hr.	Daily	Hr.	Daily	Bare Costs	Incl. O&P
1 Labor Foreman (outside)	$18.90	$151.20	$32.10	$256.80	$19.67	$33.00
1 Laborer	16.90	135.20	28.70	229.60		
1 Equip. Operator (medium)	23.20	185.60	38.20	305.60		
1 Drill Rig, Wells		2877.00		3164.70		
1 Pickup Truck, 3/4 Ton		75.80		83.40	123.03	135.34
24 L.H., Daily Totals		$3424.80		$4040.10	$142.70	$168.34

Crew B-23B	Hr.	Daily	Hr.	Daily	Bare Costs	Incl. O&P
1 Labor Foreman (outside)	$18.90	$151.20	$32.10	$256.80	$19.67	$33.00
1 Laborer	16.90	135.20	28.70	229.60		
1 Equip. Operator (medium)	23.20	185.60	38.20	305.60		
1 Drill Rig, Wells		2877.00		3164.70		
1 Pickup Truck, 3/4 Ton		75.80		83.40		
1 Pump, Cntfgl, 6"		229.80		252.80	132.61	145.87
24 L.H., Daily Totals		$3654.60		$4292.90	$152.28	$178.87

Crew B-24	Hr.	Daily	Hr.	Daily	Bare Costs	Incl. O&P
1 Cement Finisher	$22.10	$176.80	$35.65	$285.20	$20.70	$34.52
1 Laborer	16.90	135.20	28.70	229.60		
1 Carpenter	23.10	184.80	39.20	313.60		
24 L.H., Daily Totals		$496.80		$828.40	$20.70	$34.52

Crew B-25	Hr.	Daily	Hr.	Daily	Bare Costs	Incl. O&P
1 Labor Foreman	$18.90	$151.20	$32.10	$256.80	$18.80	$31.60
7 Laborers	16.90	946.40	28.70	1607.20		
3 Equip. Oper. (med.)	23.20	556.80	38.20	916.80		
1 Asphalt Paver, 130 H.P		1457.00		1602.70		
1 Tandem Roller, 10 Ton		183.60		201.95		
1 Roller, Pneumatic Wheel		244.60		269.05	21.42	23.57
88 L.H., Daily Totals		$3539.60		$4854.50	$40.22	$55.17

Crew B-25B	Hr.	Daily	Hr.	Daily	Bare Costs	Incl. O&P
1 Labor Foreman	$18.90	$151.20	$32.10	$256.80	$19.17	$32.15
7 Laborers	16.90	946.40	28.70	1607.20		
4 Equip. Oper. (medium)	23.20	742.40	38.20	1222.40		
1 Asphalt Paver, 130 H.P.		1457.00		1602.70		
2 Rollers, Steel Wheel		367.20		403.90		
1 Roller, Pneumatic Wheel		244.60		269.05	21.55	23.71
96 L.H., Daily Totals		$3908.80		$5362.05	$40.72	$55.86

Crew B-26	Hr.	Daily	Hr.	Daily	Bare Costs	Incl. O&P
1 Labor Foreman (outside)	$18.90	$151.20	$32.10	$256.80	$19.42	$32.80
6 Laborers	16.90	811.20	28.70	1377.60		
2 Equip. Oper. (med.)	23.20	371.20	38.20	611.20		
1 Rodman (reinf.)	24.85	198.80	44.50	356.00		
1 Cement Finisher	22.10	176.80	35.65	285.20		
1 Grader, 30,000 Lbs.		432.00		475.20		
1 Paving Mach. & Equip.		1580.00		1738.00	22.86	25.15
88 L.H., Daily Totals		$3721.20		$5100.00	$42.28	$57.95

Crew B-27	Hr.	Daily	Hr.	Daily	Bare Costs	Incl. O&P
1 Labor Foreman (outside)	$18.90	$151.20	$32.10	$256.80	$17.40	$29.55
3 Laborers	16.90	405.60	28.70	688.80		
1 Berm Machine		188.40		207.25	5.89	6.48
32 L.H., Daily Totals		$745.20		$1152.85	$23.29	$36.03

Crew B-28	Hr.	Daily	Hr.	Daily	Bare Costs	Incl. O&P
2 Carpenters	$23.10	$369.60	$39.20	$627.20	$21.03	$35.70
1 Laborer	16.90	135.20	28.70	229.60		
24 L.H., Daily Totals		$504.80		$856.80	$21.03	$35.70

Crew B-29	Hr.	Daily	Hr.	Daily	Bare Costs	Incl. O&P
1 Labor Foreman (outside)	$18.90	$151.20	$32.10	$256.80	$18.42	$31.07
4 Laborers	16.90	540.80	28.70	918.40		
1 Equip. Oper. (crane)	24.00	192.00	39.50	316.00		
1 Gradall, 3 Ton, 1/2 C.Y.		823.40		905.75	17.15	18.87
48 L.H., Daily Totals		$1707.40		$2396.95	$35.57	$49.94

Crew B-30	Hr.	Daily	Hr.	Daily	Bare Costs	Incl. O&P
1 Equip. Oper. (med.)	$23.20	$185.60	$38.20	$305.60	$20.27	$33.60
2 Truck Drivers (heavy)	18.80	300.80	31.30	500.80		
1 Hyd. Excavator, 1.5 C.Y.		683.40		751.75		
2 Dump Trucks, 16 Ton		953.20		1048.50	68.19	75.01
24 L.H., Daily Totals		$2123.00		$2606.65	$88.46	$108.61

Crew B-31	Hr.	Daily	Hr.	Daily	Bare Costs	Incl. O&P
1 Labor Foreman (outside)	$18.90	$151.20	$32.10	$256.80	$17.30	$29.38
4 Laborers	16.90	540.80	28.70	918.40		
1 Air Compr., 250 C.F.M.		127.60		140.35		
1 Sheeting Driver		7.20		7.90		
2-50 Ft. Air Hoses, 1.5" Dia.		10.00		11.00	3.62	3.98
40 L.H., Daily Totals		$836.80		$1334.45	$20.92	$33.36

Crew B-32	Hr.	Daily	Hr.	Daily	Bare Costs	Incl. O&P
1 Laborer	$16.90	$135.20	$28.70	$229.60	$21.63	$35.83
3 Equip. Oper. (med.)	23.20	556.80	38.20	916.80		
1 Grader, 30,000 Lbs.		432.00		475.20		
1 Tandem Roller, 10 Ton		183.60		201.95		
1 Dozer, 200 H.P.		863.40		949.75	46.22	50.84
32 L.H., Daily Totals		$2171.00		$2773.30	$67.85	$86.67

Crew B-32A	Hr.	Daily	Hr.	Daily	Bare Costs	Incl. O&P
1 Laborer	$16.90	$135.20	$28.70	$229.60	$21.10	$35.03
2 Equip. Oper. (medium)	23.20	371.20	38.20	611.20		
1 Grader, 30,000 Lbs.		432.00		475.20		
1 Roller, Vibratory, 29,000 Lbs.		436.80		480.50	36.20	39.82
24 L.H., Daily Totals		$1375.20		$1796.50	$57.30	$74.85

Crew B-32B	Hr.	Daily	Hr.	Daily	Bare Costs	Incl. O&P
1 Laborer	$16.90	$135.20	$28.70	$229.60	$21.10	$35.03
2 Equip. Oper. (medium)	23.20	371.20	38.20	611.20		
1 Dozer, 200 H.P.		863.40		949.75		
1 Roller, Vibratory, 29,000 Lbs.		436.80		480.50	54.18	59.59
24 L.H., Daily Totals		$1806.60		$2271.05	$75.28	$94.62

Crew No.	Bare Costs		Incl. Subs O & P		Cost Per Labor-Hour	
Crew B-32C	**Hr.**	**Daily**	**Hr.**	**Daily**	**Bare Costs**	**Incl. O&P**
1 Labor Foreman	$18.90	$151.20	$32.10	$256.80	$20.38	$34.02
2 Laborers	16.90	270.40	28.70	459.20		
3 Equip. Oper. (medium)	23.20	556.80	38.20	916.80		
1 Grader, 30,000 Lbs.		432.00		475.20		
1 Roller, Steel Wheel		183.60		201.95		
1 Dozer, 200 H.P.		863.40		949.75	30.81	33.89
48 L.H., Daily Totals		$2457.40		$3259.70	$51.19	$67.91

Crew B-33A	Hr.	Daily	Hr.	Daily	Bare Costs	Incl. O&P
1 Equip. Oper. (med.)	$23.20	$185.60	$38.20	$305.60	$23.20	$38.20
.25 Equip. Oper. (med.)	23.20	46.40	38.20	76.40		
1 Scraper, Towed, 7 C.Y.		170.00		187.00		
1.25 Dozer, 300 H.P.		1373.75		1511.15	154.38	169.81
10 L.H., Daily Totals		$1775.75		$2080.15	$177.58	$208.01

Crew B-33B	Hr.	Daily	Hr.	Daily	Bare Costs	Incl. O&P
1 Equip. Oper. (med.)	$23.20	$185.60	$38.20	$305.60	$23.20	$38.20
.25 Equip. Oper. (med.)	23.20	46.40	38.20	76.40		
1 Scraper, Towed, 10 C.Y.		189.10		208.00		
1.25 Dozer, 300 H.P.		1373.75		1511.15	156.29	171.91
10 L.H., Daily Totals		$1794.85		$2101.15	$179.49	$210.11

Crew B-33C	Hr.	Daily	Hr.	Daily	Bare Costs	Incl. O&P
1 Equip. Oper. (med.)	$23.20	$185.60	$38.20	$305.60	$23.20	$38.20
.25 Equip. Oper. (med.)	23.20	46.40	38.20	76.40		
1 Scraper, Towed, 12 C.Y.		189.10		208.00		
1.25 Dozer, 300 H.P.		1373.75		1511.15	156.29	171.91
10 L.H., Daily Totals		$1794.85		$2101.15	$179.49	$210.11

Crew B-33D	Hr.	Daily	Hr.	Daily	Bare Costs	Incl. O&P
1 Equip. Oper. (med.)	$23.20	$185.60	$38.20	$305.60	$23.20	$38.20
.25 Equip. Oper. (med.)	23.20	46.40	38.20	76.40		
1 S.P. Scraper, 14 C.Y.		1468.00		1614.80		
.25 Dozer, 300 H.P.		274.75		302.25	174.28	191.70
10 L.H., Daily Totals		$1974.75		$2299.05	$197.48	$229.90

Crew B-33E	Hr.	Daily	Hr.	Daily	Bare Costs	Incl. O&P
1 Equip. Oper. (med.)	$23.20	$185.60	$38.20	$305.60	$23.20	$38.20
.25 Equip. Oper. (med.)	23.20	46.40	38.20	76.40		
1 S.P. Scraper, 24 C.Y.		2297.00		2526.70		
.25 Dozer, 300 H.P.		274.75		302.25	257.18	282.89
10 L.H., Daily Totals		$2803.75		$3210.95	$280.38	$321.09

Crew B-33F	Hr.	Daily	Hr.	Daily	Bare Costs	Incl. O&P
1 Equip. Oper. (med.)	$23.20	$185.60	$38.20	$305.60	$23.20	$38.20
.25 Equip. Oper. (med.)	23.20	46.40	38.20	76.40		
1 Elev. Scraper, 11 C.Y.		805.40		885.95		
.25 Dozer, 300 H.P.		274.75		302.25	108.02	118.82
10 L.H., Daily Totals		$1312.15		$1570.20	$131.22	$157.02

Crew B-33G	Hr.	Daily	Hr.	Daily	Bare Costs	Incl. O&P
1 Equip. Oper. (med.)	$23.20	$185.60	$38.20	$305.60	$23.20	$38.20
.25 Equip. Oper. (med.)	23.20	46.40	38.20	76.40		
1 Elev. Scraper, 20 C.Y.		1597.00		1756.70		
.25 Dozer, 300 H.P.		274.75		302.25	187.18	205.89
10 L.H., Daily Totals		$2103.75		$2440.95	$210.38	$244.09

Crew No.	Bare Costs		Incl. Subs O & P		Cost Per Labor-Hour	
Crew B-34A	**Hr.**	**Daily**	**Hr.**	**Daily**	**Bare Costs**	**Incl. O&P**
1 Truck Driver (heavy)	$18.80	$150.40	$31.30	$250.40	$18.80	$31.30
1 Dump Truck, 12 Ton		325.00		357.50	40.63	44.69
8 L.H., Daily Totals		$475.40		$607.90	$59.43	$75.99

Crew B-34B	Hr.	Daily	Hr.	Daily	Bare Costs	Incl. O&P
1 Truck Driver (heavy)	$18.80	$150.40	$31.30	$250.40	$18.80	$31.30
1 Dump Truck, 16 Ton		476.60		524.25	59.58	65.53
8 L.H., Daily Totals		$627.00		$774.65	$78.38	$96.83

Crew B-34C	Hr.	Daily	Hr.	Daily	Bare Costs	Incl. O&P
1 Truck Driver (heavy)	$18.80	$150.40	$31.30	$250.40	$18.80	$31.30
1 Truck Tractor, 40 Ton		303.00		333.30		
1 Dump Trailer, 16.5 C.Y.		103.20		113.50	50.78	55.85
8 L.H., Daily Totals		$556.60		$697.20	$69.58	$87.15

Crew B-34D	Hr.	Daily	Hr.	Daily	Bare Costs	Incl. O&P
1 Truck Driver (heavy)	$18.80	$150.40	$31.30	$250.40	$18.80	$31.30
1 Truck Tractor, 40 Ton		303.00		333.30		
1 Dump Trailer, 20 C.Y.		116.40		128.05	52.43	57.67
8 L.H., Daily Totals		$569.80		$711.75	$71.23	$88.97

Crew B-34E	Hr.	Daily	Hr.	Daily	Bare Costs	Incl. O&P
1 Truck Driver (heavy)	$18.80	$150.40	$31.30	$250.40	$18.80	$31.30
1 Truck, Off Hwy., 25 Ton		920.40		1012.45	115.05	126.56
8 L.H., Daily Totals		$1070.80		$1262.85	$133.85	$157.86

Crew B-34F	Hr.	Daily	Hr.	Daily	Bare Costs	Incl. O&P
1 Truck Driver (heavy)	$18.80	$150.40	$31.30	$250.40	$18.80	$31.30
1 Truck, Off Hwy., 22 C.Y.		946.60		1041.25	118.33	130.16
8 L.H., Daily Totals		$1097.00		$1291.65	$137.13	$161.46

Crew B-34G	Hr.	Daily	Hr.	Daily	Bare Costs	Incl. O&P
1 Truck Driver (heavy)	$18.80	$150.40	$31.30	$250.40	$18.80	$31.30
1 Truck, Off Hwy., 34 C.Y.		1226.00		1348.60	153.25	168.58
8 L.H., Daily Totals		$1376.40		$1599.00	$172.05	$199.88

Crew B-34H	Hr.	Daily	Hr.	Daily	Bare Costs	Incl. O&P
1 Truck Driver (heavy)	$18.80	$150.40	$31.30	$250.40	$18.80	$31.30
1 Truck, Off Hwy., 42 C.Y.		1310.00		1441.00	163.75	180.13
8 L.H., Daily Totals		$1460.40		$1691.40	$182.55	$211.43

Crew B-34J	Hr.	Daily	Hr.	Daily	Bare Costs	Incl. O&P
1 Truck Driver (heavy)	$18.80	$150.40	$31.30	$250.40	$18.80	$31.30
1 Truck, Off Hwy., 60 C.Y.		1663.00		1829.30	207.88	228.66
8 L.H., Daily Totals		$1813.40		$2079.70	$226.68	$259.96

Crew B-34K	Hr.	Daily	Hr.	Daily	Bare Costs	Incl. O&P
1 Truck Driver (heavy)	$18.80	$150.40	$31.30	$250.40	$18.80	$31.30
1 Truck Tractor, 240 H.P.		331.60		364.75		
1 Low Bed Trailer		171.80		189.00	62.93	69.22
8 L.H., Daily Totals		$653.80		$804.15	$81.73	$100.52

Crew B-34N	Hr.	Daily	Hr.	Daily	Bare Costs	Incl. O&P
1 Truck Driver (heavy)	$18.80	$150.40	$31.30	$250.40	$18.80	$31.30
1 Dump Truck, 12 Ton		325.00		357.50		
1 Flatbed Trailer, 40 Ton		120.60		132.65	55.70	61.27
8 L.H., Daily Totals		$596.00		$740.55	$74.50	$92.57

Crew No.	Bare Costs		Incl. Subs O & P		Cost Per Labor-Hour	
Crew B-35	**Hr.**	**Daily**	**Hr.**	**Daily**	**Bare Costs**	**Incl. O&P**
1 Laborer Foreman (out)	$18.90	$151.20	$32.10	$256.80	$21.80	$36.45
1 Skilled Worker	23.05	184.40	39.05	312.40		
1 Welder (plumber)	26.15	209.20	42.90	343.20		
1 Laborer	16.90	135.20	28.70	229.60		
1 Equip. Oper. (crane)	24.00	192.00	39.50	316.00		
1 Electric Welding Mach.		77.75		85.55		
1 Hyd. Excavator, .75 C.Y.		458.80		504.70	13.41	14.76
40 L.H., Daily Totals		$1408.55		$2048.25	$35.21	$51.21
Crew B-35A	**Hr.**	**Daily**	**Hr.**	**Daily**	**Bare Costs**	**Incl. O&P**
1 Laborer Foreman (out)	$18.90	$151.20	$32.10	$256.80	$20.86	$34.87
2 Laborers	16.90	270.40	28.70	459.20		
1 Skilled Worker	23.05	184.40	39.05	312.40		
1 Welder (plumber)	26.15	209.20	42.90	343.20		
1 Equip. Oper. (crane)	24.00	192.00	39.50	316.00		
1 Equip. Oper. Oiler	20.15	161.20	33.15	265.20		
1 Welder, 300 amp		75.20		82.70		
1 Crane, 75 Ton		1217.00		1338.70	23.08	25.38
56 L.H., Daily Totals		$2460.60		$3374.20	$43.94	$60.25
Crew B-36	**Hr.**	**Daily**	**Hr.**	**Daily**	**Bare Costs**	**Incl. O&P**
1 Labor Foreman (outside)	$18.90	$151.20	$32.10	$256.80	$19.82	$33.18
2 Laborers	16.90	270.40	28.70	459.20		
2 Equip. Oper. (med.)	23.20	371.20	38.20	611.20		
1 Dozer, 200 H.P.		863.40		949.75		
1 Aggregate Spreader		41.40		45.55		
1 Tandem Roller, 10 Ton		183.60		201.95	27.21	29.93
40 L.H., Daily Totals		$1881.20		$2524.45	$47.03	$63.11
Crew B-36A	**Hr.**	**Daily**	**Hr.**	**Daily**	**Bare Costs**	**Incl. O&P**
1 Labor Foreman (outside)	$18.90	$151.20	$32.10	$256.80	$20.79	$34.61
2 Laborers	16.90	270.40	28.70	459.20		
4 Equip. Oper. (med.)	23.20	742.40	38.20	1222.40		
1 Dozer, 200 H.P.		863.40		949.75		
1 Aggregate Spreader		41.40		45.55		
1 Roller, Steel Wheel		183.60		201.95		
1 Roller, Pneumatic Wheel		244.60		269.05	23.80	26.18
56 L.H., Daily Totals		$2497.00		$3404.70	$44.59	$60.79
Crew B-36B	**Hr.**	**Daily**	**Hr.**	**Daily**	**Bare Costs**	**Incl. O&P**
1 Labor Foreman (outside)	$18.90	$151.20	$32.10	$256.80	$20.54	$34.20
2 Laborers	16.90	270.40	28.70	459.20		
4 Equip. Oper. (medium)	23.20	742.40	38.20	1222.40		
1 Truck Driver, Heavy	18.80	150.40	31.30	250.40		
1 Grader, 30,000 Lbs.		432.00		475.20		
1 F.E. Loader, crl, 1.5 C.Y.		354.40		389.85		
1 Dozer, 300 H.P.		1099.00		1208.90		
1 Roller, Vibratory		436.80		480.50		
1 Truck, Tractor, 240 H.P.		331.60		364.75		
1 Water Tanker, 5000 Gal.		118.00		129.80	43.31	47.64
64 L.H., Daily Totals		$4086.20		$5237.80	$63.85	$81.84
Crew B-36C	**Hr.**	**Daily**	**Hr.**	**Daily**	**Bare Costs**	**Incl. O&P**
1 Labor Foreman (outside)	$18.90	$151.20	$32.10	$256.80	$21.46	$35.60
3 Equip. Oper. (medium)	23.20	556.80	38.20	916.80		
1 Truck Driver, Heavy	18.80	150.40	31.30	250.40		
1 Grader, 30,000 Lbs.		432.00		475.20		
1 Dozer, 300 H.P.		1099.00		1208.90		
1 Roller, Vibratory		436.80		480.50		
1 Truck, Tractor, 240 H.P.		331.60		364.75		
1 Water Tanker, 5000 Gal.		118.00		129.80	60.44	66.48
40 L.H., Daily Totals		$3275.80		$4083.15	$81.90	$102.08

Crew No.	Bare Costs		Incl. Subs O & P		Cost Per Labor-Hour	
Crew B-37	**Hr.**	**Daily**	**Hr.**	**Daily**	**Bare Costs**	**Incl. O&P**
1 Labor Foreman (outside)	$18.90	$151.20	$32.10	$256.80	$18.12	$30.58
4 Laborers	16.90	540.80	28.70	918.40		
1 Equip. Oper. (light)	22.20	177.60	36.55	292.40		
1 Tandem Roller, 5 Ton		108.20		119.00	2.25	2.48
48 L.H., Daily Totals		$977.80		$1586.60	$20.37	$33.06
Crew B-38	**Hr.**	**Daily**	**Hr.**	**Daily**	**Bare Costs**	**Incl. O&P**
2 Laborers	$16.90	$270.40	$28.70	$459.20	$18.67	$31.32
1 Equip. Oper. (light)	22.20	177.60	36.55	292.40		
1 Backhoe Loader, 48 H.P.		207.80		228.60		
1 Hyd.Hammer, (1200 lb)		112.00		123.20	13.33	14.66
24 L.H., Daily Totals		$767.80		$1103.40	$32.00	$45.98
Crew B-39	**Hr.**	**Daily**	**Hr.**	**Daily**	**Bare Costs**	**Incl. O&P**
1 Labor Foreman (outside)	$18.90	$151.20	$32.10	$256.80	$17.23	$29.27
5 Laborers	16.90	676.00	28.70	1148.00		
1 Air Compr., 250 C.F.M.		127.60		140.35		
2 Air Tools & Accessories		22.40		24.65		
2-50 Ft. Air Hoses, 1.5″ Dia.		10.00		11.00	3.33	3.66
48 L.H., Daily Totals		$987.20		$1580.80	$20.56	$32.93
Crew B-40	**Hr.**	**Daily**	**Hr.**	**Daily**	**Bare Costs**	**Incl. O&P**
1 Pile Driver Foreman (out)	$24.45	$195.60	$43.80	$350.40	$22.16	$39.00
4 Pile Drivers	22.45	718.40	40.25	1288.00		
1 Building Laborer	16.90	135.20	28.70	229.60		
1 Equip. Oper. (crane)	24.00	192.00	39.50	316.00		
1 Crane, 40 Ton		915.20		1006.70		
1 Vibratory Hammer & Gen.		1403.00		1543.30	41.40	45.54
56 L.H., Daily Totals		$3559.40		$4734.00	$63.56	$84.54
Crew B-41	**Hr.**	**Daily**	**Hr.**	**Daily**	**Bare Costs**	**Incl. O&P**
1 Labor Foreman (outside)	$18.90	$151.20	$32.10	$256.80	$17.73	$30.01
4 Laborers	16.90	540.80	28.70	918.40		
.25 Equip. Oper. (crane)	24.00	48.00	39.50	79.00		
.25 Equip. Oper. Oiler	20.15	40.30	33.15	66.30		
.25 Crawler Crane, 40 Ton		228.80		251.70	5.20	5.72
44 L.H., Daily Totals		$1009.10		$1572.20	$22.93	$35.73
Crew B-42	**Hr.**	**Daily**	**Hr.**	**Daily**	**Bare Costs**	**Incl. O&P**
1 Labor Foreman (outside)	$18.90	$151.20	$32.10	$256.80	$19.52	$32.76
4 Laborers	16.90	540.80	28.70	918.40		
1 Equip. Oper. (crane)	24.00	192.00	39.50	316.00		
1 Welder	26.15	209.20	42.90	343.20		
1 Hyd. Crane, 25 Ton		624.20		686.60		
1 Gas Welding Machine		75.20		82.70		
1 Horz. Boring Csg. Mch.		381.60		419.75	19.30	21.23
56 L.H., Daily Totals		$2174.20		$3023.45	$38.82	$53.99
Crew B-43	**Hr.**	**Daily**	**Hr.**	**Daily**	**Bare Costs**	**Incl. O&P**
1 Labor Foreman (outside)	$18.90	$151.20	$32.10	$256.80	$17.30	$29.38
4 Laborers	16.90	540.80	28.70	918.40		
1 Drill Rig & Augers		2877.00		3164.70	71.93	79.12
40 L.H., Daily Totals		$3569.00		$4339.90	$89.23	$108.50

Crew No.	Bare Costs		Incl. Subs O & P		Cost Per Labor-Hour	
Crew B-44	**Hr.**	**Daily**	**Hr.**	**Daily**	**Bare Costs**	**Incl. O&P**
1 Pile Driver Foreman	$24.45	$195.60	$43.80	$350.40	$21.51	$37.71
4 Pile Drivers	22.45	718.40	40.25	1288.00		
1 Equip. Oper. (crane)	24.00	192.00	39.50	316.00		
2 Laborer	16.90	270.40	28.70	459.20		
1 Crane, 40 Ton, & Access.		915.20		1006.70		
45 L.F. Leads, 15K Ft. Lbs.		67.50		74.25	15.35	16.89
64 L.H., Daily Totals		$2359.10		$3494.55	$36.86	$54.60
Crew B-45	**Hr.**	**Daily**	**Hr.**	**Daily**	**Bare Costs**	**Incl. O&P**
1 Building Laborer	$16.90	$135.20	$28.70	$229.60	$17.85	$30.00
1 Truck Driver (heavy)	18.80	150.40	31.30	250.40		
1 Dist. Tank Truck, 3K Gal.		240.40		264.45	15.03	16.53
16 L.H., Daily Totals		$526.00		$744.45	$32.88	$46.53
Crew B-46	**Hr.**	**Daily**	**Hr.**	**Daily**	**Bare Costs**	**Incl. O&P**
1 Pile Driver Foreman	$24.45	$195.60	$43.80	$350.40	$20.01	$35.07
2 Pile Drivers	22.45	359.20	40.25	644.00		
3 Laborers	16.90	405.60	28.70	688.80		
1 Chain Saw, 36" Long		33.40		36.75	.70	.77
48 L.H., Daily Totals		$993.80		$1719.95	$20.71	$35.84
Crew B-47	**Hr.**	**Daily**	**Hr.**	**Daily**	**Bare Costs**	**Incl. O&P**
1 Blast Foreman	$18.90	$151.20	$32.10	$256.80	$17.90	$30.40
1 Driller	16.90	135.20	28.70	229.60		
1 Crawler Type Drill, 4"		634.20		697.60		
1 Air Compr., 600 C.F.M.		286.00		314.60		
2-50 Ft. Air Hoses, 3" Dia.		35.40		38.95	59.73	65.70
16 L.H., Daily Totals		$1242.00		$1537.55	$77.63	$96.10
Crew B-47A	**Hr.**	**Daily**	**Hr.**	**Daily**	**Bare Costs**	**Incl. O&P**
1 Drilling Foreman	$18.90	$151.20	$32.10	$256.80	$21.02	$34.92
1 Equip. Oper. (heavy)	24.00	192.00	39.50	316.00		
1 Oiler	20.15	161.20	33.15	265.20		
1 Quarry Drill		827.40		910.15	34.48	37.92
24 L.H., Daily Totals		$1331.80		$1748.15	$55.50	$72.84
Crew B-47C	**Hr.**	**Daily**	**Hr.**	**Daily**	**Bare Costs**	**Incl. O&P**
1 Laborer	$16.90	$135.20	$28.70	$229.60	$19.55	$32.63
1 Equip. Oper. (light)	22.20	177.60	36.55	292.40		
1 Air Compressor, 750 CFM		305.00		335.50		
2-50' Air Hoses, 3"		35.40		38.95		
1 Air Track Drill, 4"		634.20		697.60	60.91	67.00
16 L.H., Daily Totals		$1287.40		$1594.05	$80.46	$99.63
Crew B-47E	**Hr.**	**Daily**	**Hr.**	**Daily**	**Bare Costs**	**Incl. O&P**
1 Laborer Foreman	$18.90	$151.20	$32.10	$256.80	$17.40	$29.55
3 Laborers	16.90	405.60	28.70	688.80		
1 Truck, Flatbed, 3 Ton		160.40		176.45	5.01	5.51
32 L.H., Daily Totals		$717.20		$1122.05	$22.41	$35.06
Crew B-48	**Hr.**	**Daily**	**Hr.**	**Daily**	**Bare Costs**	**Incl. O&P**
1 Labor Foreman (outside)	$18.90	$151.20	$32.10	$256.80	$18.42	$31.07
4 Laborers	16.90	540.80	28.70	918.40		
1 Equip. Oper. (crane)	24.00	192.00	39.50	316.00		
1 Centr. Water Pump, 6"		229.80		252.80		
1-20 Ft. Suction Hose, 6"		14.50		15.95		
1-50 Ft. Disch. Hose, 6"		11.90		13.10		
1 Drill Rig & Augers		2877.00		3164.70	65.28	71.80
48 L.H., Daily Totals		$4017.20		$4937.75	$83.70	$102.87

Crew No.	Bare Costs		Incl. Subs O & P		Cost Per Labor-Hour	
Crew B-49	**Hr.**	**Daily**	**Hr.**	**Daily**	**Bare Costs**	**Incl. O&P**
1 Labor Foreman (outside)	$18.90	$151.20	$32.10	$256.80	$19.14	$32.84
5 Laborers	16.90	676.00	28.70	1148.00		
1 Equip. Oper. (crane)	24.00	192.00	39.50	316.00		
2 Pile Drivers	22.45	359.20	40.25	644.00		
1 Hyd. Crane, 25 Ton		624.20		686.60		
1 Centr. Water Pump, 6"		229.80		252.80		
1-20 Ft. Suction Hose, 6"		14.50		15.95		
1-50 Ft. Disch. Hose, 6"		11.90		13.10		
1 Drill Rig & Augers		2877.00		3164.70	52.19	57.40
72 L.H., Daily Totals		$5135.80		$6497.95	$71.33	$90.24
Crew B-50	**Hr.**	**Daily**	**Hr.**	**Daily**	**Bare Costs**	**Incl. O&P**
1 Pile Driver Foremen	$24.45	$195.60	$43.80	$350.40	$20.59	$36.02
6 Pile Drivers	22.45	1077.60	40.25	1932.00		
1 Equip. Oper. (crane)	24.00	192.00	39.50	316.00		
5 Laborers	16.90	676.00	28.70	1148.00		
1 Crane, 40 Ton		915.20		1006.70		
60 L.F. Leads, 15K Ft. Lbs.		90.00		99.00		
1 Hammer, 15K Ft. Lbs.		359.80		395.80		
1 Air Compr., 600 C.F.M.		286.00		314.60		
2-50 Ft. Air Hoses, 3" Dia.		35.40		38.95		
1 Chain Saw, 36" Long		33.40		36.75	16.54	18.19
104 L.H., Daily Totals		$3861.00		$5638.20	$37.13	$54.21
Crew B-51	**Hr.**	**Daily**	**Hr.**	**Daily**	**Bare Costs**	**Incl. O&P**
1 Labor Foreman (outside)	$18.90	$151.20	$32.10	$256.80	$17.46	$29.54
4 Laborers	16.90	540.80	28.70	918.40		
1 Truck Driver (light)	18.25	146.00	30.35	242.80		
1 Light Truck, 1.5 Ton		122.00		134.20	2.54	2.80
48 L.H., Daily Totals		$960.00		$1552.20	$20.00	$32.34
Crew B-52	**Hr.**	**Daily**	**Hr.**	**Daily**	**Bare Costs**	**Incl. O&P**
1 Labor Foreman	$18.90	$151.20	$32.10	$256.80	$19.09	$32.49
1 Carpenter	23.10	184.80	39.20	313.60		
4 Laborers	16.90	540.80	28.70	918.40		
.5 Rodman (reinf.)	24.85	99.40	44.50	178.00		
.5 Equip. Oper. (med.)	23.20	92.80	38.20	152.80		
.5 F.E. Ldr., T.M., 2.5 C.Y.		380.50		418.55	6.79	7.47
56 L.H., Daily Totals		$1449.50		$2238.15	$25.88	$39.96
Crew B-53	**Hr.**	**Daily**	**Hr.**	**Daily**	**Bare Costs**	**Incl. O&P**
1 Building Laborer	$16.90	$135.20	$28.70	$229.60	$16.90	$28.70
1 Trencher, Chain, 12 H.P.		59.40		65.35	7.43	8.17
8 L.H., Daily Totals		$194.60		$294.95	$24.33	$36.87
Crew B-54	**Hr.**	**Daily**	**Hr.**	**Daily**	**Bare Costs**	**Incl. O&P**
1 Equip. Oper. (light)	$22.20	$177.60	$36.55	$292.40	$22.20	$36.55
1 Trencher, Chain, 40 H.P.		208.60		229.45	26.08	28.68
8 L.H., Daily Totals		$386.20		$521.85	$48.28	$65.23
Crew B-54A	**Hr.**	**Daily**	**Hr.**	**Daily**	**Bare Costs**	**Incl. O&P**
.17 Labor Foreman (outside)	$18.90	$25.70	$32.10	$43.66	$22.58	$37.31
1 Equipment Operator (med.)	23.20	185.60	38.20	305.60		
1 Wheel Trencher, 67 H.P.		767.40		844.15	81.99	90.19
9.36 L.H., Daily Totals		$978.70		$1193.41	$104.57	$127.50
Crew B-54B	**Hr.**	**Daily**	**Hr.**	**Daily**	**Bare Costs**	**Incl. O&P**
.25 Labor Foreman (outside)	$18.90	$37.80	$32.10	$64.20	$22.34	$36.98
1 Equipment Operator (med.)	23.20	185.60	38.20	305.60		
1 Wheel Trencher, 150 H.P.		1326.00		1458.60	132.60	145.86
10 L.H., Daily Totals		$1549.40		$1828.40	$154.94	$182.84

Crew B-55	Hr.	Daily	Hr.	Daily	Bare Costs	Incl. O&P
1 Laborers	$16.90	$135.20	$28.70	$229.60	$17.58	$29.53
1 Truck Driver (light)	18.25	146.00	30.35	242.80		
1 Auger, 4" to 36" Dia		614.80		676.30		
1 Flatbed 3 Ton Truck		160.40		176.45	48.45	53.30
16 L.H., Daily Totals		$1056.40		$1325.15	$66.03	$82.83

Crew B-56	Hr.	Daily	Hr.	Daily	Bare Costs	Incl. O&P
2 Laborer	$16.90	$270.40	$28.70	$459.20	$16.90	$28.70
1 Crawler Type Drill, 4"		634.20		697.60		
1 Air Compr., 600 C.F.M.		286.00		314.60		
1-50 Ft. Air Hose, 3" Dia.		17.70		19.45	58.62	64.48
16 L.H., Daily Totals		$1208.30		$1490.85	$75.52	$93.18

Crew B-57	Hr.	Daily	Hr.	Daily	Bare Costs	Incl. O&P
1 Labor Foreman (outside)	$18.90	$151.20	$32.10	$256.80	$18.72	$31.54
3 Laborers	16.90	405.60	28.70	688.80		
1 Equip. Oper. (crane)	24.00	192.00	39.50	316.00		
1 Barge, 400 Ton		273.20		300.50		
1 Power Shovel, 1 C.Y.		792.00		871.20		
1 Clamshell Bucket, 1 C.Y.		44.00		48.40		
1 Centr. Water Pump, 6"		229.80		252.80		
1-20 Ft. Suction Hose, 6"		14.50		15.95		
20-50 Ft. Disch. Hoses, 6"		238.00		261.80	39.79	43.77
40 L.H., Daily Totals		$2340.30		$3012.25	$58.51	$75.31

Crew B-58	Hr.	Daily	Hr.	Daily	Bare Costs	Incl. O&P
2 Laborers	$16.90	$270.40	$28.70	$459.20	$18.67	$31.32
1 Equip. Oper. (light)	22.20	177.60	36.55	292.40		
1 Backhoe Loader, 48 H.P.		207.80		228.60		
1 Small Helicopter, w/pilot		2108.00		2318.80	96.49	106.14
24 L.H., Daily Totals		$2763.80		$3299.00	$115.16	$137.46

Crew B-59	Hr.	Daily	Hr.	Daily	Bare Costs	Incl. O&P
1 Truck Driver (heavy)	$18.80	$150.40	$31.30	$250.40	$18.80	$31.30
1 Truck, 30 Ton		215.20		236.70		
1 Water tank, 6000 Gal.		118.00		129.80	41.65	45.82
8 L.H., Daily Totals		$483.60		$616.90	$60.45	$77.12

Crew B-60	Hr.	Daily	Hr.	Daily	Bare Costs	Incl. O&P
1 Labor Foreman (outside)	$18.90	$151.20	$32.10	$256.80	$19.30	$32.38
3 Laborers	16.90	405.60	28.70	688.80		
1 Equip. Oper. (crane)	24.00	192.00	39.50	316.00		
1 Equip. Oper. (light)	22.20	177.60	36.55	292.40		
1 Crawler Crane, 40 Ton		915.20		1006.70		
45 L.F. Leads, 15K Ft. Lbs.		67.50		74.25		
1 Backhoe Loader, 48 H.P.		207.80		228.60	24.80	27.28
48 L.H., Daily Totals		$2116.90		$2863.55	$44.10	$59.66

Crew B-61	Hr.	Daily	Hr.	Daily	Bare Costs	Incl. O&P
1 Labor Foreman (outside)	$18.90	$151.20	$32.10	$256.80	$17.30	$29.38
4 Laborers	16.90	540.80	28.70	918.40		
1 Cement Mixer, 2 C.Y.		161.00		177.10		
1 Air Compr., 160 C.F.M.		82.00		90.20	6.08	6.68
40 L.H., Daily Totals		$935.00		$1442.50	$23.38	$36.06

Crew B-62	Hr.	Daily	Hr.	Daily	Bare Costs	Incl. O&P
2 Laborers	$16.90	$270.40	$28.70	$459.20	$18.67	$31.32
1 Equip. Oper. (light)	22.20	177.60	36.55	292.40		
1 Loader, Skid Steer		143.20		157.50	5.97	6.56
24 L.H., Daily Totals		$591.20		$909.10	$24.64	$37.88

Crew B-63	Hr.	Daily	Hr.	Daily	Bare Costs	Incl. O&P
5 Laborers	$16.90	$676.00	$28.70	$1148.00	$16.90	$28.70
1 Loader, Skid Steer		143.20		157.50	3.58	3.94
40 L.H., Daily Totals		$819.20		$1305.50	$20.48	$32.64

Crew B-64	Hr.	Daily	Hr.	Daily	Bare Costs	Incl. O&P
1 Laborer	$16.90	$135.20	$28.70	$229.60	$17.58	$29.53
1 Truck Driver (light)	18.25	146.00	30.35	242.80		
1 Power Mulcher (small)		103.20		113.50		
1 Light Truck, 1.5 Ton		122.00		134.20	14.08	15.48
16 L.H., Daily Totals		$506.40		$720.10	$31.66	$45.01

Crew B-65	Hr.	Daily	Hr.	Daily	Bare Costs	Incl. O&P
1 Laborer	$16.90	$135.20	$28.70	$229.60	$17.58	$29.53
1 Truck Driver (light)	18.25	146.00	30.35	242.80		
1 Power Mulcher (large)		202.20		222.40		
1 Light Truck, 1.5 Ton		122.00		134.20	20.26	22.29
16 L.H., Daily Totals		$605.40		$829.00	$37.84	$51.82

Crew B-66	Hr.	Daily	Hr.	Daily	Bare Costs	Incl. O&P
1 Equip. Oper. (light)	$22.20	$177.60	$36.55	$292.40	$22.20	$36.55
1 Backhoe Ldr. w/Attchmt.		168.40		185.25	21.05	23.16
8 L.H., Daily Totals		$346.00		$477.65	$43.25	$59.71

Crew B-67	Hr.	Daily	Hr.	Daily	Bare Costs	Incl. O&P
1 Millwright	$24.05	$192.40	$38.85	$310.80	$23.13	$37.70
1 Equip. Oper. (light)	22.20	177.60	36.55	292.40		
1 Forklift		239.80		263.80	14.99	16.49
16 L.H., Daily Totals		$609.80		$867.00	$38.12	$54.19

Crew B-68	Hr.	Daily	Hr.	Daily	Bare Costs	Incl. O&P
2 Millwrights	$24.05	$384.80	$38.85	$621.60	$23.43	$38.08
1 Equip. Oper. (light)	22.20	177.60	36.55	292.40		
1 Forklift		239.80		263.80	9.99	10.99
24 L.H., Daily Totals		$802.20		$1177.80	$33.42	$49.07

Crew B-69	Hr.	Daily	Hr.	Daily	Bare Costs	Incl. O&P
1 Labor Foreman (outside)	$18.90	$151.20	$32.10	$256.80	$18.96	$31.81
3 Laborers	16.90	405.60	28.70	688.80		
1 Equip Oper. (crane)	24.00	192.00	39.50	316.00		
1 Equip Oper. Oiler	20.15	161.20	33.15	265.20		
1 Truck Crane, 80 Ton		1020.00		1122.00	21.25	23.38
48 L.H., Daily Totals		$1930.00		$2648.80	$40.21	$55.19

Crew B-69A	Hr.	Daily	Hr.	Daily	Bare Costs	Incl. O&P
1 Labor Foreman	$18.90	$151.20	$32.10	$256.80	$19.15	$32.01
3 Laborers	16.90	405.60	28.70	688.80		
1 Equip. Oper. (medium)	23.20	185.60	38.20	305.60		
1 Concrete Finisher	22.10	176.80	35.65	285.20		
1 Curb Paver		573.60		630.95	11.95	13.15
48 L.H., Daily Totals		$1492.80		$2167.35	$31.10	$45.16

Crew B-69B	Hr.	Daily	Hr.	Daily	Bare Costs	Incl. O&P
1 Labor Foreman	$18.90	$151.20	$32.10	$256.80	$19.15	$32.01
3 Laborers	16.90	405.60	28.70	688.80		
1 Equip. Oper. (medium)	23.20	185.60	38.20	305.60		
1 Cement Finisher	22.10	176.80	35.65	285.20		
1 Curb/Gutter Paver		687.40		756.15	14.32	15.75
48 L.H., Daily Totals		$1606.60		$2292.55	$33.47	$47.76

Crew No.	Bare Costs		Incl. Subs O & P		Cost Per Labor-Hour	
Crew B-70	**Hr.**	**Daily**	**Hr.**	**Daily**	**Bare Costs**	**Incl. O&P**
1 Labor Foreman (outside)	$18.90	$151.20	$32.10	$256.80	$19.89	$33.26
3 Laborers	16.90	405.60	28.70	688.80		
3 Equip. Oper. (med.)	23.20	556.80	38.20	916.80		
1 Motor Grader, 30,000 Lb.		432.00		475.20		
1 Grader Attach., Ripper		70.60		77.65		
1 Road Sweeper, S.P.		420.00		462.00		
1 F.E. Loader, 1-3/4 C.Y.		236.20		259.80	20.69	22.76
56 L.H., Daily Totals		$2272.40		$3137.05	$40.58	$56.02
Crew B-71	**Hr.**	**Daily**	**Hr.**	**Daily**	**Bare Costs**	**Incl. O&P**
1 Labor Foreman (outside)	$18.90	$151.20	$32.10	$256.80	$19.89	$33.26
3 Laborers	16.90	405.60	28.70	688.80		
3 Equip. Oper. (med.)	23.20	556.80	38.20	916.80		
1 Pvmt. Profiler, 750 H.P.		4351.00		4786.10		
1 Road Sweeper, S.P.		420.00		462.00		
1 F.E. Loader, 1-3/4 C.Y.		236.20		259.80	89.41	98.36
56 L.H., Daily Totals		$6120.80		$7370.30	$109.30	$131.62
Crew B-72	**Hr.**	**Daily**	**Hr.**	**Daily**	**Bare Costs**	**Incl. O&P**
1 Labor Foreman (outside)	$18.90	$151.20	$32.10	$256.80	$20.30	$33.88
3 Laborers	16.90	405.60	28.70	688.80		
4 Equip. Oper. (med.)	23.20	742.40	38.20	1222.40		
1 Pvmt. Profiler, 750 H.P.		4351.00		4786.10		
1 Hammermill, 250 H.P.		1315.00		1446.50		
1 Windrow Loader		805.60		886.15		
1 Mix Paver 165 H.P.		1674.00		1841.40		
1 Roller, Pneu. Tire, 12 T.		244.60		269.05	131.10	144.21
64 L.H., Daily Totals		$9689.40		$11397.20	$151.40	$178.09
Crew B-73	**Hr.**	**Daily**	**Hr.**	**Daily**	**Bare Costs**	**Incl. O&P**
1 Labor Foreman (outside)	$18.90	$151.20	$32.10	$256.80	$21.09	$35.06
2 Laborers	16.90	270.40	28.70	459.20		
5 Equip. Oper. (med.)	23.20	928.00	38.20	1528.00		
1 Road Mixer, 310 H.P.		1610.00		1771.00		
1 Roller, Tandem, 12 Ton		183.60		201.95		
1 Hammermill, 250 H.P.		1315.00		1446.50		
1 Motor Grader, 30,000 Lb.		432.00		475.20		
.5 F.E. Loader, 1-3/4 C.Y.		118.10		129.90		
.5 Truck, 30 Ton		107.60		118.35		
.5 Water Tank 5000 Gal.		59.00		64.90	59.77	65.75
64 L.H., Daily Totals		$5174.90		$6451.80	$80.86	$100.81
Crew B-74	**Hr.**	**Daily**	**Hr.**	**Daily**	**Bare Costs**	**Incl. O&P**
1 Labor Foreman (outside)	$18.90	$151.20	$32.10	$256.80	$20.78	$34.53
1 Laborer	16.90	135.20	28.70	229.60		
4 Equip. Oper. (med.)	23.20	742.40	38.20	1222.40		
2 Truck Drivers (heavy)	18.80	300.80	31.30	500.80		
1 Motor Grader, 30,000 Lb.		432.00		475.20		
1 Grader Attach., Ripper		70.60		77.65		
2 Stabilizers, 310 H.P.		2122.00		2334.20		
1 Flatbed Truck, 3 Ton		160.40		176.45		
1 Chem. Spreader, Towed		66.00		72.60		
1 Vibr. Roller, 29,000 Lb.		436.80		480.50		
1 Water Tank 5000 Gal.		118.00		129.80		
1 Truck, 30 Ton		215.20		236.70	56.58	62.24
64 L.H., Daily Totals		$4950.60		$6192.70	$77.36	$96.77

Crew No.	Bare Costs		Incl. Subs O & P		Cost Per Labor-Hour	
Crew B-75	**Hr.**	**Daily**	**Hr.**	**Daily**	**Bare Costs**	**Incl. O&P**
1 Labor Foreman (outside)	$18.90	$151.20	$32.10	$256.80	$21.06	$34.99
1 Laborer	16.90	135.20	28.70	229.60		
4 Equip. Oper. (med.)	23.20	742.40	38.20	1222.40		
1 Truck Driver (heavy)	18.80	150.40	31.30	250.40		
1 Motor Grader, 30,000 Lb.		432.00		475.20		
1 Grader Attach., Ripper		70.60		77.65		
2 Stabilizers, 310 H.P.		2122.00		2334.20		
1 Dist. Truck, 3000 Gal.		240.40		264.45		
1 Vibr. Roller, 29,000 Lb.		436.80		480.50	58.96	64.86
56 L.H., Daily Totals		$4481.00		$5591.20	$80.02	$99.85
Crew B-76	**Hr.**	**Daily**	**Hr.**	**Daily**	**Bare Costs**	**Incl. O&P**
1 Dock Builder Foreman	$24.45	$195.60	$43.80	$350.40	$22.76	$39.69
5 Dock Builders	22.45	898.00	40.25	1610.00		
2 Equip. Oper. (crane)	24.00	384.00	39.50	632.00		
1 Equip. Oper. Oiler	20.15	161.20	33.15	265.20		
1 Crawler Crane, 50 Ton		1260.00		1386.00		
1 Barge, 400 Ton		273.20		300.50		
1 Hammer, 15K Ft. Lbs.		359.80		395.80		
60 L.F. Leads, 15K Ft. Lbs.		90.00		99.00		
1 Air Compr., 600 C.F.M.		286.00		314.60		
2-50 Ft. Air Hoses, 3" Dia.		35.40		38.95	32.01	35.21
72 L.H., Daily Totals		$3943.20		$5392.45	$54.77	$74.90
Crew B-77	**Hr.**	**Daily**	**Hr.**	**Daily**	**Bare Costs**	**Incl. O&P**
1 Labor Foreman	$18.90	$151.20	$32.10	$256.80	$17.57	$29.71
3 Laborers	16.90	405.60	28.70	688.80		
1 Truck Driver (light)	18.25	146.00	30.35	242.80		
1 Crack Cleaner, 25 H.P.		44.40		48.85		
1 Crack Filler, Trailer Mtd.		157.00		172.70		
1 Flatbed Truck, 3 Ton		160.40		176.45	9.05	9.95
40 L.H., Daily Totals		$1064.60		$1586.40	$26.62	$39.66
Crew B-78	**Hr.**	**Daily**	**Hr.**	**Daily**	**Bare Costs**	**Incl. O&P**
1 Labor Foreman	$18.90	$151.20	$32.10	$256.80	$17.30	$29.38
4 Laborers	16.90	540.80	28.70	918.40		
1 Paint Striper, S.P.		134.20		147.60		
1 Flatbed Truck, 3 Ton		160.40		176.45		
1 Pickup Truck, 3/4 Ton		75.80		83.40	9.26	10.19
40 L.H., Daily Totals		$1062.40		$1582.65	$26.56	$39.57
Crew B-79	**Hr.**	**Daily**	**Hr.**	**Daily**	**Bare Costs**	**Incl. O&P**
1 Labor Foreman	$18.90	$151.20	$32.10	$256.80	$17.40	$29.55
3 Laborers	16.90	405.60	28.70	688.80		
1 Thermo. Striper, T.M.		549.20		604.10		
1 Flatbed Truck, 3 Ton		160.40		176.45		
2 Pickup Trucks, 3/4 Ton		151.60		166.75	26.91	29.60
32 L.H., Daily Totals		$1418.00		$1892.90	$44.31	$59.15
Crew B-80	**Hr.**	**Daily**	**Hr.**	**Daily**	**Bare Costs**	**Incl. O&P**
1 Labor Foreman	$18.90	$151.20	$32.10	$256.80	$17.57	$29.83
2 Laborer	16.90	270.40	28.70	459.20		
1 Flatbed Truck, 3 Ton		160.40		176.45		
1 Fence Post Auger, T.M.		357.40		393.15	21.58	23.73
24 L.H., Daily Totals		$939.40		$1285.60	$39.15	$53.56
Crew B-80A	**Hr.**	**Daily**	**Hr.**	**Daily**	**Bare Costs**	**Incl. O&P**
3 Laborers	$16.90	$405.60	$28.70	$688.80	$16.90	$28.70
1 Flatbed Truck, 3 Ton		160.40		176.45	6.68	7.35
24 L.H., Daily Totals		$566.00		$865.25	$23.58	$36.05

Crews

Crew No.	Bare Costs		Incl. Subs O & P		Cost Per Labor-Hour	
Crew B-80B	**Hr.**	**Daily**	**Hr.**	**Daily**	**Bare Costs**	**Incl. O&P**
3 Laborers	$16.90	$405.60	$28.70	$688.80	$18.23	$30.66
1 Equip. Oper. (light)	22.20	177.60	36.55	292.40		
1 Crane, Flatbed Mnt.		201.80		222.00	6.31	6.94
32 L.H., Daily Totals		$785.00		$1203.20	$24.54	$37.60
Crew B-81	**Hr.**	**Daily**	**Hr.**	**Daily**	**Bare Costs**	**Incl. O&P**
1 Laborer	$16.90	$135.20	$28.70	$229.60	$17.85	$30.00
1 Truck Driver (heavy)	18.80	150.40	31.30	250.40		
1 Hydromulcher, T.M.		196.00		215.60		
1 Tractor Truck, 4x2		215.20		236.70	25.70	28.27
16 L.H., Daily Totals		$696.80		$932.30	$43.55	$58.27
Crew B-82	**Hr.**	**Daily**	**Hr.**	**Daily**	**Bare Costs**	**Incl. O&P**
1 Laborer	$16.90	$135.20	$28.70	$229.60	$19.55	$32.63
1 Equip. Oper. (light)	22.20	177.60	36.55	292.40		
1 Horiz. Borer, 6 H.P.		62.80		69.10	3.93	4.32
16 L.H., Daily Totals		$375.60		$591.10	$23.48	$36.95
Crew B-83	**Hr.**	**Daily**	**Hr.**	**Daily**	**Bare Costs**	**Incl. O&P**
1 Tugboat Captain	$23.20	$185.60	$38.20	$305.60	$20.05	$33.45
1 Tugboat Hand	16.90	135.20	28.70	229.60		
1 Tugboat, 250 H.P.		440.60		484.65	27.54	30.29
16 L.H., Daily Totals		$761.40		$1019.85	$47.59	$63.74
Crew B-84	**Hr.**	**Daily**	**Hr.**	**Daily**	**Bare Costs**	**Incl. O&P**
1 Equip. Oper. (med.)	$23.20	$185.60	$38.20	$305.60	$23.20	$38.20
1 Rotary Mower/Tractor		222.20		244.40	27.78	30.55
8 L.H., Daily Totals		$407.80		$550.00	$50.98	$68.75
Crew B-85	**Hr.**	**Daily**	**Hr.**	**Daily**	**Bare Costs**	**Incl. O&P**
3 Laborers	$16.90	$405.60	$28.70	$688.80	$18.54	$31.12
1 Equip. Oper. (med.)	23.20	185.60	38.20	305.60		
1 Truck Driver (heavy)	18.80	150.40	31.30	250.40		
1 Aerial Lift Truck, 80'		501.80		552.00		
1 Brush Chipper, 130 H.P.		164.80		181.30		
1 Pruning Saw, Rotary		10.55		11.60	16.93	18.62
40 L.H., Daily Totals		$1418.75		$1989.70	$35.47	$49.74
Crew B-86	**Hr.**	**Daily**	**Hr.**	**Daily**	**Bare Costs**	**Incl. O&P**
1 Equip. Oper. (med.)	$23.20	$185.60	$38.20	$305.60	$23.20	$38.20
1 Stump Chipper, S.P.		87.20		95.90	10.90	11.99
8 L.H., Daily Totals		$272.80		$401.50	$34.10	$50.19
Crew B-86A	**Hr.**	**Daily**	**Hr.**	**Daily**	**Bare Costs**	**Incl. O&P**
1 Equip. Oper. (medium)	$23.20	$185.60	$38.20	$305.60	$23.20	$38.20
1 Grader, 30,000 Lbs.		432.00		475.20	54.00	59.40
8 L.H., Daily Totals		$617.60		$780.80	$77.20	$97.60
Crew B-86B	**Hr.**	**Daily**	**Hr.**	**Daily**	**Bare Costs**	**Incl. O&P**
1 Equip. Oper. (medium)	$23.20	$185.60	$38.20	$305.60	$23.20	$38.20
1 Dozer, 200 H.P.		863.40		949.75	107.93	118.72
8 L.H., Daily Totals		$1049.00		$1255.35	$131.13	$156.92

Crew No.	Bare Costs		Incl. Subs O & P		Cost Per Labor-Hour	
Crew B-87	**Hr.**	**Daily**	**Hr.**	**Daily**	**Bare Costs**	**Incl. O&P**
1 Laborer	$16.90	$135.20	$28.70	$229.60	$21.94	$36.30
4 Equip. Oper. (med.)	23.20	742.40	38.20	1222.40		
2 Feller Bunchers, 50 H.P.		864.80		951.30		
1 Log Chipper, 22" Tree		1207.00		1327.70		
1 Dozer, 105 H.P.		434.00		477.40		
1 Chainsaw, Gas, 36" Long		33.40		36.75	63.48	69.83
40 L.H., Daily Totals		$3416.80		$4245.15	$85.42	$106.13
Crew B-88	**Hr.**	**Daily**	**Hr.**	**Daily**	**Bare Costs**	**Incl. O&P**
1 Laborer	$16.90	$135.20	$28.70	$229.60	$22.30	$36.84
6 Equip. Oper. (med.)	23.20	1113.60	38.20	1833.60		
2 Feller Bunchers, 50 H.P.		864.80		951.30		
1 Log Chipper, 22" Tree		1207.00		1327.70		
2 Log Skidders, 50 H.P.		1490.80		1639.90		
1 Dozer, 105 H.P.		434.00		477.40		
1 Chainsaw, Gas, 36" Long		33.40		36.75	71.96	79.16
56 L.H., Daily Totals		$5278.80		$6496.25	$94.26	$116.00
Crew B-89	**Hr.**	**Daily**	**Hr.**	**Daily**	**Bare Costs**	**Incl. O&P**
1 Skilled Worker	$23.05	$184.40	$39.05	$312.40	$19.98	$33.88
1 Building Laborer	16.90	135.20	28.70	229.60		
1 Cutting Machine		47.45		52.20	2.97	3.26
16 L.H., Daily Totals		$367.05		$594.20	$22.95	$37.14
Crew B-89A	**Hr.**	**Daily**	**Hr.**	**Daily**	**Bare Costs**	**Incl. O&P**
1 Skilled Worker	$23.05	$184.40	$39.05	$312.40	$19.98	$33.88
1 Laborer	16.90	135.20	28.70	229.60		
1 Core Drill (large)		106.00		116.60	6.63	7.29
16 L.H., Daily Totals		$425.60		$658.60	$26.61	$41.17
Crew B-89B	**Hr.**	**Daily**	**Hr.**	**Daily**	**Bare Costs**	**Incl. O&P**
1 Equip. Oper. (light)	$22.20	$177.60	$36.55	$292.40	$20.23	$33.45
1 Truck Driver, Light	18.25	146.00	30.35	242.80		
1 Wall Saw, Hydraulic, 10 H.P.		75.10		82.60		
1 Generator, Diesel, 100 KW		179.60		197.55		
1 Water Tank, 65 Gal.		13.60		14.95		
1 Flatbed Truck, 3 Ton		160.40		176.45	26.79	29.47
16 L.H., Daily Totals		$752.30		$1006.75	$47.02	$62.92
Crew B-90	**Hr.**	**Daily**	**Hr.**	**Daily**	**Bare Costs**	**Incl. O&P**
1 Labor Foreman (outside)	$18.90	$151.20	$32.10	$256.80	$18.95	$31.74
3 Laborers	16.90	405.60	28.70	688.80		
2 Equip. Oper. (light)	22.20	355.20	36.55	584.80		
2 Truck Drivers (heavy)	18.80	300.80	31.30	500.80		
1 Road Mixer, 310 H.P.		1610.00		1771.00		
1 Dist. Truck, 2000 Gal.		208.00		228.80	28.41	31.25
64 L.H., Daily Totals		$3030.80		$4031.00	$47.36	$62.99
Crew B-90A	**Hr.**	**Daily**	**Hr.**	**Daily**	**Bare Costs**	**Incl. O&P**
1 Labor Foreman	$18.90	$151.20	$32.10	$256.80	$20.79	$34.61
2 Laborers	16.90	270.40	28.70	459.20		
4 Equip. Oper. (medium)	23.20	742.40	38.20	1222.40		
2 Graders, 30,000 Lbs.		864.00		950.40		
1 Roller, Steel Wheel		183.60		201.95		
1 Roller, Pneumatic Wheel		244.60		269.05	23.08	25.38
56 L.H., Daily Totals		$2456.20		$3359.80	$43.87	$59.99

Crew B-90B	Bare Costs Hr.	Bare Costs Daily	Incl. Subs O & P Hr.	Incl. Subs O & P Daily	Bare Costs	Incl. O&P
1 Labor Foreman	$18.90	$151.20	$32.10	$256.80	$20.38	$34.02
2 Laborers	16.90	270.40	28.70	459.20		
3 Equip. Oper. (medium)	23.20	556.80	38.20	916.80		
1 Roller, Steel Wheel		183.60		201.95		
1 Roller, Pneumatic Wheel		244.60		269.05		
1 Road Mixer, 310 H.P.		1610.00		1771.00	42.46	46.71
48 L.H., Daily Totals		$3016.60		$3874.80	$62.84	$80.73

Crew B-91	Hr.	Daily	Hr.	Daily	Bare Costs	Incl. O&P
1 Labor Foreman (outside)	$18.90	$151.20	$32.10	$256.80	$20.54	$34.20
2 Laborers	16.90	270.40	28.70	459.20		
4 Equip. Oper. (med.)	23.20	742.40	38.20	1222.40		
1 Truck Driver (heavy)	18.80	150.40	31.30	250.40		
1 Dist. Truck, 3000 Gal.		240.40		264.45		
1 Aggreg. Spreader, S.P.		771.20		848.30		
1 Roller, Pneu. Tire, 12 Ton		244.60		269.05		
1 Roller, Steel, 10 Ton		183.60		201.95	22.50	24.75
64 L.H., Daily Totals		$2754.20		$3772.55	$43.04	$58.95

Crew B-92	Hr.	Daily	Hr.	Daily	Bare Costs	Incl. O&P
1 Labor Foreman (outside)	$18.90	$151.20	$32.10	$256.80	$17.40	$29.55
3 Laborers	16.90	405.60	28.70	688.80		
1 Crack Cleaner, 25 H.P.		44.40		48.85		
1 Air Compressor		52.40		57.65		
1 Tar Kettle, T.M.		40.95		45.05		
1 Flatbed Truck, 3 Ton		160.40		176.45	9.32	10.25
32 L.H., Daily Totals		$854.95		$1273.60	$26.72	$39.80

Crew B-93	Hr.	Daily	Hr.	Daily	Bare Costs	Incl. O&P
1 Equip. Oper. (med.)	$23.20	$185.60	$38.20	$305.60	$23.20	$38.20
1 Feller Buncher, 50 H.P.		432.40		475.65	54.05	59.46
8 L.H., Daily Totals		$618.00		$781.25	$77.25	$97.66

Crew B-94A	Hr.	Daily	Hr.	Daily	Bare Costs	Incl. O&P
1 Laborer	$16.90	$135.20	$28.70	$229.60	$16.90	$28.70
1 Diaph. Water Pump, 2"		46.20		50.80		
1-20 Ft. Suction Hose, 2"		3.55		3.90		
2-50 Ft. Disch. Hoses, 2"		6.30		6.95	7.01	7.71
8 L.H., Daily Totals		$191.25		$291.25	$23.91	$36.41

Crew B-94B	Hr.	Daily	Hr.	Daily	Bare Costs	Incl. O&P
1 Laborer	$16.90	$135.20	$28.70	$229.60	$16.90	$28.70
1 Diaph. Water Pump, 4"		87.40		96.15		
1-20 Ft. Suction Hose, 4"		7.30		8.05		
2-50 Ft. Disch. Hoses, 4"		10.50		11.55	13.15	14.47
8 L.H., Daily Totals		$240.40		$345.35	$30.05	$43.17

Crew B-94C	Hr.	Daily	Hr.	Daily	Bare Costs	Incl. O&P
1 Laborer	$16.90	$135.20	$28.70	$229.60	$16.90	$28.70
1 Centr. Water Pump, 3"		53.80		59.20		
1-20 Ft. Suction Hose, 3"		5.45		6.00		
2-50 Ft. Disch. Hoses, 3"		7.50		8.25	8.34	9.18
8 L.H., Daily Totals		$201.95		$303.05	$25.24	$37.88

Crew B-94D	Hr.	Daily	Hr.	Daily	Bare Costs	Incl. O&P
1 Laborer	$16.90	$135.20	$28.70	$229.60	$16.90	$28.70
1 Centr. Water Pump, 6"		229.80		252.80		
1-20 Ft. Suction Hose, 6"		14.50		15.95		
2-50 Ft. Disch. Hoses, 6"		23.80		26.20	33.51	36.86
8 L.H., Daily Totals		$403.30		$524.55	$50.41	$65.56

Crew B-95A	Hr.	Daily	Hr.	Daily	Bare Costs	Incl. O&P
1 Equip. Oper. (crane)	$24.00	$192.00	$39.50	$316.00	$20.45	$34.10
1 Laborer	16.90	135.20	28.70	229.60		
1 Hyd. Excavator, 5/8 C.Y.		418.00		459.80	26.13	28.74
16 L.H., Daily Totals		$745.20		$1005.40	$46.58	$62.84

Crew B-95B	Hr.	Daily	Hr.	Daily	Bare Costs	Incl. O&P
1 Equip. Oper. (crane)	$24.00	$192.00	$39.50	$316.00	$20.45	$34.10
1 Laborer	16.90	135.20	28.70	229.60		
1 Hyd. Excavator, 1.5 C.Y.		683.40		751.75	42.71	46.98
16 L.H., Daily Totals		$1010.60		$1297.35	$63.16	$81.08

Crew B-95C	Hr.	Daily	Hr.	Daily	Bare Costs	Incl. O&P
1 Equip. Oper. (crane)	$24.00	$192.00	$39.50	$316.00	$20.45	$34.10
1 Laborer	16.90	135.20	28.70	229.60		
1 Hyd. Excavator, 2.5 C.Y.		1181.00		1299.10	73.81	81.19
16 L.H., Daily Totals		$1508.20		$1844.70	$94.26	$115.29

Crew C-1	Hr.	Daily	Hr.	Daily	Bare Costs	Incl. O&P
2 Carpenters	$23.10	$369.60	$39.20	$627.20	$20.00	$33.93
1 Carpenter Helper	16.90	135.20	28.60	228.80		
1 Laborer	16.90	135.20	28.70	229.60		
32 L.H., Daily Totals		$640.00		$1085.60	$20.00	$33.93

Crew C-2	Hr.	Daily	Hr.	Daily	Bare Costs	Incl. O&P
1 Carpenter Foreman (out)	$25.10	$200.80	$42.60	$340.80	$20.33	$34.48
2 Carpenters	23.10	369.60	39.20	627.20		
2 Carpenter Helpers	16.90	270.40	28.60	457.60		
1 Laborer	16.90	135.20	28.70	229.60		
48 L.H., Daily Totals		$976.00		$1655.20	$20.33	$34.48

Crew C-2A	Hr.	Daily	Hr.	Daily	Bare Costs	Incl. O&P
1 Carpenter Foreman (out)	$25.10	$200.80	$42.60	$340.80	$22.23	$37.43
3 Carpenters	23.10	554.40	39.20	940.80		
1 Cement Finisher	22.10	176.80	35.65	285.20		
1 Laborer	16.90	135.20	28.70	229.60		
48 L.H., Daily Totals		$1067.20		$1796.40	$22.23	$37.43

Crew C-3	Hr.	Daily	Hr.	Daily	Bare Costs	Incl. O&P
1 Rodman Foreman	$26.85	$214.80	$48.10	$384.80	$21.79	$38.03
3 Rodmen (reinf.)	24.85	596.40	44.50	1068.00		
1 Equip. Oper. (light)	22.20	177.60	36.55	292.40		
3 Laborers	16.90	405.60	28.70	688.80		
3 Stressing Equipment		48.60		53.45		
.5 Grouting Equipment		78.47		86.30	1.99	2.18
64 L.H., Daily Totals		$1521.47		$2573.75	$23.78	$40.21

Crew C-4	Hr.	Daily	Hr.	Daily	Bare Costs	Incl. O&P
1 Rodman Foreman	$26.85	$214.80	$48.10	$384.80	$23.36	$41.45
2 Rodmen (reinf.)	24.85	397.60	44.50	712.00		
1 Building Laborer	16.90	135.20	28.70	229.60		
3 Stressing Equipment		48.60		53.45	1.52	1.67
32 L.H., Daily Totals		$796.20		$1379.85	$24.88	$43.12

Crew C-5	Hr.	Daily	Hr.	Daily	Bare Costs	Incl. O&P
1 Rodman Foreman	$26.85	$214.80	$48.10	$384.80	$22.39	$39.00
2 Rodmen (reinf.)	24.85	397.60	44.50	712.00		
1 Equip. Oper. (crane)	24.00	192.00	39.50	316.00		
2 Building Laborers	16.90	270.40	28.70	459.20		
1 Hyd. Crane, 25 Ton		624.20		686.60	13.00	14.30
48 L.H., Daily Totals		$1699.00		$2558.60	$35.39	$53.30

Crew No.	Bare Costs		Incl. Subs O & P		Cost Per Labor-Hour	
Crew C-6	**Hr.**	**Daily**	**Hr.**	**Daily**	**Bare Costs**	**Incl. O&P**
1 Labor Foreman (outside)	$18.90	$151.20	$32.10	$256.80	$18.10	$30.43
4 Laborers	16.90	540.80	28.70	918.40		
1 Cement Finisher	22.10	176.80	35.65	285.20		
2 Gas Engine Vibrators		52.00		57.20	1.08	1.19
48 L.H., Daily Totals		$920.80		$1517.60	$19.18	$31.62
Crew C-7	**Hr.**	**Daily**	**Hr.**	**Daily**	**Bare Costs**	**Incl. O&P**
1 Labor Foreman (outside)	$18.90	$151.20	$32.10	$256.80	$18.76	$31.40
5 Laborers	16.90	676.00	28.70	1148.00		
1 Cement Finisher	22.10	176.80	35.65	285.20		
1 Equip. Oper. (med.)	23.20	185.60	38.20	305.60		
1 Equip. Oper. (oiler)	20.15	161.20	33.15	265.20		
2 Gas Engine Vibrators		52.00		57.20		
1 Concrete Bucket, 1 C.Y.		15.60		17.15		
1 Hyd. Crane, 55 Ton		911.00		1002.10	13.59	14.95
72 L.H., Daily Totals		$2329.40		$3337.25	$32.35	$46.35
Crew C-8	**Hr.**	**Daily**	**Hr.**	**Daily**	**Bare Costs**	**Incl. O&P**
1 Labor Foreman (outside)	$18.90	$151.20	$32.10	$256.80	$19.57	$32.53
3 Laborers	16.90	405.60	28.70	688.80		
2 Cement Finishers	22.10	353.60	35.65	570.40		
1 Equip. Oper. (med.)	23.20	185.60	38.20	305.60		
1 Concrete Pump (small)		719.20		791.10	12.84	14.13
56 L.H., Daily Totals		$1815.20		$2612.70	$32.41	$46.66
Crew C-8A	**Hr.**	**Daily**	**Hr.**	**Daily**	**Bare Costs**	**Incl. O&P**
1 Labor Foreman (outside)	$18.90	$151.20	$32.10	$256.80	$18.97	$31.58
3 Laborers	16.90	405.60	28.70	688.80		
2 Cement Finishers	22.10	353.60	35.65	570.40		
48 L.H., Daily Totals		$910.40		$1516.00	$18.97	$31.58
Crew C-8B	**Hr.**	**Daily**	**Hr.**	**Daily**	**Bare Costs**	**Incl. O&P**
1 Labor Foreman (outside)	$18.90	$151.20	$32.10	$256.80	$18.56	$31.28
3 Laborers	16.90	405.60	28.70	688.80		
1 Equipment Operator	23.20	185.60	38.20	305.60		
1 Vibrating Screed		37.65		41.40		
1 Vibratory Roller		436.80		480.50		
1 Dozer, 200 H.P.		863.40		949.75	33.45	36.79
40 L.H., Daily Totals		$2080.25		$2722.85	$52.01	$68.07
Crew C-8C	**Hr.**	**Daily**	**Hr.**	**Daily**	**Bare Costs**	**Incl. O&P**
1 Labor Foreman (outside)	$18.90	$151.20	$32.10	$256.80	$19.15	$32.01
3 Laborers	16.90	405.60	28.70	688.80		
1 Cement Finisher	22.10	176.80	35.65	285.20		
1 Equipment Operator (med.)	23.20	185.60	38.20	305.60		
1 Shotcrete Rig, 12 CY/hr		227.40		250.15	4.74	5.21
48 L.H., Daily Totals		$1146.60		$1786.55	$23.89	$37.22
Crew C-8D	**Hr.**	**Daily**	**Hr.**	**Daily**	**Bare Costs**	**Incl. O&P**
1 Labor Foreman (outside)	$18.90	$151.20	$32.10	$256.80	$20.03	$33.25
1 Laborer	16.90	135.20	28.70	229.60		
1 Cement Finisher	22.10	176.80	35.65	285.20		
1 Equipment Operator (light)	22.20	177.60	36.55	292.40		
1 Compressor, 250 CFM		127.60		140.35		
2 Hoses, 1", 50'		6.90		7.60	4.20	4.62
32 L.H., Daily Totals		$775.30		$1211.95	$24.23	$37.87

Crew No.	Bare Costs		Incl. Subs O & P		Cost Per Labor-Hour	
Crew C-8E	**Hr.**	**Daily**	**Hr.**	**Daily**	**Bare Costs**	**Incl. O&P**
1 Labor Foreman (outside)	$18.90	$151.20	$32.10	$256.80	$20.03	$33.25
1 Laborer	16.90	135.20	28.70	229.60		
1 Cement Finisher	22.10	176.80	35.65	285.20		
1 Equipment Operator (light)	22.20	177.60	36.55	292.40		
1 Compressor, 250 CFM		127.60		140.35		
2 Hoses, 1", 50'		6.90		7.60		
1 Concrete Pump (small)		719.20		791.10	26.68	29.35
32 L.H., Daily Totals		$1494.50		$2003.05	$46.71	$62.60
Crew C-10	**Hr.**	**Daily**	**Hr.**	**Daily**	**Bare Costs**	**Incl. O&P**
1 Laborer	$16.90	$135.20	$28.70	$229.60	$20.37	$33.33
2 Cement Finishers	22.10	353.60	35.65	570.40		
24 L.H., Daily Totals		$488.80		$800.00	$20.37	$33.33
Crew C-10B	**Hr.**	**Daily**	**Hr.**	**Daily**	**Bare Costs**	**Incl. O&P**
3 Laborers	$16.90	$405.60	$28.70	$688.80	$18.98	$31.48
2 Cement Finishers	22.10	353.60	35.65	570.40		
1 Concrete mixer, 10 CF		105.20		115.70		
2 Concrete finisher, 48" dia		54.40		59.85	3.99	4.39
40 L.H., Daily Totals		$918.80		$1434.75	$22.97	$35.87
Crew C-11	**Hr.**	**Daily**	**Hr.**	**Daily**	**Bare Costs**	**Incl. O&P**
1 Skilled Worker Foreman	$25.05	$200.40	$42.45	$339.60	$23.47	$39.60
5 Skilled Worker	23.05	922.00	39.05	1562.00		
1 Equip. Oper. (crane)	24.00	192.00	39.50	316.00		
1 Truck Crane, 150 Ton		1610.00		1771.00	28.75	31.63
56 L.H., Daily Totals		$2924.40		$3988.60	$52.22	$71.23
Crew C-12	**Hr.**	**Daily**	**Hr.**	**Daily**	**Bare Costs**	**Incl. O&P**
1 Carpenter Foreman (out)	$25.10	$200.80	$42.60	$340.80	$22.55	$38.07
3 Carpenters	23.10	554.40	39.20	940.80		
1 Laborer	16.90	135.20	28.70	229.60		
1 Equip. Oper. (crane)	24.00	192.00	39.50	316.00		
1 Hyd. Crane, 12 Ton		612.00		673.20	12.75	14.03
48 L.H., Daily Totals		$1694.40		$2500.40	$35.30	$52.10
Crew C-13	**Hr.**	**Daily**	**Hr.**	**Daily**	**Bare Costs**	**Incl. O&P**
2 Struc. Steel Worker	$24.90	$398.40	$48.10	$769.60	$24.30	$45.13
1 Carpenter	23.10	184.80	39.20	313.60		
1 Gas Welding Machine		75.20		82.70	3.13	3.45
24 L.H., Daily Totals		$658.40		$1165.90	$27.43	$48.58
Crew C-14	**Hr.**	**Daily**	**Hr.**	**Daily**	**Bare Costs**	**Incl. O&P**
1 Carpenter Foreman (out)	$25.10	$200.80	$42.60	$340.80	$20.44	$34.66
3 Carpenters	23.10	554.40	39.20	940.80		
2 Carpenter Helpers	16.90	270.40	28.60	457.60		
4 Laborers	16.90	540.80	28.70	918.40		
2 Rodmen (reinf.)	24.85	397.60	44.50	712.00		
2 Rodman Helpers	16.90	270.40	28.60	457.60		
2 Cement Finishers	22.10	353.60	35.65	570.40		
1 Equip. Oper. (crane)	24.00	192.00	39.50	316.00		
1 Crane, 80 Ton, & Tools		1020.00		1122.00	7.50	8.25
136 L.H., Daily Totals		$3800.00		$5835.60	$27.94	$42.91

Crews

Crew No.	Bare Costs		Incl. Subs O & P		Cost Per Labor-Hour	
Crew C-14A	**Hr.**	**Daily**	**Hr.**	**Daily**	**Bare Costs**	**Incl. O&P**
1 Carpenter Foreman (out)	$25.10	$200.80	$42.60	$340.80	$22.93	$39.16
16 Carpenters	23.10	2956.80	39.20	5017.60		
4 Rodmen (reinf.)	24.85	795.20	44.50	1424.00		
2 Laborers	16.90	270.40	28.70	459.20		
1 Cement Finisher	22.10	176.80	35.65	285.20		
1 Equip. Oper. (med.)	23.20	185.60	38.20	305.60		
1 Gas Engine Vibrator		26.00		28.60		
1 Concrete Pump (small)		719.20		791.10	3.73	4.10
200 L.H., Daily Totals		$5330.80		$8652.10	$26.66	$43.26
Crew C-14B	**Hr.**	**Daily**	**Hr.**	**Daily**	**Bare Costs**	**Incl. O&P**
1 Carpenter Foreman (out)	$25.10	$200.80	$42.60	$340.80	$22.90	$39.03
16 Carpenters	23.10	2956.80	39.20	5017.60		
4 Rodmen (reinf.)	24.85	795.20	44.50	1424.00		
2 Laborers	16.90	270.40	28.70	459.20		
2 Cement Finishers	22.10	353.60	35.65	570.40		
1 Equip. Oper. (med.)	23.20	185.60	38.20	305.60		
1 Gas Engine Vibrator		26.00		28.60		
1 Concrete Pump (small)		719.20		791.10	3.58	3.94
208 L.H., Daily Totals		$5507.60		$8937.30	$26.48	$42.97
Crew C-14C	**Hr.**	**Daily**	**Hr.**	**Daily**	**Bare Costs**	**Incl. O&P**
1 Carpenter Foreman (out)	$25.10	$200.80	$42.60	$340.80	$21.65	$36.95
6 Carpenters	23.10	1108.80	39.20	1881.60		
2 Rodmen (reinf.)	24.85	397.60	44.50	712.00		
4 Laborers	16.90	540.80	28.70	918.40		
1 Cement Finisher	22.10	176.80	35.65	285.20		
1 Gas Engine Vibrator		26.00		28.60	.23	.26
112 L.H., Daily Totals		$2450.80		$4166.60	$21.88	$37.21
Crew C-14D	**Hr.**	**Daily**	**Hr.**	**Daily**	**Bare Costs**	**Incl. O&P**
1 Carpenter Foreman (out)	$25.10	$200.80	$42.60	$340.80	$22.79	$38.74
18 Carpenters	23.10	3326.40	39.20	5644.80		
2 Rodmen (reinf.)	24.85	397.60	44.50	712.00		
2 Laborers	16.90	270.40	28.70	459.20		
1 Cement Finisher	22.10	176.80	35.65	285.20		
1 Equip. Oper. (med.)	23.20	185.60	38.20	305.60		
1 Gas Engine Vibrator		26.00		28.60		
1 Concrete Pump (small)		719.20		791.10	3.73	4.10
200 L.H., Daily Totals		$5302.80		$8567.30	$26.52	$42.84
Crew C-14E	**Hr.**	**Daily**	**Hr.**	**Daily**	**Bare Costs**	**Incl. O&P**
1 Carpenter Foreman (out)	$25.10	$200.80	$42.60	$340.80	$22.14	$38.25
2 Carpenters	23.10	369.60	39.20	627.20		
4 Rodmen (reinf.)	24.85	795.20	44.50	1424.00		
3 Laborers	16.90	405.60	28.70	688.80		
1 Cement Finisher	22.10	176.80	35.65	285.20		
1 Gas Engine Vibrator		26.00		28.60	.30	.33
88 L.H., Daily Totals		$1974.00		$3394.60	$22.44	$38.58
Crew C-14F	**Hr.**	**Daily**	**Hr.**	**Daily**	**Bare Costs**	**Incl. O&P**
1 Laborer Foreman (out)	$18.90	$151.20	$32.10	$256.80	$20.59	$33.71
2 Laborers	16.90	270.40	28.70	459.20		
6 Cement Finishers	22.10	1060.80	35.65	1711.20		
1 Gas Engine Vibrator		26.00		28.60	.36	.40
72 L.H., Daily Totals		$1508.40		$2455.80	$20.95	$34.11

Crew No.	Bare Costs		Incl. Subs O & P		Cost Per Labor-Hour	
Crew C-14G	**Hr.**	**Daily**	**Hr.**	**Daily**	**Bare Costs**	**Incl. O&P**
1 Laborer Foreman (out)	$18.90	$151.20	$32.10	$256.80	$20.16	$33.16
2 Laborers	16.90	270.40	28.70	459.20		
4 Cement Finishers	22.10	707.20	35.65	1140.80		
1 Gas Engine Vibrator		26.00		28.60	.46	.51
56 L.H., Daily Totals		$1154.80		$1885.40	$20.62	$33.67
Crew C-14H	**Hr.**	**Daily**	**Hr.**	**Daily**	**Bare Costs**	**Incl. O&P**
1 Carpenter Foreman (out)	$25.10	$200.80	$42.60	$340.80	$22.53	$38.31
2 Carpenters	23.10	369.60	39.20	627.20		
1 Rodman (reinf.)	24.85	198.80	44.50	356.00		
1 Laborer	16.90	135.20	28.70	229.60		
1 Cement Finisher	22.10	176.80	35.65	285.20		
1 Gas Engine Vibrator		26.00		28.60	.54	.60
48 L.H., Daily Totals		$1107.20		$1867.40	$23.07	$38.91
Crew C-15	**Hr.**	**Daily**	**Hr.**	**Daily**	**Bare Costs**	**Incl. O&P**
1 Carpenter Foreman (out)	$25.10	$200.80	$42.60	$340.80	$21.23	$35.88
2 Carpenters	23.10	369.60	39.20	627.20		
3 Laborers	16.90	405.60	28.70	688.80		
2 Cement Finishers	22.10	353.60	35.65	570.40		
1 Rodman (reinf.)	24.85	198.80	44.50	356.00		
72 L.H., Daily Totals		$1528.40		$2583.20	$21.23	$35.88
Crew C-16	**Hr.**	**Daily**	**Hr.**	**Daily**	**Bare Costs**	**Incl. O&P**
1 Labor Foreman (outside)	$18.90	$151.20	$32.10	$256.80	$20.74	$35.19
3 Laborers	16.90	405.60	28.70	688.80		
2 Cement Finishers	22.10	353.60	35.65	570.40		
1 Equip. Oper. (med.)	23.20	185.60	38.20	305.60		
2 Rodmen (reinf.)	24.85	397.60	44.50	712.00		
1 Concrete Pump (small)		719.20		791.10	9.99	10.99
72 L.H., Daily Totals		$2212.80		$3324.70	$30.73	$46.18
Crew C-17	**Hr.**	**Daily**	**Hr.**	**Daily**	**Bare Costs**	**Incl. O&P**
2 Skilled Worker Foremen	$25.05	$400.80	$42.45	$679.20	$23.45	$39.73
8 Skilled Workers	23.05	1475.20	39.05	2499.20		
80 L.H., Daily Totals		$1876.00		$3178.40	$23.45	$39.73
Crew C-17A	**Hr.**	**Daily**	**Hr.**	**Daily**	**Bare Costs**	**Incl. O&P**
2 Skilled Worker Foremen	$25.05	$400.80	$42.45	$679.20	$23.46	$39.73
8 Skilled Workers	23.05	1475.20	39.05	2499.20		
.125 Equip. Oper. (crane)	24.00	24.00	39.50	39.50		
.125 Crane, 80 Ton, & Tools		127.50		140.25	1.57	1.73
81 L.H., Daily Totals		$2027.50		$3358.15	$25.03	$41.46
Crew C-17B	**Hr.**	**Daily**	**Hr.**	**Daily**	**Bare Costs**	**Incl. O&P**
2 Skilled Worker Foremen	$25.05	$400.80	$42.45	$679.20	$23.46	$39.72
8 Skilled Workers	23.05	1475.20	39.05	2499.20		
.25 Equip. Oper. (crane)	24.00	48.00	39.50	79.00		
.25 Crane, 80 Ton, & Tools		255.00		280.50		
.25 Walk Behind Power Tools		6.80		7.50	3.19	3.51
82 L.H., Daily Totals		$2185.80		$3545.40	$26.65	$43.23
Crew C-17C	**Hr.**	**Daily**	**Hr.**	**Daily**	**Bare Costs**	**Incl. O&P**
2 Skilled Worker Foremen	$25.05	$400.80	$42.45	$679.20	$23.47	$39.72
8 Skilled Workers	23.05	1475.20	39.05	2499.20		
.375 Equip. Oper. (crane)	24.00	72.00	39.50	118.50		
.375 Crane, 80 Ton & Tools		382.50		420.75	4.61	5.07
83 L.H., Daily Totals		$2330.50		$3717.65	$28.08	$44.79

Crew No.	Bare Costs		Incl. Subs O & P		Cost Per Labor-Hour	
Crew C-17D	**Hr.**	**Daily**	**Hr.**	**Daily**	**Bare Costs**	**Incl. O&P**
2 Skilled Worker Foremen	$25.05	$400.80	$42.45	$679.20	$23.48	$39.72
8 Skilled Workers	23.05	1475.20	39.05	2499.20		
.5 Equip. Oper. (crane)	24.00	96.00	39.50	158.00		
.5 Crane, 80 Ton & Tools		510.00		561.00	6.07	6.68
84 L.H., Daily Totals		$2482.00		$3897.40	$29.55	$46.40

Crew C-17E	Hr.	Daily	Hr.	Daily	Bare Costs	Incl. O&P
2 Skilled Worker Foremen	$25.05	$400.80	$42.45	$679.20	$23.45	$39.73
8 Skilled Workers	23.05	1475.20	39.05	2499.20		
1 Hyd. Jack with Rods		77.55		85.30	.97	1.07
80 L.H., Daily Totals		$1953.55		$3263.70	$24.42	$40.80

Crew C-18	Hr.	Daily	Hr.	Daily	Bare Costs	Incl. O&P
.125 Labor Foreman (out)	$18.90	$18.90	$32.10	$32.10	$17.12	$29.08
1 Laborer	16.90	135.20	28.70	229.60		
1 Concrete Cart, 10 C.F.		48.80		53.70	5.42	5.96
9 L.H., Daily Totals		$202.90		$315.40	$22.54	$35.04

Crew C-19	Hr.	Daily	Hr.	Daily	Bare Costs	Incl. O&P
.125 Labor Foreman (out)	$18.90	$18.90	$32.10	$32.10	$17.12	$29.08
1 Laborer	16.90	135.20	28.70	229.60		
1 Concrete Cart, 18 C.F.		74.80		82.30	8.31	9.14
9 L.H., Daily Totals		$228.90		$344.00	$25.43	$38.22

Crew C-20	Hr.	Daily	Hr.	Daily	Bare Costs	Incl. O&P
1 Labor Foreman (outside)	$18.90	$151.20	$32.10	$256.80	$18.59	$31.18
5 Laborers	16.90	676.00	28.70	1148.00		
1 Cement Finisher	22.10	176.80	35.65	285.20		
1 Equip. Oper. (med.)	23.20	185.60	38.20	305.60		
2 Gas Engine Vibrators		52.00		57.20		
1 Concrete Pump (small)		719.20		791.10	12.05	13.26
64 L.H., Daily Totals		$1960.80		$2843.90	$30.64	$44.44

Crew C-21	Hr.	Daily	Hr.	Daily	Bare Costs	Incl. O&P
1 Labor Foreman (outside)	$18.90	$151.20	$32.10	$256.80	$18.59	$31.18
5 Laborers	16.90	676.00	28.70	1148.00		
1 Cement Finisher	22.10	176.80	35.65	285.20		
1 Equip. Oper. (med.)	23.20	185.60	38.20	305.60		
2 Gas Engine Vibrators		52.00		57.20		
1 Concrete Conveyer		145.20		159.70	3.08	3.39
64 L.H., Daily Totals		$1386.80		$2212.50	$21.67	$34.57

Crew C-22	Hr.	Daily	Hr.	Daily	Bare Costs	Incl. O&P
1 Rodman Foreman	$26.85	$214.80	$48.10	$384.80	$25.10	$44.80
4 Rodmen (reinf.)	24.85	795.20	44.50	1424.00		
.125 Equip. Oper. (crane)	24.00	24.00	39.50	39.50		
.125 Equip. Oper. Oiler	20.15	20.15	33.15	33.15		
.125 Hyd. Crane, 25 Ton		78.03		85.85	1.86	2.04
42 L.H., Daily Totals		$1132.18		$1967.30	$26.96	$46.84

Crew C-23	Hr.	Daily	Hr.	Daily	Bare Costs	Incl. O&P
2 Skilled Worker Foremen	$25.05	$400.80	$42.45	$679.20	$23.26	$39.19
6 Skilled Workers	23.05	1106.40	39.05	1874.40		
1 Equip. Oper. (crane)	24.00	192.00	39.50	316.00		
1 Equip. Oper. Oiler	20.15	161.20	33.15	265.20		
1 Crane, 90 Ton		1416.00		1557.60	17.70	19.47
80 L.H., Daily Totals		$3276.40		$4692.40	$40.96	$58.66

Crew No.	Bare Costs		Incl. Subs O & P		Cost Per Labor-Hour	
Crew C-24	**Hr.**	**Daily**	**Hr.**	**Daily**	**Bare Costs**	**Incl. O&P**
2 Skilled Worker Foremen	$25.05	$400.80	$42.45	$679.20	$23.26	$39.19
6 Skilled Workers	23.05	1106.40	39.05	1874.40		
1 Equip. Oper. (crane)	24.00	192.00	39.50	316.00		
1 Equip. Oper. Oiler	20.15	161.20	33.15	265.20		
1 Truck Crane, 150 Ton		1610.00		1771.00	20.13	22.14
80 L.H., Daily Totals		$3470.40		$4905.80	$43.39	$61.33

Crew C-25	Hr.	Daily	Hr.	Daily	Bare Costs	Incl. O&P
2 Rodmen (reinf.)	$24.85	$397.60	$44.50	$712.00	$19.65	$35.48
2 Rodmen Helpers	14.45	231.20	26.45	423.20		
32 L.H., Daily Totals		$628.80		$1135.20	$19.65	$35.48

Crew C-27	Hr.	Daily	Hr.	Daily	Bare Costs	Incl. O&P
2 Cement Finishers	$22.10	$353.60	$35.65	$570.40	$22.10	$35.65
1 Concrete Saw		92.20		101.40	5.76	6.34
16 L.H., Daily Totals		$445.80		$671.80	$27.86	$41.99

Crew C-28	Hr.	Daily	Hr.	Daily	Bare Costs	Incl. O&P
1 Cement Finisher	$22.10	$176.80	$35.65	$285.20	$22.10	$35.65
1 Portable Air Compressor		18.60		20.45	2.32	2.55
8 L.H., Daily Totals		$195.40		$305.65	$24.42	$38.20

Crew D-1	Hr.	Daily	Hr.	Daily	Bare Costs	Incl. O&P
1 Bricklayer	$24.00	$192.00	$39.90	$319.20	$20.95	$34.83
1 Bricklayer Helper	17.90	143.20	29.75	238.00		
16 L.H., Daily Totals		$335.20		$557.20	$20.95	$34.83

Crew D-2	Hr.	Daily	Hr.	Daily	Bare Costs	Incl. O&P
3 Bricklayers	$24.00	$576.00	$39.90	$957.60	$21.56	$35.84
2 Bricklayer Helpers	17.90	286.40	29.75	476.00		
40 L.H., Daily Totals		$862.40		$1433.60	$21.56	$35.84

Crew D-3	Hr.	Daily	Hr.	Daily	Bare Costs	Incl. O&P
3 Bricklayers	$24.00	$576.00	$39.90	$957.60	$21.63	$36.00
2 Bricklayer Helpers	17.90	286.40	29.75	476.00		
.25 Carpenter	23.10	46.20	39.20	78.40		
42 L.H., Daily Totals		$908.60		$1512.00	$21.63	$36.00

Crew D-4	Hr.	Daily	Hr.	Daily	Bare Costs	Incl. O&P
1 Bricklayer	$24.00	$192.00	$39.90	$319.20	$18.92	$31.57
3 Bricklayer Helpers	17.90	429.60	29.75	714.00		
1 Building Laborer	16.90	135.20	28.70	229.60		
1 Grout Pump, 50 C.F./hr		110.90		122.00		
1 Hoses & Hopper		15.60		17.15		
1 Accessories		11.70		12.85	3.45	3.80
40 L.H., Daily Totals		$895.00		$1414.80	$22.37	$35.37

Crew D-5	Hr.	Daily	Hr.	Daily	Bare Costs	Incl. O&P
1 Block Mason Helper	$17.90	$143.20	$29.75	$238.00	$17.90	$29.75
8 L.H., Daily Totals		$143.20		$238.00	$17.90	$29.75

Crew D-6	Hr.	Daily	Hr.	Daily	Bare Costs	Incl. O&P
3 Bricklayers	$24.00	$576.00	$39.90	$957.60	$20.95	$34.83
3 Bricklayer Helpers	17.90	429.60	29.75	714.00		
48 L.H., Daily Totals		$1005.60		$1671.60	$20.95	$34.83

Crews

Crew No.	Bare Costs		Incl. Subs O & P		Cost Per Labor-Hour	
Crew D-7	**Hr.**	**Daily**	**Hr.**	**Daily**	**Bare Costs**	**Incl. O&P**
1 Tile Layer	$22.05	$176.40	$35.50	$284.00	$19.58	$31.50
1 Tile Layer Helper	17.10	136.80	27.50	220.00		
16 L.H., Daily Totals		$313.20		$504.00	$19.58	$31.50
Crew D-8	**Hr.**	**Daily**	**Hr.**	**Daily**	**Bare Costs**	**Incl. O&P**
3 Bricklayers	$24.00	$576.00	$39.90	$957.60	$21.56	$35.84
2 Bricklayer Helpers	17.90	286.40	29.75	476.00		
40 L.H., Daily Totals		$862.40		$1433.60	$21.56	$35.84
Crew D-9	**Hr.**	**Daily**	**Hr.**	**Daily**	**Bare Costs**	**Incl. O&P**
3 Bricklayers	$24.00	$576.00	$39.90	$957.60	$20.95	$34.83
3 Bricklayer Helpers	17.90	429.60	29.75	714.00		
48 L.H., Daily Totals		$1005.60		$1671.60	$20.95	$34.83
Crew D-10	**Hr.**	**Daily**	**Hr.**	**Daily**	**Bare Costs**	**Incl. O&P**
1 Stone Mason Foreman	$25.00	$200.00	$41.60	$332.80	$21.56	$35.77
1 Stone Mason	23.00	184.00	38.25	306.00		
2 Bricklayer Helpers	17.90	286.40	29.75	476.00		
1 Equip. Oper. (crane)	24.00	192.00	39.50	316.00		
1 Truck Crane, 12.5 Ton		482.20		530.40	12.06	13.26
40 L.H., Daily Totals		$1344.60		$1961.20	$33.62	$49.03
Crew D-11	**Hr.**	**Daily**	**Hr.**	**Daily**	**Bare Costs**	**Incl. O&P**
2 Stone Masons	$23.00	$368.00	$38.25	$612.00	$21.30	$35.42
1 Stone Mason Helper	17.90	143.20	29.75	238.00		
24 L.H., Daily Totals		$511.20		$850.00	$21.30	$35.42
Crew D-12	**Hr.**	**Daily**	**Hr.**	**Daily**	**Bare Costs**	**Incl. O&P**
2 Stone Masons	$23.00	$368.00	$38.25	$612.00	$20.45	$34.00
2 Bricklayer Helpers	17.90	286.40	29.75	476.00		
32 L.H., Daily Totals		$654.40		$1088.00	$20.45	$34.00
Crew D-13	**Hr.**	**Daily**	**Hr.**	**Daily**	**Bare Costs**	**Incl. O&P**
1 Stone Mason Foreman	$25.00	$200.00	$41.60	$332.80	$21.80	$36.18
2 Stone Masons	23.00	368.00	38.25	612.00		
2 Bricklayer Helpers	17.90	286.40	29.75	476.00		
1 Equip. Oper. (crane)	24.00	192.00	39.50	316.00		
1 Truck Crane, 12.5 Ton		482.20		530.40	10.05	11.05
48 L.H., Daily Totals		$1528.60		$2267.20	$31.85	$47.23
Crew E-1	**Hr.**	**Daily**	**Hr.**	**Daily**	**Bare Costs**	**Incl. O&P**
2 Struc. Steel Workers	$24.90	$398.40	$48.10	$769.60	$24.90	$48.10
1 Gas Welding Machine		75.20		82.70	4.70	5.17
16 L.H., Daily Totals		$473.60		$852.30	$29.60	$53.27
Crew E-2	**Hr.**	**Daily**	**Hr.**	**Daily**	**Bare Costs**	**Incl. O&P**
1 Struc. Steel Foreman	$26.90	$215.20	$51.95	$415.60	$25.08	$47.31
4 Struc. Steel Workers	24.90	796.80	48.10	1539.20		
1 Equip. Oper. (crane)	24.00	192.00	39.50	316.00		
1 Crane, 90 Ton		1416.00		1557.60	29.50	32.45
48 L.H., Daily Totals		$2620.00		$3828.40	$54.58	$79.76
Crew E-3	**Hr.**	**Daily**	**Hr.**	**Daily**	**Bare Costs**	**Incl. O&P**
1 Struc. Steel Foreman	$26.90	$215.20	$51.95	$415.60	$25.57	$49.38
2 Struc. Steel Worker	24.90	398.40	48.10	769.60		
1 Gas Welding Machine		75.20		82.70	3.13	3.45
24 L.H., Daily Totals		$688.80		$1267.90	$28.70	$52.83

Crew No.	Bare Costs		Incl. Subs O & P		Cost Per Labor-Hour	
Crew E-4	**Hr.**	**Daily**	**Hr.**	**Daily**	**Bare Costs**	**Incl. O&P**
1 Struc. Steel Foreman	$26.90	$215.20	$51.95	$415.60	$25.40	$49.06
3 Struc. Steel Workers	24.90	597.60	48.10	1154.40		
1 Gas Welding Machine		75.20		82.70	2.35	2.59
32 L.H., Daily Totals		$888.00		$1652.70	$27.75	$51.65
Crew E-5	**Hr.**	**Daily**	**Hr.**	**Daily**	**Bare Costs**	**Incl. O&P**
1 Struc. Steel Foremen	$26.90	$215.20	$51.95	$415.60	$25.02	$47.57
7 Struc. Steel Workers	24.90	1394.40	48.10	2693.60		
1 Equip. Oper. (crane)	24.00	192.00	39.50	316.00		
1 Crane, 90 Ton		1416.00		1557.60		
1 Gas Welding Machine		75.20		82.70	20.71	22.78
72 L.H., Daily Totals		$3292.80		$5065.50	$45.73	$70.35
Crew E-6	**Hr.**	**Daily**	**Hr.**	**Daily**	**Bare Costs**	**Incl. O&P**
1 Struc. Steel Foreman	$26.90	$215.20	$51.95	$415.60	$24.79	$47.01
12 Struc. Steel Workers	24.90	2390.40	48.10	4617.60		
1 Equip. Oper. (crane)	24.00	192.00	39.50	316.00		
1 Equip. Oper. (light)	22.20	177.60	36.55	292.40		
1 Crane, 90 Ton		1416.00		1557.60		
1 Gas Welding Machine		75.20		82.70		
1 Air Compr., 160 C.F.M.		82.00		90.20		
2 Impact Wrenches		28.80		31.70	13.35	14.69
120 L.H., Daily Totals		$4577.20		$7403.80	$38.14	$61.70
Crew E-7	**Hr.**	**Daily**	**Hr.**	**Daily**	**Bare Costs**	**Incl. O&P**
1 Struc. Steel Foreman	$26.90	$215.20	$51.95	$415.60	$25.02	$47.57
7 Struc. Steel Workers	24.90	1394.40	48.10	2693.60		
1 Equip. Oper. (crane)	24.00	192.00	39.50	316.00		
1 Crane, 90 Ton		1416.00		1557.60		
2 Gas Welding Machines		150.40		165.45	21.76	23.93
72 L.H., Daily Totals		$3368.00		$5148.25	$46.78	$71.50
Crew E-8	**Hr.**	**Daily**	**Hr.**	**Daily**	**Bare Costs**	**Incl. O&P**
1 Struc. Steel Foreman	$26.90	$215.20	$51.95	$415.60	$25.00	$47.67
9 Struc. Steel Workers	24.90	1792.80	48.10	3463.20		
1 Equip. Oper. (crane)	24.00	192.00	39.50	316.00		
1 Crane, 90 Ton		1416.00		1557.60		
4 Gas Welding Machines		300.80		330.90	19.51	21.46
88 L.H., Daily Totals		$3916.80		$6083.30	$44.51	$69.13
Crew E-9	**Hr.**	**Daily**	**Hr.**	**Daily**	**Bare Costs**	**Incl. O&P**
2 Struc. Steel Foremen	$26.90	$430.40	$51.95	$831.20	$24.75	$46.63
5 Struc. Steel Workers	24.90	996.00	48.10	1924.00		
1 Welder Foreman	26.90	215.20	51.95	415.60		
5 Welders	24.90	996.00	48.10	1924.00		
1 Equip. Oper. (crane)	24.00	192.00	39.50	316.00		
1 Equip. Oper. Oiler	20.15	161.20	33.15	265.20		
1 Equip. Oper. (light)	22.20	177.60	36.55	292.40		
1 Crane, 90 Ton		1416.00		1557.60		
5 Gas Welding Machines		376.00		413.60	14.00	15.40
128 L.H., Daily Totals		$4960.40		$7939.60	$38.75	$62.03
Crew E-10	**Hr.**	**Daily**	**Hr.**	**Daily**	**Bare Costs**	**Incl. O&P**
1 Struc. Steel Foreman	$26.90	$215.20	$51.95	$415.60	$25.57	$49.38
2 Struc. Steel Workers	24.90	398.40	48.10	769.60		
1 Gas Welding Machines		75.20		82.70		
1 Truck, 3 Ton		160.40		176.45	9.82	10.80
24 L.H., Daily Totals		$849.20		$1444.35	$35.39	$60.18

Crew No.	Bare Costs		Incl. Subs O & P		Cost Per Labor-Hour	
Crew E-11	**Hr.**	**Daily**	**Hr.**	**Daily**	**Bare Costs**	**Incl. O&P**
2 Painters, Struc. Steel	$21.35	$341.60	$43.00	$688.00	$19.87	$38.23
1 Building Laborer	16.90	135.20	28.70	229.60		
1 Air Compressor 250 C.F.M.		127.60		140.35		
1 Sand Blaster		15.60		17.15		
1 Sand Blasting Accessories		11.70		12.85	6.45	7.10
24 L.H., Daily Totals		$631.70		$1087.95	$26.32	$45.33

Crew E-12	Hr.	Daily	Hr.	Daily	Bare Costs	Incl. O&P
1 Welder Foreman	$26.90	$215.20	$51.95	$415.60	$24.55	$44.25
1 Equip. Oper. (light)	22.20	177.60	36.55	292.40		
1 Gas Welding Machine		75.20		82.70	4.70	5.17
16 L.H., Daily Totals		$468.00		$790.70	$29.25	$49.42

Crew E-13	Hr.	Daily	Hr.	Daily	Bare Costs	Incl. O&P
1 Welder Foreman	$26.90	$215.20	$51.95	$415.60	$25.33	$46.82
.5 Equip. Oper. (light)	22.20	88.80	36.55	146.20		
1 Gas Welding Machine		75.20		82.70	6.27	6.89
12 L.H., Daily Totals		$379.20		$644.50	$31.60	$53.71

Crew E-14	Hr.	Daily	Hr.	Daily	Bare Costs	Incl. O&P
1 Struc. Steel Worker	$24.90	$199.20	$48.10	$384.80	$24.90	$48.10
1 Gas Welding Machine		75.20		82.70	9.40	10.34
8 L.H., Daily Totals		$274.40		$467.50	$34.30	$58.44

Crew E-16	Hr.	Daily	Hr.	Daily	Bare Costs	Incl. O&P
1 Welder Foreman	$26.90	$215.20	$51.95	$415.60	$25.90	$50.03
1 Welder	24.90	199.20	48.10	384.80		
1 Gas Welding Machine		75.20		82.70	4.70	5.17
16 L.H., Daily Totals		$489.60		$883.10	$30.60	$55.20

Crew E-17	Hr.	Daily	Hr.	Daily	Bare Costs	Incl. O&P
1 Structural Steel Foreman	$26.90	$215.20	$51.95	$415.60	$25.90	$50.03
1 Structural Steel Worker	24.90	199.20	48.10	384.80		
1 Power Tool		4.40		4.85	.27	.30
16 L.H., Daily Totals		$418.80		$805.25	$26.17	$50.33

Crew E-18	Hr.	Daily	Hr.	Daily	Bare Costs	Incl. O&P
1 Structural Steel Foreman	$26.90	$215.20	$51.95	$415.60	$24.96	$46.89
3 Structural Steel Workers	24.90	597.60	48.10	1154.40		
1 Equipment Operator (med.)	23.20	185.60	38.20	305.60		
1 Crane, 20 Ton		647.45		712.20	16.19	17.80
40 L.H., Daily Totals		$1645.85		$2587.80	$41.15	$64.69

Crew E-19	Hr.	Daily	Hr.	Daily	Bare Costs	Incl. O&P
1 Structural Steel Worker	$24.90	$199.20	$48.10	$384.80	$24.67	$45.53
1 Structural Steel Foreman	26.90	215.20	51.95	415.60		
1 Equip. Oper. (light)	22.20	177.60	36.55	292.40		
1 Power Tool		4.40		4.85		
1 Crane, 20 Ton		647.45		712.20	27.16	29.87
24 L.H., Daily Totals		$1243.85		$1809.85	$51.83	$75.40

Crew E-20	Hr.	Daily	Hr.	Daily	Bare Costs	Incl. O&P
1 Structural Steel Foreman	$26.90	$215.20	$51.95	$415.60	$24.44	$45.64
5 Structural Steel Workers	24.90	996.00	48.10	1924.00		
1 Equip. Oper. (crane)	24.00	192.00	39.50	316.00		
1 Oiler	20.15	161.20	33.15	265.20		
1 Power Tool		4.40		4.85		
1 Crane, 40 Ton		772.50		849.75	12.14	13.35
64 L.H., Daily Totals		$2341.30		$3775.40	$36.58	$58.99

Crew No.	Bare Costs		Incl. Subs O & P		Cost Per Labor-Hour	
Crew E-22	**Hr.**	**Daily**	**Hr.**	**Daily**	**Bare Costs**	**Incl. O&P**
1 Skilled Worker Foreman	$25.05	$200.40	$42.45	$339.60	$23.72	$40.18
2 Skilled Workers	23.05	368.80	39.05	624.80		
24 L.H., Daily Totals		$569.20		$964.40	$23.72	$40.18

Crew E-24	Hr.	Daily	Hr.	Daily	Bare Costs	Incl. O&P
3 Structural Steel Workers	$24.90	$597.60	$48.10	$1154.40	$24.48	$45.63
1 Equipment Operator (medium)	23.20	185.60	38.20	305.60		
1-25 Ton Crane		624.20		686.60	19.51	21.46
32 L.H., Daily Totals		$1407.40		$2146.60	$43.99	$67.09

Crew E-25	Hr.	Daily	Hr.	Daily	Bare Costs	Incl. O&P
1 Welder	$24.90	$199.20	$48.10	$384.80	$24.90	$48.10
1 Cutting Torch		20.00		22.00		
1 Gases		64.80		71.30	10.60	11.66
8 L.H., Daily Totals		$284.00		$478.10	$35.50	$59.76

Crew F-3	Hr.	Daily	Hr.	Daily	Bare Costs	Incl. O&P
2 Carpenters	$23.10	$369.60	$39.20	$627.20	$20.80	$35.02
2 Carpenter Helpers	16.90	270.40	28.60	457.60		
1 Equip. Oper. (crane)	24.00	192.00	39.50	316.00		
1 Hyd. Crane, 12 Ton		612.00		673.20	15.30	16.83
40 L.H., Daily Totals		$1444.00		$2074.00	$36.10	$51.85

Crew F-4	Hr.	Daily	Hr.	Daily	Bare Costs	Incl. O&P
2 Carpenters	$23.10	$369.60	$39.20	$627.20	$20.80	$35.02
2 Carpenter Helpers	16.90	270.40	28.60	457.60		
1 Equip. Oper. (crane)	24.00	192.00	39.50	316.00		
1 Hyd. Crane, 55 Ton		911.00		1002.10	22.78	25.05
40 L.H., Daily Totals		$1743.00		$2402.90	$43.58	$60.07

Crew F-5	Hr.	Daily	Hr.	Daily	Bare Costs	Incl. O&P
2 Carpenters	$23.10	$369.60	$39.20	$627.20	$20.00	$33.90
2 Carpenter Helpers	16.90	270.40	28.60	457.60		
32 L.H., Daily Totals		$640.00		$1084.80	$20.00	$33.90

Crew F-6	Hr.	Daily	Hr.	Daily	Bare Costs	Incl. O&P
2 Carpenters	$23.10	$369.60	$39.20	$627.20	$20.80	$35.06
2 Building Laborers	16.90	270.40	28.70	459.20		
1 Equip. Oper. (crane)	24.00	192.00	39.50	316.00		
1 Hyd. Crane, 12 Ton		612.00		673.20	15.30	16.83
40 L.H., Daily Totals		$1444.00		$2075.60	$36.10	$51.89

Crew F-7	Hr.	Daily	Hr.	Daily	Bare Costs	Incl. O&P
2 Carpenters	$23.10	$369.60	$39.20	$627.20	$20.00	$33.95
2 Building Laborers	16.90	270.40	28.70	459.20		
32 L.H., Daily Totals		$640.00		$1086.40	$20.00	$33.95

Crew G-1	Hr.	Daily	Hr.	Daily	Bare Costs	Incl. O&P
1 Roofer Foreman	$21.65	$173.20	$39.65	$317.20	$18.45	$33.79
4 Roofers, Composition	19.65	628.80	36.00	1152.00		
2 Roofer Helpers	14.45	231.20	26.45	423.20		
1 Application Equipment		138.40		152.25		
1 Tar Kettle/Pot		52.50		57.75		
1 Crew Truck		109.10		120.00	5.36	5.89
56 L.H., Daily Totals		$1333.20		$2222.40	$23.81	$39.68

Crew No.	Bare Costs		Incl. Subs O & P		Cost Per Labor-Hour	
Crew G-2	**Hr.**	**Daily**	**Hr.**	**Daily**	**Bare Costs**	**Incl. O&P**
1 Plasterer	$20.95	$167.60	$34.75	$278.00	$18.63	$31.13
1 Plasterer Helper	18.05	144.40	29.95	239.60		
1 Building Laborer	16.90	135.20	28.70	229.60		
1 Grouting Equipment		110.90		122.00	4.62	5.08
24 L.H., Daily Totals		$558.10		$869.20	$23.25	$36.21
Crew G-3	**Hr.**	**Daily**	**Hr.**	**Daily**	**Bare Costs**	**Incl. O&P**
2 Sheet Metal Workers	$25.60	$409.60	$42.85	$685.60	$21.25	$35.78
2 Building Laborers	16.90	270.40	28.70	459.20		
32 L.H., Daily Totals		$680.00		$1144.80	$21.25	$35.78
Crew G-4	**Hr.**	**Daily**	**Hr.**	**Daily**	**Bare Costs**	**Incl. O&P**
1 Labor Foreman (outside)	$18.90	$151.20	$32.10	$256.80	$17.57	$29.83
2 Building Laborers	16.90	270.40	28.70	459.20		
1 Light Truck, 1.5 Ton		122.00		134.20		
1 Air Compr., 160 C.F.M.		82.00		90.20	8.50	9.35
24 L.H., Daily Totals		$625.60		$940.40	$26.07	$39.18
Crew G-5	**Hr.**	**Daily**	**Hr.**	**Daily**	**Bare Costs**	**Incl. O&P**
1 Roofer Foreman	$21.65	$173.20	$39.65	$317.20	$17.97	$32.91
2 Roofers, Composition	19.65	314.40	36.00	576.00		
2 Roofer Helpers	14.45	231.20	26.45	423.20		
1 Application Equipment		138.40		152.25	3.46	3.81
40 L.H., Daily Totals		$857.20		$1468.65	$21.43	$36.72
Crew G-6A	**Hr.**	**Daily**	**Hr.**	**Daily**	**Bare Costs**	**Incl. O&P**
2 Roofers Composition	$19.65	$314.40	$36.00	$576.00	$19.65	$36.00
1 Small Compressor		13.10		14.40		
2 Pneumatic Nailers		39.00		42.90	3.25	3.58
16 L.H., Daily Totals		$366.50		$633.30	$22.90	$39.58
Crew G-7	**Hr.**	**Daily**	**Hr.**	**Daily**	**Bare Costs**	**Incl. O&P**
1 Carpenter	$23.10	$184.80	$39.20	$313.60	$23.10	$39.20
1 Small Compressor		13.10		14.40		
1 Pneumatic Nailer		19.50		21.45	4.07	4.48
8 L.H., Daily Totals		$217.40		$349.45	$27.17	$43.68
Crew H-1	**Hr.**	**Daily**	**Hr.**	**Daily**	**Bare Costs**	**Incl. O&P**
2 Glaziers	$22.75	$364.00	$37.55	$600.80	$23.83	$42.83
2 Struc. Steel Workers	24.90	398.40	48.10	769.60		
32 L.H., Daily Totals		$762.40		$1370.40	$23.83	$42.83
Crew H-2	**Hr.**	**Daily**	**Hr.**	**Daily**	**Bare Costs**	**Incl. O&P**
2 Glaziers	$22.75	$364.00	$37.55	$600.80	$20.80	$34.60
1 Building Laborer	16.90	135.20	28.70	229.60		
24 L.H., Daily Totals		$499.20		$830.40	$20.80	$34.60
Crew H-3	**Hr.**	**Daily**	**Hr.**	**Daily**	**Bare Costs**	**Incl. O&P**
1 Glazier	$22.75	$182.00	$37.55	$300.40	$19.83	$33.08
1 Helper	16.90	135.20	28.60	228.80		
16 L.H., Daily Totals		$317.20		$529.20	$19.83	$33.08
Crew J-1	**Hr.**	**Daily**	**Hr.**	**Daily**	**Bare Costs**	**Incl. O&P**
3 Plasterers	$20.95	$502.80	$34.75	$834.00	$19.79	$32.83
2 Plasterer Helpers	18.05	288.80	29.95	479.20		
1 Mixing Machine, 6 C.F.		90.00		99.00	2.25	2.48
40 L.H., Daily Totals		$881.60		$1412.20	$22.04	$35.31

Crew No.	Bare Costs		Incl. Subs O & P		Cost Per Labor-Hour	
Crew J-2	**Hr.**	**Daily**	**Hr.**	**Daily**	**Bare Costs**	**Incl. O&P**
3 Plasterers	$20.95	$502.80	$34.75	$834.00	$20.06	$33.13
2 Plasterer Helpers	18.05	288.80	29.95	479.20		
1 Lather	21.40	171.20	34.65	277.20		
1 Mixing Machine, 6 C.F.		90.00		99.00	1.88	2.06
48 L.H., Daily Totals		$1052.80		$1689.40	$21.94	$35.19
Crew J-3	**Hr.**	**Daily**	**Hr.**	**Daily**	**Bare Costs**	**Incl. O&P**
1 Terrazzo Worker	$22.10	$176.80	$35.55	$284.40	$19.90	$32.03
1 Terrazzo Helper	17.70	141.60	28.50	228.00		
1 Terrazzo Grinder, Electric		59.55		65.50		
1 Terrazzo Mixer		121.60		133.75	11.32	12.45
16 L.H., Daily Totals		$499.55		$711.65	$31.22	$44.48
Crew J-4	**Hr.**	**Daily**	**Hr.**	**Daily**	**Bare Costs**	**Incl. O&P**
1 Tile Layer	$22.05	$176.40	$35.50	$284.00	$19.58	$31.50
1 Tile Layer Helper	17.10	136.80	27.50	220.00		
16 L.H., Daily Totals		$313.20		$504.00	$19.58	$31.50
Crew K-1	**Hr.**	**Daily**	**Hr.**	**Daily**	**Bare Costs**	**Incl. O&P**
1 Carpenter	$23.10	$184.80	$39.20	$313.60	$20.68	$34.78
1 Truck Driver (light)	18.25	146.00	30.35	242.80		
1 Truck w/Power Equip.		160.40		176.45	10.03	11.03
16 L.H., Daily Totals		$491.20		$732.85	$30.71	$45.81
Crew K-2	**Hr.**	**Daily**	**Hr.**	**Daily**	**Bare Costs**	**Incl. O&P**
1 Struc. Steel Foreman	$26.90	$215.20	$51.95	$415.60	$23.35	$43.47
1 Struc. Steel Worker	24.90	199.20	48.10	384.80		
1 Truck Driver (light)	18.25	146.00	30.35	242.80		
1 Truck w/Power Equip.		160.40		176.45	6.68	7.35
24 L.H., Daily Totals		$720.80		$1219.65	$30.03	$50.82
Crew L-1	**Hr.**	**Daily**	**Hr.**	**Daily**	**Bare Costs**	**Incl. O&P**
.25 Electrician	$26.40	$52.80	$42.95	$85.90	$26.20	$42.91
1 Plumber	26.15	209.20	42.90	343.20		
10 L.H., Daily Totals		$262.00		$429.10	$26.20	$42.91
Crew L-2	**Hr.**	**Daily**	**Hr.**	**Daily**	**Bare Costs**	**Incl. O&P**
1 Carpenter	$23.10	$184.80	$39.20	$313.60	$20.00	$33.90
1 Carpenter Helper	16.90	135.20	28.60	228.80		
16 L.H., Daily Totals		$320.00		$542.40	$20.00	$33.90
Crew L-3	**Hr.**	**Daily**	**Hr.**	**Daily**	**Bare Costs**	**Incl. O&P**
1 Carpenter	$23.10	$184.80	$39.20	$313.60	$23.76	$39.95
.25 Electrician	26.40	52.80	42.95	85.90		
10 L.H., Daily Totals		$237.60		$399.50	$23.76	$39.95
Crew L-3A	**Hr.**	**Daily**	**Hr.**	**Daily**	**Bare Costs**	**Incl. O&P**
1 Carpenter Foreman (outside)	$25.10	$200.80	$42.60	$340.80	$25.27	$42.68
.5 Sheet Metal Worker	25.60	102.40	42.85	171.40		
12 L.H., Daily Totals		$303.20		$512.20	$25.27	$42.68
Crew L-4	**Hr.**	**Daily**	**Hr.**	**Daily**	**Bare Costs**	**Incl. O&P**
1 Skilled Workers	$23.05	$184.40	$39.05	$312.40	$19.98	$33.83
1 Helper	16.90	135.20	28.60	228.80		
16 L.H., Daily Totals		$319.60		$541.20	$19.98	$33.83

Crew No.	Bare Costs		Incl. Subs O & P		Cost Per Labor-Hour	

Crew L-5	Hr.	Daily	Hr.	Daily	Bare Costs	Incl. O&P
1 Struc. Steel Foreman	$26.90	$215.20	$51.95	$415.60	$25.06	$47.42
5 Struc. Steel Workers	24.90	996.00	48.10	1924.00		
1 Equip. Oper. (crane)	24.00	192.00	39.50	316.00		
1 Hyd. Crane, 25 Ton		624.20		686.60	11.15	12.26
56 L.H., Daily Totals		$2027.40		$3342.20	$36.21	$59.68

Crew L-5A	Hr.	Daily	Hr.	Daily	Bare Costs	Incl. O&P
1 Structural Steel Foreman	$26.90	$215.20	$51.95	$415.60	$25.18	$46.91
2 Structural Steel Workers	24.90	398.40	48.10	769.60		
1 Equip. Oper. (crane)	24.00	192.00	39.50	316.00		
1 Crane,SP, 25 Ton		575.20		632.70	17.98	19.77
32 L.H., Daily Totals		$1380.80		$2133.90	$43.16	$66.68

Crew L-6	Hr.	Daily	Hr.	Daily	Bare Costs	Incl. O&P
1 Plumber	$26.15	$209.20	$42.90	$343.20	$26.23	$42.92
.5 Electrician	26.40	105.60	42.95	171.80		
12 L.H., Daily Totals		$314.80		$515.00	$26.23	$42.92

Crew L-7	Hr.	Daily	Hr.	Daily	Bare Costs	Incl. O&P
1 Carpenters	$23.10	$184.80	$39.20	$313.60	$19.54	$32.97
2 Carpenter Helpers	16.90	270.40	28.60	457.60		
.25 Electrician	26.40	52.80	42.95	85.90		
26 L.H., Daily Totals		$508.00		$857.10	$19.54	$32.97

Crew L-8	Hr.	Daily	Hr.	Daily	Bare Costs	Incl. O&P
1 Carpenters	$23.10	$184.80	$39.20	$313.60	$21.23	$35.70
1 Carpenter Helper	16.90	135.20	28.60	228.80		
.5 Plumber	26.15	104.60	42.90	171.60		
20 L.H., Daily Totals		$424.60		$714.00	$21.23	$35.70

Crew L-9	Hr.	Daily	Hr.	Daily	Bare Costs	Incl. O&P
1 Skilled Worker Foreman	$25.05	$200.40	$42.45	$339.60	$21.13	$35.59
1 Skilled Worker	23.05	184.40	39.05	312.40		
2 Helpers	16.90	270.40	28.60	457.60		
.5 Electrician	26.40	105.60	42.95	171.80		
36 L.H., Daily Totals		$760.80		$1281.40	$21.13	$35.59

Crew L-10	Hr.	Daily	Hr.	Daily	Bare Costs	Incl. O&P
1 Structural Steel Foreman	$26.90	$215.20	$51.95	$415.60	$25.27	$46.52
1 Structural Steel Worker	24.90	199.20	48.10	384.80		
1 Equip. Oper. (crane)	24.00	192.00	39.50	316.00		
1 Hyd. Crane, 12 Ton		612.00		673.20	25.50	28.05
24 L.H., Daily Totals		$1218.40		$1789.60	$50.77	$74.57

Crew M-1	Hr.	Daily	Hr.	Daily	Bare Costs	Incl. O&P
3 Elevator Constructors	$27.85	$668.40	$45.55	$1093.20	$26.46	$43.28
1 Elevator Apprentice	22.30	178.40	36.45	291.60		
5 Hand Tools		70.00		77.00	2.19	2.41
32 L.H., Daily Totals		$916.80		$1461.80	$28.65	$45.69

Crew M-3	Hr.	Daily	Hr.	Daily	Bare Costs	Incl. O&P
1 Electrician Foreman (out)	$28.40	$227.20	$46.20	$369.60	$23.82	$39.16
1 Common Laborer	16.90	135.20	28.70	229.60		
.25 Equipment Operator, Medium	23.20	46.40	38.20	76.40		
1 Elevator Constructor	27.85	222.80	45.55	364.40		
1 Elevator Apprentice	22.30	178.40	36.45	291.60		
.25 Crane, SP, 4 x 4, 20 ton		143.60		157.95	4.22	4.65
34 L.H., Daily Totals		$953.60		$1489.55	$28.04	$43.81

Crew M-4	Hr.	Daily	Hr.	Daily	Bare Costs	Incl. O&P
1 Electrician Foreman (out)	$28.40	$227.20	$46.20	$369.60	$23.66	$38.90
1 Common Laborer	16.90	135.20	28.70	229.60		
.25 Equipment Operator, Crane	24.00	48.00	39.50	79.00		
.25 Equipment Operator, Oiler	20.15	40.30	33.15	66.30		
1 Elevator Constructor	27.85	222.80	45.55	364.40		
1 Elevator Apprentice	22.30	178.40	36.45	291.60		
.25 Crane, Hyd, SP, 4WD, 40 Ton		215.30		236.85	5.98	6.58
36 L.H., Daily Totals		$1067.20		$1637.35	$29.64	$45.48

Crew Q-1	Hr.	Daily	Hr.	Daily	Bare Costs	Incl. O&P
1 Plumber	$26.15	$209.20	$42.90	$343.20	$23.53	$38.60
1 Plumber Apprentice	20.90	167.20	34.30	274.40		
16 L.H., Daily Totals		$376.40		$617.60	$23.53	$38.60

Crew Q-1C	Hr.	Daily	Hr.	Daily	Bare Costs	Incl. O&P
1 Plumber	$26.15	$209.20	$42.90	$343.20	$23.42	$38.47
1 Plumber Apprentice	20.90	167.20	34.30	274.40		
1 Equip. Oper. (medium)	23.20	185.60	38.20	305.60		
1 Trencher, Chain		1310.00		1441.00	54.58	60.04
24 L.H., Daily Totals		$1872.00		$2364.20	$78.00	$98.51

Crew Q-2	Hr.	Daily	Hr.	Daily	Bare Costs	Incl. O&P
1 Plumber	$26.15	$209.20	$42.90	$343.20	$22.65	$37.17
2 Plumber Apprentices	20.90	334.40	34.30	548.80		
24 L.H., Daily Totals		$543.60		$892.00	$22.65	$37.17

Crew Q-3	Hr.	Daily	Hr.	Daily	Bare Costs	Incl. O&P
2 Plumbers	$26.15	$418.40	$42.90	$686.40	$23.53	$38.60
2 Plumber Apprentices	20.90	334.40	34.30	548.80		
32 L.H., Daily Totals		$752.80		$1235.20	$23.53	$38.60

Crew Q-4	Hr.	Daily	Hr.	Daily	Bare Costs	Incl. O&P
2 Plumbers	$26.15	$418.40	$42.90	$686.40	$24.84	$40.75
1 Welder (plumber)	26.15	209.20	42.90	343.20		
1 Plumber Apprentice	20.90	167.20	34.30	274.40		
1 Electric Welding Mach.		77.75		85.55	2.43	2.67
32 L.H., Daily Totals		$872.55		$1389.55	$27.27	$43.42

Crew Q-5	Hr.	Daily	Hr.	Daily	Bare Costs	Incl. O&P
1 Steamfitter	$26.25	$210.00	$43.10	$344.80	$23.63	$38.78
1 Steamfitter Apprentice	21.00	168.00	34.45	275.60		
16 L.H., Daily Totals		$378.00		$620.40	$23.63	$38.78

Crew Q-6	Hr.	Daily	Hr.	Daily	Bare Costs	Incl. O&P
1 Steamfitters	$26.25	$210.00	$43.10	$344.80	$22.75	$37.33
2 Steamfitter Apprentices	21.00	336.00	34.45	551.20		
24 L.H., Daily Totals		$546.00		$896.00	$22.75	$37.33

Crew Q-7	Hr.	Daily	Hr.	Daily	Bare Costs	Incl. O&P
2 Steamfitters	$26.25	$420.00	$43.10	$689.60	$23.63	$38.78
2 Steamfitter Apprentices	21.00	336.00	34.45	551.20		
32 L.H., Daily Totals		$756.00		$1240.80	$23.63	$38.78

Crew Q-8	Hr.	Daily	Hr.	Daily	Bare Costs	Incl. O&P
2 Steamfitters	$26.25	$420.00	$43.10	$689.60	$24.94	$40.94
1 Welder (steamfitter)	26.25	210.00	43.10	344.80		
1 Steamfitter Apprentice	21.00	168.00	34.45	275.60		
1 Electric Welding Mach.		77.75		85.55	2.43	2.67
32 L.H., Daily Totals		$875.75		$1395.55	$27.37	$43.61

Crew No.	Bare Costs		Incl. Subs O & P		Cost Per Labor-Hour	
Crew Q-9	**Hr.**	**Daily**	**Hr.**	**Daily**	**Bare Costs**	**Incl. O&P**
1 Sheet Metal Worker	$25.60	$204.80	$42.85	$342.80	$23.05	$38.58
1 Sheet Metal Apprentice	20.50	164.00	34.30	274.40		
16 L.H., Daily Totals		$368.80		$617.20	$23.05	$38.58
Crew Q-10	**Hr.**	**Daily**	**Hr.**	**Daily**	**Bare Costs**	**Incl. O&P**
2 Sheet Metal Workers	$25.60	$409.60	$42.85	$685.60	$23.90	$40.00
1 Sheet Metal Apprentice	20.50	164.00	34.30	274.40		
24 L.H., Daily Totals		$573.60		$960.00	$23.90	$40.00
Crew Q-11	**Hr.**	**Daily**	**Hr.**	**Daily**	**Bare Costs**	**Incl. O&P**
2 Sheet Metal Workers	$25.60	$409.60	$42.85	$685.60	$23.05	$38.58
2 Sheet Metal Apprentices	20.50	328.00	34.30	548.80		
32 L.H., Daily Totals		$737.60		$1234.40	$23.05	$38.58
Crew Q-12	**Hr.**	**Daily**	**Hr.**	**Daily**	**Bare Costs**	**Incl. O&P**
1 Sprinkler Installer	$25.90	$207.20	$42.80	$342.40	$23.30	$38.50
1 Sprinkler Apprentice	20.70	165.60	34.20	273.60		
16 L.H., Daily Totals		$372.80		$616.00	$23.30	$38.50
Crew Q-13	**Hr.**	**Daily**	**Hr.**	**Daily**	**Bare Costs**	**Incl. O&P**
2 Sprinkler Installers	$25.90	$414.40	$42.80	$684.80	$23.30	$38.50
2 Sprinkler Apprentices	20.70	331.20	34.20	547.20		
32 L.H., Daily Totals		$745.60		$1232.00	$23.30	$38.50
Crew Q-14	**Hr.**	**Daily**	**Hr.**	**Daily**	**Bare Costs**	**Incl. O&P**
1 Asbestos Worker	$23.75	$190.00	$40.75	$326.00	$21.38	$36.68
1 Asbestos Apprentice	19.00	152.00	32.60	260.80		
16 L.H., Daily Totals		$342.00		$586.80	$21.38	$36.68
Crew Q-15	**Hr.**	**Daily**	**Hr.**	**Daily**	**Bare Costs**	**Incl. O&P**
1 Plumber	$26.15	$209.20	$42.90	$343.20	$23.53	$38.60
1 Plumber Apprentice	20.90	167.20	34.30	274.40		
1 Electric Welding Mach.		77.75		85.55	4.86	5.35
16 L.H., Daily Totals		$454.15		$703.15	$28.39	$43.95
Crew Q-16	**Hr.**	**Daily**	**Hr.**	**Daily**	**Bare Costs**	**Incl. O&P**
2 Plumbers	$26.15	$418.40	$42.90	$686.40	$24.40	$40.03
1 Plumber Apprentice	20.90	167.20	34.30	274.40		
1 Electric Welding Mach.		77.75		85.55	3.24	3.56
24 L.H., Daily Totals		$663.35		$1046.35	$27.64	$43.59
Crew Q-17	**Hr.**	**Daily**	**Hr.**	**Daily**	**Bare Costs**	**Incl. O&P**
1 Steamfitter	$26.25	$210.00	$43.10	$344.80	$23.63	$38.78
1 Steamfitter Apprentice	21.00	168.00	34.45	275.60		
1 Electric Welding Mach.		77.75		85.55	4.86	5.35
16 L.H., Daily Totals		$455.75		$705.95	$28.49	$44.13
Crew Q-17A	**Hr.**	**Daily**	**Hr.**	**Daily**	**Bare Costs**	**Incl. O&P**
1 Steamfitter	$26.25	$210.00	$43.10	$344.80	$23.75	$39.02
1 Steamfitter Apprentice	21.00	168.00	34.45	275.60		
1 Equip. Oper. (crane)	24.00	192.00	39.50	316.00		
1 Truck Crane, 12 Ton		612.00		673.20		
1 Electric Welding Mach.		77.75		85.55	28.74	31.61
24 L.H., Daily Totals		$1259.75		$1695.15	$52.49	$70.63

Crew No.	Bare Costs		Incl. Subs O & P		Cost Per Labor-Hour	
Crew Q-18	**Hr.**	**Daily**	**Hr.**	**Daily**	**Bare Costs**	**Incl. O&P**
2 Steamfitters	$26.25	$420.00	$43.10	$689.60	$24.50	$40.22
1 Steamfitter Apprentice	21.00	168.00	34.45	275.60		
1 Electric Welding Mach.		77.75		85.55	3.24	3.56
24 L.H., Daily Totals		$665.75		$1050.75	$27.74	$43.78
Crew Q-19	**Hr.**	**Daily**	**Hr.**	**Daily**	**Bare Costs**	**Incl. O&P**
1 Steamfitter	$26.25	$210.00	$43.10	$344.80	$24.55	$40.17
1 Steamfitter Apprentice	21.00	168.00	34.45	275.60		
1 Electrician	26.40	211.20	42.95	343.60		
24 L.H., Daily Totals		$589.20		$964.00	$24.55	$40.17
Crew Q-20	**Hr.**	**Daily**	**Hr.**	**Daily**	**Bare Costs**	**Incl. O&P**
1 Sheet Metal Worker	$25.60	$204.80	$42.85	$342.80	$23.72	$39.45
1 Sheet Metal Apprentice	20.50	164.00	34.30	274.40		
.5 Electrician	26.40	105.60	42.95	171.80		
20 L.H., Daily Totals		$474.40		$789.00	$23.72	$39.45
Crew Q-21	**Hr.**	**Daily**	**Hr.**	**Daily**	**Bare Costs**	**Incl. O&P**
2 Steamfitters	$26.25	$420.00	$43.10	$689.60	$24.98	$40.90
1 Steamfitter Apprentice	21.00	168.00	34.45	275.60		
1 Electrician	26.40	211.20	42.95	343.60		
32 L.H., Daily Totals		$799.20		$1308.80	$24.98	$40.90
Crew Q-22	**Hr.**	**Daily**	**Hr.**	**Daily**	**Bare Costs**	**Incl. O&P**
1 Plumber	$26.15	$209.20	$42.90	$343.20	$23.53	$38.60
1 Plumber Apprentice	20.90	167.20	34.30	274.40		
1 Truck Crane, 12 Ton		612.00		673.20	38.25	42.08
16 L.H., Daily Totals		$988.40		$1290.80	$61.78	$80.68
Crew Q-22A	**Hr.**	**Daily**	**Hr.**	**Daily**	**Bare Costs**	**Incl. O&P**
1 Plumber	$26.15	$209.20	$42.90	$343.20	$21.99	$36.35
1 Plumber Apprentice	20.90	167.20	34.30	274.40		
1 Laborer	16.90	135.20	28.70	229.60		
1 Equip. Oper. (crane)	24.00	192.00	39.50	316.00		
1 Truck Crane, 12 Ton		612.00		673.20	19.13	21.04
32 L.H., Daily Totals		$1315.60		$1836.40	$41.12	$57.39
Crew Q-23	**Hr.**	**Daily**	**Hr.**	**Daily**	**Bare Costs**	**Incl. O&P**
1 Plumber Foreman	$28.15	$225.20	$46.20	$369.60	$25.83	$42.43
1 Plumber	26.15	209.20	42.90	343.20		
1 Equip. Oper. (medium)	23.20	185.60	38.20	305.60		
1 Power Tool		4.40		4.85		
1 Crane, 20 Ton		647.45		712.20	27.16	29.87
24 L.H., Daily Totals		$1271.85		$1735.45	$52.99	$72.30
Crew R-1	**Hr.**	**Daily**	**Hr.**	**Daily**	**Bare Costs**	**Incl. O&P**
1 Electrician Foreman	$26.90	$215.20	$43.75	$350.00	$23.32	$38.30
3 Electricians	26.40	633.60	42.95	1030.80		
2 Helpers	16.90	270.40	28.60	457.60		
48 L.H., Daily Totals		$1119.20		$1838.40	$23.32	$38.30
Crew R-1A	**Hr.**	**Daily**	**Hr.**	**Daily**	**Bare Costs**	**Incl. O&P**
1 Electrician	$26.40	$211.20	$42.95	$343.60	$21.65	$35.78
1 Helper	16.90	135.20	28.60	228.80		
16 L.H., Daily Totals		$346.40		$572.40	$21.65	$35.78

Crew No.	Bare Costs		Incl. Subs O & P		Cost Per Labor-Hour	
Crew R-2	**Hr.**	**Daily**	**Hr.**	**Daily**	**Bare Costs**	**Incl. O&P**
1 Electrician Foreman	$26.90	$215.20	$43.75	$350.00	$23.41	$38.47
3 Electricians	26.40	633.60	42.95	1030.80		
2 Helpers	16.90	270.40	28.60	457.60		
1 Equip. Oper. (crane)	24.00	192.00	39.50	316.00		
1 S.P. Crane, 5 Ton		317.80		349.60	5.68	6.24
56 L.H., Daily Totals		$1629.00		$2504.00	$29.09	$44.71
Crew R-3	**Hr.**	**Daily**	**Hr.**	**Daily**	**Bare Costs**	**Incl. O&P**
1 Electrician Foreman	$26.90	$215.20	$43.75	$350.00	$26.12	$42.58
1 Electrician	26.40	211.20	42.95	343.60		
.5 Equip. Oper. (crane)	24.00	96.00	39.50	158.00		
.5 S.P. Crane, 5 Ton		158.90		174.80	7.95	8.74
20 L.H., Daily Totals		$681.30		$1026.40	$34.07	$51.32
Crew R-4	**Hr.**	**Daily**	**Hr.**	**Daily**	**Bare Costs**	**Incl. O&P**
1 Struc. Steel Foreman	$26.90	$215.20	$51.95	$415.60	$25.60	$47.84
3 Struc. Steel Workers	24.90	597.60	48.10	1154.40		
1 Electrician	26.40	211.20	42.95	343.60		
1 Gas Welding Machine		75.20		82.70	1.88	2.07
40 L.H., Daily Totals		$1099.20		$1996.30	$27.48	$49.91
Crew R-5	**Hr.**	**Daily**	**Hr.**	**Daily**	**Bare Costs**	**Incl. O&P**
1 Electrician Foreman	$26.90	$215.20	$43.75	$350.00	$22.99	$37.80
4 Electrician Linemen	26.40	844.80	42.95	1374.40		
2 Electrician Operators	26.40	422.40	42.95	687.20		
4 Electrician Groundmen	16.90	540.80	28.60	915.20		
1 Crew Truck		109.10		120.00		
1 Tool Van		129.35		142.30		
1 Pickup Truck, 3/4 Ton		75.80		83.40		
.2 Crane, 55 Ton		182.20		200.40		
.2 Crane, 12 Ton		122.40		134.65		
.2 Auger, Truck Mtd.		575.40		632.95		
1 Tractor w/Winch		254.40		279.85	16.46	18.11
88 L.H., Daily Totals		$3471.85		$4920.35	$39.45	$55.91
Crew R-6	**Hr.**	**Daily**	**Hr.**	**Daily**	**Bare Costs**	**Incl. O&P**
1 Electrician Foreman	$26.90	$215.20	$43.75	$350.00	$22.99	$37.80
4 Electrician Linemen	26.40	844.80	42.95	1374.40		
2 Electrician Operators	26.40	422.40	42.95	687.20		
4 Electrician Groundmen	16.90	540.80	28.60	915.20		
1 Crew Truck		109.10		120.00		
1 Tool Van		129.35		142.30		
1 Pickup Truck, 3/4 Ton		75.80		83.40		
.2 Crane, 55 Ton		182.20		200.40		
.2 Crane, 12 Ton		122.40		134.65		
.2 Auger, Truck Mtd.		575.40		632.95		
1 Tractor w/Winch		254.40		279.85		
3 Cable Trailers		480.75		528.85		
.5 Tensioning Rig		157.45		173.20		
.5 Cable Pulling Rig		916.00		1007.60	34.12	37.54
88 L.H., Daily Totals		$5026.05		$6630.00	$57.11	$75.34
Crew R-7	**Hr.**	**Daily**	**Hr.**	**Daily**	**Bare Costs**	**Incl. O&P**
1 Electrician Foreman	$26.90	$215.20	$43.75	$350.00	$18.57	$31.13
5 Electrician Groundmen	16.90	676.00	28.60	1144.00		
1 Crew Truck		109.10		120.00	2.27	2.50
48 L.H., Daily Totals		$1000.30		$1614.00	$20.84	$33.63

Crew No.	Bare Costs		Incl. Subs O & P		Cost Per Labor-Hour	
Crew R-8	**Hr.**	**Daily**	**Hr.**	**Daily**	**Bare Costs**	**Incl. O&P**
1 Electrician Foreman	$26.90	$215.20	$43.75	$350.00	$23.32	$38.30
3 Electrician Linemen	26.40	633.60	42.95	1030.80		
2 Electrician Groundmen	16.90	270.40	28.60	457.60		
1 Pickup Truck, 3/4 Ton		75.80		83.40		
1 Crew Truck		109.10		120.00	3.85	4.24
48 L.H., Daily Totals		$1304.10		$2041.80	$27.17	$42.54
Crew R-9	**Hr.**	**Daily**	**Hr.**	**Daily**	**Bare Costs**	**Incl. O&P**
1 Electrician Foreman	$26.90	$215.20	$43.75	$350.00	$21.71	$35.88
1 Electrician Lineman	26.40	211.20	42.95	343.60		
2 Electrician Operators	26.40	422.40	42.95	687.20		
4 Electrician Groundmen	16.90	540.80	28.60	915.20		
1 Pickup Truck, 3/4 Ton		75.80		83.40		
1 Crew Truck		109.10		120.00	2.89	3.18
64 L.H., Daily Totals		$1574.50		$2499.40	$24.60	$39.06
Crew R-10	**Hr.**	**Daily**	**Hr.**	**Daily**	**Bare Costs**	**Incl. O&P**
1 Electrician Foreman	$26.90	$215.20	$43.75	$350.00	$24.90	$40.69
4 Electrician Linemen	26.40	844.80	42.95	1374.40		
1 Electrician Groundman	16.90	135.20	28.60	228.80		
1 Crew Truck		109.10		120.00		
3 Tram Cars		340.35		374.40	9.36	10.30
48 L.H., Daily Totals		$1644.65		$2447.60	$34.26	$50.99
Crew R-11	**Hr.**	**Daily**	**Hr.**	**Daily**	**Bare Costs**	**Incl. O&P**
1 Electrician Foreman	$26.90	$215.20	$43.75	$350.00	$24.77	$40.54
4 Electricians	26.40	844.80	42.95	1374.40		
1 Equip. Oper. (crane)	24.00	192.00	39.50	316.00		
1 Helper	16.90	135.20	28.60	228.80		
1 Common Laborer	16.90	135.20	28.70	229.60		
1 Crew Truck		109.10		120.00		
1 Crane, 12 Ton		612.00		673.20	12.88	14.16
56 L.H., Daily Totals		$2243.50		$3292.00	$37.65	$54.70
Crew R-12	**Hr.**	**Daily**	**Hr.**	**Daily**	**Bare Costs**	**Incl. O&P**
1 Carpenter Foreman	$23.60	$188.80	$40.05	$320.40	$21.06	$36.18
4 Carpenters	23.10	739.20	39.20	1254.40		
4 Common Laborers	16.90	540.80	28.70	918.40		
1 Equip. Oper. (med.)	23.20	185.60	38.20	305.60		
1 Steel Worker	24.90	199.20	48.10	384.80		
1 Dozer, 200 H.P.		863.40		949.75		
1 Pickup Truck, 3/4 Ton		75.80		83.40	10.67	11.74
88 L.H., Daily Totals		$2792.80		$4216.75	$31.73	$47.92
Crew R-15	**Hr.**	**Daily**	**Hr.**	**Daily**	**Bare Costs**	**Incl. O&P**
1 Electrician Foreman	$26.90	$215.20	$43.75	$350.00	$25.78	$42.02
4 Electricians	26.40	844.80	42.95	1374.40		
1 Equipment Operator	22.20	177.60	36.55	292.40		
1 Aerial Lift Truck		250.20		275.20	5.21	5.73
48 L.H., Daily Totals		$1487.80		$2292.00	$30.99	$47.75
Crew R-18	**Hr.**	**Daily**	**Hr.**	**Daily**	**Bare Costs**	**Incl. O&P**
.25 Electrician Foreman	$26.90	$53.80	$43.75	$87.50	$20.59	$34.18
1 Electrician	26.40	211.20	42.95	343.60		
2 Helpers	16.90	270.40	28.60	457.60		
26 L.H., Daily Totals		$535.40		$888.70	$20.59	$34.18
Crew R-19	**Hr.**	**Daily**	**Hr.**	**Daily**	**Bare Costs**	**Incl. O&P**
.5 Electrician Foreman	$26.90	$107.60	$43.75	$175.00	$26.50	$43.11
2 Electricians	26.40	422.40	42.95	687.20		
20 L.H., Daily Totals		$530.00		$862.20	$26.50	$43.11

Crew No.	Bare Costs		Incl. Sub O & P		Cost Per Labor-Hour	
Crew R-21	**Hr.**	**Daily**	**Hr.**	**Daily**	**Bare Costs**	**Incl. O&P**
1 Electrician Foreman	$26.90	$215.20	$43.75	$350.00	$26.44	$43.03
3 Electricians	26.40	633.60	42.95	1030.80		
.1 Equip. Oper. (med.)	23.20	18.56	38.20	30.56		
.1 Hyd. Crane 25 Ton		57.52		63.25	1.75	1.93
32. L.H., Daily Totals		$924.88		$1474.61	$28.19	$44.96
Crew R-22	**Hr.**	**Daily**	**Hr.**	**Daily**	**Bare Costs**	**Incl. O&P**
.66 Electrician Foreman	$26.90	$142.03	$43.75	$231.00	$22.39	$36.90
2 Helpers	16.90	270.40	28.60	457.60		
2 Electricians	26.40	422.40	42.95	687.20		
37.28 L.H., Daily Totals		$834.83		$1375.80	$22.39	$36.90
Crew R-30	**Hr.**	**Daily**	**Hr.**	**Daily**	**Bare Costs**	**Incl. O&P**
.25 Electrician	$28.40	$56.80	$46.20	$92.40	$20.71	$34.43
1 Electrician	26.40	211.20	42.95	343.60		
2 Laborers, (Semi-Skilled)	16.90	270.40	28.70	459.20		
26 L.H., Daily Totals		$538.40		$895.20	$20.71	$34.43

Location Factors

Costs shown in *Means cost data publications* are based on National Averages for materials and installation. To adjust these costs to a specific location, simply multiply the base cost by the factor for that city. The data is arranged alphabetically by state and postal zip code numbers. For a city not listed, use the factor for a nearby city with similar economic characteristics.

STATE	CITY	Residential
ALABAMA		
350-352	Birmingham	.86
354	Tuscaloosa	.73
355	Jasper	.70
356	Decatur	.76
357-358	Huntsville	.84
359	Gadsden	.73
360-361	Montgomery	.75
362	Anniston	.68
363	Dothan	.74
364	Evergreen	.71
365-366	Mobile	.79
367	Selma	.72
368	Phenix City	.73
369	Butler	.71
ALASKA		
995-996	Anchorage	1.28
997	Fairbanks	1.30
998	Juneau	1.28
999	Ketchikan	1.30
ARIZONA		
850,853	Phoenix	.87
852	Mesa/Tempe	.85
855	Globe	.81
856-857	Tucson	.85
859	Show Low	.84
860	Flagstaff	.85
863	Prescott	.82
864	Kingman	.82
865	Chambers	.82
ARKANSAS		
716	Pine Bluff	.77
717	Camden	.66
718	Texarkana	.71
719	Hot Springs	.64
720-722	Little Rock	.82
723	West Memphis	.75
724	Jonesboro	.73
725	Batesville	.72
726	Harrison	.73
727	Fayetteville	.62
728	Russellville	.71
729	Fort Smith	.77
CALIFORNIA		
900-902	Los Angeles	1.07
903-905	Inglewood	1.04
906-908	Long Beach	1.03
910-912	Pasadena	1.04
913-916	Van Nuys	1.07
917-918	Alhambra	1.08
919-921	San Diego	1.03
922	Palm Springs	1.00
923-924	San Bernardino	1.01
925	Riverside	1.07
926-927	Santa Ana	1.05
928	Anaheim	1.08
930	Oxnard	1.08
931	Santa Barbara	1.07
932-933	Bakersfield	1.05
934	San Luis Obispo	1.07
935	Mojave	1.04
936-938	Fresno	1.11
939	Salinas	1.12
940-941	San Francisco	1.22
942,956-958	Sacramento	1.12
943	Palo Alto	1.17
944	San Mateo	1.21
945	Vallejo	1.14
946	Oakland	1.19
947	Berkeley	1.22
948	Richmond	1.23
949	San Rafael	1.21
950	Santa Cruz	1.15
951	San Jose	1.20
952	Stockton	1.10

STATE	CITY	Residential
CALIFORNIA (CONT'D)		
953	Modesto	1.09
954	Santa Rosa	1.15
955	Eureka	1.11
959	Marysville	1.11
960	Redding	1.12
961	Susanville	1.11
COLORADO		
800-802	Denver	.95
803	Boulder	.94
804	Golden	.91
805	Fort Collins	.90
806	Greeley	.79
807	Fort Morgan	.94
808-809	Colorado Springs	.92
810	Pueblo	.93
811	Alamosa	.90
812	Salida	.92
813	Durango	.94
814	Montrose	.89
815	Grand Junction	.93
816	Glenwood Springs	.92
CONNECTICUT		
060	New Britain	1.08
061	Hartford	1.07
062	Willimantic	1.07
063	New London	1.07
064	Meriden	1.07
065	New Haven	1.08
066	Bridgeport	1.08
067	Waterbury	1.08
068	Norwalk	1.08
069	Stamford	1.09
D.C.		
200-205	Washington	.92
DELAWARE		
197	Newark	1.01
198	Wilmington	1.01
199	Dover	1.01
FLORIDA		
320,322	Jacksonville	.78
321	Daytona Beach	.85
323	Tallahassee	.72
324	Panama City	.66
325	Pensacola	.76
326,344	Gainesville	.76
327-328,347	Orlando	.79
329	Melbourne	.86
330-332,340	Miami	.83
333	Fort Lauderdale	.84
334,349	West Palm Beach	.83
335-336,346	Tampa	.87
337	St. Petersburg	.76
338	Lakeland	.84
339,341	Fort Myers	.81
342	Sarasota	.85
GEORGIA		
300-303,399	Atlanta	.89
304	Statesboro	.67
305	Gainesville	.73
306	Athens	.74
307	Dalton	.70
308-309	Augusta	.76
310-312	Macon	.77
313-314	Savannah	.79
315	Waycross	.71
316	Valdosta	.71
317	Albany	.73
318-319	Columbus	.76
HAWAII		
967	Hilo	1.23
968	Honolulu	1.24

Location Factors

STATE	CITY	Residential
STATES & POSS.		
969	Guam	1.60
IDAHO		
832	Pocatello	.89
833	Twin Falls	.73
834	Idaho Falls	.72
835	Lewiston	.99
836-837	Boise	.90
838	Coeur d'Alene	.85
ILLINOIS		
600-603	North Suburban	1.11
604	Joliet	1.13
605	South Suburban	1.10
606	Chicago	1.15
609	Kankakee	1.03
610-611	Rockford	1.04
612	Rock Island	.97
613	La Salle	1.02
614	Galesburg	.99
615-616	Peoria	1.02
617	Bloomington	.98
618-619	Champaign	.99
620-622	East St. Louis	.99
623	Quincy	.98
624	Effingham	.99
625	Decatur	.99
626-627	Springfield	.99
628	Centralia	.97
629	Carbondale	.96
INDIANA		
460	Anderson	.92
461-462	Indianapolis	.96
463-464	Gary	1.02
465-466	South Bend	.92
467-468	Fort Wayne	.92
469	Kokomo	.94
470	Lawrenceburg	.88
471	New Albany	.86
472	Columbus	.93
473	Muncie	.92
474	Bloomington	.96
475	Washington	.91
476-477	Evansville	.91
478	Terre Haute	.91
479	Lafayette	.92
IOWA		
500-503,509	Des Moines	.93
504	Mason City	.78
505	Fort Dodge	.77
506-507	Waterloo	.81
508	Creston	.83
510-511	Sioux City	.87
512	Sibley	.74
513	Spencer	.76
514	Carroll	.76
515	Council Bluffs	.82
516	Shenandoah	.74
520	Dubuque	.86
521	Decorah	.77
522-524	Cedar Rapids	.94
525	Ottumwa	.84
526	Burlington	.88
527-528	Davenport	.98
KANSAS		
660-662	Kansas City	.95
664-666	Topeka	.78
667	Fort Scott	.85
668	Emporia	.73
669	Belleville	.74
670-672	Wichita	.81
673	Independence	.74
674	Salina	.73
675	Hutchinson	.68
676	Hays	.74
677	Colby	.76
678	Dodge City	.74
679	Liberal	.68
KENTUCKY		
400-402	Louisville	.92
403-405	Lexington	.84

STATE	CITY	Residential
KENTUCKY (CONT'D)		
406	Frankfort	.82
407-409	Corbin	.67
410	Covington	.95
411-412	Ashland	.93
413-414	Campton	.68
415-416	Pikeville	.78
417-418	Hazard	.67
420	Paducah	.89
421-422	Bowling Green	.89
423	Owensboro	.83
424	Henderson	.92
425-426	Somerset	.67
427	Elizabethtown	.88
LOUISIANA		
700-701	New Orleans	.86
703	Thibodaux	.82
704	Hammond	.80
705	Lafayette	.78
706	Lake Charles	.80
707-708	Baton Rouge	.80
710-711	Shreveport	.79
712	Monroe	.75
713-714	Alexandria	.74
MAINE		
039	Kittery	.81
040-041	Portland	.89
042	Lewiston	.89
043	Augusta	.84
044	Bangor	.87
045	Bath	.82
046	Machias	.82
047	Houlton	.87
048	Rockland	.80
049	Waterville	.76
MARYLAND		
206	Waldorf	.84
207-208	College Park	.85
209	Silver Spring	.85
210-212	Baltimore	.90
214	Annapolis	.86
215	Cumberland	.85
216	Easton	.69
217	Hagerstown	.86
218	Salisbury	.75
219	Elkton	.83
MASSACHUSETTS		
010-011	Springfield	1.05
012	Pittsfield	1.00
013	Greenfield	1.00
014	Fitchburg	1.09
015-016	Worcester	1.10
017	Framingham	1.11
018	Lowell	1.12
019	Lawrence	1.12
020-022, 024	Boston	1.17
023	Brockton	1.11
025	Buzzards Bay	1.09
026	Hyannis	1.08
027	New Bedford	1.11
MICHIGAN		
480,483	Royal Oak	1.05
481	Ann Arbor	1.05
482	Detroit	1.10
484-485	Flint	.99
486	Saginaw	.96
487	Bay City	.96
488-489	Lansing	.98
490	Battle Creek	.94
491	Kalamazoo	.93
492	Jackson	.94
493,495	Grand Rapids	.84
494	Muskegon	.90
496	Traverse City	.83
497	Gaylord	.85
498-499	Iron Mountain	.92
MINNESOTA		
550-551	Saint Paul	1.14
553-555	Minneapolis	1.18

Location Factors

STATE	CITY	Residential
MINNESOTA (CONT'D)		
556-558	Duluth	1.10
559	Rochester	1.05
560	Mankato	1.02
561	Windom	.85
562	Willmar	.86
563	St. Cloud	1.10
564	Brainerd	.98
565	Detroit Lakes	.99
566	Bemidji	.96
567	Thief River Falls	.94
MISSISSIPPI		
386	Clarksdale	.60
387	Greenville	.69
388	Tupelo	.63
389	Greenwood	.63
390-392	Jackson	.72
393	Meridian	.66
394	Laurel	.62
395	Biloxi	.75
396	McComb	.73
397	Columbus	.64
MISSOURI		
630-631	St. Louis	1.01
633	Bowling Green	.91
634	Hannibal	.88
635	Kirksville	.81
636	Flat River	.94
637	Cape Girardeau	.87
638	Sikeston	.84
639	Poplar Bluff	.85
640-641	Kansas City	1.01
644-645	St. Joseph	.95
646	Chillicothe	.86
647	Harrisonville	.96
648	Joplin	.83
650-651	Jefferson City	.89
652	Columbia	.89
653	Sedalia	.87
654-655	Rolla	.89
656-658	Springfield	.85
MONTANA		
590-591	Billings	.88
592	Wolf Point	.85
593	Miles City	.87
594	Great Falls	.89
595	Havre	.82
596	Helena	.89
597	Butte	.84
598	Missoula	.84
599	Kalispell	.83
NEBRASKA		
680-681	Omaha	.90
683-685	Lincoln	.79
686	Columbus	.69
687	Norfolk	.78
688	Grand Island	.78
689	Hastings	.76
690	Mccook	.69
691	North Platte	.75
692	Valentine	.66
693	Alliance	.65
NEVADA		
889-891	Las Vegas	1.01
893	Ely	.92
894-895	Reno	.97
897	Carson City	.97
898	Elko	.93
NEW HAMPSHIRE		
030	Nashua	.91
031	Manchester	.91
032-033	Concord	.88
034	Keene	.73
035	Littleton	.81
036	Charleston	.71
037	Claremont	.72
038	Portsmouth	.85

STATE	CITY	Residential
NEW JERSEY		
070-071	Newark	1.13
072	Elizabeth	1.15
073	Jersey City	1.12
074-075	Paterson	1.13
076	Hackensack	1.12
077	Long Branch	1.12
078	Dover	1.12
079	Summit	1.12
080,083	Vineland	1.10
081	Camden	1.11
082,084	Atlantic City	1.14
085-086	Trenton	1.12
087	Point Pleasant	1.11
088-089	New Brunswick	1.12
NEW MEXICO		
870-872	Albuquerque	.86
873	Gallup	.86
874	Farmington	.86
875	Santa Fe	.86
877	Las Vegas	.86
878	Socorro	.86
879	Truth/Consequences	.84
880	Las Cruces	.83
881	Clovis	.85
882	Roswell	.86
883	Carrizozo	.86
884	Tucumcari	.86
NEW YORK		
100-102	New York	1.37
103	Staten Island	1.30
104	Bronx	1.32
105	Mount Vernon	1.18
106	White Plains	1.21
107	Yonkers	1.22
108	New Rochelle	1.23
109	Suffern	1.15
110	Queens	1.30
111	Long Island City	1.33
112	Brooklyn	1.34
113	Flushing	1.32
114	Jamaica	1.32
115,117,118	Hicksville	1.22
116	Far Rockaway	1.31
119	Riverhead	1.23
120-122	Albany	.96
123	Schenectady	.96
124	Kingston	1.04
125-126	Poughkeepsie	1.08
127	Monticello	1.05
128	Glens Falls	.88
129	Plattsburgh	.93
130-132	Syracuse	.96
133-135	Utica	.93
136	Watertown	.92
137-139	Binghamton	.92
140-142	Buffalo	1.06
143	Niagara Falls	1.04
144-146	Rochester	.99
147	Jamestown	.91
148-149	Elmira	.89
NORTH CAROLINA		
270,272-274	Greensboro	.74
271	Winston-Salem	.74
275-276	Raleigh	.75
277	Durham	.74
278	Rocky Mount	.64
279	Elizabeth City	.62
280	Gastonia	.74
281-282	Charlotte	.75
283	Fayetteville	.72
284	Wilmington	.72
285	Kinston	.62
286	Hickory	.62
287-288	Asheville	.72
289	Murphy	.66
NORTH DAKOTA		
580-581	Fargo	.81
582	Grand Forks	.76
583	Devils Lake	.81
584	Jamestown	.75

STATE	CITY	Residential
NORTH DAKOTA (CONT'D)		
585	Bismarck	.81
586	Dickinson	.78
587	Minot	.81
588	Williston	.78
OHIO		
430-432	Columbus	.96
433	Marion	.94
434-436	Toledo	1.02
437-438	Zanesville	.91
439	Steubenville	.96
440	Lorain	1.03
441	Cleveland	1.03
442-443	Akron	1.00
444-445	Youngstown	.97
446-447	Canton	.95
448-449	Mansfield	.97
450	Hamilton	.96
451-452	Cincinnati	.96
453-454	Dayton	.93
455	Springfield	.94
456	Chillicothe	.97
457	Athens	.88
458	Lima	.91
OKLAHOMA		
730-731	Oklahoma City	.81
734	Ardmore	.79
735	Lawton	.83
736	Clinton	.78
737	Enid	.78
738	Woodward	.77
739	Guymon	.67
740-741	Tulsa	.80
743	Miami	.83
744	Muskogee	.72
745	Mcalester	.75
746	Ponca City	.78
747	Durant	.77
748	Shawnee	.77
749	Poteau	.78
OREGON		
970-972	Portland	1.02
973	Salem	1.02
974	Eugene	1.01
975	Medford	1.00
976	Klamath Falls	1.00
977	Bend	1.02
978	Pendleton	.99
979	Vale	.98
PENNSYLVANIA		
150-152	Pittsburgh	.98
153	Washington	.94
154	Uniontown	.92
155	Bedford	.88
156	Greensburg	.95
157	Indiana	.91
158	Dubois	.90
159	Johnstown	.91
160	Butler	.94
161	New Castle	.94
162	Kittanning	.95
163	Oil City	.91
164-165	Erie	.97
166	Altoona	.90
167	Bradford	.90
168	State College	.93
169	Wellsboro	.89
170-171	Harrisburg	.93
172	Chambersburg	.89
173-174	York	.89
175-176	Lancaster	.91
177	Williamsport	.86
178	Sunbury	.90
179	Pottsville	.90
180	Lehigh Valley	.99
181	Allentown	1.02
182	Hazleton	.90
183	Stroudsburg	.93
184-185	Scranton	.96
186-187	Wilkes-Barre	.92
188	Montrose	.90

STATE	CITY	Residential
PENNSYLVANIA (CONT'D)		
189	Doylestown	1.04
190-191	Philadelphia	1.13
193	Westchester	1.07
194	Norristown	1.06
195-196	Reading	.95
PUERTO RICO		
009	San Juan	.84
RHODE ISLAND		
028	Newport	1.07
029	Providence	1.07
SOUTH CAROLINA		
290-292	Columbia	.73
293	Spartanburg	.72
294	Charleston	.72
295	Florence	.67
296	Greenville	.71
297	Rock Hill	.65
298	Aiken	.84
299	Beaufort	.67
SOUTH DAKOTA		
570-571	Sioux Falls	.77
572	Watertown	.73
573	Mitchell	.75
574	Aberdeen	.76
575	Pierre	.76
576	Mobridge	.74
577	Rapid City	.76
TENNESSEE		
370-372	Nashville	.84
373-374	Chattanooga	.77
375,380-381	Memphis	.83
376	Johnson City	.72
377-379	Knoxville	.74
382	Mckenzie	.70
383	Jackson	.71
384	Columbia	.73
385	Cookeville	.68
TEXAS		
750	Mckinney	.77
751	Waxahackie	.78
752-753	Dallas	.84
754	Greenville	.70
755	Texarkana	.75
756	Longview	.69
757	Tyler	.76
758	Palestine	.69
759	Lufkin	.74
760-761	Fort Worth	.84
762	Denton	.77
763	Wichita Falls	.80
764	Eastland	.73
765	Temple	.76
766-767	Waco	.78
768	Brownwood	.69
769	San Angelo	.73
770-772	Houston	.86
773	Huntsville	.69
774	Wharton	.71
775	Galveston	.84
776-777	Beaumont	.83
778	Bryan	.74
779	Victoria	.75
780	Laredo	.73
781-782	San Antonio	.79
783-784	Corpus Christi	.78
785	Mc Allen	.76
786-787	Austin	.80
788	Del Rio	.66
789	Giddings	.69
790-791	Amarillo	.79
792	Childress	.76
793-794	Lubbock	.77
795-796	Abilene	.76
797	Midland	.77
798-799,885	El Paso	.76
UTAH		
840-841	Salt Lake City	.84

Location Factors

STATE	CITY	Residential
UTAH (CONT'D)		
842,844	Ogden	.82
843	Logan	.83
845	Price	.73
846-847	Provo	.84
VERMONT		
050	White River Jct.	.74
051	Bellows Falls	.75
052	Bennington	.74
053	Brattleboro	.75
054	Burlington	.80
056	Montpelier	.81
057	Rutland	.82
058	St. Johnsbury	.75
059	Guildhall	.74
VIRGINIA		
220-221	Fairfax	.86
222	Arlington	.87
223	Alexandria	.90
224-225	Fredericksburg	.76
226	Winchester	.72
227	Culpeper	.78
228	Harrisonburg	.68
229	Charlottesville	.73
230-232	Richmond	.82
233-235	Norfolk	.79
236	Newport News	.77
237	Portsmouth	.75
238	Petersburg	.80
239	Farmville	.69
240-241	Roanoke	.73
242	Bristol	.68
243	Pulaski	.66
244	Staunton	.69
245	Lynchburg	.70
246	Grundy	.68
WASHINGTON		
980-981,987	Seattle	1.01
982	Everett	1.03
983-984	Tacoma	.99
985	Olympia	.99
986	Vancouver	.98
988	Wenatchee	.92
989	Yakima	.95
990-992	Spokane	1.00
993	Richland	.99
994	Clarkston	.98
WEST VIRGINIA		
247-248	Bluefield	.89
249	Lewisburg	.89
250-253	Charleston	.97
254	Martinsburg	.85
255-257	Huntington	.96
258-259	Beckley	.90
260	Wheeling	.92
261	Parkersburg	.92
262	Buckhannon	.91
263-264	Clarksburg	.91
265	Morgantown	.92
266	Gassaway	.92
267	Romney	.87
268	Petersburg	.89
WISCONSIN		
530,532	Milwaukee	1.06
531	Kenosha	1.06
534	Racine	1.04
535	Beloit	1.02
537	Madison	1.01
538	Lancaster	1.00
539	Portage	.98
540	New Richmond	1.00
541-543	Green Bay	1.03
544	Wausau	.96
545	Rhinelander	.97
546	La Crosse	.96
547	Eau Claire	1.00
548	Superior	.99
549	Oshkosh	.96

STATE	CITY	Residential
WYOMING		
820	Cheyenne	.76
821	Yellowstone Nat. Pk.	.71
822	Wheatland	.72
823	Rawlins	.70
824	Worland	.69
825	Riverton	.70
826	Casper	.77
827	Newcastle	.69
828	Sheridan	.74
829-831	Rock Springs	.74
ALBERTA		
	Calgary	1.05
	Edmonton	1.05
	Fort McMurray	1.03
	Lethbridge	1.04
	Lloydminster	1.03
	Medicine Hat	1.03
	Red Deer	1.03
BRITISH COLUMBIA		
	Kamloops	1.01
	Prince George	1.02
	Vancouver	1.07
	Victoria	1.02
MANITOBA		
	Brandon	.99
	Portage la Prairie	.99
	Winnipeg	.99
NEW BRUNSWICK		
	Bathurst	.91
	Dalhousie	.91
	Fredericton	.97
	Moncton	.91
	Newcastle	.91
	Saint John	.97
NEWFOUNDLAND		
	Corner Brook	.92
	St. John's	.93
NORTHWEST TERRITORIES		
	Yellowknife	.99
NOVA SCOTIA		
	Dartmouth	.93
	Halifax	.94
	New Glasgow	.93
	Sydney	.92
	Yarmouth	.93
ONTARIO		
	Barrie	1.10
	Brantford	1.11
	Cornwall	1.10
	Hamilton	1.11
	Kingston	1.11
	Kitchener	1.05
	London	1.09
	North Bay	1.08
	Oshawa	1.10
	Ottawa	1.11
	Owen Sound	1.08
	Peterborough	1.09
	Sarnia	1.12
	Sudbury	1.03
	Thunder Bay	1.07
	Toronto	1.14
	Windsor	1.08
PRINCE EDWARD ISLAND		
	Charlottetown	.88
	Summerside	.88
QUEBEC		
	Cap-de-la-Madeleine	1.10
	Charlesbourg	1.10
	Chicoutimi	1.10
	Gatineau	1.09
	Laval	1.09
	Montreal	1.09
	Quebec	1.11

STATE	CITY	Residential
QUEBEC (CONT'D)		
	Sherbrooke	1.09
	Trois Rivieres	1.10
SASKATCHEWAN		
	Moose Jaw	.91
	Prince Albert	.90
	Regina	.91
	Saskatoon	.90
YUKON		
	Whitehorse	.89

Abbreviations

A	Area Square Feet; Ampere
ABS	Acrylonitrile Butadiene Stryrene; Asbestos Bonded Steel
A.C.	Alternating Current; Air-Conditioning; Asbestos Cement; Plywood Grade A & C
A.C.I.	American Concrete Institute
AD	Plywood, Grade A & D
Addit.	Additional
Adj.	Adjustable
af	Audio-frequency
A.G.A.	American Gas Association
Agg.	Aggregate
A.H.	Ampere Hours
A hr.	Ampere-hour
A.H.U.	Air Handling Unit
A.I.A.	American Institute of Architects
AIC	Ampere Interrupting Capacity
Allow.	Allowance
alt.	Altitude
Alum.	Aluminum
a.m.	Ante Meridiem
Amp.	Ampere
Anod.	Anodized
Approx.	Approximate
Apt.	Apartment
Asb.	Asbestos
A.S.B.C.	American Standard Building Code
Asbe.	Asbestos Worker
A.S.H.R.A.E.	American Society of Heating, Refrig. & AC Engineers
A.S.M.E.	American Society of Mechanical Engineers
A.S.T.M.	American Society for Testing and Materials
Attchmt.	Attachment
Avg.	Average
A.W.G.	American Wire Gauge
AWWA	American Water Works Assoc.
Bbl.	Barrel
B. & B.	Grade B and Better; Balled & Burlapped
B. & S.	Bell and Spigot
B. & W.	Black and White
b.c.c.	Body-centered Cubic
B.C.Y.	Bank Cubic Yards
BE	Bevel End
B.F.	Board Feet
Bg. cem.	Bag of Cement
BHP	Boiler Horsepower; Brake Horsepower
B.I.	Black Iron
Bit.; Bitum.	Bituminous
Bk.	Backed
Bkrs.	Breakers
Bldg.	Building
Blk.	Block
Bm.	Beam
Boil.	Boilermaker
B.P.M.	Blows per Minute
BR	Bedroom
Brg.	Bearing
Brhe.	Bricklayer Helper
Bric.	Bricklayer
Brk.	Brick
Brng.	Bearing
Brs.	Brass
Brz.	Bronze
Bsn.	Basin
Btr.	Better
BTU	British Thermal Unit
BTUH	BTU per Hour
B.U.R.	Built-up Roofing
BX	Interlocked Armored Cable
c	Conductivity, Copper Sweat
C	Hundred; Centigrade
C/C	Center to Center, Cedar on Cedar
Cab.	Cabinet
Cair.	Air Tool Laborer
Calc	Calculated
Cap.	Capacity
Carp.	Carpenter
C.B.	Circuit Breaker
C.C.A.	Chromate Copper Arsenate
C.C.F.	Hundred Cubic Feet
cd	Candela
cd/sf	Candela per Square Foot
CD	Grade of Plywood Face & Back
CDX	Plywood, Grade C & D, exterior glue
Cefi.	Cement Finisher
Cem.	Cement
CF	Hundred Feet
C.F.	Cubic Feet
CFM	Cubic Feet per Minute
c.g.	Center of Gravity
CHW	Chilled Water; Commercial Hot Water
C.I.	Cast Iron
C.I.P.	Cast in Place
Circ.	Circuit
C.L.	Carload Lot
Clab.	Common Laborer
C.L.F.	Hundred Linear Feet
CLF	Current Limiting Fuse
CLP	Cross Linked Polyethylene
cm	Centimeter
CMP	Corr. Metal Pipe
C.M.U.	Concrete Masonry Unit
CN	Change Notice
Col.	Column
CO_2	Carbon Dioxide
Comb.	Combination
Compr.	Compressor
Conc.	Concrete
Cont.	Continuous; Continued
Corr.	Corrugated
Cos	Cosine
Cot	Cotangent
Cov.	Cover
C/P	Cedar on Paneling
CPA	Control Point Adjustment
Cplg.	Coupling
C.P.M.	Critical Path Method
CPVC	Chlorinated Polyvinyl Chloride
C.Pr.	Hundred Pair
CRC	Cold Rolled Channel
Creos.	Creosote
Crpt.	Carpet & Linoleum Layer
CRT	Cathode-ray Tube
CS	Carbon Steel, Constant Shear Bar Joist
Csc	Cosecant
C.S.F.	Hundred Square Feet
CSI	Construction Specifications Institute
C.T.	Current Transformer
CTS	Copper Tube Size
Cu	Copper, Cubic
Cu. Ft.	Cubic Foot
cw	Continuous Wave
C.W.	Cool White; Cold Water
Cwt.	100 Pounds
C.W.X.	Cool White Deluxe
C.Y.	Cubic Yard (27 cubic feet)
C.Y./Hr.	Cubic Yard per Hour
Cyl.	Cylinder
d	Penny (nail size)
D	Deep; Depth; Discharge
Dis.;Disch.	Discharge
Db.	Decibel
Dbl.	Double
DC	Direct Current
DDC	Direct Digital Control
Demob.	Demobilization
d.f.u.	Drainage Fixture Units
D.H.	Double Hung
DHW	Domestic Hot Water
Diag.	Diagonal
Diam.	Diameter
Distrib.	Distribution
Dk.	Deck
D.L.	Dead Load; Diesel
DLH	Deep Long Span Bar Joist
Do.	Ditto
Dp.	Depth
D.P.S.T.	Double Pole, Single Throw
Dr.	Driver
Drink.	Drinking
D.S.	Double Strength
D.S.A.	Double Strength A Grade
D.S.B.	Double Strength B Grade
Dty.	Duty
DWV	Drain Waste Vent
DX	Deluxe White, Direct Expansion
dyn	Dyne
e	Eccentricity
E	Equipment Only; East
Ea.	Each
E.B.	Encased Burial
Econ.	Economy
E.C.Y	Embankment Cubic Yards
EDP	Electronic Data Processing
EIFS	Exterior Insulation Finish System
E.D.R.	Equiv. Direct Radiation
Eq.	Equation
Elec.	Electrician; Electrical
Elev.	Elevator; Elevating
EMT	Electrical Metallic Conduit; Thin Wall Conduit
Eng.	Engine, Engineered
EPDM	Ethylene Propylene Diene Monomer
EPS	Expanded Polystyrene
Eqhv.	Equip. Oper., Heavy
Eqlt.	Equip. Oper., Light
Eqmd.	Equip. Oper., Medium
Eqmm.	Equip. Oper., Master Mechanic
Eqol.	Equip. Oper., Oilers
Equip.	Equipment
ERW	Electric Resistance Welded
E.S.	Energy Saver
Est.	Estimated
esu	Electrostatic Units
E.W.	Each Way
EWT	Entering Water Temperature
Excav.	Excavation
Exp.	Expansion, Exposure
Ext.	Exterior
Extru.	Extrusion
f.	Fiber stress
F	Fahrenheit; Female; Fill
Fab.	Fabricated
FBGS	Fiberglass
F.C.	Footcandles
f.c.c.	Face-centered Cubic
f′c.	Compressive Stress in Concrete; Extreme Compressive Stress
F.E.	Front End
FEP	Fluorinated Ethylene Propylene (Teflon)
F.G.	Flat Grain
F.H.A.	Federal Housing Administration
Fig.	Figure
Fin.	Finished
Fixt.	Fixture
Fl. Oz.	Fluid Ounces
Flr.	Floor
F.M.	Frequency Modulation; Factory Mutual
Fmg.	Framing
Fndtn.	Foundation
Fori.	Foreman, Inside

Abbreviations

Foro.	Foreman, Outside
Fount.	Fountain
FPM	Feet per Minute
FPT	Female Pipe Thread
Fr.	Frame
F.R.	Fire Rating
FRK	Foil Reinforced Kraft
FRP	Fiberglass Reinforced Plastic
FS	Forged Steel
FSC	Cast Body; Cast Switch Box
Ft.	Foot; Feet
Ftng.	Fitting
Ftg.	Footing
Ft. Lb.	Foot Pound
Furn.	Furniture
FVNR	Full Voltage Non-Reversing
FXM	Female by Male
Fy.	Minimum Yield Stress of Steel
g	Gram
G	Gauss
Ga.	Gauge
Gal.	Gallon
Gal./Min.	Gallon per Minute
Galv.	Galvanized
Gen.	General
G.F.I.	Ground Fault Interrupter
Glaz.	Glazier
GPD	Gallons per Day
GPH	Gallons per Hour
GPM	Gallons per Minute
GR	Grade
Gran.	Granular
Grnd.	Ground
H	High; High Strength Bar Joist; Henry
H.C.	High Capacity
H.D.	Heavy Duty; High Density
H.D.O.	High Density Overlaid
Hdr.	Header
Hdwe.	Hardware
Help.	Helpers Average
HEPA	High Efficiency Particulate Air Filter
Hg	Mercury
HIC	High Interrupting Capacity
HM	Hollow Metal
H.O.	High Output
Horiz.	Horizontal
H.P.	Horsepower; High Pressure
H.P.F.	High Power Factor
Hr.	Hour
Hrs./Day	Hours per Day
HSC	High Short Circuit
Ht.	Height
Htg.	Heating
Htrs.	Heaters
HVAC	Heating, Ventilation & Air-Conditioning
Hvy.	Heavy
HW	Hot Water
Hyd.;Hydr.	Hydraulic
Hz.	Hertz (cycles)
I.	Moment of Inertia
I.C.	Interrupting Capacity
ID	Inside Diameter
I.D.	Inside Dimension; Identification
I.F.	Inside Frosted
I.M.C.	Intermediate Metal Conduit
In.	Inch
Incan.	Incandescent
Incl.	Included; Including
Int.	Interior
Inst.	Installation
Insul.	Insulation/Insulated
I.P.	Iron Pipe
I.P.S.	Iron Pipe Size
I.P.T.	Iron Pipe Threaded
I.W.	Indirect Waste
J	Joule
J.I.C.	Joint Industrial Council
K	Thousand; Thousand Pounds; Heavy Wall Copper Tubing, Kelvin
K.A.H.	Thousand Amp. Hours
KCMIL	Thousand Circular Mils
KD	Knock Down
K.D.A.T.	Kiln Dried After Treatment
kg	Kilogram
kG	Kilogauss
kgf	Kilogram Force
kHz	Kilohertz
Kip.	1000 Pounds
KJ	Kiljoule
K.L.	Effective Length Factor
K.L.F.	Kips per Linear Foot
Km	Kilometer
K.S.F.	Kips per Square Foot
K.S.I.	Kips per Square Inch
kV	Kilovolt
kVA	Kilovolt Ampere
K.V.A.R.	Kilovar (Reactance)
KW	Kilowatt
KWh	Kilowatt-hour
L	Labor Only; Length; Long; Medium Wall Copper Tubing
Lab.	Labor
lat	Latitude
Lath.	Lather
Lav.	Lavatory
lb.; #	Pound
L.B.	Load Bearing; L Conduit Body
L. & E.	Labor & Equipment
lb./hr.	Pounds per Hour
lb./L.F.	Pounds per Linear Foot
lbf/sq.in.	Pound-force per Square Inch
L.C.L.	Less than Carload Lot
L.C.Y.	Loose Cubic Yards
Ld.	Load
LE	Lead Equivalent
LED	Light Emitting Diode
L.F.	Linear Foot
Lg.	Long; Length; Large
L & H	Light and Heat
LH	Long Span Bar Joist
L.H.	Labor Hours
L.L.	Live Load
L.L.D.	Lamp Lumen Depreciation
L-O-L	Lateralolet
lm	Lumen
lm/sf	Lumen per Square Foot
lm/W	Lumen per Watt
L.O.A.	Length Over All
log	Logarithm
L.P.	Liquefied Petroleum; Low Pressure
L.P.F.	Low Power Factor
LR	Long Radius
L.S.	Lump Sum
Lt.	Light
Lt. Ga.	Light Gauge
L.T.L.	Less than Truckload Lot
Lt. Wt.	Lightweight
L.V.	Low Voltage
M	Thousand; Material; Male; Light Wall Copper Tubing
M^2CA	Meters Squared Contact Area
m/hr; M.H.	Man-hour
mA	Milliampere
Mach.	Machine
Mag. Str.	Magnetic Starter
Maint.	Maintenance
Marb.	Marble Setter
Mat; Mat'l.	Material
Max.	Maximum
MBF	Thousand Board Feet
MBH	Thousand BTU's per hr.
MC	Metal Clad Cable
M.C.F.	Thousand Cubic Feet
M.C.F.M.	Thousand Cubic Feet per Minute
M.C.M.	Thousand Circular Mils
M.C.P.	Motor Circuit Protector
MD	Medium Duty
M.D.O.	Medium Density Overlaid
Med.	Medium
MF	Thousand Feet
M.F.B.M.	Thousand Feet Board Measure
Mfg.	Manufacturing
Mfrs.	Manufacturers
mg	Milligram
MGD	Million Gallons per Day
MGPH	Thousand Gallons per Hour
MH, M.H.	Manhole; Metal Halide; Man-Hour
MHz	Megahertz
Mi.	Mile
MI	Malleable Iron; Mineral Insulated
mm	Millimeter
Mill.	Millwright
Min., min.	Minimum, minute
Misc.	Miscellaneous
ml	Milliliter, Mainline
M.L.F.	Thousand Linear Feet
Mo.	Month
Mobil.	Mobilization
Mog.	Mogul Base
MPH	Miles per Hour
MPT	Male Pipe Thread
MRT	Mile Round Trip
ms	Millisecond
M.S.F.	Thousand Square Feet
Mstz.	Mosaic & Terrazzo Worker
M.S.Y.	Thousand Square Yards
Mtd.	Mounted
Mthe.	Mosaic & Terrazzo Helper
Mtng.	Mounting
Mult.	Multi; Multiply
M.V.A.	Million Volt Amperes
M.V.A.R.	Million Volt Amperes Reactance
MV	Megavolt
MW	Megawatt
MXM	Male by Male
MYD	Thousand Yards
N	Natural; North
nA	Nanoampere
NA	Not Available; Not Applicable
N.B.C.	National Building Code
NC	Normally Closed
N.E.M.A.	National Electrical Manufacturers Assoc.
NEHB	Bolted Circuit Breaker to 600V.
N.L.B.	Non-Load-Bearing
NM	Non-Metallic Cable
nm	Nanometer
No.	Number
NO	Normally Open
N.O.C.	Not Otherwise Classified
Nose.	Nosing
N.P.T.	National Pipe Thread
NQOD	Combination Plug-on/Bolt on Circuit Breaker to 240V.
N.R.C.	Noise Reduction Coefficient
N.R.S.	Non Rising Stem
ns	Nanosecond
nW	Nanowatt
OB	Opposing Blade
OC	On Center
OD	Outside Diameter
O.D.	Outside Dimension
ODS	Overhead Distribution System
O.G.	Ogee
O.H.	Overhead
O & P	Overhead and Profit
Oper.	Operator
Opng.	Opening
Orna.	Ornamental
OSB	Oriented Strand Board
O. S. & Y.	Outside Screw and Yoke

Abbreviations

Abbreviation	Meaning
Ovhd.	Overhead
OWG	Oil, Water or Gas
Oz.	Ounce
P.	Pole; Applied Load; Projection
p.	Page
Pape.	Paperhanger
P.A.P.R.	Powered Air Purifying Respirator
PAR	Parabolic Reflector
Pc., Pcs.	Piece, Pieces
P.C.	Portland Cement; Power Connector
P.C.F.	Pounds per Cubic Foot
P.C.M.	Phase Contract Microscopy
P.E.	Professional Engineer; Porcelain Enamel; Polyethylene; Plain End
Perf.	Perforated
Ph.	Phase
P.I.	Pressure Injected
Pile.	Pile Driver
Pkg.	Package
Pl.	Plate
Plah.	Plasterer Helper
Plas.	Plasterer
Pluh.	Plumbers Helper
Plum.	Plumber
Ply.	Plywood
p.m.	Post Meridiem
Pntd.	Painted
Pord.	Painter, Ordinary
pp	Pages
PP; PPL	Polypropylene
P.P.M.	Parts per Million
Pr.	Pair
P.E.S.B.	Pre-engineered Steel Building
Prefab.	Prefabricated
Prefin.	Prefinished
Prop.	Propelled
PSF; psf	Pounds per Square Foot
PSI; psi	Pounds per Square Inch
PSIG	Pounds per Square Inch Gauge
PSP	Plastic Sewer Pipe
Pspr.	Painter, Spray
Psst.	Painter, Structural Steel
P.T.	Potential Transformer
P. & T.	Pressure & Temperature
Ptd.	Painted
Ptns.	Partitions
Pu	Ultimate Load
PVC	Polyvinyl Chloride
Pvmt.	Pavement
Pwr.	Power
Q	Quantity Heat Flow
Quan.;Qty.	Quantity
Q.C.	Quick Coupling
r	Radius of Gyration
R	Resistance
R.C.P.	Reinforced Concrete Pipe
Rect.	Rectangle
Reg.	Regular
Reinf.	Reinforced
Req'd.	Required
Res.	Resistant
Resi.	Residential
Rgh.	Rough
RGS	Rigid Galvanized Steel
R.H.W.	Rubber, Heat & Water Resistant; Residential Hot Water
rms	Root Mean Square
Rnd.	Round
Rodm.	Rodman
Rofc.	Roofer, Composition
Rofp.	Roofer, Precast
Rohe.	Roofer Helpers (Composition)
Rots.	Roofer, Tile & Slate
R.O.W.	Right of Way
RPM	Revolutions per Minute
R.S.	Rapid Start
Rsr	Riser
RT	Round Trip
S.	Suction; Single Entrance; South
SCFM	Standard Cubic Feet per Minute
Scaf.	Scaffold
Sch.; Sched.	Schedule
S.C.R.	Modular Brick
S.D.	Sound Deadening
S.D.R.	Standard Dimension Ratio
S.E.	Surfaced Edge
Sel.	Select
S.E.R.;	Service Entrance Cable
S.E.U.	Service Entrance Cable
S.F.	Square Foot
S.F.C.A.	Square Foot Contact Area
S.F.G.	Square Foot of Ground
S.F. Hor.	Square Foot Horizontal
S.F.R.	Square Feet of Radiation
S.F. Shlf.	Square Foot of Shelf
S4S	Surface 4 Sides
Shee.	Sheet Metal Worker
Sin.	Sine
Skwk.	Skilled Worker
SL	Saran Lined
S.L.	Slimline
Sldr.	Solder
SLH	Super Long Span Bar Joist
S.N.	Solid Neutral
S-O-L	Socketolet
sp	Standpipe
S.P.	Static Pressure; Single Pole; Self-Propelled
Spri.	Sprinkler Installer
spwg	Static Pressure Water Gauge
Sq.	Square; 100 Square Feet
S.P.D.T.	Single Pole, Double Throw
SPF	Spruce Pine Fir
S.P.S.T.	Single Pole, Single Throw
SPT	Standard Pipe Thread
Sq. Hd.	Square Head
Sq. In.	Square Inch
S.S.	Single Strength; Stainless Steel
S.S.B.	Single Strength B Grade
sst	Stainless Steel
Sswk.	Structural Steel Worker
Sswl.	Structural Steel Welder
St.;Stl.	Steel
S.T.C.	Sound Transmission Coefficient
Std.	Standard
STK	Select Tight Knot
STP	Standard Temperature & Pressure
Stpi.	Steamfitter, Pipefitter
Str.	Strength; Starter; Straight
Strd.	Stranded
Struct.	Structural
Sty.	Story
Subj.	Subject
Subs.	Subcontractors
Surf.	Surface
Sw.	Switch
Swbd.	Switchboard
S.Y.	Square Yard
Syn.	Synthetic
S.Y.P.	Southern Yellow Pine
Sys.	System
t.	Thickness
T	Temperature; Ton
Tan	Tangent
T.C.	Terra Cotta
T & C	Threaded and Coupled
T.D.	Temperature Difference
T.E.M.	Transmission Electron Microscopy
TFE	Tetrafluoroethylene (Teflon)
T. & G.	Tongue & Groove; Tar & Gravel
Th.;Thk.	Thick
Thn.	Thin
Thrded	Threaded
Tilf.	Tile Layer, Floor
Tilh.	Tile Layer, Helper
THHN	Nylon Jacketed Wire
THW.	Insulated Strand Wire
THWN;	Nylon Jacketed Wire
T.L.	Truckload
T.M.	Track Mounted
Tot.	Total
T-O-L	Threadolet
T.S.	Trigger Start
Tr.	Trade
Transf.	Transformer
Trhv.	Truck Driver, Heavy
Trlr	Trailer
Trlt.	Truck Driver, Light
TV	Television
T.W.	Thermoplastic Water Resistant Wire
UCI	Uniform Construction Index
UF	Underground Feeder
UGND	Underground Feeder
U.H.F.	Ultra High Frequency
U.L.	Underwriters Laboratory
Unfin.	Unfinished
URD	Underground Residential Distribution
US	United States
USP	United States Primed
UTP	Unshielded Twisted Pair
V	Volt
V.A.	Volt Amperes
V.C.T.	Vinyl Composition Tile
VAV	Variable Air Volume
VC	Veneer Core
Vent.	Ventilation
Vert.	Vertical
V.F.	Vinyl Faced
V.G.	Vertical Grain
V.H.F.	Very High Frequency
VHO	Very High Output
Vib.	Vibrating
V.L.F.	Vertical Linear Foot
Vol.	Volume
VRP	Vinyl Reinforced Polyester
W	Wire; Watt; Wide; West
w/	With
W.C.	Water Column; Water Closet
W.F.	Wide Flange
W.G.	Water Gauge
Wldg.	Welding
W. Mile	Wire Mile
W-O-L	Weldolet
W.R.	Water Resistant
Wrck.	Wrecker
W.S.P.	Water, Steam, Petroleum
WT., Wt.	Weight
WWF	Welded Wire Fabric
XFER	Transfer
XFMR	Transformer
XHD	Extra Heavy Duty
XHHW; XLPE	Cross-Linked Polyethylene Wire Insulation
XLP	Cross-linked Polyethylene
Y	Wye
yd	Yard
yr	Year
Δ	Delta
%	Percent
~	Approximately
Ø	Phase
@	At
#	Pound; Number
<	Less Than
>	Greater Than

RESIDENTIAL COST ESTIMATE

OWNER'S NAME: ________ APPRAISER: ________

RESIDENCE ADDRESS: ________ PROJECT: ________

CITY, STATE, ZIP CODE: ________ DATE: ________

CLASS OF CONSTRUCTION
- ☐ ECONOMY
- ☐ AVERAGE
- ☐ CUSTOM
- ☐ LUXURY

RESIDENCE TYPE
- ☐ 1 STORY
- ☐ 1-1/2 STORY
- ☐ 2 STORY
- ☐ 2-1/2 STORY
- ☐ 3 STORY
- ☐ BI-LEVEL
- ☐ TRI-LEVEL

CONFIGURATION
- ☐ DETACHED
- ☐ TOWN/ROW HOUSE
- ☐ SEMI-DETACHED

OCCUPANCY
- ☐ ONE STORY
- ☐ TWO FAMILY
- ☐ THREE FAMILY
- ☐ OTHER ________

EXTERIOR WALL SYSTEM
- ☐ WOOD SIDING—WOOD FRAME
- ☐ BRICK VENEER—WOOD FRAME
- ☐ STUCCO ON WOOD FRAME
- ☐ PAINTED CONCRETE BLOCK
- ☐ SOLID MASONRY (AVERAGE & CUSTOM)
- ☐ STONE VENEER—WOOD FRAME
- ☐ SOLID BRICK (LUXURY)
- ☐ SOLID STONE (LUXURY)

*LIVING AREA (Main Building)		*LIVING AREA (Wing or Ell) ( )		*LIVING AREA (Wing or Ell) ( )	
First Level	________ S.F.	First Level	________ S.F.	First Level	________ S.F.
Second Level	________ S.F.	Second Level	________ S.F.	Second Level	________ S.F.
Third Level	________ S.F.	Third Level	________ S.F.	Third Level	________ S.F.
Total	________ S.F.	Total	________ S.F.	Total	________ S.F.

*Basement Area is not part of living area.

MAIN BUILDING	COSTS PER S.F. LIVING AREA
Cost per Square Foot of Living Area, from Page ________	$
Basement Addition: ________ % Finished, ________ % Unfinished	+
Roof Cover Adjustment: ________ Type, Page ________ (Add or Deduct)	( )
Central Air Conditioning: ☐ Separate Ducts ☐ Heating Ducts, Page ________	+
Heating System Adjustment: ________ Type, Page ________ (Add or Deduct)	( )
Main Building: Adjusted Cost per S.F. of Living Area	$

MAIN BUILDING TOTAL COST: $ ________ /S.F. (Cost per S.F. Living Area) x ________ S.F. (Living Area) x ________ (Town/Row House Multiplier (Use 1 for Detached)) = $ ________ (TOTAL COST)

WING OR ELL ( ) ________ STORY	COSTS PER S.F. LIVING AREA
Cost per Square Foot of Living Area, from Page ________	$
Basement Addition: ________ % Finished, ________ % Unfinished	+
Roof Cover Adjustment: ________ Type, Page ________ (Add or Deduct)	( )
Central Air Conditioning: ☐ Separate Ducts ☐ Heating Ducts, Page ________	+
Heating System Adjustment: ________ Type, Page ________ (Add or Deduct)	( )
Wing or Ell ( ): Adjusted Cost per S.F. of Living Area	$

WING OR ELL ( ) TOTAL COST: $ ________ /S.F. (Cost per S.F. Living Area) x ________ S.F. (Living Area) = $ ________ (TOTAL COST)

WING OR ELL ( ) ________ STORY	COSTS PER S.F. LIVING AREA
Cost per Square Foot of Living Area, from Page ________	$
Basement Addition: ________ % Finished, ________ % Unfinished	+
Roof Cover Adjustment: ________ Type, Page ________ (Add or Deduct)	( )
Central Air Conditioning: ☐ Separate Ducts ☐ Heating Ducts, Page ________	+
Heating System Adjustment: ________ Type, Page ________ (Add or Deduct)	( )
Wing or Ell ( ): Adjusted Cost per S.F. of Living Area	$

WING OR ELL ( ) TOTAL COST: $ ________ /S.F. (Cost per S.F. Living Area) x ________ S.F. (Living Area) = $ ________ (TOTAL COST)

TOTAL THIS PAGE ________

Page 1 of 2

RESIDENTIAL
COST ESTIMATE

Total Page 1			$
	QUANTITY	UNIT COST	
Additional Bathrooms: __________ Full, __________ Half			
Finished Attic: __________ Ft. x __________ Ft.	S.F.		+
Breezeway: ☐ Open ☐ Enclosed __________ Ft. x __________ Ft.	S.F.		+
Covered Porch: ☐ Open ☐ Enclosed __________ Ft. x __________ Ft.	S.F.		+
Fireplace: ☐ Interior Chimney ☐ Exterior Chimney ☐ No. of Flues ☐ Additional Fireplaces			+
Appliances:			+
Kitchen Cabinets Adjustment: (+/−)			
☐ Garage ☐ Carport: __________ Car(s) Description __________________ (+/−)			
Miscellaneous:			+

ADJUSTED TOTAL BUILDING COST $

REPLACEMENT COST	
ADJUSTED TOTAL BUILDING COST	$ __________
Site Improvements	
(A) Paving & Sidewalks	$ __________
(B) Landscaping	$ __________
(C) Fences	$ __________
(D) Swimming Pool	$ __________
(E) Miscellaneous	$ __________
TOTAL	$ __________
Location Factor	x __________
Location Replacement Cost	$ __________
Depreciation	−$ __________
LOCAL DEPRECIATED COST	$ __________

INSURANCE COST	
ADJUSTED TOTAL BUILDING COST	$ __________
Insurance Exclusions	
(A) Footings, Site Work, Underground Piping	−$ __________
(B) Architects' Fees	−$ __________
Total Building Cost Less Exclusion	$ __________
Location Factor	x __________
LOCAL INSURABLE REPLACEMENT COST	$ __________

SKETCH AND ADDITIONAL CALCULATIONS

Page 2 of 2

C

F

G

H

M

N

O

P

Q

R

S

T

X

Y

Z

Notes

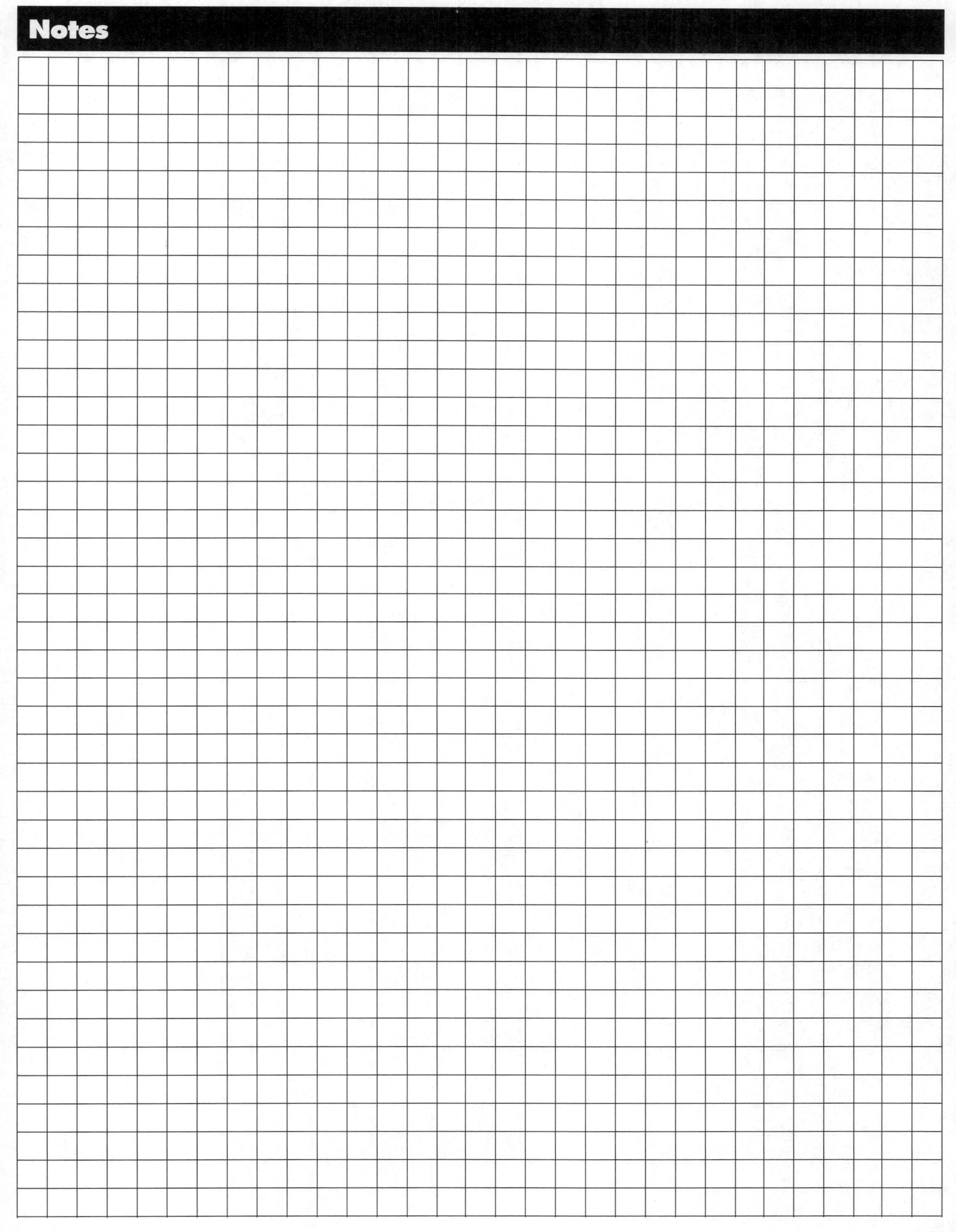

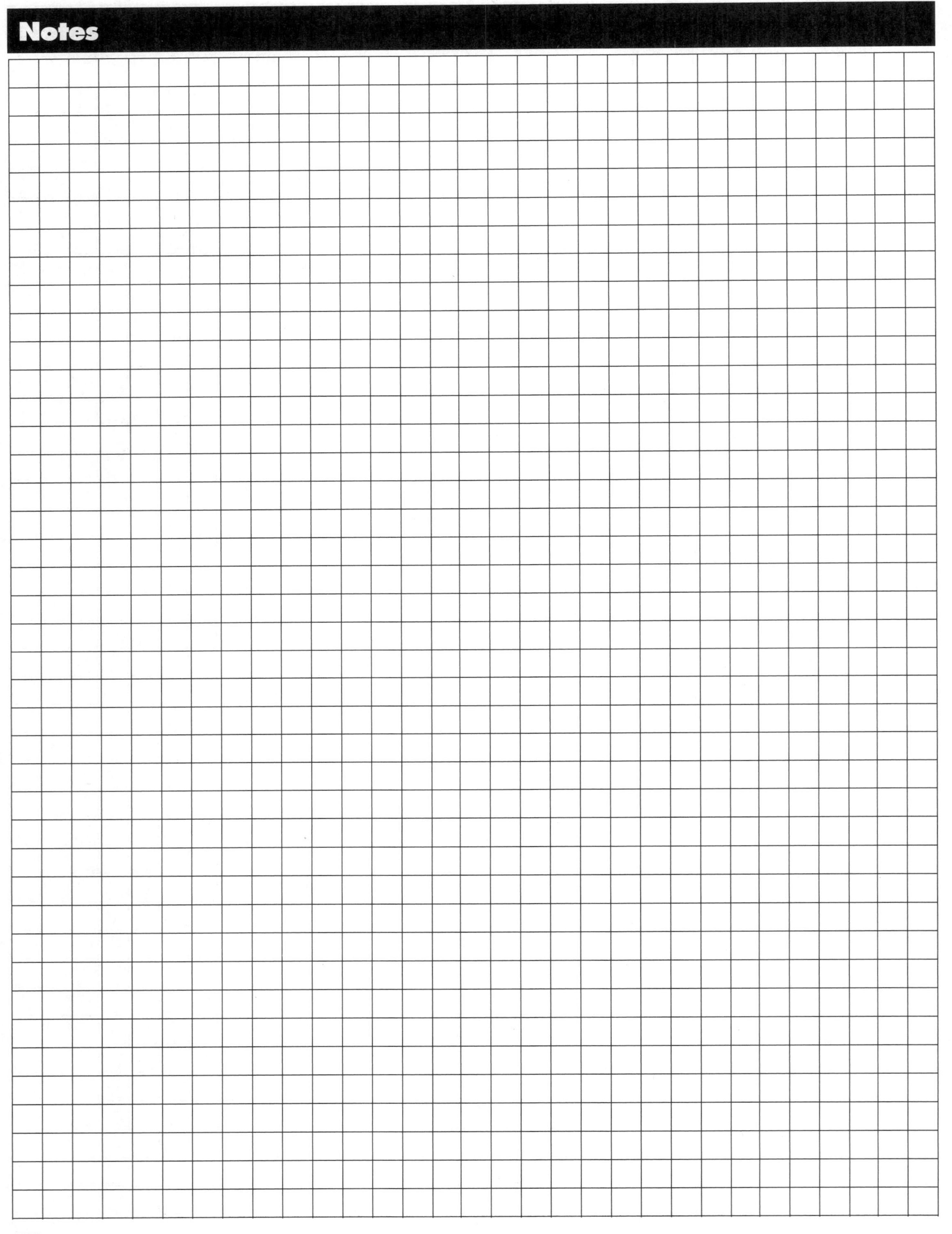

Notes

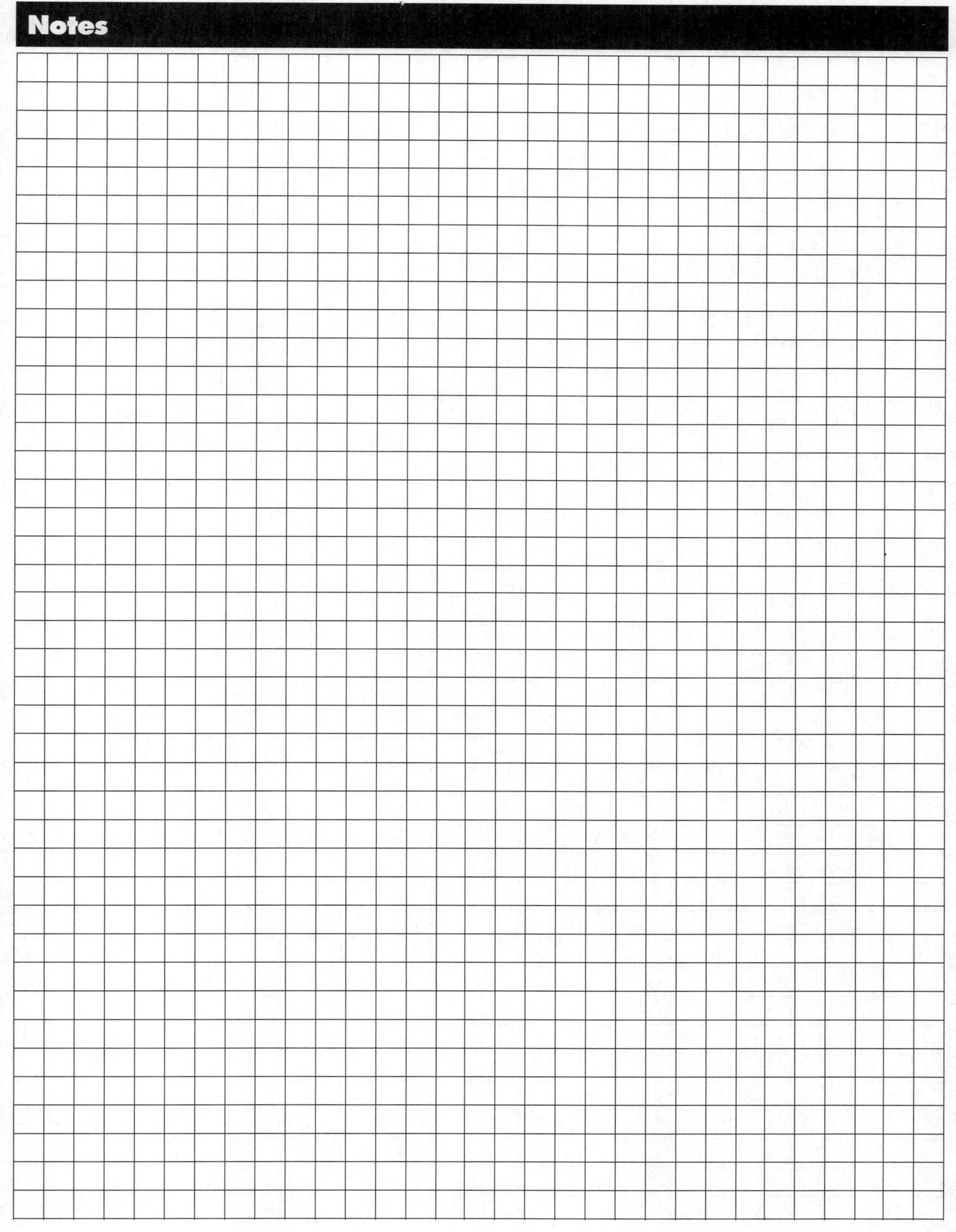

Notes

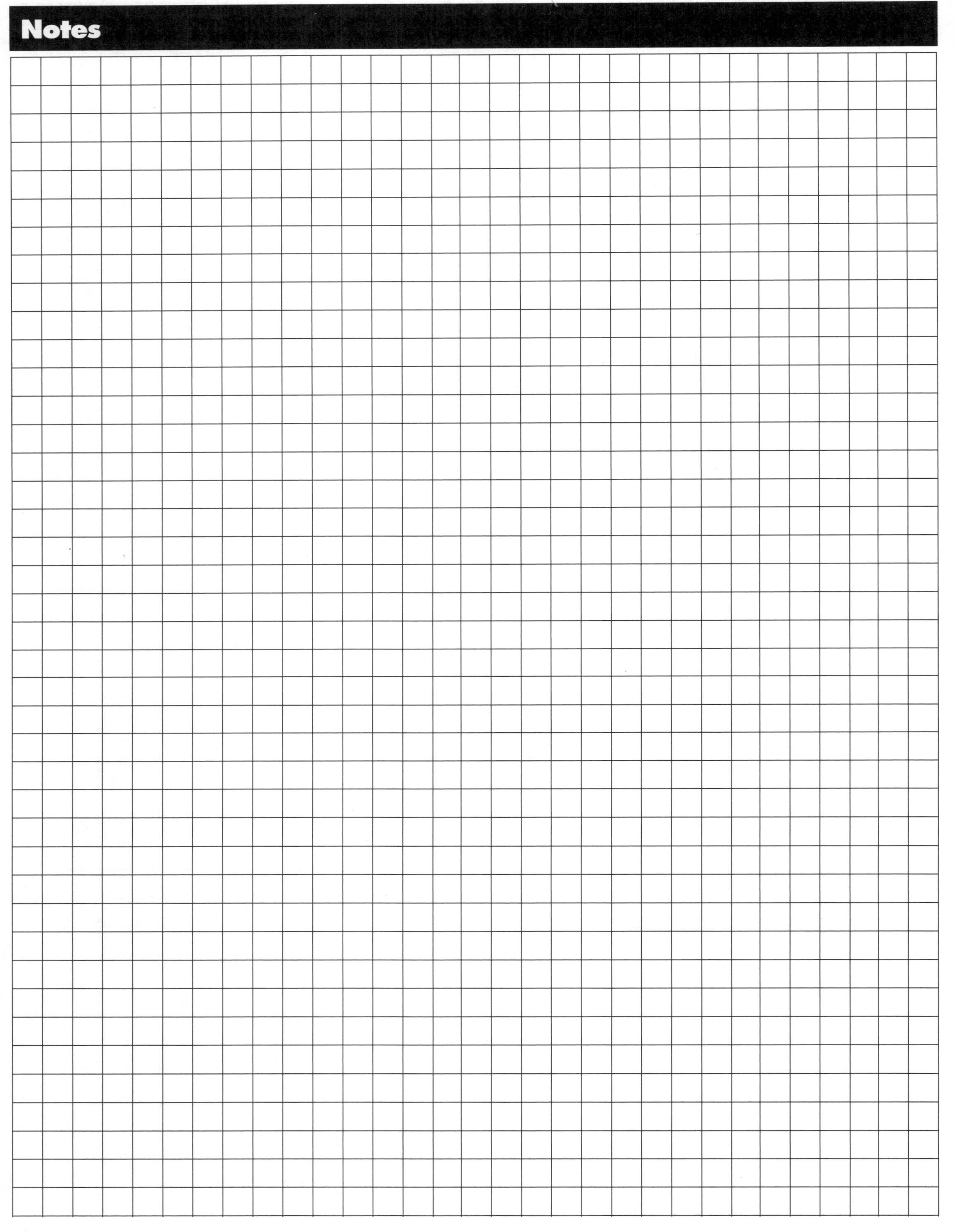

Notes

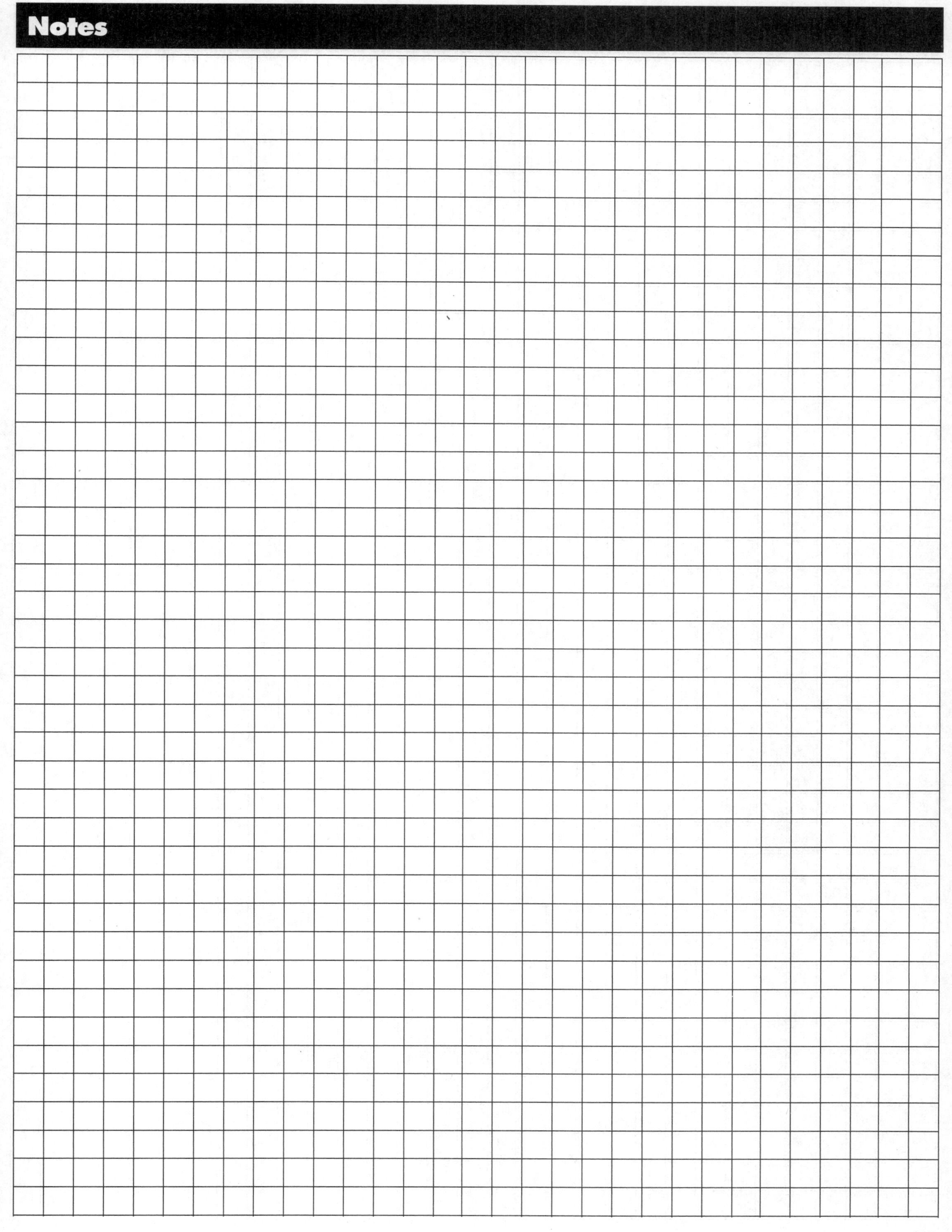

Notes

For more information visit Means Web Site at www.rsmeans.com

Reed Construction Data/RSMeans . . . a tradition of excellence in Construction Cost Information and Services since 1942.

Table of Contents

Book Selection Guide

The following table provides definitive information on the content of each cost data publication. The number of lines of data provided in each unit price or assemblies division, as well as the number of reference tables and crews is listed for each book. The presence of other elements such as an historical cost index, city cost indexes, square foot models or cross-referenced index is also indicated. You can use the table to help select the Means' book that has the quantity and type of information you most need in your work.

Unit Cost Divisions	Building Construction Costs	Mechanical	Electrical	Repair & Remodel.	Square Foot	Site Work Landsc.	Assemblies	Interior	Concrete Masonry	Open Shop	Heavy Construc.	Light Commercial	Facil. Construc.	Plumbing	Western Construction Costs	Resi-dential
1	1102	821	906	1012		1044		846	1013	1101	1050	751	1527	873	1101	709
2	2536	1320	460	1455		7430		545	1383	2495	4645	679	4194	1574	2518	758
3	1404	89	69	725		1239		204	1781	1400	1401	207	1315	47	1404	238
4	809	22	0	617		650		564	1017	785	589	373	1032	0	795	303
5	1788	212	167	916		723		909	647	1756	991	754	1770	287	1770	726
6	1232	82	78	1215		448		1216	312	1223	598	1371	1343	47	1571	1473
7	1260	155	71	1232		464		486	404	1259	346	954	1314	165	1260	746
8	1785	28	0	1862		294		1621	620	1771	49	1187	1980	0	1787	1166
9	1575	47	0	1417		234		1671	359	1532	156	1318	1792	49	1566	1206
10	858	47	25	493		194		697	179	860	0	391	879	232	858	215
11	1018	322	169	500		134		812	44	925	107	217	1077	291	925	103
12	307	0	0	46		210		1416	27	298	0	62	1420	0	298	63
13	1084	950	359	446		367		846	75	1068	258	445	1698	889	1050	193
14	342	55	0	255		36		291	0	342	30	12	341	33	340	6
15	1981	12970	617	1769		1562		1174	59	1992	1762	1198	10672	9474	2011	826
16	1282	470	9614	1003		739		1108	55	1300	760	1081	9381	415	1231	552
17	427	354	427	0		0		0	0	427	0	0	427	356	427	0
Totals	**20790**	**17950**	**12967**	**14963**		**15768**		**14406**	**7975**	**20534**	**12642**	**11000**	**42162**	**14742**	**20912**	**9283**

Assembly Divisions	Building Construction Costs	Mechanical	Elec-trical	Repair & Remodel.	Square Foot	Site Work Landsc.	Assemblies	Interior	Concrete Masonry	Open Shop	Heavy Construc.	Light Commercial	Facil. Construc.	Plumbing	Western Construction Costs	Asm Div	Resi-dential
A		19	0	192	150	540	612	0	550		542	149	24	0		1	374
B		0	0	810	2470	0	5578	333	1914		0	2020	145	0		2	217
C		0	0	635	810	0	1204	1514	145		0	710	266	0		3	588
D		1031	782	685	1814	0	2430	752	0		0	1310	1031	881		4	871
E		0	0	85	255	0	292	5	0		0	255	5	0		5	393
F		0	0	0	123	0	126	0	0		0	123	3	0		6	350
G		465	172	332	111	1836	584	0	482		432	110	87	540		7	299
																8	760
																9	80
																10	0
																11	0
																12	0
Totals		**1515**	**954**	**2739**	**5733**	**2376**	**10826**	**2604**	**3091**		**974**	**4677**	**1561**	**1421**			**3932**

Reference Section	Building Construction Costs	Mechanical	Electrical	Repair & Remodel.	Square Foot	Site Work Landsc.	Assemblies	Interior	Concrete Masonry	Open Shop	Heavy Construc.	Light Commercial	Facil. Construc.	Plumbing	Western Construction Costs	Resi-dential
Tables	129	45	85	69	4	80	219	46	71	130	61	57	104	51	131	42
Models					102							43				32
Crews	410	410	410	391		410		410	410	393	410	393	391	410	410	393
City Cost Indexes	yes	yes	yes	yes	yes	yes	yes	yes	yes	yes	yes	yes	yes	yes	yes	yes
Historical Cost Indexes	yes	yes	yes	yes	yes	yes	yes	yes	yes	yes	yes	yes	yes	yes	yes	no

Annual Cost Guides

For more information
visit Means Web Site
at www.rsmeans.com

Means Building Construction Cost Data 2004

Available in Both Softbound and Looseleaf Editions

The "Bible" of the industry comes in the standard softcover edition or the looseleaf edition.

Many customers enjoy the convenience and flexibility of the looseleaf binder, which increases the usefulness of *Means Building Construction Cost Data 2004* by making it easy to add and remove pages. You can insert your own cost information pages, so everything is in one place. Copying pages for faxing is easier also. Whichever edition you prefer, softbound or the convenient looseleaf edition, you'll be eligible to receive *The Change Notice* FREE. Current subscribers receive *The Change Notice* via e-mail.

$108.95 per copy, Softbound
Catalog No. 60014

$136.95 per copy, Looseleaf
Catalog No. 61014

Means Building Construction Cost Data 2004

Offers you unchallenged unit price reliability in an easy-to-use arrangement. Whether used for complete, finished estimates or for periodic checks, it supplies more cost facts better and faster than any comparable source. Over 23,000 unit prices for 2004. The City Cost Indexes cover over 930 areas, for indexing to any project location in North America. Order and get *The Change Notice* FREE. You'll have year-long access to the Means Estimating **HOTLINE** FREE with your subscription. Expert assistance when using Means data is just a phone call away.

$108.95 per copy
Over 650 pages, illustrated, available Oct. 2003
Catalog No. 60014

Means Building Construction Cost Data 2004

Metric Version

The Federal Government has stated that all federal construction projects must now use metric documentation. The *Metric Version* of *Means Building Construction Cost Data 2004* is presented in metric measurements covering all construction areas. Don't miss out on these billion dollar opportunities. Make the switch to metric today.

$108.95 per copy
Over 650 pages, illus., available Dec. 2003
Catalog No. 63014

For more information visit Means Web Site at www.rsmeans.com

Annual Cost Guides

Means Mechanical Cost Data 2004

• HVAC • Controls

Total unit and systems price guidance for mechanical construction...materials, parts, fittings, and complete labor cost information. Includes prices for piping, heating, air conditioning, ventilation, and all related construction.

Plus new 2004 unit costs for:
- Over 2500 installed HVAC/controls assemblies
- "On Site" Location Factors for over 930 cities and towns in the U.S. and Canada
- Crews, labor and equipment

$108.95 per copy
Over 600 pages, illustrated, available Oct. 2003
Catalog No. 60024

Means Plumbing Cost Data 2004

Comprehensive unit prices and assemblies for plumbing, irrigation systems, commercial and residential fire protection, point-of-use water heaters, and the latest approved materials. This publication and its companion, *Means Mechanical Cost Data*, provide full-range cost estimating coverage for all the mechanical trades.

$108.95 per copy
Over 500 pages, illustrated, available Oct. 2003
Catalog No. 60214

Means Electrical Cost Data 2004

Pricing information for every part of electrical cost planning: More than 15,000 unit and systems costs with design tables; clear specifications and drawings; engineering guides and illustrated estimating procedures; complete labor-hour and materials costs for better scheduling and procurement; the latest electrical products and construction methods.
- A Variety of Special Electrical Systems including Cathodic Protection
- Costs for maintenance, demolition, HVAC/ mechanical, specialties, equipment, and more

$108.95 per copy
Over 450 pages, illustrated, available Oct. 2003
Catalog No. 60034

Means Electrical Change Order Cost Data 2004

You are provided with electrical unit prices exclusively for pricing change orders based on the recent, direct experience of contractors and suppliers. Analyze and check your own change order estimates against the experience others have had doing the same work. It also covers productivity analysis and change order cost justifications. With useful information for calculating the effects of change orders and dealing with their administration.

$108.95 per copy
Over 450 pages, available Oct. 2003
Catalog No. 60234

Means Facilities Maintenance & Repair Cost Data 2004

Published in a looseleaf format, *Means Facilities Maintenance & Repair Cost Data* gives you a complete system to manage and plan your facility repair and maintenance costs and budget efficiently. Guidelines for auditing a facility and developing an annual maintenance plan. Budgeting is included, along with reference tables on cost and management and information on frequency and productivity of maintenance operations.

The only nationally recognized source of maintenance and repair costs. Developed in cooperation with the Army Corps of Engineers.

$238.95 per copy
Over 600 pages, illustrated, available Dec. 2003
Catalog No. 60304

Means Square Foot Costs 2004

It's Accurate and Easy To Use!

- **Updated 2004 price information,** based on nationwide figures from suppliers, estimators, labor experts and contractors.
- "How-to-Use" Sections, with **clear examples** of commercial, residential, industrial, and institutional structures.
- Realistic graphics, offering true-to-life illustrations of building projects.
- Extensive information on using square foot cost data, including **sample estimates** and **alternate pricing methods.**

$119.95 per copy
Over 450 pages, illustrated, available Nov. 2003
Catalog No. 60054

Annual Cost Guides

For more information
visit Means Web Site
at www.rsmeans.com

Annual Cost Guides

Means Open Shop Building Construction Cost Data 2004

The latest costs for accurate budgeting and estimating of new commercial and residential construction... renovation work... change orders... cost engineering. *Means Open Shop BCCD* will assist you to...

- Develop benchmark prices for change orders
- Plug gaps in preliminary estimates, budgets
- Estimate complex projects
- Substantiate invoices on contracts
- Price ADA-related renovations

$108.95 per copy
Over 650 pages, illustrated, available Dec. 2003
Catalog No. 60154

Means Heavy Construction Cost Data 2004

A comprehensive guide to heavy construction costs. Includes costs for highly specialized projects such as tunnels, dams, highways, airports, and waterways. Information on different labor rates, equipment, and material costs is included. Has unit price costs, systems costs, and numerous reference tables for costs and design. Valuable not only to contractors and civil engineers, but also to government agencies and city/town engineers.

$108.95 per copy
Over 450 pages, illustrated, available Nov. 2003
Catalog No. 60164

Means Building Construction Cost Data 2004

Western Edition

This regional edition provides more precise cost information for western North America. Labor rates are based on union rates from 13 western states and western Canada. Included are western practices and materials not found in our national edition: tilt-up concrete walls, glu-lam structural systems, specialized timber construction, seismic restraints, landscape and irrigation systems.

$108.95 per copy
Over 600 pages, illustrated, available Dec. 2003
Catalog No. 60224

Means Heavy Construction Cost Data 2004

Metric Version

Make sure you have the Means industry standard metric costs for the federal, state, municipal and private marketplace. With thousands of up-to-date metric unit prices in tables by CSI standard divisions. Supplies you with assemblies costs using the metric standard for reliable cost projections in the design stage of your project. Helps you determine sizes, material amounts, and has tips for handling metric estimates.

$108.95 per copy
Over 450 pages, illustrated, available Dec. 2003
Catalog No. 63164

Means Construction Cost Indexes 2004

Who knows what 2004 holds? What materials and labor costs will change unexpectedly? By how much?

- Breakdowns for 316 major cities.
- National averages for 30 key cities.
- Expanded five major city indexes.
- Historical construction cost indexes.

$237.95 per year
$59.50 individual quarters
Catalog No. 60144 A,B,C,D

Means Interior Cost Data 2004

Provides you with prices and guidance needed to make accurate interior work estimates. Contains costs on materials, equipment, hardware, custom installations, furnishings, labor costs . . . every cost factor for new and remodel commercial and industrial interior construction, including updated information on office furnishings, plus more than 50 reference tables. For contractors, facility managers, owners.

$108.95 per copy
Over 550 pages, illustrated, available Oct. 2003
Catalog No. 60094

Means Concrete & Masonry Cost Data 2004

Provides you with cost facts for virtually all concrete/masonry estimating needs, from complicated formwork to various sizes and face finishes of brick and block, all in great detail. The comprehensive unit cost section contains more than 8,500 selected entries. Also contains an assemblies cost section, and a detailed reference section which supplements the cost data.

$99.95 per copy
Over 450 pages, illustrated, available Dec. 2003
Catalog No. 60114

Means Labor Rates for the Construction Industry 2004

Complete information for estimating labor costs, making comparisons and negotiating wage rates by trade for over 300 cities (United States and Canada). With 46 construction trades listed by local union number in each city, and historical wage rates included for comparison. No similar book is available through the trade.

Each city chart lists the county and is alphabetically arranged with handy visual flip tabs for quick reference.

$239.95 per copy
Over 300 pages, available Dec. 2003
Catalog No. 60124

Residential & Light Commercial Construction Standards, 2nd Edition

By RSMeans and Contributing Authors

New, updated second edition of this unique collection of industry standards that define quality construction. For contractors, subcontractors, owners, developers, architects, engineers, attorneys, and insurance personnel, this book provides authoritative requirements and recommendations compiled from the nation's leading professional associations, industry publications, and building code organizations. This one-stop reference is enhanced by helpful commentary from respected practitioners, including identification of items most frequently targeted for construction defect claims. The new second edition provides the latest building code requirements.

$59.95 per copy
600 pages, illustrated, Softcover
Catalog No. 67322A

Designing & Building with the IBC, 2nd Edition

By Rolf Jensen & Associates, Inc.

This updated comprehensive guide helps building professionals make the transition to the 2003 *International Building Code*®. Includes a side-by-side code comparison of the IBC 2003 to the IBC 2000 and the three primary model codes, a quick-find index, and professional code commentary.

- The Quick-Find Index lets you instantly identify the IBC equivalent to the model code sections you are familiar with: the *BOCA® National Building Code, Uniform Building Code™*, or *Standard Building Code©*.
- The Code Comparison highlights the differences between the IBC 2003, IBC 2000, and the model codes — in an easy-to-read, side-by-side format.
- The Code Commentary points out issues to watch out for in making the transition to the new international code.

With illustrations, abbreviations key, and an extensive Resource section.

$99.95 per copy
Over 875 pages
Catalog No. 67328A

Means Estimating Handbook, 2nd Edition

By RSMeans

Updated new Second Edition answers virtually any estimating technical question - all organized by CSI MasterFormat. This comprehensive reference covers the full spectrum of technical data required to estimate construction costs. The book includes information on sizing, productivity, equipment requirements, code-mandated specifications, design standards and engineering factors.

This reference—widely used in the industry for tasks ranging from routine estimates to special cost analysis projects—has been completely updated with new and expanded technical information and an enhanced format for ease-of-use.

$99.95 per copy
Over 900 pages, Hardcover
Catalog No. 67276A

For more information visit Means Web Site at www.rsmeans.com

Reference Books

Value Engineering: Practical Applications

. . . For Design, Construction, Maintenance & Operations

By Alphonse Dell'Isola, PE

A tool for immediate application—for engineers, architects, facility managers, owners, and contractors. Includes: Making the Case for VE—The Management Briefing, Integrating VE into Planning and Budgeting, Conducting Life Cycle Costing, Integrating VE into the Design Process, Using VE Methodology in Design Review and Consultant Selection, Case Studies, VE Workbook, and a Life Cycle Costing program on disk.

$79.95 per copy
Over 450 pages, illustrated, Hardcover
Catalog No. 67319

Facilities Operations & Engineering Reference

By the Association for Facilities Engineering and RSMeans

An all-in-one technical referance for planning and managing facility projects and solving day-to-day operations problems. Selected as the official Certified Plant Engineer reference, this handbook covers financial analysis, maintenance, HVAC and energy efficiency, and more.

$109.95 per copy
Over 700 pages, illustrated, Hardcover
Catalog No. 67318

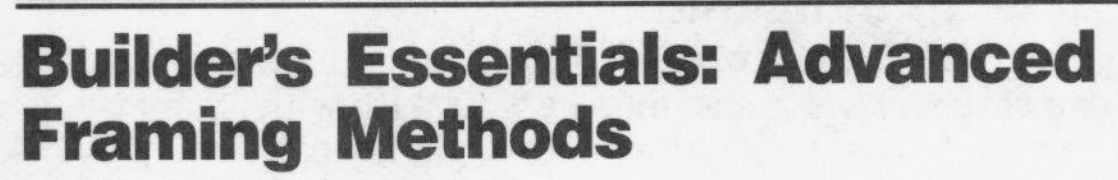

Builder's Essentials: Advanced Framing Methods

By Scot Simpson

A highly illustrated, "framer-friendly" approach to advanced framing elements. Provides expert, but easy-to-interpret, instruction for laying out and framing complex walls, roofs, and stairs, and special requirements for earthquake and hurricane protection. Also helps bring framers up to date on the latest building code changes, and provides tips on the lead framer's role and responsibilities, how to prepare for a job, and how to get the crew started.

$24.95 per copy
250 pages, illustrated, Softcover
Catalog No. 67330

HVAC: Design Criteria, Options, Selection

Expanded 2nd Edition

By William H. Rowe III, AIA, PE

Includes Indoor Air Quality, CFC Removal, Energy Efficient Systems, and Special Systems by Building Type. Helps you solve a wide range of HVAC system design and selection problems effectively and economically. Gives you clear explanations of the latest ASHRAE standards.

$84.95 per copy
Over 600 pages, illustrated, Hardcover
Catalog No. 67306

Total Productive Facilities Management

By Richard W. Sievert, Jr.

This book provides a comprehensive program to:

- Achieve business goals by optimizing facility resources
- Implement best practices through benchmarking, evaluation & project management
- Increase your value to the organization

$29.98 per copy
Over 270 pages, illustrated, Hardcover
Catalog No. 67321

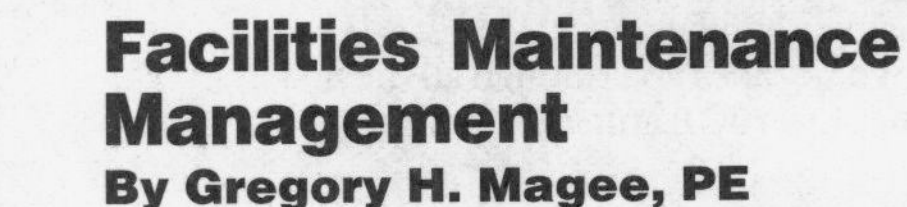

Facilities Maintenance Management

By Gregory H. Magee, PE

Now you can get successful management methods and techniques for all aspects of facilities maintenance. This comprehensive reference explains and demonstrates successful management techniques for all aspects of maintenance, repair, and improvements for buildings, machinery, equipment, and grounds. Plus, guidance for outsourcing and managing internal staffs.

$86.95 per copy
Over 280 pages with illustrations, Hardcover
Catalog No. 67249

Cost Planning & Estimating for Facilities Maintenance

In this unique book, a team of facilities management authorities shares their expertise at:

- Evaluating and budgeting maintenance operations
- Maintaining & repairing key building components
- Applying *Means Facilities Maintenance & Repair Cost Data* to your estimating

Covers special maintenance requirements of the 10 major building types.

$89.95 per copy
Over 475 pages, Hardcover
Catalog No. 67314

The Building Professional's Guide to Contract Documents

New 3rd Edition

By Waller S. Poage, AIA, CSI, CVS

This comprehensive treatment of Contract Documents is an important reference for owners, design professionals, contractors, and students.

- Structure your Documents for Maximum Efficiency
- Effectively communicate construction requirements to all concerned
- Understand the Roles and Responsibilities of Construction Professionals
- Improve Methods of Project Delivery

$64.95 per copy, 400 pages
Diagrams and construction forms, Hardcover
Catalog No. 67261A

Reference Books

For more information visit Means Web Site at www.rsmeans.com

Builder's Essentials: Plan Reading & Material Takeoff

By Wayne J. DelPico

For Residential and Light Commercial Construction

A valuable tool for understanding plans and specs, and accurately calculating material quantities.
Step-by-step instructions and takeoff procedures based on a full set of working drawings.

$35.95 per copy
Over 420 pages, Softcover
Catalog No. 67307

Planning & Managing Interior Projects

2nd Edition

By Carol E. Farren, CFM

Addresses changes in technology and business, guiding you through commercial design and construction from initial client meetings to post-project administration. Includes: evaluating space requirements, alternative work models, telecommunications and data management, and environmental issues.

$69.95 per copy
Over 400 pages, illustrated, Hardcover
Catalog No. 67245A

Builder's Essentials: Best Business Practices for Builders & Remodelers: An Easy-to-Use Checklist System

By Thomas N. Frisby

A comprehensive guide covering all aspects of running a construction business, with more than 40 user–friendly checklists. This book provides expert guidance on: increasing your revenue and keeping more of your profit; planning for long-term growth; keeping good employees and managing subcontractors.

$29.95 per copy
Over 220 pages, Softcover
Catalog No. 67329

Builder's Essentials: Framing & Rough Carpentry 2nd Edition

By Scot Simpson

A complete training manual for apprentice and experienced carpenters. Develop and improve your skills with "framer-friendly," easy-to-follow instructions, and step-by-step illustrations. Learn proven techniques for framing walls, floors, roofs, stairs, doors, and windows. Updated guidance on standards, building codes, safety requirements, and more. Also available in Spanish!

$24.95 per copy
Over 150 pages, Softcover
Catalog No. 67298A Spanish Catalog No. 67298AS

How to Estimate with Means Data & CostWorks

By RSMeans and Saleh Mubarak

New 2nd Edition!

Learn estimating techniques using Means cost data. Includes an instructional version of Means CostWorks CD–ROM with Sample Building Plans.
The step-by-step guide takes you through all the major construction items. Over 300 sample estimating problems are included.

$59.95 per copy
Over 190 pages, Softcover
Catalog No. 67324A

Unit Price Estimating Methods

New 3rd Edition

This new edition includes up-to-date cost data and estimating examples, updated to reflect changes to the CSI numbering system and new features of Means cost data. It describes the most productive, universally accepted ways to estimate, and uses checklists and forms to illustrate shortcuts and timesavers. A model estimate demonstrates procedures. A new chapter explores computer estimating alternatives.

$59.95 per copy
Over 350 pages, illustrated, Hardcover
Catalog No. 67303A

Interior Home Improvement Costs, 8th Edition

Provides 66 updated estimates for the most popular projects including estimates for home offices, in-law apartments, and remodeling for disabled residents. Includes guidance for remodeling older homes as well as attic and basement conversions, kitchen and bath remodeling, lighting and security; fireplaces; storage; stairs, new floors, and walls and ceilings.

$19.95 per copy
250 pages, illustrated, Softcover
Catalog No. 67308D

Exterior Home Improvement Costs, 8th Edition

Provides quick estimates for 64 projects including room additions and garages; windows and doors; dormers, roofs, and skylights; decks and pergolas; paving and patios; painting and siding; walls, fences, and porches; landscaping and driveways. It also contains price comparison sheets for material and equipment.

$19.95 per copy
270 pages, illustrated, Softcover
Catalog No. 67309D

Means Landscape Estimating Methods

4th Edition

By Sylvia H. Fee

This revised edition offers expert guidance for preparing accurate estimates for new landscape construction and grounds maintenance. Includes a complete project estimate featuring the latest equipment and methods, and **two chapters on Life Cycle Costing, and Landscape Maintenance Estimating.**

$62.95 per copy
Over 300 pages, illustrated, Hardcover
Catalog No. 67295B

Means Environmental Remediation Estimating Methods, 2nd Edition

By Richard R. Rast

Guidelines for estimating 50 standard remediation technologies. Use it to prepare preliminary budgets, develop estimates, compare costs and solutions, estimate liability, review quotes, negotiate settlements.

A valuable support tool for ***Means Environmental Remediation Unit Price*** and ***Assemblies*** books.

$99.95 per copy
Over 750 pages, illustrated, Hardcover
Catalog No. 64777A

For more information
visit Means Web Site
at www.rsmeans.com

Reference Books

Means Illustrated Construction Dictionary, Condensed Edition, 2nd Edition

By RSMeans

Recognized in the industry as the best resource of its kind, this has been further enhanced with updates to existing terms and the addition of hundreds of new terms and illustrations . . . in keeping with the most recent developments in the industry.
The best portable dictionary for office or field use—an essential tool for contractors, architects, insurance and real estate personnel, homeowners, and anyone who needs quick, clear definitions for construction terms.

$59.95 per copy
Over 500 pages
Catalog No. 67282A

Means Repair and Remodeling Estimating

New 4th Edition
By Edward B. Wetherill & RSMeans

Focuses on the unique problems of estimating renovations of existing structures. It helps you determine the true costs of remodeling through careful evaluation of architectural details and a site visit.
New section on disaster restoration costs.

$69.95 per copy
Over 450 pages, illustrated, Hardcover
Catalog No. 67265B

Facilities Planning & Relocation

New, lower price and user-friendly format.
By David D. Owen

A complete system for planning space needs and managing relocations. Includes step-by-step manual, over 50 forms, and extensive reference section on materials and furnishings.

$89.95 per copy
Over 450 pages, Softcover
Catalog No. 67301

Means Square Foot & Assemblies Estimating Methods

New 3rd Edition!

Develop realistic Square Foot and Assemblies Costs for budgeting and construction funding. The new edition features updated guidance on square foot and assemblies estimating using UNIFORMAT II. An essential reference for anyone who performs conceptual estimates.

$69.95 per copy
Over 300 pages, illustrated, Hardcover
Catalog No. 67145B

Means Electrical Estimating Methods

3rd Edition

Expanded new edition includes sample estimates and cost information in keeping with the latest version of the CSI MasterFormat and UNIFORMAT II. Complete coverage of Fiber Optic and Uninterruptible Power Supply electrical systems, broken down by components and explained in detail. Includes a new chapter on computerized estimating methods. A practical companion to *Means Electrical Cost Data.*

$64.95 per copy
Over 325 pages, Hardcover
Catalog No. 67230A

Means Mechanical Estimating Methods

New 3rd Edition

This guide assists you in making a review of plans, specs, and bid packages with suggestions for takeoff procedures, listings, substitutions and pre-bid scheduling. Includes suggestions for budgeting labor and equipment usage. Compares materials and construction methods to allow you to select the best option.

$64.95 per copy
Over 350 pages, illustrated, Hardcover
Catalog No. 67294A

Successful Estimating Methods: From Concept to Bid

By John D. Bledsoe, PhD, PE

A highly practical, all-in-one guide to the tips and practices of today's successful estimator. Presents techniques for all types of estimates, and advanced topics such as life cycle cost analysis, value engineering, and automated estimating. Estimate spreadsheets available at Means Web site.

$32.48 per copy
Over 300 pages, illustrated, Hardcover
Catalog No. 67287

Means Spanish/English Construction Dictionary

By RSMeans, The International Conference of Building Officials (ICBO), and Rolf Jensen & Associates (RJA)

Designed to facilitate communication among Spanish- and English-speaking construction personnel—improving performance and job-site safety. Features the most common words and phrases used in the construction industry, with easy-to-follow pronunciations. Includes extensive building systems and tools illustrations.

$22.95 per copy
250 pages, illustrated, Softcover
Catalog No. 67327

Means Productivity Standards for Construction

Expanded Edition *(Formerly Man-Hour Standards)*

Here is the working encyclopedia of labor productivity information for construction professionals, with labor requirements for thousands of construction functions in CSI MasterFormat.
Completely updated, with over 3,000 new work items.

$49.98 per copy
Over 800 pages, Hardcover
Catalog No. 67236A

Project Scheduling & Management for Construction

New 2nd Edition
By David R. Pierce, Jr.

A comprehensive yet easy-to-follow guide to construction project scheduling and control—from vital project management principles through the latest scheduling, tracking, and controlling techniques. The author is a leading authority on scheduling with years of field and teaching experience at leading academic institutions. Spend a few hours with this book and come away with a solid understanding of this essential management topic.

$64.95 per copy
Over 250 pages, illustrated, Hardcover
Catalog No. 67247A

Reference Books

For more information visit Means Web Site at www.rsmeans.com

Cyberplaces: The Internet Guide for AECs & Facility Managers
By Paul Doherty, AIA, CSI
2nd Edition
$59.95 per copy
Catalog No. 67317A

Concrete Repair and Maintenance Illustrated
By Peter H. Emmons
$69.95 per copy
Catalog No. 67146

Superintending for Contractors:
How to Bring Jobs in On-Time, On-Budget
By Paul J. Cook
$35.95 per copy
Catalog No. 67233

HVAC Systems Evaluation
By Harold R. Colen, PE
$84.95 per copy
Catalog No. 67281

Basics for Builders: How to Survive and Prosper in Construction
By Thomas N. Frisby
$34.95 per copy
Catalog No. 67273

Successful Interior Projects Through Effective Contract Documents
By Joel Downey & Patricia K. Gilbert
Now $24.98 per copy, limited quantity
Catalog No. 67313

Building Spec Homes Profitably
By Kenneth V. Johnson
$29.95 per copy
Catalog No. 67312

Estimating for Contractors
How to Make Estimates that Win Jobs
By Paul J. Cook
$35.95 per copy
Catalog No. 67160

Means Forms for Building Construction Professionals
$47.48 per copy
Catalog No. 67231

Understanding Building Automation Systems
By Reinhold A. Carlson, PE & Robert Di Giandomenico
$29.98 per copy, limited quantity
Catalog No. 67284

Managing Construction Purchasing
By John G. McConville, CCC, CPE
Now $19.98 per copy, limited quantity
Catalog No. 67302

Fundamentals of the Construction Process
By Kweku K. Bentil, AIC
Now $34.98 per copy, limited quantity
Catalog No. 67260

Means ADA Compliance Pricing Guide
$29.98 per copy
Catalog No. 67310

Means Facilities Maintenance Standards
By Roger W. Liska, PE, AIC
Now $79.95 per copy
Catalog No. 67246

How to Estimate with Metric Units
Now $9.98 per copy, limited quantity
Catalog No. 67304

Means Plumbing Estimating Methods, 2nd Edition
Now $59.95 per copy, limited quantity
Catalog No. 67283A

Preventive Maintenance for Higher Education Facilities

By Applied Management Engineering, Inc.

An easy-to-use system to help facilities professionals establish the value of PM, and to develop and budget for an appropriate PM program for their college or university. Features interactive campus building models typical of those found in different-sized higher education facilities, such as dormitories, classroom buildings, laboratories, athletic facilities, and more. Includes PM checklists linked to each piece of equipment or system in hard copy and electronic format, along with required labor hours to complete the PM tasks. Helps users select and develop the best possible PM plan within budget.

$149.95 per copy
150 pages, Hardcover
Catalog No. 67337

Preventive Maintenance Guidelines for School Facilities

By John C. Maciha

This new publication is a complete PM program for K-12 schools that ensures sustained security, safety, property integrity, user satisfaction, and reasonable ongoing expenditures.

Includes schedules for weekly, monthly, semiannual, and annual maintenance. Comes as a 3-ring binder with hard copy and electronic forms available at Means website, and a laminated wall chart.

$149.95 per copy
Over 225 pages, Hardcover
Catalog No. 67326

Historic Preservation: Project Planning & Estimating

By Swanke Hayden Connell Architects

- *Managing Historic Restoration, Rehabilitation, and Preservation Building Projects and*
- *Determining and Controlling Their Costs*

The authors explain:

- How to determine whether a structure qualifies as historic
- Where to obtain funding and other assistance
- How to evaluate and repair more than 75 historic building materials
- How to properly research, document, and manage the project to meet code, agency, and other special requirements

$99.95 per copy
Over 675 pages, Hardcover
Catalog No. 67323

Means Illustrated Construction Dictionary, 3rd Edition

Long regarded as the Industry's finest, the Means *Illustrated Construction Dictionary* is now even better. With the addition of over 1,000 new terms and hundreds of new illustrations, it is the clear choice for the most comprehensive and current information.

The companion CD-ROM that comes with this new edition adds many extra features: larger graphics, expanded definitions, and links to both CSI MasterFormat numbers and product information.

- 19,000 construction words, terms, phrases, symbols, weights, measures, and equivalents
- 1,000 new entries
- 1,200 helpful illustrations
- Easy-to-use format, with thumbtabs

$99.95 per copy
Over 790 pages, illustrated Hardcover
Catalog No. 67292A

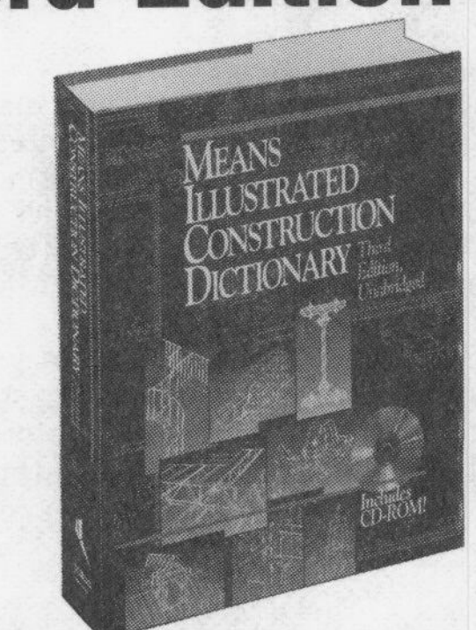

Seminars

Means CostWorks Training

This one-day seminar course has been designed with the intention of assisting both new and existing users to become more familiar with CostWorks program. The class is broken into two unique sections: (1) A one-half day presentation on the function of each icon; and each student will be shown how to use the software to develop a cost estimate. (2) Hands-on estimating exercises that will ensure that each student thoroughly understands how to use CostWorks. You must bring your own laptop computer to this course.

CostWorks Benefits/Features:
- Estimate in your own spreadsheet format
- Power of Means National Database
- Database automatically regionalized
- Save time with keyword searches
- Save time by establishing common estimate items in "Bookmark" files
- Customize your spreadsheet template
- Hot key to Product Manufacturers' listings and specs
- Merge capability for networking environments
- View crews and assembly components
- AutoSave capability
- Enhanced sorting capability

Unit Price Estimating

This interactive two-day seminar teaches attendees how to interpret project information and process it into final, detailed estimates with the greatest accuracy level.

The single most important credential an estimator can take to the job is the ability to visualize construction in the mind's eye, and thereby estimate accurately.

Some Of What You'll Learn:
- Interpreting the design in terms of cost
- The most detailed, time tested methodology for accurate "pricing"
- Key cost drivers—material, labor, equipment, staging and subcontracts
- Understanding direct and indirect costs for accurate job cost accounting and change order management

Who Should Attend: Corporate and government estimators and purchasers, architects, engineers...and others needing to produce accurate project estimates.

Square Foot and Assemblies Cost Estimating

This two-day course teaches attendees how to quickly deliver accurate square foot estimates using limited budget and design information.

Some Of What You'll Learn:
- How square foot costing gets the estimate done faster
- Taking advantage of a "systems" or "assemblies" format
- The Means "building assemblies/square foot cost approach"
- How to create a very reliable preliminary and systems estimate using bare-bones design information

Who Should Attend: Facilities managers, facilities engineers, estimators, planners, developers, construction finance professionals...and others needing to make quick, accurate construction cost estimates at commercial, government, educational and medical facilities.

Repair and Remodeling Estimating

This two-day seminar emphasizes all the underlying considerations unique to repair/remodeling estimating and presents the correct methods for generating accurate, reliable R&R project costs using the unit price and assemblies methods.

Some Of What You'll Learn:
- Estimating considerations—like labor-hours, building code compliance, working within existing structures, purchasing materials in smaller quantities, unforeseen deficiencies
- Identifies problems and provides solutions to estimating building alterations
- Rules for factoring in minimum labor costs, accurate productivity estimates and allowances for project contingencies
- R&R estimating examples are calculated using unite prices and assemblies data

Who Should Attend: Facilities managers, plant engineers, architects, contractors, estimators, builders...and others who are concerned with the proper preparation and/or evaluation of repair and remodeling estimates.

Mechanical and Electrical Estimating

This two-day course teaches attendees how to prepare more accurate and complete mechanical/electrical estimates, avoiding the pitfalls of omission and double-counting, while understanding the composition and rationale within the Means Mechanical/Electrical database.

Some Of What You'll Learn:
- The unique way mechanical and electrical systems are interrelated
- M&E estimates, conceptual, planning, budgeting and bidding stages
- Order of magnitude, square foot, assemblies and unit price estimating
- Comparative cost analysis of equipment and design alternatives

Who Should Attend: Architects, engineers, facilities managers, mechanical and electrical contractors...and others needing a highly reliable method for developing, understanding and evaluating mechanical and electrical contracts.

Plan Reading and Material Takeoff

This two-day program teaches attendees to read and understand construction documents and to use them in the preparation of material takeoffs.

Some of What You'll Learn:
- Skills necessary to read and understand typical contract documents—blueprints and specifications
- Details and symbols used by architects and engineers
- Construction specifications' importance in conjunction with blueprints
- Accurate takeoff of construction materials and industry-accepted takeoff methods

Who Should Attend: Facilities managers, construction supervisors, office managers...and other responsible for the execution and administration of a construction project including government, medical, commercial, educational or retail facilities.

Facilities Maintenance and Repair Estimating

This two-day course teaches attendees how to plan, budget, and estimate the cost of ongoing and preventive maintenance and repair for existing buildings and grounds.

Some Of What You'll Learn:
- The most financially favorable maintenance, repair and replacement scheduling and estimating
- Auditing and value engineering facilities
- Preventive planning and facilities upgrading
- Determining both in-house and contract-out service costs; annual, asset-protecting M&R plan

Who Should Attend: Facility managers, maintenance supervisors, buildings and grounds superintendents, plant managers, planners, estimators...and others involved in facilities planning and budgeting.

Scheduling and Project Management

This two-day course teaches attendees the most current and proven scheduling and management techniques needed to bring projects in on time and on budget.

Some Of What You'll Learn:
- Crucial phases of planning and scheduling
- How to establish project priorities, develop realistic schedules and management techniques
- Critical Path and Precedence Methods
- Special emphasis on cost control

Who Should Attend: Construction project managers, supervisors, engineers, estimators, contractors...and others who want to improve their project planning, scheduling and management skills.

Advanced Project Management

This two-day seminar will teach you how to effectively manage and control the entire design-build process and allow you to take home tangible skills that will be immediately applicable on existing projects.

Some Of What You'll Learn:
- Value engineering, bonding, fast-tracking and bid package creation
- How estimates and schedules can be integrated to provide advanced project management tools
- Cost engineering, quality control, productivity measurement and improvement
- Front loading a project and predicting its cash flow

Who Should Attend: Owners, project managers, architectural and engineering managers, construction managers, contractors...and anyone else who is responsible for the timely design and completion of construction projects.

For more information visit Means Web Site at www.rsmeans.com

Seminars

2004 Means Seminar Schedule

Location	Dates
Las Vegas, NV	March TBD
Washington, DC	April TBD
Denver, CO	May TBD
San Francisco, CA	June 7-10
Cape Cod, MA	September TBD
Washington, DC	September TBD
Las Vegas, NV	October TBD
Orlando, FL	November TBD
Atlantic City, NJ	November 15-18
San Diego, CA	December 13-16

Note: Call for exact dates and details.

Registration Information

Register Early... Save up to $125! Register 30 days before the start date of a seminar and save $125 off your total fee. *Note: This discount can be applied only once per order. It cannot be applied to team discount registrations or any other special offer.*

How to Register Register by phone today! Means toll-free number for making reservations is: **1-800-334-3509, ext. 5115.**

Individual Seminar Registration Fee $895. Individual CostWorks Training Registration Fee $349. To register by mail, complete the registration form and return with your full fee to: Seminar Division, Reed Construction Data, RSMeans Seminars, 63 Smiths Lane, Kingston, MA 02364.

Federal Government Pricing All Federal Government employees save 25% off regular seminar price. Other promotional discounts cannot be combined with Federal Government discount.

Team Discount Program Two to four seminar registrations: $760 per person–Five or more seminar registrations: $710 per person–Ten or more seminar registrations: Call for pricing.

Consecutive Seminar Offer One individual signing up for two separate courses at the same location during the designated time period pays only $1,435. You get the second course for only $540 **(a 40% discount)**. Payment must be received at least ten days prior to seminar dates to confirm attendance.

Refund Policy Cancellations will be accepted up to ten days prior to the seminar start. There are no refunds for cancellations received later than ten working days prior to the first day of the seminar. A $150 processing fee will be applied for all cancellations. Written notice of cancellation is required . Substitutions can be made at anytime before the session starts. **No-shows are subject to the full seminar fee.**

AACE Approved Courses The RSMeans Construction Estimating and Management Seminars described and offered to you here have each been approved for 14 hours (1.4 recertification credits) of credit by the AACE International Certification Board toward meeting the continuing education requirements for re-certification as a Certified Cost Engineer/Certified Cost Consultant.

AIA Continuing Education We are registered with the AIA Continuing Education System (AIA/CES) and is committed to developing quality learning activities in accordance with the CES criteria. RSMeans seminars meet the AIA/CES criteria for Quality Level 2. AIA members will receive (28) learning units (LUs) for each two-day RSMeans Course.

Daily Course Schedule The first day of each seminar session begins at 8:30 A.M. and ends at 4:30 P.M. The second day is 8:00 A.M.–4:00 P.M. Participants are urged to bring a hand-held calculator since many actual problems will be worked out in each session.

Continental Breakfast Your registration includes the cost of a continental breakfast, a morning coffee break, and an afternoon break. These informal segments will allow you to discuss topics of mutual interest with other members of the seminar. (You are free to make your own lunch and dinner arrangements.)

Hotel/Transportation Arrangements RSMeans has arranged to hold a block of rooms at each hotel hosting a seminar. To take advantage of special group rates when making your reservation, be sure to mention that you are attending the Means Seminar. You are, of course, free to stay at the lodging place of your choice. **(Hotel reservations and transportation arrangements should be made directly by seminar attendees.)**

Important Class sizes are limited, so please register as soon as possible.

Note: *Pricing subject to change.*

Registration Form

Call 1-800-334-3509 x5115 to register or FAX 1-800-632-6732. Visit our Web site www.rsmeans.com

Please register the following people for the Means Construction Seminars as shown here. Full payment or deposit is enclosed, and we understand that we must make our own hotel reservations if overnight stays are necessary.

☐ Full payment of $ ________________ enclosed.

☐ Bill me

Name of Registrant(s)
(To appear on certificate of completion)

____

____

____

P.O. #: ________________

GOVERNMENT AGENCIES MUST SUPPLY PURCHASE ORDER NUMBER

Firm Name ________________

Address ________________

City/State/Zip ________________

Telephone No. fax No. ________________

E-mail Address ________________

Charge our registration(s) to: ☐ MasterCard ☐ VISA ☐ American Express ☐ Discover

Account No. ________________ Exp. Date ________________

Cardholder's Signature ________________

Seminar Name City Dates

____

____

Please mail check to: Seminar Division, Reed Construction Data, RSMeans Seminars, 63 Smiths Lane, P.O.Box 800, Kingston, MA 02364 USA